# International Petroleum & Petrochemical Technology Conference 2023

Beijing, China
31 May – 2 June 2023

ISBN: 978-1-7138-8196-4
ISSN: 1742-6588

**Printed from e-media with permission by:**

Curran Associates, Inc.
57 Morehouse Lane
Red Hook, NY 12571

**Some format issues inherent in the e-media version may also appear in this print version.**



Institute of Physics
Dirac House, Temple Back
Bristol BS1 6BE UK

Phone: 44 1 17 929 7481
Fax: 44 1 17 920 0979

techtracking@iop.org

**Additional copies of this publication are available from:**

Curran Associates, Inc.
57 Morehouse Lane
Red Hook, NY 12571 USA
Phone: 845-758-0400
Fax: 845-758-2633
Email: curran@proceedings.com
Web: www.proceedings.com

# TABLE OF CONTENTS

# Preface

As an academic conference, IPPTC aims to provide an ideal opportunity for industry experts, leading engineers, researchers and technical managers as well as university scholars to share ideas and research achievements related to petroleum & petrochemical technology and discuss the practical challenges encountered and the solutions adopted. IPPTC is held annually since 2017, it becomes a platform to bridge the knowledge gap covering the whole stream in oil and gas between China and the world.

Following the success of IPPTC in previous years, this multi-disciplined conference will be taken place at the same time with China International Petroleum and Petrochemical Equipment Exhibition (CIPPE). This academic conference aims to provide an ideal opportunity for industry experts, leading engineers, researchers and technical managers as well as university scholars to share ideas and research achievements related to petroleum & petrochemical technology and discuss the practical challenges encountered and the solutions adopted. It will create a platform to bridge the knowledge gap between China and the world.

## A full list of committee

| | |
|---|---|
| **Chen Jianfeng** | China Petroleum Pipeline Bureau |
| **Chen Junbin** | Xi'an Shiyou University |
| **Cheng Zhenmin** | East China University of Science and Technology |
| **Zhu Xiaoli** | Sinopec Research Institute of Petroleum Processing Co., Ltd. |
| **Dai Baohua** | China Economics and Development Research Institute Company Ltd. |
| **Deng Wenan** | China University of Petroleum |
| **Ding Yunjie** | Dalian Institute of Chemical Physics |
| **Dong Dazhong** | Research Institute of Petroleum Exploration and Development |
| **Dong Qun** | Northeast Petroleum University |
| **Feng Zihui** | Daqing Oilfield Company PetroChina |
| **Gao Pei** | Schlumberger (Shenzhen) China |
| **Hao Yongmei** | Changzhou University |

Journal of Physics: Conference Series **2594** (2023) 011001 doi:10.1088/1742-6596/2594/1/011001

| | |
|---|---|
| **He Zhiliang** | China Petrochemical Corporation |
| **Hu Zongquan** | Sinopec Petroleum Exploration and Production Research Institute |
| **Jiang Tingxue** | Sinopec Research Institute of Petroleum Processing Co., Ltd. |
| **Li Hexing** | Shanghai University of Electric Power |
| **Li Peimin** | BGP, PetroChina |
| **Li Yujun** | Liaohe Oilfield Company PetroChina |
| **Liang Hongbao** | Northeast Petroleum University |
| **Lin Dong** | Southeast Oil and Gas Field Company |
| **Liu Keyu** | China University of Petroleum |
| **Liu Yu** | China Petroleum Pipeline Bureau |
| **Liu Zhongsheng** | Sinopec Petroleum Exploration & Production Research Institute |

Journal of Physics: Conference Series **2594** (2023) 011001 doi:10.1088/1742-6596/2594/1/011001

| | |
|---|---|
| **Luo Jianhui** | Research Institute of Petroleum Exploration and Development |
| **Meng Xianghai** | China University of Petroleum |
| **MohammadRahman** | Texas A&M University at Qatar |
| **MuhammadGibrata** | Dragon Oil, United Arab Emirates |
| **Nie Haikuan** | Sinopec Petroleum Exploration Production Research Institute |
| **Nie Maixi** | Xi'an University of Architecture and Technology |
| **Omid Pivehzhani** | University of Malaysia |
| **Pan Jinjun** | Xingjiang Oilfield Company PetroChina |
| **Peng Chong** | Sinopec Dalian Research Institute of Petroleum and Petrochemical |
| **Roberto Aguilera** | Curtin University |
| **Shabeer Ahmed** | Oil and Gas Development Company Limited Pakistan |
| **Shuai Jian** | China University of Petroleum |
| **Sui Yongli** | China Petroleum Pipeline Bureau |
| **Sun Chongzheng** | China University of Petroleum |
| **Hang Huaiyuan** | Tianjing University |
| **Wang Jingfu** | Sinopec Guangzhou (Luoyang) ing Company |
| **Wu Qing** | CNOOC Refinery Co., Ltd. |
| **Wu Lei** | Chuanqing Drilling ing Company CNPC |
| **Xia Tongqiang** | China University of Mining and Technology |
| **Xie Chaogang** | Sinopec Petroleum Exploration and Production Research Institute |
| **Xiu Guangli** | East China University of Science and Technology |
| **Xu Re** | CNOOC Huizhou Refinery Company |
| **Y. Frank Cheng** | University of Calgary |
| **Yang Zhichao** | Sinopec Beijing Research Institute of Chemical Industry |
| **YASIR M.Mukhtar** | China University of Petroleum |
| **Zhang Jingcheng** | Sinopec Research Institute of Petroleum ing Co., Ltd. |
| **Zhang Kai** | China University of Petroleum |
| **Zhang Rongjun** | Xi'an Shiyou University |
| **Zhang Tongyi** | Sinopec Research Institute of Petroleum ing Co., Ltd. |

# Peer Review Statement

All papers published in this volume have been reviewed through processes administered by the Editors. Reviews were conducted by expert referees to the professional and scientific standards expected of a proceedings journal published by IOP Publishing.

- **Type of peer review:** Double Anonymous
- **Conference submission management system:** Morressier
- **Number of submissions received:** 86
- **Number of submissions sent for review:** 86
- **Number of submissions accepted:** 86
- **Acceptance Rate (Submissions Accepted / Submissions Received × 100):** 100
- **Average number of reviews per paper:** 3
- **Total number of reviewers involved:** 258
- **Contact person for queries:**
  **Name:** Fangyuan Cheng
  **Email:** cfy_sol@126.com
  **Affiliation:** IPPTC Organizing Committee

# Improving Hydraulic Fracturing Productivity in Khalda Oil Fields (Western Desert -Egypt)

**Ahmed Abuzeid Ahmed [1,3], and Abdel Wahab Abdel Hameed Bayoumi [2,4]**
[1]Khalda Petroleum Company, Court Street, Kafr El Dawar, Beheira, Egypt
[2] Al-Azhar University, Faculty of Engineering, Mining and Petroleum Engineering Department, Cairo, Egypt.

[3]ahmed.abozaid@khalda-eg.com, ahmedabozaidoil2010@gmail.com
[4]abdewahabbayoumi@gmail.com

**Abstract**. The hydraulic fracturing technique is used to create a highly conductive path between the reservoir and wellbore at an economic rate. There are two types of hydraulic fracturing based on the pumping schedule: the first is the conventional hydraulic fracturing which uses continuous pumping schedule and forms a proppant pack inside the fracture. The second is the channel hydraulic fracturing which uses intermittent (pulsing) pumping schedule and forms proppant pillars inside the fracture. The aim of this work is to study the factors affecting the fracture conductivity of the channel fracturing technique. Technical and economical comparison has been conducted between the above-mentioned two types of fracturing using real production data. The Fracturing Computer-Aided Design and Evaluation (FracCADE) simulator was utilized for this purpose. The results indicate that the conductivity of the channel fracturing technique is greater than that of the conventional one. The pulse time and perforated cluster length were found to have a significant effect on the fracture conductivity because of their influence on the pillar spacing. The results indicate also that: the cost of the channel fracturing is greater than that of the conventional one by about **26.28 %**. However, the NPV of the produced oil by the channel fracturing is more than the same of the conventional one. In addition, the payout time of the channel fracturing is shorter which means a faster return of investments.

## 1. Introduction

In the conventional fracturing technique, the fractures are fully filled with a uniform proppant pack. Therefore, the major applications for improving the fracture conductivity include enhancing proppant roundness and strength, lowering gel loadings, and improving gel breakers. All these treatments are aimed at enhancing fluid mobility through the porous proppant pack. The porosity of the proppant pack is related to the interconnected interstitial spaces between the proppant particulates. [1]

The channel fracturing technique delinks the fracture conductivity from the proppant pack's permeability and fractures are not fully filled with uniform proppant pack leading to reducing water and proppant consumption. This technique is accomplished by forming the inhomogeneous distribution of proppant pillars and networks of open channels within hydraulic fractures. The inhomogeneous proppant placement is achieved with a pulsing a pump schedule protocol in which the proppant-laden fluid is pumped in short pulses that are alternated with pulses of proppant-free fluid. A proppant-laden pulse will create a pillar and a proppant-free pulse will create channels that are supported by pillars of proppants created during fracturing.[2,3] To make sure that the pillars will keep channels open at reservoir conditions with fluid flow through the channels, fibrous materials are added at engineered concentrations and used to :(a) minimize the dispersion of the proppant-laden pulses as they travel throughout the surface equipment and lines, along the wellbore and within the fracture; (b) enhance the proppant carrying capacity, thus boosting the proppant to transport farther within the formation; (c) mitigate proppant settling within the fracture to maintain the heterogeneous placement of proppant and deliver better vertical coverage of the fracture;[2] and (d) help to consolidate pillar structure before fractures are closed. [4] Another way to create stable and highly conductive channels within a propped fracture is by blending proppant coated with an agglomerating agent, such as a non-hardening resin, in a gel fracture fluid instead of fibers. The resin coats the proppant with a thin, tacky, non-hardening coating that dramatically increases inter grain friction and holds the proppant aggregates together as a proppant-aggregate bed thereby reducing proppant flowback, and enhances the fracture conductivity.[5] Cluster Perforation, that means the zone isn't fully perforated,[3] is preferred to separate proppant pulse into smaller slugs, so that proppant packs are distributed greater evenly and regularly along the wellbore which improves the fracture conductivity. This way, hydrocarbons can flow through the channels separating the proppant pillars rather than flowing through the proppant pack itself as in conventional fractures. Channel Fracturing is called HiWay*Flow-Channel Fracturing by Schlumberger and conductor fracturing by Halliburton.

The fracture conductivity ($F_c$) defines the conductive path provided by the proppant material to enhance deliverability and provide economic benefit when the well is placed on production. Traditionally this is measured as the fracture permeability ($K_f$) and propped fracture width ($W_f$) as shown by the following equation[6]:

$$F_c = K_f W_f$$

It is typically reported in millidarcy-feet (MD-ft) and is a key design parameter. Many factors affect the fracture conductivity such as fracture closure stress, proppant size, sorting and shape, proppant concentration, proppant strength, proppant embedment, fines migration, and fracturing fluid residue.[7,8]. In addition to some other factors in the case of channel hydraulic fractures such as pulse time, pillars radius, pillars spacing, pillars height, confining stress, and fiber concentration[9,10,11].

## 2. Formation Description

Three wells from the concession area of Khalda Petroleum Company located in Egypt Western Desert have been investigated in this study. These are: M-01X, W-10, and W-12. The well M-01X is producing from Alam Elbueib-3A (AEB-3A) while both wells W-10 and W-12 are producing from Alam Elbueib-3C (AEB-3C).

### *2.1 AEB-3A Formation for Well M-01X*

The structure contour map for the field of the well M-01X is illustrated on (Figure1) and the petrophysical data of AEB-3A formation are presented in (Table 1).

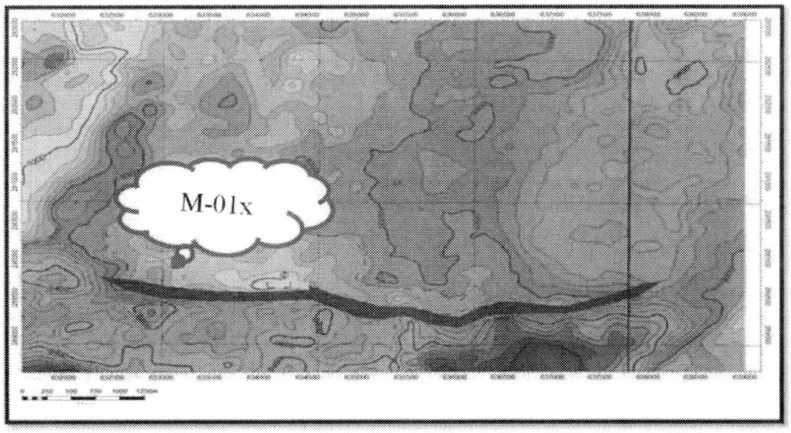


Figure 1. The structure contour map of the M-01X field

Table 1. The petrophysical data for well M-01X

| Reservoir | Subzone | NEP, ft | PHI% | Sw% | Pres, psi | Gradient Psi/ft |
|---|---|---|---|---|---|---|
| AEB-1 | | 16 | 11.3 | 36 | NA | NA |
| AEB-3A | AEB-3A1 | 8 | 13 | 30 | 4240 | 0.32 |
| | AEB-3A2 | 32 | 13 | 18 | 4270 | 0.35 |
| | AEB-3A3 | 20 | 12 | 11 | 4324 | 0.38 |
| | AEB-3A4 | 30 | 13 | 25 | 4428 | 0.27 |
| AEB-3C | Scattered sand | 33 | 11.8 | 25 | 4520 | Scattered |
| AEB-6 | | 8 | 8.7 | 26 | 5200 | 0.22 |
| Total | | 117 | | | | |

### *2.2 AEB-3C Formation for Wells W-10 and W-12*

The structure contour map for the field of wells W-10 and W-12 is illustrated on (Figure 2) and the petrophysical data of AEB-3C formation for both wells are presented in (Table 2) and (Table 3).

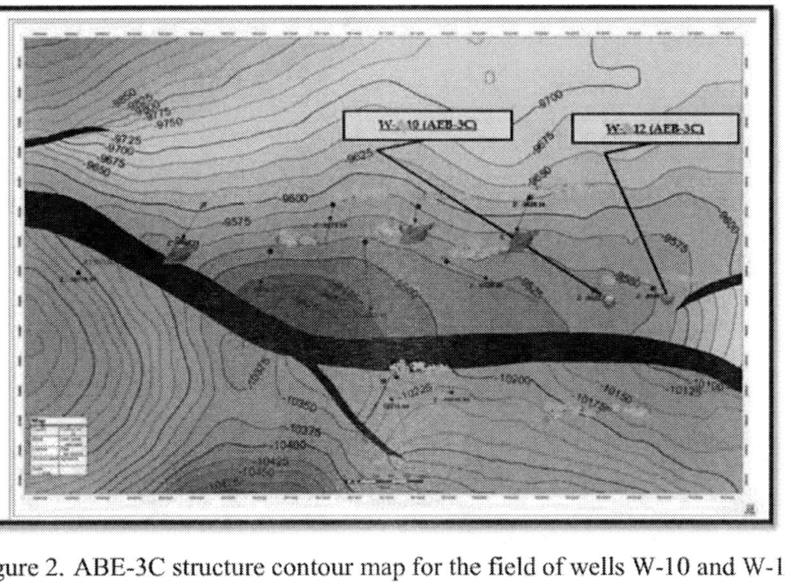


Figure 2. ABE-3C structure contour map for the field of wells W-10 and W-12

Table 2. The petrophysical data for well W-10

**Petrophysical Zone Averages Report**

**Reservoir Summary**

| Zn # | Zone Name | Top | Bottom | Gross | Net | N/G | Av Phi | Av Sw | Av Vcl Ari | Av Vsd Ari | Phi*H | PhiSo*H |
|---|---|---|---|---|---|---|---|---|---|---|---|---|
| 1 | AR -G | 5834.50 | 6078.00 | 243.50 | 17.25 | 0.071 | 0.209 | 1.000 | 0.151 | 0.538 | 3.61 | 0.00 |
| 2 | U. BAH. | 6078.00 | 6358.50 | 280.50 | 164.75 | 0.587 | 0.166 | 0.999 | 0.164 | 0.599 | 30.70 | 0.04 |
| 3 | L. BAH. | 6358.50 | 6796.50 | 438.00 | 153.25 | 0.350 | 0.123 | 0.945 | 0.160 | 0.571 | 18.80 | 1.03 |
| 4 | KHARITA | 6796.50 | 8647.00 | 1850.50 | 919.75 | 0.497 | 0.140 | 0.939 | 0.190 | 0.615 | 128.83 | 7.90 |
| 5 | DAHAB | 8647.00 | 8949.50 | 302.50 | 86.00 | 0.284 | 0.140 | 0.928 | 0.160 | 0.580 | 12.04 | 0.87 |
| 6 | ALAMEIN | 8949.50 | 8984.00 | 34.50 | 0.00 | 0.000 | --- | --- | --- | --- | --- | --- |
| 7 | AEB -1 | 8984.00 | 9334.50 | 350.50 | 67.00 | 0.191 | 0.151 | 0.888 | 0.112 | 0.665 | 10.14 | 1.14 |
| 8 | AEB -2 | 9334.50 | 9974.50 | 640.00 | 197.00 | 0.308 | 0.117 | 0.980 | 0.187 | 0.694 | 23.14 | 0.46 |
| 9 | AEB -3A | 9974.50 | 10090.50 | 116.00 | 45.50 | 0.392 | 0.116 | 0.956 | 0.064 | 0.722 | 5.29 | 0.23 |
| 10 | AEB -3C | 10090.50 | 11926.00 | 1835.50 | 262.50 | 0.143 | 0.116 | 0.604 | 0.121 | 0.683 | 30.40 | 12.04 |
| 11 | AEB -3D | 11926.00 | 12077.00 | 151.00 | 14.00 | 0.093 | 0.104 | 0.792 | 0.129 | 0.570 | 1.46 | 0.30 |
| 12 | AEB -3E | 12077.00 | 12371.00 | 294.00 | 150.00 | 0.510 | 0.102 | 0.961 | 0.060 | 0.710 | 15.28 | 0.60 |
| | **All Zones** | **5834.50** | **12371.00** | **6536.50** | **2077.00** | **0.318** | **0.135** | **0.912** | **0.160** | **0.635** | **279.68** | **24.62** |

Table 3. The petrophysical data for well W-12

**Petrophysical Zone Averages Report**

**Reservoir Summary**

| Zn # | Zone Name | Top TVDSS | Bottom TVDSS | Gross TVDSS | Net TVDSS | N/G TVDSS | Phi*H Av Phi | PhiSo*H Av Sw | Vcl*H Av Vcl | Vsd*H Av Vsd | Koil*H Av Koil | Av Kwat | Kabs*H Av Kabs |
|---|---|---|---|---|---|---|---|---|---|---|---|---|---|
| 1 | ARG | 5860.00 | 6096.00 | 236.00 | 10.50 | 0.044 | 2.57 | 0.06 | 1.23 | 5.83 | --- | | --- |
| | | 5114.77 | 5350.77 | 236.00 | 10.50 | 0.044 | 0.245 | 0.975 | 0.117 | 0.555 | --- | --- | --- |
| 2 | UBAH | 6096.00 | 6390.00 | 294.00 | 165.00 | 0.561 | 36.84 | 0.44 | 15.01 | 107.60 | 0.01 | | 64594.01 |
| | | 5350.77 | 5644.77 | 294.00 | 165.00 | 0.561 | 0.223 | 0.988 | 0.091 | 0.652 | 0.000 | 373.308 | 391.479 |
| 3 | LBAH | 6390.00 | 6804.00 | 414.00 | 52.00 | 0.126 | 8.62 | 0.05 | 6.20 | 30.03 | 0.00 | | 3037.97 |
| | | 5644.77 | 6058.77 | 414.00 | 52.00 | 0.126 | 0.166 | 0.994 | 0.119 | 0.578 | 0.000 | 57.526 | 58.422 |
| 4 | KHARITA | 6804.00 | 8630.00 | 1826.00 | 1109.75 | 0.608 | 201.53 | 5.00 | 107.04 | 715.90 | 0.00 | | 0.45 |
| | | 6058.77 | 7884.77 | 1826.00 | 1109.75 | 0.608 | 0.182 | 0.975 | 0.096 | 0.645 | 0.000 | 1.780 | 1.780 |
| 5 | DAHAB | 8630.00 | 8950.00 | 320.00 | 1.50 | 0.005 | 0.13 | 0.02 | 0.29 | 0.70 | --- | | --- |
| | | 7884.77 | 8204.77 | 320.00 | 1.50 | 0.005 | 0.086 | 0.861 | 0.193 | 0.466 | --- | --- | --- |
| 6 | ALAMEIN B | 8950.00 | 9025.00 | 75.00 | 9.00 | 0.120 | 0.62 | 0.00 | 0.19 | 1.52 | --- | | --- |
| | | 8204.77 | 8279.77 | 75.00 | 9.00 | 0.120 | 0.069 | 1.000 | 0.021 | 0.169 | --- | --- | --- |
| 7 | AEB1 | 9025.00 | 9405.00 | 380.00 | 56.50 | 0.149 | 7.63 | 0.77 | 7.33 | 32.19 | --- | | --- |
| | | 8279.77 | 8659.77 | 380.00 | 56.50 | 0.149 | 0.135 | 0.900 | 0.130 | 0.570 | --- | --- | --- |
| 8 | AEB2 | 9405.00 | 9907.00 | 502.00 | 199.00 | 0.396 | 29.74 | 1.13 | 12.03 | 110.50 | --- | | --- |
| | | 8659.77 | 9161.77 | 502.00 | 199.00 | 0.396 | 0.149 | 0.962 | 0.060 | 0.555 | --- | --- | --- |
| 9 | AEB3A | 9907.00 | 10106.00 | 199.00 | 23.25 | 0.117 | 2.89 | 0.32 | 0.73 | 16.29 | --- | | --- |
| | | 9161.77 | 9360.77 | 199.00 | 23.25 | 0.117 | 0.124 | 0.890 | 0.031 | 0.701 | --- | --- | --- |
| 10 | AEB3C | 10106.00 | 11872.00 | 1766.00 | 406.00 | 0.230 | 49.74 | 16.89 | 49.90 | 264.91 | --- | | --- |
| | | 9360.77 | 11126.77 | 1766.00 | 406.00 | 0.230 | 0.123 | 0.701 | 0.123 | 0.652 | --- | --- | --- |
| 11 | AEB3D | 11872.00 | 12011.00 | 139.00 | 22.50 | 0.162 | 2.86 | 1.41 | 2.46 | 14.06 | --- | | --- |
| | | 11126.77 | 11265.77 | 139.00 | 22.50 | 0.162 | 0.127 | 0.508 | 0.109 | 0.625 | --- | --- | --- |
| 12 | AEB 3E | 12011.00 | 12074.00 | 63.00 | 31.25 | 0.496 | 4.05 | 0.33 | 1.85 | 22.33 | --- | | --- |
| | | 11265.77 | 11328.77 | 63.00 | 31.25 | 0.496 | 0.130 | 0.918 | 0.059 | 0.715 | --- | --- | --- |
| | INJ ZONE | 10291.00 | 10351.00 | 60.00 | 20.50 | 0.342 | 2.61 | 1.18 | 1.85 | 14.40 | --- | | --- |
| | | 9545.77 | 9605.77 | 60.00 | 20.50 | 0.342 | 0.127 | 0.546 | 0.090 | 0.702 | --- | --- | --- |
| | All Zones | 5860.00 | 12074.00 | 6214.00 | 2086.25 | 0.336 | 347.23 | 24.42 | 204.27 | 1321.86 | 0.01 | | 67632.43 |
| | | 5114.77 | 11328.77 | 6213.99 | 2086.25 | 0.336 | 0.166 | 0.930 | 0.098 | 0.634 | 0.000 | 297.296 | 311.312 |

## 3. FracCADE Simulator Software

FracCADE software is used in this research. This software is useful for studying the treating fluids and proppants, pumping schedules, production forecast, and economic evaluation. It's used to estimate the effective fracture conductivity for both conventional and HiWay hydraulic fracturing techniques[12].

## 4. Results and discussion

### *4.1. Channel Fracturing of Well W-10*

Table (4) presents the general properties of well W-10 in which the channel fracturing technique has been applied. The other parameters for the same well are summarized in (Table 5). The production rate and bottom hole pressure were recorded and plotted versus time as shown on (Figure 3). The productivity index (J) was calculated to be 0.1022 BBL/D/PSI and the specific productivity index (J/h) was calculated as 0.0022 BBL/D/PSI/FT. The well was shut-in on Apr-19 to conduct work over and isolate its pay zone.

Table 4. General properties for well W-10

| Parameter | Value |
|---|---|
| Hydraulic fracture technique | Channel fracturing |
| Rock Type | Sand Stone |
| Zone name | AEB-3C |
| Pay zone depth | 10572 (ft) |
| Net pay thickness | 46 (ft) |
| Average Reservoir Pressure | 4600-4750 (psi) |

Table 5. Parameters for well W-10

| Well (W-10) Parameters | Value |
|---|---|
| Hydraulic fracture technique | Channel fracturing |
| Production starting Date | September 2015 |
| Annual Decline Rate | 60% |
| Decline Factor | 0.926 |
| Economic Limit | 5 BOPD |
| Initial oil production rate after H. F | 356 BOPD |

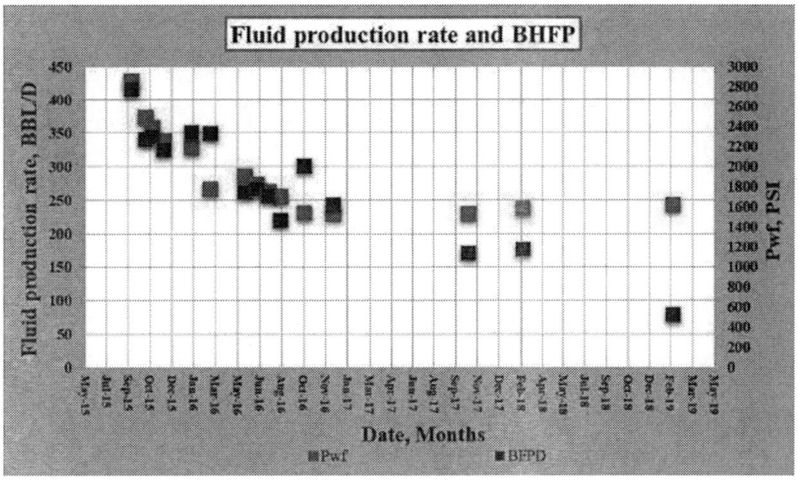


Figure 3. Fluid production rate and bottom hole flowing pressure versus time for the channel fracturing of well W-10

### *4.2. Conventional Fracturing of Well W-12*

Table (6) presents the general properties of well W-12 in which the conventional fracturing technique has been applied. The other parameters for the same well are summarized in (Table 7). The production rate and bottom hole pressure were recorded and plotted versus time as shown on (Figure 4). The productivity index (J) was calculated to be 0.0643 BBL/D/PSI and the specific productivity index (J/h) was calculated as 0.00095 BBL/D/PSI/FT. The well was shut-in on Nov-18 due to downhole pump problems, then, it was converted to a water injection well.

Table 6. General properties for well W-12

| Parameter | Value |
|---|---|
| **Hydraulic fracture technique** | **Conventional fracturing** |
| **Rock Type** | **Sand Stone** |
| **Zone name** | **AEB-3C** |
| **Pay zone depth** | **11203 (ft)** |
| **Net pay thickness** | **68 (ft)** |
| **Average Reservoir Pressure** | **4000-4300 (psi)** |

Table 7. Parameters for well W-12

| Well (W-12) Parameters | Value |
|---|---|
| **Hydraulic fracture technique** | **Conventional fracturing** |
| **Production starting Date** | **May 2015** |
| **Annual Decline Rate** | **60%** |
| **Decline Factor** | **0.926** |
| **Economic Limit** | **5 BOPD** |
| **Started oil production rate after H. F** | **243 BOPD** |

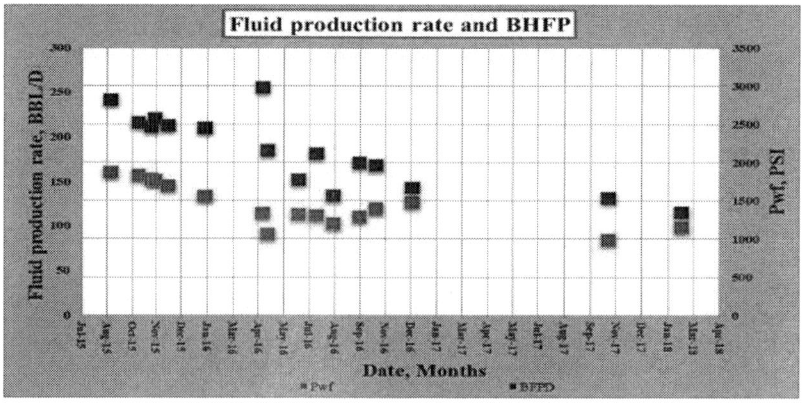


Figure 4. Fluid production rate and bottom hole flowing pressure versus time for the conventional of well W-12

### *4.3. Well M-01X*

The well M-01X was perforated through the AEB-3A4 reservoir interval (10,428-10,462 ft). After lifting and testing, the well started to produce 813 BFPD with 1 % BS&W, and 0.565 MMSCF/D gas. The memory gauges revealed a low productivity index. Therefore, it was decided to enhance reservoir productivity by hydraulic fracturing.

The simulator was run several times to obtain the fracture conductivity by the two investigated fracturing techniques. It was found that the fracture conductivity of the channel fracture [745,294 Md. Ft] is much higher than that of the conventional one [3,244 Md. Ft]. Actually, after conducting the channel fracturing, the well started to produce 1282 BFPD with 2 % BS&W, and 0.141 MMSCF/D gas.

### *4.4. Factors Affecting the Fracture Conductivity*

#### *4.4.1 Effect of Pulse Time*

The simulator was run several times to obtain the effective fracture conductivity under different pulse times and constant pump rate (20 bbl/min for every stage), same perforated cluster length (6.55 ft), and same fiber concentration (30 Ib/Mgal). The effective fracture conductivity for different pulse times is presented on (Figure 5).

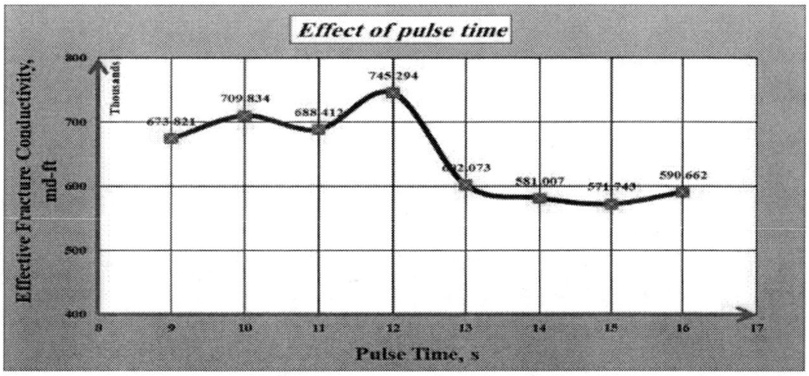


Figure 5. Pulse time effect

This Figure indicates that the maximum effective fracture conductivity (745,294 Md. Ft) is obtained at a pulse time of 12 seconds. Actually, this value is conformable with the empirical experiences and other laboratory engineering standpoint. At the beginning and during the dirty pulse stage (fluid with proppant), the increase in the pulse time results in increasing the pillar radius. This leads to a narrower horizontal pillar spacing and, thus, reduces the fracture conductivity. Near the end and during the clean stage (fluid without proppant), the increase in the pulse time results in more distribution of the proppant packs. This leads to a wider horizontal pillar spacing which tends to collapse the fracture surface and decrease the height of the pillars. Thus, reduces the fracture conductivity.

#### *4.4.2 Effect of Perforated Cluster length*

The simulator was run several times to obtain the effective fracture conductivity under different perforated cluster lengths keeping the other parameters constants. The effective fracture conductivity for different lengths of the perforated cluster is presented on (Figure 6).

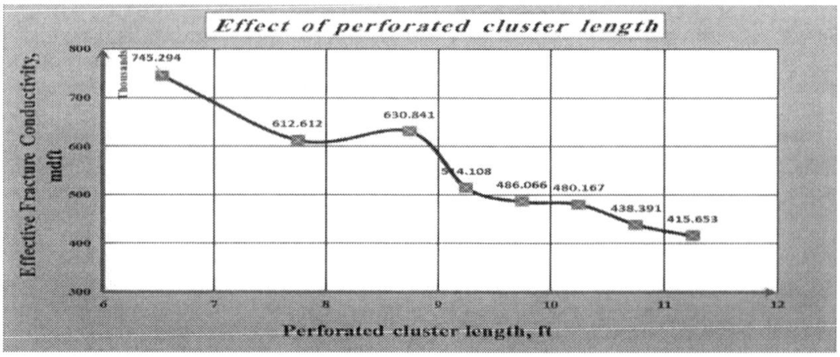


Figure 6. Perforated cluster length effect

This Figure indicates an inverse proportion between the fracture conductivity and the perforated cluster length. This is because the increase in the perforated cluster length results in increasing the number of pillars. This leads to narrow flowing channels and, thus, reduces the fracture conductivity. The maximum effective fracture conductivity (745,294 Md. Ft) is obtained at a perforated cluster length of 6.55 Ft.

#### 4.4.3 *Effect of Confining Stress*

The simulator was run several times to obtain the effective fracture conductivity under different confining stresses keeping the other parameters constants. The effective fracture conductivity for different confining stresses is presented on (Figure 7).

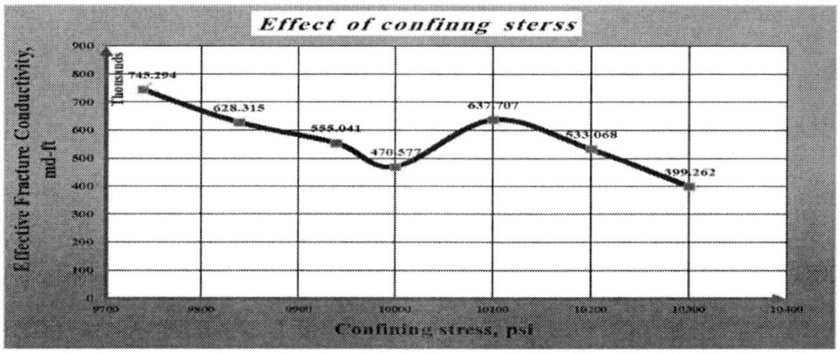


Figure 7. Confining stress effect

This Figure indicates that increasing the confining stress leads to reducing the fracture conductivity. This is because increasing the confining stresses results in decreasing the fracture width and the height of the proppant pillars. The proppant particles become more compacted and may be crushed, which impairs the permeability and reduces the fracture conductivity. The maximum effective fracture conductivity (745,294 Md. Ft) is obtained at a confining stress of 9,738 psi.

#### 4.4.4 *Effect of Fiber Concentration*

The simulator was run several times to obtain the effective fracture conductivity under different fiber concentrations keeping the other parameters constants. The effective fracture conductivity is presented on (Figure 8).

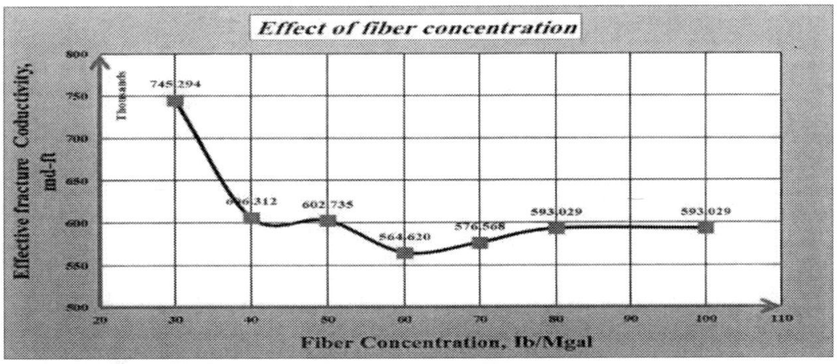


Figure 8. Fiber concentration effect

This Figure indicates that the fiber concentration has a little effect on the fracture conductivity. The fracture stimulation requires an amount of fiber to minimize the dispersion of the proppant-laden pulses and help to consolidate the pillar structure before closing the fractures. The maximum effective fracture conductivity (745,294 Md. Ft) is obtained at a fiber concentration of 30 Ib/Mgal.

*4.4.5 Effect of Proppant Concentration*

The simulator was run several times to obtain the effective fracture conductivity under different proppant concentrations, constant pump rate (20 bbl/min per stage), constant fiber concentration (30 Ib/Mgal), and same pulse time (12 sec). The effective fracture conductivity for different proppant concentrations is presented on (Figure 9).

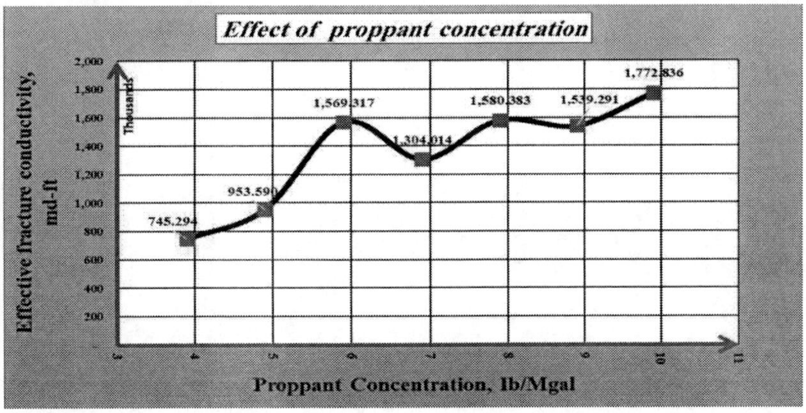


Figure 9. Proppant concentration effect

This Figure indicates that increasing the proppant concentration leads to increasing the fracture conductivity. This is because increasing the proppant concentration results in increasing the elasticity modulus of pillars. Thus, increases the fracture conductivity.

The maximum effective fracture conductivity (1,772,836 Md. Ft) is obtained at a proppant concentration of 9.9 Ib/Mgal.

*4.5. Comparison Between the Channel and Conventional Fracturing Techniques*

*4.5.1 Technical Comparison*

The simulator was run several times to obtain the effective fracture conductivity by the investigated two fracturing techniques at different conditions. The results are illustrated on Figures (10) through (12).

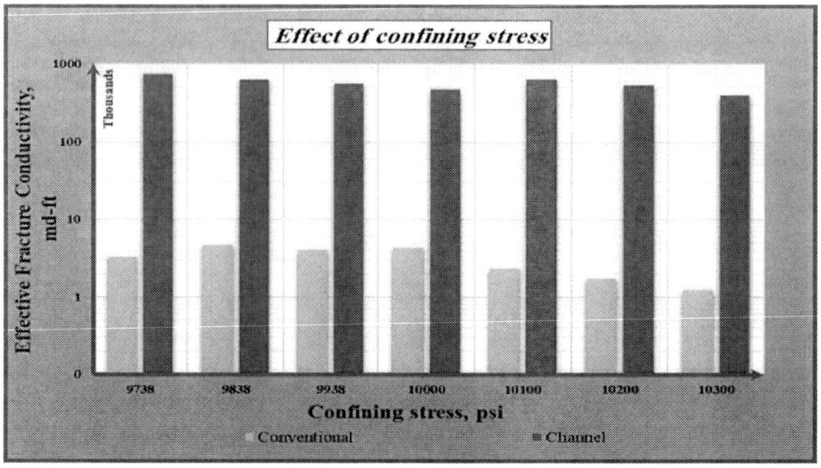


Figure 10. Comparison between the conventional and channel fracturing techniques regarding the confining stress

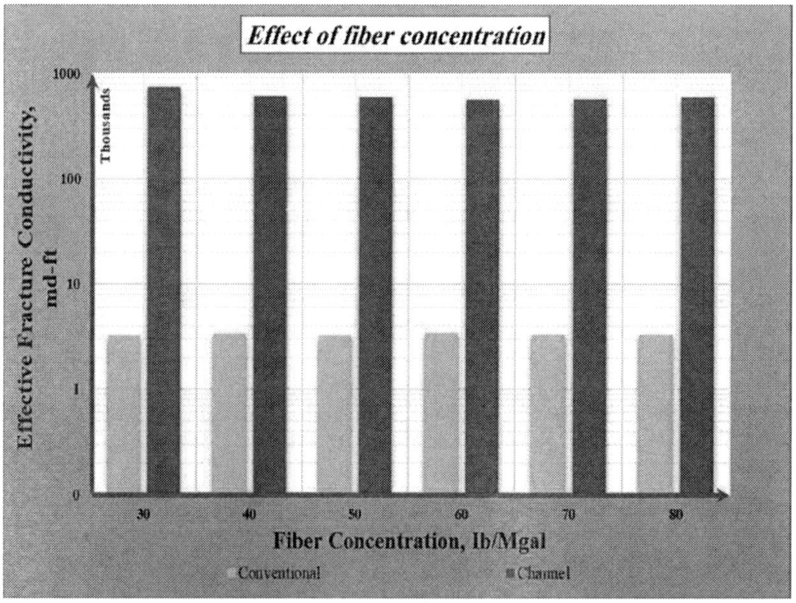


Figure 11. Comparison between the conventional and channel fracturing techniques regarding the fiber concentration

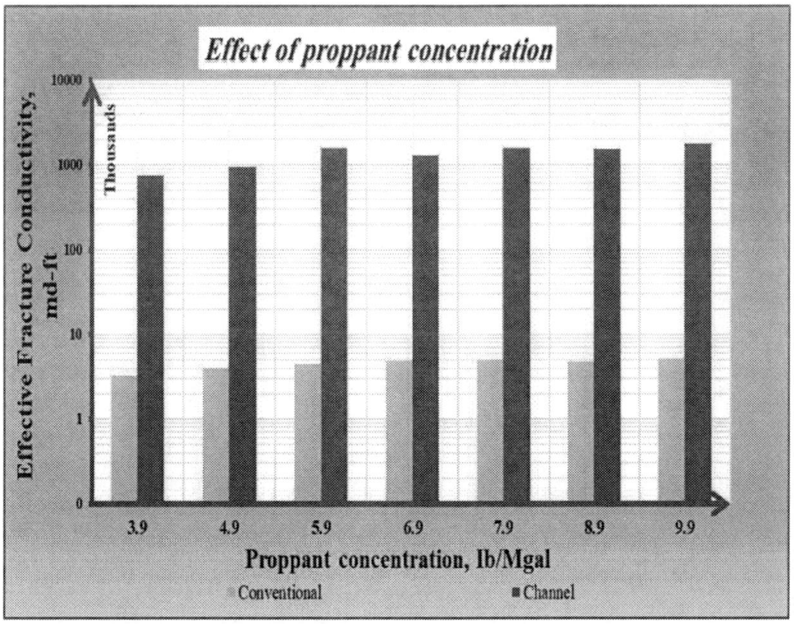


Figure 12. Comparison between the conventional and channel fracturing techniques regarding the proppant concentration

These Figures indicate that the fracture conductivity of the channel fracturing is much higher than that of the conventional one at all conditions. The logarithmic scale is used on the Y-axis because of the wide range of results.

Figure (13) illustrates the cumulative fluid production by both of the channel fracturing and conventional one. The cumulative fluid production by the channel fracturing is greater than that of the conventional one by 32%.

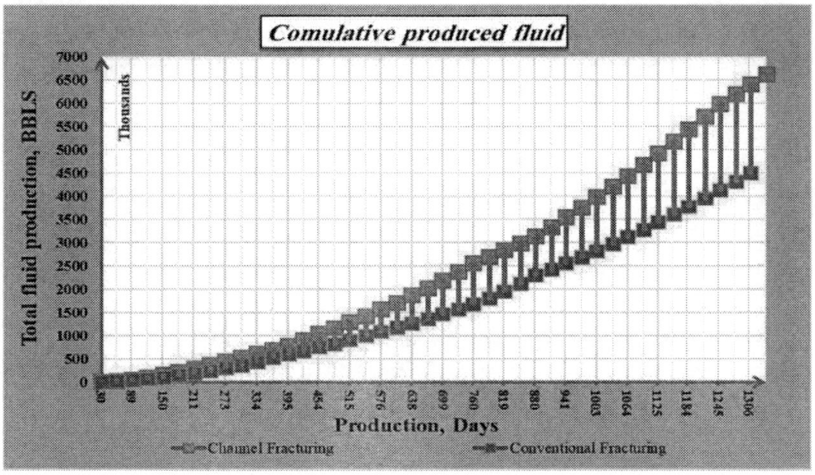


Figure 13. Cumulative fluid production by the channel fracturing technique and the conventional one

### *4.5.2 Economical Comparison*

Using the data presented in Tables (4) and (5), the net present value (NPV) of the oil produced from well W-10 by channel fracturing, has been calculated and plotted on (Figure 14). The NPV of the produced oil was found to be 5.09 $MM on 1-Dec-21 while the payout time of the channel fracturing was estimated to be **6** months.

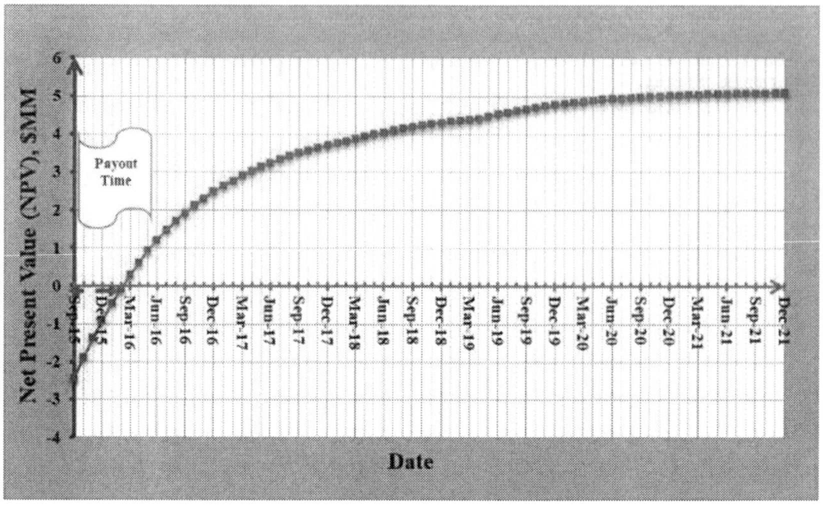


Figure 14. NPV of the produced oil and payout time by channel fracturing

The data presented in Tables (6) and (7), was used to calculate the net present value (NPV) of the oil produced from well W-12 by conventional fracturing. The results are plotted on (Figure 15). The NPV of the produced oil was found to be 2.93 $MM on 1-Aug-21 while the payout time is **10** months.

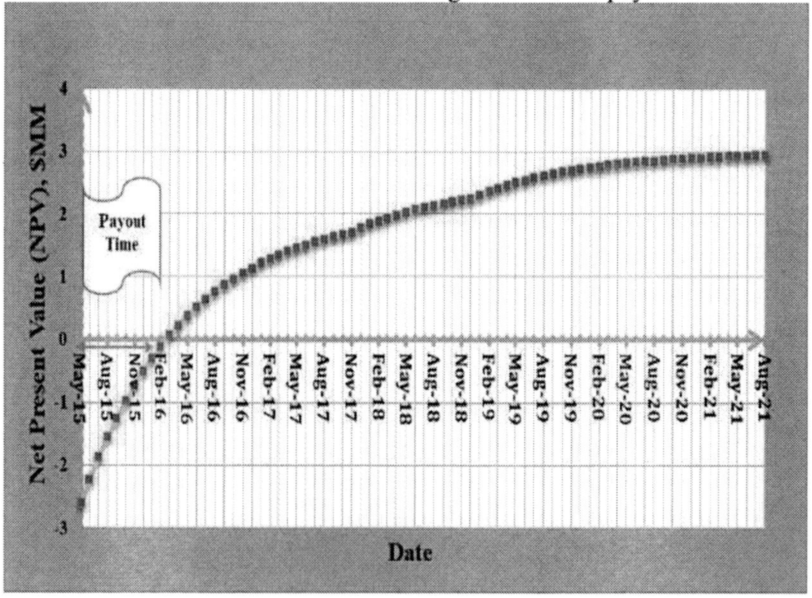


Figure 15. NPV of the produced oil and payout time by conventional one

Figure (16) illustrates the economical comparison between the channel and conventional fracturing techniques. The figure indicates that the NPV of the produced oil in the case of the channel fracturing is greater than that of the conventional one. The payout time is shorter in case of the channel fracturing.

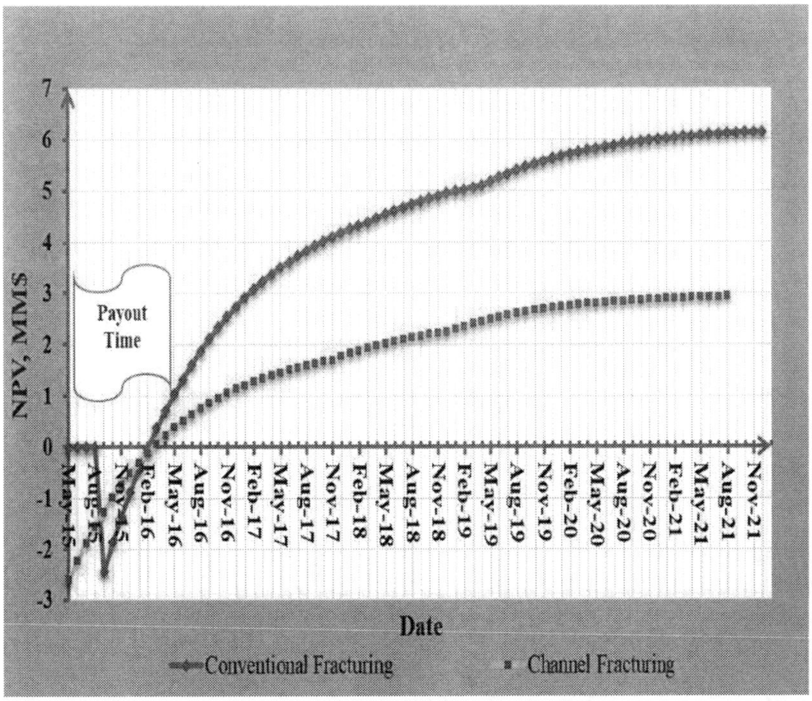


Figure 16. NPV of the produced oil and Payout time by channel and conventional techniques

## 5. Conclusions

The following conclusions may be drawn from this study:

1) The channel fracturing has a greater effect on the fracture conductivity than the conventional one. The difference is about 742,050 Md. Ft.

2) The productivity index after the channel fracturing is greater by 37.08 % while the specific productivity index is increased by 56.82 %. Besides, the cumulative fluid production by channel fracturing has been found to be greater by 32 %.

3) The pulse time, the perforated cluster length, and the confining stress have a negative effect on the fracture conductivity.

4) The fiber concentration has a little effect on the fracture conductivity.

5) The proppant concentration has a considerable effect on the fracture conductivity.

6) The cost of the channel fracturing is greater than that of the conventional one by about 26.28%; However, the NPV of the produced oil by the channel fracturing is more than the same of the conventional one. In addition, the payout time of the channel fracturing is shorter which means a faster return of investments.

## 6. Acknowledgments

I would like to express my thanks and praise to Allah and then my sincerest gratitude to my supervisor, Prof. Dr. Abdel Wahab Bayoumi. My sincere gratitude to the Petroleum Engineering staffs (Al-Azhar University). I would like to thank my parents for all the unwavering love. Finally, I am grateful to my wife and sons, Yousef, and Mohammed for all their support, patience, advice, and encouragement throughout this study.

## 7. Abbreviations

| | |
|---|---|
| AEB-3A | Alam Elbueib-3A |
| AEB-3C | Alam Elbueib-3C |

| | |
|---|---|
| BFPD | Barrel Fluid Per Day |
| BOPD | Barrel Oil Per Day |
| BBL/D/PSI | Unit of productivity index, Barrel Per Day per Pound per Inch Square |
| FracCADE | Fracturing Computer-Aided Design and Evaluation |
| Fc | Fracture Conductivity, Md. Ft. |
| HiWay | Name of simulator module that is used to generate the HiWAY pumping schedule |
| H. F | Hydraulic Fracture |
| Ib/Mgal | Unit of fiber concentration, pound per thousand gallons of fracturing fluid |
| J | Productivity index, BBL/D/PSI |
| J/h | Specific productivity index that means productivity index divided by interval length, BBL/D/PSI/Ft |
| Kf | Fracture Permeability, Md |
| md-ft | millidarcy foot is a unit of fracture conductivity |
| MMSCF/D | Million Standard Cubic Feet Per Day of gas |
| PPG | Unit for proppant concentration that means pounds of proppant added to each gallon of fracturing fluid |
| Sec | Unit of time, Second |
| Wf | Fracture width, In |

## 8. References

[1] Nguyen, Philip D., and Max L. Philippi. "Methods for forming proppant-free channels in proppant packs in subterranean formation fractures." U.S. Patent No. 9,657,560. 23 May 2017, pp. 1-7.
[2] Yan, Xia, et al. "Theoretical analysis of fracture conductivity created by the channel fracturing technique." Journal of Natural Gas Science and Engineering 31 (2016): 320-330.
[3] Barasia, A., and P. Pankaj. "Tail-in proppant and its importance in channel fracturing technique." SPE Bergen One Day Seminar. Society of Petroleum Engineers, 2014, pp. 1-12.
[4] Gillard, Matthew Robert, et al. "A new approach to generate fracture conductivity." SPE Annual Technical Conference and Exhibition. Society of Petroleum Engineers, 2010, pp. 1-13.
[5] Inyang, U. A., P. D. Nguyen, and J. Cortez. "Development and field applications of highly conductive proppant-free channel fracturing method." SPE Unconventional Resources Conference. SPE-168996-MS, Texas, USA, April 2014, PP. 1-9.
[6] Economides, Michael J., and Tony Martin. Modern fracturing: Enhancing natural gas production. Houston: ET Publishing, 2007, pp. 242-268.
[7] Gidley, John L. "Recent advances in hydraulic fracturing." (1989), pp. 116-122.
[8] SEPTAR stimulation team of the Shell Group Company. Stimulation Field Guidelines part II (Revision): Hydraulic Fracturing. RIJSWIJK: The Global EP Library Publishing, (2000),pp. 33-44.
[9] Qian, Bin, et al. "Research and practice of the impulse sand fracturing technology." Natural Gas Industry B 2.4 (2015): 334-340.
[10] Hou, Bing, et al. "Parameter simulation and optimization in channel fracturing." Journal of Natural Gas Science and Engineering 35 (2016): 122-130.
[11] Wang, Jiehao, and Derek Elsworth. "Role of proppant distribution on the evolution of hydraulic fracture conductivity." Journal of Petroleum Science and Engineering 166 (2018): 249-262.
[12] Team of Schlumberger Company. "FracCADE* fracturing design and evaluation software",2010, https://www.slb.com,(accessed 02 March 2021), pp. 1-3.

Journal of Physics: Conference Series **2594** (2023) 012002 doi:10.1088/1742-6596/2594/1/012002

# Assessment of the probability of success and risking in hydrocarbon exploration: A case study from Zindapir Anticline, Sulaiman Foldbelt, Indus Basin, Pakistan

**S. Ahmed[1], A. Nazeer[2], W Habib[1], S H Solnagi[3] and S Hameed[4]**

1 Oil and Gas Development Company Limited, Islamabad, Pakistan
shabeer.ahmed@ogdcl.com

1 Oil and Gas Development Company Limited, Islamabad, Pakistan
whabib@ogdcl.com

2 Pakistan Petroleum Limited, Islamabad, Pakistan
a_nazeer@ppl.com.pk

3 Center for Pure and Applied Geology, University of Sindh, Jamshoro, Pakistan
sarfrazsolangi60@gmail.com

4 IPR Transoil Corporation, Islamabad, Pakistan
shameed@gmail.com

**Abstract:** Role of the statistical calculation is very important in petroleum exploration business. It is used to calculate the associated risk and determines the chances of geological success prior to drilling of an exploratory well in mapped prospect area. This manuscript is focusing on the geological modelling of Zindapir Anticline, Sulaiman Fold Belt, Indus basin, Pakistan using mainly based on the concept of probability developed by Coordinating Committee for Coastal and Offshore Geoscience Program (CCOP) in East and Southeast Asia in 2000. Study area comprises of Zindapir Anticline with two abandoned wells Well X-1 and Zindapir-1 which were drilled and declared as dry holes. Detailed geological modelling indicates that Zindapir structure is still virgin to flow hydrocarbons, and there are ample chances of compartments, which were developed as result of transpressional movement at the periphery of the structure. Interpreted seismic line also supports the presence of different compartments in structure. These compartments may host hydrocarbon, generated and migrated either from Bharti Syncline and Sulaiman Foredeep or generated by indigenous source rocks possibly deposited in the south of Zindapir-1 and Well X-1. Presence of trap, hydrocarbon charge and reservoir in Zindapir structure indicates there is better probability of success for hydrocarbons, and good chances for the development of traps because of transpressional movement. Traps may contain the hydrocarbon, possibly generated from Bharti Syncline in west and Sulaiman Foredeep in east whereas southern part of Zindapir Anticline was not considered, and northern part of Zindapir Anticline was mapped with very open grid size. Present seismic data is limited which is the major reason that due to poor probability of effective trap it is recommended to acquire additional seismic data for further detailed picture structure of Zindapir Anticline. An equation for the calculation of geological chances success for new development well is also proposed in this manuscript.

## 1.Introduction

Schrödinger[1], a noble prized physicist published interesting scientific investigation between relationships of two or more than two separate associated physical systems with outcome of common representative. This common representative was called as match and geological event or factors in this manuscript. The scientific approach of statistics in Geology was initiated in the middle 50 by Baecher G.B.[2]. The period of growth of quantitative methods in geology occurred after 1950 when it was discovered by Rogerson. A[3] that both statistical and mathematical techniques are applicable to the huge quantity of geological numerical data. Coordinating Committee for Coastal and Offshore Program (CCOP) in East and Southeast Asia[4] considered the role of dependency and correlation during risking of prospects, sharing two or more than two common geological factors. Risk performance analysis in probabilistic and statistical terms in petroleum exploration estimation of risk assessment as main factor prior to the drilling of prospect Harbaugh, J.W.[5] and Salleh S.H. [6]. According to Bickel J.E[7] prospect in common basin is likely to share geological features and required optimal approach of model to understand the individual risk factor by doing the postmortem of failed well and by determining the optimal drilling strategy. Lerche I.[8] gave a review of economic risking methods commonly used in hydrocarbon exploration and described how numerical illustrations are provided to show how the Kelly method and the corporate confidence procedure operate for both single and multiple projects.

This manuscript is focused on probability-based model for hydrocarbon exploration in Zindapir structure. Two wells were drilled in Zindapir structure and declared as dry wells. Nazeer,[9] and Shazia A.[10] discussed the presence of effective petroleum system in Zindapir Structure and surrounding areas. In this manuscript, optimal approach of model will be used to delineate risk associated with individual geological factors for successful discovery of hydrocarbon in Zindapir Anticline as a part of case study. Jadoon I.A.K. and Zaib M. O.[11] worked on surface structural map of the area which is very useful to integrate extension of surface features with in the subsurface. It is important to describe geological factors for calculating effective petroleum prospect. Morosov A.L.and Bratvold R. B.[12] worked on the probability elicitation using geo-statistics in hydrocarbon exploration and proposed a novel of approach of geo-statistics to help experts elicit geological probabilities more objectively, especially useful during the exploratory phase. Abbasi S.A.[13] and Solangi S.H.[14] worked on the interpretation of seismic data and confirmed the presence of synclines and faulted anticlines in the subsurface of Sulaiman Fold Belt area Milkov A.V.[15] worked on Reporting the expected exploration outcome, when, why and how the probability of geological success and success-case volumes for the well differ from those for the prospect and concluded that and concluded that his contribution is very useful for hydrocarbon explorers to better distinguish geological PoS and success-case volumes for the prospect from those for the exploration well. This will result in clearer expectations of exploration outcomes, better drilling decisions, less disappointment with drilling outcomes, and more accurate postmortem evaluations. LaCosta William C.P. and Milkov A. V.[16] published their work on Petroleum exploration portfolios generated with different optimization approaches: Lessons for decision-makers has concluded that he formal portfolio optimization approaches to drill risky prospects to de-risk plays.

## 2. Types of Geological Events

There are three types of geological events used for probabilistic approach of assessment and discussed as below.

**2.1. Independent Events:** are those events which could not affected by other events. e.g., trapping mechanism, source rock maturity, reservoir effectiveness are independent events in effective petroleum system.

**2.2. Mutually Exclusive Events:** are defined as those events which do not occur or happen at same time. Petroleum well is either drilled with successful discovery or abandoned as dry well.

**2.3. Dependent Events:** are defined as those events which are affected by other events. Later phase of uplifting may breach the structure.

## 3. Data Control for Assessment of Probability of Success and Risking

Risking analysis in hydrocarbon exploration is dependent upon the type of geological data from prospect to basin, validity of data, available information, and level of confidence. Geologist uses three approaches to determine the probability of geological factors.

**3.1. Stochastic Approach of Probability:** It is ratio of number of events occurs and the total number of trials. Drilling of multiple wells has proved the presence of regional shale in Upper Goru (Cretaceous) reduced the risk of seal in the Lower and Punjab Platform and Sind Monocline of Indus Basin.

**3.2. Objective Approach of Probability:** This approach of probability is based on valid arguments and hypothesis. Source of data is historical and empirical and supported by analogues. Historical data clearly shows that valid structural trap in Cretaceous sands of Goru result in successful discovery in Middle and Lower Indus Basin.

**3.3. Subjective Approach of Probability:** Third approach is based on the sum of individual understanding or belief about the happening of events. Sometimes related to management strategy and sometimes imposed due to personal beliefs of Top management. Sometime based on theoretical model supported by analogues basins. In Pakistan, wells were drilled in the Punjab Platform, Central Indus Basin to support the theory of Infra Cambrian source rock after discovery of heavy oil in Rajasthan, India. If data is not supportive, third approach is not good idea for explored basin. Such approach is useful for virgin and frontier areas, but factor of risk is always high.

## 4. Probability Rules of Geological Event

Geological modelling of risking always initiated with considerable amount of uncertainty which could not be reduced without presence of valid geological data, valid arguments; scientific investigation about various geological factors which defines the geological chances of success by eliminating amount of risk. Authors observed that various approach were adopted by Petroleum explorationists (Pakistan). Most of techniques were dependent upon the corporate culture of company; sometimes imposed by personal belief of top hierarchy.

Coordinating Committee for Geoscience Programs in East and Southeast Asia (CCOP) was formed by in 1966 were the Republic of China, Japan, the Republic of Korea and the Philippines as; and currently working as s an intergovernmental organization with group of 15 countries including Japan and Indonesia. The Coordinating Committee for Geoscience Programs in East and Southeast Asia (CCOP) set the standards for probabilistic assessment of prospect and defined certain rules and definitions. The probability scale ranges from 0.0 to 1.0 where the end points of the scale; P = 1.0 means 100% certainty and P = 0.0 means 0% certainty. CCOP [4] considered four fundamental rules of probability which are used during prospect risk assessment and necessary for executive decision prior to drill. These rules are discussed below in separate headings. Probability of any geological event "P" is equal to 1 and (1- P) for risking.

**Geological Risk = 1- Geological Chances of Success**

Seismic quality evaluates the geometrical body. Uncertainty will arise if resolution of these surfaces on seismic profiles is poor. CCOP (2007) considered the seismic data as important parameter for evaluation of probability of existence of the mapped trap as a geometrical body as effective trap. Trap could not be mapped without seismic data. Absence of seismic data shares the risk more than 80%. Quality of seismic interpretation depends upon (a) seismic data quality, (b) seismic coverage (c) coverage with grid size (d) seismic interpretation /depth conversion and (e) identification of top and bottom. Seismic interpretation requires wells logs to calculated velocity control which is further used for depth identification. Absence of nearby wells and wells logs always create ambiguity and final interpretation is not absolute. Probability is not more than 0.9 if 3D seismic or closed grid 2D data is not present. Flow chart showing in **Figure 1** probability of success in drillable prospect.

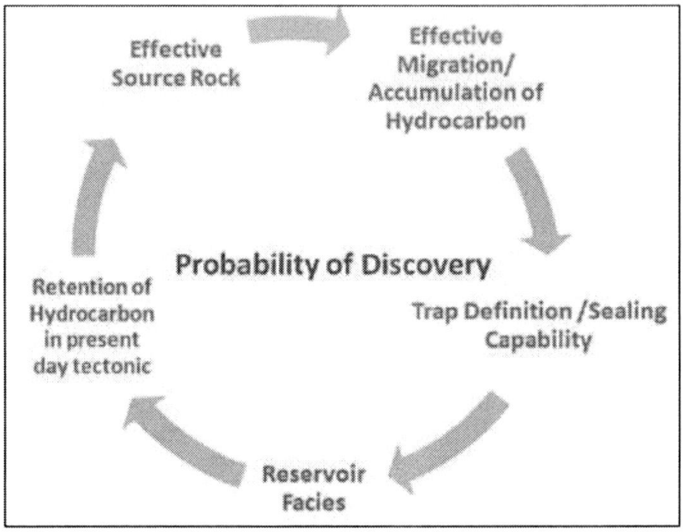


**Figure 1.** Flow chart showing probability of success in drillable prospect

Journal of Physics: Conference Series **2594** (2023) 012002 doi:10.1088/1742-6596/2594/1/012002

| Probability | Data Control | TOC | Maturation | Migration Pathway | Data quality: Trap delineation / Sealing Mecahnism: 3D | 2D grid size: Dense | 2D grid size: Open | 2D grid size: Very Open | Geological Model | Analogues | Basin Classifiaction | Prospect |
|---|---|---|---|---|---|---|---|---|---|---|---|---|
| 1 | Data control and quality is excellent | Good to very good | Matured | Proved Fault Conduits / Migration pathways | Nearby wells | | | | Absolute | Identical to discovered field | Proved basins | Drillable Prospect |
| 0.9 | | | | | | Nearby wells | | | Very likely to Absolute | | | |
| 0.8 | Data control and quality is good | Fair to good | | | Distant wells | | Nearby wells | | Very likely | Similar geological condtion tested by wells in similar petroleum system. Lateral continuity is convincing | Frontier Basin-Proved basin | |
| 0.7 | | | | | | | | | More likely | | | Workable prospect |
| 0.6 | Data control and quality is good | | | Carrier beds / Fault connectivity is poor/Long distance migration | Uncertain | Distant wells | Distant wells | Nearby wells | Likely | Geological factors are known. Limited seismic and well control. Lateral continuity is probable | | |
| 0.5 | | | | | | | | | Likely with uncertainities | | | |
| 0.4 | Data control and quality is good | Poor to fair | Unmatured | | | | | Distant wells | Unfavourable | Valid concepts but data is unconvincing on available data | Initial evaluation | |
| 0.3 | | | | | | Uncertain | Uncertain | | Questionable | | | |
| 0.2 | Geological data does not supports | | | | Absence of petroleum system | | | | Very questionable | Unlikely to questionable | | |
| 0.1 | | | | | | | | | Highly questionable | | | |

**Table 1.** The table is showing scale of probability which defined relationship between probability of geological chances of success, proposed geological model/mapping and quality of data present for evaluation purpose. Red color shows high risk, yellow color shows moderate risk good for initial screening of exploration prospect in frontier basin and proved basin. Green color shows low risk drillable prospect. (Modified after CCOP)

**Geological Chances of Success (GCS)** is probabilistic approach of scientific evaluation of drillable prospect with defined petroleum plays, defined petroleum system on the basis of geological modelling and regional knowledge which delineate the source rock facies, kitchen area, time and mechanism of charging, expulsion and accumulation of hydrocarbon in reservoir facies along vertical and lateral seal to retain the hydrocarbons in trap. Geological chances of success calculate the probability of risk associated with drilling and its chances of success which could not be described, nor understood without knowing the individual geological event and how individual geological events are evolved from geological data. Data control and type of available data is very important in making decisions. **Table-1** is showing scale of probability with seismic mapping as important parameters which controls the quality of data.

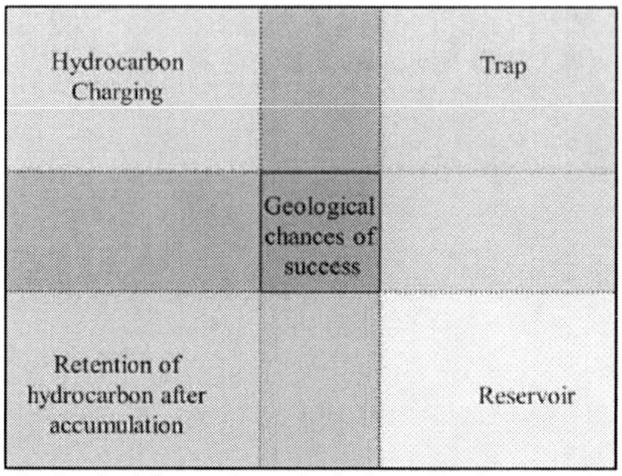


**Figure 2**. Geological factor for calculating effective petroleum system

**4.1. Multiplication Rule of Probability:** This theorem applies when probability of two events A and B or more than two events occur simultaneously and equal to product of individual events. Assessment of petroleum system is good example because all the geological factor (reservoir, trap, hydrocarbon charging and retention after accumulation) occurs simultaneously to calculate the Geological Chances Success (GCS). Fig. 2 is showing illustration of geological factor for calculating effective petroleum system as Geological Chances of success.

$$\mathbf{P_G = (P_{Reservoir})\ x(P_{Trap})\ x\ (P_{HC\ Charge})\ x\ (P_{Retention})}$$

Whereas,
HC = Hydrocarbon,
PG = Geological chances of success
PReservoir = Reservoir probability
PTrap = Trap probability
PHC Charging=Hydrocarbons charge probability
PRetention= Accumulation of retention probability

**4.2. Addition Rule of Probability:** When two events, A and B, are two geological events, the probability that A or B will occur is the sum of the probability of each event minus the probability of both will happen. If P is probability.

P(A) = Probability of Event A
P(B) =Probability of Event B
P(A ∪ B) = Probability that either A or B happens
P(A ∩ B) =Probability of A and B happening together
Then

$$\mathbf{P(A \cap B) = P(A) + P(B) - P(A \cup B)}$$

Let us consider a prospect area with the presence of Kerogen Type I and Kerogen Type II source rock in identified kitchen area; located as synclines which are separated from limbs of anticline by fault. Source rock in both syncline and indigenous source rock produces hydrocarbons. Source rock with shallower burial depth may produces oil; and similar source rock may produce gas due to elevated temperature at greater burial depth. Long route migration along faults could not be ignored and fault conduits act as migration pathway. Therefore, there is probability of producing gas condensate from kitchen area. Whereas production of oil with some gas is also possible at elevated temperature because both oil or gas condensate are two independent exclusive geological outcomes of similar source rock of different maturity levels at various depths. Sometimes such geological events may result into charging of multiple source rocks with different APIs of oil or gas condensate with different composition of inerts. Probability of oil or gas condensate depends on the evaluation of source rock in associated kitchen area. Is shown in Figure 3.

**1)For Oil Case**

(oil and gas condensate are exclusive events)

$P = P_{\text{oil}} + P_{\text{Gas Condensate}} - P_{\text{(Oil + Gas Condensate)}}$

$P = 1/2+0 - 0 = 0.5$

**2)For Gas Condensate Case**

(oil and gas condensate are exclusive events)

$P = P_{\text{oil}} + P_{\text{Gas}} - P_{\text{(Oil + Gas Condensate)}}$

$P = 1/2+ 0 - 0 = 0.5$

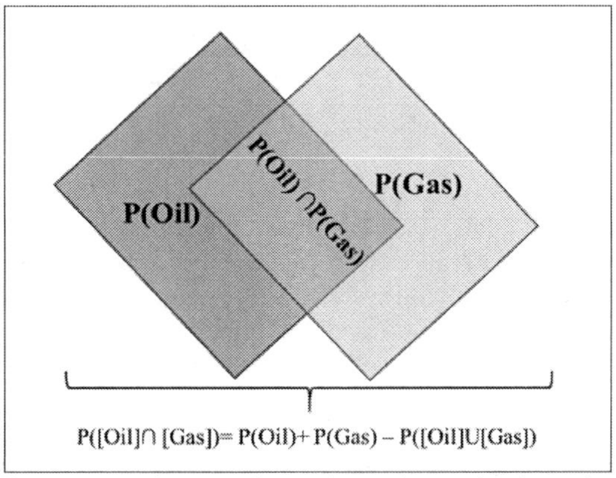


**Figure 3.** Probability of oil or gas condensate depends on the evaluation of source rock in kitchen area.

**4.2. Combination Rule of Probability:** This rule is applicable for estimation of risking associated with discovery of hydrocarbons in the two or more than two reservoirs. According to rule, all reservoirs may act as independent geological units. Therefore, the probability of occurring one or both two independent geological events (A, B) can be estimated by following mathematical relation.

$$\mathbf{P = (1- [1-P(A)] \times [1-P(B)])}$$

Zindapir Anticlinorium consist of four compartments: Rodho, Afiband, Dhodak, and Zindapir Anticline from north to south. Gas condensates were discovered from Lower Goru Sands (Cretaceous) and Chiltan Carbonate in three compartments. Probability of presence of hydrocarbons in both reservoirs is high because both reservoirs are producing hydrocarbon in other compartment of individual anticlinorium.

$$\mathbf{P= \{[1- [(1- P_{Lower\ Goru}) - (1- P_{Chiltan})]\}}$$

$$\mathbf{P = \{1-[(1-0.5)- (1-0.5)]\}=1}$$

Whereas

$P_{\text{Lower Goru}}$ = Charging of Lower Goru (Cretaceous)

$P_{\text{Chiltan}}$ = Charging of Chiltan Limestone

**4.3. Conditional Rule of Probability:** Conditional probability of event B will occur with the occurrence of event A. If event B will happen with happening of Event A then Event A is independent of B, then the probability conditional probability P (A/B) is **P(A|B) = P(A ∩ B) /P (B).** Sakesar Carbonates (Eocene) is known reservoir of Potwar Sub basin. Sakesar Carbonates (Eocene) will produce only if fracture network is present (Figure 4).

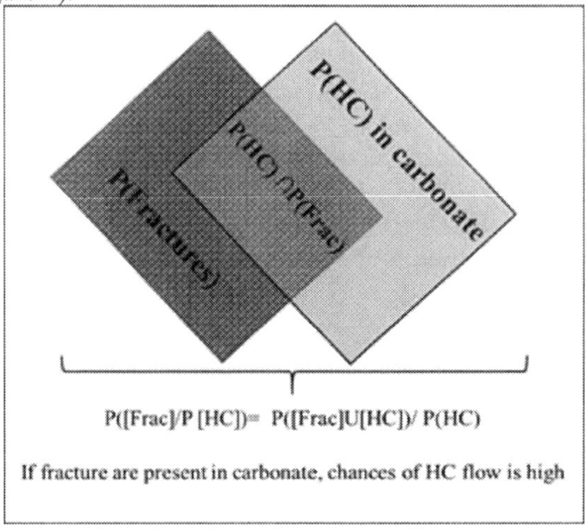


**Figure 4.** Event of hydrocarbon flows in Sakesar Limestone (Eocene) occurs when fracture network is present.

## 5.Important Geological Information

**5.1. Basin** is an area, in which sediments accumulate. Sedimentary basins are of different range in size from hundreds of meters to large parts of ocean basins. Basin may have one or more than one Petroleum play associated with petroleum system. Petroleum system deals with trap formation and process of generation of hydrocarbon from source rock., migration and accumulation of hydrocarbons in reservoir rock.

**5.2. Petroleum System** could not be defined without defining the hydrocarbon kitchen area and Petroleum System. Areas underlain by thermally mature generative source rocks in a sedimentary basin are called hydrocarbon kitchens (CCOP, 2000).

**5.3. Petroleum Play** is defined as an aerial extension of mutually related geological factors (Trilogy of source rock, presence of reservoir facies with sealing mechanism) which occurs in single petroleum system. Petroleum play is used to define the prospect with commercial and undiscovered hydrocarbons.

**5.4. Prospect** is geographical location or small mapped area with defined boundaries where the existence of oil and/or gas has not been proved, but geological studies shows that economically recoverable hydrocarbons are present (after CCOP, 2000). **Figure 5** is schematic illustration of the relation between basin, petroleum system, play and prospect. Set of basic elements required to calculate geological chances of success for drillable prospect.

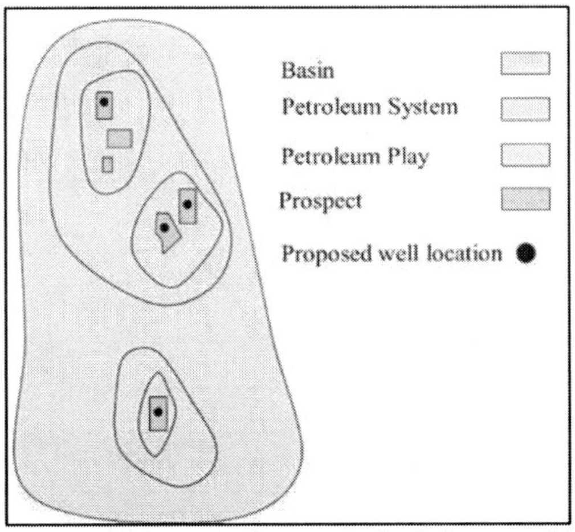


**Figure 5.** Set of basic elements required to calculate geological chances of success for drillable prospect

## 6. Tectonics, Structure and Exploration History of Area

The Zindapir Anticlinorium is located in the eastern part of Sulaiman Fold Belt (SFB), Pakistan **(Figure 6).** The older rocks have been drilled in several wells of the adjoining region of Punjab Platform and extend in the west towards Sulaiman Lobe [17] indicating the presence of Infracambrian-Permian rocks in the sub-surface of Sulaiman Fold Belt.

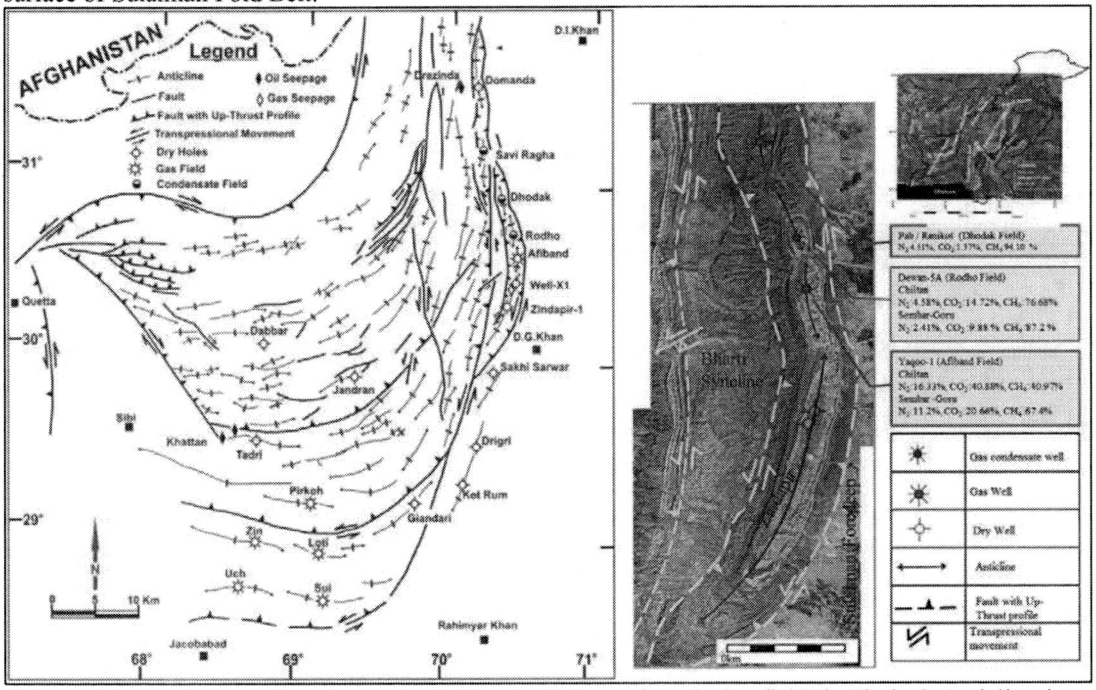


**Figure 6.** Tectonic map of study area including dry holes and producing fields in Zindapir Anticlinorium (modify after Nazeer A. et al., 2012, 2013 and 2018 and Nazeer A. Ph.D. Thesis 2019).

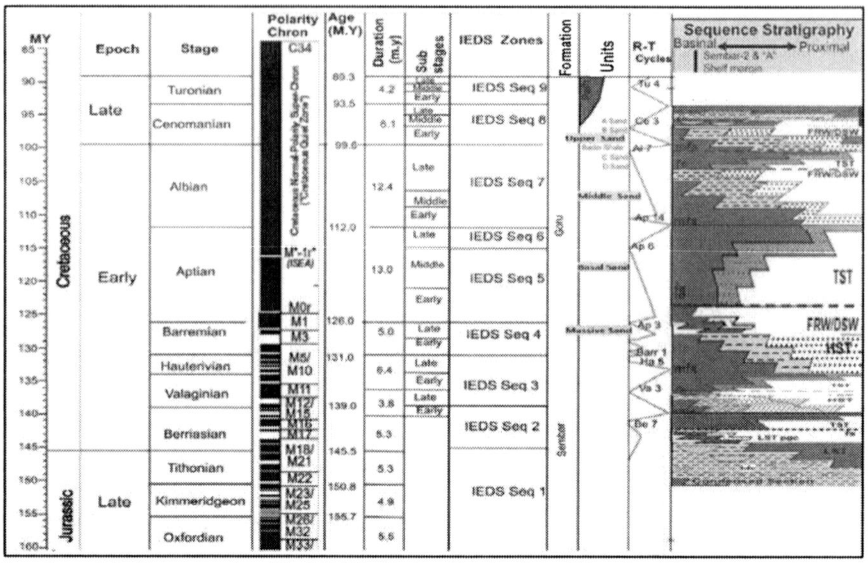


**Figure 7.** Stratigraphic framework of in Middle Indus Basin (Modify after Nazeer A., Ph.D. Thesis 2019)

The study area is located in the southern part of the Zindapir Anticlinorium, which is composed of the Rodho, Dhodak, Afiband and Zindapir Anticline. The Anticlinorium is bounded to the east by the Sulaiman depression and to the west by the Bharti Syncline and covers an area of approximately 6000 sq. km. Dhodak-1 (1976) was first commercial discovery in Dhodak structure of Zindapir Anticlinorium, which is in the frontal part of the Eastern Sulaiman Fold Belt (SFB) with gas condensate discovery in the Late Cretaceous and Paleocene reservoirs. Zindapir-1 (1986) was second well which was drilled by Oil and Gas Development Company Limited (OGDCL) to a TD of 4406 m in Alozai Formation (Triassic) of Zindapir structure with hydrocarbon shows. Afiband-1 (1986) was drilled by OGDCL in 1986 without successful discovery in Afiband structure. Dewan-1 & Dewan 5-A (2005 & 2007) encountered hydrocarbons in Goru and Chiltan formations and prove the existence of active petroleum system in the deeper horizon of Middle Jurassic–Early Cretaceous rocks. Dewan-1 (2005), Dewan 5-A (2007) were drilled in Rodho structure. Yaqoot -1(2009) was drilled by Dewan Petroleum Limited with commercial discovery of gas condensate in Afiband structure.

Zindapir Anticlinorium is associated with transpressional tectonic regime. Landsat image **(Figure 6)** is showing the style of structuration in Sulaiman Fold Belt (SFB) and their role in trap mechanism. Gas-condensate were discovered in multiple reservoirs consist of Jurassic to Paleocene siliciclastic and carbonate reservoirs; and present in the north of Zindapir Anticline as part of major anticlinorium [9].

**7.Sembar-Goru Petroleum system**

The Sembar-Goru sequence is present Indus Basin, Pakistan which extends from Kohat-Potwar basin in north to Arabian Sea in the south and form Sembar Goru Petroleum System (Upper Jurassic to Mid Cretaceous) in Middle Indus Basin and Lower Indus Basin. Sembar-Goru sequences (Late Jurassic to Late Cretaceous age) are fluvio-deltaic facies with varying lithology of siliciclastic dependent on its depositional position in the basin from shelf to slope. The Lower Goru sands are proven reservoir sands and Sembar-Goru shales are considered as principal petroleum source in Lower and Middle Basin [15]. Interbedded and overlying shales of the Sembar / Lower Goru formations are providing sealing mechanism.

The Sembar-Goru sequences are present in the subsurface of Indus Basin extending from Kohat-Potwar basin in north to Arabian Sea in the south of Pakistan. On eastern side, the Sembar-Goru sequences extend into India whereas its western limit is present in Kirthar and Sulaiman Fold Belts. First discovery was made in Lower Goru Sands reservoir of Khara Tar # 1 well which was drilled in Rajasthan, India near the eastern border of Pakistan. Union Texas Pakistan (UTP) made first discovery in Goru Sands in Southern Sindh Monocline (SSM) in 1981. By 1997 the operator had made a total of 49 Lower Goru - Sembar discoveries in the area, followed by State owned company OGDCL with number of discoveries. The Sembar-Goru Petroleum Play discoveries are located within the platform part of the Middle and Lower Indus Basin, from the Mari High all the way down to the Sindh Monocline **Figure 7** is showing Sembar-Goru sequences in Indus Basin, Pakistan.

## 8. Effective Source Rock and Hydrocarbon Charge

Quality and effectiveness of source rock is dependent on type of kerogen, level of maturity and volume of mature rock. Every petroleum prospect has been initially screening out on the basis of effectiveness of source rock and charging mechanism along trap. The probability factor for the petroleum charge system "$P_{HC\ Charge}$" is a product of the probability of effective source rock "$P_{Effective\ Source}$" and the probability of effective migration "$P_{Effective\ Migration}$".

$$P_{HC\ Charge} = (P_{Effective\ Source}) \times (P_{Effective\ Migration})$$

Results of biostratigraphy of Well-X1 and Zindapir-1 indicates that shales of Lower Cretaceous Sembar Formation which are considered to be the principal source of hydrocarbons in Lower and Middle Indus regions are absent in both wells of Zindapir Anticline (Adeel N et al., 2013) [9]. Rock pyrolysis data of Zindapir-1 was re-interpreted after integrating biostratigraphy.

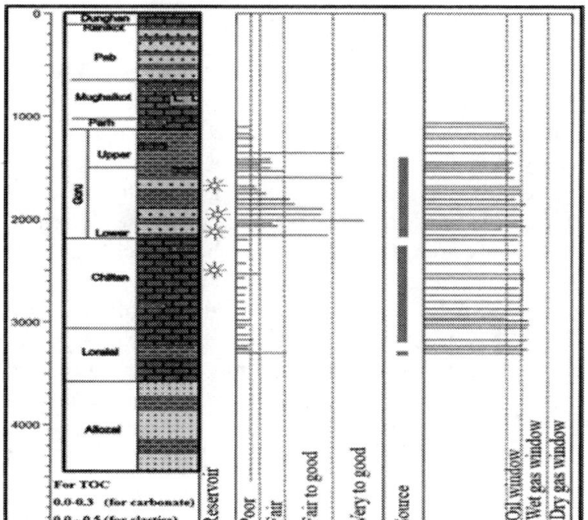


**Figure 8.** Composite petroleum system of the Zindapir-1 well which was drilled in the Zindapir Anticline

**8.1. Effective Source Rock:** The shales of Upper Goru Formation have poor to very good source rock potential (TOC between 0.25-2.25%) (Fig.8), bulk of Lower Goru shales show fair to very good source potential (TOC between 0.5-2.75%), The carbonates of Chiltan Formation have generally poor with limited fair source potential (TOC <0.3-0.45%), The Lorlai Limestone has also predominance of poor organic content with limited fair to good organic richness (TOC <0.3-1.0%). Hydrogen Index (HI) Vs Oxygen Index (OI) plot shows that the formations contain typically gas prone type III Kerogen. Vitrinite reflectance "%Ro" indicate that maturity level source rocks fall in the zone of oil and wet gas/condensate (Ro 0.5 ≈ 1.4: PI between 0.1- 0.7). Composite petroleum system of the Zindapir-1 is shown in Fig. 8. **Table. 2** indicated scheme of probability used in this manuscript. Result also includes that quality of source rock has not been reduced in the absence of Sembar facies. Therefore, probability of effective source rock "$\mathbf{P}_{\text{Effective Source}}$" is taken as 1.

| Source Rock Volume | TOC | Maturity | Probability |
|---|---|---|---|
| Suffient | Good to Very Good | Matured | 0.8-1.0 |
| Marginal | Fair to Good | | 0.6-0.8 |
| Marginal Mature | Poor to fair | Not Matured | 0.3-0.6 |

**Table. 2.** Probability scheme of effective source rock (CCOP) [4]

**8.2. Effective Hydrocarbon Migration:** Discovery of hydrocarbons of varying composition at Dhodak, Rodho and Afiband anticlines of Zindapir Anticlinorium located from north to south along the strikes indicate multiple source rocks with various level of maturity, possibly expelled and migrated through fault as conduits within the potential kitchen area in Sulaiman Foredeep in the east and Bharti Syncline in the west, due to increased overburden or being generated by indigenous potential source rocks of varying composition [9]. Probably transportation of inert gasses and hydrocarbons migrated along carrier beds and accumulated the migrated hydrocarbons in the varying quantities in deeper and shallower reservoirs. This hypothesis will be more explanatory for the development of uneven relationship of inert gases in shallower and deeper reservoirs of study area. The **Figure 9** is showing charging mechanism in Middle Indus Basin. Probability scheme of effective migration is given in **Table.3.**

Seismic Line 785-SK- 04 passing over the Zindapir Anticlinorium, flattened over the Eocene Intra Ghazij (**Figure 10** shows that structure was not present at that time[9]. Therefore, only those hydrocarbons were retained in structures which were generated / migrated at the time of structure formation and the age of structure is younger than Eocene.[16] described the younger age (Late Tertiary) of trap formation in Suleiman Fold Belt which is most possible reasons for under filled structures and even for the absence of hydrocarbon accumulation in some of the structures in Sulaiman Fold Belt.

| Migration | Time of Migration | | |
|---|---|---|---|
| Trap | Trap formed before time of HC formation | Overlaps with time Trap Formation | Trap formed when source rock is cooked |
| Local Migration | 0.9-1.0 | 0.7-1.0 | 0.1-0.4 |
| Lateral Migration with barriers | 0.8-0.9 | 0.5-0.7 | 0.1-0.3 |
| Lateral Migration without barriers | 0.5-0.8 | 0.5-0.6 | 0.1-0.3 |
| Verticle Migration with barriers | 0.7-0.9 | 0.5-0.7 | 0.1-0.3 |
| Verticle Migration without barriers | 0.5-0.6 | 0.5-0.6 | 0.1-0.2 |
| Long distance migration | 0.5-0.6 | 0.5-0.6 | 0.1-0.4 |
| Trap is in migration shadow | 0.2-0.4 | 0.1-0.3 | 0.1 |

**Table. 3.** Probability scheme of effective migration (CCOP, 2000)

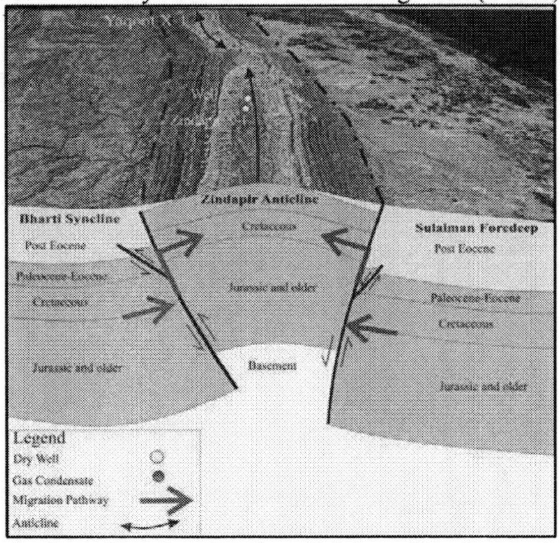


**Fig. 9.** Model showing hydrocarbon charging from Sulaiman Foredeep east and Bharti Syncline to Zindapir-1 in west. Deep seated faults conduits act as migration pathway.

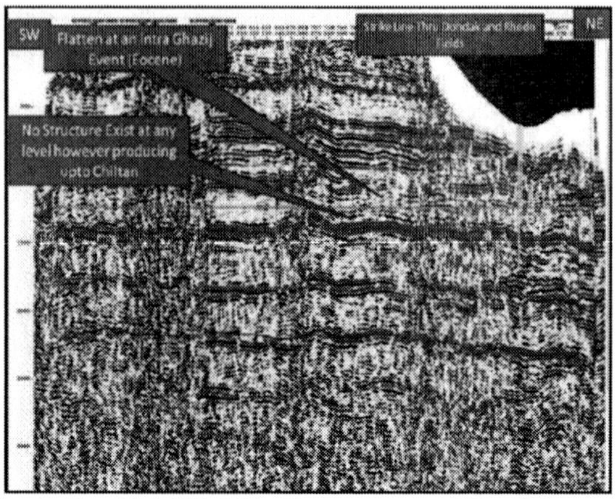


**Figure 10.** Interpreted seismic line 785-SK-85 is flattened over Ghazij level. It shows that there was no structure at the time of Middle Eocene. Structures were under filled by migrated hydrocarbons in Neogene [9].

1D Basin Model suggests (Fig. 11) that the Goru source rock is in wet gas window and it has never achieved dry gas window in its history in the Zindapir area. Goru Shales was deposited in the late to early cretaceous time and attained top oil window in late cretaceous, top condensate / wet gas window in Early to Middle Eocene time, but it has never acquired main gas generation phase. Similarly, Chiltan Limestone could not achieve dry gas window. Fig.12. is showing a plot of production index and % Ro showing production of indigenous hydrocarbons and migrated hydrocarbons.

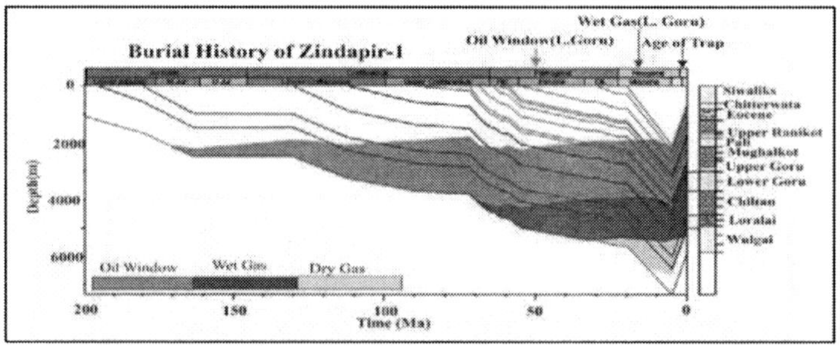


**Fig. 11.** Present day trap was not prevailing at the time of early hydrocarbon generation. Later phase of Orogeny (Post Miocene) charged the trap.

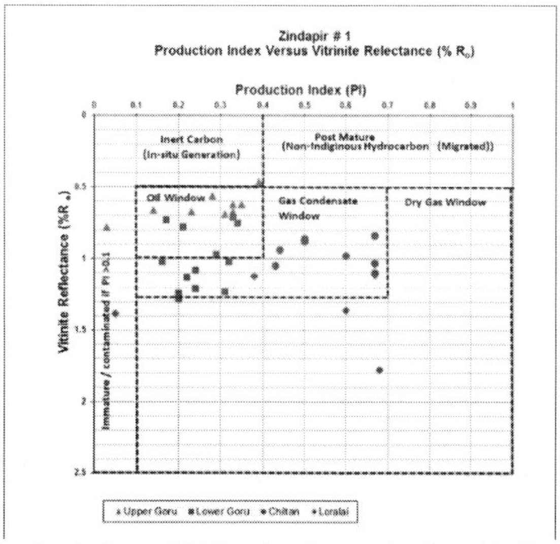


**Figure 12.** Plot of production index and % Ro showing production of indigenous hydrocarbons and migrated hydrocarbons.

Although Well X1 and Zindapir-1 and well did not flow with commercial hydrocarbon in Zindapir Anticline. But hydrocarbon discoveries in Dhodak, Rodho and Afiband prove effective source rock in Zindapir Anticlinorium, indicating Bharti Syncline in west and Sulaiman Foredeep in the east as potential kitchen area for Sembar-Goru play. Maximum burial is present in Bharti syncline and Punjab Platform. The faults seem ideal pathways. Therefore, the probability will enhance from 0.8 to 1. Probably those hydrocarbons play role in the filling of structure which were migrated later in different parts of Zindapir Anticlinorium. Retention of Hydrocarbon is not subject of hydrocarbon charging mechanism. Therefore, probability of effective migration is 1.

**8.3. Probability of Hydrocarbon Charge:** The probability factor for the petroleum charge system "$P_{HC\ Charge}$" is a product of the probability of effective source rock "$P_{Effective\ Source}$" and the probability of effective migration "$P_{Effective\ Migration}$".

$$P_{HC\ Charge} = (P_{Effective\ Source}) \times (P_{Effective\ Migration})$$

$$P_{HC\ Charge} = (1) \times (1)$$

$$P_{HC\ Charge} = 1$$

## 9.Reservoir Facies and its Effectiveness

Reservoir facies is based on distribution of reservoir facies in prospect area and thickness of net sand. The effectiveness of reservoir is based on porosity, permeability and fractures trends of study area. Fracture network is very important for carbonates rocks. Lower Goru Sands (Early Cretaceous) and Chiltan Limestone (Middle Jurassic) are proved reservoir rocks of Zindapir Anticlinorium with hydrocarbon flow from Afiband structure and Rodho structure. Both reservoirs are producing in other parts of Zindapir Anticlinorium. Facies map has been prepared to understand the presence of reservoir quality. Wireline logs of various petroleum wells were used.

**9.1. Chiltan Limestone (Jurassic) Reservoir:** Chiltan Limestone (Middle Jurassic) consists of thick to massive, dark grey carbonates with common oolitic beds. Karstification, paleosols / paleo - exposure surfaces are important geological feature at the top Chiltan Limestone. These carbonates are represented by oolitic and grain stones with shoaling upward sequences of moderate to high energy depositional environment with localized cavernous porosity (Loralai, spring, Sulaiman Fold Belt, Baluchistan, Pakistan).

Figure. 13. is showing Gamma Ray (GR) curve of Zindapir-1 well is showing shoaling up sequence in Chiltan Formation.

FMI log is not present in Zindapir -1 and Albaraka-1 well due to limitation of data in Ph.D. project. But Zindapir anticline shares similar geology throughout the Zindapir Anticlinorium and share similar geology in other parts of anticlinorium (e.g. Dhodak, Rodho and Afiband). Chiltan Limestone has flown in Rodho and Afiband has with presence of matrix porosity. Log derived facies indicate that high energy carbonates are present **(Table. 4).** Considering conditional rule of probability, we assume that permeability and fracture network is present in Afiband and Rodho structure and producing hydrocarbons.

| **Interval (m)MD** | **Thicknes (m)** | **Proisty (%)** |
|---|---|---|
| **2196.5-2207.5** | 11.1 | **3-7** |
| **2343.5-2354.6** | 11.3 | 3-7 |
| **2364.4-2369.8** | 5.4 | 2-4 |
| **2373.7-2402.1** | 28.4 | 2-8 |
| **2446-2453** | 7 | 3-4 |
| **2514-2582.6** | 68 | 2-6 |
| **2611.6-2627.9** | 16.3 | 2-3 |
| **2770-2777** | 7 | 2-4 |

**Table 4.** Log derived porosity of Chiltan Limestone

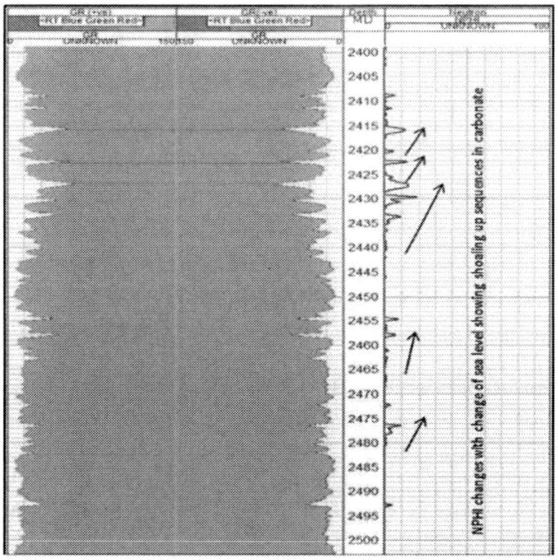


**Figure 13.** GR curve of Zindapir-1 well is showing shoaling up sequence in Chiltan Formation (Increase in neutron porosity "NPHI" shows increase in grain size) of Jurassic age [17]

Gross depositional map indicates that Zindapir Anticlinorium is sitting on regional position with higher energy carbonates. Whereas Chiltan Limestone has flown from Rodho, Afiband of Zindapir Anticlinorium. Whereas Dhodak has not been drilled to Chiltan Limestone. Regionally Chiltan Limestone has been flown from Jandran in western Sulaiman Fold Belt, Nandpur and Panjpir in Punjab Platform, central Indus Basin, Pakistan. Hence, Chiltan Limestone is proved reservoir. **Figure14.** Gross Depositional Environment (GDE) map of Chiltan Limestone in which trend of isopach map of Chiltan Limestone indicates westward increase in thickness.

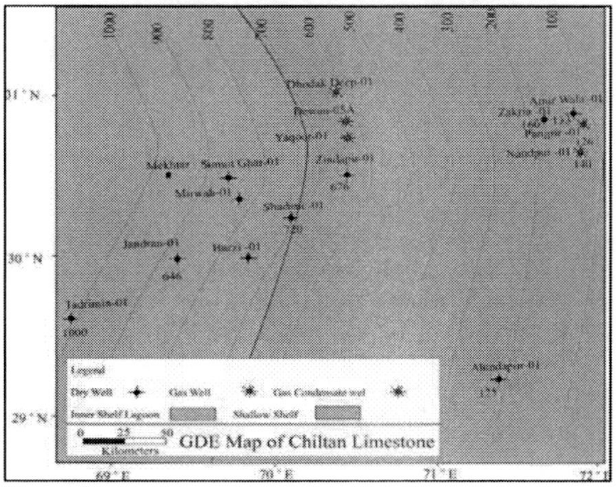


**Figure 14.** Gross Depositional Environment (GDE) map of Chiltan Limestone in which trend of isopach map of Chiltan Limestone indicates westward increase in thickness.

**9.2. Lower Goru Sands (Cretaceous) Reservoir:** Wireline correlation of Zindapir-1 and Ahmedpur-1 (**Figure 15**) shows that Sembar is present in the east of study area. The regional wireline correlation of

Sembar and Goru shows westward progradation from Punjab Platform to Sulaiman Fold Belt. The trend of the correlation pattern indicates fluvio-deltaic sands in the east (Shadani & Zindapir wells) passing westward into distal slope and basin muds (Jandran wells) across shelf edge**.** Shelf is also prograding with deposition younger progrades. Gross depositional maps clearly show that shoreface sand is present in Lower Goru which proved reservoir in Afiband and Rodho Field in Zindapir Anticlinorium. Average porosity is 6-8% which is good for gas reservoir.

Iqbal et al.[18]] described the young age (Late Tertiary) and fast uplift rate of the Sulaiman Fold Belt. The most possible reasons for under filled structures and even for the absence of hydrocarbon accumulation in some of the structures in Sulaiman Fold Belt. Seismic Line 785-SK- 04 passing over the Zindapir Anticlinorium, flattened over the Eocene Intra Ghazij. Results indicate that structure was not present at that time [17] Therefore, only those hydrocarbons were retained in structures which were generated / migrated at the time of structure formation and the age of structure is younger than Eocene. Shales of Goru are buried deep in kitchen area with more potential to generate hydrocarbons as compared to in-situ source rock of Zindapir Anticline Nazeer A. et al.,[19].

According to Peresson and Farrukh D.[20], Zindapir anticline is interpreted as the easternmost branch of a large-scale positive flower structure. In the core of the Zindapir Anticline brittle Dunghan Limestone (Paleocene) is exposed which shows synthetic shear faults on the landsat images and proves the highly left-lateral transpressive character of the fold. In such transpressional regimes, palm tree or positive flower structures have been identified using seismic data. Good quality of seismic data has ample chances to drill the well in optimistic location. Present seismic quality is very poor to infer the compartments but Landsat imagery.

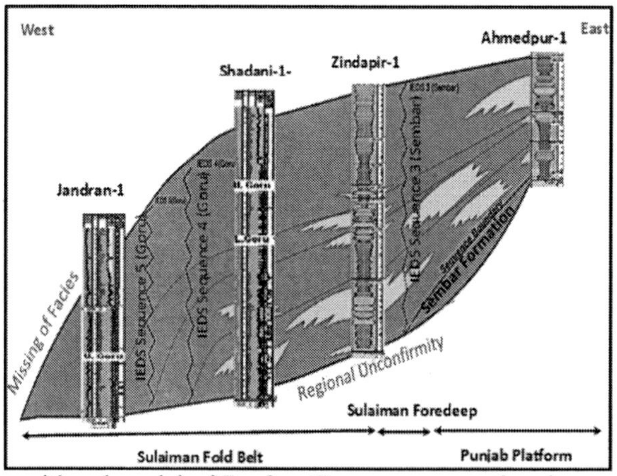


**Figure. 15.** Regional depositional model of Sembar-Goru sequences showing westward progradation of shallower facies in east passing into equivalent deep basinal setting in west.

Gross Depositional Environment (GDE) map as shown in **Figure 16** of Lower Goru Sands showing shoreface Sand is present in Zindapir-1 well drilled on Zindapir Structure.

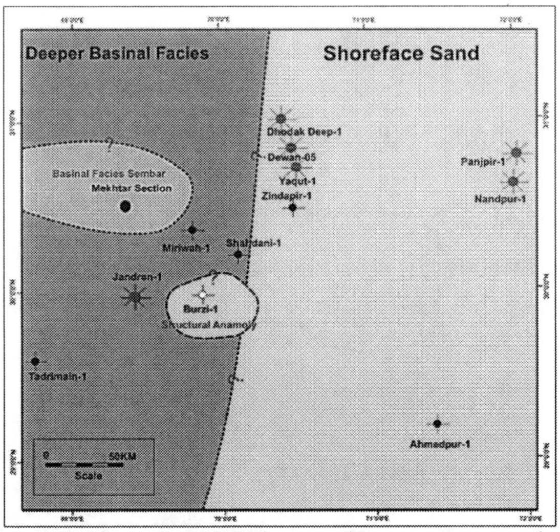


**Figure 16.** Gross Depositional Environment (GDE) map of Lower Goru sands showing shoreface sand is present in Zindapir-1 well drilled on Zindapir structure.

Table 5. is showing log derived porosity and gross thickness in Lower Goru formation.

| Interval (m)MD | Thicknes (m) | Proisty (%) |
|---|---|---|
| **1686.5-1727** | 40.5 | 3-9 |
| **1736.7-1766.7** | 30 | 3-6 |
| **1931-1980.5** | 49.5 | 1-9 |
| **1984.4-1003** | 18.6 | 5-7 |
| **2014.5-2030** | 15.5 | 1-7 |
| **2041.4-2058.4** | 17.1 | 1-8 |

**Table 5.** Log derived porosity and gross thickness in Lower Goru

## 10. Probability of Reservoir

Probability of the Reservoir "$\mathbf{P}_{\mathbf{Reservoir}}$ r" is product of $\mathbf{P}_{\mathbf{Reservoir\ Facies}}$ (reservoir facies with minimum net thickness) and $\mathbf{P}_{\mathbf{Reservoir\ Effectiveness}}$ (effectiveness of the reservoir, with respect to minimum porosity, permeability and hydrocarbon saturation).

$$\mathbf{P}_{\mathbf{Reservoir}} = (\mathbf{P}_{\mathbf{Reservoir\ Facies}}) \text{ x } (\mathbf{P}_{\mathbf{Reservoir\ Effectiveness}})$$

Table.6. is showing scheme used in the manuscript for calculation for Probability of Reservoir.

| **Uncertainity** | | **Good** | **Fair** | **Poor** |
|---|---|---|---|---|
| $P_{\text{Reservior Facies}}$ | Net Pay | 0.8-1.0 | 0.5-0.7 | 0.1-0.4 |
| $P_{\text{Reservior}}$ Effectivness | Porosity | 0.8-1.0 | 0.5-0.7 | 0.1-0.4 |
| | Permeability | 0.8-1.0 | 0.5-0.7 | 0.1-0.4 |

**Table 6.** Scheme used for calculation for Probability of Reservoir

Landsat interpretation in **(Figure 17)** showing that Zindapir anticline is a large structure, which was possibly developed because of transpressional tectonic stresses. Riedel Shear has been interpreted on the exposed surface of Dungan Carbonate (Paleocene). Presence of Riedel shear is indication of strike slip fault in wrench fold belt. In Zindapir-1, right stepping Riedel shear was interpreted on surface exposure of Paleocene limestone showing left lateral movement. The structural interpretation strongly suggests the presence of fracture network in Chiltan Limestone.

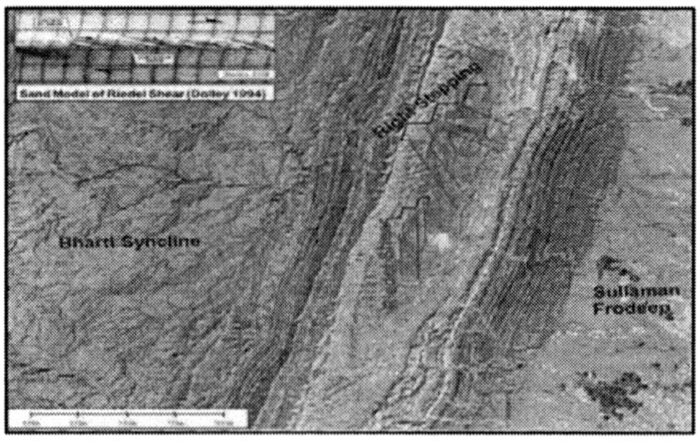


**Figure 17.** Right stepping Riedel shear in Paleocene (Dunghan) carbonate exposed in the core of Zindapir Anticline

As discussed earlier that FMI logs are not present in Zindapir-1. On the basis of Conditional Rule of Probability, it has been assumed that similar reservoir effectiveness parameters are present in in overall anticlinorium. Reservoir thickness is good. Matrix porosity is proved on wireline logs. **Table.7** is showing probability of Chiltan Limestone reservoir in Zindapir Anticline.

| $P_{\text{Chiltan Limestone Reservior}}$ | | | | | |
|---|---|---|---|---|---|
| Uncertainity | | Good | Fair | Poor | P |
| $P_{\text{Reservior Facies}}$ | Net Pay | 0.8-1.0 | 0.5-0.7 | 0.1-0.4 | 1 |
| $P_{\text{Reservior Effectivness}}$ | Porosity | 0.8-1.0 | 0.5-0.7 | 0.1-0.4 | 0.8 |
| | Permeability | 0.8-1.0 | 0.5-0.7 | 0.1-0.4 | |
| $P_{\text{Reservoir}} = (P_{\text{Reservoir Facies}}) \times (P_{\text{Reservoir Effectiveness}})$ | | | | | 0.8 |

**Table 7.** Probability of Chiltan Limestone reservoir in Zindapir Anticline

Permeability data is not usually available in exploration prospect. It may be ignored in calculation of probability. But reservoir thickness and type of reservoir could not be ignored. Shoreface sand have good reservoir effectiveness. On the basis of Conditional Rule of Probability, it has been assumed that similar reservoir effectiveness parameters are present in in overall anticlinorium. Lower Goru sands are producing in Rodho and Afiband. Probability of Lower Goru sands in Zindapir is given in Table 8

| $P_{\text{Lower Goru Sand Reservior}}$ | | | | | |
|---|---|---|---|---|---|
| Uncertainity | | Good | Fair | Poor | P |
| $P_{\text{Reservior Facies}}$ | Net Pay | 0.8-1.0 | 0.5-0.7 | 0.1-0.4 | 1 |
| $P_{\text{Reservior Effectivness}}$ | Porosity | 0.8-1.0 | 0.5-0.7 | 0.1-0.4 | 1 |
| | Permeability | 0.8-1.0 | 0.5-0.7 | 0.1-0.4 | |
| $P_{\text{Reservoir}} = (P_{\text{Reservoir Facies}}) \times (P_{\text{Reservoir Effectiveness}})$ | | | | | 1 |

**Table 8.** Probability of Lower Goru sands in Zindapir Anticline

Journal of Physics: Conference Series **2594** (2023) 012002 doi:10.1088/1742-6596/2594/1/012002

**10.Trap Mechanism**

Trap is second important factor for the assessment of effective petroleum system. Trap is a subsurface geologic feature with specific geometries to retain hydrocarbon with sealing capabilities. Every trap consists of mapped geometry in the form of shape, area and volume and enclosed by seal.

$$P_{Trap} = (P_{Effective\ Trap}) \times (P_{Sealing\ Mechanism})$$

**11.1. Definition of Effective Trap:** Seismic quality evaluates the geometrical body. Uncertainty will arise if resolution of these surfaces on seismic profiles is poor. If there is no seismic, there is no perfect definition of subsurface trap and risk is very higher. CCOP (2000) considered (1) seismic data quality, (2) aerial coverage (3) velocity control/depth conversion and (4) identification of top and bottom important parameters. Table.10. is showing the probability of effective trap (After CCOP ) [4]

| Seismic mapping | 3D Siesimic | 2D seismic | | |
|---|---|---|---|---|
| | | Dense Grid Size | Open Grid Size | Very Open Grid Size |
| Good near by wells | 0.8-1.0 | 0.5-0.9 | 0.5-0.8 | 0.3-0.6 |
| Coreelation by distant wells | 0.7-.8 | 0.4-0.7 | 0.4-0.7 | 0.2-0.5 |
| Analogue Model (uncertain) | 0.3-0.7 | 0.3-0.7 | 0.2-0.6 | 0.1-0.4 |

**Table 9.** Probability of effective trap (After CCOP (2000) [4]

Landsat Interpretation of area is shown in **Figure 18.**

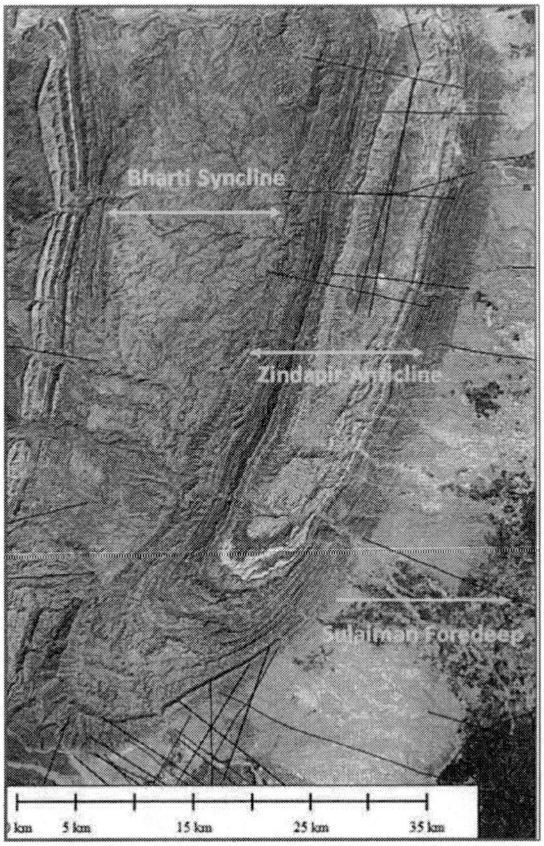

**Figure 18.** Landsat Interpretation of area

Seismic acquisition has never been acquired in Southern part of Zindapir Anticline. Central and Northern part of Zindapir Anticline was mapped with 2D seismic data very open grid size. Present seismic data is limited and not optimistic for detail structural configuration. Available seismic data is old and not reliable to map structural closure of Zindapir -1. Subsurface compartment could not be delineated with acquiring 3D seismic data. Drilling of two wells are inadequate for mapping of big size structural closure of Zindapir Anticline. Based on Landsat interpretation and available seismic data, probability of effective trap "$P_{Effective\ Trap}$" is taken 0.5.

**11.2. Sealing Mechanism:** Seal is top surface in four-way dip closure which retained the hydrocarbons in reservoir sands or limestone in geometrical shape. Shale, salt, anhydrite and carbonates have a very good sealing property (probability ranges from 0.9-1.0). Thick shales have good sealing properties (probability ranges from 0.8-1.0). Basalt has acceptable to good sealing properties (probability ranges from 0.6-0.8). Faults cutting top surface has a poor ((probability ranges from 0.4-0.6) to acceptable sealing properties and depends on sand/shale or sand/sand contact.

Intra formational Goru shales have good sealing capabilities for Lower Goru (Late Cretaceous) reservoir Sands. In addition, tight zones are also present in Chiltan carbonate (Middle Jurassic) reservoir. Geometrical shape of structure is very important to define the sealing mechanism. **Figure 9** represent schematic cross section of Zindapir Anticline. Available data does not indicate the breaching of structure. Therefore, probability of sealing mechanism "$P_{\text{Sealing Mechanism}}$" is taken as 1. Thus, probability of trap "PTrap" is 0.5.

Similarly, Chiltan limestone has tight zones, indicated by wireline logs. Such tight zones have retained hydrocarbons in Rodho and Afiband. Therefore, probability of sealing mechanism "PSealing Mechanism" is taken as 1. Thus, probability of trap "PTrap" is 0.5.

## 12. Probability of Retention After Accumulation

Three types of probabilities have been discussed in this manuscript which defines the existing petroleum system at the time of trap formation. The fourth probability is related to latest event of tectonic. As a result, the tension and compression associated with plate movement is so great that blocks of rock fracture or break apart and displacement takes place between different blocks, creating fault in nature. However, process of folding and faulting is not simple. Once the faults formed as consequences of displacement, faults and fold may occur simultaneously by changing geometry of folds. But late tectonic activity plays vital role in delineating of existing petroleum system and defined the fourth type of probability.

Fourth type of probability will be measured to assess the additional risk associated with drilling of new exploration well. This geological event deals with trap after the accumulation of hydrocarbons and based on confidence level generated by quality and quantity of seismic data and good data. Hydrocarbon retention is affected by late tectonic activity after accumulation of hydrocarbons. Thus, trap may be breached by reactivation of faults. Pathway may be blocked due to uplifting and erosion. Top point of structure may be altered, and volume of hydrocarbon changed. Compartments may be formed by reactivation of faults compression and transpression. According to CCOP, the probability of effective retention, potential fault reactivation, regional uplift (and subsequent erosion), together with tectonic and/or isostatic tilting of the trap after accumulation must be evaluated.

The present data does not confirm the breaching of structure that related to late tectonic activity. Discoveries in Dhodak, Rodho and Afiband strongly rejects the isostatic tilting of the trap after accumulation of hydrocarbons. Therefore, we assume that there is no negative impact on the volume of hydrocarbon in connection to tectonic activity after accumulation. Thus, probability of retention after accumulation is 1.

## 13.Keynotes

a)Geological Chances of success is probability based risking calculated by geologist prior to drilling.
b)Probability outcome of geological chances of success is based on confidence level over data control (Table.1).
c)Regional geology is base of prospect evaluation.
d)Regional play will be evaluated with careful attention.
e)Careful attention is required during the comparison between regional basin and prospect during probability.
f)Prospect always evaluated as individual prospect.
g)Reservoir -seal pair either exist or not exist. Always use facies mapping for the forecasting of their location in basin.
h)Avoid "double-risking.
i)Thought provoking argument is required for those factors which affect the sealing mechanism
j)Retention of hydrocarbon after accumulation will be studied with careful attention.
k)Hydrocarbon's migration may take place by following long route migration along carrier beds and fault conduit. Good example is PUNJAB Platform, PAKISTAN.
l)Presence of kitchen area with adequate overburden is very important for charging mechanism.

m)It is very important to develop some standard for probabilities ranges during assessment of risking in drillable prospect.
n)We considered CCOP (2000) as standard or this publication.

**14.Probability of Geological Success ($P_G$)**
Overall evaluation shows that Probability of Effective Trap is shown in Table 10.

| $P_{\text{Geological Chances of Success}}$ ($P_G$) | | | | | | | | |
|---|---|---|---|---|---|---|---|---|
| Formation | $P1=P_{\text{HC Charge}}$ | | $P2=P_{\text{Reservior}}$ | | $P3=P_{\text{Trap}}$ | | $P4=P_{\text{Retension after accumulation}}$ | $P_G=P1 \times P2 \times P3 \times P4$ |
| | $P_{\text{Effective Source}}$ | $P_{\text{Effective Migration}}$ | $P_{\text{Reservior Facies}}$ | $P_{\text{Reservior effectivness}}$ | $P_{\text{Effective Trap}}$ | $P_{\text{Sealing Mechanism}}$ | | |
| Lower Goru | 1 | 1 | 1 | 1 | 0.5 | 1 | 1 | 0.5 |
| Chiltan | 1 | 1 | 1 | 0.8 | 0.5 | 1 | 1 | 0.4 |

**Table 10.** Overall evaluation shows that probability of effective trap is major concern for valid geological model. This is because of limited seismic with wide area grid on Zindapir Anticline. Color Scheme is given in Table-1

**15.Screening of Risk Factor**
We have discussed earlier that probability is confidence level over data control. If data control is good, confidence level increases which further indicate that there is no uncertainty in the geological model. If data control is less, there is more uncertainty and assumption are wrong. Therefore, screening of risk factor is very important. For the screening of risk factor, we compare all the geological events used to calculate "Geological chances of success". The table is showing comparison of all geological events. This model indicates that drilling of new well without acquiring additional seismic is major risk. Probability for discovery is 0.45 and failure is 0.55. Table 8 is showing screening of risk factor. This model strongly recommends additional seismic in Zindapir Anticline prior to drilling of new well.
**15.1. Careful Approach during Risk Analysis:** Several risk assessment software is present in E & P industry with slight difference. Different excel sheets are also used by different E & P companies of PAKISTAN. Most important criteria depend upon value of assigned probability and its scientific reasoning behind it. As Experts are still susceptible to individual bias and group bias which influence the probabilities they generate Chen B.[21]. Without using some defined value of probabilities, experts are always biased due

to their individual believes and believes of their peers in company. That is why important of standardization in the assigning value of probability could not be ignored.
This manuscript is based on technique used by CCOP (2000) as standard method to calculate Geological Chances of Success. CCOP (2000) is outcome of extensive research which was carried out by intergovernmental organization of "Coordinating Committee for Geoscience Programs in East and Southeast Asia (CCOP)" in order to contribute to economic development by facilitating and coordinating the implementation of applied geoscience programs. Their work is detailed and may use as standard to define different input parameters of assigned probabilities on scientific basis.

## 16.Chances of Success for Development Wells

Author of this manuscript strongly observed that Geological Chances of Success has been calculated for every development well as practice. During process of risking, experts use those parameters during forecasting of new development well which have no use When exploration well successfully flow with hydrocarbons, question related to hydrocarbon charge source rock quality, and presence of sealing mechanism are resolved. Even effect of late tectonic has no importance. Every oil and gas field need to drill more development well to keep the economical flow of hydrocarbons. At that stage several risks are associated with new development well which are listed below.
1)Reservoir thickness
2)Reservoir effectiveness due variation in porosity and permeability.
3)Velocity control/depth conversions
4)Seismic imaging and resolution
5)Difference in seismic re-processing and its validity
6)Type of seismic vintage
7)Oil water contact
8)Blind faults
9)Drainage area
10)Distance from major faults /faults.

Risk is less than 20% if optimistic 3D seismic interpretation is present. Lack of 3D Seismic data / closed grid 2D seismic data always increased risk more than 50%. 10 % risk is always associated with new development well because we are dealing with time domain and whatever latest technology is available in world, geological input is always in time domain.
We have used Probability of well completion (PWC) for Geological chances of success for new development well and its equation is given below

$$\mathbf{P_{WC} = (P_{Reservoir}) \times (P_{Seismic\ Risk}) \times (P_{Water\ Contact}) \times (P_{Fault})}$$

Whereas
$P_{Reservoir} = (P_{Reservoir\ Facies}) \times (P_{Reservoir\ Effectiveness})$
$P_{Seismic\ Risk}$= 5% risk if interpretation quality is good
$P_{Fault}$= 5% risk of blind faults if no indication of fault
(10 % risk is always associated with new development well)
Note: - $P_{Seismic\ Risk}$ and $P_{Fault}$ are beyond the sight of seismic interpreter due to nature of data in time domain.

**17.Conclusions**

1)Probability of Hydrocarbon Charge ( $P_{HC\ Charge}$) is 1 indicating the presence of matured source to charge the reservoir through fault conduits (Fig-9)
2)Probability of Reservoir ($P_{Reservoir}$) of Goru Sand is 1 based on petrophysical interpretation, producing in Afbiand, Rodho and Dhodak from south to north.
3)Probability of Reservoir ($P_{Reservoir}$) of Chiltan Limestone is 0.8 which is based on wireline interpretation. Image log was acquired in Zindapir-1 well. Structural interpretation strongly suggests the presence of fracture network in Chiltan Limestone. GR log motifs and neutron porosity interpret the shoaling up sequence in Chiltan Formation of Zindapir-1. Moreover, Chiltan is producing from other part of Anticlinorium.
4)Zindapir anticline shares similar geology throughout the Zindapir Anticlinorium and share similar geology in other parts of anticlinorium (e.g., Dhodak, Rodho and Afiband). The present data does not confirm the breaching of structure that related to late tectonic activity. Discoveries in Dhodak, Rodho and Afiband strongly rejects the isostatic tilting of the trap after accumulation of hydrocarbons. Therefore, probability of retaining hydrocarbon is 1.
5)Presence of Hydrocarbon Charge, transpressional tectonics and reservoir and in study area indicates the potential to flow hydrocarbons having good chances of success because of transpressional movement at the periphery of the structure. These compartments may contain the hydrocarbons, generated and migrated either from Bharti Syncline in west and Sulaiman Foredeep in east (Figure 9).
6)Southern part of Anticline was not mapped where as northern part of Zindapir anticline was mapped (Fig. 18). There is major reasoning for poor Probability of Effective Trap due to low quality seismic data.

**18. Recommendations**

1)Zindapir is a large structure, which was developed because of transpressional movements. It is considered that the hydrocarbons generated from the Sulaiman depression and Bharti Syncline s might have migrated through wrench related fault conduits and trapped in structural traps (if available) in Zindapir anticline. Therefore, preferably 3D seismic with better resolution is required for detail investigation of possible structural traps.
2)Zindapir prospect is strongly recommended for further evaluation as exploration prospect after acquiring new seismic data.

**19. Success Ratio in Zindapir prospect**

The success ratio (SR) is the ratio between the number of successes (discoveries) and the number of trials (wildcat wells). Whereas the wildcat well is an exploratory well drilled in unproven area or fully exploited area have no historical history of production. Success ratio of basin is different from success ratio with proved petroleum system in mapped area.

In the case of discovery of multiple formations in limited mapped area with defined geological and geographical boundaries, success ratio of individual play will be varying and depends upon an aerial extension of mutually related geological factors (trilogy of source rock, presence of reservoir facies with sealing mechanism). Table-11 is showing Success Ratio of exploratory wells drilled for Chiltan Play in Zindapir Anticlinorium as whole petroleum system and success ratio in individual anticlines sharing similar petroleum system.

| Chiltan (Middle Jurrasic ) with target reservior in Zindapir Anticlinorium | | | | | |
|---|---|---|---|---|---|
| Anticline | Exploratory Wells drilled Chiltan Limestone | | | | |
| | Zindapir | Afiband | Rodho | Dhodak | Total Wells |
| N=No of wells | 2 | 1 | 3 | 1 | 6 |
| n = No of successes | 0 | 1 | 3 | 1 | 4 |
| n+1 | 1 | 1 | 4 | 2 | 5 |
| N+2 | 4 | 3 | 5 | 3 | 8 |
| Success Ratio (SR) | 1/4 | 2/3 | 4/5 | 2/3 | 5/8 |
| | 0.25 | 0.33 | 0.80 | 0.67 | 0.63 |
| Success Ratio = (n+1)/ (N+2) | | | | | |

**Table 11.** Success ratio in Zindapir Anticlinorium and associated anticlines

## 20. Relation between Geological chances of success and Success Ratio

Geological chances of success and Success ratio are two different things. Geological chances of success do not apply the drilling of prospect. It is applicable to validate the geological model of prospect before drilling and further indicate the risk associated with geological modelling. In the case of Zindapir-1, Geological Chance of Success for Goru Sands is 0.5 and 0.4 for Chiltan Limestone. Lower value is hit by trap definition which could not be improved with acquisition of seismic data.

The success ratio is rate of success on the basis of exploratory wells drilled with successful discoveries of hydrocarbon flow. But it never validates the geological model. Success ratio in any virgin area begin with ½. Minimum risk associated with exploration of prospect is 50% and will be improved with improvement in geological model. Success Ratio is used to rank the different basins, sub basin, system and plays during initial screening blocks offered for exploration.

## 21. Decision tree analysis

Decision tree is analytical tools to identify, describe, quantify, and explain uncertainty and its consequences during drilling of exploration well. In this paper, Geological Chances of success in Zindapir prospect will be improved trap delineation which could be valid without acquiring additional 2D seismic with close grid or 3D

seismic. Associate risk is 50% in above mention geological model with two options

**Option-1:-** Drill of well without shooting seismic two possible outcomes: (1) a 50% chance of a profitable producing well or (2) a 50% chance of a dry hole.

**Option-2:-** First shoot a seismic line with following possible outcomes: (1) a 33% chance of getting a negative seismic result (that is, structure is breached) and 2) a 67% chance of getting an encouraging seismic result and therefore drilling the well in the most favorable location.

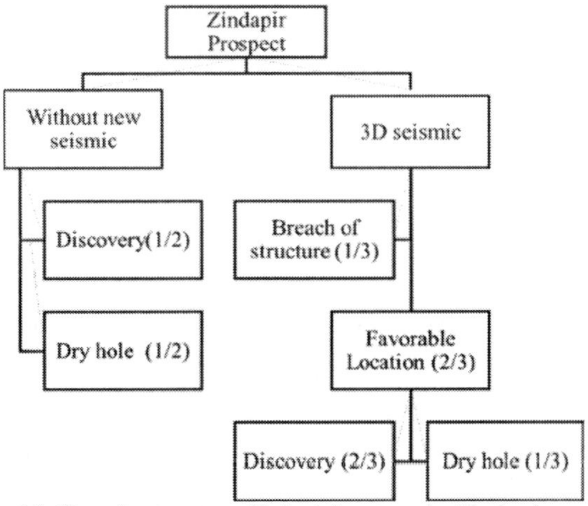


**Figure 19**. Development well decision tree in Zindapir prospect.

**References**

1. Schrödinger, E. 1935. Discussion of Probability Relations between Separated Systems. Mathematical Proceedings of the Cambridge Philosophical Society, 31 (4), 555-563. doi:10.1017/S0305004100013554
2. Baecher, G. 1975. Probability Theory in Geological Exploration. IIASA Research Memorandum. IIASA, Laxenburg, Austria. 1-18.
3. Rogerson, A. 1990. Probability, Statistic and Computer Modelling in Geology, International Conference on Teaching Statistics (ICOTS 3), Melbourne, Australia, 135-138.
4. Coordinating Committee for Coastal and Offshore Geoscience Program (CCOP) in East and Southeast Asia. 2000, The CCOP Guidelines for Risk Assessment of Petroleum Prospects: Bangkok, Thailand. 35 p.
5. Harbaugh, J.W., Davis, J.C., Wendebourg, J. 1995. Computing Risk for Oil Prospects: Principles and Programs: New York, Pergamon, 452 p.
6. Salleh, S.H., Rosales, E. and Mota., I., F. 2007. Influence of different probability based models on oil prospect exploration decision making: a case from southern Mexico, Revista Mexicana de Ciencias Geológicas, V. 24, No. 3. 306-317.
7. Bickel, J.E., Smith, J.E., and Meyer, J.L. 2008. Modeling dependence among geologic risks in sequential exploration decisions, Reservoir Evaluating and Engineering, SPE Annual Technical Conference, Texas, USA. 352-361.
8. Lerche I. 2019. A review of economic risking methods commonly used in hydrocarbon exploration. Journal of Petroleum Exploration and Production Technology. Volume 9. 1579–1591.
9. Nazeer A., Solangi, S.H., Brohi, I.A., Usmani, P., Napar, L.D., Jhangir, M., Hameed, S. and Manshoor, S.M. 2012-2013. Hydrocarbon potential of Zindapir Anticline, Eastern Sulaiman Foldbelt, Middle Indus Basin, Pakistan. Pakistan Journal of Hydrocarbon Research (PJHR). Vol. 22 & 23.

10. Shazia A., Abbasi S.A., Qureshi S. N., Solangi S.H., Siddiqui I. , and Mirza M.Q. 2015. Structural interpretation of seismic data in the south of Zindapir area, Sulaiman Fold Belt, Pakistan, Journal of Himalayan Earth Sciences. Volume 48. No. 1, 2015. 112-123.
11. Jadoon I.A.K. and Zaib M.O. 2018. Map on: Tectonic map of Sulaiman Fold Belt, Pakistan, Department of Earth Sciences, COMSATS University, Islamabad (Abbottabad Campus), Pakistan.
12. Morosov A.L. and Bratvold R. B. 2021. Probability elicitation using geo-statistics in hydrocarbon exploration. Springer Nature. https://doi.org/10.1007/s10596-021-10084-9
13. Abbasi, S. A. 2007), 2D Seismic Reflection Data Interpretation of Sulaiman Fold Belt Area Line No. 812-DA-08 and L36-91-01. M.Sc. Thesis: Quaid-e-Azam University, Islamabad, Pakistan 77-120.
14. Solangi S.H., Abbasi S.A., Ali A., Asim S., Lasharai R.A., Brohi I.A. and Siddiqui I. 2014. Study of subsurface structural trend and stratigraphic architecture using seismic data. Sindh University Research Journal (Science Series). Vol.46 (3) 373-380.
15. Milkov A.V. 2021. Reporting the expected exploration outcome: When, why and how the probability of geological success and success-case volumes for the well differ from those for the prospect. Journal of Petroleum Science and Engineering 204. (2021) 108754. 1-8.
16. LaCosta C.P., Milkov A.V. 2022. Petroleum exploration portfolios generated with different optimization approaches: Lessons for decision-makers William. Journal of Petroleum Science and Engineering 214 (2022) 110459. 1-19
17. Ali S.M., Ahmed J. and Ahmed R. 1995. Evidence of wrench tectonics in the Sulaiman Fold Belt, Pakistan and its implication for hydrocarbon prospects. Abstract and paper presented in Second South Asian Geological Congress, Colombo, Sri-Lanka.
18. IEDS (Integrated Exploration and Development Services). 1995. A Sequence Stratigraphic Study of Lower Goru-Sembar Formations of Lower and Middle INDUS Basin of PAKISTAN and Rajasthan. Multi-Client Study.
19. Iqbal M., Chaudhry M.N. and Bannert D. 2008. Hydrocarbon exploration concepts for the Eastern Frontal part of Sulaiman Fold Belt, Pakistan: Society of Petroleum Engineers (SPE)/Pakistan Association of Petroleum Geoscientists (PAPG) Annual Technical Conference, Islamabad, Pakistan.
20. Nazeer A., Abbasi, S.A. and Solangi S.H. 2018. Sedimentary facies interpretation of Gamma Ray (GR) log as basic well logs in Central and Lower Indus Basin of Pakistan. Geodesy and Geodynamics, Vol 7, No. 6, 432-443.
21. Nazeer A., Tariq T., Murteza, G., Shah S.H. and Danyal, A. 2012. An Overview of the distribution of inert gases in deeper reservoirs of the Sulaiman Fold Belt, Pakistan, Society of Petroleum Engineers (SPE)/Pakistan Association of Petroleum Geoscientists (PAPG) Annual Technical Conference, Islamabad, Pakistan.
22. Nazeer A. 2019. Facies Modelling, Source Rock Evaluation and Petroleum Potential of Sembar-Goru Sequences in Zindapir Anticline, Sulaiman Fold Belt, Pakistan. University of Sindh, Jamshoro, Pakistan, Ph.D. Thesis, 1-196.
23. http://prr.hec.gov.pk/jspui/handle/123456789/20350
24. Peresson. H, and Farrukh D. 2009. Integrating Structural Geology and GIS: Wrench Tectonics and Exploration Potential in the Eastern Sulaiman Fold Belt. Proceedings, Pakistan Association of Petroleum Geoscientists (PAPG)/Society of Petroleum Engineers (SPE) Annual Technical Conference, Islamabad, Pakistan.
25. Chen B. 2016. Research on Verification for Chance of Geologic Success. Department of Petroleum Engineering and Applied Geophysics, Norwegian University of Science and Technology. Master Thesis. 1-60.

# Application of New Technologies and Best Operational Practices to Accelerate the Learning Curve in Wells Drilled in the Ayatsil Field, Mexico

**Andrés Isaac Merchan Nájera1(⊠), Orlando José Urribarri Romero2(⊠) , Carlos Alberto García Lomelí 3(⊠), Mario Maya Perez 4(⊠), Jorge Chiw 5(⊠) ,Juan Rutilo Rayo Perez 6**

1 PEMEX Exploration and Production,
Carmen City, Mexico
2 Schlumberger Well Construction,
Carmen City, Mexico
3 PEMEX Exploration and Production,
Carmen City, Mexico
4 PEMEX Exploration and Production,
Carmen City, Mexico
5 PEMEX Exploration and Production,
Carmen City, Mexico

andres.isaac.merchan@pemex.com, oromero15@slb.com,
carlos.alberto.garcial@pemex.com, mario.maya@pemex.com,
jorge.chiw@pemex.com

**Abstract.** Today drilling wells is one of the biggest capital expenditures and is employed starting from exploration, delineation, initial wells for production, and fresh production incorporation when existing wells production have declined. The estimated cost for drilling new wells in Ayatsil field is around 25 MM$, which requires a high level of decision to achieve production goals without exceeding the budget assigned. Therefore, to make the right decision requires an integration of multidisciplinary group of specialists (geologists, geomechanics, reservoir, production and drilling engineers) from well design to execution phases.

This paper will illustrate the methodology and process applied by the operator to optimize the drilling stages and accelerate field production, as a result the operator developed the operational excellence project, that consist of five phases and being executed by a multidisciplinary project team.
The drilling team has been successful in reducing the depth versus days curves from an average of 130 days in 2017 to an average of less than 58 days in 2022. The best performance achieved till now in terms of total meterage is 4,200 meters drilled in 51 days from the surface. The continuous improvement of the Ayatsil project has resulted in world class drilling performance.

The success factors include standardized well design, performance improvement processes that was made possible by the multidisciplinary well decision team like, flat times efficiency. In addition, the outcome of this approach has resulted that 33 million dollars have been saved from the original budget in reduction of well costs and impulse the productivity of Ayatsil field.

## 1 Introduction

### *1.1 Ayatsil field*

The Ayatsil field is geographically located in the Gulf of Mexico, 130 kilometers Northwest of Ciudad del Carmen, Campeche (Figure 1).

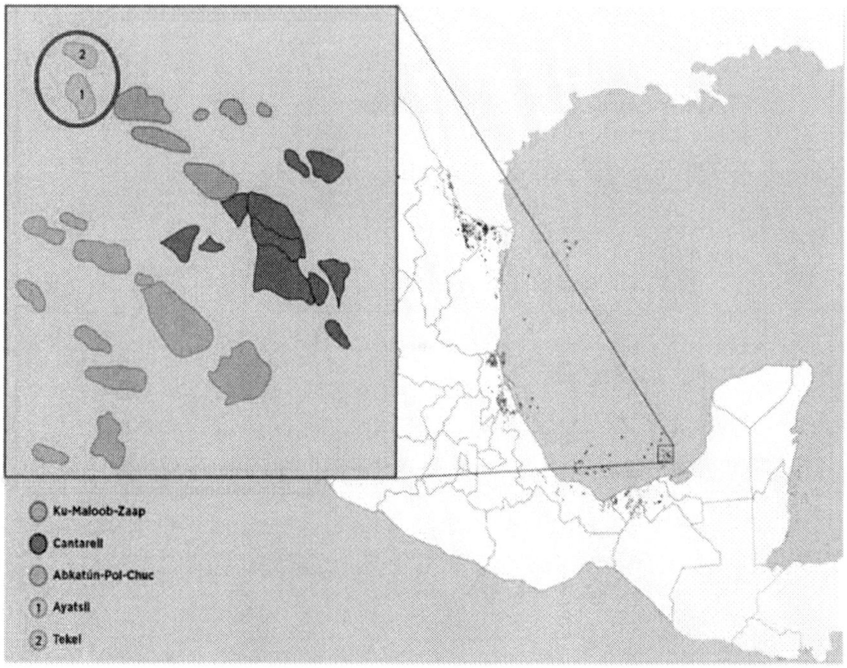


**Figure 1** *The red circle shows the location of Ayatsil field in the Gulf of Mexico.*

Ayatsil is the largest producer of Extra Heavy Oil field in Mexico, its production of extra heavy oil began at the end of 2015 and as more wells were added to the field, as a result of application the operational excellence project (Figure 2).

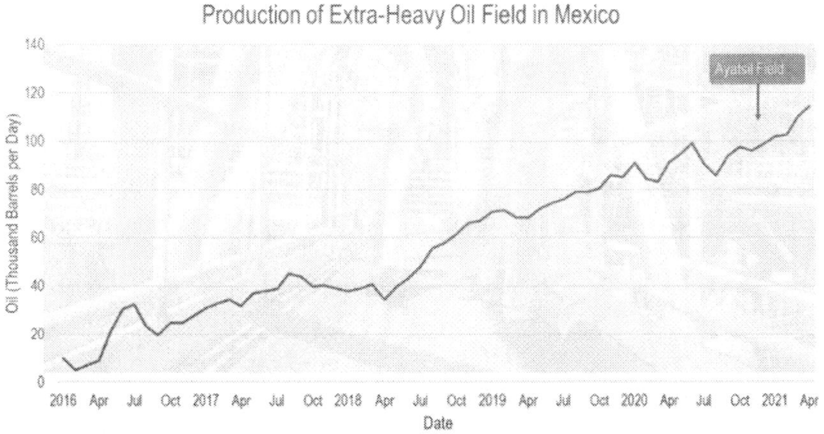


**Figure 2** *Production of Extra-Heavy hydrocarbon in the Ayatsil field* [1].

In 2019, the operator applied a set of strategic changes to optimize the drilling stages and accelerate field production. The operator implemented the policy of continuous improvement in relation to the processes, resources, and technologies available in the market, thus obtaining a notable improvement in drilling performance, resulting in more efficient drilling times and anticipated oil production.

Some of the common drilling challenges that was encountered during drilling was well control operations like kick/influx due to overpressured formation, total lost circulation and drilling through hard formation. This has significantly affected drilling times and hence well cost. Hence in order to reduce non-production time (NPT), several corrective measures were adopted. Some of the measures were improvement in the rates of penetration (ROP) and tripping times, as well as reducing flat time in the different stages during the well construction process (running casing, drilling and cementing).

To mitigate the challenges and risks described above and to reduce NPTs, the operator developed the operational excellence project (OEP) integrating 5 phases, which are listed below:

- Phase 1: Identification and evaluation of areas of opportunity
- Phase 2: Generation and selection of alternatives
- Phase 3: Development of the best solution alternative
- Phase 4: Execution of the solution alternative
- Phase 5: Evaluation of the executed solution alternative

All the above phases will include the mapping of the challenges with associated risk and mitigation, the process of reducing the NPTs, the multidisciplinary approach to optimize drilling, and the suitable application of technology to execute this new model for drilling well of Ayatsil field.

## 2 Field drilling challenges

### *2.1 Drilling campaing 2017-2019*

During the development of the field between 2017 and 2019, the total time to drill and complete the wells was averaging around 130 days due to the difficulties in drilling (Figure 3). At this time, a project was started to identify the risks associated with drilling in the field and the opportunity to

Journal of Physics: Conference Series **2594** (2023) 012003 doi:10.1088/1742-6596/2594/1/012003

mitigate the risks with the creation of the OEP that will improve in the decision making during critical activities.

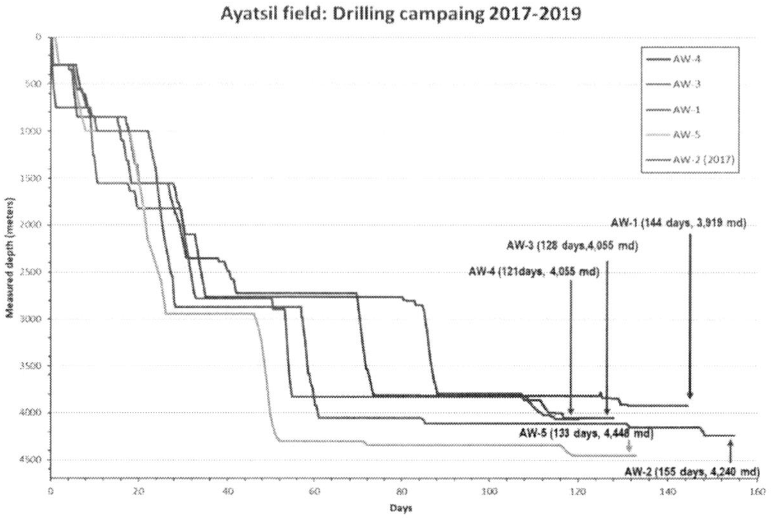


**Figure 3** *Depth versus Days Graph of Ayatsil Field drilling campaing, 2017 to 2019.*

Several of the drilling problems are associated with lithological complexity, which has the following characteristics:

- Interbedded permeable zones containing water and gas with different pressures which can cause a cross flow contaminating the drilling fluid.
- Severe loss of circulation or influx (water/gas) during drilling in these permeable sections.
- Interbedded section of naturally fractured mudstones causing breakout during drilling or Bottom Hole Assembly (BHA) trip outs.
- Total loss of circulation drilling the production zone (Cretaceous) due to its low pressure (0.5 g/cm3) depleted by production and with a high permeability.

In the diagram below, a detailed geological column with major issues which were identified and characterized to develop mitigation plan and solve the problems from the source (Figure 4).

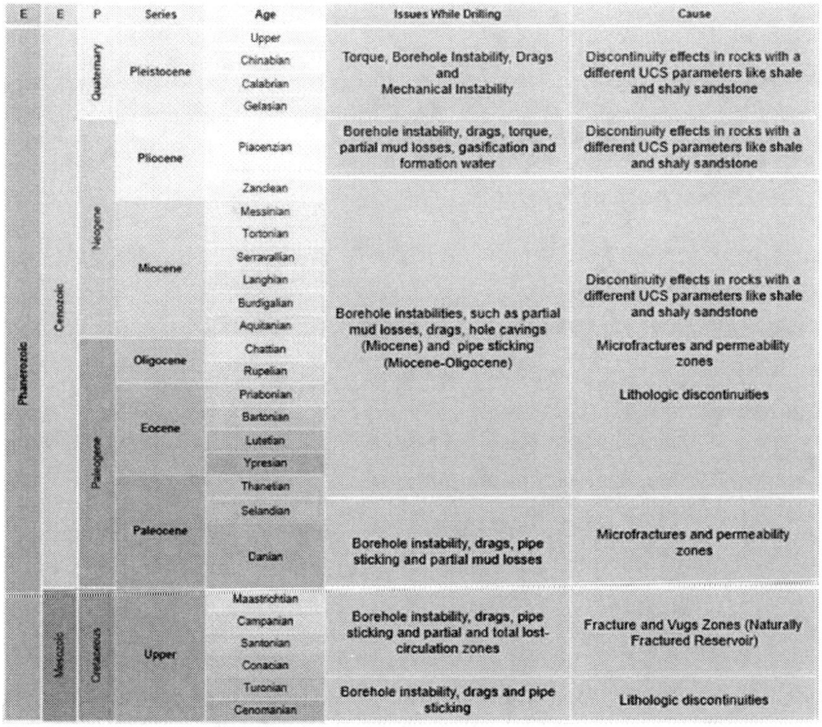

**Figure 4** *Geological prognosis a related problem.*

## 3 Operational Excellence Project (OEP)

### *3.1 Process Optimization*

OEP as a corporate operational strategy in drilling wells of Ayatsil field. The methodology used by the operator consists of 5 phases [5], which are listed below:

- Phase 1: Identification and evaluation of areas of opportunity
- Phase 2: Generation and selection of alternatives
- Phase 3: Development of the best solution alternative
- Phase 4: Execution of the solution alternative
- Phase 5: Evaluation of the executed solution alternative

The application of the 5 phases of OEP is essential for the correct selection, development, and execution of the adopted strategy for the new wells to be drilled in Ayatsil Field. OEP was applied for drilling in Ayatsil Field during the 2020 to 2022 leading to successful results. This paper emphasizes the stages of monitoring, execution, and evaluation of the methodology, which correspond to the deliverables of OEP team during the well construction process.

### *3.2 Phase 1: Identification and evaluation of areas of opportunity 8*

This phase starts with the selection of the multidisciplinary well decision team (MWDT), which is a team integrated of geologist, reservoir, geomechanics, drilling engineers and supported by services team. This multidisciplinary team work together to evaluate areas of opportunity in the well portfolio and to identify all drilling alternatives (Figure 5).

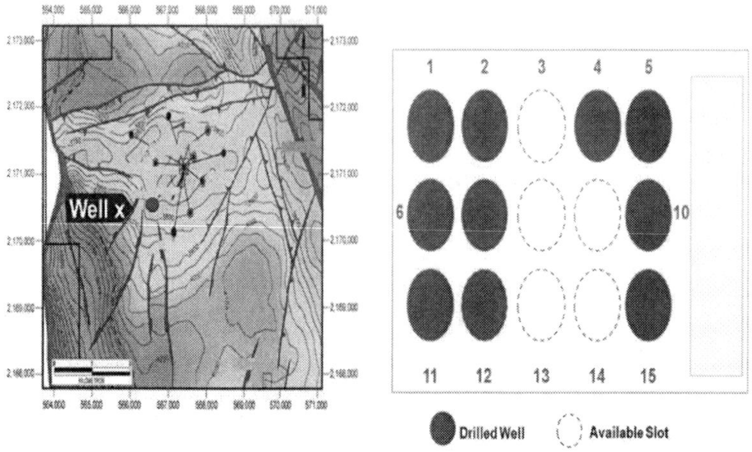


**Figure 5** *Identification and evaluation of areas of opportunity.*

The principal goal of this phase is to identify new opportunities using new technologies and the best operational practices to drill the new wells.

### *3.3 Phase 2: Generation and selection of alternatives*

The MWDT will check and identify all applicable data to suggest the best trajectory to hit the optimum target point. All plans are presented to the decision review panel (DRP) for review, validation, and suggestions (Figure 6).

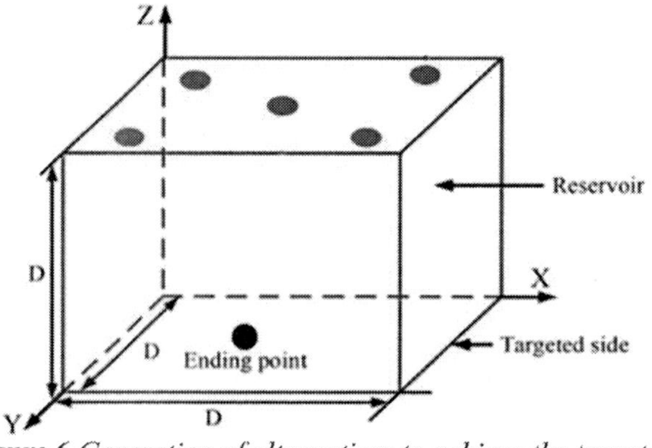


**Figure 6** *Generation of alternatives to achieve the target* [4,6].

MWDT provided the suggestions for the best direction to hit the target point and analyze alternatives with their respective risk factors.

### *3.4 Phase 3: Development of the best solution alternative*

The detailed to days vs depth graph, directional plan, expenditure days or drilling procedures, must be finalized and send to the members of DRP for approval in phase 3 (Figure 7). To complete all the potential risks, the DRP will collaborate with different teams to prepare the risk matrix at this stage of the process. This will help to identify all the potential risks identified by the MWDT, as well as the lessons learned, and best operational practices of the historical wells drilled in the field (Figure 8).

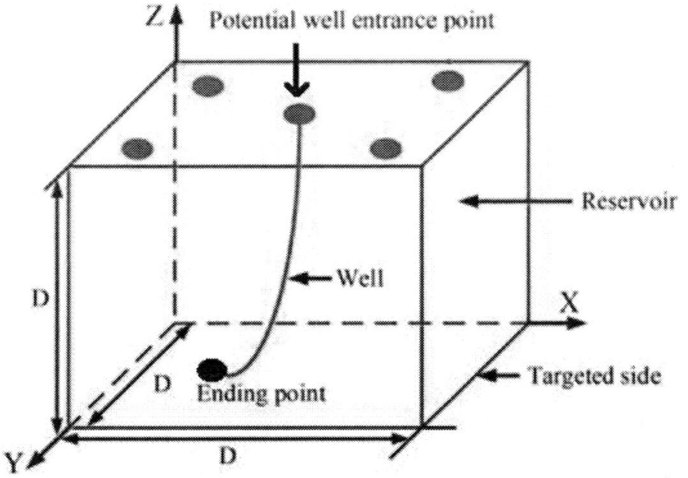


**Figure 7** *Generation of alternatives to achieve the target [4].*

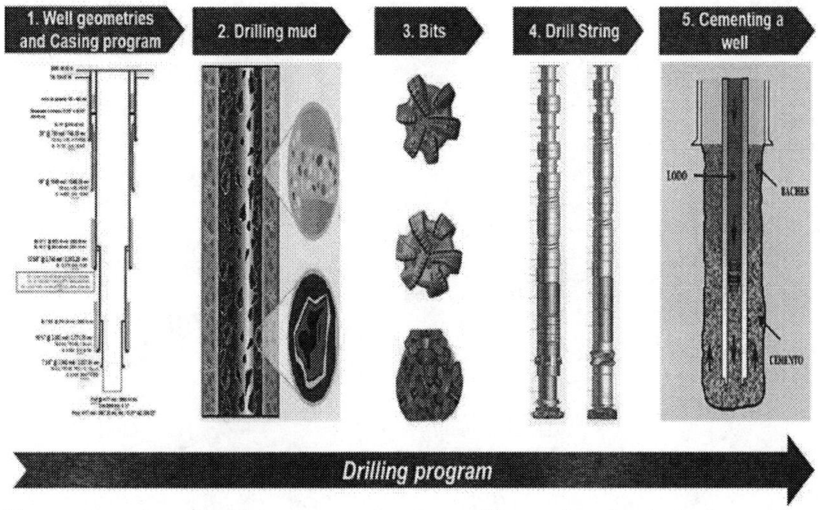


**Figure 8** *The engineering plan for constructing the wellbore, the plan includes casing programs, drilling mud, well geometries, bits, geomechanics analysis and estimations.*

This stage includes the optimization of the directional drilling target and plan, day versus depth curve including the authorization for expenditure days and perfect well days. This objective was also to develop well design assessment and drilling program.

### *3.5 Phase 4: Execution of the solution alternative*

The target of the pre job is to inform those involved in the execution of the project, the objective of the new project, risks, lesson learned, recent changes and responsibilities for the members of MWDT (Figure 9).

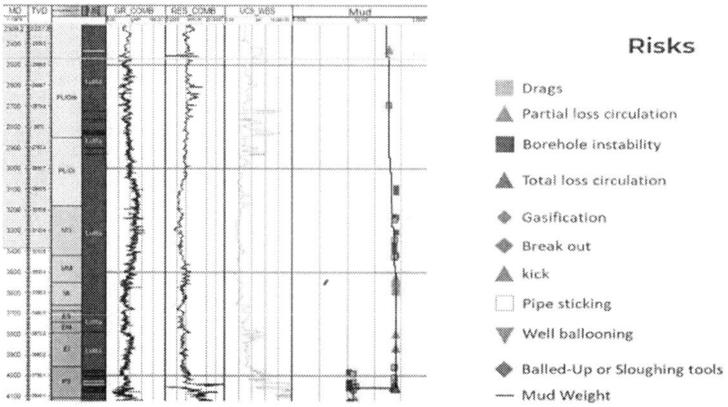


**Figure 9** *Pre job Meeting, show risks, lessons learned, recent changes and opportunities to improve the processes in the new drilling well.*

To capture the complete potential risks, the MWDT will conduct the meeting at this stage. The purpose of this meeting is to identifies all the potential and real risk based on the history of the drilled wells. In the other hand, the MWDT provides the best road map, especially for complex operations during the execution (Figure 10).

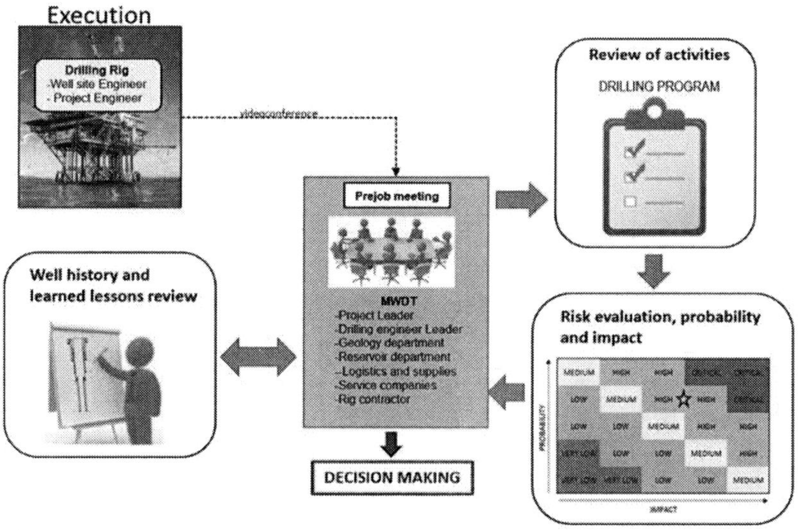


**Figure 10** *MWDT provides the best road map* [3,11].

## 3.6 *Phase 5: Evaluation of the executed solution alternative*

It is considered the last stage of the project; the objective is to evaluate the performance of the drilling and the costs of the project in relation to the premises generated in phase 1. The technical limit, the performance metrics are updated, and the non-productive times of the project are identified (Figure 11).

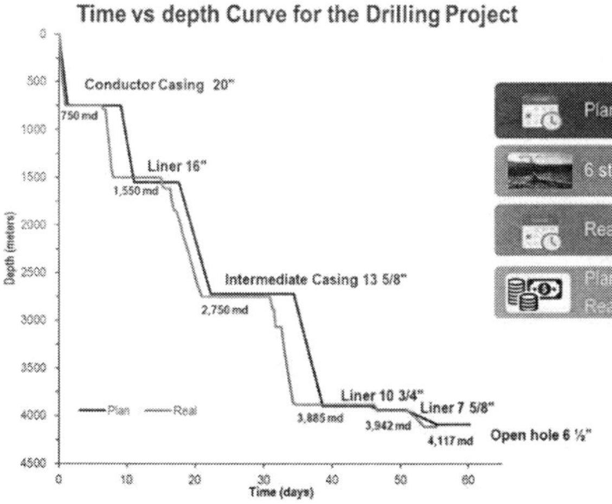


**Figure 11** *Evaluation of performance of the drilling project* [2,7,8,9,10].

The final stage captures all new situations encountered and the process improvement in the project. In this stage, the technical limit is reviewed, lost time is identified, and the performance metrics compared to the best well drilled in the field is updated.

This process is not designed to eliminate all the risk but minimizes them by reducing the level of risk during well intervention. In otherwise, this process applies a scalable risk assessment in all projects and records the causes of NPTs which are used to highlight risks and to determine the probability and magnitude of these events recurring in the drilling of future wells (Figure 12).

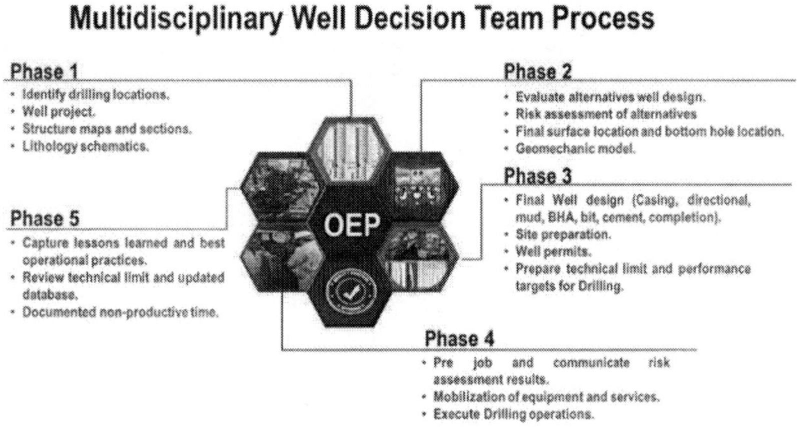


**Figure 12** *Multidisciplinary Well Decision Team Workflow* [3].

## 4 Technologies and operation optimization

### *4.1 Elimination of the 30" conductor*

At the beginning of the project, the Ayatsil wells were drilled with a 36" bit to the depth of 300 md and 30" conductor casing was run. Next was the 20" superficial casing at 1000 md. The two initial

Journal of Physics: Conference Series **2594** (2023) 012003 doi:10.1088/1742-6596/2594/1/012003

sections (36" and 26") used water base mud for drilling. An analysis was carried out with formation samples from the seabed to a depth of 1000 md where the cation exchange of the shales was analyzed to review their reactivity to water. It was possible to detect that the reaction began to increase after 650 md ( Figure 13). Based on the information analyzed, it was decided to remove the 30" pipe and start drilling with 26", then run the 20" surface casing to 700 md. Due to the fact that the drilling was without circulation, seawater was used to drill with the help of viscous pills.

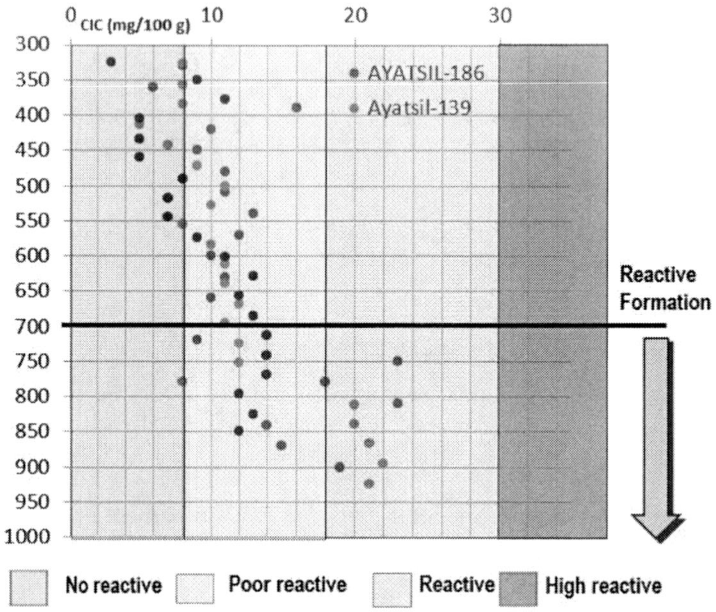


**Figure 13** *Formation (shales) reactive to water.*

Formation samples were taken from 300 to 1000 meters from three wells in the field, analyzing their reaction and based on this, it was possible to select the depth where it could be drilled with seawater (Figure 14).

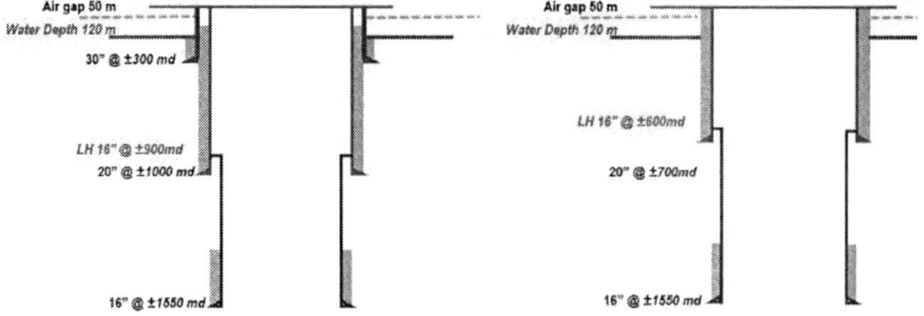


**Figure 14** *Well schematic eliminating 30" conductor.*

The benefits of this optimization were 4.1 days of rig time and a reduction in cost from the conductor pipe, accessories, casing running equipment, and cement job. To minimize risk a reamer shoe and casing running tool was used for the 20" casing.

### *4.2 Controlled mud weight increments to prevent severe lost circulation.*

In the Ayatsil field, an intermediate section (14 1/2" x 16 1/2") is drilled, which is characterized by having permeable intercalations (Figure 15) with varied range of pore pressures. The first wells drilled used a mud density to contain the highest pore pressure observing severe loss of circulation; in this sense, these permeable zones were identified to carry out a controlled increase in mud weight, minimizing fluid loss and giving time to form a strong mud cake enough to withstand the increase in density to finish the section [3].

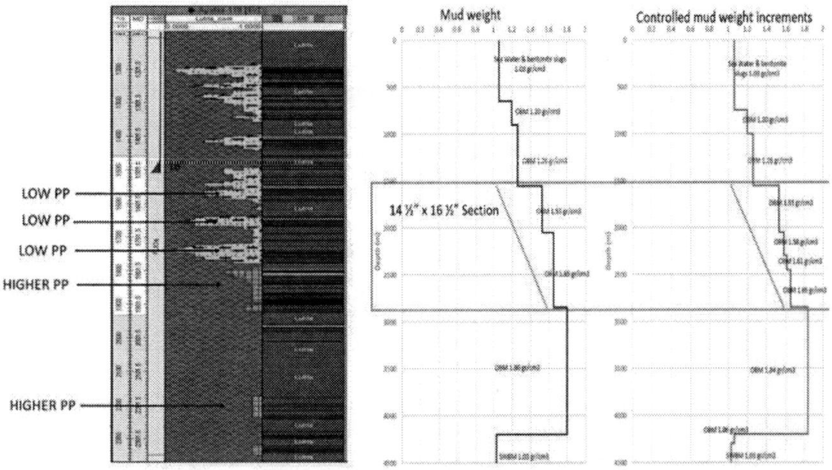


**Figure 15** *Controlled mud weight increments.*

After adjusting the program for the increase in density, it was possible to reduce drilling times since previously; there were additional times for pumping Lost Circulation Materials (LCM) and, in some severe cases, pumping cement plugs to seal this permeable zone.

### *4.3 13 5/8" swell packer to insolate annular after primary cement job*

Due the presence of interbedded sands in the intermediate section most of the 13 5/8" casing cementing jobs are done with total loss of circulation. Sometimes while waiting for the cement to set, they observed flow in the annular space between 13 5/8" – 20", hindering the operation from removing the Blow Out Preventers (BOPs) and additional time for well control activities.

The solution for this problem was to design a swell packer installed in 13 5/8" casing at 500 meters reactive to Oil-based drilling Mud (OBM) with a short time to insulate the annular with the 20" casing and proceed to remove safely the BOPs (Figure 16).

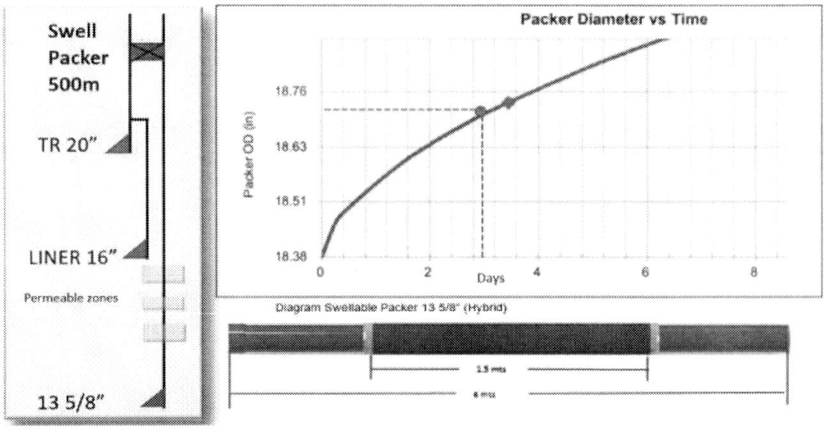


**Figure 16** *13 5/8" swell packer design.*

For the design of this packer, an oil-reactive and accelerated elastomeric compound was selected to make a seal in three days. To simulate its swelling, variables with temperature, fluid properties, and centralization was considered in the case of inclinations greater than 5 degrees. Once these packers are installed, the integrity is tested with 1000 psi in the annulus before removing BOPs. With this technology, it was possible to reduce the NPTs to control the pressures in this annular space, which in some cases reached 15 days.

### *4.4 Wellbore stability during drilling (use of asphalts)*

Another challenge was detected in the overpressured section (12 1/4") with presence of interbedded fractured carbonate, which did not represent a problem when drilled, the problem came out when this formation was hit by the continuous rotation of the BHA and start to collapses causing wellbore instability and stuck pipe problems most of them when tripping out thru this section. The image shows some of the cavings (Figure 17) recovered on the surface and their geological origin (Miocene – Oligocene).

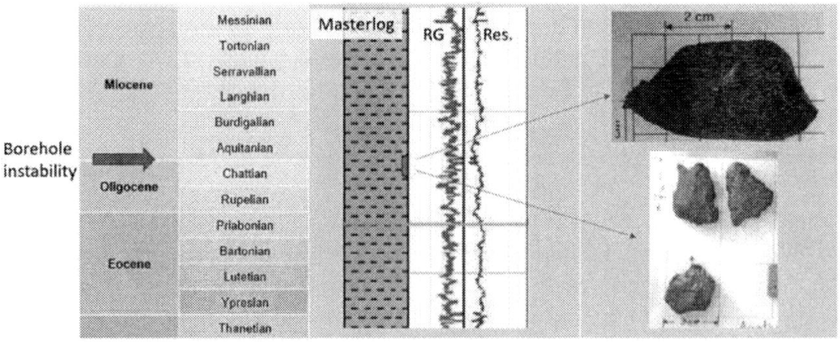


**Figure 17** *Unstable formation location and cavings.*

This unstable zone was causing continuous solids pack off in the BHA; An integrated work was carried out in conjunction with the fluid company on the problem where a change in the formulation came out, which is adding a sulfonated asphalt to improve borehole stability by plugging

Journal of Physics: Conference Series 2594 (2023) 012003 doi:10.1088/1742-6596/2594/1/012003

microfractures; This eventually would also help increasing fluid lubricity and reducing torque and drag (Table 1).

**Table 1.** *Physical properties asphalt selected.*

| **Typical physical properties** | |
|---|---|
| Physical appearance | Fine black gray powder |
| Specific gravity | 1.40- 1.65 |
| pH (2% solution) | 7- 10 |
| Solubility in water | 65% to 80% at 72 degF |

This solution helped in reducing the NPT by eliminating activities to stabilize the borehole with continuous reaming or, worst case, drilling a possible sidetrack.

### *4.5 Use of swell packers as a replacement to cementing*

After the high-pressure section, the production zone (Cretaceous) follows, which is very permeable but has very low pore pressure (0.5 g/cm3), requiring a short liner 7 5/8" (60 -90 meters) to cover the transition to the depleted reservoir. At first, attempts were made to cement this liner without success since this section had partial fluid circulation, making it difficult to place a small volume of cement. The solution was to replace the cement with swell packer (Figure 18).

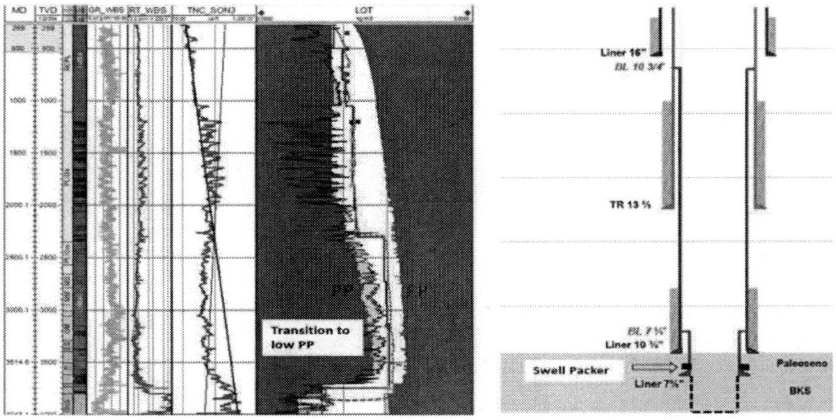


**Figure 18** *Pressures profile model and well schemati.*

A 7 5/8" swell packer to replace cement job was designed for an 8 ½" hole size, with a hybrid elastomer reactive with OBM or Water-Based Drilling Mud (WBM) due to the next section being drilled with seawater base mud (Figure 19).

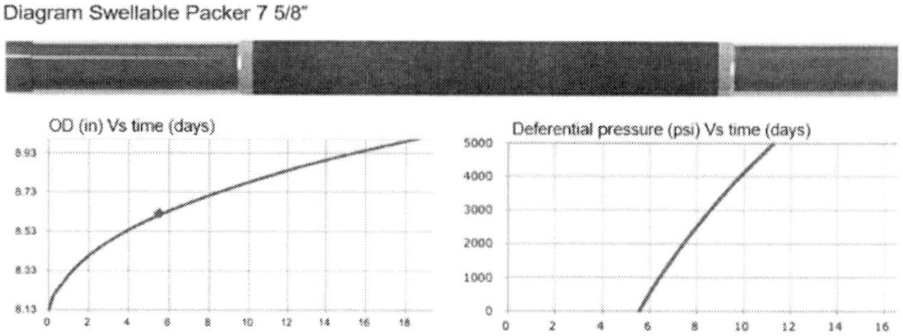


**Figure 19** *7 5/8" swell packer expansion simulation*

This solution helped to reduce the operational times by approximately one day, which eliminated the cementing job and a trip for the drill out.

*4.6 Drilling technique with total loss of circulation*

The last section (6 1/2") is the depleted reservoir zone (Cretaceous) which is characterized by low pore pressure (0.5 g/cm3) and very high permeability. The formation is mainly naturally fractured dolomite [3,11]. From the beginning of this section, drilling is carried out with a total loss of circulation Seawater-based mud is used to have a greater capacity to generate mud volumes, generally greater than 3,000 m3 (Figure 20).

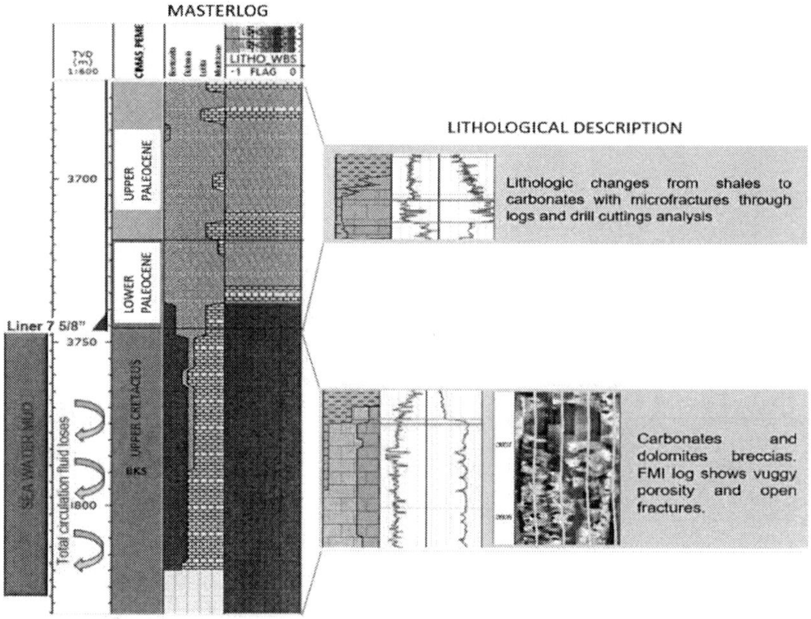


**Figure 20** *Lithological description of the reservoir section.*

Due to the fact that there is no circulation during drilling, the cleaning of the hole is poor, which means that the risk of packing the BHA with cuttings is high. Hence, to minimize this risk, the following operating practices were established:

- The Rate of Penetration (ROP) is controlled at 5 m/hr to give enough time for the cuttings to be crushed in small parts and injected into the formation through the larger fractures.

- A short trip with backreaming is made every 10 meters drilled to stabilize erratic torques and move cuttings that may be on the stabilizer.
- With the Pressure While Drilling (PWD) tool, the fluctuations of the fluid level in the annulus are monitored to check if they are plugging the fractures with the cuttings or if the admission is unrestricted to the formation.
- Every 10 meters drilled; a viscous pill (5 m3) must be pumped to help carry cuttings to the zone of mayor fractures.

These operating practices have helped to considerably reduce NPTs caused by stuck pipe in the reservoir section of the well.

## 5 Results and conclusions

### *5.1 Results*

The performance improvement process, standardized well design and equipment improvement, are proven indicators that improved drilling performance in Ayatsil field. It is important to mention, all those factors may not be new innovations but are evolving in response to the Ayatsil field new challenges. During the period 2019 to 2022 Ayatsil drilling campaign, a total 26 wells were drilled. In the end, the result was that the utilization of OEP in Ayatsil field was effective in improving drilling operations in order to achieve goals [5].

The successful performance of Ayatsil field from 2019 to 2022 can be summarized in the following:

1. By optimizing the required operations, and selecting the best technology for each problem, drilling times per well were considerably reduced in the Ayatsil field. A new benchmark of 51 days with a key performance indicator (KPIs) of 82.4 m/day (270.3 ft/day) in a field with a complicated geological structure was achieved. Before all these actions wells were drilled in 85 days with a KPI of 49.4 m/day (Figure 21).

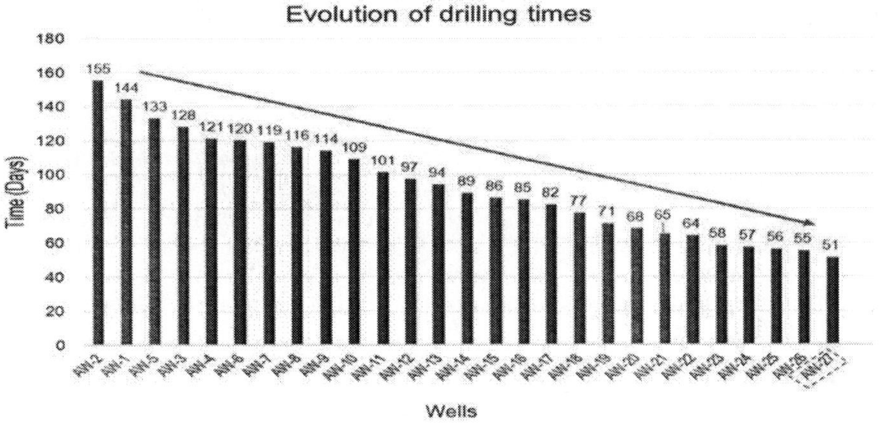


**Figure 21** *Data of day per well* [2,7,9].

2. The project was able to reach an oil production of 117,000 Barrels per day (BPD) anticipating the forecast for the field development (Figure 22).

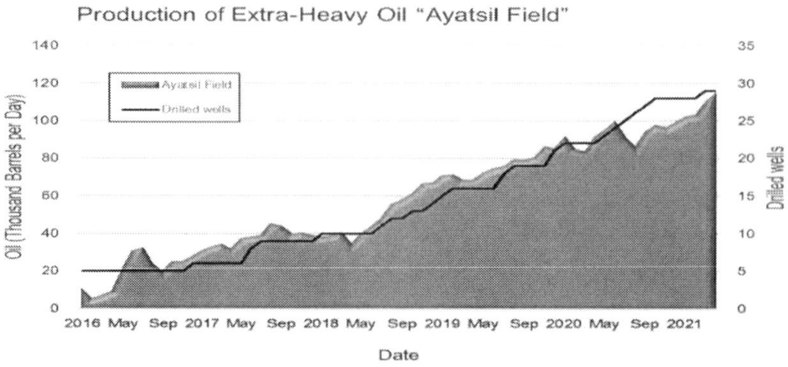


**Figure 22** *Historical production data.*

*5.2 Conclusions*

1. This paper shows that Operator has achieved its internal goal to achieve optimum drilling time. This achievement is marked by achieving a new benchmark of drilling a well in 51 days in 2022.
2. The purpose of the OEP is to provide guidance and management required to plan, design, and drilling new wells in Ayatsil field.
3. OEP is a systematic approach in drilling which has five critical areas that assesses the opportunity, defines the strategic objectives, executes the strategic plan, evaluates the strategic performance, and makes corrective adjustments in strategy.
4. The experience of the multidisciplinary team has shown that the application of new technology is necessary but not sufficient to solve the complexity of drilling operations.
5. Currently, the field continues to drill and seeks a record in each well, capitalizing on the lessons learned.

**Acknowledgments.**
The authors thank the Ku Maloob Zaap Asset for permission to publish this paper.
The authors would also like to thanks to all the people who contributed to make this publication possible.

**References**

[1].National Hydrocarbons Commision. National Hydrocarbons Commision. *National Hydrocarbons Commision.* [Online] June 19, 2023. https://produccion.hidrocarburos.gob.mx/#.

[2].*Drilling performance monitoring and optimization: a data-driven approach.* Lashari, Shan e Zehra, et al. 4, s.l. : Spinger link, April 2019, Journal of Petroleum Exploration and Production Technology, Vol. 9, pp. 2747-2756.

[3].*Real-Time Enabled Workflows Help Reduce Drilling Events, Minimize Drilling Times and Anticipate Production in Extra Heavy Oil Asset.* Carillo, Luis A., et al. s.l. : International Petroleum Technology Conference, November 2016, OnePetro.

[4].*A novel approach to optimising well trajectory in heterogeneous reservoirs based on the fast-marching method.* Lyu, Zehao, et al. s.l. : Elsevier, April 2021, Jornaul of Natural Gas Science and Engineering, Vol. 88.

[5].PEMEX E&P. *Operational Guide for Infrastructure Standardization, Monitoring Operations and Decision Making in Real Time por Drilling Activities.* GO-PP-TC-0002-2014. Carmen : PEMEX E&P, 2014.

[6].*Optimizing the separation factor along a directional well trajectory to minimize collision risk.* Mansouri, Vahid, et al. s.l. : Springer Link, 2020, Journal of Petroleum Exploration and Production Technology, Vol. 10, pp. 2113-2125.
[7].*Analysis of rate of penetration prediction in drilling using data-driven models based on weight on hook measurement.* Hashemizadeh, Abbas, et al. s.l. : Springer Link, 2022, Earth Science Informatics, Vol. 15, pp. 2133-2153.
[8].*Real-time and multi-objective optimization of rate-of-penetration using machine learning methods.* Zhang, Chengkai, et al. s.l. : Elsevier, April 2023, Geonergy Science and Engineering, Vol. 223.
[9].*Evaluation of coupled machine learning models for drilling optimization.* Hegde, Chiranth and Gray, Ken. s.l. : Elsevier, 2018, Journal of Natural Gas Science and Engineering, Vol. 56, pp. 397-407.
[10].*Improving drilling performance through optimizing controllable drilling parameters.* Bani, Ahmed Mustafa, et al. 3, s.l. : Springer link, 2021, Journal of Petroleum Exploration and Production, Vol. 11, pp. 1223-1232.
[11].*Applying a Real-Time Engineering Methodology to Drill of Potential Undesirable Events.* Perez, Carlos, et al. Houston, Texas : Offshore Technology Conference, 2012. Offshore Technology Conference.

Journal of Physics: Conference Series **2594** (2023) 012004 doi:10.1088/1742-6596/2594/1/012004

# Depositional Architecture and Reservoir Quality Assessment of Abu Roash (G) Dolomite, as a Low Resistivity Pay, Razzak Oilfield, Western Desert, Egypt.

**A Abu Shady1, Mena Fekry1, M Elsayed1, Scott Durocher1, Rushdy A Ali2**
(1) Khalda-apache Petroleum Company
(2) Premier Oil Group

Mena.fekry@khalda-eg.com rushdy.ali@permiercorex.com

**Abstract.** Abu-Roash (G) dolomite (A/R-G) is considered one of the most prolific oil-bearing reservoirs in the Western desert. In this study, Razzak Oilfield was selected to investigate the geological and petrophysical findings of these successions that were assigned stratigraphically to the upper Cretaceous (Late Cenomanian), in the Alamein basin. Sedimentologically, three main facies groups have been recognized in the studied successions of A/R-G and include dolostone, limestone and mudstone facies; and depositional; are divided into three parts: A) Back-reef to reef crest (Reefal depositional setting) - dolostone facies group; graded to B) Inner-ramp of limestone / dolomitic limestone facies types; then C) mud-flat (intertidal setting) of calcareous mudstone facies type. Petrophysically, most of the recorded pore sizes in the dolomite zone are mesopores (≈1 micron), with scattered macropores (2-10 microns), as identified by MICP on some selected 1.5" core plugs. In general; A/R-G dolomite displays low resistivity pay (high conductivity) due to: A) The heterogeneity of moldic and vuggy pore size distributions (macropores); associated with the dominant intercrystalline-dolomite mesopores; and B) Fe-rich minerals that increase the conductivity; where most of the identified dolomite crystals are secondary, late-diagenetic stage ferroan dolomite; and confirmed that petrographically by thin-section analysis.

## Introduction

The Upper Cretaceous – Late Cenomanian Abu Roash (G) Dolomite (A/R-G); Figure 1; is one of the most oil-bearing reservoirs as a low resistivity pay and it is widely distributed in the Alamein basin, Northern Western Desert, Egypt. In this study, Razzak Oilfield was selected to investigate the geological and petrophysical controls of Abu Roash (G) Dolomite. The Razzak field "complex", so named for its four separate culminations (West Razzak, Razzak Main, East Razzak and North Razzak; Figure 1). The Razzak field "complex" is developed along the northeast-southwest-trending Qattara-Alamein ridge; it is a prominent linear, structurally high trend along which the Razzak, Yidma, and Alamein fields are located (Abdin et al., 1993; and Figure 2). The Qattara-Alamein axis is situated within the Alamein basin, which appears to have had its maximum basinal development during the Early Cretaceous. The entire Razzak field complex extends for 17 km (10.6 mi); East Razzak is 3.5 km by 1.5 km (2.2 by 0.9 mi); Razzak Main is 3.0 by 1.0 km (1.9 by 0.6 mi); and West Razzak is 4.5 by 2.0 km (2.8 by 1.2 mi).

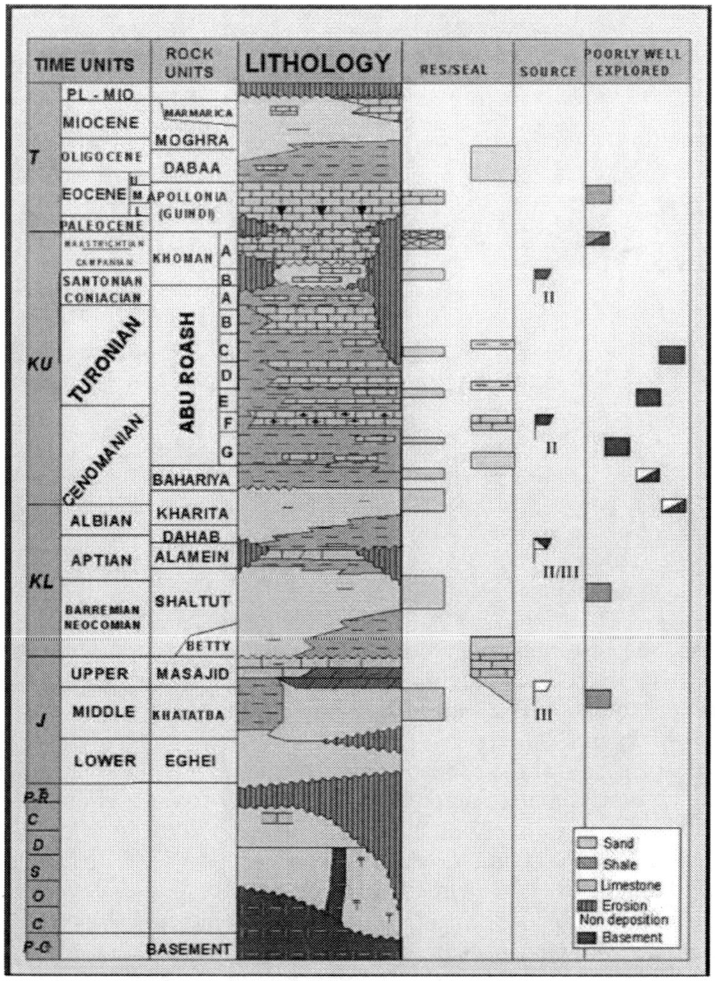


**Fig. 1.** Stratigraphic column of the Alamein Basin, Western Desert, showing the position of reservoir, seal and source rock horizons (Wahdan et al., 1996).

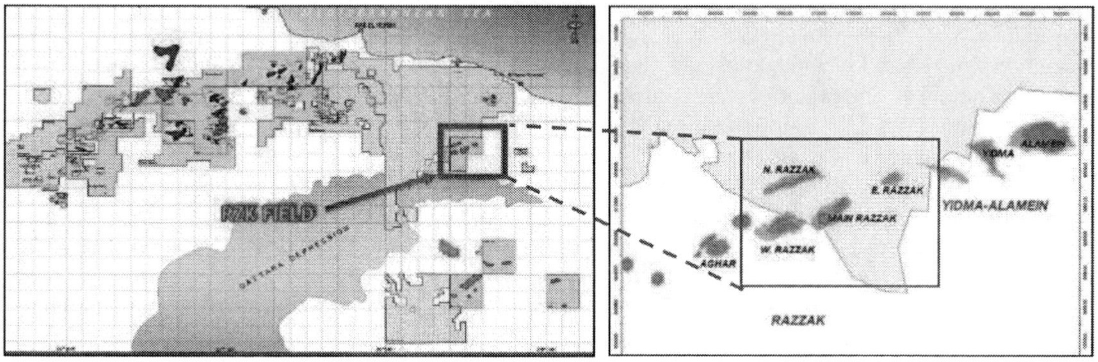


**Fig. 2.** Location Map of Razzak Oil Field; with four separate culminations - West Razzak, Razzak Main, East Razzak and North Razzak.

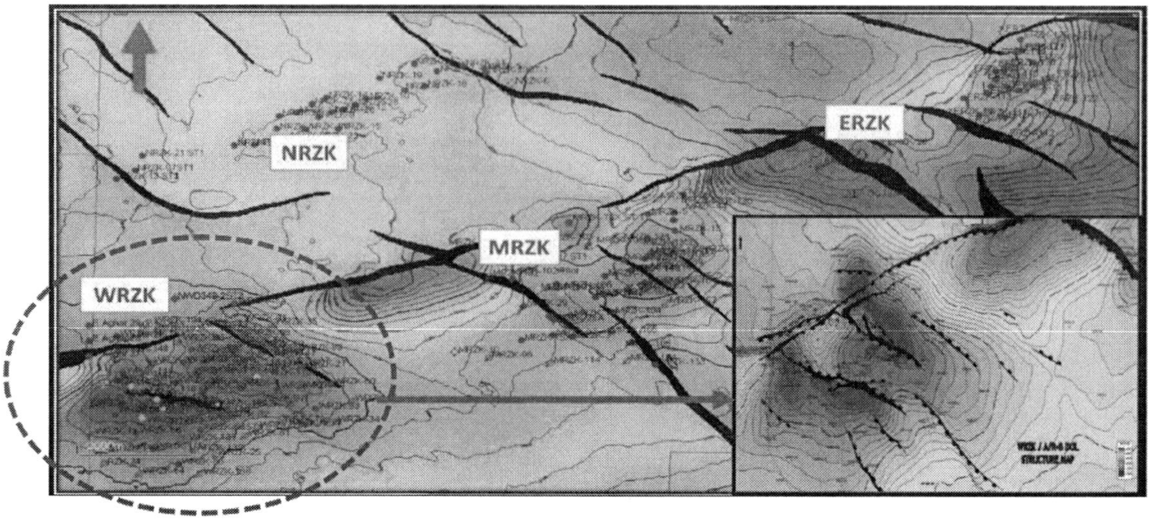


**Fig. 3.** Structure Contour Map of Abu Roash-G Dolomite – Razzak Oilfield. **Abbreviations:** WRZK-West Razzak Field; MRZK: Main Razzak Field; ERZK: East Razzak Field; and NRZK: North Razzak Field.

## Analytical Procedures and Aim of Study

Sedimentological and petrographical studies were applied on cores and cutting samples; which were selected from 11 wells over Razzak Oilfield. In addition to Mercury Injection Capillary Pressure (MICP) tests. In order to understand and interpret the obtained information with regard to reservoir quality as a low resistivity pay.
The sedimentological analysis of the available cores, which involves the identification of lithofacies types, their detailed description and interpretation of possible paleo-environmental setting; whereas; the petrographical analysis of core/drill cutting samples to identify and characterize carbonate mineralogy, texture and composition (skeletal and non-skeletal grains) for dolostone and limestone facies types, which further allows the petrographical classification of rock types and the evaluation of mineralogical controls (authigenic cement) on porosity and permeability. Also, understand the precipitation / occurrence of dolomite as primary or secondary (due to diagenesis). MICP was applied on some core plug samples; to understand the reservoir quality of the dolomite sequence (A/R-G).

## Sedimentological Core Descriptions and Depositional Settings of West Razzak Wells

Sedimentological core logging was applied for some selected core intervals throughout five wells drilled in West Razzak field (WRZK-12, -24, -60, -66 and -93 wells). Three main facies groups have been recognized and include dolostone, limestone and mudstone facies (Figures 3 and 4); as follows:

A) The most important facies is dolostone; which included seven lithofacies types; slightly argillaceous to faintly-laminated and massive-dolostone (Wackestone, Packstone and Grainstone; based on Dunham Classification; 1962). These are associated with reefal facies, which make up the bulk of the described cores. The reefal facies (based on Dunham's Classification and modified by Embry & Klovan; 1971; and James; 1984) include Floatstone with scattered coral fragments >2 mm in size, coral Bafflestone, and stromatolites/algal mats Bindstone.
B) Limestone includes slightly argillaceous to argillaceous limestone lithofacies types; massive to laminated textures, which are sometimes deformed and dolomitic due to bioturbation and dolomitisation process; respectively.
C) Mudstone includes calcareous mudstone and bioclastic mudstone lithofacies types.

Three depositional settings (Figure 5) have been recognized and fluctuated between Back-reef to reef crest (Reefal setting); as represented by the Bindstone (reef-crest) and Bafflestone to Floatstone (back-reef) dolostone lithofacies types. The depositional setting then graded to Inner-Ramp, as evidenced by the limestone/dolomitic limestone lithofacies types due to the occurrence of common amounts of peloids and benthic foram packstone to Dolo-packstone microfacies types; and confirmed that petrographically by thin-section analysis.

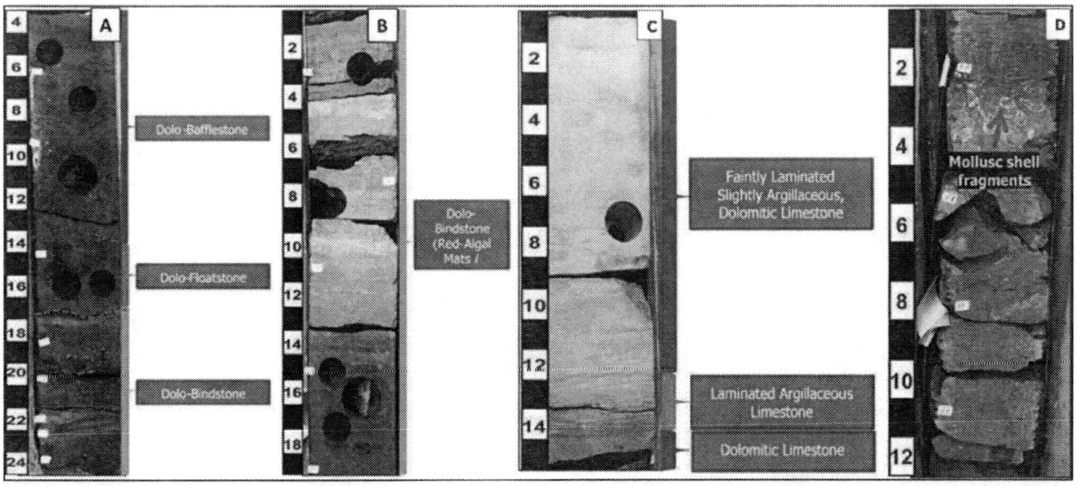


**Fig. 4.** Examples for some core photos from Well WRZK-66 represented the identified facies as follows: **Photos A and B** - Reefal facies and includes Dolo-Bafflestone, Dolo-Floatstone and Dolo-Bindstone. **Photo C** - Limestone Facies and includes Faintly Laminated Argillaceous, Dolomitic Limestone; Laminated Argillaceous Limestone and Dolomitic Limestone. **Photo D** - Calcareous Mudstone and Bioclastic (dominated by mollusc shell fragments) Calcareous Mudstone.

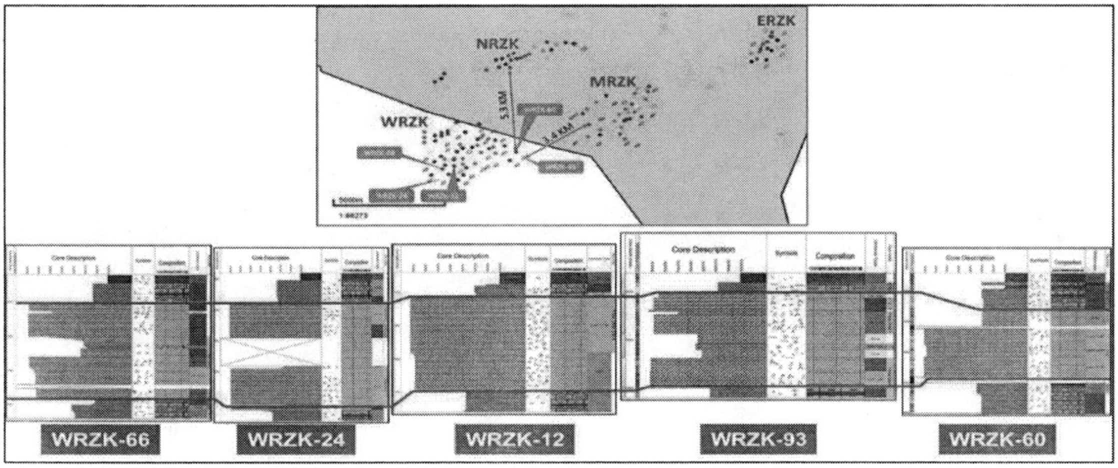


**Fig. 5.** Sedimentological core-logging correlation for dolostone facies cross the studied core wells of West Razzak Field (WRZK-66, WRZK-24, WRZK-12, WRZK-93 and WRZK-60 Wells).

Journal of Physics: Conference Series **2594** (2023) 012004 doi:10.1088/1742-6596/2594/1/012004

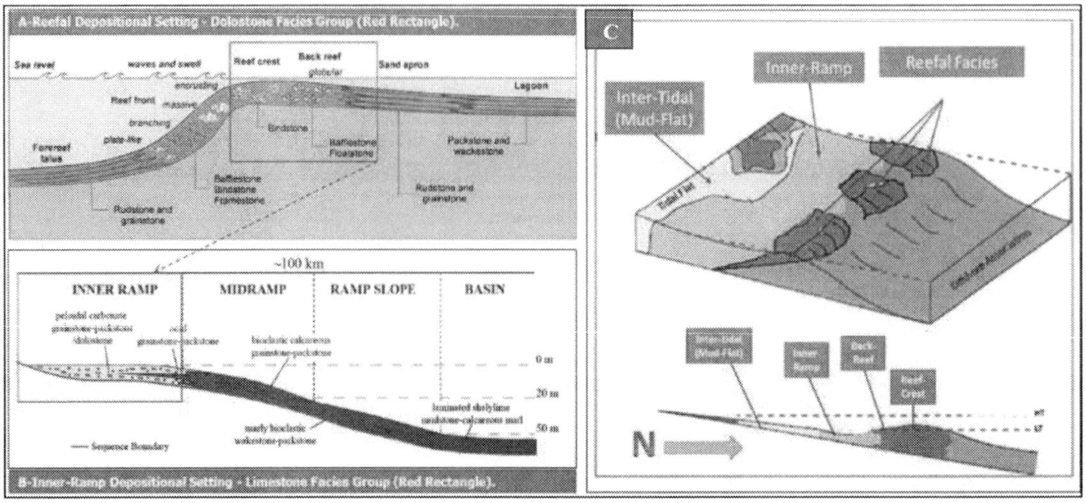


**Fig. 6.** Depositional settings of WRZK Field (A/R G Dolomite): **A)** Reefal depositional setting; **B)** Inner-Ramp depositional setting. **C)** Sketch diagrams show the distribution (fluctuation) of three depositional settings from reefal to inner ramp; then inter-tidal mud-flat (Abbreviations: HT high tide, LT low tide).

## Reservoir Quality of Abu Roash (G) Dolomite
## Methods:
### *Flow Unit Distribution R35*

vR35 (Winland's term) represents the average pore throat size that controls the fluid flow within the pore geometry.

Log R35 = 0.732 + 0.588 Log k – 0.864 log Φ

| r35 Calc / Classification | r35 | 35 = (10)^{0.732+0.588*LOG(**Permeability in (mD)**)-0.864*LOG(**Porosity in (%)**)} |
|---|---|---|
| Rock type 1 (Nano Pores) | < 0.1 | |
| Rock type 2 (Micro Pores) | 0.1 - 0.5 | |
| Rock type 3 (Meso Pores) | 0.5 - 2.0 | |
| Rock type 4 (Macro Pores) | 2.0 - 10 | |
| Rock type 5 (Mega Pores) | > 10 | |

**Correlation Based On Flow Unit distributions**

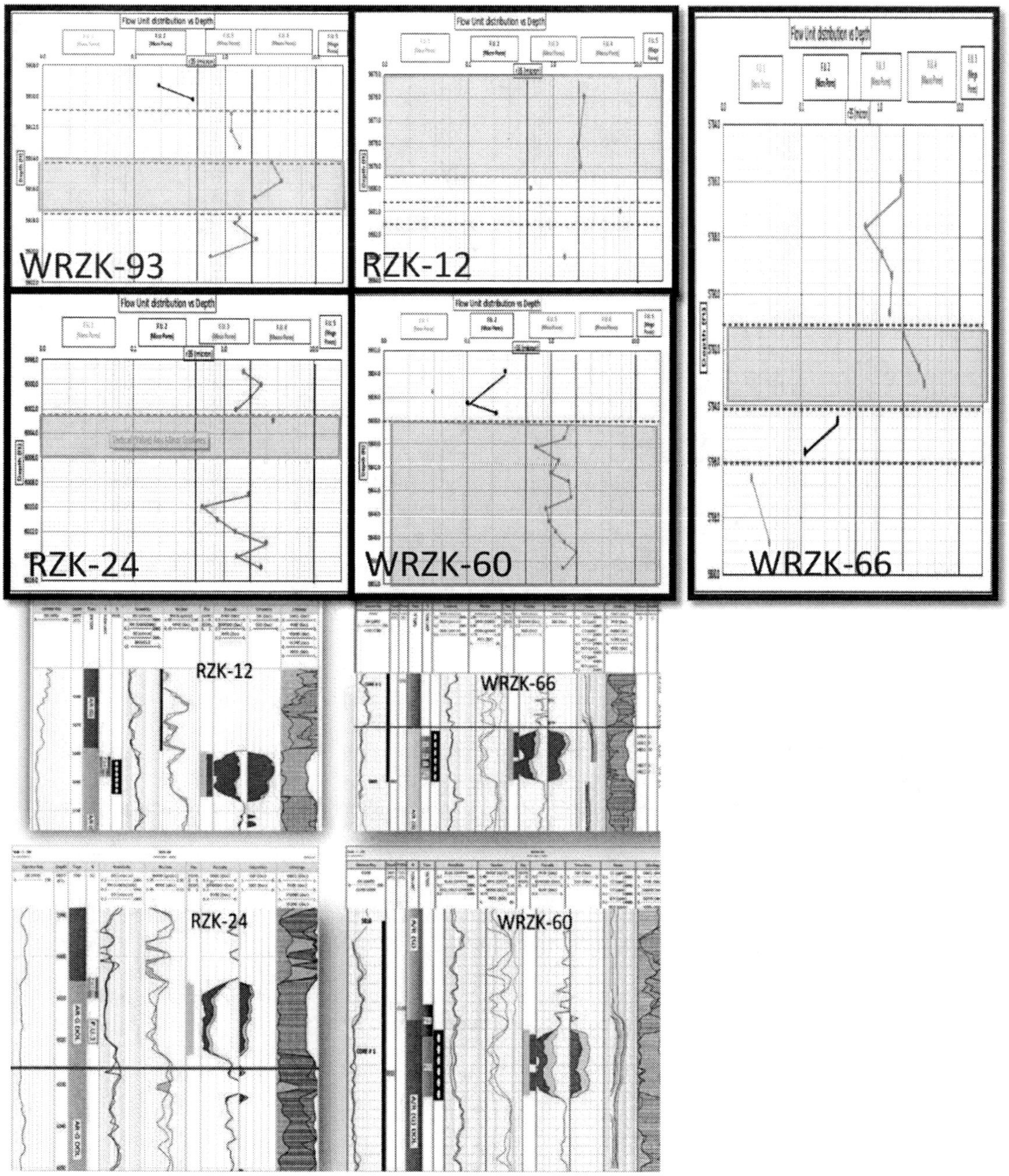


Petrophysically, the dolomite facies considers the best reservoir Petrophysically, the dolomite facies considers the best reservoir quality in the studied successions of the A/R-G; where most of the recorded pore sizes in the dolomite intervals are mesopores (≈1 micron),
with scattered macropores (2-10 microns), as identified by MICP on some selected 1.5" core plugs (Figure 6). Generally; Abu Roash - G Dolomite displays low resistivity pay (high conductivity) due to: **A)** The heterogeneity of moldic and vuggy pore size distributions (macropores); associated with the

Journal of Physics: Conference Series **2594** (2023) 012004 doi:10.1088/1742-6596/2594/1/012004

dominant intercrystalline-dolomite mesopores; and **B)** Fe-rich minerals that increase the conductivity; where the identified dolomite crystals in the successions are ferroan dolomite (secondary-dolomite) as a late-diagenetic stage; and confirmed that petrographically by thin-section analysis (Figure 6; Photos A and B)

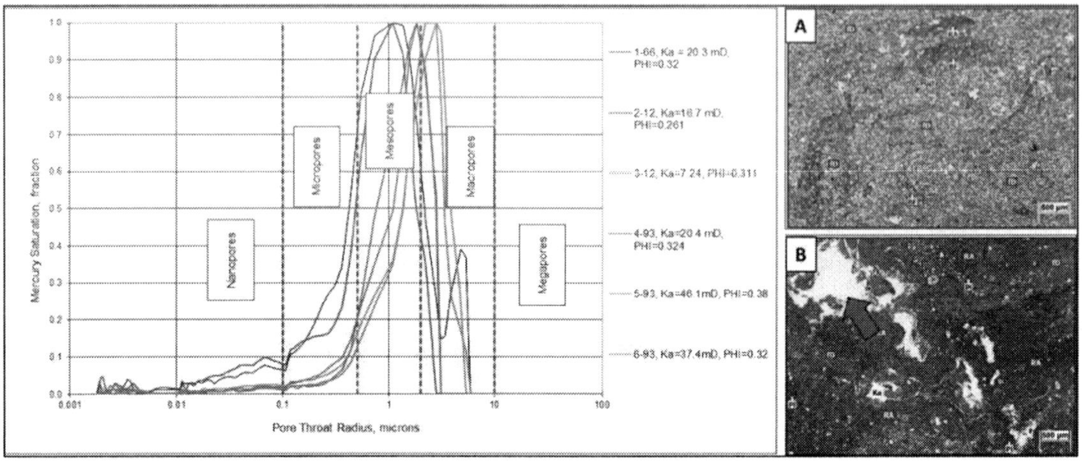


**Fig. 6.** Mercury injection capillary pressure results chart for some selected core plugs from Well WRZK-66. Mesopores (≈1 microns) are the most dominant pore sizes in the A/R-G dolomite (illustrated by thin-section photo

A; intercrystalline porosity between ferroan-dolomite crystals), with some scattered macropores (2-10 microns) represented by vuggy and moldic pores (illustrated by thin-section photo B; Red Arrow refers to vuggy pore).

**Depositional Settings of Razzak Field:**

The composition and internal organisation of the facies types and assemblages (dolostone, limestone and calcareous mudstone facies) distinguished in the cored and drill cutting successions from the studied wells in Razzak Field could have been suggest deposition in Three depositional settings:

**Back-reef to reef crest (Reefal depositional setting);** that includes the Bindstone (reef-crest) and Bafflestone to Floatstone (back-reef) – dolostone facies group; then, the depositional setting graded to

**Inner-Ramp of limestone / dolomitic limestone facies types**– due to occurrence of common amounts of peloids and benthic forams – packstone to Dolo-packstone microfacies types that observed in the petrography analysis; then the depositional-shifted to

**Mud-Flat (intertidal setting)** of calcareous mudstone (shale) facies type. The biogenic (bioclastic-fauna) content that detected macroscopically in the studied cores; include ostracods and pelecypod – mollusc shell fragments.

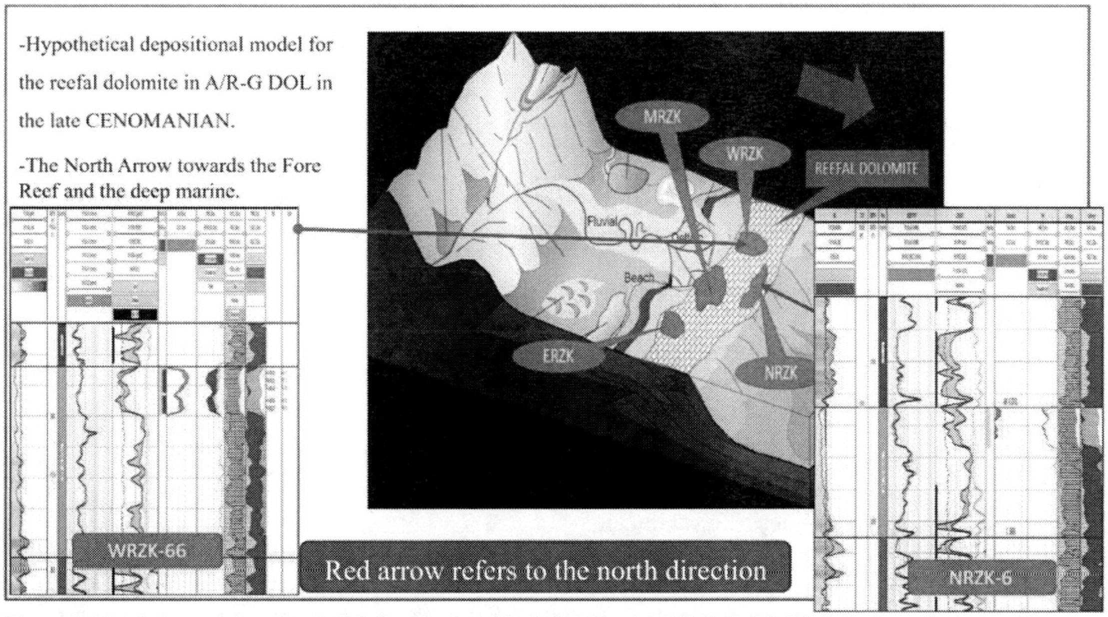


**Hypothetical depositional model for the reefal dolomite in ARG-DOL FM. In late Cenomanian in RZK Oil field, N W.DESERT, EGYPT.**

## Conclusions

Abu-Roash (G) dolomite is considered one of the most prolific oil-bearing reservoirs in the Western desert. The depositional settings of Abu Roash (G) - dolomite are divided into three parts: A) Back-reef to reef crest (Reefal depositional setting) - dolostone facies group; graded to B) Inner-ramp of limestone / dolomitic limestone facies types; then C) mud-flat (intertidal setting) of calcareous mudstone facies type. Most of the recorded pore sizes in the dolomite zone are mesopores ($\approx 1$ microns), with scattered macropores (2-10 microns), as identified by MICP analysis. Abu Roash - G Dolomite displays low resistivity pay (high conductivity) due to: A) The heterogeneity of moldic and vuggy pore size distributions (macropores); associated with the dominant intercrystalline-dolomite mesopores; and

B) Fe-rich minerals that increase the conductivity; where most of the identified dolomite crystals in thin-section analysis are secondary, late-diagenetic stage - ferroan dolomite.

## Acknowledgments

We would like to express our grateful to the Khalda Petroleum Company's management on their support and encouragement during this project and also the continuous cooperation with Premier Oilfield Group-Corex that covered the analytical tests.

## References

[1].Abdine A.S., Meshref W., Wasfi S., Shahin A.N., Aadland A., Aal A., [1993] Razzak Field-Egypt, Razzak-Alamein Basin, Northern Western Desert, *AAPG Special Volume, TR: Structural Traps VIII,* 29-56.

[2].Dunham, R.J., [1962] Classification of carbonate rocks according to depositional texture in "Classification of Carbonate Rocks" (W. E. Ham, ed.), *AAPG Memoir No. 1*, 108-121.

[3].Embry, AF, and Klovan, JE, [1971] A Late Devonian reef tract on Northeastern Banks Island, NWT: *Canadian Petroleum Geology Bulletin*, **v. 19**, 730-781.

[4].Wahdan, M., Fahmy, M., Brekelmans, H., [1996] 3D reservoir modelling and simulation in BAPETCO: case studies from the BED 1, 2 and 3 fields, Western Desert, Egypt. In: Proceedings 13th Petroleum Conference, vol. 2, EGPC, Cairo, pp. 250–260.

Journal of Physics: Conference Series **2594** (2023) 012005 doi:10.1088/1742-6596/2594/1/012005

# Exploration of the comparison method between nuclear magnetic logging porosity and core detection porosity

**Gui-pu Jiang1, Yan Fu1(✉) ,Zhong-shan Shen1,Wen Yu1, Teng-fei Cao1, Ying Du1, Min Liu1,Yu-ming Lei 1, Ming-duo Dai2, Yi-xiao Li3**
1 No.4 oil Production Plant of Daqing Oilfield Co., LTD
Daqing City, Heilongjiang Province
2Heilongjiang University
3Northeast Normal University
Fuy-cy4@petrochina.com.cn

**Abstract.** Porosity is the basis of reservoir parameter interpretation, permeability and saturation parameters are closely related to porosity, and the interpretation model of porosity has been established with the correlation between core porosity and electrical properties, and core porosity is the effective porosity measured by applying liquid saturation method. With the application of nuclear magnetic logging technology, the interpretation of porosity is more refined and can be subdivided into total porosity, effective porosity, capillary-bound water porosity, and clay-bound water porosity. In order to apply the nuclear magnetic logging technology more accurately, it is necessary to find a method to compare the two porosity detection methods. In this paper, we discuss the meaning of core porosity through the operation procedure of core detection porosity, combined with core piezometric curve and core density calculation method, get the conclusion that core detection porosity approximates total porosity, explore the comparison method between total porosity and core porosity of nuclear magnetic logging, and lay the geological foundation for the next step of applying nuclear magnetic logging data to improve porosity interpretation.

## 1 First Section

At present, the porosity electrical interpretation models commonly used in oilfield are all based on core-detected porosity as the standard, in recent years, with the application of nuclear magnetic logging technology[1], there is another reference standard and evaluation method. The core detection porosity only represents a single point, and the reference value decreases when the sampling points are small and the intra-layer variation is obvious; the nuclear magnetic logging detection porosity represents a layer segment[2], which cannot show the variation of the physical abrupt layer within the layer, so the two methods of detecting porosity can complement each other and evaluate each other, so the study of the comparison method between nuclear magnetic logging porosity and core porosity is especially important..

## 2 Definition of porosity

Quoted from "Explanation of Common Terms Used in Oil and Gas Field Development (Third Edition)", effective porosity is the ratio of connected pore volume to rock sample volume, total porosity is the sum of effective porosity and invalid porosity (not connected), also called absolute porosity. According to the definition of fine pore structure of nu
clear magnetic logging, effective porosity is the sum of capillary-bound water porosity and flowable porosity,[3] and total porosity is the sum of effective porosity and clay-bound water.

## 3 The porosity of the core test approximates the total porosityty

### *3.1 Analysis of core testing methods*

The currently used porosity electrical interpretation model comes from core analysis data, according to "GBT 29172-2012 Core Analysis Method", different determination methods are chosen to obtain the total porosity and effective porosity. The core data we applied is measured by the alcoholic liquid saturation method, which is based on Archimedes' principle, i.e.[4] the law of buoyancy, and measured as the effective porosity. According to the operation process, the rock sample should go through the process of washing, oil removal, drying and vacuum saturation[5], and the drying temperature is 105℃, according to the core analysis method 6.1.1.6, the dried rock sample will lose some or all of the clay-bound water according to the degree of drying,[6] so the effective porosity detected by the core liquid saturation method should be greater than the actual effective porosity, and infinitely close to the total porosity.[7] Similar to the washed clothes (non-waterproof type) the water excluded under human wringing pressure is similar to flowable water, the water excluded under centrifugal force is similar to capillary-bound water[8], and the water further dried out is similar to clay-bound water[6].

### *3.2 Analysis by means of density calculation*

According to the mass formula: m=ρv, the total mass of the rock sample is the sum of the masses of the various parts of the rock sample material.

$m=\rho_b V_{total} = \rho_{ma} V_{bone} + \rho_f V_{hole} == \rho_{ma} V_{total} (1 - \phi) + \rho_f V_{total}\ \phi$

Rectify that: $\phi = (\rho ma - \rho b)/(\rho ma - \rho f)$

where: $\phi$ total rock porosity %.

$\rho_{ma}$ rock skeletal density g/cm3, taken as 2.65 for general sandstone and 2.68 for calcareous sandstone.

$\rho_b$ bulk density of rock $g/cm^3$.

$\rho_f$ pore fluid density $g/cm^3$.

Rock density can be obtained by two methods, one is density logging and the other is core testing.

**Density logging method.**Density logging directly measures $\rho_b$ as the natural density[9], containing the formation fluid. Since the detection depth of density logging is shallow, about 15 cm, for high permeability formations, the detection range does not exceed the intrusion zone, so the corresponding $\rho_f$ is mud filtrate, and the density of freshwater mud filtrate is 1.0 g/cm3, but for low permeability formations, $\rho_f$ is the density of mud and formation fluid, and the calculated value of freshwater mud density alone is small, and density logging is influenced by more factors of drilling, and the calculation error is large, as follows Table, the calculated total porosity is smaller than the effective porosity from core testing.

**Table 1.** Statistical Table of Porosity Calculated from Density Logging of Well X5.

| Rock sample numberl | Effective porosity (%) | Density logging ($g/cm^3$) | Calculation of total porosity (%) |
|---|---|---|---|
| 405 | 14 | 2.4 | 15.2 |
| 410 | 17.8 | 2.36 | 17.6 |

| | | | |
|---|---|---|---|
| 415 | 15 | 2.43 | 13.3 |
| 420 | 13.9 | 2.54 | 6.7 |
| 425 | 16.2 | 2.42 | 13.9 |
| 430 | 15.4 | 2.41 | 14.5 |
| 435 | 15.5 | 2.43 | 13.3 |
| 440 | 14.8 | 2.44 | 12.7 |
| 445 | 15.3 | 2.39 | 15.8 |
| 450 | 15.6 | 2.39 | 15.8 |
| 455 | 13.7 | 2.39 | 15.8 |
| 460 | 10.7 | 2.56 | 5.5 |
| 465 | 17.6 | 2.43 | 13.3 |
| 470 | 15.2 | 2.45 | 12.1 |
| 475 | 14.2 | 2.43 | 13.3 |
| 480 | 12.3 | 2.45 | 12.1 |
| 485 | 13.7 | 2.42 | 13.9 |
| 490 | 15.7 | 2.41 | 14.5 |
| 351 | 12.7 | 2.32 | 20.0 |
| 352 | 12.6 | 2.32 | 20.0 |
| 410 | 17.2 | 2.52 | 7.9 |
| 57-3 | 17.3 | 2.36 | 17.6 |
| 59-3 | 15.9 | 2.42 | 13.9 |
| 33-3 | 26.8 | 2.24 | 24.8 |
| 38-3 | 28.2 | 2.19 | 27.9 |
| Average value | 15.9 | 2.4 | 14.9 |

**Core detection method.**The core test ρb[10], which is the dry density of the rock sample, contains pore space without oil and water, corresponding to ρf is the air density, 1.29x10-3g/cm3. 67 rock samples of various oil formation groups in high and low permeability were statistically compared, and the difference between the effective porosity explained by the core and the total porosity of the rock sample calculated from the rock skeleton density was very small, only +0.1%[11].

**Table 2. Summary of calculated total porosity and effective porosity of rock samples**

| Pound sign | Number of rock samples (block) | Dry eight (g) | Bulk density (g/cm$^3$) | Effective porosity (%) | Calculate the total porosity (%) |
|---|---|---|---|---|---|
| X5-1 | 25 | 28.193 | 2.266 | 15.2 | 14.5 |
| X5-2 | 2 | 23.174 | 1.900 | 27.5 | 28.3 |
| X2 | 35 | 23.468 | 1.949 | 26.0 | 26.5 |
| X1 | 5 | 23.221 | 1.938 | 26.4 | 26.9 |
| Average value | 67 | 25.204 | 2.065 | 22.0 | 22.1 |

### *3.3 Analysis of mercury injection experiment data*

According to "GBT 29171-2012-Determination of rock capillary pressure curve": the rock sample should also go through the process of washing, degreasing and drying, and if the determination is effective porosity, the maximum saturation of incoming mercury should be close to 100%.

**Table 3. Comparison of the distribution of permeability contribution values of core X5 wells**

| Sample number | Permeability classificationr (mD) | Microcontribution porosity (%) | No contribution porosity (%) |
|---|---|---|---|
| 38-3 | 973 | 13.3 | 3.9 |
| 33-3 | 160 | 18.1 | 15 |
| 59-3 | 4.4 | 20.1 | 18.2 |
| 445 | 2.1 | 21.7 | 13.2 |

**FIG. 1 Mercury pressure curve of rock sample 38-3 of X5 well**

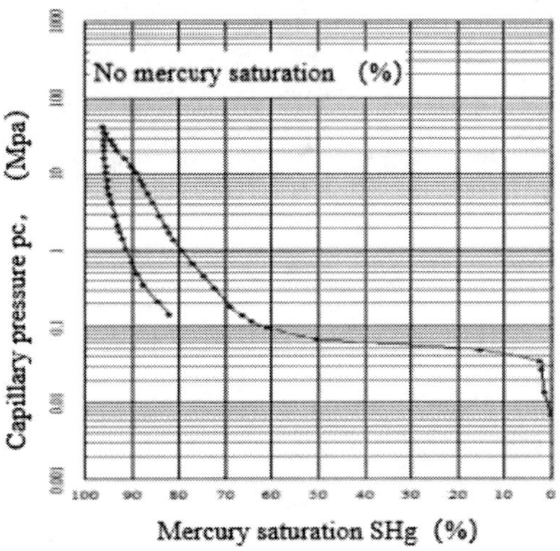


## 4 NMR logging interprets porosity

### *4.1 Nuclear magnetic logging explains the principle of porosity*

NMR logging measures the resonance relaxation characteristics of hydrogen nuclei in the formation, and probes the response of formation fluid rather than the response of the formation as a whole. This is the biggest difference between NMR logging and conventional logging. During logging, the instrument's inherent permanent magnet creates a gradient magnetic field in the radial direction of the formation to magnetize the hydrogen nuclei in the formation, and then an alternating electromagnetic field is applied to generate nuclear magnetic resonance in the formation Like. NMR logging uses spin echo method to measure the transverse relaxation time (T2) of hydrogen nuclei, which can obtain the total porosity, effective porosity, free fluid porosity, bound water porosity, pore size distribution (qualitative), permeability and other parameters of formation.[12]

The MRIL -- P instrument in use (FIG. 2) has a minimum echo interval of 0.6ms, enabling the observation of the clay water signal, thus obtaining the total porosity of the formation and the clay-bound water volume. The T2 distribution of NMR logging contains several independent relaxation times, and

T2 can be pre-selected, such as 0.5, 1, 2, 4, 8, 16, 32, 64, 128, 256, 512, 1024ms... And the corresponding porosity structure is more accurate.

In the NMR logging results, the porosity below 32ms represents capillary bound water, corresponding to the second peak of T2 spectrum. Conventional fracturing cannot be used. The area above 32ms is the effective usable porosity. Between 32ms-4ms is capillary bound water porosity. The clay bound water porosity below 4ms at T2 is the void porosity[13].

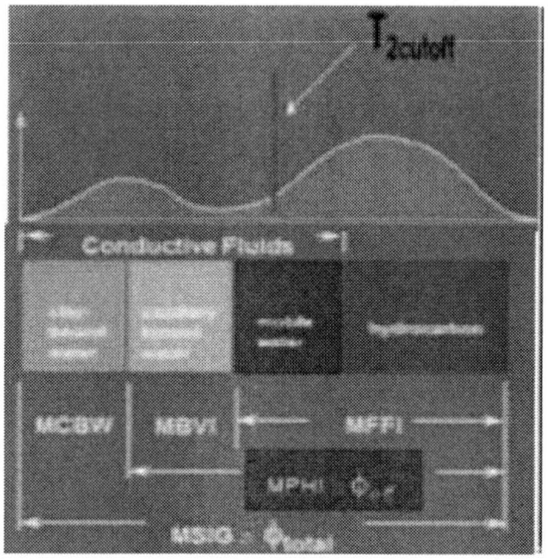


**FIG. 2 Principles of nuclear magnetic logging**

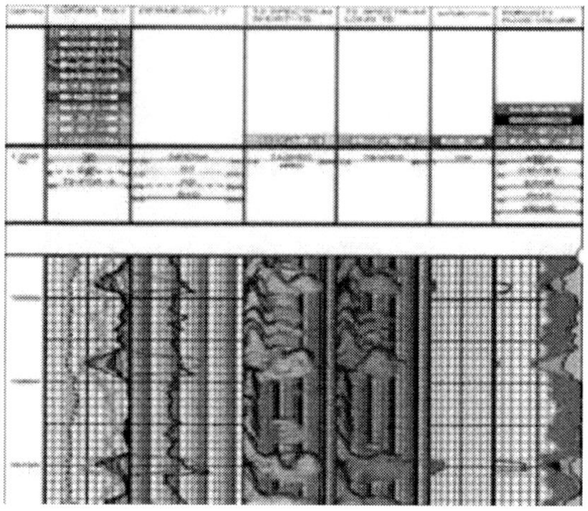

**FIG. 3 Nuclear magnetic logging interpretation results**

*4.2 Comparative analysis of nuclear magnetic logging porosity and core porosity*

**Comparison method.**NMR logging porosity is different from core porosity detection methods. NMR logging detects the overall pore structure of the sandstone group, while core porosity is the porosity at a certain depth point in the sandstone group. When core data is applied, the thickness weighted average of all rock sample data in the sandstone group should be made, and the relative thickness of the two should be equal[14-20].

**Comparison result.**According to the data of NMR logging and core synchronous detection, the absolute error between NMR logging interpretation and core interpretation effective porosity is 0.3%, and the absolute error between NMR logging interpretation effective porosity and core interpretation is 3.9%.

**Table 4.Statistical table of nuclear magnetic logging and core synchronous detection data of well X5**

| Number of rock samples (block) | NMR log interpretation results | | Core interpretation Porosity (%) | Total porosity | | Effective porosity | |
|---|---|---|---|---|---|---|---|
| | Total porosity (%) | Effective porosity (%) | | Absolute error (%) | Relative error (%) | Absolute error (%) | Relative error (%) |
| 14 | 15.2 | 11.0 | 14.9 | 0.3 | 2.0 | 3.9 | 26.2 |

## 5 Conclusion

5.1 Core detection effective porosity approximates total porosity.

5.2 The total porosity of NMR logging is the whole pore structure of sandstone group, including effective porosity and clay-bound water porosity.

5.3 When the NMR logging data is compared with the core data, the core detected porosity corresponds to the total NMR logging porosity.

5.4 The total porosity of nuclear magnetic logging corresponds to the equal thickness contrast between total porosity and core porosity, and the thickness weighted average of all rock sample data in the sandstone group should be made.

## References

[1] 1.Zhang, H.T.,2020, Applications of nuclear magnetic resonance (NMR) logging in tight sandstone reservoir pore structure characterization.In: Arabian journal of geosciences.13, pp.5-8.

[2] 2.Zhao,Y.L.,2014, How to reduce the core porosity measurement error. In: Inner Mongolia petrochemical Industry., 012,pp.70-71.

[3] 3.Murray, Christopher Jay, 1996,The use of nuclear magnetic resonance T(2) relaxation measurements in the characterization of the porous microstructure and reservoir properties of sandstones. In: University of South Carolina., pp.171.

[4] 4.Stefanidou Maria, 2010,Methods for porosity measurement in lime-based mortarst. In: Construction and Building Materials., 12,pp.2572-2578.

[5] 5.Li,G.M.,2012, Evaluation of uncertainty in core porosity measurement by liquid saturation method. In:Measurement and testing techniques., 4,pp.83-84.

[6] 6.Zhang,C.,2014, Effect and correction of clay - bound water on capillary pressure curve of mercury injection. In: Science and technology guide., 32.

[7] 7.Fu,Y.H., 2018,Preliminary study and comparison of shale core gas porosity measurement parameters. In: Special reservoir., 3,pp.144-148174.

[8] 8.Huang,K.J.,1998,Measurement accuracy of porosity. In: Natural gas exploration and development., 003,pp.75-7774.

[9] 9.Yuan,C.,The combination of formation element logging and density logging can improve the accuracy of porosity evaluation.Logging technique.,

[10] 10.C. Baiyegunhi,2014,The correlation of dry density and porosity of some rocks from the Karoo Supergroup: A case study of selected rock types between Grahamstown and Queenstown in the Eastern Cape Province, South Africas. In: International Organization of Scientific Research. 12.

[11] 11.Fu,Y.W.,2019, Analysis of pore size distribution and fractal dimension in tight sandstone with mercury intrusion porosimetry'. In: Results in Physics., 1.

[12] 12.Crow Heather L, 2020,Downhole nuclear magnetic resonance logging in glaciomarine sediments near Ottawa, Ontario, Canada. In:Near surface geophysics.pp.17.Ottawa, Ontario, Canada .

[13] 13.Son Hye-min, 2017,Pore structure classification and reservoir evaluation of pyroclastic rock based on nuclear magnetic logging'. In: Proceedings of the 2017 China Earth Sciences Joint Annual Conference (CGU2017)., 5,pp.3326-3327. Beijing.

[14] 14.Chris Petrich, 2006,Modelling the interrelationships between permeability, effective porosity and total porosity in sea ice'. In: Cold regions science and technology., 2,pp.131-144.

[15] 15.JRoshan Hamid,2016,Measurement of total porosity for gas shales by gas injection porosimetry (GIP) method'. In:Energy & fuels., 15,pp.694-707.

[16] 16.Aqel Mohammed A., 2016,permeability control on the relationship between nuclear magnetic resonance (NMR) measured porosity and routine core porosity'. In: Arabian journal of geosciences, 5,pp.5.

[17] 17.Tian,H.,2020, Reservoir Porosity Measurement Uncertainty and its Influence on Shale Gas Resource Assessment. In: Journal of petroleum Geology(English version)., 002,pp.223-242.

[18] 18.Alita Siswandani, 2016,Development of Porosity Measurement Method in Shale Gas Reservoir Rock'. In: Journal of Physics: Conference Series., 1.

[19] 19.Bagheripour P.,2013,Fuzzy ruling between core porosity and petrophysical logs: Subtractive clustering vs. genetic algorithm-pattern search. In: Journal of Applied Geophysics., 2013.

20.Janssen C., 2011,Nanoscale porosity in SAFOD core samples (San Andreas Fault)'. In:Earth and Planetary Science Letters: A Letter Journal Devoted to the Development in Time of the Earth and Planetary System.,1a2.

Journal of Physics: Conference Series **2594** (2023) 012006 doi:10.1088/1742-6596/2594/1/012006

# Analysis of the Reasons for Shutdown of Wells During the Stable Production Stage of Low-pressure and Low-permeability Gas Field Development

**Zheng Wei-shi1, 2(✉)**

1 Drilling & Production Engineering Technology Research Institute, CNPC Chuanqing Drilling Engineering Company Limited,Sichuan, China
2 National Engineering laboratory for Low-permeability oil and gas field exploration and development, Shaanxi, China

zwsxjxde@163.com

**Abstract.** With the extension of the development time of the Sulige gas field, the stage of increasing reserves and production has entered a stable production stage. The number of shut down wells is increasing, and the impact on production is also increasing, which is extremely unfavorable for the stable production of the gas field. Therefore, strengthening the management of shut down wells is becoming increasingly important. This article collects, organizes, and analyzes data from 202 shut down wells in the central region of the Sulige gas field based on geological understanding of the reservoir, fluid composition, logging data, water gas ratio, and gas production profile testing. The results show that wellbore fluid accumulation and water lock damage, wellbore scaling, energy depletion, poor reservoir properties, and throttle failure are the main factors causing shut down wells, This provides an important reference basis for formulating production resumption measures and gas increase targets for shut down wells.

## 1 Introduction

Sulige Gas Field is located in the Suligemiao area of Ordos City, Inner Mongolia. It is the first large gas field in China with a proven reserve of over one trillion cubic meters, with an exploration area of 60000 square kilometers and a total resource of 5.5 trillion cubic meters. As of the end of December 2022, the daily gas production of Sulige Gas Field has reached 105 million cubic meters, and the natural gas production has exceeded 30 billion cubic meters. More than 1200 wells have been built in the central area of Sulige Gas Field, including over 210 low production and low efficiency wells, accounting for 18% of the total, and the number is constantly increasing. In the situation of severe resource degradation, it is difficult to stabilize production, and it is necessary to conduct a systematic analysis of low yield and low efficiency wells and conduct research on measures to increase production and tap potential.

## 2 Current Situation of Long Stop Gas Wells in SuA Block

Data collection, organization, and analysis were conducted on 202 shut down wells in the research area. The results showed that wells with a single open flow rate of less than 50000 cubic meters per day accounted for about half of the total number of wells, and wells with a casing pressure of less than 4.0 MPa accounted for 40% of the total number of wells, making them the main source of long-term shutdowns. The specific situation is as follows.

### *2.1 Statistical analysis of open flow according to production testing*

(1) 101 wells with an open flow rate of less than 50000 cubic meters per day, accounting for 50%;
(2) Open flow rate greater than 100000 cubic meters/day: 72 wells, accounting for 35%;
(3) Open flow rate greater than 200000 cubic meters/day: 36 wells, accounting for 18%.

### *2.2 Statistical analysis based on the current casing pressure of shut down wells*

(1) 84 wells with casing pressure less than 4.0MPa, accounting for 42%;
(2) 50 wells with casing pressure greater than 8.0MPa, accounting for 24%;
(3) 20 wells with casing pressure greater than 12.0MPa, accounting for 10%.

## 3 Analysis on the Reasons for Long Stop Gas Well Production in SuA Block

An analysis was conducted on the reasons for the shutdowns of 202 wells. The main reasons for shutdowns include fluid accumulation shutdowns (52.0%), energy depletion (21.3%), poor reservoir properties (12.9%), wellbore failures (6.4%), and safety outsourcing (7.4%). The shutdown of accumulated liquid is the main reason for low production/shutdown, and it is also the main target for long-term well shutdown, resumption of production, and stimulation management.

### *3.1 wellbore fluid accumulation and reservoir water lock*

After the gas well enters the middle and later stages of production, the decrease in natural gas production leads to a decrease in liquid carrying capacity, which in turn leads to wellbore fluid accumulation and an increase in water saturation of near wellbore reservoirs. When the liquid column exceeds a certain height, the gas well cannot continue production due to the small pressure difference. The phenomenon of water locking and well killing caused by low production liquid accumulation is common in the development wells of the Sulige gas field. Statistics have been conducted on some typical wells that were shut down due to wellbore fluid accumulation. From the statistical results, it can be seen that wellbore fluid accumulation wells generally have high remaining recoverable reserves and are expected to restore their production capacity by eliminating wellbore fluid accumulation.

#### *3.1.1 Evaluation of Fluid Accumulation*

Gas well liquid accumulation has a serious impact on gas well production and normal production. Therefore, it is particularly important to accurately predict the critical liquid carrying flow rate of gas wells and prevent wellbore liquid accumulation during the production process of gas wells. During the production process of a water producing gas well, liquid exists in the wellbore in the form of droplets. When the production rate of the gas well is less than the critical liquid carrying flow rate, the droplets will slide to the bottom of the well, causing liquid accumulation in the wellbore.

The commonly used methods for calculating the critical liquid carrying flow rate include the Turner method and the Li Min method, which were used to calculate the critical liquid carrying flow rate of some typical wells. According to the calculation results, the actual production of each well (0.3-0.8 million cubic meters per day) is much lower than the critical liquid carrying flow rate value (Tunnel model 4.0-70000 cubic meters per day; Li Min model 1.5-25000 cubic meters per day), which is prone to wellbore fluid accumulation.

#### *3.1.4 Analysis of reservoir water lock damage caused by fluid accumulation*

A fatal weakness of tight sandstone gas reservoirs is their tendency to absorb water and cause water lock damage, while long-term wellbore fluid accumulation will cause water lock damage in the reservoir near the wellbore area. There are two main reasons for water locking, namely capillary force self suction and liquid phase retention effect.
(1) Capillary force self suction
It is a common phenomenon in low permeability and low porosity tight sandstone gas reservoirs where the initial water saturation is lower than the bound water saturation, and the reservoir is in a sub bound water state. Therefore, there is excess capillary pressure present, and when foreign fluids enter, they are easily sucked into the pore space. The pressure difference between the wetting phase and the non wetting phase on both sides of the meniscus in a capillary is generally defined as the capillary pressure, which is proportional to the interfacial tension and inversely proportional to the radius of the porous medium. Due to the small pore size of tight gas reservoirs, water lock damage is prone to occur.
(2) Liquid phase retention aggregation
The retention effect of water, also known as the aggregation effect of liquid phase, is the primary and fundamental damage factor in the operation process of tight sandstone gas reservoirs. The retention and aggregation of the liquid phase are another important factor causing water lock damage. The slow or difficult flow back of foreign fluids invading the reservoir, or even inability to flow back, can further exacerbate water lock damage. The time required to discharge a liquid column with a length of L from a capillary tube with a radius of r is:

$$t = \frac{4\mu L^2}{\Pr^2 - 2r\sigma\cos\theta}$$

In the formula, Pr is the driving pressure; σ is the interfacial tension; R is the capillary radius, L is the length of the liquid column, θ is the contact angle, μ Is the viscosity of the foreign fluid.
It can be seen that the smaller the capillary radius r, the longer the drainage time. As the drainage process progresses, the liquid is gradually discharged from the large to small capillaries, and the drainage speed decreases accordingly. The throat radius of tight gas reservoirs is small, making it difficult to discharge liquid, resulting in severe water lock damage.

### *3.2 Well bore scaling*

The wellbore is a necessary passage for the flow of natural gas. Once scaling and blockage occur in the wellbore, it will have a throttling or even cut-off effect on the flow of natural gas, leading to a decrease or shutdown of gas production.

#### *3.2.1 Type of wellbore scaling*

The types of wellbore fouling can be divided into several categories:
(1) Corrosion inhibitor residue. It is easy to adhere and package corrosion products, rock debris, sand and gravel in the underground, forming viscous blockages.
(2) Ca, Mg precipitates. The fluid produced by gas wells contains corrosive media such as highly mineralized water, H2S, CO2, etc., causing wellbore corrosion, scaling, and inorganic blockage.

(3) The foaming agent forms an insoluble viscous substance with calcium and magnesium particles in the formation water. The analysis results of solid plugging materials show that the organic content in the plugging materials is about 20%, and the main components are inorganic compound corrosion products (about 80%), water scaling products and solid sand particles, which are one of the reasons for reservoir and wellbore plugging.

### *3.2.2 Factors affecting scaling*

Well bore scaling is a quite complex process. For salt scale, the primary factor for scaling is that the solubility is in a supersaturated state. The supersaturated concentration is not only related to the solubility, but also affected by many factors such as thermodynamics, crystallization kinetics, fluid mechanics, etc. For corrosive scale, scaling is influenced by the combined effects of the conveying medium, materials, and surrounding environment.

(1) The effect of temperature on scaling

The main effect of temperature is to change the solubility of salts that are prone to scaling. Salt scale is mainly composed of carbonates. When the temperature increases, calcium bicarbonate decomposes and forms calcium carbonate scale. For salt scale mainly composed of calcium sulfate, barium sulfate, and strontium sulfate, the main reason is that the sulfate ions in the medium combine with calcium ions, barium ions, and strontium ions to form insoluble precipitates. Since most of these reactions are also endothermic reaction, with the increase of temperature, more precipitation will occur. Temperature can also affect the reproduction rate of bacteria and the rate of electrochemical reactions in steel. Due to the fact that each type of bacteria has a suitable temperature for growth, different types of bacteria have different requirements for temperature. The optimal temperature for most bacteria is 20-40 ℃, so as the temperature of the medium in contact with the wellbore changes, the breeding rate of bacteria will also change, and the corrosion on the wellbore will also change, thereby affecting the rate of corrosion scale formation.

(2) The effect of pressure on scaling

Pressure has an impact on the scaling of calcium carbonate, calcium sulfate, and barium sulfate. Calcium carbonate scaling involves gas in the reaction, and pressure has a relatively significant impact on it. The pressure decreases and the reaction proceeds to the right, which can promote scaling. For the wellbore, the pressure also decreases from the underground to the surface, so the scaling trend always increases. At the same time, due to the complex sedimentary environment of oil layers and underground scales, the pressure changes greatly, which can decrease from tens to hundreds of MPa to a few MPa or form negative pressure. The new pressure drop has increased the trend of scaling.

(3) The effect of flow rate on scaling

For various types of dirt, the growth rate of dirt decreases with increasing fluid velocity. This can be explained as although an increase in flow rate can increase the deposition rate of dirt, at the same time, the increase in erosion rate caused by an increase in flow rate is more significant, resulting in a decrease in the overall growth rate. When the flow rate decreases, the probability of solid particles and microbial excreta carried in the medium depositing increases, and the probability of wellbore scaling also significantly increases, especially in areas with structural mutations. The sudden change in flow rate can also be explained as a change in pressure. If the flow rate suddenly increases, it will cause local degassing, reduce the partial pressure of carbon dioxide, and cause calcium carbonate scaling.

(4) The effect of pH value on scaling

Research has shown that increasing the pH value of the solution, carbonate dissolution will rapidly crystallize, increasing the thermal resistance of progressive fouling, shortening the induction period of fouling formation, and promoting the growth of fouling; But if the pH value is too low, it will increase corrosion and cause corrosion scale. The determination of the pH value of the medium requires consideration of both aspects and the selection of an appropriate pH value, with a recommended range of 6.5 to 8.0.

### *3.3 Energy exhaustion*

Due to the limited controlled reserves of a single well and the influence of strong planar heterogeneity geological factors, some wells have low actual controlled reserves and are in the economic abandonment stage of their life cycle. The formation pressure is low, and the remaining recoverable reserves are small. In the actual production process, there is no significant pressure difference in the wellhead oil casing pressure. When the well is opened and closed, the oil casing pressure increases and decreases simultaneously. After a long period of shut-in, the wellhead pressure value recovers, but the pressure rapidly decreases after the well is opened for production.

### *3.4 Poor reservoir properties*

The low production caused by poor reservoir properties is characterized by low gas production from a single well, but after long-term shut-in, the wellhead pressure can recover to a higher level. Due to the fact that the production rate of a single well is much lower than the critical liquid carrying flow rate, the occurrence of wellbore fluid accumulation and reservoir water lock damage in high reservoir water saturation will further exacerbate the decline in production, which may lead to shutdown. This type of gas well is mostly distributed at the edges of sand bodies in rich areas, low structural areas, nose wings, nose depressions, and other areas. The statistical results show that this type of well has low open flow rate (basically less than 30000 cubic meters per day), low cumulative gas production (basically no more than 1 million cubic meters), and 70% tested water production.

### *3.5 Throttle failure*

Sand accumulation above the throttling device, severe fluid accumulation above the throttling device, deformation and scaling of the oil pipe, no fishing neck or broken fishing neck of the downhole throttling device, deformation or inability to lift the throttling device rubber cylinder or slip, leftover tools underground, and broken steel wire underground are all the reasons for throttling device failure. These types of wells have gradually increased in recent years, seriously affecting the later production of gas wells.

## 4 Conclusion

Based on the analysis and research of reservoir geological understanding, fluid composition, logging data, water gas ratio, and gas production profile testing of 202 shut down wells in the central area of Sulige Gas Field, it is believed that wellbore fluid accumulation and water lock damage, wellbore scaling, energy depletion, poor reservoir properties, and throttle failure have led to low production/shutdown of gas wells. In the process of increasing the potential and efficiency of gas wells, resuming production after long-term shutdowns, and increasing production management, it is necessary to verify the dynamic reserves of gas wells and predict the recoverable reserves of single wells for different types of shutdowns, clarify the distribution pattern of remaining gas in long shutdowns, analyze the potential of different types of shutdowns, and develop targeted measures for tapping potential.

## References

1. Fu Jinhua, Fan Liyong, Liu Xinshe, etc Reservoir Formation Conditions and Key Technologies for Exploration and Development of Sulige Gas Field [J]. Journal of Petroleum, 2019, 40 (2): 17
2. Tang Hai, Lv Jianjiang, Lv Dongliang, et al. Study on the influencing factors of water lock in tight and low permeability gas reservoirs [J]. Journal of Southwest Petroleum University. 2009,31 (4): 91-94

3. Liao Ruiquan, Xu Yonggao, Hu Xuebin. The damage of water lock effect on low permeability reservoirs and its inhibition and release methods [J]. Natural Gas Industry. 2002,22 (6): 87-89
4. Dong Bo, Lan Lin, Chen Zhihui, etc Study on capillary self absorption characteristics of tight gas reservoir rocks [J] Drilling and Production Technology, 2012
5. Wei Maowei, Xue Yuzhi, Li Gongrang, et al. Research progress on water lock release technology. Drilling and completion fluids [J]. 2009,26 (6): 65-68
6. Pang Zhenyu, Sun Wei, Luo Wanying, et al. Analysis of Factors Influencing Water Lock Space and Damage Degree in Low Permeability Tight Gas Reservoirs [J]. Geological Science and Technology Information. 2014,33 (3): 140-144
7. Zeng Wei, Chen Shu, Xiang Haiyang. Water lock damage in abnormally low water saturation reservoirs [J]. Natural Gas Industry. 2010,30 (7): 42-44
8. Dong Wenwu, Qu Hongjun, Wei Xinshan, et al. Analysis of water lock damage and influencing factors in the Upper Paleozoic reservoirs in the eastern Ordos Basin [J]. Journal of Yangtze University Natural Science Edition: Science and Engineering (early ten days). 2014 (11): 122-124
9. Zhang Zhenhua, Yan Jienian. Research on the influencing factors and prediction methods of water loss damage in low-permeability sandstone reservoirs [J]. Petroleum exploration and development. 2000,27 (3): 75-78
10. Tang Hai, Lv Jianjiang, Lv Dongliang, et al. Study on the influencing factors of water lock in tight and low permeability gas reservoirs [J]. Journal of Southwest Petroleum University. 2009,31 (4): 91-94
11. Lin Guangrong, Shao Chuangguo, Xu Zhenfeng, et al. Research on water lock damage and its relief methods in low permeability gas reservoirs [J]. Petroleum Exploration and Development. 2003,30 (6): 117-118
12. Liao Ruiquan, Xu Yonggao, Hu Xuebin. The damage of water lock effect on low permeability reservoirs and its inhibition and relief methods [J]. Natural Gas Industry. 2002, 22 (6): 87-89
13. Ma Hongxing, Shi Aiping, Wang Zhimin, et al. Study on water lock damage in low permeability sandstone reservoirs [J]. Petroleum drilling and production technology. 2004, 26 (4): 49-51

Journal of Physics: Conference Series **2594** (2023) 012007 doi:10.1088/1742-6596/2594/1/012007

# Discussion on Well Control Technique During Post-fracturing Tool Interval

**Xing-liang SONG1,2, Yi ZHANG 1,2(✉), Chun-hui ZHANG1,2, Xin-rong ZHOU 1,2, LI-hong KONG1,2**

1.Production Engineering & Research Institute, PetroChina Daqing Oilfield, Daqing 163453,China;
2.Heilongjiang Key Laboratory of oil-gas reservoir increase production and injection, Daqing Oilfield , Daqing 163453, China

Zhangyi001@petrochina.com.cn

**Abstract.** In the oil and gas exploration and development safety work, well control is the top priority, which goes deep into every field of oil exploration and development production. How to improve the well control consciousness of drilling engineering in the normal production process of actual drilling engineering has become an important subject faced by petroleum exploration and development. This paper focuses on the research and analysis of fracturing string internal well control problems, and puts forward a series of solutions and suggestions according to the practical experience.

## 1. Introduction

Blowout phenomenon is often encountered in the process of oil exploitation. It is a phenomenon that the fluid hidden in the stratum is expelled from the ground or infiltrated into other strata under the action of external force, which has great destruction. The production process of petroleum exploration work will be seriously affected by the surrounding environment, interfere with the oil and gas reservoir, destroy the oil resources, and even cause casualties, oil well abandonment and other huge losses. Well control is an effective way to solve these problems. It is a scientific and effective way to control the formation pressure and keep the pressure level in the well, so as to ensure oil exploration and production. In the process of oil field exploitation and exploration, the productivity and permeability will decrease at a certain period, so in order to improve the oil recovery, fracturing technology needs to be used. Tools need to be placed downhole before fracturing can be performed. As fracturing can increase pressure in the well, it is difficult to safely extract the tools from the well

[1-3]..

## 2.Well control techniques during post-frac tool run

In general, after fracturing oil and water Wells, to prevent blowout of the well, the method is to calculate the kill fluid value based on the formation pressure, use the high density kill fluid to adjust the formation pressure to maintain the pressure balance, and then use the tools in the wellhead to do the relevant operation[4-6]. However, hydraulic well has many disadvantages, such as pollution of formation environment, which reduces the service life of oil well. Due to the requirements of the job, a pressure operation will be required, with the focus on effectively connecting the tubing plug to the control device to remove the string without jetting.

## 3. Well control problems in the frac string

After fracturing, well control should be performed on the string used in production. However, due to the high pressure generated in the well, the plugging effect is ineffective and the following tools should be well controlled. Therefore, the personnel need to set a "controllable sealer" in each clamp segment in advance to effectively seal the tool segment string. The full diameter plug valve can be installed in stages in the sequence shown before fracturing. The installation of such a device enables the full-bore plug valve to be controlled at all times, or the packer can be used as a marker to indicate the installation of the full-bore plug valve. In the fracturing process, the "full-diameter plug valve" is opened. Because of its larger diameter, it does not affect the plugging effect and can make the fracturing fluid pass smoothly. After fracturing, the production personnel can conduct reasonable plugging according to the actual pressure in the well. When the pressure value in the well is small, plugging can be stopped and the whole set of pipe string can be pulled out. If the pressure in the well is larger, plugging can be done safely according to the operation process. When this set of procedures is completed, the upper pipe string can be normally pulled out. To the last frac pipe, remove the full bore plug valve, close the full bore plug valve, remove the clamp section, then remove the next full bore plug valve and close it in time. Remove all the clamp and remove the frac string at the same time according to the above procedure. If there are more than two full-diameter plug valves adjacent to each other, the operator can control the high-pressure fluid through the receiving tank of the full-diameter plug valve at the wellhead based on the actual conditions in the well. When the pressure in the well drops to a stable level, the operator can proceed with subsequent production[7-8].

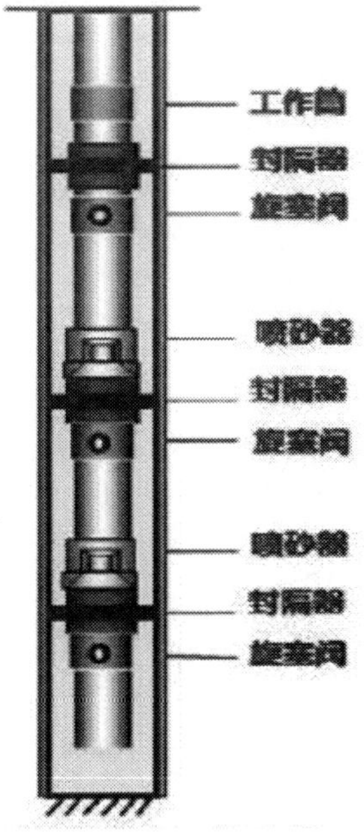


**Fig. 1.** Tool section string connection diagram

## 4. Analysis of key fracturing technologies in downhole operations

### *4.1. Separate and stratified fracturing construction technology*

At present, the separation and stratification fracturing technology is usually used in the stimulation process of oilfield fracturing. This technology is complicated in the production process of oilfield, and requires a large amount of capital cost injection. Operators can use a reasonable downhole packer according to the actual conditions of the field in production. For example, a single packer can be used to perform deep frac operations for larger Wells. The double packer has a wide range of application and can be widely applied in any oil well fracturing construction, so it has a good guarantee for the stratified fracturing construction of the reservoir. In addition, slip-sleeve packers have gained popularity in fracturing operations in the field production process. These packers are often used in deep Wells and in projects where multiple fracturing techniques have been performed[9-10].

### *4.2. flow-limiting stratified fracturing technology*

When fracturing is carried out in multiple reservoirs at the same time, the flow-limiting stratified fracturing technology can be adopted. Therefore, the operation of injection pressure can be adjusted according to different conditions of the construction process. Therefore, it is necessary for the staff to effectively control the injection value of fracturing fluid in different intervals according to the preliminary design of the reservoir and the number of related holes. Due to a lot of perforation friction generated in the construction process, the pressure data of the oil well will be significantly increased, resulting in the diversion of crack fluid after entering the reservoir. Therefore, it is necessary for the staff to adopt different reservoir treatment methods according to the difference in pressure, so as to achieve the best construction effect of stratified fracturing along the oil and gas. This fracturing technology has been widely popularized in more construction operations and has strong practicability. However, it should be noted that the longitudinal fractures formed in the construction process will produce large data differences due to the action of pressure. Therefore, in the actual construction and production process, there are clear data standards for the manufacturing accuracy of relevant fracturing equipment.

## 5. Solutions to downhole fracturing problems

### *5.1 Optimization of fracturing fluid ratio*

In the process of downhole fracturing construction, a very important operation process is the preparation of fracturing fluid. The main function of fracturing fluid is that artificial fractures will occur under the action of high pressure, and its own viscosity is large and lack of friction, so it will not pollute the oil and gas reservoir. In the present downhole fracturing construction of oil fields, active agents are usually applied on the construction operating surface. In the early stage of the project implementation, the fracturing fluid preparation should be restructured and optimized according to the actual production conditions of the oilfield, so as to effectively improve the fracturing construction effect, reduce the production cost of the oilfield and not pollute the reservoir.

### *5.2 Optimize sand ratio composition*

Sand ratio is the key to ensure the quality of downhole pressure construction projects. The ratio of total sand volume to net liquid volume is referred to as sand ratio for short. If the sand volume value is lower than the standard data in the construction process, the oil viscosity of the base will increase. In normal operation and construction, it is necessary to control the sand ratio value within an effective range, and reasonably adjust the data differences generated by repeated well states according to the actual underground mining conditions, so as to form a scientific and reasonable sand ratio. It is necessary to timely predict the fracture conductivity and design a reasonable construction scheme during the fracturing construction of polymer flooding well. Polymer flooding Wells have low fluid viscosity, so make sure the fractures are short and wide, and adjust the sand ratio to a reasonable ratio based on the actual operation. On the contrary, it is necessary to ensure that the fractures are long and narrow when the produced fluid viscosity is high, so as to avoid sand production in the production process.

## 6. Conclusion

With the continuous progress in the field of petroleum exploration and development, people pay more and more attention to the awareness of well control in the process of drilling. The current fracturing operation is more and more popular, and the pressure in the well is increasing gradually, which poses a huge challenge to the oil field exploitation industry. By adopting the above reasonable operation mode, the problem of fluid discharge in the string after fracturing can be effectively controlled, the drilling construction safety can be guaranteed, and the blowout accident can be prevented.

**References**

[1] CAI Fubao. Research on well control technology during post-fracturing tool interval [J]. Chemical Industry Management,2018(26):192.

[2] Wang Shuping, Shi Yunqing, Yan Yan, Zheng Rongchen, Liu Jiandang. Calculation method of deflating radius of fractured horizontal well in channel superposition reservoir [A]. Natural Gas Professional Committee of China Petroleum Society, Sichuan Petroleum Society and Zhejiang Petroleum Society. Proceedings of 2017 National Natural Gas Academic Annual Conference [C]. Natural Gas Professional Committee of China Petroleum Society, Sichuan Petroleum Society, Zhejiang Petroleum Society: Natural Gas Professional Committee of China Petroleum Society,2017:5.

[3] LIANG Huzhan. Discussion on the prediction method of the final deflating radius of fractured gas well in Daniudi low permeability gas Reservoir [J]. China Petroleum and Chemical Standards and Quality,2017,37(19):29-30.

[4] WANG Yousong. Research on Temporary Plugging Technology of Potential Pressure fractures in old Wells of Penglaizhen Formation Gas Reservoir in western Sichuan [D]. Chengdu University of Technology,2016.

[4] CASTRO L, WATKINS T, BEDORE B A, et al. Reducing operational time, fluid usage, hydraulic horsepower, risk, and downtime: targeted fracs using CTenabled frac sleeves [R]. SPE 154391-MS, 2012.

[5] YUAN F, BLANTON E, CHRIS P, et al. Is it possible to do unlimited multistage fracturing economically? [R]. SPE 167791-MS, 2014.

[6] YUAN F, BLANTON E, INGLESFIELD J. Single-sizeball interventionless multi-stage stimulation system improves stimulated reservoir volume and eliminates milling requirements: case studies [ R]. SPE 171183 - MS, 2014.

[7] Pei Yingjie, Ma Sihui, Li Lin, et al.Review on fracturing technology of shale gas Wells [J]. Petrochemical Technology, 2019, 26 (7): 266-267.

[8]TENG J W, LIU Y S. Analysis of shale distribution, storage potential and prospect for shale oil and gas in China [ J]. Progress in Geophysics, 2013, 28 ( 3): 1083-1108.

[9] SUN C C, TAN C D, SONG J, et al. Overview of horizontal well fracturing technology at home and abroad [ J ]. China Petroleum and Chemical Industry, 2014 (10): 50-52.

[10] ZHANG H Z, HE Y Q, LIU J, et al. Current condition and development trend of foreign phased fracturing technology for horizontal well [ J]. Petroleum Technology Forum, 2012, 31 (6): 47-52.

# Study On Fine Injection and Production in the Infill Well Area of A Transition Belt

**Chao WANG1,2(✉), Zhen-kun ZHU1,2 , Jing-qi WANG1,2 , Li-li WANG1,2 , Ying GUO1,2 , Yu-peng ZHOU1,2**

1. Daqing Oilfield Production Technology Institute, Daqing Oilfield Limited Company, Daqing, China
2. Heilongjiang Province Key Laboratory of Oil and Gas Reservoir Stimulation, Daqing, China
443696466@qq.com

**Abstract**. A transition belt of Daqing oilfield was put into operation in July 2019.The injection and production perfection degree of some well groups is low, the injection-production ratio is unbalanced, the formation energy is insufficient, the yield declines greatly, and the water cut rises quickly.In order to slow down the decline of the block and control the rising speed of water cut, the actual problems existing in the block are discussed. By drawing on the successful adjustment practice of the infill well area of the transitional belt in the B transition belt, the injection and production status and adjustment methods of the infill well area in the transitional belt in A transition belt were studied, two aspects of injection-production system and injection-production structure in block are considered. Different adjustment countermeasures were formulated for oil and water Wells in different strips, and the direction of fine injection and production adjustment was clarified. By using the well focus, structure adjustment and water injection production structure adjustment and so on, effectively reduce the block production decline, control water cut increase speed, and reduce the contradiction between the layers. To realize the purpose of further increase the level of reservoir utilization and improve the development level of A transition belt. At the same time, it provides experience for the injection and production adjustment of the infill well area for the future transition zone.

## 1. Introduction

A transition belt is located in the south of Daqing oilfield .It was developed in 1973.It has successively gone through the development stages of basic well pattern, primary infill pattern, secondary infill pattern, transition zone outspread, class I reservoir and class II reservoir polymer flooding[1-4].

In July 2019,aiming at the problems of low density of well pattern, low degree of water drive control, low rate of oil recovery, high proportion of unused thin layer and uneven use of sand body in the transitional zone[5-8],in order to improve the control degree of well pattern, further improve the injection-production relationship in block, and improve the development effect of A Transitional belt, the infill adjustment of A Transitional belt has been carried out.

## 2. Initial Problems Exist in the Infill Wells in A Transition Belt

### *2.1. Some Well Groups Have Low Injection-production Perfection and Low Injection-production Ratio*

Transition zone plan transfer scheme design 104 ports,the formulation scheme was optimized to 101 ports,the actual implementation optimization is 96 ports.After the new drilling in the transition zone was put into development, the injection of supporting oil Wells was not carried out in time. With the growth of the production time, the energy of the oil layer itself was constantly consumed, resulting in the decline of the formation fluid supply capacity.

**Table 1.** Statistical table of production status of A transition zone

| Y/d | Conversion well number(mouth) | Injection rate($m^3$) | fluid volume(t) | Oil product(t) | Flowing pressure(MPa) | Working fluid level(m) | Injection-production ratio |
|---|---|---|---|---|---|---|---|
| 201908 | / | 11642 | 15853 | 670 | 5.39 | 340.3 | 0.89 |
| 201909 | / | 11609 | 16617 | 771 | 5.35 | 378.8 | 0.84 |
| 201910 | / | 12268 | 18000 | 835 | 5.37 | 356.4 | 0.83 |
| 201911 | 7 | 13226 | 17727 | 816 | 5.09 | 422.1 | 0.91 |
| 201912 | 19 | 15518 | 16448 | 786 | 4.82 | 530.8 | 1.11 |
| total | 26 | 64263 | 84645 | 3878 | 5.22 | 393.2 | 0.92 |

Till December 2019, a total of 26 Wells were injected, and 47 new Wells in the transition zone with complete injection and production were put into production, accounting for only 24.2% of the infill Wells in the transition zone, assuming that three or more injection well points were considered perfect. The injection-production ratio of the block is 1.11, and the cumulative injection-production ratio from August to December 2019 is 0.92.

### *2.2. The Water Cut in Some Well Groups Rises Fast*

By December 2019, a total of 167 infill wells in the transition zone had been put into production, with an average water cut of 94.6%. As of December 2019, those Wells with a water cut increase of more than 2% accounted for 32.1% of the total number of new Wells.

**Table 2** Statistical table of water cut appreciation of infill Wells

| Water-based appreciation classification(percent) | December 2019 | | | | |
|---|---|---|---|---|---|
| | Open well number(mouth) | percent(%) | fluid volume(t) | Oil product(t) | Water rate(%) |
| $\triangle fw<-2$ | 28 | 20.9 | 50 | 5.5 | 89.0 |
| $-2\leq\triangle fw<0$ | 31 | 23.1 | 46 | 2.8 | 93.9 |
| $0\leq\triangle fw<2$ | 32 | 23.9 | 58 | 2.1 | 96.4 |

| | | | | | |
|---|---|---|---|---|---|
| △fw≥2 | 43 | 32.1 | 53 | 2.2 | 95.8 |
| total | 134 | 100.0 | 52 | 3.0 | 94.0 |

### *2.3. The Production of Some Well Groups Decreases Rapidly*

By December 2019, new Wells with a production decline rate of more than 20% accounted for 35.1% of the total number of new Wells, and the monthly average production decline rate for infill Wells in the transition zone reached 1.47%.

**Table 3** Classification table of infill Wells in transition zone according to oil production decline range

| Oil production decline range classification (percent) | December 2019 | | | | |
|---|---|---|---|---|---|
| | Open well number(mouth) | percent(%) | fluid volume(t) | Oil product(t) | Water rate(%) |
| △ft<0 | 74 | 55.2 | 55 | 3.8 | 93.1 |
| 1< △ft ≤20 | 13 | 9.7 | 56 | 2.7 | 95.2 |
| 20< △ft ≤50 | 22 | 16.4 | 57 | 2.7 | 95.2 |
| 50< △ft ≤70 | 15 | 11.2 | 36 | 1.2 | 96.6 |
| △ft >70 | 10 | 7.5 | 41 | 1.2 | 97.0 |
| total | 134 | 100.0 | 52 | 3.0 | 94.2 |

## 3. Study on the Adjustment Method of New Well Excavation Potential in A Transition Zone

The production decline rate and water cut rise rate of middle and high permeability oil fields in high water cut stage are controlled, and the development indicators are constantly improving [9]. In order to guarantee the development effect and ensure the uniform production of oil reservoir, according to the experience of water flooding development over the years[10-11], the injection-production structure should be adjusted according to the problems existing in the transition zone, and the injection-production relationship should be improved to slow down the production decline rate and water cut rise rate, so as to ensure the uniform production of oil reservoir and improve the development effect.

### *3.1. Optimize the Transfer Principle and Improve the Injection-production Relationship*

Considering factors such as well group perfection, injection-production relationship between old and new Wells, and production decline rate, the principle of well group with single injection-production priority, well group with relatively perfect injection-production or well group with old wells where water can be added appropriately delayed is determined.

With the continuous improvement of the injection-production relationship, the daily fluid production in the transition zone increases, the injection-production ratio increases and the formation energy recovers. For example, in well N, the daily fluid output was 58 tons at the initial stage of production, but the injection and production were not perfect, and there was only one injection well around the initial stage of production. In November 2019, the well group put in a new injection well, which increased the daily fluid output of this well by 17 tons and increased the daily oil production by 1.2 tons. In August, the second injection well was put in, which increased the daily fluid output by 11 tons and increased the daily oil production by 0.8 tons.

### *3.2. Strengthen the Adjustment of Water Injection Structure and Control the Rate of Water Cut Rise*

In order to ensure the formation energy is sufficient, the water injection structure of the transition zone is adjusted according to the pressure distribution, water cut distribution and liquid level distribution of the production well.

Is a combination of injection profile data, in view of the monolayer onrush of water injection well, according to the water flooding adjustment technology standard, fine division of conversion well injection interval, permeability differential control at less than 3.4, the variation coefficient control within 0.43, ensure water sandstone thickness ratio over 85%, the control interval injection intensity in m3/d.m around 9, to ensure the transition zone development effect. Increase the intensity of testing and adjustment to slow down the rise of water cut.

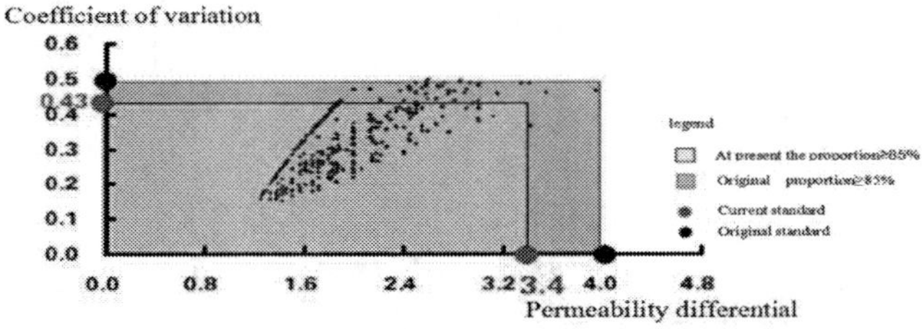


**Fig. 1.** Relationship diagram of interlayer permeability range, coefficient of variation and water absorption ratio

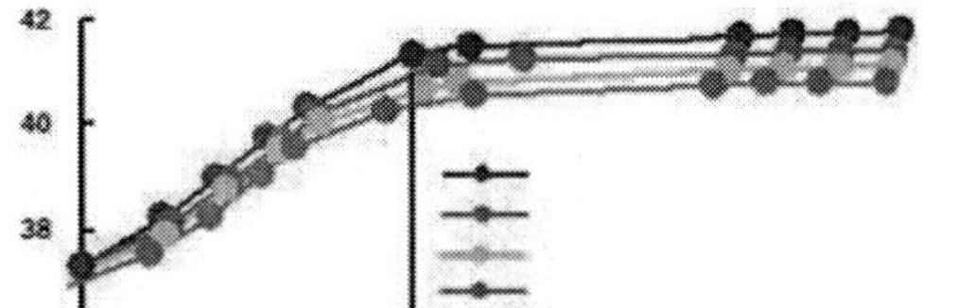


**Fig. 2.** Relationship curve between water injection intensity and recovery factor in different well patterns

Second, combined with the pressure measurement results of oil wells, the control of the reservoir with high formation pressure and high comprehensive water cut is invalid. Add effective water injection to the reservoir with low formation pressure and low comprehensive water cut. A total of 76 wells were adjusted for water flow, increasing water injection by 936$m^3$ and controlling water injection by 1988$m^3$.

### *3.3. Strengthening Measures for Oil Wells to Tap Potential and Slow Down the Rate of Production Decline*

According to the statistical results of the newly drilled water-flooded layer interpretation data in A transition zone, the proportion of no-low flooded thickness is about 27.4%, the proportion of medium water-flooded thickness is about 36.8%, and the proportion of high water-flooded thickness is about 35.8%. Preliminary determined that the potential Wells in the transition zone are mainly distributed in four belts, and the potential oil layer is the thin oil layer of A oil formation.

**Table 4** Statistical table of new well flooding in the A transition zone

| classification | project | No-flooded | | Low-flooded | | Middle flooded | | High flooded | | Total flooded | |
|---|---|---|---|---|---|---|---|---|---|---|---|
| | | Thickness (m) | Percent (%) | thickness (m) | percent (%) | thickness (m) | percent (%) | thickness (m) | percent (%) | thickness (m) | percent (%) |
| Effective thickness grade | ≥2.0m | 91.1 | 2.3 | 724.0 | 18.1 | 1474.7 | 36.8 | 1715.7 | 42.8 | 3914.4 | 97.7 |
| | 1.0m～2.0m | 69.1 | 5.7 | 296.8 | 24.3 | 482.4 | 39.6 | 370.8 | 30.4 | 1150.0 | 94.3 |
| | 0.5m～1.0m | 48.3 | 8.6 | 199.5 | 35.3 | 234.0 | 41.4 | 83.1 | 14.7 | 516.6 | 91.4 |
| | <0.5m | 95.8 | 31.1 | 143.6 | 46.6 | 56.0 | 18.2 | 12.7 | 4.1 | 212.3 | 68.9 |
| Different stripe | A zebra | 126.1 | 4.2 | 628.0 | 20.9 | 1078.2 | 35.8 | 1178.7 | 39.1 | 2884.9 | 95.8 |
| | Two, three bands | 140.3 | 5.5 | 599.0 | 23.5 | 960.5 | 37.6 | 853.8 | 33.4 | 2413.3 | 94.5 |
| | Four bands | 37.9 | 7.1 | 136.9 | 25.7 | 208.4 | 39.1 | 149.8 | 28.1 | 495.1 | 92.9 |
| Reservoir group | S I | 117.6 | 18.2 | 292.8 | 45.4 | 209.7 | 32.5 | 25.4 | 3.9 | 527.9 | 81.8 |
| | S II | 135.1 | 5.2 | 627.5 | 24.0 | 1016.5 | 38.8 | 839.0 | 32.0 | 2483.0 | 94.8 |
| | SIII | 32.5 | 3.0 | 218.9 | 20.5 | 414.2 | 38.9 | 400.5 | 37.6 | 1033.6 | 97.0 |
| | P I | 16.0 | 1.2 | 178.3 | 12.9 | 459.1 | 33.1 | 732.9 | 52.8 | 1370.3 | 98.8 |
| | P II组 | 3.1 | 0.8 | 46.4 | 12.2 | 147.6 | 38.7 | 184.5 | 48.3 | 378.5 | 99.2 |
| total | | 304.3 | 5.0 | 1363.9 | 22.4 | 2247.1 | 36.8 | 2182.3 | 35.8 | 5793.3 | 95.0 |

In view of the no-low flooded condition, the following two types of potential well groups are modified for fluid production structure: one is the oil Wells with low fluid production intensity, relatively perfect injection-production relationship and large injection-production ratio after putting into production; Second, the oil Wells with high fluid production intensity at the initial stage of production, but with a fast production decline rate due to the slow injection schedule, and with complete injection and production at present.

Fracking measures are used to release formation energy and slow production decline. In A transition zone, a total of 22 wells were implemented, with an average daily increase of 47 tons of fluid and 4.6 tons of oil in the initial measures of a single well. The comprehensive water cut decreased by 3.6 percentage points, and the reduced production decreased by 0.6 percentage points, with a total oil increase of 1.4×104t.

For example, in well N, a new injection well was injected in November 2019, and there were three injection wells in this well group. The injection-production relationship was perfect, but the fluid production intensity of this well was low and the injection-production ratio of the well group was large. Fracturing measures were modified in March 2020, and the well group increased 81 tons of fluid and 4.2 tons of oil daily in the initial stage of fracturing.

## 4. Adjustment Effect of Tapping Potential in A Transition Zone

### *4.1. The Perfection of Injection and Production in A Transition Zone Has Been Improved*

As of the end of September, a total of 67 wells had been injected in the infill well area of A transition zone. In terms of the perfection degree of well groups (those with 3 or more water injection well points are considered perfect), 132 of the 194 well groups have been perfected, accounting for 68.1% of the total well groups, an increase of 43.7 percentage points compared with last December.

### *4.2. The Reservoir Water Absorption Condition Has Been Improved*

According to the data of 55 wells water absorption profiles, the average ratio of water absorption layers in a single well is 77.6%, which is 12.6 percentage points higher than before adjustment. The average thickness ratio of sandstone absorbed by single well is 85.2%, which is 11.4 percentage points higher than before adjustment. The effective thickness ratio was 78.2%, 8.8 percentage points higher than before the adjustment.

### *4.3. The Speed of Water Cut rise and Yield Decline in The Transition Zone Are Effectively Controlled*

In September 2019, 16977 tons of liquid and 74019 tons of oil were produced daily in the region, and the comprehensive water cut was 95.64%. From December 2019 to June 2020, the monthly average water cut increased at a rate of 0.1. At present, the water cut increased and the production decline rate have been effectively controlled, and the water cut decreased to 95.6% in September.

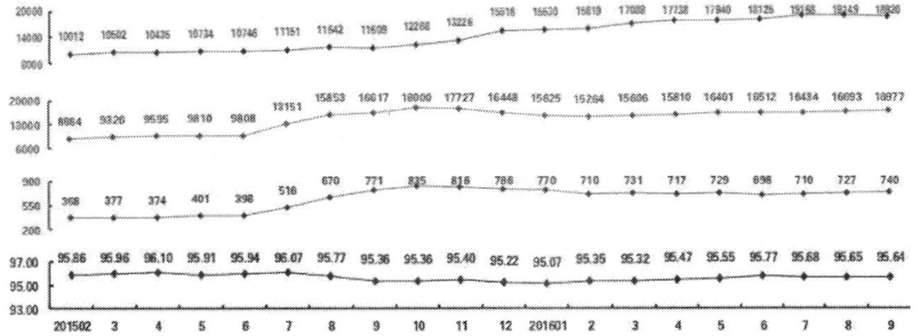


**Fig.3** Production curve of A transition zone

## 5. Conclusions

Subdivide the water injection well to control the water injection in high pressure and high aquifer secti on, and increase the water injection in low pressure and low aquifer section can effectively control the water cut rise rate. Fracturing the well group with relatively perfect injection-production relationship and sufficient water supply can effectively tap remaining oil, effectively contr ol water cut rise rate and slow down production decline rate.

## References

[1] References Chi Ming, Liu Dehua. Standard and Quality of Petroleum and Chemical I ndustry in China, "Research on Boundary of Subdivision Water Injection Technology in Ultra-high Water Cut Sandstone Oilfield", 2013(05).
[2] Nurudeen Oluwatosin Yusuf, Lynn Silpngarmlers.Application of Assisted History Match Approach in a Brown Fiedl Saves Thirty Percent in Project Time.SPE-208250-MS
[3] Zhao Hui, FU Chunbo. Fine injection-production adjustment and application of reservoir.Petrochemical Application, 2021,40:71-74.
[4] BI Yushuai. Some ideas on promoting efficient development of reservoir unit in Shaerxia reservoir in east China. Petroleum Research,2020(4).
[5] Hongfu Shi, Yifan He,Xianbo Luo,Xiao Shu.Improve the Performance and Mitigate the Risk of Brownfield Infilling Project.OTC-30767-MS.
[6] WANG Liang. Implementing fine injection-production management to continuously improve the development level of complex small fault blocks . Cleaning the W orld,2019(5).
[7] LIU Shuang. Application effect of fine injection-production technology in Sazhong Oilfield . Oil Recovery Engineering Collection,2017(2).
[8] Wang Kaiyue. Application of fine injection well testing and research in a oil production plant of Daqing Oilfield . Forei gn Logging Technology,2020(6).
[9] Liu Jintang. Analysis of measures for fine management of water injection well.Chemical Engineering and Equipment,2 019(7).
[10] Guili Zou, Chengwu Yuan, Xiangzhu Li, Qian Peng. Overall injection-production coupling parameter optimization method for high water cut reservoir.Petrochemical Application,2017(6):28 -33.
[11] Jun Wan, Xinguang Sui, Bing Liu, Yanming Bai.Reinjection of Finely Disposed Produced-water Reduces Formation Damage in Low Permeability Zones.SPE-82234-MS

Journal of Physics: Conference Series **2594** (2023) 012009 doi:10.1088/1742-6596/2594/1/012009

# Research on Fracturing Technology for Low Production and Low Efficiency Casing Damaged Wells

**Kun ZHANG[1,2(✉)],Qing-bo MAO[1,2],Zi-han MA[1,2],Su HUANG[1,2],Yang ZHANG[1,2]**

1Daqing Oilfield Production Technology Institute, Daqing Oilfield Limited Company, Daqing, China

2 Heilongjiang Provincial Key Laboratory of Oil and Gas Reservoir Stimulation, Daqing, China

zhang-kun@petrochina.com.cn

**Abstract**. With the further development of Daqing Oilfield, the proportion of casing damaged wells is increasing day by day, because the casing damage location is mostly located in the upper part of the reservoir, the conventional technology can not carry out targeted fracturing construction,the low-yield and low-efficiency wells which limits the efficient development of the oil field. In order to satisfy the needs of subdivision and large-scale fracturing construction of low-yield and low-efficiency wells with casing damage, this paper studied the fracturing design principles, determined the optimal design theory of different types of wells, and can make effective fracturing schemes for different reservoirs and well conditions. At the same time, a kind of small diameter segmented fracturing tools have been developed, which can meet the fracturing operation of wells with diameter above 104mm, so that the number of fracturing segments and the scale of reconstruction are greatly higher than those of conventional technology. 30 wells have been tested in the field, and the effect of increasing oil and injection is obvious. The results show that this process can effectively improve the production status of casing damage, low production and low efficiency wells and increase the productivity of single wells.

## 1. Introduction

Low-yield and low-efficiency wells are determined according to the benefit limit of cost calculation and the boundary of low-production and low-efficiency wells, which affects the overall development effect of the oil field to a great extent[1].

Through continuous analysis and exploration, it is possible to separate some oil wells from low-yield and low-efficiency wells. Therefore, the treatment of low-yield and low-efficiency wells is an important means for the smooth development of old oilfields[2].In the old area of Daqing Oilfield, the development time is long, the water cut is high, the proportion of low production and low efficiency wells is high, and most of them have casing damage, and the measures to increase production are seriously limited. Low production wells are generally characterized as low liquid content and high water cut, in which low liquid production wells are mainly caused by poor reservoir physical properties, formation plugging and other factors, while high water cut and low production wells are mainly caused by formation water and injected water. By analyzing the deep reasons for the formation of low production wells, the potential tapping direction and potential of single wells can be analyzed.

## 2. Principles of Fracturing Design for Low Production and Low Efficiency Wells

For low production and low efficiency wells, fracturing is a better way to achieve fluid extraction. Through the optimization of effective measures and parameters, it can ensure the high efficiency of reservoir production and improve the oil production capacity of the well. so that the production capacity of each reservoir can be brought into full play to a certain extent [3]. In view of the higher water cut before fracturing, the better connectivity of oil wells and the higher degree of recovery in some oil layers, the combination of various fracturing methods and the combination of selective fracturing with general fracturing and multi-fracture fracturing are selected to improve the permeability of thin and poor layers and tap the remaining oil at the top of thick oil layers, so as to tap the potential of oil wells. The combination of lifting and control of connected wells to control high aquifers to achieve the effect of measures to improve quality and increase efficiency. In view of the poor development of oil layer and the poor effect of measures, it is necessary to increase the scale of measures to effectively improve the connectivity effect of oil and water wells, and the corresponding fracturing transformation can be carried out in connected wells. In view of the fact that the effect of conventional fracturing measures is getting worse, the remaining oil analysis is strengthened, the principle of well and layer selection is optimized, the technology and scale of measures are optimized according to the dynamic and static data of the well area, and the new fracturing technology is actively explored to maximize the effect and benefit of the measures.

## 3. Development of Small Diameter Subdivision Fracturing Technology and Matching Tools

The development life of low-yield and low-efficiency wells is long, the casing damage rate is high, and the fracturing operation is difficult, so it is difficult to meet the requirements of subdivision, multi-stage and large-scale targeted transformation by using conventional fracturing tools. For this reason, a small diameter segmented fracturing tool is developed, and the string structure is shown in figure 1. The process adopts double-seal and single-card fracturing, equipped with small diameter expansion Packer and guided pressure sandblasting, single string can be fractured more than 5 segments, single stage sand 50m3 or more. It can be used for fracturing reconstruction of different well conditions above diameter 104mm. It has the advantages of safety, environmental protection and high efficiency, and its performance is more than 3 times higher than that of conventional small diameter fracturing tools [4].

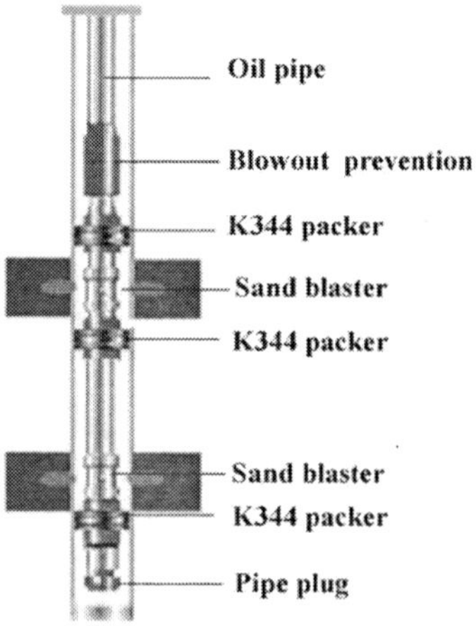


Fig. 1. Tubing string of double packers single-stage fracturing process

### *3.1. Development of Small Diameter Fracturing Packer*

The key component of packer is the rubber cylinder with sealing ability,and the high temperature mechanical property of the rubber cylinder is the main factor affecting its sealing ability[5].In order to improve the working performance,reinforcing materials can be used inside the rubber cylinder for reinforcement[6].The outer diameter of the Packer is designed according to serialization,which can be flexibly selected according to different well conditions. In order to improve the strength of rubber drum, the combination of rubber drum reinforcement cord is redesigned[7].The rubber tube adopts expanded structure, oil-resistant and heat-resistant rubber material,and the interior include steel wire and aramid fiber,which have small residual deformation and large expansion[8-9],it can ensures the pressure-bearing capacity and takes into account the safety at the same time. The overall diameter of the rubber drum is small, the passability is high, and the residual deformation after pressure is small, which increases the scope of application of the process string and reduces the risk of string blockage caused by casing deformation caused by fracturing. The Packer can meet the requirements of temperature resistance and pressure difference 70MPa, which can meet the needs of various reservoir reconstruction.

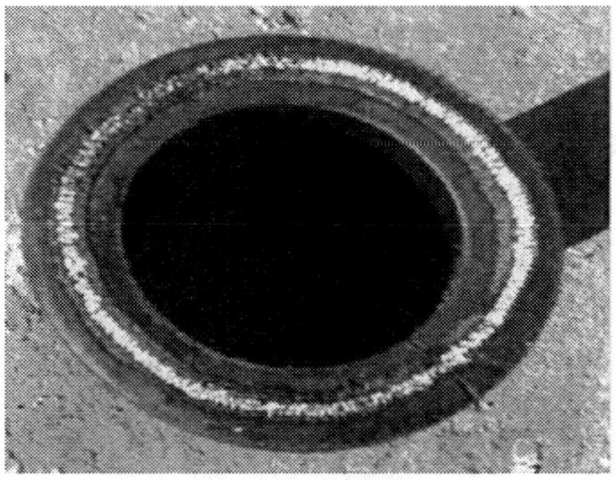

Fig. 2. Cross section of small-diameter packer rubber barrel

### *3.2 Development of Small Diameter Transmit Pressure Sand-blasting*

The outside diameter of the sandblast is 98mm, which can be applied to the staged fracturing construction in various well conditions to reduce the risk of string sticking. The integrated design of pressure guide and sand blasting is adopted.The study of nozzle performance is of great significance to the efficiency and service life of hydraulic sand jet perforator[10]. The guide channel is located inside the tool. The body is treated with local high frequency quenching to effectively improve the wear resistance. The throttle nozzle and sand outlet are made of special materials, the bushing is made of variable diameter porous design, and the wear-resistant alloy material is used. The single layer sand can reach more than 50m3, and the single sand blast can reach more than 200m3, which can meet the needs of multi-stage large-scale fracturing reconstruction.

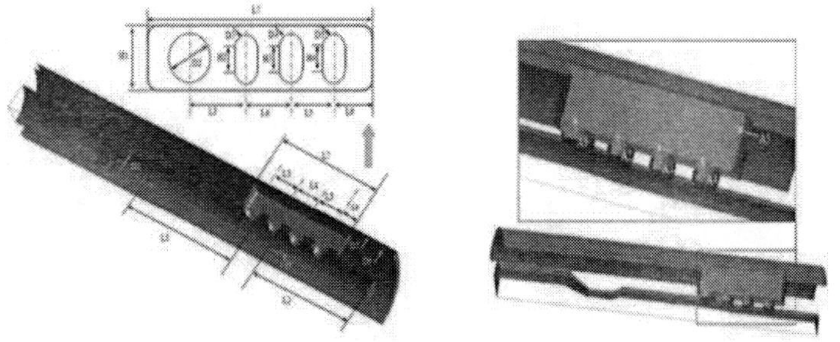

Fig. 3. Schematic diagram of flow field structure model of sand-blasting

### *3.3 Development of Tubing Blowout Preventer*
The double packers single-stage fracturing technology needs to complete the whole well construction step by step, so a small diameter tubing blowout prevention tool is designed, which can open and close the tubing channel and realize the tubing blowout prevention in the process of lifting by controlling the pressure difference of the oil jacket. to meet the requirements of the new environmental protection law. The tool has the functions of repeated opening, closing and backwashing, ensuring layer upon layer fracturing, preventing blowout layer by layer, greatly shortening the waiting time, improving the construction environment of operators, reducing labor intensity and saving construction cost. The safe, environmental protection and efficient fracturing construction of small diameter double packers single-stage fracturing technology is realized.

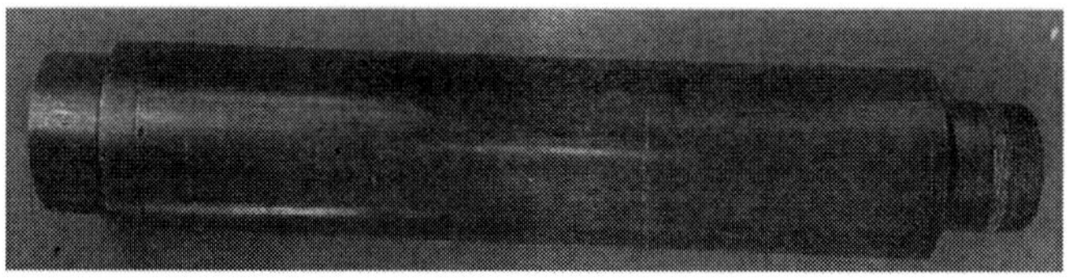

*Fig. 4. Real picture of blowout prevention tool for small diameter tubing*

## 4 Analysis of Field Application Effect
Aiming at the low production and low efficiency wells in the old area of Daqing Oilfield, the field application of 30 wells has been completed by using small diameter double sealing and single card sectional fracturing technology. The maximum single well fracturing is 7 segments, the maximum sand volume of single well is 229m$^3$, the maximum sand amount of single layer is 100m$^3$, and the minimum interval of single layer fracturing is 0.8m.Among them, the whole well fracturing 4 sections of the middle XX well, adding sand 97.5m$^3$, daily fluid production before fracturing is 9.6t, daily oil production is 1.18t, water cut is 87.7%, daily liquid production is 47.62t, daily oil production is 8.95t, water cut is 81.2%, and oil production is increased by 7.58 times.The application of this technology can effectively improve the productivity of low-yield and low-efficiency wells and provide strong technical support for accurately tapping the potential of the oil field.

## 5 conclusion
Subdivision, fine control and large-scale targeted fracturing transformation is an effective means to increase fluid production in low-yield and low-efficiency wells, which can greatly increase productivity and turn the original low-yield and low-efficiency wells into effective wells, or even high-yield wells.
The small diameter double packers single-stage fracturing tecnology can meet the fracturing needs of low production and low efficiency wells, and can also be used for segmented fracturing of casing damaged wells with different well diameters to promote the efficient development of oilfields and improve quality and efficiency.

**References**

[1] Guo Qinghua, Ma Guoliang, long Xiaoyong, Li Zhengxi. Study on comprehensive treatment technology of Low-yield and Low-efficiency Wells in Laojunmiao Oilfield. Chemical Industry Management, 2017.05:151-153
[2] Chen Qing,Yu Zhong, Zhang Chunhui, Dai Wenxin, Peng Minglan, Research on Comprehensive treatment of low yield and low efficiency Wells "Chemical Management", 2016.10:157
[3] Xue Yong, potential Analysis of inefficient Wells and Countermeasures for resuming production[J].China Chemical Trade,2017.02:230
[4] Liu Xin, Zhu Shaogong, Li Bolui, Wang Jing, Yu Zhe, Zhang Hongyan. Large-scale fracturing technology and supporting tools for casing damaged wells in Daqing Oilfield Petroleum drilling and production Technology[J]. Oil Drilling & Production Technology,2018 (40):499-502
[5] DUAN Fenghai,LI Zhen,HOU Chengxun,GUO Zhaohui,LI Fuping,DUAN Youshun.Study on the Properties of High Temperature and High Pressure Packer Cartridge[J].Oil Field Equipment,2020,49(2):66-70.
[6] TROVATTI E,CARVALHO A J F,RIBEIRO S J L,et a1.Simple green approach to reinforce natural rubber with bacterial cellulose nanofibers[J].Biomacromolecules,2013,14(8):2667-2674.
[7] Zhang Kun.Development and Application Of Small Diameter Fracturing packer in casing damaged wells[J].Oil Field Equipment,2022,1(51):105-111.
[8] BOKOBZA L.Multiwall carbon nanotube-filled natural rubber： Electrical and mechanical properties[J].Express Polymer LeRers,2012,6(3):213-223.
[9] KANNY K,MOHAN T P.Rubber nanocomposites with nano-clay as the filler[J].Progress in Rubber Nanocomposites,2017:153-177.
[10] Zhang Derong,Wu Simeng.Optimization Design of Nozzle of Hydraulic Sand Jet Perforator Based on FLUENT[J].China Petroleum Machinery

# Optimization Design And Experimental Study of Self-oscillating Cavitating Jet Descaling Nozzle Based On Cleaning Operation

**You-quan Huang[1,3], Zhong-chao Lin[2,3], Jiang Sun[2,3], Xu-yang Zhao[2,3], Ping-an Zhang[2,3]**

1. Daqing Oilfield Co., Ltd, 2.Daqing Oilfield Production Engineering Research Institute, 3. Heilongjiang Provincial Key Laboratory of Oil and Gas Reservoir Stimulation

linzhongchao@petrochina.com.cn

**Abstract.** The scale of tubing in polymer flooding and ASP flooding Wells is serious in Daqing oilfield. In order to improve the cleaning effect of tubing, the self-vibrating cavitation jet nozzle was developed and the cavitation jet technology was used to remove scale. In the design of nozzle structure, it is necessary to derive the basic relation that the structure of the oscillating cavity should satisfy when strong resonance is generated, and calculate the key length dimension $L_1$ of the oscillating cavity. The numerical simulation method the different diffusion Angle $\alpha$ and diffusion length $L_3$ nozzle wall under the condition of gas generation rate and axial velocity distribution, determine the optimum size of diffuser. The scale-removing performance test of cavitation jet verifies the performance of self-vibration cavitation nozzle and determines the optimal target distance and injection Angle. According to the reasonable dimension relation of the oscillating cavity when resonance occurs, the length $L_1$ of the oscillating cavity when strong resonance occurs is calculated as 10.4mm. Use Fluent fluid calculation software painted wall gas generating rate, the relationship between the axial velocity curve, determine the diffusion Angle $\alpha$ to 30°, spread long L3 is 2.5 mm, jet cavitation effect is best. The descaling performance test shows that under the inlet pressure of 20MPa, the target distance of 15mm and the injection Angle of 60°, the effective cleaning area is the largest within 1s. The designed nozzle has a reasonable structure and efficient descaling performance.

## 1. Introduction

Since the 1990s, the polymer flooding and ASP flooding technologies in Daqing Oilfield have entered industrial application successively, and the effect of improving oil recovery is remarkable[1-2]. However, in the field application process, the polymer flooding and ASP flooding wells have serious scaling phenomenon, resulting in frequent pump jamming, rod breakage and other conditions, which greatly affects the oil well's mechanical recovery rate. Coated tubing is used on site to reduce the scaling rate, but it cannot fundamentally solve the scaling problem[3]. Common scaling methods include chemical scaling, mechanical scaling and high-pressure water jet scaling[4-5]. In recent years, cavitation jet technology has been widely used in the field of tubing cleaning due to its simple and efficient cleaning, energy conservation and environmental protection[6]. As the key component of the cleaning equipment, the structure design of the nozzle has a direct impact on the effect of the cavitation jet. In this paper, numerical simulation and experimental research methods are combined to carry out the structural optimization design of cavitation nozzle, and to develop a high-efficiency oil tube inner wall descaling nozzle.

## 2. Technical analysis

### *2.1. Basic structure*

Self-oscillation cavitation refers to the self-excited oscillation of low pressure fluid in the resonant cavity to form large structure vortex ring and produce strong cavitation[7]. When the self- oscillation cavitation jet is used to clean the oil pipe, good cleaning effect can be achieved. At the same time, the operation time and water consumption are reduced, which has considerable social and economic benefits.

At present, the commonly used self-vibrating air jet nozzle is the organ tube cavitation nozzle. It is composed of an organ tube type resonator with length L and diameter D1 as the oscillation amplifier. The cavity inlet is connected with the inlet pipe with diameter D0, and $(D_0/D_1)^2$ forms the inlet contraction section of the cavity. The lower part of the resonant cavity is connected with a cylindrical section with a diameter of d, and $(D_1/d)^2$ forms the outlet contraction section of the resonant cavity.

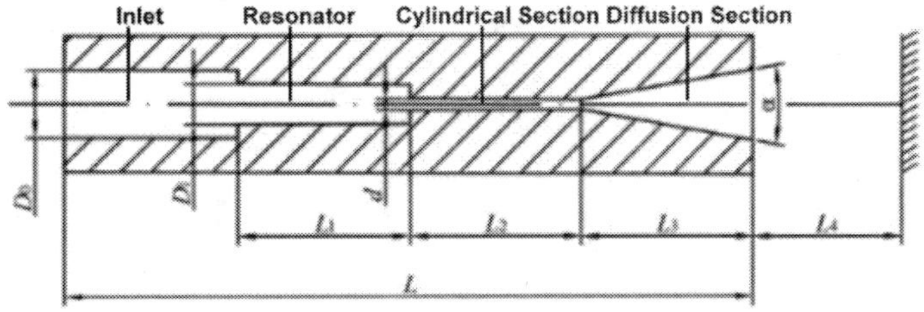


**Fig. 1.** Organ tube type cavitation nozzle

### *2.2. Working principle*

The outlet contraction section of the organ tube cavitation nozzle is both a self-exciting mechanism and a feedback mechanism. When a stable fluid passes through, its contraction surface can not only make the fluid generate initial pressure fluctuations, but also feedback the pressure fluctuations back to the resonant cavity, forming feedback pressure oscillation. According to the transient flow theory, when the frequency of the pressure fluctuation matches the natural frequency of the organ tube resonator, the feedback pressure oscillation can be amplified, thus creating fluid resonance and forming standing wave in the resonator. The vortex in the fluid boundary layer turns into a large structure separated vortex circulation and forms a low pressure region, which greatly enhances the cavitation effect.

### *2.3. Design method*

Fluid resonance is generated by self-excitation. According to the principle of underwater acoustics, the pressure fluctuation frequency $f^*$ of the fluid is determined by the critical Strouhal number Sr of the nozzle. The natural frequency of the resonant cavity depends on the contraction degree of the inlet section $(D_0/D_1)^2$ and the outlet section $(D_1/d)^2$. Therefore, the design of the organ tube resonator must first calculate the natural frequency f of the resonator, and then make the pressure fluctuation frequency $f^*$ that may be generated by the fluid self-excitation equal to the natural frequency f (resonance condition), and then design the appropriate resonator size.

According to the research of Li Gensheng et al.[8], the general expression of the natural frequency f of the organ tube is:

$$f = K_N \frac{a}{L_1} \qquad (1.1)$$

$$K_N = \frac{2N-1}{4}[(\frac{D_0}{D_1})^2 \quad 1>> \quad \frac{D_1}{d}{}^2 >> 1] \qquad (1.2)$$

Where, f is the natural frequency of the organ tube, in Hz; $K_N$ is modulus coefficient; N is the oscillation modulus; a is the sound velocity in the fluid, in m/s; $L_1$ is the length of the resonator, in m; $D_0$ is the diameter of the mouth, in m; $D_1$ is the diameter of the resonator, in m; d is the diameter of the cylindrical section, in m.
The pressure fluctuation frequency $f^*$ of the fluid is expressed by the nozzle critical Strouhal number Sr and Mach number Ma:

$$f^* = Sr\frac{Ma}{d}a \qquad (1.3)$$

According to the experimental results of S C Crow, F H Champagne [9] and V E Johnson et al [10]. By equating Equation (1.1) with Equation (1.3), we can obtain the basic relationship that the organ tube resonator structure should meet when generating the strongest resonance:

$$\frac{L_1}{d} \approx \frac{K_N}{MaSr} \qquad (1.4)$$

*2.4. Main structural parameters of self-oscillating cavitating jet nozzle*
The reduction of the length of the nozzle cylinder is beneficial to improve the cavitation effect. In combination with the actual situation on site and taking into account the machinability and accuracy of the nozzle, the diameter d of the cylindrical section is 1mm, the length $L_2$ is 1mm, the diameter $D_0$ of the inlet cavity is 6.4mm, and the diameter $D_1$ of the resonant cavity is 3.2mm. Since $(D_0/D_1)^2$ and $(D_1/d)^2$ are both>>1, taking the oscillation modulus as 1, then KN =0.25 can be obtained from equation (1.2). Taking the Mach number Ma and the critical Strouhal number Sr under strong resonance conditions, the optimum length $L_1$ of the resonator can be obtained from formula (1.4), which is 10.4 mm. The main specifications and dimensions of the nozzle are shown in Table 1:

**Table 1** Main structural parameters of cavitation nozzle

| Structural parameters | Parameter value (mm) |
|---|---|
| Inlet diameter D0 | 6.4 |
| Diameter of resonator D1 | 3.2 |
| Diameter of cylinder section d | 1.0 |
| Resonator length L1 | 10.4 |
| Cylinder length L2 | 1.0 |

The diffusion angle a and the length of the diffusion section $L_3$ of the self-oscillating cavitating jet nozzle are also the key parameters that affect the cavitation performance of the nozzle, and their reasonable values need to be optimized through numerical simulation.

## 3. Technical analysis

*3.1. Motion analysis geometry and mesh model*
According to the nozzle structure and main technical parameters of cavitation nozzle in Figure 1, the fluid domain of self-oscillating cavitation jet nozzle is established. Due to the axisymmetric characteristics of the nozzle, in order to save computational resources and improve computational efficiency, the nozzle is simulated in two-dimensional symmetry. Use ICEM software to complete the mesh generation of nozzle two-dimensional model. In order to improve the calculation accuracy, the cylindrical section and the diffusion section are densified. The mesh division of nozzle is shown in Figure 2.

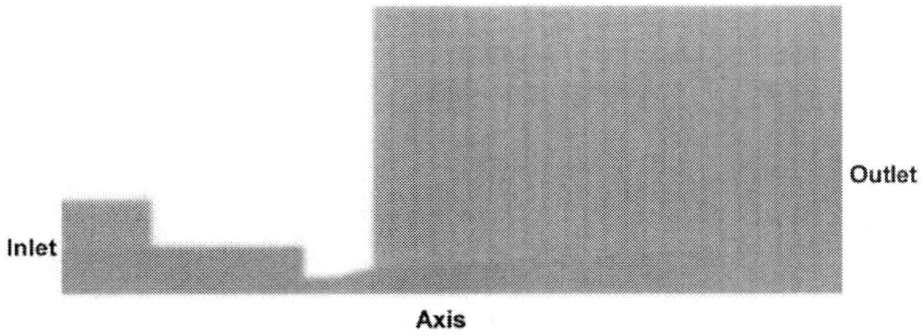


**Fig. 2.** Detail of grid division of self-oscillating cavitating jet nozzle

*3.2. Boundary conditions*
Fluent is used to carry out CFD numerical simulation of the grid model of the self-oscillating cavitating jet nozzle. The pressure inlet boundary condition is selected for the inlet, which is set as 20MPa. The pressure outlet boundary condition is selected for the outlet, which is set as standard atmospheric pressure. The non-slip adiabatic wall boundary condition is adopted for the solid wall, and the axisymmetric boundary condition is adopted for the symmetry axis.

*3.3. Flow field parameter conditions*
For the numerical model, a total of 8 numerical experiments were carried out with diffusion angle a and diffusion section length L3 as variable parameters to study the influence of diffusion angle and diffusion section length on gas phase formation rate. The specific calculation parameters are shown in Table 1.

**Table 2** Numerical simulation calculation parameters

| Diffusion angle a (°) | Diffusion section length $L_3$ (mm) |
|---|---|
| 30 | 1.5 |
| | 2 |
| | 2.5 |
| | 3 |
| 20 | 1.5 |
| 30 | |
| 40 | |
| 50 | |

### *3.4. Analysis of influence factors*

Calculate the gas phase formation rate under different influence parameters, take the nozzle wall coordinate as the abscissa, unit: mm, the gas phase formation rate as the ordinate, unit: 10-4/(kg·m-3·s-1), and draw the relationship curve between the gas phase formation rate and the nozzle wall coordinate; Calculate the axial velocity under the conditions of different influence parameters, take the nozzle axis coordinate as the abscissa, the unit is mm, the velocity is the ordinate, the unit is m·s-1, and draw the relationship curve between the velocity and the nozzle axis coordinate.

#### *3.4.1 Effect of diffusion angle*

It can be seen from the calculation results in Fig. 3 and Fig. 4 that the distribution of jet velocity with nozzle axis coordinates is almost the same when the diffusion angle is 20 ° and 40 °. The axis coordinates of the stable high-speed zone at 30 ° are the largest, and the isokinetic nucleus is the longest. After the jet passes through the nozzle, the gas phase generation rate rapidly increases to the maximum value, and the phenomenon of "isokinetic nucleus" appears at the position 8 mm away from the outlet. Then, with the gradual extension of the isokinetic nucleus, the velocity began to decrease gradually, and after a short rise at a distance of more than 9 mm, the velocity continued to decline, indicating that a strong vortex ring effect was formed in the flow field and cavitation effect was formed. When the diffusion angle is from 20° to 30°, the gas phase formation rate on the wall is the highest and the isokinetic nucleus length is the longest. When the angle continues to increase, the fluid begins to impinge on the nozzle surface, and the resistance of the nozzle to the fluid movement increases significantly, resulting in a large amount of jet energy loss. Therefore, the length of the isokinetic nucleus begins to decrease. In comprehensive consideration, when the diffusion angle a is 30 °, the spraying effect is better.

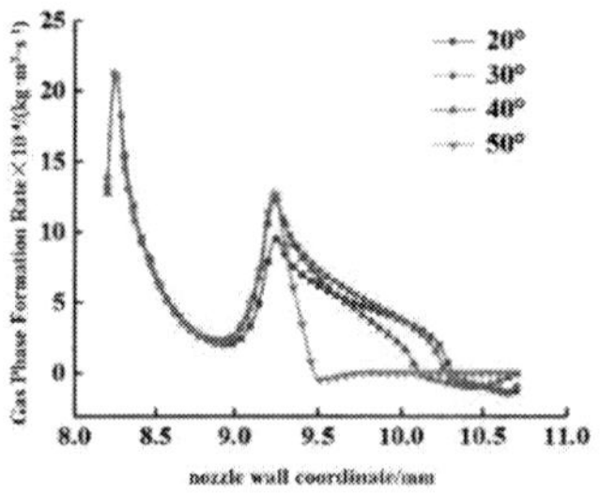


**Fig. 3.** Influence curve of diffusion angle on gas phase formation rate on nozzle wall

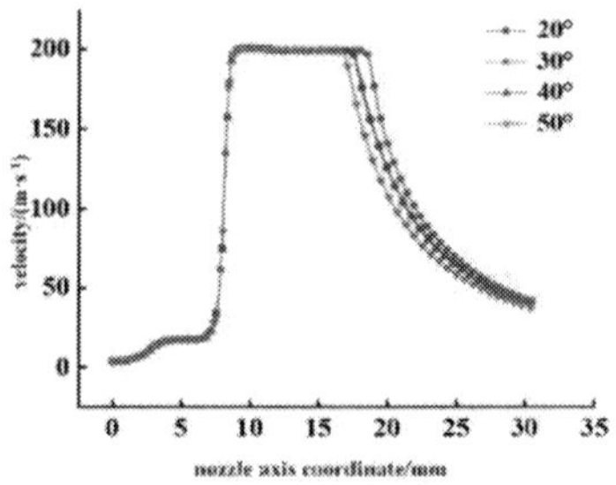


**Fig. 4.** Velocity distribution of nozzle axis

#### *3.4.2 Effect of diffusion section length*

According to the calculation results in Fig. 5 and Fig. 6, when the diffusion angle is 30 ° and the length of diffusion section increases from 1.5mm to 2.5mm, the gas phase generation rate and the length of isokinetic nucleus increase. Continue to increase the length of the diffusion section to 3mm, the gas phase generation rate and the length of the isokinetic nucleus will not increase significantly, and the length of the wall without gas phase generation will become longer. Therefore, when the length of the diffusion section is 2.5mm, the injection effect is better.

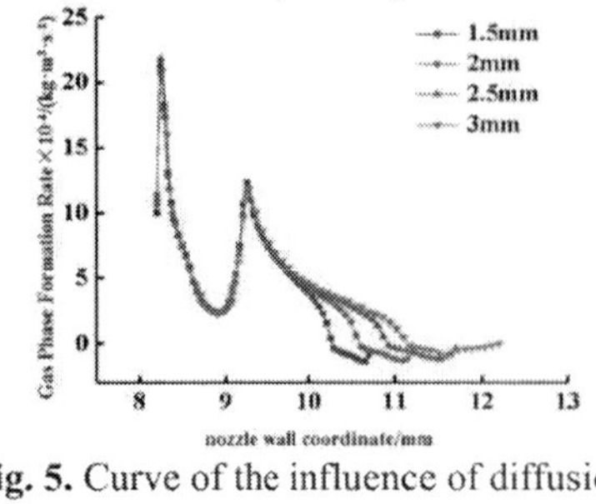


**Fig. 5.** Curve of the influence of diffusion

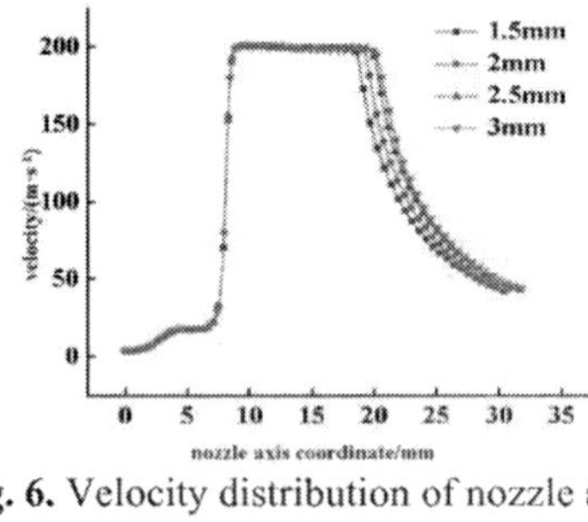


**Fig. 6.** Velocity distribution of nozzle axis

section length on the gas phase formation rate on the nozzle wall

The spray target distance and spray angle have a great impact on the scale removal effect. The best target distance and spray angle for scale removal will be determined through indoor tests.

## 4. Cavitation jet descaling performance test

### *4.1. Test platform and nozzle structure*

Figure 7 shows the nozzle descaling performance test platform, which is mainly composed of pump unit, control and flow field measurement. Its working principle is that when the working medium is delivered to the nozzle after being pressurized by the high-pressure pump, the self-oscillation cavitation phenomenon will occur after passing through the nozzle. The cavitation effect will be verified by measuring the jet velocity at the nozzle outlet. According to the nozzle structure calculation results, the test nozzle structure is processed as shown in Figure 8.

**Fig. 7.** Nozzle descaling performance test platform

**Fig. 8.** Picture of single nozzle

### *4.2. Scale removal performance test with different target distances*

Considering that the force on the hinge joint of the lifting mechanism is relatively large, it is necessary to check the strength of the padeye of the hinge joint. Next, take the padeye at the hinge of piston rod and rotary arm as an example for finite element analysis.

The tubing is the tubing with an outer diameter of F73mm pulled from the well at a ASP flooding operation site in Daqing. Although the tubing with an outer diameter of F48 mm can not be passed smoothly. Cut the tubing along the axis, and it can be seen that the scaling on the inner wall of the tubing is very obvious.

**Fig. 9.** Wax deposition inside the tubing

During the test, the pressure of the high-pressure liquid supply pump is set at 20MPa, the central axis of the nozzle outlet is designed to be perpendicular to the inner wall of the oil pipe, the test time is controlled at 1s, and the test target distance is controlled within 10-30mm. The best target distance is analyzed by measuring the impact depth and effective impact area of the cleaning area with an electronic microscope.

The test shows that when the cleaning target distance is 10-17 mm, the inner wall of the tubing is impacted with metal color, and the scale removal effect is obvious. It is proved that the nozzle structure can produce complete cavitation and can remove the hard scale on the inner wall of the tubing.

When the target distance is 10-15mm, the effective cleaning area gradually increases to 37mm2. When the cleaning target distance is greater than 15mm, the effective cleaning area decreases. When the target distance is greater than 18mm, the effective cleaning area decreases to 0. Therefore, according to the test results, the optimal target distance of nozzle descaling is initially set as 15mm.

**Table 3** Comparison of cleaning effect and effective cleaning area of nozzles with different target distances

| number | Injection angle (°) | Cleaning time (s) | Target distance (mm) | Whether metal color is exposed | Effective cleaning area ($mm^2$) |
|---|---|---|---|---|---|
| 1 | 90 | 1 | 10 | yes | 9 |
| 2 | 90 | 1 | 11 | yes | 13 |
| 3 | 90 | 1 | 12 | yes | 16 |
| 4 | 90 | 1 | 13 | yes | 25 |
| 5 | 90 | 1 | 14 | yes | 30 |
| **6** | **90** | **1** | **15** | **yes** | **37** |
| 7 | 90 | 1 | 16 | yes | 28 |
| 8 | 90 | 1 | 17 | yes | 13 |
| 9 | 90 | 1 | 18 | no | 0 |
| 10 | 90 | 1 | 19 | no | 0 |
| 11 | 90 | 1 | 20 | no | 0 |
| 12 | 90 | 1 | 21 | no | 0 |
| 13 | 90 | 1 | 22 | no | 0 |
| 14 | 90 | 1 | 23 | no | 0 |
| 15 | 90 | 1 | 24 | no | 0 |
| 16 | 90 | 1 | 25 | no | 0 |
| 17 | 90 | 1 | 26 | no | 0 |
| 18 | 90 | 1 | 27 | no | 0 |
| 19 | 90 | 1 | 28 | no | 0 |
| 20 | 90 | 1 | 29 | no | 0 |
| 21 | 90 | 1 | 30 | no | 0 |

### *4.3. Scale removal performance test with different injection angles*

Set the spray target distance of 15mm and the working time of 1s to test the cleaning effect and scale removal area under different spray angles (30 °, 45 °, 50 °, 60 °, 70 °, 80 ° and 90 °).

**Table 4** Comparison of cleaning effect and effective cleaning area of nozzles with different injection angles

| number | Injection angle (°) | Cleaning time (s) | Target distance (mm) | Whether metal color is exposed | Effective cleaning area ($mm^2$) |
|---|---|---|---|---|---|
| 1 | 30 | 1 | 15 | no | 0 |
| 2 | 45 | 1 | 15 | yes | 25 |
| 3 | 50 | 1 | 15 | yes | 31 |
| **4** | **60** | **1** | **15** | **yes** | **40** |
| 5 | 70 | 1 | 15 | yes | 28 |
| 6 | 80 | 1 | 15 | yes | 26 |
| 7 | 90 | 1 | 15 | yes | 37 |

The test results show that when the nozzle outlet pressure is 20 MPa, the target distance is 15 mm, the working time is 1 s, and the spray angle is 60°, the cleaning area is 40mm2, and the scale removal effect is the best.
Based on the comprehensive test results, the optimal scale removal target distance and spray angle of a single nozzle are determined, which provides a design basis for the development of a nozzle for high performance tubing cleaning with multiple nozzles.

## 5. Conclusion

In this paper, a self-oscillating cavitating jet nozzle is designed to solve the problem of tubing scaling. The length of nozzle resonator is designed by theoretical calculation; The optimal diffusion length and diffusion angle are determined by numerical simulation; The performance test of cavitation jet descaling is designed to determine the optimal target distance and spray angle for descaling. Through this study, the following conclusions are drawn:
(1) In view of the resonance problem of the self-resonating cavitating jet nozzle, based on the underwater acoustic principle, the basic relationship that the resonator structure should meet when generating strong resonance is established, and the optimal resonator length is determined to be 10.4 mm;
(2) In order to further enhance the cavitation effect of the self-resonating cavitating jet nozzle, based on the numerical simulation method of hydrodynamics, the geometric model and mesh model of the self-resonating cavitating jet nozzle are established, and the numerical simulation is carried out. The relationship between the angle of the diffusion section, the length of the diffusion section and the gas phase formation rate on the wall is studied. It is found that the cavitation effect is the strongest when the angle of the diffusion section is 30° and the length of the diffusion section is 2.5mm;
(3) The experimental section of self-oscillating cavitating jet nozzle is designed, and the scale removal performance test system of cavitating jet is built. The reliability of scale removal by self-oscillating cavitating jet nozzle is proved by indoor experiments.
(4) Under the experimental conditions of this paper, when the target distance is 15mm and the injection angle is 60°, the effective cleaning area is the largest in 1s, with the best descaling effect, and the goal of developing an efficient descaling nozzle for the inner wall of tubing is achieved.

### 6. References

[1] Sun Longde, Jiang Tongwen, Wang Fenglan 2021 Thinking about oilfield life. *J. Petroleum* 42 (01): 56-63(2021)

[2] Wang Yupu, Cheng Jiecheng 2003 Scaling characteristics and adaptability of mechanical recovery method during ASP flooding. *J. Daqing Petroleum Institute* 27 (2): 20-22(2003)

[3] Xu Guomin 2009 Problems and development direction of ASP flooding injection-production technology. *J. Petroleum Machinery* 37 (2): 77-80(2009)

[4] Shen Xiaoming, Li Gensheng, Ma Jiaji 1996 Research on tubing scaling mechanism and hydraulic scaling technology. *J. Petroleum Drilling Technology* 24 (3): 43-45(1996)

[5] Johnson Jr Virgil E Chahine G L Lindenmuth W T . 1982 The development of structured cavitating jet for deep hole bits SPE11060 1982

[6] Cheng Xiaorui, Zhang Shuyan, Chen Hongxing 2018 Application progress and prospective analysis of cavitation jet technology. *J. Energy and Chemical* 39 (4): 19-24(2018)

[7] Li Gensheng, Shen Zhonghou 1992 Design principle of self-oscillation cavitation nozzle for organ tube. *J. Petroleum University (Spontaneous Combustion Science Edition)* 16 (5): 35-39(1992)

[8] Yi Can, Li Gensheng, Zhang Dingguo 2005 Experimental study on cavitation initiation ability of self-oscillating nozzle. *J. China Mechanical Engineering* 16 (21): 1945-1948(2005)

[9] Crow S C and Champage F H.: Orderly structure in jet turbulence. *J.* Fluid Mechanics 48(3):547-591(1971)

[10] Johnson V E Jr,Lindenmuth W T,Conn A F 1981 Feasibility study of tuned-resonator,pulsating cavitating water jet for deep hole drilling. SAND 81-7126,1981

# Study on Economic and Efficient Completion Design of Tight Gas Reservoir

**Meng-CAI[1,2], Zhi-bo AN[1,2],Wen-hai MA[1,2], Jin-xing CHENG[1,2],Qian ZHANG[1,2]**

1 Daqing Oilfield Production Technology Institute, Daqing, China

2 Heilongjiang Provincial Key Laboratory of Oil and Gas Reservoir Stimulation, Daqing, China

Email:anzhibo@petrochina.com.cn;Da Qing,Hei Longjiang Province,China

**Abstract.** Tight gas reservoirs in Daqing Oilfield has the characteristics of deep burial, high temperature and high pressure, and the produced natural gas generally contains CO2. At present, the production casing and tubing of the tight gas reservoir are completed with 13Cr110 material, which has high cost, but the gas production is low, and it declines very fast, so the economic benefit is poor. In order to reduce cost and increase the efficiency, it is necessary to explore a new well completion method in tight gas reservoir. We study the corrosion law of different materials through laboratory experiments, and finally determine that the production casing adopts the assembly method, material of production tubing next to the production packer is 13Cr, other parts are ordinary carbon steel. The tubing and downhole tools are made of carbon steel, and the casing and tubing annulus are injected with annulus protection fluid for corrosion protection. With the comprehensive analysis, this completion method can meet the requirements of gas well corrosion prevention and production. After 20 years of development, the well construction cost of a single well has been reduced by 79% to 7.553 million RMB. The effective implementation of this design can provide reference for the subsequent development of tight gas reservoir in Daqing Oilfield.

## 1. Introduction

The deep gas well development reservoirs in Daqing Oilfield are mainly volcanic rock and tight gas reservoirs. The volcanic rock reservoirs are developed earlier and the development plan is relatively mature. The tight gas reservoirs have been gradually developed in recent years. At present, the development plan mainly refers to acidic volcanic rock reservoirs, but compared with volcanic rock, the physical properties are poor, the pressure is higher, and the development is difficult, but the production is low and the decline rate is fast. In order to reduce the cost of well construction, to improve the economic benefits of block development and achieve cost reduction and efficiency increase at the same time, it is necessary to explore a new completion method in tight gas reservoir[1-3].

## 2. Geological conditions of tight gas reservoir and difficulties in design

### *2.1. Geology of tight gas reservoir*

The gas component of tight gas reservoir in Daqing deep gas well is mainly methane, with methane content of 91.844%~91.927%, with an average of 91.886%; Water is produced generally, with carbon dioxide content of 3.625%~4.983%, with an average of 4.304%; The average relative density of natural gas is 0.6134~0.6174, with an average of 0.6154. The formation water salinity is 6370.66mg/L, bicarbonate is 2379.39mg/L, and chloride ion content is 1881.21mg/L, which is $NaHCO_3$ water type.

*2.2. Difficulties in design*

**Poor physical properties of tight gas reservoir** . The rock types of tight gas reservoirs in Daqing deep gas wells are mainly sandy conglomerate and sandstone, with large burial depth, strong diagenesis and poor reservoir physical properties. The thickness of gas layer is mainly ≤ 10m, the effective thickness is mainly ≤ 5m, and the thickness of single layer is thin; The statistical results of core analysis show that the porosity is between 2.0 and 9.0%; Permeability is 0.03-0.5D, belonging to ultra-low porosity and low permeability tight reservoir.

**High temperature and pressure of tight gas reservoir** . The central buried depth of Daqing deep gas well tight gas reservoir is more than 3500 meters, the measured formation pressure can reach 54MPa, and the pressure coefficient is 1.379, which is a normal pressure system. The measured formation temperature reaches 139.4 ℃ , and the average geothermal gradient is 3.684 ℃ /100m, all of which belong to relatively high geothermal gradient.

**The rapid production decline of tight gas reservoir** . Tight gas reservoirs are developed later than acidic volcanic reservoirs, and the production and open flow are generally low at the initial stage of production, and the production decline rate is fast. The open flow of tight glutenite reservoir and volcanic reservoir with similar formation conditions differs by 3~5 times. After two years of production, the production of gas wells in volcanic reservoir decreases by 3~5%, and the production of tight glutenite reservoir decreases by 20%. The stable production capacity is poor, and the economic benefit is poor.

## 3. Current completion of tight gas reservoir

There are five industrial gas wells in the tight gas reservoir reserve area currently developed by the atmospheric oilfield, including two vertical wells and three horizontal wells. The well bore structure adopted is shown in Fig. 1[4-5]:

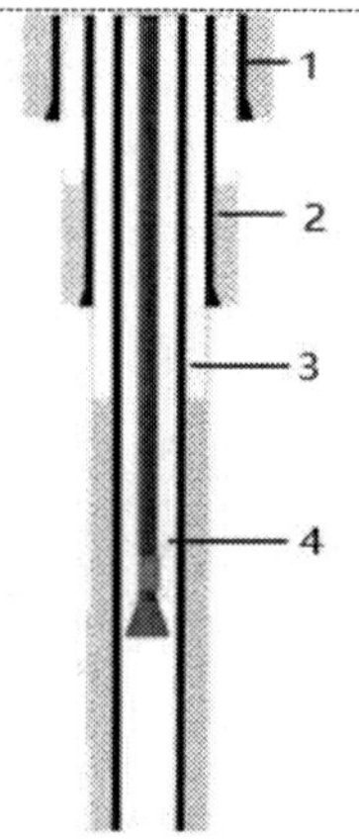


**Fig. 1.** Diagram of completion string of tight gas reservoir.

1—Surface casing J55; 2—Technical casing N80; 3—Production casing 13Cr110; 4—Tubing 13Cr110

At present, this well bore structure is adopted for the completed gas wells in tight gas reservoirs. The gas wells are in normal production, and no corrosion of casing and tubing is found.

## 4. Optimization of completion design for tight gas reservoir

*4.1. Casing material optimization*

The main corrosive gas in Daqing tight gas reservoir is $CO_2$. Because $CO_2$ local corrosion is the main form of pipe string damage, in order to reduce the cost of well construction and explore low-cost anti-corrosion technology, the local corrosion research of P110 tubing is carried out through laboratory tests, and the prediction chart of local corrosion sensitive area made of ordinary carbon steel is established.

Local corrosion law of P110 carbon steel: local corrosion sensitivity of P110 under different temperature and pressure ranges, as shown in Figure 2, P110 steel has serious local corrosion at 60~100 ℃, and tends to comprehensive corrosion at about 100 ℃. When the temperature is constant, the partial pressure of CO2 is in the range of 0.2~4MPa, and the local corrosion rate increases with the increase of the partial pressure of $CO_2$.[6-8]

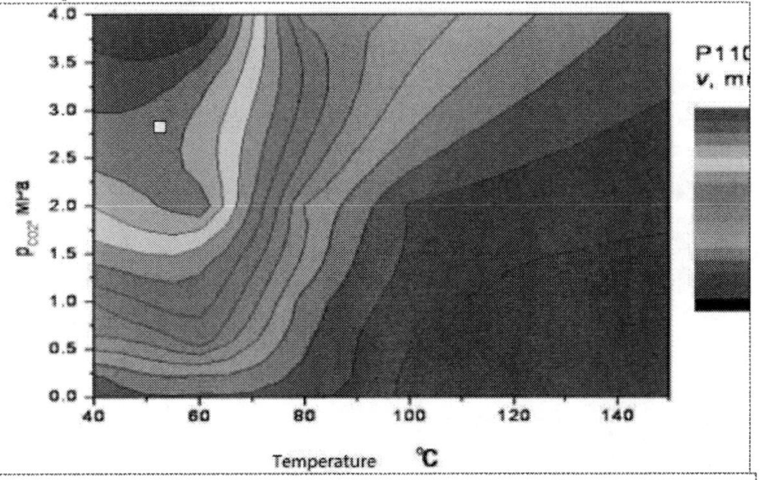


**Figure2** P110 Local Corrosion Pattern of Carbon Steel.

In view of the large local corrosion hazard, if 13Cr is used for corrosion protection in the whole wellbore, the completion cost is high. Develop low-cost local corrosion protection methods. Firstly, the local corrosion profile and galvanic corrosion rate of ordinary carbon steel tubing and casing are calculated by using the local corrosion prediction model. The casing assembly shall be carried out at the position where the uniform corrosion rate and galvanic corrosion rate are low; Complete the well with packers, use production packers and annulus chemical injection valves to achieve efficient injection of corrosion inhibitors, fully protect the oil jacket annulus, and reduce the corrosion rate[9-10].

Journal of Physics: Conference Series **2594** (2023) 012011 doi:10.1088/1742-6596/2594/1/012011

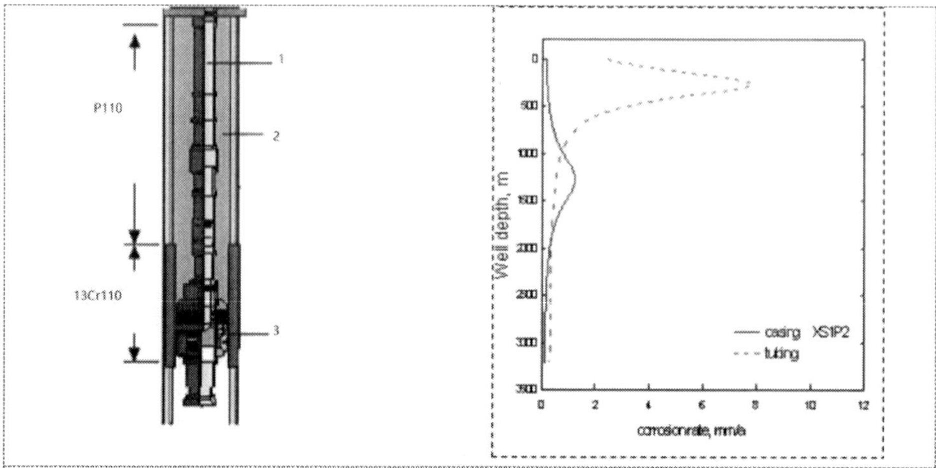


**Figure3** Schematic diagram of casing assembly.

1—Tubing; 2—Annulus protection fluid; 3—Packer

### *4.2. Tubing material optimization*

Corrosion rate of different materials is predicted based on gas production and pressure of single well predicted by gas reservoir. The corrosion rate of carbon steel and low chromium steel is higher than the severe corrosion limit, but the price of carbon steel and 3Cr pipes can reduce the cost by more than 60% compared with 13Cr pipes. Due to the special process of 9Cr materials, the market price is generally higher than 13Cr materials, so it is not considered.

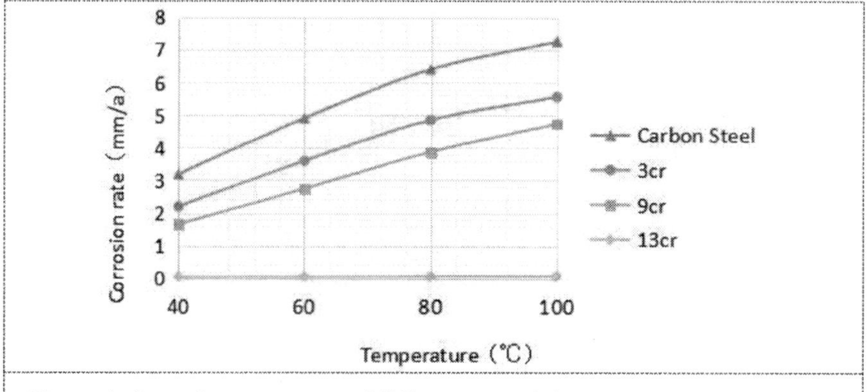


**Figure4** Corrosion rate curve of different materials at different temperatures.

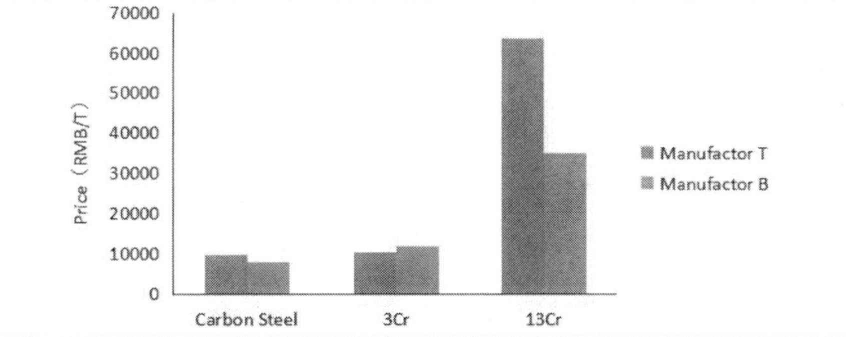


**Figure5** Comparison of prices of different pipes.

Considering the decline of gas production of new wells in the block, the corrosive environment, the anti-corrosion performance of different materials, and the purchase price, the combination plan of corrosion prevention and corrosion inhibitor of ordinary carbon steel is designed. By calculating the corrosion rate and stress curve of ordinary carbon steel tubing, the plan can meet the requirements of 20 years of safe production[11-12].

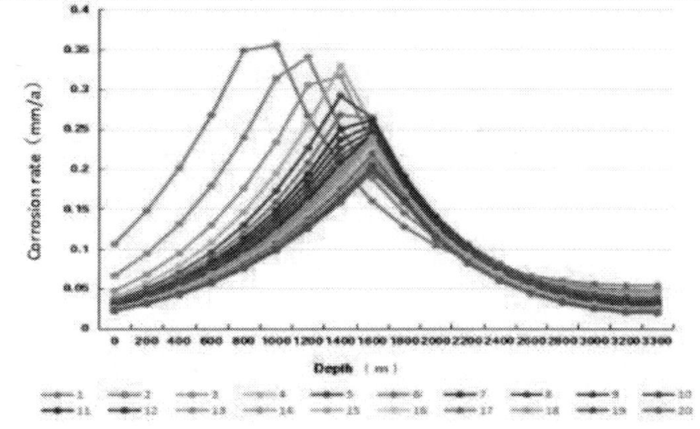


**Figure6** Corrosion rate diagram of ordinary carbon steel tubing (corrosion inhibition rate 80%).

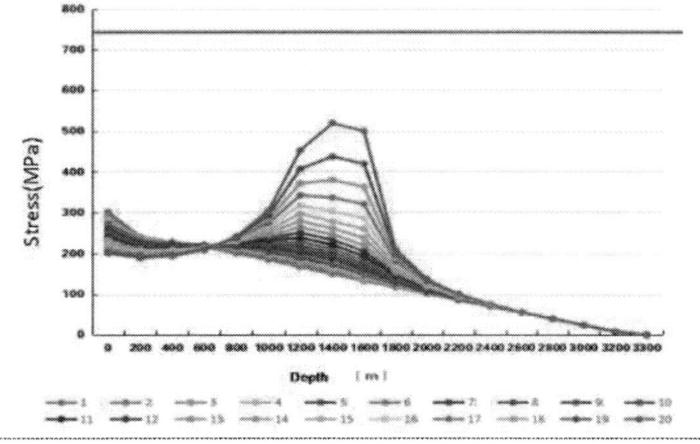


**Figure7** 20-year stress curve of ordinary carbon steel tubing (corrosion inhibition rate 80%).

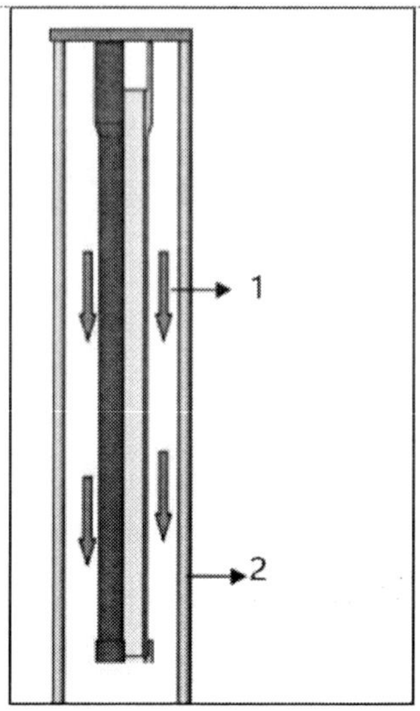


**Figure8** Diagram of inhibitor injection.

1—Corrosion inhibitor; 2—Tubing

*4.3. Mechanical check of pipe string*

The stress analysis under different working conditions is carried out by using the WELLCAT software, and the results are shown in the following figure. The calculation results show that the tubing string meets the requirements for safe production, and the safety factor remains above 1.6 under different working conditions.

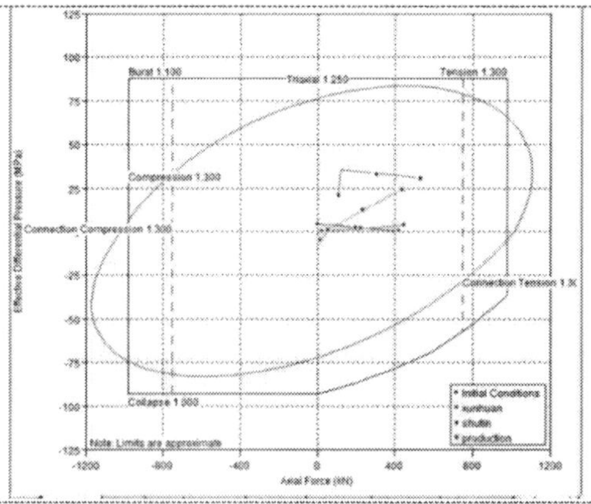


**Figure9** Full life cycle string strength.

Journal of Physics: Conference Series **2594** (2023) 012011 doi:10.1088/1742-6596/2594/1/012011

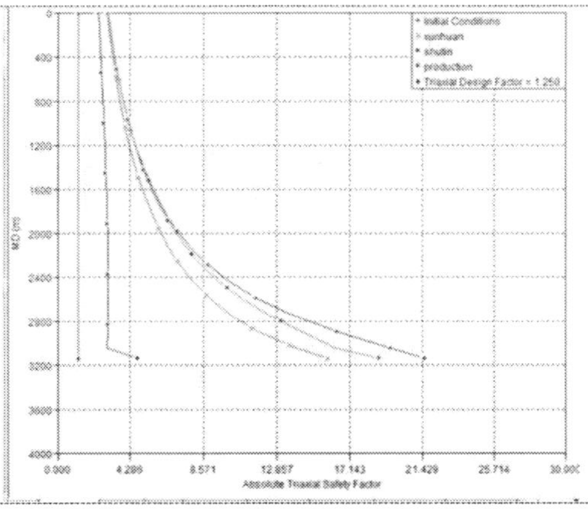


**Figure10** Safety factor of pipe string.

## 5. Comparison of economic benefits

According to the 20-year production life of the gas well and 3000m of tubing, the economic benefits of the two schemes are compared. The whole wellbore 13Cr has a high cost and is not recommended. The combined casing+ordinary carbon steel scheme can meet the 20-year safety production requirements and significantly reduce the well construction cost. Therefore, it is recommended to adopt this plan and use ordinary carbon steel tubing to complete the well. During the implementation of the plan, corrosion inhibitor shall be injected into the wellbore in a timely manner.

**Table 1.** Comparison of economic benefits of different schemes

| Year | Cost of 13Cr oil casing in the whole wellbore (10000 RMB) | Combined casing+ordinary carbon steel tubing+corrosion inhibitor | | |
|---|---|---|---|---|
| | | Total cost | Cost savings | |
| | | 10000 RMB | 10000 RMB | % |
| 1 | 951 | 135.7 | 815.3 | 85.73 |
| 4 | | 150.7 | 800.3 | 84.15 |
| 10 | | 165.7 | 785.3 | 82.58 |
| 20 | | 195.7 | 755.3 | 79.42 |

*Note: The cost of corrosion inhibitor is 50000 RMB per year for each well*

## 6. Conclusion

(1)Compared with volcanic rock reservoir, Daqing deep gas well tight gas reservoir has deep burial depth, high temperature, high pressure, poor reservoir physical properties, and greater development difficulty, but the production is low and the decline rate is fast. At present, the main completion string is the production casing and the oil casing are made of 13Cr110 material. The gas well meets the normal requirements, and no oil casing corrosion is found. However, the cost of this completion method is high, in order to achieve cost reduction and efficiency increase, Well completion mode of tight gas reservoir gas well.

(2)Through indoor experiments to study the local corrosion law of P110 carbon steel in $CO_2$ environment, optimize the casing combination structure, and finally determine to use the packer to complete the well, and inject corrosion inhibitor into the casing and tubing annulus for corrosion protection, which can ensure the overall safety and reliability of the pipe string.

(3)The oil pipe material is optimized. Considering the anti-corrosion performance and economy, the common carbon steel oil pipe+corrosion inhibitor anti-corrosion scheme is optimized. The stress analysis under different working conditions is simulated with WELLCAT software, and the strength meets the requirements.

By calculating the cost of conventional well completion oil and casing and the cost of optimized tight gas reservoir oil and casing, the cost saved at the initial stage of well construction is 85.73%, and after 20 years of development, the cost saved is about 79%, which is 7.553 million RMB..

**References**:

[1]An Zhibo.Optimization design of completion string for pilot test well of four-station gas storage group[G].Collected Works of Oil Production Engineering of Daqing Oil Field Co., Ltd. Oil Production Engineering Research Institute.2019(1).Beijing, Petroleum Industry Press,61-64.

[2].He Haijun.Prediction Model of Downhole String Corrosion Rate in Daqing Lamadian Gas Storage[G].Collected Works of Oil Production Engineering of Daqing Oil Field Co., Ltd. Oil Production Engineering Research Institute. 2017(3). Beijing, Petroleum Industry Press,73-76.

[3].Li Yufei,Ai Fangbin,She Chaoyi,Research and Application of Completion Technology for Longwangmiao High Temperature High Pressure Acidic High Production Gas Well[C],2015,2015 National Natural Gas Academic Annual Conference.

[4].Huang Weiming.Integrity evaluation technology of deep gas wells in daqing oilfield[G].Collected Works of Oil Production Engineering of Daqing Oil Field Co., Ltd. Oil Production Engineering Research Institute.2018(4).Beijing,Petroleum Industry Press,21-26.

[5].Wang Hao.Research on intelligent completion production optimization method[J].Xi'an Shiyou University,2017.

[6].Ma Wenhai, Gao Fei, Gao Chunliang.Research and application of anti-corrosion technology for casing assembly in CO2 gas wells.The Third Oil and Gas Field Development Technology Conference in 2009.

[7].De Waard.Predictive Model for CO2 Corrosion Engineering in Wet Natural Gas Pipelines[J]. Corrosion, 1991,47(12):977~985.

[8]. A Systematic Approach to Monitoring and Management of Annulus Pressures [J]. SPE-189412-MS, 2018.

[9]. Advanced Well Modeling and Diagnostics to Manage Annulus Pressure [J]. SPE-183303-MS, 2016.

[10].Xiangning Zhang; Minglu Zhang; Chunmei Ge. Successful Tight Gas Development Practice in Ordos Basin of China.OTC-26697-MS,2016.

[11].He Liu, Jianwen Yan, Zixiu Yao.A New Practical Approach to Building Tight Reservoir Development Integration Data Model in China.SPE-186983-MS,2017.

[12].Zhang Zhi,Zhang Chao.Design method of completion string for high temperature and high pressure gas well[J].Journal of Chongqing University of Science and Technology: Natural Science Edition,2017,1,67-70.

# Seismic Identification and Hydrocarbon Accumulation Characteristics of Submarine Fan in offshore Suriname

**Hai-guang Bian(✉), Zhi-xin Wen , Zhao-ming Wang, Zheng-jun He, Cheng-peng Song, Zuo-dong Liu, Xue-ling Wang, Heng-xuan Li, Tian-yu Ji**

Department of C Research Department of Oil and Gas Resource Exploration, Research Institute of Petroleum Exploration and Development, Beijing, China

bianhaiguang@petrochina.com.cn

**Abstract:** In recent years, oil and gas fields have been continuously discovered in the Suriname-Guyana Basin of South Atlantic, the reservoir of which is submarine turbidite sandstone and become one of the hot exploration objectives in the world. Based on the interpretation and analysis of 2D seismic and well data, this paper presents the integrated geological and geophysical evaluation of the Suriname-Guyana basin within Suriname and describes the identification characteristics and hydrocarbon accumulation process of submarine fan in offshore Suriname. The tectonic evolution of this region can be divided into four stages, breakup of Pangea, intracontinental rift, regional uplift and drifting over the South Atlantic. The Upper Cretaceous marks the transition of a transgressive to a regressive system with increasing sediment input from the south west, creating a broad delta plain environment across the shallow offshore area. It has been confirmed that multi-stage submarine fans have developed from Upper Cretaceous to Neogene in offshore Suriname, with overlapping characteristics of multiple erosive surfaces. It is suggested that erosive surface forming local unconformity is the important identification mark of submarine fan. Among them, the Paleogene and Upper Cretaceous submarine fans are the most widely distributed and have the most obvious characteristics on seismic profiles. The bottom of the Paleogene and Upper Cretaceous fans are identified by the erosive surfaces. The fan root has obvious characteristics of cutting valley, forming inter-bedded with the deep-sea facies toward the outer fan, along with litho-logic granularity thinning and scope expansion. The distal parts of the submarine fan are surrounded by abyssal shale, which can form a good source, reservoir and cap configuration. Therefore, turbidite fan sandstone with good source rock and cap configuration is the favorable reservoir.

## 1. Introduction

In modern continental slopes and similar ancient environments, analysis of seismic bathymetric data, drilling core data and outcrop shows large-scale collapse of massive sedimentary bodies and consequent gravity flow movements. The submarine fan is formed by the deposition of terrigenous clastic material mainly carried by the underwater gravity flow on the edge of the continental shelf, which is mainly com-posed of debris flow, turbidity current and pelagic sedimentation. A lot of research work has been done on the sedimentary model and genetic mechanism of submarine fan, also known as sea floor fan, gravity flow fan, turbidite fan, underwater alluvial fan and nearshore subaqueous fan [1-5]. In recent years, a large number of large oil and gas fields of turbidite sandstone have been discovered along the South Atlantic coast of South America. Turbidite sandstone formed by submarine fan has become one of the most favorable oil and gas reservoirs. Almost all submarine fans affect the sea-bed topography to some extent, which provides the possibility for identification on seismic profiles. Identification and search for turbidite sandstone of submarine fans play an important role in oil and gas exploration.

In this paper, Tectonic evolution and sedimentary setting of Suriname-Guyana is described, the distribution range and lithologic characteristics of the submarine fan in offshore Suriname are interpreted. Finally, hydrocarbon accumulation configuration and the controlling effect of submarine fan turbidite sandstone on hydrocarbon accumulation is studied.

## 2. Regional Geologic Setting

The offshore Suriname is a part of the Suriname-Guyana basin, covering an area of about 230,000 square kilometers (Fig .1). It can be divided into three tectonic units: the shelf, the deep sea subbasin and the Demerara Plateau. In recent years, a number of turbidite sandstone oil and gas fields have been discovered in the deep sea of the basin, and C & T heavy oil field has been discovered onshore. The exploration Wells in the shallow sea area have good oil and gas shows, which has a certain exploration potential.

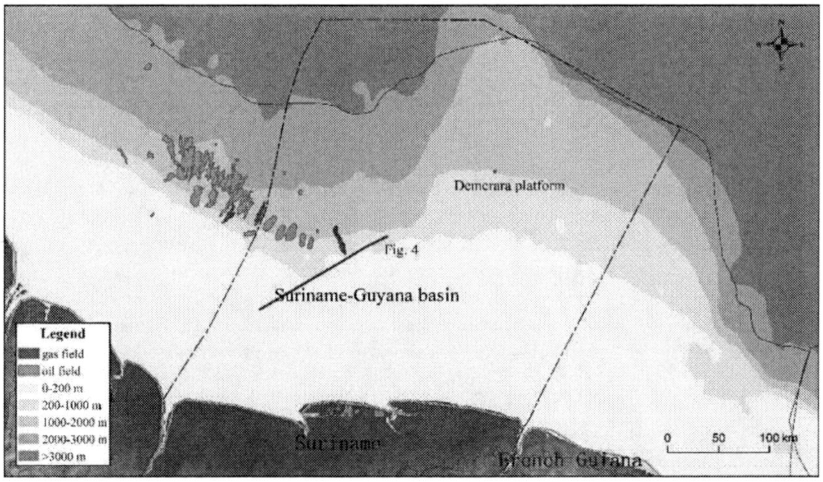


**Fig. 1.** Oil and gas fields and the location of study area in Suriname-Guyana Basin

The Demerara plateau is caused by the African Plate counterclockwise rotation and rifting away from South America, dividing the basin into the Central Atlantic passive margin to the west and the Equatorial Atlantic Transform margin to the east[6-14]. the African Plate counterclockwise rotation to South America is recorded by smaller compressional structures, resulting in up to 50 km of NE-SW shortening and 6 km of erosion.

The Suriname-Guyana basin was formed as part of the tectonic evolution of the Central, Equatorial and South Atlantic Oceans which can be subdivided into four major stages according to the research of this paper: Break up of Pangea phase, intracontinental rift phase, regional uplift phase, and drifting over the South Atlantic phase.

### *2.1. Triassic break up of Pangea phase (~200 Ma)*

The initiation of the Guiana Basin was caused by the breakup of Pangea around 200 million years ago and the subsequent drift between Gondwanaland & Laurasia which is recognized in the acreage on offer as the Nickerie Graben, part of a failed arm of the Central Atlantic [15-20]. This is reflected in the fact that the Nickerie Graben is part of the failed branch of the mid-Atlantic.

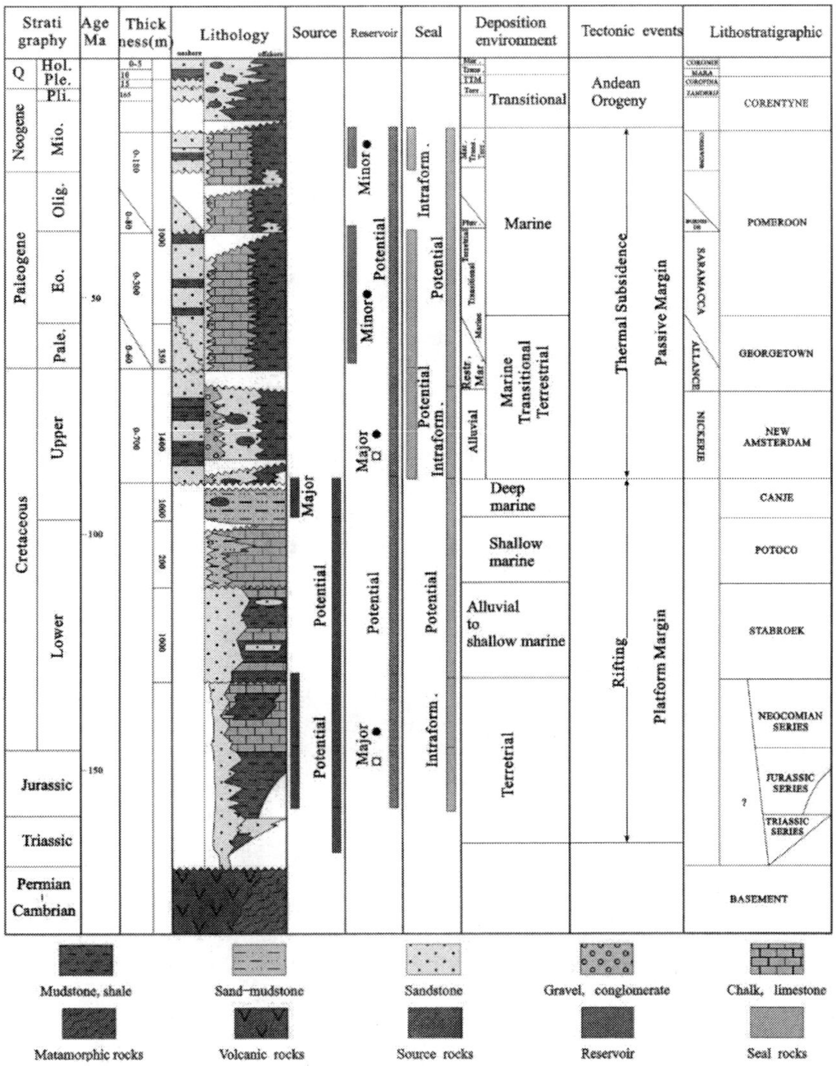


**Fig. 2.** A comprehensive stratigraphic column map of Suriname-Guyana basin

### *2.2. Mid-Late Jurassic intracontinental rift phase (200Ma-150Ma)*

This phase created a syn-rift sequence of Lower Jurassic sediments, while the Upper Jurassic drift phase created a passive margin with a moderate to high fluvial/deltaic discharge rapidly deposited in an inner neritic environment (Fig. 2). The principal structural elements of the Suriname-Guyana Basin that were created during this period include the Jurassic syn-rift grabens underlying the shelf, the shelf itself, a deep marine basin and the Demerara Plateau [21-23]. Remnants of earliest rifting are present in the form of the Nickerie and the Commewijne Grabens, and their onshore equivalent Takutu Graben on the border between Guyana and Brazil which has a proven Jurassic Petroleum System. The end of this period is marked by the Break – up Un-conformity.

### *2.3. Early-Late Cretaceous regional uplift phase (150 Ma-95 Ma)*

The South Atlantic rift and drift phase cause dramatic compression of the Jurassic section within the Suriname-Guyana basin due to rotational tectonic movement between South America and Africa which created major uplift and subsequent erosion leading to the basin-wide Break-up Unconformity [9-11]. The compressional features are recognized at the north-eastern edge of the Suriname-Guyana Basin (Fig .2). The reactivation is characterized by rather steep dipping normal faults in the Shallow offshore.

### *2.4. Late Cretaceous to recent Atlantic drifting phase (95 Ma-0 Ma)*

The fourth major phase of the development of Guiana Basin occurred following the complete rifting of Gondwanaland and the opening of the Equatorial Atlantic rift which connected the South Atlantic and Central Atlantic oceans[24-25]. Associated large igneous province volcanism and an ocean predisposed to anoxia and euxinia led to global Oceanic Anoxic Event which created the world class Cenomanian-Turonian Canje formation source rock which has been identified as one of the main sources for the oil found in the basin so far (Fig. 2). The drift phase of this tectonic chapter also marked the transition from a tectonically active basin to a passive margin.

The Equatorial Atlantic drift phase is characterized in the Guiana Basin as a passive margin sedimentary setting albeit with evidence of strike-slip tectonics observable in the sedimentary column. The transform fault systems associated with this drift can be traced on the present-day sea floor. The transform fault lineaments connecting the Demerara Plateau to Equatorial Guinea Bissau Plateau in West Africa.

## 3. The Identification and Characterization of Submarine Fan

### *3.1. Sedimentary Facies and Sequence*

The sedimentary section in the onshore to shallow offshore is underlain by the Gui-ana shield. The Guiana shield is Precambrian aged crystalline basement consisting of volcanic, granitoid and metamorphic rocks.

Permo-Triassic dykes indicate the onset of rifting within the Suriname-Guyana ba-sin. The initial provenance of the sediment of the basin is the Guiana shield which has undergone intense weathering and erosion. River systems which developed through major faults/rifts acted as the transportation and depositional systems which ultimately created the sedimentary section as described in this report. The Precam-brian basement is evident on 2D seismic data from the onshore and southern part of the shallow offshore.

The Jurassic Rift Grabens in the shallow offshore area are little explored. The rift was first penetrated in 1983 by the I/23-1X well where ~400 meters of alternating silt and volcanic material were penetrated in the Nickerie Graben. Recent evaluation indicates that the thickness of the Rift is more than 10km thick below the Cretaceous and Tertiary section. The stratigraphy of the Nikerie Rift basin is in part inferred from the Takutu Graben where lithologies include Triassic/Jurassic volcanics to Jurassic shale, salt, and sands. 40-degree API oil was tested from Jurassic fault sys-tems.

Jurassic sequence is unconformably overlain by the Berriasian-Albian sequence, where it progressively onlaps the older sequence in an up-dip southerly direction and consequently the earliest Cretaceous section is absent in the most southerly near-shore acreage. The sedimentary record of this sequence is related to tectonic evolu-tion when compression caused major inversion of the pre-Aptian/Albian sediments and created a regional peneplanation with the break in stratigraphy evident through-out the basin. The gross depositional environments transits from continental, delta plain, neritic to shelf margin in the shallow offshore. The extend in the deep water is characterized by bathyal to abyssal deposits.

Journal of Physics: Conference Series **2594** (2023) 012012 doi:10.1088/1742-6596/2594/1/012012

The Albian-Cenomanian-Turonian sequence marks the onset of the passive margin with the basin and is subdivided into two distinct stratigraphic play intervals: Albian bounded by Mid Cretaceous flooding surface and top Albian unconformity and Cenomanian-Turonian bounded by top Albian unconformity and top Turonian un-conformity. The Albian interval is deposited during the equatorial Atlantic rifting and drifting phase when the African plate was still close to South America with sed-iment input still dominantly from the east. Seismic observations also indicate well developed delta in the east of the shallow offshore area. The Cenomanian-Turonian comprises the Cenomanian and Turonian interval (source rock equivalent interval) and is also characterized by a relative sea level rise. The sequence was deposited towards the end of the Equatorial Atlantic Rifting and Drifting phase. The deposited environment reflects a retreat of the Continental and the Delta plain zone relatively to the underlying Albian interval. The south exhibits a transition from delta plain to neritic environments, while a neritic environment dominates towards the north.

The Upper Cretaceous Coniacian-Mid Campanian marks the transition of a trans-gressive to a regressive system with increasing sediment input from the south west, creating a broad delta plain environment across the shallow offshore area. A shelf margin zone could not be identified for this sequence as no clear clinoforms/foresets were evident on the seismic data, possibly because they have been removed by the Cretaceous Mega erosional event which created a cut and fill system in this area (Fig. 3). The shallow offshore area is dominated by a delta plain environment and gradually transitioning to a neritic environment. The regression continued during Mid Campanian-Maastrichtian which is characterized by tremendous sediment input in the west creating a broad delta plain across the shallow offshore area. Overall, this sequence is characterized by a transition from delta plain to neritic environments similar to the underlying Mid Campanian-Maastrichtian.

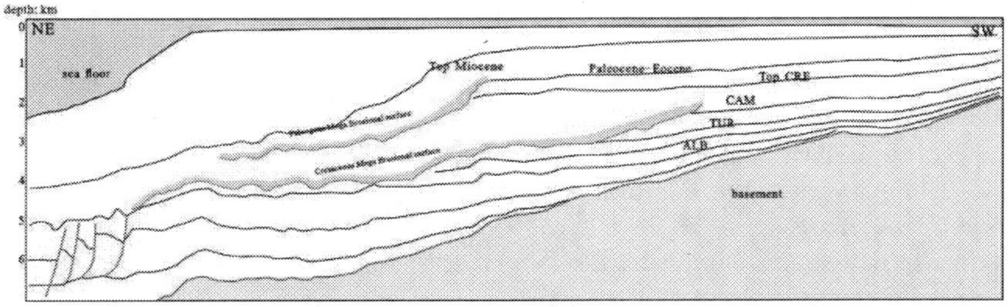


**Fig. 3.** Seismic interpretation profile and submarine fan erosion characteristics in Suriname (see seismic location in Figure 1)

From the Upper Cretaceous to present, the coast of Suriname has remained a passive margin. The basin was, however, affected during the Tertiary by the eastward movement of the Caribbean plate, which resulted in a series of strike-slip faults parallel to the Surinamese coast.

During the Miocene, the Pacific plate started to under-thrust the South American continent, causing it to tilt to the east. The Amazon river's lower tributaries conse-quently shifted to the south, beheading the big rivers that were draining in the Gui-ana basin and changed the basin from a sand dominated system to a clay-dominated system. It also ended the carbonate growth on the shelf due to the mud carried from the present Amazon river mouth along the coast of the Guyana.

### *3.2. Erosion Surface on seismic profile*

According to the drilling data and seismic interpretation section, the oldest strata in this region are determined to be the Middle Jurassic, but the existence of older strata is inferred by the association of the adjoint Atlantic margin. Five local unconformi-ties caused by erosion are identified from the seismic section: the top unconformity of the Jurassic, the top unconformity of the Aptian, the top unconformity of the Tu-ronian, the Maastrichtian unconformity and the Middle Miocene unconformity (Fig .4).

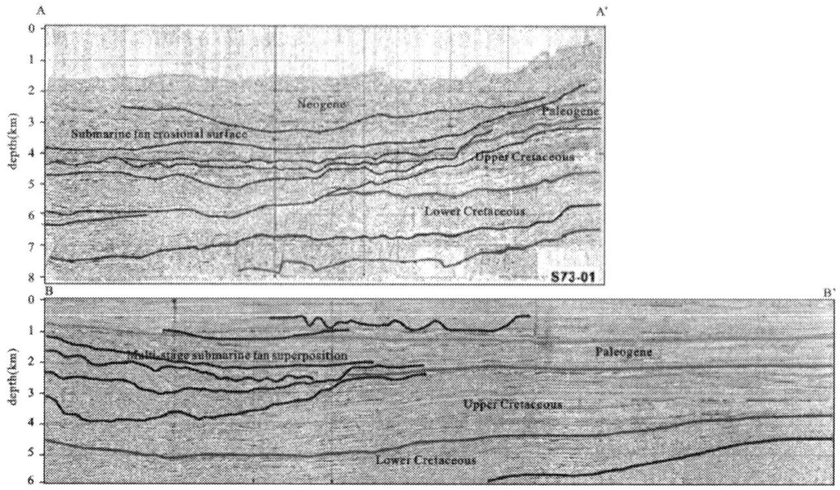


**Fig. 4.** Superposition characteristics and erosion surface identification of multi-stage submarine fans in offshore Suriname

These erosional unconformities are mainly caused by sediment gravity collapse. Several sedimentary cycles developed in the Suriname sea area, beginning with Ear-ly Cretaceous transgression followed by accretion from Aptian to Campanian. The rapid accumulation and compaction of sediments on the edge of continental shelf lead to strata collapse in some areas, and the collapsed sediments migrated and dif-fused along the continental slope to the deep sea in the form of mass transport under gravitational process. The resulting submarine fan have been widely recognized in this region, of which the erosion surface characteristics are tracked and character-ized by seismic data and are related to the Berbice Valley in Guyana.

The Cretaceous submarine fans have the characteristic of multi-stage superposition and wide distribution, and its bottom erosion surface has the characteristics of pene-trating several layers. The fan roots cut through the Campanian shallow Marine sandstone and gradually cut through the Turonian and Albian towards the fan end. Therefore, subaqueous fan erosion causes different strata loss degrees in different areas.

Compared with the Cretaceous submarine fans, the Paleogene fan has a smaller dis-tribution range. It can be seen from the seismic section that the erosion surface of the Paleogene underwater fan is shorter and the cutting stratum is less, which mainly erodes the Paleocene and Oligocene, and in some areas erodes the Upper Cretaceous. From the provenance direction, the Paleogene underwater fan was caused by several collapse at the edge of continental shelf, while the Cretaceous underwater fan was mainly formed from the same collapse point. Therefore, the fan root channel width of Cretaceous subaqueous fan is larger. Although these submarine fans are located near the continental slope, the Cretaceous submarine fan is located in the west-central part of the basin center, while the Paleogene fan is located near the Demerara Plateau to the east.

## 4. Characteristics of Hydrocarbon Accumulation

From seismic data, it can be observed that the Suriname underwater fan was caused by sediment gravity collapse. At the edge of the continental shelf, the gravitational collapse of sediment causes strata to slip and move, forming the mass transport complexes. In the local area, the strata shaped by the mass transport complexes move towards deep water to generate thrust nappe structures. The mass transport distance is more than 100km, the sediment accumulation thickness is up to hundreds of meters. The lithology of the Cretaceous submarine fan in Suriname does not change much from the fan root to the fan end, mainly middle-fine sandstone, but the grain size is finer towards the fan end. The thickness of fan body is characterized by thick layer near source and thin layer far source.

The submarine fan body is a reservoir with good physical conditions and the highest porosity can reach 25%. However, oil and gas accumulation requires good configuration of all elements. Only with good reservoir conditions and lack of source rock and cap conditions, large oil and gas accumulation cannot be formed. In the fan root and the middle part of the submarine fan, the sandstone has the best physical proper-ty, but its overlying and underlying strata are mainly shallow sea sandstone, which can not form a good reservoir cap configuration and provide hydrocarbon source. In deep-water area, the underlying and overlying strata of the fan end are dominated by deep-sea facies shale, which can form Source rock and cap rock, which is conducive to the formation of oil and gas accumulation. Therefore, the lack of mud shale in the fan root and fan of underwater fan is not conducive to oil and gas accumulation, and the fan end can form lithologic traps and close to the source rock due to the contact with deep-sea shale. The deep-sea facies fan is the most favorable exploration tar-gets.

## 5. Conclusion

The Surinamese-Guyana basin has undergone four major stages of tectonic evolution. During Early-Late Cretaceous regional uplift phase, the African Plate counterclockwise rotation to South America created compressional structures and local uplift on the Demerara plateau in the passive marginal environment, which has a great influence on the tectonic deposition and hydrocarbon accumulation in the basin.

The offshore Suriname has developed Upper Cretaceous transgression-regressive transitional sequence and Paleogene shallow Marine sequence. Under the action of sediment gravity loading, stratigraphic collapse occurred at multiple locations on the edge of continental shelf, which was initially gravity detachment structure, and then developed into thrust nappe structure towards deep water, which also led to the formation of submarine fan and turbidite sandstone.

The submarine fan can be tracked and identified by the erosion surface forming local unconformity on the seismic profile. The fan root has obvious incised valley characteristics and the fan end forms sandstone mud-stone interbedding characteristics, which shows obvious local unconformity erosion surface in the seismic section.

The submarine fan has the characteristics of multi period superposition. Five unconformity erosion surfaces have been identified on the seismic profile from the Lower Cretaceous to the Paleogene, which is mainly caused by sediment gravity collapse.

Submarine fan sandstone is an important reservoir, but good Source rock and cap rock configuration are required to form oil and gas accumulation. The turbidite sandstone at the end of the underwater fan interfingers with deep-sea shale, making it a favorable exploration target.

## References

[1] G Shanmugam and R J Moiola. Submarine fans: Characteristics, models, classification, and reservoir potential. Earth-Science Reviews 24, 383-428 (1988).

[2] TAN MingXuan, WU Feng, MA HaoRan, et al. Facies Model, Sedimentary Process and Depositional Record of Submarine Fans, and Their Implications. Acta Sedimentologica Sinica 40(2), 435-449 (2022).

[3] Richards M, Bowman M, Reading H. Submarine-fan systems i : Characterization and stratigraphic prediction. Marine and Petroleum Geology 15(7), 689-717 (1998).

[4] Hage S，Galy V V，Cartigny M J B，et al. Efficient preservation of young terrestrial organic carbon in sandy turbidity-current deposits. Geology 48(9), 882-887 (2020).

[5] Roger G. Walker. Deep-water sandstone facies and ancient submarine fans: Models for exploration for stratigraphic traps. AAPG Bulletin 62(6), 932-966 (1978).

[6] Max Casson, Jason Jeremiah, Gerome Calves, et al. Evaluating the segmented post-rift stratigraphic architecture of the Guyana continental margin. Petroleum Geoscience 27, 1-25 (2021).

[7] James Trude, Bill Kilsdonk, Tim Grow and Bryan Ott. The structure and tectonics of the

Guyana Basin. Geological Society, London, Special Publications 524, 1-20 (2022).
[8] Museur Thomas, Graindorge David, Klingelhoefer Frauke, et al. Deep structure of the Demerara Plateau: From a volcanic margin to a Transform Marginal Plateau. Tectonophysics 803, 1-22 (2021).
[9] Labails, C., Olivet, J.L., Aslanian, D. and Roest, W.R. An alternative early opening scenario for the Central Atlantic Ocean. Earth and Planetary Science Letters 297, 355–368 (2010).
[10] Loncke, L., Roest, W.R. et al. Transform marginal plateaus. Earth-Science Reviews, 203, 102940 (2020).
[11] Sibuet, J.C. and Mascle, J. Plate kinematic implications of Atlantic equatorial fracture zone trends. Journal of Geophysical Research: Solid Earth 83, 3401–3421 (1978).
[12] Greenroyd, C.J., Peirce, C., Rodger, M., Watts, A.B. and Hobbs, R.W. Demerara Plateau the structure and evolution of a transform passive margin. Geophysical Journal International 172, 549–564 (2008).
[13] Kneller, E.A., Johnson, C.A., Karner, G.D., et al. Inverse methods for modeling non-rigid plate kinematics: Application to Mesozoic plate reconstructions of the Central Atlantic. Computers & Geosciences 49, 217–230 (2012).
[14] Loncke, L., Maillard, A. e. Structure of the Demerara passive-transform margin and associated sedimentary processes. Initial results from the IGUANES cruise. Geological Society, London, Special Publications, 431, 179–197 (2016).
[15] Wenxiu Yang, Alejandro Escalona. Tectonostratigraphic evolution of the Guyana Basin. AAPG Bulletin 95 (8), 1339-1368 (2011).
[16] Benkhelil J, J. Mascle, P. Tricart. The Guinea continental margin: An example of a structurally complex transform margin. Tectonophysics 248, 117-137 (1995).
[17] Pindell J. Geologic rationale for hydrocarbon exploration in the Caribbean and adjacent regions[J]. Journal of Petroleum Geology 14, 237-257 (1991).
[18] Mann P, L Gahagan, M. Gordon. Tectonic setting of the world's giant oil and gas fields. AAPG Memoir 78, 15-105 (2003).
[19] Brouwer M J N, M M Schwander. Cenozoic carbonate banks, Foz do Amazons Basin, northeastern Brazil[J]. AAPG Studies in Geology 27(2), 174-178(1988).
[20] Campbell A. Shelf-geometry response to changes in relative sea level on a mixed carbonate-siliciclastic shelf in the Guyana Basin. Sedimentary Geology 175, 259-275 (2005).
[21] Mann P, F W Taylor, R Lawrence, et al. Actively evolving microplate formation by oblique collision and sideways motion along strike-slip faults. Tectonophysics 246, 1-69 (1995).
[22] Meyers A P, B M Stefano, F Astrid. Origins and accumulation of organic matter in expanded Albian to Santonian black shale sequences on the Deremara Rise, South American margin. Organic Geochemistry 37, 1816-1830 (2006).
[23] Gouyet S, P Unternehr, A Mascle. The French Guyana margin and the Demerara Plateau: Geological history and petroleum plays. European Association of Petroleum Geoscientists Special Publication 4, 411-422 (1994).
[24] Goss S, D Mosher, G D Wach. Continental margin development of the equatorial Atlantic gateway: Suriname, South America. Central Atlantic Conjugate Margins Conference, Halifax, 282–291 (2008).
[25] Garciacaro J E, P Mann, A Escalona. Regional structure and tectonic history of the obliquely colliding Columbus foreland basin, offshore Trinidad and Venezuela. Marine and Petroleum Geology 28, 126–148 (2011).

# Wygd Study on prediction method of drilling pressure in low permeability oilfield

**Ce Liu**

Daqing Heilongjiang
PetroChina Daqing Oilfield Co., Ltd. No.8 Oil Production Plant

Liuce11@petrochina.com.cn

**Abstract.** During the high water cut development stage of low permeability in M block, the overall recovery degree is low, there has 100 infilling wells been drilled since 2020 to improve the development effect. Before drilling, in order to avoid the risk of pressure safety, it is necessary to shut down the peripheral water injection wells to relieve the pressure, at the same time.Since the M Block reservoir is a low permeability reservoir, the pressure relief time after drilling off is long, which increases the impact on the production of developing wells and increases the fluid transportation and the related production costs, while there is no uniform standard for the time, pressure and distance of drilling off . In order to control the production loss of developing wells and save the shut-in cost .This paper applies numerical simulation and reservoir engineering method, by establishing fine geological model, optimizing simulation technology, considering the influence of start-up pressure gradient, establishing theoretical formula and mathematical model of drilling pressure, clarifying the law of drilling-off pressure relief and pressure drop, calculating the formation pressure of single well in the whole area, forming the drilling-off limit and rapid pressure prediction method, through comparison and verification, the overall deviation is 3.8%, and the method is accurate and reliable.

## 1 Introduction

Block M is a low-permeability structure-lithological reservoir, with 9 oil formation groups and 13 small layers, with an average single-layer sandstone thickness of 1.3m, an effective thickness of 0.4m, an air permeability of 44mD, and a sedimentary microfacies of delta leading facies sedimentation, narrow band-like underwater diverted channels and unstable distribution mat-like sand.The total number of wells in the block is 169 injections and 353 extraction, with a comprehensive water cut of 76.6%, oil recovery speed of 0.74%, a production degree of 11.4%, a utilization degree of only 71.3%, and a water flooding control degree of 72.3%. In order to further improve the degree of water flooding control and oil recovery speed, and improve the development effect of high water cut blocks, 60 infill wells were deployed and 50 water wells were drilled and closed. According to the requirements of reducing the wellhead pressure to 3NPa in the past, the average drilling time takes more than one year, considering the problem of formation energy deficit caused by long-term drilling and large production loss of old wells, it is necessary to optimize drilling and optimization of reservoirs with different permeability rates.

In this paper, considering the influence of multi-well interference, the pressure calculation formula of multi-source and multi-sink is established, and the pressure distribution to be drilled in different well patterns is obtained; On the basis of traditional numerical simulation, the influence of "unconnected, shale" and other ineffective layers on water absorption is deducted[1]. At the same time, the starting pressure gradient considering the oil-water phase is introduced, and the influence factors of drilling and shut-in pressure relief are clarified. The formation pressure of the whole area is obtained by using the equivalent method; On this basis, the drill-in distance, drill-in time and drill-in pressure limits under different permeability levels and well spacing conditions are given, and a multi-parameter fast pressure prediction program is compiled[12], forming a drill-in limit chart with well spacing, reservoir permeability and cumulative injection-production ratio as variables, quantifying the drill-in limit of the whole area, and achieving cost reduction and efficiency increase of the whole area.

## 2 Numerical simulation of drilling pressure relief

According to the distribution of block faults, the whole area was divided from 90 large and small faults into 21 independent fault blocks, and the 3D geological modeling of GPTModel software was adopted, the grid size was 40m×40m, and the number of grids was 262×174×13=592,600.The model reserve is 1718×104t, which is 1.8% error compared with the calibration reserve of 1750×104t, the porosity of the reservoir in Block 703 is mainly distributed between 0.18-0.24, the reservoir permeability is mainly distributed between 20mD-60mD, the net hair ratio of pure sandstone is mainly distributed in about 0.1, and the effective sandstone is mainly distributed between 0.5-0.7.

Use Eclipse software to do reservoir numberical simulation, with 13 simulated layers, 262×174 planar meshes, 40m × 40m mesh size, 329 time steps, and the overall water content fitting accuracy was 83.6%.

### *2.1 Invalid water injection treatment*

In the actual low-permeability block, due to the phenomenon of "unconnected, mudstone and other invalid layers absorbing water"[5], the injection-production ratio at the wellhead is relatively high. In the actual model, the water injection flow pressure is limited to deduct part of the water volume[25]. The formation pressure distribution of different oil layers in the well pattern is matched to obtain the formation pressure at the place to be drilled after deducting the invalid water injection and water cut matching[9].

### *2.2 Introduction of starting pressure*

Since there is a certain relationship between permeability and starting pressure gradient, the smaller the permeability, the greater the starting pressure gradient[18]. A relationship needs to be established to calculate different pressure gradients at different permeability, so as to make the productivity calculation results more accurate. Relation between single-phase oil-water starting pressure gradient and air permeability[11]:

$$G_o = 0.0454\left(\frac{k_g}{\mu_o}\right)^{-1.2738} \qquad G_w = 0.0238\left(\frac{k_g}{\mu_w}\right)^{-0.14486} \tag{1}$$

The proposed starting pressure gradient of two-phase flow is as follows:

$$G(k,S,\mu_o,\mu_w)=0.0238\left(\frac{k_g}{\mu_w}\right)^{-0.14486}\left(\frac{S_w-S_{wc}}{1-S_{wc}}\right)+0.0454\left(\frac{k_g}{\mu_w}\right)^{-0.12738}\left(1-\frac{S_w-S_{wc}}{1-S_{wc}}\right) \tag{2}$$

First, based on the historical fitting of the model, the fluid production of the oil well under the measured pressure is simulated according to the actual pressure measurement data of the oil wells around the well to be drilled. Secondly, according to the difference between the simulated fluid production and the actual fluid production of the oil wells around the well to be drilled[2], the injection pressure of the water injection well under the simulated production is calculated through comparative analysis. Finally, the calculated injection pressure is used to simulate the pressure at the point to be drilled, that is, to predict the pressure at the point to be drilled[6].
On the whole, the formation pressure of the block after drilling is mainly distributed between 3-32MPa, and the pressure in the east of the block is higher than that in the west.(see Fig. 1).

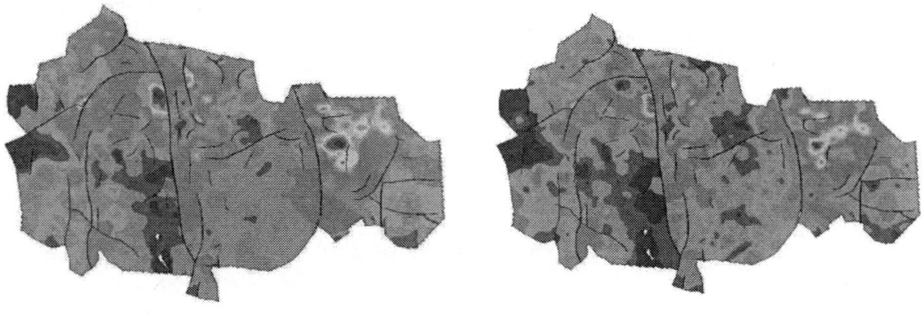

Average pressure Maximum pressure

**Fig. 1.**Pressure distribution map of the whole area

## 3 The pressure calculated using the reservoir engineering method

In order to facilitate the calculation, 69 wells are divided into 17 balance zones according to faults, and 69 wells to be drilled are divided into four situations according to the number of wells: 0 injection, 1 injection, 2 injection and 3 injection. The reservoir engineering method predicts pressure, which can be divided into three situations according to the injection and production relationship: one injection and one production, multiple well interference, and no water injection well.

### *3.1 Prediction of new well pressure in one injection and one production well pattern*

The plane radial flow pressure distribution formula is applied to calculate the pressure at the well site to be drilled in the well network, which is suitable for the situation when the well is opened and closed[15].

$$P = P_e - \frac{P_e - P_w}{\ln \frac{R_e}{R_w}} \ln \frac{R_e}{r} \qquad (3)$$

In the formula: P—pressure at any point in the formation, MPa/m; Pe—supply boundary pressure, MPa; Pw—bottom flow pressure of oil production wells, MPa; Re—supply radius, m; Rw—wellbore radius, m; r—distance between infill wells and base well networks, m.

### *3.2 Pressure prediction under multi-well interference well pattern*

For multi-well production systems, during normal injection and production, the pressure at each point in the formation is only related to the position and does not change with time[7].Multiple wells work at the same time, and the pressure value of a certain point in the formation can be calculated according to the principle of superposition of potentials; For the multi-well production system, after the water well is drilled, the multi-source multi-sink system is split into multiple one-source sink systems, and the formation pressure of multiple systems at the site to be drilled is obtained, and the maximum value is taken[10].

$$p = p_e - \frac{\mu}{2\pi} \sum_{i=1}^{n} \frac{\pm q_i B_o}{(kh)_i} \ln \frac{r_{ei}}{r_i} \qquad (4)$$

In the formula: p - the pressure of the sub-layer to be calculated at the point to be drilled; Ki - the formation permeability of the sub-layer from the ith well to the point to be drilled; Hi - effective thickness of the formation from the ith well to the point to be drilled; Qi - average daily liquid production or daily water injection of the ith well; Bo - volume coefficient of crude oil; Pe - reservoir hydrostatic column pressure.

### *3.3 Pressure prediction without water injection well*

When there are no wells in the well pattern and only other wells, the formation pressure at the place to be drilled is only affected by the wells. Take the well as the center to calculate its influence on the surrounding pressure drop[2].

$$P_e = \frac{2 \ln \frac{R_e}{R_w} \overline{P} - P_w}{2 \ln \frac{R_e}{R_w} - 1} \qquad (5)$$

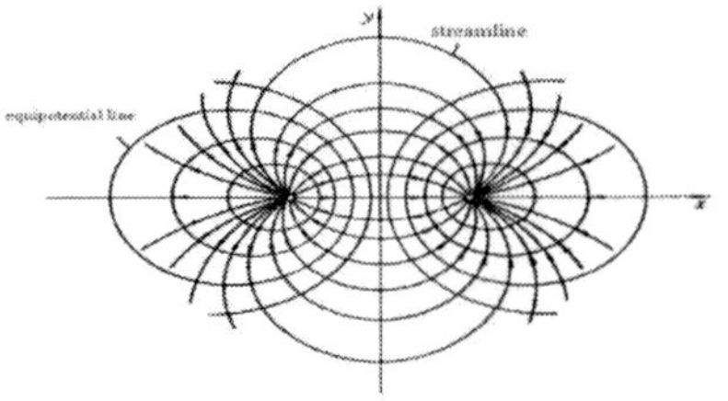


**Fig. 2.**One source one sink schematic

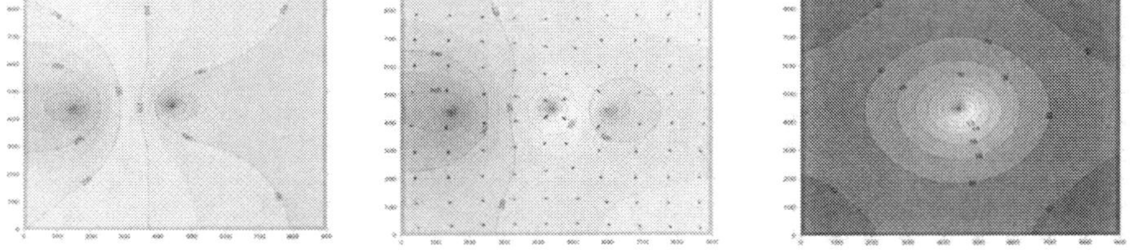

(a)One shot, one shot (b)Multi-well interference (c)No water injection well

**Fig. 3.**Pressure prediction distribution map of different injection-mining relationships

Finally, the average formation pressure of all 69 infill wells is predicted to be 19.30MPa, and the formation pressure coefficient is 1.27. There were 32 wells with a pressure of less than 3MPa, accounting for 46.4%; The pressure coefficient is less than 1.2, and 39 wells to be drilled are calculated, accounting for 56.5%.

## 4 Fast prediction method of drill-in limit and pressure

### *4.1 Drilling and closing distance limit under different injection-production*

According to the pressure distribution formula of the surface seepage of the injection well, a pressure drop distribution funnel is formed between the injection and production distances[22], and the pressure at the place to be drilled is mainly related to the injection and production distance and injection pressure[23],.Consider the drilling distance boundary at different injection and production distances[4]. Considering the upper limit of drilling fluid density, the balance between drilling fluid and formation pressure coefficient[8], a reasonable formation pressure coefficient of 1.6 is determined, and the main principle of finding the distance limit is that the formation pressure coefficient at the site to be drilled is lower than 1.6.

The reservoir engineering method and numerical simulation method were comprehensively applied to solve the pressure combined pressure coefficient[26], the pressure coefficient and injection and production distance plates were drawn[27], and the drilling distance limit was determined according to the plates[24].It can be seen from the plate that when the injection and production well distance is 200m, the minimum drilling distance limit is within the injection and production line[13], the minimum distance limit to reach the safety pressure coefficient is 100m, the maximum drilling distance is outside the injection and production line, and the maximum distance limit of the safety pressure coefficient is 300m, so the distance limit of the drilling distance of the 200m well is determined to be 100-300m, and the corresponding drilling distance limit of 300 and 400m injection wells is 150-420 and 200-550m by the same reason.

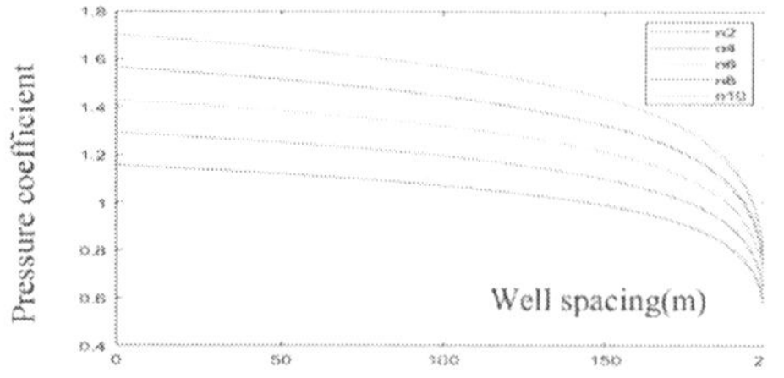


**Fig. 4.**Prediction of downforce coefficient of 200m injection production well distance

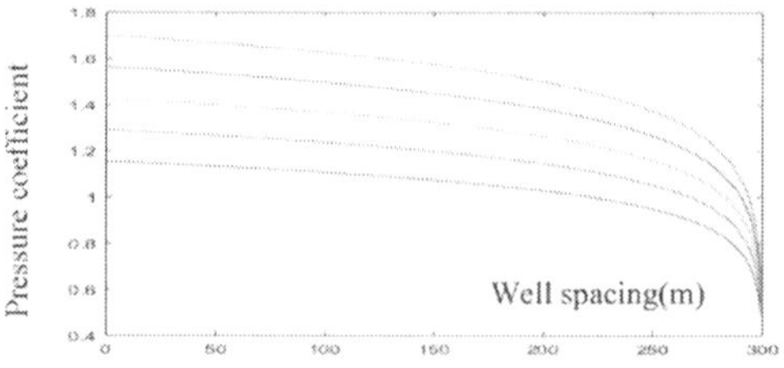


**Fig. 5.**Prediction of downforce coefficient of 300m injection production well distance

*4.2 Drilling and closing pressure limit under different injection-production well spacing and penetration rate*

The research law shows that under the same injection-production well spacing, the rate of wellhead pressure decline after the injection well shuts in is related to the permeability[14]. When the monthly decreasing amplitude of different permeability models is less than 0.5MPa, the pressure reduction rate is slow. Since the pressure drop rate decreases in the late stage of drilling and shut-in, and the pressure drop rate slows down when the drilling and shut-in time is extended[20], the wellhead pressure at the monthly decreasing amplitude of 0.5MPa is selected, and the formation pressure coefficient can be guaranteed below 1.6, so it is used as the limit of drilling and shut-in pressure.

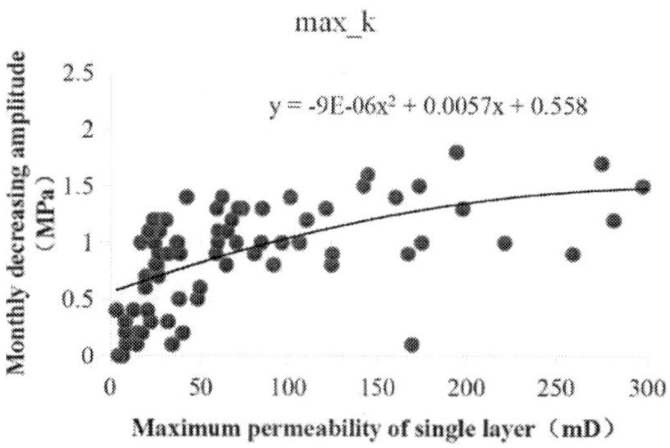


**Fig.6.**Prediction of downforce coefficient of 300m injection production well distance

**Table 1.** Lower limit of drilling and closing pressure in different formations

| Well spacing（m）/Permeability（mD） | 30 | 60 | 90 |
|---|---|---|---|
| 200 | 7 | 5 | 3 |
| 250 | 8 | 6 | 4 |
| 300 | 9 | 7 | 5 |
| 350 | 10 | 8 | 6 |
| 400 | 11 | 9 | 7 |
| 450 | 12 | 10 | 8 |

### *4.3 Drilling time limits under different injection and production distances, permeability and pre-off pressure*

Under the same injection-production well spacing and permeability, the wellhead pressure drop rate is similar after the injection well is shut in. Due to the difference of water injection pressure of single well before drilling and shut-in, the well with low pressure will reach the pressure limit faster.

**Table 2.** Lower limit of drilling and closing pressure in different formations

| Well spacing (m) | 200 | | | 250 | | | 300 | | | 350 | | | 400 | | | 450 | | |
|---|---|---|---|---|---|---|---|---|---|---|---|---|---|---|---|---|---|---|
| Pressure before shut-in (MPa) | L | M | H | L | M | H | L | M | H | L | M | H | L | M | H | L | M | H |
| Permeability (mD) | 6 | 7 | 8 | 7 | 8 | 9 | 8 | 9 | 10 | 9 | 10 | 11 | 10 | 11 | 12 | 11 | 12 | 13 |
| | 4 | 5 | 6 | 5 | 6 | 7 | 6 | 7 | 8 | 7 | 8 | 9 | 8 | 9 | 10 | 9 | 10 | 11 |
| | 2 | 3 | 4 | 3 | 4 | 5 | 4 | 5 | 6 | 5 | 6 | 7 | 6 | 7 | 8 | 7 | 8 | 9 |

### *4.4 Fast Prediction Method and Chart of Well Pressure to Be Drilled*

According to the percolation formula, the block average well spacing, well diameter, well depth, well flow pressure and other single well parameters are used to form a fast calculation chart and calculation method for formation pressure and pressure coefficient[16]. Based on the research results, a drill-in boundary chart with well spacing[17], reservoir permeability and cumulative injection-production ratio as variables is formed to calculate the distance, pressure, time and other boundaries under different parameters[19].

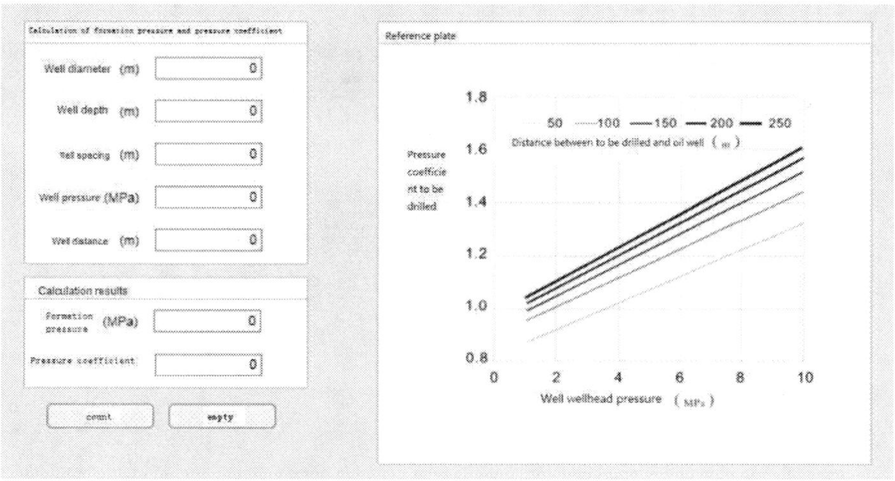


**Fig.7.**Rapid prediction method of well pressure to be drilled

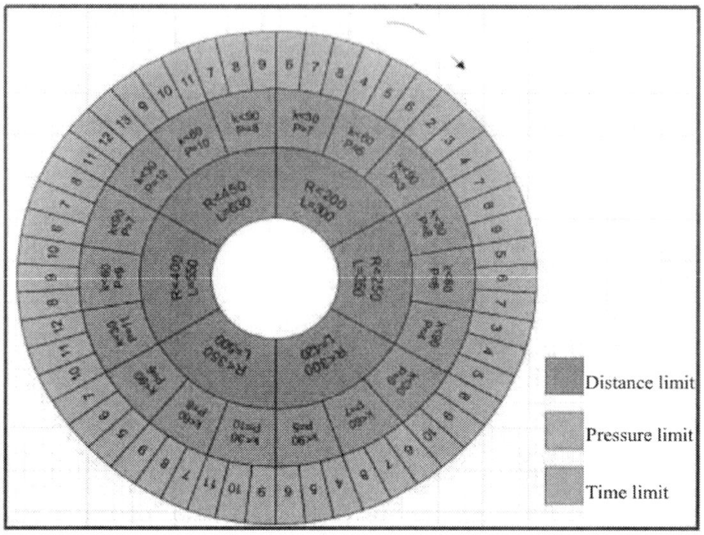


**Fig.8.**Rapid prediction chart of drilling shut-in pressure and limit

Note:

The dark blue area is the limit area of drilling and closing distance.

R-Injection-production well spacing,L-Distance limit;

The light blue area is the limit area of drilling and closing pressure.

K-Reservoir permeability,P-Pressure limit;

The light green area is the limit area of drilling and closing time.The number is the time limit from low to high pressure before closing.

## 5. Practical application

Comparing the rapid prediction method and the actual pressure measurement data of 63 wells, the overall error of formation pressure prediction is 3.8%, and the error is less than 6% in 50 wells, accounting for 80.6%, which shows that the pressure prediction method is accurate and reliable.

**Table 3.** Lower limit of drilling and closing pressure in different formations

| Error proportional classification | Number of wells | Proportion （%） |
|---|---|---|
| Below 2 | 17 | 27.4 |
| 2-4 | 23 | 37.1 |
| 4-6 | 10 | 16.1 |
| 6-8 | 5 | 8.1 |
| 8-10 | 4 | 6.5 |
| 10-12 | 3 | 4.8 |
| Total | 62 | 100 |

## 6. Conclusion

1. Block M is a typical medium-low permeability reservoir. Due to the strong plane and interlayer heterogeneity, the injection and production conditions of each well group are different. The average monthly pressure drop is about 0-4MPa when drilling and shutting down the well, and the decline is slow after 1MPa.

2. Both the theoretical model and actual data show that at the initial stage of drilling and shut-in, the wellhead pressure drops rapidly by 3-10MPa; in the middle and late stages of drilling and shut-in, the wellhead pressure drops slowly, and the pressure drop is not obvious after the monthly pressure drops below 1MPa.

3. The larger the starting pressure gradient, the more obvious the rising trend in the process of water injection; The larger the starting pressure gradient, the higher the pressure before drilling and the higher the initial pressure after drilling and closing[21]; The larger the starting pressure gradient, the more obvious the pressure drop in the later stage of drilling and shut-in, but the pressure relief is difficult.

4. Seen from the actual pressure drop data, the pressure drop in the hole has a strong correlation with the well spacing and permeability. According to reservoir engineering method and numerical simulation method, the prediction error of formation pressure is about 5%. According to the seepage formula, a fast calculation method is formed, and the error with the actual pressure measurement data is about 4%.

## References

[1] Ying-song Huang:Classification and evaluation of heterogeneous reservoirs in Zhengzhuang Oilfield of Shengli Oilfield. Petroleum exploration and development 34（6）,729-733(2008).

[2] Ya-ting Zhu:Study on subdivision water injection policy of low permeability conglomerate reservoir based on single sand body [J]. Journal of Chongqing University of Science and Technology (Natural Science Edition) 19(4),13.17(2017).

[3] Ying-fang Zhou,Yan-jun Fang,Xiao-dong Wang,et al:Study on displacement efficiency of non-piston water flooding in multi-layer reservoirs [J]. Oil and gas geology and oil recovery 16（1）,86-93(2009).

[4] De-qing Li:Study on tapping potential technology of horizontal wells in high water cut and positive rhythm thick oil reservoirs in Shengli Oilfield. Petroleum Geology and Engineering 23（3）,59-61(2009).

[5] Fu-sheng Huang,Yong-sheng Zhao:Simulation study on optimal classification of subdivided water injection in sandstone oilfield at high water cut stage. Petroleum geology and development in Daqing 17（2）,30-31(1998).

[6] Gang Hu,Xuan-hua Tian,Quan-wen Liu, et al:Study on the technical boundary of subdivision water injection in Shengli fault block reservoir. Practice and understanding of mathematics 47（3）,102-113(2017).

[7] Jian-guo Zhang,Guang-lun Lei:Seepage mechanics of oil and gas reservoirs [M]. Petroleum University Press,Dongying(1998).

[8] Shi-yi Yuan,Xin-min Song:Development technology of fractured reservoirs [M]. Petroleum Industry Press, Beijing(2004).

[9] Qiang Dai,Yong-gang Duan,Wei Chen, et al:Research status of percolation in low

permeability gas reservoirs [J]. Special oil and gas reservoirs 14(1),11-14(2007).

[10] Bing-yu Ji,Jian-fei Zhan:Calculation formula of effective time and water breakthrough time of oil wells [J]. Petroleum Geology and Development in Daqing 19(5),24- 26(2000).

[11] Ai-min Lv,Huai-ning Xing,Jian-nan Zhao, et al:Policy boundary of layered water injection technology in ultra-high water cut stage [J]. Science, technology and engineering 19(21):,155-161(2019).

[12] Wen-feng Peng,Qiao-liang Zhang,Xiao Lei, et al:Discussion on reservoir-forming model and water-cut law in early development of low-resistivity reservoirs [J]. Science, technology and engineering 16(23), 139-143(2016).

[13] Xiao-yan Yang,Kai Huang,Chao Ma,et al:Analysis of relative permeability curve under different reservoir conditions [J]. Science and technology and engineering 12(14),3340-3343(2012).

[14] Zheng-ling Geng:Study on Optimization of Vertical Balanced Displacement in Multi-layer Reservoirs at Extra-high Water Cut Stage [D]. China Shiyou University (East China),Qingdao(2013).

[15] Wen Yu:Subdivision standard of water injection in the west of Xing 6 area considering the difference of plane connectivity [D]. Petroleum Geology and Development in Daqing 38(4),96-104(2019).

[16] Min-feng Chen:Study on permeability ratio Boundary of Strata Reorganization in Severe Heterogeneous Reservoir Development [J]. Offshore Oil and Gas in China 19( 5),40-12(2017).

[17] Kao-ping Song,Yu-shu Wu,Bing-yu Ji:φ function method for predicting remaining oil saturation distribution in water drive reservoirs [J]. Acta petrolei sinica 27(3),91—95(2006).

[18] Hong-tao Liu,Hong-tao Liu,Tiany Zhao:Theoretical and Experimental Study on Nonlinear Flow Starting Pressure Gradient in Low Permeability Reservoir.IOP Conference Series: Earth and Environmental Science.vol 571, issue 1. 2020. pp 012044-.

[19] Shamsiev M. N.; Gadil'shina V. R.:Pressure-Transient Analysis of Vertical Wells in Low-Permeability Reservoirs.Lobachevskii Journal of Mathematics.vol 43, issue 5. 2022. pp 1195-1200.

[20] Yue-jun Zhao,Kao-ping Song,Guang-juan Fan:The Numerical Simulation Research of The Remaining Oil Distribution Regulation in Test Site of Low Permeability Oil Field.Acta Geologica Sinica(English Edition).vol 89, issue s1. 2015. pp 335-338.

[21] Xing-jia Liu:Discussion on the probiems and treatment methods in the development of aonan low permeability oilfield.IOP Conference Series: Earth and Environmental Science,vol 558, issue 2. 2020. pp 022054-.

[22] Nan Zhang:Development Characteristics and Development Technology Analysis of Low Permeability Oilfield.Journal of Physics: Conference Series.vol 1649, issue 1. 2020. pp 012021-.

[23] Wan-Li Kang,Bo-Bo Zhou,Issakhov Miras,Gabdullin Marabek:Advances in enhanced oil recovery technologies for low permeability reservoirs.Petroleum Science,vol 19, issue 4. 2022. pp 1622-1640.

[24] Chuan-zhi Cui,Shang-wei Wei,Zhen Wang , Yin Qian, Zhong-wei Wu:Response time of waterflooding in low-permeability reservoirs.Unconventional Resources,vol 2, 2022. pp 85-90.

[25] Wei-xia Liu:Effect evaluation method of dynamic profiling for low permeability reservoir.IOP Conference Series: Earth and Environmental Science,vol 300, issue 2. 2019. pp 022023-022023.

[26] Naser Behnam,Mehdi Hosseini,Sina Shahbazi,A Criterion for Estimating the Minimum Drilling Mud Pressure to Prevent Shear Failure in Oil Wells.Geotechnical and Geological Engineering,vol 38, issue 5. 2020. pp 227-236.

[27] Bai-song Tang:Research and Application of Drilling Formation Pressure Change Prediction Technology Based on Recorded Well Data.Journal of Research in Science and Engineering,vol 3,issue 7. 2021.

# Study of Plasma Arc Underwater Penetration Processes in Steel Structures Covered with Non-conductive Materials

**Peng LIU1, Qing-xiang WANG1, Yu-fei XU1, Yong-hong LIU1(✉), Xin-lei WU1, Chi MA1**

1 College of Mechanical and Electronic Engineering, China University of Petroleum (East China), Qingdao, Shandong, 266580, PR China.

liuyhupc@163.com

**Abstract**. Recently, plenty of offshore platforms have faced decommissioning. Offshore steel structures covered with rust, mud, and barnacles require suitable cutting methods for their removal. The non-transferred plasma arc is highly efficient in cutting both conductive and non-conductive materials and is expected to play an important role in offshore demolition operations. This work aims to investigate the underwater penetration process of non-transferred plasma arcs on steel structures covered with non-conductive substances. The workpiece consisted of a 5 mm cement layer and a 10 mm steel plate, whose removal mass and hole morphology were recorded at several moments. It was found that the hole of the cement layer was larger than that of the steel plate, while the mass removal rate of the steel plate was much higher than that of the cement layer. The hole in the cement layer gradually changes from an inverted cone to a drum, and the hole in the steel plate is always an inverted cone. A combined heat source model with time variation is proposed based on the experimental results, and a 2D axisymmetric simulation model is developed. The error between the simulated and experimental hole dimensions was within 3.1%, proving that the proposed heat source model can reflect the heat process of the workpiece well.

## 1. Introduction

As the need for modern industrial development grew, human beings built numerous production platforms offshore to access more oil resources [1]. These offshore structures have a service life of around 20-30 years [2]. Worldwide, a large number of offshore structures have reached the end of their serviceable life [3]. The decommissioned offshore platforms have to be dismantled to protect the marine environment [4]. To achieve the decommissioning of offshore structures, many technicians have tried various cutting methods to do the job. Diamond wire cutting is an external mechanical cutting method that can cut a wide range of materials, but the cutting process is subject to severe vibration [5]. Arc cutting is highly efficient for cutting metallic materials, while it cannot cut non-conductive materials [6]. Oxy-fuel cutting is capable of cutting non-conductive materials, but its thermal efficiency underwater is low, and ignition is difficult [7]. Plasma arc cutting is divided into transferred plasma arc cutting and non-transferred plasma arc cutting. The transferred plasma arc burns between the torch cathode and the sheet metal anode, mechanically compressing the arc column through the nozzle orifice, resulting in a high concentration of energy in the arc column [8]. Thus, the transferred plasma arc is extremely efficient for cutting sheet metal. The non-transferred plasma arc burns between the water-cooled cathode and anode copper nozzles. It mainly relies on the plasma jet ejected from the nozzle to achieve the thermal removal of the material [9]. Thus, the non-transferred plasma arc can cut non-conductive materials.

Most of the previous research on plasma arc cutting has been aimed at improving the cutting efficiency of metal sheets. Parameters such as operating current, gas flow rate, and ignition distance are usually treated as variable parameters [10-11]. The dimensional accuracy of the kerf, the surface quality, and the heat-affected area are often used as evaluation indicators [12-14]. Unlike production activities in the land environment, offshore decommissioning is more focused on the complete penetration of the object being removed, regardless of the quality of the kerf [15]. Due to years of immersion in seawater, the surfaces of offshore structures may be covered with rust, mud, and barnacles [16]. These surface coverings are non-conductive and can cause many difficulties in removal operations [17]. A study reports the experimental results of cutting concrete plates using plasma arc jets. It was found that the power lost to the concrete plate through heat transfer was 25-30% of the power required to melt the concrete plate, and the viscosity and surface tension were the main factors preventing the molten concrete from being discharged from the kerf [9]. Another report presents the results of laser penetration of "steel-cement-carbonate" plates. It was found that steel plates required the highest penetration specific energy, followed by carbonate plates and finally cement [18]. In summary, research on the underwater cutting of steel structures covered with non-conductive substances on the surface is very scarce.

This study aims to investigate the underwater penetration process of the non-transferred plasma arc on steel structures covered with non-conductive substances. The workpiece consisting of a 5 mm cement layer and a 10 mm steel plate was used to equate the steel structure covered with non-conductive material. First, the removal quality and hole shape of the workpiece at different moments were recorded through underwater penetration experiments. Then, a new combined heat source with time variation was proposed based on the experimental results, and the heat process of the workpiece was simulated. Finally, the results of the experiments and simulations were compared, and the conclusions of this study were drawn. This research may promote the application of the non-transferred plasma arc in offshore decommissioning.

## 2. Experimental analysis

### 2.1. Experimental setup and arrangements

The experimental setup is shown in Fig. 1. The plasma torch and workpiece are fixed in a water tank with a liquid level of 20 CM above the torch nozzle outlet. The positive pole of the power supply is connected to the copper nozzle, and the negative pole is connected to the torch cathode. Under the excitation of a high frequency and high voltage pulse, a plasma arc is ignited between the torch cathode and the nozzle, which is then blown out of the nozzle by compressed air to form a plasma arc jet. Under the continuous ablation of the high-temperature plasma arc jet, the workpiece material is removed until it is penetrated. The workpiece is used to simulate a steel structure covered with a non-conductive material, with a 5mm cement layer on its top and a 10mm steel plate on its bottom. The hole dimensions of the workpiece are shown in Fig. 1. $cd_1$ and $cd_2$ are the hole diameters on the top and bottom of the cement layer, respectively. $sd_1$ and $sd_2$ are the hole diameters on the top and bottom of the steel plate, respectively. $wd_1$ is the penetration depth of the workpiece. All operating parameters were kept stable during the experiment, with an operating current of 170 A, an operating gas of 30 L/min air, an ignition distance of 3 mm, and a nozzle diameter of 3 mm. All experiments were repeated three times, and the average of the three experiments was taken as the final result.

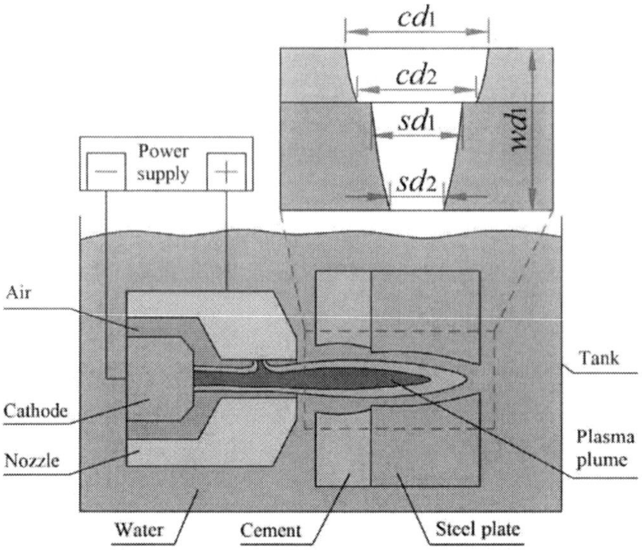

**Fig. 1.** Experimental setup

### *2.2. Experimental results*

*2.2.1. Hole shape.* The shape of the hole was recorded at different moments to study the formation process of the hole in the workpiece. Fig. 2 (a), (b), (c), and (d) show the experimental results at 1, 2, 4, and 6 s, respectively. As can be seen, the holes in both the cement layer and the steel plate gradually expand as the penetration process proceeds. The cement layer was completely penetrated after 1 s, and the steel plate was completely penetrated after 6 s. The hole in the cement layer is significantly larger than the hole in the steel plate, which gradually develops from an inverted cone to a drum shape. The hole in the steel plate always shows an inverted cone shape. It can also be seen that there is not much slag attachment to the hole surface of the workpiece, which may be due to the poor bonding of the two materials. Until the workpiece is penetrated, the slag particles produced from the steel plate can only be discharged outside through the holes in the cement layer. Due to the slag particles of the steel plates do not adhere well to the cement layer, the slag attachment of the workpiece is low.

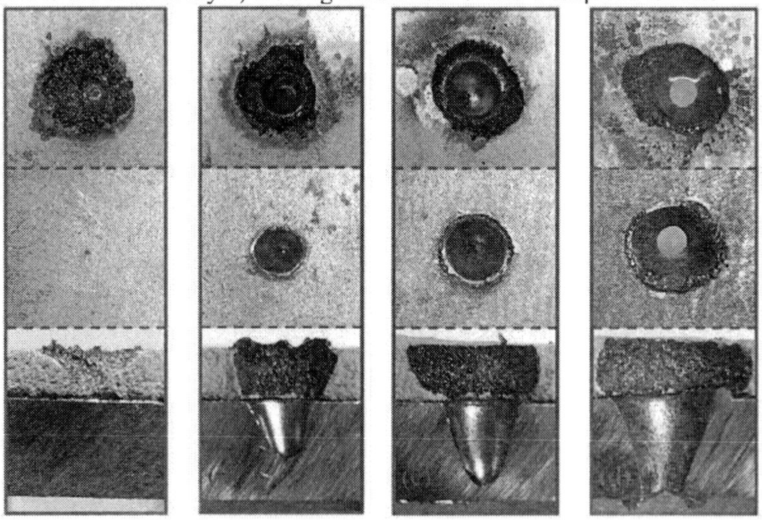
**Fig. 2.** The workpiece hole shape

*2.2.2. Experimental data.* Fig. 3 shows the experimental data recorded during the penetration process. The hole dimensions of the cement layer and the steel plate are shown in Fig. 3(a). The meaning of the variables in Fig. 3(a) has been explained in Subsection 2.1. It can be seen that the hole diameter and the hole depth of the workpiece gradually increase as time increases. The diameters of the workpiece holes are, in descending order, $cd_1$, $cd_2$, $sd_1$, and $sd_2$. It indicates that the hole in the cement layer is always larger than the hole in the steel plate. In addition, the gap between the hole diameters at the top and bottom of the cement layer gradually decreases. At the 6th second, the difference between $cd_1$ and $cd_2$ is already very small.

Fig. 3 (b) shows the expansion rate of the hole dimensions of the workpiece. $cdr_1$ and $cdr_2$ are the expansion rates of the hole diameters at the top and bottom of the cement layer, respectively. $sdr_1$ and $sdr_2$ are the expansion rates of the hole diameters at the top and bottom of the steel plate, respectively. $wdr_1$ is the expansion rate of the hole depth of the workpiece. It can be seen that the expansion rate of the hole diameter and depth basically maintain a gradual decrease. As the hole expands, the plasma arc also gradually spreads. The expansion of the diameter and length of the plasma arc jet results in severe energy attenuation. As a result, the expansion rate of the hole dimensions gradually decreases. There is a slight increase in $cdr_2$ and $wdr_1$ during the first two seconds. It may be related to the warming process of the workpiece. The overall temperature rise of the workpiece is not high at the beginning of the penetration (before 1 second). While after a period of heating (after 1 second), the overall temperature of the workpiece rises, thus causing a slight increase in $cdr_2$ and $wdr_1$.

The removal mass of the workpiece is shown in Fig. 3 (c). $mc_1$ is the removal mass of the cement layer, $ms_1$ is the removal mass of the steel plate, and $mw_1$ is the removal mass of the workpiece. As seen, the removal mass of both the steel plate and the cement layer gradually increased as the penetration progressed. As the density of the steel plate is much greater than that of the cement, more mass is removed from the steel plate, although the hole in the cement layer is greater than that in the steel plate. The mass removal rate of the workpiece is shown in Fig. 3(d). $mcr_1$ is the mass removal rate of the cement layer, $msr_1$ is the mass removal rate of the steel plate, and $mwr_1$ is the mass removal rate of the workpiece. As seen, $mcr_1$, $msr_1$, and $mwr_1$ all gradually increase. The mass removal rate is closely related to the expansion rate of the hole volume, which indicates a gradual increase in the volume expansion rate of the hole.

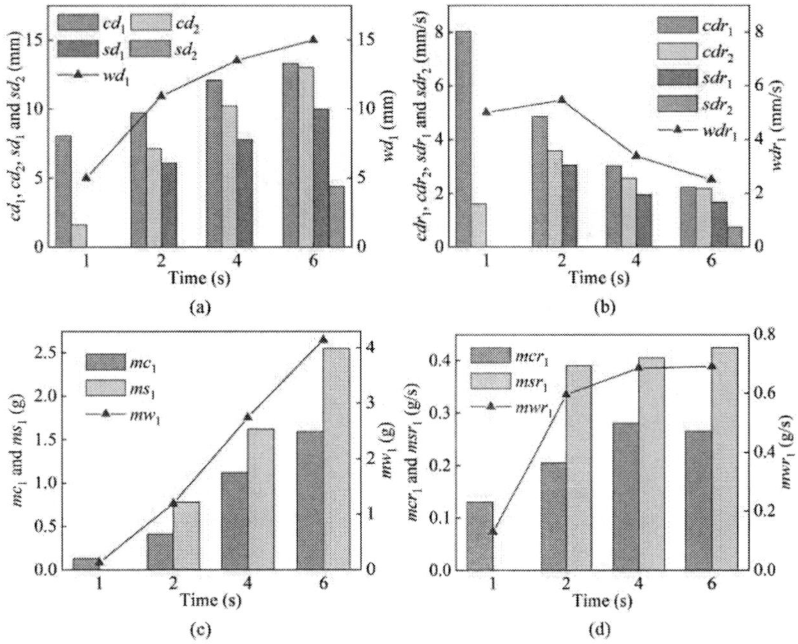


**Fig. 3.** Experimental data

## 3. Simulation analysis

### *3.1. Simulation model*

*3.1.1. Material properties.* A 2D axisymmetric simulation model was developed by COMSOL to analyze the heating process of the workpiece. The simulation model is shown in Fig. 4. In Fig. 4, OABC is the cement layer, OC is 5 mm, and OA is 20 mm. CBED is steel plate, and CD is 10mm. The variation of the physical parameters of the steel plate with temperature is considered in the model, and the latent heat of melting is ignored. In addition, it is assumed that the steel plate is removed when it exceeds its melting point (1767 K). Although there is no fixed melting point for cement, it essentially becomes molten when its temperature reaches 1674K. Therefore, it is considered that the cement layer is removed when its temperature reaches 1674 K. The material properties of the cement layers and steel plates are shown in Table 1. K1, K2 and K3 are the thermal conductivity of solid steel, molten steel and cement, respectively. Rho1, Rho2 and Rho3 are the densities of solid steel, molten steel and cement, respectively. Cp1, Cp2 and Cp3 are the specific heat capacities of solid steel, molten steel and cement, respectively.

**Table 1.** Material properties

| Solid steel | | | Molten steel | | | Cement layer | | |
|---|---|---|---|---|---|---|---|---|
| K1<br>W/(m·K) | Rho1<br>kg/m$^3$ | Cp1<br>J/(kg·K) | K2<br>W/(m·K) | Rho2<br>kg/m$^3$ | Cp2<br>J/(kg·K) | K3<br>W/(m·K) | Rho3<br>kg/m$^3$ | Cp3<br>J/(kg·K) |
| 45 | 7850 | 460 | 35 | 7000 | 660 | 1.5 | 1800 | 840 |

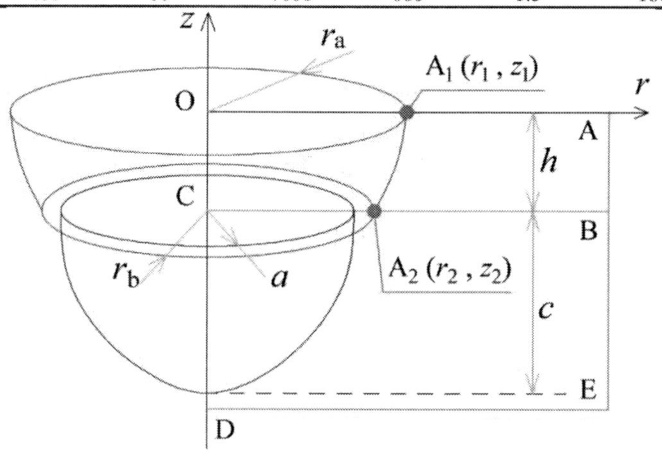


**Fig. 4.** Simulation model

*3.1.2. Boundary conditions.* The underwater processing of the plasma arc generates significant heat energy, and the temperature near the workpiece kerf is extremely high. The penetration process produces many bubbles due to the high-temperature plasma arc. Therefore, the boiling heat exchange theory of water should be used for the thermal analysis of the penetration process. In the model shown in Fig. 4, the initial temperatures of the cement and steel plates were set at 294 K. OD was the symmetry axes, set as thermally insulated boundaries. OA, AE, and DE were set as heat dissipation boundaries. The boundary heat dissipation coefficients are shown in Table 2. CB was the contact surface between the cement layer and the steel plate, set as the thermal contact boundary, ignoring radiative heat transfer. As the contact surface of the steel plate and the cement layer was filled with water at the seam, the thermal conductivity of the contact surface was taken to be consistent with Table 2.

**Table 2.** Boiling heat transfer coefficient of water

| Temperature (K) | 294 | 374 | 404 | 494 | >494 |
|---|---|---|---|---|---|
| Heat dissipation coefficient [W/(m^2*K)] | 250 | 250 | 10840 | 1030 | 1030 |

*3.1.3. Heat source model.* The simulation results for the underwater penetration process of the plasma arc are influenced by a combination of material properties, heat dissipation conditions, and the model heat source. As the hole shape of the workpiece changes during the penetration process, the heat source model also needs to change over time. In addition, as the workpiece consists of two layers of material, a single heat source cannot meet the analysis requirements. Therefore, a time-dependent combined heat source model (Eq. (1)) was proposed to analyze the heat process and penetration mechanism of the workpiece.

$$\begin{cases} Q_1 = f_1 q_1,\ f_1 = 1, & t \le 1 \\ Q_1 = f_1 q_2,\ f_1 = 0.8 - 0.1(\mathrm{t}-1), & 1 < \mathrm{t} \le 6 \\ Q_2 = f_2 q_1,\ f_2 = 0.2 + 0.1(\mathrm{t}-1), & 1 \le \mathrm{t} \le 5 \\ Q_2 = f_2 q_2,\ f_2 = 0.2 + 0.1(\mathrm{t}-1), & 5 < \mathrm{t} \le 6 \end{cases} \tag{1}$$

In Eq. (1), $q_1$ is the semi-ellipsoidal volumetric heat source, calculated according to Eq. (2). $q_2$ is the cone heat source, calculated according to Eq. (3). $Q_1$ and $Q_2$ are the heat sources for the cement layer and the steel plate, respectively, in the simulation model. $f_1$ and $f_2$ are the heat source distribution coefficients for the cement layer and the steel plate, respectively.

$$q_1 = \frac{6\sqrt{3}\eta UI}{a^2 c\pi\sqrt{\pi}} \exp\left(-\frac{3r^2}{a^2} - \frac{3z^2}{c^2}\right) \tag{2}$$

In Eq. (2), $\eta$ is the thermal efficiency of the plasma arc of 0.7. $U$ and $I$ are the voltage and current of the plasma arc, which are collected by a current sensor and oscilloscope, as shown in Fig. 5. $r$ and $z$ are the two axes of the simulation model. $a$ and $c$ are the eigenvalues of the semi-ellipsoidal volumetric heat source axis, as shown in Fig. 4. $a$ and $c$ are taken according to the hole dimensions of the workpiece measured after the experiment.

$$q_2 = \frac{9\eta UI}{\pi(1-\exp(-3))(z_1 - z_2)(r_1^2 + r_1 r_2 + r_2^2)} \exp\left(-\frac{3r^2}{r_0^2}\right) \tag{3}$$

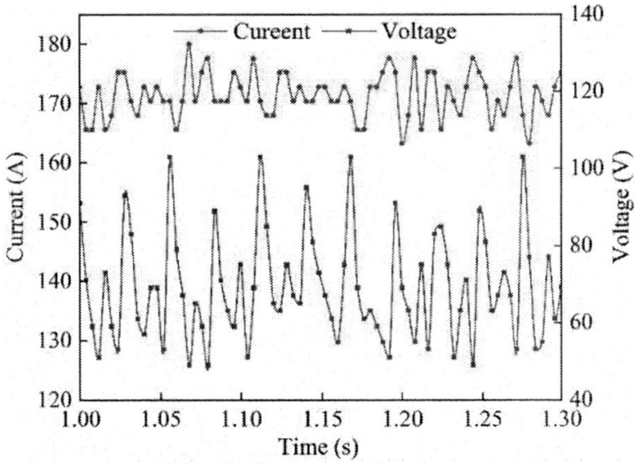


**Fig. 5.** Current and voltage

In Eq. (3), $(r_1, z_1)$ and $(r_2, z_2)$ are the coordinates of the cone heat source characteristic points, as shown in Fig. 4. $(r_1, z_1)$ and $(r_2, z_2)$ are taken according to the hole dimensions of the workpiece measured after the experiment. $r_0$ is the bus bar of the cone heat source, calculated according to Eq. (4).

$$r_0 = \sqrt{a_1^2 (1 - \frac{z^2}{b_1^2})}, \quad a_1^2 = \frac{r_2^2 z_1^2 - r_1^2 z_2^2}{z_1^2 - z_2^2}, \quad b_1^2 = \frac{r_2^2 z_1^2 - r_1^2 z_2^2}{r_2^2 - r_1^2} \tag{4}$$

### *3.2. Simulation results*

The shape of the simulated hole in the workpiece at different moments is shown in Fig. 6. The isotherms in Fig. 6 are in units of K. As seen, the hole in the cement layer is always larger than the hole in the steel plate, and its shape gradually changes from an inverted cone to a drum shape. The hole shape of the steel plate is consistently kept inverted cone. It is essentially the same shape as the experimental hole in Fig. 2. As seen from the isotherms of the cement and steel plates, the steel plate shows a wide range of temperature distribution, and its isotherms are sparse. It is mainly influenced by the thermal conductivity of the material. As the thermal conductivity of cement is much lower than that of steel plates, it is easier for heat to accumulate near a cement hole and for heat from a steel plate to be transferred from the surface of the hole to its interior. The temperature distribution in Fig. 6 shows that as time increases, the heat accumulation near the hole increases and the area of the workpiece affected by the heat becomes wide. This result can be used to explain the trend in mass removal rate in Fig. 3(d). As time increases, the hole in the workpiece becomes larger, the plasma arc becomes diffuse, and its energy gradually decays. On the other hand, the heated area increases with the hole, and the heat accumulates near the hole. As a result, the volume removal rate of the workpiece increases, and thus the mass removal rate increases.

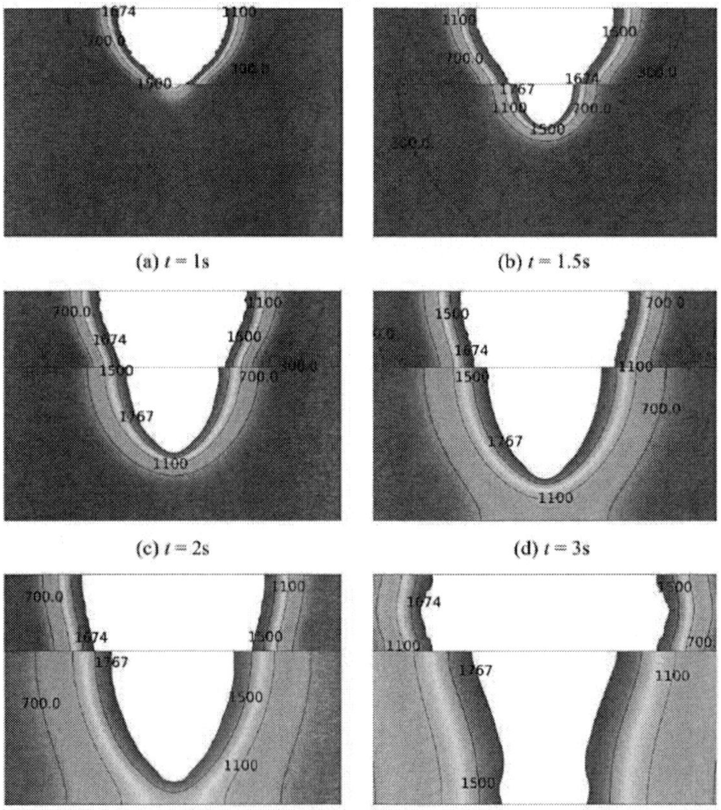


**Fig. 6.** Hole shape and temperature distribution

Fig. 7(a) shows the hole dimensions of the simulation model at different moments. $scd_1$ and $scd_2$ are the hole diameters on the top and bottom of the cement layer, respectively. $ssd_1$ and $ssd_2$ are the hole diameters on the top and bottom of the steel plate, respectively. $swd_1$ is the hole depth of the workpiece. It can be seen that the results in Fig. 7(a) and Fig. 3(a) are in general agreement. Fig. 7(b) shows the errors in the simulated and experimental hole dimensions. As seen, the error between the simulation results and the experimental results is kept below 3.1%. It indicates that the developed simulation model can reflect the penetration process of the workpiece well.

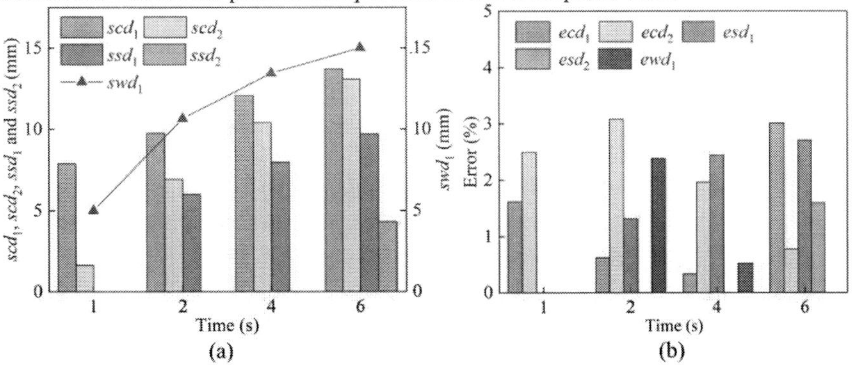


**Fig. 7.** Hole dimensions and errors

## 4. Conclusion

This work investigated the penetration process of a steel structure covered with non-conductive material through experimental and simulation methods. A double-layer workpiece consisting of a 5mm cement layer and a 10mm steel plate was successfully penetrated under a 175A non-transferred plasma arc, and the following conclusions were drawn.

(1) With the penetration time increasing, the hole of the cement layer gradually became drum-shaped from an inverted cone, and the hole of the steel plate was always an inverted cone. The hole diameter of the cement layer is greater than that of the steel plate, but its removal mass is smaller than that of the steel plate.

(2) As the penetration time increases, the hole dimension expansion rate of the workpiece gradually decreases, but its mass removal rate gradually increases. The mass removal rate of the workpiece is closely related to its heating process. The larger the hole surface area, the more heat energy accumulates and the higher the mass removal rate of the workpiece.

(3) A new combined heat source with time variation was proposed. The penetration process of the workpiece was simulated by this combined heat source, and the temperature distribution pattern inside the workpiece was obtained. The simulated and experimental hole shapes match well. The hole dimension error is below 3.1%.

## References

[1] Kaiser, M.J. and M. Liu, Decommissioning cost estimation in the deepwater US Gulf of Mexico–Fixed platforms and compliant towers. Marine structures, 2014. 37: p. 1-32.
[2] Chandler, J., et al., Engineering and legal considerations for decommissioning of offshore oil and gas infrastructure in Australia. Ocean engineering, 2017. 131: p. 338-347.
[3] Kaiser, M.J., Decommissioning forecast in the deepwater Gulf of Mexico, 2013–2033. Marine structures, 2015. 41: p. 96-126.
[4] Fam, M.L., et al., A review of offshore decommissioning regulations in five countries – Strengths and weaknesses. Ocean engineering, 2018. 160: p. 244-263.
[5] Molfino, R.M. and M. Zoppi, A robotic system for underwater eco-sustainable wire-cutting. Automation in construction, 2012. 24: p. 213-223.
[6] Tian, X., et al., Experimental Investigation on Electric Arc Cutting Process for Casing. Materials and manufacturing processes, 2014. 29(2): p. 166-174.
[7] Yan, H., L. Wu and J. Yu, The environmental impact analysis of hazardous materials and the

development of green technology in the shipbreaking process. Ocean engineering, 2018. 161: p. 187-194.

[8] Lazarević, A., Ž. Ćojbašić and D. Lazarević, Computationally intelligent modelling of the plasma cutting process. International journal of computer integrated manufacturing, 2020. 33(3): p. 252-264.

[9] Chamorro, J.C., et al., Plasma Cutting of Concrete: Heat Propagation and Molten Material Removal From the Kerf. IEEE transactions on plasma science, 2019. 47(6): p. 2859-2867.

[10] Naik, D.K. and K. Maity, Experimental analysis of the effect of gas flow rate and nature on plasma arc cutting of hardox-400. Welding in the world, 2019. 64(2): p. 345-352.

[11] Dupláк, J., et al., Prediction model of surface roughness parameters of structural steel created by plasma arc cutting via full factor experiment. Materialwissenschaft und Werkstofftechnik, 2019. 50(10): p. 1207-1220.

[12] Tekaüt, İ., Experimental investigation of the effects of cutting parameters on the cutting surface quality and heat affected area (HAZ) width in plasma arc cutting processes. Journal of the Faculty of Engineering and Architecture of Gazi University, 2020.

[13] Peko, I., et al., Modelling of Dross Height in Plasma Jet Cutting Process of Aluminium Alloy 5083 Using Fuzzy Logic Technique. Tehnički vjesnik, 2020. 27(6): p. 1767-1773.

[14] Masoudi, S., et al., Development of an intelligent model to optimize heat-affected zone, kerf, and roughness in 309 stainless steel plasma cutting by using experimental results. Materials and manufacturing processes, 2019. 34(3): p. 345-356.

[15] Liu, D., et al., Effects of processing parameters on arc stability and cutting quality in underwater wet flux-cored arc cutting at shallow water. Journal of manufacturing processes, 2018. 33: p. 24-34.

[16] Uzun, D., et al., Does the barnacle settlement pattern affect ship resistance and powering? Applied ocean research, 2020. 95: p. 102020.

[17] Petersen, D.S., et al., Competing with barnacle cement: wetting resistance of a re-entrant surface reduces underwater adhesion of barnacles. J R Soc Interface, 2018. 15(145).

[18] Da Silva MG, Braga AM, De Faria G V., Rosario FF. Application of lasers for perforation of petroleum wells. OTC Bras 2017 2017:1091–106. https://doi.org/10.4043/28067-ms.

# Electrodeposition of Double Chromium Plating to Improve Corrosion Performance of Screw Drilling in Chloride Environment

**Shuai ZHAO[1✉], Chang LIU[1], Jian GAO[1], Ming-yu BAO[1], Jing-bo WANG[1✉], Yue GAO[2], Li TANG[2], Qiu-ping TANG[3]**

1 Safety, Environment and Technology Supervision Research Institute of PetroChina Southwest Oil and Gas Field Company, Chengdu, China
2 Northeast Sichuan Gas Mine of PetroChina Southwest Oil and Gas Field Company, Dazhou Sichuan, China
3 Northwest Sichuan Gas Field of PetroChina Southwest Oil and Gas Field Company, Jiangyou, Sichuan

zhaoshuai0927@petrochina.com.cn, jingbowang@petrochina.com.cn

**Abstract.** To adapt to and satisfy rising energy demand, oil and gas exploration and production are gradually spreading to locations with poor underground conditions, and the demand for screw drilling instruments is rising as well. When underground drilling tool materials are subjected to these harsh operating circumstances, they corrode severely and may even cause drill pipe breakage and blowout incidents. Failure due to corrosion is still the most common type of failure, with drilling fluid being the leading cause of downhole drilling tool corrosion and equipment damage, particularly in high-temperature and high-pressure working conditions with higher corrosiveness, which must be avoided and controlled. Double chromium was successfully plated on the surface of the screw rotor using the electrodeposition method to extend the service life of screw drilling instruments. Electrochemical methods such as scanning electron microscopy, open circuit potential, electrochemical impedance spectroscopy, and potentiodynamic polarization curve were used to analyze the plating's morphology and corrosion resistance. The research results show that at high chloride concentrations (50g/L~100g/L), when the chloride ion concentration is 60g/L, corrosion resistance is the lowest, and that double chromium plating enhances threaded drilling tool corrosion resistance. The dual chromium plating's polarization resistance has been enhanced by 18 times. The double chrome plated milky white chrome plating functions as a physical barrier to prevent corrosive substances from diffusing to the substrate interface, corrosive ions from reaching the substrate, and considerably improving the screw rotor's anti pitting performance.

## 1. Introduction

Screw drilling tools have become increasingly popular in the construction of directional and horizontal wells as China's oil and gas development industry has grown and downhole oil and gas production technology has matured. Screw drilling is powered by a positive displacement motor [1]. Screw drilling tools frequently fail during the drilling process in the oilfield. Screw rotor surface corrosion is a common failure mode, accounting for 61.3% of total failure [2]. Fig.1 depicts the surface corrosion of a failed spiral rotor after downhole operation. Fig. 1 shows that the corrosion is local, with the corrosion pit appearing at the top of the screw rotor's spiral line. Because of its direct contact with mud and meshing with the stator during operation, the rotor is prone to corrosion and wear [3]. According to the findings of Zhang et al, screw drilling experienced fatigue, abrasive, and tribo-corrosion wears under water and mud conditions [4]. Liu proposed effective improvement measures based on design enhancement, material upgrading, process optimization, and other factors after analyzing the causes of common screw drill motor assembly failures during drilling construction [5].

**Fig.1.** Spiral drill pipe rotor surface corrosion.

Electrodeposited hard chromium (EHC) plating has been widely used in industrial fields such as construction, mining, aerospace, machine tools, automotive transport, and oilfield processing due to its high wear and corrosion resistance, excellent hardness and electrical properties, and low cost [6,7,8]. EHC refers to thick chromium layers (1-500 m in thickness) rather than thin decorative layers (0.250 m in thickness). The tungsten carbide bearing (TC bearing) is the most important factor influencing the safety and durability of a screw-drilling tool. According to the findings of Liu et al, hot cracking of YG8 cemented carbide blocks (WC-92%, Co-8%) is caused primarily by the diffusion of Co elements into the solder and the formation of W2C and Co3W3C brittle phases during heating and holding time [9]. Guo et al used high velocity oxygen fuel (HVOF) technology to deposit WC-12Co and WC-10Co-4Cr plating on the surface of a screw pump rotor substrate (45 steel). In both plating, microgalvanic corrosion occurs between the WC and the metal bonded phase, resulting in the dissolution of the bonded phase and the shedding of some WC particles, and the corrosion products can protect the plating [10]. Hard chromium plating is an effective method of increasing surface hardness [11]. The milky white chromium plating layer is milky white and lusterless, with good plating toughness, few pores, few cracks, a soft color, and good corrosion resistance [12]. Double chromium plating refers to the hard chromium plating on the surface of milky white chromium plating. It combines the properties of milky white chrome plating and hard chrome plating, and it is primarily used for plating parts that require both wear resistance and corrosion resistance [13].

In this paper, electrodeposition was used to prepare double chromium plating on the surface of 45# steel. The surface morphology was studied using SEM. The electrochemical impedance spectroscopy technique was used to analyze the film layer's ability to resist corrosion and to form protective measures against corrosion of 45# steel in a chlorine corrosion environment.

## 2. Experiment

### *2.1 Electrodeposited double chromium plating*

Table 1 shows the electroplating process parameters for this test. To ensure plating performance and improve deposition rate, two additives, LD-5680A and LD-5680B, were added to the electroplating solution of hard chromium plating. Test material using spiral rotor commonly used material 45 steel, sample size 50 mm 25mm 5mm, polishing with 180 #, 500 #, 800 #,1000 # sandpaper on the sample step by step, and then separately with acetone, anhy-drous ethanol degreasing dehydration, the cold wind blow dry.

This test employs double bath separate plating as the electroplating process. In the first stage, immerse the sample in the white chromium plating solution for electroplating, with a deposition rate of 10 m/h and an electroplating time of 2 hours. The opalescent chromium plating will be quickly placed in the hard chromium plating bath for secondary electroplating in the second stage. The hard chromium plating deposition rate is 60 m/h, and the electroplating time is 1 hour. During the electroplating process, immerse the sample in the electroplating bath for 2 minutes each time to remove the oxide film or surface defects on the sample surface, improving plating adhesion and ensuring plating thickness uniformity. Because of the high current density of chromium plating, the current density during positive plating cannot be adjusted directly and quickly from zero to the value required by the process. It is necessary to gradually increase the current density to reach the required current density process condition value in 10 times (within 3 minutes) to ensure the stability of the deposition speed and plating performance, so that the double chromium plating with high hardness and corrosion resistance is obtained.

**Table 1.** Electroplating process parameters of double chromium plating.

| | $CrO_3$ (g/L) | $SO_4^{2-}$ (g/L) | $Cr^{3+}$ (g/L) | T (℃) | i ($A/dm^2$) | LD-5680A (ml/L) | LD-5680B (ml/L) |
|---|---|---|---|---|---|---|---|
| Hard chrome plating | 220 | 3 | 2 | 60 | 60 | 10 | 10 |
| Milky white chromium plating | 250 | 2.5 | 2 | 71 | 25 | - | - |

*2.2 Analysis of the morphology of films*

The surface morphology analysis of double chrome plating was studied using a Thermo Fisher Scientific Aztec X-Max80 scanning electron microscope.

*2.3 Electrochemical test*

Electrochemical experiments were carried out on a CS 310H electrochemical workstation with a carbon rod as an auxiliary electrode, a KCl saturated Ag/AgCl as a reference electrode, and a double chrome plating sample (10 mm × 10 mm) as a working electrode. A 1 hour open circuit potential measurement was performed after that. The electrochemical impedance spectroscopy (EIS) was measured using an excitation signal of 10 $mV_{Ag/AgCl}$ in the frequency range of $10^5$~$10^{-2}$ Hz, and the results were analyzed using ZSimpWin software. The test range for potentiodynamic polarization is -0.6 $V_{Ag/AgCl}$~1.5 $V_{Ag/AgCl}$, and the scan rate is 0.5 mV/s. Test solution was 10 g/L, 20 g/L, 30 g/L, 40 g/L, 50 g/L, 60 g/L, 70 g/L, 80 g/L, 90 g/L and 100 g/L NaCl (pH 6.7). Other ionic components are: $Na^+$ concentration is 57.24g/L, $Mg^{2+}$ concentration is 0.04g/L, $HCO_3^-$ concentration is 0.75g/L, $SO_4^{2-}$ concentration is 4.00g/L, $Ca^{2+}$ concentration is 1.02g/L. Experiments were conducted at ambient temperature (293±2 K), at atmospheric pressure.

## 3. Double chromium plating experimental results analysis

*3.1 Microscopic morphology*

Fig. 2 shows that the surface of the double chromium plating contains a series of "dome-shaped crystal package particles," which could be due to the fact that at the start of the chromium electroplating, the chromium deposition begins to nucleate at some sensitive positions, such as high stress or surface defects, and grows up continuously with the increase of the electroplating time, forming a series of "dome-shaped particles" on the surface. Many tiny microcracks can also be found distributed around the "dome-shaped parti-cles," and the microcracks intersect at a point in some places. This could be due to the electroplating process's reaction of hydrogen and chromium to produce hydride, in which the hydrogen atom enters the chromium crystal and distorts the crystal structure, and under the combined action of the plating's crystal structure and the oxides, hydroxides, and hydride inclusions with high internal stress, many microcracks form on the surface or inside of the plating. The plating section micrograph shows that there is no gap between the double chromium plating and the substrate.

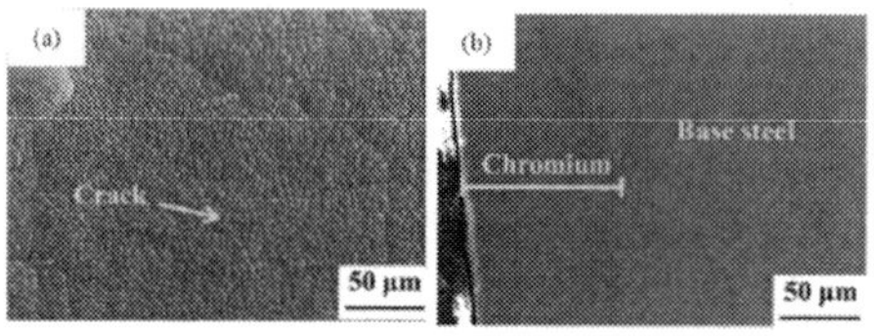


**Fig.2.** The original morphology of double chromium plating.

*3.2 The open circuit potential*

Fig. 3 depicts the trend of the open circuit potential of chromium plating with various chlorides. Fig. 3 shows that when NaCl concentration was low (10 g/L~40 g/L), the self-corrosion potential of double chromium plating first increased and then decreased with concentration increase, with a maximum value of 30 g/L. When NaCl was at a high concentration (50 g/L~100 g/L), the self-corrosion potential of the chromium plating layer increased first, then decreased, with a maximum value at 60 g/L. This is due to the metal surface producing an oxide film or corrosion product film [14], which causes the open circuit potential to rise. The opencircuit potential ($E_{ocp}$) can thermodynamically represent the material's proclivity to corrosion in solution. The $E_{ocp}$ of double chromium plating shifts as NaCl concentration increases, indicating that the thermodynamic activity or electrochemical activity of the material changed as chloride concentration increased.

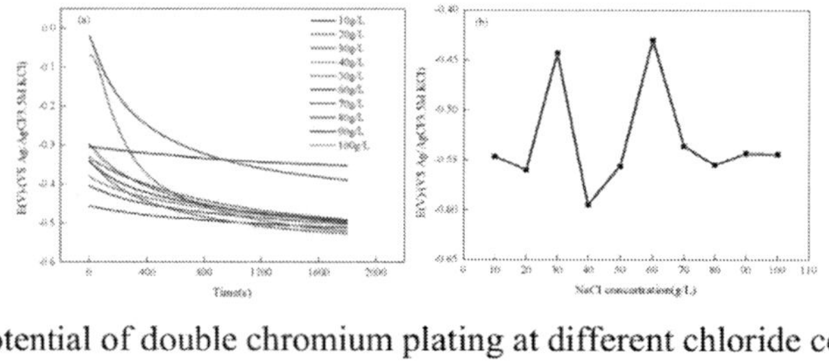

**Fig. 3.** Open circuit potential of double chromium plating at different chloride concentration.

*3.3 EIS*

Fig. 4 illustrate the Nyquist and Bode of double chromium plating at different chloride concentration. The Nyquist curve of double chromium plating in chloride solution exhibits the same features and is constituted of a single semicircular arc as shown in Fig. 4(a). The Bode-phase diagram in Fig. 4(b) comprises two overlapping time constants. Two time constants could be differentiated in the impedance spectra for all the samples (Fig. 4(b)). The first time constant appearing at high frequencies may be related to the behaviour of the electrolyte/plating interface for cracked samples and of the electrolyte/oxide layer interface for crack-free plating [15]. On the other hand, the second time constant at lower frequencies would correspond to the electrolyte/substrate interface. Furthermore, from the Bode diagrams it can be observed that the phase angle did not reach zero at high frequencies, as can be expected for a purely resistive behavior [16], suggesting the presence of defects or weak spots in the studied surfaces.

The equivalent circuit (EEC) [17] of double chromium plating under different chloride corrosion solution is determined using Zsimpwin data fitting software, as illustrated in Fig. 5. $R_s$ denotes the solution resistance, $Q_1$ represents the electric double layer capacitance, and $R_1$ indicates hard chrome plating resistance in EEC. $Q_2$ indicates the milky white chromium film's capacitive characteristics, $R_2$ represents the charge transfer resistance, and total impedance Z is computed as follows [18]:

$$Z = R_S\left(1/\left(Z_{Q_1} + 1/\left(R_1 + 1/\left(Z_{Q_2} + 1/R_2\right)\right)\right)\right) \tag{1}$$

The impedance of the electric double layer is generally related to the angular frequency of the excitation signal, and it can be represented as [19]:

$$Z_{Q_2} = R_2/\left(1 + \left(j\omega R_2 C_2\right)^{n_2}\right) \tag{2}$$

$\omega$: angular frequency; $j$: the sign of the imaginary number; $n$: The index of the constant phase element.

When n=0, it denotes resistance, and when n=1, it means capacitance. n typically ranges from 0.5 to 1 [20], and n is proportional to the roughness of the electrode surface [21]. $R_2$ represents the resistance of ions moving under the influence of an electric field. The greater the value, the stronger the corrosion resistance of the material; the greater $R_1$, the more stable the created passivation plating and the less probable it will be harmed in the medium. The stronger the protective impact [22]. The solution resistance $R_S$ is minimal in comparison to the charge transfers resistance $R_2$. As a result, the charge transfer resistance $R_2$ determines the corrosion rate. In this system, the polarization resistance $R_p$ is defined as $R_p = R_s + R_1 + R_2$, and the value of $R_p$ reflects the impedance value of double chromium plating in the simulated solution under different stresses, in see Fig. 4 (d). The polarization resistance of double chromium plating is $\Omega \cdot cm^2$, while that of single chromium plating is 3007 $\Omega \cdot cm^2$ [23], a factor of 18 times increase. According to the analysis in Fig. 4 (d), when the concentration of chloride ion is low (10 g/L~40 g/L), when the concentration of chloride ion is 30 g/L, Rp is 21777$\Omega \cdot cm^2$, which is the minimum value; when the concentration of chloride ion is high (50 g/L~100 g/L), Rp is 19532.92 $\Omega \cdot cm^2$, which is the minimum value. The results show that at low chlorine concentrations, the resistance value of the double chromium plating decreases first and then increases as the chloride ion concentration increases. The resistance value of the double chromium plating has the lowest value at 60g/L at high chlorine concentration, and its law is consistent with the conclusion of open circuit potential.

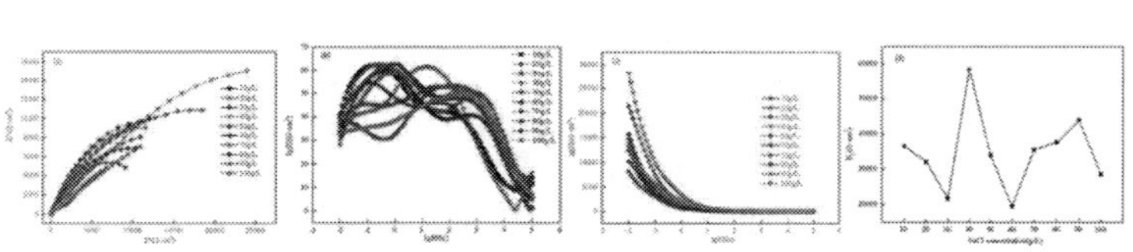

**Fig. 4.** EIS of double chromium plating at different chloride concentration.

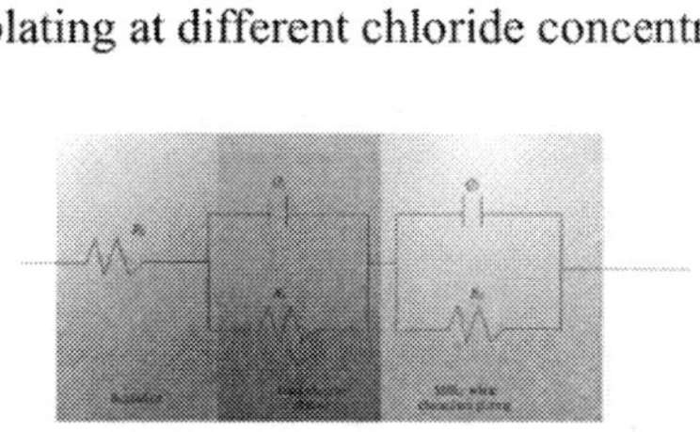

**Fig. 5.** EEC of double chromium plating at different chloride concentration.

*3.4 Potentiodynamic polarization curve*

The potentiodynamic polarization curves of double chromium plating at various chloride concentrations are shown in Fig. 6(a). As shown in Fig. 6 (a), as the concentration of NaCl increases from 10g/L to 40g/L, the cathodic current decreases near the corrosion potential and the cathodic reaction is inhibited because the increase in chloride ion concentration reduces the solubility of oxygen in solution, and because the cathodic reaction is primarily a reduction reaction of oxygen, the cathodic current density decreases with the increase in chloride ion concentration.

Fig. 6(a) depicts the potentiodynamic polarization curves of double chromium plating at different chloride concentration. It can be seen from Fig. 6 (a) that when the concentration of NaCl changes from 10g/L to 40g/L, the cathodic current decreases near the corrosion potential, and the cathodic reaction is inhibited because the increase of chloride ion concentration reduces the solubility of oxygen in solution, while the cathodic reaction is mainly the reduction reaction of oxygen, so the cathodic current density decreases with the increase of chloride ion concentration. When the concentration of NaCl in the solution changes from 10g/L to 30g/L, the current density increases, which accelerates plating dissolution. According to Fig. 6 (b), when the chloride ion concentration is 60g/L, the anode curve has a weak left shift, and the left shift is more noticeable as the potential is far from the self-corrosion potential, and the anode current increases, indicating that the self-corrosion current density is greatest at high chloride concentrations, and the chloride ion concentration is 60g/L.

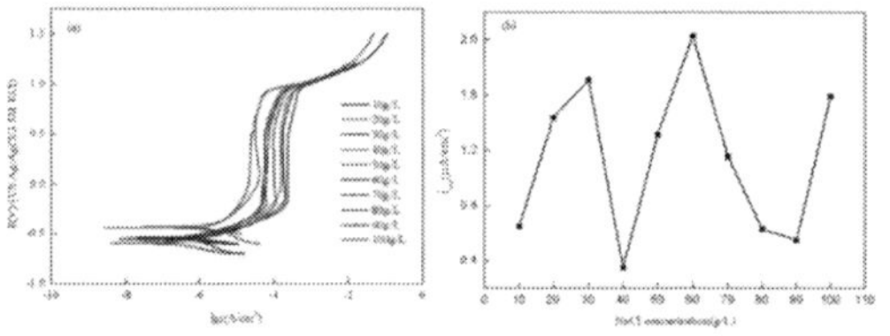

**Fig. 6.** Potentiodynamic polarization curve of double chromium plating at different chloride concentration.

## 4. Double chrome plating corrosion resistance mechanism analysis

Fig. 7 depicts the corrosion resistance mechanism of double chromium plating. Hard chromium plating has some microcracks in addition to high porosity. This is due to the fact that the decomposition of chromium hydride during the electrodeposition process results in high strength residual internal stress and microcracks. Simultaneously, the density of these microcracks in the hard chromium plating layer increases with thickness, and the density of microcracks is closely related to the plating's hardness, corrosion resistance, and wear resistance. Once there are penetrating vertical microcracks in the hard chromium plating, liquid substances can easily diffuse to the plating/substrate interface junction and promote pitting corrosion in the matrix. The cathodic reaction during the corrosion process is primarily an oxygen absorption reaction. $Cl^-$ can activate the sample surface and destroy the oxidation passivation film, increase surface defects, and promote pitting corrosion in corrosion solutions with high chlorine concentrations. Furthermore, because of its high mobility (its size is smaller than that of $NO_3^-$), they can easily penetrate the microcracks and pores of the hard chromium plating to the interface junction of the crack-free white plating. However, the crack-free milky white chromium bottom layer has fewer microcracks than the hard chromium plating on the surface layer, and the porosity is low and the air tightness is good, preventing the corrosive medium from contacting the substrate and thus preventing corrosion of the substrate.

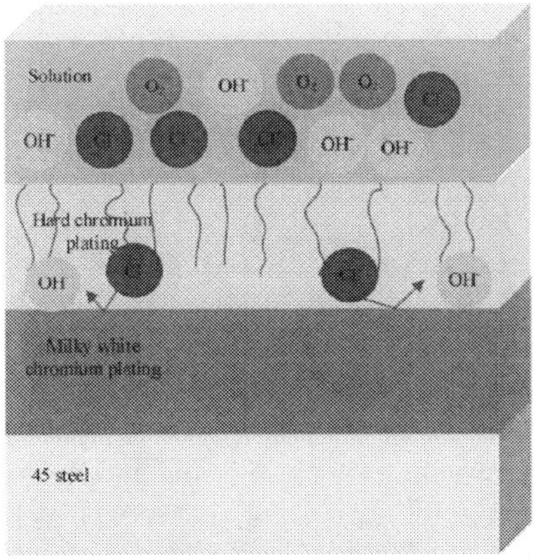


**Fig. 7.** Corrosion resistance mechanism diagram of double chromium plating.

## 5. Conclusion

(1) At low chlorine concentrations (10g/L~40g/L), the corrosion resistance of the double chromium plating decreased first and then increased with chloride concentration increase.
(2) At high chlorine concentrations (50g/L~100g/L), the corrosion resistance is the worst when the concentration of chloride ion is 60g/L.
(3) $Cl^-$ can activate the sample surface and destroy the oxidation passivation film, increase surface defects, and promote pitting corrosion in corrosion solutions with high chlorine concentrations.
(4) The crack-free milky white chromium bottom layer has fewer microcracks than the hard chromium plating on the surface layer, and the porosity is low and the air tightness is good, preventing the corrosive medium from contacting the substrate and thus preventing corrosion of the substrate.
(5) Under conditions of high chloride ion concentration, double chromium plating improves the corrosion resistance of screw drilling tools.

**Acknowledgments.** The project is supported by National Natural Science Foundation of China (Number 52174062).

## References

[1] Ranjbar, K., Sababi, M. Failure assessment of the hard chrome coated rotors in the downhole drilling motors. Engineering Failure Analysis, 20:147-155 (2012).
[2]Naiming, L., Dali, L., Zou, J. J., et al. Surface texture-based surface treatments on ti6al4v titanium alloys for tribological and biological applications: a mini review. Materials (Basel, Switzerland), 11(4):1-28 (2018).
[3] Wang, F. C. Research and application of chromium plating technology of screw drilling tool rotor. Heilongjiang Science,2022,13(12):41-43 (2022).
[4] Zhang, Z.Q., Yu, X., Zhao, W.W., et al. Exploring wear detection method for special drilling parts in liquid media. International Journal of Refractory Metals and Hard Materials, 61: 249-258 (2016).
[5] Liu, S.M. Failure cause analysis and improvement of screw drill motor assembly. Value Engineering, 41 (19): 111-113(2022).
[6] Kim, D., Kim, M., Nam, K.S., et al. Duplex coating for improvement of corrosion resistance in chromium deposit. Surface & Coatings Technology, 169:650-654 (2003).
[7] Tyler, J. M. Automotive Applications for Chromium. Metal Finishing, 93(10):11-14 (1995).
[8] Wasekar, N., Sundararajan, G. Sliding wear behavior of electrodeposited Ni–W alloy and hard

chrome coatings. Wear, 342-343:340-348 (2015).
[9] Liu, Q. C., Cui, M., He, Q. K., et al. Failure analysis of tungsten carbide bearing on screw drill. Engineering Failure Analysis, 89:37-45(2018).
[10] Guo, F.F., Zhu, C.C., Zhao, E.L., et al. Tribology and corrosion properties of HVOF-sprayed WC coatings on surface of screw pump rotor. Journal of Jiangsu University-Natural Science Edition, 43(3):332-340 (2022).
[11] Lampa. C., Smirnov, I. High speed laser cladding of an iron based alloy developed for hard chrome replacement. Journal of Laser Applications, 31(2): 022511(2019).
[12] Nieto, P., Gonzalo, E.G., Lasheras, F.S., et al. A Hybrid Predictive Approach for Chromium Layer Thickness in the Hard Chromium Plating Process Based on the Differential Evolution/Gradient Boosted Regression Tree Methodology. International Journal of Interactive Multimedia and Artificial Intelligence, 6 (4):39-48 (2020).
[13] Lasheras, F. S., Jav, V., Nieto, P., et al. The use of design of experiments to improve a neural network model in order to predict the thickness of the chromium layer in a hard chromium plating process. Mathematical and Computer Modelling, 52(7-8):1169-1176(2010).
[14] Rocchini, G. Evaluation of the electrochemical parameters by means of series expansion [J]. Corrosion Science, 36(8): 1347-1361 (1994).
[15] Imaz, N., Ostra, M., Vidal, M., et al. Corrosion behaviour of chromium coatings obtained by direct and reverse pulse plating electrodeposition in NaCl aqueous solution. Corrosion Science, 78:251-259(2014).
[16] Abreu, C.M., Cristóbal, M.J., Losada, R., et al. High frequency impedance spectroscopy study of passive films formed on AISI 316 stainless steel in alkaline medium. Journal of Electroanalytical Chemistry 572:335–345 (2004).
[17] Xiao, Y., Tang, J. L., Wang, Y. Y., et al. Corrosion behavior of 2205 duplex stainless steel in NaCl solutions containing sulfide ions. Corrosion Science, 200,110240-110256(2022).
[18] Sanchez, M., Gregori, J., Alonso, M. C., et al. Anodic growth of passive layers on steel rebars in an alkaline medium simulating the concrete pores. Electrochimica Acta,52,47-53 (2006).
[19] Macdonald, J. R. Impedance spectroscopy. Annals of Biomedical Engineering, 20,289-305(1992).
[20] Zhang, X., Pehkonen, S. O., Kocherginsky, N. et al. Copper corrosion in mildly alkaline water with disinfectant monochloramine. Corrosion Science, 44,2507-2528(2002).
[21] Cox, B., Wong, Y. M. Simulating porous oxide films on zirconium alloys. Journal of Nuclear Materials. 218,324-334(1995).
[22] Zhou, X. C., Cui, Q. Q., Jia, J. H., et al. Effect of Cl concentration on SCC behavior of 316L stainless steel in alkaline NaCl/$Na_2S$ solution. Chinese Journal of Corrosion and Protection, 37,526-532(2017).
[23] Zhou, W. Corrosion effect of cl- ions on 42crmo drilling tool steel and its double chromium plating under high temperature and high pressure. Chongqing University of Science and Technology, (2018).

Journal of Physics: Conference Series **2594** (2023) 012016 doi:10.1088/1742-6596/2594/1/012016

# The cambrian gog quartzite and deadwood formations of alberta: lessons for subsurface carbon sequestration

**Jon J Noad1,2** ✉

1 Stantec Consulting, 130-2886 Sunridge Way NE, Calgary, Alberta, Canada
2 University of Adelaide, Adelaide SA 5005, Australia
jonnoad@hotmail.com; jonathan.noad@stantec.com

**Abstract.** Injecting carbon dioxide into reservoirs has been undertaken since the 1940s, mostly as a method to enhance oil recovery. With the recognition of the impact of climate change, interest in carbon sequestration has increased exponentially, with hundreds of projects in progress around the globe. Many of these projects focus on Cambrian quartzites, which meet the requirements for carbon dioxide injection (CCS). Several CCS projects in Alberta, Canada, currently focus on using these reservoirs, including the Basal Cambrian Sandstone and Deadwood Formations, for injection and storage.

A series of spectacular outcrops of the Cambrian Gog Quartzite are available for study in the area around Lake Louise in the Canadian Rockies of Alberta. These outcrops provide compelling analogues to other clastic Cambrian formations in the province that have been utilized for CCS. As an example, the Gog Quartzite was used by Shell as a direct reservoir analogue when modeling CO2 injection into the Basal Cambrian Sandstone in the QUEST Project. The Gog Quartzite also has potential to act as a CCS reservoir in its own right and may reach 2 kilometres in thickness.

Outcrops of the Gog Quartzite were studied at Sink Lake, Spiral Tunnels and in the vicinity of Lake Louise in Alberta. They include fluvial and shallow marine, high net:gross sandstone beds with stacked trough cross-bedded sandstones. These are commonly interbedded with thin mudstone beds. Sedimentary structures include planar and trough cross-beds, wave ripples and contorted horizons interpreted as seismites with associated dewatering. Some units of the Gog Quartzite are heavily bioturbated with burrows including *Arenicolites* and *Skolithos* as well as other marine traces such as trilobite related ichnofacies.

The Cambrian Deadwood Formation is also a target for carbon dioxide sequestration and provides the main reservoir for the Aquistore Project in Saskatchewan. It has been targeted by exploration wells in eastern Alberta and a new core study has demonstrated its suitability as a CO2 reservoir. Facies ranging from regolith overlying igneous basement, through alluvial fan and tidal deposits, and out into shallow marine settings with well developed parasequences, are all seen in the cores.

The outcrop and core studies provide new data on Cambrian reservoirs which can be applied directly to potential Cambrian CCS targets in the subsurface of Alberta. The data can also be used to identify other potential Cambrian CO2 sinks around the world.

## 1 Introduction

### *1.1 Importance of CO2 injection*

There are many indications that increasing concentrations of atmospheric CO2 are contributing to global warming [1]. One way to mitigate this is to sequester CO2 (CCS) in the subsurface. Various methods are available to trap the gas, with the simplest being to utilize existing porous bedrocks in which to inject the gas. This proven technique has been utilized for more than 80 years for enhanced oil recovery (EOR), but in recent times has been employed as a stand alone option to sequester CO2. No less than 26 CCS projects are ongoing in Alberta, Canada, alone [2], with many more in development worldwide. Many of these projects are reliant on Cambrian reservoir rocks to host the sequestration of CO2, and both outcrop and subsurface core data are available as analogues in Alberta, improving our understanding of reservoir behaviour when used as CCS targets.

### *1.2 Requirements for CO2 injection*

Four main elements are required for successful CO2 injection (CCS) [3]. The first is a reservoir with sufficient porosity and permeability to allow ingress of supercritical fluids. Typical values for porosity are between 5 and 10%, while typical values for permeability range from 5 to 50 mD. Both carbonate and clastic reservoirs may exhibit such properties, with Cambrian clastic reservoirs providing some of the most suitable storage media for CCS projects in North America.

- Reservoir with sufficient porosity and permeability to allow ingress of supercritical fluids, but not so high as to allow unrestricted propagation of the injection front.
- Laterally extensive reservoir with relatively homogeneous properties.
- Depth of burial of 1 to 4 kilometres.
- Highly efficient seal, ideally with an associated "back up" secondary seal at a different stratigraphic horizon.
- Tectonically quiescent conditions.

### *1.3 Cambrian Reservoirs and CCS*

Cambrian formations such as the Basal Cambrian Quartzite [4] and the Deadwood Formation [5] meet the criteria listed above and have proven successful in projects such as Shell's QUEST (Alberta) and Aquistore (Saskatchewan) respectively. Both formations have the potential to sequester large quantities of CO2 and provide a template for the use of clastic Cambrian formations elsewhere around the globe for carbon sequestration.

Two new, complimentary studies on clastic Cambrian formations have been undertaken in Alberta to provide extra data on the depositional character and reservoir properties of such formations. The first was an outcrop study examining the Gog Quartzite in south central Alberta [6] [7], while the second examined a series of cores drilled through the Deadwood Formation in the subsurface of southeastern Alberta [8] [9]. The results are summarized below.

## 2 Outcrop study on the Gog Quartzite

The Gog Quartzite is a Cambrian clastic deposit outcropping around Lake Louise [6]. It provides an excellent analogue to the Basal Cambrian Sandstone (BCS), as demonstrated by its use as a direct analogue for modelling the behaviour of the BCS as the primary reservoir utilized for |Shell's QUEST CCS project in Alberta [10]. The BCS has excellent reservoir properties for CCS [11] including porosities up to 17% and suffient permeability to allow lateral propagation of the CO2 injection front over time [10]. While the Gog Quartzite is typically metamorphosed, leading to poorer reservoir properties, it provides an excellent Cambrian depositional analogue to other potential clastic Cambrian CCS reservoir rocks.

A series of outcrops of the Gog Quartzite were identified in south central Alberta (Figure 1) and fieldwork was undertaken to characterize the facies (Table 1) making up these potential reservoir rocks. Outcrops included those in Kicking Horse Pass (Sink Lake and Spiral Tunnels, both outcropping along Highway 1); Lake Louise; Taylor Lake; Moraine Lake and several smaller outcrops. Study of these outcrops allowed the development of a facies scheme that could then be tied directly to reservoir development.

Figure 1. View of Gog Quartzite outcrops used in this study: Moraine Lake (1); Sink Lake (2); Spiral Tunnels (3); Lake Louise (4); and Taylor Lake (5).

Table 1. Facies identified from outcrops of the Gog Quartzite.

| Facies | Interpretation | Characteristics |
|---|---|---|
| Trough cross-bedded sandstone with Arenicolites, Skolithos | Deposition in shallow marine setting as a result of longshore drift | Main reservoir facies for hosting liquid CO2 |
| Rippled sandstone | Wave rippled sandstone, | Possibly less porous |

| | | |
|---|---|---|
| | typically overlying trough cross-bedded sandstone. Part of an overall shallowing upward cycle | sandstone that may act as a baffle |
| Contorted sandstone | Response to dewatering of thick sandstone beds | Likely to act as a poor quality seal due to cementation associated with the agitation of the sandbody |
| Black shale with trilobite burrows, Rusophycos | Thin shales indicating marine flooding surfaces | Potential baffle |

The studied outcrops provided data on both the facies and facies associations (Table 1) [12]. Both fluvial and shallow marine quartzites were observed in the field, but here we will focus on the shallow marine deposits. The outcrops are typically very high net:gross with thicker, metre scale, trough cross-bedded quartzites and thin, cm scale mudstone beds. The quartzites may show evidence of dewatering and exhibit wave rippled surfaces. Trace fossils are common and are dominated by *Skolithos* and *Arenicolites* species [13] [14], confirming the shallow marine depositional setting. Shell's depositional model [10] interprets these deposits as fluvio-influenced tide-dominated bay margin, tide-dominated bay margin, proximal to distal bay and offshore deposits.

View of roadside outcrop at Sink Lake

Trough cross-bedded sandstone deposited in shallow marine conditions. Note the fracture running down the face.

Thin interbedded black mudstone beds

The excellent outcrops at Spiral Tunnels provide the opportunity to examine stacking patterns in sandstone beds.

Fig 2. Photographs showing the facies in the Gog Quartzite.

The outcrops also provided information on the structural data and how it related to the tectonic setting. Both open and closed fractures were observed, the majority of them perpendicular to bedding. In addition, several major folds were recorded in the thick, stacked quartzite beds exposed at the Spiral Tunnels.

| | |
|---|---|
|  |  |
| Large fold in stacked Gog Quartzite beds at Spiral Tunnels | Both filled and open fractures seen in outcrop |
|  | 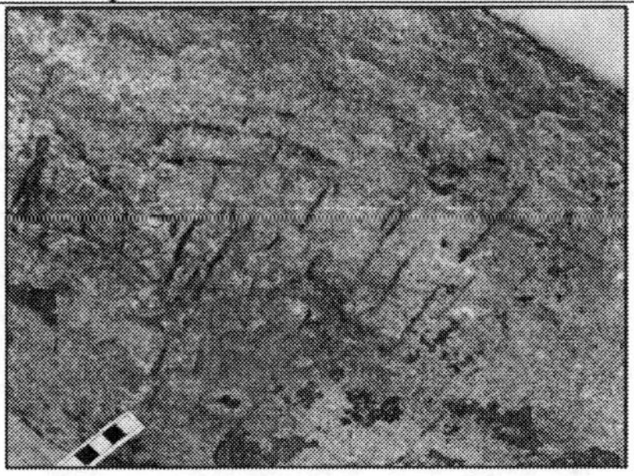 |
| Arenicolites trace fossils – lug worm burrows | Skolithos trace fossils (Pipe rock) |

Figure 3. Structural and trace fossil data in photographs from outcrops of the Gog Quartzite.

## 3 Core study on the Deadwood Formation

The Late Cambrian Deadwood Formation is a versatile reservoir, having produced oil and natural gas in North Dakota as well as being a proven helium producer in Saskatchewan. It is also a significant aquifer in some areas and has been mined for gold in the Black Hills of Dakota. It is currently being considered as a target for carbon sequestration [15].

Several deep cores were selected [8] to cover a broad spectrum of depositional settings within the Deadwood Formation in southern Alberta. The cores include the basal contact, where the Deadwood overlies regolith and underlying Precambrian, coarse grained gneiss. Other cores expose glauconitic quartz sandstone beds, varicoloured siltstones and interbedded green and purple, micaceous shales and the top contact of the Deadwood with overlying Devonian limestone beds. Deformation fabrics and fracturing were also recorded. These sediments are interpreted as having been deposited in a shallow marine to tidal setting, supported by sedimentary structures, fossil brachiopods and bioturbation including Skolithos and a variety of other burrows.

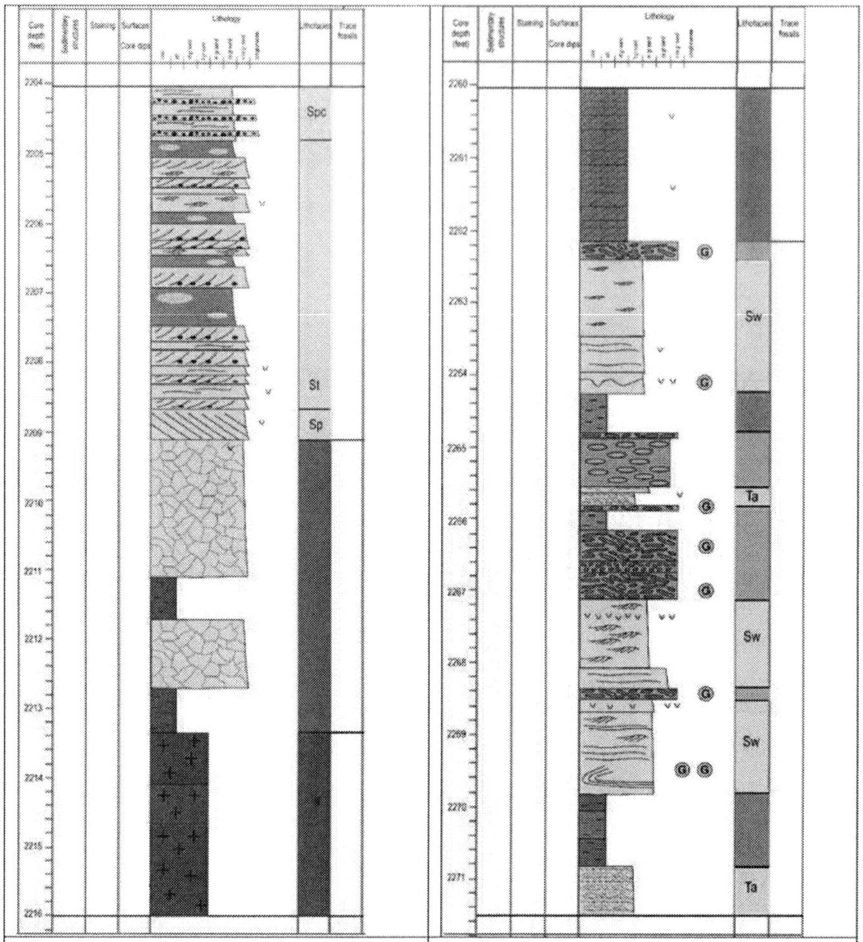


| | |
|---|---|
| Well 5-1-17-14W4 (2205 to 2218 m), interpreted as being deposited in a terrestrial setting with alluvial fans. | Well 14-35-34-17W4 (2066 to 2077 m), interpreted as being deposited in a tidal setting. |

Fig 4. Two wells cored through the Deadwood Formation showcasing different depositional settings. Facies key: St = trough xbs; Sp = planar xbs; Sw = wavy xlams; Ta = tidal algal; G = glauconite; Ig = gneiss; brown = regolith; orange = tidal, flat pebble conglomerate; light blue = Ordovician Dolomite [8].

The most proximal core (5-1-17-14W4: Figure 4) shows interbedded palaeosols and thin quartzite beds, interpreted to have been deposited in fluvial and alluvial fan settings. Core from Well 14-35-34-17W4 (Figure 4shows a variety of interpreted tidal settings. Cores from wells 9-8-39-26W4 and 1-27-60-26W4 (not shown) comprise interpreted, repeated, shallow marine parasequences, typically 1 to 2 m in thickness, passing up from mudstone to wavy laminated sandstone to trough cross-bedded sandstone. Few fractures were seen in the core.

## 4 Synthesis of outcrop and core data

Projects such as QUEST (Alberta [10] and Aquistore (Saskatchewan [17] demonstrate the efficacy of Cambrian reservoirs as targets for carbon sequestration. The analogues presented in this paper demonstrate that the reservoir facies reach 2000 m in thickness in the Gog Quartzite and 200 m thick in the Deadwood Formation. Reservoir properties are variable, with a focus on less metamorphosed deposits.

In the Gog Quartzite, the most prospective depositional settings are fluvial and shallow marine environments where the reservoir facies may form thick, stacked packages. Generally net:gross in these deposits is very high, with only very thin mudstone partings separating thick, stacked sandstone packages. Packages of trough cross-bedded sandstone may reach more than 100 m in thickness. Despite the age and tectonic overprint, the fractures in the Gog are not pervasive and may be open or closed. It is envisaged that fraccing might be necessary to open up fractures within this relatively cemented reservoir.

The Deadwood Formation is more heterogeneous [8] with a variety of depositional settings (observed in cores) ranging from alluvial fan deposits, tidal deposits and shallow marine clastic parasequences. This facies association has proven successful for CCS and both alluvial fan deposits and shallow marine deposits are considered highly prospective for sequestration. Fractures are relatively rare in the Deadwood Formation in the subsurface.

Cambrian clastic deposits have been recognized in many locations around the world including China [17] [18], France [19], South Africa [20] and Scotland [21]. In light of the success in utilization of similar sedimentary deposits in western Canada, similar reservoirs worldwide should be considered as targets for sequestration. The most prospective depositional settings are fluvial and shallow marine environments where the reservoir facies may form thick, stacked packages.

## 5 Conclusions

Numerous carbon sequestration projects are in development or already in operation in western Canada. Many of these projects are utilizing clastic Cambrian reservoirs. The Cambrian Gog Quartzite (in outcrop) and Deadwood Formation (in core) provide excellent analogues to these subsurface reservoirs in terms of these reservoir character and properties. The laterally extensive nature of the outcrops of the Gog Quartzite provides data that cannot be viewed in the subsurface, and the data available in outcrop and core has already been utilized for reservoir modeling in existing CCS projects. Finally it is strongly recommended that other coutries around the world need to revisit their own clastic Cambrian deposits as these are highly prospective as the CCS reservoirs of the future.

**Acknowledgments.** The project is supported by Stantec Consulting.

## References

[1] Gore, A.A, 2006. An Inconvenient Truth: The Planetary Emergency of Global Warming and What We Can Do About It. Published by Rodale Press, Emmaus, Pennsylvania.
[2] https://cdl.canadiandiscovery.com/corporate/products-services/carbon-capture-utilization-and-storage-ccus
[3] https://jpt.spe.org/carbon-sequestration-pick-the-right-reservoir

[4] Desjardins, P. and Smith , M. 2013 The Basal Cambrian Sandstone in the Subsurface of Alberta: Quest Carbon Capture and Storage Injection Target. Conference abstract. GeoConvention, Calgary, Alberta.
[5] Slind, O.L., Andrews, G.D., Murray, D.L., Norford, B.S. and Paterson, D.F. (1994): Middle Cambrian to Lower Ordovician Strata; in: Geological Atlas of the Western Canada Sedimentary Basin, G.D. Mossop and I. Shetsen (comp.), Canadian Society of Petroleum Geologists and Alberta Research Council.
[6] Hein, F.J. and McMechan, M.E. 1994. Proterozoic and Lower Cambrian Strata of the Western Canada Sedimentary Basin. in: Geological Atlas of the Western Canada Sedimentary Basin, G.D. Mossop and I. Shetsen (comp.), Canadian Society of Petroleum Geologists and Alberta Research Council.
[7] Noad, J.J. 2023. The Gog Quartzite: CCS Analogue Extraordinaire and Potential Target. Extended abstract. Geoconvention, Calgary, Alberta.
[8] Noad, J.J. 2022. The Mighty Deadwood: Alberta's most versatile reservoir. Extended abstract. CSPG Core Conference, Calgary, Alberta.
[9] Greggs, D. H. (2000). The stratigraphy, sedimentology, and structure of the lower paleozoic deadwood formation of western Canada (Unpublished master's thesis). University of Calgary, Calgary.
[10] Winkler, M. Generation-4 Integrated Reservoir Modeling Report: Quest CCS Project (2011).
[11] Weides, S., Moeck, I., Majorowicz, J. and Grobe, M. 2014. The Cambrian Basal Sandstone Unit in central Alberta — an investigation of temperature distribution, petrography, and hydraulic and geomechanical properties of a deep saline aquifer. Canadian Journal of Earth Sciences (2014) 51 (8): 783–796.
[12] Hein, F.J. 1987: Tidal / littoral offshore shelf deposits-Lower Cambrian Gog Group, southern Rocky Mountains, Canada. Sedimentary Geology 52, 155–182.
[13] Johnston, P.A., Collom, C.J., and Desjardins, P., 2017, Lower to Middle Cambrian of the southern Canadian Rockies, in Hsieh, J.C.C., ed., Geologic Field Trips of the Canadian Rockies: 2017 Meeting of the GSA Rocky Mountain Section: Geological Society of America Field Guide 48, p. 71–121.
[14] Desjardins, P.R., Ma´ngano, M.G., Buatois, L.A. & Pratt, B.R. 2010: Skolithos pipe rock and associated ichnofabrics from the southern Rocky Mountains, Canada: colonization trends and environmental controls in an early Cambrian sand-sheet complex. Lethaia Vol. 43, pp. 507–528.
[15] Nesheim, T.O. 2021. The Deadwood Formation: A Potential Stratigraphic Unit for CO2 Sequestration. North Dakota Geological Survey. January 2021 Newsletter. 11-13.
[16] Natural Resources Canada 2013. Aquistore Project, a Deep Saline CO2 Storage Demonstration Project. Published flyer.
[17] Gu, Z., Jian, X., Watts, A.B., Zhai, X., Liu, G. and Jiang, H. 2023. Formation and evolution of an Early Cambrian foreland basin in the NW Yangtze Block, South China. Journal of the Geological Society, 180 (3).
[18] Chough, S.K. et al. 2010. Cambrian stratigraphy of the North China Platform: Revisiting principal sections in Shandong Province, China. Geosciences Journal 14(3):235-268.
[19] Went, D. and Andrews, M. 1991. Alluvial fan, braided stream and possible marine shoreface deposits of the Lower Palaeozoic Erquy-Fréhel group, northern Brittany. Read at the Annual Conference of the Ussher Society, January 1991.
[20] Whaley, J. 2010. The Cape Peninsula. Article in Geo-Expro Magazine.
[21] Hallam, A. and Swett, K. 1996. Trace fossils from the Lower Cambrian Pipe Rock of the north-west Highlands. Scottish Journal of Geology, vol. 2, pp. 10-107.

Journal of Physics: Conference Series 2594 (2023) 012017 doi:10.1088/1742-6596/2594/1/012017

# Research on the Fault Diagnosis Method of Reciprocating Compressors Based on the Probability Diagram Model

**Wencai LIU** [1(✉)], **Xiaofei WANG** [2], **Jinxing ZHANG** [3], **Huiyong ZHENG** [1]
1 CNPC Research Institute of Safety & Environment Technology, Beijing, China
2 The 12th oil production plant of CNPC Changqing Oilfield Company, Xian, China
3 Petro China Tarim Oilfeld Company, Korla, China
zhangjx@petrochina.com.cn
4 CNPC Research Institute of Safety & Environment Technology, Beijing, China

wencailiu@cnpc.com.cn

**Abstract.** A reciprocating compressor is one of the key pieces of equipment in the petrochemical production industry. Its structure is complex and there are many sources of excitation causing vibration, making the analysis of the causes of failure challenging. Depending upon fault types, the failures in reciprocating compressors can be classified as leakage, wear, fracture, loosening, impact, blockage, etc.; depending upon where the faults occur, the failures can be classified by components affected such as a valve type, a transmission component type, a sealing component type, etc. This paper proposes a parameter self-learning method for diagnosing mechanical faults. With a probability diagram model for the "fault-sign" relationship on the equipment established according to Bayesian networks, the paper proposes a parameter self-learning method in which, based on the failure rate curve, the prior probability is determined subject to the double-parameter Weibull distribution while the conditional probability can be adjusted automatically according to the diagnosis results. After the probability diagram diagnosis model is built, it collects real-time operating data of the equipment, extracts eigenvalues, sets alarm thresholds, and enters into the diagnosis model the comparison results between the eigenvalues and alarm thresholds as the node states in the sign layer. The changes of the node states in the sign layer will change the posterior probabilities of the corresponding nodes in the fault layer so as to deliver the set of diagnosis results with the highest probability of fault occurrence, thus achieving the purpose of diagnosing the equipment fault.

## 1 Introduction

Reciprocating compressors are core equipment in oil refining, chemical and other process industries. However, because of their complex structure and multiple wearing parts, typical failures occur from time to time such as piston rod rupture, cylinder knocks, packing wear, breakage of connecting rod bolts and valve discs, etc. The event of failure may affect the stable operation of the compressor, threatening the safety of production, or even result in equipment shutdown and process interruption, causing serious economic losses[1]. Therefore, monitoring the conditions of key components of the compressor, studying the fault diagnosis, and intelligently diagnosing the failure of reciprocating compressors can effectively reduce economic losses and avoid major safety accidents, which is important for the safe production of enterprises.

## 2 Typical Failures in Reciprocating Compressors

The movement form of reciprocating compressors is relatively complex, and there exist many problems such as multiple vulnerable parts, many excitation sources that cause vibration, and the difficulty in distinguishing the failure types[2]. Table 1 lists some common failures in reciprocating compressors; Table 2 lists the fault-sign relationship for reciprocating compressors.

**Table 1.** Typical failures in reciprocating compressors.

| Fault Type | Faulty Component | Failure Name |
|---|---|---|
| Valve failure | Valve springs | Failed valve springs |
| | Valve discs | Broken valve discs |
| Transmission component failure | Crankshaft drive shaft | Imbalance |
| | Couplings and crosshead slides | Misalignment |
| | Connecting rod bolt | Broken connecting rod bolt |
| | Crosshead pins | Broken crosshead pin |
| | Crosshead fastening threads of piston rod | Loose crosshead fastening thread of piston rod |
| Piston assembly failure | Piston fastening threads of piston rod | Loose fastening threads of piston rod |
| | Support ring | Worn support ring |
| | Support ring | Fractured support ring |
| | Piston rod | Worn piston rod |
| | Piston rod | Broken piston rod |
| | Piston ring | Worn piston ring |
| | Piston ring | Broken piston ring |
| Cylinder failure | Cylinder block | Cylinder knocks |
| | Cylinder block | Cylinder scoring |

**Table 2.** Fault-sign relationship table for reciprocating compressors.

| Fault Type | Correlating Signs |
|---|---|
| Shaft misalignment | Whether the crankcase vibration RMS is alarmed; |
| | Whether the double-fold frequency of the vibration waveform is dominant; |
| Imbalance | Whether the crankcase vibration RMS is alarmed; |
| | Whether the single-fold frequency of the vibration waveform is dominant; |
| Bearing wear | Whether the crankcase vibration RMS is alarmed; |
| | Whether the mean value of the piston rod displacement is alarmed; |
| Intake valve failure | Whether the temperature of the intake valve is alarmed; |
| | Whether the crosshead vibration is alarmed at the peaks of 40°±10° and 220°±10°; |
| Exhaust valve failure | Whether the temperature of the exhaust valve is alarmed; |
| | Whether the crosshead vibration is alarmed at the peaks of 120°±10° and 300°±10°; |
| Piston assembly failure | Whether the mean value of the piston rod displacement is alarmed; |
| | Whether the peak-to-peak value of the piston rod displacement is alarmed; |
| Piston rod rupture | Whether the peak-to-peak value of the piston rod displacement is alarmed; |
| | Whether the fast change value of the piston rod displacement is alarmed; |
| | Whether the crosshead vibration is alarmed at either peak 0° or 180°; |
| Loose components | Whether the crankcase vibration RMS is alarmed; |
| | Whether the peak-to-peak value of the piston rod displacement is alarmed; |

| | |
|---|---|
| | Whether the crosshead vibration is alarmed at either peak 0° or 180°; |
| | Whether the crankcase vibration RMS is alarmed; |
| | Whether the peak-to-peak value of the piston rod displacement is alarmed; |
| Cylinder knocks | Whether the crosshead vibration is alarmed at the either peak 0° or 180°; |
| | Whether the crankcase vibration RMS is alarmed; |
| | Whether the peak-to-peak value of the piston rod displacement is alarmed; |
| Cylinder scoring | Whether the crosshead vibration is alarmed at either peak 0° or 180°; |
| | Whether the crankcase vibration RMS is alarmed; |

## 3 Fault Diagnosis Based on the Probability Diagram Model

The probabilistic graphical model is the product of a close combination of probability theory and graph theory[3]. It is a model that represents probabilistic relationships among variables in a graphical way, consisting of nodal variables and edges (correlations between variables). Probabilistic graphical models can not only succinctly represent the complex relationship among variables but also reason and learn to accurately diagnose equipment failures. Depending upon whether the edge is a one-way or two-way arrow, probabilistic graph models can be classified into two broad categories: one using directed acyclic graphs to represent the dependencies among variables, is called directed graph models (Bayesian networks), and the other is acyclic graph models. Among them, Bayesian networks, powerful in inferring uncertainties and expressing knowledge, have been successfully applied in many fields, such as fault diagnosis, reliability analysis, medical diagnosis, etc. Therefore, this paper applies the Bayesian network in probabilistic graphical models to diagnose failures in reciprocating compressors. In the Bayesian network structure, each node represents a variable, and the nodes that are correlated are connected by one-way arrows pointing from the parent node ("cause") to the child node ("effect"). In addition, conditional probability is used in the Bayesian network to represent the degree of mutual influence between these two nodes. For example, Figure 1 shows a simple Bayesian network. Due to its graphical structure, the Bayesian network can fuse information from multiple sources and represent knowledge in a concise and straightforward way[4]. Bayesian networks, powerful in addressing uncertainty problems, can effectively reduce diagnosis time and labor. For this reason, they have been applied in the fault diagnosis of various mechanical devices.

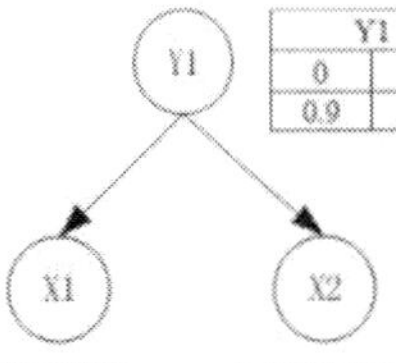


| | Y1 |
|---|---|
| 0 | 1 |
| 0.9 | 0.1 |

| | | X1 | |
|---|---|---|---|
| | | 0 | 1 |
| Y1 | 0 | 0.85 | 0.15 |
| | 1 | 0.1 | 0.9 |

| | | X2 | |
|---|---|---|---|
| | | 0 | 1 |
| Y1 | 0 | 0.6 | 0.4 |
| | 1 | 0.3 | 0.7 |

**Fig. 1.** Example of Bayesian network

Bayesian network-based reasoning refers to solving problems in real processes based on Bayesian networks, including diagnostic reasoning and causal reasoning. The former is to deduce the cause from the result, reasoning from the bottom to the top. In Bayesian networks, the prior probability of the parent node and the conditional probability of each child node need to be determined. If each element in the variable set $U = (x_1, x_2, ..., x_n)$ corresponds to a node in Bayesian networks, the joint probability density $P(x_1, x_2, ..., x_n)$ shall be:

$$P(U) = P(x_1, x_2, ..., x_n) = \prod_{i=1}^{n} (x_i|\pi_i)$$

And the marginal probability of Xi shall be:

$$P(X_i) = \sum_{expx_i} P(U)$$

The Bayesian network-based fault diagnosis method mainly refers to calculating the posterior probability of each fault based on the observed evidence obtained. The posterior probability of a fault is calculated as shown in Equation 1:

$$P(U|e) = \frac{P(U,e)}{P(e)} = \frac{P(U)P(e|U)}{P(e)} (1)$$

Finally, the fault is located by comparing the posterior probabilities of the parent nodes[5]. The two main aspects for building a Bayesian network are as follows:

(1) Determine the node information based on prior knowledge (e.g., failure cases, etc.) to establish the structure of a Bayesian network;

(2) Analyze the fault samples to calculate the parameters of the Bayesian network.

The probabilistic graphical diagnosis model this paper studies is achieved via Bayesian networks. This method, combining intelligent algorithms and traditional expert experience, can accurately diagnose the mechanical failures of equipment.

Bayesian networks, as the critical part of the fault diagnosis method based on the probabilistic graphical model, includes the fault layer, the sign layer, the prior probability set, and the conditional probability set. The probabilistic graphical model M established for the fault diagnosis in reciprocating compressors is shown in the following equation:

$$M = (F, S, E, P, C)$$

Symbol Description:

F: the set of fault nodes, $f_i$ stands for the $i$th fault node. $F = \{f_i\}_{i=1,...,M}$;

S: the set of all sign nodes, $s_j$ stands for the $j$th sign, $s_j \in S$, $S = \{s_j\}_{j=1,...,N}$ S

E: connecting lines/edges in the set of sign nodes and fault nodes, $e_{i,j}$ stands for the $i$th fault correlated with the $j$th sign.

The set of prior probabilities P: $p_i$ stands for the prior probability of the $i$th fault.

The set of conditional probabilities is C: $C_i$ stands for the conditional probability table of the $i$th fault node.

The setting of the parameters of the probabilistic graph model mainly includes the set of prior probabilities P and the set of conditional probabilities C. The value of Pi is related to the fault type, and Ci represents the reinforcement or attenuation of the signs on the occurrence of the corresponding fault, which is mainly related to the fault mechanism. The probabilistic graph model, by self-learning the fault sample data, determines the set of prior probabilities P of each fault node and the set of conditional probabilities C of its corresponding sign node.

$$P_i = \frac{P(f_i Number\ of\ fi\ occurrences\ )}{P(Total\ of\ events)}$$

The probability diagram model delivers the probability of each fault type, but for a troubleshooting engineer, only the most probable faults are needed. Therefore, the diagnosis result output module further processes the posterior probabilities of all faults delivered by the probability diagram model. The result output layer, based on the probability of each fault type delivered by the dynamic probability diagram model, adjusts the fault types that need to be shown to the relevant staff, thus assisting the personnel working for petrochemical enterprises in deciding on fault diagnosis.

During equipment fault diagnosis[6], the monitoring system monitors the real-time working conditions of each component during equipment operation and transmits the obtained parameter data to the sign extraction layer in time. Next, this layer processes and transforms the data, and enters the set of various sign states into the probability diagram model. Then the fault diagnosis model based on the probability diagram model automatically reasons and calculates the occurrence probability of each fault, and delivers the diagnosis conclusion through the diagnosis result output module. Finally, relevant personnel evaluates the diagnosis conclusion and saves the fault information into the fault storage module[7] [8]. The fault diagnosis method based on the probability diagram model combines expert knowledge and fault diagnosis rules, so the causal relationship between faults and signs is clear and explicit. Moreover, the method determines the connection strength between each fault and the sign according to the principle of probability statistics, and combines the prior knowledge with the posterior information, thus avoiding the influence of subjective bias and other factors and improving the accuracy of fault diagnosis results.

### *3.1 Parameter Setting of the Probability Map Model*

(1) Establish guidelines for setting up conditional probability tables based on correlation/association levels:

a. The conditional probability table is confirmed according to the correlation level, and the corresponding probability of each level is as follows, i.e., the probability of the corresponding fault if the sign of that level exists.

Level 1: 90%; Level 2: 80%; Level 3: 70%; Level 4: 60%

b. The two or more signs in the conditional probability table are complementary and form a sign group, i.e., a changing trend reflected. In the process of fault establishment, the fact that two or more signs occur will not enhance the probability. Its corresponding conditional probability table: the probability of the highest occurrence level in the sign group shall dominate.

c. The probability of each sign and sign group in the conditional probability table is calculated by referring to the "OR" relationship.

(2) Design of the correlation between sign characteristics and faults

The correlation/association of sign characteristics with faults is an important basis for setting/updating the parameters of the conditional probability table.

a. Correlation types

Correlation types: positive correlation, negative correlation, and levels of correlation.

b. Classification and definition of correlation/association levels

① Level 1, highly correlated. The correlation coefficient reaches: 0.9 or more

② Level 2, strongly correlated. The correlation coefficient reaches: 0.75~0.9 or more
③ Level 3, generally correlated. The correlation coefficient reaches: between 0.5 and 0.75
④ Level 4, weakly correlated. The correlation coefficient reaches: 0.5 or more

c. Principles/methods for determining the correlation

The basis for determining the level: in principle, the consistency between the sign change and the fault occurrence, or a state that reflects the presence or absence of a fault.

### *3.2 Personalized Threshold Configuration*

As the operation varies from compressor to compressor, the use of standard characteristic thresholds will affect the accuracy of diagnosis[9] [10]. Therefore, we propose a personalized method for calculating the thresholds. To calculate the eigenvalue for each normal datum of reciprocating compressors: firstly, obtain the monitoring data set R under normal operating conditions for a certain period (including 100 sets of whole-cycle monitoring data on the equipment); secondly, randomly take 10 sets of equally spaced adjacent data from the 100 sets of whole-cycle data to form the data set N; thirdly, calculate the eigenvalue for each group of data in N to obtain the eigenvalue list Q; fourthly, calculate the confidence interval of the eigenvalues based on the 3σ criterion to obtain the personalized threshold file set M for the device; finally, regard this as the standardized threshold M for each device and diagnose the fault of the device with M thereafter.

### *3.3 Derivation of Diagnostic Results with the Probabilistic Graphical Model*

The probability diagram model is used for the calculation. To illustrate the method specifically, the following will show the calculation with the example of fault A. The structure of part of the probability diagram model of the valve failure is shown in Figure 2, but in practice, it is not necessary to calculate manually in this way. Instead, after the structure of a probabilistic graphical model is constructed in the computer, the model can directly deliver the diagnostic result if a sign variable is entered.

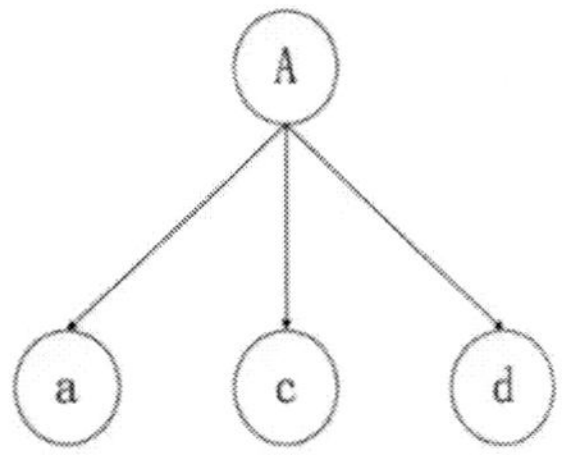


**Fig. 2.** Probabilistic model diagram for fault A

After building the above Bayesian network model, the response probability parameter values need to be filled in. Table 3 below shows the conditional probability table based on Bayesian networks for a valve failure.

**Table 3.** Conditional probability table based on Bayesian networks for valve failure.

| | a | | c | | d | |
|---|---|---|---|---|---|---|
| A | 0 | 1 | 0 | 1 | 0 | 1 |
| 0 | 0.8 | 0.2 | 0.6 | 0.3 | 0.6 | 0.3 |
| 0 | | 3 | 7 | 5 | 5 | |
| 1 | 0.1 | 0.8 | 0.2 | 0.7 | 0.2 | 0.7 |
| 2 | 8 | 3 | 7 | 1 | 9 | |

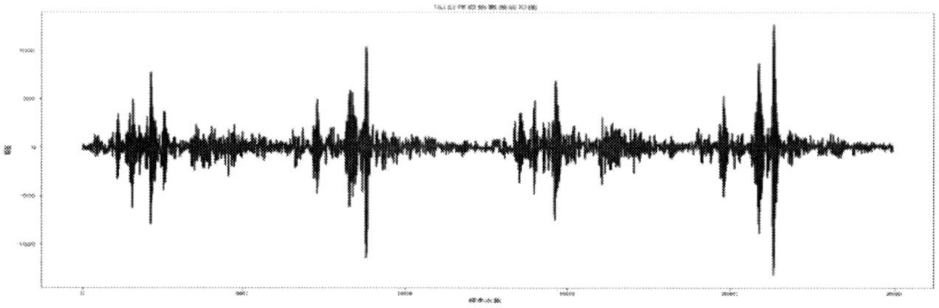

**Fig. 3.** Waveform diagram for valve failure

A Bayesian network model can be built with the Netica software. When the node state input is changed, an equipment fault can be diagnosed with the possible fault type and its probability of occurrence delivered. In Netica software when prior probability and conditional probability are entered into the Bayesian network model in Netica software to calculate the valve failure probability (as shown in Figure 3), joint probabilities of each node in the sign layer will be directly obtained. Table 4 below shows the joint probability table of the nodes in the sign layer.

**Table 4.** Joint probability table of sign nodes in Bayesian network for valve failure.

| Sign Node | Node State | Probability |
|---|---|---|
| a | a=0 | 0.644 |
| | a=1 | 0.356 |
| c | c=0 | 0.538 |
| | c=1 | 0.462 |
| d | d=0 | 0.549 |
| | d=1 | 0.451 |

From this, it can be seen that when the state of the sign node is known, the probability of its corresponding fault can be obtained, as shown in Equation 2 below. When all three sign nodes including the input valve temperature, crosshead opening angle range vibration amplitude, and crosshead closing angle range vibration amplitude all exceed their thresholds respectively, the probability of the valve failure is 86.1%.

$$P(A=1|a=1,c=1,d=1) = \frac{P(A=1)P(a=1,c=1,d=1|A=1)}{P(A=0)P(a=1,c=1,d=1|A=0)+P(A=1)P(a=1,c=1,d=1|A=1)} = \frac{0.23\times 0.88\times 0.77\times 0.79}{0.77\times 0.2\times 0.37\times 0.35+0.23\times 0.88\times 0.77\times 0.79} = 0.861 \quad (2)$$

In addition, according to the working features of reciprocating compressors, this paper takes into consideration the characteristics of vibration signals of the equipment before and after the failure during the fault diagnosis to deliver a final result by comprehensively analyzing multi-point diagnosis results.

**Table 5.** Fault diagnosis results.

| Serial No. | Case Type | Number of Cases | Accuracy of Diagnosis |
|---|---|---|---|
| 1 | Valve failure | 31 | 100% |
| 2 | Piston assembly failure | 11 | 81.81% |
| | | 6 | 83.30% |
| 3 | Crosshead wear | 7 | 85.71% |
| 4 | Bearing bush wear | 3 | 100% |
| 5 | Loose connecting rod bolt | 2 | 100% |
| 6 | Cylinder scoring | 1 | 100% |
| 8 | Piston rod rupture | 2 | 100% |
| 11 | Fluid accumulation | 2 | 100% |
| 12 | Loose piston pin | 1 | 100% |

## 4 Conclusion

This method is validated by using the data from engineering cases. The fault diagnosis method based on the probability diagram model diagnoses a variety of failures in reciprocating compressors, yielding good results. This fault diagnosis method, if applied to industrial production, can effectively guide the field production personnel in maintenance inspection and fault diagnosis, and thus maintenance and production costs can be reduced.

## References

[1]Roberts T M.,Talebzadeh M.:Acoustic emission monitoring of fatigue crack propagation [J].Journal of Constructional Steel Research,2003, 59（6）:695-712.

[2]Parker A P.:The mechanics of fracture and fatigue[J].London：E and F N Spon Ltd,1981.

[3]Gong Z,Duquesnay D L, S L.Measurement and interpretation of fatigue crack growth in7075Aluminum alloy using acoustic emission monitoring[J].Journal of Testing and Evaluation,1998,26（6）:567-574.

[4]Roberts T M,Talebzadeh M． Fatigue life prediction based on crack propagation and acoustic emission count rates [J]． Journal of Constructional Steel Research,2003,59：679-694.

[5]Lademo O G,Hopperstad O S,Langseth M . An evaluation of yield criteria and flow rules for Aluminum alloys[J]. International Journal of Plasticity,1999,15（2）:191-208
[6]Pearl J. Fusion, propagation, and structuring in belief networks[J]. Artificial Intelligence, 1986, 29(3):241-288.
[7]Pearl J. Probabilistic Reasoning in Intelligent Systems[M]. Palo Alto: Morgan Kaufmann, 1988.
[8]Ibrahim H , Patruni J R . Bayesian network-based failure analysis of fire safety barriers in floating LNG facility[J]. SN Applied Sciences, 2019.
[9]Cooper G F, Herskovits E. A Bayesian method for the induction of probabilistic networks from data[J]. Machine Learning,1992,9(4).
[10]Friedman N. Learning Belief Networks in the Presence of Missing Values and Hidden Variables[M]. Morgan Kaufmann Publishers Inc. 1997.

# Study on Simulation for Erosion-Corrosion in Petrochemical Process Piping

Wencai LIU (✉), Yajun LI, Gen TONG, Minhui ZHANG
wencailiu@cnpc.com.cn

**Abstract.** Erosion-corrosion is a damage to metal caused by relative motion at high velocity between metal surface and erosive/corrosive fluid, resulting from combined actions by mechanical erosion and electrochemical corrosion. Since petrochemical plants install numerous process pipes, especially involving complex oil refining processes and multi-phase transformation between solid, liquid and gas, the erosion-corrosion becomes one of major problems that disturb industrial work safety. So, it is important to create a mechanistic simulation model by fully considering various factors and closely simulating erosion-corrosion process.

## 1 Researches on Erosion-Corrosion

Erosion-corrosion in the bends for solid particle-containing multi-phase fluid is a very complicated process. Gas/liquid flow regimes in piping are highly variable, leading to complicated spatial distribution of solid particles. Meanwhile, motion and distribution of particles directly affect erosion-corrosion patterns in piping. In recent years, Chinese and foreign scientists have made a number of researches about erosion-corrosion in piping, focusing mainly on two respects, namely experimental study and numerical computation[1]. In the respect of experimental study, the researchers systematically investigate erosion-corrosion profiles in bends under various flow regimes including two-phase gas/solid flow, two-phase liquid/solid flow, annular flow and churn flow. In the respect of numerical computation, the scientists mainly investigate erosion-corrosion rates in the bends for two-phase fluid (gas/solid or liquid/solid) in consideration of different factors; as for the computation about multi-phase fluid, they mainly design appropriate simplification methods, allowing for simplified computation on gas and liquid phases. These researches have produced many valuable outcomes, promoting development and improvement of the theories in erosion-corrosion[2].

Numerical computation on erosion-corrosion began earlier in foreign countries. Since 1990s, computational fluid dynamics (CFD) has been applied for this purpose. Many theoretical prediction models previously created by scientists have been used for numerical computation, further advancing the theories in erosion-corrosion. Edwards et al. investigated erosion-corrosion rates in two types of 90° bend with curvatures of 1.5 and 5 respectively by using Ahlert's model, finding that elliptical erosion-corrosion pattern is generated at outer arc of bends; they believed that it is caused by intense impingement of particles onto walls. Chen et al. investigated the erosion-corrosion patterns in standard bends with curvature of 1.5 and the trajectory of solid particles in piping; they found that V-shaped pattern appears in the bends, and believed that it results from secondary impingement by solid particles. Pereira et al. calculated erosion/corrosion rates in 90° bends using the particle-wall impingement/reflecting model developed by Grant and Tabakoff; they also found V-shaped pattern in the bends, and believed it is mainly due to sliding friction between the particles at low velocity and the walls. On the foundation of Pereira et al.'s study, Duarte et al. made further research, finding that two factors, i.e. the mass flow rate of particles and how particles are coupled with gas, will notably affect erosion-corrosion patterns; in case of 1-way coupling, V-shaped pattern will appear in the bend, while in case of 4-way coupling, erosion-corrosion pattern will be not obvious; moreover, they found that as the mass flow rate increases, the pattern generated by computation based on 4-way coupling becomes vaguer; it was believed that this finding is mainly caused by the interaction between solid particles. Solnordal et al. found that for smooth walls, the erosion-corrosion pattern corresponds with the trajectory of particles. Chinese researches on erosion-corrosion in the piping for two-phase gas/solid fluid focus mainly on how different factors affect erosion-corrosion rates. Lin Nan investigated erosion-corrosion profiles in bends and tees along high-pressure gas pipelines; by assessing the effects of flow velocity, erosion angle, temperature level and particle feature on the erosion-corrosion in bends, the author identified the positions with heaviest erosion-corrosion under different conditions, thus providing guidance for optimizing piping design and mitigating erosion-corrosion. Yu Fei et al. addressed the bends for two-phase gas/solid; by using Eulerian multi-phase flow model and discrete phase model (DPM), they investigated the head loss in bends and the erosion-corrosion on walls at different parameters such as bending radius ratio, bending angle and cross-section shape; they also recommended the parameter ranges for mitigating erosion-corrosion[3]. Qian Dongliang investigated a real case – Panyu 35-2 subsea gas pipeline; the author predicted erosion-corrosion levels in tees, valves and jumpers along this pipeline by using the software Fluent, and assessed the effects of flow velocity, mass flow rate, shape factor and particle size of sand onto the erosion-corrosion through grey relational analysis.

According to above-mentioned information, in experimental studies on erosion-corrosion in the bends for two-phase gas/solid fluid, Chinese and foreign scientists mainly apply non-destructive testing and erosion-corrosion probes to determine erosion-corrosion rates at different positions. Although above-mentioned methods can discover erosion-corrosion distribution at typical positions on outer arc, it is somewhat difficult to accurately depict 3D patterns on full bend surface in consideration of different factors[4]. Due to the difference in motion status of solid particles, the erosion-corrosion patterns in bends are diverse. The research outcomes about mechanism of erosion-corrosion patterns are not enough, and no consensus has been reached. Since the motion status of solid particles, the distribution of particles and the distribution of gas/liquid phases directly affect erosion-corrosion patterns, the investigation into relation between these three factors is of great importance for discovering the mechanism of erosion-corrosion by multi-phase flow[5]. In order to address this problem, it is necessary to develop a proper, complete and accurate computational method for assessing erosion-corrosion in piping by multi-phase flow.

## 2 Erosion-Corrosion Models

Erosion-corrosion by two-phase flow in piping is a complicated process. Many variable factors may affect erosion-corrosion patterns and rates. In addition, erosion-corrosion assessment involves the interaction between liquid and solid particles and the interaction between solid particles and walls, leading to complicated computation. To accurately predict erosion-corrosion by particles in piping, we need to address two problems: (1) the trajectory of particles in piping and the velocity and angle at which particles impinge walls; (2) accurate calculation of material loss due to impingement by particles. The modelling includes step-by-step calculation as per erosion-corrosion prediction procedure; in-depth assessment on the effects of inner flow field, particle motion and different variable factors on the erosion-corrosion; development of appropriate particle impingement model for discovering erosion-corrosion mechanism; creation of the models for calculating erosion-corrosion rates and for predicting most heavily eroded/corroded positions. In existing erosion models, metal loss is quantified with erosion-corrosion rate. Erosion-corrosion rate is defined as: lost mass of wall material in unit period on unit area, expressed in $kg/m^2 \cdot s$. Overall erosion-corrosion rate is determined by calculating accumulative damage on walls by each particle:

$$E_f = \frac{1}{A_f} \sum_{\pi(f)} \dot{m}_\pi e_r \#(1)$$

$A_f$—— unit area of mesh;

$m_\pi$—— mass flow rate of particles that impinge walls;

$e_r$—— erosion-corrosion rate.

Herein, we can know that, in existing erosion-corrosion models, the damage level is calculated with erosion-corrosion rate based on mass loss in unit period on unit area. As can be understood apparently, this equation is a simplified model which only considers the effect of flow status of particles on the erosion-corrosion rate. So, other scientists make improvement based on this model, and develop different mathematic models for calculating erosion-corrosion rate by considering different factors.

DNV corrected model for erosion-corrosion is:

$$e_r = Kf(\theta)\left(\frac{u_{rel}}{u_{ref}}\right)^n \#(2)$$

$K$—— material-related coefficient;

$f(\theta)$—— impact angle function of particle; $\theta = [0, \pi/2]$, expressed in radian;

$u_{rel}$—— relative velocity between particle and wall;

$u_{ref}$—— reference velocity constant of particle;

$n$—— exponent.

In this model, we can conclude: in contrast to Ahlert's model, DNV model does not specifically consider particle shape, so that it is more simplified. In default data experiments, the coefficient for DNV corrected model is generated through the experiment about erosion on carbon steel by air-driven sand. DNV corrected model does not present the effect based on particle size.

Archard Correlation corrected model for erosion-corrosion is:

$$E_f = \frac{1}{A_f dt} \sum_{\pi(f)} e_r \#(3)$$

$$e_r = \alpha F s \#(4)$$

$dt$—— increment of flow duration;

$e_r$—— abrasion rate;

$\alpha$—— abrasion coefficient;

$F$—— normal force;

$s$—— slip distance.

This model mainly addresses the abrasion when particle touches wall in tangent direction or impinges wall at low angle. It is mainly intended to calculate abrasion rate of material eroded at low angle. In addition, this model does not consider the factors such as impact angle, particle shape and particle size; it only considers the normal force onto metal exerted by particle and the slip distance driven by this force. Herein, default value of abrasion coefficient is 0.01 kg/J, corrosion level being slight to severe.

Finnie's model for erosion-corrosion:

$$E = kV_p^n f(\gamma) \#(5)$$

$k$—— non-dimensional mass;

$V_p$—— particle impact velocity;

$f(\gamma)$—— non-dimensional function of impact angle;

$\gamma$—— radian.

Impact angle is the angle between particle impact direction and wall; in case of metal material, the exponent n is often set as 2.3~2.5. Finnie model defines erosion-corrosion rate as the function of kinetic energy of particle that impacts onto wall, setting n = 2; the model is presented as:

$$E = kV_p^2 f(\gamma) \#(6)$$

$$f(\gamma) = \begin{cases} \frac{1}{3}cos^2\gamma & tan\gamma > \frac{1}{3} \\ sin\,(2\gamma\,) - 3sin^2\gamma & tan\gamma \le \frac{1}{3} \end{cases} \#(7)$$

Erosion-corrosion on wall caused by particle is a complex function related to particle impact, particle feature and wall feature. This model assumes that for most metals, erosion-corrosion can be considered as the function between particle impact angle and velocity, omitting the factors such as particle size and shape.

The model used in research is Oka model based on Newton's second law and momentum theorem; it simulates multi-phase flow by introducing passivation, impingement/reflecting and turbulence mechanisms etc. Oka model is intended to simulate erosion-corrosion on different materials by changing their Vickers hardness. It considers the effect of particle size on erosion-corrosion rate by means of reference particle size.

$$e_r = e_{90} g(\alpha) \left(\frac{u_{rel}}{u_{ref}}\right)^{k_2} \left(\frac{D_p}{D_{ref}}\right)^{k_3} \#(8)$$

$$g(\alpha) = (sin\alpha)^{n_1}(1 + H_v(1 - sin\alpha))^{n_2} \#(9)$$

$n_1, n_2$—— exponent;
$H_v$—— Vickers hardness of material, expressed in GPa;
$u_{rel}$—— relative velocity between particle and wall;
$u_{ref}$—— reference velocity constant of particle;
$D_{ref}$—— reference radius;
$k_2, k_3$—— exponent;
$e_{90}$—— reference erosion-corrosion rate.

This model considers more complete factors. In applications, we need to adjust the parameters depending on material conditions so as to accurately calculate erosion-corrosion on material.

## 3 Simulation for Erosion-Corrosion in Special Piping

### *3.1 Infinite Element Simulation for Special Piping*

There are two CFD-based methods for numerical analysis on particle motion in flow field, namely Eulerian-Eulerian method and Eulerian-Lagrangian method. The main difference between two methods is that: Eulerian-Eulerian method assumes solid particles in piping as continuous phase, analyzes distribution and motion of particles by using Eulerian method, and cannot accurately track the angle and velocity at which particles impinge walls; whereas, Eulerian-Lagrangian method assumes solid particles as discrete phase, analyzes trajectory of particles in Lagrangian coordinate system, and can track particles by solving the equation about force exerted on particles. Since the computation on erosion-corrosion requires accurate acquisition of information about particle impingement, Eulerian-Lagrangian method is used for this purpose. The CFD-based prediction on erosion-corrosion mainly includes three steps: computation on flow field, tracking of particle trajectory, and calculation of erosion-corrosion rate. Computation on flow field is the foundation of overall computation on erosion-corrosion. On this foundation, the trajectory of particles in piping is calculated. Finally, by collecting various parameters about impingement on walls by particles, we calculate erosion-corrosion rate. In this study, all CFD-based numerical computations on erosion-corrosion are made with the software FLUENT.

To address practical applications, as shown in Picture 1, we develop simulation models for tees, crosses and bends. In consideration of calculation efficiency and erosion-corrosion features in piping, we create appropriate models. Cu-Ni alloy and Ti alloy have better performance to resist erosion. However, because of high price of Ti alloy, stainless steel is always the most widely used material to make petrochemical piping. In petrochemical processes, bends, tees and crosses are often used. So, we develop simulation models for different pipings.

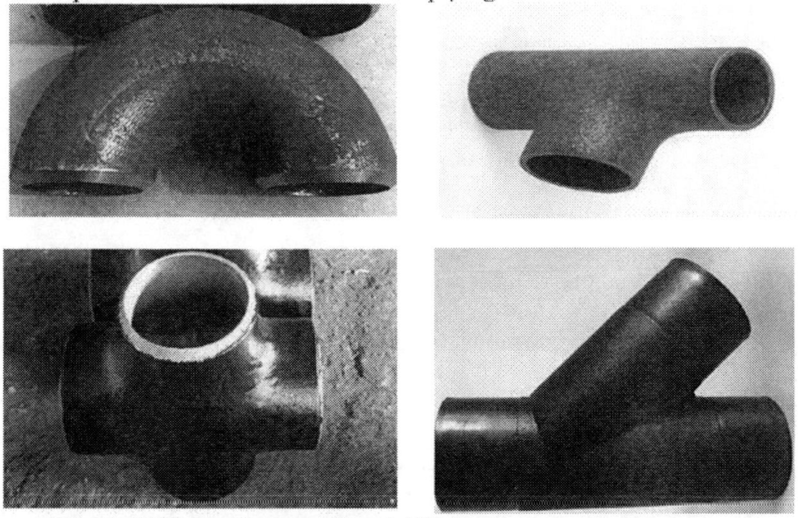

**Fig. 1.** Different Types of Special Piping

Electrochemical conductivity of piping is set as 6×10$^7$ s/m. Piping with straight pipe length L > 10D is selected. Simulated environment is normal temperature and pressure.

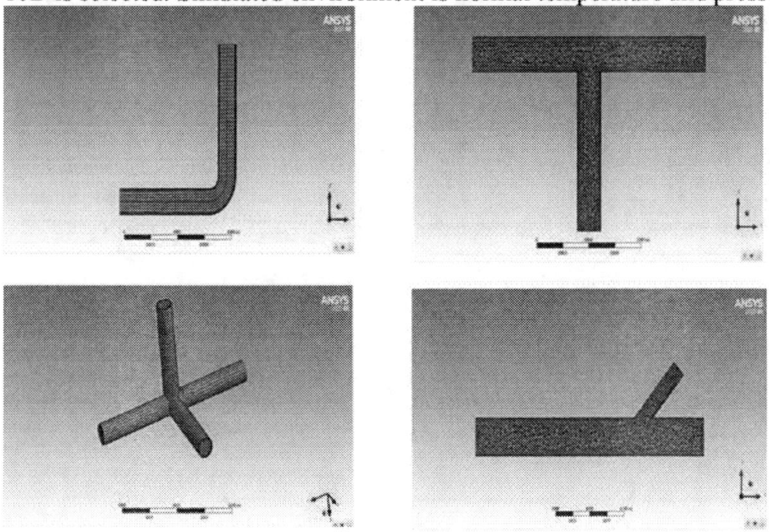

**Fig. 2.** Meshing

To facilitate analysis on erosion-corrosion rate at walls, the meshing for walls is dense, while the meshing for inner space is relatively sparse. Tet Primitive method is used to generate hexahedral mesh with tetrahedral topology. Cooper method is used to develop meshing for other geometries, as shown in Picture 2.

Many solid sands are suspended in the materials transported in petrochemical piping. Their particle sizes are mostly at micrometer level. Through complete consideration, we select the oil with density $\rho$ = 960 kg/m$^3$ as continuous phase and select the particles with density $\rho$= 1550 kg/m$^3$ as discrete phase. We assume that initial velocity of particles is same as that of oil, and set particle size to be 10 μm~1 mm, flow velocity to be 1~10 m/s, and flow rate to be 1e-8~1e-6 kg/s. Standard k-ε turbulence model is used for computation, and SIMPLE algorithm is used for pressure-velocity coupling. Continuous phase in simulation is crude oil, which is incompressible fluid. So we select the boundary conditions at inlet as Velocity Inlet. Since the pressure and velocity at outlet is unknown, we select the boundary conditions at outlet as Outflow. For walls, the boundary type is set as "wall", and the feature is set as "reflect", indicating that particles reflect after they impinge walls.

Boundary conditions for discrete phase: Escape conditions are selected at inlet and outlet in DPM model, and Reflect conditions are used for walls. We assume that particle phase is injected from surface source, where inertial partials are thrown from boundary surface at inlet and the initial velocity of particles is same as the inlet velocity of fluid. In addition, we assume that the mass of particles does not change and there is no heat transfer, so only particle density is set. In the computation, restitution coefficients from Grant and Tabakoff particle-wall impingement/reflecting model and Forder et al.'s particle-wall impingement/reflecting model are used to calculate how particles change before/after they impinge walls.

Moreover, according to Chen et al.'s research, the number of solid particles has big effect on erosion-corrosion rate. If in the computation, the number of particles "injected" into piping is small, it will lead to big calculation error. Chen et al. and Liu et al. found that, after the number of particles injected into piping reaches 20,000, as the number of injected particles increases, the erosion-corrosion rate becomes virtually constant. Therefore, in the compotation on erosion-corrosion, for each working condition, we set the particle number injected into piping to be more than 20,000, so as to eliminate the effect of particle number on the erosion-corrosion. In the computation with FLUENT, SIMPLE algorithm is used for pressure-velocity coupling, and Second Order Upwind scheme is used for dispersion of momentum and turbulent kinetic energy. In the computation on erosion-corrosion, in order to improve calculation efficiency and ensure stability and reliability of calculation, we firstly set convergence conditions under steady states (residual error is $10^{-5}$) and determine initially converged flow field of continuous phase; then we set 2-way coupling for discrete phase and continuous phase, and calculate the trajectory of discrete phase after iterative computation on continuous phase every 5 steps; after then, we introduce updated momentum and energy of discrete phase into the computation on continuous phase; afterwards, we alternately calculate the trajectory of discrete phase and the flow field of continuous phase; until steady convergence, we post-process discrete phase and observe flow field of continuous phase.

Because the computation involves multiple models, the accuracy in each step of calculation is essential to ensure reliability of final computation on erosion-corrosion. Therefore, we validate the models used in each step of calculation to ensure that the models have enough calculation accuracy.

### *3.2 Analysis on Simulation Results*

By using control variates method, this paper respectively analyzes the profiles of variable maximum erosion-corrosion rate in bends under dual-coupled actions between different factors e.g. volume fraction of moisture, particle size of sand, mass flow rate of sand and flow velocity of medium, and obtains the relation between different factors and maximum erosion-corrosion rate in bends.

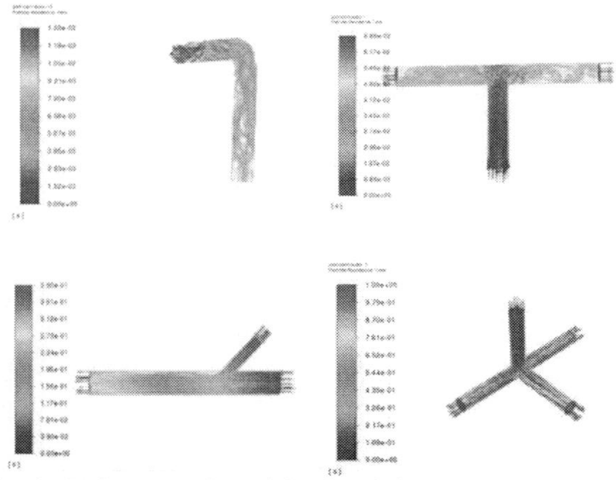

**Fig. 3.** Trajectory of Particles in Piping

The factors such as particle size, flow rate and erosion velocity produce different effects on the damage at structural positions, as shown in Picture 3. We make detailed simulation analysis by using Fluent.

The study investigates the difference in erosion-corrosion at different positions in bends through CFD simulation. The picture shows the contour of erosion rate. Erosion-corrosion results from complicated interaction between mechanical effect and electrochemical effect. However, as for stainless steel, electrochemical testing shows that the corrosion action makes small contribution to overall erosion-corrosion rate. Erosion action is dominant in the erosion-corrosion on stainless steel, i.e. the mechanical effect from fluid and sand makes great contribution to the erosion-corrosion on stainless steel. The erosion rate in bends is actually affected by sand concentration, impact velocity and impact angle. According to the simulation, as shown in Picture 4, the sand concentration is much higher in outer side than in inner side and is highest in outer outlet of bend. But, the impact velocity of sand is higher in inner side than in outer side, and the maximum impact velocity occurs in the inmost side of bend. Obviously, the sand concentration and the impact velocity have opposite effects on the erosion-corrosion in stainless steel bends. In inner side, although sand flows at high velocity, only small quantity of or even no sand impacts inner walls, resulting in low erosion rate. In outer side, although impact velocity of sand is slightly lower, frequent impact by sand leads to high erosion rate. The maximum erosion-corrosion rate occurs at outer outlet of bend, corresponding to the zone with densest sand. In the straight pipe before bend, by gravity, sand concentration is high at bottom, leading to high erosion rate.

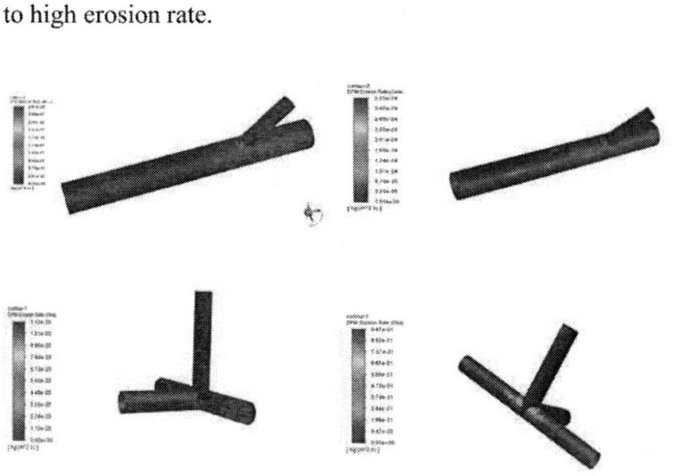

**Fig. 4.** Distribution of Erosion-Corrosion Rate in Tees and Crosses

Journal of Physics: Conference Series **2594** (2023) 012018 doi:10.1088/1742-6596/2594/1/012018

In order to analyze the effect of sand particle size on the profile of erosion-corrosion in bends, we use the coupled method between sand particle size and mass flow rate to explore the profile of sand particle size vs. erosion-corrosion in bends. According to the data about shale gas production, we set the coupled working conditions: flow velocity of medium 10 m/s, particle size of sand 0-0.5 mm, mass flow rate of sand $1\times10^{-3}$, $8\times10^{-4}$ and $5\times10^{-4}$ kg/s. Though calculation, we get the relation between maximum erosion-corrosion rate and sand particle size, as shown in Picture 5.

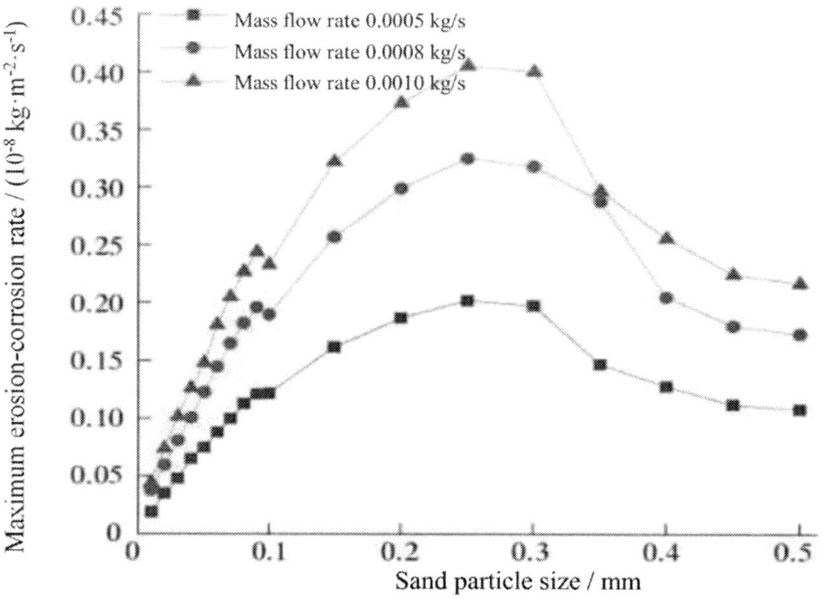


**Fig. 5.** Distribution of Erosion-Corrosion Rate in Tees and Crosses

In order to analyze the effect of medium flow velocity on the profile of erosion-corrosion in bends, we use the coupled method between sand particle size and medium flow velocity to explore the profile of medium flow velocity vs. erosion-corrosion in bends. According to the data about shale gas production, we set the coupled working conditions: flow velocity of medium 7-12 m/s, particle size of sand 0.1-0.4 mm, mass flow rate of sand $8\times10^{-4}$ kg/s. Through calculation, we get the relation between maximum erosion-corrosion rate and variable medium flow velocity, as shown in Picture 6.

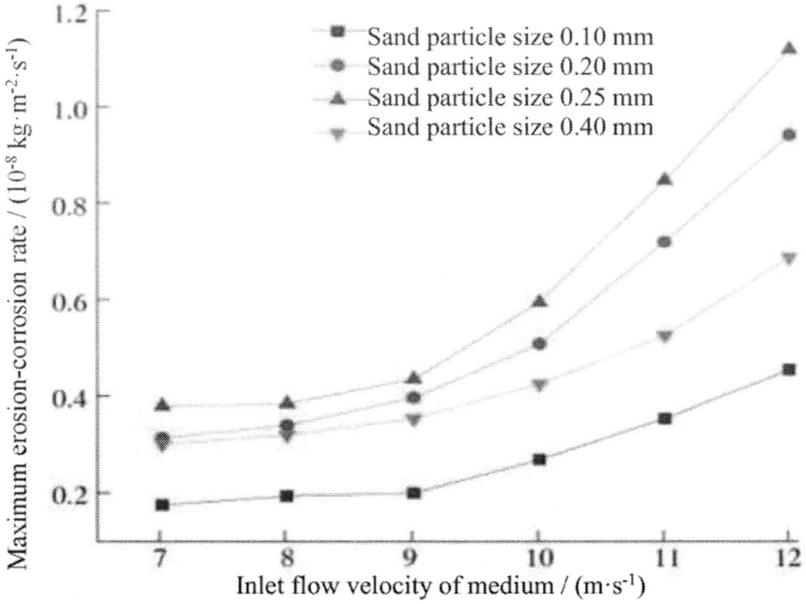


**Fig. 6.** Effect of Medium Flow Velocity on Erosion-Corrosion Rate

Under the coupled action between sand particle size and medium flow velocity, as sand particle size increases through 0.25-0.30 mm, the effect of medium flow velocity on the maximum erosion-corrosion rate gradually increases, and the slope of curve gradually increases; when medium flow velocity is high, as shown in Picture 7, the variable sand particle size has obvious effect on the maximum erosion-corrosion rate.

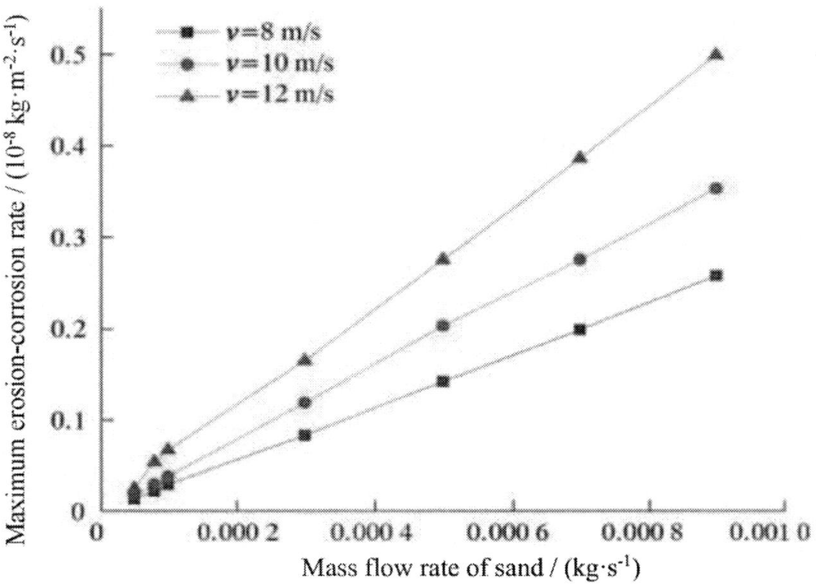


**Fig. 7.** Effect of Mass Flow Rate on Erosion-Corrosion Rate

In case of same flow velocity of medium, as the mass flow rate of sand increases, the maximum erosion-corrosion rate in bends increases linearly. This is because that as the mass flow rate of sand increases, the sand particles will impinge inner walls at increasing frequency and the piping will experience more intense erosion-corrosion by sand particles. Under the coupled action between sand flow rate and medium flow velocity, when the mass flow rate of sand is high, the flow velocity of medium has obvious effect on the maximum erosion-corrosion rate. The higher the flow velocity of medium is, the greater the slope of curve is, suggesting that when the flow velocity of medium is high, the mass flow rate of sand has obvious effect on the maximum erosion-corrosion rate.

According to simulation analysis, as the flow velocity is increasing, the quantity of solid particles carried by fluid will increase, the probability that particles impinge inner walls of bends will increase, and the erosion-corrosion rate in bends will significantly increase. Since the particles with smaller size are better coupled with the fluid medium, it allows for more sufficient impingement between particles and walls, resulting in more frequent impingement between them, leading to higher erosion-corrosion rate in bends. As the particle size is increasing, the particles are subject to greater effect of gravity and are difficultly carried by fluid medium, resulting in more frequent collision between particles and greater energy loss of particles themselves. Finally, it causes that the solid particles impinge walls at less frequency and less intensity, leading to lower erosion-corrosion rate on inner walls of bends.

We mesh the erosion-corrosion-prone positions on walls of bends and tees. Under erosion by 5000 particles, the erosion frequency at different positions in bends and tees is shown in Picture 8. It indicates how different zones are prone to erosion-corrosion.

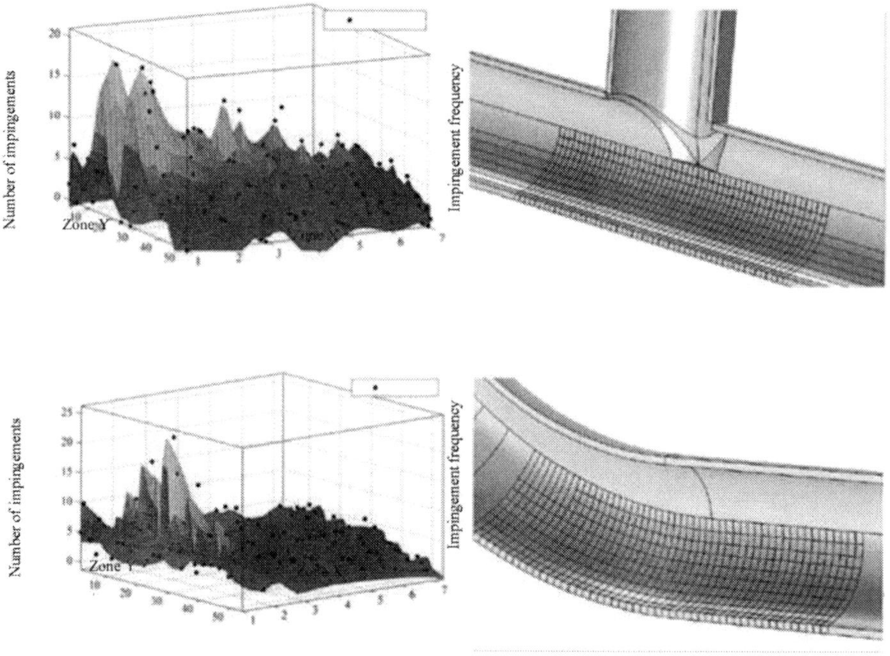


**Fig. 8.** Distribution of Particle Impingement

As can be seen, the most erosion-corrosion-prone position in bends is the interface between straight pipe and curved pipe. In addition, the bend in curved direction is more prone to erosion-corrosion, indicating the zone with electrochemical corrosion. However, a tee has two points with maximum corrosion, also indicating its structural symmetry. Therefore, the erosion-corrosion in tees should features two major zones with electrochemical corrosion.

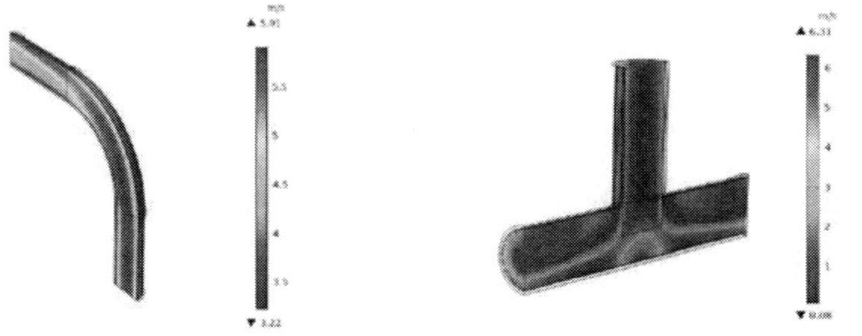

**Fig. 9.** Distribution of Turbulence Velocity

In the flow field in k-ω turbulence model, initial average velocity is set as 5 m/s. The contour of velocity in bends is shown in Picture 9. We obtain that the maximum velocity in flow field is 5.91 m/s, occurring at deformed positions of curved pipe. The minimum velocity is 3.22 m/s, occurring near outer walls of curved pipe and near inner walls of straight pipe. Also, we calculate the flow field in k-ω turbulence model for tees, and set turbulence velocity as 5 m/s. The contour of turbulence velocity is shown in Picture 10. We obtain that the maximum velocity in flow field is 6.31 m/s, occurring near the vertical changeover of flow field; the minimum velocity in flow field is 0.08 m/s, occurring at the vertical changeover of flow field. As can be found, due to deformation of flow field, the velocity near pipe walls is many times lower than that in the center. In addition, by comparing the velocity contours for two structures, it shows that the change of velocity field in tees is greater than in bends. This is because structural deformation of tees are more complicated.

As flow velocity is increasing, the quantity of solid particles carried by fluid will increase, the probability that particles impinge inner walls of bends will increase, and the erosion-corrosion rate in bends will significantly increase. Since the particles with smaller size are better coupled with the fluid medium, it allows for more sufficient impingement between particles and walls, resulting in more frequent impingement between them, leading to higher erosion-corrosion rate in bends. When particle size is smaller than certain range, as particle size increases, the abrasion caused by impingement will become greater and the erosion-corrosion rate will become higher. However, as the particle size increases further, the particles are subject to greater effect of gravity and are difficultly carried by fluid medium, resulting in more frequent collision between particles and greater energy loss of particles themselves. Finally, it causes that the solid particles impinge walls at less frequency and less intensity, leading to lower erosion-corrosion rate on inner walls of bends. According to analysis, the greatest erosion-corrosion by fluid occur centrally near the interface between bends and pipes; under the pressure 0 MPa-1.2 MPa, the erosion-corrosion rate in inner walls of piping is proportional to the flow velocity/rate at inlet; when particle size is smaller than certain range, it is proportional to particle size; at 0.26 mm, it reaches peak value; when greater than 0.26 mm, the erosion-corrosion rate is inversely proportional to particle size.

The effect of orientation on the erosion-corrosion rate is shown in Picture 10. The erosion-corrosion rate in bends is highest in horizontal-vertical (H-V) upward direction. It is mainly because that in this direction, gravity changes the motion direction of particles, and the zone area where particles impinge walls is smallest among four directions. As a result, the quantity of particles that impinge walls in unit period is most, leading to high erosion-corrosion rate. As shown in pictures, in vertical-horizontal (V-H) direction, the erosion-corrosion profiles are virtually same at different angles of bends. It is mainly because that gravity changes the motion direction of particles, thus the particles deviate from flow lines and impinge the bends at different positions.

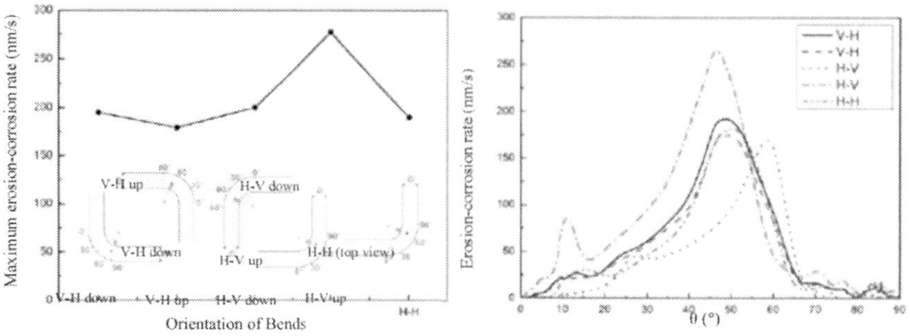


**Fig. 10.** Effect of Bend Orientation on Erosion-Corrosion Rate

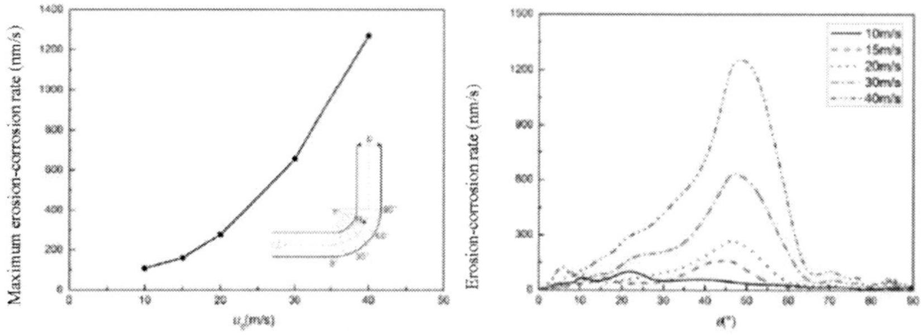


**Fig. 11.** Erosion-Corrosion Rates at Different Angles of Bends

The effect of flow velocity on the erosion-corrosion rate is shown in Picture 11. It can be known that the maximum erosion-corrosion rate increases as the flow velocity of fluid increases, presenting an exponential relation. It indicates that flow velocity has great effect on the erosion-corrosion rate. The erosion-corrosion rates at different angles are shown in the picture. When flow velocity is low (10 m/s), the maximum erosion-corrosion occurs at small angle, about 20°. When flow velocity becomes high, the maximum erosion-corrosion occurs at the angle between 45-50°. It is mainly because that when flow velocity is low, solid particles move for longer period in straight pipe at inlet, so that by gravity, the impingement is strongest and the erosion-corrosion is heaviest at small angle of bends.

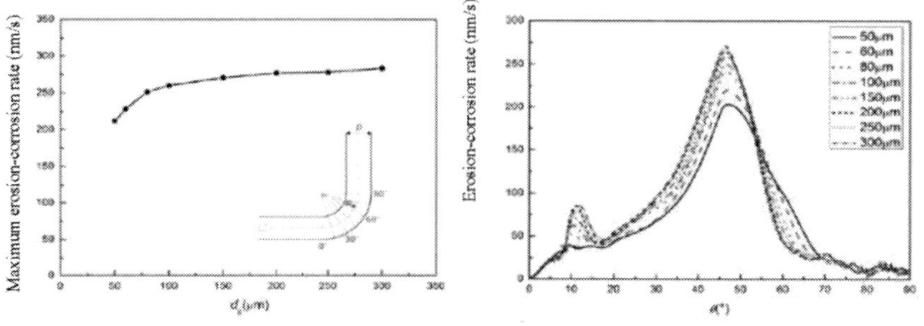


**Fig. 12.** Effect of Particle Size on Erosion-Corrosion at Different Angles

The effect of particle size on erosion-corrosion rate is shown in Picture 12. As solid particle size increases, the erosion-corrosion rate gradually increases. When particle size is greater than 100 μm, as particle size increases, the erosion-corrosion rate tends to be steady. It is mainly because that when particle phase is coupled with continuous phase, big solid particles exchange more momentum with gas. As particle size increases, the final erosion-corrosion rate tends to be steady. When solid particle size increases from 50 μm to 100 μm, the erosion-corrosion rate increases significantly. When particle size is greater than 100 μm, the erosion-corrosion rate does not increase obviously. Therefore, 100 μm is the critical diameter for variation of erosion-corrosion rate. According to pictures, at different particle sizes, the maximum erosion-corrosion rate in bends occurs at about 45°. Since small particles have better follow-up behavior, the drag force allows small-size particles to have slightly bigger impact angle, thus the angle with heaviest erosion-corrosion is slightly larger than that for big-size particles.

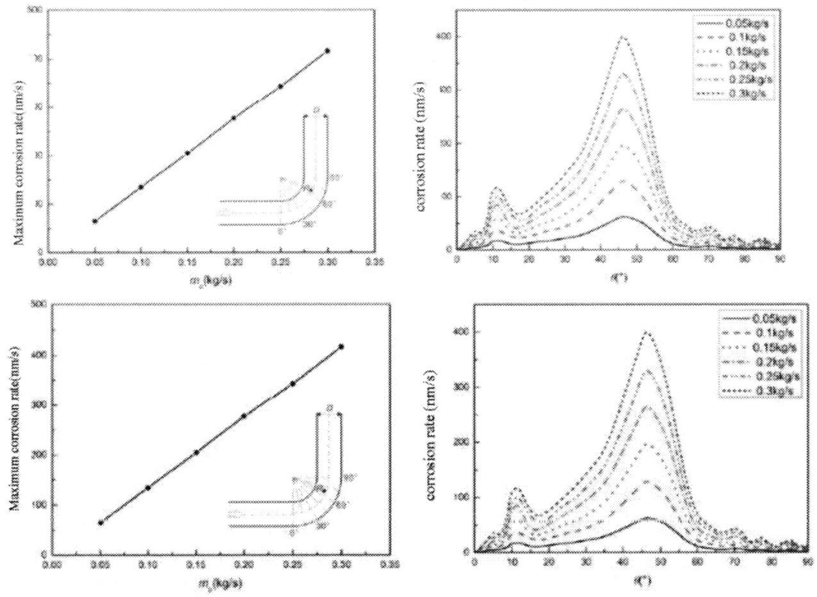


**Fig. 13.** Effect of Flow Rate on Erosion-Corrosion Angle

The effect of particle flow rate on the erosion-corrosion rate is shown in Picture 16. As particle flow rate increases, the erosion-corrosion rate in bends will increase linearly. It is mainly because that when particle flow rate increases, the quantity of particles that impinge walls in unit period will increase and the erosion-corrosion rate will become higher. The position with maximum erosion-corrosion in bends occur at 45-50°, and it does not vary with the particle flow rate.

## 4 Conclusion

This paper incorporates actual working conditions of petrochemical process piping, and carries out simulation for erosion-corrosion-prone positions such as bends, tees and crosses. It fully considers the effect of various factors, e.g. medium, particle size, flow velocity and flow rate, on the erosion process, and provides theoretical supports for addressing erosion-corrosion problems in terms of corrosion protection for processes and corrosion monitoring on equipment etc. in the future.

## References

1. YANG Hong-wei , LI Ning . Study on petroleum chemical enterprise equipment reliability management [J] . Chemical Engineering & Equipment, 2010( 8) : 252 — 253.
2. Al-Sharrah G K , Edwards D , Hankinson G . A new safety risk index for use in petrochemical planning [J] . Process Safety and Environmental Protection, 2007, 85( 6) : 533 — 540.
3. YU Yang , WANG Dong , JI Xu . Models of quality management based on knowledge for concrete enterprise [J] . Concrete, 2011( 3) : 90 — 92.
4. WU Xiao-yue , SHA Ji-chang . Study on reliability modeling of FMS [J] . Automation of Machinery Industry , 1999, 21( 2) : 23 — 25.
5. ZENG Sheng-kui , ZHAO Ting-di , ZHANG Jian-guo , et al . System reliability design and analysis course [M] . Beijing:Beijing Aeronautics and Astronautics Press, 2001: 31 — 35.

# Application of Integrated Evaluation on Seawater Injection to Improve Injectivity for Carbonate Reservoir in Offshore Oilfield

**Yuanbing Wu1, Dandan Hu1(✉), Bing He2, Songhao Hu3**

1 RIPED, PetroChina Co. Ltd, Beijing, China
2 Zhenhua Oil Company, NORINCO GROUP Beijing, China
3 Xinjiang Oilfield Company, China National Petroleum Corporation, Karamy, Xinjiang Province, China

hudandan@petrochina.com.cn

**Abstract.** Seawater flooding is popular for carbonate reservoirs in offshore oilfields and seawater is sometimes the unique choice of injection water source due to the lack of surface water supply. Seawater is injected into reservoirs to maintain reservoir pressure after water treatment for technically and economically effective field development. Uneven injection and production in B oilfield caused unbalance of pressure distribution and poor injection performance. It is essential to improve the injectivity of single well to a satisfied level through a integrated workflow on the basis of performance analysis, brine compatibility study, core flooding experiment, strategy and solution optimization. In this paper, the injection status has been analyzed and the decisive impacting factors were given including formation properties, stimulation activity, the brines compatibility and pre-treatment of injection water, which play a very important role in injection performance sustainability as the seawater injection continues. Scale modelling was done with formation water, mud filtrate, injection water and stimulation fluid for self-scaling assessment. Mixing experiment of two of the brines of four introduced at either atmospheric or reservoir condition. Core flooding experiment was carried out with formation water and filtered injection water after critical velocity for core flooding is confirmed. Then the injectivity for each wells have been analyzed at different locations of the reservoir with different formation properties, solutions have been subsequently put forward and the improvement on injectivity of injectors before and after the stimulation or even re-stimulation and the effect of cleanup after the stimulation was also compared. It was concluded that injectivity could be enhanced through stimulation and the cleanup after stimulation and the sulfate should be removed or reduced in the process of the pre-treatment of injection water from the remote terminal to ensure desirable water injection and production performance.

## 1 Introduction

Water injection is an effective method to maintain formation pressure and obtain high recovery efficiency. If seawater is properly filtered, chemically stable and free of bacteria, injection into the formation is successful and does no damage to the reservoir [1-3]. A large number of case studies have shown that seawater injection in offshore oil fields is about 5% more efficient than ordinary water injection [4-6]. In the 1970s, seawater injection was started in the Middle East to restore reservoir pressure and maintain stable oil production for giant carbonate reservoirs. Based on years of practical seawater injection experience, the challenges of seawater injection in offshore fields are: blockage caused by suspended solids and bacteria; Injected water damages the reservoir; Corrosion of the injection system by seawater; Scaling effect when seawater injection and production water re-injection are used together; And the reservoir becomes sour. The objective of this paper is: (1) to describe geological characteristics for B oilfield in offshore, analyze injection status and problems;2) to discuss the experimental research on brines compatibility study and core flooding experiment to determine the impacting factors on low injectivity;3) to put forward solutions to enhance injectivity for the current well pattern.

## 2 Description and Analysis

### *2.1 Reservoir Description*

B oilfield is a dome structure as a symmetric NE-SW trending anticlinal feature with flank and has 5 sublayers of different formation properties from top to bottom with interlayers(Fig.1). The initial reservoir pressure is about 4850 psi and the saturation pressure is about 2961 psi. The reservoir is divided into 10 subzones named from B1 to B10 with buried depth ranging from 9549ft to 10109ft. Sub layers with odd number are oil bearing of different net pay layers of dolomite or limestone with anhydrite nodules inter-bedded and sub-layers with even number are inter-layers of physical properties or lithology. Although the average values of permeability and porosity of each oil bearing sub-layer show less heterogeneity but the wide ranges of permeability and porosity of each oil bearing sub-layers indicate horizontal discrepancy of formation properties(Table 1).

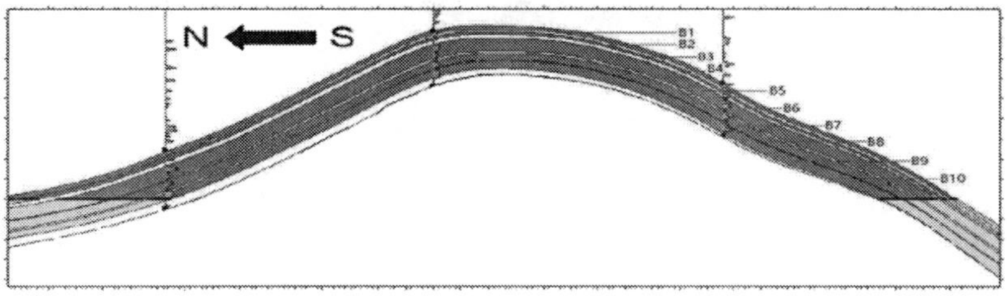


**Fig.1.** Reservoir Profile of B Oilfield

**Table 1.** Reservoir Property

| Sub-Layer | Porosity(%) | | Permeability(mD) | |
|---|---|---|---|---|
| | Range | Average | Range | Average |
| B1 | 5.3-24.9 | 12.4 | 0.18-24.9 | 5.8 |
| B3 | 4.9-15.6 | 8.8 | 0.15-71.0 | 9.8 |
| B5 | 2.5-26.5 | 11.4 | 0.01-523 | 4.3 |
| B7 | 2.5-25.8 | 10.3 | 0.01-2820 | 6 |
| B9 | 3.9-25.8 | 13.8 | 0.06-293 | 6.2 |

### 2.2 *Injection Status Analysis*

The offshore oilfield was producing oil from multiple reservoirs and mix-uneven well pattern drilled from two platforms with peripheral sea water injection was adopted for the reason of limited well slots. With PPL or PPL and ICD have been adopted as way of lower completion and swell packer as annulus isolation for comingled water injection sea water was pre-treated on one island and was injected into each injector through pipeline network and injection manifold. In order to maintain the injection capacity of the injectors regular pipeline pigging was carried out to make sure that the injection water quality met the injection quality standard.

The production of B reservoir was mainly from B5\B7\B9 layers, and the estimated recovery factor is respectively 4.46%, 3.93% and 3.98%, with the cumulative oil production accounting for 95.2%. The recovery degree of B1\B3 layer is lower, with the recovery factor of 0.74% and 0.9%, respectively, with the cumulative oil production accounting for 4.8%. The reservoir pressure was 3100psi before seawater injection, and the pressure remained 65% of the initial reservoir pressure(Fig.2). The pressure of C5\C7\C9 reservoir decreased obviously in the B-01H test of the new well of Nan platform, reflecting the good plane connectivity of the reservoir(Fig.3).

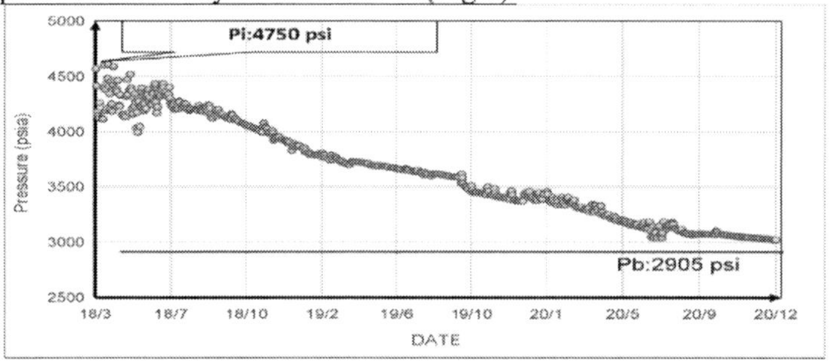


**Fig.2.** Reservoir pressure change for B oilfield

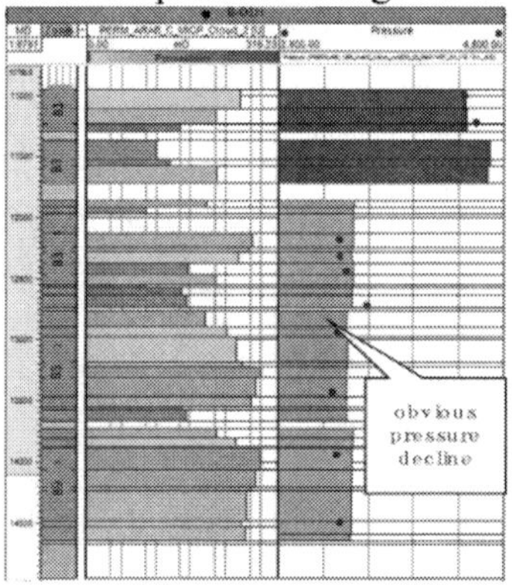


**Fig.3.** MDT for B-01H

The formation pressure ceased the downtrend and started noticeable recovery after the water injection but VRR imbalance was noticed between the north and the south regions. The VRR for north region was 1.05 with production rate of 7 MBPD and injection rate of 11.5 MBPD whereas the VRR for south region is 3.5 with production rate of 1 MBPD and injection rate of 5.5 MBPD. The VRR unbalance lead to unbalance of development of different regions and definite earlier water breakthrough at different regions and finally resulted in unbalanced reservoir development. Although regionally dynamic adjustment on production and injection has been done for balanced injection-production, this adjustment was doomed to be time-consuming and less efficient with unbalanced VRR of big difference in different regions if no effort was made from the side of injection.

Although frequent pigging for injection pipeline is periodically implemented water injection capacity of nearly all 5 injectors showed deteriorating trend as the water injection continues after the enhancement of injectivity of all the 5 injectors through the stimulation, re-stimulation and cleanup after stimulation.

Injectivity of all injectors showed impressive injection improvement after the stimulation with the same stimulation strength for each sub-layers but the injection rate cannot be stabilized and showed trend of deterioration under the same injection pressure. Cleanup was implemented on one injectors and the results indicated some improvement on the injectivity.

Different lower completion with either PPL or ICD were adopted. Two injectors with ICD showed quiet different injection performance and this difference eliminated possible ICD blockage.

## 3 Lab Experiments

### *3.1 Brines Compatibility Study*

Synthetic seawater and formation water were prepared based on the compositional data received from the field laboratory. Seawater was used as the base water or diluted as required. Synthetic formation water was used for core saturation and for compatibility tests. The screening and final selection of the composition of the modified brines were in accordance with the suggestions given by predecessors (Zhang et al. 2007; Fathi et al. 2012; Awolayo et al. 2014)[7-9]. In order to identify the root causes of deteriorating trend of injection from compatibility perspective of brines involved, formation water, injection seawater, mud filtrate and simulation fluid were introduced into the experiment for brines compatibility check [10-16]. The physical mixing experiment at different mixing ratio indicated that the formation water is incompatible with the injection water and the mud filtrate but shows no incompatibility with the simulation fluid through visual turbidity check(Fig. 4).

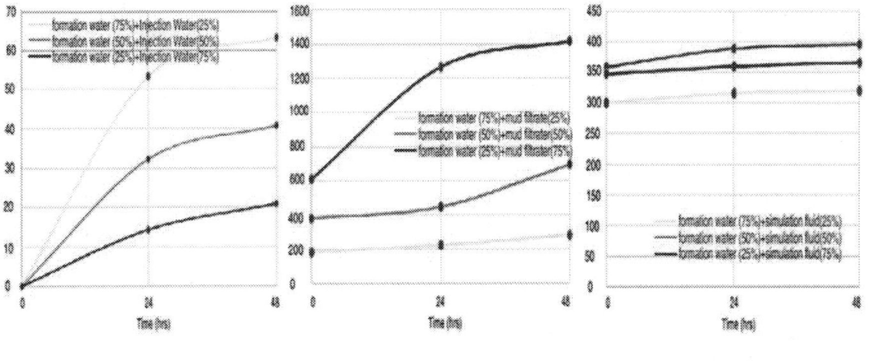


**Fig.4.** Turbidity of Blend

The physical mixing experiments showed that at room temperature and atmospheric pressure the formation water and injection water are deemed initially incompatible although decreasing scaling risk is expected with increasing volume of injection water due to a dilution effect, the mud filtrate is deemed incompatible with formation water and the simulation fluid is inconclusive regarding to compatibility with formation water(Fig.5 and Fig.6).

Journal of Physics: Conference Series **2594** (2023) 012019 doi:10.1088/1742-6596/2594/1/012019

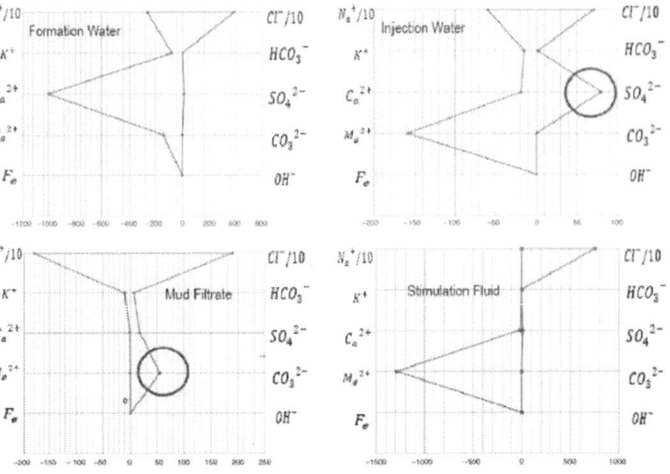


**Fig.5.** Composition of Brines

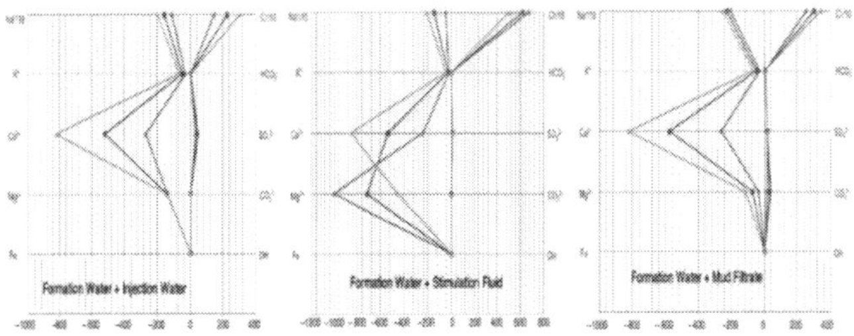

**Fig.6.** Ionic Composition of Blen

Geochemical modelling was also created with the four brines mentioned above and the result showed that the four brines have the risk of self-scaling due to calcite at both reservoir and surface conditions and the formation water presents additional self-scaling risk of anhydrite and celestite(Fig.7). At reservoir conditions, for the mixing scenarios with injection water and formation water, sulfate based mineral scales are predicted to be saturated. For the mixing scenarios with formation water and mud(Fig.8), a significant risk to the calcite scaling as well as iron carbonate was predicted. The saturation index values obtained are well above zero and precipitation is almost certain to occur.

Based on modeling outputs, at both reservoir and surface conditions, the self-scaling modeling determined that all brines (i.e., formation, injection, and mud filtrate) present some initial risk to scaling due to Calcite. The Formation water presents additional self-scaling risk due to Anhydrite and Celestite. While it does not guarantee a scale will form, it is possible.

Based on modeling outputs, at reservoir conditions, for the mixing scenarios with injection water and formation water, sulfate based mineral scales are predicted to be saturated. For the mixing scenarios with formation water and mud, a significant risk to the calcite scaling as well as iron carbonate was predicted. The saturation index values obtained are well above zero and precipitation is almost certain to occur. Based on water to water compatibility and formation damage experiments results, it is inferred that the observed damage may relate to scaling risk to calcium sulfate and calcite.

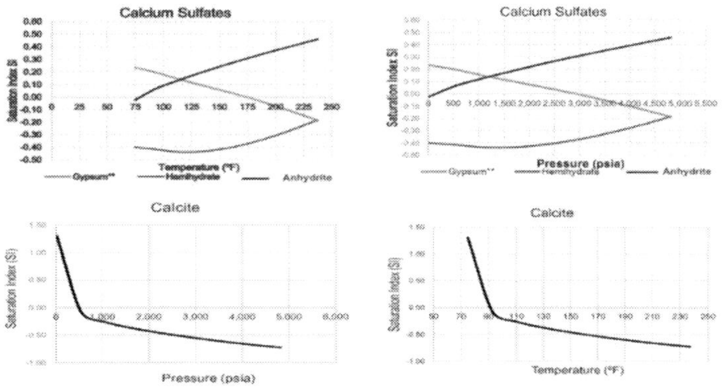


**Fig.7.** Formation Water Self-scaling Tendency

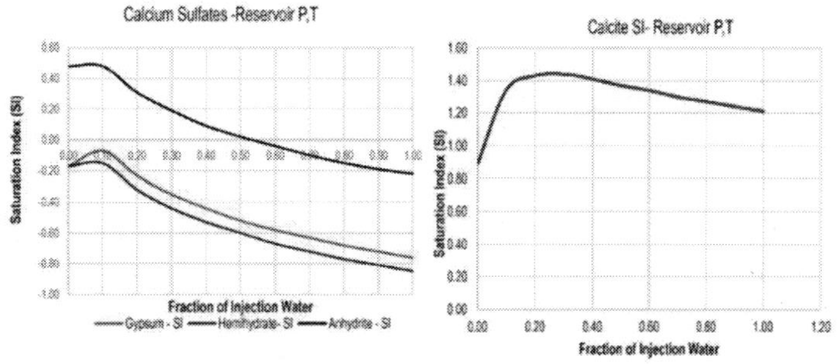


**Fig.8.** Saturation Index of Blend of Formation Water and Injection Water

### 3.2 *Core Flooding Experiment*

Three groups of cores with different levels of permeability was selected (high, medium and low permeability) to present the reservoir heterogeneity for the core flooding experiment and critical velocity tests were carried out to determine the critical velocity during the core flooding experiment to validate the result of the core flooding experiment(Fig.9).

The cores of the three level of permeability were flooded with different of brines in the order of formation water, filtered injection water, formation water, unfiltered injection water and formation water to determine the formation damage during water injection(Fig.10). For the core with lower permeability the permeability decreased from 5.19md to 4.65md after filtered injection water was injected, the permeability decreased to 4.41md after formation water was injected, the permeability decreased to 3.9md after unfiltered injection water was injected and finally the permeability reached 3.93md after formation water was injected again (Table 2). For the cores with high and medium permeability the core flooding experiment shows the same trend corresponding to the deteriorating trend of the injection capacity of the injectors. The permeability change when the cores were flooded with formation water and injection water in alternation indicates that the injection water, either filtered or unfiltered, reduces the permeability and the permeability failed to restore when formation water was injected again indicates that the formation damage caused by injection water for incompatibility is irreversible.

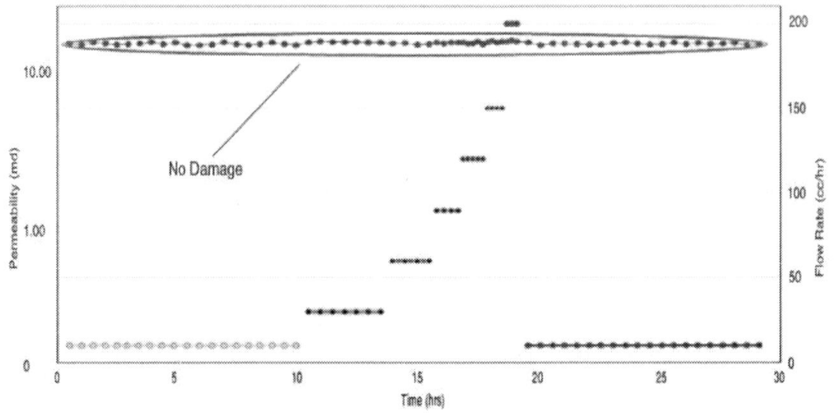


a. No Damage

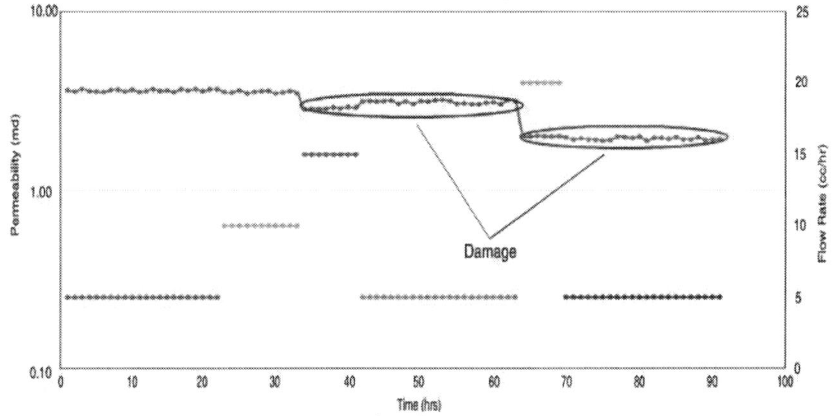


b. Damage

**Fig.9** Critical Velocity

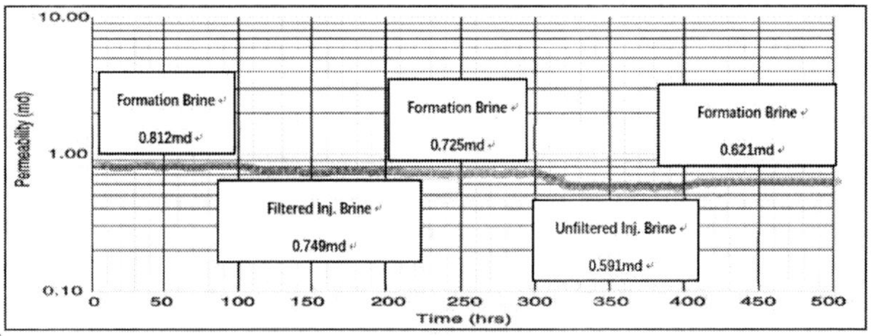


a.Lower Permeability Core

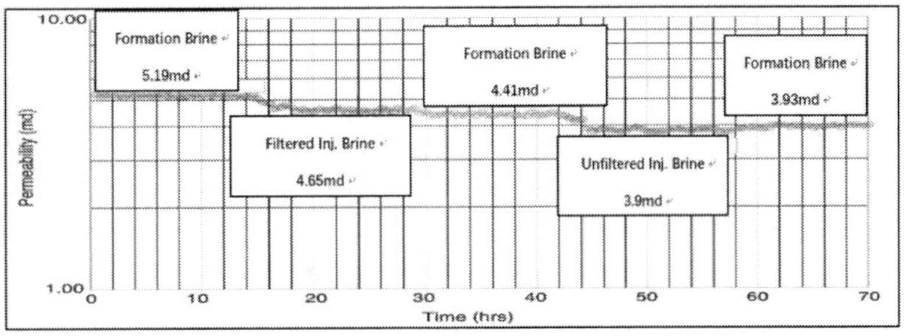


b.Medium Permeability Core

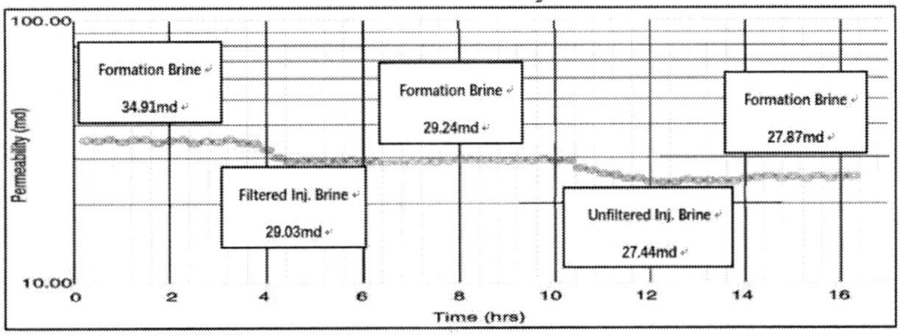


c. High Permeability Core

**Fig.10** Formation Damage from Core Flooding

**Table 2.** Permeability Reduction

| Permeability | Brine | KW(md) | Kw Reduction(%) |
|---|---|---|---|
| Low | Formation Water | 0.812 | 0 |
| | Filtered Injection Water | 0.751 | 7.49 |
| | Formation Water | 0.725 | 10.7 |
| | Unfiltered Injection Water | 0.591 | 27.19 |
| | Formation Water | 0.621 | 23.52 |
| Medium | Formation Water | 5.19 | 0 |
| | Filtered Injection Water | 4.65 | 10.48 |
| | Formation Water | 4.41 | 14.98 |
| | Unfiltered Injection Water | 3.9 | 24.95 |
| | Formation Water | 3.93 | 24.34 |
| High | Formation Water | 34.91 | 0 |
| | Filtered Injection Water | 29.03 | 16.86 |
| | Formation Water | 29.24 | 16.24 |
| | Unfiltered Injection Water | 25.33 | 27.44 |
| | Formation Water | 25.18 | 27.87 |

## 4 Integrated Analysis and Discussion

Five injectors were drilled for water injection at different location of the reservoir and showed different initial injectivity and injectivity of one injector was greatly improved after side track.
Clay-free BARADRIL-N fluid system with properly-sized acid soluble bridging material was used during the drilling process for the purpose of formation protection. The density of the drilling fluid was set to 9.2 ppg, 0.3 ppg more than the density in accordance with the formation pressure, to meet the requirement of wellbore stability(Fig.11). Acid soluble bridging material with D50 particle sizes of 25, 50 and 90 micros were added for loss control during the drilling process but still some circulation loss were observed and this indicated possible foreign particles invasion and plugging or even migration when water injection starts and continues (Table 3).

**Table 3** Injectivity before and after the stimulation

| INJECTOR | Average Perm (mD) | Injectivity Before Stimulation (bbls/d) | Injection Pressure Before Stimulation(psi) | Injectivity After Stimulation (bbls/d) | Injection Pressure After Stimulation (psi) |
|---|---|---|---|---|---|
| INJ-01 | 9.80 | 510 | 1620 | 6170 | 1463 |
| INJ-02 | 12.65 | 0 | 1956 | 5871 | 152 |
| INJ-03 | 16.07 | 1980 | 2417 | 6048 | 0 |
| INJ-04 | 41.94 | 1656 | 1741 | 6912 | 63 |
| INJ-05 | 46.88 | 4080 | 970 | 6498 | 1356 |

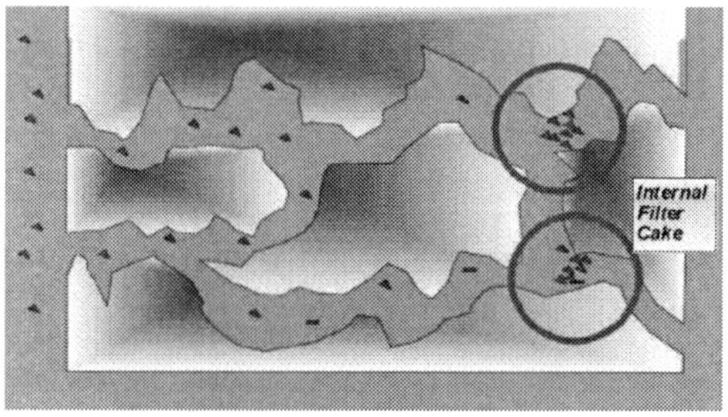


**Fig.11.** Particle Migration

Stimulation were done on all 5 injectors and re-stimulation were done on two injectors to improve the injectivity of the injectors. The stimulation on the injectors greatly enhanced the injectivity of the injectors and it is demonstrated by the impressive injectivity improvement after the stimulation compared with that before the stimulation[17-21](Fig.12). Increase on the stimulation strength to some sub-layers of good physical property indicated good result of treatment through the injectivity after the stimulation and this indicates that the sub-layers with good physical properties tends to be more easily damaged than the sub-layers of poor physical properties.

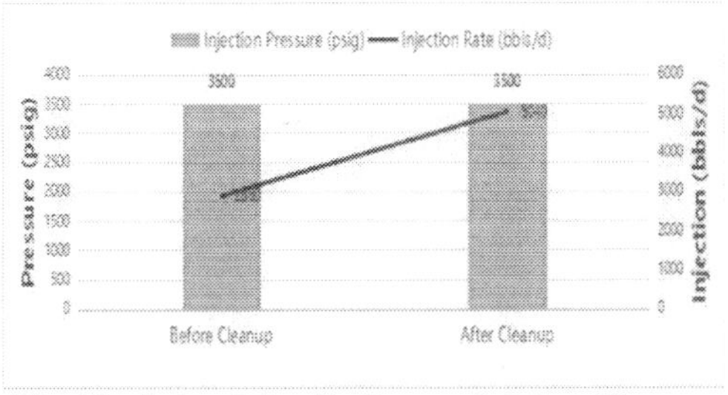


**Fig.12.** Injectivity Before and After Cleanup

The deterioration of injection was usually attributed to formation damage during drilling and completion, incompatibility between injection water and formation fluid or rapid regional pressure buildup. The injectivity improvement with the stimulation, re-stimulation and cleanup demonstrated their effectiveness and the result difference of the pressure survey from different wells eliminated the possibility of local pressure buildup so brine compatibility seems to be an answer to the deteriorating trend of injection and the solution to balanced injection and production for better reservoir management. Several different brines were involved from the drilling stage to the final injection stage therefore brines compatibility and possible formation damage should be evaluated for good water injection from the beginning when the injector was drilled till when the pre-treated sea water is continuously injected into the reservoir.

## 5 Conclusions

The discrepancy of formation properties of the injectors is decisive to the final injection performance of the injectors so optimization on well trajectory of injectors is decisive for the injection performance of the injectors.

The remarkable improvement on the injectivity of injectors through stimulation and the limited improvement of injectivity through re-stimulation with nearly the same stimulation strength validate the effect of stimulation and call for more stimulation strength if re-stimulation is planned.

It can be inferred that the observed damage may relate to scaling risk to calcium sulfate and calcite. Given that sulfate scales are mostly acid insoluble; it is suggested to remove or reduce the sulfate in the process of the pre-treatment of injection water from the remote terminal to ensure desirable water injection performance.

## References

[1]Andersen KI, Halvorsen E, Sælensminde T, Østbye NO. Water management in a closed loop—problems and solutions at brage feld. In: SPE European Petroleum Conference, 24–25 October, Paris, France; 2000. https://doi.org/10.2118/65162-MS.

[2]Awolayo A, Sarma H, AlSumaiti A. An experimental investigation into the impact of sulfate ions in smart water to improve oil recovery in carbonate reservoirs. Transp Porous Media. 2016.

[3]Graham GM, Boak LS, Hobden CM. Examination of the efect of generically diferent scale inhibitor species (PPCA and DETPMP) on the adherence and growth of barium sulphate scale on metal surfaces. In: SPE International Symposium on Oilfeld Scale, 30–31 January, Aberdeen, United Kingdom; 2001. https://doi.org/10.2118/68298-MS.

[4]Dalia Salem Abdallah, Mark Grutters, Robert Stalker, Rob Hutchison, Christopher Stewart, Samantha C. Wilson.The Impact of Sulfate Level Reduction on Seawater Injection Economics: Using Sulfate Scale Precipitation Kinetics to Our Advantage.In SPE Abu Dhabi International Petroleum Exhibition & Conference, November 15–18, 2021.https://doi.org/10.2118/207612-MS.

[5]Yisheng Hu, Eric Mackay.Reactive Transport Modelling of a Carbonate Reservoir under Seawater Injection. In SPE International Oilfield Scale Conference and Exhibition, June 20 – 21, 2018.https://doi.org/10.2118/190757-MS.

[6]Jose Plasencia, Marcelo Andreotti, Hua Wang, Zamir Alam.Subsea Seawater Treatment and Injection - Performance Tests of Filtration Membrane Modules at Simulated Subsea Conditions.In SPE Offshore Technology Conference, April 30–May 3, 2018. https://doi.org/10.4043/28810-MS.

[7]Zhang P, Tweheyo MT, Austad T. Wettability alteration and improved oil recover by spontaneous imbibition of sea water into chalk: impact of the potential determining ions Ca2+, Mg2+, and SO4 2 – . Colloids Surf A Physicochem Eng. 2007;301(1 – 3):199 – 208. https ://doi.org/10.1016/j.colsurfa.2006.12.058.

[8]Fathi SJ, Austad T, Strand S. Water-based enhanced oil recovery (EOR) by "smart water" in carbonate reservoirs. In: SPE EOR Conference at Oil and Gas West Asia, 16–18 April, Muscat, Oman; 2012. https://doi.org/10.2118/154570-MS.

[9]Awolayo A, Sarma H, AlSumaiti AM. A laboratory study of ionic efect of smart water for enhancing oil recovery in carbonate reservoirs. In: SPE EOR Conference at Oil and Gas West Asia, 31 March–2 April, Muscat, Oman; 2014. https://doi.org/10.2118/169662-MS.

[10]M. M. Jordan; I. R. Collins; E. J. Mackay. Low-Sulfate Seawater Injection for Barium Sulfate Scale Control: A Life-of-Field Solution to a Complex Challenge.In: SPE International Symposium and Exhibition on Formation Damage Control, Lafayette, Louisiana, USA, February 2006.https://doi.org/10.2118/98096-MS.

[11]Samaneh Ashoori; Mohammad Zahedzadeh; Sakineh Shokrolahzadeh; Mojgan Radmehr; Emad Roayaei; Mohammad Emadi. Efficiency Improvement of Sea Water Injection Plant-Experience from an Offshore Oil Field.In: 20th World Petroleum Congress, Doha, Qatar, December 2011.Paper Number: WPC-20-1151.

[12]M. M. Jordan, E. J. Mackay.Integrated Field Development for Effective Scale Control throughout the Water Cycle in Deep Water Subsea Fields.In: SPE Europec/EAGE Annual Conference, June 13–16, 2005. https://doi.org/10.2118/94052-MS.

[13]Eric Mackay, Trevor Jones, Bill Ginty.Oilfield Scale Management in the Siri Asset: Paradigm Shift Due to the Use of Mixed PWRI / Seawater Injection.In: SPE Europec/EAGE Annual Conference, June 4–7, 2012.https://doi.org/10.2118/154534-MS.

[14]Alotaibi MB, Nasralla RA, Nasr-El-Din HA. Wettability challenges in carbonate reservoirs. In: SPE improved oil recovery symposium.New York: Society of Petroleum Engineers; 2010. https://doi.org/10.2118/129972-MS.

[15]Aminnaji M, Fazeli H, Bahramian A, Gerami S, Ghojavand H. Wettability alteration of reservoir rocks from liquid wetting to gas wetting using nanofuid. Transp Porous Media. 2015;109(1):201–16.https://doi.org/10.1007/s11242-015-0509-6.

[16]Anderson WG. Wettability literature survey-part 2: wettability measurement. J Pet Technol. 1986;38(11):1246–62. https://doi.org/10.2118/13933-PA.

[17]Anderson WG. Wettability literature survey-part 6: the effects of wettability on water fooding. J Pet Technol. 1987;39(12):1–50. https://doi.org/10.2118/16471-PA.

[18]Austad T, Strand S, Hognesen EJ, Zhang P. Seawater as IOR fuid in fractured chalk. In: SPE international symposium on oilfeld chemistry. New York: Society of Petroleum Engineers; 2005. https://doi.org/10.2118/93000-MS.

[19]Austad T, Strand S, Madland MV, Puntervold T, Korsnes RI. Seawater in chalk: an EOR and compaction fuid. SPE Reserv Eval Eng. 2008;11(4):648–54. https://doi.org/10.2118/118431-PA.

[20]Bera A, Mandal A, Belhaj H, Kumar T. Enhanced oil recovery by nonionic surfactants considering micellization, surface, and foaming properties. Pet Sci. 2017;14(2):362–71. https://doi.org/10.1007/s12182-017-0156-3.

[21]Blake TD. The physics of moving wetting lines. J Colloid Interface Sci. 2006;299:1–13.

Journal of Physics: Conference Series **2594** (2023) 012020 doi:10.1088/1742-6596/2594/1/012020

# Application of Novel Differential Injection-production Optimization Strategy in Tight Pioclastic Limestone Reservoir

**Dandan Hu1(✉), Shu Wang1, Limin Zhao1, Bing He2, Ruicheng Ma1, Yuanbing Wu1, Songhao Hu3**

1 RIPED, PetroChina Co. Ltd, Beijing, China
2 Zhenhua Oil Company, NORINCO GROUP, Beijing, China
3 Xinjiang Oilfield Company, China National Petroleum Corporation, Karamy, Xinjiang Province, China

hudandan@petrochina.com.cn

**Abstract.** The permeability of the ultra-low permeability bioclastic limestone reservoir in Middel East A Oilfield is about 5md. The pore structure of the reservoir is in multimodal mode and has strong micro heterogeneity. The single well productivity was low and it was difficult to sustain production after the vertical well development was adopted in the early stage, and it has been gradually transformed into horizontal well pattern at the flank area. However, the reservoir pressure and production declined rapidly, and the development effect became worse. Through the combination of reservoir micro-characteristics and macro-production performance characteristics, this paper clarifies the ultra-low permeability bioclastic limestone oil displacement mechanism and remaining oil distribution law under the micro-macroscopic dual scale, and clarifies the reasonable well spacing of vertical well pattern and horizontal injection-production well pattern. Based on the differential single well performance characteristics and water injection development mechanism, the injection-production connectivity evaluation model based on rapid numerical simulation is established, the regional differential injection-production strategy is formulated, the injection-production optimization adjustment criteria are established, the establishment of effective displacement system for the reservoir is supported, and the adjustment of more than 280 producers have been carried out in 2020-2022. The daily oil production of a single well has been restored from 307 barrels/day in 2019 to 483 barrels/day in 2022, and the water cut is reduced from 27.1% to 22.1%, the reservoir pressure recovery of 496psi provides important experience guidance for the scale benefit production of such tight bioclastic limestone reservoir.

## 1 Introduction

The carbonates are one of most common reservoirs that contain hydrocarbons in Middle East. More than four-fifths of the oil reserves are produced in carbonates, and these reservoirs are characterized by high heterogeneity and complex fracture, which makes fluid movement unpredictable and causes high water cut problems. The recovery factor of the carbonates in many cases are low ($\approx 5\%$) under depletion drive. The problems occurring in tight carbonate rocks are low injectivity so that it is difficult to establish effective displacement system. But water flooding is the most efficient and cost effective IOR technique for pressure support as well as oil displacement in carbonates if properly managed. This paper takes Mishrif reservoir in A oilfield as the target, after vertical wells were put into development in the early stage, the single well productivity was low, the production decline was fast, and the recovery factor was low. However, after the combined development of horizontal well pattern and vertical well pattern was gradually turned into the later stage, the horizontal injectors had pressure suppression, slow water injection effect, and slow pressure recovery. The microscopic characteristics of the reservoir were combined with the macroscopic production performance characteristics to define the oil displacement mechanism and remaining oil distribution law of tight bioclastic limestone under the micro-macroscopic dual-scale, so that the reasonable well spacing were determined. The injection-production connectivity evaluation model has been established based on rapid numerical simulation to evaluate the connectivity between producers and injectors, and then the main control factors well group units have been divided based on static and dynamic characteristics. Moreover, the regional differential water injection development strategy has been put forward and implemented weekly period to support the establishment of effective displacement system.

## 2 Geological Characteristics and Production Challenge

Mishrif reservoir has a burial depth of 2800 meters, a reservoir thickness of 27.0~31.0 meters, an average of 29.3 meters, and is subdivided into five sub-layers, Mi4-1 to Mi4-5. Mi4-1 and Mi4-4 show uneven light color patch fabric, and the other small layers are relatively uniform. The reservoir lithology is mainly composed of green algal pelite/grain marlstone, bioclastic pelite/grain marlstone and planktonic (coccidian) grain marlstone, which is deposited in a lower energy environment. Mi4-5 is mainly composed of planktonic (coccidian) grainstone, which belongs to low energy environment. Mi4-4 and Mi4-3 are mainly composed of green algae/bioclastic grainstone, which belongs to relatively low energy environment. Mi4-2 has increased bioclastic content, mainly composed of green algae/bioclastic packstone, green algae/bioclastic grainstone, which belongs to relatively medium-low energy environment. Mi4-1 is mainly composed of green algae/bioclastic grainstone, mixed with green algae/bioclastic packstone. In general, from Mi4-5 to Mi4-1, there is a sedimentary cycle in which the water body becomes shallow upward. Core porosity ranges from 3.4% to 27.2%, with an average of 18.4%, and core permeability ranges from 0.01 to 254.14mD, with an average of 4.1mD. The permeability of the reservoir is generally low, and the physical property of the south is better than that of the north. As to the formation properties the south is superior to the north and in the vertical the altitude order is Mi4-2, Mi4-3, Mi4-4 and Mi4-1.

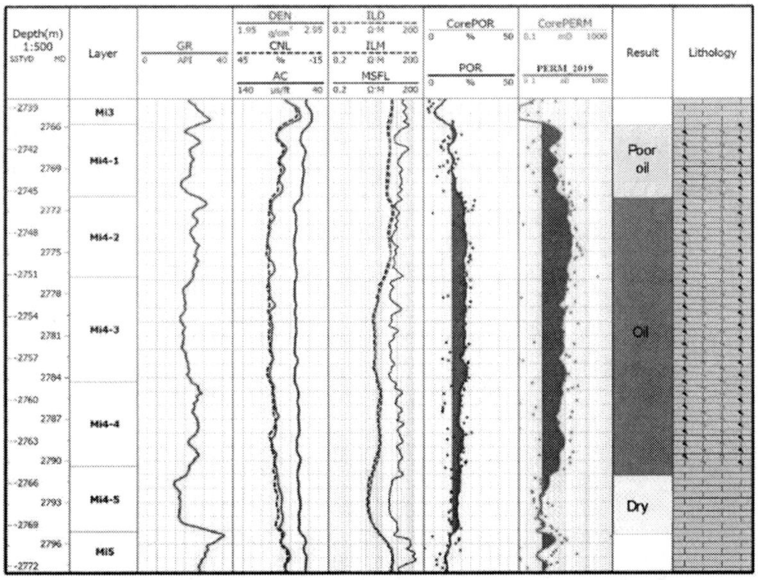


**Fig.1. Logging Interpretation Results of M Reservoir**

In the FDP plan, inverted nine-point well pattern with vertical well were jointly produced with the lower bottom water reservoir. But it was found that the combined productivity contributed little for Mishrif reservoir and the water cut increased rapidly, and the production of the vertical wells was low for wells only perforated in Mishrif reservoir. In 2017, the development mode of the horizontal well pattern was gradually started in flank area. After five years of water injection development, the well pattern in flank area has been basically improved, the reservoir pressure has been restored, and the liquid and oil production have been stabilized. However, the water injection wells have serious pressure holdup, the overall effective characteristics of water injection are not obvious, the vertical well pattern in high areas is imperfect, the pressure remains low, and a large amount of remaining oil is still enriched.

## 3 Macro-microscopic characteristics and remaining oil distribution

### *3.1 Microscopic oil displacement mechanism*

Unlike conventional sandstone, complex pore structures such as fractures and pores are widely developed in bioclastic limestone. For different bioclastic limestones, due to the influence of various factors such as sedimentary environment and diagenesis process, there will also be significant differences in pore space and throat, leading to significant differences in porosity, permeability, wettability and other physical properties.

Through the analysis of high-precision imaging results in the process of core saturation oil and water drive oil displacement, it can be seen that with the process of water drive oil, some oil in the core is driven out, oil saturation decreases, and water saturation increases. The pore structure of bioclastic limestone will affect the characteristics of phase permeability and wettability, and then affect the microscopic distribution characteristics of oil and water and the law of fluid migration. The high-resolution imaging method is used to monitor the change of fluid distribution in the micro pores of bioclastic limestone, and further clarify the micro seepage and displacement mechanism of bioclastic limestone. Taking the core of low permeability bioclastic limestone reservoir in M reservoir as an example, the pore throat structure of the core before displacement is shown in Fig. 2. It can be seen from this that the core porosity and permeability of low permeability bioclastic limestone are low and the core is dense. The statistical calculation of relevant characteristic parameters of pore structure

Journal of Physics: Conference Series **2594** (2023) 012020 doi:10.1088/1742-6596/2594/1/012020

is carried out, and the results are shown in Fig. 3. It can be seen that the pore and throat radius of the core is small, the pore branches are few, and the connectivity is poor, indicating that the tortuosity of the pore throat is small.

a. Gray image b. Core pore facies c. Core skeleton facies d. 2D image

**Fig.2. CT scan image before displacement (3D)**

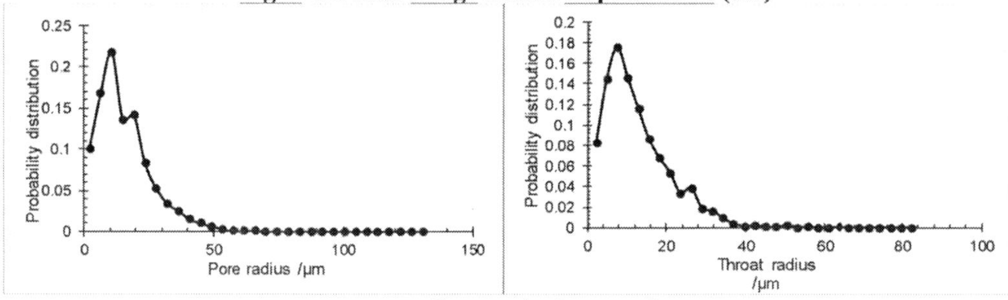


a. pore radius b. throat radius

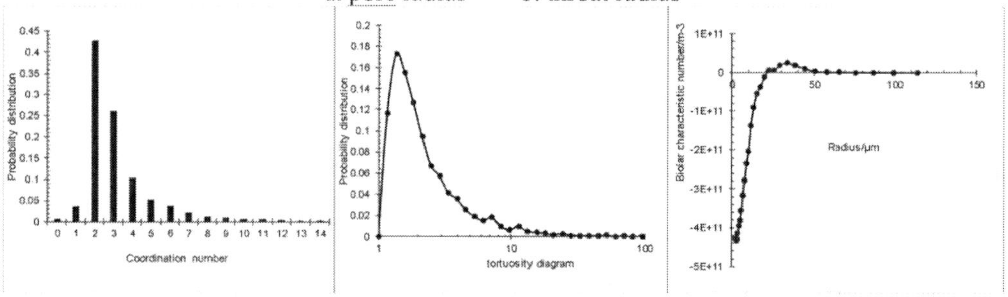


c. coordination number d. tortuosity diagram e. connectivity function

**Fig.3. Distribution of core pore structure characteristic parameters**

The remaining oil distribution types are mainly grid, multi-pore, single-hole, oil film and branching. The change of remaining oil distribution type under different displacement PV is shown in Fig. 4. It can be seen from the results that the remaining oil is mainly single-pore, and the proportion of single-pore remaining oil remains basically unchanged with the increase of displacement PV. Branched, filmy and grid-like residual oil accounts for less. The proportion of porous residual oil gradually increased. The distribution of remaining oil under different displacement PV numbers is shown in Fig. 5, and the oil saturation is shown in Fig. 6. With the continuous increase of displacement PV, the oil in the core is continuously used and the oil saturation is continuously reduced.

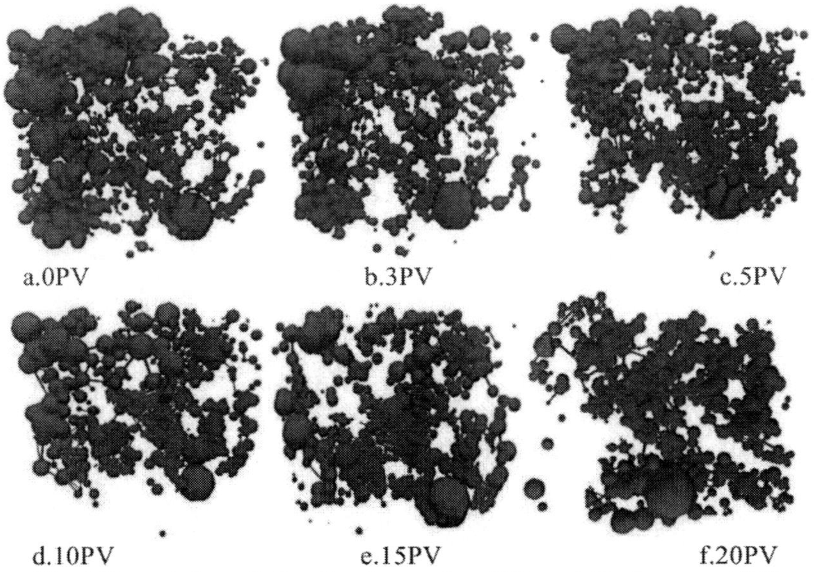


**Fig.4. Remaining oil distribution under different displacement PV**

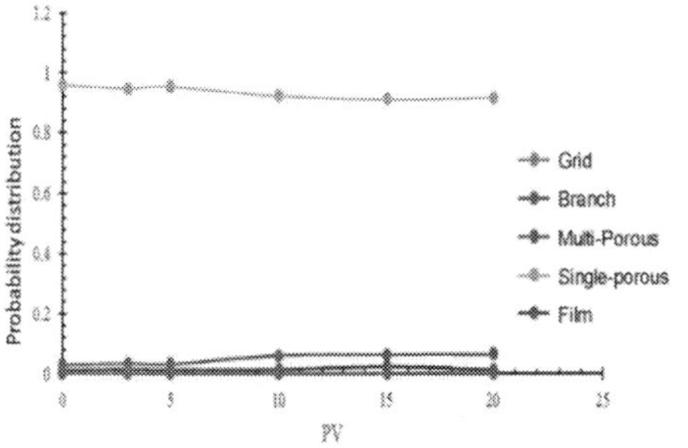


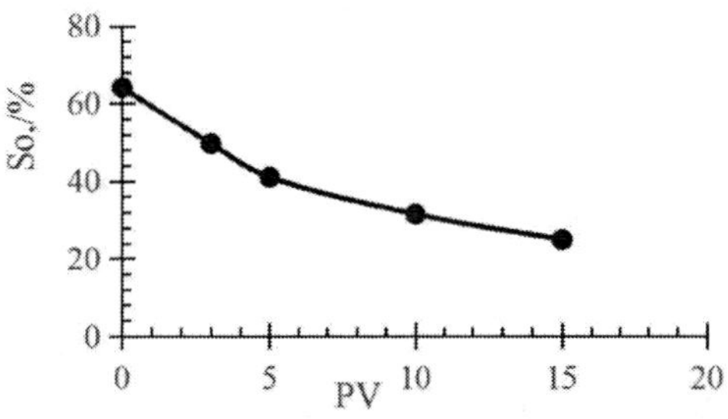


**Fig.5. The change of remaining oil distribution Fig.6. The change of oil saturation**

According to differential analysis results of the microscopic drivel, the influencing factors of residual oil saturation are: coordination number, effective connecting porosity, permeability, wetting angle, pore throat ratio, and pore size. The effect of coordination number on residual oil saturation is obvious, which indicates that the influence of pore connectivity is large.

### *3.2 Production performance and water injection effect*

There are 32 horizontal wells in Mi4 reservoir. The productivity of horizontal wells is obviously related to well trajectory through good property layer. From the perspective of productivity, the production decline of production wells is relatively slow, the production is in a continuous decline state, and the decline rate changes significantly with time. The production decline of later injectors is relatively fast during liquid drainage, and presents a two-stage change. The production decline is relatively rapid at the initial stage of liquid drainage, and the production decline rate is significantly reduced after one year of liquid drainage. For production wells, when the horizontal section is mainly located at the top of Mi4-2 sublayer, the productivity is the best. When the horizontal section is mainly located at Mi4-3 sublayer, the productivity is the second. When the horizontal section is mainly located at the middle and lower part of Mi4-2 sublayer, the productivity is relatively poor. The M0-4H well whose well trajectory crosses Mi4-2 and Mi4-3 sublayer back and forth has the worst productivity. The productivity performance of water injection wells is different from that of production wells. When the horizontal section of water injection wells is mainly located at the top of Mi4-3 sublayer, the productivity performance is the best. When the horizontal section passes through

Mi4-2 and Mi4-3 sublayer simultaneously, the productivity takes the second place. When the horizontal section passes through Mi4-3 and Mi4-4 sublayer simultaneously, the productivity is the worst. In terms of production time (including water injection horizontal wells during liquid drainage), the average flowing time of horizontal wells in Mi4 layer is about 200 days, and the flowing time of very few wells is less than one month. For horizontal wells with a flowing time of more than 800 days, the horizontal section is mainly located in Mi4-2 and Mi4-3. Among 10 horizontal wells with a flowing time of more than 800 days, 3 horizontal sections are located at the top of Mi4-2, 3 are located in Mi4-3 and 4 are located in the middle and lower part of Mi4-2.

**Table 1. Comparison of initial productivity and cum. oil of horizontal wells**

| Well Type | Well trajectory type | Initial oil rate(B/d) | Cum. Oil(Mbbls) |
|---|---|---|---|
| Producer | Mi4-2 top | 1142 | 977 |
| | Cross Mi4-2 | 791 | 864 |
| | Cross Mi4-3 | 1051 | 566 |
| Injector | Mi4-3 top | 1605 | 418 |
| | Cross Mi4-2 | 1021 | 362 |
| | Cross Mi4-3 | 1159 | 276 |
| | Cross Mi4-2 and Mi4-3 | 1270 | 280 |
| | Cross Mi4-2 and Mi4-3 many times | 632 | 93 |

The injection-production well pattern of Mi4 layer is mainly linear horizontal well pattern. The pressure in the water injection area has been significantly restored, especially around the injectors. The static pressure of the formation has reached more than 4000 psi for injectors due to the influence of water injection. The low pressure area in the reservoir is mainly located in the area where the vertical wells are not injected with water at the high crest area and the northern fracture area. The pressure at the high crest area is kept at a low level, the well pattern in the vertical well area is not perfect, and the horizontal well pattern area has the phenomenon of pressure holding, and the producer is not fully effective by waterflooding. According to the actual production performance of the effective producers, the injection water speed is 0.8-1m/d, the flow pressure and fluid volume recovery is relatively obvious, and the decline of oil production in some wells is slowed down.

### *3.3 Remaining oil distribution law*

The remaining oil in Mi4 layer is relatively rich, concentrated between the producers and the southeast of the high crest area. Relatively speaking, the recovery factor in the flank area is high, the factor in the high crest is low, the near well area is produced, and the remaining oil is relatively enriched (Fig. 7). For Mi4 layer, the remaining oil is mainly distributed at the top of the structure and between the wells, and the flooding degree of the flank area is high, indicating that the water injection development effect of the horizontal well is obviously better than that of the vertical well area. In view of the production performance that the residual oil is distributed in the vertical well area of Mi4 layer and the scope of the water injection is small by vertical well, the potential measures of the residual oil can maximize the development of water flooding to replace the residual oil and improve the reservoir ultimate recovery factor.

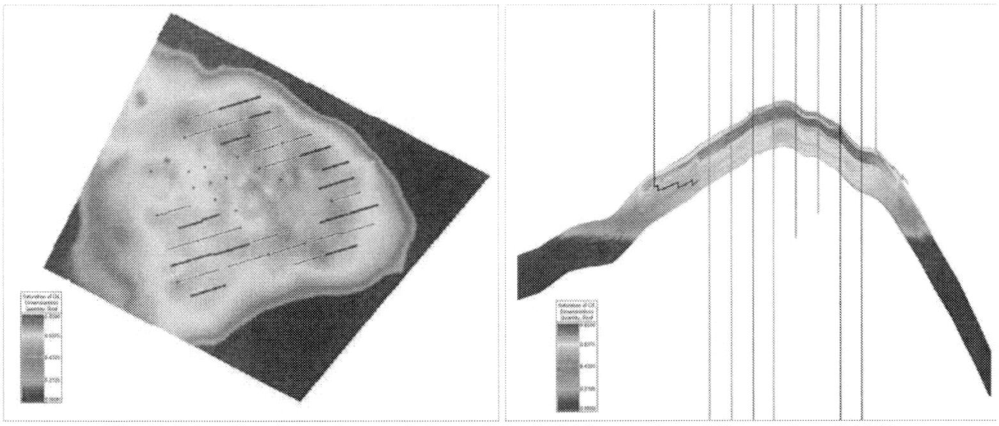

**Fig.7. Distribution of remaining oil in Mi4 layer (as of December 2021)**

### 3.4 *Rational well and row spacing*

Current row spacing of the horizontal well pattern is between 300 and 600m, and some water injection effect has been shown, which can be used as the foundation well pattern in the flank area. Vertical well area is not perfect due to low productivity, comparative analysis of five-spot, inverted nine-spot, lineal vertical well injection, horizontal well injection of four cases of prediction results, five-spot with 450 meters well distance has the best water injection effect, can keep good reservoir pressure and produce more residual oil. Otherwise, lineal vertical well injection and infilled horizontal well injection all need to drill more new wells and incremental oil shows no significant difference.

## 4 Study for water injection strategy

### 4.1 *Injection-production connectivity evaluation*

Based on the established connectivity model for the reservoirs, using water cut, daily oil rate and cum. oil production as the history matching parameters, the fitting rate of the cum. oil production was 99.7% and the water cut was 85.6%. The oil rate and the water cut match the actual value of the reservoir relatively well. The curves of the oil rate and the water cut in the later production period are basically consistent, and the connectivity evaluation results and cleavage coefficient between producers and injectors can be obtained.

Considering the influence of vertical connectivity, there are also obvious differences from the analysis of horizontal well test curve, including standard linear flow, double layer flow and transitional flow. Double layer flow type characteristics: the pressure derivative first appears a long-term 1/4 slope transition section, and then transition to 1/2, horizontal well trajectory through multiple small layers and not repeated, relatively poor connectivity. Transitional characteristics: the pressure derivative has a long-term slope of non-1/2 straight segment, layer heterogeneity and polysegment production. Standard linear flow: the pressure derivative curve has only a very short transition zone, and then maintains 1/2 slope for a long time. the horizontal well trajectory is located in the lower part of Mi4-3, with low injection capacity, well pressure and poor vertical connectivity, suspected to be affected by septal interlayer.

Combined with production characteristics, the horizontal well pattern area is divided into four kinds of main factors, respectively is double layer flow plus cross multi-sub-layer, transitional flow plus cross single sub-layer, standard linear flow plus incline single sub-layer, combined with the fault area distribution and well distance, is divided into eight units: 0-3 row, fault zone 1, fault area 2, small well distance area, 7-9 row west, 8-10 row, 8-10 row east, 8-10 row west.

**Table 2. Evaluation and analysis of well connectivity**

| Type | Well test curve | Typical well trajectory | Description and analysis |
|---|---|---|---|
| Double layer flow | 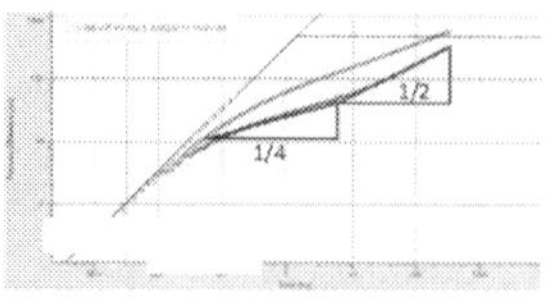  | 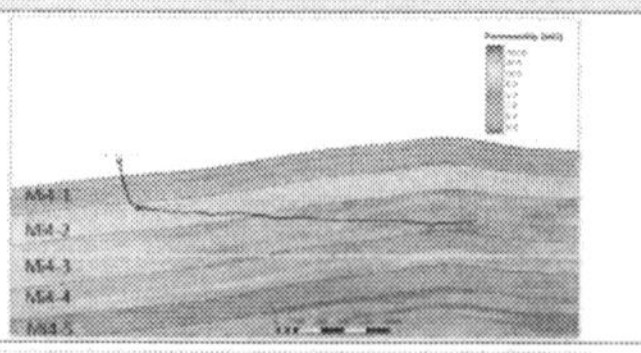 | • The pressure derivative first appears a long-term transition of 1/4 slope, and then transits to 1/2<br>• The trajectory of horizontal water injectors cross more than 2 sub-layers<br>• Obvious high injection pressure and poor injection-production connectivity |
| Transitional flow | 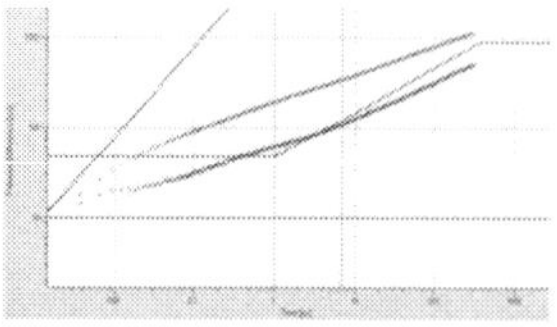 | 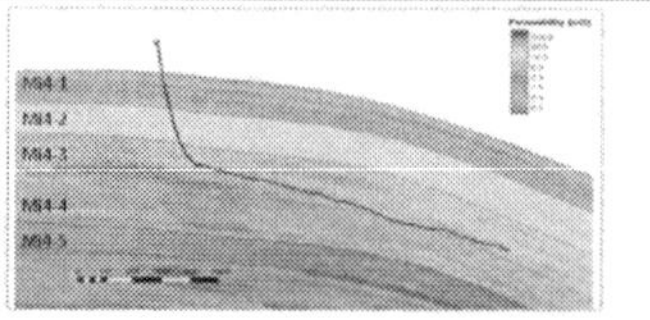 | • The pressure derivative has a long-term slope non-1/2 linear segment<br>• The trajectory of horizontal wells is mainly inclined to penetrate in the same layer<br>• Higher well injectivity and good injection-production connectivity |
| Standard linear flow | 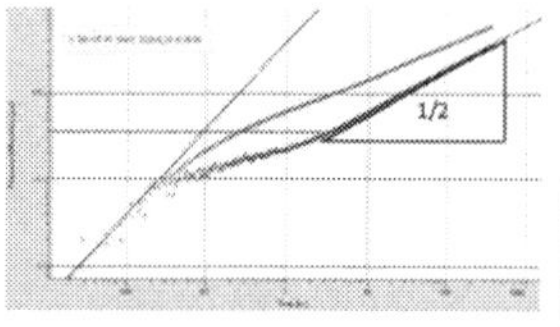  | 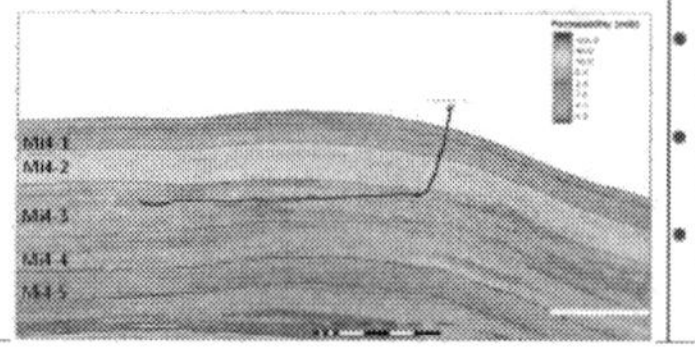 | • The pressure derivative curve only has a very short transition zone, and then maintains a slope of 1/2 for a long time<br>• The trajectory of horizontal water injectors traverses horizontally, located in the middle and lower part<br>• Lower well injectivity with high injection pressure and poor injection-production connectivity |

*4.2 Regional differential water injection strategy*

The numerical simulation results of the full reservoir show that after the reservoir pressure level reaches 0.80 of initial reservoir pressure, the rational reservoir pressure is 3600 psi, the injection-production ratio in the pressure recovery stage is suitable for 1.2-1.4, and 1.0 injection-production ratio is suitable in the later stage of water injection. From the perspective of actual production performance, the VRR reaches 0.95 when the pressure recovery for some area. According to the relationship of VRR of main controlling area, pressure restoring rate and pressure build-up of injectors, rational VRR for different main controlling area are given to ensure the stable production rate and the safety of water injection well in the water injection pipeline. Moreover, the stimulation activities, such as acid washing, deep acidification and surfactant depressurization and injection increase, can use to improve water injection effect.

After two years of injection and production optimization, the oil production for single well increased from 307b/d in 2019 to 610b/d in 2022, the water cut decreased from 27.1% to 22.1%, the average reservoir pressure restored by 331psi, and the total incremental oil reached by 520,000 tons.

**Table 3. Rational VRR for different main controlling area and static pressure of single well**

| The main controlling factor | Region | Current reservoir pressure(psi) | Well connectivity | BHP of injectors (psi) | Static pressure of single well | | |
|---|---|---|---|---|---|---|---|
| | | | | | ≥3600 | 3600-3300 | <3200 |
| Transitional flow+cross single sub-layer | Row 0-3 | 3461 | good | 4415 | 0.95-1.05 | 1.05-1.20 | 1.20-1.40 |
| | Row 5-7 west | 3389 | good | 4167 | | | |
| Standard linear flow +incline single sub-layer | Small well spacing | 3537 | fair | 4468 | - | 1.10-1.25 | - |
| | Row 8-10 middle | 3633 | fair | 4680 | 1.00-1.05 | 1.10-1.20 | 1.20-1.30 |

| | | | | | | | |
|---|---|---|---|---|---|---|---|
| Double layer flow+cross multi-sub-layer | row8-10 east | 3459 | poor | 5269 | 1.00-1.05 | 1.05-1.08 | 1.15-1.25 |
| | Row 8-10 west | 3533 | poor | 5279 | | | |
| Fracture area | Fracture area 1 | 3054 | poor | 5478 | | - | |
| | Fracture area 2 | 3647 | good | 4230 | 0.60 | - | - |

## 5 Conclusion

1)Core pore and throat radius is small, pore branches less, poor connectivity, pore throat twists and tortuous degree is small, the remaining oil mainly single pore, and with the increase of displacement for PV, single hole remaining oil proportion basically remain unchanged, residual oil saturation factors from large to small in order: coordination number, effective connection porosity, permeability, wetting angle, throat ratio, pore size.
2)Through the combination of micro characteristics and macroscopic production dynamic characteristics, the development effect of horizontal well development is influenced by well trajectory, inter-well connectivity and other factors, and there are obvious differences in the natural flowing period, cum. oil production and pressure recovery after water injection. The remaining oil is mainly concentrated between the production wells and the southeast of the high crest area. The producing degree in the flank area is relatively high, and the degree in high crest area is relatively low. The joint development mode of linear horizontal well pattern and vertical well pattern is adopted, and the row distance of the horizontal well is 300-600m, and the well spacing is 450m for vertical wells.
3)Based on the differentiated dynamic characteristics of the four main controlling factors and the water drive mechanism, the evaluation method and model of injection and production connectivity were established based on rapid numerical simulation, the differentiated water injection strategy was formulated, and the establishment of effective displacement system of tight pioclastic limestone reservoir was supported through real-time optimization of injection and production.

## Reference

[1]Roberts, D.G. Sedimentary Basins and Petroleum Geology of the Middle East. Mar. Pet. Geol. 1997, 16, 379.
[2]Wayne, M.A. 2008. Geology of Carbonate Reservoirs, first edition. Hoboken, New Jersey: John Wiley & Sons.
[3]Alyan M., Martin, J., and Irwin D. 2015. Field Development Plan Optimization for Tight Carbonate Reservoirs. Presented at the Abu Dhabi International Petroleum Exhibition and Conference, Abu Dhabi, UAE, 9-12 November. SPE-177695-MS. https://doi.org/10.2118/177695-MS.
[4]Bigno, Y., Al-Bahry, A., Melanson, D.D. et al 2001. Multilateral Waterflood Development of a Low-Permeability Carbonate Reservoir. Presented at the SPE Annual Technical Conference and Exhibition, New Orleans, Louisiana, 30 September – 3 October. SPE-71609-MS. https://doi.org/10.2118/71609-MS.
[5]Dismuke, C.T., Jokhio, S.A., Mubarak, H.K. et al 2010. Optimized Field Development of an Offshore Carbonate Reservoir in Saudi Arabia. Presented at the SPE/DGS Annual Technical

Symposium and Exhibition, Al-Khobar, Saudi Arabia, 4-7 April. SPE-136949-MS. https://doi.org/10.2118/136949-MS.

[6]Ghauri, W.K., Osborne, A.F. and Magnusen, W.L. 1974. Changing Concepts in Carbonate Waterflooding - West Texas Denver Unit Project – An Illustrative Example. SPE J Pet Technol 26 (06): 595–607. SPE-4683-PA. https://doi.org/10.2118/4683-PA.

[7]Hognesen, E.J., Strand, S., and Austad, T. 2005. Waterflooding of Preferential Oil-Wet Carbonates: Oil Recovery Related to Reservoir Temperature and Brine Composition. Presented at the SPE Europe/EAGE Annual Conference, Madrid, Spain, 13-16 June. SPE-94166-MS. https://doi.org/10.2118/94166-MS.

[8]Mahmood, S., Salazar, P., Zhao, X. et al 2017. Waterflooding in Giant Carbonate Reservoir; Successes and Challenges. Presented at the Abu Dhabi International Petroleum Exhibition and Conference, Abu Dhabi, UAE, 13-16 November.SPE-188532-MS. https://doi.org/10.2118/188532-MS.

[9]Wei, C., Li, Y., Song, B. et al 2014. Waterflooding Surveillance and Optimization for a Super Giant Carbonate Reservoir.Presented at the SPE Annual Technical Conference and Exhibition, Amsterdam, the Netherlands, 27-29 October. SPE-170621-MS. https://doi.org/10.2118/170621-MS.

[10]Mukanov, A., Bigeldiyev, A., Batu, A. et al 2020. Features of Field Development with Tight Carbonate Reservoirs by Waterflooding. Paper presented at the SPE Annual Caspian Technical Conference in Nur-Sultan, Kazakhstan. October 21 – 22, 2020. SPE-202534-MS. https://doi.org/10.2118/202534-MS.

[11]Twartz, S., Gorjy, F., and Milne, I. et al 1998. A Multiple Realisation Approach to Managing Uncertainty In The North Rankin Gas Condensate Field, Western Australia. Paper presented at the 1998 SPE Asia Pacific Oil & Gas Conference and Exhibition held in Perth, Australia, 12-14 October. SPE-500078. https://doi.org/10.2118/50078-MS.

[12]Briceno, M.; Summers, L.E.; Quijada, O.J. Horizontal Wells Pressure Build Up Analysis in Orinoco Heavy Oil Belt. In Proceedings of the SPE International Thermal Operations and Heavy Oil Symposium and International Horizontal Well Technology Conference, Calgary, AB, Canada, 4–7 November 2002.

[13]Feng, W.C.; Nurafza, P.; Al-Shamma, B.; Gringarten, A. Bilinear Flow in Horizontal Wells in a Homogeneous Reservoir:Huntington Case Study. In Proceedings of the European Conference 2015, Madrid, Spain, 1–4 June 2015.

[14]Baba, A.; Azzouguen, A.; Mazouzi, A.; Tiab, D.; Aissaoui, K. Determination of the Controlling Factors and Origins of the Bilinear Flow from Horizontal Well Transient Responses. In Proceedings of the Canadian International Petroleum Conference, Calgary,AB, Canada, 11–13 June 2002.

[15]Yonglv, J.; Xuewen, Y.; Jianguo, G. Study on Variable Well Storage in Well Test Analysis. Xinjiang Pet. Geol. 1998, 19, 4.

[16]Shiqing, C.; Weirong, X.; Kangliang, G.; Jianlin, L. Typical Curve Characteristics and Case Analysis of Variable Wellbore Storage Effect. Well Test. 1992, 4, 9.

[17]Keyi, Z.; Lize, L.; Liping, J.; Kangliang, G.; Fei, L.; Haiping, H.; Xin, L. Pressure buildup curve behavior of pore type carbonate reservoir in AH Oilfield of Iraq. Fault-Block Oil Gas Field 2014, 21, 762–766. [CrossRef]

[18]Liu, Y.; Wu, X.; Wang, Y.; He, F.; Yuan, Y. Build a bridge between well testing theory and application. Well Test. 2006, 15, 5.

[19]Enjie, H.; Lize, L.; Keyi, Z.; Xinmin, W.; Liping, J.; Zhaolei, S.; Kai, H. Water Production Characteristics and Control Measures of Porous Carbonate Reservoir in Iraq. J. Yangtze Univ. 2015, 12, 1–6.

[20]Yong, L.; Qi, Z.; Chenji, W.; Benbiao, S. Dynamic Characterization of Different Reservoir Stacked Patterns for a Giant Carbonate Reservoir in Middle East. In Proceedings of the SPE Reservoir Characterisation and Simulation Conference and Exhibition, Abu Dhabi, United Arab Emirates, 5 May 2017.

[21]Gringarten, A.C.; Ramey, H.J.J. The Use of Source and Green's Functions in Solving Unsteady-Flow Problems in Reservoirs. Soc.Pet. Eng. J. 1973, 13, 285–296.
[22]Xiangyan, K. Advanced Seepage Mechanics, 2nd ed.; University of Science and Technology of China Press: Hefei, China, 2010;pp. 203–233.
[23]Nengqiang, L. Practical Modern Well Test. Interpretation Method, 4th ed.; Petroleum Industry Press: Beijing, China, 2009; pp. 149–153.
[24]Xiaoping, L.; Liehui, Z.; Qiguo, L. Well Test. Analysis Method, 1st ed.; Petroleum Industry Press: Beijing, China, 2008; pp. 50–53.
[25]Suzuki, K.; Nanba, T. Horizontal Well Pressure Transient Behavior in Stratified Reservoirs. In Proceedings of the SPE Annual Technical Conference and Exhibition, Dallas, TX, USA, 6–9 October 1991.
[26]Xiaoyun, L.; Yongqiang, W.; Chong, C.; Lianchao, J.; Lu, L. Theory and Application of Unstable Well Test Analysis for Horizontal Wells. Well Test. 2012, 21, 5–10.
[27]Liang, W.; Ruiting, B. Well Test Analysis and Application of Horizontal Well In Low Permeability Reservoir. West.-China Explor.Eng. 2010, 22, 44–47.
[28]Xin, L.; Baozhen, J.; Zheng, C.; Yang, W.; Ting, Z. Application of Combination of Dynamic Analysis and Numerical Well Test In Solving Multi-solution Problem of Well Test. Neijiang Technol. 2014, 35, 3.

Journal of Physics: Conference Series **2594** (2023) 012021 doi:10.1088/1742-6596/2594/1/012021

# Integrated study of structural styles and their suitability for hydrocarbon entrapment, Southern Kirthar Fold Belt, Pakistan

**A A Halepoto[1*], S Ahmed[2], M H Agheem[1], A A A D Hakro[1], R A Lashari[1] and S B Ahmedani[1]**

[1]Centre for Pure and Applied Geology, University of Sindh, Jamshoro, Pakistan

[2]Oil and Gas Development Company Limited (OGDCL), Islamabad, Pakistan

[*]Corresponding Author's Email: aijaz.halepoto@usindh.edu.pk

**Abstract.** Present study is concerned with the interpretation of satellite imagery and 2D seismic profiles to understand the structural styles and petroleum system elements of the Southern Kirthar Fold Belt of Pakistan. Southern Kirthar Fold Belt is the part of regional Himalaya-Suleiman and Kirthar fold belt chain, which is produced in the Cenozoic due to collision and under-thrusting of the Indian Plate with Eurasian Plate. The exploration activities are being carried out in the study area since 19$^{th}$ century and number of gas discoveries are made, but many structures are still barren of hydrocarbons. Present study identifies four structural trends, which are 0, 350,330 and 020 in relation with 010 trending transpressional Ornach-Nal plate boundary. Three different styles of the studied anticlines are identified on the basis of seismic interpretation, which are anticlines cut by both fore-thrusting and back-thrusting reverse faults, anticlines cut by only fore-thrusting reverse faults and anticlines not cut by faults. Sufficient throw along the faults is identified in the seismic profiles, which can provide the juxtaposition trapping as well as recharging medium from the source to trap. Gas fields are discovered from the four out of six studied anticlines irrespective of style.

## 1. Introduction

Almost entire discovered hydrocarbons of the world are produced from the subsurface traps developed in the geological structures, therefore, their study is exceptionally imperative for the hydrocarbon exploration [1] [2]. Recognition of structural traps is one of the decisive and vital task in hydrocarbon exploration [3]. Fold belts developed in contractional regimes consists of the structural traps for a large part of the most important petroleum provinces of the world such as Middle East and Venezuela [2]. Fold belts are active and conspicuous tectonic features of convergent continental plate boundaries, and are very important areas for the exploration of hydrocarbons [4]. They contain more than 700 billion barrels out of 2800 billion barrels of oil of known oil reserves [5]. Kirthar Fold Belt is the part of an active regional foreland fold belt chain, known as Himalaya-Suleiman and Kirthar fold belt (figure 1), which developed on the north western boundary of the Indian Plate due to the collision and under-thrusting of Indian Plate with different tectonic blocks of Eurasian Plate during the Tertiary and Quaternary [6] [7] [8] [9] [10]. Present study area is located in the southern extreme of this active regional foreland fold belt chain and it is an emerging hydrocarbon province of Pakistan. Number of gas fields such as Mazarani (1959), Sari (1966), Hundi (1971), Bhit (1997), Badhra (2000) and Rehman (2015) etc. has been discovered in the Kirthar Fold Belt from shallow reservoirs, generally not more than 2000m deep (figure 2).

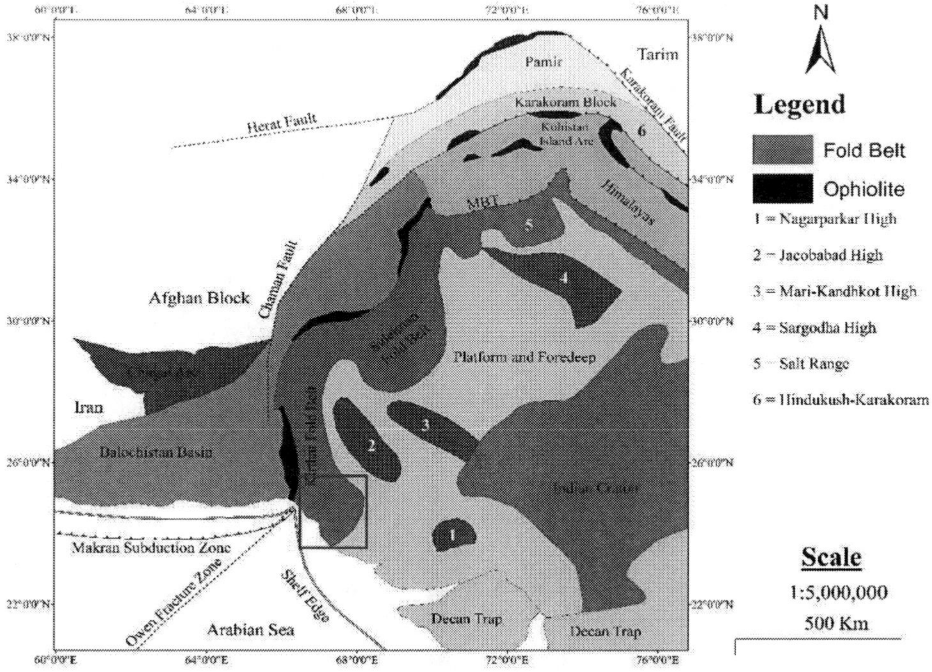


**Figure 1.** Map of regional tectonic framework showing study area in red rectangle (modified after [11] [6] [12] [4]).

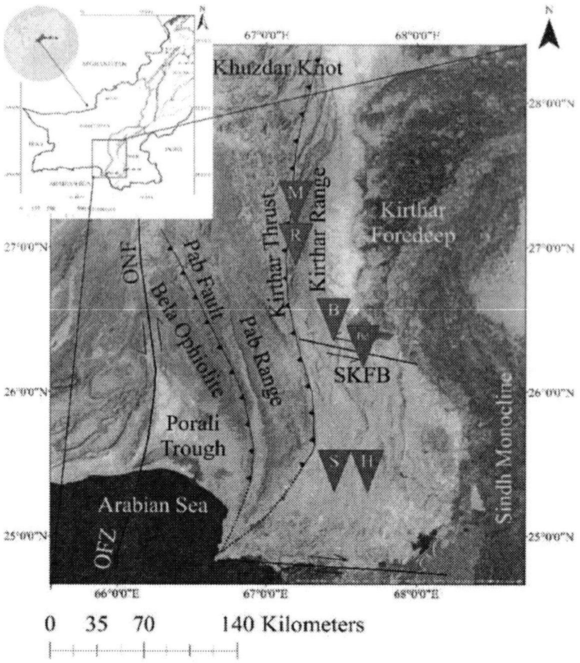


**Figure 2.** Landsat 8 True Colour Composite of Southern Kirthar Fold Belt, showing location of discovered gas fields in filled red triangles (modified after [13] [28] [29] [12] [7]). M=Mazarani, Z=Zamzama, Bh=Bhit, Ba=Badhra, S=Sari, H=Hundi.

The stratigraphy, petroleum system elements and main tectonic episodes of the Kirthar Fold Belt are shown in figure 3. The Precambrian basement rocks are exposed in the southeast of study area. However, Paleozoic sequences are not yet reported from the study area, but well developed very thick Mesozoic and Cenozoic sequences exist in the study area.

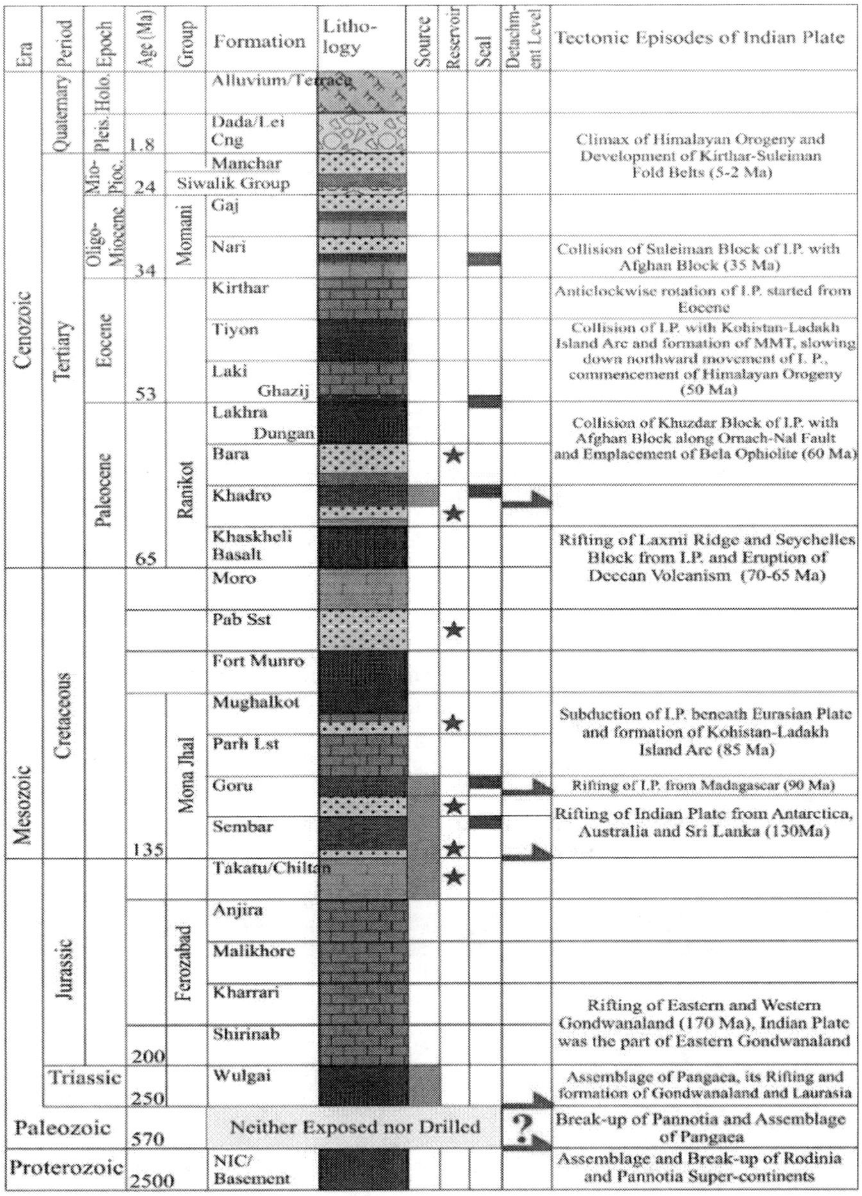


**Figure 3.** Stratigraphy, Petroleum System Elements, and Detachment Levels of the SKFB in relation with tectonic episodes of the Indian Plate (modified after [13] [6] [14] [15] [16] [17] [18] [19] [20] [21] [22] [23]).

The source rock potential has been reported from the Late Jurassic, Early Cretaceous and Early Paleocene rocks. The Late Cretaceous Pab Sandstone and Paleocene Ranikot Group are chief producing reservoirs in the Kirthar Fold Belt. The deeper reservoirs such as Jurassic carbonates and sand bodies of different Cretaceous strata such as Goru and Mughalkot formations are not yet proved proliferous. The exploration activities in the Southern Kirthar Fold Belt (henceforth SKFB) started in the early of the 20th century. Geological Survey of India and Burma Oil Company Limited (BOCL) carried out reconnaissance fieldwork and geological mapping of the study area in the last quarter of 19th and first half of the 20th century respectively. The earliest exploratory well named Drigh Road-1

Journal of Physics: Conference Series **2594** (2023) 012021 doi:10.1088/1742-6596/2594/1/012021

for the search of hydrocarbons was drilled by British Oil Company (BOC) in 1939 in the Karachi city (table 1). They observed some oil shows in the Oligocene Nari Formation and Miocene Gaj Formation. Although well does not yield any recoverable hydrocarbons, hence abandoned. The first gas discovery in the Kirthar Fold Belt was made by Pakistan Petroleum Limited (PPL) in the Mazarani Folded Zone in Kirthar Range during 1958. Pak Hunt Oil Company drilled Phulji-1 in 1958 (Kirthar Foredeep), which proved dry hole, but on the same prospect BHP drilled Zamzama-1 in 1998 and discovered hydrocarbon reserves of 9.5 MMBO and 1,755 BCFG. Pak Hunt Oil Company also drilled Sari and Badhra anticlines of the SKFB, but failed to discover the hydrocarbons. However, later gas reserves were discovered in both these structures when Oil and Gas Development Company Limited (OGDCL) and LASMO drilled these structures in 1966 and 1998 respectively. The history of exploratory wells drilled in the SKFB is tabulated in table 1, which shows the success ratio of gas discoveries has increased in recent times. But there are many structures in the SKFB, where no hydrocarbon discovery is yet made. It indicates some sort of deficit in the petroleum system of the study area, which might be poor or absence of charging system and deterioration and leakage of traps. These exploration risks are associated with structural styles of the study area, therefore present work is aimed to evaluate the subsurface structural styles through seismic interpretation to understand the presence of effective charging system, trap integrity, timing of hydrocarbon generation and trap development.

## 2. Data and Methodology

Satellite image (754-RGB) of Landsat 8 satellite, 2D seismic profiles and well data is used for the present study. The surface structures are marked on the satellite image to understand the regional structural trend and inter-relationship of the structures and plate boundary. Total six 2D seismic profiles of the study area are interpreted, four digitally and two manually, in terms of subsurface stratigraphy, folds and fault pattern to understand the availability of effective charging system and intact traps. PetroMod (Petroleum System Modeling) of the SKFB is developed by [24] is adopted in present work to understand the maturation, generation and migration history of hydrocarbons. Deformation episodes of the study area summarised in figure 3 and correlated with PetroMod to understand the history of trap development and hydrocarbon entrapment in these traps.

**Table 1.** Exploration History in the SKFB, discoveries are highlighted in green (source: Pakistan Energy Yearbook, 2021).

| Well Name | Company | Year | Total Depth | Formation | Structure Drilled | Status |
|---|---|---|---|---|---|---|
| Drigh Road-1 | BOC | 1939 | 515m | Nari Formation | No Surface structure | Abandoned |
| Lakhra-1B | BOC | 1948 | 3860m | Sembar | Lakhra anticline | Abandoned |
| Karachi-1 | PPL | 1956 | 3036m | Korara Shale | No Surface structure | Abandoned |
| Sari Singh-1 | Pak-Hunt | 1957 | 945m | Upper Ranikot Group | Sari anticline | Abandoned |
| Lakhra-1 | Pak-Hunt | 1957 | 3625m | Goru Formation | Lakhra anticline | Abandoned |
| Sumbakh-1 | Pak-Hunt | 1957 | 1933m | Upper Ranikot Group | Sumbakh anticline | Abandoned |
| Mazarani-1 | PPL | 1958 | 3084m | Fort Munro Formation | Kirthar Range | Gas |
| Badhra-1 | Pak-Hunt | 1958 | 4373m | Pab Sandstone | Badhra anticline | Abandoned |
| Karachi-2 | PPL | 1959 | 3947m | Korara Shale | No Surface structure | Abandoned |

| | | | | | | |
|---|---|---|---|---|---|---|
| Sari-1 | OGDCL | 1965 | 1480m | Lower Ranikot Group | Sari anticline | Gas |
| Benir-1 | OGDCL | 1967 | 2348m | Ranikot Group | Benir anticline | Abandoned |
| Kand-1 | OGDCL | 1968 | 2400m | Korara Shale | Kand anticline | Abandoned |
| Hundi-1 | OGDCL | 1969 | 2405m | Ranikot Group | Hundi anticline | Gas |
| Gadularo-1 | OGDCL | 1971 | 1860m | Lower Ranikot Group | Gadularo anticline | Abandoned |
| Khothar-1 | OGDCL | 1972 | 1901m | Lower Ranikot Group | Khothar anticline | Abandoned |
| Surjan-1 | OGDCL | 1979 | 3616m | Goru Formation | Surjan anticline | Abandoned |
| Sann-1 | OGDCL | 1985 | 3770m | Chiltan Formation | Lakhra anticline | Abandoned |
| Bhit-1 | LASMO | 1996 | 1839m | Pab Sandstone | Bhit anticline | Abandoned |
| Bhit-2 | LASMO | 1996 | 2067m | Fort Munro Formation | Bhit anticline | Gas |
| Badhra-2 | LASMO | 1998 | 3496m | Goru Formation | Badhra anticline | Gas |
| Lalian-1 | Tullow | 2001 | 3215m | Fort Munro Formation | Lakhra anticline | P & A |
| Zirkani-1 | PKP | 2003 | 2816m | Pab Sandstone | Kirthar Range | Gas |
| Chung-1 | PKP | 2004 | 2026m | Mughalkot Formation | Chung anticline | P & A |
| Maliri-1 | Premier | 2005 | 2989m | Pab Sandstone | Maliri anicline | P & A |
| POGC Rehman-1 | POGC | 2009 | 2801m | Mughalkot Formation | Kirthar Range | Gas |
| Ludali-1 | ENI | 2013 | 2660m | Mughalkot Formation | No Surface structure | Gas |
| Rizq-1 | POGC | 2014 | 3174m | Mughalkot Formation | Kirthar Range | Gas |
| Kotri North X-1 | PPL | 2015 | 3651m | | Lakhra anticline | Temp. Suspended |
| Unarpur-1 | UEP | 2018 | 3940m | Goru Formation | Lakhra anticline | Gas |
| Unarpur-2 | UEP | 2019 | 4137m | Goru Formation | Lakhra anticline | Gas |
| Rayyan-1 | POGC | 2023 | | Pab Sandstone | Kirthar Range | Gas |

## 3. Results and Discussion

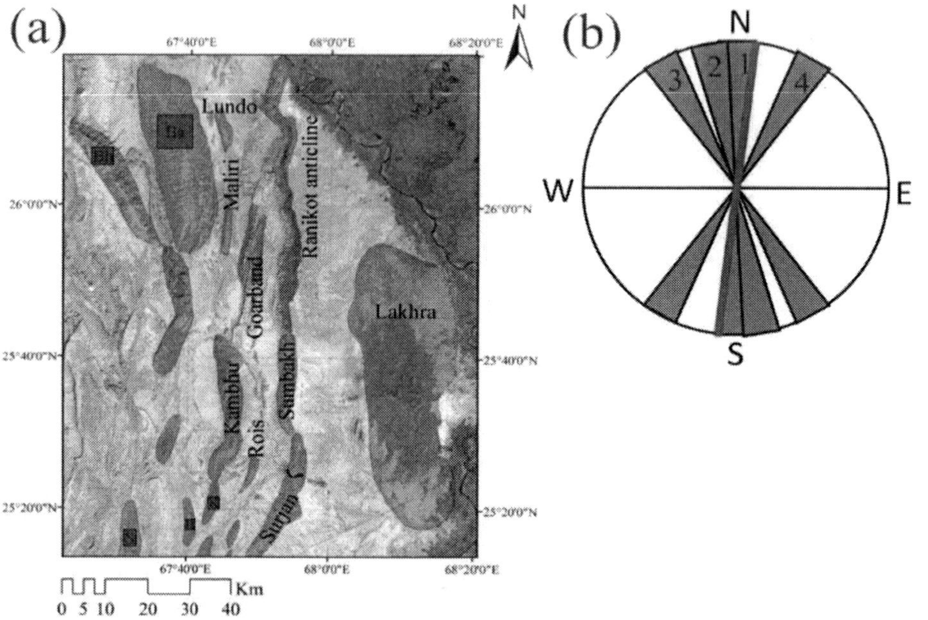


**Figure 4.** (a) Landsat 8 (OLI TIRS) 754-RGB image of the study area showing the locations of some prominent anticlines (highlighted in light blue colour). (b) Rose diagram showing the trends of the structures in the SKFB, Pakistan. Bold red line in rose diagram represents the trend of transpressional plate boundary (Image source: earthexplorer.usgs.com).

The conspicuous anticlineal structures of the SKFB are outlined on Landsat 8 (OLI TIRS) 754-RGB image (figure 4a). Image shows that most of the exposed structures trends NS direction. But, other structural trends are also identified in the SKFB as shown in rose diagram (figure 4b). Rose diagram indicates four different structural trends. Azimuthally, these trends are 0, 350,330 and 020 on the basis of their relative dominance. The bold red line in figure 4b trends 010, which shows the trend of the transpressional Ornach-Nal plate boundary in the west. Most of the structural trends are parallel to sub-parallel with the trend of plate boundary, therefore it is interpreted that these structures are produced due to the compressional forces generated along the Ornach-Nal plate boundary. Some structural trends deviate to nearly oblique to the plate boundary such as 330 and 020 in the study area (figure 4 a&b), which might be due to either subsurface thrust trend or movement on the oblique ramp or concealed tear fault.

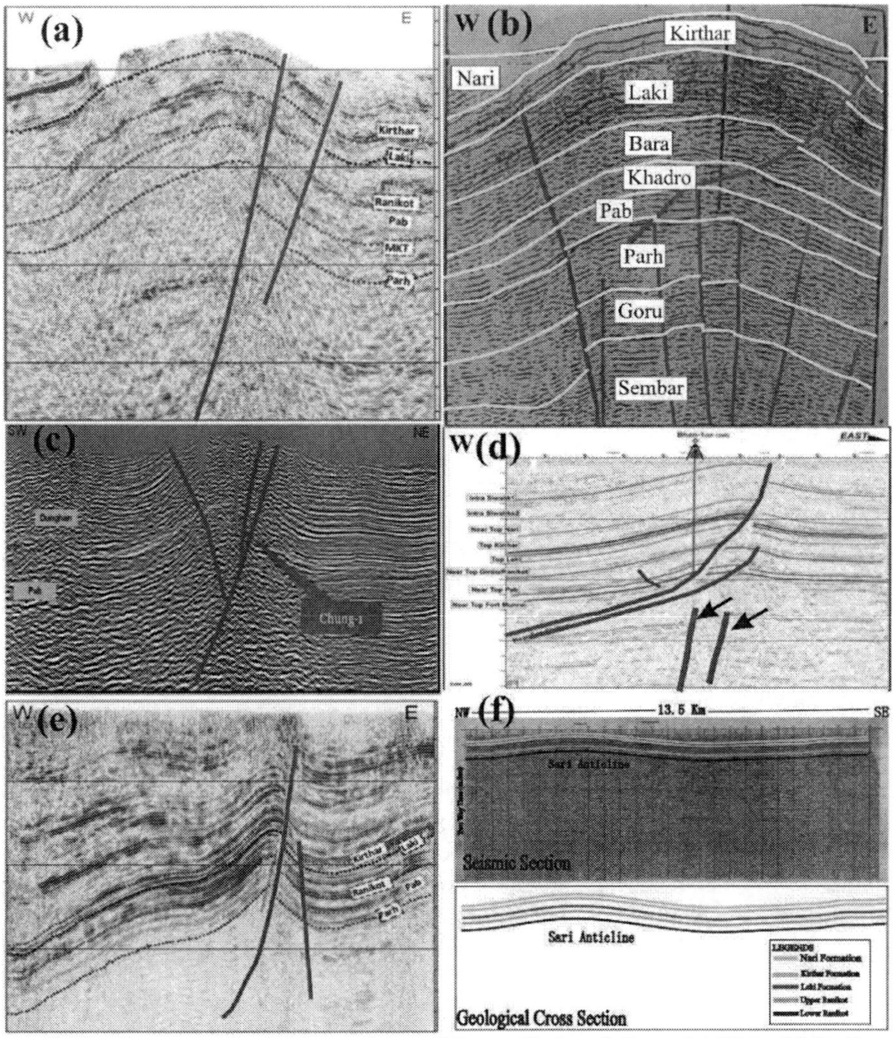


**Figure 5.** The interpreted seismic sections of the studied anticlines, (a) Bhit, (b) Badhra, (c) Chung, (d) Bhan. (e) Zamzama and (d) Sari anticline. Faults are shown by bold red lines, deeper faults in infra-Cretaceous sequence of the Bhan anticline are shown with black arrows (modified after [9] [25]).

The seismic sections of Bhit, Badhra, Chung, Bhan, Zamzama and Sari anticlines are interpreted. Gas field was discovered by ENI LASMO from Bhit anticline in 1997. Seismic interpretation of the Bhit anticline indicates that structures is cut by steeply west dipping, emergent as well as blind reverse faults (figure 5a). Stratigraphy of Mid-Cretaceous to Recent is identified in the structure, with producing tight gas reservoir in the Upper Cretaceous Pab Sandstone. The seismic section of Badhra anticline shows that anticline is cut by listric, emergent fore-thrusting and steeply back-thrusting blind faults. Vertical to sub-vertical concealed faults are also observed in the Mesozoic and older sequence, which are believed to be inverted normal faults of Mesozoic rifting (figure 5b). Stratigraphic sequence from Early Cretaceous to Recent is identified in the structure, with producing reservoir in the Upper Cretaceous Pab Sandstone. Gas field was discovered by ENI LASMO from this anticline in 1998. The seismic section of the Chung anticline shows the anticline is cut by steep fore-thrusting and back-thrusting emergent reverse faults (figure 5c). The anticline was drilled by PKP in 2004 down to the

Mid-Cretaceous Mughalkot Formation penetrating through Paleocene Ranikot Group and Pab Sandstone reservoirs, but no discovery is made from this anticline. The seismic section of Bhan anticline shows that the anticline is bounded by listric fore-thrusting blind reverse faults (figure 5d). The stratigraphy from Jurassic Chiltan Limestone to Recent is identified in the seismic interpretation of the Bhan anticline. Sub-vertical, steeply foreland dipping inverted faults are also present in the Jurassic and older rocks as identified in the seismic section. The seismic profile of the Bhan anticline shows that anticline is detached from the top Jurassic Chiltan Limestone, which suggest that most of the structures in the SKFB might have detached from the top Chiltan Formation. Detachment lies much deeper in the rest of studied structures and couldn't be identified in the interpreted seismic profiles. The anticline was drilled down to the Mid-Cretaceous sequence, but no discovery is made from this anticline. The seismic profile of the Zamzama anticline shows that the forelimb of the anticline is cut by steeply west dipping, fore-thrusting blind reverse fault as well as east-dipping back-thrusting reverse fault (figure 5e). Stratigraphic sequences from Mid-Cretaceous Parh Limestone to Recent is identified in the structure, with gas producing reservoir in the Upper Cretaceous Pab Sandstone and Paleocene sands of the Ranikot Group. This structure was earlier drilled by Pak-Hunt in 1958 without discovery by BHP re-drilled it in 1998 and discovered commercial quantity of gas reserves. The Sari anticline is the least deformed structure that is not cut by any emergent fault (figure 5f). Pak Hunt drilled the wildcat well Sari Singh-1 on this structure in 1957 down to the Upper Paleocene Dunghan Formation, but failed to produce hydrocarbons. However, eight years later, similar prospect was drilled by OGDCL down to the Early Paleocene Khadro Formation and made gas discovery from the sands of Khadro and Bara Formations of Paleocene Ranikot Group in 1966.
Three different structural styles are identified in the study area on the basis of seismic interpretation, which are:

1. Anticlines cut by both fore-thrusting and back-thrusting reverse faults, such as Badhra and Chung anticlines
2. Anticlines cut by only fore-thrusting reverse faults, such as Bhit, Bhan and Zamzama anticlines
3. Anticlines not cut by faults, such as such as Sari anticline

Seismic interpretation indicates that there is sufficient throw along the faults, which can provide the juxtaposition trapping in the reservoir. These faults can also act as recharging medium for hydrocarbon migration from the source to trap. Deeper faults present in the Mesozoic and older rocks, as identified in the Badhra and Bhan anticlines, therefore, there are bright chances that traps of Chiltan and Goru formations might have charged for hydrocarbons through these faults.
PetroMod (figure 6) indicates that the oil generation in the SKFB has been started in Mid Cretaceous, much before the deposition of reservoirs, such as Pab Sandstone, Ranikot Group and development of compressional structural traps. Therefore, it could be postulated that most of the oil generated has been leaked. Early compressional structural traps were developed in the Late Paleocene, when Indian Plate collided with the Afghan Block of Eurasian Plate along the Ornach-Nal transpressional boundary and Bela-Muslimbagh Ophiolites were emplaced [14] [9]. Other episode of compressional forces occurred during Mio-Pliocene and Pleistocene [26] [27] [7]. Gas started to generate from the Eocene and onwards, when reservoirs were deposited and compressional traps were present in the study area. Therefore, all the discoveries in the fold belt are gas-prone.

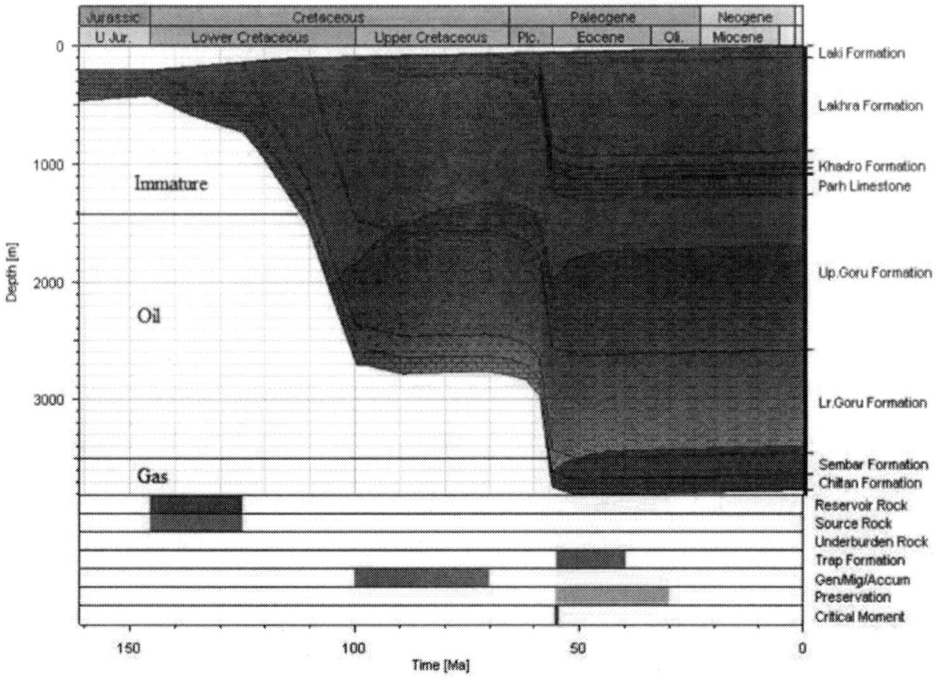


**Figure 6.** PetroMod 1D showing burial history for Well X in SKFB showing Petroleum System Element chart and Hydrocarbon zone for the Sembar Formation (Adopted after [24]).

## 4. Conclusions

1. Four structural trends, three types of structural styles and stratigraphy from Jurassic to Recent is identified during present study
2. Gas fields are discovered from four (Bhit, Badhra, Zamzama and Sari) out of six studied anticlines, irrespective of the structural style
3. Deep reservoirs (Jurassic-Chiltan and Cretaceous-Goru, Mughalkot formations) in these structures might contain commercial hydrocarbon quantities, as indicated by the presence of juxtaposition trapping and charging medium along the faults

## Acknowledgments

Authors are highly indebted to the Technical Committee of the International Petroleum and Petrochemical Technology Conference (IPPTC)-2023 for providing the chance to present our work on a very prestigious professional forum. The authors are highly thankful to University of Sindh, Jamshoro, Pakistan for providing facilities to carry out the research work for present study. We are also indebted to the Director of Centre for Pure and Applied Geology for providing technical support and cooperation to present this work in the IPPTC-2023.

## References

[1] T. P. Harding and J. D. Lowell, "Structural Styles, Their Plate-Tectoic Habitats, and Hydrocarbon Traps in Petroleum Provinces," *American Association of Petroleum Geologists Bulletin*, vol. **63**, no. 7, pp. 1016–1058, 1979, doi: 10.1306/2f9184b4-16ce-11d7-8645000102c1865d.

[2] J. Gluyas and R. Swarbrick, *Petroleum Geoscience.* Blackwell Science Ltd, 2004.

[3] G. V. Chilingar, L. A. Buryakovsky, N. A. Eremenko, and M. V. Gorfunkel, *Development in Petroleum Science 52: Geology and Geochemistry of Oil and Gas*, First Edit. Elsevier, 2005.

[4] M. Humayon, R. J. Lillie, and R. D. Lawrence, "Structural Interpretation of the Eastern

Suleiman Foldbelt and Foredeep, Pakistan," *Tectonics*, vol. 10, no. **2**, pp. 299–324, 1991.

[5] D. Roeder, "Fold–thrust belts at Peak Oil," in *Hydrocarbons in Contractional Belts*, G. P. Goffey, J. Craig, T. Needham, and R. Scott, Eds. London: The Geological Society, London, 2010, pp. 7–31.

[6] C. M. Powell, "A Speculative Tectonic History of Pakistan and Surroundings: Some constraints from Indian Ocean," in *Geodynamics of Pakistan*, A. Farah and K. A. De Jong, Eds. Quetta: Geological Survey of Pakistan, 1979, pp. 5–24.

[7] D. Bannert, A. Cheema, A. Ahmed, and U. Schaffer, "The Structural Development of the Western Fold Belt, Pakistan," *Geologisches Jahrbuch Reihe*, vol. B80, pp. 3–60, 1992.

[8] P. J. Treloar and C. N. Izatt, "Tectonics of the Himalayan collision between the Indian Plate and the Afghan Block: A synthesis," *Geological Society Special Publication*, vol. **74**, pp. 69–87, 1993, doi: 10.1144/GSL.SP.1993.074.01.06.

[9] R. Hedley, J. Warburton, and J. Smewing, "Sequence Stratigraphy and Tectonics in the Kirthar Foldbelt, Pakistan," *PAPG/SPE Annual Technical Conference, 7-8 November 2001, Islamabad, Pakistan*, no. November 2001, 2001.

[10] R. Hinsch, C. Asmar, M. Nasim, M. A. Abbas, and S. Sultan, "Linked thick-to thin-skinned inversion in the central Kirthar Fold Belt of Pakistan," *Solid Earth*, vol. **10**, no. 2, pp. 425–446, 2019, doi: 10.5194/se-10-425-2019.

[11] R. Stonely, "Evolution of the continental margin bounding a former Tethys," in *The Geology of continental margins*, C. L. Drake and C. A. Burke, Eds. Springer-Verlag, New York, 1974.

[12] A. H. Kazmi and R. A. Rana, "Tectonic Map of Pakistan," 1982.

[13] Hunting Survey Corporation, *Reconnaissance Geology of Part of West Pakistan*. Toronto, Canada: A Colombo Plan Co-operative Project, 1960.

[14] F. Allemann, "Time of Emplacement of the Zhob Valley Ophiolites and Bela Ophiolites, Baluchistan (Preliminary Report)," in *Geodynamics of Pakistan*, A. Farah and K. A. De Jong, Eds. Quetta: Geological Survey of Pakistan, 1979, pp. 215–242.

[15] E. J. Barron and C. G. A. Harrison, "An analysis of past plate motions; the South Atlantic and Indian oceans," in *Mechanisms of continental drift and plate tectonics*, P. A. Davies and S. K. Runcorn, Eds. Academic Press: London, 1980, pp. 89–109.

[16] H. A. Raza, S. M. Ali, and R. Ahmed, "Petroleum Geology of Kirthar sub-Basin and Part of Kutch Basin," *Pakistan Journal of Hydrocarbon Research*, vol. **2**, no. 1, pp. 29–73, 1990.

[17] P. S. Plummer and E. R. Belle, "Mesozoic Tectono-stratigraphic Evolution of the Seychelles Micro-continent," *Sedimentary Geology*, vol. **96**, pp. 73–91, 1995.

[18] I. B. Kadri, *Petroleum Geology of Pakistan*. Karachi: Pakistan Petroleum Limited, 1995.

[19] F. K. Bender and H. A. Raza, Eds., *Geology of Pakistan*. Gebruder Borntraeger, 1995.

[20] A. H. Kazmi and M. Q. Jan, *Geology and Tectonics of Pakistan*. Karachi: Graphic Publishers, 1997.

[21] S. M. I. Shah, Ed., *Stratigraphy of Pakistan*, Mem. 22. Quetta: Geological Survey of Pakistan, 2009.

[22] N. K. Siddiqui, *Petroleum Geology, Basin Architecture and Stratigraphy of Pakistan*. Published by the Author, 2016.

[23] S. Ahmed, S. H. Solangi, M. S. K. Jadoon, and A. Nazeer, "Tectonic evolution of structures in Southern Sindh Monocline, Indus Basin, Pakistan formed in multi-extensional tectonic episodes of Indian Plate," *Geodesy and Geodynamics*, vol. **9**, no. 2, pp. 358–366, 2018.

[24] N. Sheikh and P. H. Giao, "Evaluation of shale gas potential in the Lower Cretaceous Sembar Formation, the Southern Indus Basin, Pakistan," *Journal of Natural Gas Science and Engineering*, vol. **44**, pp. 162–176, 2017, doi: 10.1016/j.jngse.2017.04.014.

[25] N. Ahmad, Y. Zeb, and N. Ul-Haq, "Key to Success for Exploration of Pab Formation in Kirthar Foldbelt," *PAPG/SPE Annual Technical Conference, November 2015, Islamabad, Pakistan*, pp. 1–7, 2015.

[26] C. J. Banks and J. Warburton, "'Passive-roof' Duplex Geometry in the frontal structures of the

Kirthar and Sulaiman Mountain belts, Pakistan," *Journal of Structural Geology*, vol. **8**, no. 3, pp. 229–237, 1986.

[27] G. Sarwar, "Tectonic setting of the Bela Ophiolites, southern Pakistan," *Tectonophysics*, vol. **207**, pp. 359–381, 1992, doi: 10.1016/0040-1951(92)90396-N.

[28] G. Sarwar and K. A. De Jong, "Arcs, Oroclines, Syntaxes: the Curvatures of Mountain Belts in Pakistan," in *Geodynamics of Pakistan*, A. Farah and K. A. De Jong, Eds. Quetta: Geological Survey of Pakistan, 1979, pp. 341–349.

[29] A. H. Kazmi, "Active Fault Systems in Pakistan," in *Geodynamics of Pakistan*, A. Farah and K. A. De Jong, Eds. Quetta: Geological Survey of Pakistan, 1979, pp. 285–304.

Journal of Physics: Conference Series **2594** (2023) 012022 doi:10.1088/1742-6596/2594/1/012022

# The effect of hydraulic fracturing length to the well production

**Jabrayil Eyvazov, Ph.D, Assoc. Prof, Natig Hamidov**

E-mail: jabrayil.eyvazov88@gmail.com

**Abstract.** The lowering of the original or natural permeability of the reservoir rock close to the well bore is referred to as "formation damage." Additionally, it can be described as any kind of procedure that lowers the productivity of an oil, water, or gas bearing formation. Production, drilling, hydraulic fracturing, and work over operations are just a few of the steps of recovering oil and gas from reservoirs that can result in formation damage, an unfavorable operational and financial issue. A well stimulation method used to raise well productivity is hydraulic fracturing. This method helps to transfer the crude oil and natural gas through the pores and artificially develops channels to the production well because they are stuck at specific locations in the rocks. The oil and gas are extracted from the well and delivered to the surface for further utilization. Fracturing fluid, also known as pumping fluid, typically created by mixing water and sand and injecting it under high pressure into the formation when using the hydraulic fracturing process. There are certain activities that cause damage near the wellbore which lead to reduction in productivity. The drilling operation itself causes damage to the formation as the solids and fluids of the drilling fluids, as well as the fines produced due to the drilling operation, invade the formation and reduce the formation permeability and the ability of the oil and gas to flow to the wellbore. Hydraulic fracturing improves the productivity by creating a conductive path into the reservoir. The length of hydraulic fracturing influences to the production of well. As well as length of hydraulic fracturing increases, oil and gas production from well is increasing. After some point of length, it does not influence to the production of well.

## 1. Introduction

By applying high pressure to tight formations like shale, a fluid, proppant, and additives are used to either create new cracks or widen existing ones, allowing hydrocarbons to flow more easily into the well bore and ultimately to the surface facilities.

Fracking, as hydraulic fracturing is more popularly known, nowadays, primarily utilized to produce hydrocarbons. The proppant in the hydraulic fracturing fluid makes sure that once fractures made, they do not seal immediately, allowing the hydrocarbon to flow out of the tight formation gradually. The additives are made up of several chemical types, and each one of them improves a particular quality of the fluid needed for the hydraulic fracturing process to be successful.

Following are some of the causes of hydraulic fracturing:

In order to avoid well bore damage that reduces productivity, hydraulic fracturing performed. Damage to the area close to the wellbore is a result of drilling particles getting into the formation and drilling fluids not working well with the formation of interest chemically. This can remedied chemically by

hydraulic fracturing or matrix treatments to restore conductivity between the well bore and the formation.
To increase productivity, hydraulic fracturing facilitates the development of conductive hydrocarbon channels into the rock. According to Darcy's law, hydraulic fracturing increases the formation's permeability, fracture height, and coverage while boosting or maintaining reservoir pressure to increase well productivity.
The purpose of hydraulic fracturing process optimization is to maximize gas and oil production by increasing the amount of the cracked reservoir rock. The model of fracture propagation in elastic media brought on by viscous fluid injection serves as the foundation for hydraulic fracturing optimization. The surface of the cavity in an infinite elastic medium, the fluid pumping pressure that initiates and propagates the fracture, and the elastic media characteristics are the input parameters for the fracture propagation model. The fracture surface, fracture width distribution, and fracture front propagation speed are the model's output properties. The direct problem of fracture propagation is the calculation of the output characteristics using the input parameters. By figuring it out, one may forecast how the fracture will occur, how much hydrocarbon will be extracted from the fracture, how much this procedure will cost, etc.
To improve the hydraulic fracturing treatment, the ratio of the fracture length to the drainage radius must be optimized. By estimating flow rate vs. time as a function of fracture length and drainage radius for blanket reservoirs, the ideal fracture length and drainage radius can be found. The drainage radius in lenticular reservoirs is a constant value independent of the extent of the fracture treatment. The geology investigations of the region yield the drainage radius that is most likely. The engineer can optimize propped fracture half-length by adjusting ratio after obtaining a likely drainage radius value.
A proper fracture half-length and fracture conductivity are the two most crucial characteristics for optimal good performance by a fracture stimulation program.
Since,

$$F_{cd} = \frac{k_f \cdot W}{k \cdot L_f} \qquad (1)$$

Where,
$F_{cd}$ ≡ Fracture conductivity
$k_f$ ≡ Fracture permeability
W ≡ Conductive fracture width
k ≡ Formation permeability
$L_f$ ≡ Conductive fracture single wing length
Post fracture temperature logs are used in the measuring of fracture height after hydraulic fracturing. It involves determination of the prefracture temperature profile by logging followed by at the least two postfracture temperature logs. The process entails running a pre fracture temperature log so as to determine the temperature gradient in the reservoir formation and thereafter temperature logs are run after the hydraulic fracturing stimulation process. Heat transfer in the surrounding layers is by conduction whereas in the fractured payzone it is by convection. On interpreting the temperature logs, a temperature anomaly is noticed due to the different temperature recover rates after pumping hence identifying the fractured zone.

## 2. Experiment part

This study conducted on unconventional reservoir. The permeability is in range of 5-13 mD. The main purpose of the research project to define the optimal hydraulic fracture length and fracture width for defined well.
It done sensitivity analysis on hydraulic fracturing length.

In first case, length of horizontal section was 150 m. Several cases simulated based on hydraulic fracture length and total gas production results compared. It defined that, 125 m hydraulic fracture length is the best result for this well.
In second case, length of horizontal section was 200 m. Several cases simulated based on hydraulic fracture length and total gas production results compared. It defined that, 125 m hydraulic fracture length is the best results for this well, too.

**Table 1.The effect of hydraulic fracturing length within 50-150 metres to the well production**

| Date | Base Case | Fracture length, m | | | | |
|---|---|---|---|---|---|---|
| | | 50 | 75 | 100 | 125 | 150 |
| 2023 | 846.45 | 1015.74 | 1100.38 | 1185.03 | 1269.67 | 1227.35 |
| 2024 | 958.04 | 1149.64 | 1245.45 | 1341.25 | 1437.06 | 1389.15 |
| 2025 | 1071.78 | 1286.14 | 1393.31 | 1500.49 | 1607.67 | 1554.08 |
| 2026 | 1188.60 | 1426.33 | 1545.18 | 1664.04 | 1782.90 | 1723.47 |
| 2027 | 1267.11 | 1520.60 | 1647.27 | 1773.95 | 1900.67 | 1837.31 |
| 2028 | 1288.27 | 1546.00 | 1674.78 | 1803.58 | 1932.41 | 1867.99 |
| 2029 | 1296.25 | 1555.57 | 1685.09 | 1814.75 | 1944.38 | 1879.56 |
| 2030 | 1299.55 | 1559.52 | 1689.35 | 1819.37 | 1949.33 | 1884.35 |
| 2031 | 1301.09 | 1561.37 | 1691.34 | 1821.53 | 1951.64 | 1886.58 |
| 2032 | 1301.91 | 1562.35 | 1692.41 | 1822.67 | 1952.87 | 1887.77 |
| 2033 | 1302.34 | 1562.87 | 1692.95 | 1823.28 | 1953.51 | 1888.39 |
| 2034 | 1302.57 | 1563.14 | 1693.25 | 1823.60 | 1953.86 | 1888.73 |
| 2035 | 1302.69 | 1563.29 | 1693.41 | 1823.77 | 1954.04 | 1888.90 |
| 2036 | 1302.76 | 1563.37 | 1693.50 | 1823.86 | 1954.14 | 1889.00 |
| 2037 | 1302.80 | 1563.42 | 1693.56 | 1823.92 | 1954.20 | 1889.06 |
| 2038 | 1302.84 | 1563.47 | 1693.60 | 1823.98 | 1954.26 | 1889.12 |
| 2039 | 1302.87 | 1563.50 | 1693.64 | 1824.02 | 1954.31 | 1889.16 |
| 2040 | 1302.89 | 1563.53 | 1693.67 | 1824.05 | 1954.34 | 1889.19 |
| 2041 | 1302.91 | 1563.55 | 1693.69 | 1824.07 | 1954.37 | 1889.22 |
| 2042 | 1302.92 | 1563.56 | 1693.71 | 1824.09 | 1954.38 | 1889.23 |
| 2043 | 1302.93 | 1563.58 | 1693.72 | 1824.10 | 1954.40 | 1889.25 |
| 2044 | 1302.93 | 1563.59 | 1693.73 | 1824.10 | 1954.40 | 1889.25 |

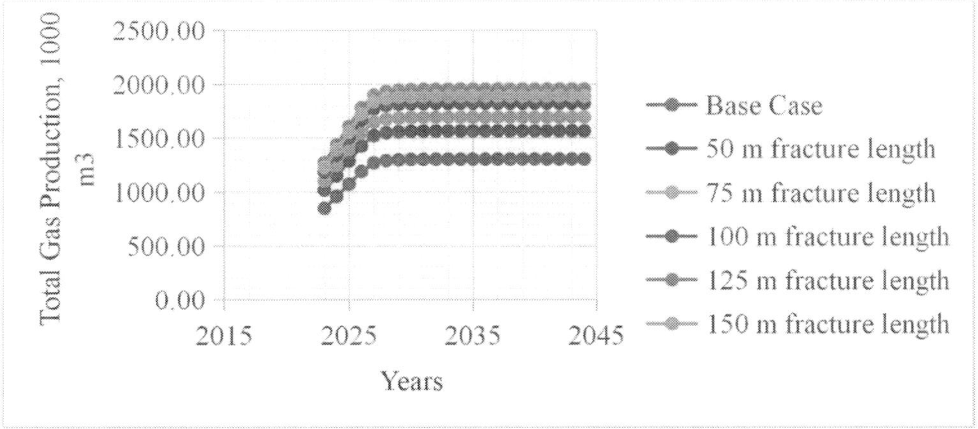


**Figure 1. The effect of hydraulic fracturing length within 50-150 metres to the well production**

**Table 2 .The effect of hydraulic fracturing length within 100-200 metres to the well production**

| Date | Base Case | Fracture length, m | | | | |
|---|---|---|---|---|---|---|
| | | 100 | 125 | 150 | 175 | 200 |
| 2023 | 846.45 | 1185.03 | 1269.67 | 1227.35 | 1185.03 | 1142.70 |
| 2024 | 958.04 | 1341.25 | 1437.06 | 1389.15 | 1341.25 | 1293.35 |
| 2025 | 1071.78 | 1500.49 | 1607.67 | 1554.08 | 1500.49 | 1446.90 |
| 2026 | 1188.60 | 1664.04 | 1782.90 | 1723.47 | 1664.04 | 1604.61 |
| 2027 | 1267.11 | 1773.95 | 1900.67 | 1837.31 | 1773.95 | 1710.60 |
| 2028 | 1288.27 | 1803.58 | 1932.41 | 1867.99 | 1803.58 | 1739.16 |
| 2029 | 1296.25 | 1814.75 | 1944.38 | 1879.56 | 1814.75 | 1749.94 |
| 2030 | 1299.54 | 1819.36 | 1949.31 | 1884.33 | 1819.36 | 1754.38 |
| 2031 | 1301.09 | 1821.53 | 1951.64 | 1886.58 | 1821.53 | 1756.47 |
| 2032 | 1301.91 | 1822.67 | 1952.87 | 1887.77 | 1822.67 | 1757.58 |
| 2033 | 1302.34 | 1823.28 | 1953.51 | 1888.39 | 1823.28 | 1758.16 |
| 2034 | 1302.57 | 1823.60 | 1953.86 | 1888.73 | 1823.60 | 1758.47 |
| 2035 | 1302.69 | 1823.77 | 1954.04 | 1888.90 | 1823.77 | 1758.63 |
| 2036 | 1302.76 | 1823.86 | 1954.14 | 1889.00 | 1823.86 | 1758.73 |
| 2037 | 1302.80 | 1823.92 | 1954.20 | 1889.06 | 1823.92 | 1758.78 |
| 2038 | 1302.84 | 1823.98 | 1954.26 | 1889.12 | 1823.98 | 1758.83 |
| 2039 | 1302.87 | 1824.02 | 1954.31 | 1889.16 | 1824.02 | 1758.87 |
| 2040 | 1302.89 | 1824.05 | 1954.34 | 1889.19 | 1824.05 | 1758.90 |
| 2041 | 1302.90 | 1824.06 | 1954.35 | 1889.21 | 1824.06 | 1758.92 |
| 2042 | 1302.92 | 1824.09 | 1954.38 | 1889.23 | 1824.09 | 1758.94 |
| 2043 | 1302.93 | 1824.10 | 1954.40 | 1889.25 | 1824.10 | 1758.96 |
| 2044 | 1302.93 | 1824.10 | 1954.40 | 1889.25 | 1824.10 | 1758.96 |

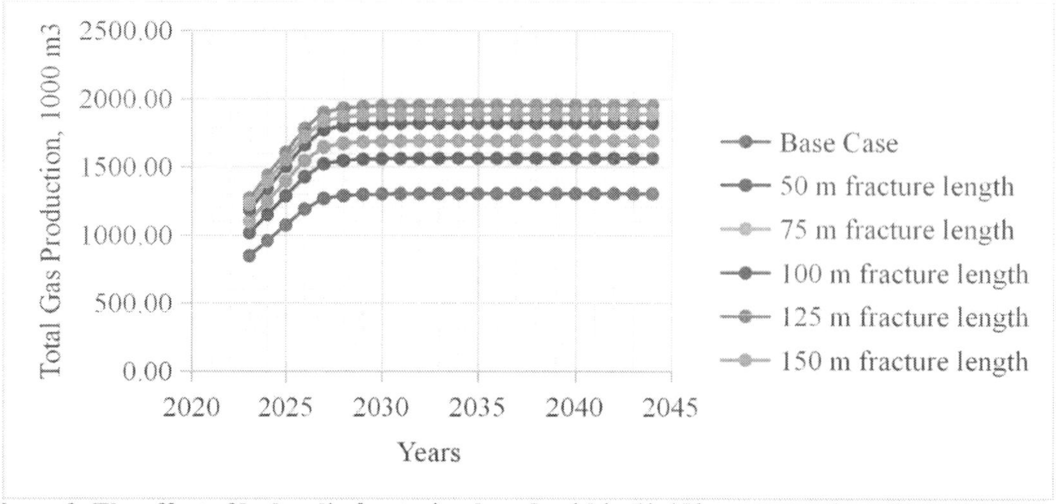


**Figure 2. The effect of hydraulic fracturing length within 50-150 metres to the well production**

## Conclusion

According to the study, it defined that, the highest fracture length does not mean highest production. In this study, for this well, the optimal fracture length is 125 m. Afterwards reservoir model simulated in various hydraulic fracture width for this well in 125 m of hydraulic fracturing length.

## References

[1]. Amer S. A. Al Jabri* and Girma T. Chala. A Review of Challenges in Hydraulic Fracturing Operation. Vol.2020:018

[2]. Dongming Zhi , Xuyang Guo, Wei Wang, Yan Jin, Caiguang Liu, Gang Chen, Zhenlin Wang. Fracturing and production analysis of the efficacy of hydraulic fracture stage reduction in the improvement of cost- effectiveness in shale oil development: A case study of Jimsar shale oil, China. DOI: 10.1002/ese3.940.2021

[3]. Hassan Z.Harraz. Hydraulic Fracturing Process. November 2014.
https://www.researchgate.net/publication/326977263

[4]. Ali Daneshy, Recent Advances in Hydraulic Fracturing for Enhanced Well Productivity

[5]. Zhai, F, A and C. A new tool for multi & multi-well hydraulic fracture modeling, SPE-173367-MS (2015)

[6]. Ahmed U, Thompson T.W, Kelkar S.M, Strawn J.A, Veghte R. and Hathaway S: "Optimizing Hydrulic Fracture Designs in Formations with Poor Containment," SPE 13375, presented in Eastern Regional Meeting, Charleston, West Virginia, 31 October – 2 November 1984.

[7]. Economides, M. and Nolte, K.G. 2000. Reservoir Stimulation, 3rd edition, New York:
John Wiley & Sons, Ltd.

[8]. Ion Pană, Iuliana Veronica Ghe,tiu, Ioana Gabriela Stan,*, Florinel Dinu , Gheorghe Brănoiu and Silvian Suditu. The Use of Hydraulic Fracturing in Stimulation of the Oil and
Gas Wells in Romania. Sustainability 2022, 14,5614. https://doi.org/10.3390/su14095614

[9]. Yehia F*, Shazly T, Noah AZ, Abou Heleik MM and Kabel KI. Hydraulic Fracturing Process Systems and Fluids: An Overview. Petroleum & Petrochemical Engineering Journal. Volume 6 Issue 3. DOI: 10.23880/ppej-16000306

[10]. Shihao Wang *, Xiangyu Yu, Philip H. Winterfeld and Yu-Shu Wu. Real-Time Simulation of Hydraulic Fracturing Using a Combined Integrated Finite Difference and Discontinuous
Displacement Method: Numerical Algorithm and Field Applications. Water 2023, 15, 938. https://doi.org/10.3390/w15050938

# Flow characteristics around the crack in pipeline

**Jafar A Ali[1] and Payman S Ahmed[2]**

[1]Department of Petroleum Engineering, Faculty of Engineering, Koya University, Koya KOY45, Kurdistan Region - F.R. Iraq

[2]Department of Manufacturing Engineering, Faculty of Engineering, Koya University, Koya KOY45, Kurdistan Region - F.R. Iraq

[1] Jafar.dalo@koyauniversity.org

**Abstract.** Pipelines transport fluids especially petroleum and its products, the transportation process may face some problems due to the defects that can occur in the pipes. Cracks are the main defects that attack the pipes and reduce their performance. It is known that the flow around cracks in pipelines are complicated and affected by various factors, it worth to be considered before decision making on the pipeline integrity. In this study flow characteristics around the crack inside the pipeline has been investigated. The part of the pipe within the crack is the crack zone, where the flow behaviour is different with the rest of the pipe. The work carried out at the laboratory using pipes with different crack length and crude oil with known properties to observe the flow at the crack zone. Then the work theoretically analysed with computational fluid dynamics to model the flow. It is found that the fluid at the crack zone is going to be more viscous lead to accumulation and changes in the characteristics of the flow. The results revealed that the flow characteristics in cracked pipeline with no leakage depend on several factors; the size of the crack, temperature, pressure, velocity and fluid viscosity.

## 1. Introduction

Pipelines are transferring fluids long distance, they are subject to variety of risks such as; human activities, proximity to rivers and roads, slope, land surface temperature and aspect [1]. Study of the fluid dynamics inside pipes are important due to its wide application in industries [2]. Pipelines are usually lying on ground and would be suffer from the risk of cracks because of the applied loads while transferring fluids [3]. Crack in a pipeline can be a serious issue as it can lead to leaks or even a pipeline failure. The severity of the issue depends on various factors such as the size and location of the crack, the type of fluid being transported, the pressure and temperature of the pipeline, and the material of the pipeline. Cracks in pipelines have a great influence on hydraulic efficiency of the system, they are reducing the pressure and velocity of the fluid pass through [4]. The flow profile inside the pipeline is one of the main factors affect the crack. It has internal loads on the pipe walls which initiate the crack, enlarge it and increase the viscosity at the crack zone [5]. This is due to reduction in the velocity of the fluid at that part of the pipe [3]. Related to the studies in pipelines particularly those involved with fractured pipes and losses, different models were developed as function of fluid flow and density [6] as well as the pipe diameter and thickness. [7] Found that the viscosity of the fluid will not influence pressure inside the pipe, but the head loss will increase somehow depends on the formation of the core flow of the viscous oil. Yoon and Son are two researchers investigating crack propagation and how affects by the flow characteristics inside the pipe.

They studied larger crack in simply supported pipes and how relate to the fluid flow behavior [8]. So, very rare studies are existing on investigating the cracked pipelines and the flow characteristics of the fluid passing through. Researchers did not consider the effects of the crack parameters on the critical flow and fluid properties particularly the viscosity which are our main objective in this paper. The paper aims at finding the characteristic of the flow at the crack zone inside the cracked pipeline.

## 2. Materials and experimental procedure

In this section we describe the experimental procedures and components. The schematic of the apparatus which consists of, tanks, pipes, valves, pumps, flowmeters and gauges are illustrated in 'figure 1'. The rig was designed in laboratory at an environment reflects the real pipeline operation. The pipe material is carbon steel with properties illustrated in Table. 1, pipes are including cracks in different sizes. The fluid used in the experiment is crude oil, the physical properties of the crude oil are presented in Table 2. The crude oil was discharging from the right-hand side tank to left hand side through the cracked pipe. The viscosity of the oil around the crack was taken every several trials.

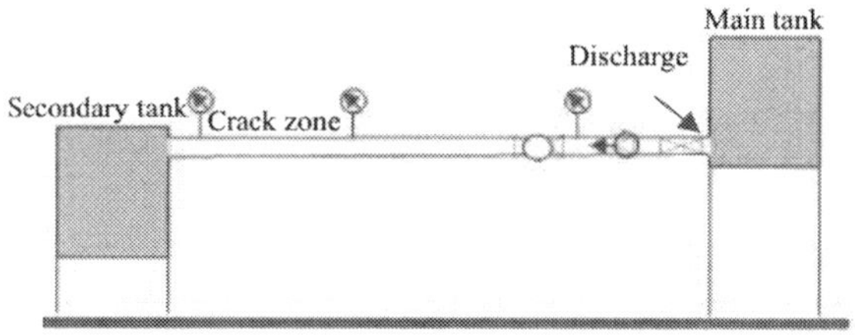


**Figure 1.** Experimental diagram [9]

## 3. Modelling and analysis

Theoretical analysis was carried out using CFD package ANSYS, Fluent tool in the Software used to solve the equations of motion and model the flow in the pipe [5]. SolidWork Software used to create the pipe 'figure 2'. The material of the pipe and its size are illustrated in Table 1. The parameters of the crack are shown in Table 3, the crack is shown in 'figure 3'. A meshgrid about 50000 cells was created on the model, the crack zone in the model is shown in 'figure 4' with mesh.

**Table 1.** Pipe metal properties.

| | |
|---|---|
| Outer radius | 30mm |
| Inner radius | 28mm |
| Pipe length | 150mm |
| Thickness | 2mm |
| Pressure | 20MPa |
| Modulus of elasticity | 205000Pa |
| Poisson's ratio | 0.28 |

**Figure 2.** Modeled pipe using (SolidWorks Software)

**Table 2.** Crude oil properties [10]

| Crude Properties | Values |
|---|---|
| Density (kg/m$^3$) | 841 |
| Viscosity (kg/m.s) | 0.003704 |
| A (m$^2$) | 0.000314 |
| Pressure (Pa) | 16500.42 |
| Enthalpy (J/kg) | 74100 |
| Temperature (K) | 290 |

**Table 3.** Crack parameters

| | |
|---|---|
| Major radius | **1.2mm** |
| Minor radius | 0.6mm |
| Largest contouring radius | 0.75mm |
| Mesh method | Tetrahedrons |
| Mesh contours | 6 |
| type | Semi elliptical crack |

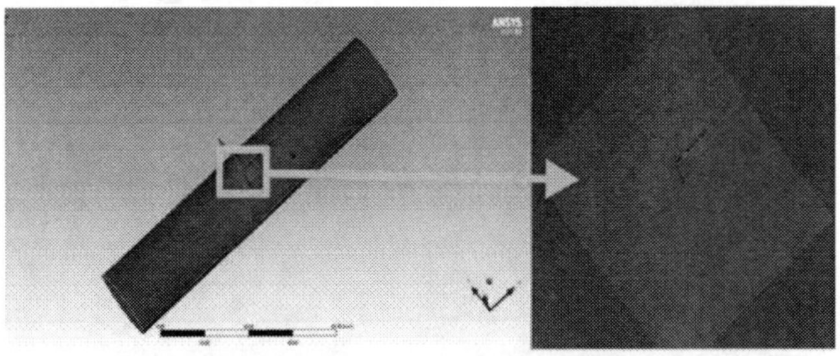

**Figure 3.** Direction of the crack

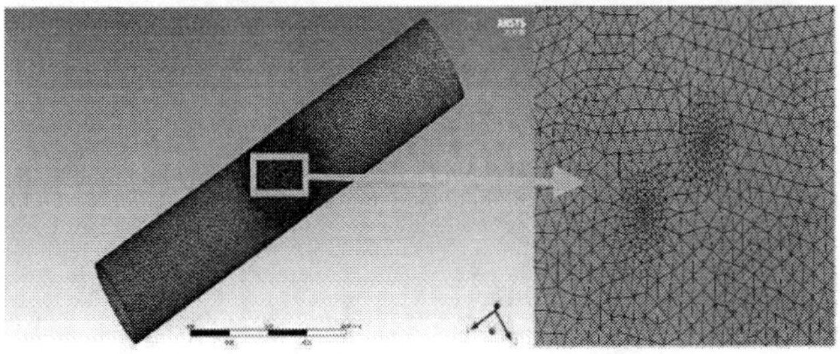

**Figure 4.** The typical mesh of the pipe and also face of the crack is appeared

Block strategy was used to generate the mesh in simulation [11]. The properties of the crude oil which have been measured in the laboratory input to the model. Crack added to the pipe using smart crack growth with two different directions; circumferential and longitudinal. The program is analyzing the flow behavior and the internal pressure on crack initiation and growth. The flow of the crude was affected by the crack as illustrated in 'figure 5'.

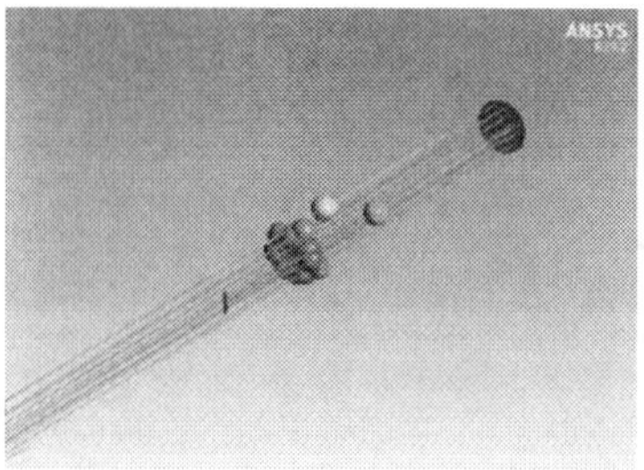

**Figure 5.** Oil particles near the crack.

The deformation of the crack under continuous load is shown in 'figure 6' while crack propagation is illustrated in 'figure 7'. When the problem continues without fixing it raises to fracture see 'figure 8'.

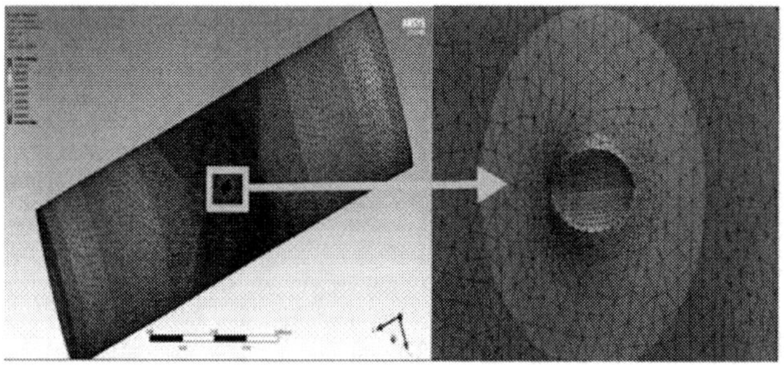

**Figure 6. Crack in pipeline**

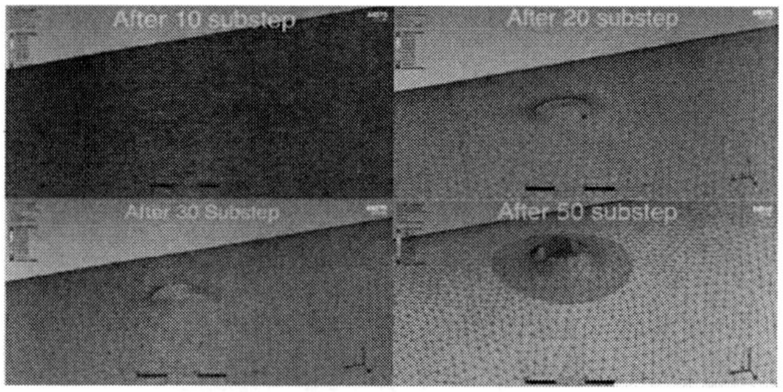


**Figure 7. Crack propagation**

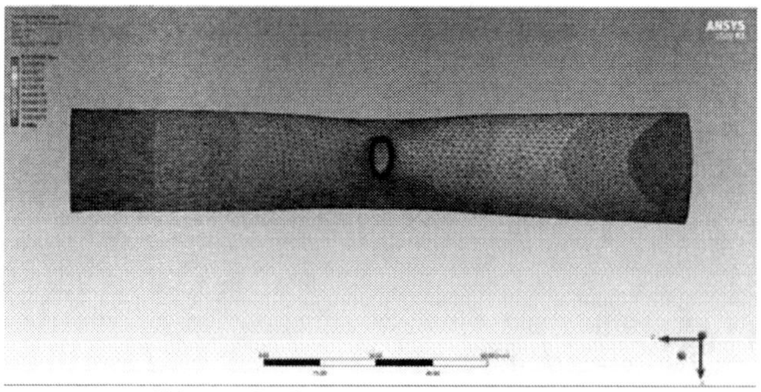

**Figure 8. Crack changed to fracture**

## 4. Mathematical analysis

Fluid flow inside a pipe is affected by factors such as the pressure drop along the pipe, the roughness of the pipe surface, and the presence of bends or other obstructions in the pipe [10]. These factors can cause changes in the velocity and pressure of the fluid as it moves through the pipe. The viscosity of fluid flowing through a crack or a leakage in a pipeline depends on various factors such as the type of fluid, the temperature of the fluid, and the size and geometry of the crack. In a crack pipeline, the viscosity of crude oil will change and move slower in the area around the crack. This can be analyzed using the newton law of viscosity equation (1).

$$\tau = \mu \frac{du}{dr} \tag{1}$$

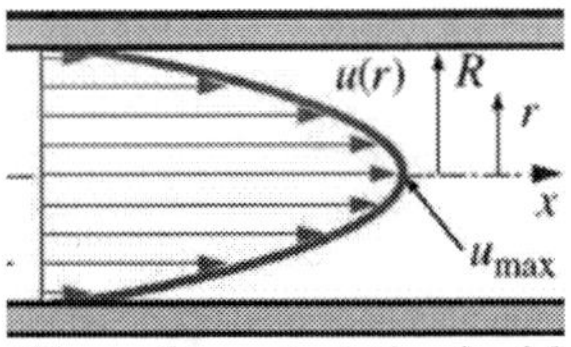


**Figure 9.** Flow characteristics inside the pipe

Where, τ is the shear stress, *u* is fluid velocity and *r* is the fluid thickness inside the pipe. Note that the thickness of the fluid in 'figure 9' is 2*r*, but for analysis 2 is cancelled. While viscosity μ:

$$-\mu = \frac{Fr}{Au} \tag{2}$$

What is observed form 'figure 10' that the thickness *r* of the fluid at the crack zone gets bigger. The bigger value for *r* increases the value of viscosity equation (2).

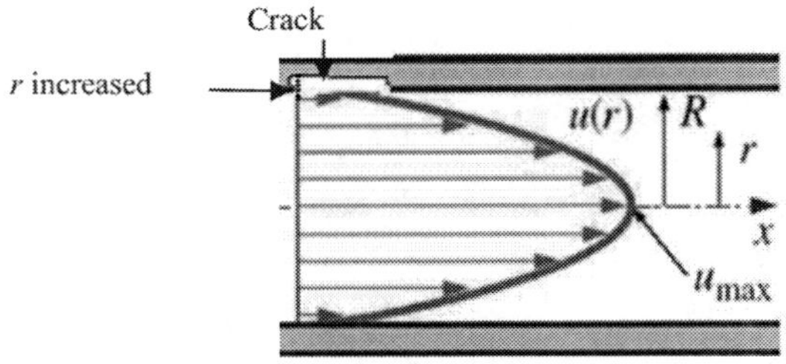


**Figure 10.** Flow characteristics inside the pipe (crack zone)

## 5. Results and discussion

Pipeline cracks pose significant risks to safety, the environment, and the economy. It is crucial for pipeline operators to implement robust safety and maintenance practices to prevent pipeline cracks and minimize the risks associated with them. Leakage in pipeline is the main cause of the risks which affect the human lives and environment [12]. As shown in the computational analysis 'figure 8' crack will change to fracture and leakage if not fixed in time. The results obtained from the mathematical analysis show that the viscosity around the crack will increase. The increase in viscosity lead to reduction in flow in the crack zone 'figure 10' i.e. crude oil flow more slowly through the crack. The reduction in velocity causes some accumulation inside the crack 'figure 11'.

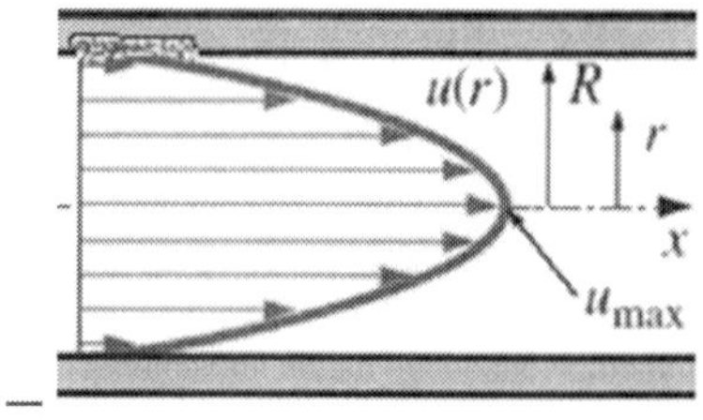


**Figure 11.** Accumulation around the crack

It is noticed that the flow at the crack will be reduced and accumulation occur. This accumulation will make the liquid with time more viscus. Accumulation of water under hydrocarbons inside the crack in pipelines, provides an ideal habitat for the growth of microbes. Microbes discharge gases such as H2S and CO2, these gases encouraging microbial induced corrosion MIC. Then the microbial corrosion or internal stress will change the crack to fracture (pipeline failure) and leakage.

## 6. Conclusion

A crack in a pipeline can lead to a fracture if the crack propagates or grows under the influence of stress or pressure. A fracture can occur when the crack reaches a critical size and the pipeline material can no longer sustain the stress, leading to a complete separation or rupture of the pipeline. In this study the following were concluded; accumulation of the crude oil at the crack zone. Fluid viscosity at the crack zone will increase. Microbial growth inside the crack to increase the probability of microbial corrosion in the pipe.

## References

[1] Ali J A, Khodakarami L and Ahmed M S 2022 Pipeline risk assessment using GIS. *UKH Journal of Science and Engineering.* **6(2)** 15-25.
[2] Paidoussis M P 1998 *Fluid-Structure Interactions. Academic Press.*
[3] Eslami G, Maleki V A and Rezaee M 2016 Effect of Open Crack on Vibration Behavior of a Fluid-Conveying Pipe Embedded in a Visco-Elastic Medium. *Latin American Journal of Solids and Structures.* **13** 135–154.
[4] Sousal C A and Romero O J 2017 Influence of oil leakage in the pressure and flow rate behaviors in pipeline. *Latin American Journal of Energy Research.* **4** 17–29.
[5] Cox A 2005 What are communities of practice? A comparative review of four seminal works. *Journal of Information Science.* **31** 527-540.
[6] Falconi B M, Tamayo E, Laurencio A H, Vega J, Gualotuña E, Grijalva E and Campana L 2018 Model of pressure losses in pipes during the transport of heavy oil with 11 API gravity. *International Journal of Mechanics.* **12** 8–13.
[7] Kumar R, Gudala M and Rao R M 2016 Studies on Emulsified Crude Oil Flow through Horizontal Pipeline. *Journal of Modern Chemistry & Chemical Technology.* **7** 82–90.
[8] Son I and Yoon H 2008 Stability analysis of pipe conveying fluid with crack and attached masses. *Journal of the Korean Society for Precision Engineering.* **25**.
[9] Ali J A 2023 *Proceedings of the 2022 International Petroleum and Petrochemical Technology Conference*, Jia'en Lin, China. Springer Nature.
[10] Kamal K and Ali J A 2021 Pressure drop in cracked pipeline with no leakage. *Journal of Mechanical Engineering Research and Developments.* **44** 410 – 417.
[11] Park J, Cho Y, Kim S and Lee J 2015 Estimation of leak rate through circumferential cracks in pipes in nuclear power plants. *Nuclear Engineering and Technology.* **47** 332–339.
[12] Golmohamadi M 2015 Pipeline leak detection, Master thesis, Missouri University of Science and Technology.

# Function Optimization of Self-shutoff Protection System for -Valve Chambers of Long-distance Natural Gas Pipelines

**Yiran GAO1(✉), Yundong YUAN1,Xiang ZHOU1,Jingwei YANG1,Qi Lin1**

1 Sine-pipeline International Company, Beijing

Yiran.gao@cnpc.com.cn

**Abstract.** The rate of pressure drop set value of shut-off valve in gas pipeline is uncertain to a certain extent, and the function of self-shutoff protection may disable under some leakage conditions. A gas pipeline leakage model is established using simulation software, to simulate the leakage, flow rate reduction, and crossover, and obtain the pressure changes in upstream and downstream shut-off valve chambers under different working conditions[1]-[3]. The rate of pressure drop is calculated by using the rate of pressure drop algorithm adopted by Line Guard 2300 system. Simulation results indicate that the gas transmission volume has little influence on the rate of pressure drop, and the leakage diameter has great influence on the rate of pressure drop, showing a positive correlation. When the pipeline runs in parallel, it is necessary to avoid opening the crossover valve under large pressure difference. Finally, the recommended rate of pressure drop threshold at the trunk line shut-off valve chamber is given, and the leakage condition of 1/2 diameter at 32km upstream can be detected under this recommended value.The simulation software is first applied in multinational pipelines, which can effectively protect pipeline defect points. Reduced methane cold venting, greatly improving the contribution to achieving emission reduction goals.

## 1. Introduction

At present, the self-shut-off protection value of valve chambers along the pipeline is generally set using unified rate of pressure drop set value and duration set value (0.15 MPa/min, 120s) [4]-[6] according to the industry experience. Excessive rate of pressure drop may lead to some accident conditions that cannot be stopped in time. while too small rate of pressure drop may lead to false shutdown under non-leakage conditions, which will interfere with the normal operation of the pipeline to a certain extent. Therefore, the appropriate rate of pressure drop set value and duration set value should be determined according to the specific working conditions of each pipeline. In addition, because the valve chamber protection system of the trunk line of the China-Central Asia Natural Gas Pipeline has been in the wild environment for a long time, it is easily affected by complex factors such as lightning strike, extremely cold/high temperature weather, stray current of cathodic protection and signal grounding. In this case, there is a risk of false shutdown of the valve chamber under non-leakage conditions. Therefore, it is of great significance to study the rate of pressure drop set value of trunk shut-off valve under various operating conditions for the gas transmission system in the

China-Central Asia Natural Gas Pipeline [7]-[14].

## 2. Calculation method of rate of pressure drop

The self-shutoff protection system of the gas pipeline valve chamber generally consists of a shut-off valve body, a pneumatic-hydraulic actuator, and an electronic control unit. The electronic control unit is composed of a power supply, CPU, solenoid valve, pressure sensor, and terminal block. The pressure sensor at the cut-off valve can convert the pipeline pressure into electrical signals and send them to the shut-off valve control system. The pressure signals are usually collected every 5 seconds. To convert the pressure value at each time point into the rate of pressure drop, the rate of pressure drop algorithm needs to be written into the microprocessor of the shut-off valve control system. At present, there are three widely used rate of pressure drop algorithms, including the traditional algorithm, the algorithm used in domestic gas pipeline and the algorithm used in Line Guard 2100/2200/2300 system [15].

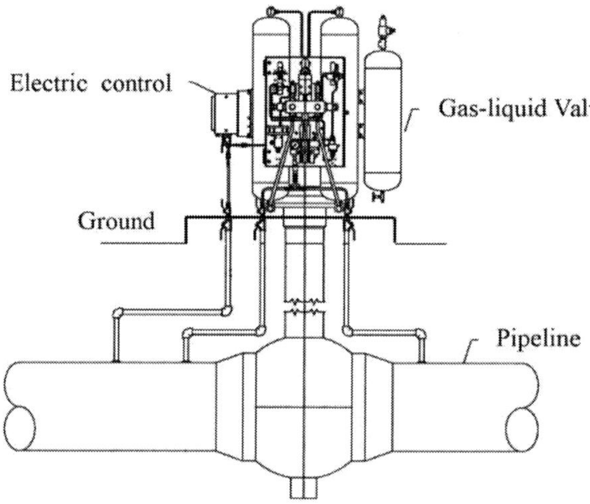


**Fig. 1.** Schematic diagram of the valve and electric control of the cut-off valve chamber

### 2.1 *Traditional algorithm*

If the pressure obtained at *T* second is *P1* and the pressure obtained at T-5 second is *P2*, the rate of pressure drop at *T* second is[16]:

$$ROD = (P_1 - P_2) \times 12\# \text{ (2.1)}$$

Where: *ROD*-the rate of pressure drop at the T second, MPa/min.

$P_1$--Pressure at T-5 second, MPa.

$P_2$--Pressure at T second, MPa.

### 2.2 *Algorithm commonly used in China*

If the pressure at 5th second, 10th second... 60th second are P1, P2... P12, and the rate of pressure drop at 35th second, 40th second... 60th second are *ROD*7, *ROD*8... *ROD*12, then[17]:

$$ROD_7 = (P_1 - P_7) \times 2$$
$$ROD_8 = (P_2 - P_8) \times 2$$
$$\cdots$$
$$\cdots$$
$$ROD_{12} = (P_6 - P_{12}) \times 2$$

The average of *ROD*7, *ROD*8... ROD12 are taken as the rate of pressure drop at 60th second, and its calculation method is:

$$ROD = \frac{\sum_{i=7}^{12} ROD_i}{6} = \frac{\sum_{i=1}^{6} P_i - \sum_{i=7}^{12} P_i}{3} \# \ (2.2)$$

### 2.3 *Line Guard 2100 2200 2300 System Algorithm*

The LineGuard electronic control unit of Shafer gas-over-oil actuator records the pressure value every 5 seconds, as shown in Fig.2.

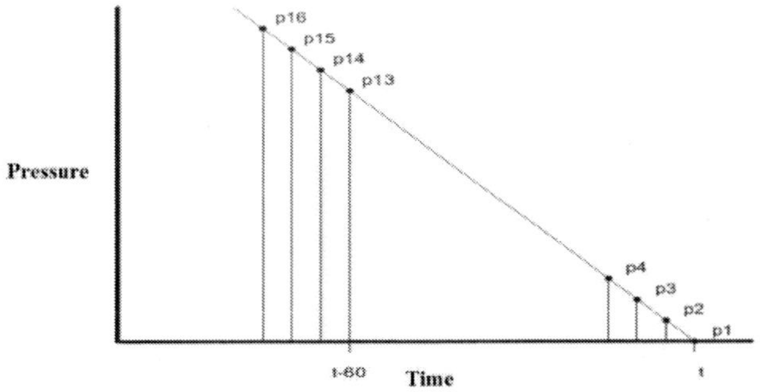


**Fig. 2.** Schematic Diagram of LineGuard 2100 2200 2300 Pressure Data Sampling

The LineGuard 2100 system takes the pressure value every 5 seconds, and the calculated rate of pressure drop is also obtained every 5 seconds. For example, a rate of pressure drop value is calculated at p1 point, and then a rate of pressure drop value is calculated after 5 seconds. When simulation is conducted with SPS software, the time step can be set for calculation to obtain the pressure data value at each moment, and the rate of pressure drop value can be updated to calculate the rate of pressure drop.

If the pressure at 5th second, 10th second … 80th second is *P*16, **P**15 … *P*1, the average pressure at 20th seconds is:

$$P_{avg,20} = \frac{P_{16} + P_{15} + P_{14} + P_{13}}{4} \# \ (2.3)$$

The average pressure value at the 80th second is:

$$P_{avg,80} = \frac{P_4 + P_3 + P_2 + P_1}{4} \# \ (2.4)$$

The rate of pressure drop at the 80th second is:

$$ROD = P_{avg,20} - P_{avg,80} = \frac{(P_{13} - P_1) + (P_{14} - P_2) + (P_{15} - P_3) + (P_{16} - P_4)}{4} \# \ (2.5)$$

The gas over oil actuator of domestic gas transmission pipeline is usually Shafer actuator produced by American Shafer Company, and the automatic control system is the Line Guard 2100/2200/2300 system matched with Shafer actuator. Hence, the voltage drop rate algorithm used in this study is Line guard 2100/2200/2300 system algorithm.

## 3. Functional simulation analysis of valve chamber self-shutoff protection system

In this paper, the Synergi Pipeline Simulator (SPS) software is used to carry out the simulation calculation and analysis of the shut-off valve chamber of the gas transmission pipeline. The SPS software is a professional pipeline simulation software owned by DNV GL. The software is suitable for hydraulic and thermal calculations of oil and natural gas pipelines, and can be operated in both online and offline modes. SPS simulation software has nearly 50 years of simulation engineering experience, and has almost become the industry standard in the field of hydraulic simulation in China and even in the world.

In order to verify the effectiveness of the software, the transient operating condition of valve chamber of the Central Asia gas pipeline was selected as an example (see Fig.3), and the simulated data and actual data on site were compared and analyzed. The valve chamber's self-shutoff protection function is triggered due to the rapid change of the pressure drop rate detected by the pressure transmitter due to the ice blockage of the pressure detection tube. After the abnormal shut-off event of the valve occurred, the upstream and downstream jumper pipes of the valve chamber were first opened in sequence. Second, the valve chamber by-pass pipe was opened for equalizing pressure upstream and downstream of the main valve. After the pressure difference between the upstream and downstream of the main valve dropped to 0.45 MPa, the Control Center remotely opened the main valve, and finally the upstream and downstream jumper pipes were closed, and the pipelines resumed normal operation. As shown in Fig.4, the simulation is carried out according to the adjustment of the operating conditions at the same time. The pressure change of the upstream and downstream of the valve chamber calculated by the simulation is basically consistent with the actual monitoring data on site, and the final pressure difference of the by-pass pipe is also basically the same. This verification shows that the use of SPS to carry out the simulation and analysis of valve chamber self-shutoff protection system has a high accuracy[18]-[19].

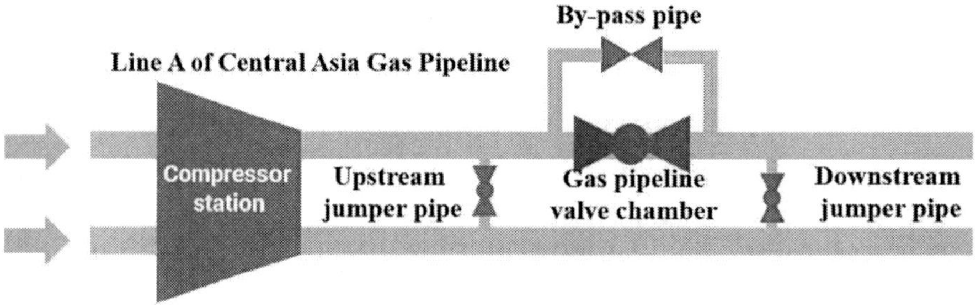


**Fig. 3.** Schematic diagram of the false shut-off process of gas pipeline valve chamber

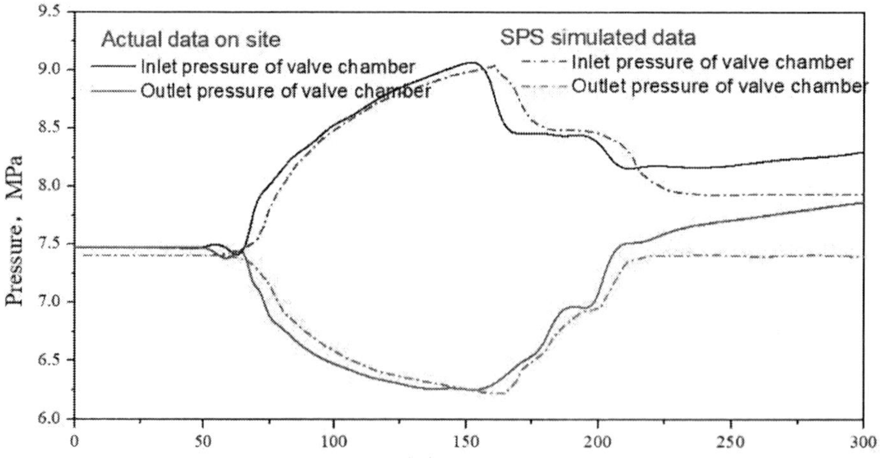


**Fig. 4.** Validation of SPS simulation analysis

### *3.1 Leakage operating condition*

Basic parameters of the pipeline: pipeline diameter 1067mm, wall thickness 15.6 mm, valve chamber spacing 32km, pipe section between stations 132km, starting point operating pressure 9MPa, flow rate 130 x 108 m 3/a, leakage formation 10s.

**Design of leakage simulation:**

Distance from upstream valve chamber: 0, 16, 32km.
Different leakage diameters: D1067 and D533.5
Total number of working conditions: 6
**The leakage point is 0L** away from the upstream valve, and the leakage diameter is D, 1/2 D.
**The leakage point is 1/2 L** away from the upstream valve, and the leakage diameter is D and 1/2 D.
**The leakage point is L** away from the upstream valve, and the leakage diameter is D, 1/2 D.
Under the set value of 120s rate of pressure drop duration, the maximum rate of pressure drop that can be detected under each leakage condition is shown in Fig. 5-1 - Fig.10.

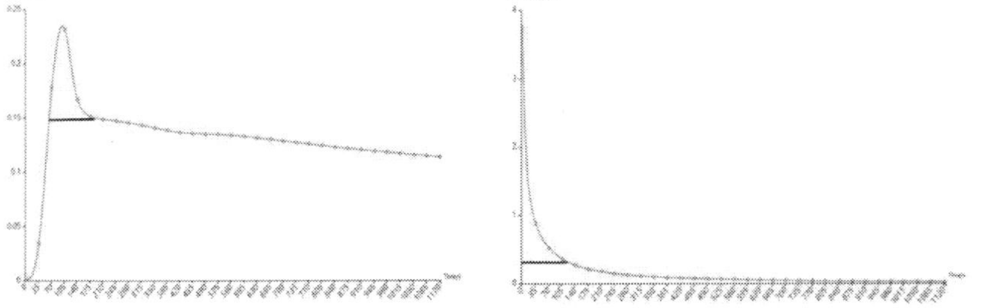

**Fig. 5.** Detection Value of the rate of pressure drop at 0km from the Leakage Point of Leakage Diameter 1067 to the Upstream Valve Chamber (Upstream and Downstream Valve Chambers)

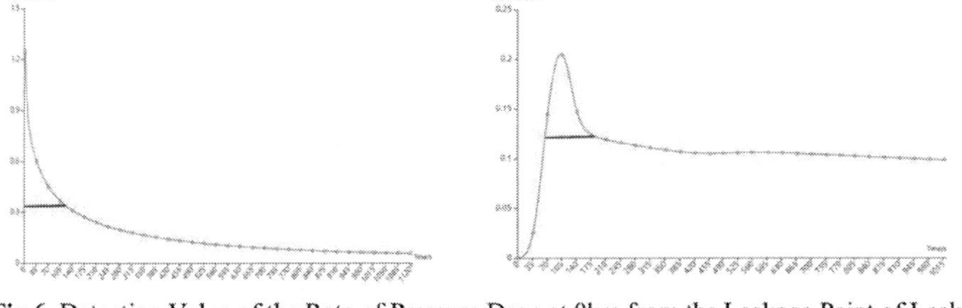

**Fig.6.** Detection Value of the Rate of Pressure Drop at 0km from the Leakage Point of Leakage Diameter 533.5 to the Upstream Valve Chamber (Upstream and Downstream Valve Chambers)

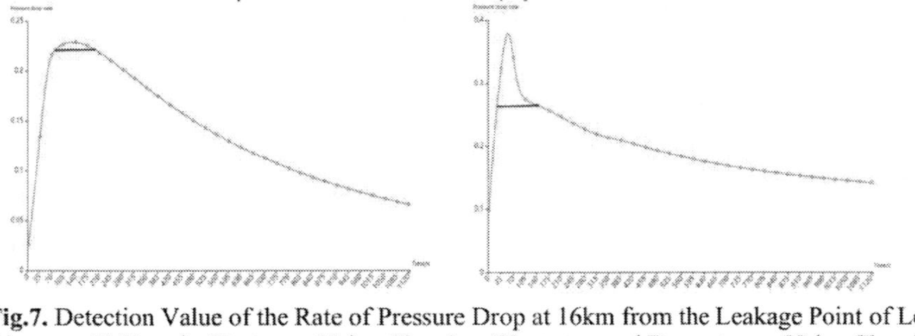

**Fig.7.** Detection Value of the Rate of Pressure Drop at 16km from the Leakage Point of Leakage Diameter 1067 to the Upstream Valve Chamber (Upstream and Downstream Valve Chambers)

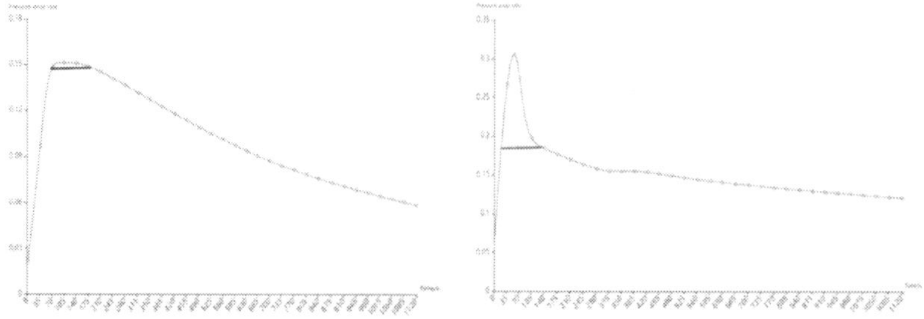

**Fig.8.** Detection Value of the Rate of Pressure Drop at 16km from the Leakage Point of Leakage Diameter 533.5 to the Upstream Valve Chamber (Upstream and Downstream Valve Chambers)

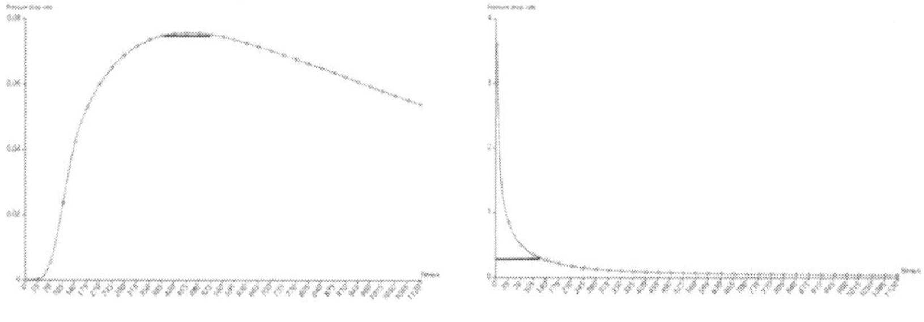

**Fig.9.** Detection Value of the Rate of Pressure Drop at 32km from the Leakage Point of Leakage Diameter 1067 to the Upstream Valve Chamber (Upstream and Downstream Valve Chambers)

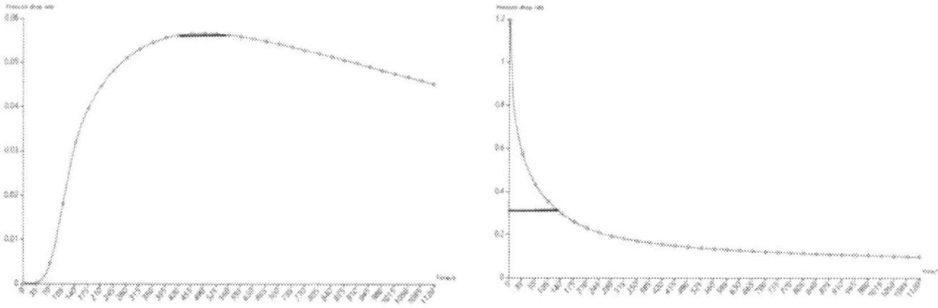

**Fig.10.** Detection Value of the Rate of Pressure Drop at 32km from the Leakage Point of Leakage Diameter 533.5 to the Upstream Valve Chamber (Upstream and Downstream Valve Chambers)

Under the 120s rate of pressure drop duration, the maximum rate of pressure drop that can be detected under each working condition is shown in Table1.

**Table1.** Maximum Detection Value of Leakage Rate of Pressure Drop under Various Working Conditions (MPa/min)

| Leakage diameter (mm) | Distance from upstream valve/km (Upstream valve chamber) | | | Distance from dwonstream valve/km (Downstream valve chamber) | | |
|---|---|---|---|---|---|---|
| | 0km | 16km | 32km | 0km | 16km | 32km |
| D533.5 | 0.332 | 0.144 | 0.056 | 0.127 | 0.178 | 0.308 |
| D1067 | 0.392 | 0.223 | 0.072 | 0.146 | 0.272 | 0.339 |

Table1 shows that by simulating different leakage conditions, the recommended rate of pressure drop threshold of upstream and downstream shut-off valves at leakage points under various leakage conditions are given to detect the benchmark of 1/2 diameter leakage conditions at 32km upstream, and the recommended duration of 120s is 0.127 MPa/min.

For 1067mm/9MPa pipeline system, when the delay time of rate of pressure drop is 120s, it can be seen that the larger the leakage diameter, the easier it is to detect the leakage, and vice versa. For a rate of pressure drop set value, the leakage caused by the same leakage caliber can be detected differently. The closer the location of the leakage point (the distance between the leakage point and the upstream valve chamber) is to the upstream valve chamber, the easier it is to detect the leakage, and vice versa.

### *3.2 Flow rate reduction condition*

Basic parameters of the pipeline: the diameter of Line A of China-Central Asia Natural Gas Pipeline is 1067mm, the wall thickness is 15.6 mm, the starting point operating pressure is 9MPa, and the designed flow rate is 130 × 108 m 3/y.

**The change of rate of pressure drop of the first shut-off valve when the distance is 9.6 km** from the upstream and the flow rate is reduced by 50%.

**The change of rate of pressure drop of the first shut-off valve when the distance is 4 km** from the upstream and the flow rate is reduced by 50%.

**The change of rate of pressure drop of the first shut-off valve when the distance is 9.6 km** from the upstream and the flow rate is reduced by 20%.

**The change of rate of pressure drop of the first shut-off valve when the distance is 4 km** from the upstream and the flow rate is reduced by 20%.

Under the set value of 120s rate of pressure drop duration, the maximum rate of pressure drop set value that can be detected under each flow rate reduction condition is shown in Fig.11. - Fig.14.

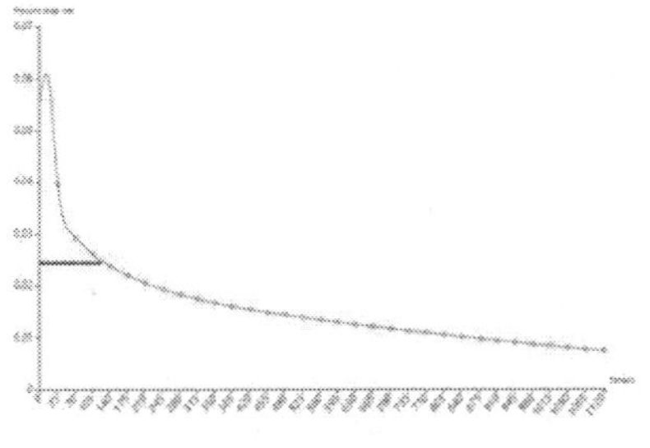

**Fig.11.** 9.6 km from the upstream, flow rate reduction by 50%

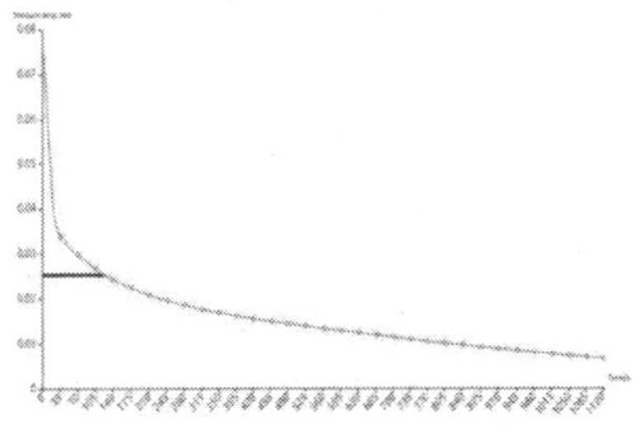

**Fig.12.** 4 km from the upstream, flow rate reduction by 50%

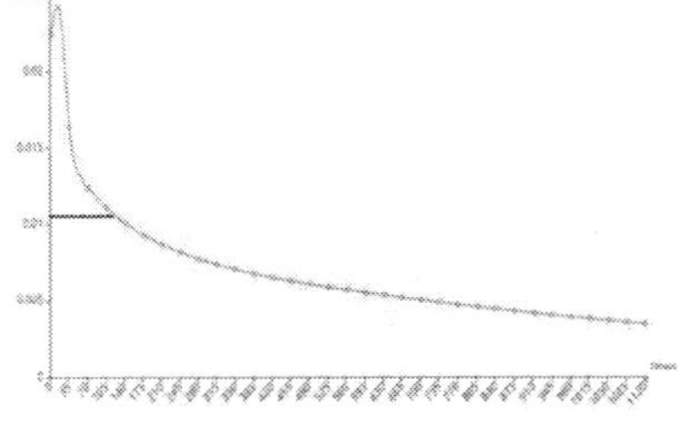

**Fig.13.** 9.6 km from the upstream, flow rate reduction by 20%

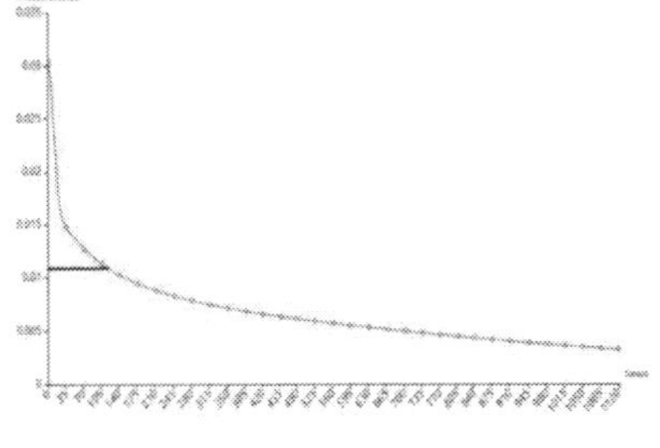

**Fig.14.** 4 km from the upstream, flow rate reduction by 20%

Under the set value of 120s rate of pressure drop duration, the maximum rate of pressure drop that can be detected under each working condition is shown in Table 2.

**Table 2.** Detection Value of rate of pressure drop under Flow Rate Reduction Condition

| Flow rate reduction | Distance | Rate of pressure drop MPa/min |
|---|---|---|
| 50% flow rate reduction | 9.6km | 0.023 |
| | 4km | 0.024 |
| 20% flow rate reduction | 9.6km | 0.0106 |
| | 4km | 0.0108 |

Table 2 shows that when the duration of rate of pressure drop is 120s, the rate of pressure drop of the first trunk shut-off valve is less than 0.025 MPa/min, indicating that there is obvious difference between the rate of pressure drop under flow rate reduction condition and leakage condition, and the valve will not be shut off by mistake.
In case of the same flow rate reduction in the upstream, for flow rate reduction conditions at different distances, the closer to the upstream, the greater the rate of pressure drop of the trunk shut-off valve. When the distance is the same, the more the reduced flow rate, the greater the rate of pressure drop.

### *3.3 Crossover operating condition*

A crossover valve is set between Line B and Line C of China-Central Asia Natural Gas Pipeline to connect the two pipelines:
**The starting point pressure of line B and Line C is set as 9MPa and 8.085 MPa,** respectively, and the upstream and downstream pressure difference of crossover valve after steady operation is 0.5 MPa.
**The starting point pressure of line B and Line C is set as 9MPa and 7.2 MPa,** respectively, and the upstream and downstream pressure difference of crossover valve after steady operation is 1 MPa.
Under the set value of 120s rate of pressure drop duration, when the upstream and downstream pressure difference of the crossover valve is 0.5 MPa, the maximum set value of the rate of pressure drop that can be detected by the upstream and downstream valves of Line B of China-Central Asia Natural Gas Pipeline is shown in Fig.15 - Fig.16.

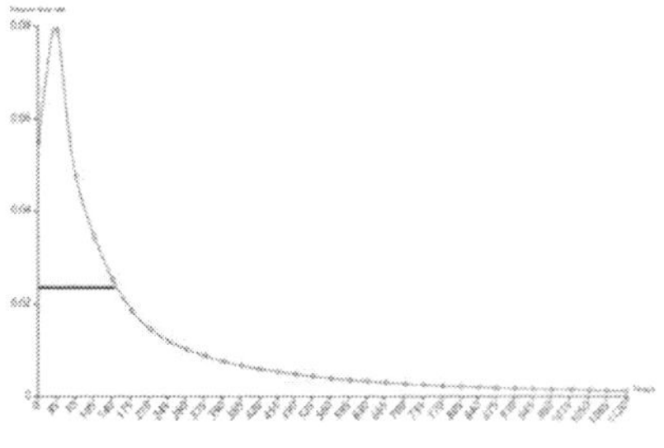

**Fig.15.** Rate of Pressure Drop of Upstream Valve of Line B of China-Central Asia Natural Gas Pipeline

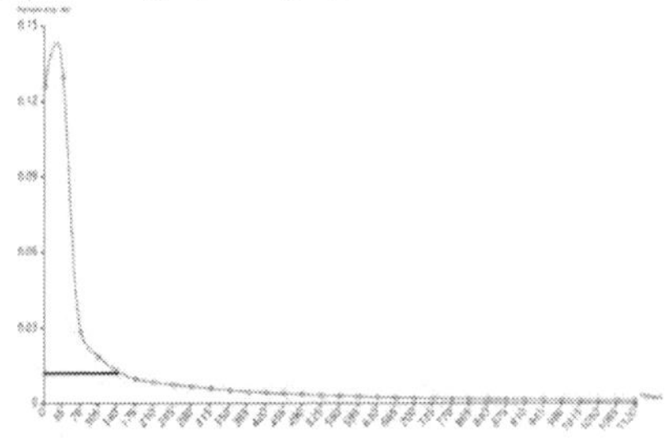

**Fig.16.** Rate of Pressure Drop of Downstream Valve of Line B of China-Central Asia Natural Gas Pipeline

Under the set value of 120s rate of pressure drop duration, when the upstream and downstream pressure difference of the crossover valve is 1 MPa, the maximum set value of the rate of pressure drop that can be detected by the upstream and downstream valves of Line B of China-Central Asia Natural Gas Pipeline is shown in Fig.17 - Fig.18.

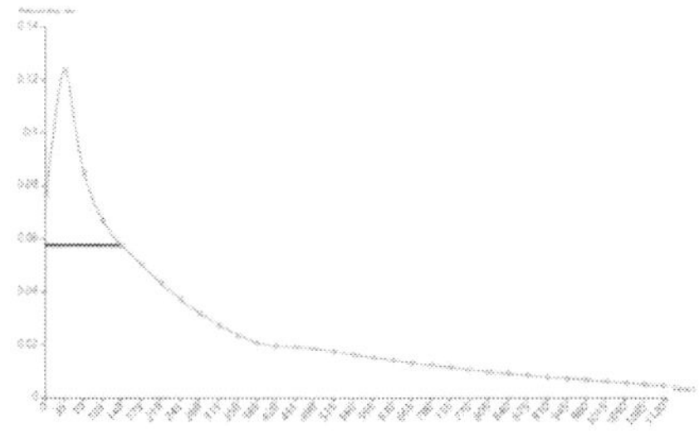

**Fig.17.** Rate of Pressure Drop of Upstream Valve of Line B of China-Central Asia Natural Gas Pipeline

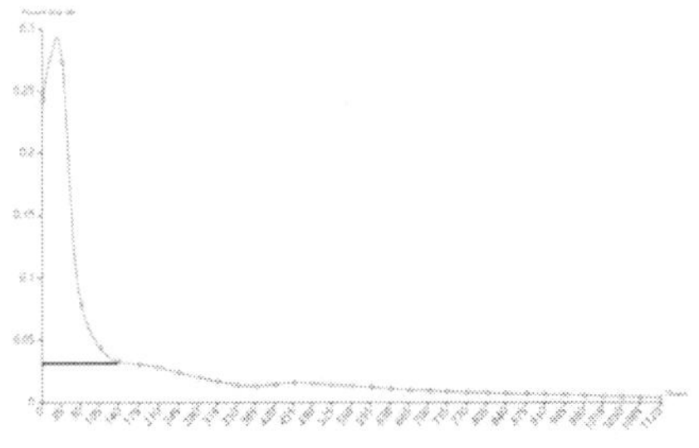

**Fig.18.** Rate of Pressure Drop of Downstream Valve of Line B of China-Central Asia Natural Gas Pipeline

Under the set value of 120s rate of pressure drop duration, the maximum rate of pressure drop that can be detected under each working condition is shown in Table 3.

**Table 3.** Detection Value of Rate of Pressure Drop after Crossover Valve Is Connected

| Upstream and downstream pressure difference after the crossover valve is connected | Valve | Rate of pressure drop MPa/min |
|---|---|---|
| 0.5 MPa | Upstream of Line B | 0.030 |
| | Downstream of Line B | 0.016 |
| 1Ma | Upstream of Line B | 0.056 |
| | Downstream of Line B | 0.035 |

After the crossover valve is connected to the two pipelines, if the upstream and downstream pressure difference is 1MPa, the maximum rate of pressure drop that can be detected is 0.056 MPa/min, which is easy to cause false shut-off of the shut-off valve. In the process of pipeline operation, it is necessary to avoid opening the crossover valve with large pressure difference.

### *3.4 Compressor startup and shutdown operating condition*

When the gas pipeline is running normally, the rate of pressure drop at the valve will change when the compressor starts and shuts down. To avoid the false shut-off of the shut-off valve, the rate of pressure drop set value must be greater than the maximum rate of pressure drop caused by non-accident conditions under the same duration.

The diameter of Line A of China-Central Asia Natural Gas Pipeline is 1067mm, the wall thickness is 15.6 mm, the starting point operating pressure is 9MPa, and the designed flow rate is 130 × 108 m 3/y. Four shut-off valves are set, the first upstream valve Blkv001 is 16km away from the compressor, the second is the upstream station valve Blkv002, the third is the downstream station valve Blkv003, and the fourth shut-off valve Blkv004 is 16km downstream of the compressor. Fig.19. is the schematic diagram of compressor startup and shutdown condition model:

**Fig.19.** Compressor Startup and Shutdown Condition Model

The compressor shutdown condition is simulated at 30s, and attention should be paid to the rate of pressure drop changes of downstream shut-off valve Blkv004 and downstream station valve Blkv003.

3.4.1 Statistical analysis of rate of pressure drop detection values under various simulated working conditions with duration of 120s.
Under the set value of 120s rate of pressure drop duration, the maximum rate of pressure drop set value that can be detected under each compressor shutdown condition is shown in Fig.20 - Fig.21.

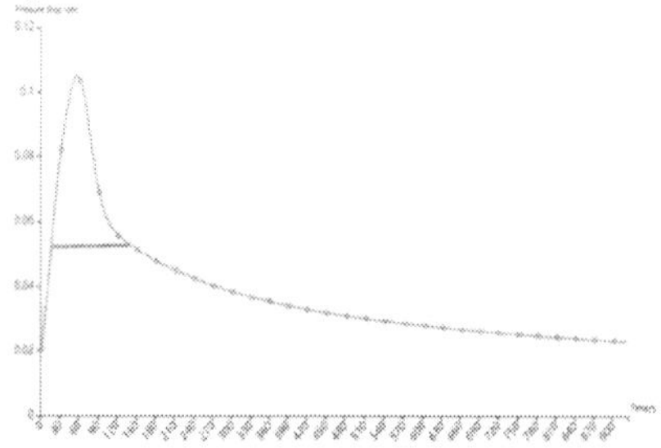

**Fig.20.** Rate of Pressure Drop of Downstream Shut-off Valve

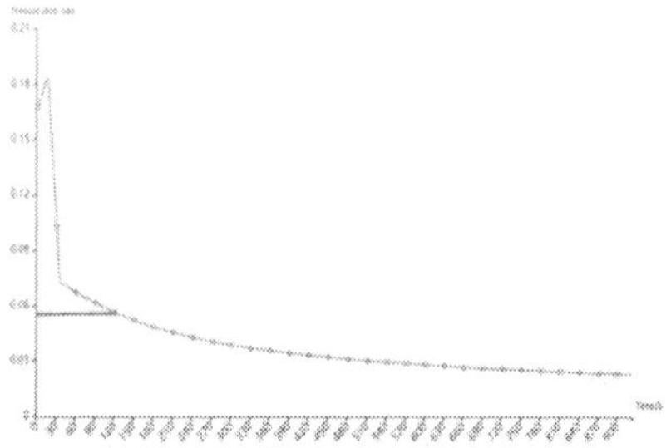

**Fig.21.** Rate of Pressure Drop of Downstream Station Valve

Under the set value of 120s rate of pressure drop duration, the maximum rate of pressure drop that can be detected under each compressor shutdown condition is shown in Table 4.

**Table 4.** Detection Value of Rate of Pressure Drop after Compressor Shutdown

| Working condition | Valve position | Rate of pressure drop MPa/min |
|---|---|---|
| Compressor shutdown | 16km downstream | 0.053 |
| | Valve in downstream station, 0.05 km away from compressor | 0.056 |

Table 4 shows that when the rate of pressure drop duration is 120s, the recommended rate of pressure drop threshold at downstream shut-off valve is 0.056 MPa/min. it can also be seen from the above figure that under the compressor shutdown condition, the rate of pressure drop of the valve in the downstream station and the shut-off valve at 16km downstream is large, which is easy to cause false shut-off of the shut-off valve, so the compressor shutdown operation should be avoided as much as possible during operation.
Finally, the compressor startup condition is simulated at 150s, and attention should be paid to the rate of pressure drop changes of upstream shut-off valve Blkv001 and upstream station valve Blkv002.

3.4.2 Statistical analysis of rate of pressure drop detection values under various simulated working conditions with duration of 120s.
Under the set value of 120s rate of pressure drop duration, the maximum rate of pressure drop set value that can be detected under the compressor startup condition is shown in Fig.22 - Fig.23.

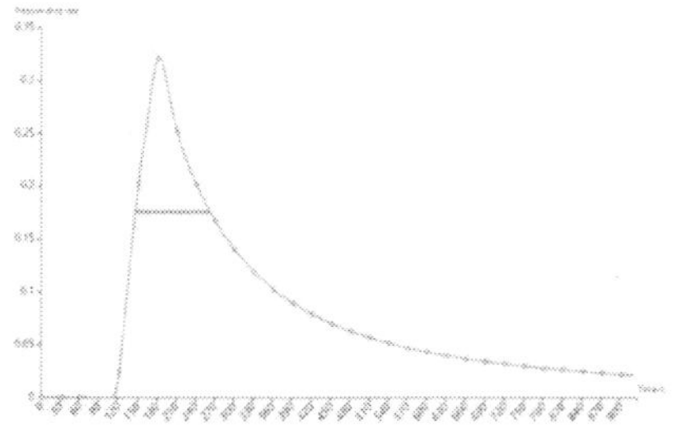

Fig.22. Rate of Pressure Drop of Upstream Shut-off Valve

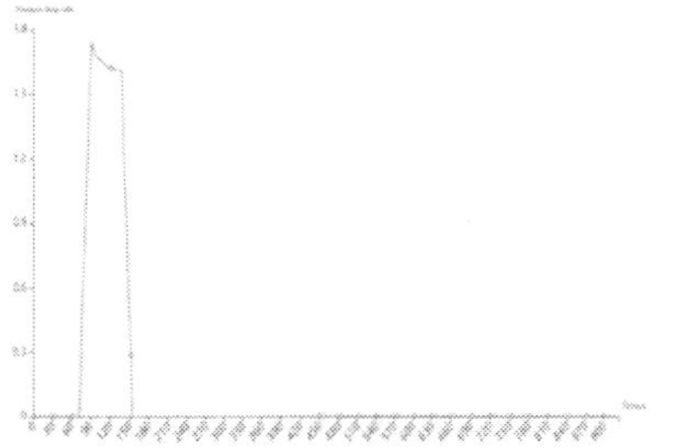

Fig.23. Rate of Pressure Drop of Upstream Station Valve

Under the set value of 120s rate of pressure drop duration, the maximum rate of pressure drop that can be detected under the compressor startup condition is shown in Table 5.

**Table 5.** Detection Value of Rate of Pressure Drop after Compressor Startup

| Working condition | Valve position | Rate of pressure drop MPa/min | Compressor pressure ratio | Compressor power/MW |
|---|---|---|---|---|
| Compressor startup | 16km upstream | 0.173 | 1.3 | 31.01 |
| | Valve in upstream station, 0.1 km away from compressor | / | | |

Table5 shows that when the rate of pressure drop duration is 120s, the recommended rate of pressure drop threshold of upstream shut-off valve is 0.173 MPa/min. it can also be seen from the above figure that under the compressor startup condition, if the duration is set to 120s, it is difficult to cause false shut-off of the valve in the station, but it is easy to cause false shut-off of the shut-off valve at 16km upstream. Therefore, the rate of pressure drop set value should be greater than 0.173 MPa/min.

Since the rate of pressure drop set value of upstream valve is usually less than 0.15 MPa/min in actual situation, it is considered that the pressure ratio of compressor has certain influence on the maximum rate of pressure drop detected by upstream valve under startup condition, so the pressure ratio is set as 1.2.

**Statistical analysis of rate of pressure drop detection values under various simulated working conditions with duration of 120s.**

Under the set value of 120s rate of pressure drop duration, the maximum rate of pressure drop set value that can be detected under the compressor startup condition is shown in Fig.24 - Fig.25.

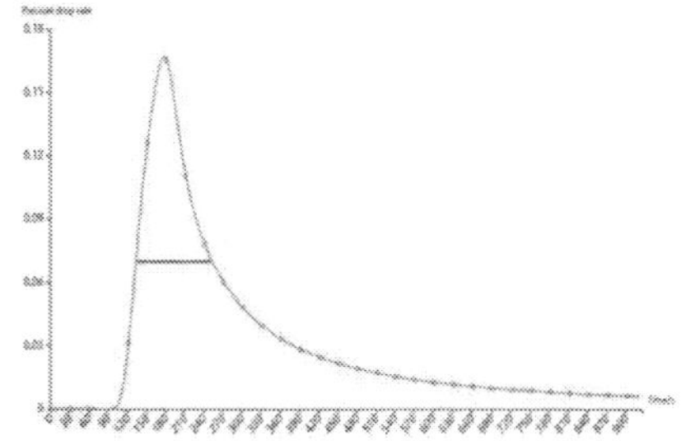

Fig.24. Rate of Pressure Drop of Upstream Shut-off Valve

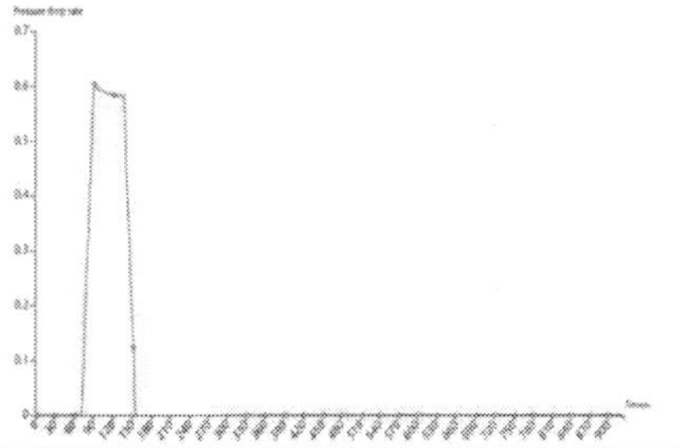

Fig.25. Rate of Pressure Drop of Upstream Station Valve

Under the set value of 120s rate of pressure drop duration, when the pressure ratio is 1.2, the maximum rate of pressure drop that can be detected under the compressor startup condition is shown in Table 6.

**Table 6.** Detection Value of Rate of Pressure Drop after Compressor Startup

| Working condition | Valve position | Rate of pressure drop MPa/min | Compressor pressure ratio | Compressor power/MW |
|---|---|---|---|---|
| Compressor startup | 16km upstream | 0.072 | 1.2 | 12.62 |
| | Valve in upstream station, 0.1 km away from compressor | / | | |

Table 6.3-1 shows that when the rate of pressure drop duration is 120s and the pressure ratio is 1.2, the recommended rate of pressure drop threshold of the upstream shut-off valve is 0.072 MPa/min, that is, the rate of pressure drop set value under this working condition should be greater than 0.072 MPa/min.

From the comparison of the above two startup conditions, it can be seen that under a certain duration and a certain flow rate, the maximum rate of pressure drop detected by the upstream valve of the compressor is closely related to the pressure ratio of the compressor, so the rate of pressure drop should be set according to the actual situation in operation.

## 4. Conclusion

*4.1* By simulating different leakage conditions, the recommended rate of pressure drop thresholds of upstream and downstream shut-off valves at leakage points under various leakage conditions are given to detect the benchmark of 1/2 diameter leakage conditions at 32km upstream, and the recommended duration of 120s is 0.127 MPa/min.

*4.2* By simulating the starting point flow rate reduction condition, the rate of pressure drop at the first shut-off valve is below 0.025 MPa/min in the case of 120s duration, which will not lead to false shut-off.

*4.3* When the pressure difference between the two ends of the crossover valve reaches 1MPa, the rate of pressure drop may reach 0.056 MPa/min in the 120s duration by simulating the opening of the crossover valve with large pressure difference, so the opening of the crossover valve with large pressure difference should be avoided as much as possible.

As a conclusion, with the help of simulation methods, combined with the historical data of operating conditions (such as the maximum pressure drop rate of daily operating conditions), through rationalized and differentiated quantitative adjustment of the pressure drop rate and delay time setting value of valve chambers in different regions, the self-shutoff protection function of the valve chamber can be better utilized, and the reliability of the entire pipeline can be improved in the end.The simulated results can provide reference for the optimization setting of the self-shut-off protection function of the valve chamber of the long-distance natural gas pipeline. For the first time, simulation software is used in multinational pipelines (the Central Asia gas pipeline) to effectively protect pipeline defects. The new values of self-shut-off protection help the pipeline operation to be more stable and safer, reduce the false shutdown caused by pressure drop, and reduce methane cold venting, greatly increasing the contribution to achieving emission reduction goals[20]-[23].

**References**

[1] Sun Fangfang, Ma Liang, Wang Fei, et al. Benchmarking of Spacing Design for Domestic and Foreign Gas Pipeline Valve Chamber [J]. Oil and Gas Field Surface Engineering, 2020, 39 (02): 60-63.
[2] Sun Xiao, An Yao, Li Wang, et al. Study on Protection Ability of Shut-off Valve in A Long-distance Natural Gas Pipeline [J]. Journal of Beijing Institute of Petrochemical Technology, 2018, 26 (03): 36-41.
[3] Lu Meitong, Luo Yexin, He Jiahuan. Protection Simulation of Gas Pipeline Shut-off Valve Burst [J]. Natural Gas and Petroleum, 2015, 33 (05): 14-17.
[4] Wang Guanghui, ZhangWenfei. Determination Method of Rate of Pressure Drop Set

Value of Quick Shut-off Valve on Trunk Line [J]. Oil and Gas Storage and Transportation, 2004, 23 (11): 37-39.
[5] Qi Jianbo, Luo Lihua, Kang Zhuo. Calculation and Analysis of Rate of Pressure Drop of Shut-off Valve on the Second West-East Gas Pipeline [J]. Pipeline Technology and Equipment, 2015 (4): 7-8.
[6] WANG Weilin, GAO Yonghe, LAI Jianbo, et al. Rate of Pressure Drop Set Value of Gas Pipeline Shut-off Valve Burst Detection System [J]. Gas and Heat
[7] Xu Jiashuang, Qiu Xingdong, Li Hairun, et al. Analysis of Rate of Pressure Drop of Damaged Long-distance Gas Pipeline Based on SPS [J]. Natural Gas and Petroleum, 2015, 33 (03): 5-10.
[8] Li Genglian, Ren Jijuan. Study on Shut-off Threshold of Shut-off Valve of Gas Pipeline Trunk Line [J]. Oil and Gas Field Surface Engineering, 2018, 37 (09): 47-50.
[9] Zhang Shuanglei, Chen Feng, Sun Zairong, et al. Study on the Set Value of Rate of Pressure Drop of Gas Pipeline Shut-off Valve [J]. Natural Gas and Petroleum, 2015, 33 (04): 7-11.
[10] ZUO L, JIANG F, JIN B, et al. Influences on the Rate of Pressure Drop in Automatic Line Break Control Valves on a Natural Gas Pipeline [J]. Pipelines: 1489-1499.
[11] LIlI Z, FANGMEI J, BIN J, et al. Value setting for the rate of pressure drop of automatic line-break control valves in natural gas pipelines [J]. Journal of Natural Gas Science and Engineering, 2015, 26.
[12] Dai Xiaohua, Yu Bin, Li Yan, et al. Study on the Rate of Pressure Drop Set Value of Line D of China-Central Asia Natural Gas Pipeline [J]. Oil and Gas Field Surface Engineering, 2018, 37 (4): 8-12.
[13] Fan Zhenning, Liang Haining. Selection of Rate of Pressure Drop of Shut-off Valve in Case of Natural Gas Pipeline Burst [J]. Oil and Gas Storage and Transportation, 2015, 34 (4): 422-425.
[14] Phan T T, Sawin A J. Automatic Linebreak Control Valve Case Study [J]. Psig Annual Meeting, 2012.
[15] Wang Chizi, Qi You, Chen Qinghai, et al. Application of A New Rate of Pressure Drop Algorithm in Interlocking Protection of Natural Gas Pipelines [C] China Oil and Gas Field Surface Engineering Technology Exchange Conference. 2013.
[16] H. Prashanth Reddy, Shankar Narasimhan, S. Murty Bhallamudi, S. Bairagi. Leak detection in gas pipeline networks using an efficient state estimator. Part II. Experimental and field evaluation[B].Computers & Chemical Engineering, Volume 35, Issue 4, 2011:662-670.
[17] Zuo Li-Li,Jin An-Tai, Chen Qian, Dong Qin-Long, Li Yuan-Xin, Zhao Si-Rui.Simulation on venting process and valve opening control method for gas trunk pipelines. Petroleum Science.Volume 19, Issue 6.2022: 3016-3028.
[18] Maciej Chaczykowski, Paweł Zarodkiewicz. Simulation of natural gas quality distribution for pipeline systems[J].Energy.Volume 134.2017: 681-698.
[19] Helgaker, J. F., Müller, B., and Ytrehus, T. (April 1, 2014). "Transient Flow in Natural Gas Pipelines Using Implicit Finite Difference Schemes." ASME[J]. Offshore Mech. Arct. Eng. August 2014; 136(3): 031701.
[20] Mohammad Soltanieh, Angineh Zohrabian, Mohammad Javad Gholipour, Eugenia Kalnay.A review of global gas flaring and venting and impact on the environment: Case study of Iran[J]. International Journal of Greenhouse Gas Control, Volume 49, 2016:Pages 488-509.
[21] Sydney Thomas, Richard A Dawe. Review of ways to transport natural gas energy from countries which do not need the gas for domestic use[J].Energy.Volume 28, Issue 14,2003: 1461-1477.
[22] H. Prashanth Reddy, Shankar Narasimhan, S. Murty Bhallamudi, S. Bairagi. Leak detection in gas pipeline networks using an efficient state estimator. Part-I: Theory and simulations. Computers & Chemical Engineering.Volume 35, Issue 4.2011:651-661.
[23] Johannes H. Andersen, Roger S. Powell.Implicit state-estimation technique for water network monitoring[J]. Urban Water.Volume 2, Issue 2,2000:123-130.

# An XFEM-based CZM Numerical Strategy for Modeling Hydraulic Fracture Propagation under Different Fracturing Schemes

**Hao-yu ZHANG1, Jun-bin CHEN1, (⊠), Yang WANG2**

1 College of Petroleum Engineering, Xi'an Shiyou University,
Xi'an, China
2 No.6 Oil Production Plant, PetroChina Changqing Oilfield Company,
Xi'an, China

chenjb@xsyu.edu.cn

**Abstract.** Segmented and cluster fracturing can improve the efficiency of volume stimulation in shale reservoirs and reduce construction costs. It is important to clarify the propagation characteristics of hydraulic fractures and stress interference under different fracturing techniques to optimize the process of clustered and staged fracturing. For this purpose, we have developed a 2D XFEM-based CZM hydraulic fracturing model. The capability of this model was validated by analytical solutions and then used to study the propagation paths of hydraulic fractures and the characteristics of stress interference under simultaneous fracturing, sequential fracturing, zipper fracturing and multi-cluster fracturing. The results show that in simultaneous fracturing, the middle cluster is compressed by the external position clusters, and the opening width of hydraulic fractures is reduced. In sequential fracturing, the fracture that first initiates create an additional stress field that inhibits the propagation distance of subsequent fractures and the propagation path of hydraulic fractures is also affected in the stress shadow region. Zipper fracturing can effectively alleviate stress interference between multiple fractures, and internal fractures can also propagate a certain distance. In multi-cluster fracturing, the fluid rate into the internal fractures may be limited, and fracture propagation is also limited by stress interference. Therefore, it is necessary to optimize the parameters of clusters to ensure that all clusters can initiate fractures normally. The research results are important for the parameter optimization of clustered and staged fracturing, especially for well factory fracturing mode.

## 1. Introduction

Hydraulic fracturing has proven to be an effective method for enhancing the production of both conventional and unconventional oil and gas reservoirs through the expansion of drainage zones. Currently, multi-fracturing techniques for horizontal wells are becoming increasingly popular in the fracturing community (Pope et al. 2010; Chaudhri 2012; Yu et al. 2014). In order to improve the efficiency of transformation and reduce operation costs, the "well factory" model is often used to develop shale oil and gas (Ladlee and Jacquet 2011). In the early period, the fracturing sequence of an individual well was changed from sequential fracturing to jump fracturing to increase fracture interference and fracture complexity. (Soliman et al. 2010; East et al. 2011) With the promotion of "well factory" model, synchronous fracturing techniques with multiple wells alternating fracturing have emerged. Based on this, conventional and improved zipper fracturing techniques have been developed by optimizing fracturing sequences and well spacing, which further promotes the formation of complex fracture networks. Furthermore, the dominant trend involves the utilization of multi-cluster hydraulic fracturing technology in horizontal wells using the well factory fracturing mode. Within the realm of hydraulic fracturing, simultaneous creation of multiple fractures is a common strategy aimed at cost reduction. This technique involves dividing the fractured interval using a bridge plug positioned within the casing of a horizontal well. Each fractured interval encompasses 3 to 12 clusters, each comprising 6 to 30 shot holes (Bunger et al. 2014; Lecampion et al. 2015). Despite the cost-effectiveness of this multi-stage, multi-cluster fracturing method, wells stimulated using this approach frequently fall short of anticipated production levels. Field data from oil and gas monitoring indicate that the production increase along the wellbore is often inconsistent, with actual stimulation being significantly less pronounced than indicated by microseismic mapping. These observations suggest that a significant portion of hydraulic fractures may be contributing less efficiently to production outcomes (Miller et al. 2011; Sookprasong et al. 2014; Ugueto C et al. 2016).

In recent times, various scholars have employed diverse modeling techniques to systematically examine the simultaneous progression of hydraulic fractures. Certain findings indicate that the interplay of stress among fractures and the subsequent impeding influence on specific fracture growth constitute the primary catalysts for notable alterations in production rates within stimulated wells (Nagel and Sanchez-Nagel 2011; Roussel and Sharma 2011; Li et al. 2017). Guo et al. (Guo et al. 2015) formulated a 3D model for fracture propagation, utilizing the finite element method to simulate segmented multi-cluster fracturing. Focusing on a study involving three clusters, they devised an optimization technique for cluster spacing, which takes into account the geometric attributes of each cluster and the stress field created amidst fractures. When fractures are evenly distributed in a three-cluster configuration, the dominant influence resides with the external fractures, prompting the central fractures to propagate linearly. Conversely, if the spacing between the three-cluster fractures varies, the irregularity in the propagation of multiple fractures becomes more prominent (Yun et al. 2016). Krzaczek et al. (Krzaczek et al. 2020) integrated the discrete element method (DEM) with computational fluid dynamics (CFD) to create a numerical model for assessing the response of fluid-induced fractures in rock formations. Similarly, Yang et al. (Yang et al. 2021) constructed a discrete element model (DEM) to study the simultaneous hydraulic fracturing of paired wells. Their findings emphasized the significance of azimuth and spacing on fracture progression and pressure, highlighting the susceptibility of hydraulic fracture migration due to stress shadow effects between the two wells. However, establishing a correlation between rock attributes (such as Young's modulus and permeability) and DEM parameters presents a challenge. Moreover, conducting field-scale DEM simulations encounters the hurdle of substantial computational expenses. Another frequently employed approach for hydraulic fracturing modeling is the cohesive zone model (CZM). However, CZM imposes limitations on fracture propagation paths, making the representation of stress interference issues between fractures complicated.

The extended finite element method (XFEM) has gained prominence in the fracturing field, with numerous researchers using it to simulate hydraulic fracturing procedures (Sheng et al. 2018; Wang et al. 2018; Cruz et al. 2019). Nonetheless, there is a scarcity of documented investigations focusing on the propagation behavior of hydraulic fractures across varied fracturing modes using XFEM. There is also a lack of in-depth studies on the comparative effects of stress interference on hydraulic fractures under different fracturing schemes. For this purpose, the XFEM-based cohesive zone method (XFEM-based CZM) was used to develop a simultaneous fracturing model, a sequential fracturing model, a zipper fracturing model and a multi-cluster fracturing model to compare the effect of stress interference on fracture propagation. By analyzing the fracture propagation morphology of each model, we hope to illustrate the role of stress disturbances in different fracturing schemes so that the fracturing community can better optimize fracturing operation parameters. The structure of this paper is as follows. Section 2 outlines the methodology employed in numerical simulation. Subsequently, section 3 validates the efficacy of the model proposed. In section 4, the impact of distinct fracturing strategies on fracture morphology is expounded. A concise summary based on the numerical results is given in section 5.

## 2. Modeling Theory

While hydraulic fracturing primarily occurs within three-dimensional formations, the focus of this study is to evaluate the free propagation of hydraulic fractures under different fracturing strategies. Therefore, we take a 2D plane containing a fractured interval as the study area (Fig. 1). In order to provide the numerical model with the capabilities mentioned above, three primary methodologies are employed in this study: (i) XFEM-based CZM for simulating unhindered hydraulic fracture propagation and stress interactions among fractures; (ii) traction-separation law for identifying criteria governing fracture opening damage; (iii) fluid pipe element for emulating pressure drop losses in perforated clusters during fracturing fluid flow. Elaboration on these theories is provided comprehensively in the subsequent subsections. These theories are described in detail in the following subsections.

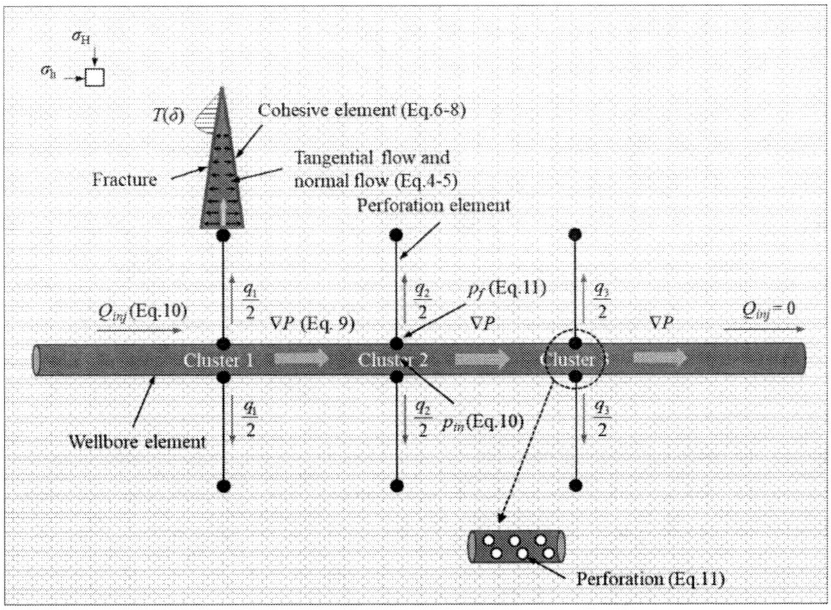


**Fig. 1.** Sketch of hydraulic fracturing operation.

## *2.1. XFEM-based CZM*

To avoid the mesh reconstruction problem in the simulation of hydraulic fracturing process using the conventional finite element method, Moes and Belytschko (Moës et al. 1999) proposed a mesh-independent XFEM fracture propagation model based on the partition of unity method (PUM). Fries and Belytschko (Fries and Belytschko 2010) proposed a displacement model for XFEM by incorporating a crack tip asymptotic an asymptotic displacement field function ($F_l(x)$) at the crack tip and a jump function ($H(x)$) that can reflect the crack surface, and incorporating discontinuous geometry, fracture and discontinuity fields, see Eq. 1,

$$\boldsymbol{u}_{xfem}(x) = \sum_{i \in I} N_i(x)\left[\boldsymbol{u}_i + H(x)\boldsymbol{a}_i + \sum_{l=1}^{4} F_l(x)\boldsymbol{b}_i^l\right] \tag{1}$$

where $I$ is the number of standard FE nodes, $N_i(x)$ is the standard FE shape function for ordinary element nodes, $\boldsymbol{u}_i$ is the displacement of the ordinary element node, $H(x)$ is the Heaviside enrichment function, which represents the nodes cut by the fracture interior, $\boldsymbol{a}_i$ is the crack penetration element node degree of freedom, $\boldsymbol{b}_i^l$ is the enrichment element node degree of freedom at the crack tip, $F_l(x)$ is the asymptotic crack tip function and is used to detail the nodes which are only cut by the fracture tip.

The $H(x)$ function is given as 1 for above the crack and -1 for below the crack. The expression is written as Eq. 2,

$$H(x) = \begin{cases} 1 & \text{if } \left(x - x^*\right) \cdot \boldsymbol{n} \geq 0 \\ -1 & \text{else} \end{cases} \tag{2}$$

where $x$ is a Gaussian point, $x^*$ is the point on the crack closest to $x$, and $\boldsymbol{n}$ is the unit vector normal to the crack at $x^*$.

The four enhanced nodes of the crack tip using the asymptotic solution proposed by Williams (Williams 1959) are defined in the local crack tip polar coordinate system as Eq. 3,

$$\{F_i(r,\theta)\}_{i=1,2,3,4} = \left\{\sqrt{r}\sin(\frac{\theta}{2}), \sqrt{r}\cos(\frac{\theta}{2}), \sqrt{r}\sin(\frac{\theta}{2})\sin(\theta), \sqrt{r}\cos(\frac{\theta}{2})\sin(\theta)\right\} \tag{3}$$

where $r$ and $\theta$ are polar co-ordinates with origin at the crack tip.

In recent years, ABAQUS has implemented XFEM-based CZM using the "phantom node" method. An enriched pressure diffusion/stress element is replicated internally by adding the angle phantom node.

As depicted in Fig. 2, solid blue circles represent the original real nodes, while hollow circles represent the corner phantom nodes. Before damage initiation, only real nodes are in play within this element, with corner phantom nodes attached to the original nodes. Once damage begins, the displacement and pore pressure degrees of freedom of corner phantom nodes become operational. At this point, the real and corner phantom nodes of the element are allowed to deform independently and the pore pressure can diffuse independently. The fracture evolution occurs after the corner phantom nodes and their corresponding real nodes generate separation displacements. A new cohesive element is created between the two enriched elements separated by the deformation through interpolation to represent the discontinuous interface behavior (shown as red hollow circles in Fig. 2). The interface strength follows to the traction separation law (TSL) as described in the cohesive zone model. Therefore, fluid pressure driven fracture propagation behavior can be simulated using XFEM-based CZM.

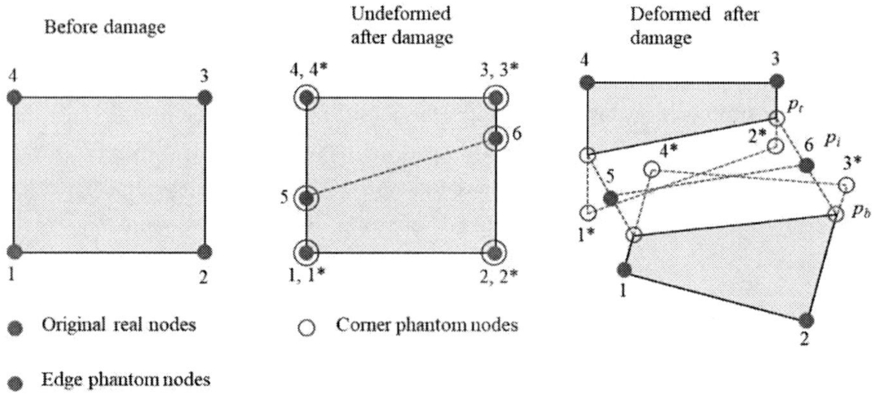


**Fig. 2**. Implementation of the XFEM with real and phantom nodes on an enriched element before and after damage.

### *2.2. Fluid Flow Within the Cohesive Zone*

Assuming that all fluids within the fracture are Newtonian, and the tangential flow equation (Boone and Ingraffea 1990) can be described as Eq. (4),

$$q = \frac{w^3}{12\mu}\nabla p \tag{4}$$

where $q$ is the tangential flow in the fracture, $w$ signifies the fracture width, $\mu$ is the fluid viscosity, and $\nabla p$ indicates the fluid pressure gradient along the direction of fracture propagation.
The fracturing fluid has the ability to infiltrate the rock through the upper and lower surfaces within the fracture. This filtration loss phenomenon can be characterized by Eq. (5).

$$\begin{cases} q_t = c_t\left(p_i - p_t\right) \\ q_b = c_b\left(p_i - p_b\right) \end{cases} \tag{5}$$

In this equation, $q_t$ and $q_b$ denote the volume flow rates per unit time on the upper and lower fracture surfaces, respectively. The terms $c_t$ and $c_b$ correspond to the leak coefficients on the upper and lower surfaces of the fracture. Similarly, $p_t$ and $p_b$ represent the pore pressures on the upper and lower fracture surfaces, while $p_i$ signifies the fluid pressure within the fracture.

### 2.3. Failure Mechanism

The progression of fracture is described by the concept of stiffness decay, with Eq. (6) taking the form:

$$\begin{aligned} t_n &= \begin{cases} (1-D)\bar{t}_n, & \bar{t}_n \geq 0 \\ \bar{t}_n & \text{otherwise (no damage to compressive stiffness)} \end{cases} \\ t_s &= (1-D)\bar{t}_s \\ t_t &= (1-D)\bar{t}_t \end{aligned} \tag{6}$$

where $\bar{t}_n$ , $\bar{t}_s$ and $\bar{t}_t$ denote the stress components in the normal and tangential directions of the element prior to damage, given the existing strain conditions.
The interface damage is represented in the numerical model by a scalar damage variable $D$. In this process, the damage variable monotonically evolves from 0 to 1 with the loading of the traction force. The damage variable $D$ can be expressed in Eq. 7 as,

$$D = \begin{cases} 0, & \delta_m \leq \delta_m^0 \\ \dfrac{\delta_m^f\left(\delta_m - \delta_m^0\right)}{\delta_m\left(\delta_m^f - \delta_m^0\right)}, & \delta_m^0 < \delta_m < \delta_m^f \\ 1, & \delta_m \geq \delta_m^f \end{cases} \tag{7}$$

where $\delta_m^0$ and $\delta_m^f$ signify the separation displacements at the initiation and full development of damage within the element.
The results of Wang (Wang et al. 2016) show that the maximum principal stress criterion is applicable to XFEM-based CZM fracture initiation, expressed as Eq. (8),

$$f = \left\{\frac{\langle \sigma_{\max} \rangle}{\sigma_{\max}^0}\right\} \tag{8}$$

where $\sigma_{\max}^0$ is the maximum principal stress allowed. The symbol $\langle \ \rangle$ denotes the Macaulay bracket, conventionally interpreted to indicate that a purely compressive stress state does not induce damage. It is assumed that damage is initiated when the maximum principal stress ratio reaches 1.

### *2.4. Fluid Flow Within a Wellbore*

The fluid pipe element has two pore pressure nodes, which create a certain pressure drop when fluid passes through the element. Therefore, the pressure drops of fluid passing through a wellbore or perforation during multi-cluster hydraulic fracturing using some fluid pipe elements. The pressure loss equation can be written as Eq. 9,

$$
\begin{aligned}
&\nabla P - \rho g \Delta Z = \left(C_L + K_t\right)\frac{\rho v^2}{2}, \\
&C_L = \frac{fL}{D_h}, \\
&f = 8\left[\left(\frac{8}{\mathrm{Re}}\right)^{12} + \frac{1}{\left(A+B\right)^{1.5}}\right]^{1/12}, \\
&A = \left[-2.457\ln\left(\left(\frac{7}{\mathrm{Re}}\right)^{0.9} + 0.27\left(\frac{K_s}{D_h}\right)\right)\right]^{16}, \\
&B = \left(\frac{37350}{\mathrm{Re}}\right)^{16}
\end{aligned}
\tag{9}
$$

where $\nabla P$ denotes the pressure disparity at the cluster node, while $\Delta Z$ represents the vertical variation of the node elevation. The variable $v$ signifies fluid velocity within the horizontal wellbore, and $\rho$ corresponds to fluid density. $G$ is the acceleration due to gravity, $C_L$ is the loss coefficient, and $f$ accounts for frictional force along the wellbore. $L$ is the length of the wellbore, $K_t$ is the loss term in fixed direction, $K_s$ is the roughness of wellbore, $D_h$ is the diameter of the wellbore, and $\mathrm{Re}$ denotes the Reynolds number of fluid within the wellbore.

The fluid in the pipe network satisfies mass conservation equation and energy conservation equation, and the expression is written as Eq. 10,

$$
\begin{aligned}
&Q_{inj} = \sum_{i=1}^{n} q_i \\
&p_{wellbore} = p_{in} + p_f + p_s
\end{aligned}
\tag{10}
$$

where $Q_{inj}$ is the total injection flow rate, $q_i$ is the flow rate into the specific cluster labeled as $i$, $n$ is the number of clusters, $p_{wellbore}$ is the fluid pressure at the injection point, $p_{in}$ represents the fluid pressure at the cluster, $p_f$ is the pressure decrement across the cluster, $p_s$ is the pressure resulting from fluid-wellbore friction.

The perforation friction can be calculated using the pressure drop model of Bernoulli equation, and its expression is as follows,

$$
p_f = 0.807249 \times \frac{\rho}{n^2 D_p^4 C^2} q_i \tag{11}
$$

where $n$ is the number of perforation in each cluster, $D_p$ is the perforation diameter, $C$ is the coefficient affecting perforation flow, generally 0.56~0.9 (Modeland et al. 2011), in this paper, the value is 0.6.

## 3. Model Validation

To validate the credibility of our fracture propagation model, we conduct a comparison with a solitary fracture propagation model, as outlined by Geertsma and De Klerk (Geertsma and De Klerk 1969). This validation assumes a homogeneous reservoir for the hydraulic fracturing process, where the fracturing fluid does not contain granular proppant, and no proppant transport or settling occurs. The analytical formulas for fracture width ($w_0$), fracture length ($L$), and net pressure ($p$) at the injection point can be explained by the following equations:

$$w_0(t) = 2.14\left[\frac{(1-v)\mu Q^3}{E}\right]^{\frac{1}{6}} t^{\frac{1}{3}} \tag{12}$$

$$L(t) = 0.65\left[\frac{EQ^3}{(1-v)\mu}\right]^{\frac{1}{6}} t^{\frac{2}{3}} \tag{13}$$

$$p(t) = 1.97\left[\frac{EQ^3\mu}{L^2(1-v)^3}\right]^{\frac{1}{4}} + \sigma_h \tag{14}$$

where $Q$ is the fluid injection rate, $\mu$ is the fluid viscosity, maintained at 1 cp in this study, $t$ is the fluid injection time, $E$ is the Young's modulus, $v$ indicates the Poisson's ratio, and $\sigma_h$ corresponds to the minimum principal stress.

(a) (b) (c)

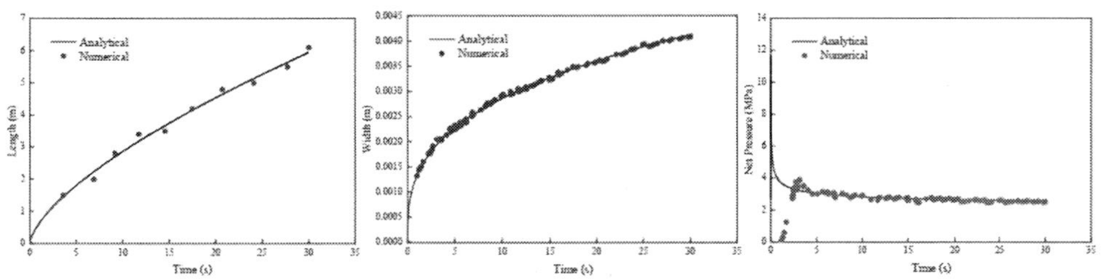


**Fig. 3.** Time evolutions of (a) length of fracture, (b) fracture width at injection point, and (c) net pressure.

## 4. Results and Discussion

Based on the XFEM-based CZM approach, we developed hydraulic fracturing models with different schemes by controlling injection point positions and initiation sequence of clusters (Fig. 4). Table 1 lists the main parameters used in the numerical simulations.

(a) (b)

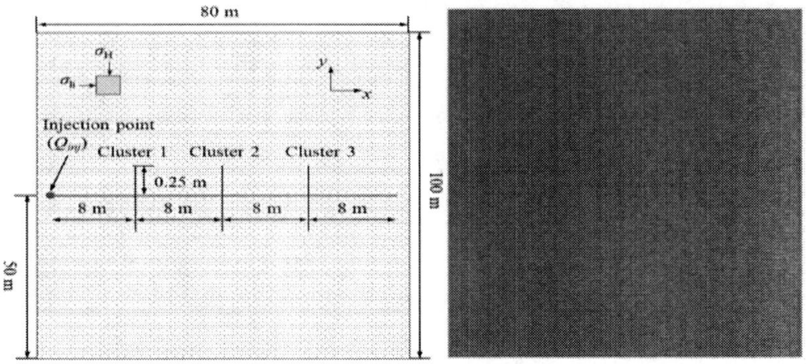


**Fig. 4.** The numerical model of hydraulic fracturing used in this paper: (a) sketch of geometric model; (b) meshing scheme.

**Table 1.** Input parameters for simulating hydraulic fracturing.

| Categories | Input Parameters | Value |
|---|---|---|
| **Rock matrix parameters** | Elastic modulus | 30 GPa |
| | Poisson's ratio | 0.25 |
| | Permeability coefficient | $1\times10^{-7}$ m/s |
| | Specific weight of fluid | 9.8 kN/m$^3$ |
| | Porosity | 0.1 |
| | Initial pore pressure | 20 MPa |
| | Maximum horizontal stress | 10 MPa |
| | Minimum horizontal stress | 8 MPa |
| **Cohesive zone parameters** | Normal stress expected value, $t_n$ | 12 MPa |
| | 1st shear nominal stress expected value, $t_s$ | 20 MPa |
| | 2nd shear nominal stress expected value, $t_t$ | 20 MPa |
| | Failure displacement, $\delta_m^f$ | 0.001 m |
| **Wellbore parameters** | Casing size | 5½-inch |
| | Pipe roughness, $K_s$ | $0.015\times10^{-3}$ mm |
| | Number of perforations, $n$ | 16 |
| | perforation diameter | 0.01 m |
| **Fracturing fluid parameters** | Fluid viscosity | 4 cp |
| | Formation leak-off coefficient, $k$ | $1\times10^{-14}$ m/( Pa • s) |
| | Injection rate per cluster | 0.02 m$^3$/s |
| | Injection time | 30 s |

### *4.1. Fracture Propagation During Simultaneous Fracturing*

Simultaneous fracturing means each perforated cluster is fractured simultaneously at the same flow rate. Therefore, there are three injection points in this model, which are located at the cluster locations, and the three clusters are fractured simultaneously by setting the injection time at the same moment. Each perforated cluster is injected at a flow rate of 0.02 m$^3$/s and the injection time is 30 s. Fig. 5a shows the pore pressure map of simultaneous fracturing. As can be seen from the figure, stress interference phenomenon is obvious during the simultaneous fracture initiation of the three clusters. The middle cluster is severely deformed by compression of external clusters, and the hydraulic fractures close to wellbore show shrinkage and a severe reduction in fracture width. In this situation, proppant inflow will become difficult and the middle fracture will risk being plugged with proppant. Fig. 5b shows a vector plot of the maximum principal stress direction at the end of fracturing. It can be seen from the figure that the maximum principal stress perturbation is most evident at the tip of hydraulic fractures. The direction of maximum principal stresses in the region outside cluster system changes sharply and covers almost the entire domain. The maximum principal stress direction even reverses to a horizontal direction near the tip of fractures. This indicates that the stress interference generated by cluster systems is not only present among fractures, but also has a very strong effect on regions outside the fracture system.

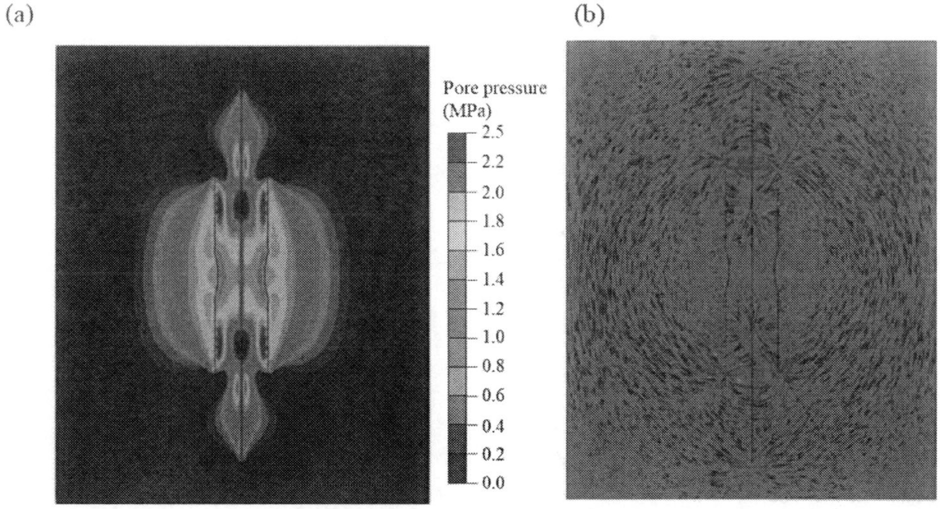


**Fig. 5.** Morphology of hydraulic fracture propagation in simultaneous fracturing: (a) fractures propagation paths; (b) maximum principal stress direction.

*4.2 Fracture Propagation During Sequential Fracturing*

We modeled sequential hydraulic fractures by varying the initiation times of each cluster. In this study, each cluster was fractured for 10 seconds with a 1 second interval and an injection rate of 0.02 $m^3/s$. Figure 6a shows the propagation pattern of the hydraulic fractures after fracturing. As can be seen from the figure, the fractures that initiated earlier caused stress interference on the subsequent clusters. Cluster 1 had the longest propagation distance, followed by cluster 2, and cluster 3 had the shortest propagation distance. This may be because the stress shadow created by the earlier initiated fractures altered the original stress field, and the later initiated fractures not only had to overcome the fracture pressure of the rock, but also had to overcome the additional residual stress field of the earlier fractures. This undoubtedly increased the energy required for subsequent cluster initiation. Therefore, under the same injection conditions, the propagation distance of the later initiated fractures was reduced. In addition, the propagation path of the hydraulic fractures was also affected in the stress shadow region. In the stress concentration area, the hydraulic fractures deviated from the original propagation path and could even maintain a planar double-wing fractures shape. Under the influence of stress interference, hydraulic fractures may deviate to one side of the horizontal borehole. Figure 6b shows the vector map of the maximum principal stress. From the figure it can be seen that the stress interference region is mainly concentrated around the fractures system. Compared with simultaneous fracturing, the range of stress interference is significantly reduced, which indicates that the order of hydraulic fracture initiation controls the range of stress interference effects.

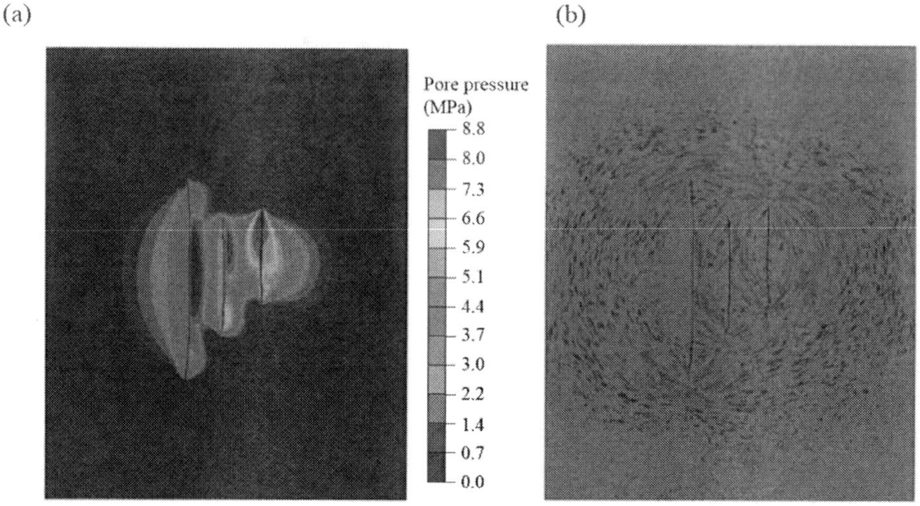


**Fig. 6.** Morphology of hydraulic fracture propagation in sequential fracturing: (a) fractures propagation paths; (b) maximum principal stress direction.

### *4.3 Fracture Propagation During Zipper Fracturing*

Zipper fracturing is a technique in which after completing the fracturing of cluster 1 and cluster 3, the fracturing operation of cluster 2 is performed. The characteristic of this fracturing scheme is that it can be combined with adjacent stimulation wells, allowing the order of fracture initiation to jump between wells. In this study, we only simulated the zipper fracturing operation of a single well. Similarly, each cluster was fractured for 10 seconds with a 1 second interval and an injection rate of 0.02 $m^3/s$. Figure 7 shows the results of our numerical simulation. It can be seen that the length of each hydraulic fracture has been increased in the zipper fracturing scheme compared to sequential fracturing. Especially for cluster 3, the propagation distance of the hydraulic fracture has increased significantly. This may be because the stress shadow area after cluster 1 initiation does not completely cover the area where cluster 3 is located, and therefore the additional residual stress field that cluster 3 has to overcome during initiation is smaller. In addition, due to the presence of normal flow at the fracture opening, the fluid in cluster 1 is filtered into the formation over time during the long construction operation, resulting in a decrease in pressure within the fracture. The perturbation in the direction of the maximum principal stress in the operated area occurred during the fracturing of cluster 2, which may have prevented cluster 2 from being affected by the stress interference caused by cluster 1 during the last fracturing period. Therefore, the propagation length of the entire hydraulic fracture system is better than that of the sequential fracturing scheme.

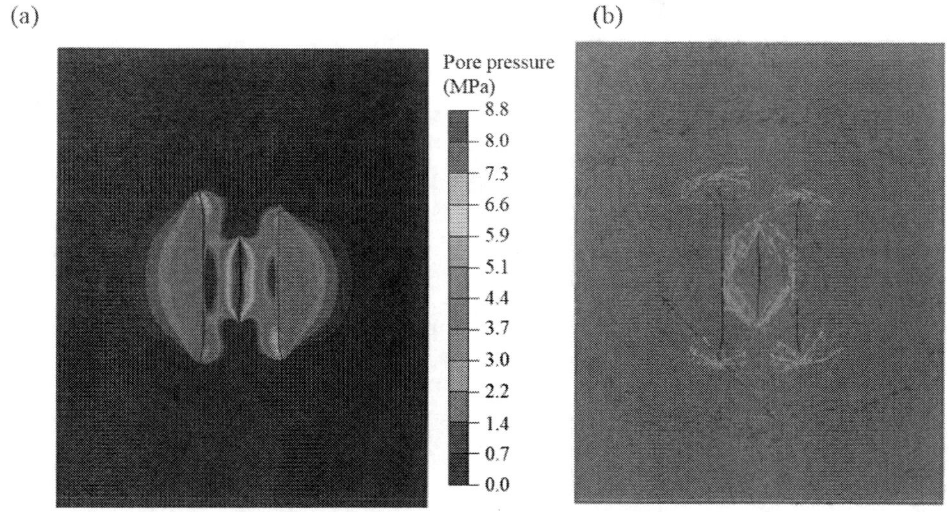


**Fig. 7.** Morphology of hydraulic fracture propagation in zipper fracturing: (a) fractures propagation paths; (b) maximum principal stress direction.

*4.4 Fracture Propagation During Multi-Cluster Fracturing*

The characteristic of multi-cluster fracturing is that fluid is injected from the wellhead and then enters the formation through the perforation clusters in the horizontal wellbore. The high-speed flow of fluid causes the formation to open up and form macroscopic hydraulic fractures. Controlling the fluid flow rate into the perforation clusters is the most interesting parameter optimization for the fracturing community. Fig. 8 shows the propagation morphology of hydraulic fractures with a cluster spacing of 8 m, injection velocity of 0.06 $m^3/s$, and injection duration of 30 s. It can be seen from the figure that the most significant feature of multi-cluster fracturing compared to synchronous fracturing is the inhibitory effect of external fractures on the middle fracture. When the cluster spacing is small, the middle fracture cannot even initiate under the influence of external fractures, which is consistent with the microseismic monitoring data at the fracturing site (Yoshida et al. 2018; Wang et al. 2022). We plotted the flow rates through the three clusters as a curve against time (Fig. 9). It can be seen from the figure that the flow rate into each cluster is dynamically unevenly distributed, with the middle cluster receiving the least amount of flow rate under the influence of stress interference. This is also the reason why the hydraulic fracture in the middle has stopped propagating. The fracturing community always expects all clusters to open smoothly, so it is necessary to optimize the operating parameters of internal clusters during multi-cluster hydraulic fracturing to ensure that all clusters can initiate successfully. Fig. 10 shows a vector map of the maximum principal stress direction, which indicates that the stress interference of multi-cluster fracturing is mainly concentrated between fractures, manifested in the inhibitory effect of external fractures on internal fractures. The stress interference outside the fractures is relatively weak and mainly concentrated in the fracture tip area.

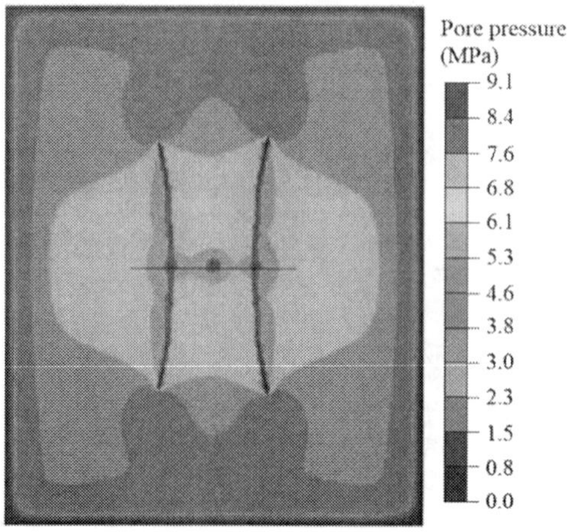


**Fig. 8.** Morphology of hydraulic fracture propagation in multi-cluster fracturing.

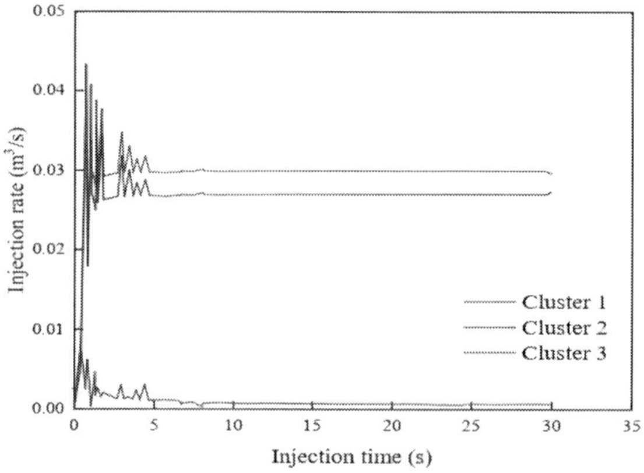


**Fig. 9.** Curve of flow rate into each cluster over time.

**Fig. 10.** Vector diagram of maximum principal stress direction at the end of multi-cluster fracturing.

## 5. Conclusions

This paper presents a numerical simulation method for hydraulic fracturing based on XFEM-based CZM and compares the characteristics of hydraulic fracture propagation paths and stress interference effects during simultaneous fracturing, sequential fracturing, zipper fracturing and multi-cluster fracturing. The main conclusions of this study are as follows:

(1) In simultaneous fracturing, medial fractures can be compressed by external fractures, resulting in a severe reduction in fracture width. Although the intermediate fracture may have a longer propagation distance under compression, the narrow opening width is not conducive to proppant deposition.

(2) Due to the stress concentration phenomenon at the tip of the hydraulic fracture as it propagates, the initial fractures in sequential fracturing can impose significant residual stresses on subsequent fractures. This additional stress field increases the energy required for the propagation of subsequent fractures and inhibits their propagation.

Zipper fracturing technique can effectively reduce the negative effects of stress interference. In this case, the stress interference on the medial hydraulic fracture is reduced, which improves the propagation distance.

(4) In multi-cluster fracturing, the medial fracture is strongly influenced by the external fractures and the fluid rate into the middle cluster is also limited, making it difficult to initiate fracturing when the cluster spacing is small.

(5) To ensure successful initiation of all clusters, stress interference should be used to encourage fracture growth, or the perforation parameters should be optimized to increase perforation friction.

## Acknowledgments

The project is supported by National Natural Science Foundation (Number 52274040).

## References

[1] Boone, T. J. and A. R. Ingraffea (1990). A numerical procedure for simulation of hydraulically - driven fracture propagation in poroelastic media. International Journal for Numerical and Analytical Methods in Geomechanics 14(1): 27-47. DOI: https://doi.org/10.1002/nag.1610140103.

[2] Bunger, A. P., R. G. Jeffrey and X. Zhang (2014). Constraints on simultaneous growth of hydraulic fractures from multiple perforation clusters in horizontal wells. Spe Journal 19(04): 608-620. DOI: https://doi.org/10.2118/163860-PA.

[3] Chaudhri, M. M. (2012). Numerical Modeling of Multi-Fracture Horizontal Well for Uncertainty Analysis and History Matching: Case Studies from Oklahoma and Texas Shale Gas Wells. SPE Western Regional Meeting, OnePetro.

[4] Cruz, F., D. Roehl and E. do Amaral Vargas Jr (2019). An XFEM implementation in Abaqus to model intersections between fractures in porous rocks. Computers and Geotechnics 112: 135-146. DOI: https://doi.org/10.1016/j.compgeo.2019.04.014.

[5] East, L., M. Y. Soliman and J. Augustine (2011). Methods for enhancing far-field complexity in fracturing operations. SPE Production & Operations 26(03): 291-303. DOI: https://doi.org/10.2118/133380-PA.

[6] Fries, T. P. and T. Belytschko (2010). The extended/generalized finite element method: an overview of the method and its applications. International journal for numerical methods in engineering 84(3): 253-304. DOI: https://doi.org/10.1002/nme.2914.

[7] Geertsma, J. and F. De Klerk (1969). A rapid method of predicting width and extent of hydraulically induced fractures. Journal of petroleum technology 21(12): 1571-1581. DOI: https://doi.org/10.2118/2458-PA.

[8] Guo, J., Q. Lu, H. Zhu, Y. Wang and L. Ma (2015). Perforating cluster space optimization method of horizontal well multi-stage fracturing in extremely thick unconventional gas reservoir. Journal of Natural Gas Science and Engineering 26: 1648-1662. DOI: https://doi.org/10.1016/j.jngse.2015.02.014.

[9] Krzaczek, M., M. Nitka, J. Kozicki and J. Tejchman (2020). Simulations of hydro-fracking in rock mass at meso-scale using fully coupled DEM/CFD approach. Acta Geotechnica 15: 297-324. DOI: https://doi.org/10.1007/s11440-019-00799-6.

[10] Ladlee, J. and J. Jacquet (2011). The implications of multi-well pads in the Marcellus Shale. DOI: https://hdl.handle.net/1813/56079.

[11] Lecampion, B., J. Desroches, X. Weng, J. Burghardt and J. E. Brown (2015). Can we engineer better multistage horizontal completions? Evidence of the importance of near-wellbore fracture geometry from theory, lab and field experiments. SPE Hydraulic Fracturing Technology Conference, OnePetro.

[12] Li, Y., J. Deng, W. Liu, W. Yan, Y. Feng, W. Cao, P. Wang and Y. Hou (2017). Numerical simulation of limited-entry multi-cluster fracturing in horizontal well. Journal of Petroleum Science and Engineering 152: 443-455. DOI: https://doi.org/10.1016/j.petrol.2017.03.023.

[13] Miller, C., G. Waters and E. Rylander (2011). Evaluation of production log data from horizontal wells drilled in organic shales. North American Unconventional Gas Conference and Exhibition, OnePetro.

[14] Modeland, N., D. Buller and K. K. Chong (2011). Statistical analysis of the effect of completion methodology on production in the Haynesville shale. North American Unconventional Gas Conference and Exhibition, OnePetro.

[15] Moes, N., J. Dolbow and T. Belytschko (1999). A finite element method for crack growth without remeshing. International journal for numerical methods in engineering 46(1): 131-150. DOI: https://doi.org/10.1002/(SICI)1097-0207(19990910)46:1<131::AID-NME726>3.0.CO;2-J.

[16] Nagel, N. B. and M. Sanchez-Nagel (2011). Stress shadowing and microseismic events: a numerical evaluation. SPE Annual Technical Conference and Exhibition, OnePetro.

[17] Pope, C. D., T. T. Palisch, E. Lolon, B. Dzubin and M. A. Chapman (2010). Improving Stimulation Effectiveness-Field Results in the Haynesville Shale. SPE Annual Technical Conference and Exhibition, OnePetro.

[18] Roussel, N. P. and M. M. Sharma (2011). Optimizing fracture spacing and sequencing in horizontal-well fracturing. SPE Production & Operations 26(02): 173-184. DOI: https://doi.org/10.2118/127986-PA.

[19] Sheng, M., G. Li, D. Sutula, S. Tian and S. P. Bordas (2018). XFEM modeling of multistage hydraulic fracturing in anisotropic shale formations. Journal of Petroleum Science and Engineering 162: 801-812. DOI: https://doi.org/10.1016/j.petrol.2017.11.007.

[20] Soliman, M. Y., L. East and J. Augustine (2010). Fracturing design aimed at enhancing fracture complexity. SPE EUROPEC/EAGE Annual Conference and Exhibition, OnePetro.

[21] Sookprasong, P., R. S. Hurt, C. C. Gill and R. LaFollette (2014). Fiber optic DAS and DTS in multicluster, multistage horizontal well fracturing: Interpreting hydraulic fracture initiation and propagation through diagnostics. SPE Annual Technical Conference and Exhibition, OnePetro.

[22] Ugueto C, G. A., P. T. Huckabee, M. M. Molenaar, B. Wyker and K. Somanchi (2016). Perforation cluster efficiency of cemented plug and perf limited entry completions; Insights from fiber optics diagnostics. Spe hydraulic fracturing technology conference, OnePetro.

[23] Wang, B., F. Zhou, D. Wang, T. Liang, L. Yuan and J. Hu (2018). Numerical simulation on near-wellbore temporary plugging and diverting during refracturing using XFEM-Based CZM. Journal of Natural Gas Science and Engineering 55: 368-381. DOI: https://doi.org/10.1016/j.jngse.2018.05.009.

[24] Wang, J., Y. Tan, M. Rijken, X. Liu, A. Singh and Y. Li (2022). Observations and Modeling of Fiber Optic Strain on Hydraulic Fracture Height Growth in Hydraulic Fracturing Test Site 2 (HFTS-2). Spe Journal 27(02): 1109-1122. DOI: https://doi.org/10.2118/209216-PA.

[25] Wang, X., C. Liu, H. Wang, H. Liu and H. Wu (2016). Comparison of consecutive and alternate hydraulic fracturing in horizontal wells using XFEM-based cohesive zone method. Journal of Petroleum Science and Engineering 143: 14-25. DOI: https://doi.org/10.1016/j.petrol.2016.02.014.

[26] Williams, M. (1959). The stresses around a fault or crack in dissimilar media. Bulletin of the seismological society of America 49(2): 199-204. DOI: https://doi.org/10.1785/BSSA0490020199.

[27] Yang, W., S. Li, Y. Geng, Z. Zhou, L. Li, C. Gao and M. Wang (2021). Discrete element numerical simulation of two-hole synchronous hydraulic fracturing. Geomechanics and Geophysics for Geo-Energy and Geo-Resources 7(3): 55. DOI: https://doi.org/10.1007/s40948-021-00257-2.

[28] Yoshida, N., A. D. Hill and D. Zhu (2018). Comprehensive modeling of downhole temperature in a horizontal well with multiple fractures. Spe Journal 23(05): 1580-1602. DOI: https://doi.org/10.2118/181812-PA.

[29] Yu, W., Z. Luo, F. Javadpour, A. Varavei and K. Sepehrnoori (2014). Sensitivity analysis of hydraulic fracture geometry in shale gas reservoirs. Journal of Petroleum Science and Engineering 113: 1-7. DOI: https://doi.org/10.1016/j.petrol.2013.12.005.

[30] Yun, X., C. Ming, W. Qi, L. Deqi, Y. Nengyu, W. Dingwei and G. Baoshan (2016). Stress interference calculation model and its application in volume stimulation of horizontal wells. Petroleum Exploration and Development 43(5): 849-856. DOI: https://doi.org/10.1016/S1876-3804(16)30101-X.

# Analysis of Well Performance Based on Bottom Hole Pressure Calculations for Gas Lift Application

**Nader Mohamed1, Elhassan Mostafa. Abdallah1(✉), Faleh H. Almahdawi2, Yasir M. F. Mukhtar3,4(✉)**

1Faculty of Engineering Technology and Built Environment, UCSI University, Kuala Lumpur, Malaysia.
2Petroleum Engineering Department, College of Engineering, University of Baghdad, Iraq
3China University of Petroleum, Beijing
4College of Petroleum Eng. & Mining, Sudan University of Science & Technology, Khartoum

elhassan@ucsiuniversity.edu.my, nxnorth90@gmail.com yasir@cup.edu.cn

**Abstract.** This paper focuses on optimizing gas lift techniques in four wells located in the North Sea to overcome the common issue of low reservoir pressure. Gas lift is a common artificial lift technique used in wells with low reservoir pressure where the natural pressure of the reservoir is insufficient to bring the oil to the surface. The success of gas lift optimization depends on identifying the deepest injection point, gas injection rate, and pressure of the gas injected. The paper proposes the deepest injection points for wells A-03, A-09A, A-10T2, and A-12BT2 as 3809.38 ft, 6649.7 ft, 3559.88 ft, and 3979.04 ft, respectively. The injection rates analyzed in this paper range from 1-10 mmscf/day, and injection pressures of 500 psia, 1000 psia, and 1500 psia were used. The response of each well to gas lift was analyzed, and increasing injection rate and pressure were found to increase oil production rate. The optimal gas injection rate and pressure for wells A-03, A-09A, and A-12BT2 were found to be 10 mmscf/day and 1500 psia, while for well A-12BT2, the best oil recovery was obtained at an injection rate of 3 mmcsf/day and injection pressure of 1500 psia. Tubing size within the range of 1-5.5 inch was examined, and it was found that increasing tubing size enhances oil flow rate. The optimal tubing size for the wells was found to be 5.5 inch, taking into consideration factors such as flow rate, pressure, temperature requirements, and fluid type.

## 1 Introduction

The importance of oil in our daily lives cannot be overstated, as it is a vital resource for transportation, electricity generation, and the production of consumer goods. The availability and affordability of oil play a crucial role in shaping global economies and ensuring our daily needs are met. The oil reservoirs that produce oil can experience decline where the oil becomes trapped and is no longer able to be produced. This can occur due to a variety of reasons, including low permeation pressure, the type of oil, or geological characteristics of the reservoir [1]. By analyzing the production system and identifying potential bottlenecks, operators can implement changes to increase production, reduce costs, and improve efficiency. This includes examining factors such as flow rates, reservoir characteristics, and artificial lift systems to optimize the overall well production system.
One popular artificial lift technique used to increase oil production is gas lift, which involves injecting gas into the well to reduce the density of the fluids and increase flow. There are two main types of gas lift systems: continuous and intermittent, which are selected based on well and reservoir characteristics [2]. Continuous gas lift systems are characterized by a continuous injection of high-pressure gas into

the wellbore. This gas mixes with the well fluids, which reduces the pressure gradient in the flow, ultimately reducing the flow bottom-hole pressure (BHFP) to achieve the desired production rate [1]. Continuous gas lift systems consume the greatest amount of energy, mainly due to the continuous flow of gas. However, they are preferred for wells that require high production rates, as they can provide a steady and constant lift to the well fluids [2]. Although they require high compression costs, continuous gas lift systems can optimize well performance and increase the overall recovery rate.

Intermittent gas lift systems, also known as cyclic gas lift systems, are characterized by periodic injections of high-pressure gas into the wellbore. The injection of gas is intermittent, which results in a pulsating flow of well fluids to the surface. This pulsating flow can help to prevent sand deposition, reduce liquid loading, and enhance oil recovery [1]. Intermittent gas lift systems are typically more cost-effective than continuous gas lift systems, as they require lower compression costs. However, they are not suitable for wells that require a steady flow of fluids, as the production rate is not constant [2].

Proper tubing size selection can minimize pressure drop and optimize flow rates in gas lift systems, which can improve overall well performance. Studies have shown that selecting an optimal tubing size can increase oil production rates by up to 15% [3]. It is essential to consider the well and reservoir characteristics when selecting tubing size, including reservoir pressure, oil viscosity, and fluid properties. Additionally, the use of advanced analytical tools, such as computational fluid dynamics, can aid in the accurate prediction of pressure loss and the optimization of tubing size for maximum oil production [4]. Therefore, optimizing tubing size can lead to cost savings and improved well performance, making it an essential factor in gas lift system design and operation.

## 2 Methodology

Utilizing PIPESIM software, gas lift optimization in oil and gas wells was analyzed based on data presented in Table 1 [5]. The results provide significant insights into the effectiveness of gas lift optimization techniques, which could potentially improve well performance and increase recovery factor. Future research could further validate these findings and explore additional factors impacting gas lift optimization, including wellbore geometry and fluid properties.

**Table 1.** North Sea wells.

| Parameter | Well | | | |
|---|---|---|---|---|
| | A-03 | A-09A | A-10T2 | A-12BT2 |
| $D_{eq}$ (ft) | 0.3983 | 0.4049 | 0.4012 | 0.3825 |
| $f_w$ (fraction) | 0.8542 | 0.5174 | 0.8047 | 0.6107 |
| $h$ (ft) | 9,686 | 9,589 | 9,646 | 9,633 |
| $h_A$ (ft) | 8,819 | 8,906 | 9,137 | 7,758 |
| $J_L$ (STB/d/psi) | 5.9 | 2.84 | 5.93 | 13.35 |
| $L$ (ft) | 11,032 | 10,240 | 11,453 | 10,264 |
| $L_A$ (ft) | 9,967 | 9,512 | 9,885 | 8,082 |
| $Ms$* (lb/1,000 STB liquid) | 0 | 0 | 0 | 0 |
| $P_r$ (psia) | 4,395 | 2,611 | 4,389 | 3,476 |
| $P_{wh}$ (psia) | 435 | 249 | 377 | 450 |
| $Q_{gi}$ (MMSCF/d) | 7.54 | 3.01 | 5.23 | 6.91 |
| $R_p$ (SCF/STB) | 797 | 797 | 797 | 797 |
| $T_{wf}$ (°F) | 260.6 | 260.6 | 260.6 | 260.6 |
| $T_{wh}$ (°F) | 206.6 | 163.9 | 213.8 | 208.4 |
| $\rho_s$ (lb/ft$^3$) | 165 | 165 | 165 | 165 |
| $\Upsilon_{gi}$ (air = 1) | 0.937 | 0.937 | 0.937 | 0.937 |
| $\Upsilon_{gn}$ (air = 1) | 0.937 | 0.937 | 0.937 | 0.937 |
| $\Upsilon_o$ (freshwater = 1) | 0.844 | 0.844 | 0.844 | 0.844 |
| $\Upsilon_w$ (freshwater = 1) | 1.161 | 1.161 | 1.161 | 1.161 |
| Measured $Q_{LS}$ (STB/d) | 9,793 | 3,441 | 9,950 | 10,289 |
| Model $Q_{LS}$ (STB/d) | 10,039 | 3,950 | 9,834 | 10,604 |
| Deviation (%) | 2.5 | 14.8 | 1.2 | 3.1 |

Journal of Physics: Conference Series **2594** (2023) 012026 doi:10.1088/1742-6596/2594/1/012026

### *2.1 Gas Lift Installation Process*

To optimize the performance of a gas lift system in oil and gas wells, the injection point depth will be optimized using the software. The location of the deepest injection point can be determined based on factors like fluid density, production rate, and pressure and temperature conditions. The software offers tools such as nodal analysis to model and analyze the gas lift system's performance under different injection point scenarios. Using PIPESIM, engineers can simulate various scenarios such as changes in injection rate or pressure to obtain valuable insights into the gas lift system's behavior. This information can be used to maximize production rate, improve efficiency, and enhance the recovery of oil from the reservoir. Designing a gas lift system using PIPESIM requires a thorough understanding of the reservoir and wellbore properties and the gas lift system's operating requirements.

### *2.2 Gas Lift Selection*

Gas lift is a complex process that involves injecting gas into the wellbore to enhance production rates. The selection of the appropriate gas lift application is a critical aspect of gas lift design and optimization. The selection process depends on various factors, such as the production rate, well depth, and reservoir characteristics. Different gas lift applications are available, including continuous gas lift and intermittent gas lift, each with its unique advantages and disadvantages.

Understanding the different types of gas lift applications and their suitability based on the oil rate is essential to ensure effective gas lift operations.

### *2.3 Pressure Loss Through Vertical Components*

In a gas lift system, pressure loss can occur through the vertical components, such as tubing and casing, which can affect the efficiency and gas consumption. Frictional resistance between the fluid and the walls of the vertical components is the main cause of pressure loss and increases with fluid velocity and surface roughness. Changes in elevation and obstructions can also contribute to pressure loss. Accurate calculation of pressure loss using the software is crucial in optimizing gas lift systems to improve their efficiency.

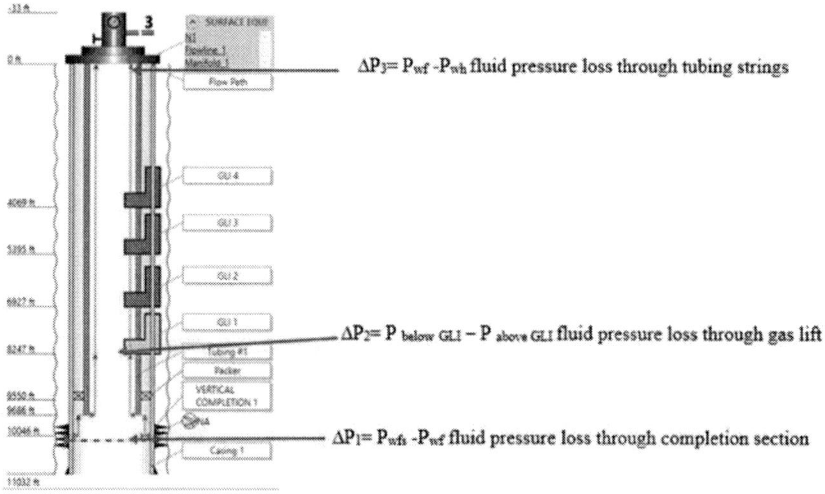


**Fig. 1.** Pressure loss through vertical components.

### *2.4 Optimum Tubing Size*

Selecting the optimal tubing size in gas lift systems is critical for achieving efficient performance and improving well economics. PIPESIM software is a commonly used simulation tool to determine the optimum tubing size by modelling gas and liquid flow. The simulation enables engineers to analyse how different tubing sizes affect system performance. Presenting an overview of the use of PIPESIM and the factors considered in simulation. The research utilized tubing sizes ranging from 3-5.5 inches with incremental changes of 0.5 inches.

## 3 Results and Discussion

### *3.1 Wells Performance*

The graphical results illustrated in figure below for each well in the North Sea field show that there are challenges in correlating inflow from reservoir formation and outer flow from vertical components in each well. As a solution, gas lift is suggested to improve well performance and increase the recovery factor. Gas lift works by injecting gas into the wellbore to reduce the fluid's density and facilitate its flow to the surface. However, a feasibility study and optimization analysis must be conducted to determine the most effective gas lift design and operating condi-tions. Several parameters, such as gas injection rate, wellhead and bottomhole pressures, and the number and size of gas lift valves, need to be evaluated for the optimization process. Careful planning, design, and optimization are critical to achieving optimal results. Overall, gas lift is an effective technique for improving well performance and increasing recovery factor, particularly in situations where traditional production methods are inadequate.

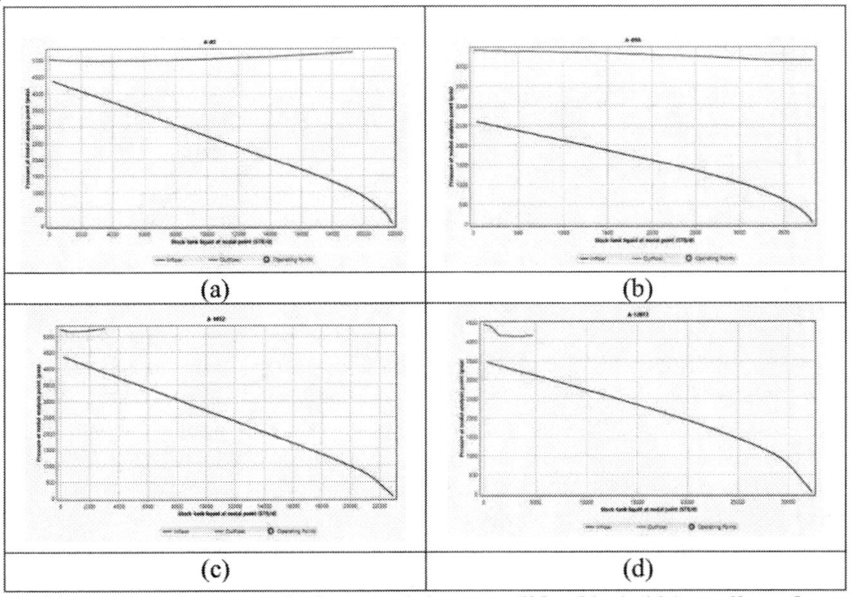


**Fig. 2.** (a) A-03 well performance before applying gas lift, (b) A-09A well performance before applying gas lift, (c) A-10T2 well performance before applying gas lift, (d) A-12BT2 well performance before applying gas lift.

### *3.2 Gas Lift Installation*

**Deepest Injection Point**

The depth of the injection point in gas lift systems plays a crucial role in lifting well fluids to the surface, and its location is determined by various well characteristics. The deeper the injection point, the better the lifting effect. However, the depth is subject to practical limitations depending on factors such as the available gas supply, tubing size, and well completion design. Multiple injection points at different depths can offer more flexibility in optimizing lift performance. The specific design and operation of the gas lift system should be based on individual well conditions and production goals. The deepest injection points for wells A-03, A-09A, A-10T2, and A-12BT2 were calculated using PIPSIM simulator and are 3809.38 ft, 6649.7 ft, 3559.88 ft, and 3979.04 ft, respectively, due to different well head pressures as shown in Fig. 3.

Journal of Physics: Conference Series **2594** (2023) 012026 doi:10.1088/1742-6596/2594/1/012026

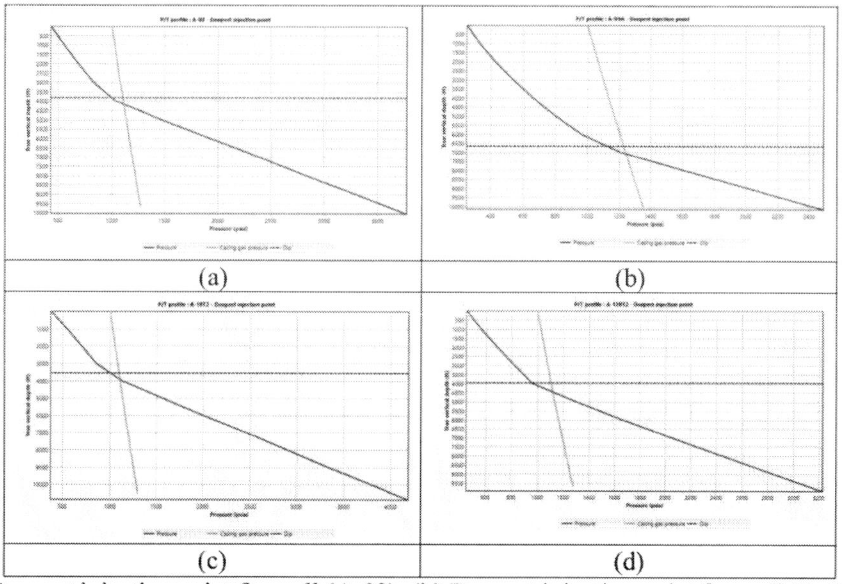


**Fig. 3.** (a) Deepest injection point for well (A-03), (b) Deepest injection point for well (A-09A), (c) Deepest injection point for well (A-10T2), (d) Deepest injection point for well (A-12BT2).

**Gas Lift Responses**

Fig. 4 below shows the gas lift response in well A-03 for injection pressures of 1000 and 1500 psia and injection rates ranging from 1 to 10 mmscf/day. Increasing the injection pressure and rate leads to a higher oil production rate and deepest injection point. The optimum injection rate and pressure for well A-03 are 10 mmscf/day and 1500 psia, respectively, using continuous gas lift. The oil rate increases up to 6407 stb/day by applying the selected injection rate and pressure. While well A-09A has the lowest bottom hole pressure and has the highest vertical length, increasing gas injection rate and pressure from 6-10 mmscf/day and from 1000-1500 psia, do not have any impact on increasing the oil production rate. However optimum injection rate and pressure for well A-09A are 5.5 mmscf/day and 1000 psia, respectively. For well A-10T2, the applied injection rate was from the range of 1-10 mmscf/day and the injection pressure was from 1000 psia and 1500 psia, while increasing the injection rate more than 3 mmscf/day leads to decrease the oil rate. Therefore, the optimum injection rate and pressure for well A-10T2 will be 3 mmscf/day and 1500 psia respectively. Moreover, the deepest injection point where the gas left valve should be placed on depth of 5359 ft. Applying injection rate from range 1-10 mmscf/day and injection pressure of 1000 psia and 1500 psia.for well A-12BT2. Increasing gas injection pressure increases oil production rate, but injection rates above 8 mmscf/day decrease oil rate slightly. Optimum injection rate and pressure are 8 mmscf/day and 1500 psia, respectively, leading to a maximum oil production rate of 6690 stb/day. Continous gas lift is recommended.

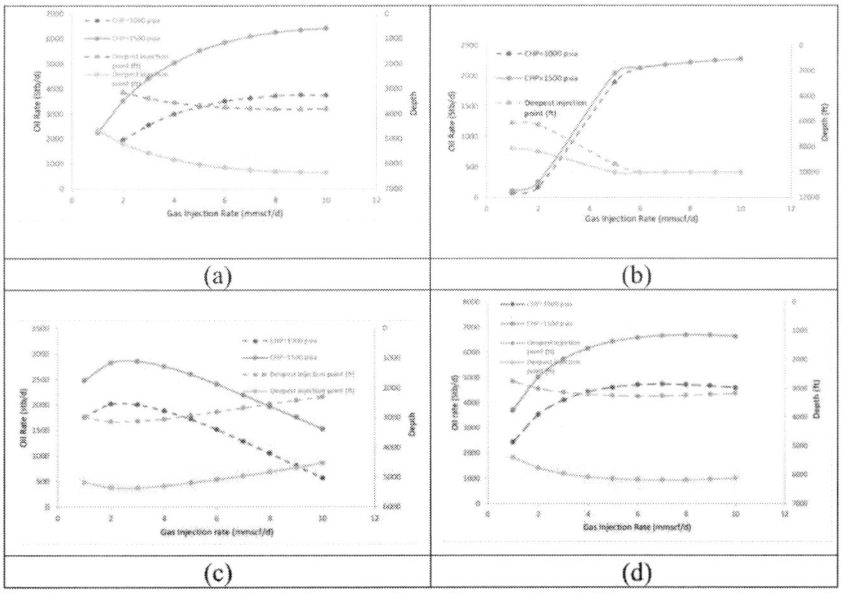


**Fig. 4.** (a) Gas Lift response obtained from well A-03, (b) Gas Lift response obtained from well A-09A, (c) Gas Lift response obtained from well A-10T2, (d) Gas Lift response obtained from well A-12BT2.

**Gas Lift Design**

Gas lift design involves determining optimal injection gas rate, injection depth, and other parameters for desired production rate. It analyzes well characteristics like reservoir pressure, temperature, fluid properties, and completion design. Conventional gas lift design is based on descended liquid column depth and maximum working gas pressure [6]. Fig. 5 below shows the gas lift design for well A-03, based on the calculations using PIPESIM three gas lift valves should be installed in this well at depth of 5000ft, 4070ft and 2400 ft to increase the production rate. However, this calculation was based on the when injecting gas with rate at 10 mmscf/day and 1500 psia. Based on the simulation results obtained using PIPESIM, well A-09A should implement 8 valves at depth of 1616.445 ft, 3111.236 ft, 4356.217 ft, 5419.157ft, 6302.404 ft, 7016.142 ft, 7578.171 ft, and 8002.937 ft. However, these valves should increase the productivity in this well. Moreover, the calculation was obtained when the injection gas rate and pressure with 5.5 mmscf/day and 1000 psia respectively. Fig. 5 (b) illustrate gas lift design on well A-09A. For A- 10T2, the design was obtained when applying 3 mmscf/day and 1000 psia rate and pressure respectively. However, four gas lift valves should be implemented at depth of 2557.252 ft, 4049.912 ft, 4792.646 ft, and 5157.663 ft. implementing intermittent gas lift application should increase the productivity in this well, but it will increase the gas injection rate. When injecting gas rate with 8 mmscf/day at pressure of 1500 psia as artificial gas lift, well A-12BT2 should has four gas lift valves at different depth. However, the setting depth of these valves should be placed withing these depth 2298.739 ft, 3836.075 ft, 4729.178 ft, and 5259.007 ft. The suggested gas lift valves should increase the production rate in this well. However, further investigations will be discussed to improve the productivity in each well.

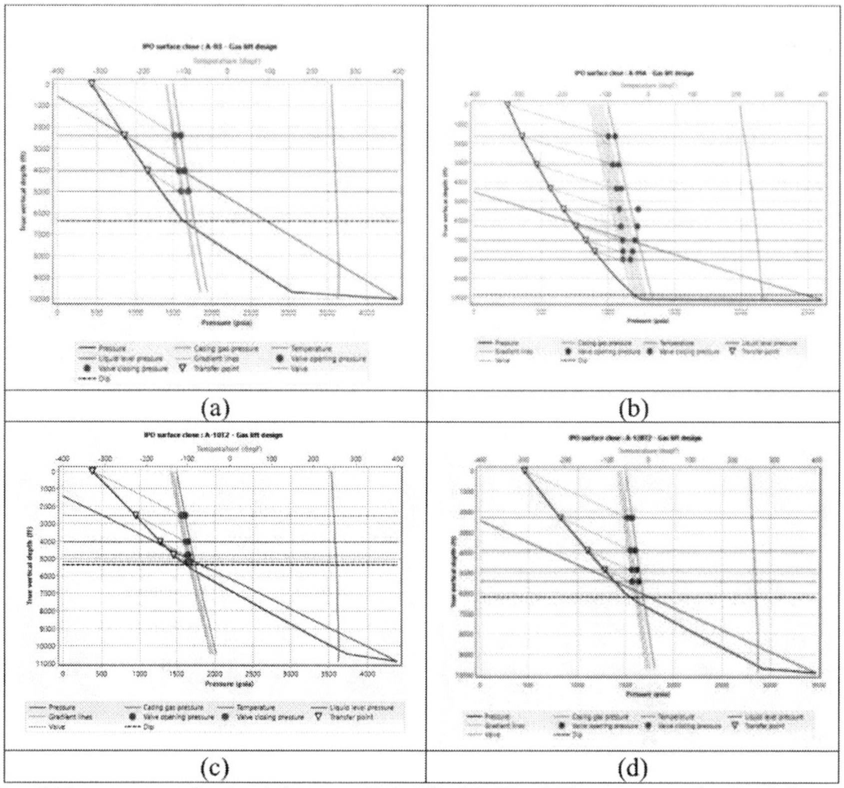


**Fig. 5.** (a) Gas Lift response obtained from well A-03, (b) Gas Lift response obtained from well A-09A, (c) Gas Lift response obtained from well A-10T2, (d) Gas Lift response obtained from well A-12BT2.

### *3.3 Gas Lift Implementation*

**Intermittent Gas Lift**

Intermittent gas lift was applied in the well A-03 using the gas lift valves that was mentioned in the previous section. However, Fig. 6 (a) illustrating the vertical components in well A-03 and nodal analysis after applying intermittent gas lift application. As a result of injecting gas rate at 8 mmscf/day and 1500 psia pressure, the liquid production rate was obtained to have a value of 5737.613 stb/day, while the outlet oil production rate was calculated to be 836.544 stb/day. Gas lift valves, as discussed earlier, were utilized to implement intermittent gas lift in well A-09A. However, Fig. 6 (b) showing the well's vertical components and nodal analysis after the intermittent gas lift was applied. The application of intermittent gas lift at a gas injection rate of 6 mmscf/day and a pressure of 1000 psia yielded a liquid production rate of 1601.468 stb/day and an outlet oil production rate of 772.8685 stb/day. Installing intermittent gas left resulted in increasing the liquid production rate 2296.88 stb/day and increases the oil production rate 448.5807 stb/day. However, Fig. 6 (c), illustrate the well vertical components, as showing four gas lift volves were placed at different depth based on gas lift response results. In well A-12BT2, the gas lift valves mentioned earlier were utilized to implement intermittent gas lift. Fig. 6 (d) displays the vertical components in the well where 4 gas lift valves were implanted at different elevation, while nodal analysis after applying the intermittent gas lift method. With a gas injection rate of 8 mmscf/day and a pressure of 1500 psia, the liquid production rate was determined to be 6118.307 stb/day, while the outlet oil production rate was computed to be 2381.857 stb/day.

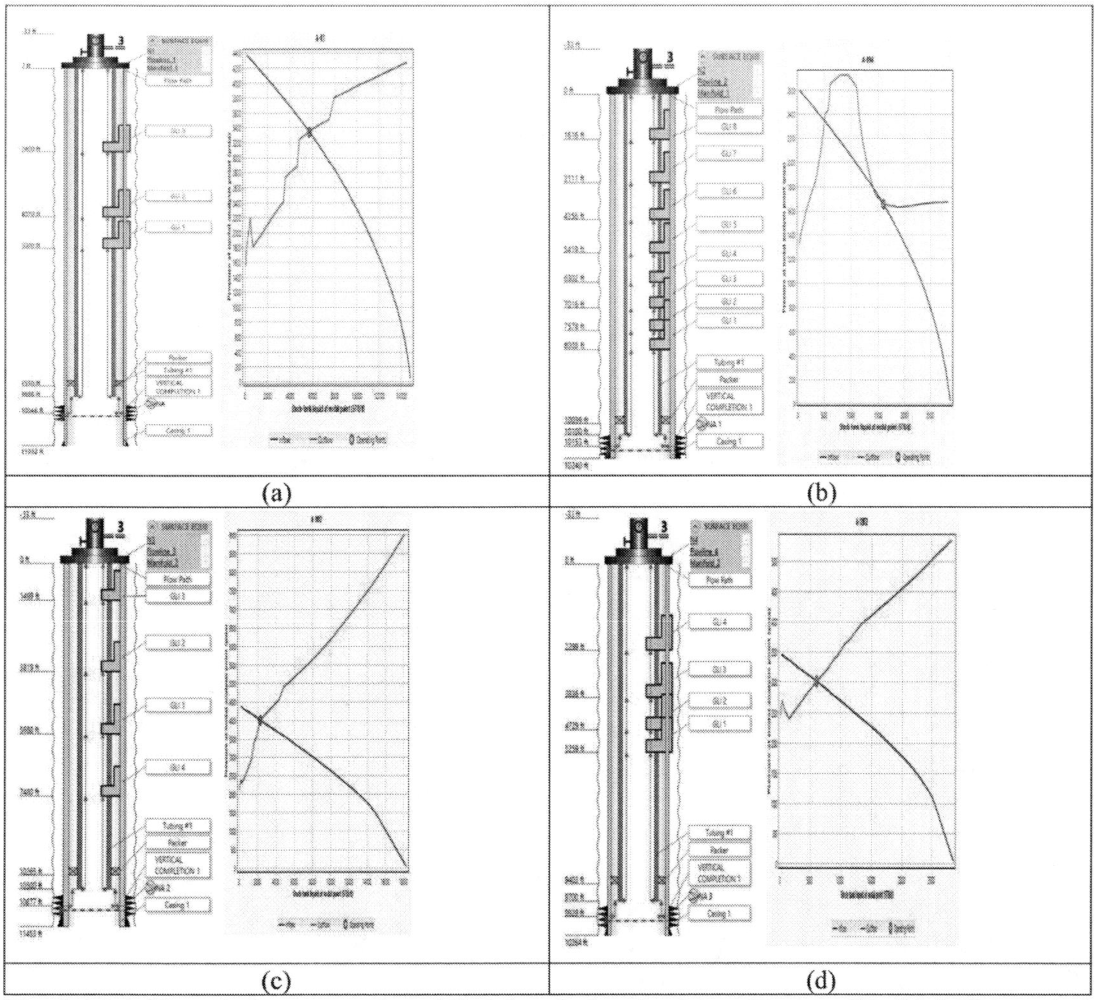


**Fig. 6.** (a) A-03 vertical components and Nodal analysis using intermittent gas lift, (b) A-09A vertical components and Nodal analysis using intermittent gas lift, (c) A-10T2 vertical components and Nodal analysis using intermittent gas lift, (d) A-12BT2.

### *3.4 Continues Gas Lift*

Implementing continues gas lift in well A-03 leads to enhance the liquid production rate as well the oil production rate 7326.322 stb/day and 1068.178 stb/day respectively. Moreover, in this scenario one valve were placed at the bottom of the well at depth of 5000 ft while the injection rate and pressure were 8 mmscf/day and 1500 psia respectively. Fig. 7 (a) illustrating well vertical components and the nodal analysis using continues gas lift applied in well A-03. The continuous gas lift implementation in Well A-09A resulted in an increase in both the liquid and oil production rates. The liquid production rate increased to 1601.401stb/day, while the oil production rate increased to 772.8362 stb/day. To achieve this, a valve was placed at the bottom of the well at a depth of 8002.937 ft, and an injection rate of 6 mmscf/day and pressure of 1000 psia were applied. Fig. 7 (b) displays the well's vertical components and the nodal analysis using continuous gas lift applied in Well A-09A. Well A-10T2 resulted in a noticeable improvement in both liquid and oil production rates. The liquid production rate rose to 3441.53 stb/day, while the oil production rate increased to 672.1309 stb/day. To achieve this, a valve was installed at a depth of 7479.901ft in the well, and a gas injection rate of 3 mmscf/day and pressure of 1500 psia were utilized. Fig. 7 (c) illustrates the vertical components of the well and the nodal analysis following the implementation of continuous gas lift in Well A-10T2. Well A-12BT2 resulted in a significant improvement in both liquid and oil production rates. The liquid production rate increased to 6105.183 stb/day, while the oil production rate increased to 2376.748 stb/day. To achieve this, a valve was installed at a depth of 5259.007 ft in the well, and a gas injection rate of 8 mmscf/day and pressure of 1500 psia were utilized. Fig. 7 (d) depicts the vertical components of the well and the nodal analysis after the implementation of continuous gas lift in Well A-12BT2.

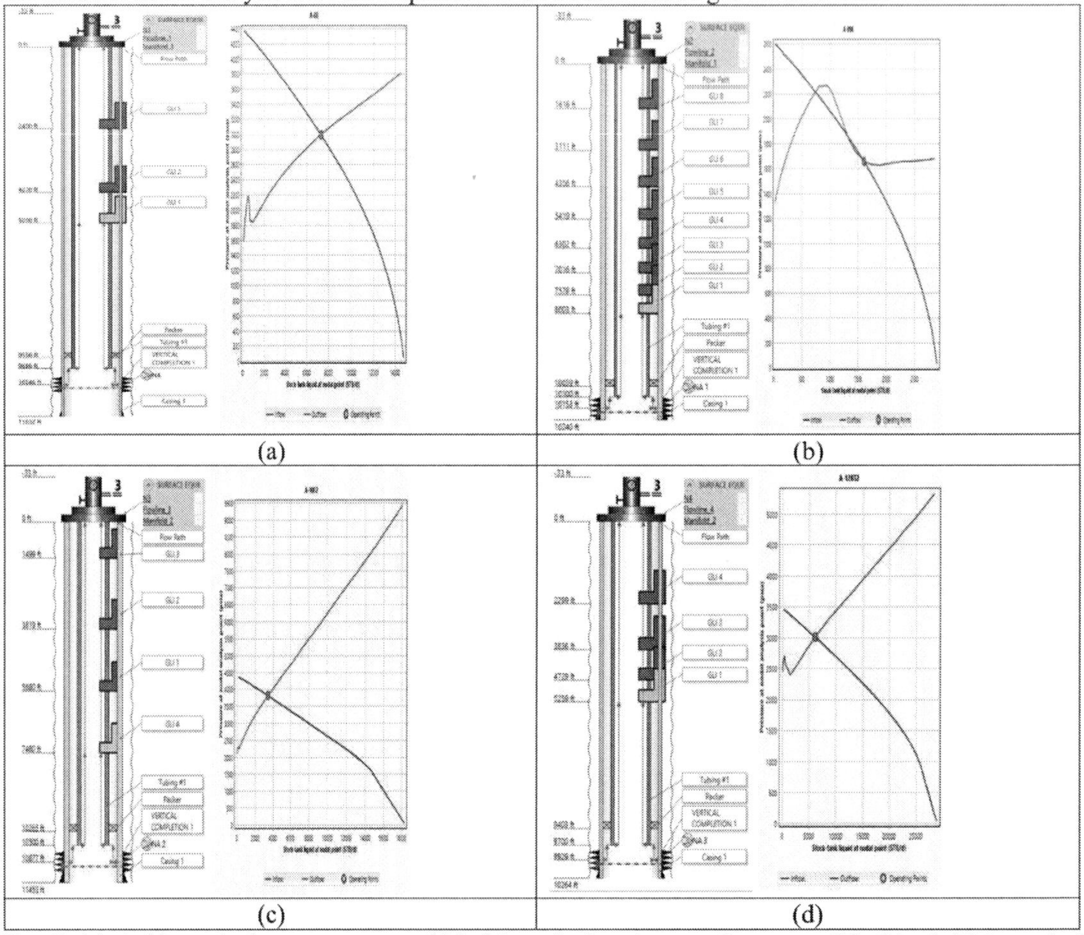


**Fig. 7.** (a) A-03 vertical components and Nodal analysis using continues gas lift, (b) A-09A vertical components and Nodal analysis using continues gas lift, (c) A-09A vertical components and A-09A Nodal analysis using continues gas lift, (d) A-12BT2 vertical components and Nodal analysis using continues gas lift.

**Table 2.** Comparison oil rate utilizing continues and intermittent gas lift application.

| Well | Intermittent Gas Lift Oil Production Rate (stb/day) | Continues Gas Lift Oil Production Rate (stb/day) |
|---|---|---|
| A-03 | 836.544 | 1068.178 |
| A-09A | 772.8685 | 772.8362 |
| A-10T2 | 448.5807 | 672.1309 |
| A-12BT2 | 2381.857 | 2376.748 |

### *3.5 Pressure Loss Through Vertical Components*

When implementing gas lift, there is pressure loss that occurs through the vertical components of the well. This pressure loss is due to frictional losses that occur as the gas moves through the tubing, and it can result in a decrease in the injection pressure available at the bottom of the well. However, in this section discuss the pressure loss through the vertical components in each well of the data provided from North Sea field. Due to the data limitation, only pressure loss through vertical completion is calculated. These components including the completion equipment, gas lift components, and the wellhead are the three-component used to analyse the pressure loss in each well. However, the mathematical equations below are utilized.

Fluid pressure loss through tubing strings.

$$\Delta P3 = Pwf - Pwh \quad (1)$$

Fluid pressure loss through gas lift

$$\Delta P2 = P\ below\ GLI - P\ above\ GLI \quad (2)$$

Fluid pressure loss through completion section.

$$\Delta P1 = Pwfs - Pwf \quad (3)$$

Based on the equations above pressure loss through vertical components in each well was determined as showing in **Table 3.**

**Table 3.** Comparison oil rate utilizing continues and intermittent gas lift application.

| Well | ΔP1 (psia) | ΔP2 (psia) | ΔP3 (psia) |
|---|---|---|---|
| A-03 | 1403.633 | 427.103 | 2555.772 |
| A-09A | 956.401 | 350.0445 | 1405.974 |
| A-10T2 | 584.424 | 422.954 | 3427.897 |
| A-12BT2 | 469.158 | 312.924 | 2556.11 |

### *3.6 Pressure Loss Through Vertical Components*

Selecting the optimum tubing size is an important consideration in many fluid transport systems. It is critical to ensure that the tubing size is appropriate for the fluid flow requirements to ensure efficient operation of the system. Choosing the wrong tubing size can result in problems such as excessive pressure drop, inefficient operation, and even system failure.

In order to determine the optimal tubing size, a range of tubing sizes from 3-5.5 inches were tested. Increasing the tubing size resulted in an increase in production rate across all wells, with the largest tubing size of 5.5 inches exhibiting the highest production rate. Therefore, the optimal tubing size for selection is 5.5 inches. The nodal analysis for the selection of tubing size for each well in the North Sea field is illustrated in the figures below.

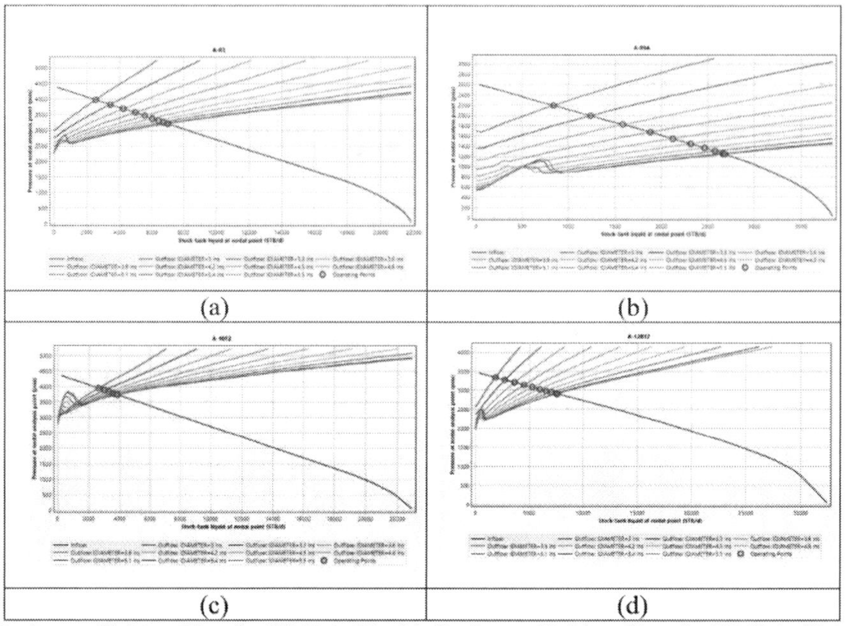

**Fig. 8.** (a) Optimum Tubing Size for Well A-03, (b) Optimum Tubing Size for Well A-09A, (c) Optimum Tubing Size for Well A-10T2, (d) Optimum Tubing Size for Well A-12BT2.

## 4 Conclusion

The use of gas lift systems, simulated by PIPESIM software, can significantly improve the performance of oil and gas wells, especially in challenging situations where traditional production methods may be limited. Continuous gas lift is the most suitable method for most of the wells, as shown by the oil production rate data obtained from both continuous and intermittent gas lift methods for each well. The analysis of pressure loss in gas lift systems and the selection of the optimal tubing size are crucial factors in maximizing oil production rates. Increasing the tubing size from 3-5.5 inches led to an increase in production rate across all wells, with the largest tubing size of 5.5 inches exhibiting the highest production rate. It is important to consider other factors such as well depth, fluid properties, and production goals when selecting the tubing size.

**Acknowledgments.** The project is supported by National Natural Science Foundation (Number *********).

## References

[1]Al-Shammasi, M., & Al-Marhoun, M. (2018). Impact of Reservoir Heterogeneities on the Performance of Gas Lift Wells. Journal of Petroleum Science and Engineering, 166, 715-725. doi: 10.1016/j.petrol.2018.03.015.

[2]Chacón-Nava, J. G., Vásquez-Cruz, A., & De la Cruz-Álvarez, H. (2018). Comparison of Continuous and Intermittent Gas Lift Systems for Increasing Oil Production. Journal of Petroleum Science and Engineering, 163, 662-672. doi: 10.1016/j.petrol.2018.01.009.

[3]Shokir, E. M., Abdelghany, M. M., & Abu El-Ata, M. I. (2019). Prediction of Gas-Lift Performance in Vertical and Deviated Wells. Journal of Petroleum Science and Engineering, 178, 1011-1025.

[4]He, Y., Huang, Z., Ma, Y., & Zhang, Y. (2019). A novel method for optimizing the operation of continuous gas-lift systems in offshore oilfields. Journal of Natural Gas Science and Engineering, 67, 24-35.

[5]Liu, Q., Tang, J., Ke, W., Wang, H., & Orivri, U. D. (2023). Case Study: Successful Application of a Novel Gas Lift Valve in Low Pressure Wells in Fuling Shale Gas Field. Processes, 11(1), 19.

[6]Yadua, A. U., Lawal, K. A., Eyitayo, S. I., Okoh, O. M., Obi, C. C., & Matemilola, S. (2021). Performance of a gas-lifted oil production well at steady state. Journal of Petroleum Exploration and Production Technology, 11(6), 2805-2821.

# Analysis of Lens Surface Shape Based on Theory of Plates and Shells

**Yu-chen JIN**(✉)

College of Pipeline and Civil Engineering, China University of Petroleum (East China),Qingdao, China

627155057@qq.com

**Abstract**. The purpose of this article is to improve the accuracy and expand the force analysis of the lens by using the theory of plates and shells and reciprocal theorem. The first one concerns the deflection expression of a circular lens of equal thickness supported at several uniform points. The second calculates the deflection of the lens under its own weight, and then starting from known Zernike coefficient, PV value and RMS value in optics, the first 37 Zernike coefficients are obtained. The third compares the Zernike coefficients and lens center deflections of partic-ular cases. The derivation process based on plates and shells theory and reciprocal theorem is given in details. As validation, the lens deflec-tion expression is applied to calculate 4 points support, the predicted re-sults illustrate great agreements with the finite element simulation solu-tion. Finally, the proposed method has shown great potential for engi-neering analysis and may also bring inspirations for optical field, as well as other plate and shell theory system.

## 1. Introduction

Lithography machine is an indispensable equipment in chip manufacturing and 3D printing technology[1, 2]. As the most crucial part, the objective lens can seriously af-fect the working accuracy of the lithography machine[3-5].Researches on objective lens are mostly based on material selection and equipment manufacturing process[6, 7], but few researchers have conducted the deformation analysis under load.

It is clearly to find by observing the objective lens of the lithography machine and referring to other optical lenses that the axial thickness is much smaller than the lens diameter, and the degree of deformation of the objective lens under its own weight is also much smaller than the lens thickness. Furthermore, the macroscopic distribution of the supporting points, and its size is fully consistent with the definition of thin plate[8]. The lens can be abstracted as a circular thin plate model. Normally, the lens is supported by uniform points along the boundary.

Bassali[9, 10] was the first to use the complex variable function method to study the deformation problem of a circular plate of equal thickness supported at several points. Although an explicit calculation formula for the deflection, bending moment, torsion moment and shear stress of the circular plate at any point was given, but there is no numerical solution. Subsequently, in the case of multi-point support with a symmetrical load applying to the thin circular annulus, the solution of deflection was obtained[11-13]. When the load distribution area becomes a concentric ellipse, Anwar[14] and Bassali[15] put forward the corresponding solution. After that, Bassali[16] turned his attention to clamped thin isotropic plates. Timoshenko[8] gave an analytical ex-pression for the case where a circular thin plate has two points of support in the di-ameter direction. This expression is only a special case, and the solution cannot be generalized to a general form, or applied to the case of non-uniform point support. Yu and Pan[17] deduced the deflection expression of a circular plate under uniform load. The supporting points of the circular plate were located at discrete points which distributed equidistantly along the circumference of a concentric circle. Vaughan[18] derived the deflection equation of the point-support thin circular plate, which was used to predict the deflection of the circular plate under uniform loading. Li Nong[19] firstly applied the reciprocal theorem to analyze the deflection surface of the elastic thin circular plate. The deflection expressions for the case of several points of support were given, but there were still some errors in the derivation process, which affect the accuracy of the final results.

This study focuses on revisiting the plate and shell theory, and Zernike polynomials to combine both optical model and mechanical analysis is proposed to evaluate the overall deflection of the lens. The outline of the present work is given briefly in the following. Section 2 presents the theory of elastic circular plate in polar coordinate system and the reciprocal theorem, as well as Zernike fitting. Section 3 is dedicated to the numerical examples aspects in which several-points supported lens are given to demonstrate for the accuracy.

## 2. The reciprocal theorem and Zernike fitting

*2.1 Circular plate theory in polar coordinates*

The differential equation of an elastic surface in polar coordinates can be written as

$$D\left(\frac{\partial^2}{\partial r^2}+\frac{1}{r}\frac{\partial}{\partial r}+\frac{1}{r^2}\frac{\partial^2}{\partial \theta^2}\right)\left(\frac{\partial^2 w}{\partial r^2}+\frac{1}{r}\frac{\partial w}{\partial r}+\frac{1}{r^2}\frac{\partial^2 w}{\partial \theta^2}\right)=q \tag{1}$$

where $D$ and $q$ are the bending stiffness and uniform load of the plate, and $w$ represents the displacement expression.

When the boundary of a circular plate is axisymmetric around the Z-axis, and the transverse load on it is also axisymmetric around the Z-axis, then the above formula can be simplified to

$$D\left(\frac{\partial^2}{\partial r^2}+\frac{1}{r}\frac{\partial}{\partial r}\right)\left(\frac{\partial^2 w}{\partial r^2}+\frac{1}{r}\frac{\partial w}{\partial r}\right)=q \tag{2}$$

Now the solution to this differential equation are presented as

$$w=C_1\ln r+C_2r^2\ln r+C_3r^2+C_4+w_1 \tag{3}$$

where $w_1$ is any particular solution.

When the load received by the circular plate is distributed continuously along the radial direction, the load can be approximately expressed as a polynomial

$$q=A_0+A_1r+A_2r^2+\cdots \tag{4}$$

And then we can take the particular solution as

$$w_1=r^4(B_0+B_1r+B_2r^2+\cdots) \tag{5}$$

In a cross section with a constant pole diameter, the expressions of $M_r$, $M_{r\theta}$ and $F_{Sr}$ are as follows

$$M_r=-D[\frac{\partial^2 w}{\partial r^2}+\mu(\frac{1}{r}\frac{\partial w}{\partial r}+\frac{1}{r^2}\frac{\partial^2 w}{\partial \theta^2})]$$
$$M_{r\theta}=-D(1-\mu)[\frac{1}{r}\frac{\partial^2 w}{\partial r\partial\theta}-\frac{1}{r^2}\frac{\partial w}{\partial\theta}] \tag{6}$$
$$F_{Sr}=-D\frac{\partial}{\partial r}\nabla^2 w$$

where $\nabla^2 w=\frac{\partial^2 w}{\partial r^2}+\frac{1}{r}\frac{\partial w}{\partial r}+\frac{1}{r^2}\frac{\partial^2 w}{\partial \theta^2}$, and $D=\frac{Ed^3}{12(1-\mu^2)}$.

*2.2 The reciprocal theorem*

Assuming the system with a unit concentrated load is the first elastic body, and its displacement component can be expressed as $u_1$ , $v_1$ and $w_1$ , its surface force component can be written as $P_{x1}$ , $P_{y1}$ and $P_{z1}$ . The real system is regarded as the second elastic body, whose displacement

Journal of Physics: Conference Series **2594** (2023) 012027 doi:10.1088/1742-6596/2594/1/012027

component can be written as $u_2$ , $v_2$ and $w_2$ , its surface force component is given as $P_{x2}$ , $P_{y2}$ and $P_{z2}$ , and physical force component are $X_2$ , $Y_2$ and $Z_2$ . By using $\Delta_2$ to represent the displacement of the second elastic body, the following formula can be obtained by using the theorem of reciprocity

$$\Delta_2 = \iiint_{\Omega} \left( X_2 \ {}_1 + Y_2 v_1 + Z_2 \ {}_1 \right) \mathrm{d}\Omega + \iint_S \left( P_{x2} \ {}_1 + P_{y2} \ {}_1 + P_{z2} \ {}_1 \right) \mathrm{d}S$$
$$- \iint_S \left( P_{x1} u_2 + P_{y1} v_2 + P_{z1} w_2 \right) \mathrm{d}S \tag{7}$$

*2.3 Derivation of lens deflection*

The first elastic body is a circular thin plate with fixed periphery, acting on a unit concentrated load at a certain flow coordinate $(\xi, \varphi)$ ; the second elastic body is a uniform point support lens model, which can be abstracted as the same size as the first elastic body. The expression of the surface deflection of the lens model can be obtained by applying the reciprocal theorem between the two elastomers

$$\omega_2(\xi,\varphi) = \iint_A q\omega_1(\xi,\varphi,r,\theta) r dr d\theta + \int_0^{2\pi} M_{1r}\theta_a d\theta + \int_0^{2\pi} V_{1r} f_a d\theta \tag{8}$$

where $\omega_1(\xi,\varphi,r,\theta)$ indicates the displacement in the coordinate $(r,\theta)$ , and the parameter $(\xi,\varphi)$ represents where the unit of concentrated load acts. The gravity of the lens is equivalent to a uniform load and record it as $q_0 = \rho dg$ . The deflection and rotation angle of the lens at the boundary are expressed as $f_a$ and $\theta_a$ :

$$f_a = \sum_{m=0,k,2k}^{\infty} f_m \cos m\theta \qquad \theta_a = \sum_{m=0,k,2k}^{\infty} \theta_m \cos m\theta \tag{9}$$

The result is got:

$$\omega = \frac{q_0 \left( a^2 - \xi^2 \right)}{64D} + \sum \frac{\theta_m}{2a} \left( \frac{\xi}{a} \right)^m \left( a^2 - \xi^2 \right) \cos m\varphi$$
$$+ \sum \frac{f_m}{2a^2} \left( \frac{\xi}{a} \right)^m \left( a^2 - \xi^2 \right) \left[ (m+2) a^2 - m\xi^2 \right] \cos m\varphi \tag{10}$$

## 3. Numerical example

Considering a circular objective lens of equal thickness, the parameters are collected by Table 1.

**Table 1. Material parameters**

| Young's modulus | Poisson ration | Density | Radius | Thickness |
|---|---|---|---|---|
| 70000 | 0.15 | 2200e-12 | 65 | 5 |

The lens is uniformly supported at four points and five points respectively, the schematic is displayed as

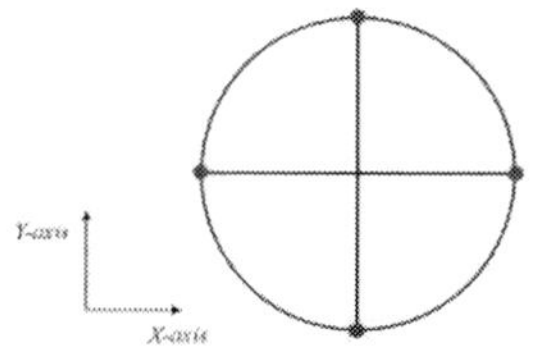


**Fig. 1.** Example diagrams of four-point support

Then apply the reciprocal theorem mentioned in this article to calculate the models and take the four point's situation as example. The finite element method is used to perform finite element simulation on the above model to obtain the coordinates of each grid point on the upper surface of the lens, and to get the numerical solution of the deflection corresponding to each coordinate. The selected coordinate points are the same as those used in the finite element analysis. After calculation, analytic solutions of

Journal of Physics: Conference Series **2594** (2023) 012027 doi:10.1088/1742-6596/2594/1/012027

the deflection of each point on the upper surface of the lens are obtained. The operations for five-point support are the same as four points.

The deflection of the center point of the circular lens is compared, and the results are presented in Table 2.

**Table 2. Error of the deflection of the center point between proposed and n solution**

| | Deflection with 4 points($10^{-4}$) | Deflection with 5 points($10^{-4}$) |
|---|---|---|
| Proposed solution | 2.24 | 2.02 |
| Numerical solution | 2.29 | 2.08 |
| Error(%) | 2.18 | 2.88 |

Table 2 indicates that the errors between two solutions are less than 3%, which clarifies the accuracy of the proposed method. All solutions are both fitted to Zernike, the clear aperture is selected to be 90% of the diameter of the lens. For the four points case, the wavefront aberration and some of the first 37 Zernike coefficients are expressed in Fig.2 and Fig.3 and Table 3, respectively.

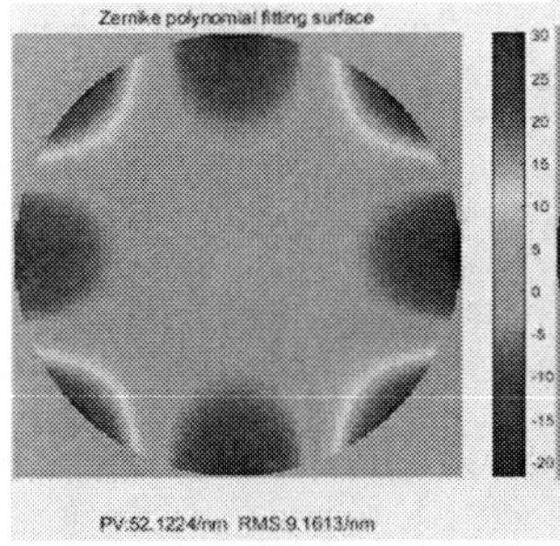


**Fig. 2.** Wavefront aberration of four-point supported lens with proposed solution

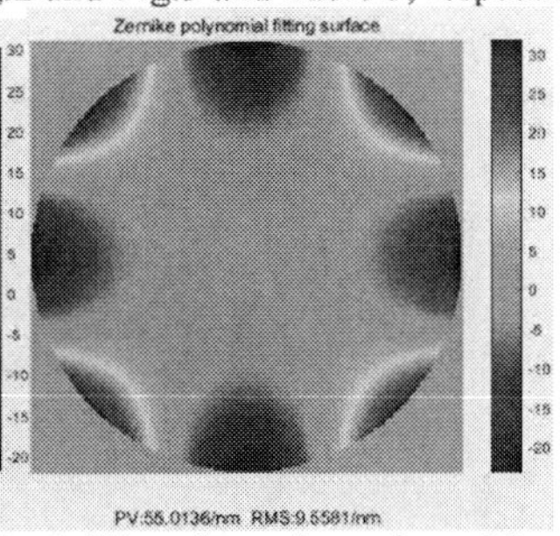


**Fig. 3.** Wavefront aberration of four-point supported lens with numerical solution

**Table 3. Errors of some indicators between proposed and n solution**

| | $Z_1$ | $Z_3$ | $Z_4$ | $Z_9$ | $Z_{17}$ | $Z_{28}$ | PV | RMS |
|---|---|---|---|---|---|---|---|---|
| Proposed solution | 615.66 | 54.05 | -7902.52 | 454.22 | -2925.57 | 206.10 | 55.12 | 9.16 |
| Numerical solution | 599.88 | 53.35 | -7789.46 | 451.34 | -3060.40 | 188.17 | 55.01 | 9.56 |
| Error(%) | 2.63 | 1.31 | 1.45 | 0.64 | 4.41 | 9.53 | 0.20 | 4.18 |

The trend of the four-point support between our proposed solution and numerical solution is exactly the same. Considering the Zernike coefficient, the first, third, fourth, ninth, seventeenth and twenty-eighth items are non-zero items, errors of the first five items are less than 5%, the twenty-eighth item reaches to 9.53%, which is also acceptable. The accuracy of PV value and RMS value is also very gratifying.

## 4. Conclusion

In this article, the lithography machine objective lens is reframed as an elastic circular thin plate model for the purpose of enhancing the parsing accuracy of the general circular lens. Combined with the reciprocal theorem, theory of plates and shells is applied to study the deformation of objective lens, and Zernike coefficient in the field of optics is adopted to characterize the overall accuracy of the lens. The deflection analytical solution of a uniform thickness lens under the support of several uniform points is deduced. The proposed solutions is compared with those obtained by finite element solutions, including the deflection of the lens center point, the lens PV value, the RMS value and the Zernike coefficient.

The calculation example shows that in the case of uniform support at multiple points, 3% error of the lens center point deflection is achieved using the proposed solution, and the errors of the PV value and the RMS value are all less than 5%, as for the majority of 37 Zernike coefficients, the error reaches to 6%, which is still small enough to be reasonable.

The research can help understand the deformation mechanism of circular lens under load and is beneficial to improve the accuracy of finite element analysis. Also, our findings may extremely enrich the optical analysis.

**References**

[1]J. U. Kim, S. Lee, and T. I. Kim, "Recent Advances in Unconventional Lithography for Challenging 3D Hierarchical Structures and Their Applications," J. Nanomater. 2016, 17 (2016).
[2]M. Colburn, T. Bailey, B. J. Choi, J. G. Ekerdt, S. V. Sreenivasan, and C. G. Willson, "Development and advantages of step-and-flash lithography," Solid State Technology 44, 67 (2001).
[3]T. R. Corle, G. S. Kino, and S. M. Mansfield, "Lithography system employing a solid immersion lens," (1992).
[4]S. Dong, K. Guo, X. Li, H. Chen, and M. Ni, "Design and Analysis of a Precise Adjustment Mechanism for Lithography Objective Lens Based on Slit Diaphragm Flexure," in Mechanism and Machine Science, X. Zhang, N. Wang, and Y. Huang, eds. (2017), pp. 403-414.
[5]Y. Sohda, N. Saitou, H. Itoh, and H. Todokoro, "An objective lens system for e-beam cell projection lithography," Microelectronic Engineering 23, 73–76 (1994).
[6]K. H. Schuster and W. Clauss, "Microlithography projection objective with crystal lens," (2010).
[7]S. Dong, K. Guo, X. Li, H. Chen, and M. Ni, "Design and Analysis of a Precise Adjustment Mechanism for Lithography Objective Lens Based on Slit Diaphragm Flexure," (2017).
[8]S. P. Timoshenko and S. Woinowsky-Krieger, Theory of plates and shells (McGraw-hill, 1959).
[9]W. A. Bassali and H. G. Eggleston, "The transverse flexure of thin perforated elastic plates supported at several points," Mathematical Proceedings of the Cambridge Philosophical Society 53, 744 (1957).
[10]W. A. Bassali, "Thin circular plates supported at several points along the boundary," Mathematical Proceedings of the Cambridge Philosophical Society 53, 525-535 (1957).
[11]W. A. Bassali, "Deflection surface of a symmetrically loaded thin circular annulus on multipoint supports," Arab Gulf Journal of Scientific Research 2(1984).
[12]W. A. Bassali, "Normally loaded thin circular annulus of four point supports," (1992).
[13]W. A. Bassali, "Symmetrically loaded thin annular plate supported at the four corners of a concentric rectangle," Indian Journal of Pure and Applied Mathematics 21(1990).
[14]M. N. Y. Anwar and W. A. Bassali, "Bending of a thin circular plate uniformly loaded over a concentric ellipse and symmetrically supported at four points on its axs," Journal of the University of Kuwait 15(1988).
[15]W. A. Bassali and M. N. Y. Anwar, "Thin circular plate under normal paraboloidal loading over a concentric ellipse and symmetrically supported at four points," (1988).
[16]W. A. Bassali, "Some Problems in the Small Deflexions of Clamped Thin Isotropic Plates," ZAMM Journal of applied mathematics and mechanics: Ztschrift fur angewandte Mathematik und Mechanik 40, 493-507 (2010).
[17]J. C. L. Yu and H. H. Pan, "Uniformly loaded circular plate supported at discrete points," International Journal of Mechanical ences 8, 333-340 (1966).
[18]H. Vaughan, "Deflection of uniformly loaded circular plates upon equispaced point supports," Journal of Strain Analysis for Engineering Design 5, 115-120 (1970).
[19]Li, NongFu, and Bao-lian, "The symmetrical bending of an elastic circular plate supported at K internal points," Applied Mathematics & Mechanics (1991).

# Investigation and Application of Reversible Temporary Plugging Steering Agent for Acidification

**Gong-shuai Guan1(✉), Jing-wen Wang1, Yong-zhuo Shan1, Chao Tang1, Wen-ping Sun1**

1 Oil Production Engineering Research Institute of PetroChina Daqing Oilfield Co., Heilongjiang Provincial Key Laboratory for Increasing Oil and Gas Reservoir Yield
guangongshuai@petrochina.com.cn

**Abstract.** In order to solve the problems of insolubility of conventional plant gum type solid steering agents and the need to return after stratified acidification to unblock the formation, two chemically reversible temporary plugging agents were developed in this paper. The variation of viscosity of liquid steering agent and the dissolution of particles of solid-liquid combination steering agent were evaluated indoors. The liquid steering agent single-pipe core blocking experiment and double-pipe core splitting rate experiment were carried out, and the indoor over-hole plate pressure measurement experiments of different types of steering agents were conducted. And the two steering agents were applied in the field. The results show that the viscosity of liquid steering agent can change with the pH value, showing the characteristics of becoming viscous in acid and decomposing in water, which can increase the injection pressure by 1.95MPa after plugging the core in the indoor experiment, and can increase the construction pressure by 1.4MPa in the field application. What's more, the pressure of solid-liquid combination steering agent indoor over pore plate is 3 times of liquid temporary plugging agent, and the construction pressure can be increased by 3.0 MPa. Both of the above temporary plugging agents can be dissolved by themselves in the process of water injection after the measure, and the plugging of the reservoir can be released without returning to the drainage. Meanwhile, it can also meet the needs of different pressure difference layer section for diversion acidification.

## 1 Introduction

Acidizing is one of the most economical and effective means to unblock reservoirs and improve injection and recovery capacity during oilfield development. Due to the non-homogeneity of the reservoir, acidizing fluid is more likely to acidify the high permeability layer and intensify the permeability difference between reservoirs. Therefore, the field construction process requires stratified acidizing in order to achieve uniform unblocking of reservoirs with different permeability[1-22]. In the past, solid steering agent was used to plug the high permeability layer, so that the subsequently injected acid could enter the relatively low permeability reservoir for unblocking and achieve the purpose of uniform acidification in layers. However, the main component of solid steering agent is plant gum, which has poor solubility and must be returned to drainage after acidification. Generally, the formation pressure is relied on to remove the solid steering agent blocked in the borehole to unblock the wellbore and restore the water injection capacity of the blocked section, and even, in some wells, the formation pressure is insufficient, and the solid steering agent needs to be carried out of the wellbore by gas lift. Both of these methods require the return of the formation fluid containing the residual acid solution to the surface, which is difficult and expensive to handle, and also poses a potential risk of environmental pollution during transport and disposal. For this reason, the field needs a reversible temporary plugging agent that can dissolve itself after acidizing, which can be unblocked without returning to the surface after the measure and reduce the risk of safety and environmental protection.

## 2 Research and evaluation of reversible temporary blocking steering agent

In view of the problem that conventional plant gum type solid steering agent does not dissolve and must be returned to the drainage, two kinds of agents, low strength liquid reversible temporary plugging agent and high strength solid-liquid combination reversible temporary plugging agent, have been developed to realize diversion acidification and uniform unblocking of different pressure difference layer sections and improve the effect of measures.

### *2.1 Low strength liquid reversible temporary plugging steering agent*

An environmentally friendly surfactant is used, which has a viscosity comparable to water in aqueous solution with neutral pH, but undergoes a conjoining reaction in the acid system and transforms from a monomer to a giant micelle structure, which increases the viscosity of the solution substantially and achieves plugging of the reservoir rock pores. During the normal production process after acidizing measures, after the high viscosity residual acid comes into contact with the injected water, the giant micelle structure will be decomposed into single spheres and the viscosity will drop rapidly, which can automatically unblock the rock pores without returning to the drainage.

#### *2.1.1 Experiment to Evaluate the Change of Viscosity of Steering Agent Solution with pH Value*

Different concentrations of steering agent solutions were prepared with water and the initial viscosity of the solutions was determined. The pH of the solution was adjusted to 1 using 12% hydrochloric acid by mass fraction, and the viscosity of the solution was measured at this time. Then the pH of the solution was adjusted to 7 with 12 wt% of NaOH and the viscosity of the solution was measured. The test results are shown in **Fig. 1**.

From **Fig. 1**, it can be seen that the viscosity increased significantly as the solution changed from neutral to acidic, and after adjusting to neutral again, the viscosity could be restored to the initial value. As the concentration of the steering agent increases, the viscosity of the solution in acidic environment also gradually becomes larger. Take 5% concentration of steering agent as an example, the initial viscosity of the solution is 1.5 mPa·s. When the pH value is adjusted to 1, the viscosity of the solution increases to 324.3 mPa·s. When the pH value is adjusted to 7 by adding NaOH solution, the viscosity of the solution decreases to 2.2 mPa·s.

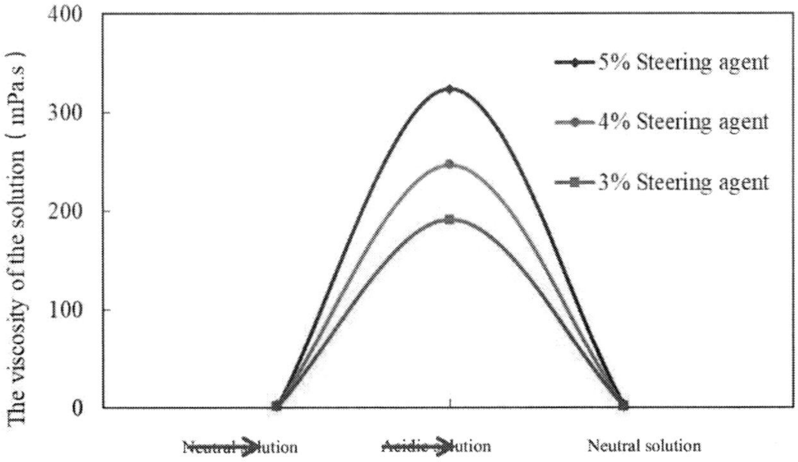


**Fig. 1.** Variation curve of viscosity of steering agent solution with pH

*2.1.2 Single Pipe Core Plugging Performance Evaluation Experiment*

The core was vacuum saturated with water and placed in a core holder. The simulated formation water was displaced into the core at a constant flow rate and the permeability of the water phase was measured. Then the liquid temporary plugging agent solution with a steering agent concentration of 5% and HCl concentration of 12% was displaced by 0.3 PV, and then the simulated formation water was displaced and the injection pressure was measured at this time.

The experimental results showed that the initial injection pressure was 0.02 MPa when replacing the simulated formation water, and the injection pressure increased to 1.97 MPa after replacing the liquid temporary plugging agent, and the injection pressure increased by 1.95 MPa, indicating that the core was effectively plugged.

*2.1.3 The Evaluation Experiment of Double Tube Core Fractionation Rate*

Two cores with large differences in permeability were selected, and the aqueous-phase permeability of the two cores was first measured separately using simulated formation water, and then a double-pipe parallel core replacement test was conducted using simulated formation water to measure the water flowing through the two cores before blocking. The cores were then plugged with liquid temporary plugging agent, and the simulated formation water was replaced again to determine the water flow through the two cores after plugging with liquid temporary plugging agent. Finally, the hydraulic permeability of the two cores was measured separately using the simulated formation water. The experimental results are shown in **Fig. 2**.

**Fig. 2** shows that the water-phase permeability before sealing of the low-permeability and high-permeability cores were $67.0\times10^{-3}um^2$ and $214.2\times10^{-3}um^2$, respectively, and the corresponding water diversion ratios were 28.6% and 71.4%, respectively. The water phase permeability after sealing of low permeability and high permeability cores were $57.4\times10^{-3}um^2$ and $36.1\times10^{-3}um^2$, respectively, corresponding to the diversion water ratio of 69.6% and 30.4%, respectively. From the experimental results, it can be seen that due to the difference in permeability, the liquid temporary plugging agent is more likely to enter the high permeability core, which causes a significant decrease in permeability by 83.0%, while the permeability of the low permeability core only decreases by 14%, and the permeability of the high permeability core is lower than that of the low permeability core after plugging instead. From the change of diversion water ratio before and after blocking, it can be seen

Journal of Physics: Conference Series **2594** (2023) 012028 doi:10.1088/1742-6596/2594/1/012028

that the diversion water of high permeability cores decreased by 57%, and the diversion water of low permeability cores increased by 143%, which achieved the purpose of diversion.

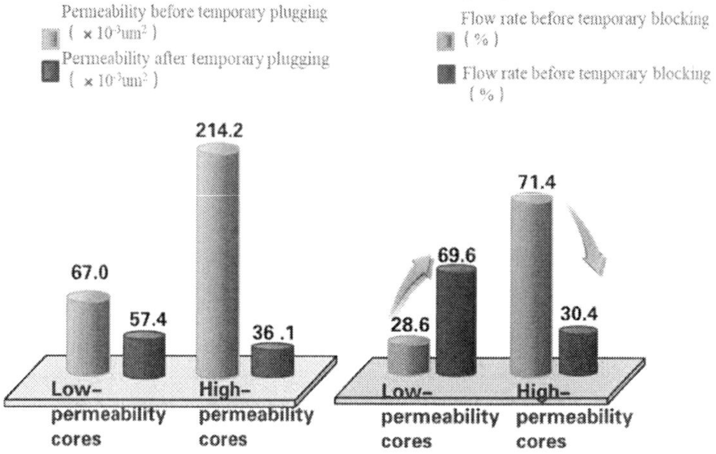


**Fig. 2.** Results of dual-tube parallel core repulsion experiments

### *2.2 High Strength Solid-liquid Combination Reversible Temporary Plugging Steering Agent*

Due to the limited plugging strength of liquid temporary plugging agent, a high-strength solid-liquid combination of reversible temporary plugging steering agent was studied to meet the needs of reservoir diversion and acidification with large pressure difference between layers. A kind of solid particles slowly dissolved in the liquid steering agent is preferred to take advantage of the high viscosity of the liquid steering agent to carry the solid particles, which not only maintains the characteristics of the liquid steering agent in the layer with high viscosity blocking, but also utilizes the principle of "bridging blocking" of the solid particles at the gunkhole and core end face. The high strength solid-liquid combination of reversible temporary plugging steering agent can achieve the purpose of further improving the plugging strength and enabling the subsequent injection of acid into the high-pressure layer section to unblock.

#### *2.2.1 The Evaluation Experiment Solid Particles Dissolution Performance*

The solid particles were added into three systems of water, acid solution and liquid temporary plugging agent to observe the dissolution and settlement at different times, and the results are shown in **Fig. 3**.

**Fig. 3** shows that the solid particles dissolve faster in the aqueous solution, with a dissolution rate of 35.7% in 15min, and slower in the liquid temporary plugging agent, with only 3.6% dissolved in 15min, which is 10% of the dissolved amount in water. It shows that the solid particles can achieve the purpose of temporary plugging of high permeability layer during the construction process, while the injection of water after construction can ensure rapid unblocking. In addition, the solid particles can be suspended and dispersed in the liquid steering agent, and it takes 24h to completely settle to the bottom of the container, but only 3s in water to sink to the bottom, which can meet the demand of carrying injection and sealing the reservoir on site.

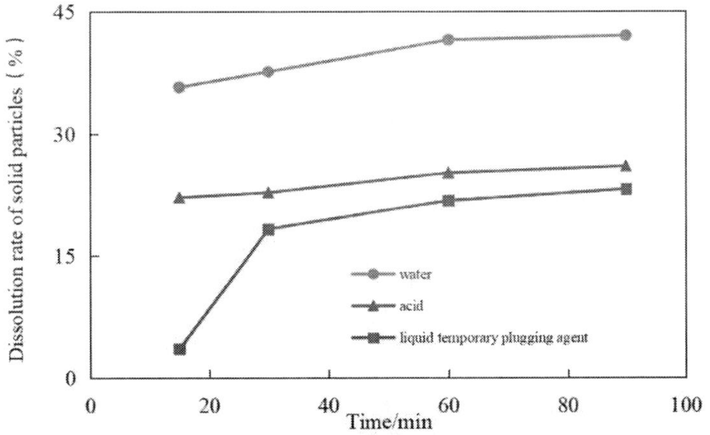


**Fig. 3.** Dissolution rate of solid particles in different liquids

### *2.2.2 Solid-liquid Combination Reversible Temporary Plugging Steering Agent Blocking Strength Evaluation Experiment*

The indoor simulation of the solid particle bridge plugging principle was used to evaluate the plugging strength of the steering agent by measuring the pressure across the orifice plate, and the pressure of the temporary plugging agent with different concentrations of solid-liquid ratio through the orifice plate of 0.15 mm diameter was tested. The results are shown in **Fig. 4**.

**Fig. 4** shows that the pressure of the solid-liquid temporary plugging agent through the pore plate is significantly higher than that of the single liquid temporary plugging agent. For example, the pressure of liquid plugging agent containing 20% solid particles is three times higher than that of single liquid plugging agent when the amount of plugging agent is 50 ml. In addition, with the increase of solid particle content, the over-hole plate pressure also increases gradually. With the increase of temporary plugging agent dosage, the pressure of over-hole plate also increases gradually. Thus, it can be seen that the solid-liquid combination of temporary plugging agent can significantly improve the plugging strength, which is suitable for the diversion acidizing of large pressure difference layer section.

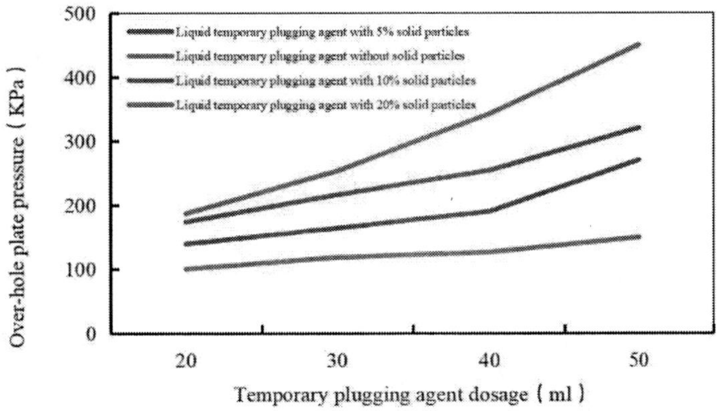


**Fig. 4.** Different types of temporary plugging agent over the orifice plate pressure curve

## 3 Field test effect and analysis

Two agents, low-strength liquid reversible temporary plugging agent and high-strength solid-liquid combination reversible temporary plugging agent, were used to carry out diversion acidizing field tests and compared with the pressure change of cage acidizing wells during construction. The pressure variation curves are shown in **Fig. 5**.

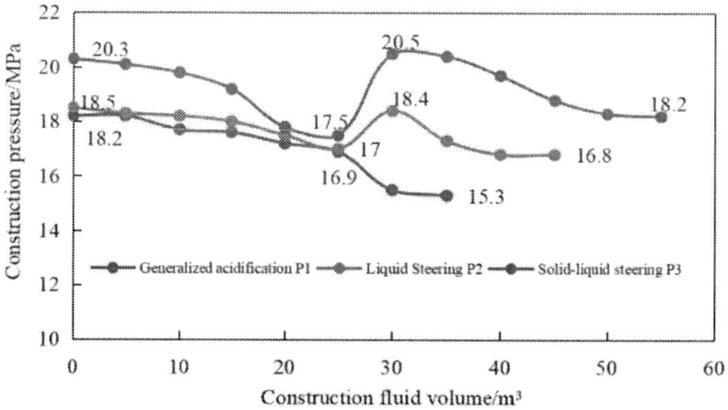


**Fig. 5.** Liquid reversible temporary plugging steering agent sealing before and after the pressure change curve

As shown in **Fig. 5**, the results of the field test showed that the pressure of the caged acidizing well gradually decreased during the construction process, and the pressure dropped by 1.6 MPa at the end of construction.

In the stratified acidizing well using liquid steering agent, the pressure dropped first during the construction process, indicating that the injected acid played an unblocking role. After the injection of liquid steering agent, the construction pressure rose by 1.4 MPa, proving that the high permeability layer section was effectively blocked, so that the subsequently injected acid entered the low permeability layer section for unblocking, and finally the construction pressure dropped by 1.6 MPa, which realized the stratified acidizing of different layer sections.

In the stratified acidizing well using solid-liquid combination steering agent, the construction pressure of the first layer section decreased by 1.8 MPa, and after injecting solid-liquid combination steering agent, the construction pressure increased by 3.0 MPa, and its sealing strength was much higher than that of single liquid steering agent. The solid-liquid combination steering agent can significantly increase the construction pressure and achieve the plugging of the layer section with large pressure difference.

## 4 Conclusion

In order to solve the problem of insoluble traditional plant-based solid guiding agents and the need for backflow after layered acidification to clear the formation, two chemical reversible temporary plugging agents were developed in this paper, and the following conclusions were obtained:

(1) The results show that the viscosity of liquid steering agent can change with the pH value, showing the characteristics of becoming viscous in acid and decomposing in water.

(2) The plugging effect of liquid reversible temporary plugging steering agent on core is obvious in the indoor test, which can increase the injection pressure by 1.95MPa, and the construction pressure can be increased by 1.4MPa during the field test to meet the need of diversion acidizing in small pressure difference layer section.

(3) The blocking strength of the solid-liquid combination of reversible temporary plugging steering agent is significantly higher than that of the single liquid temporary plugging agent,

and its indoor over-hole plate pressure is three times that of the single liquid temporary plugging agent, which can increase the construction pressure by 3.0 MPa during the field test to meet the needs of diversion acidizing of large differential pressure layer section.

(4) The two chemically reversible temporary plugging agents can be dissolved by themselves in the process of water injection after the measure, and the plugging of the reservoir can be released without returning to the drainage.

(5) The two reversible temporary plugging steering agents researched solve the problems of insolubility of solid steering agents with plant gum as the main component and the need to return to drainage after acidification, and realize the full coverage of diversion acidification for different pressure difference layer sections.

**Acknowledgments.** The project is supported by National Science and Technology Major Project "Demonstration Project of Tight Oil Development in Songliao Basin" (Number 2016ZX05071).

## References

[1] Shi L., Zhang W., Wei X. et al. Application of temporary plugging and acidizing technology in oil fanzhuang oilfield[J]. Break Block Oil and Gas Field, 2008, No. 90(06): 114-116.
[2] Zhong S., Ren S., Huang Y., et al. Research and application of uniform acid distribution technology for ultra-deep horizontal wells in Yuanba marine phase[J]. Petroleum Drilling and Production Technology, 2013,35(06):83-86.
[3] Qing X. Research and application of temporary plugging and acidizing technology in Jing'an oilfield[D]. Xi'an University of Petroleum, 2014.
[4] Liu P. Research on temporary plugging acidizing process in Chengdao oilfield[D]. China University of Petroleum (East China), 2019.DOI:10.27644.
[5] Zhang X.F., Ou T.H., Zhao Y.P. Experimental research and application of temporary plugging and acidizing process technology[J]. Daqing Petroleum Geology and Development, 2000(01):44-46+54.
[6] Chen XY, Zhang ZQ, Li D. Experimental study on chemical splitting acidification process[J]. Modern Chemical, 2016,36(08):67-70+72.
[7] Wang W., He H., Liu C., et al. Research and application of temporary plugging and acidizing process technology for low permeability reservoirs[J]. Science, Technology and Engineering, 2010,10(34):8533-8535.
[8] Zhu B. Research and application of layered acidizing technology for immobile pipe column in low permeability oilfield[J]. Inner Mongolia Petrochemical, 2016,42(06):116-118.
[9] Xu J. Application of stratified acidification process technology for water injection wells in the present river[J]. Inner Mongolia Petrochemical, 2013,39(01):107-109.
[10] Chen L. Application of water-soluble acidizing temporary plugging process technology in isolated oil fields[J]. Neijiang Science and Technology,2015,36(02):36-37.
[11] Zou H. Y., Peng Y. K. Adaptation study of phase conversion acid in Gaas oilfield[J]. Inner Mongolia Petrochemical, 2013,39(10):148-149.
[12] Zhang N. L., Li Y., Zhang H. G., et al. Research and effect evaluation of layered temporary plugging acidizing technology[J]. Petrochemical Applications, 2017,36(08):24-29.
[13] Zhong S , Pan B , Wang X , et al. Research and application of acidification temporary plugging agent in Yuanba gas field[J]. Reservoir Evaluation and Development, 2017.
[14] Feng P Y , Wang Z S , Cui B , et al. Research on Emulsion Temporary Plugging and Acidification Experiment in Carbonate Reservoir[J]. Science Technology and Engineering, 2015.
[15] Deng Z , Ji Z , Zhang S . New Online Shunt Acidification for Water Injection Increasing

Technology and Its Application in Huanjiang Oilfield[J]. Journal of Chemistry, 2021(6):2021.
[16] Li Q S . Development and application of WLD temporary blocking a-gent[J]. Petroleum Geology & Oilfield Development in Daqing, 2004.
[17] He J , Okere C J , Su G , et al. Formation damage mitigation mechanism for coalbed methane wells via refracturing with fuzzy-ball fluid as temporary blocking agents[J]. Journal of Natural Gas Science and Engineering, 2021, 90(5).
[18] Yang X , Xu X , Wang X , et al. Protective Workover Fluid for Biological Limestone Reservoir in Dagang Oilfield Technology Research[J]. IOP Conference Series Earth and Environmental Science, 2021, 632:022058.
[19] Cheng W , Lu C , Feng G , et al. Ball sealer tracking and seating of temporary plugging fracturing technology in the perforated casing of a horizontal well[J]. Energy Exploration & Exploitation, 2021(6).
[20] Deng Z , Ji Z , Zhang S . New Online Shunt Acidification for Water Injection Increasing Technology and Its Application in Huanjiang Oilfield[J]. Journal of Chemistry, 2021(6):2021.
[21] Wang J , Wei W , Nan M , et al. Technical Study on the Treatment of Offshore Oilfield Acidification Flow-back Fluid[J]. Journal of Yangtze University(Natural Science Edition), 2017.
[22] Li H , Shi Y . Study on the Performance Degradation of Sandstone under Acidification[J]. ACS Omega, 2020, 5(43):28333-28340.

# AE-BEGAN based Synthetic Data Augmentation for Sample-Limited High-Dimensional Problems with Application to NIR Spectral Data

**Xin-Yue Luo1,2, Xing-Rong Fan1,2(✉), Xian-Ming Zhang1(✉), Tian-Yu Chen2, Chun-Jie Huang3**

1 Engineering Research Centre for Waste Oil Recovery Technology and Equipment, Ministry of Education, Chongqing Technology and Business University, Chongqing, China
2 Chongqing Key Laboratory of Intelligent Perception and Block Chain Technology, Chongqing Technology and Business University, Chongqing, China
3 Chongqing Intelligent Supply Chain Engineering & Technology Research Center, Chongqing Jiaotong University, Chongqing, China
4 Chongqing Haolongyu Technology Co., Ltd., Chongqing, China

zxm215@126.com, fanxingrong@ctbu.edu.cn

**Abstract**. Synthetic data augmentation holds substantial research and application value in scenarios characterized by limited samples and high dimensions. It enhances the analytical ability and efficiency of spectral analysis models. This paper proposes Autoencoder-Combined Boundary Equilibrium Generative Adversarial Networks (AE-BEGAN) as a new method for augmenting synthetic data in scenarios with limited samples and high dimensions, with a specific emphasis on near-infrared (NIR) spectral data. The spectral data first undergoes pre-processing procedures that encompass advanced noise reduction algorithms and techniques for removing abnormal samples, guaranteeing elimination of unwanted disturbances and outliers. Then, the pre-processed data is utilized to train the AE-BEGAN model, which generates augmented synthetic samples. Finally, real NIR spectral data obtained from lubricant samples exhibiting different water contents were employed to validate and test the performance of the model. The experimental results demonstrate that the AE-BEGAN model outperforms other GANs in generating synthetic data of high quality and diversity, as quantified by two evaluation metrics, α-Precision and β-Recall with scores of approximately 0.86 and 0.28, respectively. The application case study confirms that the AE-BEGAN model exhibits the capability to generate derived NIR spectra and expand the number of spectra in scenarios with limited samples and high dimensions.

## 1. Introduction

Many studies have reported that abundant, diverse, and well-proportioned samples can significantly improve the efficiency of models in spectroscopic analysis, both qualitatively and quantitatively [1-2]. Moreover, with the emergence of the big data era, deep learning (DL) algorithms have been widely applied to spectral analysis, posing a greater challenge to the number of spectral samples due to the data sensitivity of these algorithms [3-5].

However, some samples are expensive and difficult to obtain, particularly in the biology and medicine fields. For example, in the process of NIR spectroscopy analysis, due to the complex component content of some tested materials (e.g. waste oil, drugs), measurement is time-consuming and laborious, making it difficult to acquire a large amount of data. This affects the adequacy of the number of modeled samples, thus reducing the efficiency and analytical ability of the established models [6-7]. Consequently, exploring effective methods of synthesizing high-quality spectra to supplement training sets is of great importance.

A multitude of data augmentation techniques have been developed to tackle the challenges posed by sample-limited problems. Moreover, in high-dimensional contexts featuring a considerable number of variables or features, synthetic data augmentation serves as an effective means to alleviate the risk of overfitting and bolster the generalization capability of models. These techniques can be broadly classified into oversampling methods, data transformations, and generative models. Generative Adversarial Networks (GANs) [8] and their variants, such as DCGAN [9], WGAN [10], WGAN-GP [11], and BEGAN [12], are among the most widely used generative models for synthetic data augmentation. Teng et al. extended the spectral database of laser-induced breakdown spectroscopy based GAN, as their original infrared spectral data were abundant yet had insufficient dimensions, making them unsuitable for scenarios with limited samples and high dimensions [13]. Zhu et al. extended the spectra of corn and tablets using BEGAN, but did not use evaluation indicators to assess the quality and diversity of synthetic spectra [14].

Limited attention has been given to the synthesis of spectral samples that are both limited and high-dimensional, leading to the development of models that yield uncertain or unstable predictions when trained on small sample sets. To address this gap, this article presents a novel approach known as Autoencoder-Combined Boundary Equilibrium Generative Adversarial Networks (AE-BEGAN) for augmenting synthetic data in scenarios characterized by limited samples and high dimensions. To demonstrate the practical application of AE-BEGAN, real NIR spectral data obtained from lubricant samples with varying water content is employed as a case study. The main contributions are summarized as follows:

(1) We propose AE-BEGAN, a general framework for synthetic data augmentation, specifically designed for high-dimensional problems with limited samples.

(2) To assess the quality and diversity of the synthetic data generated by trained models, we employ two evaluation metrics, namely $\alpha$-Precision and $\beta$-Recall.

(3) Through an application case study, we validate the capabilities of AE-BEGAN in generating derived NIR spectra and effectively expanding the quantity of spectra in scenarios characterized by limited samples and high dimensions.

## 2. Methodology

In this section, the model architecture of AE-BEGAN is proposed, followed by a detailed description of the specific implementation method for generating NIR spectral samples. Finally, two evaluation metrics for characterizing the quality and diversity of synthesized spectral samples and NIR spectral datasets of lubricant samples with varying water content are presented.

### *2.1. Proposed model architecture*

The architecture of AE-BEGAN is depicted in Fig. 1. It consists of a deep autoencoder (DAE) with a symmetric encoder and decoder, which serves as the discriminator. The generator has the same structure as the decoder in the discriminator. The generator takes random noise as input and produces synthetic data samples $G(z)$ as output. The discriminator takes both real data $X$ and generated data $G(z)$ as input and outputs $D(x)$ and $D\big(G(z)\big)$ with the main features of the real and generated data sample $G(z)$ extracted by the autoencoder. Unlike traditional GANs, which directly match the data distribution, AE-BEGAN aims to match the loss distribution of the DAE. This loss is calculated using the Wasserstein distance function. The discriminator loss for real data is defined as in *Eqn* (1):

$L(x) = |D(x) - x|^2 \# (1)$

where $x$ is real data, and $D(\cdot)$ is the output of the autoencoder discriminator. The
loss of the discriminator for the synthetic data is defined as in *Eqn* (2):

$$L\big(G(z)\big) = \big|D\big(G(z)\big) - G(z)\big|^2 \#(2)$$

The proposed generator loss $L_G$ is defined as in *Eqn* (3):

$$L_G = L\big(G(z)\big) \#(3)$$

The discriminator aims to minimize $L(x)$ and maximize $L\big(G(z)\big)$. Thus, the proposed discriminator loss $L_D$ is defined as in *Eqn* (4):

$$L_D = L(x) - k_t \cdot L\big(G(z)\big) \#(4)$$

where $k_t \in [0,1]$ is the control parameter that maintains the equilibrium between $L(x)$ and $L\big(G(z)\big)$, and $k_t$ is updated as follows:

$$k_{t+1} = k_t + \lambda_k \Big(\gamma L(x) - L\big(G(z)\big)\Big) \#(5)$$

where $\lambda_k$ is the learning rate of $k_t$, set to 0.001; the hyperparameter $\gamma \in [0,1]$ is the diversity ratio. The higher the value of $\gamma$, the better the diversity. The index $M_{global}$ to judge the convergence of the model is defined as in *Eqn* (6):

$$M_{global} = L(x) - \big|\gamma L(x) - L\big(G(z)\big)\big| \#(6)$$

This metric can be used to determine when the network reaches its final state or whether the model has crashed.

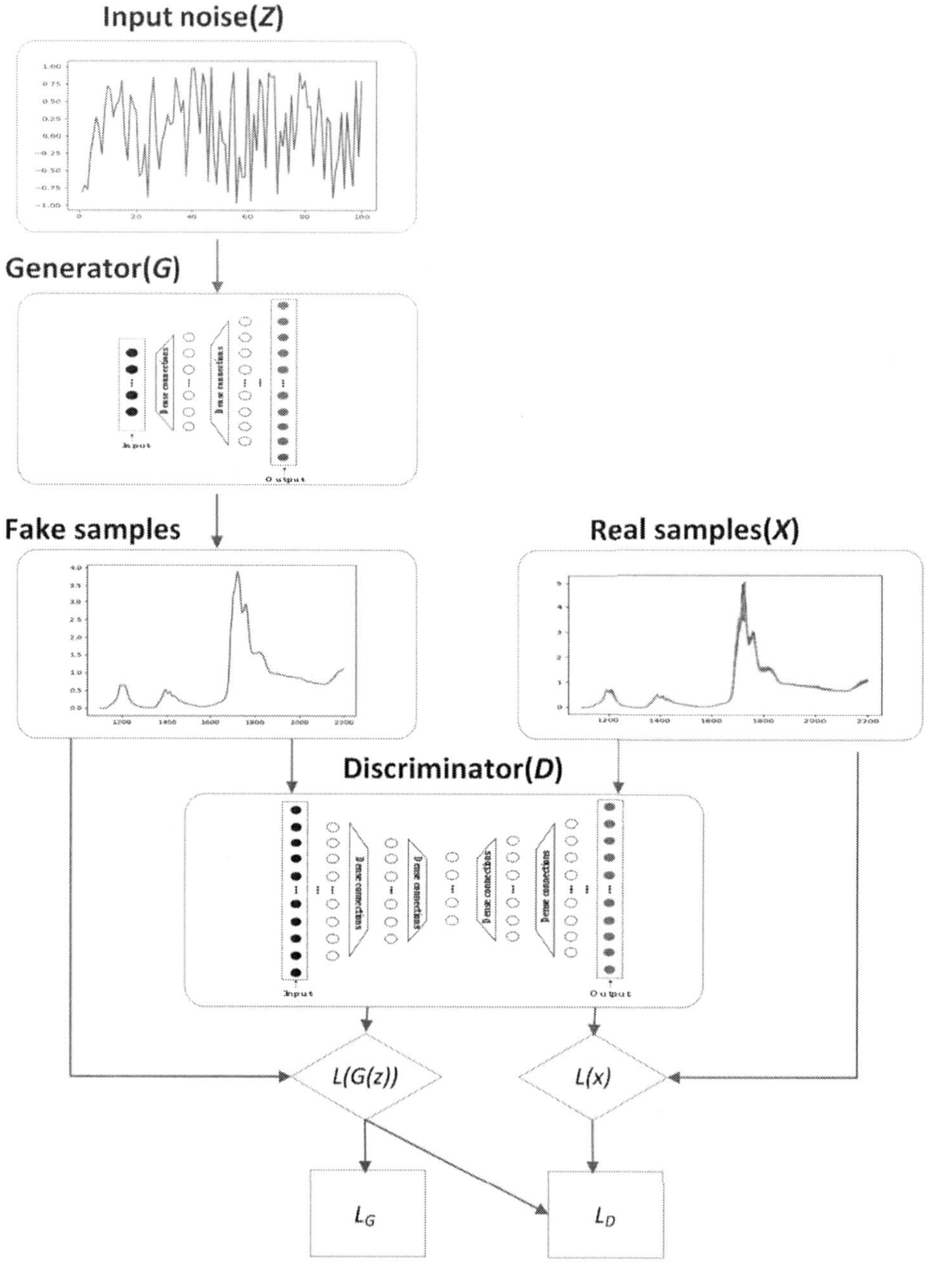


**Fig. 1.** Network architecture of Autoencoder-Combined Boundary Equilibrium Generative Adversarial Networks (AE-BEGAN).

### *2.2. Implementation*

The model was implemented and trained in PyTorch v1.12.1 with Python v3.7.16. The overall network architecture is depicted in Fig.2. In theory, a trained AE-BEGAN generator is identical to the decoder of an autoencoder and produces a realistic spectrum from a random input latent vector. We take real NIR spectral data of lubricant samples with varying water content as a case study for the application of AE-BEGAN. Therefore, the first step is to input the noise vector and upscale the network to an array that can hold the dimensions of each NIR spectrum (*i.e.*, the increasing wavelength).

For the generator, we used a network of fully connected layers with ELU (Exponential Linear Unit) active function and Batch Normalization to speed up the training process and increase the robustness of the network [15]. The generator's input is a 1×150-dimensional uniform random sample distribution with values ranging from -1 to 1. After four layers of full connected networks, the dimensions of the synthetic spectral data increased to 500, 500, 900, and finally 1100. The discriminator is an auto-encoder with a fully connected network structure. The discriminator was fed 1×1100-dimensional spectral data as input and output 1×1100-dimensional data after encoding and decoding. The input and hidden layers use the ELU activation function with a dropout function while the output layer uses the CELU (continuously-differentiable exponential linear unit) activation function [16].

Besides, we normalize the spectral samples to [0,1] using MinMaxScaler normalization and use grid search to optimize the learning rate, batch size, activation function and other parameters by maximizing the $IP_\alpha$ value (see Section 2.3).

The second step of our AE-BEGAN methodology was to develop a process to maintain fake spectra with appropriate noise and variance in the final synthetic datasets. Algorithm 1 gives the learning process of AE-BEGAN. To this end, the AE-BEGAN was trained with the Adam optimizer using an initial learning rate of 0.0001 [17]. After 20000 training iterations, the learning rate was reduced to 0.00001 and the model was trained for an additional 5000 iterations. The batch size of each epoch was set to 8 [18], and the diversity ratio γ was set to 0.3, 0.5, 0.7 and 0.9 respectively [12].

**Algorithm 1 The learning process of AE-BEGAN**

**Input:** dataset $X$, hyperparameter $\gamma$, $\lambda_k$, and $k_t$
**Output:** generator $G$ and discriminator $D$

Initialize generator $G$ and discriminator $D$
**for** batch of real samples $x$ **do**
  $z \sim N(-1,1)$;
  $L(x) \leftarrow |D(x) - x|^2$;
  $L(G(z)) \leftarrow |D(G(z)) - G(z)|^2$;
  $L_G \leftarrow L(G(z))$
  $L_D \leftarrow L(x) - k_t L_G$;
  Update $D$ with Adam optimizer on $L_D$;
  Update $G$ with Adam optimizer on $L_G$;
  Update $k_{t+1} \leftarrow k_t + \lambda_k(\gamma L_X - LG(z))$;
  Update $M_{global} \leftarrow L(x) + |\gamma L(x) - L(G(z))|$;
**end for**

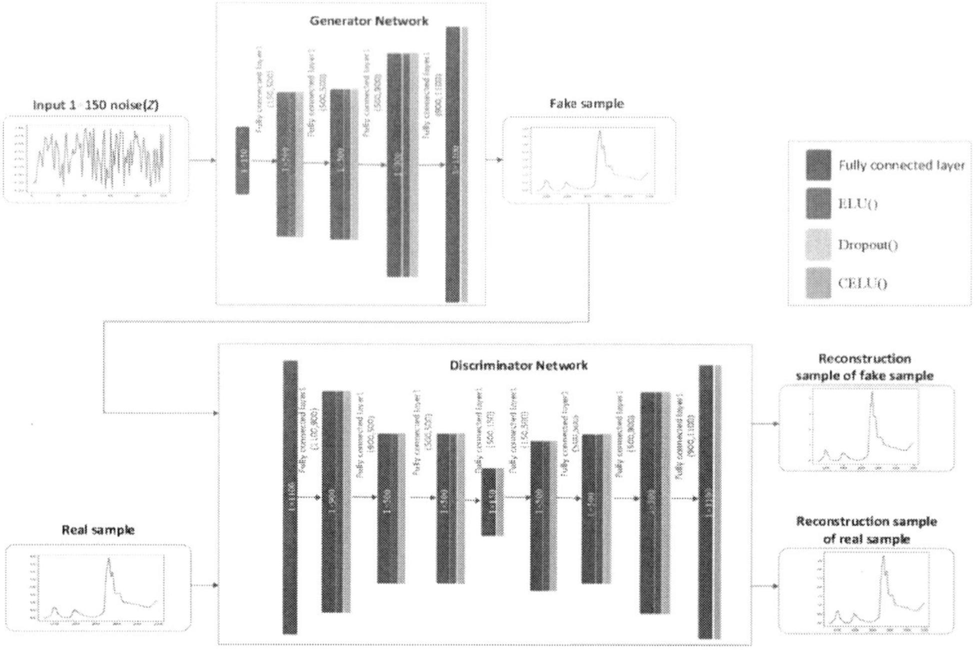


**Fig. 2.** Network architecture of AE-BEGAN for NIR spectral data of lubricant samples.

### 2.3. *Criteria for evaluating the quality and diversity of synthetic data*

Many evaluation metrics exist for generative models, but most of them are related to image data [19]. In our paper, we introduced two evaluation metrics (α-Precision, β-Recall) [20] for assessing the quality and diversity of synthetic samples by different models. α-Precision measures the similarity between synthetic samples and real samples. β-Recall measures the proportion of the synthetic sample covering the real sample.

The α-Precision ($P_\alpha$) and β-Recall ($P_\beta$) curve provides detailed views on model's quality and diversity, but it is often more convenient to summarize performance in a single number. Therefore, the integrated $P_\alpha$ and $P_\beta$ metrics is defined as in *Eqns* (6) and (7):

$$IP_\alpha = 1 - 2\Delta P_\alpha \#(6)$$

$$IP_\beta = 1 - 2\Delta P_\beta \#(7)$$

where $\Delta P_\alpha$ ($\Delta P_\beta$) is the mean absolute deviation of $P_\alpha$ ($P_\beta$), defined as

$$\Delta P_\alpha = \int_0^1 \mid P_\alpha - \alpha \mid \mathrm{d}\alpha \#(8)$$

$$\Delta P_\beta = \int_0^1 \mid P_\beta - \beta \mid \mathrm{d}\beta \#(9)$$

Where $\Delta P_\alpha \in [0,1/2]$ and $\Delta P_\beta \in [0,1/2]$ quantify the extent to which $P_\alpha$ and $P_\beta$ deviate from their optimal value. Only when the distribution of synthetic data is consistent with the distribution of real data, $IP_\alpha = IP_\beta = 1$. For further description of these metrics, please refer to Alaa's papers [20].

### 2.4. *NIR Spectral datasets*

The original dataset comprised NIR spectral data from 39 lubricating oil samples, each measured in the wavelength range of 700-2499 nm, resulting in a total of 10,284 channels. After excessive noise was removed and the data was smoothed using a Savitzky-Golay filter [21], the final dataset comprised 1,100 channels in the wavelength range of 1100 to 2199 nm, with an interval step of 1 nm (see Table 1 and Fig.3).

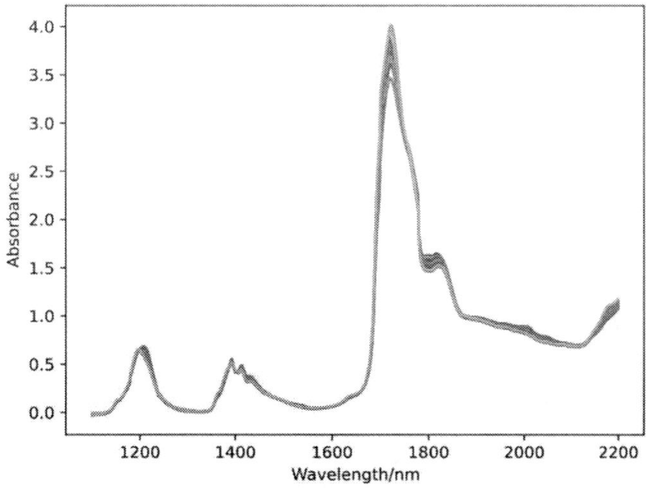


**Fig. 3.** The data of 39 NIR spectral samples of lubricating oil with varying water content.

**Table 1.** Spectral measurement data of lubricating oil samples with varying water content.

| No. | Wavelength/nm | | | | | |
|---|---|---|---|---|---|---|
| | 1100 | 1101 | 1102 | ... | 2198 | 2199 |
| 1 | -0.0167 | -0.0167 | -0.0167 | ... | 1.1748 | 1.1805 |
| 2 | -0.0267 | -0.0267 | -0.0267 | ... | 1.154 | 1.1596 |
| ... | | | | ... | | |
| 38 | -0.0091 | -0.0091 | -0.0092 | ... | 1.0782 | 1.0853 |
| 39 | -0.0126 | -0.0127 | -0.0126 | ... | 1.0676 | 1.0742 |

## 3. Results and Discussion

### *3.1. Training process of AE-BEGAN*

In order to effectively train a discriminator, it is recommended to initially use real samples in multiple training iterations. It is advised to avoid alternating between real and synthetic samples on the discriminator until it has attained a sufficient level of discrimination ability. Once this level has been reached, the real and synthetic samples can be trained alternately. To achieve optimal model convergence during the training process, it was necessary to set the initial learning rate to 0.0001 and closely monitor the global convergence metric $M_{global}$. However, after 20,000 training iterations, it was observed that the curve fluctuated without improvement. Therefore, the learning rate was decreased to 0.00001 and training was continued for an additional 5,000 epochs until the model achieved complete convergence, as shown in Fig.4. The detailed training time of the model is provided in Table 2.

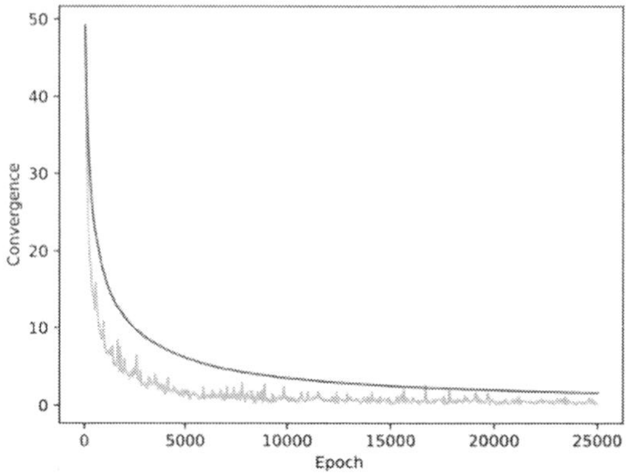


**Fig. 4.** The measure of convergence.

**Table 2.** Time spent in model training (unit: second/s).

| Gamma | The first 20000 epochs | The last 5000 epochs | Sum of all epochs | Average of four models |
|---|---|---|---|---|
| 0.3 | 174.6397 | 43.5321 | 218.1718 | 234.1269 |
| 0.5 | 182.4732 | 43.3520 | 225.8252 | |
| 0.7 | 182.8187 | 42.8128 | 225.6315 | |
| 0.9 | 213.1370 | 53.7381 | 266.8751 | |

*3.2. Comparing synthetic spectra with real spectra*

It is assumed that the average real spectrum is used as a reference and the average synthetic spectrum is employed as a sample. Subtracting the average synthetic spectrum data from the average real spectrum data yields the corresponding difference spectra, which are depicted in Fig. 5. If the synthesized sample is similar to the real sample, their mean difference should be close to zero, indicating that the synthesized sample is consistent with the real sample distribution.

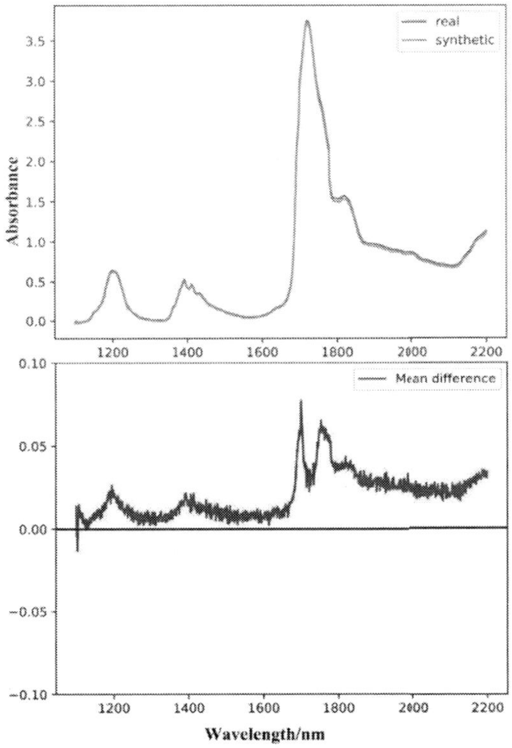


**Fig. 5.** Real and synthetic lubricating oil spectrum (upper part) and their the difference spectra (lower part).

### 3.3. NIR spectra diversity and quality

The diversity ratio of AE-BEGAN regulates the balance between the diversity and quality of the generated samples. To assess the effect of the parameter γ on the synthetic spectrum, the experiment set γ to 0.3, 0.5, 0.7 and 0.9, respectively, to conduct synthetic training on the lubricating oil spectrum. After convergence, $IP_\alpha$ and $IP_\beta$ were calculated. The Principal Component Analysis (PCA) decomposition of the synthetic spectrum and the dispersion of sample points in the PCA score diagram were observed to determine their diversity. Fig. 6 shows that, as the γ value increases, the data points represented by the first and second principal components of the synthetic spectrum become more dispersed.

Table 3 shows that the larger the $\gamma$ value, the smaller the $IP_\alpha$ value and the greater the $IP_\beta$ value of the composite data. Fig. 6 and Table 3 demonstrate that the diversity of synthetic data has a negative correlation with its quality; the greater the γ value, the better the diversity of synthetic data, but the lower the quality, and the smaller the γ value, the worse the diversity of the synthesized data, but the higher the quality.

**Table 3.** The effect of diversity ratio γ on the quality and diversity of generated samples.

| γ | $IP_\alpha$ | $IP_\beta$ |
|---|---|---|
| 0.3 | 0.87 | 0.10 |
| 0.5 | 0.79 | 0.29 |
| 0.7 | 0.71 | 0.33 |
| 0.9 | 0.63 | 0.35 |

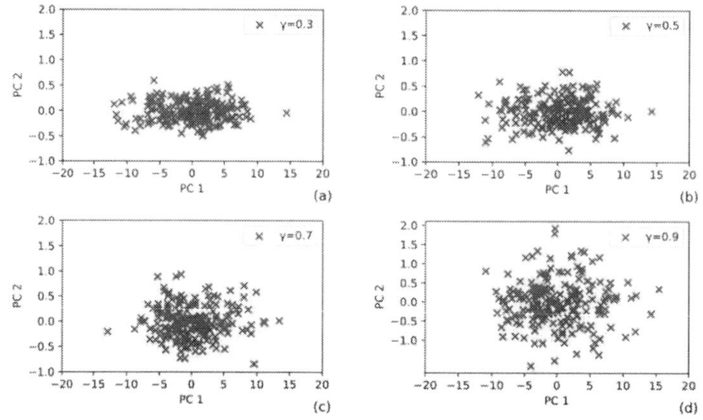

**Fig. 6.** The effect of diversity ratio γ on the diversity of synthetic spectra samples.

### 3.4. Comparison experiments

In comparison with the existing DCGAN, WGAN, WGAN-GP and the proposed AE-BEGAN, the values of $IP_\alpha$ and $IP_\beta$ are assessed respectively (Table 4). As shown in Table 4, the proposed AE-BEGAN outperforms other GANs in generating synthetic data of high quality and diversity, as quantified by two evaluation metrics, α-Precision and β-Recall with scores of approximately 0.86 and 0.28, respectively. Thus, the application case study confirms that the AE-BEGAN is capable of generating derived NIR spectra and expanding the number of spectra in sample-limited, high-dimensional scenarios.

**Table 4.** Results of model performance evaluation for different GANs under the and α-Precision ($IP_\alpha$) and β-Recall ($IP_\beta$) metrics.

| Method | $IP_\alpha$ | $IP_\beta$ |
|---|---|---|
| DCGAN | 0.50 | 0.21 |
| WGAN | 0.53 | 0.23 |
| WGAN-GP | 0.57 | 0.25 |
| AE-BEGAN | **0.86** | **0.28** |

## 4. Conclusion

The present study conducts a comparison of the mean difference between artificial and real samples, thereby demonstrating the effectiveness of AE-BEGAN in accurately capturing the distributional characteristics of real samples and generating corresponding synthetic samples. The results unequivocally validate that the distribution of artificial samples generated by AE-BEGAN closely aligns with that of the real samples. Additionally, this study explores the impact of diversity ratio on the quality and diversity of the generated samples within the AE-BEGAN framework, uncovering a negative correlation between quality and diversity. Furthermore, a comprehensive evaluation is carried out by comparing the $IP_\alpha$ and $IP_\beta$ values of synthetic spectra produced by AE-BEGAN with those of alternative GANs. The findings consistently demonstrate that AE-BEGAN outperforms other GANs, exhibiting significantly higher $IP_\alpha$ and $IP_\beta$ scores, thereby indicating its superior suitability for generating synthetic spectra. Moving forward, we intend to delve into the specific applications of AE-BEGAN in spectroscopy synthesis and thoroughly analyze the effectiveness of this synthesis technique.

**Acknowledgment**

This work is supported in part by the Science and Technology Research Program of Chongqing Municipal Education Commission (KJQN201900833), the Research Project of Chongqing Technology and Business University (1952001), the Scientific research platform open project of Chongqing Technology and Business University (KFJJ2018059), and the Natural Science Startup Foundation of Chongqing Technology and Business University (2056019). We are also very grateful to the anonymous reviewers and editors for the comments on the manuscript.

**References**

[1] Zhou L, Zhang C, Taha M F, et al. Determination of leaf water content with a portable NIRS system based on deep learning and information fusion analysis[J]. Transactions of the ASABE, 2021, 64(1): 127-135.
[2] Chu X, Huang Y, Yun Y H, et al. Chemometric methods in analytical spectroscopy technology[M]. Springer, 2022.
[3] Neo E R K, Low J S C, Goodship V, et al. Deep learning for chemometric analysis of plastic spectral data from infrared and Raman databases[J]. Resources, Conservation and Recycling, 2023, 188: 106718.
[4] Praveen B, Menon V. Dual-Branch-AttentionNet: A Novel Deep-Learning-Based Spatial-Spectral Attention Methodology for Hyperspectral Data Analysis[J]. Remote Sensing, 2022, 14(15): 3644.
[5] Zhang X, Lin T, Xu J, et al. DeepSpectra: An end-to-end deep learning approach for quantitative spectral analysis[J]. Analytica Chimica Acta, 2019, 1085(13):48-57
[6] Pavlou E, Kourkoumelis N. Deep adversarial data augmentation for biomedical spectroscopy: Application to modelling Raman spectra of bone[J]. Chemometrics and Intelligent Laboratory Systems, 2022, 228: 104634.
[7] Yang J, Xu J, Zhang X, et al. Deep learning for vibrational spectral analysis: Recent progress and a practical guide[J]. Analytica Chimica Acta, 2019, 1081(12):6-17
[8] Goodfellow I, Pouget-Abadie J, Mirza M, et al. Generative adversarial networks[J]. Communications of the ACM, 2020, 63(11): 139-144.
[9] Radford A, Metz L, Chintala S. Unsupervised representation learning with deep convolutional generative adversarial networks[J]. arXiv preprint arXiv:1511.06434, 2015.
[10] Arjovsky M, Chintala S, Bottou L. Wasserstein generative adversarial networks[C]//International conference on machine learning. PMLR, 2017: 214-223.
[11] Gulrajani I, Ahmed F, Arjovsky M, et al. Improved training of wasserstein gans[J]. Advances in neural information processing systems, 2017, 30.
[12] Berthelot D, Schumm T, Metz L. Began: Boundary equilibrium generative adversarial networks[J]. arXiv preprint arXiv:1703.10717, 2017.
[13] Teng G E, Wang Q Q, Kong J L, et al. Extending the spectral database of laser-induced breakdown spectroscopy with generative adversarial nets[J]. Optics Express, 2019, 27(5): 6958-6969.
[14] Zhu, D., Xu, L., Chen, X., Yuan, L. M., Huang, G., Li, L., Xi, C., Shi, W.: Synthetic spectra generated by boundary equilibrium generative adversarial networks and their applications with consensus algorithms[J]. Optics Express, 2020, 28(12):17196-17208.
[15] Ioffe S, Szegedy C. Batch normalization: Accelerating deep network training by reducing internal covariate shift[C]//International conference on machine learning. pmlr, 2015: 448-456.
[16] Srivastava N, Hinton G, Krizhevsky A, et al. Dropout: a simple way to prevent neural networks from overfitting[J]. The journal of machine learning research, 2014, 15(1): 1929-1958.
[17] Kingma D P, Ba J. Adam: A method for stochastic optimization[J]. arXiv preprint arXiv:1412.6980, 2014.
[18] Masters D, Luschi C. Revisiting small batch training for deep neural networks[J]. arXiv preprint arXiv:1804.07612, 2018.
[19] Borji A. Pros and cons of gan evaluation measures[J]. Computer Vision and Image Understanding, 2019, 179: 41-65.
[20] Alaa A, Van Breugel B, Saveliev E S, et al. How faithful is your synthetic data? sample-level metrics for evaluating and auditing generative models [C]//International Conference on Machine

Learning. PMLR, 2022: 290-306.
[21] Press W H, Teukolsky S A. Savitzky - Golay smoothing filters[J]. Computers in Physics, 1990, 4(6): 669-672.

# Dessert Optimization and Oil Potential Analysis of M9 Block in W Oilfield

**Fu-ji Yu**

No.8 Oil Production Plant of Daqing Oilfield Co., Ltd.
M17713390006@163.com

**Abstract**. This article first explains the fine structure of the M9 block, and the encryption equal value line is finely constructed into a diagram to identify the micro -amplitude structure. Use the block drilling to complete the inner plane and vertical distribution statistics of the block. Essence Using the seismic section interpretation technology, multi -attributes and detailed portrayal of the broken layer, implement the overall structural characteristics and TEZHEN characteristics of the block, carry out multi -attribute analysis and research on the oil layer sandstone group, preferably seismic attributes sensitive to the development of sandstone, and obtain each through the calculation of the well -shaking combination calculation. Unit sandstone thickness quantitative prediction diagram, analyze the characteristics of the oil and water distribution of the block, combine the favorable positions of the well deployment, combine the characteristics of the structure, the break, and the sandy body distribution, and prefer the dessert area. The original large sedimentation law understands that the "single structure favorable oil control zone" is required to re -evaluate the high -risk zone, and combine the fine structure interpretation with the sand body reservoir prediction. Improve the success rate of the evaluation control well in the dessert area, and follow the deployment of the block well after the guidance.
Keywords: Evaluation effect,Structural strip zone,Tibetan type,

## 1. Introduction

The M9 block structure is located in the northwest of the depression in the center of the center of the northern part of the Songliao Basin. In the continuous improvement of the theory of Chengzang, the continuous renewal of the research technology, the continuous expansion of the evaluation area, the proven reserves area of the P oil layer has basically covered the main potential area of the main potential area of the non -proven area. The M9 block blank area is naturally the main target area of the next exploration and development. The geological conditions of the target area are very complicated. In the past, the evaluation control well had many defeats. After the drilling evaluation control well controlled 42 ports, 17 oil test wells, 8 industrial oil flow wells, large high -risk areas, the original large deposition laws knew the general understanding of general, But risks also mean potential. As the successful deployment experience is continuously enriched, it will be of great significance to expand the reserves of the target area in the next step.

Through the explanation of the fine structure of the block, identify the micro -amplitude structure, implement the overall structural characteristics and fracture characteristics of the block, carry out multi -attribute analysis and research on the oil layer sandstone group, preferably seismic attributes sensitive to the development of sandstone, and obtain each through the calculation of the integration of the well seismic. Unit sandstone thickness quantitative prediction diagram, analyze the characteristics of the oil and water distribution of the block, and combine the characteristics of the structure, the break, and the distribution of the sand body, which can prefer the dessert area.

## 2. Fine geological research level

### *2.1. Layer explanation*

There are 202 drilled ports in the M9 area, including 23 scratches, 127 oil wells, 35 wells, 10 air wells, and 7 scrap wells. Careful sorting out the data in the area, there are 190 wells in 202 wells with sound wave curves, with the condition of synthetic record calibration. Based on the completion of the 190 well synthesis record calibration and the combination of well -shakes, the shaft layer and earthquake reflection are achieved. The features are further unified. Select the results of the typical well -complex seismic records, compare multiple wells, analyze the seismic section of the well, and analyze the profile. The impact of division is equal, phase stability and continuity are general, and the reflection characteristics of the earthquake reflection on the top surface of each unit inside the P oil layer are not obvious.

According to the characteristics of its earthquake, first determine the section of the maniform grid section explained in the region, comprehensively synthesize the calibration geological layered position, and explain and close the grid section one by one. Based on this In areas where the axis continuous and stable, the three -dimensional automatic tracking method is used to explain. In the position with poor phase stability, the combination of automatic tracking and manual interpretation of the section is used to finally complete the two layers of the M9 block P top surface.

### *2.2. Fault explanation*

The M9 blockbuster layer has a small vertical span and a relatively simple structure. The main seismic characteristics of the faults are reflected waves and the shafts of the same phase shaft. By view of the attribute body that can reflect the characteristics of the boundary characteristics, It is easy to establish the spatial concept of the fracture layer, such as the scale, location, direction, combination and other information. Browse and positioning by slicing and positioning of the animation and horizontal animation, the characteristics of the comparative faults on different data can be more accurately captured by the true information of the fault, and reasonably reproduce the original appearance of the fault.

### *2.3. Oil and water distribution*

The M9 block can divide the blocks in the direction of the east and west directions in a structural form into several narrow strips near the north and south. On the basis of the above -mentioned constructive unit, the statistical block has been drilled 183 in the block, and the P oil layer is divided into 9 small layers and 13 time units, and the oil levels and vertical distribution of each unit are analyzed. The PX area is mainly distributed in the southwest nasal structure belt, the central area of the central area, the northwest nasal constructor belt, and the central part of the central part of the central part. In addition to the individual layers of the northern part of the work area, the central area, and the northeast nasal structure belt of the northeast, there is basically no oil and gas enrichment. Therefore, it can be roughly determined that the oil and gas distribution of the research area has the characteristics of the south and the north. Affected by the "separation groove" effect of the ground band, there is basically no oil and gas enrichment in the central ground belt and northern nasal structure zone in the central area, and it also shows that the oil and gas distribution of the work area has the characteristics of west good.

The WX region is located in the slope area of the northern depression in S. From the vertical oil and gas distribution reflected by the section of the oil collection, the overall manifestation is the influence of the oil and the water, the broken layer cutting and the rocky changes. A single bureau's circles are mainly controlled, and the types of oil collection are mostly fault -rock composite collection.

It can be seen through the vertical oily statistics of each unit, which can be seen that the area has the characteristics of oil and water. The upper part of the P oil layer is mainly based on the oil layer and the same layer of oil and water. The vertical distribution of the oil layer in the WX area has the characteristics of double peaks. On the upper part of the oil layer, the highest number of oil and water is Ps52 and Ps8 units; the water layer gradually increases from top to bottom.

## 3. Analysis of deployment results and oil content potential

The area is broken and developed in the area, and the structure and development of the structure and the development of the four types of structure are closed, namely the nasal breaking structure, the block structure, the broken back oblique structure, and the micro -amplitude structure. Oil and gas collection is mainly controlled by the sedimentary belt and closed development, forming a composite collection. Combining structure, fracture, sandy body matching conditions, and oil and gas distribution rules, the type of P oil layer can be divided into 3 categories: the fault-rock composite collection, the structure-rock composite oil collection, and the rock collection.

The oil and gas currently discovered is mainly concentrated in the northwest nasal structure belt, the nasal nasal structure zone in the southwestern part, the medium -slope zone in the middle, and the southeast side of the central area.

In accordance with the laws of oil and water distribution and the distribution of oil collection in various districts, comprehensive use of research results such as deposition, reservoir, structure, anti - discovery, well measurement explanation, and seismic attributes, and comprehensively evaluate the P oil layer. Evaluation control well deployed in the western near -nasal structure zone and the in - favorable area in the middle of the medium -slope zone.

The H9 well area, the type of block oil is mainly rock-constructed, and the high parts of the high parts are constructed in the low range, the prediction of the sand body development area deployment evaluation control well F6, F6 oil test oil 35.604t/d , F7 well drill 25m in sandstone, and the drill is effective 5.5m.

The H3 well area, the type of oil collection is mainly rock-constructing oil collection. In the constructing high parts, the back of the rock development zone 23-1, the single well diamond encounters a valid thickness of 2.6m/1 layer, the oil trial oil Nissan 11.082T, according to the experience of typical success wells, deploy G1 wells in sandstone development and relatively favorable parts (expanded outside the proven area), drill the valid thickness of 2.7 meters/1 layer, the test oil Nissan 16.38t.

In the H10 well, the block is adjacent to the west to oblique. The top sandstone group occurs mostly on the oil layer. The H10 well test oil in the micro -amplitude constructed ring is high -yield. Therefore, finding a favorable constructor in the area is the main direction of deploying wells. The two micro -amplitude structures that are preferably constructed with relatively favorable positions and sandy body development are closed, and 2 portfolio evaluation control well A1 and B2, control the oil content 0.80km2, A1 oil Nissan 15.7t.

The C3 well area is located in the middle of the ground, and the overall opening is low -end opening. The east, west, and north are surrounded by faults. There are multiple micro -amplitude structures in the area. The validity thickness is 1.8m, Nissan oil is 19.0T/d, Nissan water 2.8m3, the block ring is closed D4, the PI3 layer is valid 1.7m, the Nissan oil is 18.1 t/d, preferably, micro - amplitude structure ring closed, in PI1, in PI1 -3, deploy 2-port evaluation control wells on the better part of the PI6-9 layer and the tap cut sand body area, the closed range of 8m, control the oil-containing area of 2.07km2, the geological reserves of 260,000 tons, the E5 drill is effective, the drill is effective The thickness is 2.7m/1 layer.

## 4. Conclusion

Multi -attribute comprehensive detailed resolution of the broken layer, seismic section interpretation technology, and encrypted equal value lines are finely constructed as diagram. Identification of micro - amplitude structure ring is an important direction for fine structure interpretation.

The M9 block oil and gas distribution has the characteristics of the north -good north and the west good east. Gradually increase.

The original large sedimentation law understands the general understanding. It is necessary to find a "single structure favorable oil control zone" to re -evaluate the high -risk zone. Combining the fine structure interpretation and the prediction of the sand storage layer will be of great significance to the deployment of the evaluation well.

**References**

[1] Deng Yunhua. Fracture-Sand Body forms a "transit station" mode of oil and gas movement [J]. China Petroleum Exploration, 2005, 6: 14-17.

[2] Liu Zerong, Xin Quanlin, Deng Junguo, etc. The formation mechanism and structural mode of the flock group oil gas collection [M]. Beijing: Petroleum Industry Press, 1998, 16-23.

[3] Cao Hui. Some of the understanding of the application of earthquake attributes [J]. Explore the physical progress of the earth, 2002,25 (5): 18-22.

[4] Zhu Xiaomin, Liu Changli, Zhang Yina, et al. The application of earthquake deposits in the prediction of the sand body prediction of the Lake Basin Delta [J].

[5] Ethics - Science and Engineering Ethics; Reports Outline Science and Engineering Ethics Findings from University of Malaya (Ethical Concerns About Human Genetic Enhancement in the Malay Science Fiction Novels)[J]. Journal of Engineering.

[6] T. Malay Peninsula &c.[J]. Harris Will. Ambit.

[7] Sex-dependent differences in the stress mitigating and antidepressant effects of selective aryl hydrocarbon receptor modulators.[J]. Madison Caitlin A,Debler Roanna A,Vardeleon Nathan I,Hillbrick Lauren,Jayaraman Arul,Safe Stephen,Chapkin Robert S,Eitan Shoshana. Journal of affective disorders.

[8] Alfred Russel Wallace: letters from the Malay Archipelago[J]. Choice Reviews Online.

[9] Petronas Kicks Off US$5bn North Malay Basin Project[J]. Asia Pacific oil and gas insight Group. Asia Pacific oil and gas insight.

[10] Application of Mango's C 7 hydrocarbon parameters to the correlation of oil in the western Niger Delta, Nigeria[J]. Chidi Eneogwe,Olusegun Ekundayo. Journal of Mining and Geology.

[11] Evaluation of source rock potential and hydrocarbon composition of oil sand and associated clay deposits from the Eastern Dahomey Basin, Nigeria[J]. Saeed Mohammed,Mimonitu Opuwari,Salam Titinchi,Timothy Bata,Mohammed B. Abubakar. Journal of African Earth Sciences.

[12] A target-oriented integrated workflow for seismic delineation of thin tight reservoirs in the eastern Sichuan Basin, China[J]. Luo Xin,Chen Xuehua,Zhang Jie,Li Kangyi,Lv Bingnan,Wen Huaguo. Journal of Petroleum Science and Engineering.

[13] Fluid energy characteristics and development potential of coalbed methane reservoirs with different synclines in Guizhou, China[J]. Zhengguang Zhang,Yong Qin,Zhaobiao Yang,Jun Jin,Congcong Wu. Journal of Natural Gas Science and Engineering.

[14] Geochemical features and genesis of the natural gas and bitumen in paleo-oil reservoirs of Nanpanjiang Basin, China[J]. MengJun Zhao,ShuiChang Zhang,Lin Zhao,Jiang Da. Science in China Series D: Earth Sciences.

[15] Depositional environment and hydrocarbon source potential of the Oligocene Ruslar Formation (Kamchia Depression; Western Black Sea)[J]. R.F. Sachsenhofer,B. Stummer,G. Georgiev,R. Dellmour,A. Bechtel,R. Gratzer,S. Ćorić. Marine and Petroleum Geology.

[16] Differential hydrocarbon enrichment in deep Paleogene tight sandstones of the Dongpu Depression in Eastern China[J]. Zhang Yuanyuan,Ren Zhanli,Jiang Youlu,Liu Jingdong. Energy Exploration & Exploitation.

[17] Study on fault-controlled hydrocarbon migration and accumulation process and models in Zhu I Depression[J]. Zhu Wenqi,Wu Keqiang,Ke Ling,Chen Kai,Liu Zhifeng. Acta Oceanologica

Sinica.
[18] Identification of effective source rocks in different sedimentary environments and evaluation of hydrocarbon resources potential: A case study of paleogene source rocks in the Dongpu Depression, Bohai Bay Basin[J]. Zhu Chenxi,Jiang Fujie,Zhang Pengyuan,Hu Tao,Liu Ying,Xu Tianwu,Zhang Yun-xian,Deng Qian,Zhou Yongshui,Xiong Hang,Song Zezhang. Journal of Petroleum Science and Engineering.
[19] The potential of shear - wave VSPs for monitoring recovery: A letter to management[J]. Stuart Crampin. The Leading Edge.
[20] Long-Term Production Testing Improves Reservoir Characterization in the Oseberg Field[J]. Fjerstad, Paul A.,Flaate, Svein A.,Jens, Hagen,Steinar, Pollen. Journal of Petroleum Technology.

# Research and application of anti - corrosion cement slurry in formation water corrosion environment

**Jiang-ben Min(✉), Jian Zhang, Xiang-chen Wan, Xiao-rong Chen, Wen-bin Wang, Kai Tang**

1. CCDC Drilling Engineering Technology Research Institute Xi'an, Shaanxi
2. National Engineering Laboratory for Exploration and Development of Low Permeability Oil and Gas Fields, Xi'an, Shaanxi

447559798@qq.com

**Abstract.** Corrosion of cementing cement ring causes the cement ring permeability to increase, resulting in serious casing corrosion. The casing corrosion perforation leads to an increase in water content of oil wells, low driving use of water injection wells, and imbalance of injection and production well patterns. Therefore, on the basis of conventional cementing cement, epoxy resin emulsion non-penetrating agent and corrosion inhibitor composed of rust and antibacterial materials are added to reduce the corrosion and permeability of cementing cement and reduce casing corrosion. The cement sample permeability experiment shows that adding 10% non-permeability agent and 2% corrosion inhibitor to low-density cement reduces the cement stone sample permeability from 0.075mD to 0.021mD, a decrease of 3.57 times. In the conventional density cement, adding 10% non-penetrating agent and 2% corrosion inhibitor, the permeability of the cement stone sample decreased from 0.041mD to 0.0009mD, a decrease of 4.56 times. The evaluation experiment of cement stone anti-corrosion performance shows that adding 10% non-penetrating agent and 2% corrosion inhibitor to cement can increase the corrosion resistance of cement samples by 6 times. At the same time, the thickening time and compressive strength of cement stone samples remain unchanged compared with conventional cement. Optimizing the technological measures of supporting anti-corrosion tools, and summing up a set of anti-corrosion cementing technology through the field application of anti-corrosion cement slurry, which provides strong technical support for the development of cementing technology under the corrosive environment of formation water in the oil field.

## 1 Introduction

With the continuous exploration and development of oilfields, the problem of casing damage in oil and water wells caused by formation water corrosion has become more and more serious, and the number of casing damage wells has continued to increase. According to statistics, there are more than 500 new casing corrosion damage wells in Changqing Oilfield each year. The casing service life is only 9.7 years. Casing damage leads to an increase in water cut of oil wells, low driving use of water injection wells, and imbalance in the injection-production well pattern, which seriously affects the development effect of the oilfield and the degree of reservoir production [10]. The casing is corroded seriously. After the casing is broken, the formation water flows back to form a flooded well. In addition, the water-type incompatibility is mixed with the formation water and scales to form a blockage, which increases the oil flow resistance and causes isolation failure, and the oil well is reduced or even shut down. In order to increase crude oil production and restore single well productivity, major oilfield companies have implemented casing subsidies[1], chemical squeezing[2], small casing secondary cementing[3] and other treatment measures for casing damaged wells. , Field tests show that casing subsidies have problems such as long construction period, high operating cost, and reduced wellbore diameter. Chemical extrusion has the problems of low extrusion success rate and short effective period of measures. Although small casing secondary cementing can be achieved Wellbore rebuilding, but the quality of cementing with narrow annulus gaps is difficult to effectively guarantee. Other related measures and technologies are limited to the anti-corrosion protection of newly run casings and the repair of old casings after damage, and the extent of improving casing corrosion is limited.

On the other hand, engineers and technicians have done a lot of research work on the management of casing damaged wells. In addition to the improvement of the above-mentioned engineering and technical measures, they also consider improving the cementing quality and improving the cement performance to enhance the sealing of the cement ring. Integrity to resist erosion by formation water. The CNPC Drilling Institute conducts mechanical analysis on the casing-cement ring-stratum combination, and uses whether the maximum differential stress experienced by the cement ring exceeds the compressive strength of the cement stone as the condition for determining the seal failure. On this basis, the seal safety factor method is proposed. The sealing ability evaluation of cement ring provides a quantitative method. Evaluation and analysis show that under the premise of ensuring the strength of cement stone, reducing the elastic modulus can effectively improve the sealing ability of the cement ring. If the strength is too low, even if the elastic modulus is very low, it cannot meet the sealing requirements of the cement ring. The strength is high enough to meet the sealing requirements even if the elastic modulus is high. Therefore, the sealing ability of cement stone depends on its sealing safety factor, rather than a single strength or elastic modulus index. Therefore, "high strength low modulus" is proposed as the guiding idea of toughness transformation [4].

Schlumberger developed the Flex Stone flexible cement slurry system in order to prevent micro-annular gaps from downhole stress factors. The system is based on the principle of solid phase particle grading, combined with flexible particles and expansion agents, etc., the Young's modulus can be reduced to 1.38GPa, the linear expansion rate can reach 2%, the applicable temperature range is

40℃~150℃, and the density range is 1.2~2.2g/cm$^3$, the system has good application effects in foreign underground gas storage, multi-level perforation and fracturing stimulation wells, and steam flooding heavy oil thermal recovery wells. Daqing Drilling and Drilling Research Institute independently developed ductile cement. The indoor evaluation of 5cm cement column gas channeling pressure is 0.51MPa, liquid channeling pressure is 4.88MPa, 10cm cement column gas channeling pressure is 0.89MPa, liquid channeling pressure is greater than 7MPa, compressive strength It has reached 44MPa, the elastic modulus is 3.8GPa, and the strain amount is 1.82%. The field test has been completed and the effect is good. CNPC Drilling Institute has developed a medium and high temperature ductile expansion cement slurry system, which can reduce the Young's modulus by more than 30% while ensuring compressive strength. This system has been applied in gas storage, shale gas horizontal wells, high pressure natural gas wells, etc. More than 50 wells have good results. CNPC Boxing has developed an elastic cement slurry system. The 7d compressive strength of the system is 22-30MPa, the tensile strength is 1.8-2.2MPa, and the Young's modulus is 5.0-6.0GPa. Chuanqing Drill Probe has developed a flexible self-stressing cement slurry system. The applicable temperature of this system is 50℃~150℃. The Young's modulus of cement stone is reduced by 20% and the compressive strength is reduced by 20% at 90℃ and 7d. , And the interface bonding strength is significantly improved. The application effect of this system in Xiangguosi gas storage is good. Sinopec Engineering Research Institute has developed an elastic-flexible cement slurry system. The flexural strength of the system can be increased by more than 40% and the bending toughness can be increased by more than 90% compared with the original slurry. Currently, more than 100 wells have been applied in Fuling shale gas horizontal wells with good results. Schlumberger developed FUTUR self-healing cement, which has a temperature range of 20℃~138℃ and a density of 1.44~1.92g/cm3. It can react with oil and methane whose concentration is lower than 95% to repair microcracks. , To prevent oil and gas from going up. The Life Cem TM cement slurry system developed by Halliburton Company contains activating components of hydrocarbons, which can react with hydrocarbon fluids in the formation. It has a self-expansion capacity of 1% to 2% and can repair microstructures of 100 to 250 μm. Cracks and micro annulus [5-9].

In summary, a large number of studies have focused on improving the sealing integrity of the cement ring from the direction of improving the mechanical properties of cement, but there are few reports on the corrosion resistance of cement itself. Therefore, according to the corrosion mechanism of formation water on the casing, this article proposes a new method to improve cementing cement ring corrosion and improve its own corrosion resistance to reduce casing damage. Aiming at the need for cement ring corrosion protection technology, on the basis of conventional cementing cement, Carry out the research and development of oil well cement non-penetrating agent, corrosion inhibitor key admixture, formed a set of anti-corrosion cement slurry system suitable for formation water corrosion environment, and effectively improved the problem of casing damage caused by cement ring corrosion.

## 2 Corrosion mechanism of casing and cement ring

### *2.1 Analysis of formation water corrosion environment*

Taking Changqing Oilfield as an example, there are corrosive water layers in Yijun, Luohe, Huachi, and Huanhe River from bottom to top. The main corrosive water layers have large water volumes and long water layer spans. The largest, most corrosive, rich ion content, high salinity, there are a lot of $SO_4^{2-}$, $Cl^-$, $HCO^{3-}$, $Na^+$, $K^+$, $Ca^{2+}$, $Mg^{2+}$ plasma, see Table 1, which shows that it is in the Luohe layer and moving fluid The corrosive environment between the surface layers is very serious.

Table 1 Luohe Water Quality Analysis Table of Changqing Oilfield

| Water source | pH | Cation index（mg/L） | | | Anion index（mg/L） | | | Total salinity（mg/L） | $CO_2$（mg/L） | Sulfate reducing bacteria（A/L） |
|---|---|---|---|---|---|---|---|---|---|---|
| | | $Na^++K^+$ | $Ca^{2+}$ | $Mg^{2+}$ | $Cl^-$ | $SO4^{2-}$ | $HCO_3^-$ | | | |
| Luohe Formation | 6.65～7.0 | 557.8～1721.6 | 30.46～415.8 | 206.1～1800 | 273～1228 | 677～1924 | 131.15～329.4 | 4918～6562 | 24～68 | 92～105 |

*2.2 Cause analysis of cement ring corrosion*

There are two main reasons for the corrosion of cement ring by formation water: water dissolution and ion exchange. Reason 1: The cement ring acts as an external barrier to support the protective sleeve. It is in contact with formation water for a long time, and the cement stone is eroded by formation water for a long time to cause dissolution; Reason 2: Formation water is rich in corrosive ions, these ions will be replaced with $Ca(OH)_2$ in cement[5], $Ca(OH)_2$ crystals will dissolve, cement stone structure will be destroyed, and the higher the salinity of the formation water , The lower the cement stone permeability, the faster the cement dissolution rate [11].

*2.3 Cause analysis of casing corrosion*

There are many reasons for casing corrosion. From the perspective of cementing, the reasons for corrosion can be divided into the following two points: ① Cementing quality factors, the Luohe formation has low pressure bearing capacity, and leakage is prone to occur during cementing, which often results in insufficient cement slurry return. As a result, there is no cement filling protection in the upper casing section, the annulus is not filled with cement, and the outer wall of the casing is exposed in the formation water rich in corrosive ions, which directly corrodes the casing; ②Cement factor, cement can return to the ground to isolate the Luohe water However, due to the cement slurry system, the cement stone formed after cement solidification has low compressive strength, poor compactness, and high permeability. Over time, the cement stone structure is damaged, and the high-salinity formation water passes through the cement ring and contacts the casing to corrode .

The formation water has a complex corrosion environment and multiple corrosive effects coexist. The formation water contains a lot of corrosive ions. When the cement ring is structurally damaged, the outer wall of the casing directly contacts the formation water, which provides conditions for electrochemical corrosion. At the same time, the water also contains a small amount of $CO_2$. And sulfate-reducing bacteria accelerate the biochemical reaction of casing corrosion, which is an important reason for the violent flooding caused by the Luo River water inversion after casing corrosion damage. In order to solve the problem of formation water casing damage, one of the important tasks is to solve the problem of cement ring corrosion.

## 3 Anticorrosive cement slurry design

### *3.1 Design method of anticorrosive cement slurry*

According to the previous research and analysis on the corrosion mechanism of casing and cement ring by formation water, combined with the cementing "compression stability and leakage prevention" criterion, the anti-corrosion cement slurry design method mainly considers the following aspects: ①In order to prevent cementing leakage, effectively reduce the collar For the density of cement slurry, a closed-cell anti-channeling perlite and hollow glass microspheres are selected as the composite lightening agent in the room, and the low-density and high-strength cement slurry basic system is designed according to the performance requirements of the cement slurry; ②The thickening time of the tail slurry is reasonably controlled to improve the cement The early strength of the cement paste can effectively improve the anti-water invasion ability of the cement paste and realize the high-quality cementation of cement paste; ③The key to the anticorrosion of cement paste is to improve the compactness and corrosion resistance of cement paste, without reducing the basic performance of cement paste. The introduction of self-developed non-permeable agents and corrosion inhibitors, non-permeable agents can effectively improve the cement stone matrix anti-corrosive ion migration performance, reduce the ability of corrosive ions to penetrate the cement stone, and form an isolation barrier between the casing and the formation; The corrosion inhibitor in the agent can be adsorbed on the surface of the metal to inhibit the anode and cathode reactions separately or at the same time, thereby reducing the corrosion current in the corrosion process, and achieving the purpose of corrosion inhibition; ④The Luohe water inverted after the cover is broken, and the lower formation water Mixed, the lower casing and cement also face the risk of corrosion, so it is necessary to also carry out anti-corrosion design for the sealing cement of the lower well section.

### *3.2 Optimization of non-penetrating agent for anticorrosive cement slurry*

Under the premise of not reducing the basic performance of cement slurry, by adding epoxy resin, which is a material that improves the density of cement stone, is a non-penetrating agent. This type of material has a certain film-forming property and can effectively improve the density of the matrix, so it is corroding In terms of prevention and treatment, it can effectively prevent the penetration of corrosive medium fluids [14]. Through the addition of epoxy resin（Figure 4）, the epoxy resin cement slurry is cross-linked to form a three-dimensional structure, which gives the epoxy resin cement slurry system excellent mechanical properties. The particles of the thermosetting resin cement slurry solidified body are tightly surrounded and improve the cement compactness At the same time, it can also give toughness to the cement stone of the oil well, reduce the damage caused by the stress concentration of the cement ring after the impact force, and improve the treatment and development life of casing damaged wells.

Figure 1 Unsaturated polyester

Figure 2 Urea formaldehyde resin

Figure 3 Phenolic Resin

Figure 4 Epoxy resin

*3.3 Optimization of cement slurry corrosion inhibitor for anticorrosive cementing*

Through investigation and analysis, it is found that the current anti-corrosion performance of cement slurry is mainly improved in the following three aspects: ①reducing the content of alkaline substances in the cement stone; ②restricting the penetration process to inhibit corrosion chemical reactions; ③improving the original compressive strength [14-16]. And most of the reported antiseptic materials are designed for cement slurry anti-corrosion in $CO_2$ or $H_2S$ corrosive environment, while the corrosion environment of Luohe water layer in Changqing Oilfield is mainly high salinity formation water and a small amount of $CO_2$ and sulfate reducing bacteria. The mechanism of action is different, therefore, the selection and design of anti-corrosion materials that need to be carried out are also different.

In this paper, based on the theory of electrochemical corrosion inhibition and film formation, drawing on the anticorrosion technology of reinforced concrete and introducing corrosion inhibitors [18], a composite oil well cement corrosion inhibitor (a low-density cementing slurry system for anticorrosion and its production

Preparation method, patent number: 201710200346.X), which is composed of three parts: anode type corrosion inhibitor, cathode type corrosion inhibitor and sulfate reducing bacteria antibacterial components, specifically sodium nitrite, calcium nitrite, sodium phosphate and One or several combinations of sodium benzoate. The mechanism of action is as follows: ① Anode-type corrosion-inhibiting materials and cathode-type corrosion-inhibiting materials act on the anode area and the cathode area respectively, and increase the polarization by forming a passivation film or generating substances that are difficult to dissolve in water, thereby increasing the cathode , The resistance between the anodes inhibits the electrochemical reaction and achieves the purpose of corrosion protection. ② Sulfate-reducing bacteria (SRB) is the main corrosive microorganism. Its main characteristic is that it multiplies under the sediment and scale in the pipeline, causing serious local corrosion of the pipe string [19]. The antibacterial components of sulfate-reducing bacteria can inhibit the growth of bacteria by interfering with the permeability of microbial cell membranes, hindering the adsorption of amino acids, and reducing the activity of enzymes to achieve the purpose of inhibiting the corrosion of sulfate-reducing bacteria.

*3.4 Evaluation of basic performance of anticorrosive cement slurry system*

According to the anti-corrosion cement slurry design method, combined with the research and development of key admixtures, and based on the principle of close packing, the particle gradation is reasonably controlled, and the cement slurry rate, compressive strength, stability, rheology, and permeability are comprehensively considered through indoor experimental research. The basic performance of the cement slurry is determined by determining the amount of the key admixtures to form a basic formulation system of low-density and conventional-density cement slurry for anticorrosion cementing. The basic performance of the cement slurry system is shown in Table 2.

Table 2 Performance table of basic formula of anticorrosive cement slurry

| formula | Water-cement ratio | density (g/cm$^3$) | Water separation (%) | Compressive strength 45℃ /MPa | Permeability mD | Rheological parameters | | Thickening time60℃/30MPa | |
|---|---|---|---|---|---|---|---|---|---|
| | | | | | | n | K($Pa\cdot s^n$) | First thick (BC) | time/min |
| 1 | 0.65 | 1.25 | 0 | $P_{24h}$-8.3; $P_{48h}$-11.8 | 0.075 | 0.76 | 0.39 | 14 | 176 |
| 2 | 0.44 | 1.88 | 0 | $P_{4h}$-5.8; $P_{24h}$-26.5 | 0.041 | 0.85 | 0.32 | 17 | 55 |

formula1: G+Compound lightening agent+Micro silicon+Superfine cement+Reinforced material+Fluid loss agent

formula2: G+Fluid loss agent+Early Strength Agent+Anti-channeling agent+Dispersant

It can be seen from Table 2 that the basic formulation of the anticorrosive cement slurry is a composite hollow glass microbead cement slurry system and a fluid loss-reducing early-strength anti-channeling cement slurry system. This research introduces key anticorrosive materials on the basis of the formulations 1 and 2 in Table 2 , That is, non-penetrating agent and corrosion inhibitor, the comprehensive cost and the amount of anti-corrosion material will affect the basic performance of cement slurry. The recommended amount of non-penetrating agent and corrosion inhibitor is 10% and 2% respectively. The performance of the formed anti-corrosion cement slurry system is shown in the table 3 and Table 4, the system is stable without water separation, high compressive strength, dense and impermeable cement, and adjustable thickening time, which can meet the cement anti-corrosion requirements of on-site cementing construction.

Table 3 Anticorrosive low cement slurry formula debugging performance table

| formula | Water-cement ratio | density (g/cm³) | Compressive strength 45℃/MPa | Permeability mD | Thickening time60℃/30MPa | |
|---|---|---|---|---|---|---|
| | | | | | First thick (BC) | time/min |
| formula1+5%non-penetrating agent | 0.65 | 1.25 | $P_{24h}$-8.0; $P_{48h}$-11.2 | 0.056 | 13 | 179 |
| formula1+10%non-penetrating agent | 0.65 | 1.24 | $P_{24h}$-7.8; $P_{48h}$-10.7 | 0.021 | 12 | 185 |
| formula1+1%Corrosion inhibitor | 0.65 | 1.25 | $P_{24h}$-8.2; $P_{48h}$-11.7 | 0.075 | 14 | 177 |
| formula1+2%Corrosion inhibitor | 0.65 | 1.25 | $P_{24h}$-8.0; $P_{48h}$-11.1 | 0.075 | 14 | 180 |
| formula1+10%non-penetrating agent+1%Corrosion inhibitor | 0.65 | 1.24 | $P_{24h}$-7.5; $P_{48h}$-10.3 | 0.021 | 12 | 188 |

Table 4 Anticorrosive conventional cement slurry formula debugging performance table

| formula | Water-cement ratio | density (g/cm³) | Compressive strength 45℃/MPa | Permeability mD | Thickening time60℃/30MPa | |
|---|---|---|---|---|---|---|
| | | | | | First thick (BC) | time/min |
| Formula2+5%non-penetrating agent | 0.44 | 1.88 | $P_{4h}$-5.7; $P_{24h}$-26.0 | 0.023 | 17 | 56 |
| Formula2+10%non-penetrating agent | 0.44 | 1.87 | $P_{4h}$-5.5; $P_{24h}$-25.2 | 0.009 | 16 | 59 |
| Formula2+1%Corrosion inhibitor | 0.44 | 1.88 | $P_{4h}$-5.6; $P_{24h}$-26.3 | 0.041 | 17 | 56 |
| Formula2+2%Corrosion inhibitor | 0.44 | 1.88 | $P_{4h}$-5.3; $P_{24h}$-25.8 | 0.041 | 17 | 58 |
| Formula2+10%non-penetrating agent+1%Corrosion inhibitor | 0.44 | 1.87 | $P_{4h}$-5.1; $P_{24h}$-25.0 | 0.009 | 16 | 60 |

It can be seen from Table 3 and Table 4 that the key admixture of anticorrosive cement slurry, and the main component of non-penetrating agent is epoxy resin emulsion, which is compatible with water and has strong adhesion, and the solidified body formed by combining with cement has good stability , Low permeability, acid and alkali corrosion resistance. The corrosion inhibitor used is a composite corrosion inhibitor, which is composed of rust-inhibiting materials and anti-bacterial materials. It has little effect on cement thickening time and compressive strength. Its corrosion-inhibiting mechanism is mainly formed in the anode or cathode. The chemical film is adsorbed on its surface and acts separately or at the same time to increase the resistance between the anode and the anode and reduce the electrochemical reaction speed; at the same time, add sulfate-reducing bacteria antibacterial ingredients to destroy the growth environment of sulfate-reducing bacteria and inhibit the growth of bacteria. Effectively improve the ability of cement to resist biological corrosion.

## 4 Evaluation of anticorrosion performance of anticorrosive cement slurry system

The anticorrosion performance evaluation of the anticorrosive cement slurry system takes the anticorrosive low density cement slurry system as an example to study. The composite hollow glass bead low-density cement slurry system contains a large amount of micron $SiO_2$ and ultrafine Portland cement particles. The average particle size of these particles is about 0.15-10 $\mu m$ , and they are evenly distributed in the gaps between the cement particles. Reduce the generation of capillary channels, form impermeable cement, introduce non-permeable agents and corrosion inhibitors on the basis of the existing low-density cement slurry, and further increase the compactness and corrosion resistance of cement.

### *4.1 Evaluation of Corrosion Inhibition Performance of Anticorrosive Low Density Cement Slurry*

The test adopts the classic weightlessness method, using the formation water samples taken from the Luohe Formation on site, and treating the N80 steel sheet according to the SY/T 5273-2000 method for evaluating the performance of corrosion inhibitors for oil field produced water and weighing and recording [20], the N80 test Place the sheet on two rubber strips with a height of 1 cm, slowly pour the cement paste into the cement stone mold, put it in the mold and the curing conditions meet the requirements of GB/T 19139-2012 [21], after curing fo r 2 days, it will be molded and demolded. Put the cement stone in a 1000ml sealed glass container containing 500g Luohe water corrosive solution for 28 days, take out the N80 steel sheet, dry and degrease dehydrated, weigh it, and evaluate the anticorrosive low density cement slurry by calculating the mass loss rate of the N80 steel sheet Anti-corrosion performance of the system （Figure5-8）. Table 5 shows the anti-corrosion performance of anticorrosive low-density cement slurry in different corrosive media.

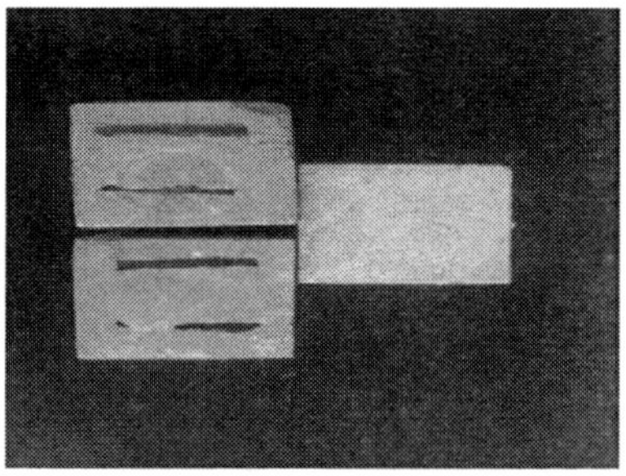

Figure5 Corrosion performance evaluation mold photo Figure6 After curing, there are pictures of coupons and test blocks without coupons

Journal of Physics: Conference Series **2594** (2023) 012031 doi:10.1088/1742-6596/2594/1/012031

Figure7 Pouring process

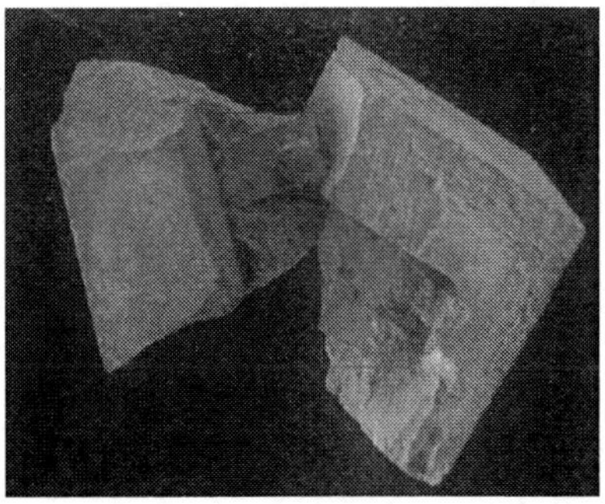

Figure8 N80 test piece in the broken cement stone

Table 5 Corrosion inhibition rate of anticorrosive low-density cement slurry in different corrosive media

| NO. | Media type | Corrosion inhibitor% | Steel sheet quality of N80 g | | | Corrosion inhibition rate% |
|---|---|---|---|---|---|---|
| | | | Before the test | After the test | Quality loss | |
| 1 | Tap water | 0 | 8.3316 | 8.3261 | 0.0055 | / |
| 2 | Luohe Water | 2 | 8.3946 | 8.3938 | 0.0008 | 84.45 |
| 3 | 10 times Luo River water | 2 | 8.4127 | 8.4118 | 0.0009 | 83.63 |
| 4 | 20 times Luo River water | 2 | 8.3245 | 8.3235 | 0.0010 | 81.81 |

The corrosion inhibition rate of corrosion inhibitor $\lambda$ is calculated using the following formula:

$$\lambda = \frac{\Delta m_1 - \Delta m_2}{\Delta m_1} \times 100\% \quad (1)$$

Among $\Delta m_1$ is the mass loss of N80 steel sheets cured in Luohe water when no corrosion inhibitor is added, g; when the corrosion inhibitor is added to 2%, $\Delta m_2$ the mass loss of N80 steel sheets cured in Luohe water with different concentrations, g.

It can be seen from Table 5 that in the low-density cement without corrosion inhibitor, N80 steel sheet has the largest quality loss. After adding 2% corrosion inhibitor, the corrosion inhibition effect is obvious, the quality loss is greatly reduced, and the quality loss is reduced by 6 times. The corrosion resistance is increased by 6 times. The anticorrosive low-density cement slurry with corrosion

Journal of Physics: Conference Series **2594** (2023) 012031 doi:10.1088/1742-6596/2594/1/012031

inhibitor added to Luohe water of 10 times and 20 times concentration can still maintain a high corrosion inhibition efficiency.

*4.2 Evaluation of compressive strength and permeability of anticorrosive low-density cement stone*

In order to understand the changes in the compressive strength and permeability of the anticorrosive low-density cement stone under long-term soaking conditions, the anti-corrosion low-density cement stone without N80 steel flakes was put into Luohe water with different concentrations and cured for 7d, 14d, 28d, and tested. The changes in compressive strength and permeability, the test results are shown in Figure 9 and Figure 10.

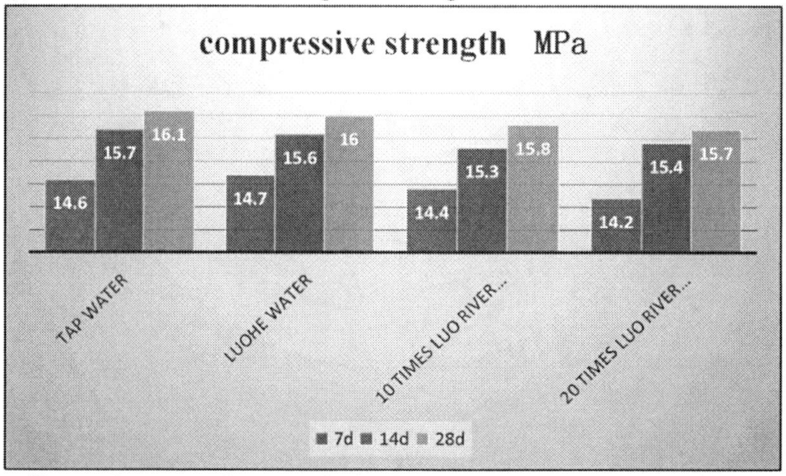


Figure9 Changes in compressive strength of anticorrosive low-density cement slurry in different corrosive media

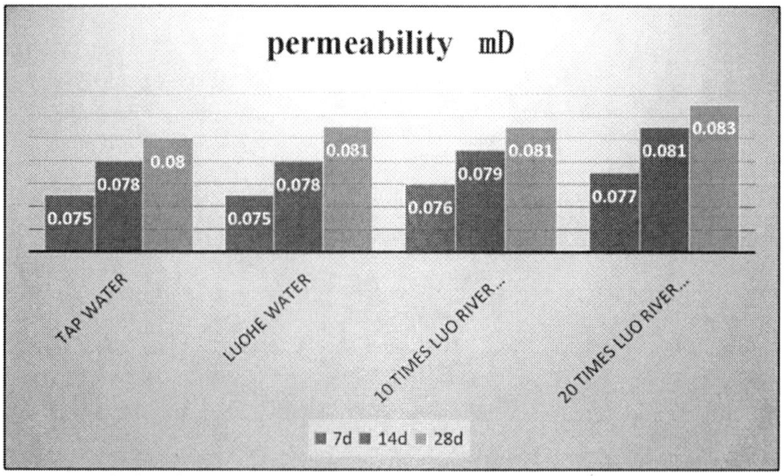


Figure 10 Permeability changes of anticorrosive low-density cement slurry in different corrosive media

It can be seen from Figure 9 and Figure 10 that with the extension of time, the compressive strength of the anticorrosive low-density cement stone has not decayed, but is still gradually increasing, and the permeability is still maintained at a low state. Dissolution occurs when the concentration of the corrosive medium changes, the $Ca(OH)_2$ crystals are well protected, and the cement stone structure is dense and stable.

Figure 11 The SEM photographs of blank cement stone and anti-corrosion cement

It can be seen from Figure 11 that the surface of the pure cement stone matrix is rough, with obvious pore structure, loose texture, and a large number of crystal-like and gel-like cement hydration products with clear morphology are interlaced and grown to form a complete network interweaving structure; The anti-corrosion cement stone matrix has a smooth surface, and no obvious pore structure and clear hydration products are observed. Instead, a clear film-like substance is found to cover the cross section of the cement stone matrix and the cement stone matrix is dense.

**5 Anticorrosive cementing supporting technology**

At present, the oilfield mainly adopts the epoxy coating outside the casing plus sacrificial anode technology and the modified epoxy phenolic spray thermal curing coating protection technology for the new casing anti-corrosion in the casing. The anti-corrosion design strictly implements relevant standard requirements， Q/SY CQ 3548-2015 《Technical requirements for external anti-corrosion technology of oil and water well casing epoxy coating plus sacrificial anode》 and SY/T 6717-2016 《Tubing and Casing Internal Coating Technical Conditions》. On this basis, the self-developed new anode material and the conventional material of the casing centralizer are integrated into a casing centralizer with anti-corrosion function to extend the service life of oil and gas well casings.

The self-developed anti-corrosion centralizer (Figure 12) (a kind of elastic skeleton centralizer for the outer wall of the casing, patent number: CN201922125330.4) includes an anti-corrosion cylinder, an upper spiral centralization clamping end and a lower spiral centralization clamping end , The upper screw-righting clamping end and the lower screw-righting clamping end are respectively sleeved on the outer walls of the two ends of the anti-corrosion cylinder. The anti-corrosion cylinder is made of composite sacrificial anode material, which has high hardness, wear resistance and corrosion resistance. Good performance. It can not only ensure the middle degree of casing in cementing construction, improve cementing displacement efficiency, but also effectively delay casing corrosion after commissioning and increase the service life of oil and gas wells. The main body of the upper and lower screw righting and clamping ends is made of highly wear-resistant material, with high tensile strength, and it is not easy to break off and trip. The anti-corrosion cylinder is clamped between the upper and lower screw-righting clamping ends, which is convenient for installation and replacement. After replacement, the anti-corrosion cylinder can be reused, saving production costs. The specific performance parameters are shown in Table 6.

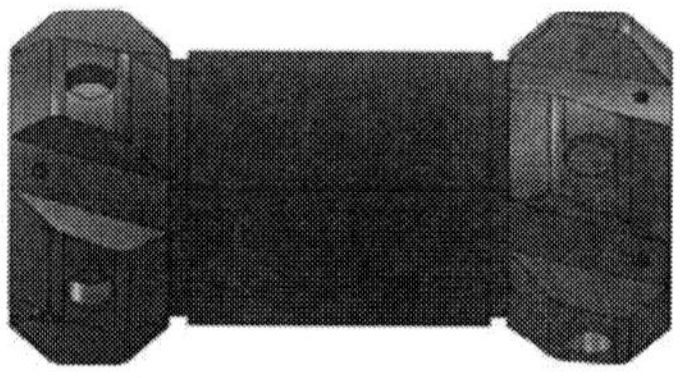

Figure 12 Diagram of anti-corrosion centralizer

Table 6 Anticorrosive centralizer performance index parameters

| performance | parameter（215.9*177.8mm） | parameter（215.9*139.7mm） |
|---|---|---|
| Axial tensile strength | 50T | 70T |
| Radial compressive strength | 12T | 18T |
| Corrosion potential | -0.85V | -0.85V |

## 6 Application of Anticorrosive Cementing Cement for Luohe Water Layer in Changqing Oilfield

In 2017, the Longdong block of Changqing Oilfield tested anti-corrosion cementing cement slurry and completed 6 wells. The Luohe layer of 6 wells was between 474-1031m. The low-density and conventional-density cement slurry were designed for anti-corrosion, and new casing was installed. The epoxy coating on the outside of the casing plus the sacrificial anode technology and the modified epoxy phenolic spray thermal curing coating protection technology for the inside of the casing are treated before. The new anti-corrosion centralizer is added to the entire well section, and the anti-corrosion cement slurry has good fluidity during on-site construction. The cement is returned to the ground, the cementing qualification rate is 100%, and the cementing quality rate is greater than 93%. So far, no casing corrosion has been found in the Luohe layer of the block.

## 7 Conclusion

1. Carry out research on anticorrosive cement system in formation water corrosion environment, prevent casing damage from the perspective of improving cementing cement ring, and provide a new idea for casing corrosion damage.

2. Experiments show that adding 10% epoxy resin emulsion non-penetrating agent and 2% corrosion inhibitor composed of rust and antibacterial materials to cement slurry can achieve a cement stone permeability of 3.57-4.56 times. The corrosion resistance is increased by 6 times, while the compressive strength of cement stone samples remains unchanged compared with conventional cement, providing a new formulation for the development of dense and low-permeability anticorrosive cement systems.

3. On the basis of ensuring the effective sealing of cement, the anti-corrosion ability of the casing body and the cement ring itself is improved, the "internal and external repair" of anti-corrosion

cementing is realized, the service life of the casing is extended, and the development of anti-corrosion cementing technology under the conditions of formation water corrosive environment is established basis.

**Acknowledgments** The project is supported by CNPC Foundation,Research on anti - corrosion cementing technology of Luohe water layer and secondary cementing technology of old well treatment in Changqing Oilfield (Number :CQ2017B-12-3-3).

**References**

[1] WANG Hai, XU Xingping, WEN Shengkui, et al.Research on Subsidy Repair Technology of Large Diameter Expansion Thin-walled Pipe in Casing Damage[J].Petroleum machinery, 2017, 45(10): 98-102.

[2]FAN Zexia , REN Shaoran , WangJiexiang , et al.Indoor Research and Field Application of Compound Squeeze Injection Technology for Oil Well Chemical Scale Prevention[J].Oil drilling technology, 2007(04):84-86.

[3]LIU Zhixiong.Cementing technology of sidetracking slim hole liner with casing window in WangCe 14-15[J].Drilling technology, 2019:42(2):121～123.

[4] LIU Shuoqiong,LI Deqi,YUAN Jinping , et al.Study on cement ring integrity of shale gas well[J].Natural gas industry, 2017, 37（7）: 76～82.

[5]LIU Jian,DU Jianbo,ZHANG Xingguo,et al.Development and application of high temperature and high pressure cement ring integrity evaluation device[C].Proceedings of Symposium on Cementing Technology of 2017,Beijing:Petroleum Industry Press, 2017: 195, 2017, 37（7）: 76～82.203.

[6]TAO Qian,ZHOU Shiming,LIU Rengguang,et al.Analysis of long-term sealing failure mechanism and preventive measures of cement ring in high pressure gas wellProceedings of Symposium on Cementing Technology of 2017,Beijing:Petroleum Industry Press, 2017: 3, 2017, 37（7）: 76～82.11.

[7] HOU Pan,GAO Ya,ZHU Zhongxi,Research and application of Flex STONE elastic cement in underground gas storage[J].Natural gas technology and economics, 2014, 8（4）: 25～27.

[8] ZHANG Huanzhi,HE Yanqing,ZHANG Huazhen.Halliburton Tight Gas Development Technology Series[J].Petroleum Technology Forum, 2013, 32（4）: 43～48.

[9]YU Yongjin,Jinjianzhou,QI Fengzhong.Research progress of functional cementing working fluid[J].Drilling fluid and completion fluid, 2016, 33（4）: 1～7.

[10]LI Xiaobing.Research on improvement of repairing damaged well technology with small casing in Longdong Oilfield[D].Master Thesis of Xi'an Shiyou University,2009:7～10.

[11]YANG Yuanguang, LIAO Gang, GUO Xiaoyang, et al.Research on anti-corrosion low-density cement slurry system[J].Drilling fluid and completion fluid,1999(04):3-5.

[12]YUE Yuezhan.Research on Pressure Reducing Material and Low Density Cement Slurry System[D].Master Thesis of Southwest Petroleum University, 2017.

[13]ZHANG Hanzhi , ZHAI Xiaopeng,LOU Yishan.Development status and prospects of China's continental shale oil drilling technology[J].Oil drilling technology, 2019,41(03):265-271.

[14]ZHANG Yihang, SONG Xuhui, XU Mingbiao, et al.Research Progress on Corrosion Prevention of Cement Stone[J].Applied Chemical Industry,2019,48(10):2450-2455.

[15]ZHAO Jun, XU Bihua, QIU Huiyang, et al.Development and application of high temperature resistant and $CO_2$ and $H_2S$ corrosion resistant cement slurry system[J].China Offshore Oil and Gas,2017,29(3):91-94.

[16]GU Tao.Research on Corrosion Damage and Prevention of Cementing Cement Stone under the Condition of $CO_2$ Geological Storage[D].Master Thesis of Southwest Petroleum University, 2017.

[17]LIU Chongjian, HUANG Baizong, XU Tongtai, et al.Cement injection theory and application in oil and gas wells[M].Petroleum Industry Press, 2001.

[18]ZHANG Tiansheng.Corrosion inhibitor[M].Chemical Industry Press, 2002.

[19]YI Shaojin, SHE Yuehui, Petroleum and Environmental Microbiology Technology[M].China University of Geosciences Press, 2002.

[20]Performance Evaluation Method of Corrosion Inhibitor for Oilfield Produced Water, SY/T 5273-2000 [S].1～12

[21] Oil well cement test method, GB/T 19139-2012 [S].14～18

[22]Zhang J , Wang C , Peng Z .Corrosion integrity of oil cement modified by environment responsive microspheres for $CO_2$ geologic sequestration wells[J].Cement and Concrete Research, 2021, 143(2):106397.

[23]Mohammadreza B, Seyed M, Eshmaiel G.Parametric study on the integrity of wellbores in $CO_2$ storage sites[J].Construction and Building Materials, 2021, 268(3):121060.

[24]B.L.S. Costa, J.C.O. Freitas, R.G.S. Araujo, Y.H. Oliveira, R.C. Santiago, F.S. Oliveira. Analysis of different oil well cement slurry formulations exposed to a CO2-rich environment[J]. Journal of $CO_2$ Utilization,2021,51(1):101636.

[25]Xingguo Zhang, Youzhi Zheng, Ziming Guo, Yong Ma, Yao Wang, Tao Gu, Tao Yang, Libin Jiao, Kaiqiang Liu, Zhongzhi Hu. Effect of $CO_2$ solution on Portland cement paste under flowing, migration, and static conditions[J]. Journal of Natural Gas Science and Engineering,2021,95(4):104179.

[26]Jinliang Liu, Yanmin Jia.Experimental investigation on durability of cement-steel pipe for wellbores under $CO_2$ geological storage environment[J].Construction and Building Materials,2020,236(6):117589.

[27]Ahmed Abdulhamid Mahmoud, Salaheldin Elkatatny.Mitigating CO2 reaction with hydrated oil well cement under geologic carbon sequestration using nanoclay particles[J],Journal of Natural Gas Science and Engineering,2019,68(2),102902.

Journal of Physics: Conference Series **2594** (2023) 012032 doi:10.1088/1742-6596/2594/1/012032

# Division Method of Dynamic Flow Units in Fluvial Reservoir at Ultra-high Water Cut Period

**Yingfei SUI1, Chuanzhi CUI1(✉), Yin QIAN1, Peifeng JIA1,2, Shuiqingshan LU1, Mengxin CHEN1**

1 College of Petroleum Engineering, China University of Petroleum (East China),
2 Petroleum Engineering Research Institute of Sinopec Shengli Oilfield Company.

ccz2008@126.com

**Abstract**. The heterogeneity of fluvial sedimentary reservoir is serious. The division of flow units in the early stage of development can provide a good guidance in the well pattern distribution and reservoir production. Due to the huge changes in the oil-water system in the long-term development reservoir, the flow units divided at the initial stage are difficult to guide the production when entering the ultra-high water cut period. Therefore, it is necessary to redefine the division indicators and methods of flow units, so as to provide support for the formulation of efficient regulation and control technology. Starting from the definition of flow unit, this paper establishes the percolation coefficient as the characteristic index of percolation capacity. Combined with the percolation characteristics of fluvial reservoir, the dynamic flow unit division method with percolation coefficient and water cut as division parameters is established. Through the division results of typical models of fluvial reservoir, it can be seen that the four types of flow units divided by the dynamic flow unit method in this paper can play a great guiding role in the later formulation of regulation and control measures. The verification of reservoir examples shows that the division results of dynamic flow units of fluvial facies can well reflect the remaining oil enrichment areas in the reservoir and classify the remaining oil enrichment areas. At the same time, the dynamic flow unit changes continuously with the development process, which can more clearly show the producing degree of each remaining oil enrichment area after the regulation and control measures. This provides a key support for formulating targeted and efficient reservoir development technique.

## 1. Introduction

Due to the influence of reservoir heterogeneity and low-level percolation barrier (sedimentary facies interface), the dynamic heterogeneity of flu-vial sedimentary reservoir is serious, and the percolation capacity of it varies greatly. The proposal of flow unit and its research method provide an effective means to understand the heterogeneity of reservoir and re-veal the distribution law of remaining oil. Flow unit refers to the reservoir with similar petrophysical characteristics and percolation characteristics that are continuously distributed in space [1-3] and it is a comprehensive reflection of reservoir lithology, physical properties and seepage characteristics [4-6]. Owing to the limitations of specific geological conditions and actual data and the different starting point of the research problem, the understanding and research methods of the flow unit are not completely consistent. Flow unit research methods are mainly divided into two types. One is the flow unit division method based on geological research, such as sedimentary facies method [7], analytic hierarchy process [8] and heterogeneous comprehensive index method [9]. The other is the reservoir parameter analysis method based on mathematical means such as flow stratification index FZI method [10], pore throat geometry R35 method [11], production dynamic parameter method, and multi parameter comprehensive method [12].

In recent years, a large number of scholars have carried out more in-depth research on the division of flow units based on these methods. Zahra [13] compared four petrophysical methods including Rock Fabric Number (RFN), Winland porosity-permeability plot (R35), Reservoir Quality Index/Flow Zone Indicator (RQI/FZI), and Bulk Volume Water (BVW), and BVW is the best method for the division of flow units in carbonate reservoirs. Rebelo [14] found that Aguilera R35 has the highest correlation with permeability by comparing three classification methods, including FZI, Winland R35 and Aguilera R35, which proved to be a reliable method for application in complex carbonate reservoirs. In addition, some scholars have further improved the existing methods. Osama [15] done through differentiation of all samples that lie on the unit slope trend line and those lie scattered on both sides of it, into three sub-flow units regardless their slopes and the use of (FZI) arithmetic average for each sub-unit instead of that at ($\phi Z$) = 1 and permeability prediction has been improved after applying the new additional processing. Oliveira [16] have found multiple hydraulic flow units directly from the dataset without resorting to usual conversion formulae, by setting up constrained clustering approaches with different must-link hard constraints through using the so called flow zone indicator (FZI) as underlying parameter. Ha [17] used unsupervised learning to automatically cluster hydraulic flow units based on flow zone indicators (FZI) in a core plug dataset.

After the reservoir development enters the ultra-high water cut period, the oil-water system in the reservoir has changed greatly compared with the initial stage of development, and the percolation capacity of oil-water is also different from the initial stage. The existing classification and evaluation parameters of flow units are mostly static parameters, and less dynamic parameters are used, which is not suitable for the research of dynamic character of flow units. After years of development, most of the reservoirs have carried out numerical simulation research. Therefore, based on the results of reservoir numerical simulation calculation, this paper establishes the characterization index and division method of dynamic flow unit of fluvial reservoir in ultra-high water bearing period, considering the geological and development dynamic parameters such as reservoir permeability and corresponding relative permeability curve, water saturation, viscosity, etc., so as to provide support for formulating targeted and efficient regulation technology in the later stage of development

## 2. Characterization index and calculation method of percolation capacity in ultra-high water cut period

The flow unit is defined as a reservoir with similar seepage characteristics, so the percolation coefficient is proposed as the characterization index of flow capacity. The percolation coefficient is defined as the oil flow, water flow or total flow of oil and water passing through the unit area under the unit pressure gradient, corresponding to the oil phase percolation coefficient, water phase percolation coefficient and two-phase percolation coefficient.

For two-phase percolation coefficient, its expression is:

$$Z_l = \frac{q_o + q_w}{A \cdot (\Delta p / \Delta x)} = \frac{kk_{ro}}{\mu_o} + \frac{kk_{rw}}{\mu_w} = \left( \frac{k_{ro}(k, S_w)}{\mu_o} + \frac{k_{rw}(k, S_w)}{\mu_w} \right) k \tag{1}$$

The expressions of oil phase percolation coefficient and water phase percolation coefficient are respectively:

$$Z_o = \frac{q_o}{A \cdot (\Delta p / \Delta x)} = \frac{kk_{ro}}{\mu_o} = \left( \frac{k_{ro}(k, S_w)}{\mu_o} \right) k \tag{2}$$

$$Z_w = \frac{q_w}{A\cdot(\Delta p/\Delta x)} = \frac{kk_{rw}}{\mu_w} = \left(\frac{k_{rw}(k,S_w)}{\mu_w}\right)k \tag{3}$$

Oil phase relative permeability ($k_{ro}$) and water phase relative permeability ($k_{rw}$) are functions of permeability ($k$) and water saturation ($S_w$), and the commonly used expression [18-20] is:

$$k_{ro} = k_{roi}\left(\frac{1-S_w-S_{or}}{1-S_{wc}-S_{or}}\right)^m \tag{4}$$

$$k_{rw} = k_{rwi}\left(\frac{S_w-S_{wc}}{1-S_{wc}-S_{or}}\right)^n \tag{5}$$

where $k_{roi}$ is the relative permeability of oil under irreducible water saturation, which is generally the benchmark permeability and its value is 1, $k_{rwi}$ is the relative permeability of water under residual oil saturation, $S_w$ is water saturation, $S_{wc}$ is the irreducible water saturation, $S_{or}$ is residual oil saturation.

The relative permeability curves of different rock pore structures are different, which can be shown in the macroscopic view that the relative permeability curves of different permeability are different. There are great differences in physical properties in the same fluvial reservoir, so it is necessary to establish a calculation method of relative permeability under any reservoir permeability, so as to obtain a more accurate percolation coefficient.

According to Eq. (4) and Eq. (5), there are mainly five characteristic parameters characterizing a relative permeability curve, including irreducible water saturation ($S_{wc}$), residual oil saturation ($S_{or}$), water phase power index ($n$), oil phase power index ($m$), and residual oil water phase relative permeability ($k_{rwi}$). By regressing the relative permeability curves of many fluvial reservoirs, the relationship between the characteristic parameters of relative permeability curves and permeability is established, and the relative permeability expression corresponding to any permeability can be obtained. Taking 101 relative permeability curves of fluvial facies sedimentary units in Shengli Oil Region as the research object, the regression collation is carried out, and the relationship formula between characteristic parameter values of relative permeability curves and permeability is established as follows [21]:

$$S_{wc} = 0.3236 - 0.0327\ln(k) \tag{6}$$

$$S_{or} = -0.0131\ln(k) + 0.1946 \tag{7}$$

$$n = 0.0056k + 1.9146 \tag{8}$$

$$m = -0.2479\ln(k) + 2.8019 \tag{9}$$

$$k_{rwi} = -0.038\ln(k) + 0.2594 \tag{10}$$

The applicable range of permeability K of Eq. (6) to Eq. (10) is 0.06 ~ 6.3$\mu m^2$, and the relative permeability curve corresponding to any permeability within this range can be calculated, so as to obtain the corresponding percolation coefficient.

## 3. Selection of dividing indexes for dynamic flow units and their dividing limits

Dynamic flow unit division method is based on reservoir numerical simulation, so it is necessary to consider the embodiment of the definition characteristics of flow unit in numerical simulation parameters. Firstly, the continuous distribution of flow units in space is mainly manifested in the closed impermeable boundary and the argillaceous interlayer between each sublayer in the numerical simulation. Secondly, the flow unit has similar petrophysical characteristics, which are mainly reflected by the absolute permeability and porosity of the grid in the numerical simulation. Finally, the flow unit has similar percolation characteristics, which can be reflected in the numerical simulation through percolation capacity and development performance. The absolute permeability and percolation capacity of the grid can be characterized by percolation coefficient, and the development dynamics can be characterized by selecting a dynamic parameter. Then the dynamic flow unit is divided by these two characterization parameters.

Comprehensively considering the goal of tapping the potential of remaining oil in the ultra-high water cut period, the characteristics of strong heterogeneity of fluvial reservoir, and the nature of flow unit characterizing percolation capacity, the oil phase percolation coefficient is selected as the division index of static percolation capacity, and the grid water cut is selected as the division index of current development dynamics.

### *3.1. Dividing boundary of oil phase seepage coefficient*

According to the expression of oil phase percolation coefficient, it can be seen that the oil phase percolation coefficient represents the percolation capacity of remaining oil, that is, the fluidity of remaining oil under a certain driving pressure difference, which is a function of remaining oil saturation and reservoir permeability. Taking this as the division index of dynamic flow unit, it fully considers the reservoir physical parameters based on numerical simulation and development dynamic parameters, and also corresponds to the goal of tapping the potential of remaining oil in ultra-high water cut period.

Considering the low economic benefit and high development cost of reservoirs in ultra-high water cut stage, the boundary of dividing oil-phase percolation coefficient based on economic benefit potential is established according to the principle of profit and loss balance. The oil phase percolation coefficient corresponding to the economic limit oil production of a single well under the condition of single well control area is used as the standard for dividing high and low percolation capacity.

The formula of economic limit oil production of single well in ultra-high water cut period is [22]:

$$Q_o = \frac{C_G \times 10^4 / 365}{(P - R_T) - R_o} \tag{11}$$

where $Q_o$ is the economic limit oil production of a single well, $C_G$ is the annual fixed cost of a single well, $P$ is the price of crude oil, $R_T$ is the tax per ton of oil, $R_o$ is the variable cost per ton of oil.

Assuming that the actual well spacing of the reservoir is $B$, the average thickness of the reservoir is $h$, and the production pressure difference of the reservoir is $\Delta p$, the boundary of dividing oil-phase percolation coefficient can be obtained from Eq. (11).

$$Z_{tc} = \frac{Q_o}{B \cdot h \cdot (\Delta p / B)} = \frac{Q_o}{\Delta p \cdot h} \tag{12}$$

The oil-phase percolation coefficient of each grid in the actual reservoir can be directly obtained from the numerical simulation results and the expression of oil-phase percolation coefficient. Firstly, the characteristic parameters of the relative permeability curve are calculated according to the grid permeability $k$. Then, the relative permeability ($k_{ro}$) of grid oil phase is calculated according to the characteristic parameter value of relative permeability curve and grid water saturation Sw. Finally, the value of oil phase percolation coefficient ($Z_o$) of the grid is calculated according to Eq. (2).

According to the calculated grid oil phase percolation coefficient and Eq. (12), if $Z_o > Z_{tc}$, the actual reservoir grid attribute is high percolation capacity; if $Z_o < Z_{tc}$, the actual reservoir grid attribute is low percolation capacity.

### *3.2. Dividing boundary of grid water cut*

The physical meaning of grid water cut is the water production rate of an oil well that is supposed to be put into production in this grid at this time. According to the definition of water cut, the expression of grid water cut is as follows:

$$f_w = \frac{Q_w}{Q_w + Q_o} = \frac{1}{1 + \frac{\mu_w}{\mu_o} \frac{k_{ro}(k, s_w)}{k_{rw}(k, s_w)}} = \frac{Z_w}{Z_o + Z_w} \tag{13}$$

According to Eq. 16, the grid water cut can be regarded as the ratio of water phase seepage coefficient to two-phase seepage coefficient in the grid in addition to having the meaning of general water cut, that is, the proportion of water phase percolation ability in oil-water two-phase percolation ability. In this way, the influence of reservoir physical properties ($k$) is removed, but the reservoir fluid characteristics under time accumulation are reflected. Considering that the oil-water system of the reservoir changes greatly in the ultra-high water cut period, it is more appropriate to select the grid water cut as the division index.

The current overall water cut of the reservoir ($f_{wtc}$) is taken as the standard for dividing the grid with high water cut and low water cut. The water cut ($f_w$) of each grid in the actual reservoir can be directly obtained from the numerical simulation results and the expression of grid water cut. Then, if $f_w > f_{wtc}$, the actual reservoir grid attribute is high water cut; if $f_w < f_{wtc}$, the actual reservoir grid attribute is low water cut.

## 4. Division and types of dynamic flow units of fluvial facies

Based on the numerical simulation results, two flow unit division indexes are used to make each grid of the reservoir have two attributes, namely, oil phase percolation capacity and grid water cut. According to this, the dynamic flow unit of fluvial reservoir is divided into four types: E, G, M and P, as shown in Table 1.

**Table 1.** Classification of dynamic flow units.

| Type | Characteristic |
|---|---|
| E | This type of flow unit mainly shows the characteristics of low water cut and high oil phase percolation capacity. This flow unit mainly refers to the positions where water flooding is not effective, such as inter well detention and residual oil controlled by faults. It is necessary to improve the well pattern or change the direction of flow field to achieve the purpose of effective displacement. |
| G | This type of flow unit mainly shows the characteristics of high water cut and high oil phase percolation capacity. This flow unit is mainly the area with good development effect and being affected by displacement, so no measures are required. |
| M | This type mainly shows the characteristics of low water cut and low oil phase percolation capacity. This flow unit is mainly in the area with low permeability and poor displacement effect. There are many remaining oil here, but it is difficult to use. Therefore, radial drilling and increasing displacement pressure difference shall be taken. |
| P | This type of flow unit mainly shows the characteristics of high water cut and low oil phase percolation capacity. This flow unit is mainly the area that has been fully displaced, with high water cut and low oil saturation. Ineffective circulation of water drive mostly exists in this area, and profile control can be taken when necessary. |

## 5. Division and types of dynamic flow units of fluvial facies

A reservoir of Gudao Oilfield in Shengli Oilfield is taken as an example, which is a typical fluvial sedimentary reservoir. At present, the comprehensive water cut of the reservoir has reached 97.6%, which belongs to the late stage of ultra-high water cut. It is the applicable reservoir for the dynamic flow unit division method in this paper.

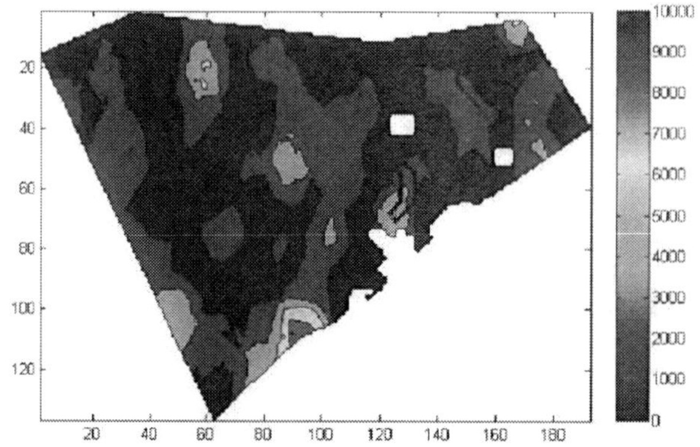


**Fig. 1.** Permeability distribution in the middle of the main production layer

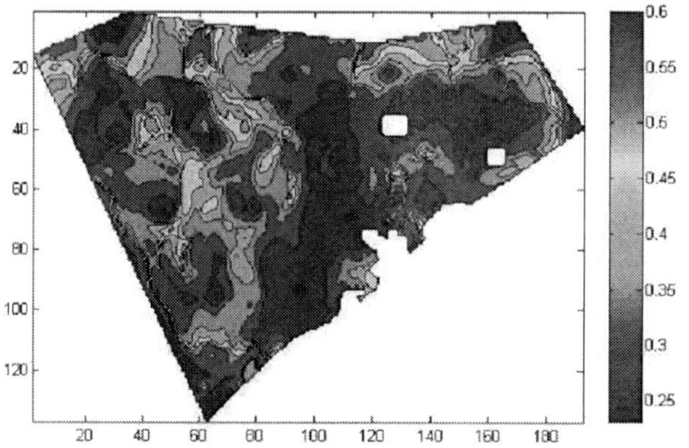


**Fig. 2.** Residual oil distribution in the middle of the main production layer.

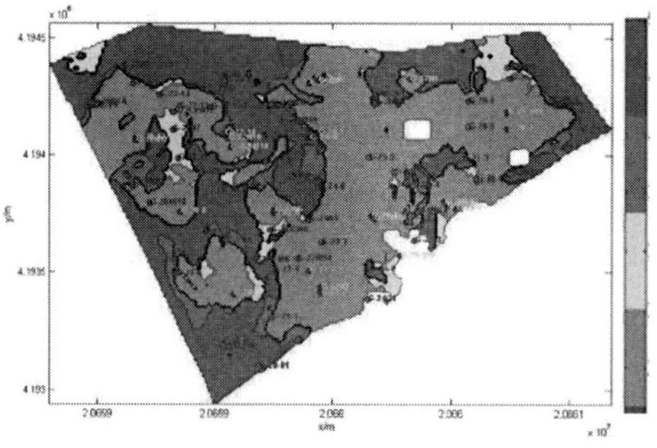

**Fig. 3.** Distribution of dynamic flow units in the middle of the main production layer. Red is Type E, pink is Type G, yellow is Type M, and green is Type P

The main production layer of the example reservoir is selected to divide the dynamic flow unit. First of all, the limit of oil-phase percolation coefficient of the reservoir is calculated according to the economic limit oil production of a single well. Assuming that the current crude oil price is \$60, the fixed cost ($C_G$) of a single well is \$86,000, the tax ($R_T$) per ton of oil is \$47.28, and the variable cost ($R_O$) per ton of oil is \$121.64, the economic limit oil production ($Q_o$) of a single well is calculated to be 0.859t/d; If the production pressure difference of the reservoir is 2 MPa and the average thickness of the reservoir is 600 cm, the division limit of the oil phase seepage coefficient can be obtained as 0.0027 according to Eq.15. Then, the abandoned water content of the oilfield ($f_{wtc}$ = 0.98) is taken as the dividing limit of the grid water cut. Finally, according to the oil phase percolation coefficient ($Z_o$) and water cut ($f_w$) of each grid, as well as the division boundaries ($Z_{tc}$ = 0.0027 and $f_{wtc}$ = 0.98), the attributes of each grid are determined, and the division of dynamic flow units is completed. Fig. 3 shows the flow unit division results when the comprehensive water cut of the numerical model of the example reservoir is 90%.

The division results show that the flow units in the middle of the main production layer are mainly Type E and Type P (see Fig. 3). Type E represents the remaining oil enrichment area and there is no effective displacement position at this stage. Type P represents the high water cut area that has been fully displaced at this stage. Type E and Type P flow units are more, indicating that there is a large difference in water drive displacement at this position in the small layer. Therefore, the large proportion of Type E and Type P flow units indicates that the water flooding displacement difference at this position of the main production layer is large. In addition, Type G flow units are few and scattered, indicating that there are few effective displacement areas at the main layer, and development measures need to be adjusted to make Type E flow units effectively displace, gradually becoming Type G, and finally becoming Type P, so as to tap the potential of remaining oil enrichment area. Finally, Type M flow units are also scattered. Type M is a low permeability area after a certain degree of water flooding displacement, but still has a large amount of remaining oil. Type M flow units are difficult to develop due to low permeability, and need greater displacement pressure difference to product the remaining oil.

### *5.1. Differences in the division of dynamic flow units in the vertical direction of reservoirs*

The division results of flow units at the upper, middle and lower parts of the main production layer are shown in Fig. 4, Fig. 3 and Fig. 5. The proportion of each flow unit in the upper, middle and lower parts of the main production layer is calculated as shown in Table 2.

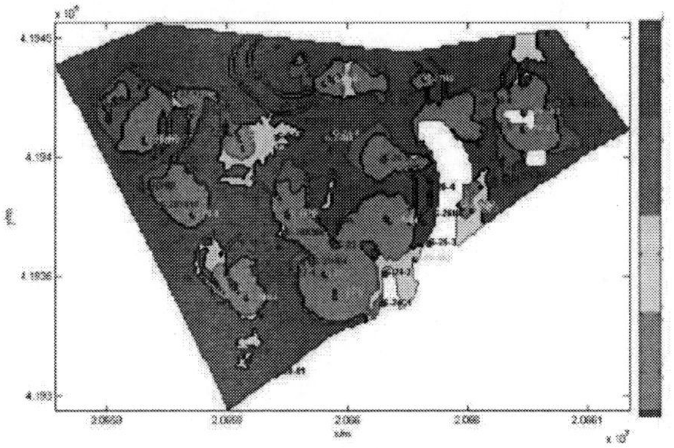

**Fig. 4.** Distribution of dynamic flow units in the upper part of the main production layer. Red is Type E, pink is Type G, yellow is Type M, and green is Type P.

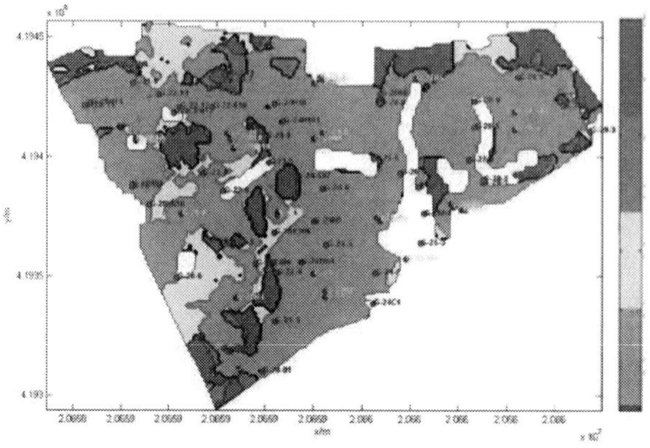

**Fig. 5** Distribution of dynamic flow units in the lower part of the main production layer. Red is Type E, pink is Type G, yellow is Type M, and green is Type P

According to the data in Table 2, the upper part of the reservoir is dominated by Type E flow units, and the lower part of the reservoir is dominated by Type P flow units, which is the inevitable result of the difference in oil and water gravity. The proportion of Type G flow units in the overall reservoir is very small. Type G flow units represent the part of the reservoir being displaced. Type G is a necessary process for Type E flow units in the main enrichment area of remaining oil to be displaced into Type M and Type P, and the lack of Type G indicates that the reservoir development effect is poor, and the area being affected by water flooding is small.

**Table 2.** Proportion of flow units in each part of the main production layer

| Proportion of flow units /% | Upper part of layer | Middle part of layer | Lower part of layer |
|---|---|---|---|
| E | 60.86 | 40.31 | 14.32 |
| G | 7.38 | 6.59 | 4.01 |
| M | 8.37 | 6.60 | 14.20 |
| P | 23.38 | 46.50 | 67.47 |

*5.2. Dynamic flow unit change during reservoir development*

When the comprehensive water cut of the target reservoir numerical model is 90%, 95% and the current water cut (97.6%), the flow units are divided, showing the evolution process of each flow unit in the middle of the main layer at different water cut stages, and the boundaries of the dividing indicators are consistent ($Z_{tc} = 0.0027, f_{wtc} = 0.98$).

It can be seen from Fig. 8, Fig. 9 and Fig. 10 that a large number of remaining oil enrichment areas, i.e. Type E flow units, have gradually become Type G after development, and finally become ineffective water drive circulation areas, i.e. Type P flow units, realizing effective development of reservoirs, taking the area of Pictorial Circle as an example. However, Type M flow units in the reservoir are reduced, but the change is not obvious.

According to the data in Table 3, during the development of the middle part of the main layer, Type E flow units decrease rapidly, while Type P flow units grow slowly first and then rapidly. This is because Class E flow units need to go through the process of becoming Type G to finally become Type P. Therefore, Type G flow units are an important indicator for evaluating the development effect of the reservoir under the current conditions. Type G flow units increased rapidly and then decreased slightly, indicating that the reservoir has gone through an effective development process and the good development effect has continued to the present. Type M flow units have decreased to a certain extent, indicating that the low permeability area has been developed to a certain extent. Up to the current development stage, Type P flow units account for about 74%, indicating that the development of this part has entered the late stage, and there are many invalid development areas. If further tapping the potential of remaining oil is required, the next step of flow unit division is required, giving the new boundary of flow unit division index of the reservoir, so as to provide new guidance for the next step of remaining oil tapping measures.

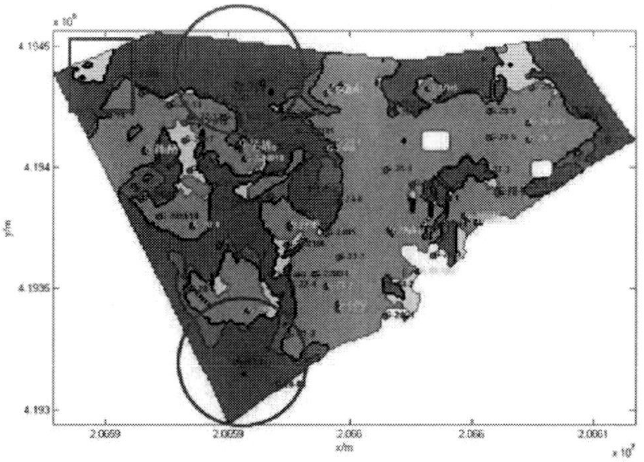


**Fig. 6** Distribution of flow units in the middle of main layer with 90% water cut. Red is Type E, pink is Type G, yellow is Type M, and green is Type P

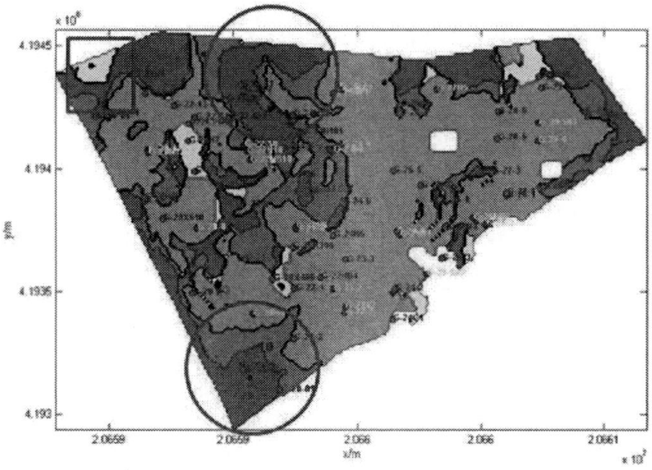


**Fig. 7** Distribution of flow units in the middle of main layer with 95% water cut. Red is Type E, pink is Type G, yellow is Type M, and green is Type P

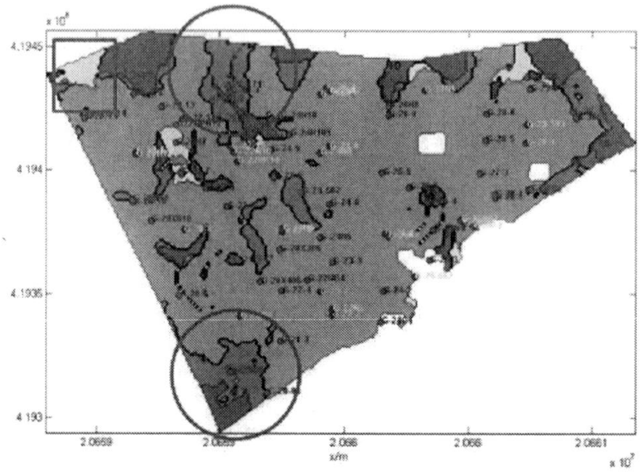

**Fig. 8** Distribution of flow units in the middle of main layer with 97.6% water cut. Red is Type E, pink is Type G, yellow is Type M, and green is Type P

**Table 3.** Proportion of flow units in various development periods in the middle of main layer

| Proportion of flow units /% | 90% water cut | 95% water cut | 97.6% water cut |
|---|---|---|---|
| E | 40.31 | 23.30 | 10.41 |
| G | 6.59 | 17.33 | 11.93 |
| M | 6.60 | 4.66 | 3.41 |
| P | 46.50 | 54.70 | 74.24 |

## 6. Conclusions

Based on the above research, the following conclusions can be drawn:

(1) Starting from the definition of flow unit, combined with the development characteristics of river facies reservoir in the ultra-high water cut period, the percolation coefficient is established as the characterization index of percolation capacity.

(2) The oil phase percolation coefficient and grid water cut are selected as the division index of dynamic flow unit.

(3) Considering the economic limit production of a single well and the water cut of the reservoir, the dynamic flow unit is divided into four types: E, G, M and P, and the corresponding control policies for each type of flow unit are determined.

(4) The verification of reservoir examples shows that the division results of dynamic flow units of fluvial facies can well reflect the remaining oil enrichment areas in the reservoir and classify the remaining oil enrichment areas.

(5) The dynamic flow unit changes continuously with the development process, which can more clearly show the producing degree of each remaining oil enrichment area after the regulation and control measures. This provides a key support for formulating targeted and efficient reservoir development technique.

**Acknowledgments.** The project is supported by Natural Science Foundation of China with No. 51974343 and Project funded by China Postdoctoral Science Foundation with NO. 2021M703588.The first author would like to acknowledge College of Petroleum Engineering at China University of Petroleum (East China) to provide experimental equipment.

## References

[1] Ebank W. J.: Flow unit concept-integrated approach to reservoir description for engineering projects. AAPG Bulletin 71(5), 551-552 (1987).
[2] Guangming T., Ogbe D. O., Munly Walt.: The use of flow units as a tool for reservoir description: a case study. SPE Formation Evaluation 10(2), 122-128 (1995).
[3] Jiyu L.: Research Progress of Flow Units. Progress in Earth Science 2000(03), 303-306 (2000).
[4] Yang L.: Reservoir flow unit model and remaining oil distribution law. Journal of Petroleum 03, 52-55 (2003).
[5] Fengchun M. A., Jincheng L. I., Yanxiong W. U.: Calculation and evaluation method of reservoir parameters in multiple oil reservoirs based on flow unit. Geoscience 34(2), 370–377 (2020).
[6] Xincai C., Bin Z., Chuqiao G.: Reservoir Permeability Calculation under Flow Unit Control. Energies 15(20), 7637 (2022).
[7] Hamlin H. S., Dutton S. P., Tyler N.: Depositional controls on reservoir properties in a braid delta sandstone, Tirrawarra oil field, South Australia. AAPG Bulletin 80(2), 139-156 (1996).
[8] Shimi P., Zhijun Y., Xue C.: A new method for quantitative study of flow units in continental reservoirs. Petroleum Exploration and Development 28(5), 68-70 (2001).
[9] Zubing L., Minggao L., Bin M.: Heterogeneous composite index method for dividing reservoir flow units. Xinjiang Geology 23(2), 203-206 (2005).
[10] Yanxin Z., Chengyan L., Li Z.: Discussion on the division of flow units by FZ1. Petroleum Exploration and Development 31(5), 130-1320 (2004).
[11] Alden J. M., Stepphen T. S., Dan J. H.: Characterization of petrophysical flow units in carbonate reservoir. AAPG Bull 81(5), 734-7560 (1997).
[12] Fa Y., Chengyan L., Xiuju Z.: Discussion on the method of multi parameter identification of flow unit. Journal of Sedimentology 24(1), 91-95 (2006).
[13] Riazi Z.: Application of integrated rock typing and flow units identification methods for an Iranian carbonate reservoir. Journal of Petroleum Science and Engineering 160, 483–497 (2018).
[14] Rebelo T. B., Batezelli A., Mattos N.: Flow units in complex carbonate reservoirs: A study case of the Brazilian pre-salt. Marine and Petroleum. Geology 140, 105639 (2022).
[15] Elnaggar M. O.: A new processing for improving permeability prediction of hydraulic flow units, Nubian Sandstone, Eastern Desert, Egypt. Journal of Petroleum Exploration and Production Technology 8(3), 677-683 (2018).
[16] Oliveira G. P., Santos M. D., Roque W. L.: Constrained clustering approaches to identify hydraulic flow units in petroleum reservoirs. Journal of Petroleum Science and Engineering 186(C), 106732 (2020).
[17] Quang H. M., Huy D. H., Duy K. T.: Hydraulic Flow Unit Classification and Prediction Using Machine Learning Techniques: A Case Study from the Nam Con Son Basin, Offshore Vietnam. Energies 14(22), 7714 (2021).
[18] Jiqiang W., Shengjie Y., Menggao Z.: A new method for calculating the relative permeability curve of oilfield in ultra-high water cut period. Petrochemical Application 36(8), 62-64 (2017).
[19] Nutao W., Hao C., Tao W.: Calculation of reservoir relative permeability curve using production data. Journal of Southwest Petroleum Institute 27(5), 36-38 (2005).
[20] Xiangfeng H., Guangming Z., Junlei G.: A new method for calculating the relative permeability curve of oil reservoirs and its application. Petroleum Geology and Engineering 27(1), 53-55 (2013).
[21] Chuanzhi C., Zhengling G., Yanzhong W.: Dynamic distribution calculation model and application of permeability in high water cut stage of water drive reservoir. Journal of China University of Petroleum (Natural Science Edition) 36(04), 118-122 (2012).
[22] Deshuo T.: Study on economic limit liquid production of reservoirs in ultra-high water cut stage. Petrochemical Technology 22(04), 156 (2015).

Journal of Physics: Conference Series **2594** (2023) 012033 doi:10.1088/1742-6596/2594/1/012033

# Research and Practice of Intelligent Separate Production Technology with Cable in Huabei Oilfield

**Ya-jun CHEN1, Zhi-yuan XU2, Dong-wen GE3, Xu ZHANG4,Bo-hua HUANG5, Tao WANG6**

1 No. 1 Oil Production Plant, Huabei Oilfield Company of PetroChina, Renqiu, He Bei Province, China;
2 Engineering Technology Research Institute, Huabei Oilfield Company of PetroChina,Renqiu, He Bei Province, China;
3 No. 1 Oil Production Plant, Huabei Oilfield Company of PetroChina, Renqiu, He Bei Province, China;
4 Engineering Technology Research Institute, Huabei Oilfield Company of PetroChina,Renqiu, He Bei Province, China;
5 Engineering Technology Research Institute, Huabei Oilfield Company of PetroChina,Renqiu, He Bei Province, China;
6 No. 1 Oil Production Plant, Huabei Oilfield Company of PetroChina, Renqiu, He Bei Province, China;
E-mail:cy1_chenyj@petrochina.com.cn

**Abstract:** In order to improve the separate production effect of sandstone reservoir in Huabei Oilfield and give full play to the production capacity of medium and low permeability layers, the intelligent separate production technology with cable is designed and developed. The technology uses cable packer to seal each layer in the well, and a set of intelligent separate production device is connected to the corresponding string position of each target layer. During the production process, the intelligent separate production device can send instructions according to the ground control center Open or close the oil layer to realize the exploitation of any layer and multi-layer co production of any layer. It can accurately measure the liquid production and properties of each section of the well, and calculate the formation permeability, skin factor and other parameters by sections, so as to realize the continuous testing and monitoring function.In the later stage, the switch opening can be adjusted by remote control to adjust the output of each layer in time to meet different production allocation requirements and achieve the purpose of increasing oil and reducing water, which is of great significance to realize the accurate development of the reservoir.

## 1. Introduction

At present, Huabei Oilfield is mainly developed in sandstone blocks. With the development of the oilfield entering the middle and late stage, it has now entered the high water cut stage, and almost all the high permeability oil reservoirs have become high aquifers. With the increase of water cut, the contradictions between oil reservoirs are becoming increasingly prominent[1-2].Due to the influence of interlayer interference, oil reservoir recovery is low. How to balance interlayer pressure, inhibit high aquifer, exploit low aquifer, and improve oil reservoir recovery is a long-term technical problem in the process of oil field development.

Since 2014, the pressure controlled intelligent separator has been introduced in Huabei Oilfield, and the preliminary test of intelligent production has been carried out.The separator performs downhole switching by receiving ground pressure instructions, and separately switches and controls layers with different permeability. However, pump truck is needed for each switch, and the construction procedure is complicated. Moreover, its pressure data records are stored and cannot reflect downhole pressure and flow changes in real time[3-5].Since 2017, Huabei Oilfield has been gradually exploring and implementing the intelligent separate production technology with cable, which connects each layer packer and intelligent separate production with cables. Downhole switch changes can be realized at any time by the command sent from the ground, and pressure data can be transmitted to the ground system in real time. With the network, remote data transmission and remote switch control functions can also be realized.

## 2. The principle of the intelligent separate production technology with cable

The intelligent separate production technology with cable is mainly composed of wireline packer, intelligent separator , ground control cabinet and remote control center.The technological principle is shown in Figure 1.

The technology adopts a one-run string structure. The target strata are equipped with intelligent separator with built-in flowmeters, temperature, pressure sensors and downhole switches. Connected to the ground control cabinet through cables, the intelligent separators switch is controlled by instructions sent from the ground, so as to realize the separate mining and stratified testing of any interval.

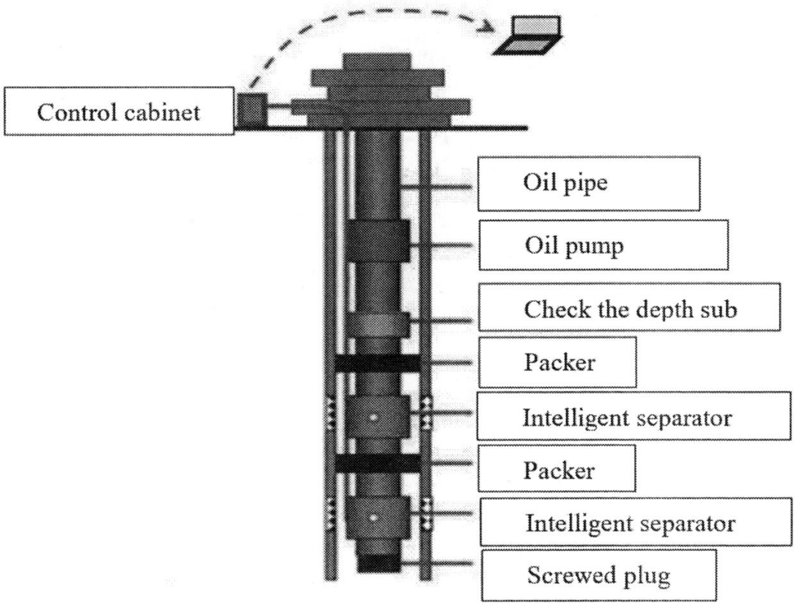


**FIG 1 .** Schematic diagram of intelligent separate production technology with cable

## 3. Key tool research

It innovatively developed the step by step unsealing packer over the cable and the intelligent separator, improved the wall-hook packer over the cable, sealing joint and other supporting tools, and formed the downhole separate production technology string with cable, which met the technical requirements of the three-stage and above intelligent separate production in the field.

### 3.1 Wall-hook packer over the cable

The intelligent separate production technology with cable requires the use of packer over the cable.For this reason, the wall-hook packer over the cable(see Figure 2) is innovatively developed, and an independently developed cable sealing joint is installed on the upper and lower ends of the ordinary packer, so as to solve the problem of packer crossing the cable.The sealing joint adopts high and low pressure double cavity design and three-stage multiple sealing mode, which effectively avoids the contact between the liquid inside the tool and the cable core and ensures the sealing property of the cable joint.The packer is connected in series through a cable joint to enable power and data transmission throughout the string.

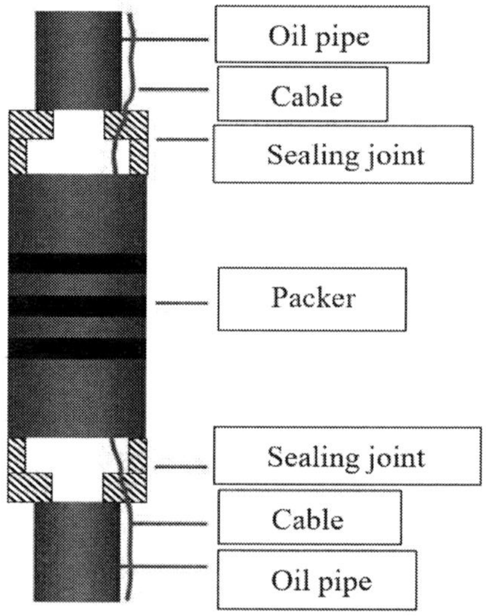


**FIG 2.** Packer crossing the cable diagram

### 3.2 Step by step unsealing packer over the cable

In the production process of multilayer production string, due to the influence of interzonal pressure difference, the packer is easy to be unsealed in advance, which affects the life of string [6-8]. Therefore, a step by step unsealing packer over the cable with pressure difference of 50MPa and temperature resistance of 150°C is designed. As shown in Figure 3, the packer is composed of setting mechanism, step by step release mechanism, pressure balancing mechanism and locking mechanism. With step by step release function, the packer can meet the technical requirements of intelligent separate production of more than three levels in the field.

**FIG 3.** Step by step unsealing packer

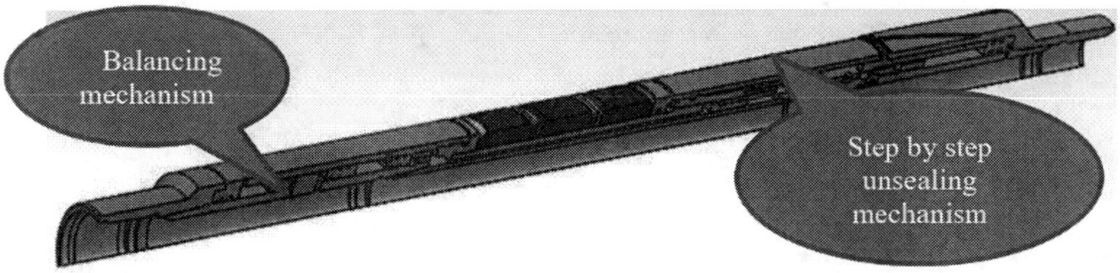


**FIG 4.** Schematic diagram of the step by step unsealing packer

The technical principle is shown in Figure 4. The upper part of the packer is designed with a two-stage balanced piston, which can fully balance the lower thrust caused by the pressure recovery in the non-producing layer and the upper pressure difference caused by the influence of the dynamic fluid level, so as to maintain the balanced force of the packer and avoid premature release due to excessive upper pressure difference.The lower part is designed as a step-by-step release mechanism. During the process of release, the pipe string is lifted up, and the inner and outer center pipes have a relative displacement and shear the release shear nails. The packer rubber cylinder loses pressure and returns to its original state to realize release [9-10]. It can be used in series in separate production wells.

### 3.3 Intelligent separator

The key tool of intelligent separate production technology with cable is intelligent separator, as shown in Figure 5. It uses pressure, temperature, downhole switches, water-holding flowmeters and ground systems. Due to the special needs, each instrument needs a special design, using a single core cable control mode, the instrument power supply for DC, signal transmission using pulse coding mode, the well switch using a negative voltage working mode, other instruments using a positive voltage working mode. Each instrument is addressed, and the ground system sends instructions to control downhole instrument operations or send data.

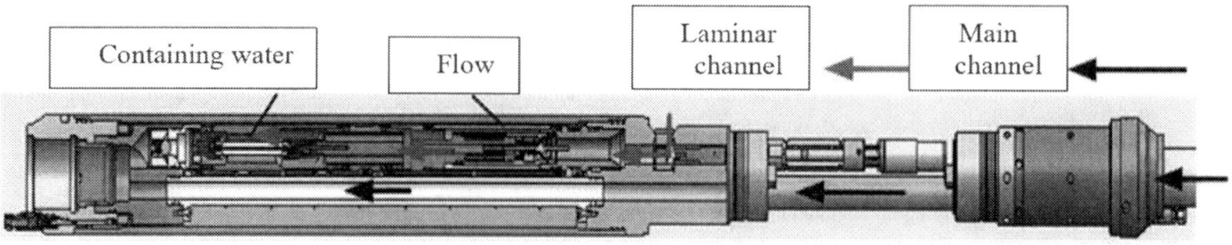


**FIG 5.** Structure diagram of intelligent separator

The strain type manometer uses the bridge composed of strain resistance as the sensor. When the pressure changes, the signal output by the sensor is processed by the circuit into the corresponding pressure data [11-12].The temperature instrument uses a platinum resistance sensor. When the temperature changes, the resistance value of the platinum resistance will change. The circuit also carries out digital processing on it.The strain-type pressure and temperature instrument circuit has an address, and when the instrument is addressed by the ground instrument, it sends pressure and temperature data to the ground system.

The downhole switch panel has 4-bit addresses, ranging from 0000 to 1111, and up to 16-bit addresses.The ground control panel supplies power to the downhole switch. If the ground instrument sends an "on" command, the motor turns forward and the ball screw turns the switch on. If the ground instrument sends a "off" command, the motor reverses and the ball screw turns the switch off.

Water-holding flowmeter , which is divided into water-holding rate and turbine flow rate two parts, the water -holding rate sensor used in this design is capacitive.When the water-holding rate of the fluid near the sensor is different, the capacitance of the water-holding rate sensor also changes. This is the working principle of water holdup sensor to measure water holdup.The circuit further processes the signals from the turbine sensor and the water holdup sensor, waits for the ground system to address them, and finally sends them to the ground system through the cable.

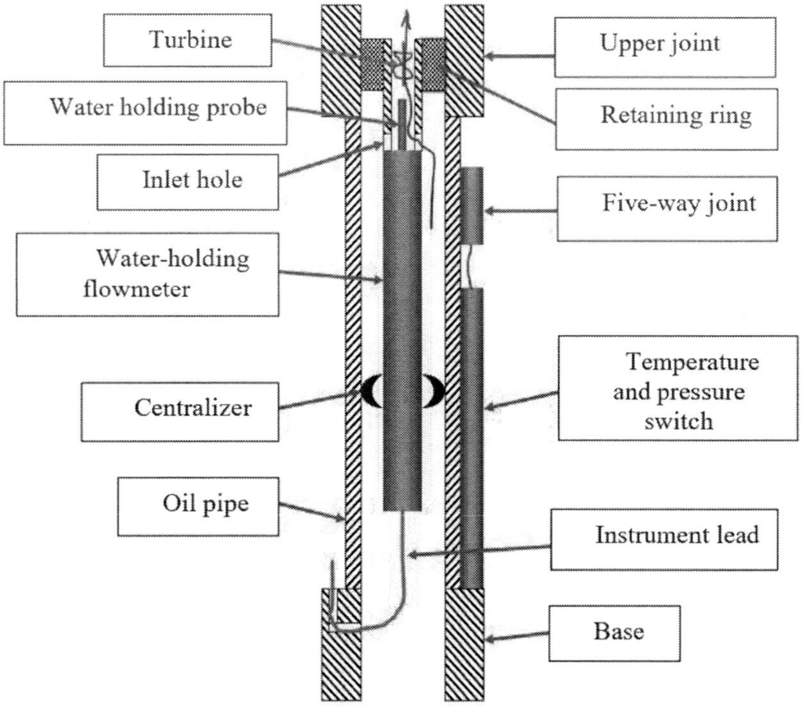


**FIG 6.** Installation diagram of water-holding flowmeter

Remote control process: First use the software of the remote control center to set the basic working parameters, and then input the corresponding instrument address of each level, set the information of each level;After the work information and parameters are set, software is used to send commands to the ground system. The ground system supplies power to the intelligent separator through the cable and sends commands. The intelligent separator starts to work for data acquisition.After the data collected by the downhole intelligent separator is transmitted to the ground system through the cable for signal processing and then transmitted to the remote control center, and then calculated and analyzed by the software, it can accurately observe the production situation of each layer in the downhole.

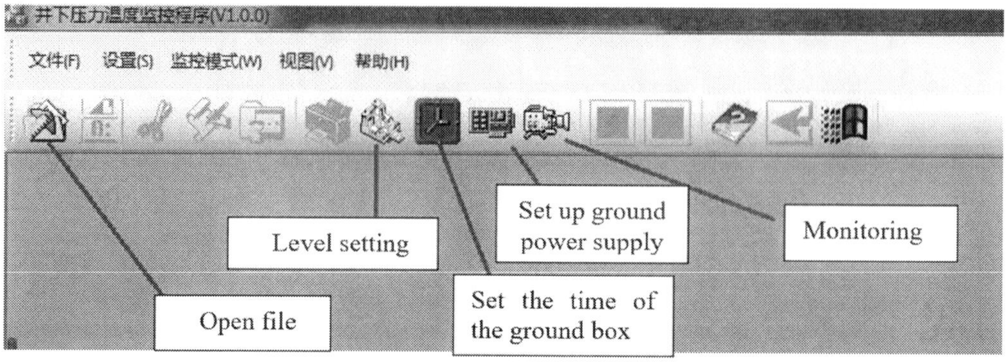


**FiG7.** Software interface of remote control center

## 4. Intelligent separate production technology construction process

(1) Run the separate production string.According to the geological parameters and production distribution requirements of the well, the separate production string was designed, and each layer was isolated with wireline packer, and an intelligent separator was placed at the corresponding position of each production layer.In the process of running the separate production string, the cable is run at the same time, and a cable protector is installed on each tubing collar.

(2) Set the packer.After the string is lowered to the design position, the magnetic positioning is used to calibrate the depth, and the position of the packer clamp point is adjusted according to the calibration data.The ground slowly pumped water into the tubing in small batches to fully set the packer.

(3) Check the packer.The wellhead cable is connected to the surface control cabinet. By injecting water into the stratum, the intelligent separator data of each layer can be read to determine whether the packer is well set.

(4) Run the rod pump barrel and rod string.According to production requirements, the ground controls the switch state of the intelligent separator at the corresponding level for production.

(5) After production for a period of time, according to downhole flow, pressure and other data recorded by the ground control cabinet, the production dynamic analysis of the oil well is carried out again, and the production layer is timely adjusted.

## 5. Field application analysis

As shown in Table 1, the intelligent separate production technology with cable has been applied to a total of 5 wells in central Hebei region of Huabei Oilfield.

**Table 1.** Application of intelligent separate production technology with cable

| Well number | Before the measures | | | After the measures | | |
|---|---|---|---|---|---|---|
| | Monthly liquid production/t | Monthly oil production/t | Moisture content/% | Monthly liquid production/t | Monthly oil production/t | Moisture content/% |
| Cha19-104 | 799 | 47.4 | 94.1 | 671 | 74.3 | 89.0 |
| Cha15-112 | 813 | 8 | 99.0 | 98 | 11.3 | 88.4 |

| | | | | | | |
|---|---|---|---|---|---|---|
| Ma95-28 | 554 | 101 | 81.8 | 222 | 58.4 | 73.7 |
| Liu107-32 | 675 | 7 | 98.9 | 325 | 7 | 97.8 |
| Wen118-10 | 996 | 15 | 98.5 | 269 | 3.3 | 98.8 |

Application Example 1: The intelligent separate production technology with cable test was carried out on well Cha 19-104. The string diagram is shown in Figure 8, and the production situation is shown in Table 2.

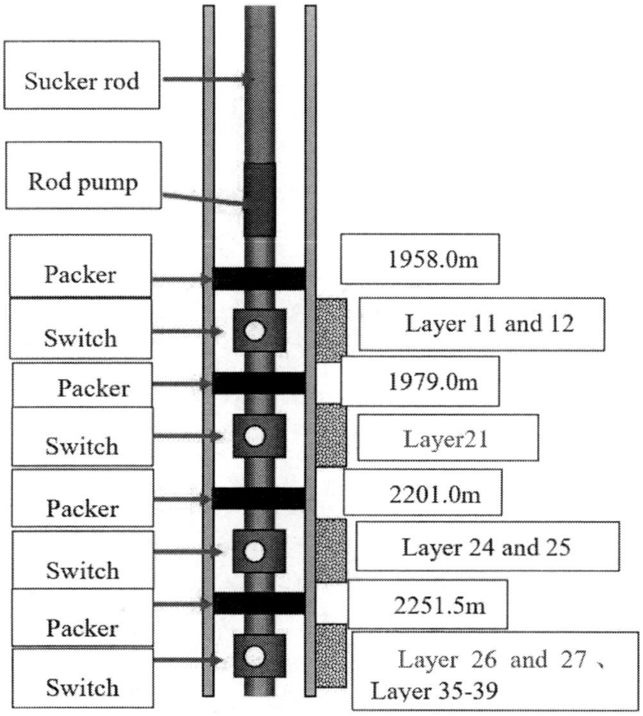


**FIG 8.** Separate layer production string diagram of well Cha 19-104

**Table2 .** Production of well Cha 19-104

| Date | Section1 | Section2 | Section3 | Section4 | Daily fluid production($m^3$/d) | Moisture content (%) | Daily oil production ($m^3$/d) | Cumulative fluid production($m^3$/d) | Cumulative oil production($m^3$/d) |
|---|---|---|---|---|---|---|---|---|---|
| 21.10-21.12.1 | ON | ON | ON | ON | 26.7 | 96 | 1.19 | | |

| | | | | | | | | | |
|---|---|---|---|---|---|---|---|---|---|
| 21.12.26 -22.1.7 | ON | ON | ON | ON | 16.62 | 100 | 0 | 166.17 | 0 |
| 22.1.8-22.2.2 | OFF | OFF | OFF | ON | 22.76 | 98 | 0.5 | 591.73 | 11.43 |
| 22.2.3-22.3.1 | OFF | OFF | OFF | ON | 21.31 | 97 | 0.6 | 2354.89 | 210.65 |
| 22.3.1-22.5.20 | OFF | OFF | OFF | ON | 19.13 | 87 | 2.55 | | |

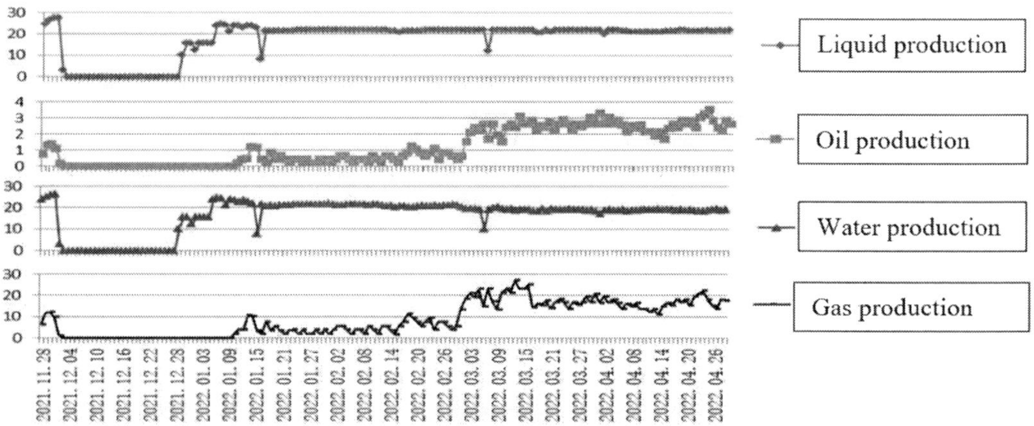


**FIG 9.** Production curve of well Cha 19-104

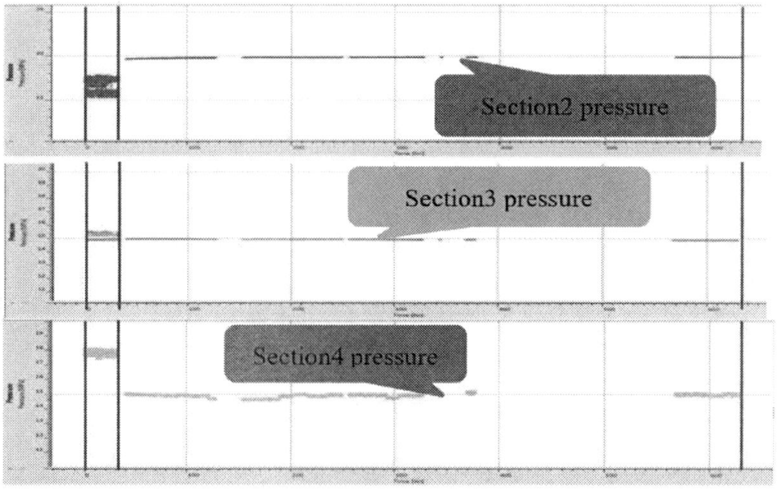


**FIG10.** Pressure test of each layer in well Cha 19-104

The well had been produced separately on December 26, 2021. It's divided into four sections.Before the separate production,fluid daily production is 26.7 $m^3/d$ and oil daily production is 1.19 $m^3/d$.During the single production of the 4th member, the daily wellhead oil production increased from $0.5m^3/d$ in the combined production period to $2.55m^3/d$ in the 4th member, indicating that there was residual potential in the 4th member. Therefore, the fourth member was single production. From March 1 to May 20, the accumulative oil increase was $201.48m^3$, and the accumulative water production was $288.00m^3$ less.

On January 7, 2022, the ground storage terminal began to record pressure data. According to the analysis of the data obtained from the test site, the downhole intelligent separator worked normally, truly recorded the dynamic changes of the bottom hole pressure and temperature, obtained a clear and complete measured curve, and obtained the formation pressure, temperature, formation parameters and other reservoir information. Verify the reliability of separate layer production controlled by intelligent separator.

## 6. Conclusion

(1) Intelligent separate layer production with cable can obtain pressure data at different stages in real time due to the separate pressure and temperature test probes in each section. It has continuous testing and monitoring functions, which is conducive to timely adjustment of reservoir development dynamics and the integration of engineering geology.

(2) According to the field application results of Huabei Oilfield, the cable controlled separate production and measurement technology can realize the function of remote data transmission and remote control switch, which plays the role of accurately measuring the liquid production quantity and properties of each section, plugging high aquifer and improving oil recovery efficiency. Compared with the previous separate layer production , it has obvious advantages.

## References

[1] Liu-ren LI, Yan-yan ZHAO. Quantitative principles and methods for classification and recombination of development strata in waterflood oilfield [J]. Journal of Xi 'an Shiyou University (Natural Science Edition),2016 (06).
[2] Ming-qiang CHEN, Jia-li GE. Fuzzy cluster analysis method for classification of oilfield development strata [J]. Journal of Southwest Petroleum Institute,1987 (02).
[3] Yong-gang GONG, et. First application of intelligent production string in Bohai oilfield [J]. Scientific Management,2018.
[4] Ying-zi CHENG, et .A new intelligent production by layer and testing technology for oil Wells [J]. Drilling & Production Technology,2015,38(3) : 46-48.
[5] Zu-qing HE, Cheng-chun LIANG, Han-xiu PENG, Ming ZHU , Tong HE. Research and experiment on intelligent production technology of horizontal well in tight reservoir in southern Ordos Basin [J]. Petroleum Drilling Technology,2017.
[6] Yi-fan WU. Research and application of multi-layer separation production techn ology in oil well [J]. Chemical Industry Management,2013 (12).
[7] Da-jian LI , Cai-yun NIU, Miao HE , Bai WANG, Hong-zheng ZHU, Wen-hao CUI . Discussion on technology of multilayer production separation in Changqing Oilfield [J]. Petrochemical Application,2013 (09).
[8] Shengli Oilfield completed the first intelligent production field test [J]. Petroleum Drilling Techniques,2009 (03).
[9] Min GAO. Contact Stress Analysis and Structure Optimization of Compression Packer [D]. China University of Petroleum (East China),2014.
[10] Xing CHEN. Design and Analysis of Sealing Mechanism of Compression Packer [D]. China

University of Geosciences (Beijing),2015.
[11] Jie ZHAO. Research on Key Technologies of Direct-reading Downhole Pressure Gauge [D]. Huazhong University of Science and Technology,2016.
[12] Chang-ling HU, Hua ZHANG, Peng ZHANG, Hai-tao SONG. Research on high precision realization method of logging pressure gauge [D]. Petroleum Instruments,2009.

# Physical Simulation of Steam Absorption Characteristics in Horizontal Wells of Super Heavy Oil Reservoir

**Yu-junLI1,2(✉)**

1. National Energy Heavy (Super Heavy) Oil Exploitation Research and Development Center, Panjin 124010
2. PetroChina Liaohe Oilfield Company, Panjin 124010

jun_liyu@126.com

**Abstract:** In the process of dual Horizontal well SAGD development of super heavy oil, the steam absorption balance of upper horizontal steam injection well is an important factor to determine the development effect of SAGD. The factors affecting the steam absorption of upper horizontal wells involve reservoir permeability, steam injection technology, steam injection parameters, etc. The physical simulation method is used to study the influence degree of each specific factor and improvement measures. Based on the similarity criterion, the reservoir scale model is designed and made according to the oilfield prototype, and the simulation experiment is carried out on the steam absorption characteristics and influencing factors along the horizontal wellbore. The experimental results show that the steam injection rate, non condensable gas, the location of the steam injection string and the reservoir permeability have great effects on the steam absorption of the reservoir. The heating uniformity along the horizontal well can be improved by properly increasing the steam injection speed, adding a certain amount of non condensable gas to the steam, and the position of the steam injection string end in the relatively low permeability area of the reservoir. This achievement has a strong guiding significance for improving the steam absorption balance of horizontal wells and improving the development level of dual level SAGD.

## 1 Introduction

Double horizontal well SAGD technology can significantly improve the development of super heavy oil reservoir. In recent years, with the continuous improvement of drilling directional technology, the double horizontal well trajectory control technology has been developed unprecedentedly, the double horizontal SAGD technology has been widely used in the development of super heavy oil. The steam absorption along the upper horizontal steam injection well bore of the double-horizontal SAGD is uneven affected by the steam injection process, steam injection parameters and reservoir heterogeneity, which resulting in the uneven expansion of the steam cavity which seriously affects the development level of the reservoir; It is a difficult problem for the research of steam injection horizontal wells to make the reservoir uniformly heated along the horizontal well.
The development process of steam injection in horizontal wells of super-heavy oil reservoirs is simulated by the physical simulation device of horizontal wells of super-heavy oil with different

reservoir characteristics under different steam injection conditions. According to the experimental results, specific steam injection methods for optimizing horizontal wells are proposed.

## 2 Physical model establishment

### *2.1 Experimental device*

The experimental device is a device designed to simulate the steam absorption condition along the wellbore of horizontal wells for heavy oil thermal recovery. The main instruments and equipment used in the experiment are: simulation device for steam absorption along the wellbore of heavy oil thermal recovery horizontal well, data acquisition device, six-way check valve, crude oil input device, steam generation device, ISCO high-pressure pump, gas flow controller, etc.

The process of the experiment is as follows ： First, the ISCO high-pressure pump is used to inject crude oil into the model body, and the injection pressure refers to the simulated reservoir pressure. At this time, the sand pipe of the model body is filled with crude oil, which can simulate the original reservoir state before steam (gas) injection; Then, according to the actual experimental needs, inject steam and other gases or mixed gases to be tested; Under the model body, connect the one-way pressure control valve and the vapor liquid collection device to regulate the "reservoir" pressure and prevent the "reservoir" from being distorted due to the pressure surge caused by the injection of steam and other gases; On the model body, install temperature and pressure sensors, and use the data acquisition system to record and monitor the temperature and pressure of the "reservoir" in real time (Figure 1).

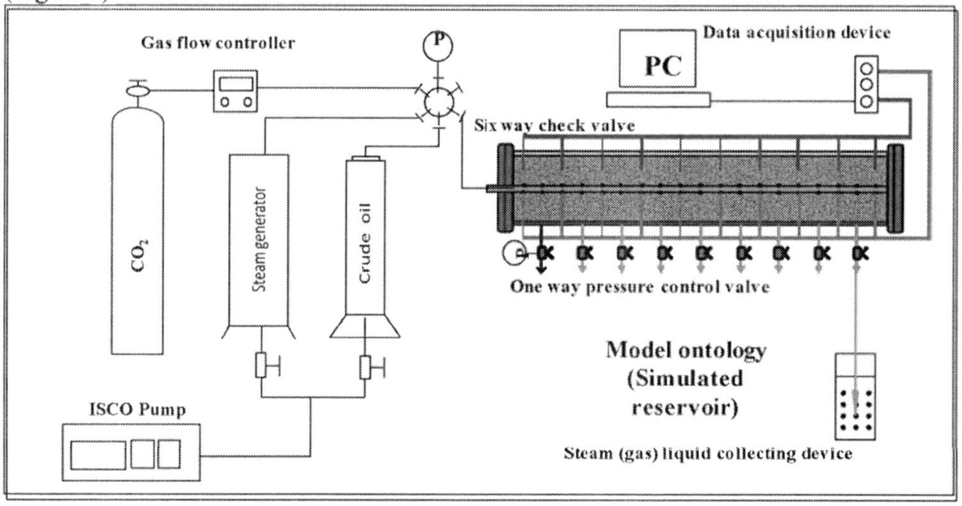


Fig. 1 Flow chart of simulation test for steam absorption along the wellbore of heavy oil thermal recovery horizontal well

### *2.2 Parameter settings related to physical model*

The high temperature and high pressure proportional physical simulation system device is used in the experiment to ensure the reliability of the physical simulation and make the simulation results could play a guiding role in the field development,. The artificial core of the simulated reservoir is made of quartz sand, whose thermal physical parameters are basically the same as that of the quartz sandstone in the field reservoir; The crude oil in the simulated reservoir in the model body is the crude oil collected on the oil field; The steam produced by the steam generator is deionized water for steam injection collected on the oil field also, and the temperature is adjusted to the actual conditions of the reservoir; This ensures the progressiveness, rationality and reliability of the physical model simulation results.

The similarity criteria commonly used in physical simulation are used to determine these parameters for reservoir physical properties (mainly permeability and porosity), injection-production time, horizontal well section length, and perforating parameters of horizontal wells cannot be consistent with the actual field conditions.

According to the specific problems to be solved in this experiment, the number of similarity criteria that play a leading and decisive role in the results is determined, and the following six main similarity criteria are selected as the number of similarity criteria for this physical simulation. See Table 1 for its physical meaning and the physical quantity to be simulated.

Table 1 Physical significance of main similarity criteria parameters of physical simulation

| Similarity criteria data | Physical significance | Analog parameter |
|---|---|---|
| $\frac{K\Delta\rho g t}{\mu L}$ | Ratio of gravity to viscous force | permeability |
| $\frac{\alpha_o}{vL}$ | Ratio of heat conduction to convection | Injection rate |
| $\frac{K\Delta p}{vL\mu}$ | Ratio of driving force to viscous force | Injection-production pressure difference |
| $\frac{vt}{L}$ | Ratio of flow scale to spatial scale | time |
| $\frac{\rho v D}{\mu}$ | The ratio of inertial force to viscous force of fluid flow in horizontal wellbore | Horizontal wellbore radius |
| $\frac{n r_p^4 L}{r_w q_m}$ | Perforation parameters, ratio of perforating flow to axial flow | bore diameterand number of perforations per unit length |

In the table, K is absolute permeability, μ m2; Δρ Is the density difference of water and oil, kg/m3; G is the acceleration of gravity, m/s2; T is time, s; μ Is the fluid viscosity, mPa · s; L is the length, m; α0 is the thermal diffusion coefficient of oil layer, m2/s; V is the seepage velocity, m/s; Δ P is the injection-production pressure difference, Pa; ρ Is the fluid density, kg/m3; D is the inner diameter of horizontal well, m; N is the number of perforations per unit length of horizontal section, number/m; Rp is the diameter of perforation, m; Rw is the radius of horizontal shaft, mm; Qi is the volume flow of fluid in the wellbore, m3/s.

Through calculation and according to the actual situation, the parameter values of model design are shown in Table 2.

Table 2 Model design parameter values

| parameter | Reservoir thickness(H), m | Horizontal well length(L), m | Permeability (K), μm2 | Wellbore radius(RW), mm | Equivalent hole radius(rp), mm |
|---|---|---|---|---|---|
| Prototype value | 50 | 196 | 1.5 | 48 | 1.8 |
| Actual value | 0.25 | 0.98 | 300 | 3.4 | 0.9 |

## 3 Experimental scheme design and steps

The experimental device uses the same N80 steel pipe with a radius of 3.4mm as the field to simulate the horizontal wellbore, and uses the equivalent hole radius to perforate the horizontal wellbore.
In the simulation of steam injection system, the main way is to change the steam injection pressure by adjusting the pressure of ISO pump, so as to simulate different steam injection speed and steam injection volume in the field; CO2 injection rate and gas injection volume are controlled by gas flow controller; At the same time, observe and correct the injection pressure and reservoir pressure through the pressure gauge on the six-way valve. In order to facilitate the simulation of the influence of injection mode on the heating along the horizontal well, the pipeline connection with a radius of 1.89mm of J55 steel grade high temperature and high pressure resistance is used between the six-way valve and the model body.
According to the field steam injection control mode, the following four different steam injection parameters and experimental schemes under reservoir conditions are designed to study the steam absorption along the horizontal well bore:
(1) Change the injection steam speed under homogeneous reservoir: 50mL/min, 75mL/min and 90mL/min;
(2) Change the injection medium under homogeneous reservoir: add CO2 in water steam and water steam (in gaseous state, the content is 50% respectively);
(3) Change injection mode under homogeneous reservoir: heel, toe, middle and both ends of steam injection horizontal well;
(4) Change reservoir conditions: homogeneous reservoir, heterogeneous reservoir with different properties;
According to the design scheme of physical simulation, the model body is designed according to the reservoir before the experiment; After completing the design of the model body, CO2 is used to verify the sealing of the whole experimental process. After verifying that there is no problem, high-temperature steam is injected into the model body for more than 40 minutes. At the same time, the water in the reservoir is saturated, and the model body is fully heated, so that the crude oil can be fully saturated when injected.
After the completion of the above basic work, carry out experiments according to different schemes, collect data and analyze the results.

## 4 Analysis of sensitive factors of steam absorption along the wellbore of horizontal wells.

In the process of thermal recovery and development of heavy oil, the injection parameters (mainly the injection rate of fluid), the composition of the injected fluid, the injection mode (the position of the steam outlet of the steam injection string), the heterogeneity of the reservoir and other conditions of the steam injection horizontal well have a great impact on the pressure drop and temperature transmission process along the wellbore of the steam injection horizontal well.

### *4.1 Effect of steam injection speed*

Reservoir simulation device under test is set as a homogeneous reservoir (permeability of 300 Darcy), and the steam injection rate is divided into 50mL/min, 75mL/min and 90mL/min at the heel of the steam injection horizontal well, and the water steam is continuously injected into the homogeneous model. According to the temperature and pressure data extracted from the data acquisition device, the characteristic curves of pressure and temperature change along the horizontal wellbore with different steam injection rates under the condition of homogeneous reservoir are drawn (as shown in Fig. 2 and Fig. 3).

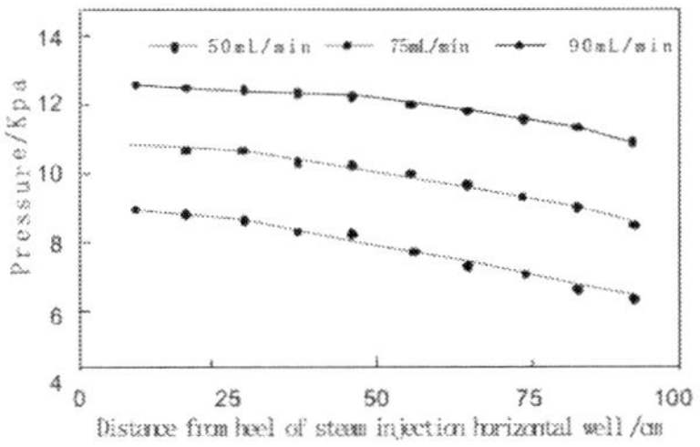


Fig. 2 Variation characteristics of steam injection rate and wellbore pressure

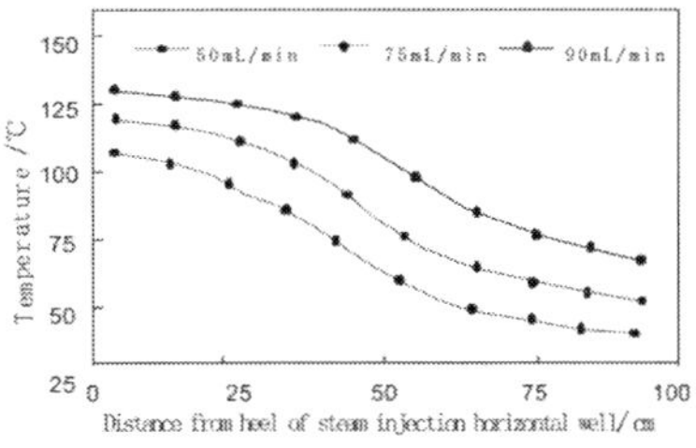


Fig. 3 Characteristics of steam injection speed and wellbore temperature

It can be seen from Fig. 2 and Fig. 3 that with the increase of steam injection rate, the pressure loss along the way decreases and the temperature decrease decreases under the homogeneous reservoir model (that is, without considering the influence of reservoir heterogeneity): when the steam injection rate is 50mL/min, the pressure at the steam injection foot is 8.8KPa and the temperature is 107 ℃; The pressure at the toe is 6.3KPa and the temperature is 31 ℃, in which the pressure drops by 2.5 KPa and the temperature drops by 76 ℃; When the steam injection speed is 75mL/min, the pressure at the steam injection foot is 11.4KPa, the temperature is 122 ℃, the pressure at the tip is 8.2KPa, the temperature is 61 ℃, the pressure decreases 2.2 KPa, and the temperature decreases 61 ℃; When the steam injection speed is 90mL/min, the pressure at the steam injection foot is 12.8KPa, the temperature is 128 ℃, the pressure at the tip is 10.9KPa, the temperature is 73 ℃, the pressure drop is 1.9 KPa, and the temperature drops 53 ℃.

According to the above data, the steam injection pressure loss and heat loss can be significantly reduced with the increase of steam injection speed, the heat utilization rate can be improved. That is to say, when formulating the steam injection plan for horizontal wells, increasing the steam injection rate can reduce the heat loss, improve the heat utilization efficiency, and also have a certain impact on the heating uniformity of the horizontal section of the reservoir.

### 4.2 *Effect of injected fluid*

Under the condition of homogeneous reservoir (the reservoir permeability is 300 Darcy), follow injection is adopted to inject steam and 50% steam+50% $CO_2$ into the reservoir model at the injection rate of 75 mL/min. According to the 10 pairs of temperature and pressure data extracted from the data acquisition device, the pressure and temperature changes at different locations along the horizontal wellbore of a homogeneous reservoir when injecting different fluids are plotted as shown in Figure 4 and Figure 5.

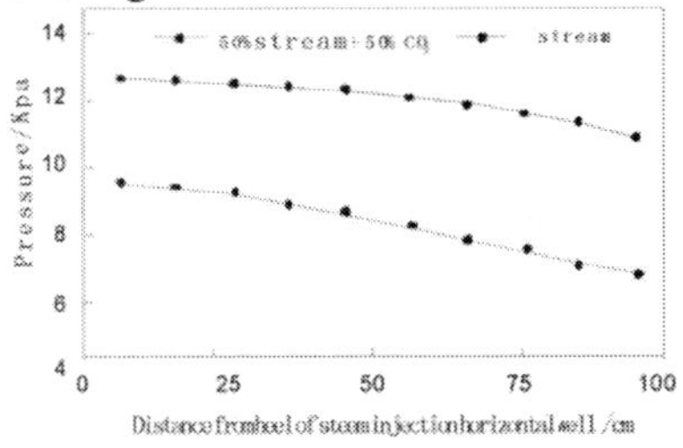


Fig. 4 Pressure variation characteristics in wellbore with different injection fluids

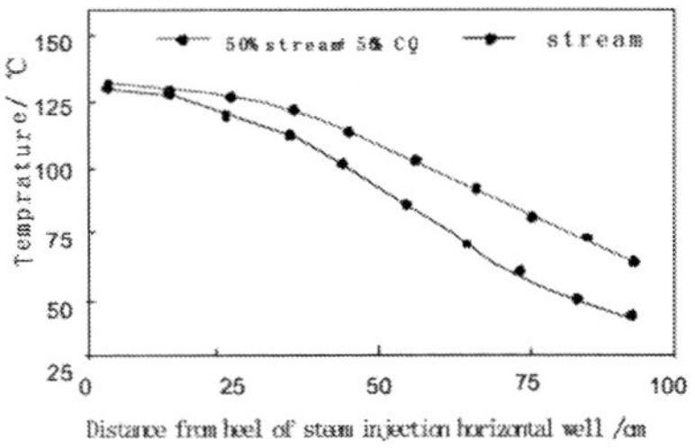


Fig. 5 Characteristics of wellbore temperature along different injection fluids

It can be seen from the figure that compared with the injection of pure steam, adding a certain amount of non-condensable gas (CO2 is mixed in this experiment) to the steam, the pressure drop loss and heat loss along the horizontal well bore are smaller, and the heating uniformity along the horizontal well section is more obvious. The main reason is that a certain amount of non-condensable gas is added to the steam. Due to the partial pressure effect of non-condensable gas, the water vapor pressure in the mixed steam decreases, the temperature of water saturated steam decreases, and the proportion of steam condensed into water decreases. Therefore, the density and viscosity of the injected fluid can be significantly reduced, and the pressure loss along the horizontal well can be reduced.

*4.3 Influence of injection mode*

Under the condition of homogeneous reservoir (reservoir permeability is 300 Darcy), steam is injected into the model by four injection methods: horizontal well heel injection, middle end injection, toe end injection and two end injection, with the injection rate of 75mL/min. According to the collected temperature and pressure data, the characteristic curves of pressure and temperature changes along the injection horizontal wellbore under different injection modes are drawn as shown in Fig. 6 and Fig. 7.

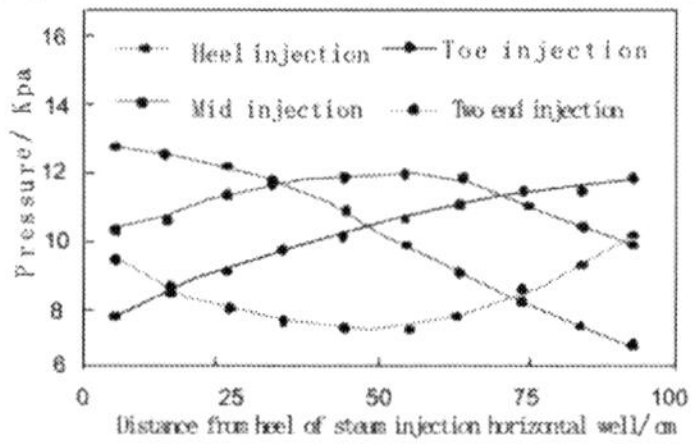


Fig. 6 Characteristics of pressure change in wellbore with different injection methods

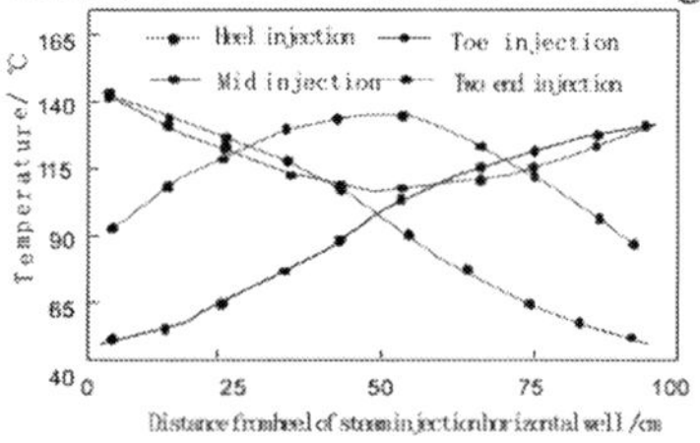


Fig. 7 Characteristics of wellbore temperature along different injection methods

It can be seen from Figure 6 that under the condition of homogeneous reservoir, no matter which injection method, the pressure at the steam outlet of the steam injection string is the largest, and the pressure decreases as the distance from the steam injection port increases. The order of pressure drop from large to small is: heel injection mode=toe injection mode>two end injection mode>middle end injection mode.

For the temperature along the path (Fig. 7), the reservoir temperature of the injection method at both ends is the highest, that is, the thermal efficiency is the best, and the temperature along the path is relatively uniform; The mid-end injection method takes the second place, while the heel and toe injection methods have the opposite characteristics of reservoir temperature rise although the thermal efficiency is basically one to one.

To sum up, the following conclusions can be drawn: different injection methods (different injection port locations) have significantly different characteristics of pressure and temperature changes, so you can change the injection method (changing the steam injection port) to make the reservoir evenly heated, and reduce the steam consumption and preheating time during preheating of SAGD horizontal wells.

*4.4 Steam absorption characteristics of different reservoir conditions*

In order to simulate the heating along the steam injection horizontal well under different reservoir conditions, the model body (artificial reservoir) is divided into three parts according to equal length, and the reservoir permeability of each part is set according to the experimental needs.

In this experiment, three kinds of reservoirs with different properties are set up: homogeneous reservoir (reservoir permeability is 320μm2)； Permeability in the middle of reservoir 430μm2, the permeability of other parts is 180μm2 and middle reservoir permeability 180μm2, 430 for other partsμ m2 。 The injection method is heel injection, and the steam injection speed is 75mL/min. The temperature and pressure data under different reservoir conditions are extracted respectively, and the characteristics of pressure and temperature changes along the horizontal wellbore are drawn (Fig. 8, Fig. 9).

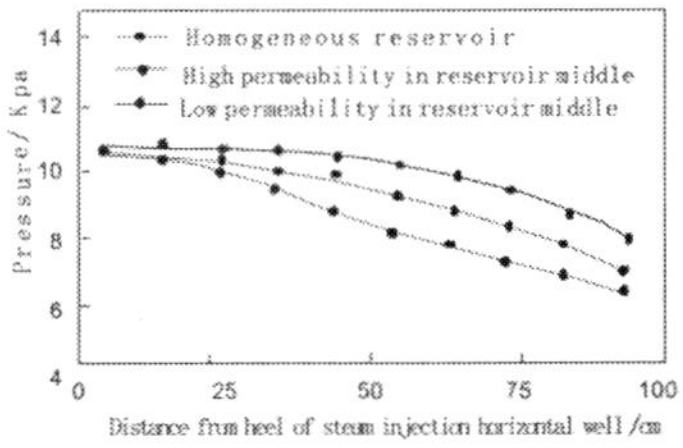


Fig. 8 Pressure variation characteristics in wellbore under different reservoir conditions

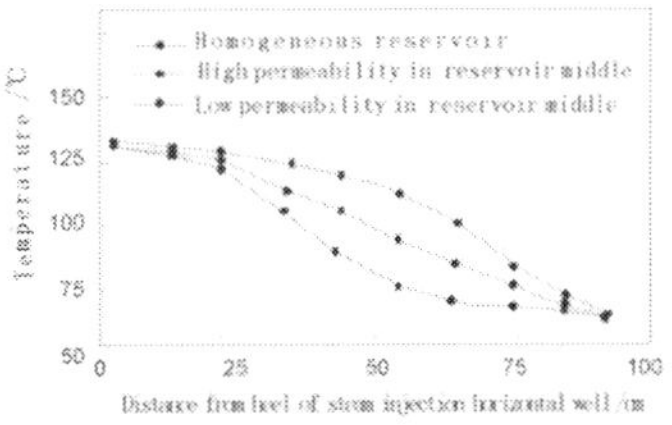


Fig. 9 Temperature characteristics along the wellbore under different reservoir conditions

It can be seen from the figure that the effect of steam injection in horizontal wells on heating heavy oil reservoirs is greatly affected by the permeability of reservoirs along the horizontal wells. In terms of the pressure along the way, the pressure drop of the middle high permeability reservoir is significantly lower than that of the homogeneous reservoir and the middle low permeability reservoir. In terms of the temperature along the way, the middle high permeability reservoir is not only high in temperature, but also more evenly heated.

The above results show that the steam injection point is located in the relatively low permeability area of the reservoir, which can improve the utilization rate of steam heat and make the heating of the reservoir more uniform.

## 5 Conclusion

(1) Applying physical simulation experimental technology to simulate the steam absorption along the wellbore during steam injection in heavy oil horizontal wells according to similarity criteria can provide theoretical basis for SAGD preheating and improving the uniform production of reservoirs.
(2) Increasing the steam injection rate and mixing non-condensable gas in the steam can reduce the heat loss along the way, and the heating uniformity of the reservoir is good, and the steam injection effect is improved.
(3)Adding a certain amount of non condensable gas (mixed with $CO^2$ in this experiment) to the steam can reduce the pressure drop and heat loss along the horizontal wellbore, and the heating uniformity along the horizontal wellbore is more obvious.
(4) When the injection mode is changed (the position of steam injection port), the characteristics of pressure and temperature change along the horizontal well are obvious; The reservoir permeability has a great influence on the heating effect of the reservoir.
(5) For reservoir permeability, placing the steam injection point in the low permeability area can improve the steam absorption profile of horizontal wells along the way, so that the horizontal wells are evenly heated along the way, thus improving the development effect of super heavy oil.

**Acknowledgments.** The project is supported by National Major Science and Technology Project "Key technologies for heavy oil/super heavy oil development"(Number：2016ZX05012).

**Reference**

[1] Butler R M. Rise of interfering steam chambers [J].,Journal of Canadian Petroleum Technology, 1987，26( 3):70 75.
[2] He Z, Murtaza M， Dehghanpour H. Modeling three-phase flow during steam chamber rise-impact of water drainage on oil production rate[C]//SPE Heavy Oil Conference-Canada. Calgary: One Petro, 2013: 9-16.
[3] Azad A, Chalaturnyk R J.An improved SAGD analytical simulator: circular steam chamber geometry[J].Journal of Petroleum science &Engineering, 2012, 82-83( 2):27-37.
[4] Chung K H, Butler R M. Geometrical effect of steam injection on the formation of emulsions In

the steam-assisted gravity drainage process[J].Journal of Canadian Petroleum Technology, 1988，27(1):36-42.
[5] Kimber K D, Puttagunta V R, Ali S. New scaling criteria and their relative merits for steam recovery experiments[J].Journal of Canadian Petroleum Technology, 1988，27( 4):42-57.
[6] Kimber K D，Ali S. Verification of scaling approaches for steam injection experiments[J].Journal of Canadian Petroleum Technology , 1989，28( 1):33-40.
[7] YUAN Hemin, HAN Dehua, ZHANG Weimin. Heavy oil sands modeling during thermal production and its seismic response[J].Geophysics. 2016. 81(11: 57-70.
[8] Doan Q，Ali S M F, George E A. Scaling criteria and model experiments for horizontal well[J].Journal of Canadian Petroleum Technology, 2012，31( 9): 57-65.
[9] LI Yu-jun, REN Fan-xiang, YANG Li-qiang et al. Description of steam chamber shape in heavy oil recovery using 4D microgravity measurement technology [J]. Petroleum Exploration and Development, 2013, 40(3):381-384
[10] CHENG-hao, SU Zhu-liu, YAN Ling-jun et al. Applications of Transient Electromagnetic Method in Dynamic Monitoring of SAGD for Ultra Heavy Oil in Shu 1 Area [J].Journal of Oil and Gas Technology, 2011, 33(7):104-107.
[11] ZHAO Qing-hui, LIU Qi-cheng, YUTao. Research On The Characteristics Of Steam Chamber Development Of Steam-Assisted Gravity Drainage [J]. Journal of Southwest Petroleum University(Science & Technology Edition),2008,30(04):123-126.
[12] ButlerRM. Steam-assisted gravity drainage: concept, development, performance and future [J]. Journal of Canadi-an Petroleum Technology, 1994, 33(2): 44-50.
[13] ZHANG Xing-yuan, LI Ting XIE Yang-bo, et al. Impacts of inter layers on SAGD[J].Special Oil &Gas Reservoirs, 2017,24(5):120-125.
[14] YANG Zhi, Meng Xiang-bing, Wu Yong-bin, Zhao Hui-long, Luo Chi-hui, Gan Shang-shan. Dual horizontal Wells in Shallow Super-heavy Oil Reservoir in Fengcheng Oilfield, [J].Special Oil &Gas Reservoirs, 2021,28(1):92-97.
[15] CHEN Hao, Su Zhu liu, YAN liangJun et al. Applications of Transient Electromagnetic Method in Dynamic Monitoring of SAGD for Ultra Heavy Oil in Shu 1 Area [J]. Journal of Oil and Gas Technology, 2011, 33(7):104-107.
[16] LIN Riyi, Qi Shangchao, Shen Wenli, et al. Rules of steam injection parameters in the preheating stage of cycle [J]. Joural of China University of Petroleum (NATURAL SCIENCE EDITION), 2018, 42 (1): 134-140.
[17] LI Jingling, Zhu Zhihong, Dou Shengjun, et al. Calculation of preheating heat transfer and analysis of influencing factors in double horizontal wells [J]. Xinjiang Petroleum Geology, 2014, 35 (1): 82-86
[18] LIU Peng, Xia Yinchu. A new steam injection method for heavy oil horizontal well [J]. Science and technology and engineering, 2011, 11 (08): 1786-1789.
[19] CHEN Zhijun, Huang Xiaodong, Peng Hui, et al. Research and application of uniform steam injection technology in heavy oil horizontal wells [J]. Xinjiang Petroleum Technology, 2010,20 (03): 9-11.
[20] YIN Hu,LIU Hui,LI Qian, et al.Technical Research on Improving Efficiency of Steam Injection in Horizontal Wells for Heavy Oil Reservoirs[J].Journal of Oil and Gas Technology,2011,33(12):119-123
[21] HAN Dehua, YAO Qiuliang, ZHAO Huizhu, et al. Challenges inheavy oil sand measurements[R]. Houston, USA: Sponsored Program Annual Meeting for Fluid/DHI, 2007.
[22] LI Hui, ZHAO Luanxiao, HAN Dehua, et al. Elastic properties of heavy oil sands: Effects of temperature, pressure, and microstructure[J].Geophysics, 2016, 81(4): 453-464.
[23] ASHRAFI O, NAVARRI P, HUGHES R, et al. Heat recovery optimization in a steam-assisted gravity drainage (SAGD) plant[J].Energy, 2016, 111:981-990.
[24] GAO Yunfeng, FAN Ting'en, GAO Jinghuai, et al. Monitoring of steam chamber in steam-assisted gravity drainage based on temperature sensitivity of oil sand[J]. Petroleum Exploration

and Development, 2021, 48(6): 1224-1231.
[25] Zhu Yunpeng, Yan Gengcheng, Zheng Gang, et al. Principle and application of numerical simulation analysis of vertical well and horizontal well assisted gravity drainage [J] Science Technology and Engineering, 2021,21( 30):12916-2924.

Journal of Physics: Conference Series **2594** (2023) 012035 doi:10.1088/1742-6596/2594/1/012035

# Nitrogen Foam Alternating Flooding Experiment in The Offshore Strong Bottom Water Sandstone Reservoir

**Hui CAI, Zhang ZHANG, Jie TAN(✉) , Chun-Yan LIU, Wen-tong ZHANG**

Tianjin Branch of CNOOC Ltd., Tianjin, China

75657043@qq.com

**Abstract**. The CFD oilfield is located in the Bohai Sea area and is a typical offshore sandstone reservoir with strong bottom water. The characteristic of the rapid increase in water content is that it quickly enters a highly high water content period, and there currently needed to be more potential tapping measures affected by offshore production facilities. The development mode should be changed to explore the development effect of nitrogen foam flooding. This study conducted two experiments. The first experiment is a gas injection expansion experiment, which studied the changes in bubble point pressure, fluid volume, density, and viscosity under different gas injection conditions. The second experiment used a high-temperature and high-pressure oil displacement device and a long core for oil displacement experiments. Two typical sand bodies were selected for the $N_2$ foam oil displacement experiment. The experimental results show that nitrogen injection has little effect on the properties of reservoir crude oil, with a viscosity reduction effect of 12.4% for heavy oil and 20.0% for thin oil. The viscosity of crude oil significantly impacts the selection of alternative media. The channel was prone to blockage and low oil displacement efficiency when injecting a single gas medium into a heavy oil system. The displacement efficiency of thin oil nitrogen foam is 73.28%. The displacement efficiency of nitrogen foam flooding for heavy oil is 43.09%. This study has a guiding significance for developing development plans for offshore sandstone reservoirs.

## 1. Introduction

The CFD oilfield is located in the Bohai Sea area, a typical offshore sandstone reservoir with strong bottom water. There are many sets of oil-water systems from top to bottom, the viscosity range is 30~350cp, and the fluid properties change significantly. The water content rises very fast, and it would soon enter the ultra-high water content period. Affected by offshore production facilities, there are few potential tapping measures at present. It has been developed for more than ten years, facing low oil recovery and high water cut problems. It is necessary to study effective EOR technology.

Higher water cut is an essential factor hindering the improvement of oil and gas field recovery. This phenomenon has appeared in oil fields developed for many years. Petroleum workers actively deal with the production problems and give a series of practical methods[1]. In 2005, a rapid increase in water cut-rate appeared in well 28-8 in two areas of the Gudao area in the Shengli oil field. Zhang Xiansong and others proposed that nitrogen foam should be used to enhance oil recovery after polymer flooding[2]. The enhanced foam flooding indoor flooding experiment was carried out, and the quartz sand filling model with a 1500mD penetration rate was used. It was found that when the injection flow rate was 0.5mL/min, the backpressure was 6MPa, the simulated oil was 74cp, the sealing capacity of the foam was the strongest when the gas-liquid ratio was 1:1, and the stability of the foam system was the best when the best polymer concentration was 1800mg/L[3]. At the same time, the laboratory test results were applied on site, and the breakthrough of Nissan oil from 10m3/d to 43$m^3$/d was achieved. Water cut also decreased from 95% to 71.5%[4].

The Jia Min effect in foam flooding can improve the displacement fluid's swept volume and washing efficiency[5]. A pressure drop funnel would form near the well so that the remaining oil on both sides of the main diagonal line would be used to form an oil wall extending from the mainline to the two sides, forming a longer water-bearing low-value production stage[6]. Nitrogen foam is mainly concentrated near the wellbore. During the subsequent water flooding process, foam forces the liquid flow to turn, thus reducing the water content of the reservoir[7]. The following year, the injection mode was optimized for the same block, and it was found that the gas-liquid alternate injection mode could effectively control the gas channeling phenomenon[8].

Polymer is usually used for mobility control in the chemical flooding EOR method, but for some high temperature and high salinity reservoirs, the mobility control ability of polymer would be limited. In these unique reservoirs, foam is used instead of polymer for fluidity control to ensure recovery of chemical flooding[9]. The decrease of foam mobility is mainly divided into two stages. The initial fluidity of foam decreases by 1 order of magnitude. The decline of bubble mobility is mainly divided into two stages: the initial stage of bubble propagation and the decrease of mobility by 1 order of magnitude. At the same time, with the liquid dilution, the surfactant concentration increases, and the fluidity continues to decrease by 1~2 orders of magnitude. The gas channeling phenomenon in the thermal recovery wells of the Girasol oilfield in Columbia has caused the high water cut problem of the reservoirs[10]. Zhu Ming and other researchers have developed a high-temperature foam system with a temperature above 300℃ for the particularity of the thermal recovery wells, which has become the research focus. The YL-10 foam agent suitable for this type of oil reservoir was screened out through a comprehensive evaluation of foaming, dynamic, plugging, and oil displacement performance[11].

Research institutions and oil fields have successfully carried out foam flooding research and field tests. Field tests have achieved a good oil displacement effect, and oil recovery has been dramatically improved[12]. However, for these oilfields, the subsequent promotion of foam flooding could be more satisfactory [13], so further theoretical and experimental studies are still needed. With the development of nitrogen foam flooding technology, some problems have also been brought about. There are three common problems. One is that crude oil has a certain degree of influence on the stability of the foam. There is a phenomenon of oil defoaming. The two is that most nitrogen foam flooding experiments are carried out under normal pressure. There is no simulation of the actual situation of the reservoir, and the three is that the core is prone to gas channeling, and the experimental results are not accurate. Therefore, this paper uses high-temperature and high-pressure oil displacement devices and long core to effectively weaken gas channeling to conduct oil displacement experiments and analyze and study different oil reservoirs and oil viscosities. It guides the design of the development scheme of offshore strong bottom water reservoirs.

## 2. Reservoir physical properties and fluid properties

In this study, two typical reservoirs in the CFD oilfield, N and G, were used to simulate nitrogen reservoir foam displacement using high pressure and high-temperature equipment to simulate reservoir conditions. See Table 1 for reservoir and fluid properties.

Table 1 Reservoir and fluid properties of two typical reservoirs

| Reservoir name | Fluid properties | | Physical properties | | | lithology | Reservoir temperature ℃ | Reservoir pressure MPa | GOR $m^3/m^3$ | Bubble point pressure MPa | Bubble point pressure crude oil viscosity cp |
|---|---|---|---|---|---|---|---|---|---|---|---|
| | Viscosity mPa·s | Density g/cm³ | K mD | Porosity % | Thickness m | | | | | | |
| the N reservoir | 142 | 0.97 | 2600 | 27 | 32 | Sandstone | 65 | 10 | 14 | 5.87 | 125.7 |
| the G reservoir | 30 | 0.94 | 1600 | 24 | 41 | Sandstone , middle interlayer | 84 | 13 | 6 | 2.96 | 23.1 |

## 3. Gas injection expansion test

The expansion experiment aims to study the changes of reservoir fluid parameters under different gas injection conditions, such as bubble point pressure, fluid volume, density, viscosity, and other main parameters. The reservoir fluid test sample is prepared in Table 4, and the injected gas is $N_2$.

### 3.1. Experimental process and steps

The experimental study was also completed in the JEFRI mercury-free high temperature and high-pressure reservoir fluid analyzer with an observation window developed and produced by the DBR company of Canada. The experimental process is shown in Fig. 1.

Experimental steps:

(1) Transfer an appropriate amount of prepared reservoir fluid sample into PVT instrument and stabilize it at reservoir temperature of 65 ℃ for 2 hours;

(2) Inject an appropriate amount of associated pressurized gas into the reservoir crude oil sample in the PVT instrument, and thoroughly stir for 2 hours to make the sample homogeneous and single-phase;

(3) Slowly reduce the pressure of oil sample in PVT cylinder, measure its bubble point, and conduct a single stripping test to test the dissolved gas volume of crude oil and the density and viscosity of the fluid

(4) After the test, continue to add nitrogen into the oil sample according to the above steps 2~3, pressurize the sample to make it homogeneous and single-phase, then depressurize and measure its bubble point and conduct a single degassing experiment. Experiment in turn and inject associated gas several times in succession.

### 3.2. Comparative analysis of results

See Fig. 1 for the test results of gas injection expansion of the N and G reservoirs. The maximum corresponding to the N and G reservoirs oil samples is $32.30m^3/m^3$ and $16.83m^3/m^3$, respectively. Compared with the original gas-oil ratio, the growth multiples of the N and G reservoirs gas oil ratio are 2.3 and 3.5 times, respectively. Therefore, the viscosity reduction effect of the G reservoir $N_2$ injection is better than the N reservoir. It can be seen from the saturation pressure of the injected gas that with the increase of the injected gas volume, the bubble point pressure of the fluid increases linearly, and the gas dissolution is complex. At the same time, because the reservoir pressure of the N and G reservoirs is low, it is challenging to realize miscible flooding under the current conditions. Therefore, the unfavorable mobility ratio may lead to easy channeling when injecting single medium nitrogen. It can be seen from the changes of volume coefficient of different injected gas and expansion coefficient of different injected gas that due to the significant increase of dissolved gas, the expansion capacity of the G reservoir after gas injection is stronger than that of the N reservoir. Compared with the N reservoir, $N_2$ injection is more likely to lead to gas channeling. It can be seen from the changes of different injected gas densities and viscosity that the viscosity reduction effect of the N reservoir is

12.4%, and that of the G reservoir is 20%. The G reservoir sample's viscosity reduction and density reduction effect after injection is better than the N reservoir.

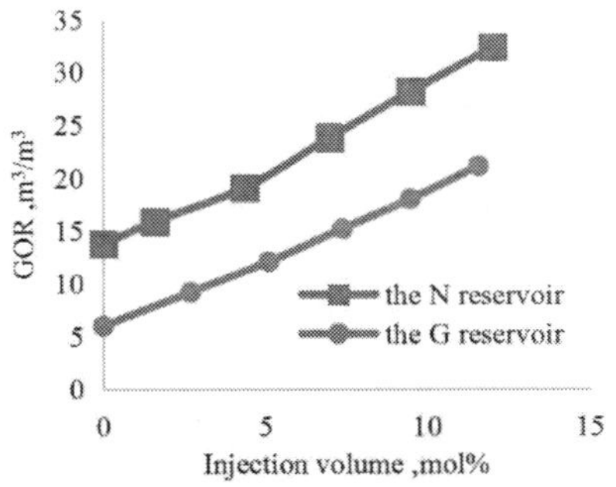


a. Variation of gas-oil ratio under different injected gas content

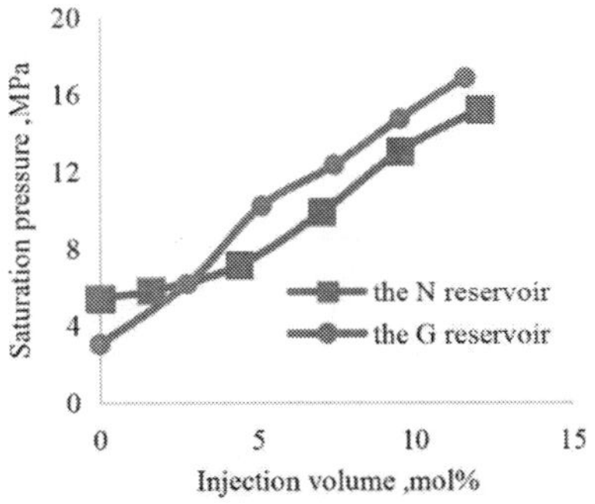


b. Variation of saturation pressure under different injected gas content

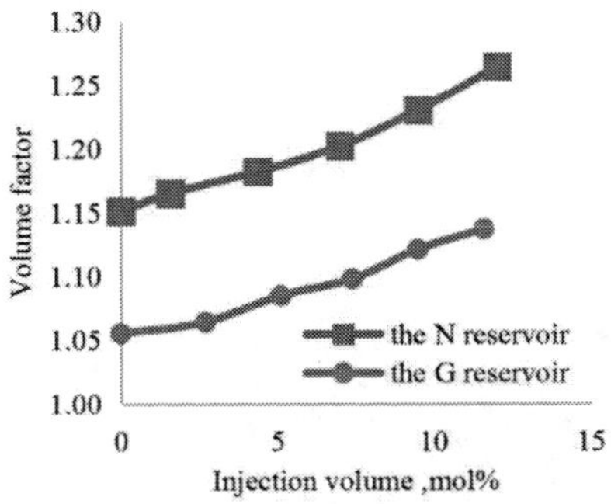


c. Variation of volume coefficient under different injected gas content

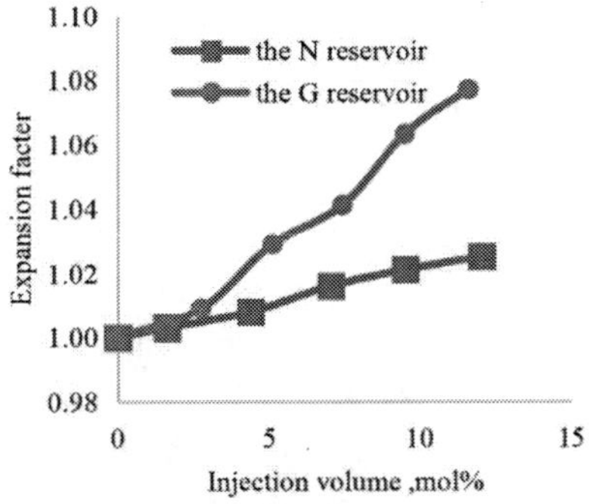


d. Variation of expansion coefficient under different injected gas content

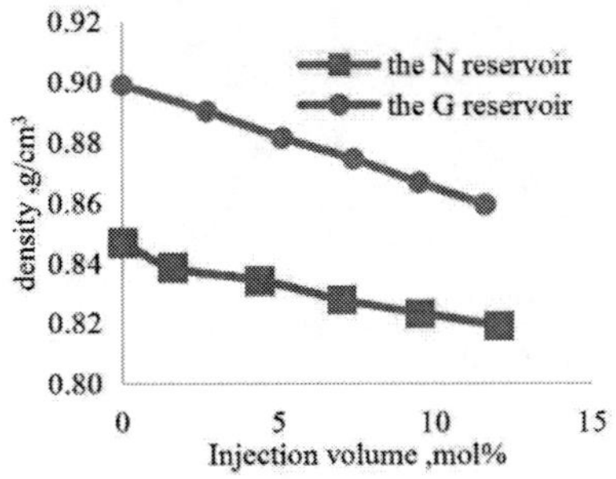


e. Density change under different injected gas content

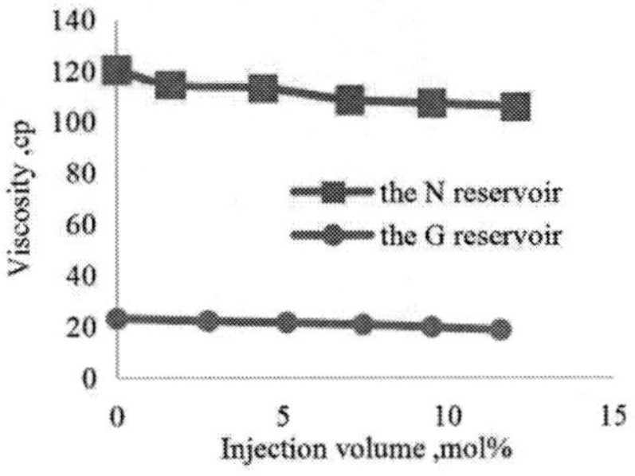


f. Viscosity change under different injected gas content

Fig. 1 Changes of main parameters in $N_2$ injection expansion test of the sample

## 4. Long core displacement experiment with different injection media

### *4.1. Experimental process and steps*

The long core experimental research was completed on Hycal long core displacement devices in Canada. The experimental process is shown in Fig. 2.

Journal of Physics: Conference Series **2594** (2023) 012035 doi:10.1088/1742-6596/2594/1/012035

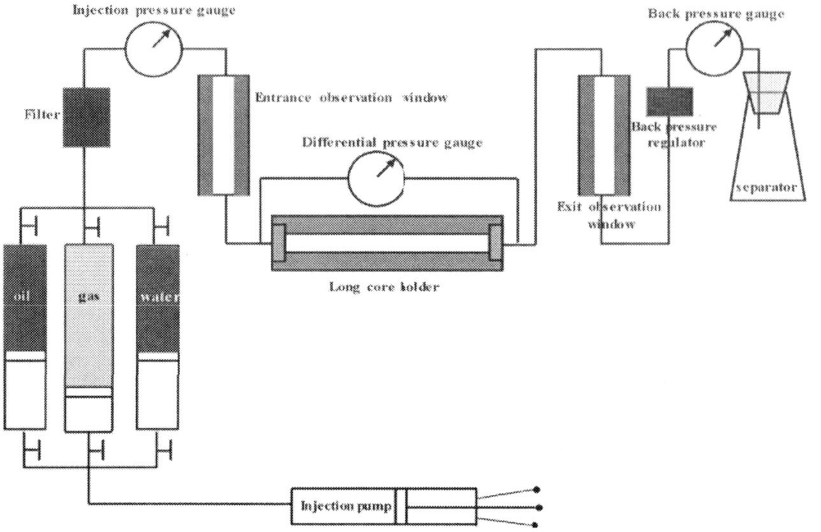


Fig. 2. Flow chart of long core experiment

This process comprises a water body, injection pump system, long core holder, back pressure regulator, differential pressure gauge, temperature control system, liquid fraction collector, gas meter, and gas chromatography. The crucial part of the long core displacement device is the one-meter triaxial long core holder (Fig. 3).

Fig. 3. The long core displacement device

### *4.2. Core and fluid sample preparation*

Due to the limitation of on-site core samples, artificial cores are used in this experiment. The lithology is feldspathic sandstone, with 14 pieces of the N and G reservoirs rock core, and the cores are relatively homogeneous. Among them, the core porosity of the N reservoir is 32.7%, the permeability is 2600mD, the core porosity of the G reservoir is 29.3%, and the permeability is 1600mD, which is in line with the reservoir porosity of the N reservoir 32~36%, the average value is 32.7%, and the average value of permeability is 2600mD. The reservoir porosity of the G reservoir is 28%~34%, the average porosity is 29.3%, and the average permeability is 1600mD. The lithology of the artificial core is rock and relatively homogeneous, so the core is packed with a shrinkable heat tube.

The permeabilities of the selected 13 cores from the outlet end to the inlet end are gradually arranged from small to large. Each core is loaded into the long core holder for use according to the core sequence in the table. The total length of the N reservoir long core is 94.961cm, and the pore volume is 155.961cm$^3$. The total length of the G reservoir long core is 94.882cm, and the pore volume is 139.689cm$^3$.

Table 2 Analysis of reservoir water quality

| Reservoir | Ion concentration, PPm | | | | | | | | Total salinity |
|---|---|---|---|---|---|---|---|---|---|
| | $Na^+$ | $K^+$ | $Mg^{2+}$ | $Ca^{2+}$ | $Cl^-$ | $SO_4^{2-}$ | $HCO_3^-$ | $CO_3^{2-}$ | |
| N | 1437.02 | 14.16 | 20.44 | 91.98 | 1964.8 | 0 | 441.73 | 14.96 | 3985.1 |
| G | 1642.53 | 45.28 | 19.07 | 71.56 | 2167.06 | 15.56 | 589.56 | 24.77 | 4575.4 |

According to the PVT analysis report of reservoir crude oil of the N reservoir and typical wells of the G reservoir, reservoir crude oil samples are prepared. The prepared the N and G reservoirs crude oil single degassing oil ratio, bubble point pressure, and reservoir condition crude oil viscosity is consistent with the sample preparation standard. According to the analysis report of reservoir water quality, see Table 2.

Injection fluid: the nitrogen concentration for displacement is 99.99%, and the foam mass concentration is 0.35%.

### *4.3. Experimental conditions and methods*

The original reservoir condition is established. The original reservoir pressure of the N reservoir is 10MPa, and the reservoir temperature is 65 degrees. The initial reservoir pressure in the G reservoir is 13MPa, the reservoir temperature is 84 degrees, and the irreducible water saturation is 20%. Under these conditions, the $N_2$ foam alternating flooding experiment is carried out under this condition, the mass concentration of the foaming agent is 0.35%, the displacement of the slug is 0.1 HCPV, and the ratio of nitrogen to foam is 1:1.

1) Prepare oil samples and instruments

Prepare the long core holder, put the core into the long core holder according to the required sequence, calibrate, clean, and blow various instruments, test temperature, and pressure, then evacuate, and keep the constant temperature to the required value for the experiment.

2) Establish bound water

After vacuumizing the long core, saturate the reservoir water, establish the Eastbound water after degassing oil displacement, record the injected water volume and water output in this process, and displace the water until the irreducible water saturation reaches 20%

3) Saturation of oil sample

The system pressure is established with degassed crude oil, and the prepared oil sample is used for displacement several times the pore volume. When the gas composition at the outlet is the same as that of the associated gas, and the gas-oil ratio is the same, the sample is saturated and forms the original state.

4) Displacement experiment

After the sample is saturated, gas bubbles alternate drive, slug size 0.1 HCPV, gas water ratio 1:1, every 0.1 HCPV after recording data.

5) Core cleaning

After the experiment, clean the core with petroleum ether and absolute alcohol. Petroleum ether mainly cleans the oil in the rock core, and absolute alcohol cleans the water in the rock core.

### *4.4. Experimental conditions and methods*

*4.4.1. Recovery degree of nitrogen bubble alternative displacement in the N reservoir.* Nitrogen foam alternative oil displacement experiment was carried out for the N reservoir under original formation pressure of 10MPa and formation temperature of 65 ℃. When 0.7HCPV is injected cumulatively, the gas-oil ratio suddenly increases, and the gas breakthrough occurs. The breakthrough point pressure difference is 0.2MPa, and the cumulative recovery rate is 21.86%. When 0.8HCPV is injected cumulatively, water begins to flow, with a pressure difference of 5.6MPa and a cumulative recovery rate of 29.31%. After the injection medium breaks through, there is still a certain amount of oil output due to the foam's profile control and displacement. After the breakthrough, the differential pressure of foam displacement is kept between 2.9-3.5MPa, and that of nitrogen displacement is kept between 0.2-0.6MPa. The differential pressure is relatively stable. When injecting 1.9HCPV, the gas-oil ratio significantly increases, the water content reaches more than 80%, and the degree of oil recovery slows down. When injecting 3.2HCPV, displacement ends, and the cumulative oil recovery rate reaches 43%. The change curves of displacement pressure difference, gas oil ratio, water content, and recovery efficiency in the displacement experiment are shown in Fig. 4.

*4.4.2. The G reservoir $N_2$ foam alternating oil recovery degree.* Under formation pressure of 13MPa and formation temperature of 84 ℃ , the G reservoir carried out a nitrogen foam alternative displacement experiment. When 0.8HCPV is injected accumulatively, water begins to flow out, and the differential pressure is 0.1MPa, and the cumulative recovery degree is 39.47%. When 1.7HCPV is injected accumulatively, the gas-oil ratio suddenly becomes larger, and the gas breakthrough occurs. The cumulative recovery degree at the breakthrough point is 53.78%, and the differential pressure is 1.1MPa. After the breakthrough, the differential pressure of foam displacement is kept between 1.1-1.3, the differential pressure of nitrogen displacement is kept between 0.6~1.0MPa, and the differential pressure is relatively stable. When 0.9HCPV is injected, the water cut is more than 70%, and oil recovery has slowed. When injecting 4.5HCPV, the displacement is completed, and the cumulative oil recovery rate reaches 74.28%. The change curves of displacement pressure difference, gas oil ratio, water content, and recovery efficiency in the displacement experiment are shown in Fig. 4.

*4.4.3. Comparative analysis of experimental results.* See Fig. 4 for a comparison of the recovery degree of gas foam alternative displacement experiment in N and G reservoirs. It can be seen from the figure that the crude oil viscosity of the N reservoir is high (142cp), and the displacement pressure difference is more significant than that of the G reservoir. The maximum foam displacement pressure difference of the N reservoir can reach 5.6MPa, while the maximum foam displacement pressure difference of the G reservoir is 1.2MPa: after gas breakthrough, due to the limited profile control and displacement of the N reservoir, the gas oil ratio increases significantly, and the gas phase seepage is dominant, and the pressure difference is less than that of G reservoir. When water breakthrough occurs in the N reservoir and G reservoir, the injection volume is around 0.8HCPV. However, there is a significant difference in injection volume between the two when a gas breakthrough occurs. When the N reservoir accumulates an injection of 0.7HCPV, gas breakthrough occurs; when the G reservoir accumulates an injection of 1.7HCPV, gas breakthrough occurs. Analysis suggests that there are two main reasons. One is the high viscosity of crude oil in the N reservoir, with a large difference in gas and oil phase mobility ratio, resulting in early gas breakthroughs. Second, for N reservoir heavy oil, the foam profile control and displacement effect is limited, and the gas-oil ratio rises step by step after gas channeling. However, the foam slug is injected after gas channeling in the G reservoir. The gas-oil ratio decreases to the level before gas channeling, and the foam profile control and displacement effect are significant. Due to the difference in the breakthrough volume of injected media, there is a significant difference in the recovery degree between the N reservoir and G reservoir. The recovery degree of the N reservoir is 43.0%, and the recovery degree of the G reservoir can reach 74.28%. From the long core displacement experiment, it can be seen that for the N reservoir heavy oil, the difference between the gas phase and oil phase mobility ratio is too significant, the injection of a single gas medium is easy to channelize, and foam flooding can delay the gas breakthrough to a certain extent: for G reservoir, the nitrogen foam alternative flooding foam system has significant profile control and displacement functions, and the oil displacement efficiency is high.

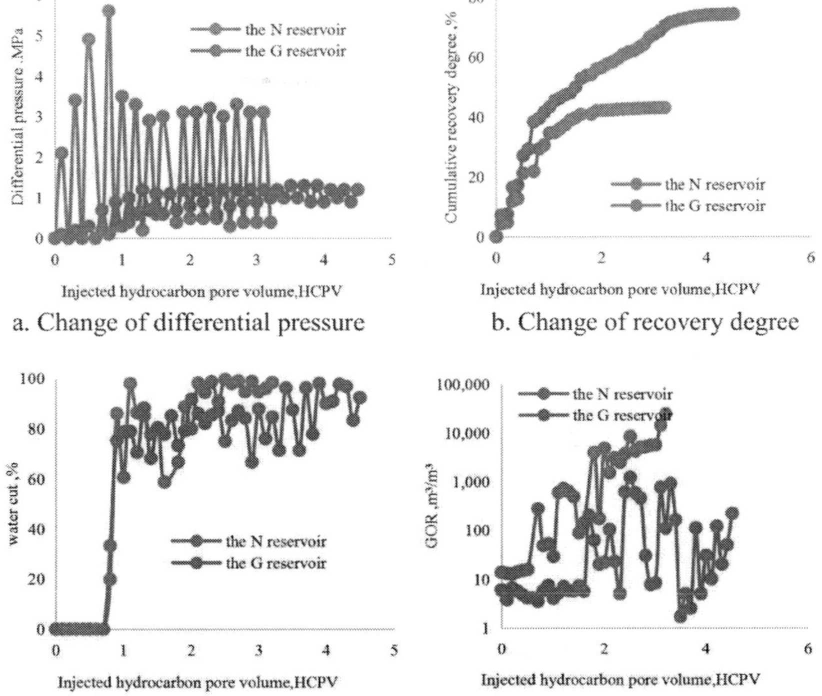


a. Change of differential pressure b. Change of recovery degree

c. Change of water cut d. Change of GOR

Fig. 4 Change of dynamic parameters with injected hydrocarbon pore volume

## 5. Conclusions

(1)The experimental results of gas injection expansion of crude oil in the N and G reservoirs show that, on the whole, nitrogen injection has little effect on the properties of reservoir crude oil due to the high viscosity and low reservoir pressure of the reservoir crude oil. The maximum gas-oil ratio of crude oil in the N and G reservoirs is $32.3m^3/m^3$ and $16.83m^3/m^3$, respectively. The growth multiples of the N and G reservoirs gas oil ratio are 2.3 and 3.5 times, respectively. The lower the viscosity, the greater the growth factor of the gas-oil ratio.

(2)It can be seen from the changes in the volume coefficient of different injected gas and the expansion coefficient of different injected gas that due to the significant increase of dissolved gas, the expansion capacity of the G reservoir after gas injection is stronger than that of the N reservoir. Those with lower viscosity have stronger expansion ability.

(3)After gas injection expansion, the viscosity reduction effect of the N reservoir is 12.4%, while that of the G reservoir is 20.0%. Lower viscosity has a better viscosity reduction effect.

(4)From the gas foam alternative displacement experiment in N and G reservoirs. The higher the viscosity of crude oil, the higher the displacement pressure difference. The injection volume of different viscosity water breakthroughs is the same, all around 0.8HCPV. But the higher the viscosity, the earlier the gas breakthrough.

(5)The long core displacement experiment results show that the crude oil viscosity greatly influences the selection of displacement medium. The single gas medium injected into the heavy oil system is easy to channel and has low efficiency in oil displacement. The oil displacement efficiency of the G reservoir is 73.28%. For heavy oil with the N reservoir, the displacement efficiency of $N_2$ foam flooding is 43.09%. The lower the viscosity, the higher the oil displacement efficiency.

This study guides the design of the offshore sandstone reservoir development scheme.

## References

[1] Kun L. (2018) The Mechanism and Parameter Optimization of Water Control with Nitrogen Foam in Bottom Water Reservoir. Dissertation, China University of Petroleum (East China)

[2] Kim, I., Worthen, A. J., Johnston, K. P., et al. (2016) Size-dependent properties of silica nanoparticles for Pickering stabilization of emulsions and foams. J. Nanopart. Res., 18, 82.

[3] Worthen, A. J., Bagaria, H. G., Chen, Y., et al. (2013) Nanoparticle-stabilized carbon dioxide-in-water foams with fine texture. J. Colloid Interface Sci., 391, 142–151.

[4] Han Q.(2010). Nitrogen foam flooding mechanism research and experiment. Dissertation, Northeastern petroleum university.

[5] Puchun Z., Hongtao M., Hui T., et al. (2015) Application of nitrogen foam in sandstone reservoirs with strong bottom water in the Tahe oilfield. Sino-Global Energy.,20(12): 38-43.

[6] Clark J. A.; Santiso E. E. (2018) Carbon sequestration through CO2 foam-enhanced oil recovery: a green chemistry perspective. Engineering, 4, 336–342.

[7] Hasannejada R., Pourafsshary P., Vatani A., et al. (2017) Application of silica nanofluid to control initiation of fines migration. Petroleum Exploration and Development, 10 (6):1-9.

[8] Emrani A. S., El-Din H. A. (2017) Stabilizing CO2 foam by use of nanoparticles. SPE 174254.

[9] AlYousef, Z., Almobarky, M., Schechter, D. (2017) Enhancing the Stability of Foam by the Use of Nanoparticles. Energy Fuels, 31, 10620–10627.

[10] Zhao J.; Torabi F. (2020) Experimental investigation and modelling of CO2-foam flow in heavy oil systems. Can. J. Chem. Eng., 98, 147–157.

[11] Xu, Z.; Li, Z., Jing, A.; Meng, F. et al. (2019) Synthesis of Magnetic Graphene Oxide (MGO) and Auxiliary Microwaves To Enhance Oil Recovery. Energy Fuels, 33, 9585–9595.

[12] Hurtado, Y., Beltrán, C., Zabala, R. D., et al. (2018) Effects of surface acidity and polarity of SiO2 nanoparticles on the foam stabilization applied to natural gas flooding in tight gas-condensate reservoirs. Energy Fuels, 32, 5824–5833.

[13] Li, S.; Wang, Q.; Li, Z. (2020) Stability and Flow Properties of Oil-Based Foam Generated by CO2. SPE J., 25, 416–431.

# Research and Practice of Heavy Oil Hydraulic Pump Lifting Technology in Bohai Bay K Oilfield

Zu-feng ZHANG1(✉),,Yan-feng SHAO1, Ying-bin ZHU2,Ling-ling LI1,Xue-wei LI1,Guo-peng LI1
c4_zhangzf @ petrochina.com.cn

**Abstract.** In the early stage of heavy oil reservoir development in Bohai Bay K Oilfield, crude oil viscosity is high, conventional lifting processes have poor adaptability, and production energy consumption is high; Entering the stage of ultra-high water cut development, the company implemented carbon dioxide huff and puff, the difficulties in lifting matching brought about by the integrated wellbore requirements of injection and production, high gas-liquid ratio lifting, and enhanced corrosivity of the produced fluid. It conducted research on concentric dual/single pipe hydraulic lifting technology for heavy oil reservoirs, and conducted field practice. The research results and field practice show that the hydraulic pump lifting technology has good adaptability for both heavy oil sand production wells and heavy oil carbon dioxide huff and puff wells, effectively reducing the energy consumption of heavy oil lifting, significantly extending the pump inspection cycle of heavy oil sand production wells, significantly improving the lifting adaptability of carbon dioxide huff and puff wells, and has high promotion value in similar reservoirs and single wells.

## 1 Preface

Bohai Bay K Oilfield is located on the northeast slope of the Kongdian uplift structural belt, which is an anticlinal structure cut by faults. Around the main structure of K Oilfield, some small fault blocks such as K101, K102, K104, and K85X1 have also been developed, with a structural area of 30km², an oil bearing area of 12.7km², and a geological reserve of 2354 × $10^4$t.

The K102 fault block is located in the north of the first fault block of K Oilfield (see Fig. 1). The reservoir development is mainly Guantao Formation (see Fig. 2). The buried depth of the reservoir is 1340-1359m, the average porosity is 31%, the permeability is 777.9-909.2mD, the geological reserves are 1,854,800 tons, and the crude oil viscosity (80 ℃) is 633-8559mPa.s[1-5]. The surface crude oil density is 0.97g/cm³, belonging to super heavy oil bottom water reservoir.Although the fault block K102 was discovered in 1996, it has not been put into development due to the factors of high oil viscosity, poor technological adaptability and high energy consumption.In 2019, effective utilization of the block was achieved through the promotion and application of horizontal wells for well construction, supporting carbon dioxide huff and puff in the reservoir, using hydraulic pumps to lift the wellbore, and adopting centralized degassing pipeline transportation on the ground. However, the average monthly power consumption of a single well in this fault block reached 56000 kWh[6-7].

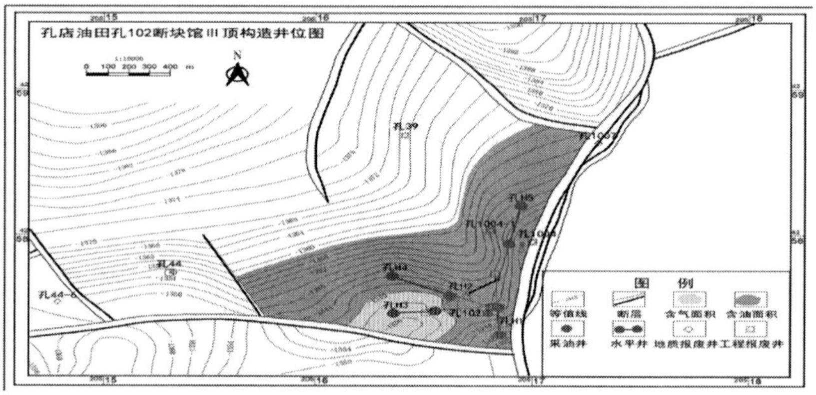


**Fig. 1.** Well Location Map of K102 Fault Block.

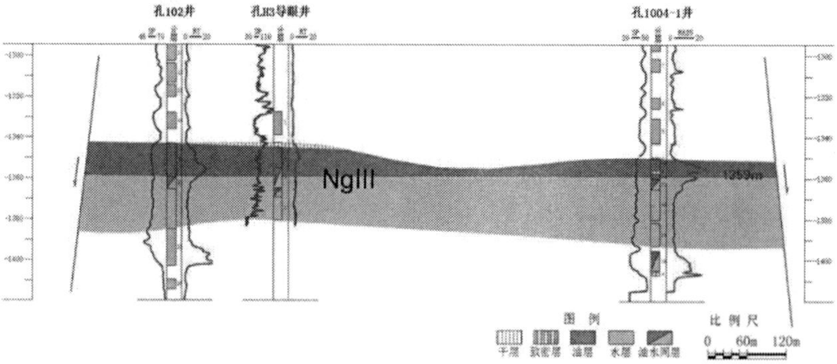


**Fig. 2.** Reservoir Profile of K102 Fault Block.

The K104 fault block is located on the south side of Block 2 in K Oilfield (see Fig. 3). The reservoir is mainly developed in the Guantao Formation (see Fig. 4), with a buried depth of 1336-1351m, an average porosity of 33%, a permeability of 1893mD, a geological reserve of 1273500 tons, a surface crude oil viscosity of 2354.5-3885.4mPa. s, and a colloidal asphalt content of 49.5%. It belongs to a heavy oil bottom water reservoir[8-9].This block was put into development in 2003, with a maximum daily production of 67 tons. In 2008, it entered the ultra-high water cut development stage, with a comprehensive water cut of 96.35%. The daily production of the block decreased to 19.13 tons. In 2020, the block implemented carbon dioxide huff and puff, with a maximum daily production of 41.8 tons. However, the average pump inspection cycle of the oil well was only 238 days, far below the average level of 1867 days before the huff and puff of the fault block[10-11].

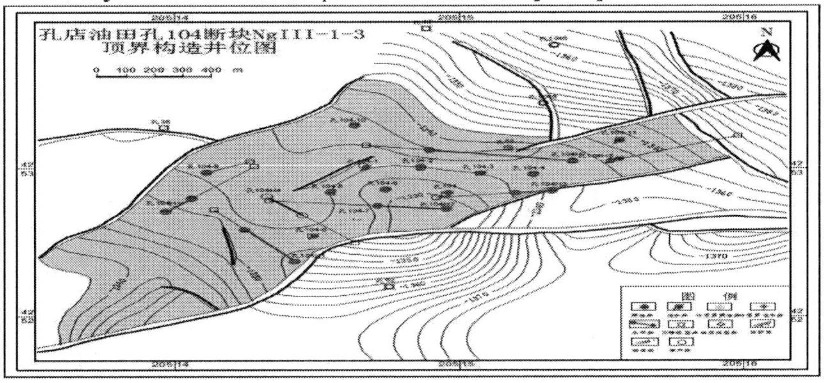


**Fig. 3.** Well Location Map of K104 Fault Block.

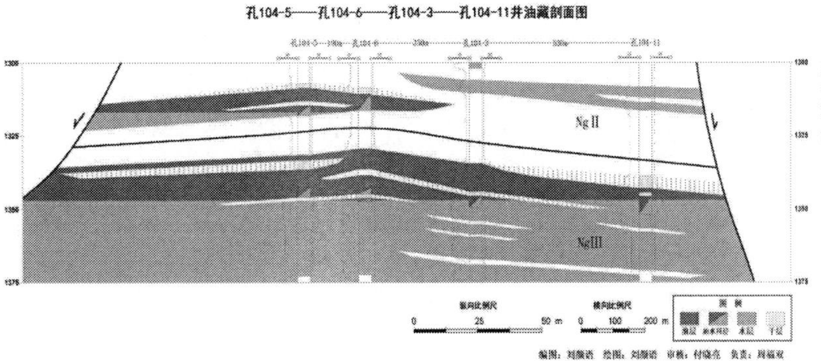


**Fig. 4.** Reservoir Profile of K104 Fault Block.

## 2 Difficulties in lifting heavy oil reservoirs in K oilfield

### *2.1 High energy consumption of heavy oil rod lift process*

**Poor adaptability of lifting process.** Heavy oil is lifted with a pump, which has a large flow resistance of crude oil in the tubing, and the operating load of surface production equipment is high when the sucker rod goes up; It is difficult to lower the sucker rod, the sucker rod is prone to bending, and the eccentric wear of the rod and tube in the wellbore is prominent. Due to the influence of crude oil viscosity, during the production process, when the oil pump goes up, the traveling valve closes slowly, while the fixed valve opens slowly. When the oil pump goes down, the traveling valve opens slowly, and the fixed valve closes slowly, reducing the pump's filling coefficient and reducing pump efficiency[12-15].

**High energy consumption of lifting process.** Heavy oil has a strong sensitivity to viscosity and temperature. When lifting with a pumping unit, in order to reduce wellbore lifting resistance and ensure normal production of the oil well, it is usually accompanied by wellbore electric heating for temperature rise and viscosity reduction lifting. The lifting of super heavy oil sometimes requires a combined viscosity reduction process of bottomhole chemical viscosity reduction and wellbore heating viscosity reduction. The installed power of heavy oil wellbore electric heating is 100KW or more, and the monthly power consumption for single well production is more than 35000 kWh, resulting in high operating energy consumption[16] (see Table 1).

**Table 1.** Statistical table of production energy consumption of heavy oil electric heating rod lift wells.

| Order numbe r | Well No | Production situation | | | Monthly power consumption（Kilowatt hour） |
|---|---|---|---|---|---|
| | | Daily liquid production（$m^3$） | Daily oil production（t） | containing water（%） | |
| 1 | KH3 | 6.7 | 3.82 | 42.99 | 42964 |
| 2 | K85-18 | 23.6 | 14.2 | 39.83 | 33714 |
| 3 | K1077H | 18.19 | 6.34 | 65.15 | 45130 |
| 4 | K59-17 | 20.75 | 9.43 | 54.55 | 31579 |
| 5 | K104-12K | 9.11 | 4.86 | 46.65 | 29874 |
| 6 | K85-26 | 8.45 | 3.25 | 61.54 | 24113 |
| 7 | K85-29H | 8.34 | 4.62 | 44.6 | 43111 |

| | | | | |
|---|---|---|---|---|
| Comprehensive average | 13.59 | 6.65 | 51.1 | 35784 |

### 2.2 *Difficulty in matching lifting technology for ultra high water cut huff and puff wells*

**Short wellbore corrosion cycle.** After carbon dioxide huff and puff production in heavy oil wells with extremely high water content, free carbon dioxide and formation production fluid are simultaneously lifted to the ground, as part of carbon dioxide is dissolved in the production fluid to generate carbonate ions. During the production process, carbonate ions have a certain degree of corrosion on the wellbore rods and tubes, resulting in the failure of the lifting process, especially for the rod lifting process, because the eccentric wear of the reciprocating movement of the rods and tubes will exacerbate the failure of the wellbore lifting process[17-18](see Table 2).

**Table 2.** Comparison of pump inspection cycles before and after carbon dioxide huff and puff in pumping wells.

| Order number | Well No | Pump inspection cycle before throughput (d) | Pump inspection cycle after huff and puff (d) | Days comparison (d) |
|---|---|---|---|---|
| 1 | K104-2 | 2389 | 86 | 2303 |
| 2 | K104-5 | 652 | 53 | 599 |
| 3 | K104-9 | 649 | 63 | 586 |
| 4 | K104H11 | 2074 | 796 | 1278 |
| 5 | K104H2 | 3272 | 441 | 2831 |
| 6 | K104H3 | 2966 | 84 | 2882 |
| 7 | K104-6 | 1543 | 250 | 1293 |
| 8 | K104-7 | 935 | 179 | 756 |
| 9 | K104H7 | 2323 | 189 | 2134 |
| Comprehensive average | | 1867 | 238 | 1629 |

**Poor adaptation to high gas-liquid ratio.** After carbon dioxide huff and puff is carried out and put into production in heavy oil wells with extremely high water content, the injected liquid carbon dioxide will be lifted to the ground in a gaseous form along with the formation produced liquid. Theory indicates that gas is one of the main factors affecting the pump efficiency of the oil well pump. For oil wells with high gas-liquid ratio, it is very easy to generate the phenomenon of gas lock in the oil well pump. According to metering statistics, at the initial stage of a huff and puff well, the gas-liquid ratio at the wellhead reaches above 120, and gas locks frequently occur in the production process of the oil well pump, resulting in abnormal fluid production of the oil well and low production time rate of the oil well.

**Difficulty in lifting due to sand control failure.** Heavy oil unconsolidated sandstone reservoirs are prone to sand production, and in order to ensure normal production of oil wells, mechanical sand control is generally adopted before production. After implementing carbon dioxide, some carbon dioxide will be dissolved in formation water to generate carbonate ions, which will cause corrosion to the bottom hole sand control assembly (see Fig. 5&Fig. 6). At the same time, gaseous carbon dioxide increases the flow rate of the formation produced fluid and fine powder particles, exacerbating the flushing of the fluid on the bottom hole sand control assembly, leading to failure of the sand control process and difficulty in lifting the supporting equipment.

**Fig. 5.** Failure Diagram of Sand Control Screen Pipe in Well K104-9.

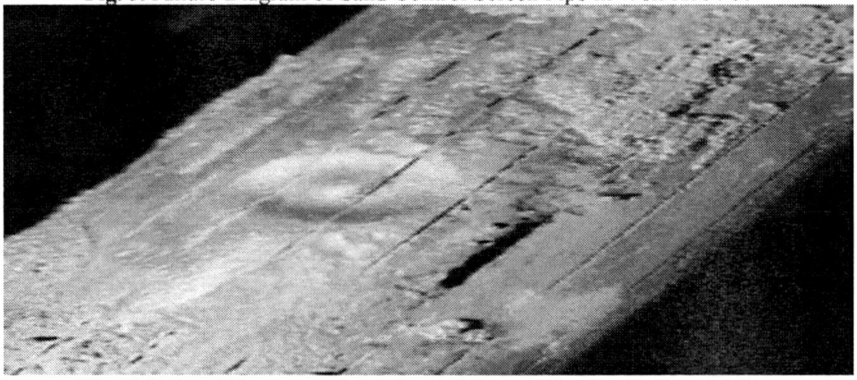

**Fig. 6.** Failure Diagram of Sand Control Screen Pipe for Well K104-2.

## 3 Hydraulic pump lifting principle and application ideas

### *3.1 Working principle of hydraulic pump*

Hydraulic pump is a lifting process that continuously sucks the formation, mainly composed of a high-pressure chamber, a mixing chamber, and a low-pressure chamber. When the power fluid passes through the high-pressure chamber nozzle and the mixing chamber throat, a negative pressure is formed in the low-pressure chamber to suck the formation fluid. The formation fluid and the power fluid are mixed and pressurized in the mixing chamber and then lifted to the ground(see Fig. 7). By improving the ground power fluid process, it is possible to replace the pump core without moving the pipe string, At the same time, it meets the requirements of integrated huff and puff injection production string[19].

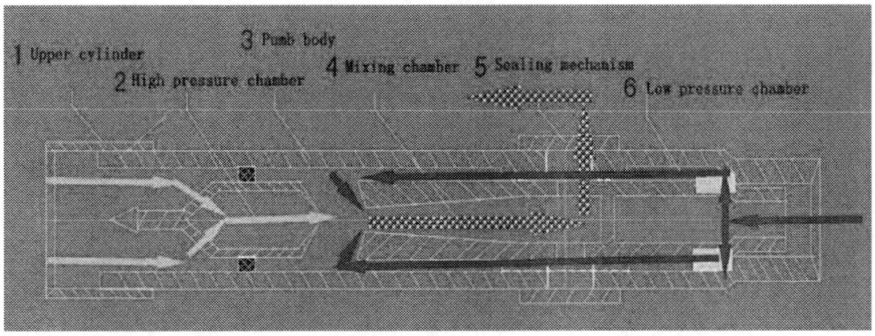


**Fig.** 7. Schematic Diagram of Hydraulic Pump Lifting Principle.

### *3.2 Matching Technology and Application Thinking of Concentric Double Tube Hydraulic Pump*

**Matching Process of Concentric Double Tube Hydraulic Pump.** The concentric double pipe hydraulic pump lifting wellbore string is mainly composed of a 48mm center pipe and a 73mm outer pipe. The power fluid drives the bottom hole jet pump to pump the formation through the center pipe, and the formation fluid and power fluid are mixed and lifted to the ground through the center pipe and outer pipe annulus(see Fig. 8). The wellhead is equipped with high-pressure power fluid and metering process, single well liquid gathering process, and casing water make-up regulation process.

Journal of Physics: Conference Series **2594** (2023) 012036 doi:10.1088/1742-6596/2594/1/012036

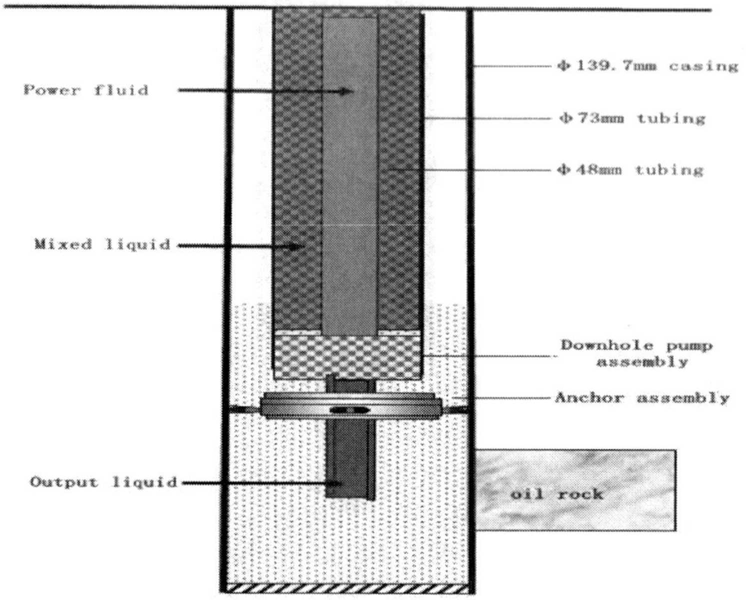


**Fig.** 8. Structural Diagram of Concentric Double Tube Wellbore String.

**Application Thinking of Concentric Double Tube Hydraulic Pump.** Research shows that when the liquid velocity at any cross-section in the drainage annulus is more than twice the settling velocity of the sand, the formation sand can be transported and discharged from the bottom of the well to the surface along with the produced fluid. Due to the small annular control area of concentric dual pipe drainage, the lifting of sand producing wells can be achieved by designing an annular mixed liquid discharge velocity greater than 2 times the sand settling velocity. At the same time, the power fluid and formation fluid are mixed and lifted to the ground, and the power fluid has a good dilution and viscosity reduction effect on heavy oil. This process is mainly used for lifting ordinary heavy oil or ordinary heavy oil sand production wells[20-21](see Fig. 9).

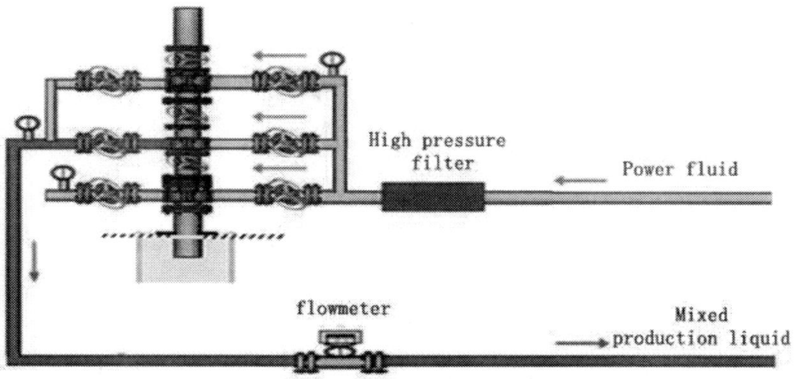


**Fig. 9.** Concentric Double Pipe Ground Process Diagram.

### *3.3 Matching Technology and Application Thinking of Single Tube Hydraulic Pump*

**Matching process of single pipe hydraulic pump.** The single pipe hydraulic pump lifts the wellbore string mainly by φ73mm center tube and Composed of φ139.7mm casing, the power fluid is pumped through the center pipe to drive the bottom hole jet pump, and the formation fluid and power fluid are mixed and lifted to the ground by the center pipe and casing annulus(see Fig. 10). The wellhead is equipped with high-pressure power fluid, metering process, and single well liquid gathering process.

**Application Thinking of Single Tube Hydraulic Pump.** Research shows that with the increase of water content in crude oil, the viscosity of the oil water mixture first shows an upward trend, and when the water content reaches 40%, the highest value appears, and then gradually decreases, with a relatively large decline. By designing the power fluid volume and adjusting the water content of the mixed liquid, the flow resistance during lifting can be reduced. At the same time, due to the special structure and suction principle of the jet pump, pump air lock phenomenon will not occur during production, and it has good adaptability for high gas-liquid ratio lifting. This process is mainly used for lifting heavy oil or high gas-liquid ratio wells after huff and puff(see Fig. 11).

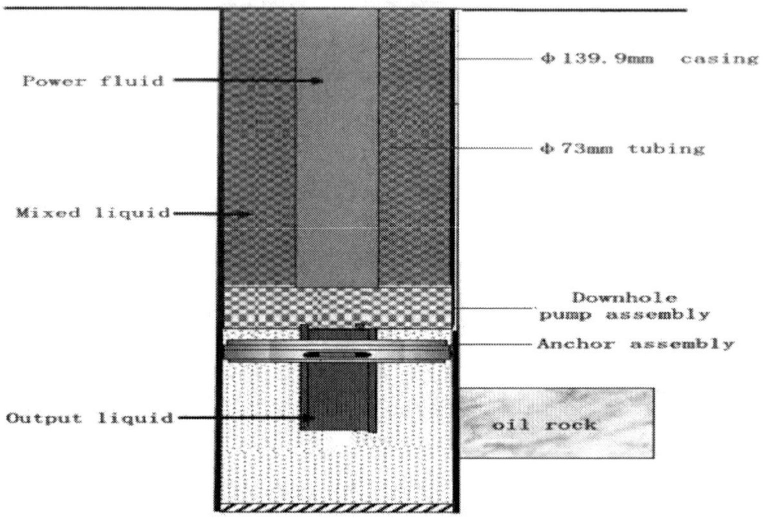


**Fig.** 10. Structural Diagram of Single Tube Wellbore String.

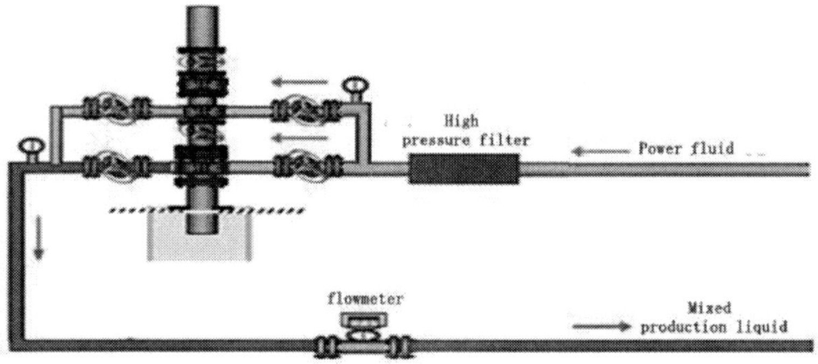


**Fig.** 11. Single pipe ground process diagram.

## 4 Field application effect

### *4.1 Lifting energy consumption of heavy oil hydraulic pumping*

KH3 belongs to a horizontal well in K102 fault block, with a crude oil viscosity of 8630 millipascals per second (50 degrees Celsius). In 2020, the new well was equipped with a pumping unit to electrically lift the wellbore. After measurement, the monthly power consumption was as high as 43000 kWh. In June 2022, a single tube hydraulic pump was used for lifting conversion. This fault block water injection system was used as the power fluid source. After implementation, the monthly power consumption of a single well was only 8800 kWh, reducing energy consumption by 79.5%. In the heavy oil wells of K Oilfield, a single tube/double tube hydraulic pump was used to lift 7 wells. Compared with the electric heating lifting of the pumping unit, the average energy consumption decreased by 73.3%(see Table 3).

**Table 3.** Comparison of energy consumption reduction effects before and after hydraulic pump lifting conversion in heavy oil wells.

| Order number | Well No | Monthly power consumption before conversion (kw·h) | Monthly power consumption after conversion (kw•h) | Power saving rate (%) |
|---|---|---|---|---|
| 1 | KH3 | 42964 | 8826 | 79.5 |
| 2 | K59-17 | 31579 | 9345 | 70.4 |
| 3 | K1037-4 | 29876 | 11682 | 60.9 |
| 4 | K1037-3 | 26452 | 8459 | 68 |
| 5 | K104-9 | 31573 | 7684 | 75.7 |
| 6 | K104-7 | 26745 | 6879 | 74.3 |
| 7 | K104 | 41257 | 8753 | 78.8 |
| Comprehensive average | | 32921 | 8804 | 73.3 |

### 4.2 *Hydraulic pump lift prolongs sand production well cycle*

K1037-3 is a vertical well in the K'ernan fault block. In March 2013, the new well was put into production with a screw pump, mainly producing the NgII-3 formation. The pump inspection cycle for the sand producing oil wells in the formation is only 21 days short. Due to the high shale content of the reservoir and the small median diameter of the sand particles in the formation, there is a risk of formation plugging when conducting mechanical sand control. In September 2013, a concentric double tube hydraulic pump was used for lifting conversion, and the fault block water injection system was used as a power fluid source, After implementation, the single well pump inspection cycle was extended to 2181 days. In Kongdian Oilfield, a heavy oil sand production well was lifted 5 times by using a concentric dual pipe hydraulic pump, with an average pump inspection cycle of 1640 days(see Table 4).

**Table 4.** Statistical Table of the Effect of Applying Hydraulic Pump to Lift and Convert Heavy Oil Sand Production Wells in the Month Before Conversion.

| Order number | Well No | Pump inspection cycle before switching (d) | Pump inspection cycle after conversion (d) | Periodic comparison (d) |
|---|---|---|---|---|
| 1 | K1037-3 | 38 | 2181 | 2143 |
| 2 | K1034-4 | 169 | 1976 | 1807 |

| | | | | |
|---|---|---|---|---|
| 3 | K38-18 | 26 | 515+ | 489 |
| 4 | Y3H19 | 43 | 1345 | 1302 |
| 5 | K104 | 26 | 633+ | 607 |
| Comprehensive average | | 60 | 1330 | 1270 |

### 4.3 *Adaptability of CO2 huff and puff hydraulic pump lift*

The overall carbon dioxide huff and puff of K104 fault block was implemented in 2019. In order to meet the requirements of injection-production integration, the wellbore was lifted with a pumping unit and a hydraulic feedback pump. After the implementation of huff and puff in oil wells, the produced liquid has a high gas-liquid ratio, resulting in frequent gas locks in the oil well pump. At the same time, due to the high gas-liquid ratio, the produced liquid carries a small amount of mud and silt to wash the bottom hole sand control screen, leading to the failure of early sand control processes in some wells. The average pump inspection cycle of the oil wells is only 90 days, and the low production time rate is only 65%. In 2021, gradually apply single/double pipe hydraulic pumps for lifting process conversion, while improving the fault block high-pressure water system to provide power source. After implementation, the average pump inspection cycle for huff and puff wells reached 181 days, including 637 days for K104, and all of them remained continuously effective. The average production time rate of oil wells increased to 99.1%, increasing by 29.9 percentage points(see Table 5).

**Table 5.** Statistical Table of the Effect of Applying Hydraulic Pump to Lift and Convert Heavy Oil Sand Production Wells in the Month Before Conversion.

| Order number | Well No | Pumping unit lifting after huff and puff | | Hydraulic pump lifting after huff and puff | | Comparison of production hours and rates (%) |
|---|---|---|---|---|---|---|
| | | Pump inspection cycle (d) | Production hour rate (%) | Pump inspection cycle (d) | Production hour rate (%) | |
| 1 | K104 | 26 | 54.5 | 637+ | 99.2 | 44.7 |
| 2 | K104-7 | 179 | 82.8 | 217+ | 98.5 | 15.7 |
| 3 | K104H1 | 86 | 68.7 | 47+ | 99.3 | 30.6 |
| 4 | K104H3 | 84 | 64.6 | 56+ | 99.1 | 34.5 |
| 5 | K104-6 | 250 | 86.4 | 62+ | 99.2 | 12.8 |
| 6 | K104-5 | 53 | 58.2 | 67+ | 99.3 | 41.1 |
| Comprehensive average | | 130.4 | 69.2 | 181+ | 99.1 | 29.9 |

## 5 Conclusion and understanding

(1) Single tube hydraulic pump can effectively reduce the energy consumption of heavy oil well lifting. Compared with traditional electric heating lifting, the energy-saving and consumption reduction effect can reach more than 60%, and the energy-saving effect of using oilfield water injection system as power source is more prominent;
(2) Concentric dual pipe hydraulic pump has strong lifting adaptability for heavy oil sand production wells, and the key is that the designed mixed liquid flow rate must be more than twice the sand setting speed;
(3) Single pipe or concentric double pipe hydraulic pump can meet the requirements of integrated injection production string in huff and puff wells. At the same time, it has strong lifting adaptability for high gas liquid ratio and sand control failure after huff and puff, which can significantly improve the production time rate of huff and puff wells;
(4) Single tube and concentric double tube hydraulic pumps belong to a rodless lifting technology that can adapt to deviated deep pumping in large wells, eliminate wellbore eccentric wear from the source, and effectively extend the pump inspection cycle in heavy oil wells.

## References

1. Meng, L. X. , and B. S. Ren . "THE RESEARCH OF OPTIMAL METHOD TO AD]UST THE DEVELOPMENT OF HEAVY OIL BOTTOM WA-TER RESERVOIR." Journal of Southwest Petroleum Institute (2006).
2. Ren, W. . "The Research of Water Influx Law for Edge-Bottom Water Heavy Oil Reservoir." 1(2015).
3. Gomes, E. , and A. K. Ambastha . "Transient Pressure Behaviour Of A Heavy Oil Reservoir Undergoing Steam Injection In The Presence Of A Bottom-Water Zone." Journal of Canadian Petroleum Technology 35.04(1996):17-24.
4. Yue-Hai, L. U. . "Water Invasion Control and Safeguard Measures Resereh of Heavy Oil Reservoir." International journal of petroleum science and technology 7.2(2013).
5. Pang, Z. , et al. "Study on nitrogen foam anti-water-coning technology for conventional heavy oil reservoir." Acta Petrolei Sinica 28.5(2007):99-103.
6. Abdreshov, S. A. , et al. "TECHNOLOGY OF WATER LIFTING FROM WELLS USING AN IMPROVED WATER JET PUMP INSTALLATION." Journal of electronics, communication and instrumentation engineering research 1(2020):10.
7. Li, A. , et al. "Optimized Design of Jet Pump Lifting Parameters in High Wax Content Oil Wells." Journal of Yangtze University(Natural Science Edition) (2019).
8. Tinghui, H. U. , et al. "Study and Application of Diagnosis Curves of Water Channeling Patterns for Horizontal Well in Bottom-water Heavy Oil Reservoir." 1(2019).
9. Meng, B. , et al. "Experimental Study on the Mechanism of Nitrogen Foam to Improve the Recovery of Bottom-Water Heavy Oil Reservoir." Energy & Fuels 7(2022):36.

10. Legarre, H. A. , S. J. Johnson , and M. A. Mccaffrey . "Improved reservoir management of heavy oil assets using biomarker variability in sidewall cores and produced oils: An example from the Cymric Field, Kern County, California." (1996).
11. Tabibi M , Mirkalaei S M . A Successful Case Study on Development of a Giant Highly Fractured Carbonate Heavy Oil Reservoir in Iran[M].
12. Worth, D. , et al. "Assessment of artificial lift methods for a heavy oil field in Kuwait." Journal of Petroleum Science and Engineering 180(2019):835-843.
13. Yan, Y. , et al. "Experimental Study on Lifting Technology of Hydraulic Driving Screw Pump." IOP Conference Series: Earth and Environmental Science 769.4(2021):042066 (6pp).
14. Li, Y. , et al. "Viscosity profile prediction of a heavy crude oil during lifting in two deep artesian wells." Chinese Journal of Chemical Engineering (2017).
15. Li, S. , et al. "Modeling of lifting heavy oil assisted by enclosed thermal fluid circulation in hollow rod." Journal of Petroleum Science & Engineering 75.1-2(2010):135-142.
16. Minav, T. A. , L. Laurila , and J. J. Pyrhonen . "Analysis of electro-hydraulic lifting system's energy efficiency with direct electric drive pump control." Automation in Construction 30.MAR.(2013):144-150.
17. Ke-Xing, L. I. , P. U. Wan-Fen , and D. Song . "Research progress on corrosion,scaling and asphalt deposition in heavy oil CO_2 flooding." Special Oil & Gas Reservoirs (2007).
18. Zhang, Gcyf . "Effect of O-2 on down-hole corrosion during air-assisted steam injection for heavy oil recovery." Corrosion Engineering, Science and Technology 54.3a4(2019).
19. Amani, M. . "Production of heavy oil with a hydraulic gas pump." (1995).
20. Xue, Q. , et al. "Drainage gas recovery technology of concentric double-tube hydraulic pump." Well Testing (2013).
21. Chen, R. . "Design and application of concentric double-tube hydraulic pump sand carrying oil production diagnosis and optimal design system." Petroleum Geology and Engineering (2017).

# Research and Application of HSE Control System for High Consequence Area Water Injection Pipeline

**Shen QI1,Xue-mei WANG1, Peng ZHOU1, Jia-xiu JI1, Li-juan ZHAO1 (✉),Wen-bin DENG1**

1Daqing Oilfield Supervision Center, Ranghulu District, Daqing City, Heilongjiang Province, China

412715742@qq.com

**Abstract.** In view of the fact that the safety management technology prevention and control capabilities are weak in the safety management technology such as long operating life, high pipeline operating pressure, large control area, difficulty in detecting abnormal phenomena, slow valve opening and closing speed, and difficulty in quickly handling emergencies in oilfields, the HSE control system of water injection pipelines in high consequence areas is developed and designed with the help of technical means such as Internet of Things and edge gateway computing, which realizes real-time monitoring of water injection pipeline pressure in high consequence areas, remote control of shut-off valves, timely elimination of hidden dangers, and timely control of dangerous situations. Purpose of expeditious disposal. This research project was tested in the field of a branch of a water injection line in a typical high-consequence area. The verification results show that the system has the functions of real-time monitoring of water injection pressure and remote control of shut-off valve, which is of great significance for improving the HSE technical defense system of oilfield and improving the ability of oilfield to deal with emergencies.

## 1 Introduction

With the deepening of oilfield development and the development and changes of urban construction, the original water-injection pipeline route is gradually surrounded by shopping malls, factories, houses, highways, and other buildings (structures), forming a high-consequence area [9-11] .

At present, there are the following difficulties in the management of water injection pipe network in high consequence areas: First, the service time of the water injection pipeline is long, the average operating life is 20.3 years, the pipeline operating pressure is high, the average operating pressure is 13.6MPa, and the risk of pipeline perforation is large. Second, the control area is large, it is difficult to detect abnormal phenomena in time, and when emergencies such as pipeline perforation occur, it is difficult to quickly arrive at the scene and complete the disposal, so as to miss the golden processing time of the emergency. Third, the emergency shut-off valve in use is a manual valve, the valve switching speed is slow, and it takes about 2-3 workers to operate for about 15 minutes to complete a switch, which is difficult to quickly deal with emergencies.

## 2 Architecture and Working Principle of HSE Remote Control System of Water Injection Pipeline

### 2.1 *Overall System Architecture*

The remote-control system of the water injection pipeline has four layers: data acquisition layer, data storage layer, application layer and terminal layer. [1] (See Fig. 1)

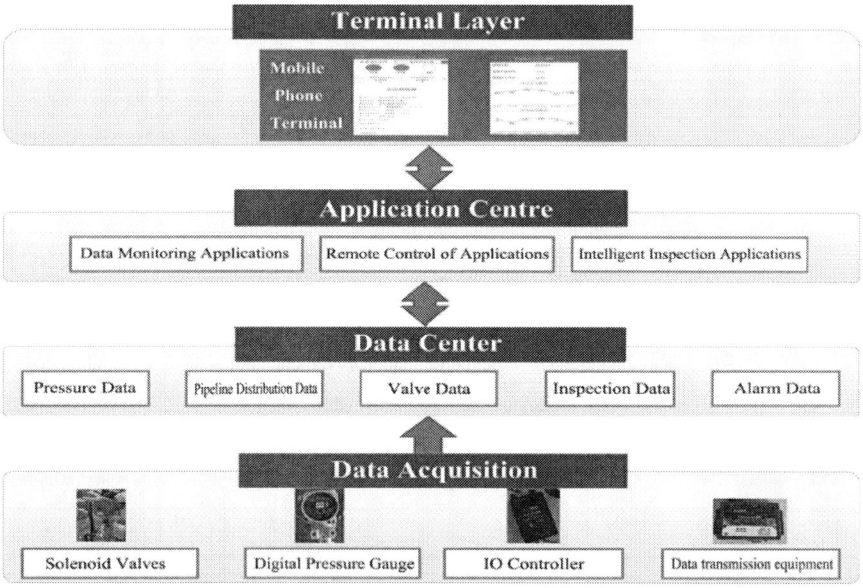


**Fig. 1.** Schematic diagram of the overall system architecture.

#### *2.1.1 Data Acquisition Layer*

The data acquisition layer is the lowest layer structure, and the equipment data information is obtained through the underlying device with 485 interface communication function, such as the switch status information of the intelligent shut-off valve, pipeline pressure data information, etc. [2] The obtained data, after encoding, is transmitted to the upper layer through the Data Transfer unit (hereinafter referred to as DTU) data transparent transmission device.

#### *2.1.2 Data Storage Layer*

The data storage layer receives the data information uploaded by the underlying layer, decodes, and analyzes it, extracts the data and stores it in the database.

#### *2.1.3 Application Layer*
Through the processing of data, the APP side data monitoring application, remote control application, intelligent inspection application and other functions can be realized[7-8] .

#### *2.1.4 Terminal Layer*
Use personal mobile phones as terminal control systems.

### *2.2 Working Principle*

#### *2.2.1 Information Collection and Transmission*
Using hardware topology, we have designed the information collection and transmission system composed of electromagnetic shut-off valve, digital remote pressure gauge and data transparent transmission equipment. [3] (See Fig. 2)

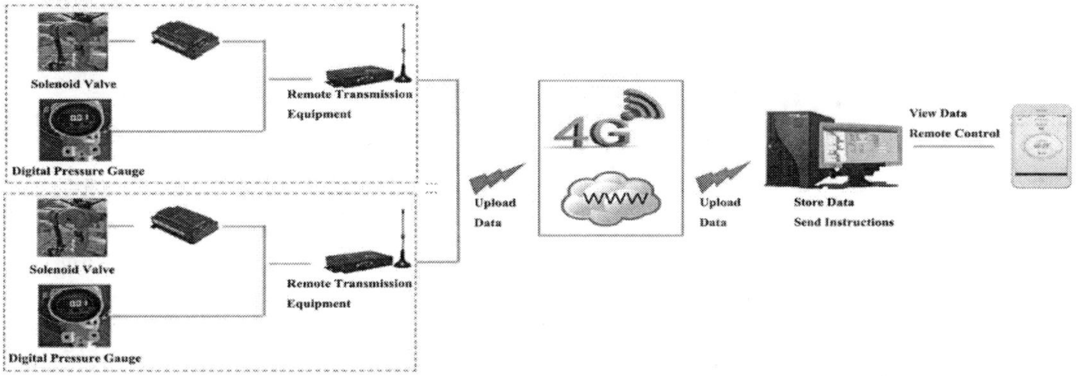


**Fig. 2.** Schematic diagram of the working network topology.

The connection mode of each hardware of the system is as follows:

The electromagnetic shut-off valve is connected with the IO controller, and the system sends switching instructions to the electromagnetic shut-off valve through the IO controller.

IO controller, digital pressure gauge and DTU data transmission device are connected. When uploading data, the DTU transmits the data to the data monitoring server remotely, parses the data and saves it to the database of the server. The upper program can complete the reading and remote control of the data information of the bottom device by sending the corresponding instructions to the DTU.

The mobile APP communicates with the server through the Internet, receives monitoring data from the server, and can issue instructions to open and close the valve in the APP.

#### *2.2.2 Data Concurrent Multi-Point Real-Time Processing*

The remote-control system needs to collect real-time data from multiple points synchronously. Due to the network bandwidth and time-consuming data processing, it is easy to produce data delay, "packet loss", and even system crash. (See Fig. 3)

Therefore, the packet queue and asynchronous thread working mode are designed. [4] When data receiving and processing, the data receiving thread, the data parsing thread and the instruction monitoring thread run at the same time, and each thread is responsible for processing different transactions to ensure that the data packet can be transmitted in real time and stably in the process of multi-bit concurrent data transmission.

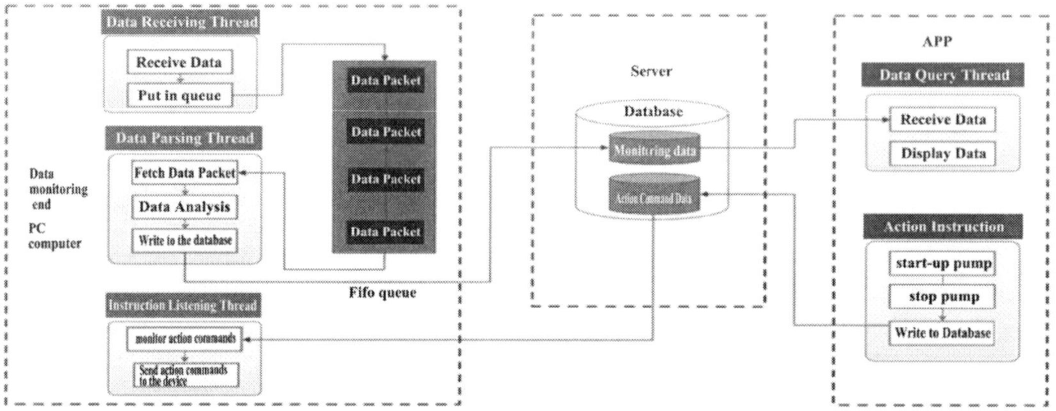


**Fig. 3.** Schematic diagram of the working principle of the control system

## 3 Core Hardware Equipment of the System

The core hardware equipment of the system is mainly composed of shut-off valve, pressure gauge, DTU, IO controller, and meet the following conditions: First, keep the equipment working normally under low temperature conditions (-30℃ to -40℃); The second is to ensure the safe operation of the system in adverse weather conditions; Third, when the network breaks down suddenly, the valve can still achieve emergency shut off.

### *3.1 Working Parameters of System Core Components*

All hardware equipment meets the requirements of industrial standards, suitable for low-temperature outdoor working environment. The following Table 1 gives working parameter table of each core component.

**Table 1.** Working parameter table of each core component.

| serial number | Hardware Name | Operating temperature | Working accuracy | Explosion-proof rating |
|---|---|---|---|---|
| 1 | Shut-off valve | -40℃～+425℃ | —— | |
| 2 | Pressure gauge | -40℃～+85℃ | Level 0.5 | Exd II BT4 |
| 3 | DTU | -40℃～+75℃ | —— | |
| 4 | IO controller | -40℃～+85℃ | —— | |

### *3.2 Backup of Manual Operation of the Shut-Off Valve*

The valve has a manual control function to ensure that the valve can still be emergency shut off in the event of an abnormality in network communication or the valve's electrical control system.

When manual operation is required on site, the manual lever can be rotated to the highest position, at which time the manual mode takes effect, and the operation panel above the valve can be turned to complete the manual opening and closing valve operation. (See Fig. 4)

**Fig. 4.** Manually operated lever for shut-off valves

## 4 Test Application of HSE Remote Control System for Water Injection Pipeline

### *4.1 Overview of The Test Area*

An oil production team is located in a prosperous commercial area, and has successively built more than 10 trading markets, 5 residential areas, 3 schools, 1 kindergarten, 1 nursing home and other old goods, grain and oil, agricultural trade, flowers, birds and fish, etc., with a total of 371 houses, with an average daily population flow of nearly 10,000 people and more than 8,000 round-trips by vehicles. The block adopts the mining method of internal water injection to maintain pressure, with a total of 48 water injection wells, an average operating life of 17.58 years and an average operating pressure of 14.6MPa.The water injection pipeline network has a complex environment and high water injection pressure, which belongs to a typical high consequence area. [9-12] .

The test point is an X-water 135 water injection branch line of an encrypted trunk line, the branch pipe diameter φ65mm, which is a typical branch water injection pipeline and is widely representative.

### *4.2 Experimental Application Effect*

#### *4.2.1 Data transmission stability test*

Data transmission stability at different acquisition frequencies. Adjust the pressure acquisition time interval (≥1s, the collection frequency can be adjusted according to the actual working conditions), observe the stability of system data transmission at different acquisition frequencies, and verify the accuracy and stability of system data transmission. The following Table 2 gives system stability test at different acquisition frequencies.

**Table 2.** System stability test at different acquisition frequencies.

| serial number | Pressure Acquisition Frequency(s) | Valve status acquisition frequency(s) | Test duration (min) | System performance |
|---|---|---|---|---|
| **1** | 1 | 1 | 60 | Stable |
| **2** | 2 | 2 | 60 | Stable |
| **3** | 3 | 3 | 60 | Stable |
| **4** | 5 | 5 | 60 | Stable |
| **5** | 10 | 5 | 60 | Stable |
| **6** | 15 | 5 | 60 | Stable |
| **7** | 20 | 10 | 60 | Stable |
| **8** | 30 | 10 | 60 | Stable |

Taking the 30-second acquisition frequency as an example, the pressure data fluctuated slightly between 14.70-14.72MPa during the process of lasting about half an hour, and no transients occurred. (See fig. 5)

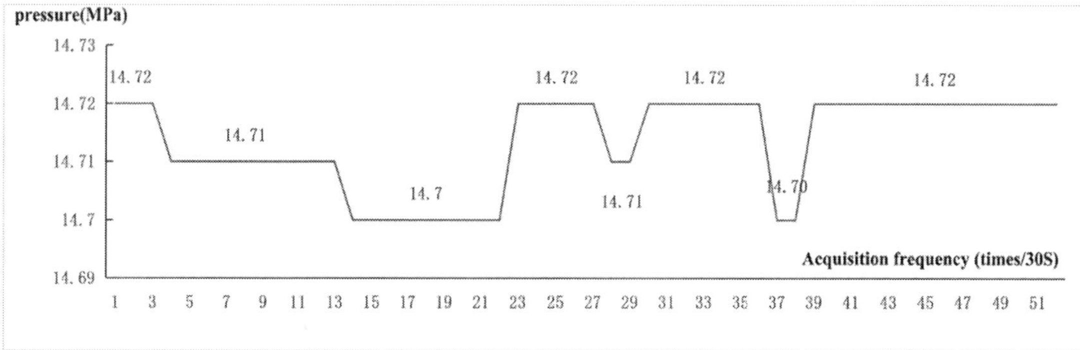


**Fig. 5.** Pressure data acquisition curve at a frequency of 30 seconds

The test results show that the data acquisition and transmission performance of the system remain stable under different acquisition frequency combinations, and no abnormal packet drops and errors occur.

Multiple acquisition frequencies and data transmission stability in scenarios. Combine a variety of acquisition frequencies and scenarios in any way, including operating time intervals and different operating states of valves, and observe the stability of data acquisition. The following Table 3 gives data transmission stability test under various acquisition frequencies and scenarios.

**Table 3.** Data transmission stability test under various acquisition frequencies and scenarios.

| serial number | Acquisition frequency (s) | Operation interval time (min) | Valve state | Real-time pressure data (MPa) |
|---|---|---|---|---|
| **1** | 1 | 1 | open | 14.17 |
| **2** | 1 | 1 | close | 7.56 |
| **3** | 1 | 60 | open | 14.17 |
| **4** | 1 | 60 | close | 7.67 |
| **5** | 3 | 720 | open | 14.18 |
| **6** | 3 | 720 | close | 7.64 |
| **7** | 3 | 1440 | open | 14.18 |
| **8** | 3 | 1440 | close | 7.64 |
| **9** | 10 | 1 | open | 14.17 |
| **10** | 10 | 1 | close | 7.69 |
| **11** | 10 | 60 | open | 14.18 |
| **12** | 10 | 60 | close | 7.85 |
| **13** | 30 | 720 | open | 14.18 |
| **14** | 30 | 720 | close | 7.86 |
| **15** | 30 | 1440 | open | 14.20 |
| **16** | 30 | 1440 | close | 7.84 |

The experimental results show that the data transmission remains stable under the arbitrary combination of different acquisition frequencies and operation scenarios, and there is no abnormal packet loss and error.

### *4.2.2 Remote control stability test*

Stability test of mobile phone controlled shut-off valve switch. Under different switch states of the valve, the mobile phone operation APP switch the valve to verify the stability of the system operation. [5] The test results show that the accurate switch of the valve can be realized by using the mobile APP. The following Table 4 gives stability test of mobile phone operated shut-off valve switch.

**Table 4.** Stability test of mobile phone operated shut-off valve switch.

| serial number | Valve state | APP test operation | Expected system response | Test results |
|---|---|---|---|---|
| **1** | open | open | valve does not operate | normal |
| **2** | open | close | Valve close | Normal shut-off |
| **3** | close | open | valve does not operate | normal |
| **4** | close | close | Valve open | Normal open |

Mobile phone remote control off valve switch stability test. In the test under different operating distances (the distance between the mobile phone operator and the shut-off valve), the remote control

Journal of Physics: Conference Series **2594** (2023) 012037 doi:10.1088/1742-6596/2594/1/012037

of the shut-off valve is carried out through the mobile phone APP to verify the accuracy and stability of the remote operation function of the system. [6] The following Table 5 gives stability test of mobile phone operated shut-off valve switch.

**Table 5.** Stability test of mobile phone operated shut-off valve switch.

| serial number | Distance (m) | APP test operation | Expected system response | Test results |
|---|---|---|---|---|
| **1** | 0 | open | valve open | normal |
| **2** | 0 | close | valve close | normal |
| **3** | 5 | open | valve open | normal |
| **4** | 5 | close | valve close | normal |
| **5** | 1000 | open | valve open | normal |
| **6** | 1000 | close | valve close | normal |
| **7** | 5000 | open | valve open | normal |
| **8** | 5000 | close | valve close | normal |
| **9** | 1200000(Weihai, China) | open | valve open | normal |
| **10** | 1200000(Weihai, China) | close | valve close | normal |

The test results show that the mobile phone APP can realize remote control of the valve within a range of 1200 kilometers, and the operation is stable, and the valve opening and closing time is only 15 seconds. (See Fig. 6)

**Fig. 6.** Mobile phone remote control valve switch state

### *4.2.3 Pipeline leakage alarm test*

When perforated leakage occurs in the water injection pipeline, it is divided into three types: small-hole, medium-hole and large-hole leakage according to the severity of the leakage. When the pipeline has abnormal conditions such as perforation leakage, the system recognizes the abnormal pressure signal in real time, and automatically sends an alarm signal to the management personnel through mobile phone SMS, APP push notification and other forms, and the management personnel do not have to arrive at the scene, they can remotely open or close the shut-off valve through the mobile phone APP.

Combined with the actual production situation, three thresholds were set in the simulation: 13.80MPa, 13.20MPa, 10.20MPa, corresponding to three types of perforation leakage. When the pressure drops to 13.8-13.20MPa, it is small hole leakage; when the pressure is between 13.2 and 10.2 MPa, it is medium hole leakage; when the pressure is lower than 10.2 MPa, it is large hole leakage. Verify whether the system can alarm normally under different thresholds.

The test results show that, according to the set threshold, when the real-time pressure exceeds the corresponding threshold, the system can accurately identify and correctly alarm. The following Table 6 gives Test whether the system can alarm normally under different thresholds.

**Table 6.** Test whether the system can alarm normally under different thresholds.

| serial number | Pressure threshold (MPa) | Real-time pressure data | Expectation | Test results |
|---|---|---|---|---|
| **1** | 13.80 | 14.12 | No alarm | No alarm |
| **2** | 13.80 | 13.78 | Small hole alarm | Small hole alarm |
| **3** | 13.80 | 13.36 | Small hole alarm | Small hole alarm |
| **4** | 13.20 | 13.10 | Medium hole alarm | Medium hole alarm |
| **5** | 13.20 | 12.46 | Medium hole alarm | Medium hole alarm |
| **6** | 13.20 | 11.92 | Medium hole alarm | Medium hole alarm |
| **7** | 10.20 | 10.06 | Large hole alarm | Large hole alarm |
| **8** | 10.20 | 9.88 | Large hole alarm | Large hole alarm |
| **10** | 10.20 | 5.94 | Large hole alarm | Large hole alarm |

## 5 Conclusion

The HSE control system of the waterflood pipeline achieves the goal of real-time monitoring and remote control of the waterflood system in the high-consequence area, timely control and prompt disposal of dangerous situations, improves the prevention, control and disposal ability of the oilfield for emergencies, and is of great significance for improving the HSE technical defense system of the oilfield and enhancing the essential safety management level of the oilfield pipeline.

It is recommended to promote the application of HSE control system of water injection pipelines in the high-consequence area of oilfields, and add a video surveillance subsystem to the system, so as to have a more intuitive and efficient global grasp of the status of the water injection pipe network. When an emergency such as pipeline leakage occurs, it can automatically and quickly check the relevant related points, realize linkage and joint control, and timely and effectively control the dangerous situation.

## References

[1] Bai Xue-jie, Guo Lei-gang, Jiang Li-ge.: Research on the application of Internet of Things

technology in smart grid. Internet of Things Technology, 12(03): 83-85, 88(2012).

[2] CAO Xu, WANG Ru-jun, WEI Li-jun et al.: Research on system architecture of Industrial Internet & Oil and Gas Pipeline Safety Production. Science and Technology of Work Safety in China, 17(S1): 5-9(2021).

[3] Hua-bin WEN.: Research and implementation of mobile collaborative computing platform based on Cloudlet three-layer structure model. Harbin Institute of Technology, (2015).

[4] Zi-Ming ZHAO, Fang LIU, Zhi-Ping CAI, et al.: Edge computing: platform, application and challenge. Journal of Computer Research and Development, 55(2),327-337 (2018).

[5] Dong Jie, Xin Ji-tao, Lian Jie.: Research on H.264 live Video technology Based on Android System. Television Technology, 39(04): 11-15(2015).

[6] Chen Zhi-hong, Wang Ming-xiao.: Application of computer vision in intelligent security. Telecommunication Science, 37(08): 142-147(2021).

[7] Z. H. Liu, J. Y. Wang and Y. P. Li. Quantized image patches co-occurrence matrix: a new statistical approach for texture classication using image patch exemplars.In: Proceedings of 3rd International Conference on Digital Image Processing, Chengdu, China, 2011, pp. 80092P-80092P-5.

[8] Liu Guo-xiu.: Construction of Six God Qu Production Process Tracing System Based on Internet of Things Technology. Chinese Journal of Medicina Guide, 2022-01-15.

[9] NACE SP0208—2008, Internal Corrosion Direct As sessment Methodology for Liquid Petroleum Pipelines[S].

[10] NACE SP0110—2010, Wet Gas Internal Corrosion Direct Assessment Methodology for Pipelines[S].

[11] NACE SP0206—2016, Internal Corrosion Direct As sessment Methodology for Pipelines Carrying Normally Dry Natural Gas[S].

[12] Yang Chun-lin, Yuan Jing-qing. Identification and Integrity Evaluation Method for High Consequence Areas of Existing Pipelines in Oil and Gas Fields. Equipment environmental engineering, 10.7643/ issn.1672-9242.2020.01.018.

# The Water Problems and Treatment Strategies of Volcanic Gas Reservoir in Daqing

**Qiang LI1,2(⊠) Guo-hong LIU1,2 Zhi-rui WANG1,2 Wen-hai MA1,2 Pin-gang MA1,2 Hao-yang LI1,2**

1. Daqing Oilfield Oil Production Engineering Research Institute, Daqing, Heilongjiang Province, China
2. Heilongjiang Provincial Key Laboratory of Oil and Gas Reservoir Stimulation, Daqing, Heilongjiang Province, China
johnlee2013@126.com

**Abstract:** In order to realize the development strategy of "stabilizing oil and increasing gas and simultaneously boosting oil and gas", Daqing Oilfield has continuously strengthened the exploration and development of deep natural gas resources since the discovery of Xushen gas Field. The volcanic gas reservoir is the main gas producing area of Songliao Basin. Through "internal excavation and external expansion", natural gas production has maintained a trend of growth for 12 consecutive years. However, due to the development of fractures and bottom water in volcanic gas reservoir, the development contradiction caused by gas well water is increasingly prominent as the gas field development enters the end of stable production stage. Therefore, on the basis of the understanding of four types of effluent water in volcanic rock reservoir, this paper carries out the analysis of the production dynamic characteristics of typical gas wells, and identifies the difficulties faced by gas wells with different types of effluent water. Through field tests, economic and feasible treatment measures are explored, including continuous circulation gas lift of strong water breakthrough gas well in fracture vertical direction, intelligent foam drainage gas production of weak water breakthrough gas well in fracture vertical direction, and deep water unlocking of gas well in pore vertical water cone and gas reservoir with high water saturation gas reservoir effluent water, which provides technical support for the sustainable and efficient development of volcanic gas reservoir.

## 1 Introduction

Natural gas resources in Daqing Songliao Basin are mainly distributed in volcanic rocks of the first member of Yingcheng Formation of the Lower Cretaceous and sandy conglomerate of the fourth member of Yingcheng Formation [1-2]. In 2002, the first exploration well in volcanic rocks reservoir, the Xushen X well was able to produce gas with a daily output of one million cubic meters, which initiated the development of volcanic gas reservoirs in Xushen gas field [3]. The volcanic gas reservoir of the first member of Ying Formation has a depth of 3200~4300m and poor reservoir physical property. The reservoir space consists of pores and fractures, with an average porosity of 6.57% and an average permeability of $0.43\times10^{-3}\mu m^2$, belonging to the low porosity and low permeability fractured gas reservoirs [4-5]. With the difficulty of external exploration and expansion of high-quality reservoirs increasing, internal potential exploration of developed blocks is the most important thing to maintain stable gas production in Songliao Basin. However, due to the development of edge and bottom water and high-angle structural fractures in volcanic gas reservoirs, formation water rapidly cones into the bottom hole along the fractures, resulting in water being seen in most gas wells at the early stage of production [6,7]. Gas well water has become a major problem restricting the internal potential exploration of volcanic gas reservoir. After water intrudes into the gas reservoir, it first affects the gas phase permeability in the reservoir. Due to the low permeability of volcanic rock reservoir and strong hydrophilicity of rock, formation water enters micro-pores and micro-fractures under the action of capillary force, which increases the water saturation of near-well reservoir and easily leads to reservoir water lock damage[8,9].At the same time, the gas wells water also increases the energy consumption of natural gas ejection-lifting in the wellbore, and the decrease of liquid carrying capacity leads to the accumulation of liquid at the bottom of the well, which increases the back pressure at the bottom of the well and may even "backflow" into the reservoir, resulting in flooding of the gas well and production suspension, affecting the final recovery of the gas reservoir by more than 30% [10]. "Water control" has become the key technology for the stable production and control decline of the Xushen gas field.

In this paper, according to the four effluent water types of volcanic rock reservoir in Daqing Songliao Basin, namely vertical strong water breakthrough in fractures, vertical weak water breakthrough in fractures, vertical water coning in pores and high water saturation gas reservoir effluent water, the production characteristics and hazards of different effluent water types of gas wells are analyzed, and continuous circulation gas lift of strong water breakthrough gas well in fracture vertical direction, intelligent foam drainage gas production of weak water breakthrough gas well in fracture vertical direction, and deep water unlocking of gas well in pore vertical water cone and gas reservoir with high water saturation gas reservoir effluent water treatment measures are proposed. To achieve comprehensive water control from the reservoir to the wellbore, to meet the current stage of Xushen gas field water treatment requirements.

## 2 Characteristics and hazards of volcanic gas reservoir effluent water in Xushen Gas Field

Due to the influence of volcanic structure and reservoir heterogeneity, the gas-water relationship of volcanic gas reservoir is complicated, showing an "upper gas and lower water" distribution on the whole. The high part of the structure is pure gas layer, and the low part of the structure is the same layer of gas-water or water layer (Figure 1).Due to poor reservoir physical property and low natural productivity, Daqing oilfield has realized effective use of volcanic rock reservoir through reservoir reconstruction technologies such as multi-layer fracturing of vertical wells and staged fracturing of horizontal wells in open holes, but also shortened the water-free gas production period of gas reservoirs. About 65% of gas wells have begun to produce water in the stage of gas test and production after pressure. Gao Tao et al., according to parameters such as gas-water ratio and water production of gas wells, divided the effluent water types of Xushen gas field into four types: vertical strong water breakthrough in fractures, vertical weak water breakthrough in fractures, vertical water cone in pores and effluent water from gas reservoirs with high water saturation [11]. Different types of effluent water from gas wells show different production dynamic characteristics [12].

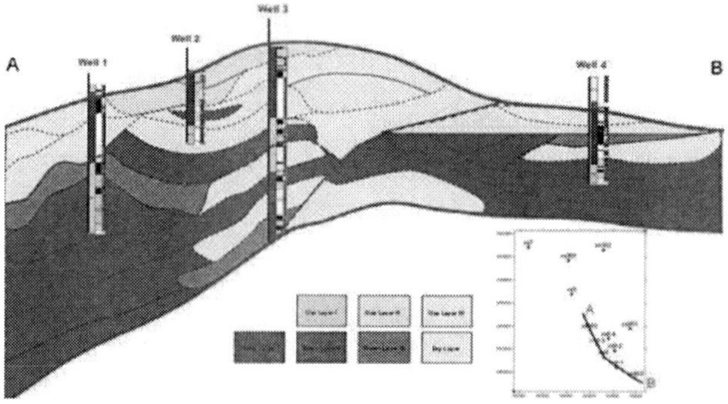

**Fig. 1.** Gas reservoir profile of Xushen gas field

Gas wells with vertical strong water breakthrough in fractures are generally located in the structural low part of volcanic rock mass with high porosity and permeability, permeability greater than 1mD, and daily gas output generally greater than $5\times10^4m^3$. As the active bottom water is directly communicated by artificial fractures, water will be seen immediately after production of this type of gas well, and the daily water output is generally greater than $50m^3$. High water production increases the flow friction in the wellbore, resulting in large oil sleeve pressure difference. At the same time, with the continuous coning or ridge of bottom water, the water saturation of the near-well reservoir keeps increasing, the gas phase permeability decreases, the tubing casing pressure shows a synchronous and rapid decline trend, and the gas well self-injection period is short. If no auxiliary drainage measures are taken, it is prone to violent flooding and stop production (Figure 2a).

Gas wells with vertical weak water breakthrough in fractures are generally located in the high structural position of the volcanic rock mass with high porosity and permeability, which is a kind of gas wells with good development effect, and the daily gas production is generally greater than $5\times10^4m^3$.Water production in gas wells is mainly due to artificial fractures communicating edge water or bottom water through natural micro-fractures. This kind of gas wells also see water immediately after being put into operation, but the daily water volume is generally less than $20m^3$.Due to weak water intrusion and abundant formation energy supply, the stable production period of gas well is long. In the later stage of production, with the energy attenuation of the bottom layer, the tubing casing pressure shows a increase trend. Through regular drainage measures, a larger degree of recovery can be obtained (Figure 2b).

The pore vertical water-cone gas well is generally located in the low part of the structure where natural fractures do not develop. The pore permeability is less than 1mD and the conductivity is less than the fracture. The daily gas production is generally less than $5\times10^4m^3$ and the daily water production is less than $5m^3$.This type of gas well is typically characterized by a fault-like decline in gas production after prolonged shut-in. This is due to the accumulation of fluid in the wellbore after shut-in, and the formation water gathering in the near-wellbore reservoir under the action of capillary force, forming water lock damage, resulting in no recovery of gas production after shut-in until the well is shut down (Figure 2c).

Gas wells with high water saturation gas reservoir effluent water are generally located in the same layer of gas and water with poor reservoir physical property. Bottom water breakthrough is caused by capillary pressure to form a transition zone between gas cap and bottom water, and water saturation is usually greater than 50%. This kind of gas well has high pressure and high yield in the early stage of production, but due to the influence of poor reservoir physical property and insufficient formation energy supply, the pressure and production decline quickly. In stable production, the daily gas production is generally less than $5\times10^4m^3$ and the daily water production is generally less than $2m^3$.Compared with the pore vertical conical gas well, this kind of gas well is more sensitive to shut-in operation because of poor reservoir physical property and easy to produce water lock damage. Prolonged shut-ins tend to destroy the established flow pressure field in the reservoir, and after re-opening the well, the production pressure differential is not sufficient to overcome the start-up pressure, and the well is almost unable to resume continuous and stable production (Figure 2d).

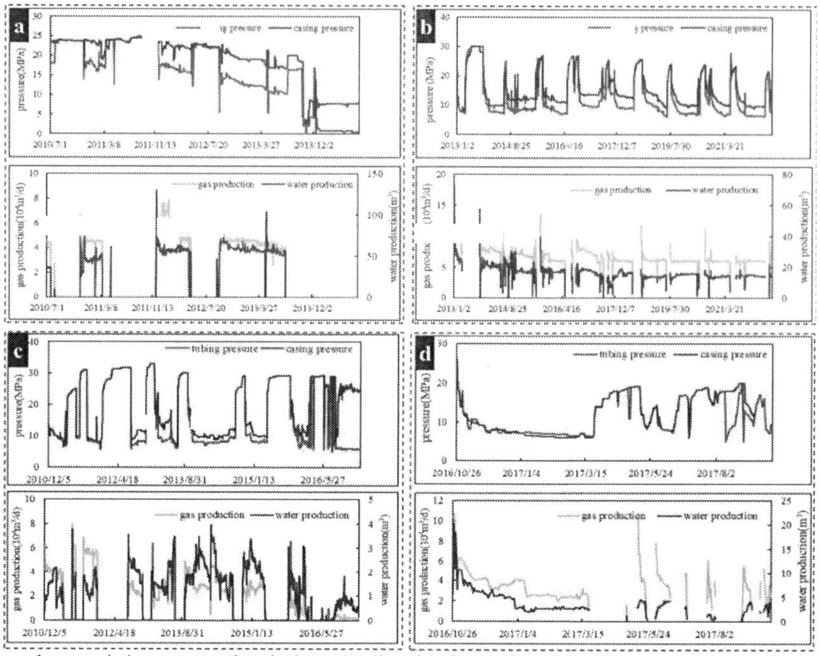


**Fig. 2.** Production characteristics curves of typical gas Wells with four effluent water types (a. Gas wells with strong water breakthrough in vertical fractures; b. Gas wells with weak water breakthrough in vertical fractures; c. Pore vertical water cone type gas well; d. gas well with high water saturation gas reservoir effluent water)

## 3 Main water control measures

### *3.1 Continuous cycle gas lift technology*

For drowned well caused by vertical strong water breakthrough in fractures, the conventional gas lift process using nitrogen and other external gas sources can discharge the fluid in the wellbore, but due to the high process cost (20,000 to 40,000 yuan per day), it can only be used for short-term strong drainage. Once the gas injection is stopped, this kind of gas well cannot rely on its own energy drainage, facing the risk of flooding and production again. The valve gas lift technology is feasible, but it requires pressure to drive down the string, and the operation cost alone is more than 1 million yuan/well. For this reason, Daqing Oilfield has developed a gas cycle gas lift process for drowned well, which is mainly composed of natural gas compressor, gas generator, gas water separator and ground mixed transport pump (Figure 3).

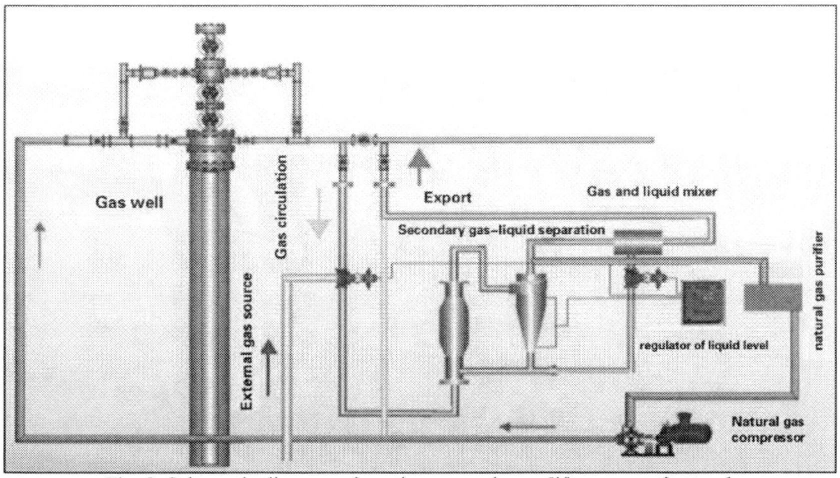


**Fig. 3.** Schematic diagram of continuous cycle gas lift process of natural gas

According to the source of natural gas, continuous cycle gas lift construction can be divided into two stages: recovery and stable production. In the stage of gas well recovery, natural gas collected and transported into the station from adjacent wells is used as the gas source. According to the surface pipeline process, there are roughly two kinds of recovery processes: ①If there are production pipelines and gas lift pipelines from the wellhead to the gas collecting station, the gas lift pipelines will be used to return the natural gas to the wellhead for continuous circulation and gas lift prying equipment. After gas-liquid separation and pressurization, the gas will be re-injected from the tubing casing annulus to the bottom of the well, and the injected gas will be discharged from the wellbore and near-well reservoir fluid through the tubing shoes to reduce the bottom-hole flow pressure, activate the gas well productivity, and restore the gas well to production. The gas and water produced by the oil pipe enter the gas collection station through the production line. ②If there is only one production pipeline from the wellhead to the gas collecting station, the natural gas in the station will be returned to the wellhead through the production pipeline for continuous circulation gas lift prying equipment, and the gas and water produced by the tubing will be recovered and treated through the blow-off process. After the gas well resumes production, according to the drainage demand of the gas well, part of the natural gas produced by the well is used as the gas source. After separation and pressurization, it is continuously injected into the tubing shoe through the oil sleeve to assist the drainage of the gas well and ensure the long-term and stable production of the gas well. The process principle is simple, the equipment composition is simple, easy to operate, the use of natural gas as the gas source, the process cost is reduced significantly (0.3~ 5,000 yuan per day), to solve the drowned well recovery and high-water production well drainage needs.

### *3.2 Intelligent foam drainage technology*

Gas wells with vertical weak water breakthrough in fractures, due to the large gas production and moderate water production, foam drainage technology can be implemented regularly to meet the demand of gas well drainage. However, due to the relatively stable daily production data of this type of gas well, it is difficult to judge the intervention time of foam drainage gas production. However, Turner[13] and Li Min[14] et al., based on the droplet model under annular (fog) flow state, calculated the critical liquid carrying flow rate of gas well with a large difference from the actual field situation. The timing of gas well fluid accumulation determined by Turner model is too early. The critical liquid carrying velocity of gas well determined by Li Min model is inconsistent with wellbore flow pattern. In addition, due to the large water production of this kind of gas well, the amount of foam drainage agent in a single well is about 50~100kg/d, coupled with the scattered distribution of gas wells, the workload of artificial foam drainage maintenance is large, and the implementation of dosing system is subject to external factors such as personnel and weather. For this reason, Daqing Oilfield has developed a high temperature foam discharge agent with a temperature resistance of 150℃[15] and an intelligent foam discharge technology system, which consists of automatic injection device, gas-liquid two-phase flow monitoring device, wellhead pressure transmitter and intelligent controller, etc. (Figure 4).

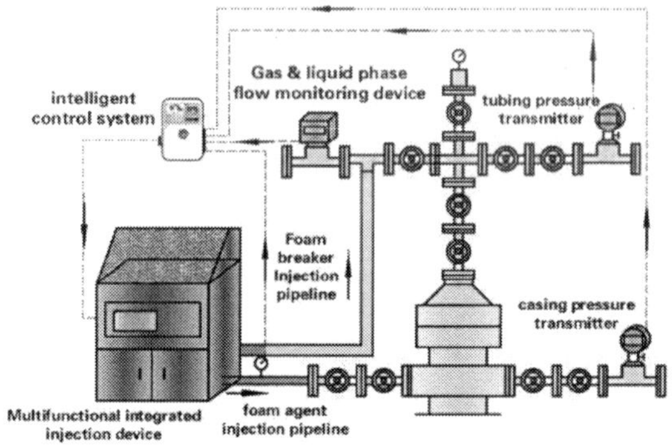


**Fig. 4.** Schematic diagram of intelligent foam drainage technology

With the decrease of liquid carrying capacity of gas well, the gas-liquid two-phase flow in wellbore gradually changes from annular fog to slug flow. After years of field exploration, it is summarized the understanding of the best efficient of foam drainage and liquid carrying in volcanic rock reservoir gas well under slug flow. In order to accurately monitor slug flow pattern, Daqing Oilfield puts forward a practical method to judge slug flow pattern of gas well by using real-time monitoring data of wellhead oil pressure. As shown in Figure 5, this set of oil pressure data is real-time data recorded by the wellhead pressure transmitter every 5 minutes. Before the slug flow occurs, the oil pressure drops slowly, and the relationship between oil pressure and time is approximately linear. When the slug flow pattern occurs in the wellbore, the oil pressure fluctuates periodically. When the liquid slug reaches the wellhead, the oil pressure rises rapidly until the liquid column passes through the wellhead. Then the oil pressure drops until the next liquid column reaches the wellhead.

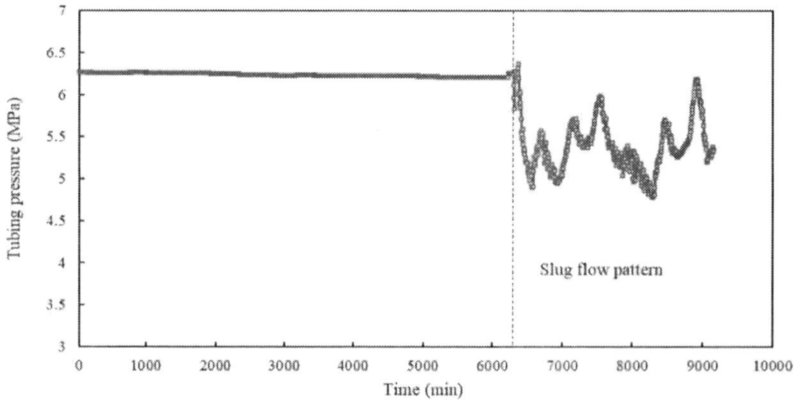


**Fig.5.** The curve of determining slug flow pattern of gas well based on real-time oil pressure and time relationship

Therefore, when the system detects the slug flow pattern characteristics of oil pressure fluctuation and the water production monitored by the two-phase flowmeter, the system automatically calculates the filling amount of foam discharge agent and automatically starts the foam discharge agent injection pump to inject the foam discharge agent through tubing casing annulus. After foam discharge agent injection, it generally takes 4 to 5 hours for foam to arrive at the wellhead, at which time the fluctuation range of oil pressure will decrease. When the system detects this characteristic or a specified time interval, the foam breaking agent injection pump will be started, and the foam breaking agent will enter the production line through the tubing pressure gauge Cock, realizing the functional requirements of automatic injection of the foam discharge agent and the foam breaking agent.

### *3.3 Deep water unlocking technology of volcanic rock reservoir*

For gas wells with longitudinal water cone and high-water saturation gas reservoir water outflow, resulting in reservoir water lock damage, drainage and gas production measures can no longer meet the technical requirements, and the conventional surfactant type water lock remover has a short effective period. In order to remove the "water seal gas" in the reservoir, dredge the gas flow channel and release the gas well productivity, Daqing Oilfield has formed the deep-water lock removal technology of volcanic rock reservoir. First of all, we independently developed a super-hydrophobic nano-water unlocking agent system, which is mainly composed of super hydrophobic nano-fluorine emulsion [16] and surfactant. By changing the rock wettability and reducing the surface tension, capillary pressure can be reduced to relieve reservoir water blocking damage. Among them, the size of super-hydrophobic nano-particles is distributed in the range of 80~90nm (Figure 6a), which can be uniformly adsorbed on the rock surface. By increasing the roughness of the rock surface, the hydrophobic type containing fluorine groups can modify the hydrophilic rock surface into vapor phase wetting. Different from hydrophobic materials such as super-hydrophobic nano-silica, the nano-particle has good dispersion in water and does not need organic solvents such as methanol as dispersant [17] (Fig. 6b), and it has good scour resistance and can effectively prolong the validity period of water-blocking effect. In addition, the gas-water surface tension can be reduced to less than 20mN/m by the combination of hydrocarbon and fluorocarbon surfactants.

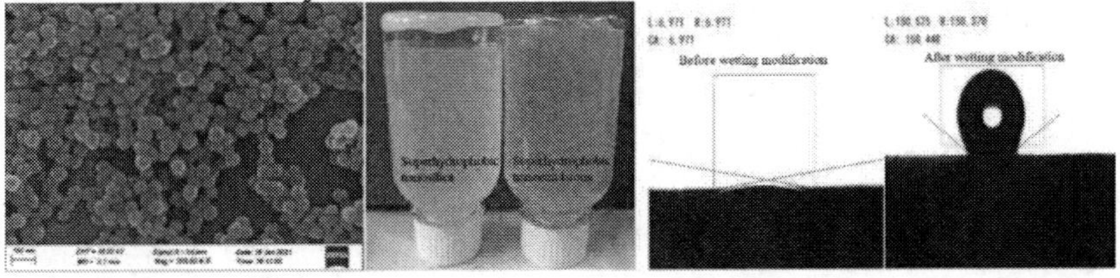


**Fig.6.** Properties of super-hydrophobic nano-emulsion (a. Particle size; b. Comparison of dispersibility between super-hydrophobic nano-silica and super-hydrophobic nano-emulsion; c. Contact Angle of rock surface before and after wetting modification)

Secondly, the slug injection process of water unlocking agent and liquid carbon dioxide was explored and formed. The high-pressure pump truck was used to inject water unlocking agent and liquid carbon dioxide successively. When the temperature in the wellbore exceeds 31.1 ℃ and the pressure exceeds 7.38 MPa, the carbon dioxide is in a supercritical state with low viscosity and strong diffusion ability, which can effectively promote the water lock release to the deep part of the reservoir. In addition, the surface tension of supercritical carbon dioxide is zero, which also has the effect of water blocking. For the near-wellbore formation where "water seal gas" blocks the energy transmission, carbon dioxide plays the role of increasing energy and assisting drainage [18]. After 2 to 4 days of shut-in reaction, the well is opened and discharged back, and the water-unlocking agent and formation water after the $CO_2$ assisted reaction are discharged to the surface.

Finally, since the pore longitudinal water cone and high-water saturation gas wells are very sensitive to pressure changes, management should be strengthened. In addition to in-station maintenance and other factors, the well should be avoided for a long time, and drainage measures such as bubble drainage should be adopted in time when liquid accumulation is found to avoid water lock injury again.

## 4 Field application

The above three measures realize the comprehensive treatment of four types of gas wells from the formation to the wellbore, basically meeting the demand for water control at the present stage. Since the "13th Five-Year Plan", more than 60 wells have been implemented, with a cumulative gas increase of $15.7\times10^8 m^3$. The drainage and gas increase effect are obvious, which provides strong engineering technical support for the stable production of Xushen gas field.

The continuous cycle gas lift was carried out in 13 flooded wells, with a recovery success rate of 100 %, and an average daily gas increase of $2.8\times10^4$ $m^3$ per well, cumulative gas increase $8.3\times10^8$ $m^3$。 The example analysis is shown in Figure 7. After the well was flooded, the oil pressure dropped to 0.6 MPa. In 2020, the intermittent nitrogen gas lift was used to discharge the accumulated liquid in the wellbore, and the oil and casing pressure recovered to balance. After the well was opened, the nitrogen gas lift was continued to be used, realizing the recovery of the gas well for 85 days and the gas increase of $210\times10^4$ $m^3$, but the construction cost was close to 3 million yuan. Considering the economic cost, the well was flooded again after the nitrogen gas lift was stopped. In 2021, continuous cycle gas lift was carried out to reactivate the well. By adjusting gas injection, the differential pressure between the oil and casing gradually decreased, daily gas production recovered to more than $3\times10^4$ $m^3$, daily water production remained more than 100 $m^3$, and daily maintenance cost was reduced by more than 70.

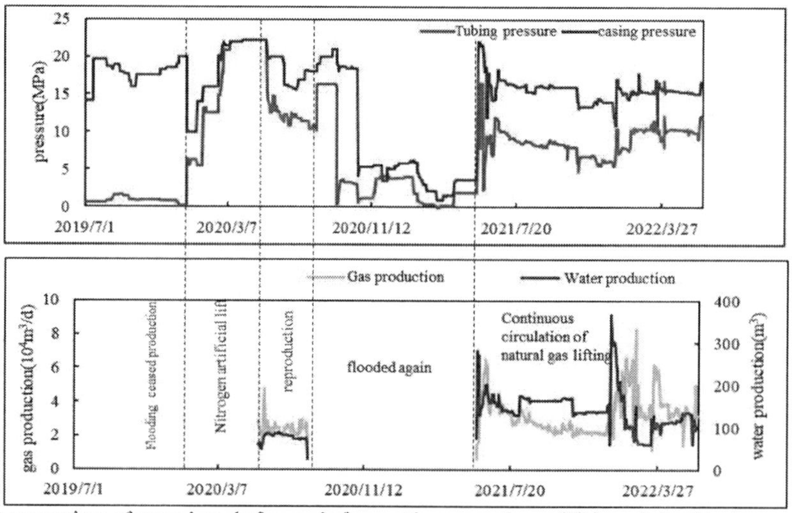


**Fig. 7.** Effect comparison of natural gas before and after continuous cycle gas lift in well A with strong water channeling in fracture longitudinal direction.

Intelligent foam drainage gas production was carried out in 39 wells, with a single injection volume of 15 L~40 L of foaming agent and a daily injection frequency of 1-2 times. The foaming agent supplement cycle of the automatic injection device was 7-15 days, greatly reducing the labor intensity, with an average daily gas increase of $0.7\times10^4$ $m^3$ per well, cumulative gas increase $5.4\times10^8$ $m^3$.Taking Figure 8 as an example, before the implementation of intelligent bubble drainage, the pressure and production of the well basically remained stable, and the average daily gas production was $2.2\times10^4$ $m^3$, daily water yield of 3.4 $m^3$, oil jacket pressure difference of about 3 MPa, with obvious characteristics of fluid accumulation. After the implementation of intelligent bubble drainage, the casing pressure decreased significantly, and the oil pressure increased slightly. At the beginning of bubble drainage, the maximum water yield reached 19.5 $m^3$, and then the gas and water production basically remained stable, and the average daily gas production increased by $0.8\times10^4$ $m^3$, daily water production increased by 7.9 $m^3$, and the differential pressure of oil jacket basically remained at 1~1.5 MPa.

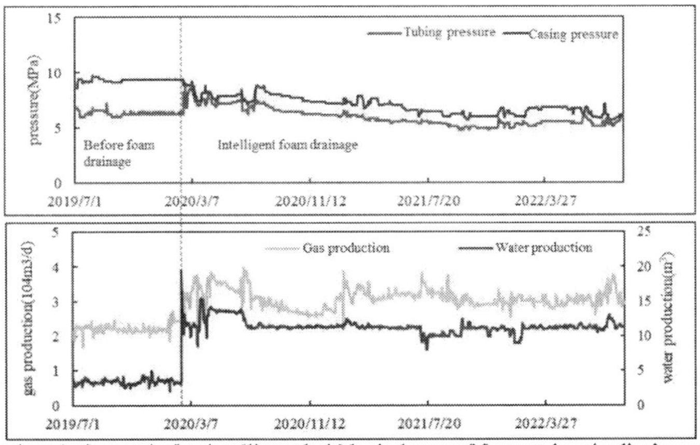


**Fig. 8.** Effect comparison before and after intelligent bubble drainage of fracture longitudinal weak water channeling gas well B

11 Wells of deep volcanic rock reservoir were descaled, with an average of 10~20 t of descaling agent and 20~40 t of liquid carbon dioxide injected into each well. After removing water lock damage, gas production can recover to more than 80% of the damage before water lock. As shown in Fig.9, Before the implementation of water release lock construction, the differential pressure between oil and casing was close to 4 MPa, which has shown the characteristics of fluid accumulation, and the gas production showed a downward trend, with an average daily gas production of $3.9 \times 10^4$ m$^3$, daily water yield 4.5 m$^3$, In April 2020, the well was shut down for 284 days due to in-station maintenance, renovation and other reasons. After the well was opened again, the pressure of the oil jacket synchronously and rapidly decreased to about 1 h, and the production could only be maintained. After the well was shut in, the wellhead pressure could not recover to more than 20 MPa. The water lock release construction was carried out from March to May in 2021, and the wellhead shut-in pressure gradually recovered to about 20 MPa. After the well was opened, the gas production and water production gradually increased with the release of the water lock injury, average daily gas production restored to $3.2\times10^4$ m$^3$, with daily water production rising to 7.0 m$^3$.

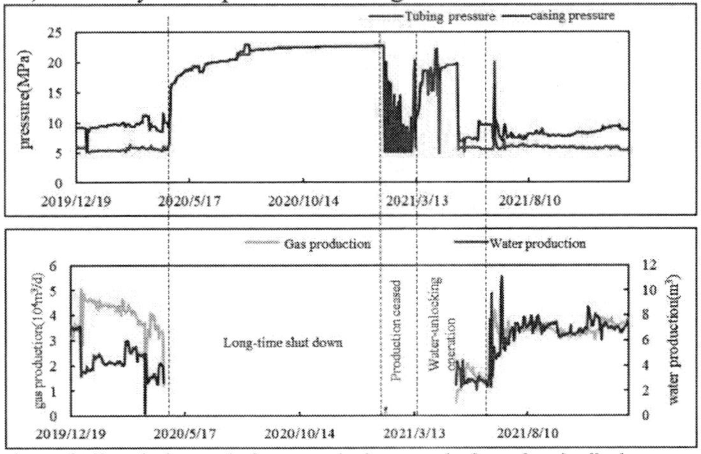


**Fig. 9.** Comparison of effects before and after water lock removal of pore longitudinal water cone gas well C

## 5 Conclusion

(1) According to the effluent characteristics of volcanic rock reservoir in Daqing, this paper analyzes the production characteristics of four types of typical gas wells with effluent problems, identifies the

potential effluent hazards, and provides guidance for the identification and judgment of the effluent risks of the same type of gas wells.
(2) According to the reservoir characteristics and the relative relationship between the gas well and the edge and bottom water, the treatment strategy of "one type, one policy" is proposed for different types of gas wells.
(3) It is necessary to strengthen the combination of geology and engineering, and comprehensively control the volcanic water outflow from various aspects such as "control", and "drainage", so as to provide new ideas for the water control of volcanic gas reservoirs.

**References**

[1] Zhang Xiaoye. Comprehensive geological study of Yingcheng Formation gas reservoir in Xushen 1 well area of Daqing Oilfield [J]. Western Exploration Engineering, 2022,34 (04): 69-72+75.
[2] Xu Zhengshun, Wang Yuming, Pang Yanming, Shu Ping, Gao Xiang, Ai Xingbo. Identification and evaluation of volcanic gas reservoirs in Daqing Xushen Gas Field [J]. Petroleum Exploration and Development, 2006 (05): 521-531.
[3] Wang Yongzhuo, Zhou Xuemin, Yin Changhai, Shu Ping, Cao Baojun, Zhu Yingkang. Reservoir forming conditions and key technologies for exploration and development of Xushen gas field [J]. Journal of Petroleum, 2019, 40 (07): 866-886.
[4] Xu Zhengshun, Fang Baocai. Characteristics and development countermeasures of volcanic gas reservoirs in Xushen Gas Field [J]. Natural Gas Industry, 2010,30 (12): 1-4+119.
[5] Liu Lukun Study on pore characteristics and evolution of volcanic reservoir in Yingyi section of Xingcheng area [D]. Northeast Petroleum University, 2017.
[6] Tan Xianchun, Shao Rui, Qiu Hongfeng. Difficulties and countermeasures for efficient development of volcanic gas reservoirs in Xushen Gas Field [J]. Natural Gas Industry, 2009,29 (08): 72-74+140-141.
[7] Gao Tao, Wang Gaowen. Identification of early water production of gas wells after fracturing by flowback rate combined with water-gas ratio and water property analysis - taking volcanic gas reservoirs in Xushen Gas Field as an example [J]. Natural Gas Industry, 2009,29 (08): 89-91+143.
[8] Shang Yao. Water lock effect of volcanic gas reservoirs in Xushen gas field [J]. Daqing Petroleum Geology and Development, 2015,34 (03): 53-56
[9] Hu Yong, Zhu Huayin, Wan Yujin, Han Yongxin, Wang Yufeng. Pore structure and gas-water seepage characteristics of Daqing volcanic rocks [J]. Journal of Southwest Petroleum University, 2007 (05): 63-65+89+200.
[10] Gao Shusheng, Yang Minghan, Ye Liyou, Liu Huaxun, Zhu Wenqing. Water invasion dynamic simulation experiment of low permeability bottom water gas reservoir and its impact on oil recovery [J]. Natural Gas Industry, 2022,42 (03): 61-70.
[11] Gao Tao. Using the combination of dynamic and static methods to classify the types of water from gas wells in Xushen Gas Field [J]. Natural Gas Industry, 2010,30 (09): 42-45+123.
[12] Liu Miao, Zhang Yongping, Zhang Feng, Wang Zhizhi. Research on adaptability of drainage gas production process in the early stage of development of Daqing volcanic gas reservoir [J]. Natural Gas Exploration and Development, 2010,33 (04): 57-60+65+95-96.
[13] Turner R G, Hubbard M G, Dukler A E. Analysis and Prediction of Minimum Flow Rate for the Continuous Removal of Liquid from Gas Wells[J]. Journal of Petroleum Technology, 1969, 21(11):1475-1482.
[14] Li M, Sun L, Li S. New View on Continuous-removal Liquids from Gas Wells[C]// SPE Permian Basin Oil and Gas Recovery Conference. 2001.
[15] Ma Pinggang, Zhang Yuguang, Xu Dekui Application of foam drainage gas production technology in high-temperature volcanic gas reservoirs in Daqing Oilfield [C]//. Proceedings of the 2015 National Natural Gas Academic Annual Conference [Publisher unknown], 2015:877-

881.
[16] Li Q, Fan Z, Liu Q, Ma W, Li J, Li N, Ma P, Zhang H. Synthesis of a Superhydrophobic Fluorinated Nano-Emulsion and Its Modification on the Wettability of Tight Sandstone. Materials. 2022; 15(11):4015.
[17] Li Yonglian, Yan Jiawei, Zhuang Yingfen. Study on preparation and properties of superhydrophobic surface based on nano-silica [J]. Journal of Mianyang Normal University. 222,41 (08), 7-11.
[18] Gao Yuequan, Hao Hongfeng, Jiang Lan. Application of carbon dioxide energy enhancement and plugging removal technology in Xiasiwan Oil Production Plant [J]. Guangzhou Chemical Industry. 2009, 37 (04).
[19] Yong D , Enbo Q , Xinpu S ,et al. Water-Yielding Mechanism and Water Control Measures in Kelameili Volcanic Gas Field, Junggar Basin[J].Xinjiang Petroleum Geology, 2014.
[20] Bennion B D , Thomas F B , Bietz R F ,et al.[Petroleum Society of Canada Annual Technical Meeting - Calgary, Alberta (1996-06-09)] Annual Technical Meeting - Remediation of Water And Hydrocarbon Phase Trapping Problems In Low Permeability Gas Reservoirs[J].[2023-06-18].DOI:10.2118/96-80.
[21] Wang Y , Kang Y , Wang D ,et al. Liquid phase blockage in micro-nano capillary pores of tight condensate reservoirs[J].Capillarity, 2022, 5(1):12-22.DOI:10.46690/capi.2022.01.02.

Journal of Physics: Conference Series **2594** (2023) 012039 doi:10.1088/1742-6596/2594/1/012039

# Lithology and Lithofacies Characteristics of Deeply Buried Clastic Reservoirs in Alternating Wet-arid Climate and its Hydrocarbon Significance

**Jian-ping LI1, Zi-ying QIU1(✉), Hong-yi LI1**

1 Exploration and Development Research Institute, China National Offshore Oil Corporation Research Institute Ltd.,
Beijing, CHINA,
1 Exploration and Development Research Institute, China National Offshore Oil Corporation Research Institute Ltd.,
Beijing, CHINA,
1 Exploration and Development Research Institute, China National Offshore Oil Corporation Research Institute Ltd.,
Beijing, CHINA,
lihy12@cnooc.com.cn

**Abstract:** There have been series of HPHT (high temperature and high pressure) reservoirs with good quality and high productivity discovered in the Triassic Skagerrak Formation in the Central Graben of the UK North Sea. The bottlenecks for further E&P activity are the complicated depositional facies and high reservoir heterogeneity. The main reservoir space is a mixed porosity network composed of primary pores, micro-pores, secondary pores, secondary micro-pores and micro-fractures. Reservoirs of the Skagerrak Formation have greatly varied lateral connectivity as well as poor vertical connectivity. In the case of sufficiently high pressure differentials, hydrocarbon can be fed from high-potential area to low-potential area. There is no simple linear relationship between depth and reservoir quality. The primary depositional texture (grain size/sorting/clay content, etc. ) linked to facies has a fundamental control on reservoir quality. Clean and well sorted cross laminated sandstone has the best reservoir quality, while the reservoir quality of planar laminated sandstone decreased due to increased degree of compaction caused by mica drapes. The reservoir quality of the fine-grained ripple cross laminated sandstone is even poorer than that of the planar laminated sandstone, due to abundant mica and clay drapes. Mottled bioturbated sandstone is fine grained with high clay content, which is more muddy because of containing burrows and insect larva. Pedoturbated sandstone has the poorest reservoir quality due to severe dolomite cementation. The dominant detrital material of the Skagerrak Formation is feldspar that provides abundant secondary porosity by alteration and dissolution. Chlorite coats are commonly developed and stained with residual oil or bitumen, which clearly inhibits quartz cementation and feldspar overgrowth. The presence of these clay coats contributes to the preservation of anomalously high porosity at depth in the cleaner and coarser grained fluvial channel facies. The Skagerrak formation facies is a dryland fluvial–lacustrine depositional system which is controlled by fluctuations of depositional base level due to salt mobilisation. The Skagerrak formation consists of several cycles that are vertically separated by lacustrine mudstone. The reservoir permeability system is composed of active fluvial channel sandstones, terminal splay sandstones, abandoned fluvial channel sandstones, minor channel sandstones connecting fluvial channel and terminal splay, as well as low-permeability faults, while the active fluvial channel sandstone is the "sweet spot". The analysis of controls on deeply buried reservoir quality is significant to precisely predict "sweet spot" in the Skagerrak Formation.

## 1 Introduction

The Skagerrak Formation sandstone reservoir has long been recognized as important target for exploration in the UK Central North Sea, where multiple oil fields have been discovered. These sandstone reservoirs develop excellent reservoir properties, with local porosity exceeding 25% and permeability reaching several darcys. However, due to their deep burial depth, high temperature and pressure, and complexity of their stratigraphic sequence and internal structures, the reservoir heterogeneity is particularly strong. These oil reserves are typically located at depths below 3500 meters, and are characterized as high-temperature and high-pressure reservoirs, with pressures up to 12900psi (890bar) and temperatures up to 350 degrees Fahrenheit (180 degrees Celsius). Their lateral connectivity varies greatly, while vertical connectivity is poor or even absent in some areas. In the case of sufficiently high pressure differentials, major faults may exhibit leakage, and short-term production and testing results may not be representative of the long-term performance of oil fields [1-3]. To better guide exploration and development, it is necessary to further clarify the sedimentary and diagenetic factors that contribute to the strong heterogeneity of these reservoirs.

The sand-prone intervals dominated by fluvial and lacustrine facies of the Skagerrak Formation usually have good lateral reservoir connectivity but poor vertical connectivity, and their internal sedimentary microfacies exhibit complex variation. It can be difficult to interpret sedimentary environments based solely on seismic interpretations and wireline log patterns without core constraint. This paper conducts a main controlling factors analysis of the reservoir quality of the Skagerrak Formation in the North Sea Triassic system by combining lithofacies, sedimentary facies and wireline logs.

## 2 Regional Geological Setting

The Triassic of the Central North Sea (Fig. 1) consists of two diachronous lithostratigraphic units: the lower Smith Bank Formation dominated by red and variegated (grey and green) mudstone, and the upper Skagerrak Formation dominated by sandstone. The thickness of each unit is greater than 500 meters, and the thickness variations are mainly controlled by the mobilisation of the underlying Permian salt. The Skagerrak Formation can be further divided into six members(in ascending order: Judy, Julius, Joanne, Jonathan Josephine and Joshua[4]), three of which(Julius、 Jonathan and Joshua) are regionally correlative lacustrine mudstones(Fig. 2).

Journal of Physics: Conference Series **2594** (2023) 012039 doi:10.1088/1742-6596/2594/1/012039

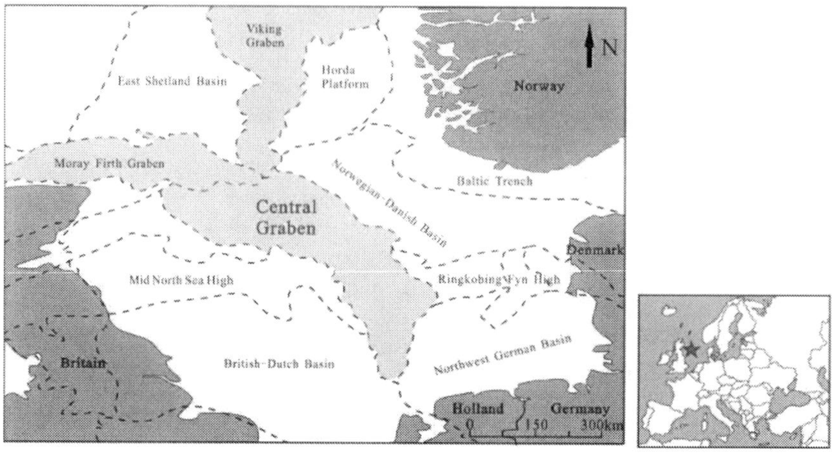


**Fig. 1.** Location map of the UK Central North Sea

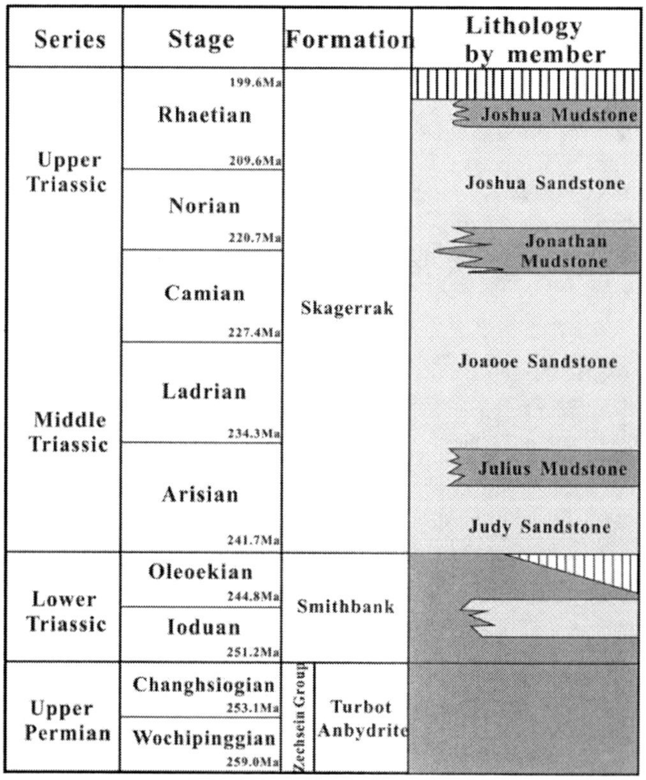


**Fig. 2.** Triassic stratigraphic column of the UK Central North Sea

The Triassic of the North Sea was distributed in a series of fluvial-lacustrine arid basins controlled by faults. The main fluvial system runs from north to south axially and sediment supply is controlled by rifting. In the Central North Sea area the accommodation change is also related to the mobilisation of the underlying Permian salt. Fluvial systems are sourced from the Scottish Highlands several hundred kilometers to the northwest and flows into a series of ephemeral to semi-permanent lakes (playas). The development of these lakes or rivers is influenced by surface runoff and fluctuations of lake levels. During the Late Triassic, fluvial channels were relatively small and bifurcated with small to medium scale cross lamination, and dolocrete conglomerates from the floodplains are developed at the channel bottom. In general, floodplain facies and in-situ soils can not be well preserved. Channel fillings, terminal splays, and shoeface sheetsands are the most important depositional facies in the Central North Sea.

The Skagerrak Formation is composed of multiple repeated sedimentary cycles, each of which is several hundred meters thick and separated by regionally correlatable lacustrine mudstones. Each cycle consists of playa mudstones (lacustrine dominatd), unconfined sheet sandstones, terminal splays and confined coarse-grained fluvial facies from bottom to top. These large-scale cyclic sediments indicate periodic expansion and contraction of the North Sea Triassic fluvial system, controlled by external factors such as regional base level, sediment supply, and climate change[5-6].

## 3 Facies and Depositional Environment

Previous studies have suggested that the Triassic sandstones in the Central North Sea were deposited in an arid to semi-arid environment and represent a fluvial-lacustrine depositional system affected by frequent fluctuations of depositional base level due to salt mobilisation. According to the biostratigraphy and lithostratigraphy characteristics, the Skagerrak Formation can be divided into six members. And mudstones are main correlation markers that can be applied to seismic interpretation and wireline log correlation. The mudstone units consist of two types: one is red-grey or grey with ripple cross-stratification, desiccation cracks, bioturbation, rootlets, etc., which is considered to be periodic floodplain deposits in the ephemeral lake setting; the other is black laminated shales with various carbonized rootlets, layered calcrete nodules, and occasional limestone, which is considered to be swamp to slightly saline lake deposits. Between these two mudstones develops braided fluvial unconfined channel filling sandstones and terminal splay sandstones as well as levees, crevasse splay and floodplain deposits varied greatly in thickness and distribution. Channel filling sandstones are mainly distributed in low-lying areas with rapid subsidence, interspersed with calcrete nodules palaeosoil, while few floodplain argillaceous deposits are remained. The sandstones are interbedded with a small amount of mud shales and cementation layers, such as calcrete nodules palaeosoils and channel lag deposits, which constitute interlayers with limited distribution.

Based on extensive observations of rock cores, the Skagerrak Formation is a series of multi-cycle sediments that were deposited in the semi-arid setting. The depositional environment varied from permanent lakes to marshes or even soil, depending on fluctuations of base level. Lacustrine mudstones commonly develop ripple cross laminated and bioturbated structures, with algal laminations and rootlets visible. Mottled mudstones are gray and rich in rootlets, rhizomes, bioturbation traces mainly left by insects (rather than invertebrate burrows). Calcic paleosols are red and formed during arid periods, while dolocretes nodules present in situ or re-deposited indicate pedoturbation processes in arid environments. At the end of each cycle, unconfined flood events carrying a large amount of debris were deposited in shallow and wide basins[7].

### *3.1 Cored Facies*

The Triassic Skagerrak Formation in the North Sea is mainly composed of sandstones, conglomerates, and various mudstones. Sandstones (Fig. 3) include cross laminated sandstone, planar laminated sandstone, ripple laminated sandstone, bioturbated sandstone, and pedogenically modified sandstone. The grain size of these sandstones ranges from silt to fine grain, with roundness from moderate to good.

The cross laminated sandstone is fine grained sandstone mainly developed in fluvial channels with scattered mica drapes. Localised pedoturbation can be seen at the top, indicating that it was sometimes in slack water of abandoned fluvial channels and sometimes exposed to the surface. The reservoir properties of this kind of sandstone vary greatly. The porosity of the cross laminated sandstone in the fluvial channel filling is generally about 25%, with a maximum of 30.3%, and the permeability is generally more than 50 mD, with a maximum of 2020 mD. The properties of the cross laminated sandstone in the terminal splay minor channels are poorer, with a porosity of up to 25.6% but a permeability of only 1.88 mD.

The planar laminated sandstone was sheetflow sedimentation deposited in slack water with abundant mica drapes . The reservoir properties of this kind of sandstone also vary greatly, and the overall properties are inferior to those of cross laminated sandstone. The porosity of the planar laminated sandstone in the fluvial channel filling is generally about 20%, and the permeability is generally 15 mD, with a maximum of 75 mD. The properties of the planar laminated sandstone in the terminal splay minor channels are even worse, with a porosity of about 15% and a permeability of 0.65-5.08 mD.

The ripple laminated sandstone is fine grained with climbing ripple lamination and mica drapes in ripple troughs. It was deposited in low-energy flow with sufficient sediment supply (climbing ripple lamination)[8]. The reservoir properties of this kind of sandstone are even poorer, with a porosity of about 15% and a permeability of only 0.3 mD. Mica drapes parallel to the lamination are often seen in planar and ripple laminated sandstones.

The bioturbated sandstone is fine to very fine grained sandstone with abundant bioturbation traces left mainly by insects. It is often developed on the edge of lakes or abandoned fluvial channels during relatively humid periods. Due to the presence of burrows and insect larvae in the lamination, this kind of sandstone is mud-prone and is generally not served as reservoir.

The pedoturbated sandstone is characterized by the presence of a large amount of calcrete or dolocrete nodules, evaporites, rootlets and rhizomes, which was deposited under arid conditions with soil formation. This kind of reservoir is generally not served as a reservoir.

During the sedimentation of the Skagerrak Formation, sandstones were arkosic with K-feldspar accounting for the dominant feldspar type. With the transformation of diagenesis, feldspar has significantly decreased. Therefore, sandstones are mainly subarkosic at present. The local abundance of mica and the increase amount of feldspar are reflected as high gamma responses on the wireline log[9].

The conglomerate (Fig. 4) is mainly composed of reworked dolomites, which can distinguish from well-rounded intraformational mudclasts. Pebbles can range up to 3cm diameter with a fine grained sand matrix and subhorizontal lamination, which is typical channel lags deposition. Dolocretes and calcretes conglomerates are mainly from floodplain, and mudclasts are from reworked deposits of lake margin, which is common in terminal splay minor channels. The reservoir properties of dolocretes conglomerate are good, with a porosity of about 20% and a maximum permeability of 200 mD. The reservoir properties of mudclasts conglomerate are poor, with a porosity of 15% and a permeability of only 0.24 mD.

The mudstone (Fig. 4) is characterized by containing silt and mica. According to lamination, bioturbation and pedoturbation, the mudstone can be divided into three types. All kinds of mudstones have a certain amount of dolomites. Mudstone units thickness ranges from 1 to 100 feet. The relatively thick mudstone units can represent regional lake flooding events of stratigraphic significance. The laminated mudstone is dark grey mudstone with interbedded thin sandstone and rare desiccation cracks, which is deposited in slack water. The mottled mudstone and muddy sandstone are grey, red and green variegated silty claystones with bioturbation traces, which represent ephemeral lacustrine deposition of aridity alternations. Insects activities occurs during periods of humidity[10]. The pedoturbated mudstone and muddy sandstone are green, grey and red variegated silty claystones and muddy sandstones with rootlets, rhizomes and calcrete or dolocrete nodules. They were deposited in arid climate with lowering water tables. The dolocrete nodules were developed during the process of soil formation[11].

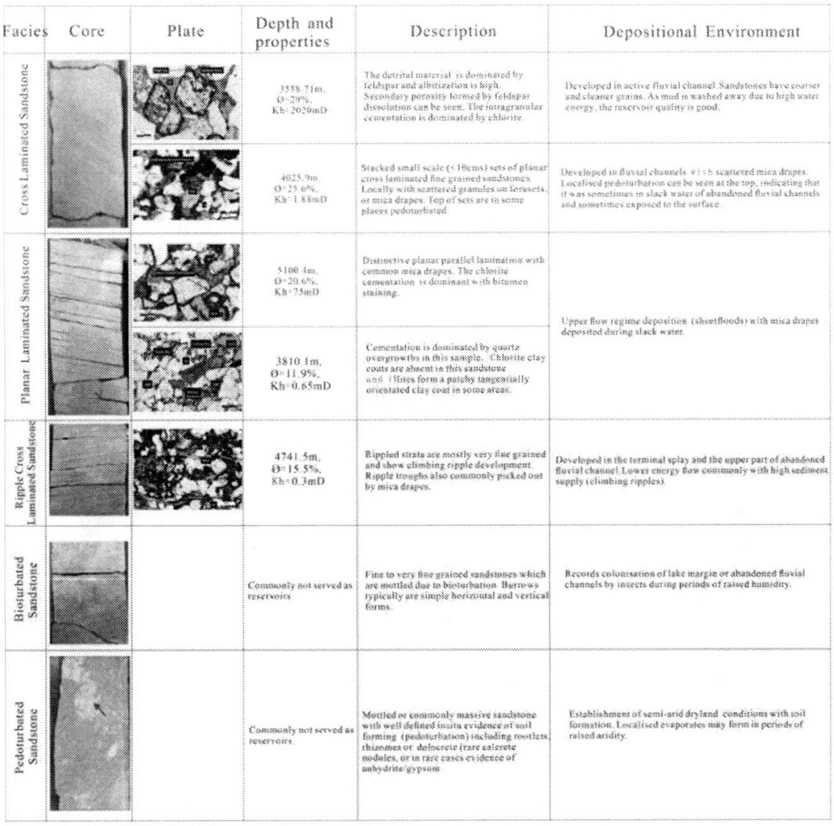


**Fig. 3.** The core facies and interpretation of the Triassic Skagerrak Formation（1）

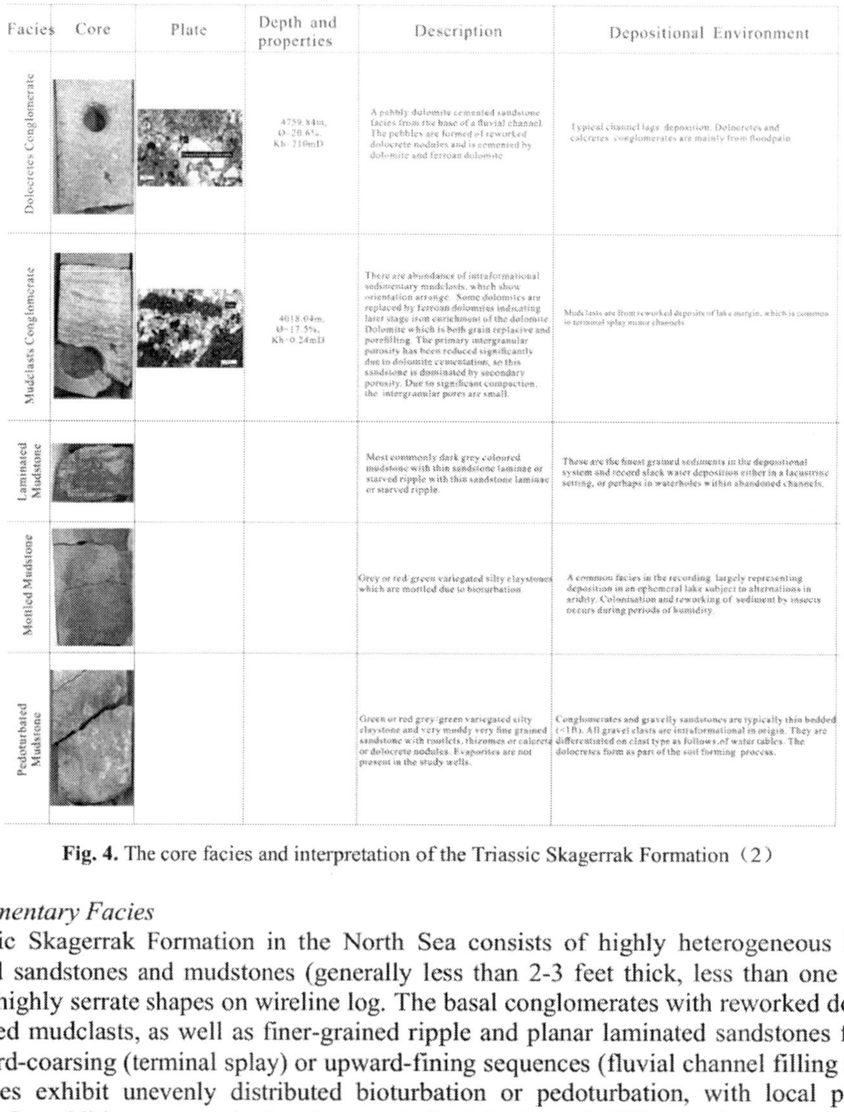


**Fig. 4.** The core facies and interpretation of the Triassic Skagerrak Formation（2）

### *3.2 Sedimentary Facies*

The Triassic Skagerrak Formation in the North Sea consists of highly heterogeneous layers with interbedded sandstones and mudstones (generally less than 2-3 feet thick, less than one meter) that appears as highly serrate shapes on wireline log. The basal conglomerates with reworked dolomite and well-rounded mudclasts, as well as finer-grained ripple and planar laminated sandstones form small-scale upward-coarsing (terminal splay) or upward-fining sequences (fluvial channel filling sediments). These cycles exhibit unevenly distributed bioturbation or pedoturbation, with local parts red or variegated. In addition to terminal splays and fluvial channel filling sediments, the dominant sedimentary facies also include floodplains and ephemeral lacustrine sediments (Table 1).

**Table 1.** Characteristic Wireline Log Response of Sedimentary Facies

| Depositional Facies | | | Typical Wireline Log Response |
|---|---|---|---|
| **Fluvial** | **Fluvial Channel** | **Active** | Fluvial elements are typically sharp based and show a more blocky but overall upward increasing GR profile reflecting increasing clay content in finer grained abandonment facies. Abandonment units may be sharp based. Enhanced cementation at base and top of fluvial units is reflected on porosity logs. |
| | | **Abandoned** | |

| | | | |
|---|---|---|---|
| | **Floodbasin** | | These units are difficult to pick on wireline without core constraint. They tend to be more muddy and dolomite cemented which would indicate higher GR and distinct porosity log response. |
| **Lacustrine** | **Terminal Splay** | **Minor Channel** | Terminal Splay Complexes form larger scale composite upward cleaning cycles on gamma ray (typically 10-20ft thick) amalgamating into packages up to 200ft thick. Overall GR tends to be hotter and more serrate reflecting the higher mica and mud content, as well as the thinly bedded character of these facies. The smaller scale terminal splay lobes form smaller scale composite upward cleaning cycles (<5ft). Minor distributary channels form small scale sharp based upward 'fining' units on gamma profile and are generally <5ft thick, but may be multistorey in architecture. |
| | | **Lobe** | |
| | **Ephemeral Lake-Playa** | | High GR response or GR peak reflecting high clay content in these muddy facies. Commonly form the base of upward coarsening units. |

*Fluvial Channel Filling Deposits.* Fluvial channel filling deposits are composed of a group of upward-fining sequences with erosional bases overlain by conglomerates. They are characterized by multistorey filling and thickness of over ten feet. These sequences consist of cross laminated and low-angle laminated sandstone, and contain abundant mudclasts of lag deposits and re-deposited calcareous paleosols (Fig. 5).

Fluvial deposits are largely controlled by incised valleys and localised microtopography, which may be influenced by salt and rifting activities. During the humid periods of climate cycle, the sediment supply increased and the fluvial channel filling deposits indicate semi-perennial fluvial environment. The lower part of the channels consist of conglomerates and fine to medium grained sandstones cemented by dolomite, which include pebbles with a diameter of 5-25mm dominated by dolocrete nodules. Fine to medium sandstones are overlain with moderate sorting and small-scale cross lamination. Mica drapes and burrows are rare in sandstones, while localised dolocrete nodules are developed. Pedogenesis and dolocrete nodules can be seen at the top of the channel. Overlapping fluvial gravel deposits may be influenced by valley incision during lake level fluctuations [12]. While fluvial systems in arid environments are dominated by ephemeral regime with high variation in flow and deposited on floodplains that are heavily eroded with rare vegetation cover. The channels change frequently and the unchanneled areas are often covered by sheetflow deposits. This fluvial system contains a large amount of bedload sand or gravel and suspended load (mud). It is dominated by braided fluvial systems which are easily eroded along the banks, and the proportion of ephemeral bedload is high with a high width-depth ratio [13]. The upper part of channels are fine grained and belong to abandoned channel deposits with higher clay content.

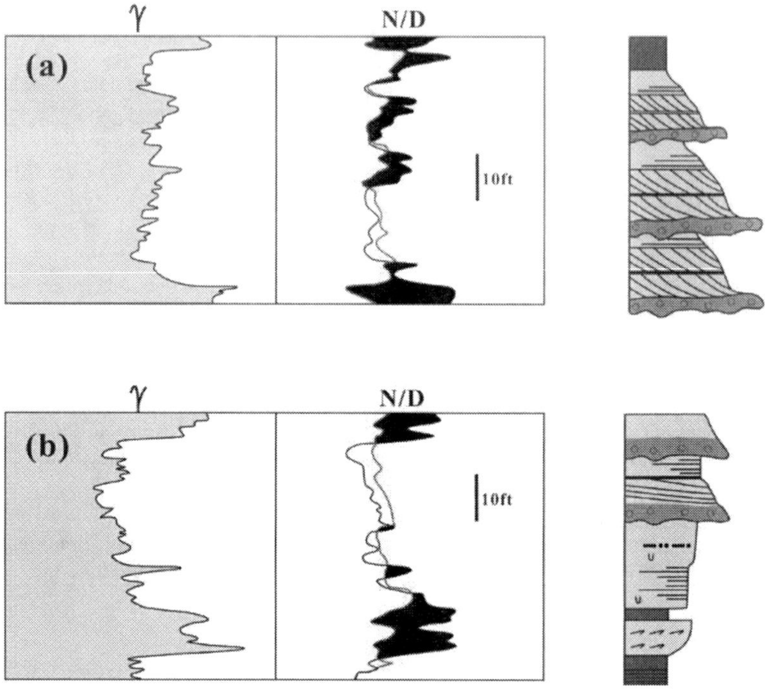


**Fig. 5.** Two basic sedimentary facies of the Triassic Skagerrak Formation(modified after T.McKIE（2005）

*Terminal Splay.* The terminal splay is one of the most important sedimentary facies types of the Skagerrak Formation (Fig. 6). Vertically, it consists of multiple upward coarsening cycles with a thickness of generally 7-15 feet and composed of planar laminated and ripple cross laminated sandstone with rich mica. Thin mudstone layers with mud cracks developed are often observed at the base of the cycle, while burrows are rare.

The terminal splay is in contact with the overlying sandstone with either abrupt transition(mudclasts or mud cracks) or gradual transition. It represents fan-shaped sediments at the terminus of a river in a lake delta or a dry environment. The sedimentary characteristics reflect the background of intermittent aridity of the lake, rapid changes in flow sedimentary conditions and reduced channel confinement. Sheet-flow is the dominant sedimentary facies. Its cored facies is dominated by planar laminated, ripple cross laminated and bioturbated sandstones with mica coatings. Climbing ripple lamination may develop under the condition of abundant sediment supply as well as slow flow velocity.

The terminal splay consists of terminal splay channels and lobes. Compared with the fluvial channel facies dominated by cross lamination, the terminal splay channel is dominated by planar lamination with thinner thickness (<3-5 feet) than that of the fluvial channel[14]. The terminal splay channels are similar to distributary channels, with mudclast conglomerates at the bottom and dolocrete or calcrete nodules are rare. The terminal splay lobes are distributed at the terminus of the terminal splay channel and are fan-shaped as a whole. The modern example that can be used for analogue is Lake Eyre in Australia (Fig. 6). According to Lake Eyre, the morphology of terminal splay lobes is controlled by topography and varies in shape [18]. Vertically, the terminal splay is often interbedded with ephemeral lacustrine mud sedimentation, which is characterized by abundant drying features such as pedogenic mudstones, desiccation cracks, rootlets and rhizomes.

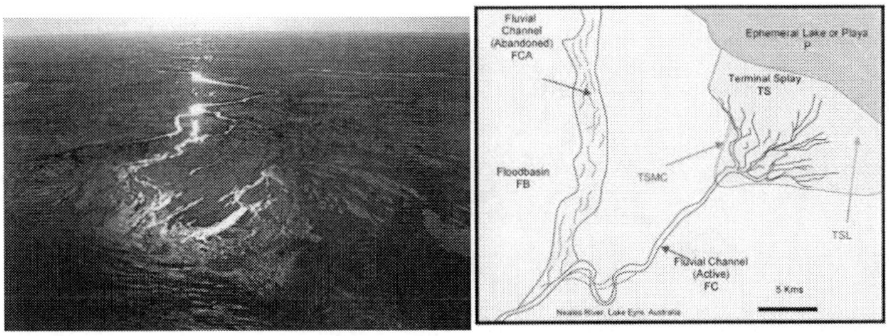


**Fig. 6.** The Lake Eyre of Australia

*Floodplain Sedimentation.* Floodplain facies are characterized by the presence of pedoturbation, especially dolomite developed in the semi-arid environment. The floodplain sedimentation developed between fluvial channels is rarely preserved during the deposition period of the Skagerrak Formation , indicating a low-sinuosity fluvial system with channels constantly modified and laterally shifted on floodplains and valley floor.

*Ephemeral Lake Sedimentation.* Ephemeral lake sedimentation is mainly composed of muddy sediments with interbedded claystones. The thickness ranges from less than 1 feet to 100 feet, and colors are more red or variegated than sandstones. Lacustrine sedimentation includes laminated mudstones, bioturbation-mottled mudstones, or pedoturbated mudstones. Sedimentary structures indicate sedimentation of very shallow water, periodic aridity or desiccation environments. The main indicators include ripple cross lamination, desiccation cracks, insect burrows, rootlets, rhizomes or pedogenesis. Thicker lacustrine sediments represent sedimentation deposited during the high stand regional lake level period and can be used as regional correlative markers. However the thickness of lacustrine sediments does not necessarily indicate lake depth, as many shallow water sedimentary textures are present, which reflect constant lake developments during periods of relatively wet climate. Thick muddy lacustrine sediments represent the basin can accommodate continuously increasing amounts of mud brought by flowing water. Based on a large number of core, wireline and seismic data, a lithofacies paleogeographic map of the Skagerrak Formation with three sandstone-dominated cycles has been compiled, such as the Judy Member lithofacies paleogeographic map (Fig. 7).

Journal of Physics: Conference Series **2594** (2023) 012039 doi:10.1088/1742-6596/2594/1/012039

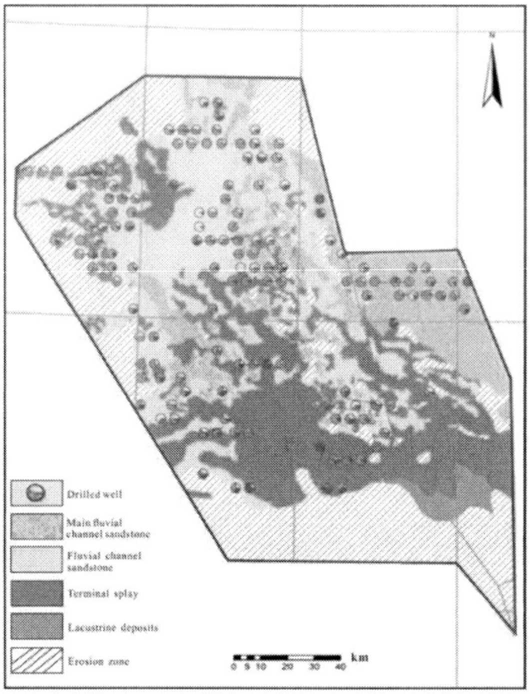


**Fig. 7.** The lithofacies paleographic map of the Judy Member of the Skagerrak Formation

## 4 Sandstone Petrography

The Triassic Skagerrak Formation in the North Sea is mainly composed of fine-grained sandstone, with feldspar as the dominant detrital component. The sedimentary facies strongly control the primary texture and mica content of the sandstone, which significantly affects the diagenetic history and reservoir quality. In general, sandstones in high energy fluvial channels are relatively clean, dominated by medium to fine sand with good sorting (Table 2-1) and contain basal conglomerates. The terminal splay is mainly composed of fine-grained sandstone with high mica content. Ductile particles have experienced early compaction, causing a significant reduction in reservoir properties (Table 2-2).

The Skagerrak Formation sandstones have undergone significant diagenetic transformation, resulting in particle dissolution, replacement, and intergranular cementation. The preserved pore system is a complex network composed of primary intergranular pores, secondary pores and micropores. This diagenetic process significantly affects the relationship between porosity, permeability and depth. Sandstones with well-connected primary intergranular pores exhibit significantly better permeability than sandstones with more isolated secondary pores and tortuous throat micropores [15].

### *4.1 Detrital Material*

The Skagerrak Formation sandstone is mainly subarkosic, containing abundant intraformational blocks such as dolocrete nodules and mudclasts. Due to replacement of kaolinite, albite, illite and dolomites or variable dissolution of feldspar and the secondary porosity, the detrital composition of sandstone has undergone significant changes.

During the period of deposition, the Skagerrak Formation was rich in feldspar. While during the diagenetic process, about one-third of the original detrital feldspar disappeared or modified due to replacement and dissolution, and the K-feldspar was often replaced by more stable albite after deep burial. Other forms of K-feldspar such as perthite and microcline are partly altered by illite and kaolinite. The content of monocrystalline quartz in the detrital component is high, and the presence of monocrystalline quartz and polycrystalline quartz indicates that the detrital material comes from a mixed metamorphic rock series. The detrital grain boundaries are generally well-rounded, indicating high-maturity source or long-distance transportation. Mica is mainly composed of muscovite with a small amount of biotite and localised pyrite. Muscovite is partly replaced by clay minerals (illite, kaolinite) during diagenesis. Rock fragments are predominantly composed of quartzose fragments from metamorphic rocks (high-grade phyllite). Heavy minerals mainly include magnetite, garnet, epidote and zircon.

### *4.2 Diagenetic Minerals*

A large number of authigenic minerals are found in the sandstones of the Triassic Skagerrak Formation, including quartz, feldspar overgrowth, chlorite, kaolinite, illite, dolomite, siderite, pyrite, and albite. The distribution of most minerals is uneven, with dolomite, chlorite, and quartz cements being locally developed.

Quartz forms euhedral syntaxial prismatic overgrowths on detrital grains, showing distinct angular shapes to detrital grains (Table 2-3). The boundary of quartz overgrowth can be identified locally by dust rims between the edge of detrital grains. Chlorite coatings cover about 80% of the detrital grain surface and significantly constrain quartz overgrowth[16].

Albite replaced K-feldspar in situ (Table 2-4). Albite is the stable form of K-feldspar, which is difficult to further be replaced or dissolved.

Dolomite is more common and its edges are often stained with iron (Table 2-5). It is a common diagenetic mineral in the Triassic sandstones of the North Sea, either as intergranular cement or as a product of grain replacement. Dolomite can be rhombic spar cement filling pore space or grain replacement. Dolomite cements often come from various sedimentary materials that are either in situ or resedimented. Basal conglomerates deposited in fluvial channels can often be strongly cemented by dolomite.

Pyrite (Table 2-6) is present in most sandstone samples in a little amount and mainly appears in the late stage of replacement . In some samples, pyrite replaces the edges of rock fragments and alters clay detrital grains. In some samples, pyrite is related to bitumen stained fractures.

Kaolinite appears either as grain replacement of feldspar or intergranular cement in trace amounts. Kaolinite only appears in a few samples, characterised by its pseudohexagonal shapes and forms distinctive booklets.

Illite appears either as clay coatings or replacement of feldspar in trace amounts, which is present in a few samples

Chlorite (Table 2-7) can be seen in most samples and appears as grain coatings covering 20%-80% of grains. In thin section, clay has a good radial flake-like crystalline form with severe bitumen staining. As mentioned above, chlorite clay coatings significantly constrain the overgrowth of quartz.

**Table 2.** Petrological characteristic chart of sandstone

| Number | Plate | Description |
| --- | --- | --- |
| 1 | 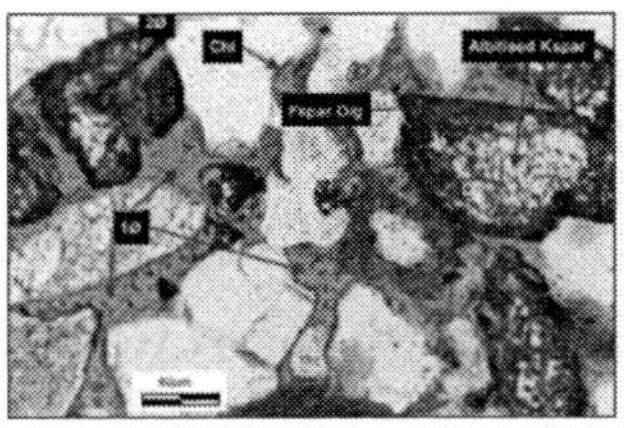 | Depth: 3590.28m, Facies: active fluvial channel, Lithology: cross laminated sandstone, Porosity：30.3%, Permeability：1350mD.<br>The sandstone is subarkosic. The dominant feldspar is K-feldspar and has a high degree of albitisation. Feldspar dissolution is developed, resulting in the generation of secondary porosity. Porosity is dominantly by primary porosity, which shows well connection due to a low degree |

| | | |
|---|---|---|
| | | of compaction and cementation as well as the absence of clay matrix. Coarser grain size has enhanced the permeability. |
| 2 | 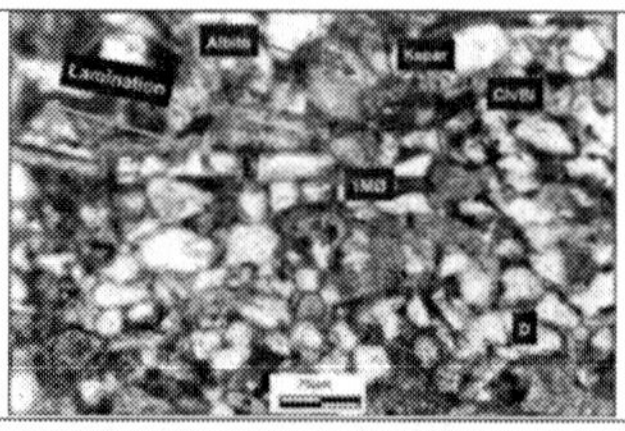  | Depth:12279ft(4029m), Facies: terminal splay minor channel, Lithology: planar laminated sandstone, Porosity: 17.7%, Permeability: 5.08mD. The orange stain is identified as k feldspar, the patchy white area is identified as albite suffering partial diagenetic replacement. Mica is abundant in this sample and shows a parallel alignment. |
| 3 |   | Depth:15613ft(4320m), Facies: fluvial channel, Lithology: planar laminated sandstone, Porosity: 11.9%, Permeability: 0.65mD. Cementation is dominated by quartz overgrowths in this sample. Chlorite clay coats are absent in this sandstone and illites form a patchy tangentially orientated clay coat in some areas. |
| 4 | 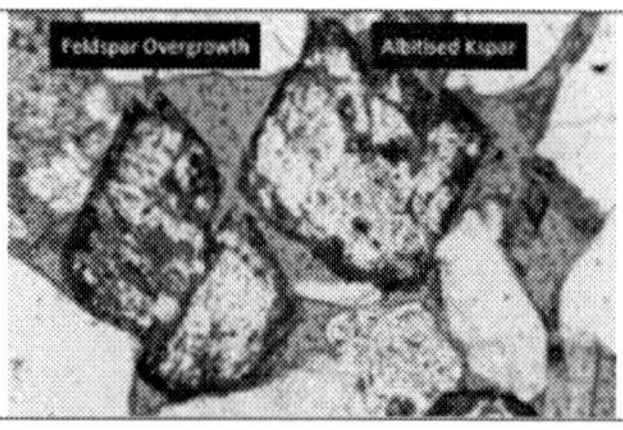  | Depth:12279ft(4029m), Facies: terminal splay minor channel, Lithology: planar laminated sandstone, Porosity: 17.7%, Permeability: 5.08mD. The orange stain is identified as k feldspar, the patchy white area is identified as albite suffering partial diagenetic replacement. Mica is abundant in this sample and shows a parallel alignment. |
| 5 |   | Depth:12247ft(4018m), Facies: terminal splay minor channel, Lithology: mudclast conglomerate, Porosity: 17.5%, Permeability: 0.24mD. There are abundance of intraformational sedimentary mudclasts, which show orientation arrange. Some dolomites are replaced by ferroan dolomites indicating later stage iron enrichment of the dolomite. Dolomite which is both grain replacive and porefilling. The primary intergranular porosity has been reduced significantly due to dolomite cementation so this sandstone is dominated by secondary porosity. Due to significant compaction, the intergranular pores are small. |
| 6 |  | Depth:12271ft(4026m), Facies: terminal splay minor channel, Lithology: cross laminated sandstone, Porosity: 25.6%, Permeability: 1.88mD. In core the sandstone shows a light oil stain (bitumen) and is also cross cut by fractures which are locally associated with pyrite cementation with framboidal shape. |
| 7 | 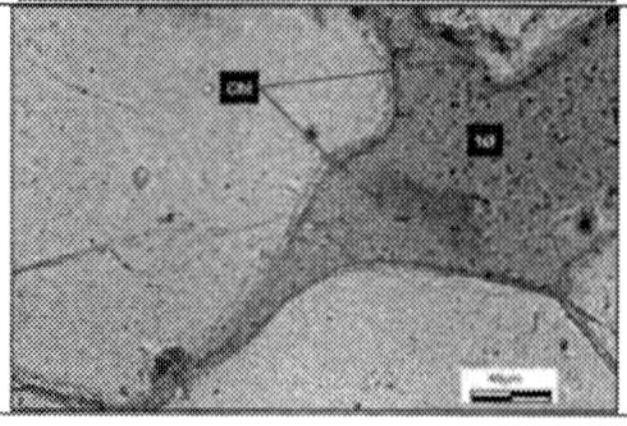 | Depth: 3558.71m, Facies: fluvial channel, Lithology: cross laminated sandstone, Porosity: 29%, Permeability: 2020mD. Occurs as coatings of microcrystalline clay, often stained with bitumen. |

### 4.3 *Rock Structure and Intergranular Volume (IGV)*

The Skagerrak Formation sandstones are characterized by fine grains (silt to medium sand) with moderately well sorting. The diameter of conglomerates in the core can range up to 3 cm, with an average of 5-10 mm. Grains are generally sub-rounded in shape.

The content of clay matrix is controlled by sedimentary facies. The cleanest sandstones are fluvial channel sandstones with coarse grains and cross lamination. Fine-grained sandstones with bioturbation tend to be more muddy due to containing burrows and insect larvae in the bedding planes . Planar and ripple cross laminated sandstones contain mica or clay drapes. These fine-grained facies are mainly deposited in terminal splay and minor channels, which have undergone varying degrees of mechanical compaction. The intergranular volume (IGV) method is used to calculate the porosity loss caused by mechanical compaction (Paxton et al., 2002). The formula for IGV is:

$$IGV = intergranular\ porosity + intergranular\ cement + intergranular\ matrix$$

For the cleanest sandstones, the primary intergranular porosity is about 40%, and the porosity loss due to compaction is about 40%-IGV.

Paxton et al. (2002) found that the porosity of clean, well-sorted, quartz-rich sandstones at depth of 3000 meters is about 26%, indicating a porosity loss rate of 14% due to mechanical compaction. Measured IGV values of these sandstones range from 12% to 26%. Assuming an primary sedimentary porosity of 40%, the porosity loss due to mechanical compaction varies from 28% to 14%. Sandstones with the highest intergranular porosity are those cleanest and coarsest grained, buried at shallower depths. The clay coatings on dissolved feldspar grains can serve as evidence for late-stage compaction. In thin sections, petrographic evidence of chemical pressure solution formed after deep burial such as quartz grain contact sutures are almost absent. Burial diagenesis has extensively altered the porosity system of the Skagerrak Formation sandstones. The porosity during deposition period was about 40%, mainly consisting of primary intergranular pores which have been downgraded to micropores by the precipitation of chlorite microcrystals and detrital clays. Secondary porosity formed by feldspar and mica dissolution also accounts for a significant proportion of the total porosity, which results in secondary porosity and secondary microporosity.

Therefore, we classify the porosity types as primary intergranular porosity, microporosity, secondary porosity and secondary microporosity. Degradation of porosity to microporosity significantly reduces permeability, as it increases the tortuosity and deformation of the porosity network. Grain dissolution can improve porosity, resulting in locally high porosity. The connection between secondary and primary porosity networks makes an effective contribution to the permeability of sandstones.

## 5 Analysis of Sedimentary and Diagenetic Factors Controlling Reservoir Quality and Performance

The samples for this study were collected from depths ranging from 11600 feet (about 3806 meters) to 15500 feet (5085 meters). Although reservoirs of best quality occur at the shallowest depth and reservoirs of poorest quality occur at the deepest depth, there is no simple linear relationship between depth and physical properties [16]. Reservoir properties are not only controlled by depth, but also by changes in sedimentary facies and regional diagenetic trends. The primary depositional texture (grain size, sorting, clay content, etc.) related to sedimentation has a fundamental control on porosity evolution and reservoir quality changes after burial. Subsequent diagenesis further modifies the porosity network. In some samples, secondary porosity and microporosity related to feldspar alteration and dissolution form a mixed porosity network. The development of microcrystalline cements such as the presence of chlorite and detrital clay matrix can cause a significant degradation of primary porosity to microporosity. By combining conventional core analysis with thin section data, it is found that the increase in quartz and doloimite cementation during diagenesis with the lack of chlorite coatings and bitumen staining will lead to low permeability, while fine grain is not the main reason. Chlorite coatings and bitumen staining inhibit the overgrowth of quartz and feldspar, which is the main reason for the high permeability of the reservoir.

### *5.1 Depositional Facies Controls on Reservoir Quality*

Core analysis shows that the basic primary textural characteristics of rock types and sandstones have a strong influence on reservoir quality. Well-sorted and clean cross laminated sandstone reservoirs have good quality, but those with high clay and mica content have significantly poorer reservoir quality. Planar laminated sandstones have slightly lower reservoir quality due to mica drapes on lamination which increases the degree of compaction. Ripple cross laminated sandstones have significantly lower reservoir quality as they are fine graind and have mica or clay drapes. Bioturbated sandstones have poorer reservoir quality due to finer grain size, higher clay content and containing burrows or insect larvae in lamination. Pedoturbated sandstones have poor reservoir quality due to high dolomite cementation which is detrimental to physical properties[17]. Conglomerates generally have poor reservoir quality due to compactional deformation of the clay clasts and high dolomite cementation.

The control of sedimentary environment on reservoir quality is also evident. Fluvial channel sandstones have coarser and cleaner grains. As mud is washed away due to high water energy, the reservoir quality is good. The middle of the channel sandbodies has the best reservoir as basal conglomerates are significantly dolomite cemented and tops are locally pedoturbated. Abandoned channels have poorer reservoir quality as they contain more mud due to lower water energy. If burial depth is shallow and diagenesis has little impact, minor channels connecting the fluvial channels and terminal splays will have a better reservoir quality. However, terminal splay sandstones have finer grains, relatively high clay content, more mica drapes and more distal depositional position in the overall system, which lead to poor reservoir quality, particularly permeability. There is an evident relationship between reservoir quality, burial depth, depositional facies and sedimentary environment. Deeply buried fluvial channel sandstones have better reservoir quality than shallow terminal splay sandstones, which indicates that the influence of depositional facies on reservoir quality is more significant than that of diagenesis. However, for some deeply buried sandstones, the superposition of diagenesis can have a large impact on reservoir quality.

### *5.2 Controls of Diagenesis on Reservoir Quality*

During the depositional period, the Skagerrak Formation was rich in feldspar. But due to grain compaction and dissolution during diagenesis, about one-third of the original feldspar fragments disappeared or were altered. The widely distributed dark gray laminated mudstones and mottled mudstones were deposited in slack water such as lakes or lagoons, and produced a large amount of organic acids. These organic acids entered the primary pores and dissolved feldspar. Fluvial channel sandstones are relatively coarse and clean with strong resistance to compaction, so the primary porosity is well preserved. Good primary porosity is both conducive to the entry of organic acids and to the removal of dissolved grains, promoting the development of dissolution porosity.

During the depositional period of the Skagerrak Formation, open-flow environments allows dissolved grains to migrate smoothly, which is conducive to the preservation and formation of chlorite and inhibits diagenesis. Chlorite coating is a key factor in the preservation of anomalously high porosity in deeply buried thick sandstones as chlorite is often stained by bitumen and significantly inhibits quartz overgrowth. In sandstones where chlorite coatings and bitumen staining are absent, the quartz overgrowth will lead to poor reservoir quality, and anomalously high porosity may be preserved if chlorite coatings and bitumen staining are present. If the primary intergranular porosity is high and chlorite coatings inhibit quartz cementation, a porosity of 30% can still be achieved with a permeability exceeding 1 Darcy even at depths of 11600-12100 feet (about 3806-3970 meters).

### *5.3 Impacts and Controls of Reservoir Quality on Oil Reservoir Performance*

Through studies above, it was clarified that active fluvial channels, abandoned fluvial channels, minor channels, terminal splays, and shoreface sheetsands can serve as reservoirs. Production and testing data from multiple oil fields have shown (Fig. 8) that the oil reservoir permeability system is composed of the above sedimentary facies and low-permeability faults[19], among which active fluvial channels play a key role. Connecting reservoirs consist of high porosity and permeability sandstones in active fluvial channels, low-permeability terminal splay sandstones, abandoned fluvial channels and minor channel sandstones. The initial inflow comes from high-permeability fluvial channel sandstones near the wellbore which depletes quickly, but is constantly fed by low-permeability sandstones. When the pressure differentials exceed 5000 psi, flow can go across faults[1].

It is effective to drill wells on active fluvial channel sandstones , so finding active fluvial channel facies is significant to exploration and development.

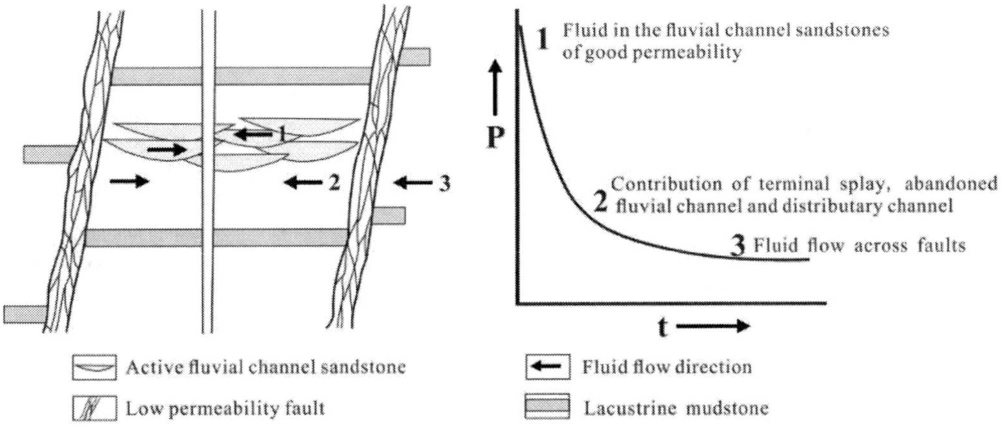


**Fig. 8.** Main factors controlling oil reservoir performance(modified after T.McKIE（2005）

## 6 Conclusions

(1) The Triassic Skagerrak Formation in the North Sea is a cyclic sedimentation of alternate fluvial-lacusrine deposits in the arid to semi-arid flashy depositional setting. Fluvial facies include active fluvial channels, abandoned fluvial channels and intervening floodplains. Lacustrine facies include terminal splay minor channels, terminal splay lobes and playas.

(2) The Skagerrak Formation sandstones are typically fine to very fine grained with medium to good sorting and are characterized by cross lamination, planar lamination, ripple cross lamination, bioturbation or pedoturbation. Burrows are dominated by insect activity traces and soils are characterised by dolocrete nodules developed in the continental semi-arid setting. Cross laminated sandstones especially active fluvial channels sandstones have good reservoir quality. Planar laminated and ripple cross laminated sandstones generally contain mica or clay drapes, which decreases the reservoir quality. Bioturbated sandstones have poor reservoir quality as they are more muddy due to burrows. Pedoturbated sandstones also have poor reservoir quality due to high dolomite cementation. Conglomerates and gravel sandstones are thin bedded and dominated by reworked dolomite and mudclasts. Some dolocrete conglomerates have good reservoir quality.

(3) Thin-section analysis shows that the detrital material of the Skagerrak Formation is mainly composed of feldspar. The significant feldspar alteration and dissolution have resulted in a large amount of secondary porosity. The widely distributed dark grey laminated shale and mottled mudstone were deposited in slack water such as lakes and lagoon, generating a large amount of organic acids. These organic acids enter the primary pores and dissolve feldspar. The sandstone grains of active

fluvial channels are relatively coarse and clean with strong resistance to compaction, thus the primary porosity can be well preserved. Good primary porosity not only facilitates the entry of organic acids but also promotes the development of dissolution pores by carrying away dissolution products.

(4) Conventional core analysis data shows that the primary sedimentary texture (grain size, sorting, and clay content) plays a fundamental control role in the reservoir quality. However, the superimposed effect of diagenesis will be evident as the burial depth increases. Although physical property data shows that reservoirs at shallow burial depth have better quality than those at deeper burial depths, reservoir quality is not simply controlled by depth. Indeed sandstones of similar sedimentary facies have similar porosity and permeability at the same burial depth, the lack of chlorite coatings will significantly reduce reservoir quality due to diagenesis (quartz overgrowth). The calculation of intergranular pore volume shows that changes in compaction degree also have a significant impact on reservoir quality. Chlorite coatings are well-developed and are often stained with residual oil and bitumen. These coatings inhibit quartz overgrowth, thus contributing to the preservation of reservoirs with anomalous high porosity under deep burial conditions.

(5) The reservoir permeability system is composed of high-permeability active fluvial channel sandstones, low-permeability terminal splay sandstones, abandoned fluvial channels, and minor channel sandstones connecting fluvial channel and terminal splay, as well as low-permeability faults. Among them, the active fluvial channel facies plays a key role. The initial inflow comes from high-permeability fluvial channel sandstones near the wellbore which depletes quickly, but is constantly fed by low-permeability sandstones. When the flow pressure differential is greater than 5000 psi, the flow can go across faults. Wells are only effective when they are drilled on active fluvial channel sandstones, so finding active fluvial channel facies is critical to exploration and development.

## Reference

[1]T. MCKIE and P. AUDRETSCH Depositional and structural controls on Triassic reservoir performance in the Heron Cluster, ETAP, Central North Sea Petroleum Geology, 285–297. 2015

[2]John A. Fisher1* & Gary J. Nichols Interpreting the stratigraphic architecture of fluvial systems in internally drained basins Journal of the Geological Society, London, Vol. 170, 2013, pp. 57–65.

[3]N. A Hodgson, 1'2 J. Farnsworth & A. J. Fraser Salt-related tectonics, sedimentation and hydrocarbon plays in the Central Graben, North Sea, UKCS Exploration Britain: Geological insights in the next decade Geological Society Special Publication No. 67, pp 31~38.

[4]T. MCKIE1*, S. J. JOLLEY1,2 & M. B. KRISTENSEN1 Stratigraphic and structural compartmentalization of dryland fluvial reservoirs: Triassic Heron Cluster, Central North Sea n. Geological Society, London, Special Publications, 165–198. 2010

[5]DAVID M. CLEVELAND, STACY C. ATCHLEY, AND LEE C. NORDT CONTINENTAL SEQUENCE STRATIGRAPHY OF THE UPPER TRIASSIC (NORIAN–RHAETIAN) CHINLE STRATA, NORTHERN NEW MEXICO, U.S.A.: ALLOCYCLIC AND AUTOCYCLIC ORIGINS OF PALEOSOL-BEARING ALLUVIAL SUCCESSIONS Journal of Sedimentary Research, 2007

[6]Alonso-Zarza, A.M., 2003. Palaeoenvironmental significance of palustrine carbonates and calcretes in the geologic record, Earth Science Review p.261-298.

[7]Bloch, S., Lander, R.H., Bonnell, L., 2002. Anomalously high porosity and permeability in deeply buried sandstone reservoirs: origin and predictability. Am. Assoc. Pet. Geol. Bull. 86, 301– 328

[8]Clemente, P. and Pérez-Arlucea, M., 1993. Depositional Architecture of the Cuerda Del Pozo Formation, Lower Cretaceous of the Extensional Cameros Basin, North-Central Spain. Jour. Sed. Pet., V. 63, pp. 437-452..

[9]Deibert, J.E. and Camilleri, P.A., 2006. Sedimentologic and tectonic origin of an incised valley fill sequence along an extensional marginal-lacustrine system in the Basin and Range province, United Sates: Implications for predictive models of the location of incised valleys. AAPG v.90, p.209-235.

[10]Ehrenberg, S. N., 1993. Preservation of anomalously high porosity in deeply buried sandstones by graincoating chlorite: Examples from the Norwegian Continental Shelf. AAPG Bulletin, v.77, No. &, P. 1260-1286.

[11]Fisher, J.A., Krapf, C.B.E., Lang, S. C., Nichols, G. J., and Payenberg, T. H. D., 2008. Sedimentology and architecture of the Douglas Creek terminal splay, Lake Eyre, central Australia. Sedimentology 55, 1915–1930.

[12]Hodgson, N, A., Farnsworth, J. and Fraser, A.J., 1992. Salt related tectonics, sedimentation and hydrocarbon plays in the Central Graben, North Sea UKCS: In Hardman R.F.P. (ed) Exploration Britain: Geological Insights for the Next Decade. Geol. Soc. London Spec. Publ. No 67, p.31-63.

[13]Goldsmith P.J., Rich, B., and Standring, J., 1995, Triassic correlation and stratigraphy in the South Central Graben, UK North Sea, in Boldy, S.A.R., ed., Permian and Triassic Rifting in Northwest Europe: Geological Society of London, Special Publication 91, p. 123–143.

[14]Goldsmith, P.J., Hudson, G., AND Van Veen P., 2003, Triassic, in Evans, D., Graham, C., Armour, A., and Bathurst, P., eds, The Millenium Atlas: Petroleum Geology of the Central and Northern North Sea: Geological Society of London, p. 105–127.

[15]Jeans, C.V., Reed, S.J.B., and Xing, M. 1993. Heavy Mineral Stratigraphy in the UK TRias: Western Approaches, Onshore England and Central North Sea, p. 609-624. In Pet. Geol. NW Europe: 4th Conf. Parker JR Editor Geological Society of London.

[16]Kent, D.V., Olsen, P.E. and Witte, W.K. 1995. Late Triassic-Early Jurassic geomagnetic polarity sequence and paleolatitudes from drill cores in the Newark rift basin (eastern North America). Journal Geophysical Research, Vol. 100, p. 14965-14998.

[17]Kessler, J. L. P., G. S. Soreghan, and H. J. Wacker, 2001. Equatorial aridity in western Pangea: Lower Permian loessite and dolomitic palaeosols in northeastern New Mexico, U.S.A.: Journal of Sedimentary Research, v. 71, p. 817– 832

[18]Lang, S.C., Payenberg G, T.H.D., Reilly M.R.W., Kassan, J., Hicks, T., and Benson, J., 2004, Modern analogues for dryland sandy fluvial–lacustrine deltas and terminal splay reservoirs: Australian Petroleum Production and Exploration Association, Journal, v. 44, p. 329–356

[19]Henrik Grob, Elisabeth Seidel, Christian Hübscher, 2020. Seismic amplitude and attribute data from Mesozoic strata in the Skagerrak(Danish-Norwegian North Sea): Indicators for fluid migration and seal bypass systems.

# Ni/ZnO@ZSM-5 Reaction Adsorption Desulfurization Catalyst for Octane Number Improvement

**Ze-kun Guan1, Yu Zhang1, Ting Zhao1, Yun-qi Liu1(✉)**

1 State Key Laboratory of Heavy Oil Processing, College of Chemistry and Chemical Engineering, China University of Petroleum (East China), Qingdao, China
liuyq-group@upc.edu.cn

**Abstract.** Reaction adsorption desulfurization technology is one of the main means of gasoline desulfurization. In this work, ZnO nanoparticles were prepared by pyrolytic oxidation using ZIF-8 as precursor. ZnO spheres with particle size of 5 nm can be formed after the pyrolysis of Zif-8. $Ni/ZnO_{5nm}$ prepared by equal volume impregnation method can achieve 99.2% desulfurization rate at 390°C, which has more excellent desulfurization performance compared to $Ni/ZnO_{100nm}$. Ni/ZnO@ZSM-5 was synthesized by in-situ growth-pyrolysis oxidation-equal volume impregnation method. In the reaction adsorption desulfurization process of FCC gasoline, the desulfurization rate reached 99.5%, and the aromatic content in the product increased by 10.7 percentage points, showing excellent octane number improvement. This work provides new ideas for the synthesis of ZnO nanoparticles as well as novel reactive adsorbent desulfurizers, and promotes the development of S-Zorb process.

## 1 Introduction

$SO_x$ is one of the main pollutants in the atmosphere. The emission of a large amount of $SO_x$ will cause acid rain, photochemical smog and other disasters. The combustion of fossil fuels is the main emission source of $SO_x$[1]. With the increasingly stringent environmental protection law, the sulfur content in gasoline is constantly restricted. At present, the sulfur content in vehicle gasoline is limited to 10 μg/g worldwide. Reactive adsorption desulfurization technology is an effective means of gasoline desulfurization and is widely developed. At present, the reaction adsorption desulfurization catalyst is mainly composed of Ni, ZnO, $Al_2O_3$, $SiO_2$, etc[2-6]. Ni is the active center of desulfurization, which is responsible for the activation of C-S[7], ZnO provides sulfur capacity[8, 9]. The combination of Ni and ZnO can realize the transfer of S from Ni to ZnO to ensure the activity of Ni[10]. $Al_2O_3$ and SiO2 are the carriers of the catalyst, providing support and pores for the catalyst[11-15].

At present, the S-Zorb process based on the reaction adsorption desulfurization principle has been put into production in large quantities for FCC gasoline desulfurization[16]. Although the process can reduce the sulfur content of gasoline to 10 μg/g, there is a problem that olefin saturation leads to a decrease in the octane number of the product. The decrease of octane number is not conducive to the production of refined oil, and it is necessary to add high octane number components in the subsequent process to reconcile. Therefore, it is of great significance to develop a new type of reactive adsorption desulfurizer to ensure the octane number of the product while desulfurization. ZSM-5 has unique pore structure and large specific surface area, so it has good coke resistance and shape selective catalytic function, which is a common aromatization catalyst[17-22]. The introduction of ZSM-5 can make the reaction adsorption desulfurization catalyst have aromatization performance, convert some olefins into aromatic hydrocarbons, and achieve the goal of improving the octane number of the product.

In this work, ZnO was prepared by using ZIF-8 as precursor, and the effect of ZnO particle size on desulfurization reaction was investigated. Ni/ZnO@ZSM-5 was synthesized by in-situ growth-pyrolysis oxidation-equal volume impregnation method, and the desulfurization performance and aromatization performance of the catalyst were investigated.

## 2 Experimental section

### *2.1 Materials*

The chemical reagents (Grade AR) and ZSM-5 ($SiO_2/Al_2O_3$=25) used in the experiments were purchased from Aladdin Chemical Co., Ltd. and Nankai University Catalyst Plant, respectively.

### *2.2 Preparation of catalysts*

5.6 g of $Zn(NO_3)_2 6H_2O$ and 6.2 g of 2-methylimidazole were dissolved in 150 ml of methanol, respectively. The two solutions were mixed and stirred at room temperature for 24 h and then centrifuged to obtain ZIF-8. ZIF-8 was calcined in air at 300, 350, 400°C for 3 h to obtain $ZnO_{300℃}$, $ZnO_{350℃}$, $ZnO_{400℃}$, respectively. Loading 8% Ni on $ZnO_{400℃}$ and $ZnO_{100nm}$ (purchased) by equal volume impregnation method were named as $Ni/ZnO_{5nm}$ and $Ni/ZnO_{100nm}$, respectively.

ZIF-8@ZSM-5 can be obtained by adding 2 g ZSM-5 to the above ZIF-8 synthesis solution. ZnO@ZSM-5 was obtained by calcining ZIF-8@ZSM-5 in air at 400°C for 3 h. Loading 8% Ni on ZnO@ZSM-5 by equal volume impregnation method was named as Ni/ZnO@ZSM-5.

### *2.3 Characterization of catalysts*

The crystal structure of the synthesized product was characterized by X-ray diffractometer (XRD, X' Pert PRO MPD, The Netherlands). The light source adopts Cu target, Ka radiation source ($\lambda$=0.15406 nm), and the test angle is 5~80°. The morphology of the catalysts were observed by scanning electron microscope (SEM, S-4800, Japan) and transmission electron microscope (TEM, JEM-2100, Japan).

### *2.4 Evaluation of catalyst activity*

A 10 ml micro-reaction device was used to evaluate the activity of the catalyst, loading 2 g catalyst (20-40 mesh), reaction pressure of 1 MPa, $H_2$/Oil=100:1. The catalyst needs to be reduced with $H_2$ at 400°C and then adjusted to the reaction temperature for experiments. The activity of Ni/ZnO was evaluated by using n-heptane and thiophene to configure model oil with S content of 1000 μg/g as raw material, and the reaction temperature was 360, 390, 420, 450 and 480°C. The catalytic activity of Ni/ZnO@ZSM-5 was evaluated using FCC gasoline with S content of 1186 μg/g as raw material at 390°C.

## 3 Results and discussion

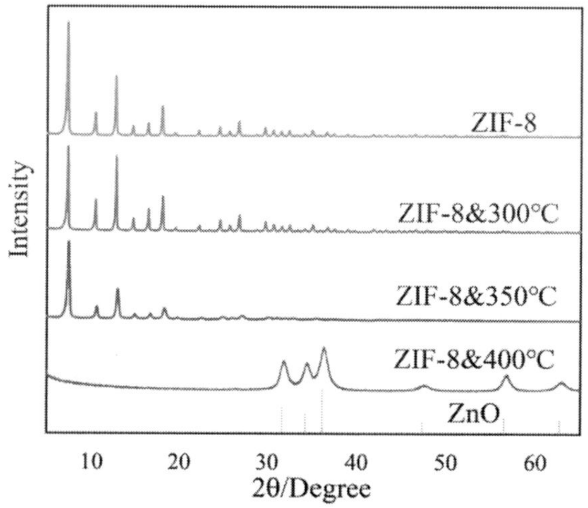


Fig. 1 XRD patterns.

From the XRD pattern (Fig. 1), it can be seen that the characteristic peaks of ZIF-$8_{300°C}$ are similar to the characteristic peaks of ZIF-8, while the lack of some peaks corresponding to ZIF-8 in ZIF-$8_{350°C}$ indicates that the crystal structure is destroyed since ZIF-8 begins to decompose at 350°C. The characteristic peak of ZIF-8 completely disappeared at 400°C, and some new characteristic peaks corresponding to ZnO appeared, indicating that ZIF-8 was completely decomposed to form ZnO.

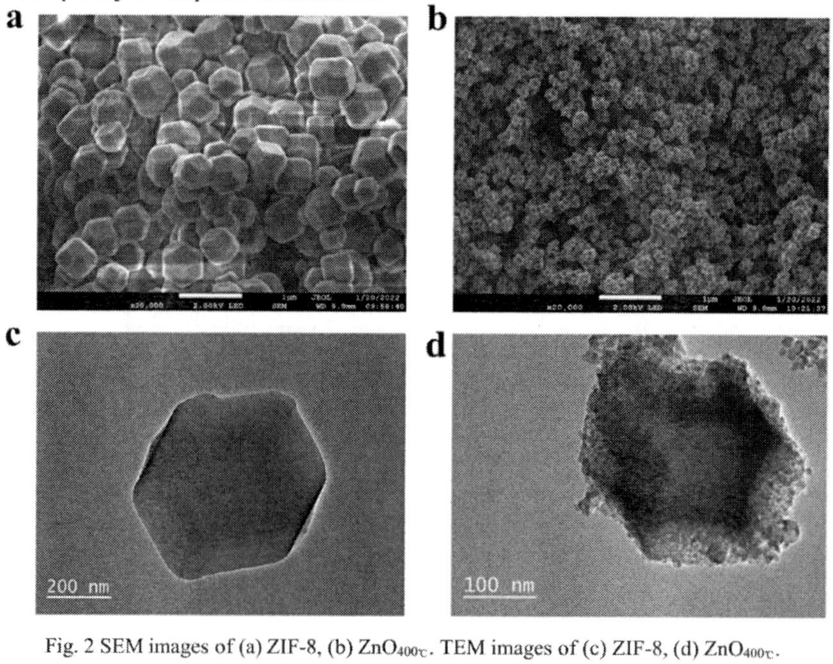


Fig. 2 SEM images of (a) ZIF-8, (b) $ZnO_{400°C}$. TEM images of (c) ZIF-8, (d) $ZnO_{400°C}$.

From the SEM diagram (Fig. 2a), it can be seen that ZIF-8 has an dodecahedral structure with a particle size of about 200 nm. The ZnO formed after pyrolysis still retains the original dodecahedral structure of ZIF-8, but shrinks significantly in size compared to ZIF-8, with a particle size of 100 nm (Fig. 2b). From the TEM diagram (Fig .c and d), it can be seen that the dodecahedral ZnO formed by the pyrolysis of ZIF-8 is composed of ZnO spheres of about 5 nm.

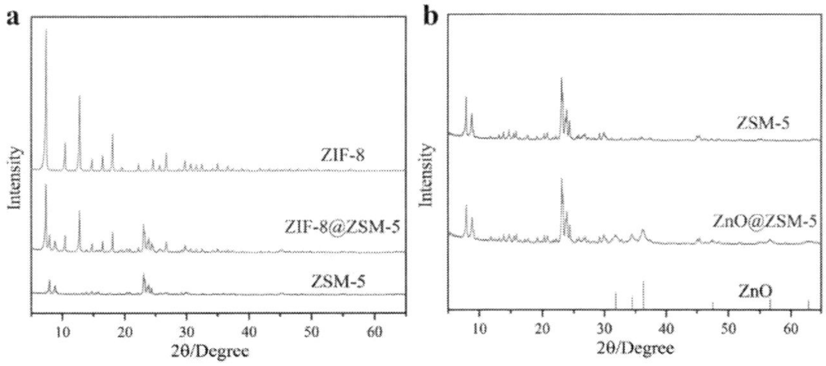


Fig. 3 XRD patterns.

In the XRD pattern of ZIF-8@ZSM-5 (Fig. 3a), the characteristic peaks of ZSM-5 and ZIF-8 appeared at the same time, indicating that the product was a composite material containing ZIF-8 and ZSM-5. The intensity of the characteristic peak of ZIF-8 in ZIF-8@ZSM-5 is significantly higher than that of ZSM-5, indicating that the content of ZIF-8 in ZIF-8@ZSM-5 is higher. Comparing the characteristic peaks of ZSM-5, it can be found that the characteristic peaks of ZSM-5 in ZIF-8@ZSM-5 did not change significantly, indicating that the original crystal structure of ZSM-5 was not destroyed during the synthesis of ZIF-8. After oxidative pyrolysis, the characteristic peak of ZIF-8 in ZIF-8@ZSM-5 disappeared completely, and the characteristic peak corresponding to ZnO appeared, indicating that ZIF-8 was completely transformed into ZnO, and the characteristic peak of ZSM-5 did not change significantly during this process, indicating that ZnO@ZSM-5 was successfully synthesized (Fig. 3b).

Fig. 4 SEM images of (a)ZSM-5, (b) ZIF-8@ZSM-5, (c) ZnO@ZSM-5.

It can be seen from SEM diagram (Fig. 4b) that a large amount of ZIF-8 is generated on the surface of ZSM-5. This is due to the weak negative charge on the surface of ZSM-5, which will attract $Zn^{2+}$ to induce ZIF-8 to nucleate on the surface of ZSM-5. After pyrolysis, ZnO is significantly reduced relative to ZIF-8 and uniformly attached to the surface of ZSM-5 without shedding, indicating that ZnO@ZIF-8 is successfully synthesized (Fig. 4c).

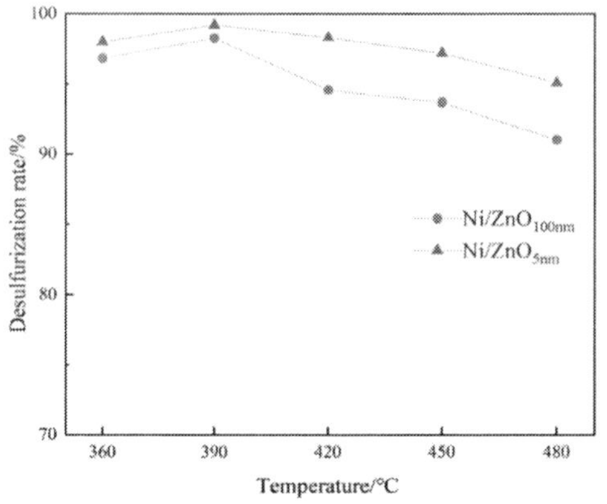


Fig. 5 Desulfurization rate of catalyst.

The desulfurization rate of $Ni/ZnO_{5nm}$ and $Ni/ZnO_{100nm}$ catalysts increased first and then decreased with the reaction temperature from 360°C to 490°C, and showed the highest desulfurization rate at 390°C. The desulfurization rate of $Ni/ZnO_{5nm}$ is obviously higher than that of $Ni/ZnO_{100nm}$, reaching 99.2%,indicating that smaller ZnO particle size is beneficial to the desulfurization of Ni/ZnO catalyst.

Table 1 PONA analysis data of products (wt.%)

| Items | P | N | O | A |
|---|---|---|---|---|
| Feed | 39.41 | 7.12 | 29.22 | 24.25 |
| Ni/ZnO@ZSM-5-$SiO_2$ | 38.41 | 8.83 | 39.92 | 12.87 |

In the reaction adsorption desulfurization process of FCC gasoline, the desulfurization rate reached 99.5%. It can be seen from Table 1 that the olefin content in the product decreased by 11.38 percentage points, and the aromatic content increased by 10.7 percentage points, indicating that the introduction of ZSM-5 promoted the aromatization reaction of olefins and had obvious octane number improvement effect.

## 4 Conclusion

ZnO nanospheres with particle size of 5 nm were prepared using ZIF as precursor. $Ni/ZnO_{5nm}$ and $Ni/ZnO_{100nm}$ catalysts were prepared by equal volume impregnation method using 5 nm and 100 nm ZnO as raw materials. The desulfurization rate of the catalyst increased first and then decreased with the reaction temperature from 360°C to 490°C, and showed the highest desulfurization rate at 390°C. The desulfurization rate of $Ni/ZnO_{5nm}$ is obviously higher than that of $Ni/ZnO_{100nm}$, indicating that smaller ZnO particle size is beneficial to the desulfurization of Ni/ZnO catalyst. Ni/ZnO@ZSM-5 was synthesized by in situ growth-pyrolytic oxidation-equal volume impregnation. In the reaction adsorption desulfurization process of FCC gasoline, the desulfurization rate reached 99.5%, and the aromatic content in the product increased by 10.7 percentage points, showing excellent octane number improvement. This paper provides new ideas for the synthesis of ZnO nanoparticles as well as novel reactive adsorbent desulfurizers, and promotes the development of S-Zorb process.

**Acknowledgments.** The present work is financially supported by National Natural Science Foundation (Number 2217080278).

## References

[1] Brunet S., Mey D., Pérot G., Bouchy C., Diehl F.: On the hydrodesulfurization of FCC gasoline: a review. Applied Catalysis A: General. 278. 143 (2005).

[2] Li L., Ju F., Ling H.: Reactivation of $NiSO_4/ZnO-Al_2O_3-SiO_2$ adsorbent for reactive adsorption desulfurization. Fuel. 339. 127411 (2023).

[3] Sun Y., Li L., Ju F., Ling H.: Evolution of nickel species in reactive adsorption desulfurization of benzothiophene. Separation and Purification Technology. 283. 120204 (2022).

[4] Ju F., Wang M., Luan H., Du P., Tang Z., Ling H.: Reactive adsorption desulfurization of NiO and Ni0 over NiO/ZnO-$Al_2O_3$-$SiO_2$ adsorbents: role of hydrogen pretreatment. RSC advances. 8. 33336 (2018).

[5] Li L., Ju F., Sun Y., Pan H., Ling H.: Self-sulfidation adsorbent for reactive adsorption desulfurization. Fuel. 313. 122696 (2022).

[6] Ju F., Wang M., Wu T., Ling H.: The Role of NiO in Reactive Adsorption Desulfurization Over NiO/ZnO-$Al_2O_3$-$SiO_2$ Adsorbent. Catalysts. 9. 79 (2019).

[7] Qiu L., Zou K., Xu G.: Investigation on the sulfur state and phase transformation of spent and regenerated S zorb sorbents using XPS and XRD. Applied Surface Science. 266. 230 (2013).

[8] Huang L., Yan L., Tang M., Wang G., Qin Z., Ge H.: Effect of Pretreatment on the Adsorption Performance of Ni/ZnO Adsorbent for Dibenzothiophene Desulfurization. ACS Omega. 3. 18967 (2018).

[9] Liu Y., She N., Zhao J., Peng T., Liu C.: Fabrication of hierarchical porous ZnO and its performance in Ni/ZnO reactive-adsorption desulfurization. Petroleum science. 10. 589 (2013).

[10] Qiu L., Zou K., Xu G.: Investigation on the sulfur state and phase transformation of spent and regenerated S zorb sorbents using XPS and XRD. Applied Surface Science. 266. 230 (2013).

[11] Bezverkhyy I., Safonova O.V., Afanasiev P., Bellat J.P.: Reaction between Thiophene and Ni Nanoparticles Supported on $SiO_2$ or ZnO: In Situ Synchrotron X-ray Diffraction Study. The Journal of Physical Chemistry C. 113. 17064 (2009).

[12] Ullah R., Bai P., Wu P., Zhang Z., Zhong Z., Etim U.J., Subhan F., Yan Z.: Comparison of the Reactive Adsorption Desulfurization Performance of Ni/ZnO–$Al_2O_3$Adsorbents Prepared by Different Methods. Energy & Fuels. 30. 2874 (2016).

[13] Ullah R., Zhang Z., Bai P., Wu P., Han D., Etim U.J., Yan Z.: One-Pot Cation–Anion Double Hydrolysis Derived Ni/ZnO–$Al_2O_3$Absorbent for Reactive Adsorption Desulfurization. Industrial & Engineering Chemistry Research. 55. 3751 (2016).

[14] Ju F., Liu C., Li K., Meng C., Gao S., Ling H.: Reactive Adsorption Desulfurization of Fluidized Catalytically Cracked (FCC) Gasoline over a Ca-Doped Ni-ZnO/$Al_2O_3$–$SiO_2$ Adsorbent. Energy & Fuels. 30. 6688 (2016).

[15] Meng X., Huang H., Shi L.: Reactive Mechanism and Regeneration Performance of NiZnO/$Al_2O_3$-Diatomite Adsorbent by Reactive Adsorption Desulfurization. Industrial & Engineering Chemistry Research. 52. 6092 (2013).

[16] Lyu Y., Sun Z., Xin Y., Liu Y., Wang C., Liu X.: Reactivation of spent S-Zorb adsorbents for gasoline desulfurization. Chemical Engineering Journal. 374. 1109 (2019).

[17] Zhou Q., Chen Y., Fan S., Wang S., Qin Z., Dong M., Wang J., Fan W.: Development and catalytic mechanism of a highly efficient Pt/Kβ catalyst for n-Heptane aromatization. Fuel. 337. 126874 (2023).

[18] Zhang L., Zhang H., Chen Z., Ning Q., Liu S., Ren J., Wen X., Li Y.: Insight into the impact of Al distribution on the catalytic performance of 1-octene aromatization over ZSM-5 zeolite. Catalysis science & technology. 9. 734 (2019).

[19] Song S., Li T., Ju Y., Li Y., Lv Z., Zheng P., Duan A., Wu P., Wang X.: Lanthanum/Gallium-Modified Zn/ZSM-5 Zeolite for Efficient Isomerization/Aromatization of FCC Light Gasoline. Industrial & Engineering Chemistry Research. 61. 9667 (2022).

[20] Chen G., Fang L., Li T., Xiang Y.: Ultralow-Loading Pt/Zn Hybrid Cluster in Zeolite HZSM-5 for Efficient Dehydroaromatization. Journal of the American Chemical Society. 144. 11831 (2022).

[21] Xu G., Zhu X.: A core-shell structured Zn/SiO2@ZSM-5 catalyst: Preparation and enhanced catalytic properties in methane co-aromatization with propane. Applied Catalysis B: Environmental. 293. 120241 (2021).

[22] Pan T., Ge S., Yu M., Ju Y., Zhang R., Wu P., Zhou K., Wu Z.: Synthesis and consequence of Zn modified ZSM-5 zeolite supported Ni catalyst for catalytic aromatization of olefin/paraffin. Fuel. 311. 122629 (2022).

Journal of Physics: Conference Series **2594** (2023) 012041 doi:10.1088/1742-6596/2594/1/012041

# Effect Analysis of Demagnetization of Permanent Magnet Synchronous Motor for Electric Drilling Tool on Air-gap Flux Density

**Qing-xue ZHANG1, Jun-guo CUI2(✉),Wen-sheng XIAO3,Lian-peng MEI4, Li-ping TAN5**

1 College of Mechanical & Electronic Engineering, China University of Petroleum (East China), Qingdao, China,
2 College of Mechanical & Electronic Engineering, China University of Petroleum (East China), Qingdao, China,

sdust_zqx93@163.com, cuijunguo@upc.edu.com,

**Abstract.** Electric drilling tools have been developed in the field of drilling because of their high efficiency. As the core component of electric drilling tool, permanent magnet synchronous motor(PMSM) provides power for drilling tool. However, in the process of drilling, the permanent magnet of PMSM will appear demagnetization.The value of air-gap flux density affects the performance and efficiency of the motor directly. Demagnetization faults are divided into uniform demagnetization and local demagnetization.In order to study the influence of demagnetization of PMSM on the air-gap flux density, mechanism of permanent magnet is analyzed Firstly, and then establishing numerical model with analytical method. Finally, simulation motor demagnetization using finite element method(FEM) based on Maxwell. No-load and load operating state are analyzed under the two fault modes are considered in this paper, and compared with the healthy state of the motor. The analytical results show that when the demagnetization fault occurs in the motor permanent magnet, the air-gap flux density will decrease accordingly.The FEM method results show that with the increase of uniform demagnetization rate, the flux density of the air-gap almost decreases in proportion to the increase of demagnetization rate. The demagnetization of single pole permanent magnet has little effect on the magnetic density of the nearby permanent magnet, but it will lead to the imbalance of the whole electromagnetic field of the motor. The air-gap flux density can be used as a characteristic index to judge the fault of demagnetization.

## 1 Introduction

Compared with traditional screw and turbodrill, electric driling tool has the advantages of high efficiency and easy control, and it has been applied in oil drilling. PMSM has been applied in electric drilling tools due to its advantages such as high efficiency, low noise and high torque current ratio [1]. Permanent magnet can provide excitation field for permanent magnet motor. At present, the most used permanent magnet material is NdFeB. However, due to high temperature, external magnetic field or vibration and other external influences, permanent magnet will appear demagnetization phenomenon. Demagnetization is mainly caused by armature reaction. During normal operation of the motor, the adverse magnetic field generated by stator current resists the residual magnetic induction in the permanent magnet, especially at high torque, which will generate more intense adverse magnetic field. This phenomenon occurs repeatedly, leading to local or uniform demagnetization of the permanent magnet [2]. High temperature will also lead to demagnetization of magnets. Short circuit between stator turns may also cause partial demagnetization of permanent magnets, which will lead to unbalanced magnetic tension, noise and mechanical vibration of the motor.

The demagnetization of PMSM will lead to complex changes in the electromagnetic field. The study on the influence of permanent magnet demagnetization can provide a basis for early fault detection and subsequent fault maintenance diagnosis [3]

## 2 Study on Demagnetization Mechanism of Permanent Magnet

### *2.1 Temperature demagnetization*

When the drive system of permanent magnet motor is overloaded or heat dissipation is not timely, the internal temperature of the motor is high, the electron arrangement in permanent magnet will begin to change with the interference of temperature, leading to the change of its magnetization ability. Changes in magnetic properties include reversible changes and irreversible changes[4]. After returning to the original temperature, the partial change of magnetic recovery is reversible, which is usually expressed by the temperature coefficient of remanent flux and the temperature coefficient of intrinsic coercivity, as shown in formula (1).

$$\begin{cases} \alpha_{B_r} = \dfrac{B_r - B_r'}{B_r'(t-t_0)} \times 100\% \\ \alpha_{H_{cj}} = \dfrac{H_{cj} - H_{cj}'}{H_{cj}'(t-t_0)} \times 100\% \end{cases} \quad (1)$$

Where $B_r$, $H_{cj}$ are respectively the remanent flux density and intrinsic coercivity of the permanent magnet at temperature $t$; $B_r'$, $H_{cj}'$ are respectively the remanent flux density and intrinsic coercivity of the permanent magnet at temperature $t_0$. When the temperature rises to a certain level, it will cause irreversible demagnetization of the permanent magnet. In order to maintain constant load operation after demagnetizing occurs, winding current increase and copper consumption increase, leading to further increase of permanent magnet operating temperature, and finally worse demagnetizing level, forming a closed-loop vicious cycle [5]. Fig. 1 shows the dynamic demagnetizing process of permanent magnet at high temperature.

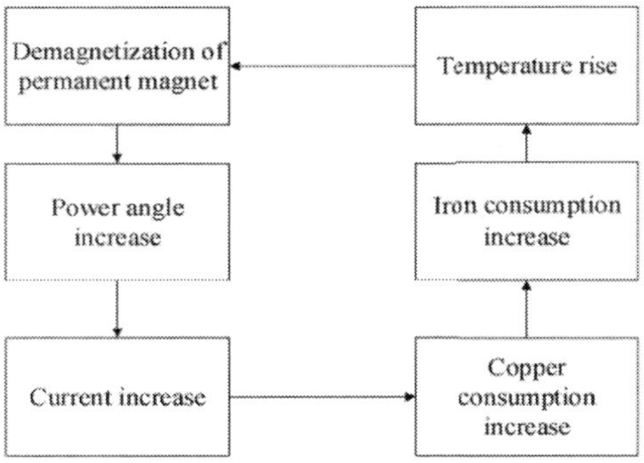


**Fig. 1.** Dynamic demagnetizing process of permanent magnet at high temperature.

*2.2 Demagnetization due to external magnetic field*

When the external magnetic field acts on the permanent magnet and reaches a certain limit, the magnetic domain inside the permanent magnet will turn the direction of rotation into demagnetization [6]. During the operation of motor, permanent magnet is subject to the AC reverse magnetic field generated by armature, as shown in Fig.2. When the permanent magnet works above knee point A, it is in a normal state. After the external magnetic field disappears, it will revert to $B_r$; if the external magnetic field exceeds point A, it works at point D. Removing external magnetic field, the remanent flux density can only revert to $B_d$[7].

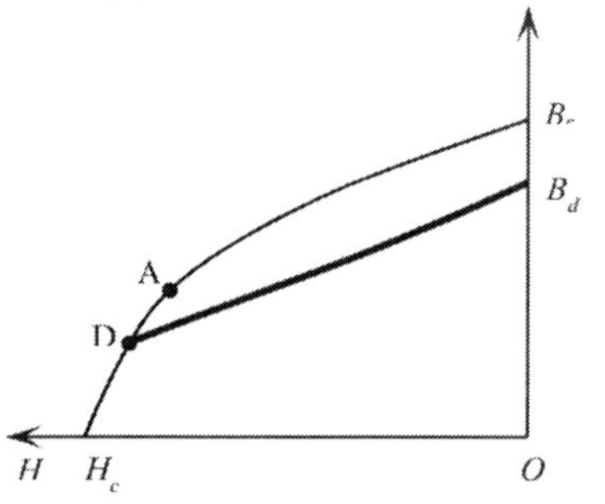


**Fig. 2.** Schematic diagram of demagnetization curve of permanent magnet

## 3 Analytical Model of Demagnetization for PMSM

According to Chapter 2, the demagnetization characteristic of permanent magnet after demagnetization is remanent flux density $B_r$, and it is mainly generated by permanent magnet when the motor in no load state. The air-gap flux can be obtained by the following formula:

$$\Phi_{PM} = \frac{B_r P_c A_m}{\sigma_0} \times 10^{-4} \tag{2}$$

Where $\Phi_{PM}$ is air-gap flux provided by permanent magnet; $P_c$ is the magnetic dense operating point of the permanent magnet; $A_m$ is the magnetic flux area, $\sigma_0$ is the magnetic leakage coefficient in no load[8].

It is assumed that the motor winding length is $l_w$, pole distance is $\tau_1$, pole arc coefficient is $\alpha_i$, the relationship between no-load air gap magnetic density $B_0$ and $B_r$ is:

$$B_0 = \frac{\Phi_{PM}}{\alpha_i \tau_1 l_w} \times 10^{-4} = \frac{B_r P_c A_m}{\sigma_0 \alpha_i \tau_1 l_w} \tag{3}$$

When the load current flows into the armature winding, the air-gap flux density is composed of two parts, one is generated by the permanent magnet, and the other is generated by the winding current [9]. The combined action of direct axis current $I_d$, AC current $I_q$ and back electromotive force $E_0$ results in the synthesized electromotive force $E_\delta$ of the air gap fundamental wave:

$$E_\delta = \sqrt{(E_0 - I_d X_{ad})^2 + (I_q X_{aq})^2} \tag{4}$$

$X_{ad}$ is the direct axis armature reaction reactance and $X_{aq}$ is the quadrature axis armature reaction reactance. The back electromotive force $E_0$ in no-load formed by air-gap fundamental wave is:

$$E_0 = 4.44 f k_{dp} N k_\phi \frac{B_r P_c A_m}{\sigma_0} \times 10^{-4} \tag{5}$$

Where $f$ is the rated frequency, $N$ is the number of series turns per phase winding, $k_{dp}$ is the winding factor, $k_{dp}$ is the waveform coefficient of the air-gap flux, and the air-gap flux $\Phi_\delta$ in load is:

$$\Phi_\delta = \frac{E_\delta}{4.44 f k_{dp} N k_\Phi} \tag{6}$$

Combined with the above formula, the relationship between air-gap flux density $B_L$ and remanent magnetic density $B_r$ under load can be obtained as follows:

$$B_L = \frac{B_r A_m}{\alpha_i \tau_1 l_w} \sqrt{[b_{mn} - (1 - b_{mn})\lambda_\sigma]^2 + (\frac{kP_c}{\sigma_0})^2} \tag{7}$$

$B_{mn}$ is the per-unit value of flux density at the rated load operating point of permanent magnet, $\lambda_\sigma$ is the per-unit value of magnetic leakage conductivity, and $k$ ranges from 0.35 to 0.85.

It can be seen from equations (3) and (7) that, in no-load and load conditions, the air-gap flux density and remanent flux density are proportional to each other under certain conditions, that is, when the demagnetization fault occurs in the motor permanent magnet, the air-gap flux density will decrease accordingly.

## 4 Finite Element Simulation Analysis of Demagnetion

### *4.1 FEM Modeling of PMSM*

In order to analyze the influence of permanent magnet demagnetization on the electromagnetic parameters of the motor, two-dimensional finite element demagnetization simulation is performed. The motor is a 14 poles, 12 slots surface-mounted PMSM with rated speed of 300rpm and rated torque of 168N·m. The stator and rotor material are set silicon steel DW465_50, and the permanent magnet adopts NdFeB. Fig.3 shows the simulation model of PMSM and the number of permanent magnet, where N/S represents the N pole and S pole of permanent magnet.

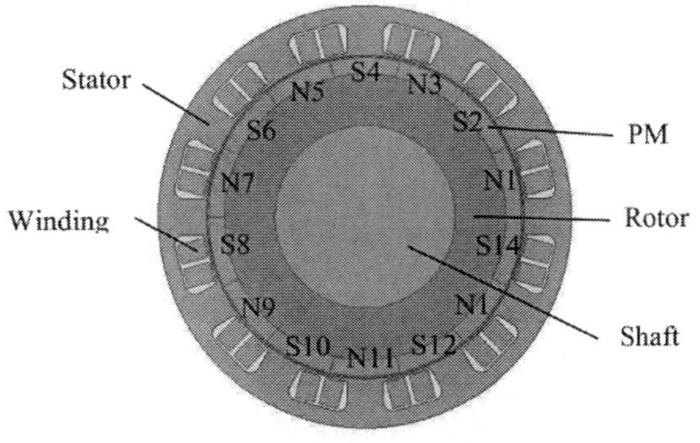


**Fig. 3.** Simulation model of PMSM and the number of permanent magnet

### *4.2 Effect Analysis of uniform demagnetizing on air-gap flux density*

This study researchs the uniform demagnetization of permanent magnets in health, 10%, 30% and 60% under no-load conditions, and simulates different degrees of demagnetization of permanent magnets by changing the parameters of remanent magnetic density and coercivity [10-11]. Fig. 4 shows the variation of air-gap flux density with the circumference under the radius of 65.05mm. It can be seen from the figure that the air-gap flux density changes approximately sinusoidal along with the circumference, and the peak value of air-gap flux density decreases with the increase of the uniform demagnetization ratio.

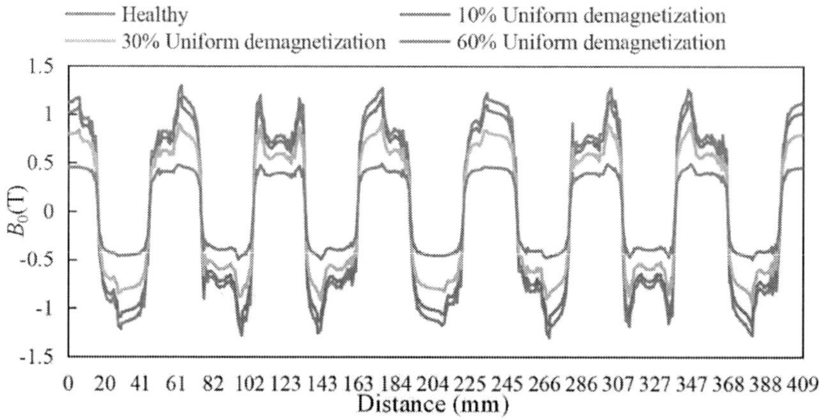


**Fig. 4.** Air-gap flux density with different uniform demagnetizing ratio under no-load

The air-gap flux density is the result of the joint action of stator current and permanent magnet under load, which is more complex than that under no-load. Fig. 5 shows that when stator current generates excitation field, the maximum of air-gap flux density slightly increases compared with no-load.

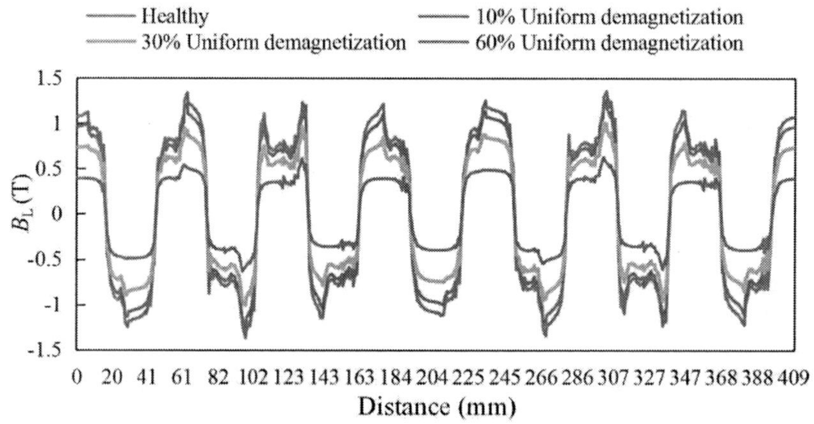


**Fig. 5.** Air-gap flux density with different uniform demagnetizing ratio under rated-load

Table 1 shows the maximum air-gap flux density under different demagnetizing ratio. It can be seen from the table that the maximum of air-gap flux density under no-load is almost proportional to the remanent flux density.

**Table 1.** Maximum air-gap flux density values for different demagnetization ratios with/without load.

| Demagnetization ratio(%) | Remanent flux density(T) | Maximum of air-gap flux density without load | Maximum of air-gap flux density with load |
|---|---|---|---|
| 0 | 1.3 | 1.298 | 1.363 |
| 10 | 1.17 | 1.178 | 1.261 |
| 30 | 0.91 | 0.974 | 1.004 |
| 60 | 0.52 | 0.492 | 0.629 |

It can be seen from Fig. 6 that the air-gap flux density decreases more slowly with demagnetization ratio under load than without load.

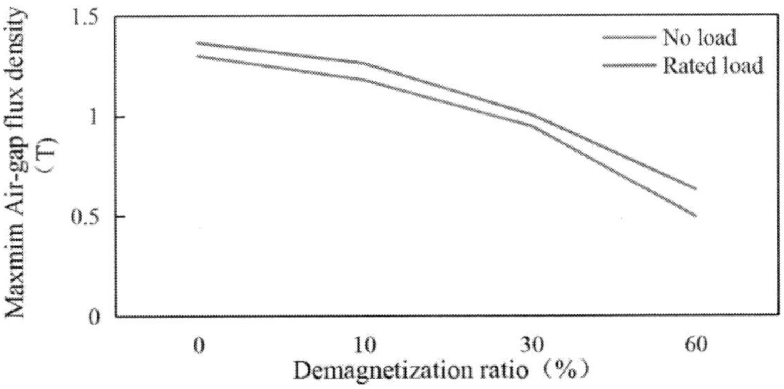


**Fig. 6.** Trend diagram of the maximum air-gap flux density

### *4.3 Effect Analysis of partial demagnetizing on air-gap flux density*

This study researches the partial demagnetization of a single permanent magnet, and N1 permanent magnets are demagnetized by 10%, 30% and 60% respectively. It can be seen from Fig. 7 and Fig. 8 that, no matter in no-load or load state, the peak value of air-*gap* flux density at the demagnetization part decreases obviously with the increase of demagnetization ratio, and the air-gap flux density curve of the remaining undemagnetized permanent magnets basically coincides. It shows that the effect of single pole demagnetization is small. Demagnetizing permanent magnet leads to unbalanced distribution of air-gap magnetic field, which is bound to cause current fluctuation.

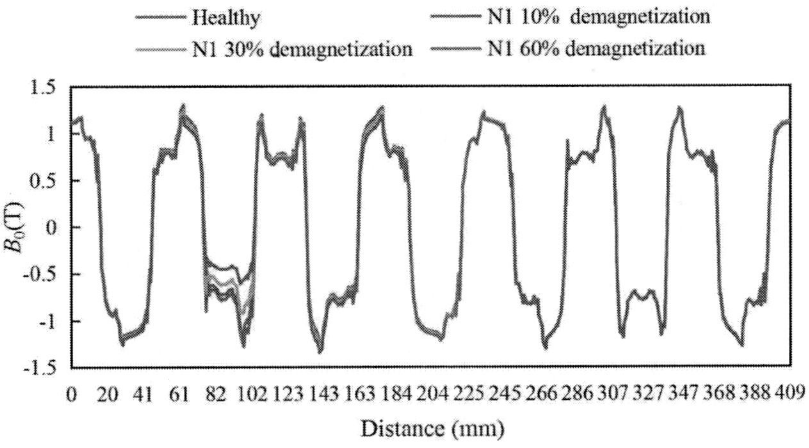


**Fig. 7.** Air-gap flux density with different partial demagnetizing ratio under no-load

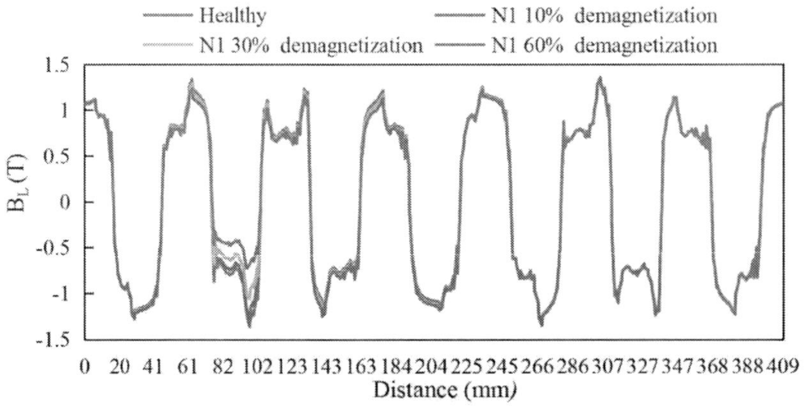


**Fig. 8.** Air-gap flux density with different partial demagnetizing ratio under rated-load

Table 2 shows the maximum air-gap flux density r under different partial demagnetization ratios. It can be seen from the table that the maximum magnetic density of air-gap flux density is almost proportional to the remanent flux density of the motor no matter without load or with load.

**Table 2.** Maximum air-gap flux density values for different demagnetization ratios without/with load.

| Demagnetization ratio of N1(%) | Remanent flux density(T) | Maximum of air-gap flux density without load | Maximum of air-gap flux density with load |
|---|---|---|---|
| 0 | 1.3 | 1.279 | 1.363 |
| 10 | 1.17 | 1.171 | 1.282 |
| 30 | 0.91 | 0.922 | 1.05 |
| 60 | 0.52 | 0.611 | 0.723 |

## 5 Conclusions

Analytical model and finite element model were used respectively to analyze the effect of permanent magnet demagnetizing on air-gap flux density in this study, and compared with the motor in a healthy state, the following conclusions were obtained:

After uniform demagnetizing of permanent magnet under no-load, the air-gap flux density is almost linear, because the air-gap magnetic density under no-load is mainly generated by permanent magnet. Under rated load, due to the effect of stator current, the air-gap flux density decreases slowly with the demagnetization ratio. For single pole partial demagnetization, it has little influence on the flux density of the nearby permanent magnet, but destroys the balance of the overall electromagnetic force. It can be seen from the above that the air-gap flux density can be used as the fault characteristic quantity in the diagnosis of demagnetization faults.

**Acknowledgments.** The project is supported by National Key Research and Development Program Project(2021YFB3401400, 2021YFB3401402).

## References

[1] Xue, Q B., Zhang, J. Advances in inteligent automatic driling technologies and euipment[J]. Exploration Engineering (Rock & Soil Drilling and Tunneling), 47(4):9-14(2020).

[2] Toliyat, H., Nandi, S., Choi, S., Homayoun M-K. Electric Machines, Modeling, Condition Monitoring, and Fault Diagnosis[M], ChinaM achine Press, BeiJing: (2014).

[3] Ding, S C., He, W., Hang, J., Tang D W. Uniform Demagnetization Fault Diagnosis of PMSM Based on Radial Air-gap Flux Density and Stator Current[J]. Proceedings of the CSEE, https://kns.cnki.net/kcms/detail/11.2107.TM.20221014.1745.011.html.(2022)

[4]Sun, Z K. Performance Analysis on the Permanent Magnet Demagnetization Fault of PMSM[J]. Electrical Machinery Technology, 6:53-56 (2022)

[5] Zhang, Y C. Research on comprehensive diagnosis method of electrical anddemagnetization faults of PMSM system in vehicle[D], JiangSu Univesity,(2021).

[6] Hu, T X. Research on Demagnetization Fault of Surface-mountPermanent Magnet [D], Synchronous Motor Guangdong University of Technology(2019).

[7] Song, J C. Demagnetization fault recognition and classificationresearch of double stator air-core permanent m agnetsynchronous linear motor[D],AnHui Univesity(2020).

[8] Sun, Y Ling., Liu Y., Lin W Y. The Selection of Magnet Optimum Operating Point for EV's PMSM [J]. Auto Electric Parts, 4:12-16(2011).

[9] Andrew, S B. Condition Monitoring and Analysis of a Permanent Magnet Synchronous Machine Drive System [D] Michigan State University(2014).

[10]Mohamed, B K., Ahmed, H B., Ameur, F A., Azzedine, B., Zakaria, G. Diagnosis and discernment between eccentricity and demagnetization faults In PMSM drives[J]. Journal of Power Electronics, 21:563-573(2021).

[11]Miguel D P., Antonio G E, Jordi-Roger R R E, Juan A O. Feature Extraction of Demagnetization Faults in Permanent-Magnet Synchronous Motors Based on Box-Counting Fractal Dimension[J].58(5) 1594-1605(2011).

# Development of Predictive Model for Wax Formation in Deep-water Pipeline Using Machine learning

**Farhana Sultana1,Elhassan Mostafa Abdallah1(✉), Shaohua Dong2, Yasir M. F. Mukhtar2,3(✉)**

1Faculty of Engineering Technology and Built Environment, UCSI University, Kuala Lumpur, Malaysia.
2Pipeline Technology & Safety Research Center, China University of Petroleum, Beijing
3College of Petroleum Eng. & Mining, Sudan University of Science & Technology, Khartoum

elhassan@ucsiuniversity.edu.my, yasir@cup.edu.cn

**Abstract.** Wax deposition in deep sea pipelines can affect oil and gas production. It happens when the pipeline temperature drops below the wax appearance temperature, causing the wax to solidify and stick to the pipeline walls. This paper discusses the limitations of current models in predicting wax deposition in pipelines and proposes the development of more complex, intelligent models that are accurate, resilient, and adaptable to novel input-output scenarios. It also highlights various methods for preventing wax deposition, including the use of chemical inhibitors and thermal remediation techniques. The paper aims to develop a neural network modeling approach for predicting wax deposition potential in producing reservoirs to prevent accumulation before removal. The parameters that were used are wax molecule concentration gradient at pipe wall, temperature gradient at pipe wall, reported wax deposition rate, the pipe wall shear stress, and crude oil viscosity. The radial basis function neural network (RBFNN), is based on the machine learning technique. RBFNN is a good choice for building a predictive wax deposition rate model because it is highly accurate, robust to noise and outliers, computationally efficient, can model complex nonlinear relationships, suitable for real-time applications and requires less training data. The RBF neural network is constructed using MATLAB. The outcomes and experimental data are consistent with a 1.5 percent relative inaccuracy when compared to prior studies. The R^2 value was 0.9906, R value was 0.9953 and the RMSE value was 0.8598. The RBF neural network-based approach for predicting wax deposition rate was successful.

## 1. Introduction

Wax accumulation can impede oil flow, resulting in pressure anomalies and an artificial barrier that can restrict or halt output. In severe cases, however, this might lead to the closure of a manufacturing facility or pipeline. When wax solid precipitates, the formation around the wellbore is damaged, permeability is diminished, and the composition and rheology of the reservoir fluid are altered. Moreover, the wax deposit is a gel made of solid wax crystals and trapped liquid, as opposed to solid wax. Over time, the deposit matures and gets tougher. The precipitation of wax components from crude oil results in several waxy crude oil features, such as oil gelation and enhanced viscosity. When the temperature falls below the wax appearance temperature (WAT), the first wax crystals begin to form

in the crude oil, and wax might precipitate throughout the cooling process. When the input coolant temperature falls below the point at which wax will emerge, crude oil paraffin components may get deposited in the wall of the cold pipeline, resulting in wax. [5]
Due to their low temperature, subsea pipes are particularly sensitive to the wax deposition process. Deep-water oil production requires avoidance of wax accumulation. Using chemical, mechanical, and thermal remediation approaches, wax deposition in crude oil production systems can be decreased or prevented. Since mechanical and thermal remediation approaches have become too expensive for wax deposition inhibitors, which are chemical additives, are gaining popularity. Hot oil and steam are circulated via bottom-hole heaters, heat-retention equipment, and other active heating systems, including thermal insulation. Scrappers and pigs are inserted into the holes and pipelines, respectively, to remove the debris mechanically. Chemical inhibitors include PPDs, among others. Wax deposition and wax precipitation are sometimes used interchangeably, although they really relate to two distinct ideas. When crude oil comes into contact with surfaces, wax deposition occurs. Wax deposits can develop from either a previously precipitated solid phase (wax) or dispersed wax molecules that have been diluted by molecular diffusion. [9]
The deposition of wax in pipelines is influenced by several parameters such as crude oil composition, temperature, flow rate, pressure, and time. As the temperature differential between the wax solution and the cold pipe increases, more wax is deposited on the pipe wall. Wax deposition increases with longer operating times and decreases with higher flow rates. Wax deposition can cause lower production rates, equipment obstructions, and complete shutdown of the production process. Mechanical methods like scrapers and pigs are used to remove wax deposits, but they are impractical for deep waters due to the pipeline's length. Chemical additives are becoming more common as wax deposition inhibitors. Solvents can also be used to remove wax buildup, and the formation can be repaired. [6]

## 2. Methodology and Experimental Setup

### *2.1 RBF Neural Network*

The optimum RBF neural network structure was determined based on a statistical study of its training, testing, and validation error, resulting in the simplest structure. This study employs the root square (R2), the root means square error (RMSE), the mean square error (MSE), The mean absolute error (MAE), and the coefficient factor for statistical analysis (R). The statistical analysis equations are provided as follows in Eqs. (1) - (4). [1] The Coefficient of Determination (R^2) was determined using the equation:

$$R^2 = 1 - \frac{\sum_{i=1}^{n} (\text{Ytest} - \text{Ypred})^2}{\sum_{i=1}^{n} (\text{Ytest} - \text{mean}(\text{Ytest}))^2} \quad (1)$$

The Root Mean Square (RMSE) was determined using the equation : (Xie & Xing, 2017a)

$$\text{RMSE} = \sqrt{\frac{1}{n}\sum_{i=1}^{n} (\text{Ytest} - \text{Ypred})^2} \quad (2)$$

The Coefficient Factor for statistical analysis (R) was determined using the equation : (Xie & Xing, 2017a)

$$R = \sqrt{1 - \frac{\sum_{i=1}^{n} (\text{Ytest} - \text{Ypred})^2}{\sum_{i=1}^{n} (\text{Ytest} - \text{mean}(\text{Ytest}))^2}} \quad (3)$$

The Mean Square Error (MSE) was determined using the equation: (Xie & Xing, 2017a)

$$\mathrm{MSE} = \frac{1}{n}\sum_{i=1}^{n} (\mathrm{Ytest} - \mathrm{Ypred})^2 \quad (4)$$

The Mean Absolute Error (MAE) was determined using the equation: (Xie & Xing, 2017a)

$$\mathrm{MAE} = \left(\frac{1}{n}\right) * \mathrm{sum}\big(\mathrm{abs}(\mathrm{Ytest} - \mathrm{Ypred})\big) \quad (5)$$

Where:

- n is the number of samples in the dataset.
- Ytest is the actual or true target values in the test set.
- Ypred is the predicted target values by the model.
- abs() is the absolute function, which returns the positive value of a number regardless of its sign.
- sum() is the summation function, which adds all the values in a given vector or array. [2]

The network is then trained using these instances after being given a set of import. A network's actual export will match its ideal output when training is complete. The weights are modified, if this is not the case. [2]

### 2.2 *The Radial Basis Function*

The radial basis function is nonnegative, nonlinear, and has a symmetrical structure. The import of a local range is taken into account. The vector separation between the import vector X and the weight vector C1 of the radbas transfer function, multiplied by the bias, is the function's net import. In this image, a dot product box allows users to insert import vectors and weight matrices to create the dot product.) [1]

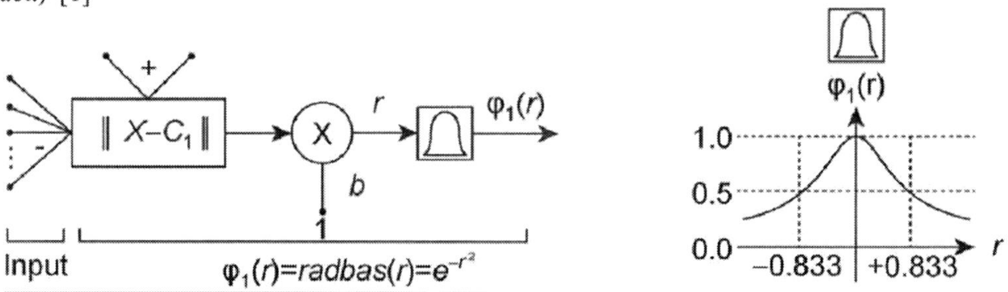


**Fig. 1** The radbas transfer function and a radial basis network with n imports. [5]

The maximum value of the function is 1 for radial basis, which signifies that the import is a value of 0. The export will increase as C1 and X get closer to one another. Since radial basis neurons can produce 1 if the input X and weight vector C1 are the same, they can be compared to detectors. The sensitivity of the radbas neuron can be changed as a result of the bias b. [10]

$$\varphi_i = \exp\left(-\frac{\|X - c_i\|^2}{2s^2}\right) \quad (6)$$

The equation above represents the import vector X, the weight vector C1, and the bias vector b. [10]

*2.3 Predictive Wax Deposition Rate Model Creation in MATLAB Using An RBF Neural Network*

Step 1. Preprocessing The Data: Each element in the training sample has a variety of units. To make computations easier and take account of dimensions, each sample's element should be standardised to the range [0, 1]. [5]

Step 2. Network Creating: Calling the function newrb (P, T, GOAL, SPREAD). P and T, the import and target vector matrices, are normalised training samples in this study with GOAL = 0.01 and SPREAD = 2. [8]

Step 3. Predictions based on a mathematical model: By using the sim function, the trained network predicted the 106 datasets in Table 3.1 as shown in Figure 2.1. At this stage of the prediction process, anti-normalization is necessary. [3]

## 3. Results and Discussion

### *3.1 The Export Calculation*

The RBF equation that was generated after building the model in MATLAB was in this form :

$$Y = \sum_{j=1}^{j} w_i \times \varphi_i + b \qquad (7)$$

The equation (7) represent the input features phi(X,c,s) which is written in the equation as are calculated using some function that takes input data 'X' and some parameters 'c' and 's' as inputs. The weights determine the contribution of each input feature to the output value, while the bias term 'b' shifts the output value up or down, depending on its value. [5]

There are n imported points that are then concealed from the neural network's outputs, and p hidden points that are also hidden from the outputs. The learning sample in this instance is (X,Y), where = ( 1, 2, 3 .... ) represents the import samples and = ( 1, 2, 3 .... ) represents the needed export samples. [11] The following formula is used to compute the export value of hide tier point I :

$$Y = \sum_{j=1}^{n} w_i \times \varphi(||X - c_i||) + b \qquad 1 \leq i \leq p; 1 \leq j \leq n \qquad (8)$$

The equation (8) was generated after substituting equation (6) and (7). For each centroid vector c i, the model computes the weighted sum of the Gaussian function outputs. During model training, the weights w i and bias term b are learnt to minimize the gap between the anticipated output Y and the actual output for a given input X. [13] The obtained parameters are substituted into the equation (8) and the following equation was generated:

$$Y = \sum_{j=1}^{84} 0.83255 \times \varphi(||X, c_i, 1) + 59.4209 \qquad 1 \leq i \leq 1 \qquad (9)$$

The final equation (9) represents a prediction model that takes an input vector X and produces an output value Y. The model is defined as a weighted sum of Gaussian functions, where each function is centered at a predefined centroid vector "c_i,1" and the weight are fixed at 0.83255. The bias term is also included and has a fixed value of 59.4209. The components of the equation are as follows:

- Y: This is the output value of the model, which represents the predicted value for the given input vector X. [5]
- X: This is the input vector to the model, which contains a set of features or variables that are used to make a prediction. [12]

- c_i,1: This is a centroid vector, which is a predefined vector that represents the center of the Gaussian function used in the model. [9]
- ||X - c_i,1||: This is the Euclidean distance between the input vector X and the centroid vector c_i,1, which measures how far apart they are in multi-dimensional space. [5]
- φ( ): This is the Gaussian function used to compute the similarity between the input vector X and the centroid vector c_i,1. The function takes the Euclidean distance ||X - c_i,1|| as input and produces a scalar output value that represents the similarity or weight of the input vector with respect to that centroid vector. The Gaussian function is defined as φ(||X - c_i,1||) = exp(-||X - c_i,1||^2 / (2s^2)) where s is a parameter that determines the width of the Gaussian function.[1]
- 0.83255: This is the weight assigned to each Gaussian function output value. The weight determines the contribution of each Gaussian function to the final prediction and is fixed at a value of 0.83255.
- 59.4209: This is the bias term, which is a constant value added to the sum of the weighted Gaussian function outputs to shift the prediction up or down. The bias term is also fixed at a value of 59.4209.
- j: This is the index of the summation, which ranges from 1 to 84. The summation symbol ∑ indicates that the terms inside the brackets are added together, where j is the index that iterates through the values 1, 2, 3, ..., 84. However, since the weight and bias term are fixed, the summation simply represents the sum of 84 identical terms of the form 0.83255×φ(||X,c_i,1) + 59.4209. [5]

This equation represents a type of machine learning model called a radial basis function (RBF) network, which uses a single Gaussian function to compute the similarity between an input vector and a predefined centroid vector. The model is commonly used for tasks such as function approximation, classification, and regression. The output value of the model is the weighted sum of the Gaussian function output and a bias term, where the weights and bias term are fixed at specific values that were learned during the training of the model. [3]

*3.2 RBF Neural Network Model Generated Results*

**Table 1** Comparison Of Each RBF Neural Network Model For Training Data.

| **Number** | **Parameters** | **RBF neural network model** | **Coefficient Factor (R)** | **Root Square (R^2)** | **RMSE** |
|---|---|---|---|---|---|
| 1 | , _ , dc/dt , dT/dr , W | 5-25-1 | 0.94712 | 0.89705 | 2.85099 |
| 2 | , _ , dc/dt , dT/dr , W | 5-84-1 | 0.000 + 0.3159i | -0.09984 | 9.31858 |
| 3 | , _ , dc/dt , dT/dr , W | 5-85-1 | 0.95849 | 0.91871 | 2.53345 |
| 4 | , _ , dT/dr , W | 4-83-1 | 0.97047 | 0.94182 | 2.14322 |
| 5 | , _ , dT/dr , W | 4-84-1 | 0.99531 | 0.99064 | 0.85983 |
| 6 | , _ , dT/dr , W | 4-74-1 | 0.96687 | 0.93484 | 2.26822 |
| 7 | , _ , dT/dr , W | 4-85-1 | 0.99521 | 0.99044 | 0.86890 |
| 8 | , dT/dr , W | 3-84-1 | 0.68565 | 0.47011 | 6.46811 |
| 9 | , dT/dr , W | 3-85-1 | 0.99369 | 0.98741 | 0.99696 |

The Table 1 provides information on nine models of RBF neural networks, with different combinations of input parameters and architecture. Model Number 1 used all five input parameters and had a relatively good fit with a coefficient factor (R) of 0.947 and a root square (R^2) of 0.897. Model Number 2 had an imaginary coefficient factor and a negative R^2, indicating a poor fit. Model Number 3 used the same input parameters as Model 1 but with a larger hidden layer size and had a good fit with an R value of 0.958 and an R^2 value of 0.919. Model Number 4 used a smaller set of input parameters and had a good fit with an R value of 0.970 and an R^2 value of 0.942. Model Number 5 had the same input parameters as Model 4 but with a larger hidden layer size and had a very good fit with an R value of 0.995 and an R^2 value of 0.991. Model Number 6 was the same as Model 4 but with a smaller hidden layer size and had a good fit with an R value of 0.967 and an R^2 value of 0.935. Model Number 7 was the same as Model 5 but with a larger hidden layer size and had a very good fit with an R value of 0.995 and an R^2 value of 0.990. Model Number 8 used only three input parameters and had a poor fit with an R value of 0.686 and an R^2 value of 0.470. Model Number 9 was the same as Model 8 but with a larger hidden layer size and had a very good fit with an R value of 0.994 and an R^2 value of 0.987. Finally, Model Number 5 was chosen with 4 input parameters, 84 neurons in 1 hidden layer and 1 output, due to its high R^2 value.

**Table 2** RBF neural network weight bias from first 5 neuron in hidden layer to output layer.

| i | w | | | | | | b |
|---|---|---|---|---|---|---|---|
| 1 | 0.8326 | -0.2163 | -0.1079 | -0.1068 | 1.7281 | 26.1211 | 59.4209 |
| 2 | 0.8326 | -0.2634 | 0.1925 | 2.2204 | 0.9044 | 0.9977 | |
| 3 | 0.8326 | -0.3626 | -0.1866 | 0.8345 | 2.9217 | -5.6091 | |
| 4 | 0.8326 | 0.5965 | -0.0554 | -0.1843 | 0.3749 | 8.5627 | |
| 5 | 0.8326 | 1.7935 | 1.7645 | -0.8608 | 0.2648 | -5.8441 | |

Table 2 shows that an RBF model used 84 radial basis functions, each with a center point and spread value, to model the relationship between input and output variables. Weight values were used to scale the contribution of each radial basis function to the final output. The bias term had a different value for each radial basis function, allowing the RBF model to capture complex relationships between input and output variables. The optimal spread value was chosen through a trial-and-error process, and the number of radial basis functions impacted the model's ability to capture complex relationships. It was important to train the RBF model on a diverse and representative dataset and use regularization techniques to prevent overfitting. While analyzing the weight biases provided insights into the RBF model's behavior, other factors such as the spread value, number of radial basis functions, and dataset quality were also important for optimizing its performance.

With 106 dataset the RBF neural network model was built, where 85 data-set were used to train the model and 21 dataset were used to test the model. Many graphs were used to test the predicted wax deposition rate model developed with machine learning in RBF Neural network. In this part, we will detail all the graphs derived from this model, including their function, benefits, and downsides.

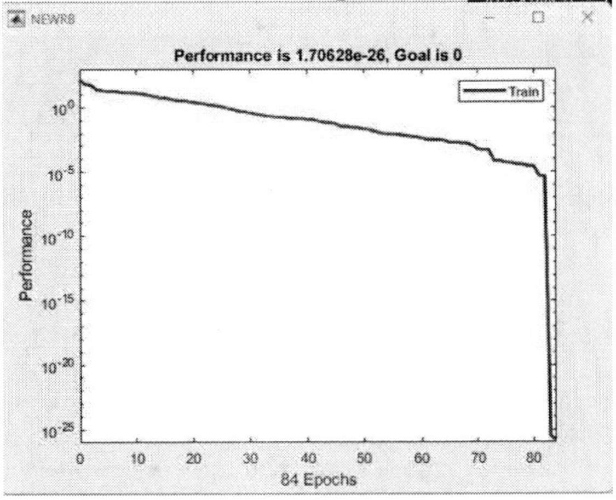


**Fig. 2** Permormance vs 84 Epochs graph

*3.3 Accuracy Analysis*

In the past, the oil and gas sector faced significant problems with wax deposition in deep-sea pipes. To address this issue, a prediction model was developed using machine learning techniques, specifically using MATLAB and a radial basis function (RBF) neural network. The dataset used for the model had 106 values with five input variables, and nine examples with varying input and neuron counts were generated, with Case 5 selected for further analysis.

Various measures, including R, R-squared, root mean squared error (RMSE), mean absolute error (MAE), and mean squared error (MSE), were used to assess the efficacy of the model. The obtained values for R, R-squared, and RMSE were 0.9953, 0.9906, and 0.85983, respectively, indicating a highly accurate model. Additionally, scatter plots, histograms, density plots, and quadratic model equations were used to analyze the performance of the model, revealing a good correlation between predicted and actual wax deposition rates on both the testing and training sets.

To test the model's ability to predict new data, input variables were removed and new data was loaded, with predictions made using the RBF model. The expected wax deposition rates based on the new data were shown, and a scatter plot was created to demonstrate the model's ability to predict wax deposition rates for fresh data.

The weight, bias, and critical value of the hidden layer were gathered to understand how the neural network produced its predictions, as well as expected wax deposition rate values and residual values, allowing for comparison between predicted and actual results. One suggestion for improving the model's accuracy was to expand the dataset, investigate additional input variables, and explore other neural network topologies and training procedures.

The MATLAB-created RBF neural network model proved to be a reliable and efficient tool for estimating the wax deposition rate in deep-sea pipes, with potential applications in pipeline design and operation optimization. It highlights the effectiveness of machine learning in addressing complex challenges in the oil and gas industry. [5]

| **Number** | **pipe wall shear stress,** | **crude oil viscosity,** | **wax molecule concentration gradient at pipe wall, dc/dt** | **temperature gradient at pipe wall, dT/dr** | **Reported wax deposition rate, W, (g/m^2 h)** | **Predictive wax deposition rate** | **Relative Tolerance %** |
|---|---|---|---|---|---|---|---|
| 1 | 11.299 | 54.2 | 0.214 | 0.583 | 11.9 | 11.82 | 0.5996 |
| 2 | 6.195 | 25.2 | 0.0167 | 1.046 | 9.5 | 9.26 | 2.4432 |
| 3 | 7.585 | 33.1 | 0.0787 | 1.041 | 9.19 | 9.03 | 1.7244 |
| 4 | 9.046 | 41.4 | 0.5073 | 1.051 | 5.57 | 5.59 | 0.4373 |
| 5 | 11.299 | 54.2 | 0.214 | 1.647 | 6.77 | 6.76 | 0.0720 |

**Table 3** Testing Model Dataset With Predictive Wax Deposition Rate And Relative Tolerance Percentage.

Table - 3 presents the results of testing a predictive model for wax deposition rate, few data-set were showed in the table. The dataset used in the testing contains 21 samples, each with five input variables, namely pipe wall shear stress, crude oil viscosity, wax molecule concentration gradient at the pipe wall, temperature gradient at the pipe wall, and reported wax deposition rate. The predictive model was used to generate the estimated wax deposition rate for each sample. The predicted wax deposition rates were then compared with the actual reported wax deposition rates to calculate the relative tolerance percentage.

The results show that the predicted wax deposition rates were close to the actual reported rates, with a mean relative tolerance of 1.47%. Only two samples had a relative tolerance greater than 5%, with the highest being 13.17% for sample 21. On the other hand, four samples had a relative tolerance percentage of less than 1%, indicating that the model was highly accurate in predicting the wax deposition rate for those samples. The mean relative tolerance was found 1.5015%.
The results of the testing demonstrate that the predictive model for wax deposition rate is reasonably accurate and can be used for practical applications in the oil and gas industry.

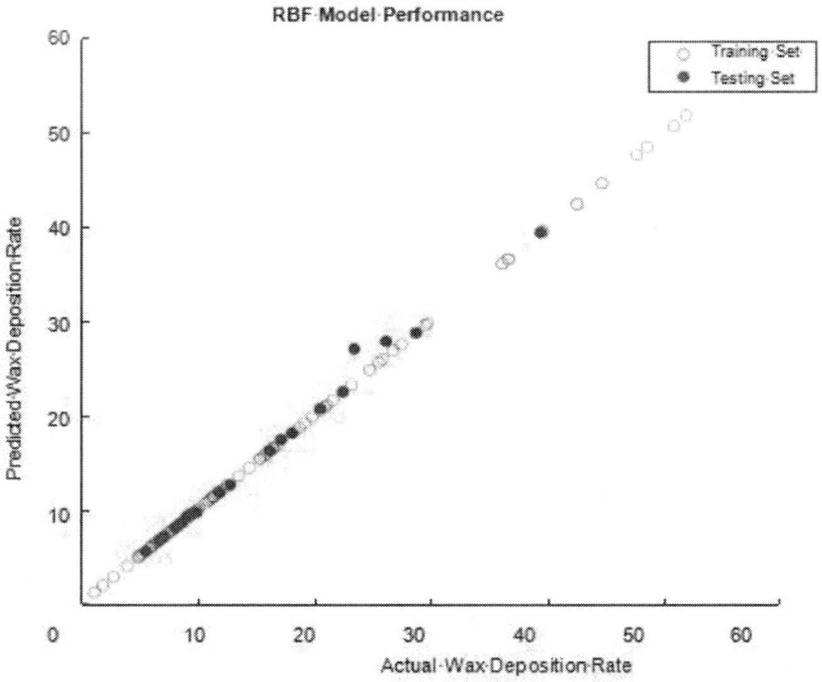


**Fig. 3** RBF Model Performance comparison between training set and Testing set graph on Predicted Reported Wax Deposition Rate vs Actual Reported Wax Deposition Rate

In Figure 4 the scatter plot of predicted vs actual wax deposition rate for the testing set and training set provided a visual representation of how well the RBF neural network model performed. The x-axis represented the predicted wax deposition rate, while the y-axis represented the actual wax deposition rate obtained from the dataset. The scatter plot showed a cluster of points that followed a linear trend, indicating that the predicted wax deposition rates were highly correlated with the actual wax deposition rates. The red markers on the scatter plot indicated the testing set, while the blue markers represented the training set.
The linear trend in the scatter plot suggested that the model effectively captured the underlying patterns in the dataset and accurately predicted the wax deposition rates. The clustering of points around the linear trend indicated that the model was not overfitting or underfitting the data and was accurately predicting the wax deposition rates for new data. The scatter plot also revealed some outliers, which were points that fell far away from the linear trend. These outliers may have been due to factors that were not accounted for in the model, or may have been due to errors or noise in the dataset. The scatter plot of predicted vs actual wax deposition rate for the testing set showed that the RBF neural network model was a highly accurate and reliable tool for predicting wax deposition rates in deep sea pipelines. The model could be used to optimize pipeline design and maintenance, leading

to cost savings and improved safety in the oil and gas industry. [1]
The graph demonstrates that the projected values were rather close to the actual values, except for a few outliers. With a slope of roughly 0.9, the line of best fit indicates that the projected values are somewhat lower than the actual values. The R-squared score of 0.9953 shows that the model explains 99.53 percent of the variance in the data, which is an excellent indication of its correctness. This graph was important for evaluating the accuracy of the model on the training set and spotting any data outliers. This graph offers a visual representation of the performance of the model on the training set, which may be used to evaluate the accuracy of the model. This graph may not be appropriate for huge datasets since it might be difficult to read the findings when there are numerous data points.

## 4. Conclusion

Machine learning may be used to forecast the rate of wax deposition in deep-sea pipes, which is advantageous for the oil and gas business. A dataset of 106 values and five input variables was used in this study to train and test the RBF neural network model, a supervised learning approach. The model's ability to forecast the wax deposition rate was determined to be reliable and accurate, as shown by the high values of several performance measures. The relationship between input data and the predicted wax deposition rate is crucially understood by the model's equation, which can be used to pinpoint the input parameters that have the highest impact on the rate of wax deposition. With the addition of new machine learning algorithms and input variables, the predictive model created using machine learning has the potential to be an effective tool for predicting the wax deposition rate in deep-sea pipelines. The paper emphasizes how crucial precise predictions are to ensuring the efficiency and security of oil and gasoline pipelines.

**Acknowledgement**
The authors completed a challenging assignment successfully with the guidance of their supervisors including professor Shaohua Dong from China university of petroleum in Beijing, and the evaluator Mr. Sami A. Musa Yagoub, as well as the support and motivation of the Petroleum Engineering department at UCSI University.

**References**

[1] Abdelrahman Musa, S., Abdelrahman Musa Yagoub, S., Eldwin Pradipta, G., & Mohammed Yahya, E. (2021). *Prediction of bubble point pressure for Sudan crude oil using Artificial Neural Network (ANN) technique Progress in Energy and Environment Prediction of bubble point pressure for Sudan crude oil using Artificial Neural Network (ANN) technique* (Vol. 15).
[2] Baghban, A., Hekmati, R., Hajiali, M., Janghorban Lariche, M., & Kamyab, M. (2018). Application of MLP-ANN as novel tool for estimation of effect of inhibitors on asphaltene precipitation reduction. *Petroleum Science and Technology*, *36*(16), 1272–1277. https://doi.org/10.1080/10916466.2018.1468777
[3] Chen, M., Sasanipour, J., Kiaian Mousavy, S. A., Khajeh, E., & Kamyab, M. (2018). Estimating phase behavior of the asphaltene precipitation by GA-ANFIS approach. *Petroleum Science and Technology*, *36*(19), 1582–1588. https://doi.org/10.1080/10916466.2018.1493503
[4] Chu, Z. Q., Sasanipour, J., Saeedi, M., Baghban, A., & Mansoori, H. (2017a). Modeling of wax deposition produced in the pipelines using PSO-ANFIS approach. *Petroleum Science and Technology*, *35*(20), 1974–1981. https://doi.org/10.1080/10916466.2017.1374405
[5] Chu, Z. Q., Sasanipour, J., Saeedi, M., Baghban, A., & Mansoori, H. (2017b). Modeling of wax deposition produced in the pipelines using PSO-ANFIS approach. *Petroleum Science and Technology*, *35*(20), 1974–1981. https://doi.org/10.1080/10916466.2017.1374405
[6] Jalalnezhad, M. J., & Kamali, V. (2016). Development of an intelligent model for wax deposition in oil pipeline. *Journal of Petroleum Exploration and Production Technology*, *6*(1), 129–133. https://doi.org/10.1007/s13202-015-0160-3
[7] Joshi, R. R. (n.d.). *Modeling and Simulation of Wax Deposition in Crude Oil Pipeline.*

http://www.irphouse.com

[8] Nait Amar, M., Jahanbani Ghahfarokhi, A., & Shang Wui Ng, C. (2022). Predicting wax deposition using robust machine learning techniques. *Petroleum*, *8*(2), 167–173. https://doi.org/10.1016/j.petlm.2021.07.005

[9] Sousa, A. M., Matos, H. A., & Pereira, M. J. (2019). Modelling Paraffin Wax Deposition Using Aspen HYSYS and MATLAB. In *Computer Aided Chemical Engineering* (Vol. 46, pp. 973–978). Elsevier B.V. https://doi.org/10.1016/B978-0-12-818634-3.50163-6

[10] Xie, Y., & Xing, Y. (2017a). A prediction method for the wax deposition rate based on a radial basis function neural network. *Petroleum*, *3*(2), 237–241. https://doi.org/10.1016/j.petlm.2016.08.003

[11] Xie, Y., & Xing, Y. (2017b). A prediction method for the wax deposition rate based on a radial basis function neural network. *Petroleum*, *3*(2), 237–241. https://doi.org/10.1016/j.petlm.2016.08.003

[12] Yu, Z., Ruoyu, W., & Xue, W. (2021). Prediction Method of Wax Deposition Rate in Crude Oil Pipeline Based on RBF Neural Network and Support Vector Machine. *E3S Web of Conferences*, *271*, 04007. https://doi.org/10.1051/e3sconf/202127104007

[13] Zabihi, R., Mowla, D., & Karami, H. R. (2019). Artificial intelligence approach to predict drag reduction in crude oil pipelines. *Journal of Petroleum Science and Engineering*, *178*, 586–593. https://doi.org/10.1016/j.petrol.2019.03.042

# Exploratory Test of Horizontal Drilling Plugs and Water Shutoff Technology in X Horizontal Gas Well

**Si-fan WANG1(⊠),3, AN-kang ZHANG 2,3, Kai-Bin HU2,3**

1 CCDC Drilling and Production Engineering Technology Research Institute, CNPC. Xi'an City, Shaanxi Province, China.
2 Oil and Gas Technology Research Institude of Petrochina Changqing Oilfield Company. Xi'an City, Shaanxi Province, China
3 National Engineering Laboratory of Low-permeability Oil & Gas Exploration and Development. Xi'an City, Shaanxi Province, China

532295652@qq.com

**Abstract**. In order to solve the problem that the tight gas X horizontal gas well could not be put into production due to the accumulation of fluid in the wellbore during the drainage stage after fracturation, based on the geological understanding of water flowing from the far end of the reservoir, and the actual situation of staged fractured with drillable bridge plug, water blocking has become the exploration direction of potential excavation and resuming production. Through the analysis of the type, structure, setting principle and unsealing principle of drillable bridge plug, put forward the technical idea of horizontal drilling plugs and mechanical water plugging. And determine the treatment method of drilling-grinding and fishing large inner diameter bridge plugs, and design the construction process. Moreover combined with theoretical calculation, the relationship between the debris carrying velocity and displacement in horizontal wells, the size of the debris and the angle of inclination was optimized. Besides that, the milling method that uses a wellhead rotating device to drive the drill pipe and grinding shoes for drilling and grinding was proposed. In the field test, successfully drill to the third large inner diameter bridge plug position according to the design requirements. And successfully put in the fishable bridge plug to achieve mechanical water blocking. After commissioning, the daily gas production volume reached 11000 $m^3$/d. The test results show that: 1) It is feasible to use the wellhead rotating drive drill pipe and grinding shoes to drill and remove the steel drillable bridge plugs in the horizontal section of the horizontal well; 2) It is also feasible to use the fishable bridge to plug in the horizontal section of the horizontal gas well with drillable bridge plug staged fractured, and the plugging effect is good. In summary, it is shown that this technology test achieves the purpose of potential excavation and resuming production of the X horizontal gas well, which has good reference value.

## 1. Introduction

Currently, there are more than 2500 horizontal gas wells in Changqing Oilfield, including hundreds of horizontal wells with low production and low efficiency due to wellbore fluid accumulation, which has seriously affected the productivity of horizontal gas wells. How to deal with wellbore fluid accumulation has become the core recovery problem for horizontal gas wells. However, currently, nitrogen gas lift, compressor gas lift, plunger drainage[1-3], bubble drainage[4], suction, and velocity string drainage processes cannot fundamentally solve the problem[5-6]. Therefore, the problem of water plugging in horizontal gas wells is extremely urgent. The current horizontal well water plugging technology mainly focuses on oil well horizontal wells [7-9], while the exploration of gas well horizontal wells is seriously insufficient, and there are few domestic documents mentioning relevant cases and countermeasures[10-15]. Through analyzing the fluid accumulation situation in the wellbore of horizontal gas well X, the wellbore condition of the horizontal gas well that can be drilled with bridge plug staged fracturing is fully evaluated. However, due to the limitations of horizontal well fracturing technology, there is currently a lack of experience in such bridge plug treatment. Therefore, based on the need to tap potential and resume production, the technical idea of horizontal drilling plug and mechanical water plugging was proposed, and the recovery of horizontal drilling plug and water plugging for tight gas X horizontal wells was successfully achieved. This expanded a new idea for the treatment of wellbore fluid accumulation in gas field horizontal wells, and provided a technical reference for the recovery of water plugging and potential tapping for similar gas well horizontal wells.

## 2. Analysis of the accumulation liquid

The horizontal section of tight gas X horizontal well has a length of 1000 m, and the drilling encounter rate of the reservoir is 67.3%. During the fluid discharge stage after fracturing, it did not rely on its own energy to spray through. Instead, it has been using the on-site nitrogen lift method for fluid discharge, with the degree of fluid discharge reaching 92.3%. During the pressure recovery process after gas lift, the oil casing pressure was not available, the oil pressure was 0 MPa, and the casing pressure was 24 MPa. In the later stage of fluid discharge, the water property analysis was performed: PH: 6-7, r: 1.02 g/ml, CL: 21185 mg/L. Judging based on the above situation and geological understanding, the remote end of the reservoir in this well is flooded by water, making it difficult to exploit and utilize it. Therefore, it is necessary to conduct water sealing operations on the two remote reservoir segments to restore the well's productivity.

## 3. Water shutoff idea

### *3.1. Water shutoff design*

X horizontal gas well has a vertical depth of 4343.91 m and a horizontal section length of 1000 m. After staged fracturing test using drillable bridge plugs and soluble balls, Φ60.3 mm tubing was lowered under pressure to above the horizontal section, with a horizontal gas reservoir casing of Φ114.3 mm. According to the original wellbore structure of X horizontal gas well (Fig.1), in order to achieve reliable water sealing of the reservoir on the right side of the drillable bridge plug at 5287 m, using a fishable bridge plug to set and seal the reservoir on the left side of 5287 m (Fig. 1) has good technical advantages.

(1) Compared to conventional cement compaction, due to the quality of cement retainers, it is first possible to avoid incorrect plugging of the left reservoir of the drillable bridge plug at 5287 m by cement, as well as permanent plugging of the right reservoir. If it is necessary to exploit the potential of plugging the reservoir in the future, cement compaction is not advisable. Secondly, it is necessary to avoid cement drilling operations, which will prolong the construction time and inevitably increase the loss of kill fluid and reservoir damage.

(2) Compared to other liquid plugging agents, it can avoid the problems of poor plugging reliability and difficulty to control the plugging time.

(3) The released bridge plug can release and seal the reservoir through fishing the bridge plug, allowing its productivity to be brought into play.

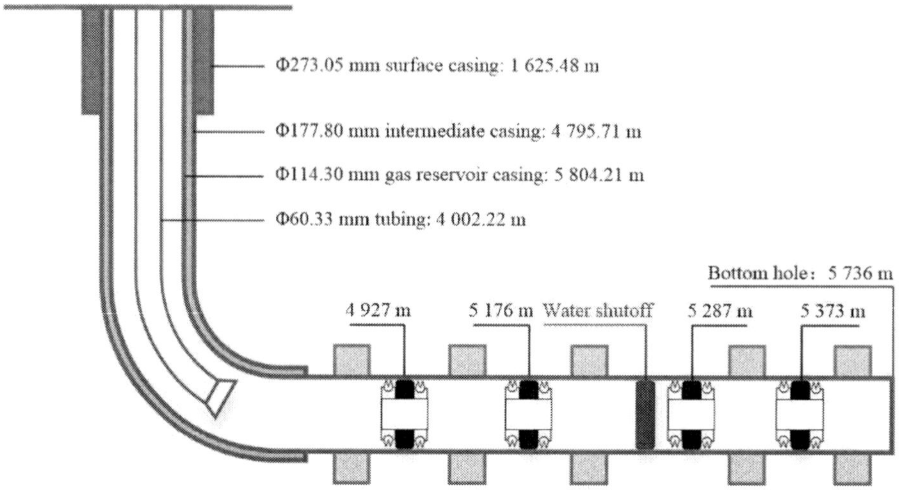


**Fig.1.** Well bore structure and water shutoff design for X horizontal gas well

### 3.2. *Technical difficulties*

To carry out mechanical water shutoff, it is first necessary to clean the wellbore, such as pulling out Φ60.3 mm tubing, purging the well, etc. However, the existence of two drillable bridge plugs at 4927 m and 5176 m in the horizontal section of Well X hinders the mechanical water shutoff operation at the distal end of the reservoir (Fig. 1). Due to the limitations of the staged fracturing and gas testing process for horizontal gas wells, the lack of relevant experience in handing drillable bridge plugs undoubtedly increases technical difficulties, especially for such steel bridge plugs in horizontal wells.

### 3.3. *Technical countermeasure*

X horizontal gas well has a vertical depth of 4343.91 m and a horizontal section length of 1000 m. After staged fracturing test using drillable bridge plugs and soluble balls, Φ60.3 mm tubing was lowered under pressure to above the horizontal section, with a horizontal gas reservoir casing of Φ114.3 mm. According to the original wellbore structure of X horizontal gas well (Fig.1), in order to achieve reliable water sealing of the reservoir on the right side of the drillable bridge plug at 5287 m, using a fishable bridge plug to set and seal the reservoir on the left side of 5287 m (Fig. 1) has good technical advantages.

*3.3.1. Structural principle of drillable bridge plug.* The Baker Hughes drillable bridge plug structure for X horizontal gas well includes expansion pipe, upper shear pin, sealant cylinder, expansion cone, lock ring, slip, slip seat, and lower shear pin[16], as shown in Figure 2. Its working principle: Firstly, connect the stress pull rod in the setting tool to the inner cavity of the bridge plug through upper and lower shear pins, and contact the end face of the push cylinder in the setting tool with the upper end face of the expansion pipe; When setting is required, the length of the stress pull rod remains unchanged. Under the action of setting (gunpowder or hydraulic) thrust, the push cylinder pushes the expansion pipe downward along the upper inclined conical surface of the expansion cone, forcing the expansion ring at the lower end of the expansion pipe to expand. At the same time, the expansion cone is pushed to cut the upper shear pin, and then the lower inclined conical surface of the lower end of the expansion cone is forcibly inserted into the integral slip, causing the slip to be anchored on the inner wall of the casing along the segmented expansion fracture, At this time, the rubber cylinder also fully expands under the action of the expansion ring and adheres to the inner wall of the casing, sealing the bridge plug and the casing annulus. Later, as the setting thrust gradually increases the lower shear pin is cut off, realizing the release of the setting tool.

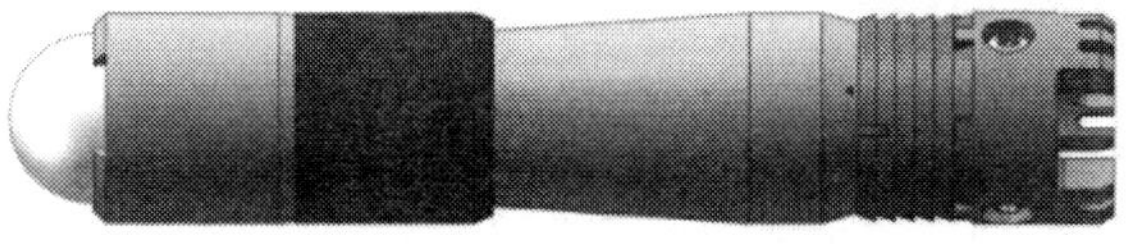

**(a)** Drillable bridge plug and soluble ball

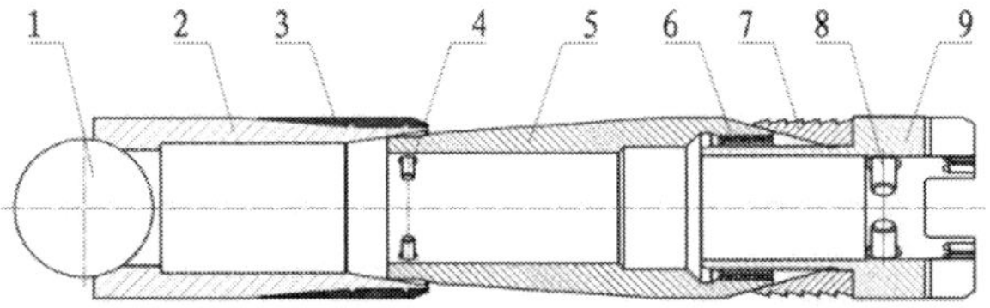


1.Soluble sphere 2.Expandable tube 3.Sealant cartridge 4.Upper shear pin 5. Expansion cone
6. Clamping ring 7.Slip 8.slip seat 9. Lower shear pin
**(b)** Structure diagram of drillable bridge plug

**FIG. 2.** Baker Hughes drillable bridge plug

*3.3.2. Countermeasure discussion.* The According to the structure of the Baker Hughes drillable bridge plug, it is a single slip structure, which only prevents the bridge plug body from flowing downward during fracturing. However, due to the constraints of the locking mechanism of the locking ring and the tooth buckle, as well as the influence of the expansion tube and rubber cylinder that have expanded and adhered to the inner wall of the casing, it is not feasible to directly apply lifting force to the inner cavity of the bridge plug through the internal fishing spear. On the one hand, the bridge plug is difficult to shrink and unlock, and on the other hand, the resistance between the expansion tube and the rubber cylinder and the inner wall of the casing is large, The lifting force of the fishing string in the horizontal section is limited, and direct fishing can only be attempted when the bridge plug slips have been damaged and dropped. The expansion tube of the drillable bridge plug is made of steel, the rubber cylinder is made of rubber, and the other parts are made of iron. Therefore drilling and fishing has become the main technical route.

For the first drillable bridge plug at 4927 m, in principle, it must be thoroughly removed to avoid affecting the drilling and grinding treatment of the second bridge plug. When continuing to process the second bridge plug in principle, removing it is the first option. If the operation is complex and difficult, you can choose to push it down to a deeper area than the water plugging position.

## 4. Construction technology

### *4.1. Construction process*

The construction process of mechanical water shutoff for X horizontal gas well is shown in Figure 3. First, use hydraulic well killing, disassemble the wellhead Christmas tree, install blowout preventers, and perform pressure testing. Then, pull out the original well tubing string, and conduct through-hole cleaning. Then, explore the sand surface and flush the sand to the drillable bridge plug position at 4927 m, it is recommended to reverse cycle sand flushing. Then, run a grinding shoe or sleeve milling barrel to drill and grind the bridge plug. According to the situation, the first bridge plug shall be removed by drilling and fishing, and the second bridge plug shall be removed by the same method. Then, the casing shall be scraped at the water shutoff position. Then, the salvageable bridge plug shall be placed and sealed at the water shutoff position, and the sealing inspection string shall be run for pressure test and sealing inspection. After the sealing inspection is qualified, the smooth tubing completion string shall be run, the gas production tree shall be replaced, the ground process shall be connected, and the pressure test shall be qualified. Then, nitrogen gas lift and liquid drainage shall be conducted, and the well shall be finally completed.

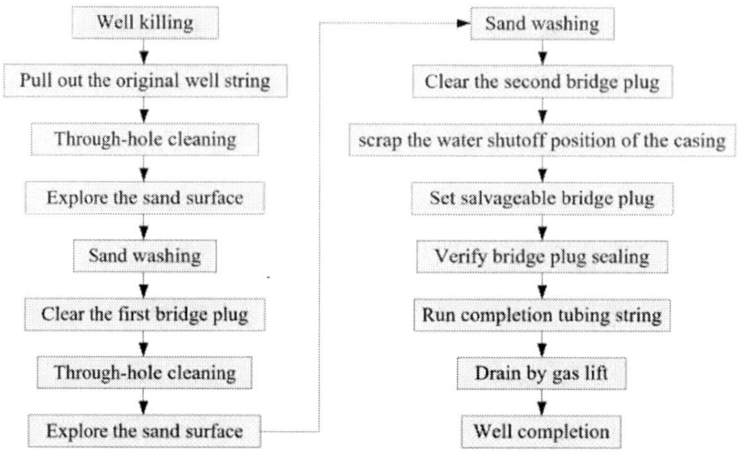


**Fig. 3.** Drilling plug and water plugging construction process for X horizontal well

### *4.2. Drilling and grinding displacement*

Based on the Newton-Rettinger calculation method, the formula for the free settling terminal velocity of spherical debris in the drilling and grinding fluid in the vertical section is formula (1), considering only the drilling and grinding fluid, debris density, and debris shape [17-23]:

$$v_e = \left(\frac{8}{3} d \frac{\rho-\rho_0}{\rho_0} g\right)^{\frac{1}{2}} \tag{1}$$

In the formula (1), The parameter $v_e$ is the free setting end velocity of the abrasive debris in the milling fluid (m/s); $d$ is the diameter of the spherical grinding (m); $\rho$ is the abrasive density (kg/m$^3$); $\rho_0$ is the density of drilling and grinding fluid (kg/m$^3$); $g$ is the acceleration of gravity (m/s$^2$).

Due to the fact that the density of iron chips is much greater than that of rock cuttings, the existing annular stopping flow rate formula for deviated well sections is limited[18-23]. Based on the experimental fitting established by Si-rui ZENG and others, the calculate formula[24] for the critical sand carrying flow rate of horizontal wells and different well inclination angles (0-90°) is:

$$v = (-0.0009(90-\mathrm{a})^2 + 0.0706(90-\mathrm{a}) + 1.7558)v_e \tag{2}$$

In the formula (2): The parameter $v$ is the critical flow rate of drilling and grinding fluid carrying cuttings in different deviated well sections (m/s); $a$ is the well inclination angle (°); $v_e$ is the free settling end velocity of the abrasive debris in the cleaning fluid (m/s).

In the formula (2): when the deviation angle of the well is ≤2°, $v < v_e$ which is not consistent with the current situation. Therefore, the condition for formula (2) to hold is that the well inclination angle is >2°.

In order to meet the actual requirements for the return of cuttings from drilling and grinding fluids, the minimum pumping flow rate is chosen to be 2 times the critical flow rate of cutting [25]. Therefore, the formula for calculation the minimum displacement required for debris to return in from the annulus is:

$$Q_z = 2\pi({r_1}^2 - {r_2}^2)v_e \quad (0 \le a \le 2°) \tag{3}$$

$$Q_z = 2\pi({r_1}^2 - {r_2}^2)v \quad (2° < a \le 90°) \tag{4}$$

In the formula (3) and (4), The parameter $Q_z$ is the minimum displacement pumped from the drilling and grinding pipe string ($m^3/s$); $r_1$ is 1/2 of the outer diameter of the annulus (m); $r_2$ is 1/2 of the inner diameter of the annulus (m); $v_e$ is the free settling end velocity of the abrasive debris in the abrasive fluid (m/s); $v$ is the critical flow rate of the drilling and grinding fluid carrying cutting in different deviated well sections (m/s).

The formula for calculating the minimum displacement required for debris to return from the drill string is:

$$Q_f = 2\pi r^2 v_e \qquad (0 \le a \le 2°) \tag{5}$$

$$Q_f = 2\pi r^2 v \qquad (2° < a \le 90°) \tag{6}$$

In the formula(5)and(6): The parameter $Q_f$ is the minimum displacement pumped from the annulus ($m^3/s$); $r$ is the 1/2 of the inner diameter of the drilling string (m); $v_e$ is the free setting end velocity in the mill (m/s); $v$ is the critical flow rate of the drilling and grinding fluid carrying cuttings in different deviated well sections (m/s).

Assuming that the abrasive debris is spheres, and that all the pumped abrasive fluid returns form the annulus (positive circulation) or, drill string (reverse circulation) between the Φ114.3 mm casing and the Φ60.3 mm drill string to the wellhead, the diameter of abrasion debris is 0.1-3.0 mm, a density is 7850 $kg/m^3$, and the drilling and grinding fluid is clean water , its density is 1000 $kg/m^3$, bring the above data into equation (1) - (6) to calculate the relationship between the size of wear debris, well inclination, and the critical flow rate and minimum displacement of cutting.

Fig 4 shows that the critical flow rate of drilling and grinding fluid carrying debris always increases with the increase of the size of the debris under the same well inclination angle. However, under the same size condition, the critical flow rate of carrying debris starts to increase with the increase of the well inclination angle from 0°, until the flow rate in the zone with the well inclination angle between 30° and 50° increases to the peak, and the peak value is more than 3.1 times of the corresponding vertical well section free sedimentation terminal velocity, and then decreases. The final horizontal critical value is 1.7 times or more of the final free settlement velocity of the corresponding vertical well section.

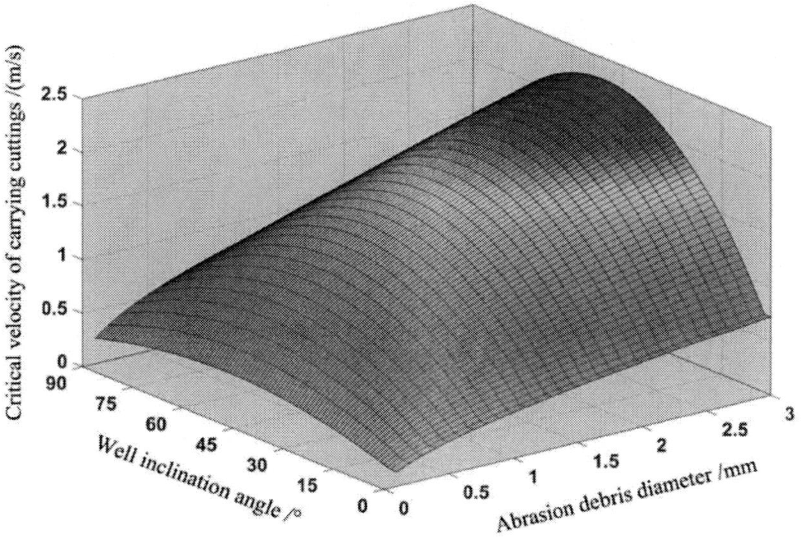


**Fig. 4.** The relationship between the critical velocity of carrying cuttings in drilling and grinding fluid and the well inclination angle and the size of abrasive particles

The relationship between the theoretically calculated minimum pumping displacement and the well inclination angle, as well as the size of abrasive particles, is shown in Figure 5. It can be seen that the displacement required for reverse circulation is far lower than that for positive circulation, which is more than three times the amount required under the same abrasive size and well inclination angle conditions. In addition, the larger the abrasive diameter, the larger the maximum displacement required, and the reddish color indicate an excessively high minimum displacement which may be difficult to achieve due to pump pressure limitations. In order to prevent larger abrasive particles from impeding drilling and grinding, on the one hand, it is recommended to control the generation of large wear debris, and on the other hand, it is recommended to conduct a reverse circulation flushing every footage and distance. As iron chips are generally in the form of wire strips and sheets, compared to spheres, they have a larger surface area under the same mass, and therefore receive stronger liquid transport under the same displacement. If the parameters such as viscosity and consistency of the drilling and grinding fluid itself are considered, the calculated minimum displacement can also be optimized and reduced.

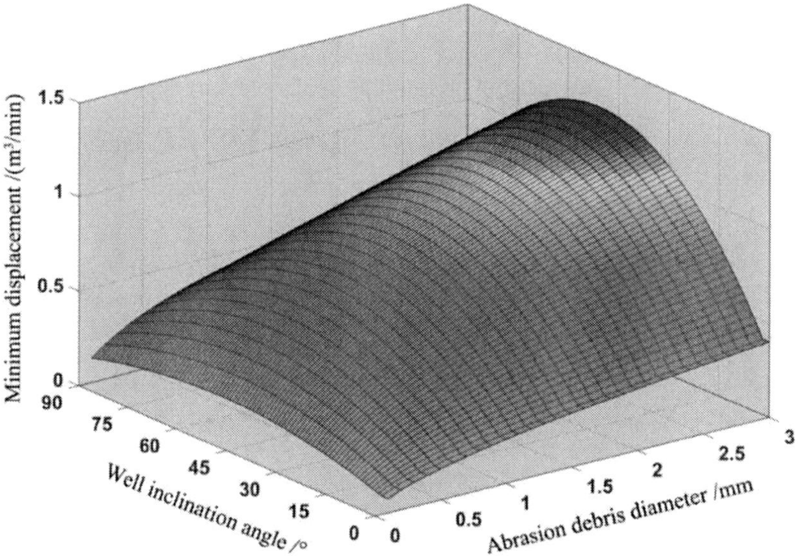


**(a)** Positive circulation

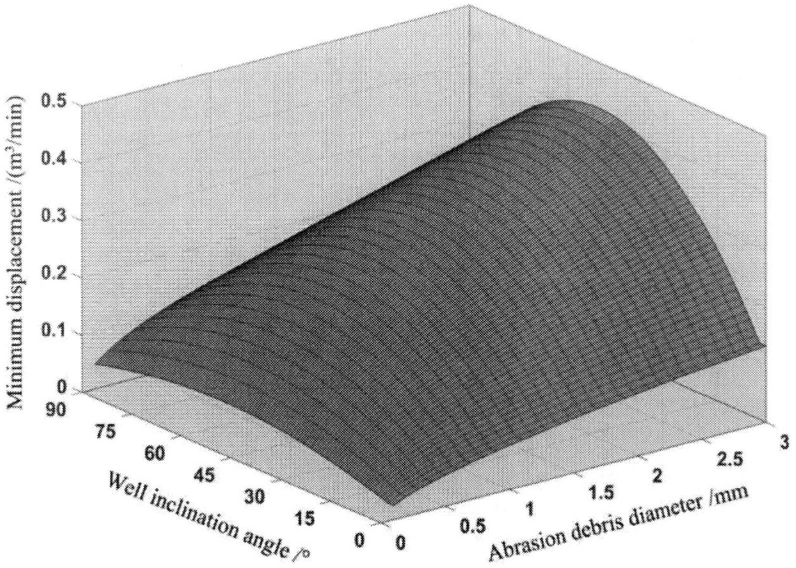


**(b)** Reverse circulation

**Fig .5.** The relationship between the minimum displacement and the well inclination angle and the size of abrasive particles

### *4.3. Drilling and grinding the pipe string*

According to the current practice of fishing bridge plug grinding and milling technology for gas exploration wells[25], it is recommended to use wellhead rotary equipment( top drive, drill floor rotary table)to drive drill pipe and grinding shoes for drilling and grinding. Compared to tubing connected screw drills, the advantages are:

1) Greater weight on bit can be applied.

2) Reverse circulation well flushing is available to remove larger sized abrasive debris.

3) Prevent reverse torsion of drilling and grinding pipe string, which may cause danger, and can apply large torque.

4) At the same displacement, the pump pressure is low.

5) Improve drilling and grinding efficiency, safety, efficiency and reliable.

## 5. Field test

### *5.1. Test overview*

After pulling out the production tubing in X well on site, the drilling and grinding of two drillable bridge plugs 4927 m and 5176 m, were successfully achieved using flat bottom grinding shoes until the third bridge plug position 5286.38 m, was reached and the drilling and grinding was stopped. The construction parameters are shown in Table 1, and the cumulative drilling and grinding time is 38 hour. The internal fishing spear was used to fish the drillable bridge plug, failed. After drilling and grinding, a reverse circulation large displacement well flushing was conducted to ensure that the wellbore above 5286.38 m was clean, and then a fishable bridge plug was inserted to send and seal the pipe string. The ball was placed at 5278.03 m to seal the distal end of the reservoir, and the production pipe string was inserted. The daily recovery gas production reached 11000 $m^3/d$.

**Table. 1.** Drilling-grinding construction parameters of drillable bridge plug

| Bridge plug position (m) | Drilling-grinding tool string | Pumping pressure (MPa) | Torque (N·m) | Rotation speed (r/min) | Drilling pressure (kN) | Displacement (L/min) | Drilling-grinding time (h) | Remarks |
|---|---|---|---|---|---|---|---|---|
| 4927 | Φ60.3 mm tubing + screw drill + flat bottom grinding shoes | 26-28 | / | / | 5-15 | 450-500 | 12 | Drilling depth 0.03 m |
| | Top drive：Φ60.3 mm drill pipe + flat bottom grinding shoes | 12 | 1500-2800 | 60-80 | 20-30 | 400 | 14 | To 5169.67 m |
| 5176 | Top drive：Φ60.3 mm drill pipe + flat bottom grinding shoes | 14-20 | 1500-3000 | 60-90 | 20-50 | 450-600 | 12 | To 5286.38 m |

### *5.2. Testing process*

After successfully pulling out the Φ60.3 mm original tubing string in the well, first use the Φ60.3 mm tubing connecting screw and flat bottom grinding shoes to conduct purging of the well. After purging to a well depth of 4927.06 m, pressurize it with 20 kN, conduct repeated exploration three times, and the depth remains unchanged. It is determined that it is the first large diameter bridge plug. Therefore, start the pump for drilling and grinding, with a pump pressure of 26-28 MPa, a displacement of 450-500 L/min, and a cumulative drilling and grinding drilling depth of 0.03 m for 4 hours. There are fresh iron chips on the magnet at the position where the drilling and grinding fluid returns. After drilling and grinding for 8 hours, there is no drilling depth, after that the Φ60.3 mm drill pipe connecting flat bottom grinding shoe was replaced and the top drive was used to drill and grind using the parameters in Table 1. After a period of 22 min, the first bridge plug was ground, it was analyzed that the bridge plug slip failed to release due to the impact of vibration drilling and grinding, resulting in the bridge plug being pushed down by the drilling punch until it reached 5169.67 m. The cumulative drilling and grinding time was 6 hours without drilling depth. After that, the free hanging weight was lowered to 475 kN to 0, without holding the pump or jumping the drill. It was determined that the bridge plug was at the second 5176 m position.

Exploring the use of a retractable fishing spear to fish the drillable bridge plug from the inside, pressurizing it to 30 kN, and lifting it up failed to fish, resulting in a failure to fish. After that, continue to drill and grind with flat bottom grinding shoes, with a displacement of 500 L/min, weight on bit of 20-50 kN, torque of 1500-2800-3000 N·m, rotational speed of 80-90 r/min, a small amount of fine sand, fresh iron chips and scale mixture are returned from the outlet, lasting for 4 hours, with a drilling depth of 20 cm. Then, use a displacement of 600 L/min, pressure of 30-50 kN, torque of 2000-2500 N·m, and rotational speed of 80-90 r/min to drill and flush, with a depth of 5286.38 m (which has met the depth requirement of 5277 m for bridge plug insertion), and continue to drill and flush for 2 hours without drilling depth, It is judged that the third bridge plug has been reached at 5287 m, and the accumulated drilling wear time during the whole process has been nearly 40 hours. Then carry out reverse circulation well flushing, flush with large displacement until the water quality at the inlet and outlet is consistent, and scrape the casing at the pre-set position of the water shutoff bridge plug. After that, the bridge plug was lowered to send the sealing string, and a ball was thrown to set the Fuxin retrievable bridge plug at 5278.03 m. The sealing inspection string was lowered to 5273.06 m for sealing inspection. After that, the production string was lowered, gas lift drainage was performed, and the well completion structure was shown in Figure 6, The goal of water plugging in the far end of the horizontal section of the X horizontal.

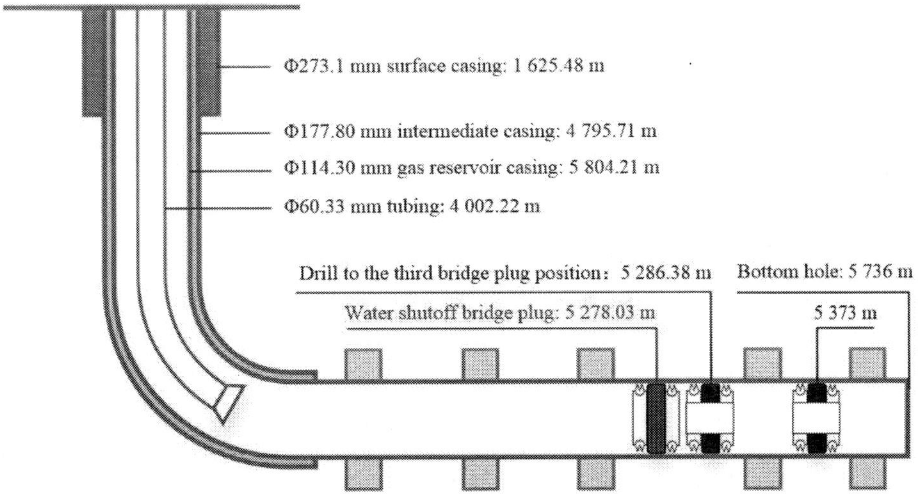


**Fig.6.** Well completion structure diagram

## 6. Conclusion

The following conclusions were made on the basis of field test and theoretical analysis:

1) Field test show that the technology of mechanical water plugging in the horizontal section of a gas well with drillable bridge plug staged fracturing is feasible and has good plugging effects.

2) The horizontal section drilling plug test shows that the technically of using wellhead rotary drive drill rod and grinding shoe drilling to treat the drillable bridge plugs in the horizontal section a horizontal wells is feasible, with lower pump pressure and higher displacement compared to screw drilling.

3) In order to continue to improve milling efficiency and avoid large size debris impeding milling, it is recommended to increase the drilling and milling displacement by more than 1.0 m3/min, and use large displacement reverse circulation flushing to remove the drilling tool up and down every milling distance to remove large size iron debris.

4) It is recommended to use soluble bridge plug for staged fracturing of horizontal wells to avoid complexity caused by subsequent horizontal wellbore operation.

## References

[1] Liu Peng, Chen Man, He Zhiguo,, et al. Study on a Gas Plunger Lift Model for Shale Gas Wells and Its Effective Application[J]. Fluid Dynamics & Materials Processing, 18(4), (2022). doi:10.32604/FDMP.2022. 019736.

[2] Yang Zhi, Yang Jian, Wang Qiang, et al. Research and Application of Plunger Airlift Draining Gas Recovery Intelligent Control Technology in Shale Gas Horizontal Wells[J]. Chemistry and Technology of Fuels and Oils, 58(3), (2022). doi: 10.1007/S10553-022-01419-Y.

[3] Li Nan, Cao Guangqiang, Jiang Yixin, et al. Numerical computational fluid dynamics (CFD) simulation study based on plunger gas lift test[J]. Frontiers in Earth Science, (2023): doi:10.3389/feart.2023.1100882.

[4] Etemad Sahand, Kantzas Apostolos, Bryant Steven. A systematic analysis of foam drainage: Experiment and model[J]. Results in Engineering, 15. 2022. doi:10.1016/J.RINENG.2022. 100551.

[5] Che Mingguang, Liu Jie, Bu Chengzhong, et al. Tubing Depth Optimization Study of Horizontal Well Drainage and Gas Production in WeiYuan Shale Gas Field. Journal of Physics[J]. Conference Series, 2458(1), (2023). doi:10.1088/1742-6596/2458/1/012032.

[6] Liu Chengting, He Liang. Atomized droplet size prediction for supersonic atomized water drainage and natural gas extraction[J]. Scientific Reports, 12(1), (2022). doi:10.1038/S41598-022-26597-X.
[7] Hu Aiguo, Li Kezhi, Feng Yunhui, et al. Laboratory Evaluation and Field Application of a Gas-Soluble Plugging Agent: Development of Bottom Water Plugging Fracturing Technology[J]. Energies, 15(18), (2022). doi:10.3390/EN15186761.
[8] Tong Le, Fan Ronggui, Yang Shuangchun, et al. Experimental Study of Oilfield Water Plugging with Organic Chromium Gel[J]. Chemistry and Technology of Fuels and Oils, 57(6), (2022). doi: 10.1007/S10553-022-01336-0.
[9] Li Rui. Study on Water Plugging Technology Outside Drivepipe of Oil Well[J]. IOP Conference Series: Earth and Environmental Science, 804(2), (2021). doi: 10.1088/1755-1315/804/2/022048.
[10] Wang Jiexiang, Wu Jingwen, Kang Hongshuai, et al. Study on Water plugging Technology of Self - Generating Gel-Foam in Offshore Oilfield[J]. IOP Conference Series: Earth and Environmental Science, 170(2),(2018). doi:10.1088/1755-1315/170/2/022139.
[11] Zhang Tao, Huang Wei, Cheng chen, et al. Research and test of chemical fixed point water shutoff technology for horizontal well in low permeability oilfield[C]. 2019 international Field Exploration and Development Conference, Xi'an (2019).
[12] Zhao Yunfeng. Research and application of chemical segment water plugging of horizontal wells in Gaosheng oilfield[J]. Petroleum Geology and Engineering, 26(03), 126-127+130+141 (2012).
[13] Hou Mei, Song Gencai, Xiang Mingguang, et al. Development and application of pipe string for water plugging technology in horizontal Wells [J]. Journal of Oil and Gas Technology, (06),787-789 (2005).
[14] Gao Chaoli, Liang Feng, Li Yang, et al. Research on water-seeking technology of horizontal wells and suggestions for water plugging technology of Wuqi Oilfield[J]. Unconventional Oil & Gas,6(01), 44-50 (2019).
[15] Chen Weiyu, Meng Kequan, Zhu liguo. Progress in research on water shutoff for horizontal well[J]. Petroch- emical Industry Application, 33(2), 1-4 (2014).
[16] Li Pengfei, Liu Xuhui, Wang Jie. Development and application status of non drilling type bridge plug with large internal diameter[J]. Oil Filed Equipment, 45(11), 93-97 (2016).
[17] Dou Guoren. Incipient motion of coarse and fine sediment[J]. Journal of Sediment Research, (6), 1-9 (1999).
[18] Pang Rende, Cui Shasha, Han Jiyong, et al. Research and Application of Drilling, Milling-Grinding Techniques for Drilling out Composite Bridge Plugs in Coiled Tubing in Horizontal Wells[J]. Petroleum Drilling Techniques, 44(01), 57-62 (2016).
[19] Song Xianzhi, Li Gensheng, Wang Mengshu, et al. Numerical Simulation on Cuttings Carrying Regularity for Horizontal Wells Drilled with Coiled Tubing[J]. Petroleum Drilling Techniques, 42(2), 28-32 (2014).
[20] Lai Fengpeng, Li Zhiping, Cen Fang, et a1. Optimum working parameters for hydraulic sand washing in horizontal wells[J], Petroleum Drilling Techniques, 35(1), 69-71 (2007).
[21] WILLIAMS C E J, BRUCE G H. Carrying capacity of drilling muds[J]. Journal of Petroleum Technology, 3(4), 111-120 (1951).
[22] Duan Mingqin, MISKA S, Yu Mengjiao, et al. Transport of small cuttings in extended-reach drilling[J]. SPE Drilling& Completion, 23(3), 258-265 (2008).
[23] Li Hongqian, Gao Chuanliang, Ren Yaoxiu, et a1. Calculations of annular slip velocity in the 2nd cleaning area during horizontal drilling[J]. Petroleum Drilling Techniques, 23(supplement 1), 27-29+85 (1995).
[24] Zeng Sirui, Dong Changyin, Wei Ran, et al. Comparing experiments and their mathching application of the borehole critical sand-carrying velocity in the horizontal well with different pitching angles[J]. Petroleum Geology and Oilfield Development in Daqing, 37

(06), 57-62 (2018).
[25] Wang Sifan, Huang Yongzhang, Hu Dongfeng. Study and application on the fishing and milling technology of layer-sealing bridge plugs in gas exploration wells. China Petroleum Machinery, 49(1), 41-46+52 (2021).

Journal of Physics: Conference Series **2594** (2023) 012044 doi:10.1088/1742-6596/2594/1/012044

# Corrosion Analysis and Prevention of Tubing Nipples in Four Wells

**Fang-ting HU1,1✉, Hong-tao LIU1,2✉,Mi-feng ZHAO1,3✉,Hai-long GENG1,4✉,Mao-xian XIONG1,5✉, Er-biao Lou1,6✉**

1. Tarim Oilfield Branch Oil and Gas Engineering Research Institute, Korla 841000, China,

1.2450126336@qq.com
2.liuhongtao-tlm@petrochina.com.cn
3.zhaomf-tlm@petrochina.com.cn
4. genghailong-tlm@petrochina.com.cn
5. xiongmx-tlm@petrochina.com.cn
6. Leb-tlm@petrochina.com.cn

**Abstract:** A high-temperature and high-pressure gas field in an oil field requires the use of super 13 chromium material due to corrosive environments. In the recent production process, there have been four consecutive cases of corrosion, fracture, and perforation of tubing nipples, which have brought enormous economic losses to the oilfield. This article analyzes the failure causes of these four tubing nipples and proposes preventive measures. In this paper, the macroscopic morphology, chemical composition, mechanical properties, metallography, and microscopic morphology of four failed tubing nipples have been analyzed. Combined with the field service conditions and string structure of the tubing nipples, it is believed that the failure of the tubing nipples is mainly due to the connection of carbon steel and stainless steel materials, forming a typical galvanic corrosion galvanic cell. The geometric and environmental factors of the materials exacerbate this corrosion. Eventually, severe galvanic corrosion occurred on the inner surface of the tubing nipple, resulting in severe wall thinning or partial perforation, and four tubing nipples failed. The failed tubing nipples in Wells A, B, and C have erosion marks, which are caused by the action of high pressure and high flow rate fluids in the well; The inner surface of the tubing nipple in Well D is rusty yellow, mainly due to the corrosion of its inner wall caused by the acid fracturing fluid injected earlier. At the same time, the string structure of four wells shows that the tubing nipple is located at the variable diameter node of the string. When the fluid in the well flows from the large hole into the smaller hole, the fluid in the well will change there, and its flow rate and pressure will also increase, further exacerbating the erosion in this area. With the development of deep wells containing carbon dioxide in oil fields and the application of carbon dioxide injection secondary oil recovery technology, the use of 13Cr martensitic stainless steel is also increasing. In order to avoid the occurrence of similar accidents that will bring huge economic losses to the oil field, it is recommended that the material of the pipe string (including accessories) of such wells be confirmed using a portable direct reading spectrometer before entering the well.

## 1. Introduction

The average well depth of a gas field in the oilfield is 6800m~7000m, the formation temperature is 152 ℃~188 ℃, the formation pressure is 113MPa~130MPa, and the partial pressure of carbon dioxide is 0.86MPa~2.56MPa, without hydrogen sulfide.According to literature [1], the gas field belongs to high-temperature high-pressure ultra deep gas well.In this case. In this environment where the partial pressure of carbon dioxide is between 0.21Mpa and 10MPa, the formation temperature is greater than 150 ℃ and less than 180 ℃, and there is no hydrogen sulfide, the super 13Cr material shall be used for the tubing according to the oil field material selection standard.

Recently, there have been four consecutive corrosion fracture and perforation of tubing nipple in the gas field (see Table 1), which has brought huge economic losses to the oilfield. The four oil pipe nipples are of the same specification. The nominal outer diameter of the coupling is 135.23mm, the nominal outer diameter of the pipe body is 114.30mm, and the nominal wall thickness is 12.7mm. See Table 1 for the failure details.

**Table 1.** Statistics of Fracture Failure Data of Four Tubing Nipples

<table>
<tr><th>Well No</th><th>Upper connecting string</th><th>Lower connecting string</th><th>Position of tubing nipple(m)</th><th>Length of tubing nipple(m)</th><th>Failure position of tubing nipple</th><th>Well depth (m)</th><th>Anomaly before failure</th></tr>
<tr><td>Well A</td><td rowspan="3">Φ114.30mm S13Cr110 special thread oil pipe (wall thickness 12.7mm)</td><td rowspan="3">Φ114.30mm S13Cr110 special thread oil pipe (wall thickness 9.65mm)</td><td>1509.51</td><td>1.03</td><td>External thread corrosion and fracture</td><td>7635</td><td>A Annulus pressure is equal to oil pressure</td></tr>
<tr><td>Well B</td><td>1610.55</td><td>1.04</td><td>Corrosion perforation of tubing nipple's body</td><td>7777</td><td>Abnormal fluctuation of oil pressure with amplitude of 12MPa</td></tr>
<tr><td>Well C</td><td>1299.24</td><td>1.02</td><td>External thread corrosion and fracture</td><td>7720</td><td>Annulus pressure A is basically consistent with oil</td></tr>
</table>

| | | | | | | | |
|---|---|---|---|---|---|---|---|
| | | | | | | | pressure |
| Well D | | | 661.80 | 1.04 | External thread corrosion and fracture | 6382 | Annulus pressure A is basically consistent with oil pressure |

From Table 1, it can be seen that the four failed tubing nipples are located in the upper position of the string. And the upper parts of these four oil pipe short joints are connected by Φ 114.30mm × 12.70mm S13Cr110 special threaded oil pipe (wall thickness 12.7mm), connected at the lower part Φ 114.30mm S13Cr110 special threaded oil pipe (wall thickness 9.65mm). All four tubing nipples have undergone corrosion, with many similarities. Therefore, it is of great significance to analyze the four failures together, find the causes, and propose preventive measures to prevent similar failures from reoccurring, leading to workover operations or accidents in ultra-deep and high-pressure wells.

The failure analysis in this article includes macroscopic morphology analysis, chemical composition analysis, mechanical property testing, metallographic analysis, and microscopic morphology analysis, as shown below.

## 2. Macro morphology analysis

### *2.1 Macro morphology analysis of Well A*

The outer surface of the failed tubing nipple in Well A is smooth, and the overall appearance is yellowish brown, without obvious corrosion pits. The macro morphology is shown in Figure 1.

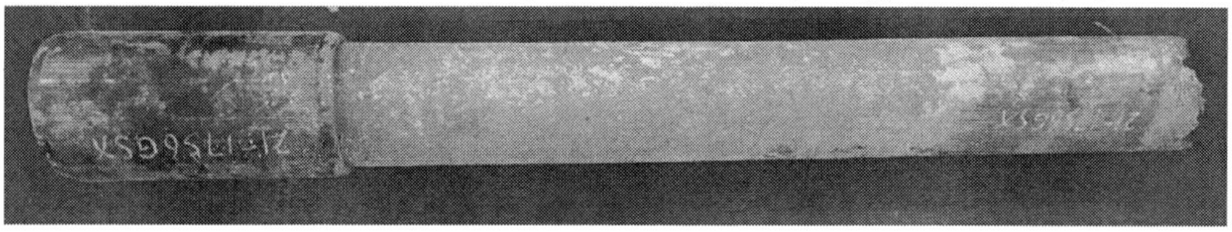

**Fig. 1.** Macromorphology of the outer surface of well A tubing nipple

The external thread end of the tubing nipple is serrated, with slight necking and severe wall thickness reduction. The thickness of the thinnest area is about 1mm, and about 70% of the thread length disappears, as shown in Figure 2.

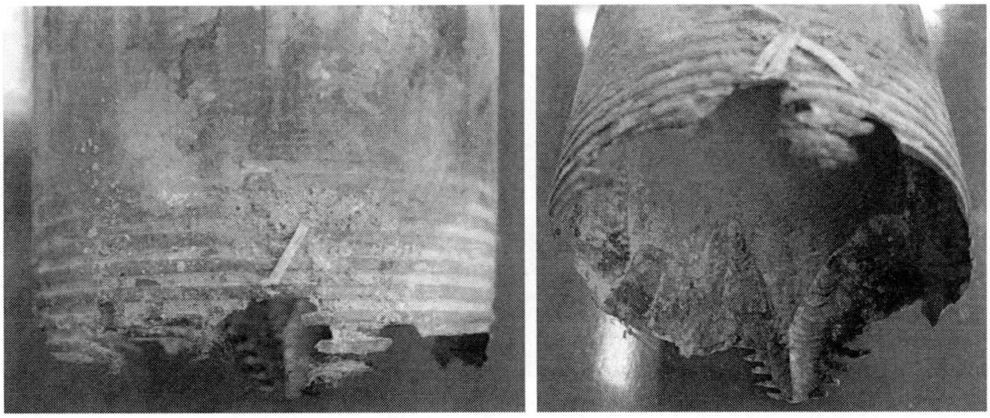

**Fig. 2.** Macro morphology of the external thread end of the tubing nipple failure in Well A

After the tubing nipple is longitudinally cut open, there is a "V" shaped wall thickness thinning area at the end of the failed external thread, and the boundary between the thinning area and the non thinning area is obvious, with clear edges and corners, and there are obvious erosion marks (there is obvious beach pattern in the middle of the wall thickness thinning area). The longitudinal section of the thinning area shows that the thinner the end of the external thread is, the more obvious the thinning is, and it is linear, as shown in Figure 3.

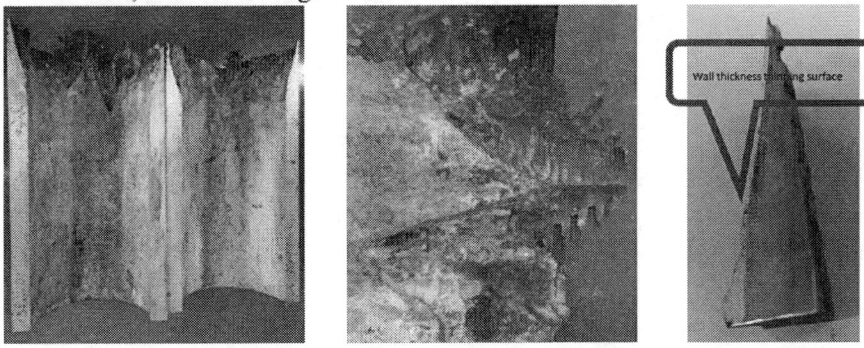

a.Longitudinal splitting morphology at the end b. Beach pattern in the thinning area c. Wall thickness thinning morphology

**Fig. 3.** Wall thickness thinning morphology of external thread end of well A tubing nipple

At the same time, it can be observed that the inner wall of the tubing nipple is covered with a dense layer of dirt in color. Using mechanical methods to remove the cover, it can be seen that the inner surface of the tubing nipple body area is smooth without obvious pitting traces, and the inner surface of the coupling is copper yellow without obvious damage and corrosion traces, as shown in Figure 4.

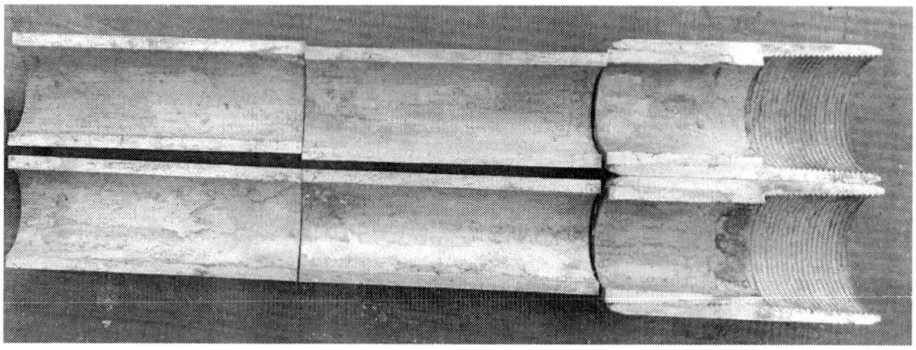

Fig. 4. Corrosion morphology of inner surface of well A tubing nipple

There are obvious corrosion pits on the inner wall of the external thread at the other end of the tubing nipple and the area adjacent to the coupling torque step. The maximum depth of the corrosion pits exceeds 5 mm, and there are also significant wall thickness thinning features on the section of this area, as shown in Figure 5.

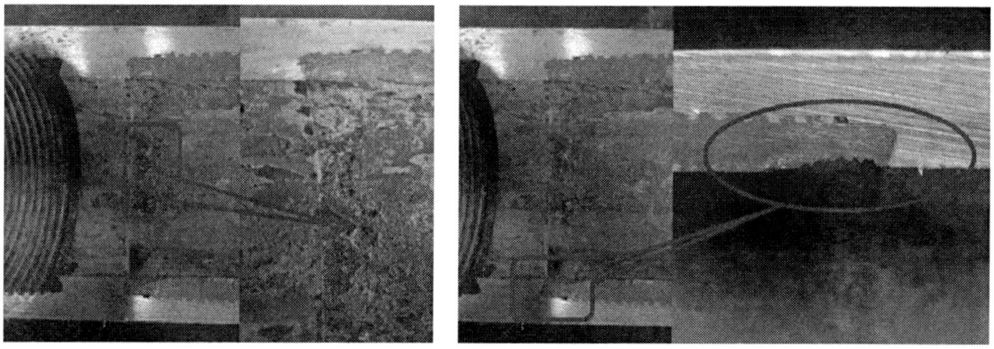

a. Corrosion pit b. Wall thickness reduction

**Fig. 5.** Morphology of inner wall of external thread and coupling area at the other end of well A tubing nipple

*2.2 Macro morphology analysis of Well B*

The failed tubing nipple of Well B is dark black on the whole, without obvious corrosion trace. A circular perforation with a hole diameter of about 10mm can be seen in the middle. The macroscopic morphology is shown in Figure 6.

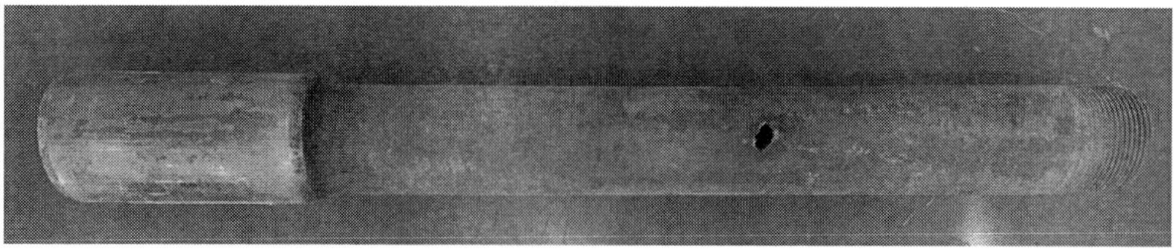

**Fig. 6. Macromorphology of the outer surface of the tubing nipple of Well B**

After the tubing nipple is longitudinally cut open, it can be seen from Figure a in Figure 7 that there is no obvious covering on the inner wall surface, and local areas present metallic luster. At the connection with the coupling (left end of Figure a in Figure 7), the corrosion pits on the inner surface of the external thread end are obvious and distributed in a circumferential direction. The corrosion pits in local areas have penetrated the wall thickness, and the internal thread is visible (Figure b in Figure 7); There are many deep and large corrosion pits near the perforation (Figure 7, Figure a), with a maximum depth of 10mm; There is no obvious corrosion on the inner surface of the external thread at the other end (the right end of a in Figure 7), but the number of external threads has decreased from 18 to 12, with 6 threads disappearing due to corrosion (Figure c in Figure 7).

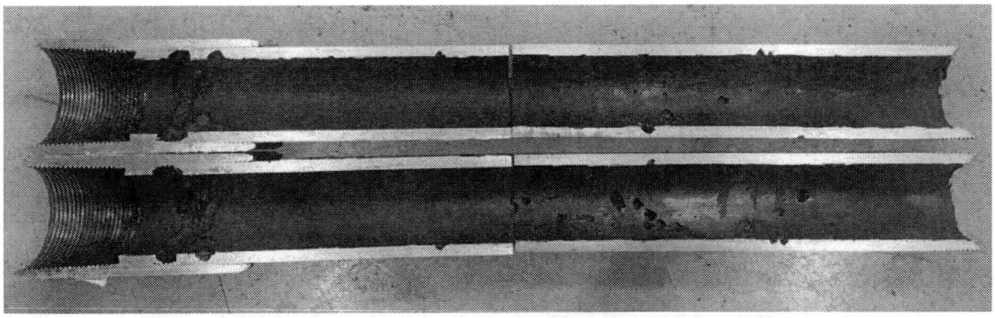

a. Corrosion morphology of internal surface

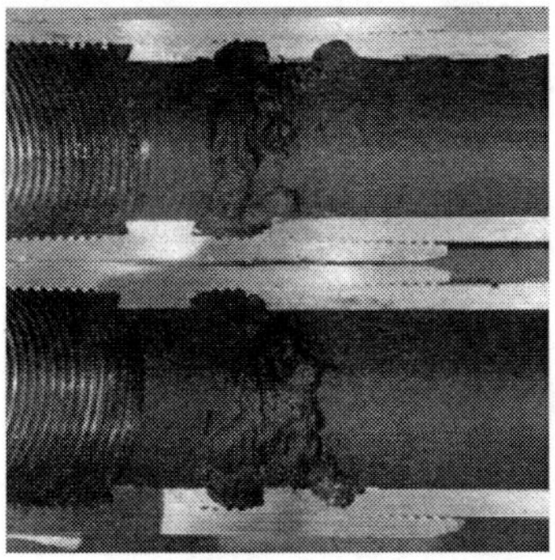

b.Corrosion morphology of coupling joint c.External thread end and inner wall at the other end

**Fig. 7. Appearance of inner wall and end face of external thread at the other end of well B tubing nipple**

*2.3 Macro morphology analysis of Well C*

There is obvious scale on the outer surface of the tubing nipple in Well C, which is generally yellowish brown. After removing the cover, no obvious corrosion pits are found. As shown in Figure 8, the right external thread end is serrated, with about 70% of the threads disappearing. At the same time, the remaining thread wall thickness is severely reduced, with the thickness of the thinnest area being about 1mm:

a. Macromorphology of the outer surface of the tubing nipple

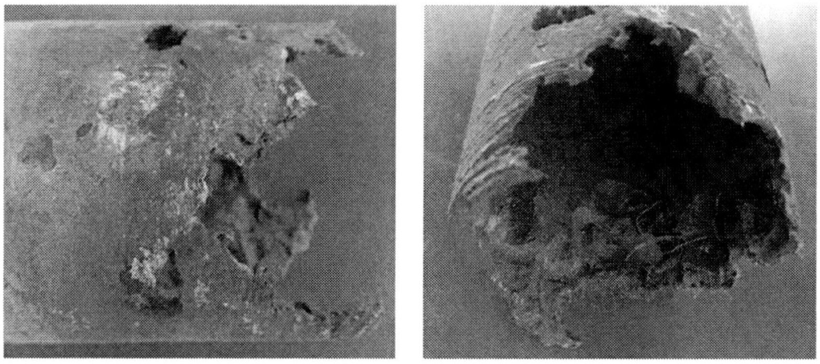

b. Corrosion morphology of external thread end of tubing nipple

**Fig. 8.** Macroscopic and External Thread Corrosion Morphology of Well C Tubing Nipple

Cut the short joint longitudinally. From Figure 9, it can be seen that there is no obvious covering on the surface of the inner wall, and some areas show metallic luster. There are many deep and large corrosion pits in some areas, up to 7mm deep. There are also obvious corrosion pits on the inner wall of the external thread connected with the coupling, with the maximum depth of more than 5mm (in the red box in Figure 9). At the same time, there are also features of wall thickness reduction:

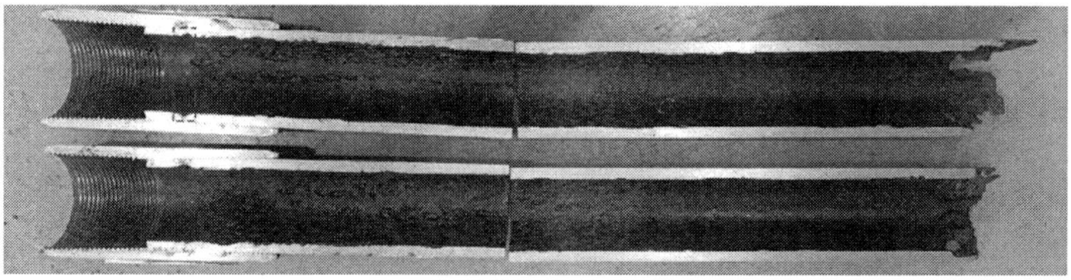

**Fig. 9.** Internal surface morphology of well C tubing nipple

### *2.4 Macro morphology analysis of Well D*

The macroscopic appearance of the failed tubing nipple in Well D is shown in Figure 10, as shown in Figure a, and it is generally dark black (covered with oil stain on the surface), without obvious corrosion pits; Visible perforations can be seen in the area about 100mm away from the external threaded joint (Figure 10, Figure b), and the outer surface of the perforations is smooth without pitting marks; The external thread end is in a zigzag shape. Due to corrosion, the thread has disappeared, and the wall thickness has been severely reduced. The thickness of the thinnest area is about 1mm (Figure 10, Figure c).

a. Macromorphology of external surface

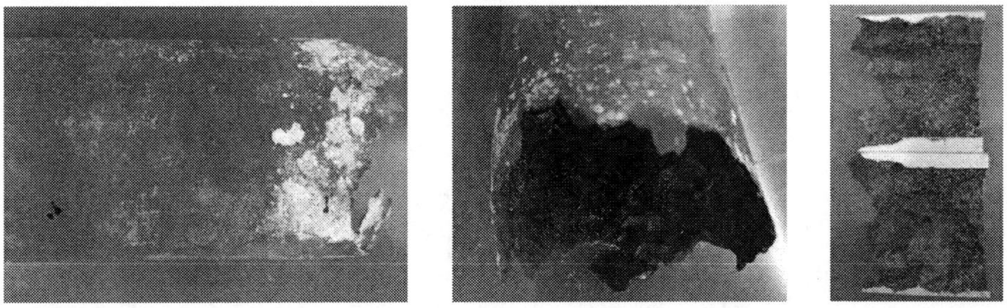

b.External surface perforation morphology c. External thread end corrosion morphology

**Fig. 10. Macro morphology of well D tubing nipple**

After longitudinally splitting the tubing nipple, it can be seen that large and deep corrosion pits are distributed on the entire inner wall (especially at the outer thread end of the nipple). The bottom of the corrosion pit is loose and rusty yellow, with a small amount of black due to the oil stain on the surface. The inner thread of the nipple coupling is intact and copper yellow, without obvious damage and corrosion traces, as shown in Figure 11.

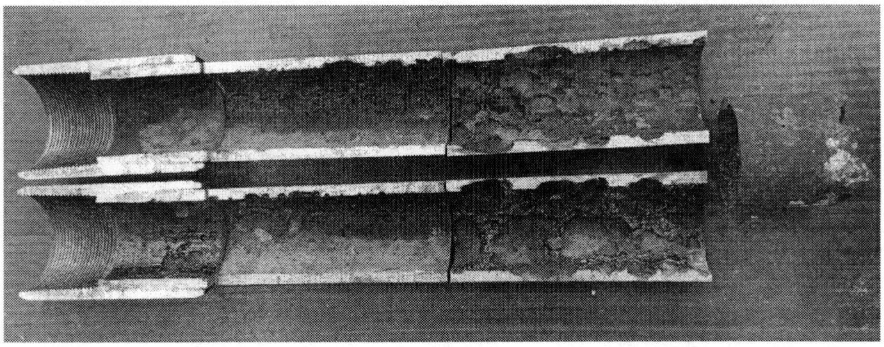

**Fig. 11.** Corrosion Morphology of the Inner Surface of Well D Tubing Nipple

## 3. Chemical composition analysis

Samples were taken from the tubing nipple bodies and couplings of the above four wells. According to the GB/T 4336-2016 standard[2], carbon sulfur analyzer and ARL 4460 direct reading spectrometer were used to analyze the chemical composition of the samples. The results are shown in Table 2. From the data in Table 2, it can be seen that the body material of the tubing nipple is ordinary carbon steel, and the coupling material is super 13Cr. Referring to the requirements of the oil field for tubing materials P110 and S13Cr, the S and P contents of the tubing nipple body and the coupling meet the requirements.

**Table 2.** Chemical Composition Analysis Results (wt%)

| Sample \ Element | C | Si | Mn | P | S | Cr | Mo | Ni | Nb | Cu |
|---|---|---|---|---|---|---|---|---|---|---|
| Body of well A tubing nipple | 0.24 | 0.21 | 1.66 | 0.0094 | 0.0048 | 0.043 | 0.014 | 0.036 | 0.0012 | 0.036 |
| Body of well B tubing nipple | 0.27 | 0.22 | 1.68 | 0.0085 | 0.0037 | 0.046 | 0.014 | 0.036 | 0.0012 | 0.037 |
| Body of well C tubing nipple | 0.26 | 0.21 | 1.66 | 0.0095 | 0.0041 | 0.036 | 0.019 | 0.038 | 0.0013 | 0.052 |
| Body of well D tubing nipple | 0.24 | 0.21 | 1.65 | 0.0097 | 0.0045 | 0.043 | 0.015 | 0.035 | 0.0009 | 0.037 |
| Oilfield requirements for P110 | / | / | / | ≤0.015 | ≤0.010 | / | / | / | / | / |
| Coupling of well A tubing nipple | 0.035 | 0.18 | 0.39 | 0.019 | 0.005 | 12.95 | 2.16 | 5.60 | 0.03 | 0.066 |
| Coupling of well B tubing nipple | 0.031 | 0.16 | 0.37 | 0.014 | 0.003 | 12.93 | 2.16 | 5.53 | 0.03 | 0.063 |
| Coupling of well C tubing nipple | 0.030 | 0.18 | 0.39 | 0.016 | 0.002 | 13.05 | 2.12 | 5.59 | 0.03 | 0.030 |
| Coupling of well D tubing nipple | 0.033 | 0.17 | 0.37 | 0.017 | 0.005 | 13.21 | 2.22 | 5.79 | 0.03 | 0.040 |
| Oilfield requirements for S13Cr110 | / | / | / | ≤0.020 | ≤0.005 | / | / | / | / | / |

## 4. Mechanical property test

### *4.1 Tensile test*

Take tensile samples from the tubing nipple body and coupling of four wells, and conduct room temperature tensile test according to ASTM A370 standard [3]. The test results are shown in Table 3. The tensile properties of the tubing nipple body and coupling meet the requirements of the oil field.

**Table 3.** Tensile test results at room temperature

| Sample | Yield strength (MPa) | Tensile strength (MPa) | Elongation after fracture (%) |
|---|---|---|---|
| Average value of well A tubing nipple body | 852 | 940 | 24.5 |
| Average value of well B tubing nipple body | 903 | 949 | 23.7 |
| Average value of well C tubing nipple body | 871 | 945 | 27.0 |
| Average value of well D tubing nipple body | 913 | 961 | 21.0 |
| Oilfield requirements for P110 | 758~965 | ≥862 | ≥16 |
| Average value of tubing nipple coupling in Well A | 858 | 909 | 22.5 |
| Average value of tubing nipple coupling in Well B | 882 | 952 | 23.3 |
| Average value of tubing nipple coupling in Well C | 858 | 922 | 22.0 |
| Average value of tubing nipple coupling in Well D | 862 | 945 | 20.8 |
| Oilfield requirements for S13Cr110 | 758-896 | ≥827 | ≥16 |

### *4.2 Hardness test*

Take transverse hardness samples from the body and coupling of the tubing nipple of four wells, and use KB30BVZ-FA testing machine to test the hardness of the samples according to ASTM E92 standard [4]. See Table 4 for the results. The coupling hardness of the tubing nipple meets the oilfield requirements.

**Table 4.** Rockwell hardness test results

| Sample | Detection value |
|---|---|
| Body of tubing nipple in Well A | 30.4、30.5、30.5 |
| Body of tubing nipple in Well B | 30.4、30.7、30.9 |
| Body of tubing nipple in Well C | 29.1、28.6、28.6 |
| Body of tubing nipple in Well D | 30.5、29.5、29.4 |
| Oilfield requirements for P110 | / |
| Coupling of tubing nipple in Well A | 28.7、29.8、29.6 |
| Coupling of tubing nipple in Well A | 31.4、31.4、31.3 |
| Coupling of tubing nipple in Well A | 30.5、30.7、30.2 |
| Coupling of tubing nipple in Well A | 28.5、28.9、29.0 |
| Oilfield requirements for S13Cr110 | Maximum hardness 32 |

### *4.3 Charpy impact energy*

Take impact samples from the tubing nipple body and coupling of four wells, and conduct Charpy impact test on the samples with the impact tester of PIT302D impact tester according to the GB/T 229 standard[5]. See Table 5 for the test results. The impact performance of the tubing nipple body and coupling meets the requirements of the oil field.

**Table 5.** Impact test results

| Sample | Sample specification (mm) | Notch shape | temperature (℃) | Impact absorbed energy (J) | | | average value (J) |
|---|---|---|---|---|---|---|---|
| Longitudinal sampling of tubing nipple body in Well A | 10×10×55 | V type | -10 | 120 | 122 | 95 | 112 |
| Longitudinal sampling of tubing nipple body in Well B | | | | 136 | 128 | 128 | 131 |
| Longitudinal sampling of tubing nipple body in Well C | | | | 120 | 125 | 121 | 122 |
| Longitudinal sampling of tubing nipple body in Well D | | | | 95 | 101 | 100 | 99 |
| Oilfield requirements for P110 | | | | Longitudinal minimum impact energy ≥ 80 | | | |
| Longitudinal sampling of coupling of tubing nipple in Well A | | | | 188 | 191 | 193 | 191 |
| Transverse sampling of coupling of tubing nipple in Well B | | | | 184 | 192 | 173 | 183 |
| Transverse sampling of coupling of tubing nipple in Well C | | | | 200 | 205 | 221 | 209 |
| Longitudinal sampling of coupling of tubing nipple in Well D | | | | 159 | 166 | 149 | 162 |
| Oilfield requirements for S13Cr110 | | | | Minimum transverse impact energy ≥120; Longitudinal minimum impact energy≥140 | | | |

## 5. Metallographic analysis

The transverse and longitudinal metallographic analysis samples are taken from the main body and coupling of the tubing nipple by the wire cutting method. According to the GB/T 13298-2015 standard[6], the microstructure, grain size and inclusion grade of the samples are analyzed by using the OLS 4100 laser confocal microscope, MEE4M metallographic microscope, image analysis system and Smartzoom 5 ultra-depth digital microscope. The results show that the metallographic structures of the tubing nipple body and coupling are tempered sorbite and tempered martensite, respectively. The grain size and inclusion results are shown in Table 6.

**Table 6.** Metallographic analysis results of tubing nipple and coupling

| Sample | Nonmetallic inclusion | Organization | Grain size |
|---|---|---|---|

| | A | B | C | D | | |
|---|---|---|---|---|---|---|
| | thin | thin | thin | thin | | |
| Body of tubing nipple in Well A | 0.5 | 1.0 | 0 | 1.0 | S | 9.0 |
| Body of tubing nipple in Well B | 0.5 | 0.5 | 0 | 1.0 | S | 10.0 |
| Body of tubing nipple in Well C | 0.5 | 0.5 | 0 | 1.0 | S | 10.0 |
| Body of tubing nipple in Well D | 0.5 | 1.0 | 0 | 0.5 | S | 9.0 |
| Coupling of tubing nipple in Well A | 0.5 | 0.5 | 0 | 1.0 | M | 7.0 |
| Coupling of tubing nipple in Well B | 0.5 | 0.5 | 0 | 1.0 | M | 8.0 |
| Coupling of tubing nipple in Well C | 0.5 | 0.5 | 0 | 1.0 | M | 8.0 |
| Coupling of tubing nipple in Well D | 0.5 | 0.5 | 0 | 0.5 | M | 7.0 |

Note: S in the table represents tempered sorbite; M stands for tempered martensite.

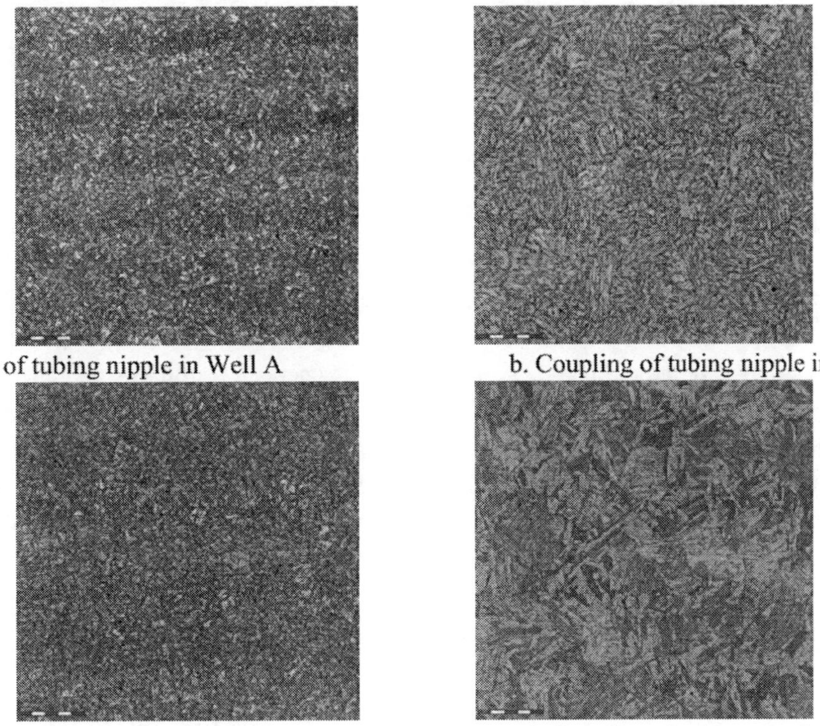

a. Body of tubing nipple in Well A

b. Coupling of tubing nipple in Well A

c. Body of tubing nipple in Well B

d. Coupling of tubing nipple in Well B

e. Body of tubing nipple in Well C

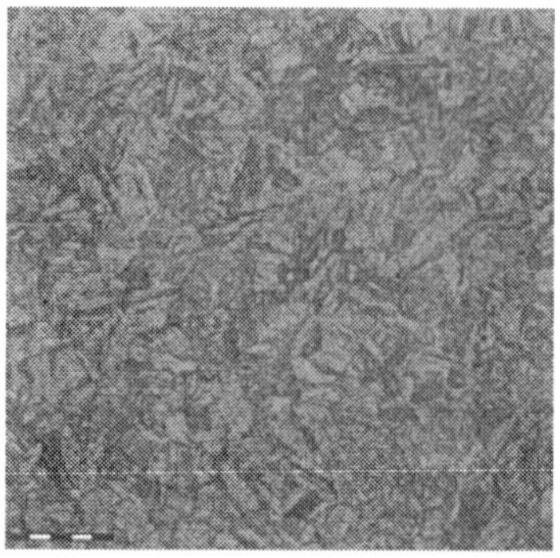

f. Coupling of tubing nipple in Well C

g. Body of tubing nipple in Well D

h. Coupling of tubing nipple in Well D

**Fig. 12.** Metallographic structure of tubing nipple body and coupling in four wells

## 6. Micromorphology analysis

Take samples from the failed tubing nipples of four wells, after ultrasonic cleaning with acetone reagent and anhydrous alcohol, use TESCAN VEGA3 XMU scanning electron microscope and energy spectrometer to observe the morphology and analyze the energy spectrum of the inner surface of the samples, and the results are as follows:

The micromorphology of the tubing nipple in these four wells has similarities. The micromorphology of the corrosion pits on the inner surface is uniform in color, and the microstructure is relatively loose, in the form of cotton wool. The corrosion products in the flat area of the inner wall are loose, in the form of particles. Slightly different is that the corrosion pits at the low magnification of the tubing nipple in Well B are bowl shaped, with cracks visible in some areas. The corrosion pits on the inner surface of the tubing nipple in Well C are connected by multiple corrosion pits, and the bottom of the corrosion pit is soil gray, with obvious cracks visible in most areas, as shown in Figure 13.

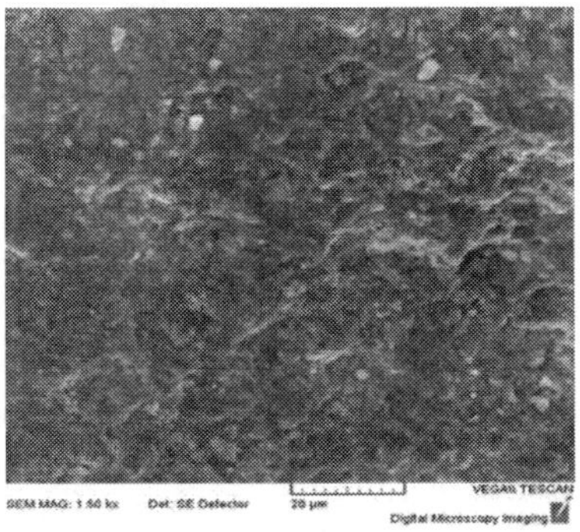

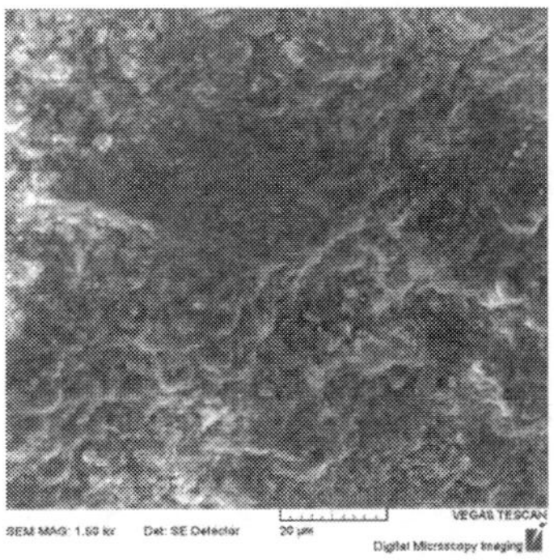

a. Morphology of corrosion pits on the inner surface of Well A sample b. Appearance of flat area on the inner surface of well A sample

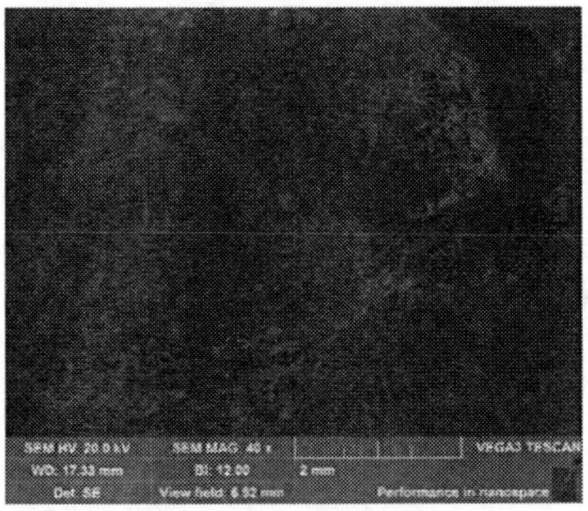

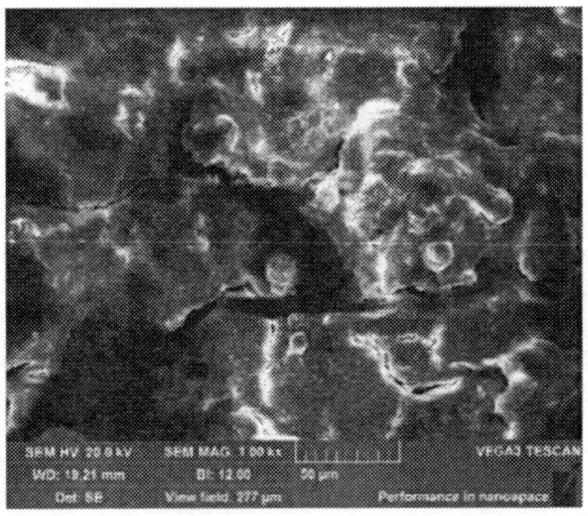

c. Macroscopic appearance of corrosion pit of well B sample d. Appearance of corrosion pit bottom of well B sample

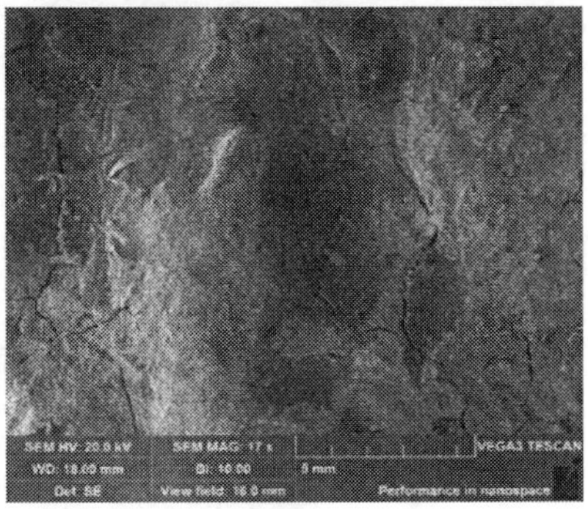

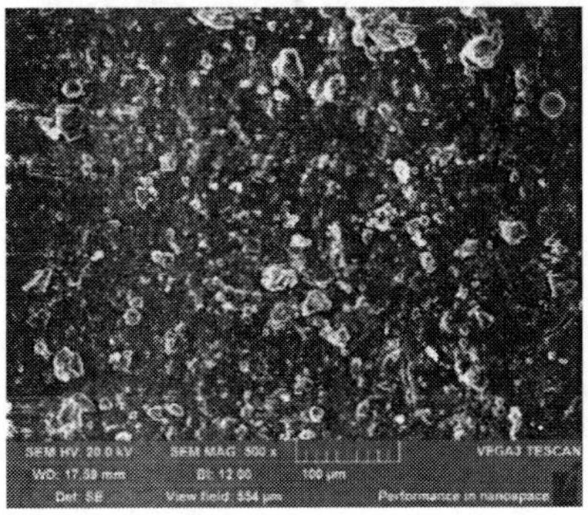

e. Macroscopic crack morphology of corrosion pit of well C sample f. Granular morphology at the bottom of corrosion pit of well C sample

Journal of Physics: Conference Series **2594** (2023) 012044 doi:10.1088/1742-6596/2594/1/012044

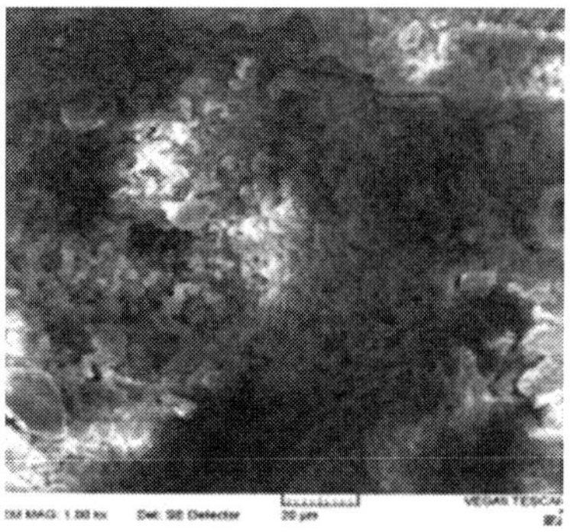

g. Corrosion pit morphology on the inner surface of well D sample

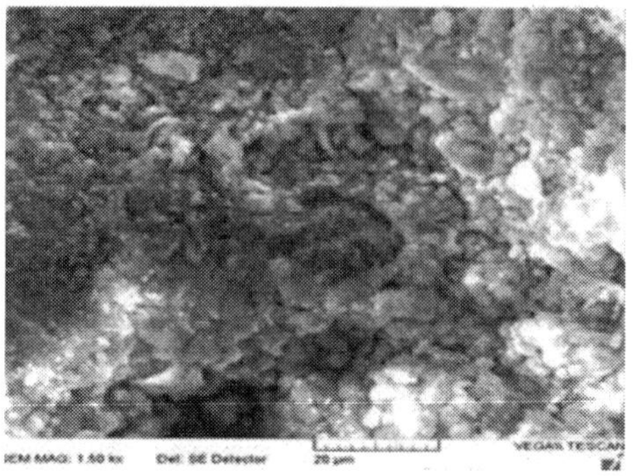

f.Morphology of flat area on the inner surface of well D sample

**Fig. 13.** Corrosion morphology of tubing nipple in four wells

The INCA350 energy spectrum analyzer was used to analyze the corrosion products in the flat area of the inner wall of the tubing nipple and at the bottom of the corrosion pit. The results showed that the corrosion products in Well A mainly contain elements such as S, O, Fe, P, K, and Ca; The corrosion products on the inner wall of the tubing nipple in Well B at the bottom of the pit mainly contain elements such as S, O, Fe, Ca, and Ba, and some areas contain more Cl elements; The corrosion pit bottom cover on the inner surface of the tubing nipple in Well C mainly contains S, O, Ca, Si, Fe, Ba, and a small amount of K, Al elements. The corrosion products at the bottom of the corrosion pit in Well D are O, Cl, and Fe elements, as shown in Figure 14:

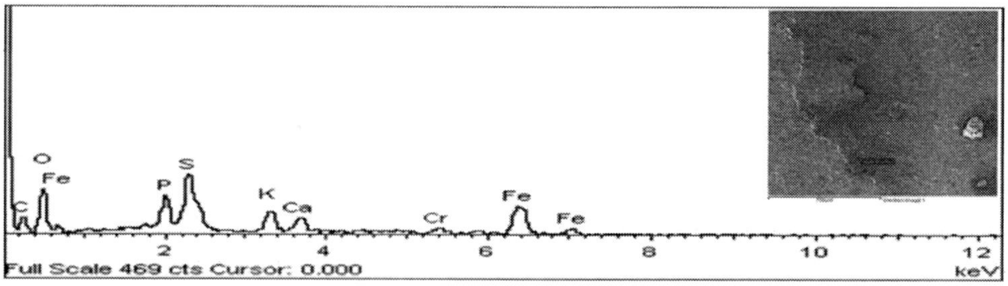


a. Energy spectrum analysis results of corrosion pit of tubing nipple in Well A

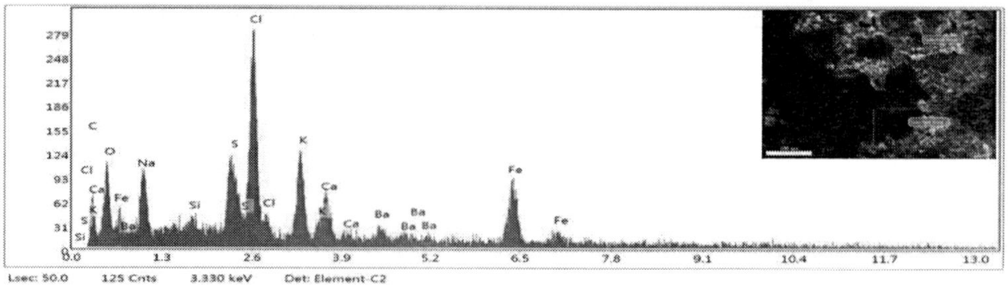


b. Energy spectrum analysis results of corrosion pits on the inner wall of well B tubing nipple

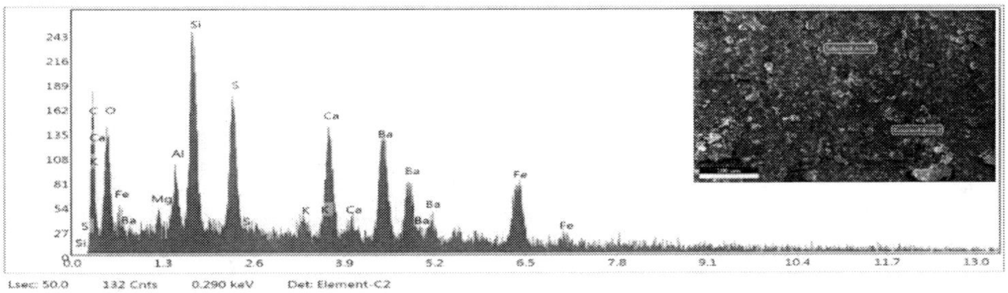


c. Energy spectrum analysis results of corrosion pit of tubing nipple in Well C

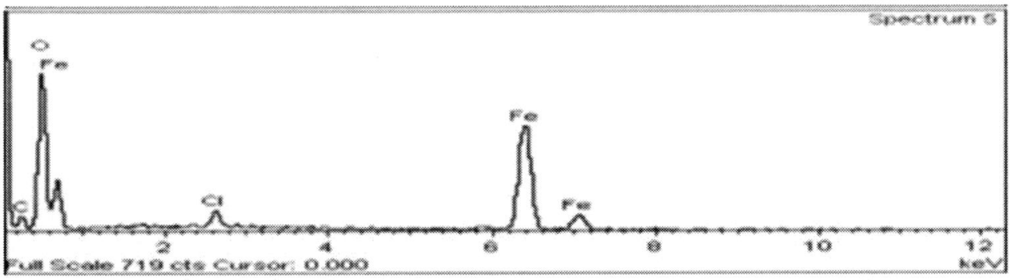


d. Energy spectrum analysis results of corrosion pit of tubing nipple in Well D

**Fig. 14.** Energy spectrum analysis results of corrosion pit of tubing nipple in four wells

## 7. Comprehensive analysis

According to the physical and chemical performance testing results of the four tubing nipples, the material of the nipple body is ordinary low-carbon steel, and the coupling material is super 13Cr. The chemical composition, tensile properties, hardness, and impact energy of the nipple body and the coupling meet the performance requirements of the oilfield for P110 and S13Cr, respectively. From the metallographic analysis, the metallographic structure of the nipple body is tempered sorbite, with a grain size grade of 9.0-10.0. The metallographic structure of the coupling is tempered martensite, with a grain size grade of 7.0-8.0. No abnormality is found. Based on the macroscopic characteristics of the failure locations of the four tubing nipples and in combination with their on-site service conditions, it is believed that the failure causes are as follows.

Firstly, the outer surfaces of the four tubing joints are smooth, without obvious uniform corrosion and pitting traces, which can eliminate the possibility of tubing joints in the four wells failing due to external wall corrosion. The wall thickness of the external thread at one end of Well A decreases linearly in the longitudinal direction, and significant corrosion can be seen at the connection between the internal surface of the external thread at the other end and the coupling torque step; There are many deep and large corrosion pits near the perforation in the central area of the inner wall of Well B, especially at the connection between the torque step of the tubing joint coupling and the joint body, where the entire circumferential corrosion pattern is concentrated; In Well C, there are many deep and large corrosion pits near the contact steps between the tubing joint and the coupling, with corrosion traces. The external thread at the other end is severely thinned due to corrosion, presenting a honeycomb shape, and partially corroded to form holes; There are obvious corrosion pits on the inner wall of the tubing joint in Well D, with greatly reduced wall thickness and obvious signs of tensile deformation (slight necking). The external thread at the other end of the D well tubing joint is serrated and disappears due to corrosion. Therefore, it is not difficult to see that all four wells have failed due to corrosion, with severe corrosion at the threaded connections. The tubing joint bodies of these four wells are all made of ordinary carbon steel, and the joint material is over 13Cr, which does not meet the requirements of the oilfield for using over 13Cr for all completion strings.

Some data indicate that significant galvanic corrosion occurs when carbon steel metals and stainless steel metals are in the same electrolyte. Galvanic corrosion is a typical type of contact corrosion, which refers to electrochemical corrosion caused by two different metals contacting each other while simultaneously in the dielectric. The two metals form a spontaneous battery, and the more active metal acts as the anode of the battery, causing corrosion. It is a form of corrosion that has extremely widespread hazards and may cause serious losses [7]. In these four wells, only the tubing nipple body is made of carbon steel, and the rest are made of stainless steel. They are located in the same downhole produced fluid medium, and typical galvanic corrosion occurs.

In order to verify the above conclusions, we connect stainless steel and ordinary carbon steel materials with a clamp slot. When in the same corrosion medium, ordinary carbon steel is subject to galvanic action, resulting in its corrosion rate increasing by 9.68% compared with normal corrosion. Super 13Cr stainless steel is subject to galvanic action, resulting in its corrosion rate decreasing by 4.79% (as shown in Figures 15 and 16).

**Fig.15.** Couple corrosion test connection sample of super 13Cr stainless steel and ordinary carbon steel

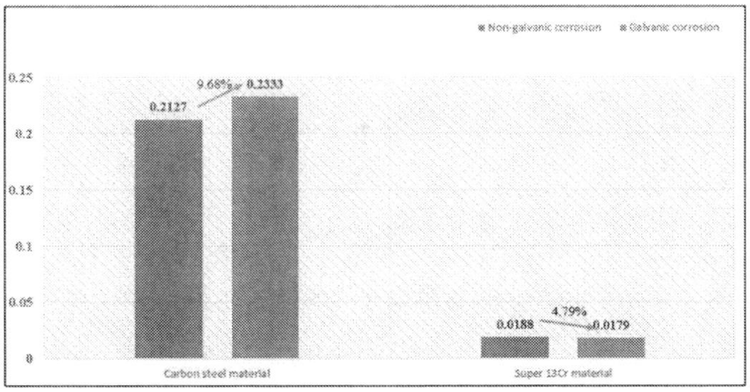


**Fig. 16.** Comparison of galvanic corrosion rate between super 13Cr stainless steel and ordinary carbon steel

In literature [8], there is also a similar conclusion that there is a significant difference in the corrosion potential between super 13Cr and P110 steel, and the open circuit potential of super 13Cr steel is significantly higher than that of P110 steel. Therefore, galvanic corrosion occurs when the two are connected, and super 13Cr steel is protected as the cathode of the galvanic couple, while P110 steel is accelerated as the anode of the galvanic couple. The corrosion products on its surface are composed of FeO. This type of galvanic corrosion can also occur between nickel based alloys and carbon steel materials [9], and the corrosion of carbon steel will also increase, with the same corrosion principle. Therefore, the selection of pipe materials is very important, which directly affects the integrity of the gas well string.

There are many factors affecting galvanic corrosion[10]. In addition to the characteristics of the material mentioned above, there are also geometric factors of the material. Here, it mainly refers to the specific area of the cathode and anode of the galvanic material. The area of the super 13Cr used as the cathode material in the pipe string is very large, and almost all of the whole pipe string, while the P110 steel used as the anode material is only the body part of the oil pipe nipple, so the area ratio of the cathode material to the anode material is very large, The phenomenon of galvanic corrosion will be more obvious.

Finally, there are environmental factors that affect galvanic corrosion. Local areas of the inner surface of the tubing nipples in wells A, B and C are shiny, with erosion morphology, especially in well C. This is because the wellbore pressure and daily output of the three wells are relatively high. Under high pressure and high flow rate, the inner wall of the wellbore will have relatively serious erosion, especially at the corrosion pit caused by galvanic corrosion between the external thread end face of the tubing nipple and the coupling torque step. Well D tubing nipple is slightly different from the other three wells. The inner surface is rusty yellow. The corrosion products at the bottom of the corrosion pit are O, Cl and Fe. According to the main components of the corrosion products, it can be inferred that the corrosion of the inner wall of the tubing nipple is mainly caused by Cl and O elements. Well D was acidified in the early stage. During the acidification process, oxygen in the air will enter the wellbore with the fracturing fluid, resulting in the following reactions, causing the inner wall of the tubing nipple to be continuously corroded and forming large and deep corrosion pits.

$$Fe + 2HCl = FeCl_2 + H_2\uparrow \quad (1)$$

$$4FeCl_2+O_2+4HCl=4FeCl_3+2H_2O \quad (2)$$

Finally, the string structure of four wells shows that the tubing nipple connects two kinds of tubing, and is at the pipe string diameter change node (the inner diameter decreases from 95.0 mm to 88.9 mm from the bottom of the well to the wellhead). When the fluid in the well flows from the large hole to the smaller hole, the fluid in the well will change there, and its flow rate and pressure will also increase, further intensifying the erosion in this area.

At the same time, the string structure of the four wells shows that the tubing nipple connects two types of tubing, and is at the pipe diameter node (the inner diameter decreases from 95.0 mm to 88.9 mm from the bottom of the well to the wellhead). When the fluid in the well flows from the large hole to the smaller hole, the fluid in the well will change there, and its flow rate and pressure will also increase, further intensifying the erosion in this area.

Based on the above test results and analysis, it can be seen that the failure of the oil pipe nipple is mainly due to the connection of carbon steel and stainless steel materials, forming a typical galvanic corrosion primary battery. The geometric and environmental factors of the material exacerbate this corrosion. Eventually, severe galvanic corrosion occurred on the inner surface of the tubing nipple, resulting in severe wall thinning or partial perforation, leading to failure.

## 8. Conclusions

(1) The material of the tubing nipple body of the four wells is ordinary carbon steel, and the coupling material is super 13Cr. Its chemical composition, tensile property, hardness and impact energy meet the performance requirements of P110 and S13Cr of the oil field respectively; The metallographic structure of the four oil pipe nipple bodies is tempered sorbite with grain size grade of 9.0-10.0. The metallographic structure of the coupling is tempered martensite with grain size grade of 7.0-8.0. No abnormality is found.

(2) The galvanic corrosion caused by the connection between carbon steel and stainless steel is the main reason for the short joint failure.

(3) Local areas on the inner surface of the tubing nipple of Well A, Well B and Well C are shiny, which is due to the serious erosion of the inner wall of the wellbore caused by high pressure and high flow rate fluid; The inner surface of the tubing nipple in Well D is rusty yellow, mainly because the acid fracturing fluid injected in the early stage aggravated the corrosion of the inner wall of the tubing nipple.

(4) With the development of deep wells containing carbon dioxide in oil fields and the application of $CO_2$ injection secondary oil recovery technology, the use of 13Cr martensitic stainless steel is also increasing. There are many protective measures for galvanic corrosion [11], but the most direct measure to mitigate galvanic corrosion is to avoid galvanic pairs prone to galvanic corrosion in terms of material selection, while avoiding the occurrence of the "large cathode small anode" situation [10]. Therefore, it is recommended to confirm the material of the pipe string (including accessories) using a portable direct reading spectrometer before entering the well.

(5) It is recommended that the oil field strengthen the management of pipe column attachments.

**References**

1. Wang Haige, F., Li Wanping, S.,Guo Xiaoxia, T.: Development of drilling and completion technology at high temperature and high pressure[R], Beijing: China Petroleum Economic and Technological Research Institute,2009,1-9.

2. GB/T 4336,Determination of multielement content of carbon steel and medium low alloy steel, China, China Iron and Steel Industry Association, 2016.

3. ASTM A370, Standard Test Methods and Definitions for Mechanical Testing of Steel Products, American, American Society for Testing and Materials,2021.

4. ASTM E92, Standard Test Methods for Vickers Hardness and Knoop Hardness of Metallic Materials, American, American Society for Testing and Materials,2017.

5.GB/T 229,Metallic materials Charpy pendulum impact test, China, China Iron and Steel Industry Association,2020.

6. GB/T 13298,Inspection methods of microstructure for metals, China, China Iron and Steel Industry Association,2015.

7. Chen Xingwei, F., Wu Jianhua, S., Wang Jia, T., etc: Research progress of influencing factors of galvanic corrosion, Corrosion science and protection technology,4(22),363-366(2010).

8. Yao Xiaofei, F., Tian Wei, S., Xie Faqin, T., etc: Study on the galvanic corrosion behavior of super 13Cr and P110 tubing steel in NaCl solution, Material Guide A: Overview,6(31),166-169(2017).

9. Zhu Dajiang, F., Fan Yu, S., Zhang Huali, T., etc: Study on galvanic corrosion behavior of corrosion-resistant alloy tubing and high grade carbon steel casing in drilling fluid, Material protection,9(55), 27-33(2022).

10. Wang Xiaohuan, F., Zhang Shenghan, S., Di Jie ,T. :Review of factors affecting galvanic corrosion of metals, Guangdong Chemical,19(49),142-147(2022) .

11. Zhang Wenyu, F.,Research progress of galvanic corrosion and protection, Overall corrosion control,12(32),51-56(2018).

# Dynamics Analysis of Droplets Collision with the Wall Driven by Buoyancy

**Feng Rong1, Li-min He1, Yu-ling Lü1(✉), Ce Wang1, Yi-shuo Han1**

1 College of Pipeline and Civil Engineering, China University of Petroleum Huadong, Qingdao, China
lylchuyun@163.com

**Abstract.** Dynamics analysis of the oil droplet in water impacting on a rigid wall in the range of Re=9.3 ~ 231.39, We=0.003 ~ 0.637 was carried out by combining microscopic experiment with theoretical study. In the oil droplet's force equation, we consider the added mass force and the water film induced force, in addition to the conventional forces, i.e. buoyancy and resistance to flow. Stokes-Reynolds drainage equation (i.e. SR model) and Young-Laplace equation (i.e. YL model) are combined with the force equation to numerically solve the motion law of oil droplets near the wall for different control parameters. The study indicates that coupled model would expect oil droplet velocities, motion trajectories and capture the kinetic behaviors of the water-film drainage evolution. The interfacial deformation of oil droplets causes changes in the pressure and the pressure distribution pattern has an important relationship with the shape of the oil droplet interface. These two forces play a dominant role in droplet motion during the collision and the maximum value of these two forces increases with increasing droplet size and Re number. Future research should shift towards complex oil-water systems and consider the effects of chemicals on oil droplets so that the application of the model will be more generalisable. At the same time, computer technology should be used to transfer the two-dimensional view to a three-dimensional view to analyse the deformation of the oil-water interface and to better understand the drainage process of the water film.

## 1. Introduction

At present, the combined water content of recovered fluids in most of China's oil fields already exceed 90% [1,2], resulting in a large amount of oil-bearing wastewater with micron-sized oil droplets, requiring efficient treatment. Gravity separators as oilfield commonly used oil-water separation equipment, the separation space utilizes gravity to separate the droplets, so the selection of the separation space height is quite important. If the height is too low, the separation is not sufficient. If the height is too high, it increases the total volume of the oil-water separation equipment [3,4]. Generally, the height of the separation space is closely related to the movement of the droplets [5,6]. The oil removal process by coalescing plates involves the adhesion, spread-ing, migration, and shedding of oil droplets on the plates [7,8,9], the interaction between the two preferentially determines the efficiency of the whole oil removal process.

On the other hand, to reduce the cost of oilfield development, for produced liquids with high water content, technicians frequently adopt the transport technology at low temperatures [10,11,12], but it is prone to causing the phenomenon of oil droplets with the small size condensation in water and to adhere on the wall during transportation, and even block the pipeline in serious cases [13,14], the oil droplet-wall interaction in this process affects the transport efficiency of the mixed transportation pipeline. The scientific problem involved in the above process is that oil droplets in water rise and collide with the solid wall. Although this is a fundamental problem, it occurs widely in liquid-fluid systems in the petrochemical industry. It is important to clarify the physical mechanism in this process to improve oil-water separation in agglomerated plates and optimize low-temperature gathering technology.

In recent years, Liu et al. [15] proposed the phase field method to numerically simulate the motion of individual colloidal droplets in water. An analysis suggests that the adhesion process of gelled crude oil droplets is dominated by the buoyancy, drag and flow shear forces and provides the trajectory of the gelled crude oil droplets. Similarly, Zhang et al. [16] maintained that during crude oil gathering at low temperatures. The gravity, buoyancy, drag and flow shear control the movement and adhesion of oil droplets. They considered the flow field, but ignored the interaction of the droplet with the surrounding fluid during acceleration process.

On the other hand, based on the gravitational oil-water separation process, Lu et al. [17,18] considered the mechanical characteristics of a single droplet during the ascending or descending stage and established a dynamic equilibrium equation according to the gravity, flow resistance, and buoyancy acting on the droplet during its movement. However, they did not consider the effect of additional force on the droplets, which results in the model's prediction accuracy of the movement trajectory is largely limited. Meon et al. [19] argued that in the oil removal process by coalescing plate packings, the collision process of oil droplets mainly relies on the effects of gravity, buoyancy, and flow resistance, and based on this, a motion model of the oil droplet was established to further analyze the droplet's dynamics properties. Similar to the study of Lu et al. [17] the study still ignores the added mass force, while not considering the effect of the near-wall on the motion of oil droplets.

In summary, there is a relative lack of research into the mechanic properties under the influence of walls, and few researchers have used staged mechanical analysis to illustrate droplet-wall collisions. And researchers always ignore the interaction be-tween the droplet and the surrounding fluid and the wall's effect. However, the force mechanism of droplets near the wall is precisely the key to solving practical engineering problems. The adhesion process of oil droplets is dynamically coherent and the speed of movement varies from moment to moment, i.e. there are inertial forces, the analysis process needs to consider the rising phase and the interaction phase with the wall to ensure the continuity and integrity of the droplet motion.

Therefore, to address the above problems, this paper adopts a research method combining microscopic experiments and theoretical analysis to conduct an in-depth dynamics study on the oil droplets in water colliding vertically with a rigid wall. And we analyzed the variation laws of oil droplet velocity, motion trajectory, and mechanical properties, et al., under different control parameters, which can not only reveal the physical mechanism in the process, but also provide a theoretical basis for the analysis of oil droplet adhesion, migration and shedding, and provide a sub-model reference for complex oil-water systems.

Nomenclature

| Physical Parameters | Specific meaning |
|---|---|
| $\rho_d$ | Dispersed phase density, kg/m$^3$ |
| $\rho_c$ | Continuous-phase density, kg/m$^3$ |
| R | Droplet radius, m |
| $C_D$ | Drag coefficient |
| $C_m$ | Virtual mass coefficient |
| $r_{max}$ | Maximum radial position, m |
| v | Initial velocity, m/s |
| $\mu_c$ | Dynamic viscosity, pa·s |
| $\sigma$ | Interfacial tension, N/m |

## 2. Experimental parts

### *2.1. Experimental equipment and materials*
We have set up an experimental system for the microscopic visualization of oil drop-let collisions with walls in water, as given in Figure 1. The complete system consists of the following parts: a self-made sample tank, micro-sized droplet preparation system and high-speed microscope camera with lenses of different magnifications. For ease of observation and photography, the sample tank is made of Plexiglas panels. We used silicone oil as the dispersed phase with a density and viscosity of 961 kg/m$^3$ and 10mpa·s, respectively, and deionized water as the continuous phase with a density of 997 kg/m$^3$ and a viscosity of 0.897mpa·s. The interfacial tension between oil and water is 0.032 N/m.

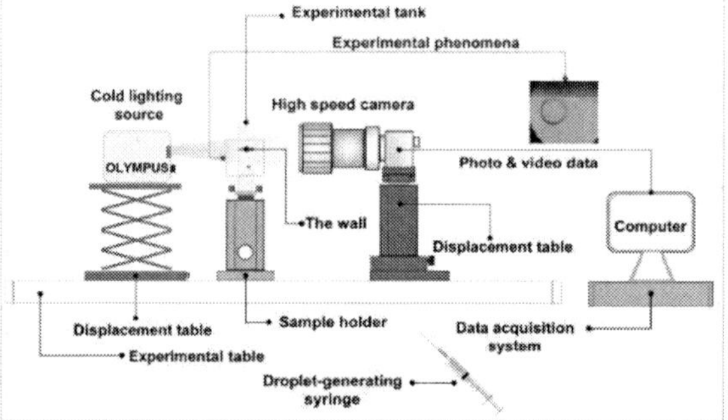


Fig. 1. Schematic diagram of the visualized experimental setup

The different volumes of microdroplets prepared, converted into oil droplet radii using the volume equation, are expressed as follows.

$$R=\sqrt[3]{\frac{3V}{4\pi}}\#(1)$$

### *2.2. Feature parameters*
For better identification and analysis of the movement of oil droplets, we define two control parameters, Reynolds number (Re) and Weber number (We) of the droplet as a way to describe the initial impact state of the droplet. In addition, we calculate the speed of movement from the coordinates of the position of the oil droplets in two pictures at a certain time interval.

$$Re=\frac{\rho_c v D_0}{\mu_c}\#(2)$$

$$We=\frac{\rho_c v^2 D_0}{\sigma}\#(3)$$

### *2.3. Experimental image of the droplet collision*
Figure 2 shows the movement of oil droplets rising and colliding with the wall. Due to the gradual increase in the droplet size, the observed field of view was not conducive to recording the motion of the droplets and we changed the lens during the experiment. In Figure 2(a) and 2(b), the oil droplets did not deform and bounce during the whole collision process. While as the droplet size increases to a certain degree, i.e., in Figure 2(c) and 2(d), the bounce phenomenon of the droplet occurs twice to consume the excessive initial kinetic energy. This phenomenon would be more pronounced as the size increases. Furthermore, if the initial position of the droplet is kept the same, the larger the droplet size, the shorter its approach time, and the longer the drainage time instead.

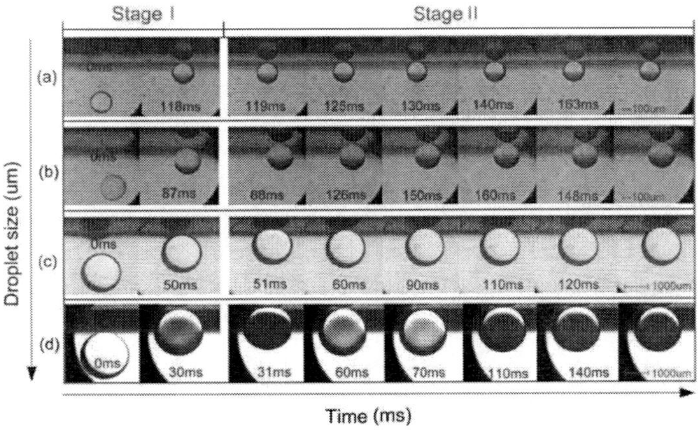


Fig. 2. Movement of oil droplets driven by buoyancy. (a) R=0.363mm.(b) R=0.457mm.(c) R=0.894mm.(d) R=1.06mm

### 2.4. *Motion stage analysis*

We separate the collision process into the approach stage (Stage Ⅰ) and the drainage stage (Stage Ⅱ) according to the variation law of barycenter position of the droplet with R=0.894mm as an example, as shown in Figure 3. In the Stage Ⅰ, the droplet maintains the same velocity and approaches the wall, which is not affected by the wall at this time, and the buoyancy and flow resistance maintain equilibrium together to regulate the droplet's motion. In the Stage Ⅱ, due to the obstruction by the wall, the force equilibrium state of the oil droplet is broken, the effect of drainage resistance and added mass force is highlighted, and the velocity of the droplet gradually decreases to 0.

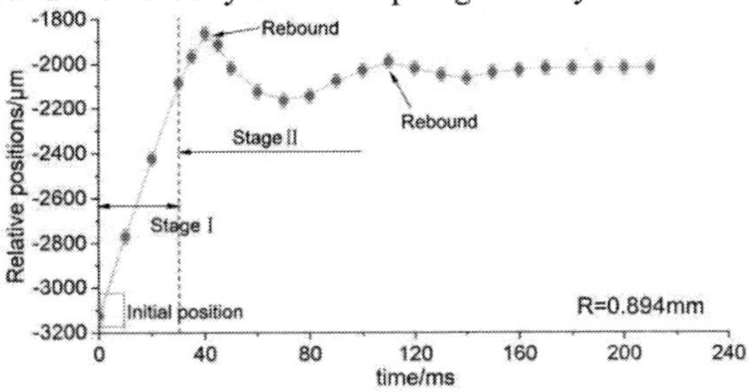


Fig. 3. Relative position of oil droplets with R=0.894mm during collision with the wall

## 3. Mathematical description

To provide a better mathematical description of this process. We establish a coordinate system. In this coordinate system, the oil droplets collide vertically with the wall at a terminal velocity. We also labelled the parameters involved in the numerical solution process as follows. h(r,t) denotes the water film distribution, $u_r$ is the radial velocity distribution, p(r,t) is the pressure distribution within the water film, $r_{max}$ is the maximum radial position of the water film distribution, and Zd is the barycenter of the oil droplet.

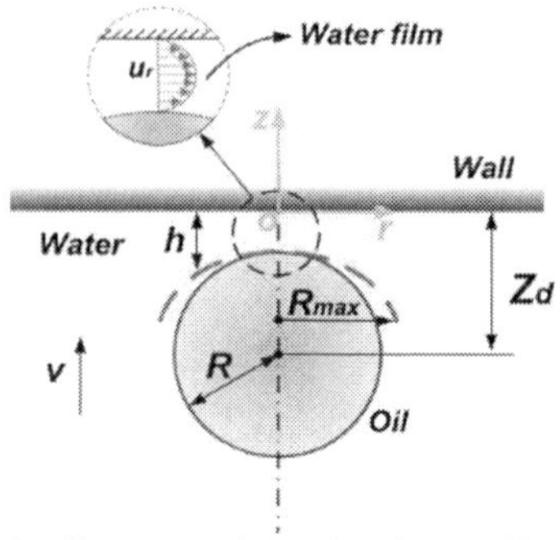


Fig. 4. Schematic diagram of the droplet collision with the wall

### 3.1. *Force model of the droplet*

First, we perform a detailed physical analysis of the process of oil droplet sticking to the wall. We assume that the initial velocity of the droplet is close to zero when it is released from the inlet and that it accelerates under the action of buoyancy and gravity, while the resistance to flow of the droplet gradually increases and the combined inertial force on the droplet gradually decreases. When the velocity increases to a certain value, the acceleration decreases to zero and the velocity of the droplet starts to remain constant. In the meanwhile, we regard the phase of uniform motion as the starting point for our force analysis. The kinetic equilibrium equation of the droplet is built based on Newton's second law. Due to the presence of gravity, the process of oil droplet sticking to the wall is first driven by a buoyancy force. This is associated with the droplet volume and density difference [20].

$$\overrightarrow{F_B}=\frac{4}{3}\pi R^3(\rho_c - \rho_d)g\#(4)$$

In order for the movement of the droplet to be in a state of equilibrium of forces. There must exist another force, namely the resistance. This force is in connection with the viscosity of the continuous phase and takes the following form [21]. In the equation below. We assume the resistance of the droplet movement process and the resistance of the spherical particles, both of which have similarities. The exact equation after collation is therefore shown below.

$$\overrightarrow{F_D}= - 6\pi(1+0.15Re^{0.687})\mu_c RV\#(5)$$

The inertia of the droplet during its ascent, impeded by the wall, means that an additional force acts on the droplet, which takes the following form [22,23].

$$\overrightarrow{F_A}=\frac{4}{3}\pi R^3\rho_c C_m\frac{dV}{dt}\#(6)$$

During the adhesion, the water film and the surface tension are the main factors influencing the drainage process. The mutual compression between the droplets and the walls causes a sudden rise in film pressure, which creates a force. Essentially, the force acts in a similar form to a resistance, hence the name of the force as drainage resistance. It has the following form [24]. Here take $r_{max} = 1.01R$.

$$\overrightarrow{F_F}= - \int_0^{r_{max}} 2\pi rp\cdot dr\ \#(7)$$

### 3.2. *Stokes-Reynolds-Young-Laplace Model*

The evolution of water films and changes in film pressure are the theoretical basis for revealing the droplet's collision dynamics. Therefore, based on the theory of thin film, we adopt the Stokes-Reynolds drainage equation to mathematically describe water film drainage [25], i.e. Equation (8).

Along with the water film drainage, the film pressure changes from moment to moment. We can describe this variation using the following equation [26].

$$\frac{\partial h}{\partial t}=\frac{1}{12\mu_c r}\frac{\partial}{\partial r}\left[rh^3\left(\frac{\partial p}{\partial r}\right)\right]\#(8)$$

$$p=\frac{2\sigma}{R}-\frac{\sigma}{r}\frac{\partial}{\partial r}\left(r\frac{\partial h}{\partial r}\right)\#(9)$$

### *3.3. Boundary conditions and model calculation*

At the beginning of the movement, no interaction between the droplet-wall occurs and no film pressure exists. When the droplet is in the vicinity of the wall, the interaction between the two is generated and the film pressure rises sharply. Combined with the coordinate system, at the initial moment, the local geometry of the droplet interface should satisfy the equation "$h(r,0)=-(H0+0.5r^2/R)$" holds. H0 indicates the initial position of the oil droplet at the initial moment. Normally, we would choose a location where the oil droplets are not affected by the wall, and also determine whether this location meets the experimental requirements. If a deformation of the droplet occurs during collision, the droplet's barycenter should satisfy the equation "$Zd=-[\ h(r_{max},0)+R-0.5r_{max}^2/R]$" holds.

At r=0, the satisfaction of $\partial h/\partial r=0$ and $\partial p/\partial r=0$ holds. We assume that the speed at which the thickness of the oil film changes is approximately equal to the velocity of the oil drop, so the boundary conditions, hence the boundary conditions, At $r = r_{max}$, the satisfaction of $\partial h/\partial t = v$, $p = 0$ holds.

The overall solution idea is to transform the partial differential equations into ordinary differential equations, which are then combined with the force equations for the oil droplets to form a system of equations, and then combined with boundary and initial conditions. The detailed steps are as follows: we first bring equation (9) into equation (8) and then discretize the spatial term on the right-hand side of the water film evolution equation using a second-order central difference format with a spatial step $\Delta r = r_{max}/N$, keeping the differential form of the time term on the right-hand side of the equation unchanged.

## 4. Results and Discussion

### *4.1. Kinetic analysis of the droplet*

Figure 5 below represents the variation law of velocity and motion trajectory of oil droplets during the rising and impacting the wall under different control parameters in deionized water. With the help of the visual experimental system, the motion trajectories of oil droplets of different sizes were photographed and the velocity in this process was recorded, while the results were compared with the theoretical calculation results, as shown in Figure 5(a) and 5(c). It can be seen that the theoretical calculation results are in good agreement with the experimental calculation results, which indicates that the dynamics characteristics of oil droplets hitting the wall can be precisely captured by the theoretical model.

For small-sized droplets, the terminal velocity of the oil droplet in the approach stage decays rapidly to 0 due to the constraint of the horizontal wall, while the distance between the oil droplet and the wall also shortens to 0, and finally reaches Zd=R, as shown in Figure 5(a) and 5(b).

For large-sized droplets, the terminal velocity (positive direction) decays rapidly to 0 after the first collision of oil droplets with the wall and then increases in the reverse direction, with the increase of Re and We numbers, the reverse maximum velocity (rebound velocity) and reverse motion amplitude also become larger. And the velocity of the oil droplet changes more rapidly from the maximum terminal velocity when it approaches the wall to the maximum rebound velocity when it starts to move away from the wall. This is since the larger the droplet size is, the stronger the constraint of the wall on the oil droplet, and the stronger the dominant effect of the added mass force and the water film-induced force on the droplet motion, as shown in Figure 5(c) and 5(d).

Moreover, further observation of the trajectory of the oil droplet reveals that its decay form is similar to viscous damping decay, but the rebound amplitude is larger than that in the viscous environment. This is because the medium damping in the deionized water environment is much less than that in the viscous environment, so the energy dissipation in the decay process is bound to be smaller than that in the viscous medium, and the interface deformation of the oil droplet collision can also store part of the kinetic energy to provide energy for the oil droplet bounce in the next moment.

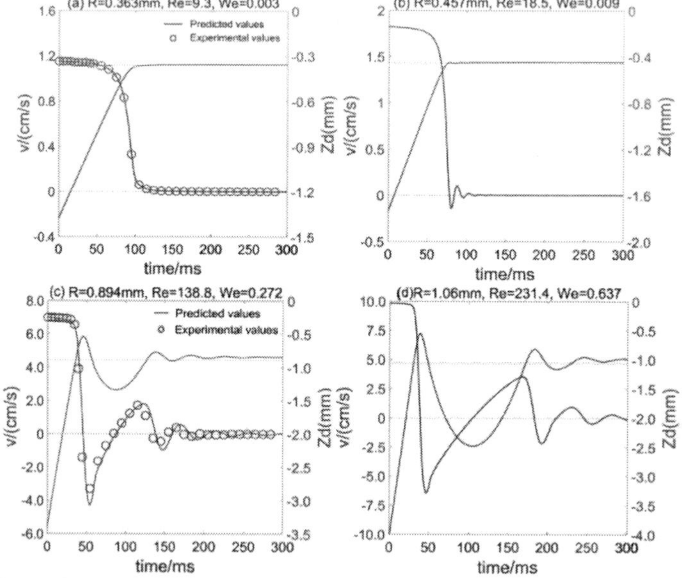


Fig. 5. Velocity and motion trajectory during the impact under different control parameters

### 4.2. *Forces on oil droplets*

Figure 6 shows the forces variation of the droplet during the impinging the wall under different control parameters, where the oil droplets are subjected to following forces, namely $F_B$ (buoyancy force), $F_D$ (flow resistance), $F_A$ (added mass force), and $F_F$ (film-induced force).In the approach stage, the oil droplet moves at terminal velocity, when the buoyancy and flow resistance are balanced. And the remaining two forces on the oil droplet as it moves away from the wall are zero. When the interaction occurs, these two forces increase dramatically, and the droplet's motion is mainly controlled by these two forces at this time. Meanwhile, the larger the Re number and We number of the drop-let, the larger the maximum film-induced force and added mass force it receives.

Specifically, at Re=9.3, Re=18.5, these two forces reach their maximum values first, and then the added mass force rapidly drops to 0, while the film-induced force falls to a level comparable to the buoyancy. When the motion tends to be stable, both two forces show small periodic changes, which is because the interface will oscillate slightly during the droplet impinging the wall, resulting in smaller fluctuations in the force, but not enough to change the droplet's motion state.

And unlike smaller droplets, at Re=138.8 and Re=231.4, both of these forces will also drop sharply after reaching the maximum value during the impact, but a local reverse change appears, which is due to the local negative pressure area near the center of the water film when the droplet bounces. When the oil droplet no longer bounces, the added mass force is 0, and film induced force decreases and equilibrates with buoyancy.

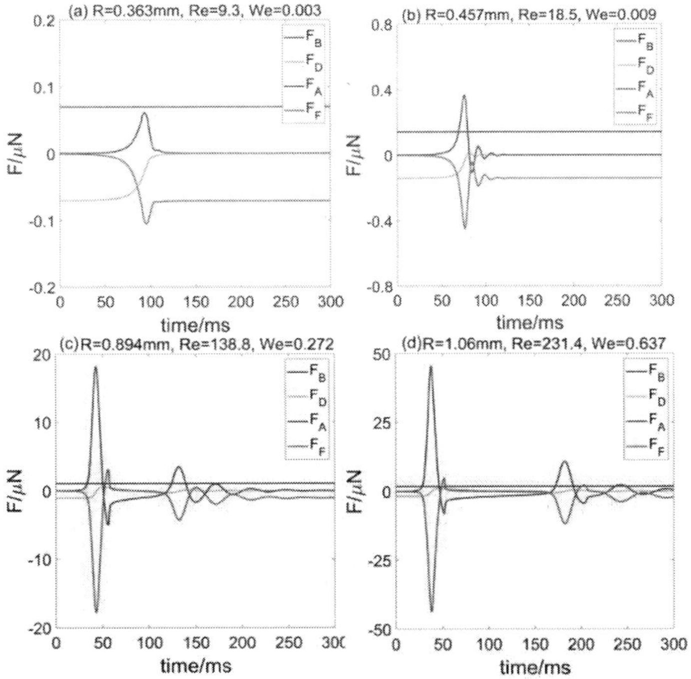


Fig. 6. Forces variation in the collision process with the wall under different control parameters

### 4.3. *Interface shape and film pressure*

Based on experimental images, when the size of the droplet reaches a certain level, the oil droplet will rebound, otherwise it will not rebound. Therefore, two representative droplet sizes, namely R=0.363mm and R=1.06mm, were selected for analysis. Figure 7 shows the interface shape and the film pressure with 80ms ~ 90ms for the droplet with R=0.363mm. During this period, the upper interface of the oil droplet does not deform and is always parabolic. This is due to the small size of the droplet, although driven by buoyancy, it does not gain very high initial kinetic energy. There-fore, when the interaction occurs, all the initial kinetic energy is dissipated due to the viscosity of the surrounding continuous phase. Meanwhile, the pressure distribution during this time period is shown on the right of Figure 7. The pressure in the film gradually increases over time, and the overall pressure distribution is in a normal curve, i.e. it has the maximum pressure at the center and gradually decreases at the sides.

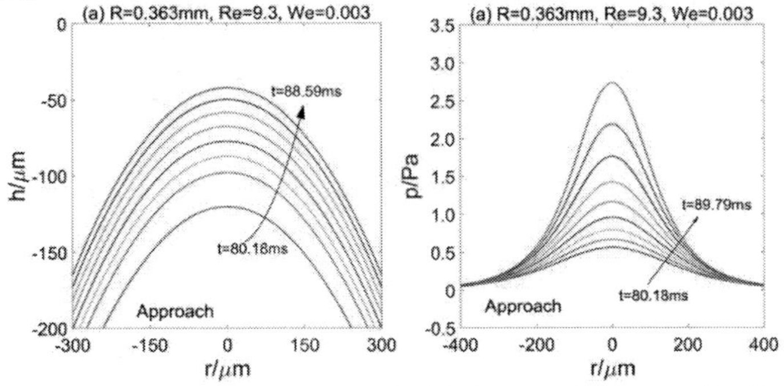


Fig. 7. In the approach stage(80ms~90ms), interfacial shape and pressure distribution near the wall for droplet size R=0.363mm

Journal of Physics: Conference Series **2594** (2023) 012045 doi:10.1088/1742-6596/2594/1/012045

Figure 8 shows the interface shape and the pressure distribution with 30ms ~ 60ms for the droplet with R=1.06mm. The barycenter of the droplet reaches the maximum deviation position as the dividing point, we separate the evolution process of the water film into the approach stage and retraction-rebound stage. As can be seen from Figure 8(A), during the period of 30ms ~ 40ms, the upper inter-face of the droplet has deformed and become very flat, and then the curvature of the interface reverses and presents a "dimple". This is because at this time, the droplet size is large and it has high initial kinetic energy. After the drainage stage, part of the energy is still not consumed, so it is gradually converted to the interface deformation energy to provide energy for the rebound at the next moment. Corresponding to the pressure distribution in the film, when the oil droplet gradually approaches the wall, the pressure in the center of the film rapidly increases and decreases along the radial direction. As the interface presents a liquid dimple, the pressure in the central position of the film starts to remain unchanged after rising rapidly. And the pressure in the film decreases along the radial direction of the water film.

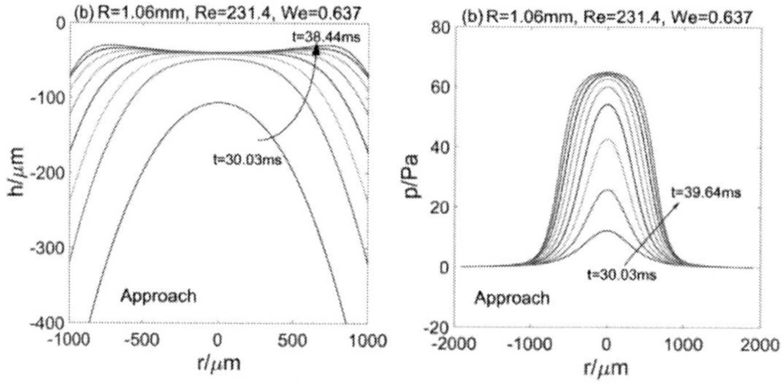


(A) In the approach stage (30ms～40ms)

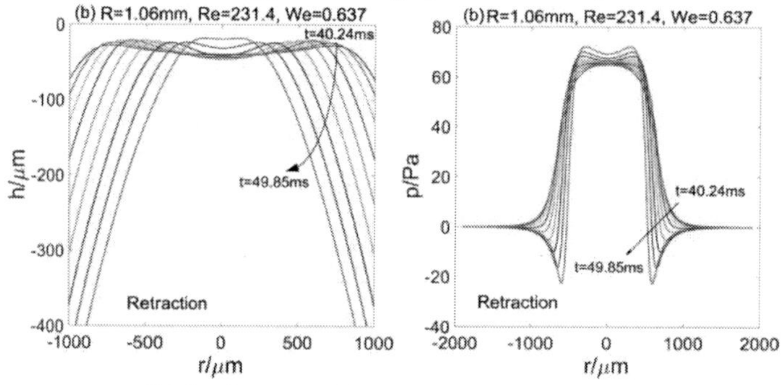


(B) In the retraction stage (40ms～50ms)

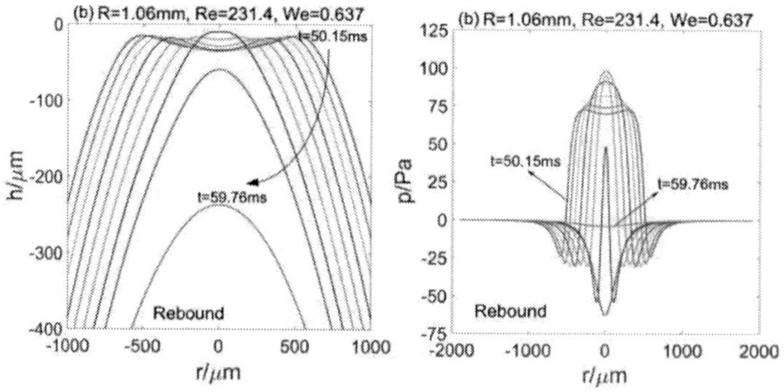


(C) In the rebound stage (50ms～60ms)

Fig. 8. Interfacial shape and pressure distribution near the wall for droplet size R=1.06mm

As can be seen from Figure 8(B), during the period of 40ms ~ 50ms, the droplet inter-face still presents a liquid dimple. Although in the retraction stage, the oil droplet centroid gradually moves away from the wall, the interface still moves toward the wall, and the radial position of the bulge at the edge of the liquid dimple moves toward the center of the water film. Furthermore, the maximum pressure will increase, and the pressure will accelerate in the radial direction. At this time, the maximum pressure was 65.28pa, which exceeded the Laplace pressure of the oil droplet (about 60.377pa). The root cause was the interfacial liquid dimple formed when the oil drop-let interacted with the wall.

As depicted in Figure 8(C), during the period (50ms ~ 60ms), when the interface at the center of the water film develops to the nearest to the wall, the oil droplet inter-face begins to move toward the barycenter, the liquid nest gradually disappears, ac-companied by the formation of a parabolic interface again, and the oil droplet re-bounds quickly. When the oil droplet rebounds for the second time, the evolution law of the upper interface of the droplet is similar to that of the first time, only the position and deformation degree of the liquid dimple will be different, which will not be repeated here. During this period, the interaction between droplet and wall surface gradually disappears, the interface returns to its initial shape, and the pressure distribution will also drop to 0. The pressure distribution law of oil droplets in the second rebound is similar to that in the first rebound, except that the pressure distribution position will change with the change in the position of the liquid dimple.

In addition, during the period of 40ms ~ 60ms, it can also be found from the pressure distribution, there will be negative pressure areas in the location of the liquid dimple and some areas outside. The negative pressure zone will develop towards the center of the water film with the rebound of the oil droplets, producing an attractive effect and elongating the oil droplets, so as to make the interface close to the wall, which is also the fundamental reason for the reverse direction of the film induced force at some moments.

## 5. Conclusion

In this paper, we study the vertical collision of the droplet with the rigid plate (Re=9.3 ~ 231.39 and We=0.003 ~ 0.637) by employing microscopic experiments and theoretical analysis, and analyze the dynamics of droplets under different control parameters.

(1) The coupling model composed of the SRYL model and the force model of the oil droplet can predict the velocity and motion trajectory of the oil droplet in the process of hitting the wall under different Re number and We number. The calculated results are in good agreement with the experimental data, and can capture the dynamics behavior of the oil droplet near the wall.

(2) When $Re \leq 27.8$, $We \leq 0.019$, the oil droplets do not bounce, the collision process interface is always parabolic in shape, the pressure takes the shape of a normal curve.

(3) When $Re \geq 46.3$ and $We \geq 0.044$, the oil droplets bounce back and the inter-face takes the shape of a dimple during collision, and the pressure distribution pattern has an important correlation with the shape of the interface.

(4) The collision process is divided into Stage Ⅰ and Stage Ⅱ. In the Stage Ⅰ, buoyancy and flow resistance together dominate the droplet motion. In the Stage Ⅱ, added mass force and the induction force play a dominant role, and as the droplet size and Re number increase, the maximum value of these two forces increases, eventually reaching a balance between the buoyancy and film induced force.

(5) In industrial processes, chemicals are commonly found in oil-water systems and their influence on the collision process of liquid droplets can't be ignored, especially due to a series of problems caused by concentration, such as interfacial concentration and interfacial viscoelasticity, which should be introduced into the new theoretical model in order to better guide actual production.

**Acknowledgments.** This work was supported by the National Natural Science Foundation of China "Mechanism of Oil Droplet Adhesion and Water Film Drainage in near-wall Flow Field" (No. 52174067).

**References**

[1] Frising T, NoïkC, Dalmazzone C. The Liquid/Liquid Sedimentation Process: From Droplet Coalescence to Technologically Enhanced Water/Oil Emulsion Gravity Separators: A Review[J]. Journal of Dispersion Science & Technology, 2006, 27(7):1035-1057.
[2] Fahim M A, Akbar A M. Removal of fine oily hazes from wastewater using deep fibrous bed Coalescer[J]. Environmental Letters, 1984, 19(3):299-319.
[3] Huang W, He X, Deng C, et al. Study on the Intensification Mechanism of Oil-water Separation Process by Using Inclined Plate Pack[J]. Advanced Engineering Sciences, 2017.
[4] Wang G, He L, Lu Y, et al. Study on oil-water separating behavior of gravity separator[J]. Acta Petrolei Sinica, 2006, 27(6):112-115.
[5] Zhao C, Zhou J, Yan Y, et al. Application of coagulation/flocculation in oily wastewater treatment: A review[J]. Science of The Total Environment, 2021, 765: 142795.
[6] Zhang Y, Huang Q, Cui Y, et al. Estimating wall sticking occurrence temperature based on adhesion force theory[J]. Journal of Petroleum Science and Engineering, 2020, 187: 106778.
[7] Han Y, He L, Luo X, et al. A review of the recent advances in design of corrugated plate packs applied for oil-water separation[J]. Journal of Industrial and Engineering Chemistry, 2017, 53: 37-50.
[8] Cao J, Wei L I, Zheng Y. Study on Separation Function of New Oil-water Separator with Corrugated Plates packing[J]. Filter & Separator, 2001.
[9] Chen J, Zhou Y, Zhou C, et al. A durable underwater superoleophobic and underoil superhydrophobic fabric for versatile oil/water separation[J]. Chemical Engineering Journal, 2019, 370: 1218-1227.
[10] Xu P, He L, Yang D, et al. Blocking characteristics of high water-cut crude oil in low-temperature gathering and transportation pipeline[J]. Chemical Engineering Research and Design, 2021(3–4).
[11] Gao W, Cheng Q, Huang Z, et al. Research and application on the gathering and transportation radius chart of multi-phase pipelines in high water cut stage[J]. Case Studies in Thermal Engineering, 2020, 21: 100654.
[12] Zhao Y, Yao L, Liu L, et al. Optimization of energy saving gathering and transportation mode in a block of Tahe oilfield[J]. Case Studies in Thermal Engineering, 2019, 13: 100378.
[13] Zhao C, Zhou J, Yan Y, et al. Application of coagulation/flocculation in oily wastewater treatment: A review[J]. Science of The Total Environment, 2021, 765: 142795.
[14] Zhang Y, Huang Q, Cui Y, et al. Estimating wall sticking occurrence temperature based on adhesion force theory[J]. Journal of Petroleum Science and Engineering, 2020, 187: 106778.
[15] Liu X Y, Zhao C C, Xiao-Qing L I, et al. Numerical Simulation on Influence of Interfacial Tension Coefficient of Water-Oil Two-Phase Flow on the Gelled Crude Oil Particle Motion State[J]. Contemporary Chemical Industry, 2014(2):301-304.
[16] Zhang Y, Huang Q, Cui Y, et al. Estimating wall sticking occurrence temperature based on adhesion force theory[J]. Journal of Petroleum Science and Engineering, 2020, 187: 106778.
[17] Lu Y. Droplets Dynamics Analyses in Gravity Oil-Water Separation[J]. OIL-GASFIELD SURFACE ENGINEERING, 1998, 000(004):1-5.
[18] Lu Y, Pan Y, Xue D. Factors Affecting Droplet Motion in Gravity Oil-Water Separation Devices[J]. Oil-Gasfield Surface Engineering, 1993, 000(003):1.
[19] Meon W, Rommel W, Blass E. Plate separators for dispersed liquid—liquid systems: hydrodynamic coalescence model[J]. Chemical Engineering Science, 1993, 48(1): 159-168.
[20] Luo S, Yong Yu. Force Balance Model and Bubble Rebound Law for Bubble-Wall Collision[J]. Physics of Gases, 2019.
[21] Yong Yu, Luo S. Study of Dynamic model for Bubble impact and Bounce From rigid wall[J]. Transactions of Beijing Institute of Technology, 2020.
[22] Kharlamov A A, Chara Z, Vlasak P. Hydraulic formulae for the added masses of an impermeable sphere moving near a plane wall[J]. Journal of Engineering Mathematics, 2008,

62(2):161-172.
[23] Manica R, Klaseboer E, Chan D. Force Balance Model for Bubble Rise, Impact, and Bounce from Solid Surfaces[J]. Langmuir, 2015, 31(24):6763-72.
[24] Kumar K, Nikolov A D, Wasan D T. Effect of Film Curvature on Drainage of Thin Liquid Films[J]. Journal of Colloid & Interface Science, 2002, 256(1):194-200.
[25] Zhang X, Manica R, Tchoukov P, et al. Effect of Approach Velocity on Thin Liquid Film Drainage Between an Air Bubble and a Flat Solid Surface[J]. Journal of Physical Chemistry C, 2017, 121(10):5573-5584.
[26] Chan D, Klaseboer E, Manica R. Theory of non-equilibrium force measurements involving deformable drops and bubbles[J]. Advances in Colloid & Interface Science, 2011, 165(2):70-90.

# Differential Diagenesis and Pore Evolution Characteristics of Different Sedimentary Microfacies Reservoirs of Bioclastic Limestone: A Case Study of MB2-MC1 Section in X Oilfield in The Middle East

**Yi-chang YU1,2(✉), Zhi-wei LI3, Bo-heng SHEN2, Rui-rui WANG1, Shi-qi SONG1, Rui GUO2, Feng-feng LI2, Jiu-zhou WANG1**

1 China National Oil and Gas Exploration and Development Company Ltd, Beijing, China
2 PetroChina Research Institute of Petroleum Exploration & Development, Beijing, China
3 Research Institute of Petroleum Processing, Sinopec, Beijing, China

yuyichang@petrochina.com.cn

**Abstract**. In order to clarify the differential diagenesis of different sedimentary microfacies of bioclastic limestone in Cretaceous M Formation in the Middle East and systematically analyze the characteristics of reservoir pore evolution, the data of regional geology, well logging, core, cast thin section, scanning electron microscope, three-dimensional CT and conventional porosity and permeability test are comprehensively used, taking MB2-MC1 section in X oilfield as an example, starting from the diagenesis and pore evolution of bioclastic-arene shoal, the characteristics of diagenesis of each sedimentary microfacies in different environments are analyzed, and the differential diagenesis, pore evolution process and its influence on physical properties are studied. It is showed that the reservoirs of MB2-MC1 section are mainly developed in the ramp shoal and platform margin shoal. The shoal facies can be further subdivided according to the type of bioclastic debris. The sedimentary microfacies determine the original fabric and bioclastic type of M Formation, and then control the diagenetic sequence and strength. Each sedimentary microfacies mainly experiences different degrees of mud crystallization and burrows in seawater environment, dissolution and cementation in atmospheric freshwater environment, compaction, pressure solution and dolomitization in burial environment. High energy microfacies have the characteristics of non-selective dissolution, mainly developing intergranular pores and intergranular dissolved pores. Medium energy microfacies have the characteristics of selective dissolution, mainly developing mould pores and intragranular pores. The medium-low energy microfacies have the characteristics of moderate cementation, moderate compaction and moderate-weak dolomitization, mainly developing intragranular pores and visceral foramen pores. The low energy microfacies have the characteristics of strong cementation, strong compaction and moderate dolomitization, mainly developing micropores, intercrystalline pores and visceral foramen pores, which are non-reservoirs.

**Keywords**. The Middle East, Bioclastic limestone, Sedimentary micro-facies, Differential diagenesis, Pore evolution

## 1. Introduction

With the continuous growth of China's oil and gas demand and the progress of technical measures, the domestic development has gradually transformed to the deep onshore, carbonate rocks and unconventional petroleum, and the scale of cooperation in overseas oil and gas exploration and development has also been expanding [1-5]. In China, high-quality reservoirs under the overall low-permeability tight background in deep clastic rocks and carbonate rocks are favorable targets for exploration and development. One of the research focuses is to restore the formation process and pore evolution characteristics of reservoirs based on laser U-Pb dating technology [6-8]. In oversea, the Middle East has become China's most important international production base. Different from China 's fractured-vuggy carbonate reservoirs and deep dolomite reservoirs, the carbonate particles are debris after the death and fragmentation of different organisms. The original fabric is well preserved, and the reservoir space are mainly various pores with relatively good physical properties [9-13]. However, the hydrodynamic conditions of different sedimentary environments of bioclastic limestone are quite discrepant, the types of microfacies are complex and diverse, and the diagenesis characteristics and strength of different sedimentary microfacies are inconsistent. Therefore, to clarify the characteristics of differential diagenesis and pore evolution is necessary [14-16].

The researches on diagenesis and pore evolution mainly focus on deep clastic rocks, especially the pore evolution characteristics and the genesis and preservation mechanism of high porosity reservoirs in tight sandstone and shale [17-23]. For carbonate reservoirs, especially bioclastic limestone in the Middle East, due to the limitation of research data, there is a lack of research on pore evolution. In addition, the study of reservoir differential diagenesis and pore evolution characteristics of different sedimentary microfacies is rarely involved [24-26]. Therefore, based on the abundant well logging, core, casting thin section, scanning electron microscope, three-dimensional CT and conventional porosity and permeability test data in X oilfield in the Middle East, the diagenesis characteristics, differential diagenesis and pore evolution process of each sedimentary microfacies in different diagenetic environments of MB2-MC1 section with the best physical properties in M Formation are systematically studied, in order to provide guidance to the distribution prediction and development optimization adjustment of reservoirs.

## 2. Geological Background

X oilfield is located in Missan Province, southeastern Iraq. Since the Cambrian, the area where the X oilfield is located has been situated in the north-ern margin of the ancient Gondwana continent for a long time, where platform sediments are mainly developed. The tectonic activity in Cretaceous is generally weak, and it is in a warm and humid low-latitude area. The climate is sunny, high in oxygen content, good in seawater circulation, and suitable for biological growth. It mainly includes benthic foraminifera, non-fixed bivalves, rudist and echinoderms. Its unique external morphology, internal structure and particle size are the most important signs for identifying bioclastic types [27-28]. After the death and fragmentation of these organisms, they are elutriated and sorted by seawater to become carbonate particles, forming a large-scale bioclastic limestone of M Formation, which is in integrated contact with the underlying mudstone of R Formation and unconformity with the overlying K Formation (see Fig. 1).

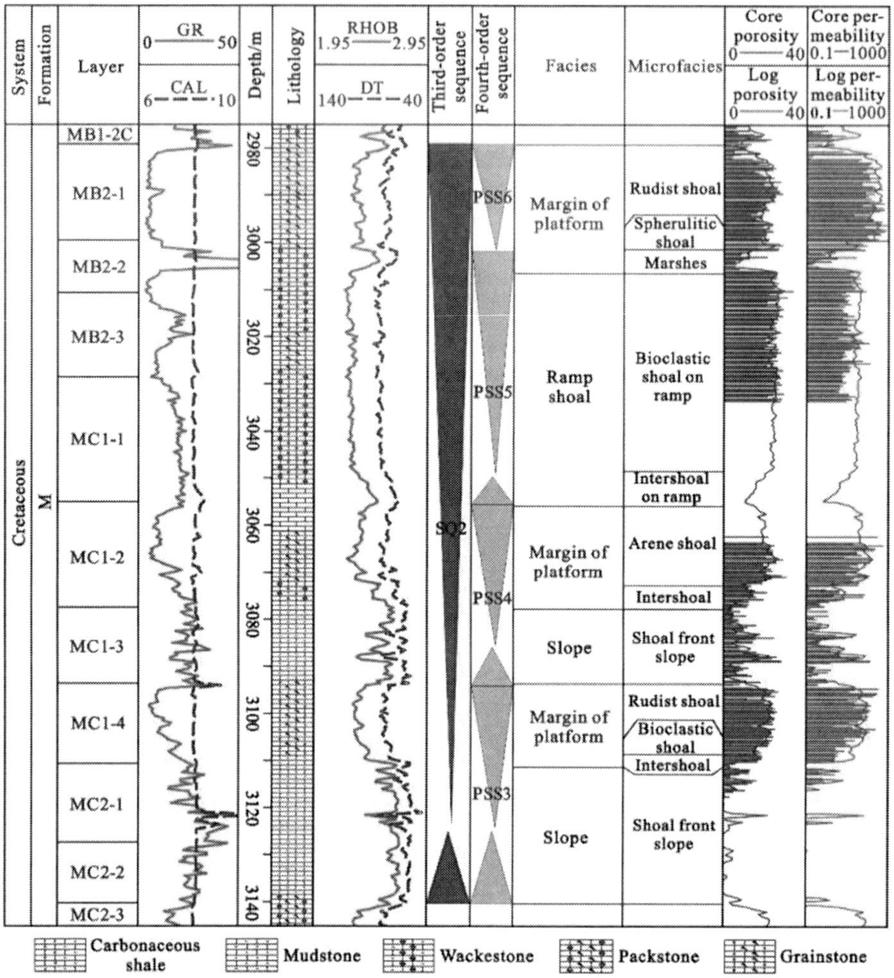


**Fig. 1.** Single well sequence sedimentation and physical properties of MB2-MC1 section in X oilfield.

The reservoir physical properties of MB2-MC1 section in X oilfield have obvious facies-controlled characteristics, and the migration and evolution of different sedimentary microfacies make the distribution of reservoir physical properties vary greatly. In addition, the diagenesis characteristics of each sedimentary microfacies are different. Under the dual influence of sedimentary process and diagenesis, the reservoirs have the characteristics of diverse pore types and complex pore throat structures. The study area is a typical porous reservoir, including primary pores such as micropores, visceral foramen pores, biological framework pores, intergranular pores (see Fig. 2a~Fig. 2d) and secondary pores such as intercrystalline pores, intergranular dissolved pores, intragranular dissolved pores, mould pores, dissolved vugs (see Fig. 2e~Fig. 2h). Core 3D CT scan shows that the main reservoir space is pores of different genesis, and microfractures account for a small proportion (see Fig. 2i).

## 3. Diagenesis and pore evolution characteristics of bioclastic-arene shoal

### *3.1. Characteristics of diagenesis of bioclastic-arene shoal*

Compared with the short sedimentary period, the diagenesis process is long and complex. Almost all high-quality reservoirs are inseparable from pore-increasing diagenetic transformation. Even for facies-controlled reservoirs, a large number of secondary pores are also formed in diagenesis, but the main controlling factor determining the distribution of reservoirs is sedimentary facies. With the increasing reserves and production of carbonate reservoirs, the research technology of carbonate diagenesis has been continuously innovated, and the research methods have also transformed from static observation statistics to dynamic restoration simulation, and the research accuracy and fidelity have been continuously improved.

**Fig. 2.** Reservoir pore types and characteristics of MB2-MC1 section in X oilfield.

The diagenesis types of bioclastic limestone in MB2-MC1 section of M Formation are diverse, and the pore-increasing diagenesis includes burrows, neomorphism and dissolution. Pore-reducing diagenesis consists of mud crystallization, cementation, compaction, pressure solution and dolomitization (see Fig. 3).

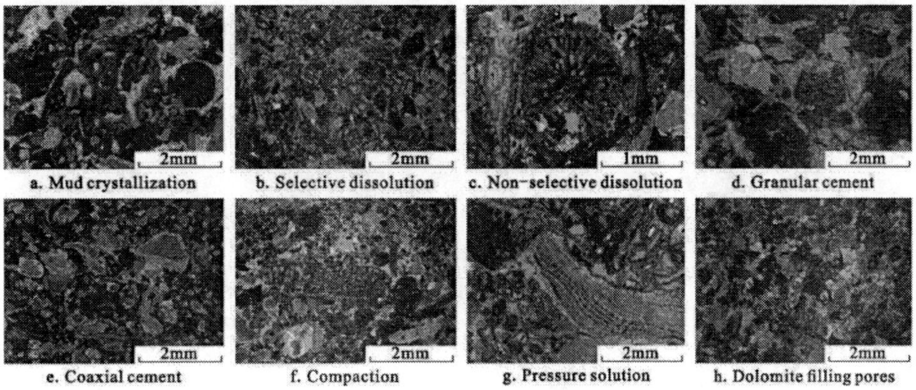


**Fig. 3.** Types and characteristics of diagenesis of bioclastic-arene shoal of MB2-MC1 section in X oilfield.

The predecessors have systematically studied the sedimentary characteristics, diagenesis evolution and the controlling effect of sedimentary diagenesis on pore evolution of the rudist shoal on the margin of platform in X oilfield [25]. This paper takes the bioclastic-arene shoal as an example to analyze its diagenesis characteristics in sea-water environment, atmospheric freshwater environment, shallow burial environment and medium burial environment. In the seawater environment, the pore fluid of bioclastic-arene shoal is normal or modified seawater, which is supersaturated to most carbonate minerals. It is characterized by biological activities and seawater cements, mainly mud crystallization, burrows and seawater cementation (see Fig. 3a). In the atmospheric freshwater environment, diagenesis is characterized by neomorphism, dissolution and cementation. When unstable bioclastics coexist with stable bioclastics, unstable bioclastics are preferentially dissolved, occurring selective dissolution (see Fig. 3b).

With the process of dissolution, after the mineral stabilization, the atmospheric freshwater expands the original pore and non-selective dissolution occurs (see Fig. 3c). The dissolution forms secondary pores, which also takes away the $Mg^{2+}$. The dissolution products are precipitated in the freshwater hyporheic zone, as a result, the equiaxed granular cements are formed in the pores (see Fig. 3d). In addition, coaxial cements are developed on the surface of echinoderm debris and arene debris (see Fig. 3e). In the burial environment, the free exchange between the diagenetic fluid and the chemical active gas in the atmosphere such as $CO_2$ is cut off, and the temperature and pressure gradually increase. Therefore, compaction occurs, and the long biological debris will form some microcracks (see Fig. 3f). With the raise of burial depth, the compaction continues, the stress concentration at the contact of particles occurs pressure solution, and the dissolved substances precipitate on the surface of other particles (see Fig. 3g). In addition, due to the influence of $CO_2$ in the carbonate system, the increase of temperature will reduce the solubility of carbonate minerals, which is conducive to dolomitization in the burial environment (see Fig. 3h).

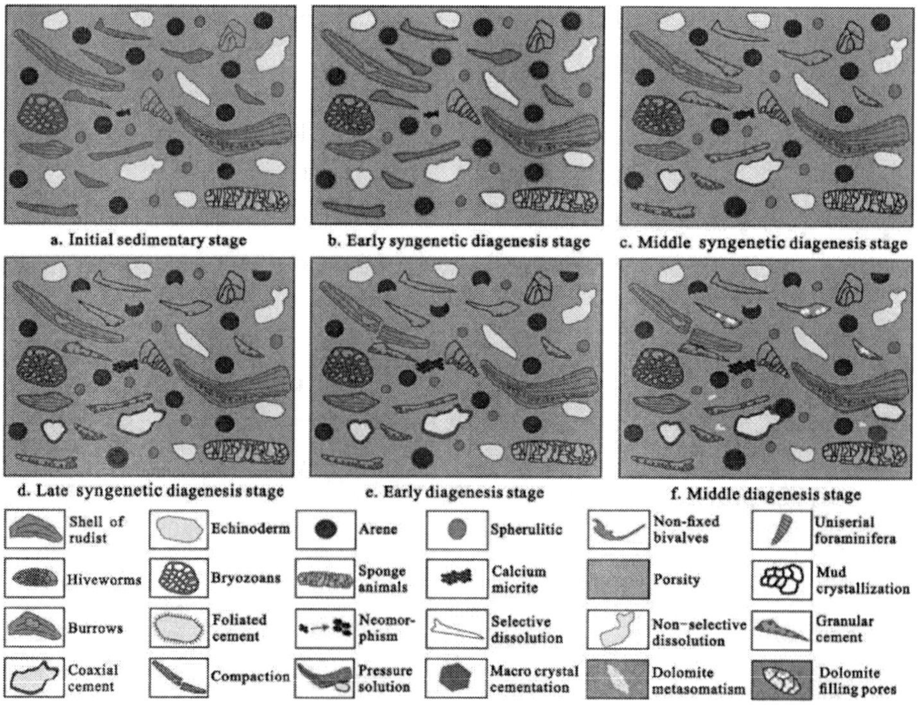


**Fig. 4.** Diagenesis evolution pattern of bioclastic-arene shoal of MB2-MC1 section in X oilfield.

### 3.2. *Pore evolution characteristics of bioclastic-arene shoal*

After the deposition of bioclastic-arene shoal, it experiences syngenetic stage, early diagenesis and middle diagenesis stages. In the initial sedimentary stage, the bioclastic-arene shoal has low micrite content, coarse grain size, a large number of primary pores, and strong fluid seepage capacity (see Fig. 4a), resulting in significant diagenesis in the syngenetic stage. In addition, in the whole diagenesis stage, the regional structure activity of X oilfield is relatively stable and the faults are not developed.

During the early syngenetic diagenesis stage, in the seawater environment, the microorganisms repeatedly perforate the carbonate shell from the outside to the inside, and then are filled with micrite, which slightly reduces the porosity. Because the micrite sleeve has a smaller particle size than the original micrite, it is darker under the microscope. It may serve as a barrier for fluid flow during the later diagenesis process, so that the internal structure of the shell can be preserved. In addition, microorganisms drill the shell of the rudist to form a small number of isolated intragranular pores. In the bioclastic-arene shoal area with strong hydrodynamic conditions, waves can promote the rapid flow of supersaturated seawater. Due to the continuous contact between sediments and seawater, the surface of the shell and particles forms fibrous and bladed rimmed cements, which block pores and reduce porosity (see Fig. 4b, Fig. 5).

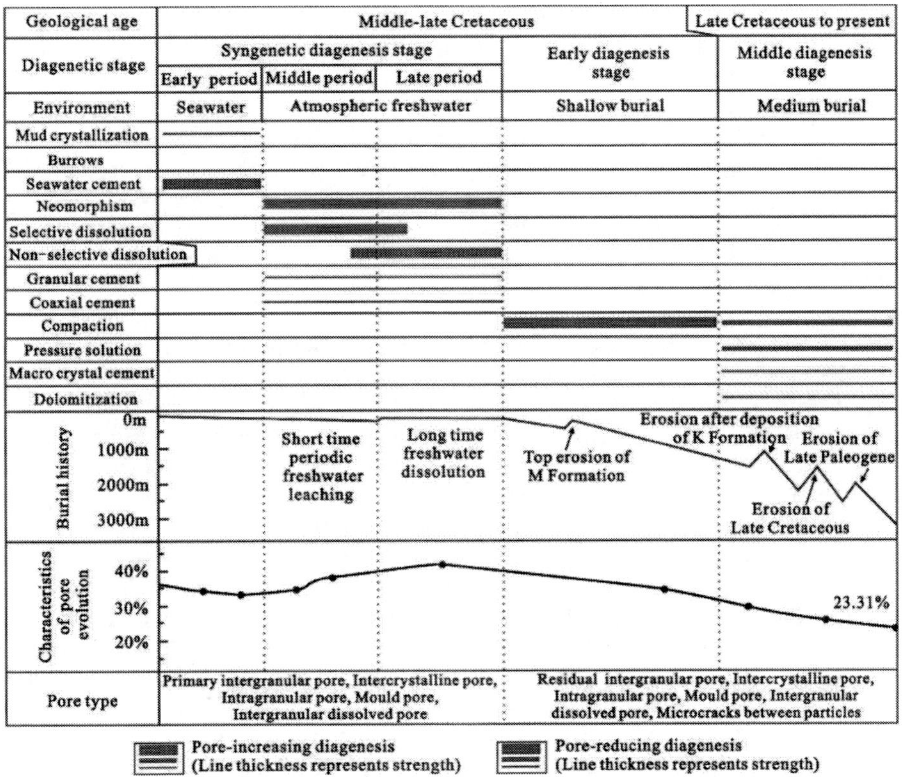


**Fig. 5.** Diagenesis sequence and pore evolution characteristics of bioclastic-arene shoal of MB2-MC1 section in X oilfield.

During the middle syngenetic diagenesis stage, since the bioclastic-arene shoal accumulates rapidly, the sediments are easily exposed to the atmospheric freshwater environment, which is subjected to short-term periodic leaching and neomorphism. The aragonite non-fixed bivalves are selectively dissolved to form mould pores. At the same time, the whole aragonite fibrous cements and most of the bladed cements formed in the seawater environment are dissolved, as a result, the porosity is greatly increased. The influence depth of dissolution in this stage is relatively limited, which has a lager influence on the upper part of the bioclastic-arene shoal. At the same time, the dissolution products provide the material basis for the precipitation in the freshwater undercurrent zone. Therefore, a small number of pores are filled with equiaxed granular cements. The echinoderm debris and arene debris are mostly calcite single crystals, thus the cement grows coaxially around its outer ring, reducing reservoir porosity (see Fig. 4c, Fig. 5).

During the late syngenetic diagenesis stage, the sea level decline cycle ends, and after the overall sedimentation of bioclastic-arene shoal, it experiences a long period of exposure. The dissolution of bladed cements continues. At this time, the stable rudist debris and echinoderm debris are dissolved, the grain boundary is uneven, and the dissolution harbor appears, which expands the original pores and forms intergranular dissolved pores, resulting the significant increase of porosity. The depth of dissolution in this stage is controlled by the exposure time. The longer the exposure time is, the more sufficient the atmospheric freshwater leaching is, and the greater the influence depth is. The calcium carbonate produced by dissolution precipitates elsewhere (see Fig. 4d, Fig. 5).

During the early diagenesis stage, the content of cement in the bioclastic-arene shoal reservoir is low, and it is difficult to weaken the compaction. In addition, some of the rudist de-bris and sponge animals are long strips, and non-selective dissolution forms irregular boundaries. Under the load pressure of the overlying strata, the pore fluid is reduced, the particles are mostly in line contact or point-line contact, the porosity is reduced, and the long strip rudist debris and sponge animals are cracked at the point with a smaller thickness, but its pore-increasing effect is limited. The compaction reduces the porosity of the bioclastic-arene shoal to a certain extent (see Fig. 4e, Fig. 5).

During the middle diagenesis stage, the compaction continues under the stress of the overlying strata, but the compaction strength is smaller than that of the shallow burial environment. The stress at the contact point of the particles is locally concentrated, thus the pressure solution occurs. The particles are in concave-convex or suture contact, the porosity is reduced, and coarse grain cementation and dolomitization are developed in some pores. The diagenesis of this stage is mainly pore-reducing, while the overall diagenesis is not strong, thus the porosity decline is limited (see Fig. 4f, Fig. 5).

## 4. Diagenesis characteristics of each sedimentary microfacies in different environments

### *4.1. Characteristics of initial sedimentary state*

The original fabric and bioclastic type, content and composition of different sedimentary microfacies are quite different. At the same time, the medium-high energy microfacies located at the high part of the palaeogeomorphology during the sedimentary period are also located at the high part of the geomorphology during the diagenetic period, which are more likely to accept the leaching of atmospheric freshwater, and the leaching products can be rapidly discharged to the surrounding area. However, the medium-low energy microfacies locate at the lower part of the palaeogeomorphology during the sedimentary period, and the atmospheric freshwater leaching is weak, and will accept the precipitation of the dissolution products form the high part of the palaeogeomorphology, thus cementation occurs. These comprehensive factors make the sedimentary microfacies have obvious differential diagenetic characteristics in seawater environment, atmospheric freshwater environment and burial environment. Like the bioclastic-arene shoal, the diagenesis sequence and pore evolution of other sedimentary microfacies are studied from the whole evolution process of sedimentation-diagenesis. In order to avoid the repetition of each microfacies, the diagenesis characteristics of different microfacies are summarized according to the initial sedimentary state, seawater diagenesis environment, atmospheric freshwater diagenesis environment, shallow burial diagenesis environment and medium burial diagenesis environment (see Fig. 6~ Fig. 10).

In the initial sedimentary stage, from the high energy microfacies included rudist shoal on the margin of platform, bioclastic-arene shoal and bioclastic-spherulitic shoal (see Fig. 6a, Fig. 6b), to the medium energy microfacies included rudist bioclastic shoal on ramp, bioclastic shoal (see Fig. 6c~Fig.

Journal of Physics: Conference Series **2594** (2023) 012046 doi:10.1088/1742-6596/2594/1/012046

6e), to the medium-low energy microfacies included intershoal (see Fig. 6f, Fig. 6g), and finally to the low energy microfacies included fine-grained deposition on shoal front slope and incision filling (see Fig. 6h, Fig. 6i), the bioclastic content gradually decreases, and the major bioclastic types change from rudist and echinoderms to non-fixed bivalves and benthic foraminifera. The particle composition is from medium-low magnesium calcite to high magnesium calcite and aragonite. The content of micrite gradually increases. The original pores are dominated by intergranular pores to visceral foramen pores and micropores. The degree of pore connectivity gradually decreases, and the seepage capacity of diagenetic fluid becomes worse (see Fig. 6).

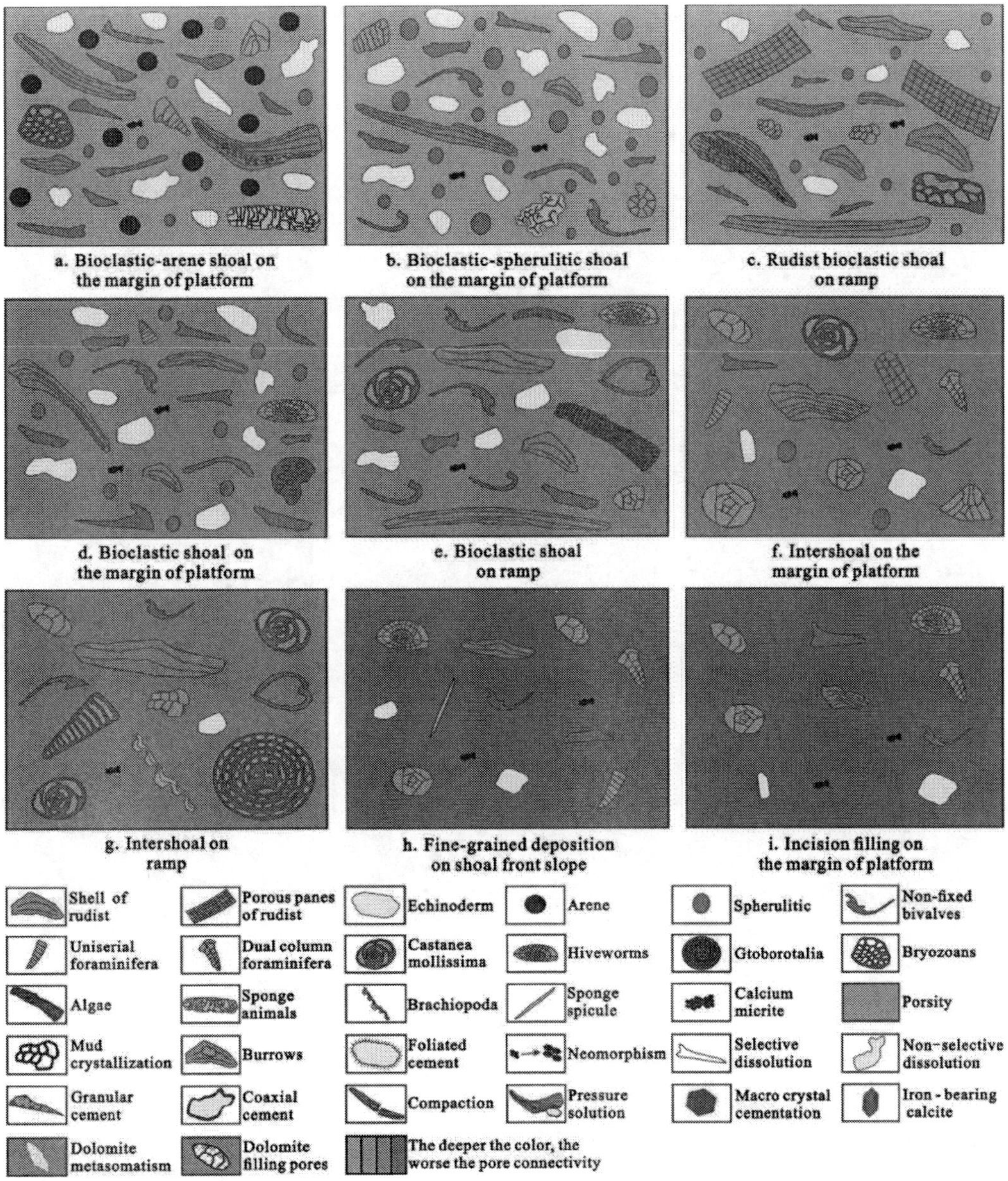


**Fig. 6.** Initial sedimentary state of each sedimentary microfacies of MB2-MC1 section in X oilfield (The legends of Fig. 7 to Fig. 10 are consistent with this).

### *4.2. Diagenetic characteristics of seawater environment*

During the early syngenetic diagenesis stage, in the seawater environment, the mud crystallization leads to a slight decrease in porosity, and the isolated pores formed by biological drilling result in a slight increase in porosity, while both of which have little effect on the reservoirs of different sedimentary microfacies.

In addition, the original pores of the medium-high energy microfacies included ramp shoal and platform margin shoal are developed better, the flux of the supersaturated seawater is larger, thus the seawater cementation is relatively strong (see Fig. 7a~Fig. 7e). However, the original pores of the medium-low energy microfacies included intershoal, shoal front slope and incision filling deposition are developed worse, the flux of seawater is smaller, thus the seawater cementation is relatively weak (see Fig. 7f~Fig. 7i). Due to the relatively frequent exposure in the study area, the unstable seawater cement on the surface of the particles is more prone to dissolution, which can be observed only in a small number of thin sections and has little effect on the reservoir. On the whole, the seawater diagenesis environment has little effect on the reservoir diagenesis difference and pore evolution.

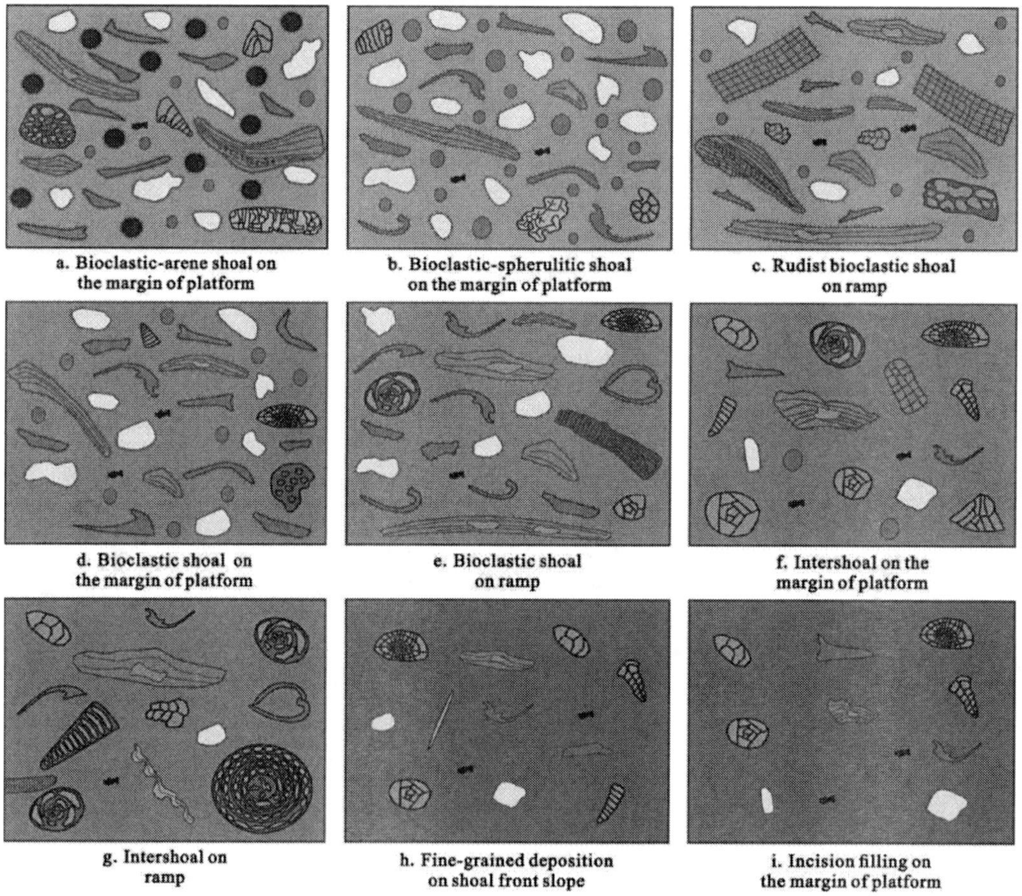


**Fig. 7.** Diagenetic characteristics of seawater environment of each sedimentary microfacies of MB2-MC1 section in X oilfield.

### *4.3. Diagenetic characteristics of atmospheric freshwater environment*

During the late syngenetic diagenesis stage, in the atmospheric freshwater environment, the water depth of the ramp shoal and platform margin shoal are shallow, the accumulation speed is fast. As a result, they are closer to the sea level, frequently exposed, and more easily subjected to leaching. At the same time, these facies have higher porosity and permeability, and the dissolution effect is obvious. The platform margin shoal is mainly low-medium magnesium calcite bioclastic debris with more sufficient leaching, thus the non-selective dissolution is developed, and the edge of the particles presents a harbor-like feature (see Fig. 8a~Fig. 8c). The ramp shoal includes aragonite, high magnesium calcite bioclastic and low-medium magnesium calcite bioclastic debris, and the leaching degree is lower than that of the platform margin shoal. Therefore, the selective dissolution is developed, and only the stable bioclastic shell is retained (see Fig. 8d, Fig. 8e). Under the action of dissolution, aragonite and high magnesium calcite dissolve and transform into low-medium magnesium calcite, completing the process of mineral stabilization. At the same time, the calcium carbonate and $Mg^{2+}$ produced by dissolution can be discharged to the surrounding area quickly, thus the reservoir physical properties become better.

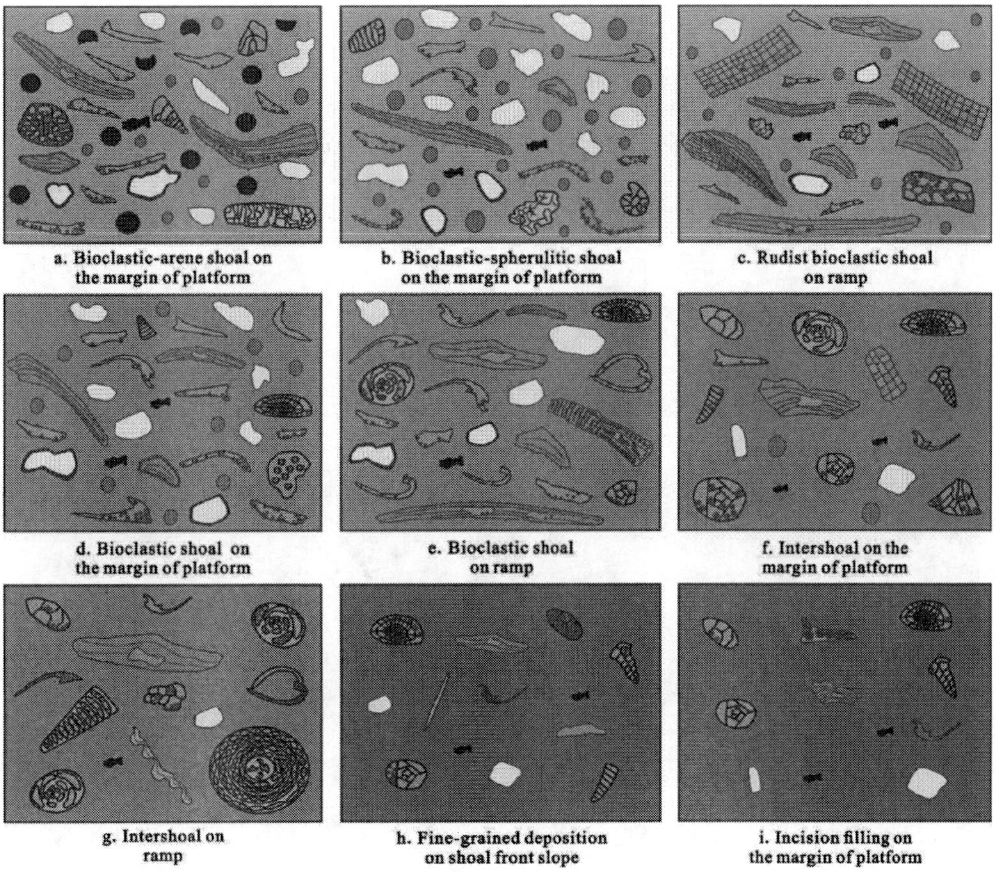


**Fig. 8.** Diagenetic characteristics of atmospheric freshwater environment of each sedimentary microfacies of MB2-MC1 section in X oilfield.

In contrast, the water depth of intershoal, shoal front slope and incision filling deposition are deep, the accumulation speed is slow. As a result, they are far from the sea level, the probability of exposure is small, and the atmospheric freshwater leaching is weak. At the same time, the original porosity and permeability of these micro-facies are low, the dissolution effect is poor, thus it is difficult to remove $Mg^{2+}$ from bioclastic debris. The intershoal in the middle-low parts of the paleogeomorphology will accept the dissolution products from the high paleogeomorphology, so that the calcium carbonate supersaturated precipitation forms cement, and the content of $Mg^{2+}$ is further increased (see Fig. 8f, Fig. 8g). In addition, the content of cement and $Mg^{2+}$ in the shoal front slope and incision filling deposition at the lower part of the palaeogeomorphology is higher, and the physical properties of the reservoir become worse (see Fig. 8h, Fig. 8i). The atmospheric freshwater diagenetic environment has the greatest impact on the differential diagenesis of reservoirs. Otherwise, the upper part of the shoal in the vertical seepage zone has the longer exposure time in the atmospheric freshwater environment, and its dissolution is more obvious, while the lower part of the shoal in the horizontal subsurface flow zone has more significant cementation, so the distribution of cement in the lower part of the shoal is more ubiquitous (see Fig. 8a~Fig. 8f).

### *4.4. Diagenetic characteristics of shallow burial environment*

During the early diagenesis stage, a large amount of water existed in original pore is discharged in the shallow burial environment, and the particles are compacted and accumulated with deformation and microfractures, which greatly reduces the porosity. The above factors make the compaction the most typical diagenesis in this stage, and its influence on the pores in the diagenetic evolution is quite different. Concretely speaking, If the pores formed in the early stage are filled with fluid or cement, or the particle content is high, which increases the support capacity of the reservoir skeleton, the later compaction has little effect on the pores. If the early dissolution is strong, the particle brittleness is strong, or the micrite content is high, thus the later compaction can greatly destroy the early pores.

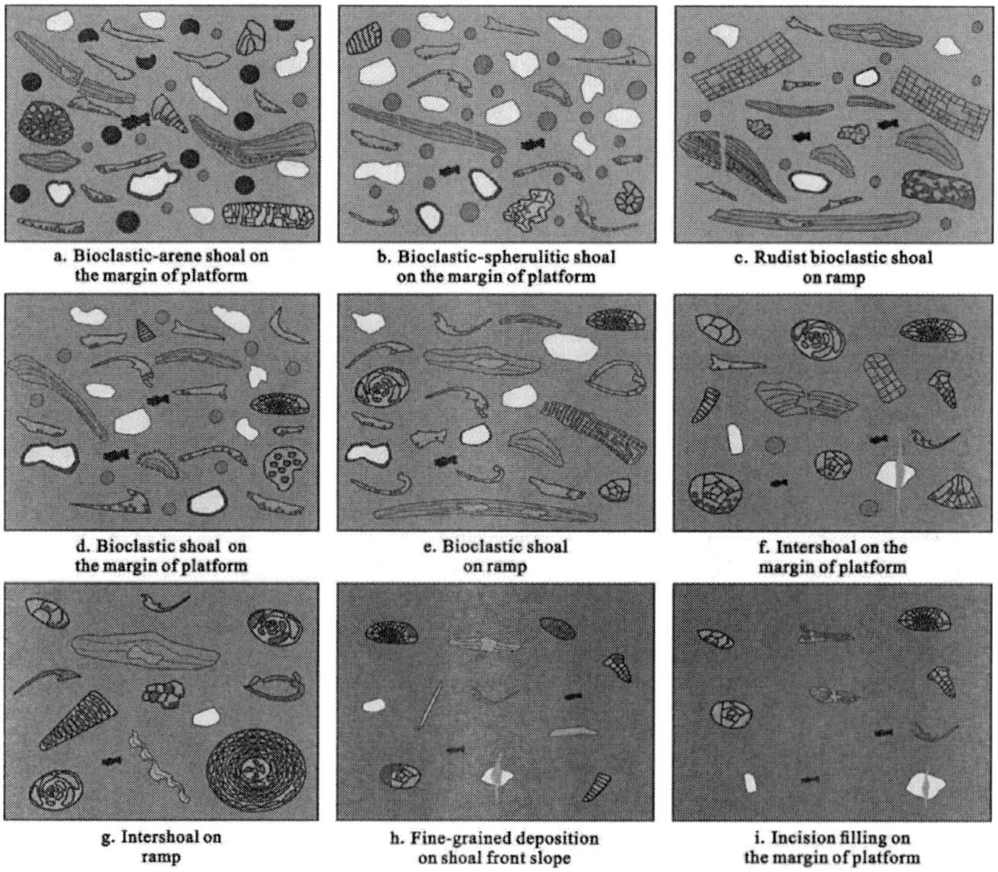


**Fig. 9.** Diagenetic characteristics of shallow burial environment of each sedimentary microfacies of MB2-MC1 section in X oilfield.

The rock types of ramp shoal and platform margin shoal are mainly grainstone and packstone with particle support structure. The compaction resistance is strong, and the pore-reducing effect of compaction is weak. Especially, the bioclastic-arene shoal with equiaxed granular arene as the main particle has the strongest compaction resistance. However, the rudist debris in the shoal facies is relatively special. Its unique shape and strong dissolution and weak cementation characteristics make its anti-compaction ability weak, and it is easy to rupture and form intergranular micro-fractures (see Fig. 9a~Fig. 9e). In contrast, the rock types of intershoal, shoal front slope and incision filling deposition are mainly mudstone and wackstone supported by micrite, with strong plasticity, more sensitive to pressure, weaker compaction resistance and more obvious porosity decline (see Fig. 9f~Fig. 9i).

### *4.5. Diagenetic characteristics of medium burial environment*

During the middle diagenesis stage, in the middle burial environment, the carbonate sediments establish the grain framework of the micrite level or the particles level. The continuous bury will increase the elastic stress between the particles, increase the solubility, and the dissolved material will precipitate again away from these contact surfaces.

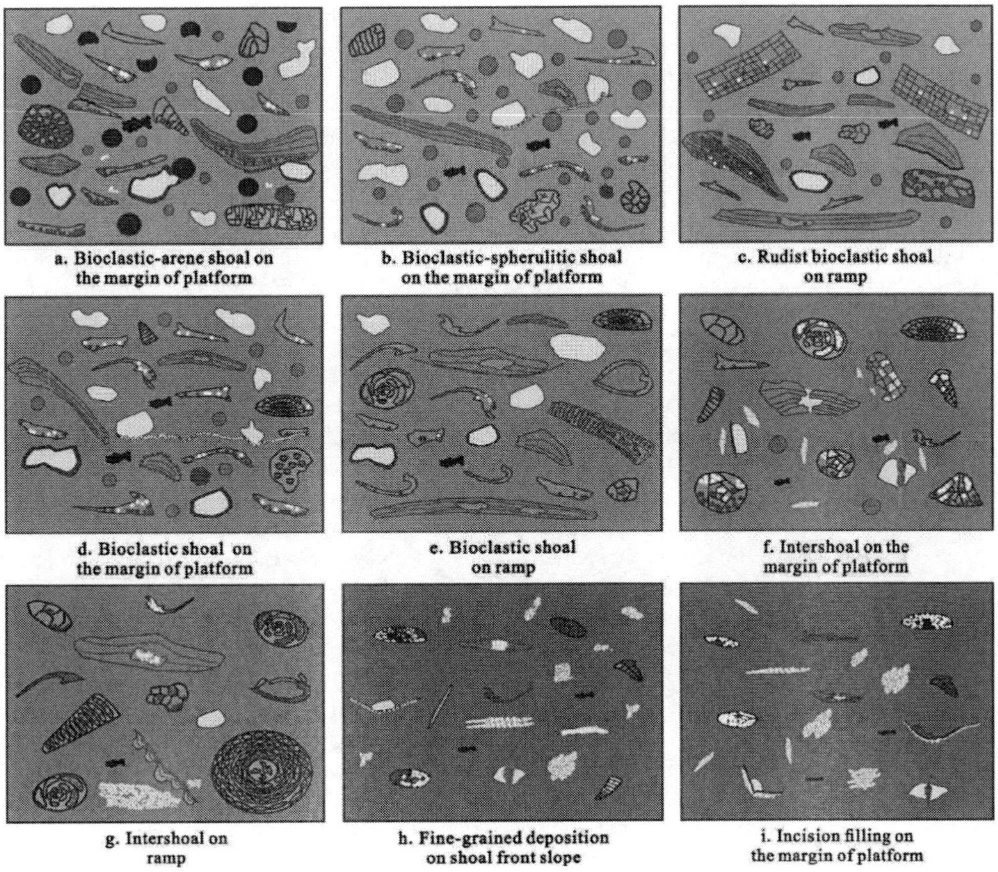


**Fig. 10.** Diagenetic characteristics of medium burial environment of each sedimentary microfacies of MB2-MC1 section in X oilfield.

The ramp shoal and platform margin shoal are particle support structures. The contact surface between particles is large, and the pressure solution effect is relatively strong. Especially for the bioclastic-arene shoal of single crystal calcite, the pressure solution effect is the most obvious (see Fig. 10a~Fig. 10e). In contrast, the intershoal, shoal front slope and incision filling deposition are micrite supporting structures, and the pressure solution is relatively weak (see Fig. 10f~Fig. 10i).

In the middle burial environment, dolomitization is the most prominent diagenesis. The ramp shoal and platform margin shoal have completed the mineral stabilization process in the atmospheric freshwater environment. The content of $Mg^{2+}$ is low, thus the degree of dolomitization is low (see Fig. 10a~Fig. 10e). In contrast, the mineral stabilization process of the medium-low energy microfacies included intershoal in the atmospheric freshwater environment is not finished completely, and the $Mg^{2+}$ dis-charged from the high part is accepted, which needs to be stabilized by micro-dissolution and micro-precipitation. During this process, $Mg^{2+}$ will be released, so that dolomitization will occur at the enrichment area of $Mg^{2+}$. At the same time, $Mg^{2+}$ will be discharged during the process of pressure dissolution, resulting in the distribution of dolomite along the suture line (see Fig. 10f, Fig. 10g). The $Mg^{2+}$ is higher in the low energy microfacies shoal front slope and incision filling deposition, and dolomitization is more developed (see Fig. 10h, Fig. 10i). In addition, since dissolution and precipitation occur in a closed system, the development degree of burial cementation in various sedimentary microfacies is low.

## 5. Diagenesis sequence and pore evolution characteristics of different sedimentary microfacies

### *5.1. Differential diagenesis and pore evolution characteristics*

The sedimentary microfacies determine the material basis of diagenesis and the characteristics of paleogeomorphology, and then control the sequence and intensity of diagenesis and pore evolution. The high energy microfacies are dominated by pore-increasing diagenesis, and the low energy microfacies are dominated by pore-reducing diagenesis, which leads to the polarization diagenesis. The rudist shoal is characterized by non-selective dissolution and medium-weak compaction. The bioclastic-arene shoal has the characteristics of non-selective dissolution and moderate-weak pressure dissolution. The bioclastic-spherulitic shoal and the rudist bioclastic shoal have the selective dissolution and moderate-weak non-selective dissolution. The bioclastic shoal on the margin of platform and on ramp are characterized by selective dissolution (see Fig. 11).

The reservoir properties of bioclastic limestone depend on the original sedimentary characteristics, and diagenesis has the dual characteristics of inheritance and development for reservoir reconstruction. Among them, the atmospheric freshwater environment and burial environment have a profound impact on differential diagenesis and pore evolution (see Fig. 11). High energy microfacies such as rudist shoal, bioclastic-arene shoal and bioclastic-spherulitic shoal are leached by freshwater rich in $CO_2$ in the atmospheric freshwater environment, non-selective dissolution occurs, and seawater cements basically disappear. Therefore, various pores of different sizes and shapes are formed, and the porosity increases. In the burial environment, these microfacies are characterized by weak compaction, pressure solution and dolomitization, and the porosity decreases slightly. Medium energy microfacies such as rudist bioclastic shoal on ramp, bioclastic shoal are leached, but the leaching is not sufficient, and the exposure frequency and time are less than those of high energy microfacies. Therefore, selective dissolution occurs, forming a large number of mould pores and intragranular dissolution pores, and porosity increases. In the burial environment, these microfacies are also characterized by weak compaction, pressure solution and dolomitization, and the porosity decreases relatively slightly.

Medium-low energy microfacies such as intershoal on the margin of platform and on ramp are subject to leaching and the dis-solution effect is poor. In addition, they accept calcium carbonate and $Mg^{2+}$ from the high geomorphological area, so that the supersaturated precipitation of calcium carbonate forms some cements and the porosity decreases. In the burial environment, the overlying strata pressure acts on the micrite, and compaction, pressure solution and dolomitization are developed, thus the porosity decreases by a moderate degree. Low energy microfacies such as fine-grained deposition on shoal front slope and incision filling are difficult to accept atmospheric freshwater leaching, and also accept calcium carbonate and $Mg^{2+}$ from other regions, which makes calcium carbonate supersaturated precipitation form a large number of cements, and the porosity decreases greatly. In the burial environment, the plasticity of micritic is strong, therefore, the compaction effect is significant. The Mg/Ca ratio is high, the dolomitization is obvious, and the porosity is greatly reduced again.

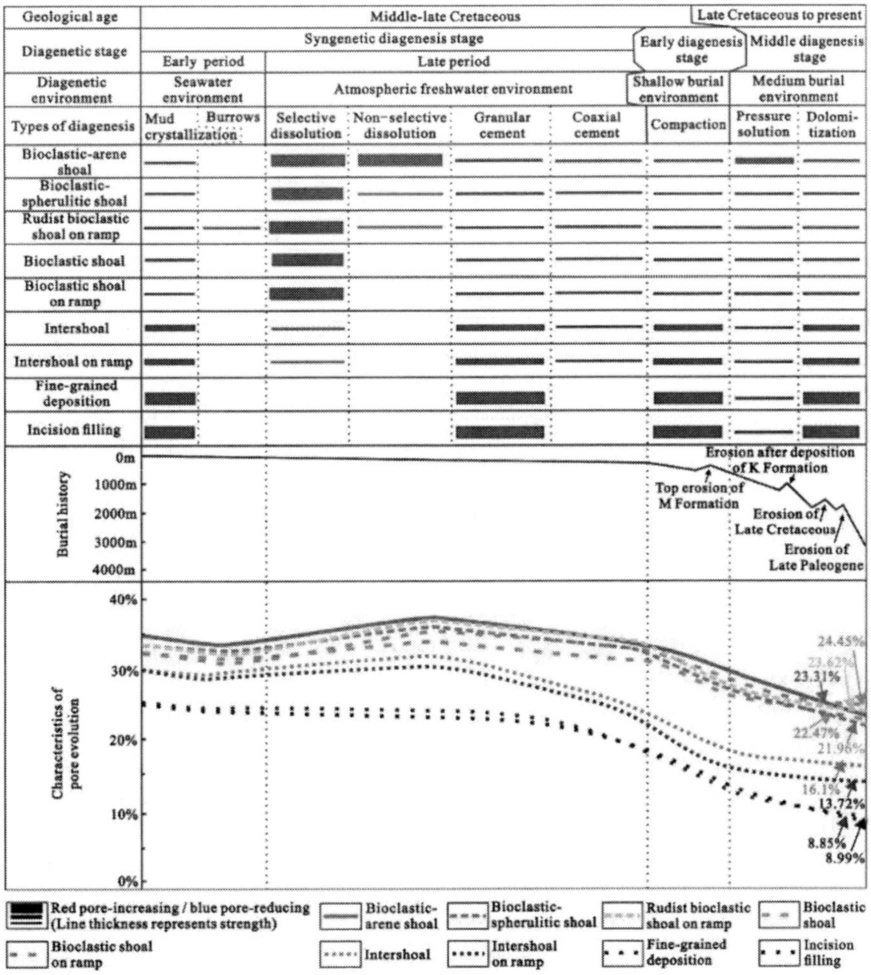


**Fig. 11.** Diagenesis sequence of MB2-MC1 section in X oilfield.

## 5.2. *Physical properties*

On the basis of the main control of facies, diagenetic evolution further enhances the difference of reservoir physical properties of various sedimentary microfacies. High energy microfacies are dominated by intergranular pores and intergranular dis-solved pores. Medium energy microfacies are dominated by mould pores and intragranular pores. The medium-low energy microfacies are dominated by intragranular pores and visceral foramen pores. The low energy microfacies are dominated by micropores, intercrystalline pores and visceral foramen pores, which are non-reservoirs. In addition, the porosity and permeability characteristics of different sedimentary microfacies are not consistent (see Table 1).

**Table 1.** Pore types and reservoir physical properties of each sedimentary microfacies of MB2-MC1 section in X oilfield.

| Sedimentary microfacies | Major pore types | Average porosity/% | Average Permeability /$10^{-3}\mu m^2$ |
|---|---|---|---|
| Rudist shoal | Intergranular pores, intergranular dissolved pores, dissolved vugs | 22.51 | 187.01 |
| Bioclastic-arene shoal | Intergranular pores, intergranular dissolved pores | 23.31 | 136.01 |
| Bioclastic-spherulitic shoal | Mould pores, intragranular pores, intergranular pores | 21.96 | 39.36 |
| Rudist bioclastic shoal on ramp | Mould pores, intragranular pores, biological framework pores | 23.62 | 27.79 |
| Bioclastic shoal | Mould pores, intragranular pores, visceral foramen pores | 22.47 | 19.26 |
| Bioclastic shoal on ramp | Mould pores, intragranular pores, visceral foramen pores | 24.45 | 10.15 |
| Intershoal | Intragranular pores, visceral foramen pores | 16.1 | 3.66 |
| Intershoal on ramp | Intragranular pores, visceral foramen pores | 13.72 | 1.31 |
| Fine-grained deposition | Visceral foramen pores, micropores, intercrystalline pores | 8.85 | 0.52 |
| Incision filling | Micropores, intercrystalline pores | 8.99 | 0.14 |

## 6. Conclusion

The carbonate particles of the shoal facies are the debris after the death and elutriation of organisms such as rudist, echinoderms, non-fixed bivalves and benthic foraminifera. The shoal facies can be further subdivided according to bioclastic types. There are various types of diagenesis in bioclastic limestone. The original fabric, pore development degree, type, content and composition of bioclastics in different sedimentary microfacies are quite inconsistent, and the paleo-geomorphological positions they are located in are different, which makes each sedimentary microfacies have obvious differential diagenesis characteristics in seawater, atmospheric freshwater and burial diagenesis environment. The rudist shoal, bioclastic-arene shoal and bio-clastic-spherulitic shoal have the characteristics of non-selective dissolution. The rudist bioclastic shoal, bioclastic shoal have the characteristics of selective dissolution.

The intershoal on the margin of platform and on ramp have the characteristics of moderate cementation, moderate compaction and moderate-weak dolomitization. The fine-grained deposition on shoal front slope and incision filling have the characteristics of strong cementation and moderate dolomitization.

The sedimentary microfacies determine the material basis of diagenesis and paleogeomorphology, and then control the process and amplitude of pore evolution. The porosity of high energy microfacies in atmospheric freshwater environment increases, then decreases slightly in burial environment, mainly developing intergranular pores and intergranular dissolved pores. The porosity of medium energy microfacies increases in the atmospheric freshwater environment and decreases relatively slightly in the burial environment, mainly developing mould pores.

The porosity of medium-low microfacies decreases in the atmospheric freshwater environment, then it continues decrease by a moderate degree in the burial environment, mainly developing intra-granular pores and visceral foramen pores. The porosity of low microfacies decreases in the atmospheric freshwater environment, then it decreases greatly in the burial environment, mainly developing micropores, intercrystalline pores and visceral foramen pores, forming non-reservoirs.

**References**

[1] Cao, Y.C., Yuan, G.H., Yang, H.J., Wang, Y.Z., Liu, K.Y., Zan, N.M., Xi, K.L., Wang, J.: Current situation of oil and gas exploration and research progress of the origin of high-quality reservoirs in deep-ultra-deep clastic reservoirs of petroliferous basins. Acta Petrolei Sinica 43(1), 112-140 (2022).

[2] Xu, H., Liu, M.J., Zhang, Z., Ye, S.J., Yang, Y.T., Wu, L., Zhang, L., Nan, H.L., Tan, X.C., Zeng, W., Lian, C.B.: Diagenesis and porosity evolution of the 3rd member of Xujiahe Formation tight sandstone reservoir in the Western Sichuan Depression, Sichuan Basin. Natural Gas Geoscience 33(3), 344-357 (2022).

[3] Qiang, S.T., Shen, P., Zhang, J., Xia, M.L., Feng, M.Y., Xia, Q.S., Tao, Y.Z., Xia, Z.Q., Lin, Y.: The Evolution of Carbonate Sediment Diagenesis and Pore Fluid in Dengying Formation, Central Sichuan Basin. Acta Sedimentologica Sinica 35(4), 797-811 (2017).

[4] Luo, Q.Q., Liu, B., Jiang, W.M., Yu, J.X., Liu, S.Q., Wang, Y.C., Wei, L.B., Cai, Z.H.: Di-agenesis and pore evolution of dolomite reservoir in the 5th member of the Ordovician Majiagou Formation, central Ordos Basin. Oil & Gas Geology 41(1), 102-115 (2020).

[5] Shi, B.Q., Wang, Z.M., Wen, Z.X., Wan, L.K., Xu, N., Liu, X.B.: The global oil and gas exploration situation in 2020 and the outlook for 2021. International Petroleum Economics 28(3), 29-35 (2020).

[6] Yang, J.Q., Ji, Y.L., Wu, H., Meng, L.J.: Diagenesis and Porosity Evolution of Deep Reservoirs in the Nanpu Sag: A case study of Sha 1 Member of the Paleogene in No. 3 structural belt. Acta Sedimentologica Sinica 40(1), 203-216 (2022).

[7] Zhang, X.F., Shi, K.B., Liu, B., Yang, Y.K., Wang, J.Q.: Retention Processes and Porosity Preservation in Deep Carbonate Reservoirs. Geological Science and Technology Information 33(2), 80-85 (2014).

[8] Zhou, J.G., Yu, Z., Wu, D.X., Ren, J.F., Zhang, D.F., Wang, S.Y., Yin, C., Liu, Y.X.: Restoration of formation processes of dolomite reservoirs based on laser U-Pb dating: A case study of Ordovician Majiagou Formation, Ordos Basin, NW China. Petroleum Exploration and Development 49(2), 285-295 (2022).

[9] Wang, Y.L., Song, X.M., Wang, G.H., Wang, H., Li, Z.M., Zhang, H., Nie, Z.: Key technologies and practices for rapid and large - scale production increase in cooperation oil and gas fields of the Middle East. Acta Petrolei Sinica 41(12), 1633-1642 (2020).

[10] Khawaja, A.M., Thabit, J.M.: Predicting the Depositional Environments of Mishrif For-mation from Seismic Isopach Map in the Dujaila Oil Field, Southeast-Iraq. Iraqi Journal of Science 62(6), 1943-1955 (2021).

[11] Al-Banna, A.S., Nassir, N.A., Al-Sharaa, G.H.: A Comparison of Mishrif Formation Characteristics in Kumait and Dujaila Oil Fields, Southern Iraq, Using Seismic Inversion Method and Petrophysical Properties Analysis. Iraqi Journal of Science 61(12), 3294-3307 (2020).

[12] Jassim, A.A., Al-Dabaj, A.A., Al-Adili, A.S.: Water Injection for Oil Recovery in Mishrif Formation for Amarah Oil Field. Iraqi Journal of Chemical and Petroleum Engineering 21(1), 39-44 (2020).

[13] Grotzinger, J., Al-Rawahi, Z.: Depositional facies and platform architecture of microbialite-dominated carbonate reservoirs, Ediacaran-Cambrian Ara Group, Sultanate of Oman. AAPG Bulletin 98(8), 1453-1494 (2014).

[14] Battiau-Queney, Y., Préat, A., Trentesaux, A., Recourt, P., Bout-Roumazeilles, V.: Late Mississippian limestone sedimentary environment in southern Pembrokeshire (Bulls laughter Bay, Wales): evidence of meteoric diagenesis and hypersaline features. Geological Magazine 158(5), 759-786 (2021).

[15] Periere, D., Durlet, C., Vernnin, E., Caline, B., Boichard, R., Meyer, A.: Influence of a major exposure surface on the development of microporous micritic limestones - Example of the Upper Mishrif Formation (Cenomanian) of the Middle East. Sedimentary Geology 353(15), 96-113 (2017).
[16] Mahdi, T.A., Aqrawi, A.A.M.: Sequence stratigraphic analysis of the mid-Cretaceous Mishrif Formation, southern Mesopotamian Basin, Iraq. Journal of Petroleum Geology 37(3), 287-312 (2014).
[17] Tian, B., Zheng, Y.W., Zhao, J.M., Zhu, X.F., He, T., Luo, X.F.: Research Development in Anomalously High Porosity of Deep Clastic Reservoirs. Science Technology and Engineering 22(22), 9456-9465 (2022).
[18] Li, L.P., Liang, J.T., Liu, S.B., Guo, Y.B., Li, K.Y., He, Y., Jin, J.X.: Diagenesis and pore evolution of dolomite reservoirs of Cambrian Xixiangchi Formation in central Sichuan Ba-sin. Lithologic Reservoirs 34(3), 39-48 (2022).
[19] Nohl, T., Wetterich, J., Fobbe, N., Munnecke, A.: Lithological dependence of aragonite preservation in monospecific gastropod deposits of the Miocene Mainz Basin: Implications for the (dia-)genesis of limestone–marl alternations. Journal of Sedimentary Research 90(11), 1500-1509 (2020).
[20] Sobczak, K., Holl, H.G., Garnett, A.: Estimating porosity and permeability in the Springbok Sandstone, Surat Basin (Queensland), using new wireline log-based workflow. The APPEA Journal APPEA J 61(2), 720-725 (2021).
[21] Roberts, N., Rasbury, E.T., Parrish, R.R., Smith, C.J., Horstwood, M., Condon, D.J.: A calcite reference material for LA-ICP-MS U-Pb geochronology. Geochemistry, Geophysics, Geosystems 18(7), 2807-2814 (2017).
[22] Godeau, N., Deschamps, P., Guihou, A., Leonide, P., Girard, J.P.: U-Pb dating of calcite cement and diagenetic history in microporous carbonate reservoirs: Case of the Urgonian Limestone, France. Geology 46(3), 247-250 (2018).
[23] Paxton, S.T., Szabo, J.O., Ajdukiewicz, J.M.: Construction of an intergranular volume compaction curve for evaluating and predicting compaction and porosity loss in rigid-grain sandstone reservoirs. AAPG Bulletin 86(12), 2047-2067 (2002).
[24] Goodner, H.M., Rankey, E.C., Zhang, C., Watney, W.L.: Rock fabric controls on pore evolution and porosity–permeability trends in oolitic grainstone reservoirs and reservoir analogs. AAPG Bulletin 104(7), 1501-1530 (2020).
[25] Yu, Y.C., Sun, L.D., Song, X.M., Guo, R., Gao, X.J., Lin, M.J., Yi, L.P., Han, H.Y., Li, F.F., Liu, H.Y.: Sedimentary diagenesis of rudist shoal and its control on reservoirs: A case study of Cretaceous Mishrif Formation, H Oilfield, Iraq. Petroleum Exploration and Development 45(6), 1007-1019 (2018).
[26] Langhorne, B., Smithe, J., Graham, R.D.: Structurally controlled hydrothermal alteration of carbonate reservoirs: Introduction. AAPG Bulletin 90(11), 1635-1640 (2006).
[27] Mahdi, T.A., Aqrawi, A.A.M., Horbury, A.D., Sherwani, G.H.: Sedimentological characterization of the mid-Cretaceous Mishrif reservoir in southern Mesopotamian Basin, Iraq. Geoarabia 18(1), 139-174 (2013).
[28] Jacob, D.E., Soldati, A.L., Wirth, R., Huth, J., Wehrmeister, U., Hofmeister, W.: Nanostructure, composition and mechanisms of bivalve shell growth. Geochimica Et Cosmochimica Acta 72(22), 5401-5415 (2008).

# Reservoir Analysis for A Low Recovery Factor Mature Reservoir with A Gas Cap and Large Aquifer: A Proposed Coning Correlation

**Xuan Yi1(✉), Zhao-ran Liu2**

1 Department of Petroleum Engineering, University of Houston,
Houston, TX, USA
2 CNPC Research Institute of Engineering Technology
Tianjin, China

meredithyi1125@gmail.com, 610390030@qq.com

**Abstract.** With the development of reservoir engineering and management, enhanced oil recovery (EOR) techniques are becoming increasingly sophisticated. Generally, the recovery factor (RF) for a conventional reservoir can range from around 10% to 60% after four decades of production, with an average recovery factor of around 35%. While the RF of the NGJ field with high porosity (0.25) and permeability (300 to 600 md) is 6.11% after producing for over 40 years, which is rare in such good reservoir conditions. In this work, some classic techniques and several advanced methods were utilized to investigate the reason behind the low RF for this reservoir.

A dynamic reservoir simulation was constructed on the basis of a statics petrophysical model. Based on the preliminary simulation study, it was found that gas coning is a common problem in the reservoir. Therefore, a concept of coning correlation function (CCF) is proposed, and the procedures for developing the CCF are explained. The critical oil rate was estimated using the Addington correlation and modified through the proposed coning correlation that varies as a function of time. The estimated critical rate through CCF is used to control flowing bottom hole pressure (FBHP) in the prediction scenarios.

The study showed that the critical rate exponentially decreased as production time increases for a single well. A larger well perforation interval would result in a lower critical rate, while a greater horizontal permeability would lead to a larger critical rate. Applying the correlation in the case of a previously perforated well, the critical oil rate was determined to provide supporting information on the optimized FBHP. In an infill well, based on an optimal critical oil rate, the distance between gas-oil contact (GOC) and the top of perforations was estimated. Eight prediction schemes were proposed based on the analysis results. The prediction scenario that involved a combination of workovers and infill drilling would bring a RF of 8.44% and the highest benefits of $50.35 MM in the next ten years.

The improved correlation was generated by a multiple-well numerical model instead of a one-well radial model. The production history was taken into account to adjust dimensionless parameters in the correlation. Compared with the Addington correlation, the modified one improves the accuracy of critical oil rate estimation.

## 1. Introduction

Gas and water coning are phenomena that occur in partially penetrated producing wells in a reservoir with gas cap and aquifer. When the well is producing, the gas oil contact (GOC) moves down, while oil water contact (OWC) rises up to the wellbore in a conical shape because of the pressure drawdown around the well. As the production rate increases, the cone height increases until it reaches the perforation. When the apex of the cone reaches the perforation, gas breakthrough or water breakthrough occurs. The oil production rate when gas breakthrough or water breakthrough happens is called the critical rate.

The presence of gas and water coning in production wells reduces the oil rate. From both economic and operational perspective, this condition is undesirable for several reasons:

- Gas price is much lower than oil price
- The production of gas and water results in an accelerated reservoir pressure depletion, which results in a lower oil production rate
- The well with gas and water coning may have to be abandoned early

The gas and water coning problem has been studied for more than several decades. Many correlations have been developed to discuss and predict some parameters of gas and water coning, such as critical rate, breakthrough time, and production after breakthrough. Gas and water coning problem was first described by Muskat and Wycokoff [1]. They presented a static conical water coning model in which, while the oil zone is partially penetrated, the water remains in the lower portion of the pay horizon. They described the movement of water related with the density difference between oil and water, the slope of the stratum, and regional pressure distribution. They pointed out that an increase of oil rate would break the balance of the low-height water cone, driving the water into the perforation. After Muskat and Wyckoff, several models were generated to predict critical rate, breakthrough time, and GOC and OWC movement after breakthrough.

For analytical approaches, Chierici, Ciucci and Pizzi [2] followed the Muskat and Wyckoff study. A potentiometric model was presented and solved for the potential function of critical rate by equating gravity and viscous forces. They developed a series of curves to predict critical rate for different reservoir characteristics. The method can be used to predict the length and position of perforation interval. Wheatley [3] pointed out some deficiencies of previous theories. The presence of the cone in pressure distribution calculation was taken into account. The model was derived by solving a potential function using Laplace's equation for pure radial flow condition. The results were modified to satisfy other conditions to make the equation as accurate as possible. Chaperon [4] studied the coning problem for steady-state and pseudo-steady state flow in anisotropic formation. Correlations were developed to calculate critical rates for vertical and horizontal wells with very short intervals. By comparing both types of wells, it was concluded that the "horizontal wells allow higher critical rates, but this advantage lessens as vertical permeability decreases." Guo and Lee [5] constructed a radial/spherical/combined (RSC) 3D flow field model, accounting for the effect of limited wellbore penetration on the oil productivity, following Abass and Bass [6] study. A new equation was derived in relation to the oil zone thickness, wellbore radius and drainage area radius etc. Their results showed that the optimum completion interval for water-coning should be less than one third from top of the total thickness of oil layer. Tabatabaei, Ghalambor and Guo [7] followed Guo and Lee's steps, assuming the similar RSC 3D model. Tabatabaei et al. model is most suitable for low-pressure-gradient case, while Guo and Lee model is applicable for high-pressure-gradient case. The results indicate that the optimum completion interval for water-coning should be less than almost half from top of the total thickness of oil layer.

The first empirical model was constructed by Schols [8], who developed it using laboratory experiment. Addington [9] generated gas-coning correlations by a one-well model constructed on an implicit radial simulator. An oil material balance equation was developed to calculate the gas oil contact (GOC) movement during production. Sensitivity to oil rate, horizontal permeability, the ratio of vertical permeability and horizontal permeability $(k_v/k_h)$ , perforation thickness, average oil viscosity and well spacing were also investigated by utilizing the semi-log plot of Gas Liquid Ratio vs. Average Oil Column Height above Perforations. A general correlation was obtained to calculate the critical rate. Yang and Wattenbarger [10] followed Addington method to estimate the critical rate and water oil ratio (WOR) movement for water coning in vertical and horizontal wells. An r-z radial model was run for vertical wells and an x-z model for horizontal wells. Water coning correlations were generated to predict critical rate, breakthrough time and WOR after breakthrough. Similarly, Benamara and Tiab [11] followed Addington's steps, studying the gas coning problem. Gas cap expansion was taken into account in their theory and new correlations were proposed for critical rate, breakthrough time and GOR after breakthrough in vertical and horizontal wells.

In recent years, researchers combine analytical and empirical models (i.e., semi-analytical models) to overcome the deficiencies of purely theoretical models. Many analytical models are derived from simplified assumptions, which cannot reflect underground complexity and variety. Empirical models are generated from laboratory and simulators, lacking the support of physical principles. Semi-analytical models are based on simplified force interaction, but also consider the ignored physical effects by fitting the discrepancy statistically with empirical data. The processes of estimating some values of analytical model's parameters by experimental data are called statistical model calibrations[12-17].This kind of model increases the accuracy of the prediction and reduces bias. Prasun and Wojtanowicz [17] generated a semi-analytical model to study water coning problem for a more specific scenario of naturally fractured reservoirs (NFR). They simplified an NFR mechanistic model and applied Chaperon's correlations in their model by replacing the permeability of matrix with an equivalent permeability of NFR. Then they developed a dual porosity/dual permeability (DPDP) model in a reservoir simulator for NFR experiments and calibrated the ignored parameters using a series of simulated experiments.

In this work, the workflow of developing a CCF based on Addington correlation is presented to help control free gas production. Subsequently, the CCF was implemented in reservoir simulations to support the proposal of various predictive scenarios.

## 2. Field Overview

GJF is located in the UA Basin, Northeast India. The basin is bounded by the eastern Himalayas in the north and by the NP hill ranges in the south and southeast (Figure 1). There are seven main structures in GJF. NGJ, as one of the seven structures, is located in the northeast part of GJF. It is separated with the other structures by two faults. The sands in NGJ structure from stratigraphically deeper to shallower depths are BE, BT, B5, B4, LT, ML and UT. The sand with largest stock tank oil original in place (STOOIP) is UT, which is the object studied in this paper.

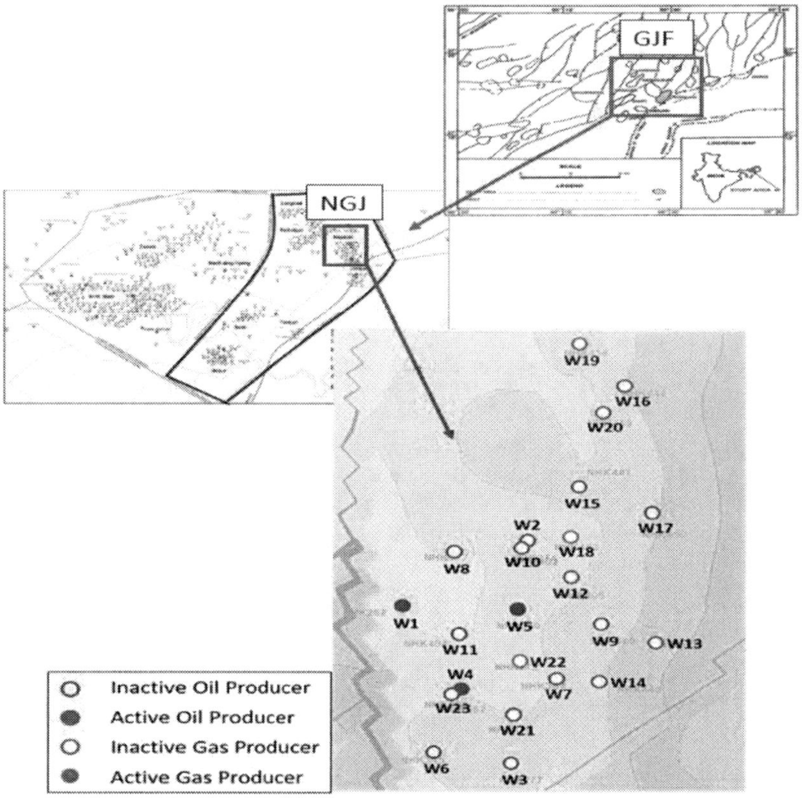


**Figure 1.** Geographical location of NGJ and wells in the field.

Table 1 gives reservoir and fluid properties for the NGJ field. The UT in NGJ Field starts from -2140 m (-7019 ft). The gross thickness of the formation is 39 m (127.92 ft). It needs to be noticed that the reservoir initially contains $2.2 \times 10^9$ $m^3$ of gas ($m = 0.4$) and a groundwater volume in aquifer which is more than 100 times STOOIP. While initial reservoir pressure is 230.97 bar (3350 psi), which is higher than 224.08 bar (3250 psi) of the bubble point pressure (BPP) at the same reference depth. The dominant primary drive mechanisms, according to the material balance analysis, are water influx and gas cap expansion.

**Table 1**. NGJ field reservoir and fluid properties

| NGJ field reservoir and fluid properties | |
|---|---|
| Formation | UT |
| Lithology | sandstone |
| Depth, m | 2286 |
| Gross thickness, m | 39 |
| Average porosity, % | 25 |
| Average permeability, md | 300 to 600 |
| Initial water saturation, % | 35 |
| Average original pressure, bar | 230.97 |
| Bubble point pressure, bar | 224.08 |
| Stock-tank oil gravity, °API | 25.6 |
| Oil viscosity, cp | 3 |
| FVF, $m^3/sm^3$ | 1.214 |
| Gas Gravity | 0.6246 |

## 3. Low Recovery Factor Reservoir

Production in NGJ field started in 1973, and the current production streams result from the effect of gas cap expansion and water influx. Usually, the average efficiency of primary recovery for the field with gas cap expansion is 30%, and 50% in case of presence of a strong water influx[18]. Total production of the field is 1.13 $MMsm^3$ (7.11 MMSTB) after over four decades, which yields a RF of 6.11% with a STOOIP of 18.5 $MMsm^3$ (116.36 MMSTB) despite the fact that there is a strong aquifer support and gas cap expansion in this reservoir.

## 4. Gas Coning Problem

Although the reported initial reservoir pressure (3340 psi) is higher than the bubble point pressure (3250 psi), presence of an initial gas cap is evident from production data. Some simple calculations can be performed to confirm the presence of an initial gas cap in this reservoir. As shown in Figure 2, gas volume produced from the oil zone is the sum of the solution gas (*B*) and free gas (*A*) liberated from solution gas. The gas volume produced from gas cap is free gas (*C*). The free gas rate in the oil zone (*A*) can be estimated by:

$$A = q_o \cdot \frac{k_{rg}/\mu_g B_g}{k_{ro}/\mu_o B_o} \quad [1]$$

Also, the sum of free gas rate from gas cap and oil zone (*A*+*C*) is:

$$A + C = q_o R_p - q_o R_s = q_o R_p - B \quad [2]$$

Figure 3 compares gas production from the oil column (A+B) and total produced gas (A+B+C). As it is clear from the plot, produced gas is much higher than the calculated gas production from the oil column all over the field production history. This indicates that an initial gas cap does exist and also, a considerable portion of gas production is from the gas cap during this time.

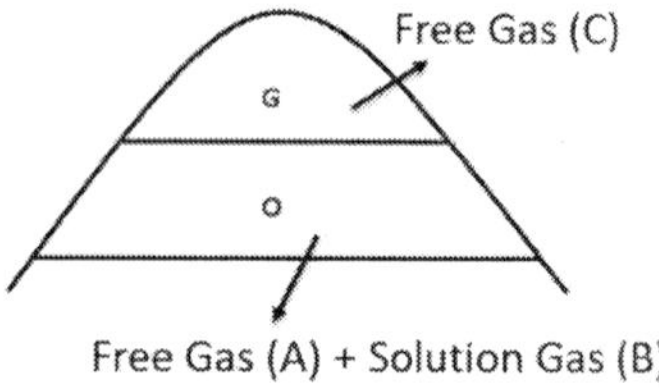

**Figure 2**. Illustration of free gas from gas cap and oil zone.

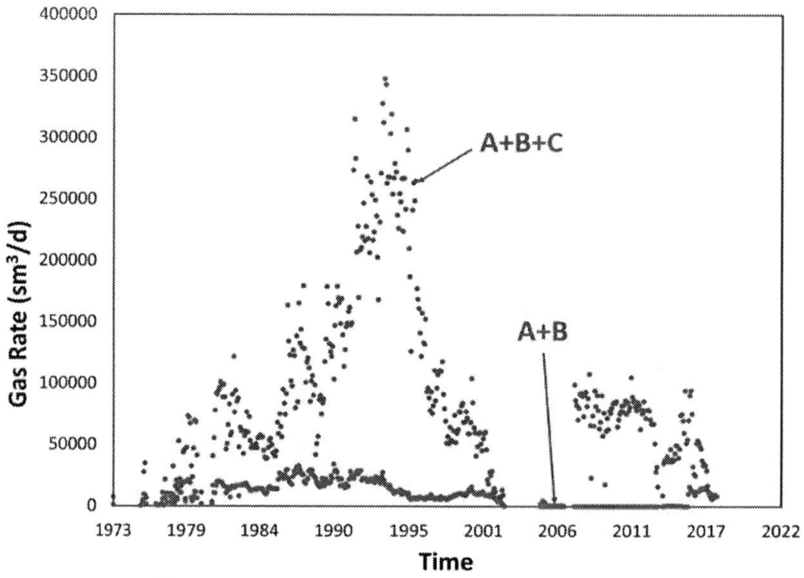

**Figure 3**. Gas production rate comparison.

## 5. Coning Correlation Development

From theorical perspective, critical rate can be developed and solved based on the equilibrium conditions of viscous forces and gravity forces. There are several analytical methods available with several assumptions to simplify the model equations. This results in an inaccurate estimation of critical rate due to the complex and changing underground conditions. Empirical correlations, based on the lab experiments or computer simulation, reflect the flow behavior around a well. Addington [9] developed an empirical correlation based on a closed outer boundary problem that never reaches steady-state conditions for reservoirs without aquifer support. In this work, Addington correlation is expanded to calculate the critical rate for reservoirs with a large aquifer volume and complex primary drive mechanisms.

### *5.1 Addington Correlation*

While testing the effects of several parameter on the gas coning phenomenon, Addington [9] observed a linear relation between gas oil ratio (GOR) and the average column height above perforation after gas breakthrough on a semi-log plot.

Based on this observation, the correlation for critical rate was generated as below

$$h_{ab} = 137.9(P)^{0.429} \quad [3]$$

where

$$P = \frac{Q_A B_o \left(\frac{k_v}{k_h}\right)^{0.1} \mu_o F_1 F_2}{k_h \sqrt{h_p}} \quad [4]$$

$$F_1 = \frac{h_{ap} + h_p}{h} \quad [5]$$

$$F_2 = \left(\frac{well\ spacing}{647 \cdot 10^3}\right)^{0.1} \quad [6]$$

In Equations [ 3 ] through [ 6 ], $h_{ab}$ is average oil column height above perforations at gas breakthrough (ft); $Q_A$ is critical oil rate calculated by Addington correlation (STB/d); $B_o$ is oil formation volume factor (RB/STB); $k_v$ is vertical permeability (mD); $k_h$ is horizontal permeability (mD); $\mu_o$ is average oil viscosity (cP); $F_1$ is geometric factor; $F_2$ is well spacing factor; $h$ is pay thickness (ft); $h_p$ is perforation thickness (ft); $h_{ap}$ is average oil column height above perforations (ft).

The correlations were applied to calculate the critical oil rate for each well in NGJ field. The results showed that the correlations do not work for wells in which gas coning happens several times during the production history.

### *5.2 Improved Addington Correlation*

Initially, Addington's approach was applied to the wells in the reservoir under study, but it was found that the GOC cannot be estimated precisely because of the strong aquifer support in NGJ. After many attempts using several variables, a new semi-log plot, *GOR vs. oil rate,* was generated (Figure 4.a)). From this plot, the produced GOR equals to the solution GOR if the oil rate is smaller than critical oil rate. Otherwise, if the oil rate exceeds the critical oil rate, the produced GOR would be larger than solution GOR which indicates contribution of gas from the gas cap. The turning point is the estimated critical oil rate. Several wells in this reservoir were selected with relatively continuous production history (W3, W4, W5, W6, W8, W11 and W19) to develop the proposed correlation. GOR was plotted against oil rate on a semi-log plot for each well. For most of the plots, as shown in Figure 4.b), the estimated critical rate is not a single value and it varies in a certain range. Some high oil rate points corresponds to low GOR (e.g., $oil\, rate = 100\, sm^3/d$ ; $GOR = 107\ sm^3/sm^3 = solution\, GOR$ ), while partial low oil rate points corresponds to high GOR (e.g., $oil\, rate = 67.5\, sm^3/d$; $GOR = 896\ sm^3/sm^3$). By analyzing wells production data, it was found that the high oil rate with low GOR usually happened in the early period, and relatively low oil rate with high GOR occurred late into the producing phase. The critical oil rate decreases as production time increases.

Therefore, based on the gas coning occurrence time, well production was divided into several periods, for which GOR was plotted against oil rate. The results for W3 shown in Figure 5 indicate a strong relationship between oil rate and GOR for each period. Then the trend lines were drawn, and the oil rates and corresponding time in days were recorded in Table 2.

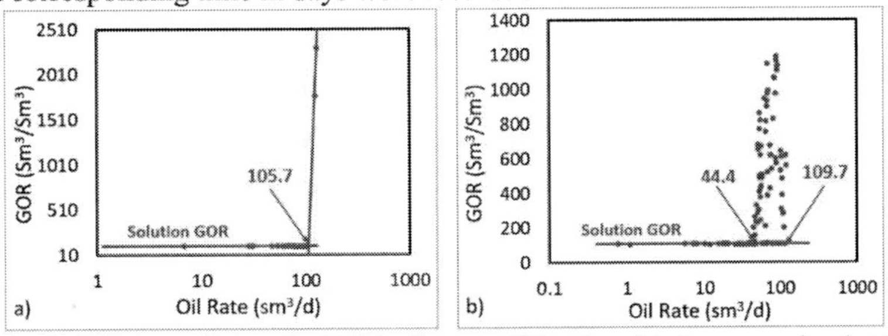


**Figure 4.** a) GOR vs. oil rate for W4 before modification. b) An example plot for W3.

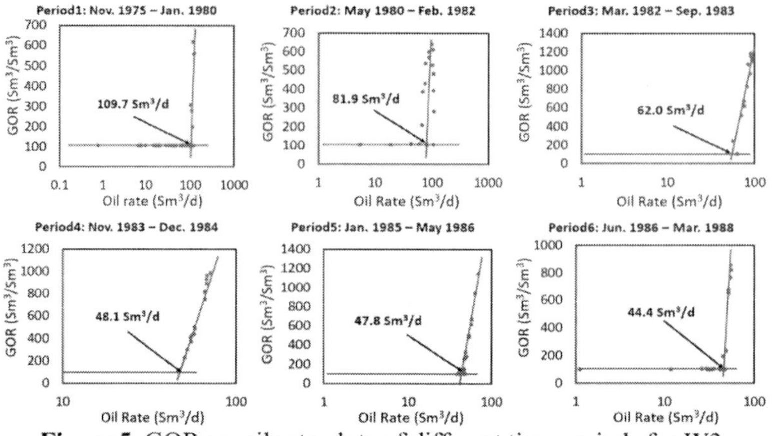


**Figure 5**. GOR vs. oil rate plots of different time periods for W3.

**Table 2.** Summary of reservoir characteristics, results from Addington's method, real critical rate, and corresponding time for wells

| Well | Period | $k_h$ | $k_v$ | $h_{ap}$ | $h_p$ | Time | Addington $Q_A$ | Estimated $Q_C$ |
|---|---|---|---|---|---|---|---|---|
| | | (mD) | (mD) | (ft) | (ft) | (days) | (sm³/d) | (sm³/d) |
| W4 | | 1250 | 500 | 91.2 | 9.84 | 2738 | 116.3 | 105.7 |
| W3 | 1 | 902 | 360 | 88.54 | 16.4 | 757 | 97.4 | 109.7 |
| | 2 | 902 | 360 | 88.54 | 16.4 | 1184 | 97.4 | 81.94 |
| | 3 | 902 | 360 | 88.54 | 16.4 | 1488 | 97.4 | 61.97 |
| | 4 | 902 | 360 | 88.54 | 16.4 | 2097 | 97.4 | 48.1 |
| | 5 | 902 | 360 | 88.54 | 16.4 | 2859 | 97.4 | 47.76 |
| | 6 | 902 | 360 | 88.54 | 16.4 | 3344 | 97.4 | 44.4 |
| W5 | 1 | 500 | 200 | 81.672 | 19.68 | 150 | 50.7 | 50.32 |
| | 2 | 500 | 200 | 81.672 | 19.68 | 759 | 50.7 | 37.32 |
| | 3 | 500 | 200 | 81.672 | 19.68 | 1185 | 50.7 | 28 |
| W8 | 1 | 375 | 150 | 67.568 | 9.84 | 1249 | 22.6 | 47.1 |
| | 2 | 375 | 150 | 67.568 | 9.84 | 1676 | 22.6 | 41 |
| | 3 | 375 | 150 | 67.568 | 9.84 | 2496 | 22.6 | 35 |
| W11 | 1 | 1250 | 500 | 49.2 | 16.4 | 182 | 54.9 | 72.29 |
| | 2 | 1250 | 500 | 49.2 | 16.4 | 456 | 54.9 | 48 |
| W19 | 1 | 1250 | 500 | 49.2 | 16.4 | * | 13.7 | 54 |
| | 2 | 1250 | 500 | 49.2 | 16.4 | * | 13.7 | 42 |
| | 3 | 1250 | 500 | 49.2 | 16.4 | * | 13.7 | 11.12 |
| W6 | 1 | 250 | 100 | 82 | 19.68 | * | 25.5 | 25 |
| | 2 | 250 | 100 | 82 | 19.68 | 393 | 25.5 | 23.06 |
| | 3 | 250 | 100 | 82 | 19.68 | * | 25.5 | 20 |
| | 4 | 250 | 100 | 82 | 19.68 | 1096 | 25.5 | 11 |

Note:
Kv/kh = 0.4; h = 127.95 ft; Oil viscosity = 2.7 cP; Porosity = 0.25; $F_2$ = 0.855
* Time cannot be determined from the plot and production data.

To find out the relationship between critical rate and time, a plot of critical rate vs. time was generated. As shown in Figure 6, the critical oil rate decreases exponentially as production time increases for a single well. As seen in the figure, the curves from different wells do not overlap with each other, so it can be concluded that the results are influenced by the unique properties of the wells.

After several attempts, a parameter group, $\frac{Q_c}{Q_A}\left(\frac{h_p}{h}\right)^2$, was plotted against well production time as shown in Figure 7. As a result, a correlation was developed for estimating critical oil rate for this reservoir. The estimated critical rate can be developed by

$$Q_c = Q_A \cdot CCF \quad [7]$$

where

$$CCF = \frac{0.023e^{-0.00055t}}{\left(\frac{h_p}{h}\right)^2} \quad [8]$$

In Equations [ 7 ] and [ 8 ], $Q_C$ is the critical rate (STB/D); $Q_A$ is the critical rate calculated by Addington's correlation (STB/D); $h_p$ is the perforation length (ft); $h$ is the oil zone thickness (ft); $t$ is time (d).

The generalization format of CCF can be written as

$$CCF = \frac{F_3 e^{-F_4 t}}{\left(\frac{h_p}{h}\right)^2} \quad [9]$$

where $F_3$ and $F_4$ are the parameters dependent on the reservoir conditions.

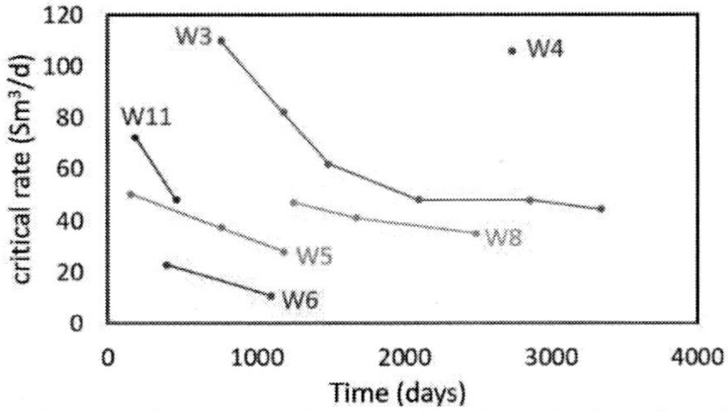


**Figure 6.** Relationship between critical rate and producing time for each well.

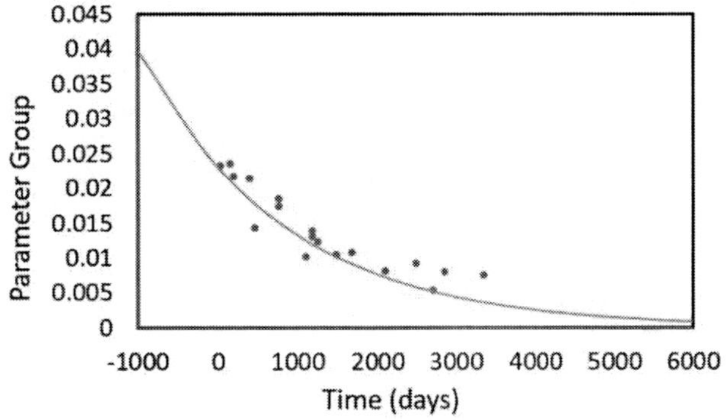


**Figure 7.** Critical rate and production time correlation.

W19 production data was not taken into account in the generation of the proposed correlation. It was used to validate the modified correlation. Well production performance for W19 (Figure 8) was plotted, in which GOR is larger than the solution GOR when the oil rate is larger than the critical rate, and GOR decreases and reaches the solution GOR when the oil rate is smaller than the critical rate as expected. The modified correlation can be used for prediction cases to avoid or minimize gas coning for proposed workover wells or infill drilling wells.

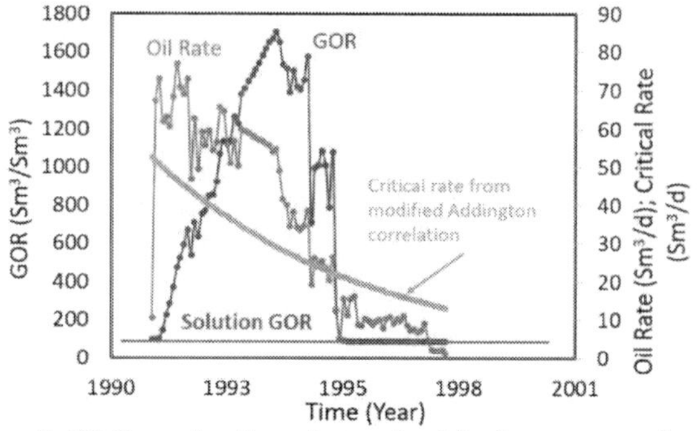


**Figure 8.** Well production plot and critical rate curve for W9.

*5.3 Discussion and comparison with other work*

Figure 9 and Table 3 compare critical rates computed by several gas coning correlations for W4. Schols [8] and Addington [9] are empirical correlations. Chaperon [4] is an analytical correlation. The result of Chaperon [4] is very high, which might due to the large density difference between oil and gas. Chaperon [4] correlation maybe more useful to estimate critical oil rate for water coning, although the paper mentioned "Water cones (vertical wells) follow the same equations as gas." Estimated critical oil rate for improved correlation is a function of time $(452.6e^{-0.00055t})$. The critical rate listed in Table 3 is estimated for May 1988 $(t = 2738\ d)$.

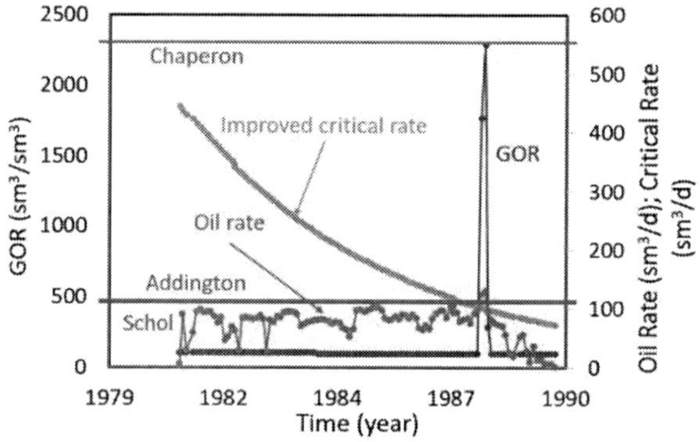


**Figure 9.** Gas coning correlation comparison.

**Table 3**. Critical rate calculated using several gas coning correlations

| Correlations | Estimated Qc (sm³/d) |
|---|---|
| Schols (1972) | 115 |
| Addington (1981) | 116.3 |
| Chaperon (1986) | 560.4 |
| Improved correlation | 100.4 |

In both Addington[9] correlation and the improved correlation, the value of the well radius and drainage radius do not significantly affect critical rate. The Schols[8] and Chaperon[4] theories show a dependence of critical rate on drainage radius. Schols[8] correlation also presents the effect of well radius on critical rate. However, Schols[8] theory applies to a fully penetrating and partially completed well rather than a partially penetrating well. The difference between these two situations should be very small when the penetration interval is much smaller than the well radius. Chaperon[4] generated the mathematical formulation for gas coning analysis by assuming no aquifer support and no flow boundary condition at initial gas oil contact. These assumptions make the application of the model very limited, which is not applicable for most reservoir conditions. Similar with Chaperon[4], Addington[9] correlation is also only applicable for the reservoir with limited gas cap and no aquifer zone.

In summary, the improved correlation described in this paper gives more accurate results for estimating critical rate. Predicted trends mostly agree with previous work that critical rates decrease with increasing well penetrations and decreasing horizontal permeability. For a single well, the critical rates decrease exponentially with time regardless of gas breakthrough occurrence as the underground reservoir conditions keep changing in production processes. The improved correlation is mainly applicable for most fields with both gas cap and aquifer zone.

## 6. Application to the Simulation Study

A 3D-simulation model of NGJ UT was constructed based on the geomodel using the PETREL commercial simulator. The well critical rates were estimated to help control the flowing bottom hole pressure. The model contains a medium-sized (m = 0.4) gas cap and a large aquifer volume ($1.04 \times 10^9$ $m^3$). The reservoir properties including PVT data, relative permeability curves and capillary pressure curve are shown in Appendix A. Table 4 gives the fluid saturation and the depth of Gas-Oil Contact (GOC) and Oil-Water Contact (OWC).

**Table 4**. Fluid saturation, GOC and OWC

| | |
|---|---|
| **Connate water saturation, fraction** | 0.35 |
| **Residual oil saturation, fraction** | 0.2 |
| **Critical gas saturation, fraction** | 0.02 |
| **GOC depth, m** | 2171 |
| **OWC depth, m** | 2210 |

The model was validated through pressure and production history match (Figure 10). Then five scenarios for several wells were designed to compare GOR performance before and after using the CCF. Table 5 presents the details of five cases and the results of the numerical tests. The results show that the producing GORs were controlled as low as 85 $m^3/m^3$ with utilization of CCF. While the GORs could reach to over 1000 $m^3/m^3$ without using CCF control.

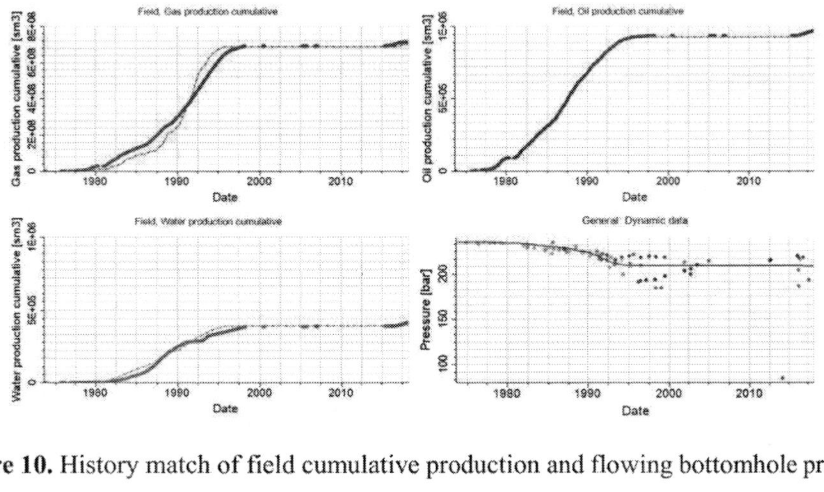


**Figure 10.** History match of field cumulative production and flowing bottomhole pressure.

**Table 5**. Numerical test results

| Cases | Actions | Without using CCF | | Using CCF | | |
|---|---|---|---|---|---|---|
| | | FBHP (bar)* | GOR (m3/m3) | Estimated critical rate (m3/d) | FBHP (bar) | GOR (m3/m3) |
| **Case1** | W11: reperforated at 2191.5 m to 2197.5 m | 189.76 | 1101.8 | 72.2 | 202 | 85.69 |
| **Case2** | W21: reperforated at 2182.2 m to 2188.2 m | 189.76 | 1699.7 | 28.6 | 205 | 85.75 |
| **Case3** | Infill 1: perforated at 2190 m to 2199 m | 189.76 | 1208.5 | 70.8 | 203 | 85.5 |
| **Case4** | Infill 2: perforated at 2183 m to 2189 m | 189.76 | 1686.1 | 64.8 | 205 | 85.71 |
| **Case5** | Infill 3: perforated at 2183m to 2189m | 189.76 | 1862.3 | 27.3 | 200 | 85.83 |

Note: *Average flowing bottom hole pressure for active wells at the end of April 2018.

Then eight prediction scenarios for the most economically advantageous situation are shown. The economic estimation is achieved by net present value (NPV) calculation (Appendix B). Figure 11 shows the comparison of field oil rate of each case. The predicted oil rate of Case1 follows the previous trend, and the field stops production in three years because of the oil rate reaching zero. In prediction case 2, assuming the casing condition was good, i.e., not requiring repairs, the workover actions were to squeeze the old perforation and perforate a new interval at -2387 to -2393 m as Action 1 in Table 5. Case 3, Case 5, Case 6, and Case 7 applied Action 2 to 5 respectively. Case 4 utilized the combination of Action 1 and 2. And Case 8 performed the combination of Action 1 to 3. Case 8 delivered the highest cumulative oil production of 1.56 MMSKL (9.81 MMSTB), leading to a RF of 8.44% in the next two decades. The economic estimation is summarized in Table 6.

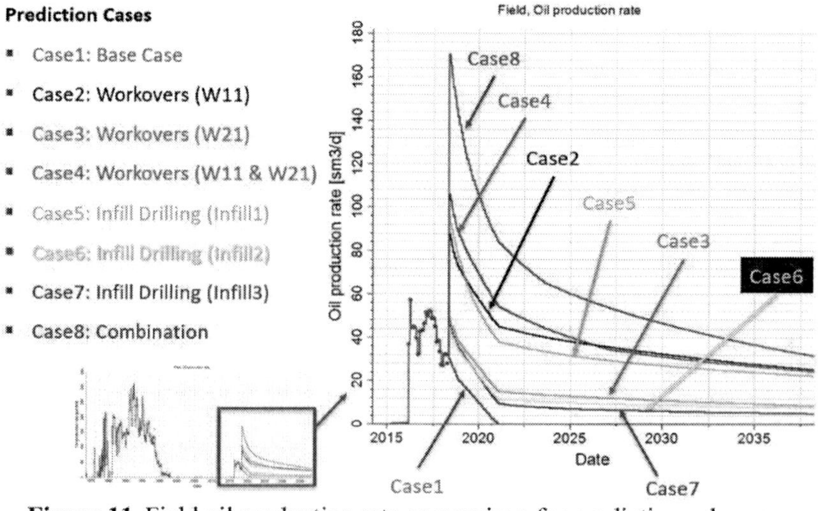


**Figure 11.** Field oil production rate comparison for prediction schemes.

**Table 6**. Summary of prediction results and economic estimation (start from April 2018 and end in April 2038).

| Cases | Np (MMSKL) | STOOIP (MMSKL) | RF | Net Present Value (NPV) |
|---|---|---|---|---|
| End of History | 1.1314 | 18.5 | 6.12% | - |
| Case1 | 1.14173 | 18.5 | 6.17% | 2.4 |
| Case2 | 1.40228 | 18.5 | 7.58% | 28.42 |
| Case3 | 1.23198 | 18.5 | 6.66% | 7.06 |
| Case4 | 1.41898 | 18.5 | 7.67% | 31.39 |
| Case5 | 1.36798 | 18.5 | 7.39% | 21.8 |
| Case6 | 1.21388 | 18.5 | 6.56% | 1.59 |
| Case7 | 1.19728 | 18.5 | 6.47% | 0.04 |
| Case8 | 1.56108 | 18.5 | 8.44% | 50.35 |

## 7. Conclusions

On the basis of the analyses and results of the work performed in this paper, the following conclusions can be made:

1. Critical rate exponentially decreases as production time increases for a single well (e.g., W3).
2. Based on the proposed gas coning correlation, a larger well perforation interval results in a lower critical rate.
3. Based on the improved gas coning correlation, a larger horizontal permeability results in a larger critical oil rate.
4. Producing GOR can be minimized from over 1000 $m^3/m^3$ to approximate 85 $m^3/m^3$ by using the improved gas coning correlation.
5. The prediction Case 8 that involved a combination of workovers of W11 and W21, and infill drilling of Infill1 would bring a RF of 8.44% and the highest benefits of $50.35 MM in the next two decades

It is worth mentioning that the oil rate dramatically reduced and reservoir pressure dropped rapidly after the gas coning occurrence. It can be concluded that the low recovery factor might be due to the frequent gas coning problems in the field.

## Acknowledgement

The authors would like to appreciate the support and permission from UH-EIP to publish this paper. The geological, production/injection, and other engineering data provided by our partner is highly appreciated.

## References

[1] Muskat, M., and Wycokoff, R. D., 1935, "An Approximate Theory of Water-coning in Oil Production," Transactions of the AIME, 114(01), pp. 144-163.
[2] Chierici, G. L., Ciucci, G. M., and Pizzi, G., 1964, "A Systematic Study of Gas and Water Coning By Potentiometric Models," Journal of Petroleum Technology, 16(08), pp. 923-929.
[3] Wheatley, M. J., "An Approximate Theory of Oil/Water Coning," Proc. SPE Annual Technical Conference and ExhibitionSPE-14210-MS.

[4] Chaperon, I., "Theoretical Study of Coning Toward Horizontal and Vertical Wells in Anisotropic Formations: Subcritical and Critical Rates," Proc. SPE Annual Technical Conference and ExhibitionSPE-15377-MS.
[5] Guo, B., and Lee, R. L. H., 1993, "A Simple Approach to Optimization of Completion Interval in Oil/Water Coning Systems," SPE Reservoir Engineering, 8(04), pp. 249-255.
[6] Abass, H. H., and Bass, D. M., "The Critical Production Rate in Water-Coning System," Proc. Permian Basin Oil and Gas Recovery ConferenceSPE-17311-MS.
[7] Tabatabaei, M., Ghalambor, A., and Guo, B., 2012, "An Analytical Solution for Water Coning in Vertical Wells," SPE Production & Operations, 27(02), pp. 195-204.
[8] Schols, R. S., 1972, "Water coning. An empirical formula for the critical oil-production rate," Erdoel-Erdgas Z. (Germany, Federal Republic of), 88(1).
[9] Addington, D. V., 1981, "An Approach to Gas-Coning Correlations for a Large Grid Cell Reservoir Simulator," Journal of Petroleum Technology, 33(11), pp. 2267-2274.
[10] Yang, W., and Wattenbarger, R. A., "Water Coning Calculations for Vertical and Horizontal Wells," Proc. SPE Annual Technical Conference and ExhibitionSPE-22931-MS.
[11] Benamara, A., and Tiab, D., "Gas Coning in Vertical and Horizontal wells, a Numerical Approach," Proc. SPE Rocky Mountain Petroleum Technology ConferenceSPE-71026-MS.
[12] Cameron, I. T., and Hangos, K., 2001, Process modelling and model analysis, Elsevier.
[13] Aashto, 2010, Guide for the Local Calibration of the Mechanistic-empirical Pavement Design Guide, American Association of State Highway and Transportation Officials.
[14] Amisigo, B. A., 2005, Modelling riverflow in the Volta Basin of West Africa: A data-driven framework, Cuvillier Verlag.
[15] Clarkson, C. R. R., and Qanbari, F., 2015, "A Semianalytical Forecasting Method for Unconventional Gas and Light Oil Wells: A Hybrid Approach for Addressing the Limitations of Existing Empirical and Analytical Methods," SPE Reservoir Evaluation & Engineering, 18(01), pp. 94-110.
[16] Prasun, S., and Ghosh, S., 2018, "A new analytical model of ultimate water cut for light oil reservoirs with bottom-water," J Oil Gas Petrochem Sci, 1(3), pp. 74-81.
[17] Prasun, S., and Wojtanowicz, A. K., 2019, "Semi-analytical prediction of critical oil rate in naturally fractured reservoirs with water coning," Journal of Petroleum Science and Engineering, 180, pp. 779-792.
[18] Thakur, G. C., and Satter, A., 1998, Integrated waterflood asset management.

## Appendix A

PVT data was obtained from three wells: W3, W4 and W8. Figure 12 shows the PVT curves of the oil phase. The data from W3 is the most reliable and acceptable among data from the three wells. The reservoir started production in 1973, and W3 PVT data was measured in 1977, which is closest to the first production date. The reservoir pressure during these four years has not changed significantly, so the W3 PVT data is considered representative of the original sample. Figure 13 shows the gas compressibility factor, gas density, gas formation volume factor, gas viscosity changing as pressure increases.

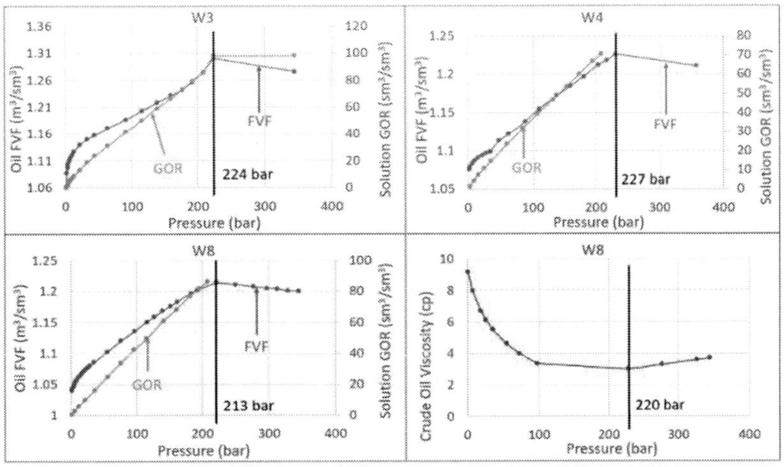


**Figure 12**. Oil phase PVT curves for W3, W4 and W8.

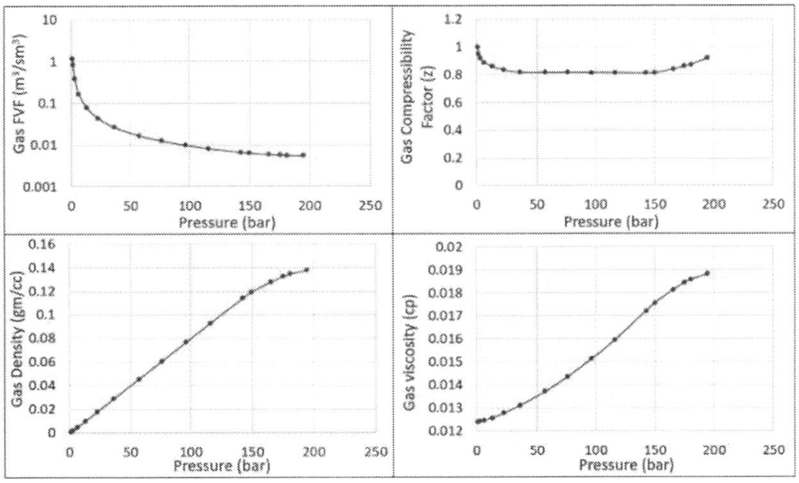


**Figure 13**. Gas property variation with pressure.

Core data was obtained from the core analysis report. Eight core samples from different depths were selected to generate the relative permeability curves and capillary pressure curve. The final curves after averaging and extrapolating are shown in Figure 14. The multicolored dots in Figure 14. c) represent lab data. The black curve is the final curve after averaging the multicolored dot values. The connate water saturation is about 35%, which is the same as the initial water saturation obtained from relative permeability properties.

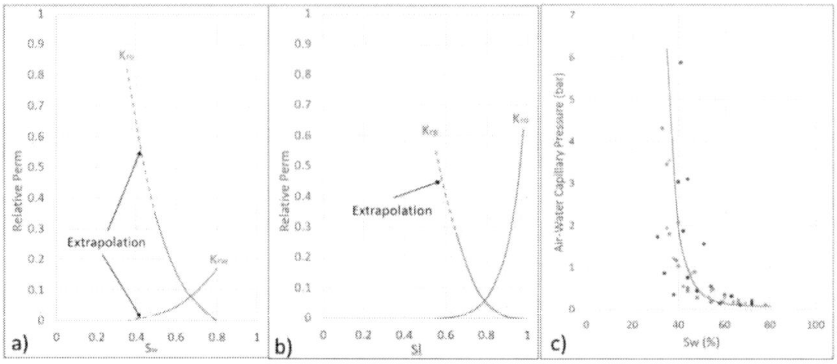


**Figure 14**. a) Oil-water relative permeability curve. b) Gas-oil relative permeability curve. c) Capillary pressure curve.

Appendix B

The NPV for incremental production is calculated by

$$NPV = \sum_{t=1}^{n} \frac{R_t}{(1+i)^t} \qquad [10]$$

where $R_t$ is the net cash flow during a single period; i is the discount rate; and t is the number of time periods.

The equations to calculate $R_t$ are presented as

$$R_t = NR + CE + OE + PT \qquad [11]$$

$$NR = GR \cdot (1 - R_{oy}) \qquad [12]$$

$$GR = N_i \cdot OP + G_i \cdot GP \qquad [13]$$

where NR is the net revenue; CE is the capital expense (used as a negative number); OE is the operating expense (used as a negative number); PT is the production taxes (used as a negative number); GR is the gross revenue; $R_{oy}$ is the royalties; $N_i$ is the oil production during a single period in STB; OP is the oil price in $/STB; $G_i$ is the gas production during a single period in MSCF; and GP is the gas price in $/MSCF.

Table 7 shows the parameters used in NPV calculation in prediction schemes.

**Table 7.** Oil/gas prices, expenses parameters used in NPV calculation.

| | |
|---|---|
| **Oil Price** | $50/stb |
| **Gas price** | $2.6/Mscf |
| **Operation expenses** | $0.49 MM for the first year with an increasing rate of 3% per year |
| **Royalties** | 12.5% of gross revenue |
| **Production taxes** | 5% of net revenue |
| **Workover expenses** | $1 MM/well |
| **Infill drilling expenses** | $4 MM/well |

# Wetting Behavior of $CO_2$ Droplets on Smooth Solid Surface: Molecular Simulation Perspective

**Heng-guang CAO, Xue-wen CAO(✉), Xiang-yang ZHAO, Yong-qing MA, Mei-qin DONG, Hao LI, Jiang BIAN**

College of Pipeline and Civil Engineering, China University of Petroleum (East China),
Qingdao, China
caoxw@upc.edu.cn

**Abstract.** The wettability of droplets on solid surfaces is important for accurately revealing the microscopic mechanisms of gas condensation nucleation and droplet growth. During the contact condensation of $CO_2$ gas on the heat exchanger surface in the pressurized liquefied natural gas technology, the wettability of $CO_2$ droplets on the heat exchanger surface directly affects the heat transfer thermal resistance of the heat exchanger, which then affects the heat transfer efficiency of methane and ethane in the heat exchanger. Therefore, molecular dynamics simulations were used to study the spreading process and wetting patterns of nanoscale $CO_2$ droplets on different energy surfaces. The results show that as the potential well depth $\varepsilon$ of the wall atoms increases, the intensity of the solid-liquid interaction increases and the corresponding surface energy increases accordingly, showing different droplet spreading rates and wetting characteristics. Unlike the interfacial characteristics of macroscopic droplets, there are significant fluctuations at the gas-liquid interface of droplets on the molecular scale, but microdroplets can still form a specific contact angle after spreading on different energy surfaces in a statistical sense, and this contact angle decreases with increasing intensity of solid-liquid interaction. The low-energy surface at potential well depths $\varepsilon$ less than 266 $J \cdot mol^{-1}$ exhibits a $CO_2$-phobicity, and the surface becomes $CO_2$-philic as the potential well depth continues to increase. The trend of the contact angle of $CO_2$ droplets affected by temperature is the same as that of the center-of-mass height, which characterizes the spreading morphology of $CO_2$ droplets. As the temperature increases, the contact angle decreases due to the further spreading and wetting of droplets on different energy surfaces. As the $CO_2$-philicity of the surface gets higher, the contact angle decreases to a greater extent.

## 1 Introduction

Floating liquefied natural gas system (FLNG) is an emerging offshore gas field development technology that combines the functions of natural gas extraction, processing, liquefaction, storage and handling, with the advantages of the short construction cycle, reusability and flexibility, and is the best choice for deep offshore gas field development [1]. However, the current natural gas pretreatment process is complex, in which different process technologies are used for dehydration [2], sour gas removal [3], and heavy hydrocarbon removal [4], which have the disadvantages of complex systems, high energy consumption, and high investment and operation costs, and are difficult to meet the strict requirements for compactness and reliability of the process flow imposed by FLNG [5]. Therefore, it is important to explore and develop new FLNG technologies for the development of the offshore oil and gas industry. Pressurized liquefied natural gas (PLNG) technology is an emerging natural gas liquefaction technology in recent years [6] that allows natural gas liquefaction at moderate pressures and higher temperatures, thus significantly reducing the facilities required for processing and liquefaction. The introduction of PLNG technology into the offshore natural gas liquefaction field can reduce the cooling capacity required for liquefaction to save energy and reduce the $CO_2$ pre-treatment requirements and facility footprint.

PLNG technology allows natural gas to contain up to 3-6% $CO_2$, thus condensation [7], crystallization [8] and frost [9] phase change precipitation of $CO_2$, heavy hydrocarbons and other impurity gases can occur during the natural gas cooling process. During the contact condensation of $CO_2$ gas on the heat exchanger surface, the wettability of $CO_2$ droplets on the heat exchanger surface directly affects the heat transfer thermal resistance of the heat exchanger [10], which then affects the heat transfer efficiency of methane and ethane in the heat exchanger. Wettability represents the degree of affinity and adhesion of the liquid on the wall surface and is measured by the contact angle [11]. The factors affecting wall wettability include both wall material and surface microstructure [12], and the difference in wettability directly affects the formation location and nucleation rate of liquid nuclei and crystal nuclei [13]. Studying the wetting state of droplets on different energy surfaces is essential for accurately predicting the nucleation rate and revealing the droplet growth mechanism.

In petroleum and natural gas engineering, $CO_2$ is often applied to enhance oil and gas recovery [14],[15], so some studies on $CO_2$ condensation have focused on the wettability of different sealing reservoirs in $CO_2$ environments. Yan et al. [16] used non-equilibrium molecular dynamics simulations to study the repulsion process of supercritical $CO_2$ ($scCO_2$) and water injection into adsorbed oil reservoirs in nanochannels, revealing the mechanism of enhanced recovery by $CO_2$ water alternating-gas injection technique. Tenney and Cygan [17] investigated the behavior of $scCO_2$ and aqueous solutions on the surface of kaolinite hydrophilic and hydrophobic substrates. When a large amount of aqueous phase is present, $scCO_2$ forms non-wetting droplets on the hydrophilic surface of kaolinite, while when a large amount of $scCO_2$ is present, the non-wetting droplets interact with the hydrophobic surface of kaolinite by adsorbing a mixture of $CO_2$ and $H_2O$ molecules. It was also noted in the literature [18] that the $CO_2$ fluid environment can effectively modify the wetting behavior of water droplets on solid surfaces. It can be seen that most simulation studies add $CO_2$ as a guest molecule and discuss the effect of the degree of $CO_2$ enrichment on the wettability of the original system, or observe the difference in wettability of the mixed system compared to the original system, and do not directly derive the wettability characteristics of pure $CO_2$ droplets. Furthermore, fundamental studies on the wetting and condensation behavior of $CO_2$ on heat transfer surfaces in a gaseous atmosphere are largely limited [19],[20], but this is essential to study the gas-liquid phase change precipitation of $CO_2$ in the PLNG process. Experimental conditions that maintain the stable presence of $CO_2$ droplets while also measuring their contact angle are not easily achieved, and experimental studies on the wettability of pure $CO_2$ droplets have hardly been reported so far.

Molecular dynamics (MD) simulations, with time scales on the order of nanoseconds and tens of nanoseconds and spatial scales on the order of nanometers and tens of nanometers, have provided new technical ideas for solving many problems that are not experimentally possible under existing conditions [21] and have become indispensable tools for studying condensation, wetting, and nucleation processes of condensable gases on structural surfaces [22]. Therefore, in this paper, MD simulations are used to study the wetting characteristics of pure $CO_2$ droplets on a smooth solid wall surface. Due to the lack of experimental measurement data on the contact angle of $CO_2$ droplets on different material surfaces, the interaction between $CO_2$ and wall atoms can be changed by adjusting the potential well parameters of the wall atoms. In the paper, the number of simulated droplet molecules and the number of atomic layers on the wall are first examined and determined, then the spreading process of $CO_2$ droplets on different energy surfaces is studied, the interfacial characteristics and contact angles of droplets on the surface are analyzed, and finally the effects of changes in potential well depth and temperature on the wetting characteristics of $CO_2$ droplets are discussed.

## 2 Simulation methods

### *2.1 Simulation model*

The droplet spreading and contact angle measurement simulation system is shown in Fig. 1, where the simulation box consists of a solid wall surface and a $CO_2$ droplet. The simulated system has lengths of 144.588, 144.588 and 200.00 Å in the x, y and z directions, respectively, and the system applies periodic boundary conditions in the x and y directions and non-periodic boundary conditions in the z direction with a rebound wall at the boundary. The number of molecules of $CO_2$ droplets gradually varies from 800, 1800, 2800, 3800, 4800 and 5800, while the number of atomic layers on the wall varies sequentially from 2, 4, 6, 8, 10 and 12. To simulate the wetting of 4800 molecular droplets on an 8-layer atomic wall surface, for example, the solid surface consists of 25600 atoms arranged in face-centered cubic form with dimensions of 144.588 Å×144.588 Å×14.4588 Å. A spherical droplet consisting of 4800 $CO_2$ molecules is placed above the surface and allowed to evolve, with a spherical radius of 40 Å and a density of 1165.8 kg/m$^3$. The horizontal projection of the droplet center is located at the center of the wall, its vertical height is 50.5412 Å from the wall, and the distance between the lowest position of the droplet and the wall is 10.5412 Å.

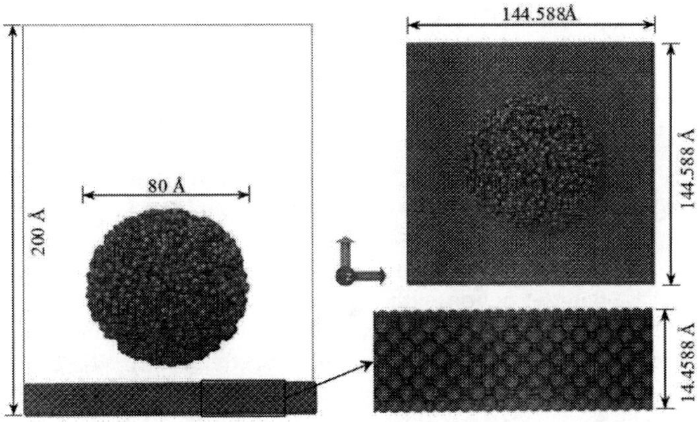


**Fig. 1** Model of $CO_2$ droplet wetting on the wall.

Considering the higher accuracy in describing the gas-liquid equilibrium characteristics of $CO_2$ and physical parameters such as surface tension, the interaction between $CO_2$ molecules is characterized using the TraPPE_flex molecular model [23]. According to this model, the interaction potential between $CO_2$ molecules is expressed as:

$$U_{ij}=4\varepsilon_{ij}\left[\left(\frac{\sigma_{ij}}{r_{ij}}\right)^{12}-\left(\frac{\sigma_{ij}}{r_{ij}}\right)^{6}\right]+\frac{q_iq_j}{4\pi\varepsilon_0 r_{ij}}+\frac{1}{2}k_r(r-r_0)^2+\frac{1}{2}k_\theta(\theta-\theta_0)^2 \quad (1)$$

where $r_{ij}$, $\varepsilon_{ij}$, $\sigma_{ij}$, $q_i$, $q_j$ and $\varepsilon_0$ are the distances of atoms to $i$ and $j$, respectively, the LJ potential trap depth, the distance between atoms where the LJ potential is 0, the charge carried by the atoms and the vacuum dielectric constant.

The forces between the wall atoms are represented by the LJ potential function [24], where the fundamental parameter is $\sigma = 2.616$ Å. The potential well depth $\varepsilon$ is adjusted to obtain different surface energies based on the LJ potential function parameter of the $CO_2$ molecule. The forces between the wall atoms and $CO_2$ molecules are calculated by adopting the Lorentz-Berthelot mixing principle, where $\sigma_{ij} = (\sigma_i + \sigma_j)/2$ , $\varepsilon_{ij} = \sqrt{\varepsilon_i \varepsilon_j}$ . For common metals such as Ag, Al, Au, Cu, Pt, etc., the potential well parameter $\varepsilon$ varies greatly among pure metals, while the distance parameter $\sigma$ is very close [24], so the parameter affecting the interaction between wall atoms and $CO_2$ molecules is mainly the potential well depth $\varepsilon$. In this paper, the simpler FCC atomic wall surface with different potential well depths is used to represent the actual wall surface with different energies. The higher the value of the potential well depth $\varepsilon$, the higher the strength between the wall atoms and the $CO_2$ molecules, and the corresponding increase in surface energy.

### *2.2 Simulation details*

The open-source software GROMACS (version 2019) was used for MD simulations, and the simulation results were visualized using VMD software. The whole wetting simulation process includes three stages: energy minimization, equilibrium and kinetic calculation. The initial model was constructed by the Packmol tool and VMD software and optimized using the conjugate gradient method. The equilibrium phase is performed in the NVT tether with a simulation time of 10 ns. The temperature is controlled by a Nosé-Hoover thermostat, the initial velocity of the particles is generated by the Maxwell-Boltzmann distribution according to the desired temperature, and the equations of motion are integrated using the leap-frog algorithm with a step size of 1 fs. The kinetic calculations are performed under the NVE system synthesis with a simulation time of 2 ns. The temperature control of the $CO_2$ droplet is withdrawn and the results of droplet wetting are sampled and analyzed. The truncation radius taken in the calculation of the LJ action is 10 Å. The PME method is adopted to sum the long-range effects of the electrostatic force, and the time step is chosen to be 1 fs.

### *2.3 Model Validation*

In order to verify the validity of the TraPPE_flex model, the gas-liquid interface of $CO_2$ was investigated. A rectangular simulation box of 40 × 40 × 140 Å$^3$ was built, and 1000 $CO_2$ molecules were randomly placed in the middle of the box with two vacuum layers on both sides of the $CO_2$ layer. Then, the NVT system synthesis is used for equilibration, and the system can achieve a stable liquid film and gas phase equilibrium. As shown in Fig. 2(a), the region with a dense molecular number indicates the liquid phase and the region with a sparse molecular number indicates the gas phase. As shown in Fig. 2(b), the mass density distribution of $CO_2$ in the temperature range of 228.15-288.15 K in the equilibrium state of the gas-liquid phase along the Z-axis direction is counted. As the temperature increases, the density and the molecular number of liquid-phase $CO_2$ decrease, while the density and the molecular number of gas-phase $CO_2$ increase. This indicates that the $CO_2$ vapor tends to condense on the liquid film to maintain the gas-liquid phase equilibrium during the cooling process. Fig. 2(c) compares the MD simulation results of the density at $CO_2$ gas-liquid coexistence with the experimental data from NIST. The simulated data at low temperatures have slight errors with the experimental data, which indicates that the TraPPE_flex model can be used for the gas-liquid phase equilibrium simulation of $CO_2$, and it can predict the gas-liquid interface behavior of $CO_2$ molecules better.

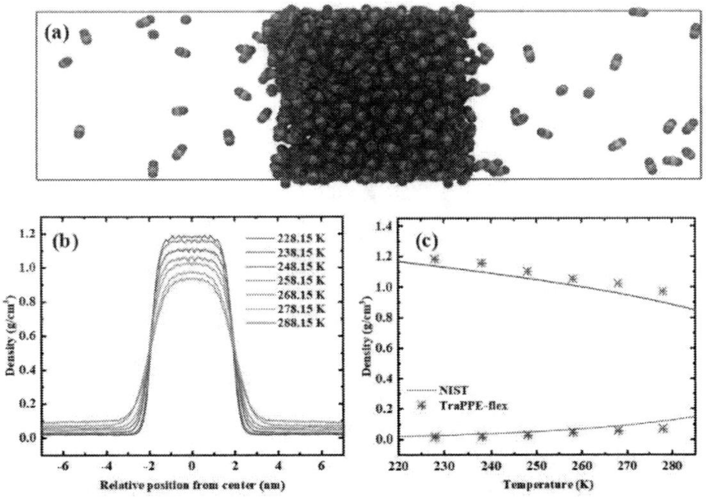


**Fig. 2** Carbon dioxide gas-liquid coexistence system. (a) Snapshot of the interface of $CO_2$ at a temperature of 228.15 K. (b) Density distribution of $CO_2$ along z-direction at different temperatures. (c) Comparison of simulated data with experimental data.

## 3 Results and Discussion

### *3.1 Calculation of contact angle*

As shown in Fig. 3, the density of $CO_2$ in the simulation box was sampled every 10 ps during the kinetic calculation stage, and the contact angle of the droplet was finally calculated using the density profile method. From Fig. 3(a), it can be seen that the $CO_2$ molecules in the center of the droplet are obviously dense, indicating that the molecules in this part are relatively stable, while the $CO_2$ molecules at the position of the gas-liquid interface are relatively loose, showing that the probability of the liquid-phase molecules in this region is relatively low, which is caused by the fluctuation of the molecules at the gas-liquid interface. Nevertheless, from a statistical point of view, the obtained density image interface is smooth enough to be approximated as a circle. And the contact angle obtained by this image measurement can be considered as the equilibrium contact angle of this surface. As shown in Fig. 3(b), the contact angle can be obtained from the following geometric relationship:

$$\theta = 90^\circ + \arcsin\left(\frac{h-r}{r}\right) \qquad (2)$$

where $h$ and $r$ are the height and radius of the droplet, respectively, and $r^*$ is the radius of the contact circle surface. As shown in Fig. 3(c), the contact angle of the droplet on the surface is calculated to be 142°, which far exceeds the neutral standard of 90°, so it can be indicated that this surface is a $CO_2$-phobic surface.

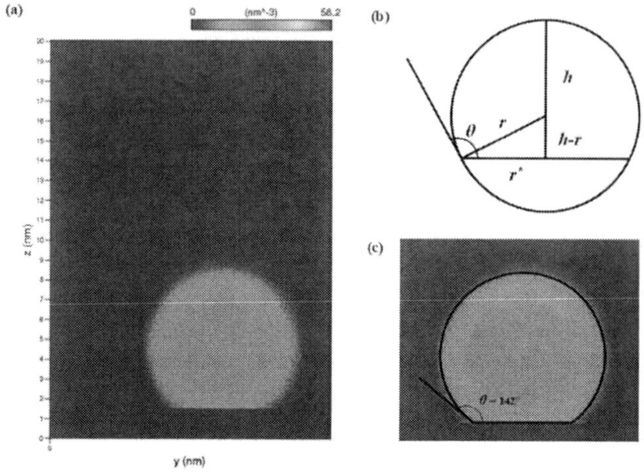


**Fig. 3** (a) Density distribution of liquid droplets on a smooth surface. (b) Schematic diagram of contact angle calculation. (c) Simulated value of contact angle.

### *3.2 Determination of droplet molecular number and wall thickness*

The number of droplet molecules and the number of atomic layers on the wall are examined and determined in the simulations on the $CO_2$-phobic, neutral, and $CO_2$-philic walls, respectively. Different types of wetted walls can be simulated by varying the potential well depths $\varepsilon$, which are 57, 266, and 500 $J{\cdot}mol^{-1}$ for sparse $CO_2$, neutral, and pro-$CO_2$ walls, respectively.

Fig. 4 gives the calculated values of contact angles for pure $CO_2$ droplets with the different numbers of molecules, on the $CO_2$-phobic, neutral, and $CO_2$-philic walls. When the number of molecules is small, the contact angle values of the droplets fluctuate widely. When the number of molecules gradually increases and reaches 4800, the values of contact angles of droplets on the hydrophobic $CO_2$, neutral and pro-$CO_2$ walls all stabilize. When the number of droplet molecules is small, the gas-liquid interface fluctuates drastically, leading to large deviations in multiple simulations. As the number of molecules increases, the gas-liquid interface becomes stable and the fluctuations decrease gradually. Therefore, the number of droplet molecules was set to 4800 in the subsequent simulations.

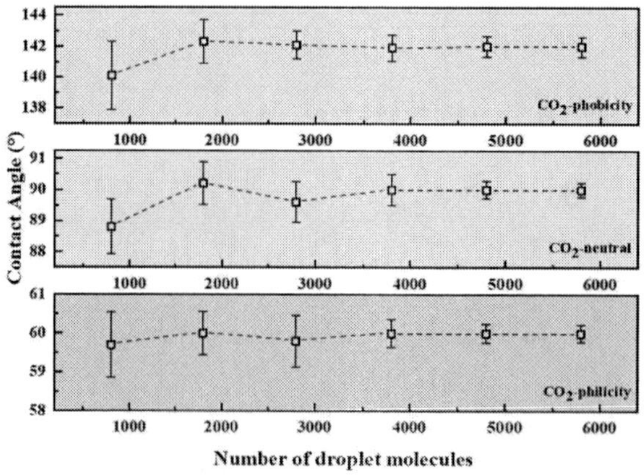


**Fig. 4** Calculated values of contact angle for different droplet molecular numbers on different wetting types of walls.

The calculated values of the contact angle of pure $CO_2$ droplets (molecular number 4800) on the $CO_2$-phobic, neutral, and $CO_2$-philic walls with the different number of atomic layers on the walls are shown in Fig. 5. The contact angle does not show a large gap with the increasing number of atomic layers on the wall surface, and it is stable around the same level value on the $CO_2$-phobic, neutral, and $CO_2$-philic walls. However, in the model setup, it should be ensured that the distance between the atom of any $CO_2$ molecule and any atom of the bottommost layer of the wall is larger than the interaction truncation radius to make the simulation setup more reasonable. Therefore, the thicker the wall, the more reasonable the motion of the atoms in the wall that can interact with the $CO_2$ molecule. However, the higher the number of atomic layers, the steeper the demand on the computational performance. Therefore, weighing the calculation result error and the calculation performance condition, a wall with 8 atomic layers was determined for the $CO_2$ wetting simulation, corresponding to a wall thickness of 14.4588 Å.

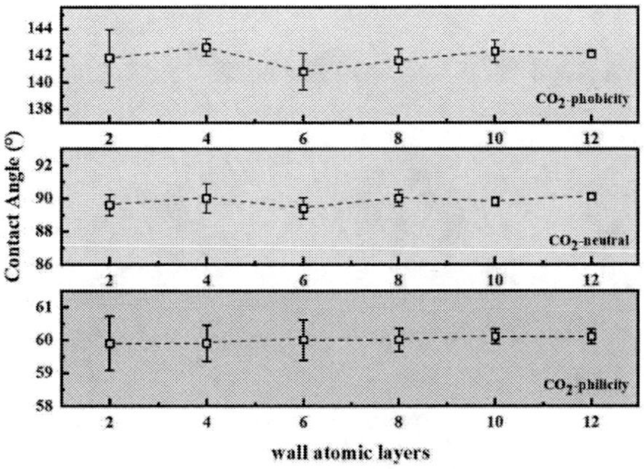


**Fig. 5** Calculated values of contact angle for different numbers of atomic layers on different wetting types of walls.

### *3.3 Spreading process of $CO_2$ droplets on surfaces with different energies*

After determining the number of molecules and the number of atomic layers on the wall of the wetting droplets, the study of the wetting process of $CO_2$ nanodroplets on different energy surfaces was carried out by varying the potential well depth $\varepsilon = 20, 57, 133, 266, 333, 433, 533$, and 1000 J·mol$^{-1}$ for the LJ potential energy of the wall atoms. The MD simulation results of the droplet wetting process on different surfaces are shown in Fig. 6. The initial $CO_2$ droplets evolve into gas-liquid mixed droplets or gas-liquid mixed films, depending on the wettability of the surface. Except for the higher energy surface with $\varepsilon = 1000$ J·mol$^{-1}$, the droplets gradually change to the spherical deficient morphology. For the lower energy surfaces with $\varepsilon$=20, 57 and 133 J·mol$^{-1}$, the liquid-wall contact line appears to retract to a certain extent and shows $CO_2$-phobic characteristics due to the weak interaction between the wall atoms and $CO_2$ molecules. As the surface energy increases, for the intermediate energy surfaces with $\varepsilon$=266 and 433 J·mol$^{-1}$, the liquid-wall contact lines expand outward rapidly, and the liquid begins to stabilize after spreading to a certain extent and shows a certain contact angle characteristic. When the surface energy increases to a higher level ($\varepsilon = 1000$ J·mol$^{-1}$), the droplet spreading characteristics appear to change significantly again. Due to the short-range characteristic of LJ action, the wall atoms act very strongly on the neighboring $CO_2$ molecules, attracting more $CO_2$ molecules to the solid surface, thus leading to the rapid spreading of $CO_2$ molecules near the wall in the plane direction, and finally, the droplets evolve into a liquid film on the wall, completely wetting the surface.

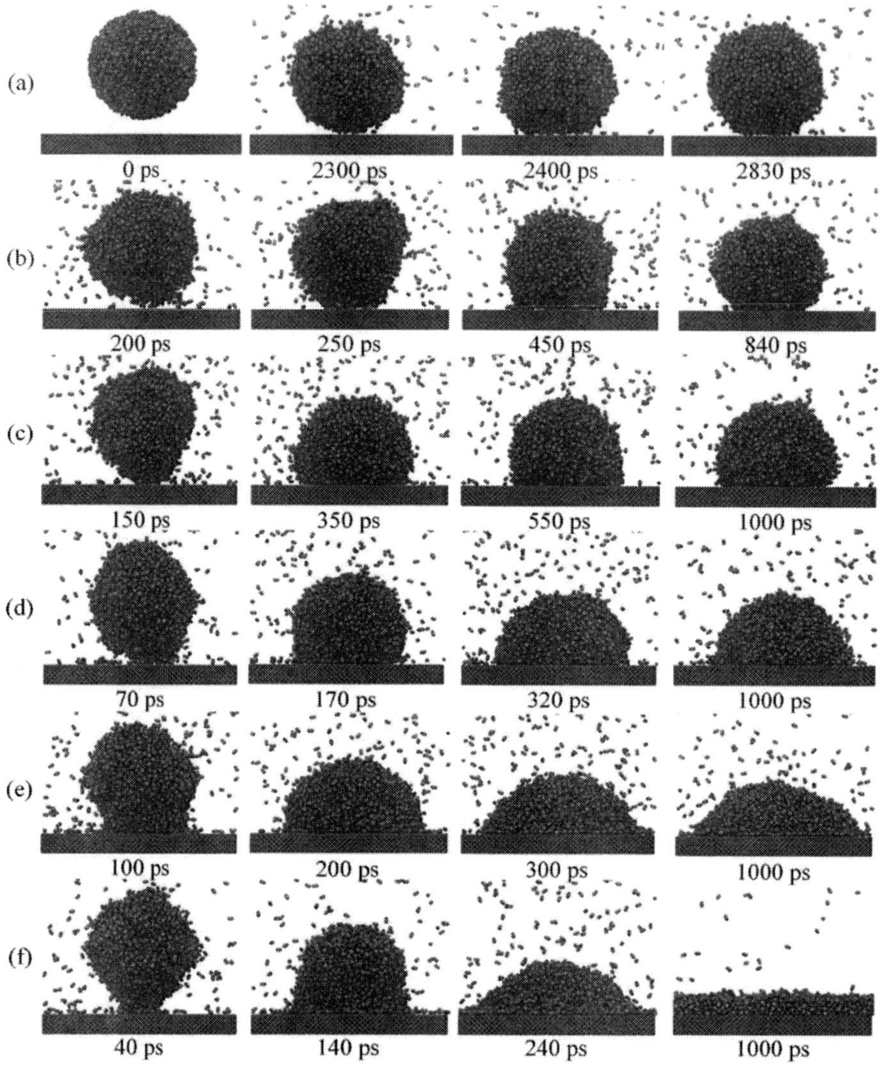


**Fig. 6** Wetting processes of droplets on surfaces with different energies (a) $\varepsilon = 20$ J·mol$^{-1}$. (b) $\varepsilon = 57$ J·mol$^{-1}$. (c) $\varepsilon = 133$ J·mol$^{-1}$. (d) $\varepsilon = 266$ J·mol$^{-1}$. (e) $\varepsilon = 433$ J·mol$^{-1}$. (f) $\varepsilon = 1000$ J·mol$^{-1}$.

In terms of droplet spreading occurrence rate, droplet spreading occurs relatively fast on high-energy surfaces. For higher energy surfaces, the force between wall atoms and $CO_2$ molecules is very strong, so $CO_2$ molecules will spread in the direction of the surface plane at a very fast rate, which is also conducive to reaching the wetting equilibrium state more quickly. For low-energy surfaces, the force between wall atoms and $CO_2$ molecules is relatively weak, so the wetting evolution of droplets on the wall surface is mainly determined by the force of $CO_2$ molecules themselves, and the morphological transition occurs more slowly considering the low force and surface tension of $CO_2$ molecules themselves.

During the wetting process, the closer the droplet is to a sphere, the larger the contact angle is, and the more it tends to spread flat on the wall to be wetted, the smaller the contact angle is. The change in the position of all $CO_2$ molecules in the droplet leads to a change in the morphology of the droplet as a whole, so the position of the droplet center of mass can be used to represent the change in the morphology of the droplet. Since smooth walls are used in the simulations, the coordinate component com[Z] of the droplet center of mass in the vertical direction can reflect the spreading of the droplet. To further investigate the wetting process of the droplet, the position change of the droplet center of mass along the z-axis with time t on three different energy surfaces was calculated. As shown in Fig. 7, the com[Z] of the droplet gradually decreases as it gradually spreads on the wall from the initial spherical shape. The higher the wall energy, the smaller the com[Z] of the droplet center of mass position. The droplet takes a longer time to reach the wetting equilibrium state on the hydrophilic wall than on the hydrophobic wall, and the droplet will be more stable after equilibrium on the hydrophilic wall, and the center-of-mass position does not fluctuate much.

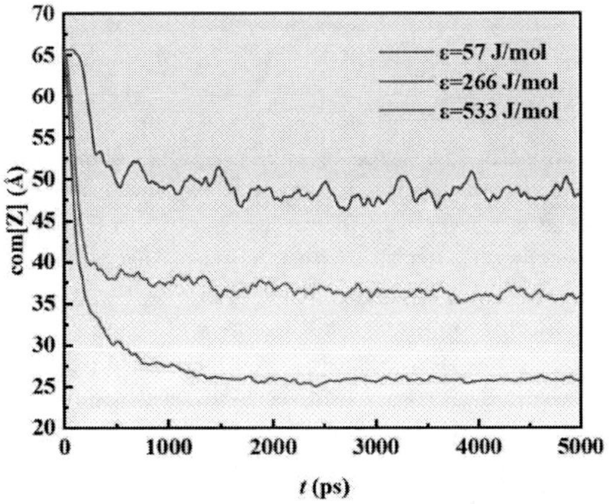


**Fig. 7** Variation of the position of the center of mass of a liquid drop along the z-axis

### 3.4 *Interface characteristics and contact angles of the droplet on the surface*

When the simulation proceeds to $1.5\times10^6$ steps, the droplet spreading reaches a stable state on most of the surfaces, which can be judged by the variation of the droplet center-of-mass position along the z-axis. However, from the droplet images shown in Fig. 8, even in the relatively stable spreading state, there is a certain degree of fluctuation at the gas-liquid interface of the $CO_2$ droplet. That is, the gas-liquid interface on the molecular scale is not smooth and stable, but there is a certain degree of molecular density rise and fall. From the gas-phase perspective, this is because the droplet is in the nanometer scale, where the gas-phase molecules no longer satisfy the continuous medium assumption. The forces at the gas-liquid interface caused by random collisions of gas-phase molecules are no longer symmetric, resulting in a large number of random fluctuations at the gas-liquid interface, which makes the gas-liquid interface not smooth. Therefore, the random fluctuations at the gas-liquid interface of nanoscale microdroplets become an important feature to distinguish their wettability from that at a macroscopic scale.

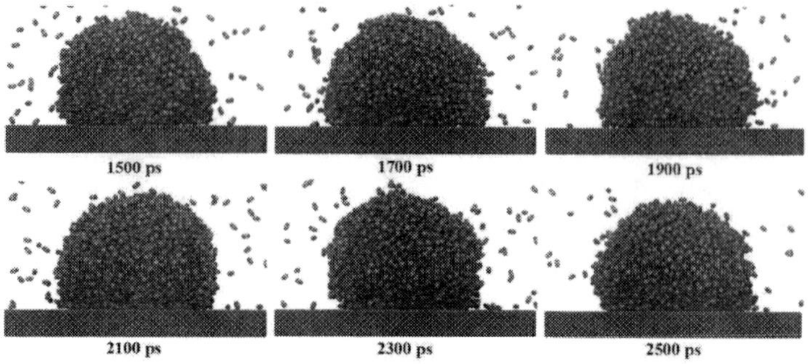


**Fig. 8** Characteristics of fluctuations at the gas-liquid interface of droplets on the surface at different moments ε = 133 J·mol$^{-1}$

Fig. 9 (a) shows the two-dimensional density cloud map of $CO_2$ droplets on different energy surfaces, which can clearly identify the shape of $CO_2$ droplets on the solid surface. With the increase of the wall metal atom potential well depth $\varepsilon$, the liquid $CO_2$ molecules are more easily attracted to the pro-$CO_2$ solid surface, and the surface shifts from $CO_2$-phobic to $CO_2$-philic. When $\varepsilon$ = 1000 J·mol$^{-1}$, the solid surface exhibits perfect $CO_2$-philic. property, resulting in the droplets forming film-like condensation on the solid surface more easily. Fig. 9(b) shows the variation curve of the $CO_2$ droplet contact angle with the wall atomic potential well depth ε. The larger the wall atomic potential well depth $\varepsilon$, the stronger the interaction between the wall particles and the droplet particles, which will promote the interaction between the droplet and the wall and reduce the contact angle. ε increases from 20 J·mol$^{-1}$ to 533 J·mol$^{-1}$, and the contact angle decreases monotonically from 158° to 60°, and the trend and magnitude of the droplet contact angle decrease both slow down gradually. In particular, a wall atomic potential trap depth $\varepsilon$ of J·mol$^{-1}$ produces a $CO_2$ droplet contact angle of about 90°, which is the critical point for the transition from $CO_2$-phobic to $CO_2$-philic. The inhomogeneity of the $CO_2$ density in the droplet is determined, especially near the droplet and the solid surface. To quantitatively distinguish the number of adsorbed layers of $CO_2$ molecules on solid surfaces with different $\varepsilon$, the density distribution curve of $CO_2$ is plotted in Fig. 9(c). The number of peaks increases significantly with the increase of $\varepsilon$ intensity. When $\varepsilon$ = 57 J·mol$^{-1}$, only one peak can be observed, while when $\varepsilon$ = 500 J·mol$^{-1}$, four peaks can be observed. These peaks reflect the adsorption layer of $CO_2$. The higher the energy of the solid surface, the higher the density of the adsorption plane and the layer closest to the solid surface shows a greater density than the other layers.

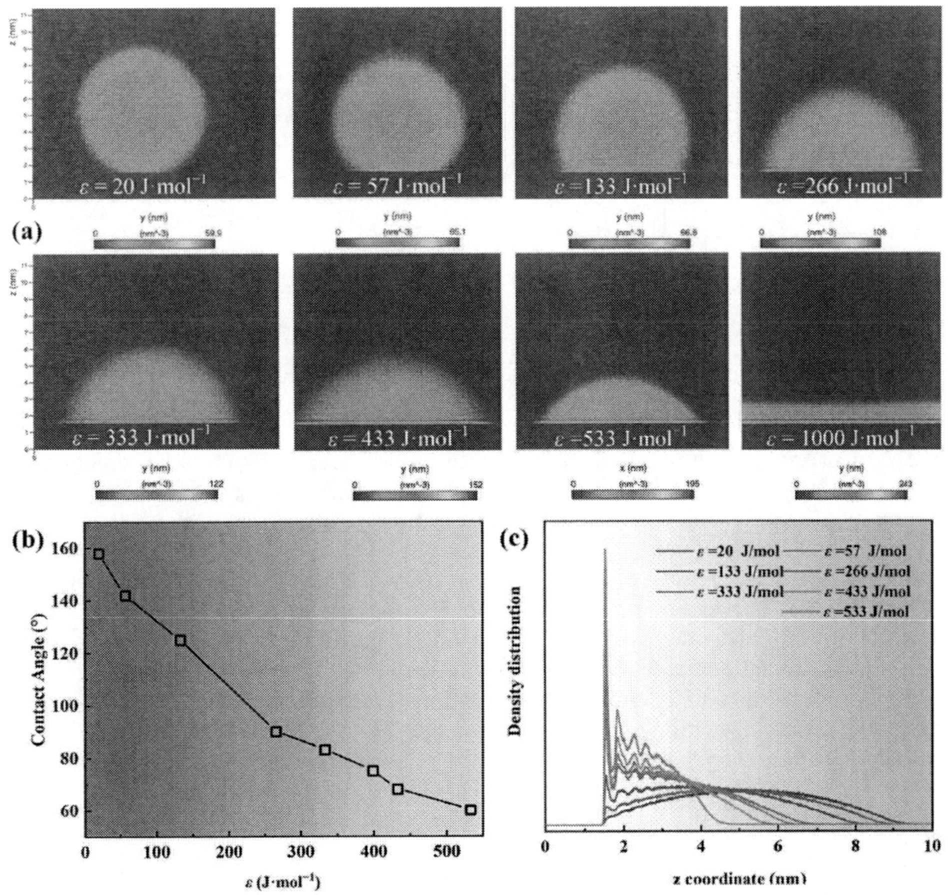


**Fig. 9** (a) Density clouds of $CO_2$ droplets on different energy surfaces. (b) Contact angles of $CO_2$ droplets on different energy surfaces. (c) Relative density distribution of $CO_2$ along the z-direction from the solid surface to the top of the simulation box.

### *3.5 Effect of temperature on the wettability of $CO_2$ droplet*

The change in temperature triggers a corresponding change in the gas-liquid interface. By changing the temperature of $CO_2$ droplets, the wetting simulation was performed on three types of walls, namely, $CO_2$-phobic, neutral and $CO_2$-philic, to obtain the wetting state of $CO_2$ droplets on the solid surface. As shown in Fig. 10, the results show that the contact angle of all three surfaces will decrease when the temperature increases and the higher the surface $CO_2$-philic, the greater the contact angle decrease. The increase in temperature leads to the decrease of gas-liquid surface tension, and as the temperature increases, more and more liquid $CO_2$ molecules evaporate and convert to a gas state. The classical Yong equation gives the relationship between the contact angle $\theta$ and the solid-gas surface tension $\gamma_{SV}$, solid-liquid surface tension $\gamma_{SL}$, and liquid-gas surface tension $\gamma_{LV}$: $\gamma_{SV} = \gamma_{SL} + \gamma_{LV}\cos\theta$. According to the Yong equation, $\cos\theta$ should increase with decreasing gas-liquid surface tension, and its corresponding contact angle $\theta$ decreases. Therefore, the contact angle decreases slightly with the increase in temperature, but the overall change is not significant in the simulation range.

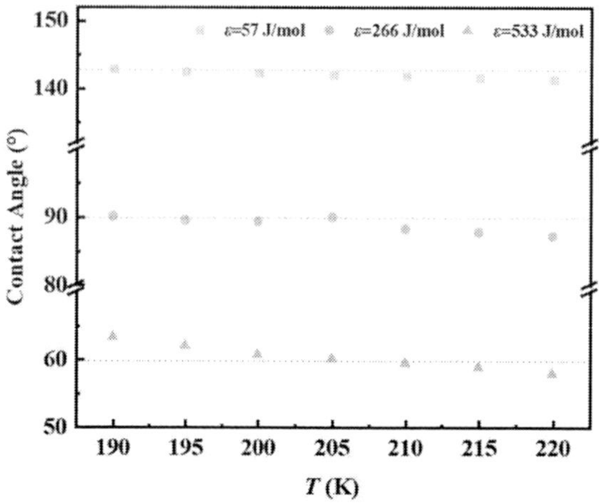


**Fig. 10** Contact angles of droplets on surfaces with different wetting types at different temperatures

The morphological changes of droplets at different temperatures can be compared by counting the com[Z] of $CO_2$ droplets at steady state form. As shown in Fig. 11, the com[Z]_Ave at each temperature is the average value of the droplet com[Z] at the kinetic calculation stage. The com[Z]_Ave of the droplet on the wall surface gradually decreases with increasing temperature, indicating that the droplet gradually spreads flat on the infiltrated wall surface as the temperature increases, so the droplet contact angle continues to decrease. That is, the molecular motion of $CO_2$ droplets intensifies as the temperature increases, and the morphological changes tend to strengthen on the walls with different wetting spreading tendencies. For the spreading result of droplets on the wetted wall, the morphological characteristic of the contact area between the outermost edge of the droplet and the wall is the contact angle, so the changing trend of the contact angle of $CO_2$ droplets affected by temperature is the same as the changing trend of com[Z]_Ave which characterizes the spreading morphology of $CO_2$ droplets.

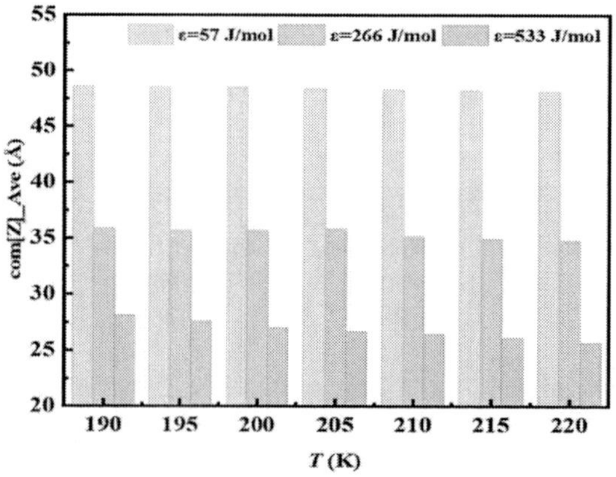


**Fig. 11** com[Z]_Ave values of droplets on walls with different wetting types at different temperatures

## 4 Conclusions

Understanding the wetting and condensation behavior of $CO_2$ on the cooling surface of a heat exchanger is a necessary starting point for studying the technology of pressurized liquefaction of natural gas with $CO_2$ impurities. In this paper, molecular dynamics methods were used to simulate and calculate the spreading process and wetting morphology of nanoscale droplets on different energy surfaces, and the following conclusions were obtained:

(1) The number of molecules of pure $CO_2$ droplets (4800) and the number of atomic layers on the wall (8 layers) for subsequent simulations when the gas-liquid interface is more stable are determined from the trend and deviation of the contact angle of $CO_2$ droplets.

(2) The gas-liquid interface forces caused by random collisions of gas-phase molecules are no longer symmetrical, making the gas-liquid interface fluctuate to some extent even in the relatively stable spreading state of $CO_2$ droplets, which is significantly different from the macroscopic droplet morphology. Under the statistical significance, the density image of the gas-liquid interface can still be obtained smoothly enough to accurately calculate the droplet contact angle using the density profile method.

(3) The contact angle of $CO_2$ droplets decreases with the increase of the wall atomic potential well depth $\varepsilon$. The wetting contact angle of $CO_2$ droplets on the wall with higher affinity for $CO_2$ molecules ($CO_2$-philicity) is smaller, and the trend and magnitude of the decrease of the contact angle of $CO_2$ droplets are gradually slowed down. In particular, the wall atomic potential trap depth $\varepsilon$ of 266 $J \cdot mol^{-1}$ produces a $CO_2$ droplet contact angle of about 90°, which is the critical point for the transition from $CO_2$-phobic to $CO_2$-philic.

(4) The trend of the $CO_2$ droplet contact angle affected by temperature is the same as that of com[Z]_Ave, which characterizes the spreading morphology of $CO_2$ droplets. As the temperature increases, the contact angle decreases due to the further spreading and wetting of droplets on different energy surfaces. As the $CO_2$-philicity of the surface gets higher, the contact angle decreases to a greater extent.

**Acknowledgments.**
The project is supported by National Natural Science Foundation (Number U21B2087).

**References**

[1]Van Nguyen, T., de Oliveira Júnior, S.: System evaluation of offshore platforms with gas liquefaction processes. Energy 144, 594–606 (2018).
[2]Liu, Y., Cao, X., Yang, J., Li, Y., Bian, J.: Energy separation and condensation effects in pressure energy recovery process of natural gas supersonic dehydration. Energy Conversion and Management 245, 114557 (2021).
[3]Cao, X., Guo, D., Sun, W., Zhang, P., Ding, G., Bian, J.: Supersonic separation technology for carbon dioxide and hydrogen sulfide removal from natural gas. Journal of Cleaner Production 288, 125689 (2021).
[4]Ogunronbi, K.E., Sepehri, A., Chen, B., Wyslouzil, B.E.: Vapor phase nucleation of the short-chain n -alkanes (n -pentane, n -hexane and n -heptane): Experiments and Monte Carlo simulations. Journal of Chemical Physics 148, 144312 (2018).
[5]Wen, C., Li, B., Ding, H., Akrami, M., Zhang, H., Yang, Y.: Thermodynamics analysis of $CO_2$ condensation in supersonic flows for the potential of clean offshore natural gas processing. Applied Energy 310, 118523 (2022).
[6] Xiong, X., Lin, W., Gu, A.: Design and optimization of offshore natural gas liquefaction processes adopting PLNG (pressurized liquefied natural gas) technology. Journal of Natural Gas Science and Engineering 30, 379–87 (2016).
[7]Cao, H., Cao, X., Chen, J., Zhao, X., Ding, G., Guo, D., Liu, Y., Li, H., Bian, J.: Molecular dynamics simulation of the transport properties and condensation mechanism of carbon dioxide. Journal of Natural Gas Science and Engineering 105, 104692 (2022).
[8]Guo, D., Cao, X., Ding, G., Zhang, P., Liu, Y., Bian, J.: Crystallization and nucleation mechanism of heavy hydrocarbons in natural gas. Energy 239, 122071 (2022).

[9]Nasrifar, K., Moshfeghian, M.: Prediction of carbon dioxide frost point for natural gas and LNG model systems. Journal of Natural Gas Science and Engineering 76, 103206 (2020).
[10]Snustad, I., Ervik, Å., Austegard, A., Brunsvold, A., He, J., Zhang, Z.: Heat transfer characteristics of $CO_2$ condensation on common heat exchanger materials: Method development and experimental results. Experimental Thermal and Fluid Science 129, 110440 (2021).
[11]Snustad, I., Røe, I.T., Brunsvold, A., Ervik, Å., He, J., Zhang, Z.: A review on wetting and water condensation - Perspectives for $CO_2$ condensation. Advances in Colloid and Interface Science 256, 291–304 (2018).
[12]Wu, J., Snustad, I., Ervik, Å., Brunsvold, A., He, J., Zhang, Z.: $CO_2$ wetting on pillar-nanostructured substrates. Nanotechnology 31, 245403 (2020).
[13]Xu, W., Lan, Z., Peng, B., Wen, R., Ma, X.: Molecular dynamics simulation on the wetting characteristic of micro-droplet on surfaces with different free energies. Wuli Xuebao/Acta Physica Sinica 64, 216801 (2015).
[14]Liu, Y., Rui, Z.: A storage-driven $CO_2$ EOR for a net-zero emission target. Engineering 18, 79-87 (2022).
[15]Yu, X., Li, J., Chen, Z., Wu, K., Zhang, L., Hui, G., Yang, M.: Molecular dynamics computations of brine-$CO_2$/$CH_4$-shale contact angles: Implications for $CO_2$ sequestration and enhanced gas recovery. Fuel 280, 118590 (2020).
[16]Yan, Y., Li, C., Dong, Z., Fang, T., Sun, B., Zhang, J.: Enhanced oil recovery mechanism of $CO_2$ water-alternating-gas injection in silica nanochannel. Fuel 190, 253–259 (2017).
[17]Tenney, C.M., Cygan, R.T.: Molecular simulation of carbon dioxide, brine, and clay mineral interactions and determination of contact angles. Environmental Science and Technology 48, 2035–2042 (2014).
[18]Yan, M., Yang, X., Lu, Y.: Wetting behavior of water droplet on solid surfaces in solvent environment: A molecular simulation study. Colloids and Surfaces A: Physicochemical and Engineering Aspects 429, 142–148 (2013).
[19]Horsch, M., Vrabec, J., Bernreuther, M., Grottel, S., Reina, G., Wix, A., Schaber, K., Hasse, H.: Homogeneous nucleation in supersaturated vapors of methane, ethane, and carbon dioxide predicted by brute force molecular dynamics. Journal of Chemical Physics 128, 164510 (2008).
[20]Snustad, I., Hollund, L., Ervik, Å., Austegard, A., Brunsvold, A., He, J., Zhang, Z.: Condensation heat transfer of $CO_2$ on Cu based hierarchical and nanostructured surfaces. International Journal of Heat and Mass Transfer 175, 121367 (2021).
[21]Cao, H., Cao, X., Zhao, X., Guo, D., Liu, Y., Bian, J.: Molecular dynamics simulation of wax molecules aggregational crystallization behavior during cooling of crude oil mixture. Case Studies in Thermal Engineering 37, 102298 (2022).
[22]Huang, D., Quan, X., Cheng, P.: An investigation on vapor condensation on nanopillar array surfaces by molecular dynamics simulation. International Journal of Heat and Mass Transfer 98, 232–238 (2018).
[23]Cadena, C., Anthony, J.L., Shah, J.K., Morrow, T.I., Brennecke, J.F., Maginn, E.J.: Why is $CO_2$ so Soluble in Imidazolium-Based Ionic Liquids?. Journal of the American Chemical Society 126, 5300–5308 (2004).
[24]Heinz, H., Vaia, R.A., Farmer, B.L., Naik, R.R.: Accurate simulation of surfaces and interfaces of face-centered cubic metals using 12-6 and 9-6 lennard-jones potentials. Journal of Physical Chemistry C 112, 17281–17290 (2008).

Journal of Physics: Conference Series **2594** (2023) 012049 doi:10.1088/1742-6596/2594/1/012049

# A 3D Geological Modeling Method for Grain-size Lithofacies of Tight Sandstone Gas Reservoirs: A Case Study of Xujiahe Formation Gas Reservoir in Sichuan Basin, China

**Xiao-fei SHANG1(✉) Yun-fei XIANG1 Zhong-qun LIU1**

1 Petroleum Exploration and Production Research Institute, SINOPEC, Beijing, China

shangxf.syky@sinopec.com

**Abstract.** Tight sandstone gas reservoir has strong heterogeneity, and it is difficult to control the change of internal storage and permeability properties by simple sedimentary facies. It is necessary to explore a geological modeling method that can more accurately characterize the distribution of high-quality reservoirs. Taking the Xujiahe Formation gas reservoir in Xinchang area, Sichuan Basin, China as an example, this paper introduces the modeling method of three orders: sand-mudstone facies, sedimentary facies and grain-size lithofacies, so as to realize the spatial characterization of high-quality tight sandstone reservoirs with strong heterogeneity. In this technical process, the sand-mudstone model is first established. Based on the sand and mudstone model, four sedimentary types (or sedimentary facies) of distributary channel, channel edge, interdistributary bay and mouth bar are further divided. The sedimentary facies model is constructed by multi-point geostatistical modeling method. The quantitative relationship between sedimentary facies and grain-size lithofacies is linked by using argillaceous content as a "bridge". The spatial distribution probability of grain-size lithofacies is constrained by the neural network clustering of argillaceous content and natural gamma-ray data in three-dimensional space. By controlling the types, ratios, and boundaries of grain-size lithofacies through sedimentary facies, and combining probabilistic bodies to cooperate with constraints, precise simulation of lithofacies can be achieved. The grain-size lithology lithofacies model established by this method follows the depositional law in space and has more reasonable contact relations between various sand bodies. The anastomosis rate of the model reached 85% with the new drilling test. This paper provides a new modeling idea for quantitative characterization and prediction of high-quality tight sandstone gas reservoirs, and provides a more accurate model basis.

## 1. Introduction

With the development of global natural gas industry, the field and scale of tight sandstone gas reservoir development are expanding rapidly [1-2]. The gas in tight sandstone gas reservoir occurs in the rocks with low porosity and permeability [3-6]. The key to the exploration and development of this kind of gas reservoir is to accurately depict the parts with good relative physical properties under the generally tight background (i.e. high-quality reservoir geological modeling).

Facies-controlled modeling takes facies (such as lithofacies, sedimentary facies, petrophysical facies, etc.) as constraint conditions to analyze and simulate various attribute models of reservoir physical properties, oil & gas properties, etc. In conventional sandstone reservoirs, the distribution of high-quality reservoirs is usually related to the sedimentary facies. However, in tight sandstone reservoirs, complex densification processes such as diagenesis often occur after the initial deposition, so the controlled response of reservoir pore structure and reservoir physical properties (porosity, permeability, etc.) to sedimentary facies decreases. It is difficult for simple sedimentary facies to control the changes of internal reservoir storage and permeability properties [7-8]. Therefore, for tight sandstone gas reservoirs, it is necessary to construct a lithology or lithofacies model that is more sensitive to the control of reservoir physical property.

The Sichuan Basin of China is rich in natural gas resources and has many gas reservoirs [9-10], among which the second member of Xujiahe Formation in Xinchang Gas field is a typical tight sandstone gas reservoir. The proven reserves of natural gas in this reservoir have exceeded $1600\times10^8$ $m^3$, but the actual recovery rate is less than 10% due to the highly heterogeneous reservoir. The main reason is that the spatial distribution of high-quality reservoir is not accurately described. Although traditional stochastic simulation method can directly build a 3D lithofacies model from Wells, it is difficult to accurately represent the distribution probability of lithofacies between Wells in the modeling process due to the strong heterogeneity of tight sandstone reservoirs. As a result, the final model lacks the geological meaning of high-quality reservoirs, so it has no predictive significance.

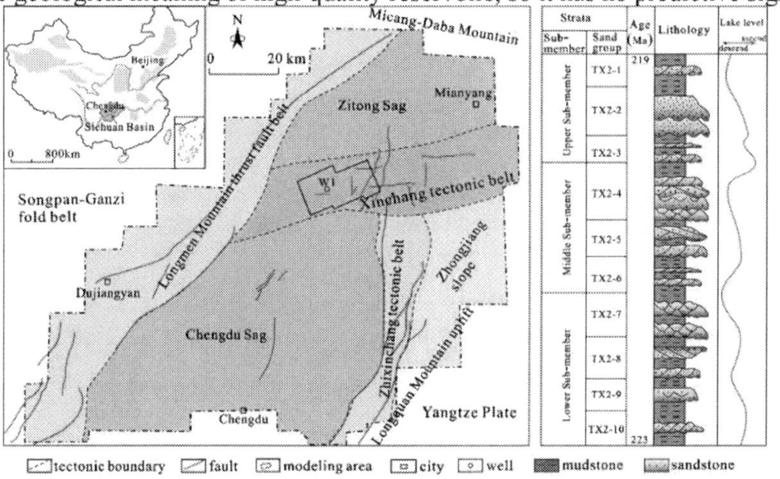


**Fig. 1.** Geological structure characteristics of Xinchang tectonic belt and sedimentary strata of Xujiahe Formation in western Sichuan Basin, China.

In this paper, the Xinchang gas field in Sichuan Basin is taken as an example. On the basis of sedimentary facies modeling, the sand body quality (such as good sand, poor sand, etc.) is further simulated more accurately, and a more reasonable and accurate lithofacies modeling method is introduced, which provides a new idea for the spatial prediction and quantitative characterization of high-quality tight sandstone gas reservoirs.

## 2. Geological setting

The west Sichuan depression is located in the west of Sichuan Basin. It is a long strip extending northeast in plane, bounded by the Longmen Mountain tectonic belt in the west and extended to the Kunlun and Qinling tectonic belt in the northeast, covering an area of about 50,000 $km^2$. In the Late Triassic, the western Sichuan depression was located at the junction of several plates and formed an overlying basin due to frequent tectonic activities [11]. The Longmen Mountain and Micang-Daba Mountains tectonic movements affect the secondary structural morphology and sedimentary filling evolution of the western Sichuan depression. Xinchang area is located in the middle part of the western Sichuan depression, formed in the early Indochinese movement, and tectonically, it is an almost east-west trending anticline (see Figure 1).

In the early stage of the Triassic, with the bi-directional subduction of the Yangtze plate to the Qiangtang and Kunlun Blocks, the Micang-Daba Mountains ancient landmasses appeared in the northeast of the Sichuan Basin, and the Longmen Mountain began to thrust and uplift in the west. During the sedimentary period of Xujiahe Formation, the tectonic movement in the east-west direction was further intensified, and the western Sichuan area subsided greatly, forming a depression along the front of Longmen Mountain [12-13]. The second member of Xujiahe Formation was tectonically active, and Longmen Mountain, Micangshan-Daba Mountains all provided source sand bodies to the depression through water system. The sedimentary system mainly developed in Xinchang area is braided river delta, which is an important reservoir of Xinchang gas field with large thickness and wide distribution.

The reservoir of the second member of Xujie Formation in western Sichuan Depression is generally gas-bearing. Previous studies have shown that the extensive development of reservoirs with good porosity and permeability (high-quality sand bodies) is the basis of stable oil and gas production and the formation of gas reservoirs with high economic production value [14-15]. In order to establish property parameter models such as porosity and permeability, it is necessary to have a facies model to measure the quality of sand body and control its 3D spatial distribution.

## 3. Method and process

According to the results of core and well logging interpretation, the average porosity of coarse sandstone is 4.8%, middle sandstone is 4.2%, and that of fine sandstone is only 3.3%. The porosity of different grain size lithology/lithofacies is obviously different. According to the interpretation of log grain size, it can be divided into four lithofacies types: medium-coarse sandstone facies, fine sandstone facies, siltstone facies and mudstone facies. The medium-coarse-grained sandstone is the favorable rock facies, and the fine sand siltstone has obvious deterioration in physical properties.

Since there is no relevant understanding of the distribution of different grain size lithofacies, if the traditional sequential indication simulation (SIS) is used to establish the lithofacies model, the randomness is too strong, and there will be a lot of "noise", and the distribution of lithofacies is difficult to guarantee. To establish a grain-size lithofacies model, we need to make it clear that the spatial distribution of sand bodies with different grain-size is controlled by sedimentary facies. For example, the grain-size of channel is coarse at the bottom and fine at the top. However, such as qualitative understanding cannot meet the needs of quantitative modeling. For example, in the channel sand body, grain size changes from coarser to finer from bottom to top, but the specific locations of different grain size lithofacies cannot be determined (see Figure 2).

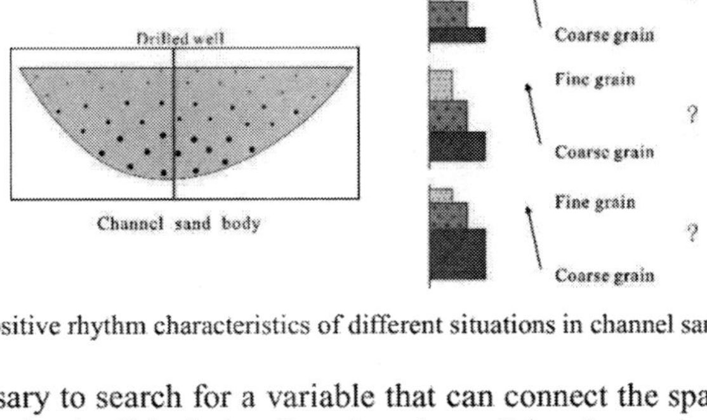


**Fig. 2.** Positive rhythm characteristics of different situations in channel sand bodies.

Therefore, it is necessary to search for a variable that can connect the spatial relationship between sedimentary facies and grain-size lithofacies. Based on the well logging grain size interpretation model, the argillaceous content in sandstone with different grain size varies significantly, in other words, the grain size of sandstone can be determined by the argillaceous content to a certain extent. For example, in the channel sedimentary facies, from the bottom of the channel upward, the sandstone grain size gradually changes from coarse sandstone to fine sandstone and siltstone. Correspondingly, the argillaceous content in sandstone changes from less to more, while the mouth bar sand body is on the contrary. In this paper, the argillaceous content is introduced as the "bridge" between sedimentary facies and lithofacies to quantitatively reflect the spatial distribution of lithofacies and accurately simulate them.

There are two basic points to establish a reliable grain-size lithofacies model. One is to accurately characterize the spatial distribution of argillaceous content. Second, the correlation between grain-size lithofacies and argillaceous content (and other factors) should be clarified. Through drilling analysis, we can basically define the distribution range and average value of argillaceous content of different grain size lithofacies. Therefore, an automatic classification method of grain-size lithofacies can be explored by using neural network clustering analysis in combination with argillaceous content and sensitive logging (such as Gamma-Ray logging) (see Figure 3). It can then be extended to 3D space to help build a grain-size lithofacies geological model.

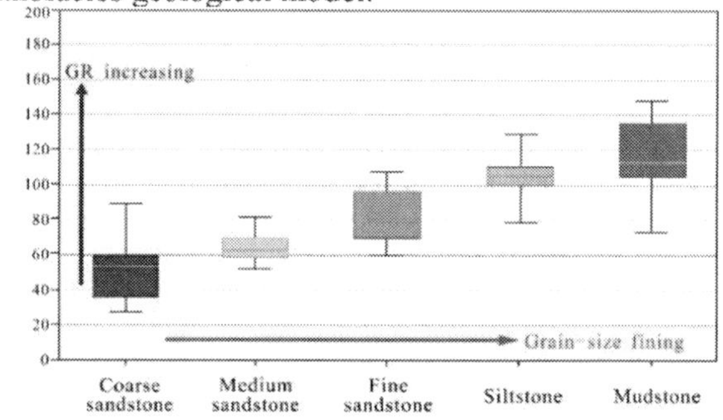


**Fig. 3.** Gamma-ray differentiation of sand bodies with different grain sizes.

In this paper, we introduce a modeling idea of grain-size lithofacies. In this modeling process, it mainly includes two parts: one is the modeling of argillaceous content controlled by sedimentary facies; The second is grain size lithofacies modeling based on argillaceous content neural network clustering (see Figure 4).

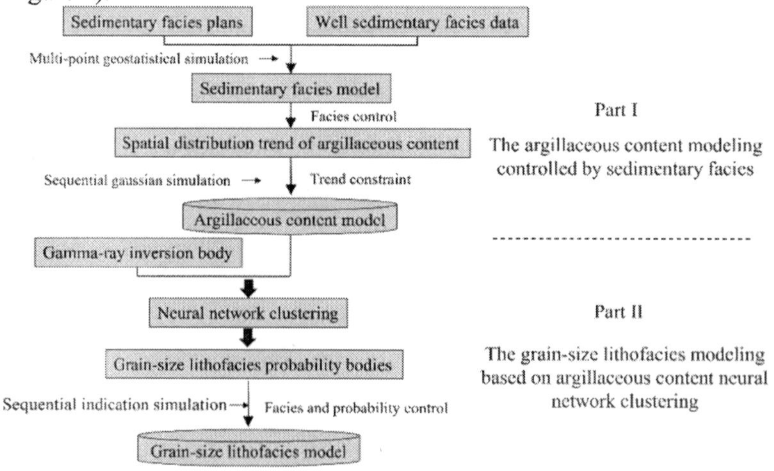


**Fig. 4.** Construction process of grain-size lithofacies model of tight sandstone.

## 4. Sedimentary facies modeling

### *4.1. Division of sedimentary sand bodies*

According to the characteristics of petrology, lithofacies association, biological fossils, and sequence structure, etc., the reservoir of the second member of the Xujiahe Formation is further divided into four sedimentary facies (or sand body sedimentary types), namely distributary channel, channel edge, interdistributary bay, and mouth bar. Among them, distributary channel in the braided river delta plain/front is the most developed sand body deposition in the study area.

*4.1.1. Distributary channel facies.* This kind of sand bodies belongs to delta plain or front. It is mainly composed of fine-medium grained sandstone, followed by coarse sandstone. In the longitudinal direction, multi-stage channel overlaps each other. The compositional maturity and texture maturity of the sandstone in the sedimentary facies are relatively low. Large plate cross-bedding, trough cross-bedding and parallel bedding are developed in the sandstone, and there are obvious scour-filling structure at the bottom of the sand body, and mud gravels can be seen on it.

*4.1.2. Channel edge.* This sand body is located at the edge of the above/underwater distributary channel. It is mainly composed of grey and fine-grained sediments. Its sedimentary texture is similar to that of the distributary channel, with tabular cross-bedding and trough cross-bedding.

*4.1.3. Interdistributary bay.* The interdistributary bay belongs to delta front, and the lithology is mainly a set of silty mudstone and a small amount of siltstone of fine suspension origin, with wavy sand bedding and horizontal bedding.

*4.1.4. Mouth bar.* The mouth bar is mainly composed of grey and dark grey medium to thick layered fine-grained deposits. The single sand layer is thicker and has the anti-rhythm characteristic of upward coarser grain size. After wave washing for a long time, the sand body has good sorting and roundness, and sedimentary texture such as collapse deformation and coiling bedding can be seen.

### *4.2. Construction of sedimentary facies model*

The sedimentary facies model was established on the basis of the sand and mudstone facies model in order to avoid the "conflict" between the two models of sand bodies. Therefore, the general idea of simulation is to adopt multi-order facies control in the sandstone model, and take different sedimentary types of sand bodies as the second facies for simulation.

The multi-point statistical geological modeling method uses "training image" instead of the variation function in the traditional statistical modeling method to represent the spatial structure and change of geological variables [16-18]. It can overcome the deficiency that the traditional two-point statistical geological modeling method can not reproduce the spatial geometric form of geological bodies well. In addition, the multi-point statistical geological modeling method adopts the sequential simulation process based on pixel rather than the iterative trial-and-error simulation process of the object-based stochastic simulation method, which can easily condition well data and other geological information and improve the computational efficiency. Therefore, in order to establish a more reasonable sedimentary sand body type model, we adopted the multi-point geostatistics method for simulation.

Taking the distribution of sand body types in each member as the training image, the quantitative distribution model of sand body types was established based on the multi-point geostatistical method of PVD-Sim algorithm [19]. The results show that the model is consistent with the interpretation of single well sand body and the spatial structure between different sand bodies expressed in the training image, and the sedimentary continuity is maintained longitudinally (see Figure 5).

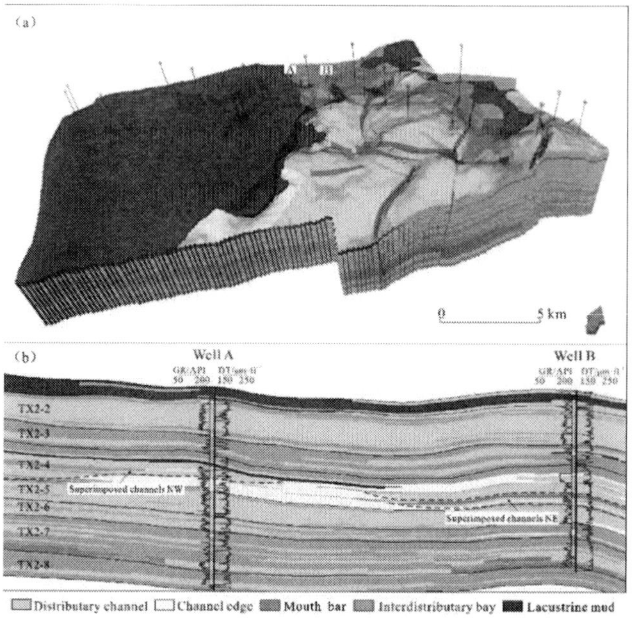


**Fig. 5.** Sedimentary facies model, wherein (a) is a three-dimensional model of sedimentary facies, and (b) is a section presentation of the model.

## 5. Argillaceous content modeling

It is obvious that argillaceous content can distinguish different lithofacies types, but the spatial distribution of argillaceous content still needs to find a related attribute (i.e. the second variable) to increase the control conditions of argillaceous content simulation. Based on the interpretation and model characterization of sedimentary facies, the distribution of argillaceous content in different sedimentary facies was further analyzed, and the controlling effect of sedimentary relative argillaceous content was analyzed from both plane and vertical aspects. Finally, the sedimentary facies control was used to construct the argillaceous content distribution model.

Firstly, the distribution range and data pattern of argillaceous content can be controlled according to different sedimentary facies. For example, in the interdistributary bay facies, the average argillaceous content is 0.32 , while the average argillaceous content in the sedimentary facies of distributary channel is only 0.09 (see Figure 6). In the whole delta front depositional environment, from the interdistributary bay to the channel edge and then to the distributary channel, the argillaceous content decreases gradually.

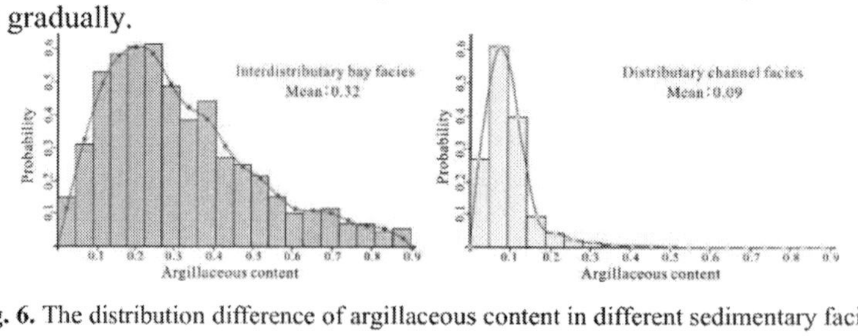


**Fig. 6.** The distribution difference of argillaceous content in different sedimentary facies.

Secondly, the sedimentary facies can control the longitudinal distribution trend of argillaceous content. For example, in the channel facies, the sand body is in a positive rhythm and the argillaceous content increases from bottom to top. Based on the sedimentary facies of the sand body such as each single channel or mouth bar, the 3D geological model of sedimentary facies is used to constrain the longitudinal distribution of argillaceous content through the corresponding longitudinal variation trend (see Figure 7).

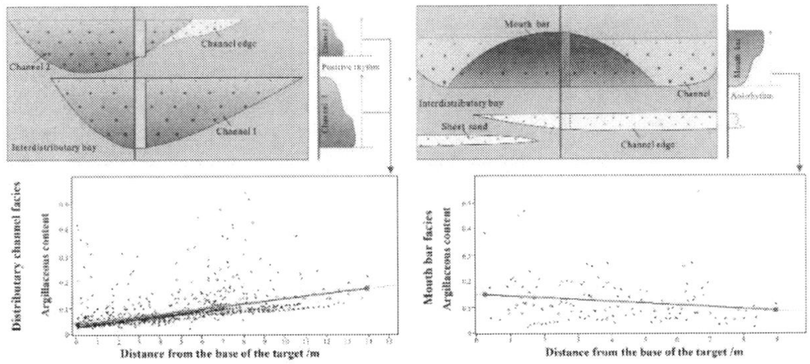


**Fig. 7.** The longitudinal distribution of argillaceous content in different sedimentary facies.

Variation function is the most important parameter in geostatistical simulation, which determines the distribution direction and scale of the simulation target. In the analysis of argillaceous content data, the variational functions were fitted from the three directions of principal, secondary and vertical variables respectively for parameter input of simulation. Under the control of sedimentary facies model, the model of argillaceous content distribution was established by SGS (see Figure 8). In the process of simulation, it is necessary to adjust the overall proportion of argillaceous content in each layer, so as to ensure that the spatial distribution of argillaceous content is consistent with the lithologic variation trend reflected by sediment.

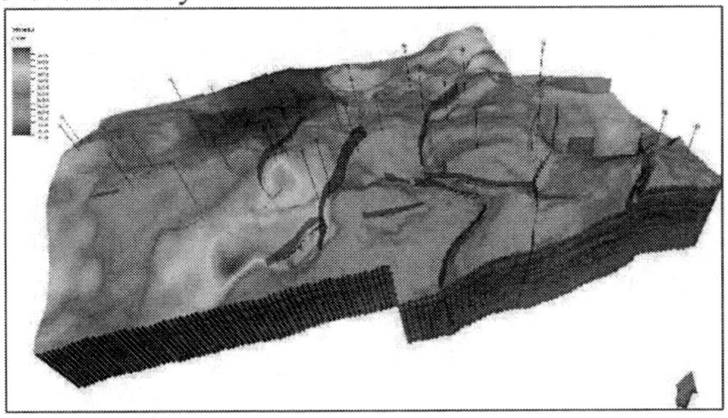

**Fig. 8.** Three-dimensional distribution model of argillaceous content.

## 6. Grain-size lithofacies modeling

Based on the continuous variable of argillaceous content, the neural network clustering method can effectively correspond to the discrete data of sandstone grain size, so as to quantify the probability of grain-size lithofacies. For example, when the argillaceous content is 0.3, the probability that the sand body is coarse sandstone facies, medium sandstone facies or fine sandstone facies can be obtained. In the neural network clustering analysis, the more data samples involved in the training, the closer the clustering result is to the objective reality and the higher the accuracy. In addition, it is also necessary to consider that the data samples involved in the training can satisfy the data distribution between Wells, so that the neural network clustering can be implemented in three-dimensional space to obtain the distribution state of the probability of each lithofacies in three-dimensional space, so that the 3D model of lithofacies can be established.

### *6.1. Neural network clustering analysis*

The concept of neural network algorithm is to simulate the rapid recognition (training process) of multiple phenomena and signals captured by the human brain to achieve fuzzy classification and estimation, such as the distinction of similar colors and shapes of various objects [20-22]. A single seismic attribute often reflects the reservoir lithology and physical properties in a biased way, which can be reduced by multi-attribute analysis. In the process of logging interpretation, natural gamma-ray data is of great reference value to the division of grain-size lithofacies, and the three-dimensional distribution data of natural gamma-ray can also be obtained by seismic inversion technology. Therefore, argillaceous content data and natural Gamma-Ray data were combined to conduct neural network clustering analysis.

*6.1.1. Single well neural network clustering.* The feasibility of using neural network to analyze grain-size lithofacies is tested in single well data. The participating Wells were randomly selected, and the well argillaceous content and gamma-ray data were selected as samples to conduct neural network training and learning. The output results were set as four discrete data. During the training, the interpreted grain-size lithofacies classification information (i.e. coarse-grained lithofacies, medium-grained lithofacies, fine-grained lithofacies and mudstone facies) of participating Wells were supervised.

The grain-size lithofacies data obtained by single well neural network clustering was compared with the data from logging interpretation. The matching degree between the two was compared by selecting the values in the middle of the lithofacies. For participating Wells and non-participating Wells, the similarity between the two grain-size lithofacies reaches more than 70%. These results indicate that the neural network clustering of grain-size lithofacies based on argillaceous content and Gamma-Ray data is effective, and confirm the feasibility of this method.

*6.1.2. 3D model neural network clustering.* The neural network analysis method based on single well is extended to three-dimensional space. With argillaceous content model and Gamma-Ray inversion model as input data, the supervised neural network learning training was also carried out with the grain-size lithofacies data obtained from all Wells in the model as supervision. The output results of training are also set as four discrete data corresponding to four lithofacies types (coarse-grained lithofacies, medium-grained lithofacies, fine-grained lithofacies and mudstone facies), as well as the probability spatial distribution of each facies.

### *6.2. Grain size lithofacies spatial distribution probability acquisition*

Based on the 3D model of argillaceous content and Gamma-Ray inversion body, the corresponding "threshold window" of argillaceous content and Gamma-Ray in each lithofacies was obtained through their distribution. Through the cross-verification of the lithofacies and argillaceous content and gamma-ray, the argillaceous content model and Gamma-Ray inversion body were clustered by neural network, and finally the spatial probability distribution of four types of lithofacies was obtained (see Figure 9).

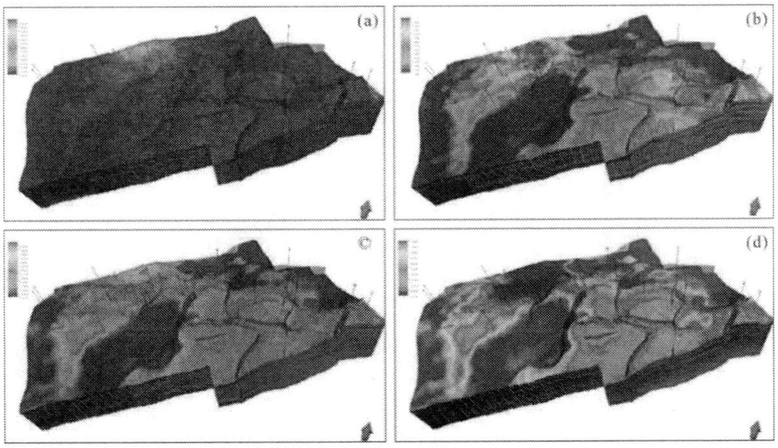

**Fig. 9.** Distribution probability of grain-size lithofacies, where (a) is coarse-grained lithofacies probability; (b) is the probability of medium-grained lithofacies, (c) is the probability of fine-grained lithofacies, and (d) is the probability of mudstone.

### *6.3. Establishment of grain-size lithofacies model*

The conditional data used for grain-size lithofacies modeling is the sampled discrete lithofacies data of grain-size lithofacies interpreted from each single well after sampling by the model grid. Lithofacies data analysis is an important basis for determining the simulation method and controlling the simulation results, which usually needs to be analyzed statistically from multiple perspectives.

*6.3.1. Longitudinal ratio distribution of grain-size lithofacies.* In the whole second member of Xujiahe Formation, the distribution of the vertical proportion of lithofacies is mainly to reflect the difference of sedimentary environment of each layer. According to the analysis of reservoir characteristics of thin section data in Xinchang area, it is found that the middle and upper members are mainly medium and coarse grains, accounting for more than 85% of the sandstone. Meanwhile, compared with the upper member, the coarse-grained lithofacies in the middle member accounts for more, about 12%, while the upper member only accounts for 2.47% of the total sandstone. The lower member is dominated by medium and fine grained lithofacies, in which fine grained sandstone accounts for 55% (see Figure 10).

Within the member, the vertical proportion of lithofacies can reflect the rhythm of vertical deposition. For example, the sedimentary evolution analysis shows that the lake plane was raised during the deposition of the TX2-4 sand formation, and the delta sand bodies in the early stage were extremely developed and mainly deposited by distributary channel. The sand bodies were thick and gradually thinned upward. At the top of the sand formation, there was a period of rapid lacustrine flooding, characterized by relatively stable mudstone deposits, which were basically distributed throughout the region. Therefore, the longitudinal variation of argillaceous content in the whole sedimentary process needs to be reflected in the 3D lithofacies modeling.

Journal of Physics: Conference Series **2594** (2023) 012049 doi:10.1088/1742-6596/2594/1/012049

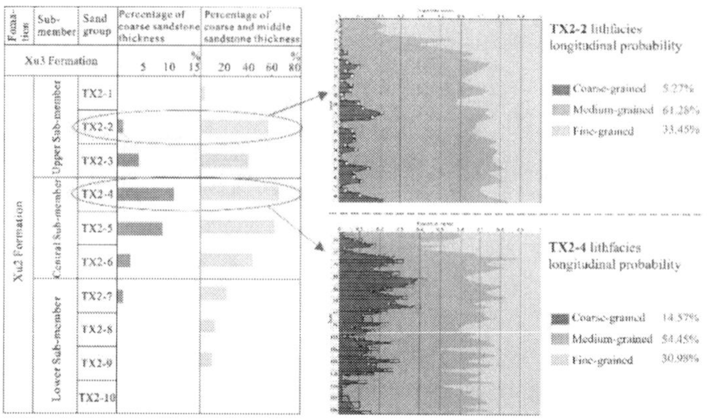


**Fig. 10.** The distribution proportion of various grain-size lithofacies in different intervals.

*6.3.2. The thickness distribution statistical characteristics of lithofacies.* The statistics of the thickness distribution of lithofacies deposits are designed to master the thickness ranges of different lithofacies, and these quantitative parameters are used as constraints in the simulation process. Different lithofacies have different thickness distributions. For example, in delta front deposits, the thickness of medium-coarse-grained lithofacies is mainly deposited in distributary channels and mouth bars, so the thickness is larger, while the thickness of fine-siltstone lithofacies and mudstone facies is mainly deposited in interdistributary bay and sheet sand, and this lithofacies are smaller.

*6.3.3. Distribution of constraint variables for lithofacies distribution.* The constraint data of grain-size lithofacies distribution are mainly reflected in the distribution trend of each grain-size lithofacies between Wells. The spatial probability data of each grain-size lithofacies obtained through the clustering of argillaceous content and Gamma-Ray neural network can be used as a good constraint variable to guide the distribution trend of each grain-size lithofacies in the simulation process.

*6.3.4. Analysis results of variational functions controlling the direction and scale of lithofacies distribution.* Each grain-size lithofacies still needs a corresponding variation function to control the continuity of its distribution. Based on the grain-size lithofacies data at drilling, the variational function fitting of each lithofacies was carried out to obtain the variable direction, range, and vertical variational parameters of each lithofacies in the reservoir.

In this study, a grain-size lithofacies model which can better characterize the reservoir quality was established for the tight sandstone gas field of Xujiahe Formation. Firstly, according to the results of drilling, seismic and geological analysis, a lithology model was established to divide the geological body into two parts: sandstone and mudstone. In sandstone reservoirs, there are good and bad quality. Three lithofacies, coarse-grained lithofacies, medium-grained lithofacies and fine-grained lithofacies, can be divided according to the grain-size of sand body which can better reflect the reservoir quality. Based on the identification results of logging lithofacies, neural network clustering method was adopted according to the internal relationship between argillaceous content and gamma-ray parameters and grain-size lithofacies. Each lithofacies was spatially constrained by corresponding probability body. By using the sequential indication simulation (SIS) method, and using the spatial probability of each grain-size lithofacies as the second variable constraint, combined with the variational function and the control of vertical distribution ratio, the 3D distribution model of grain-size lithofacies was finally simulated (see Figure 11).

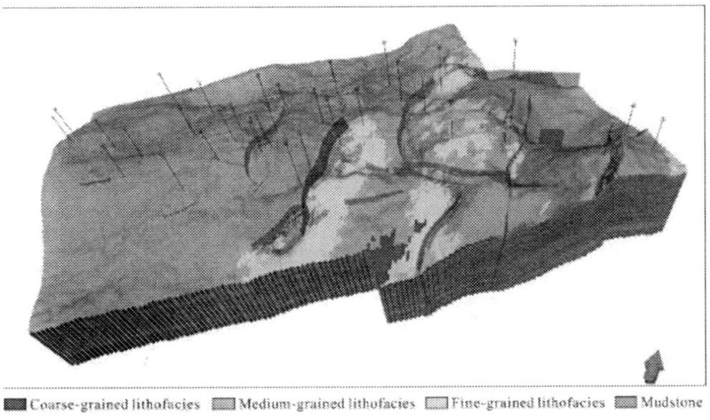


**Fig. 11.** Grain-size lithofacies 3D distribution model.

## 7. Model test and application effect

The established grain-size lithofacies model needs to be tested for its distribution in three-dimensional space. The rationality test mainly includes two aspects: first, whether the distribution of each sandstone grain-size lithofacies is distributed in the sandstone sedimentary facies; Second, whether the distribution of each grain-size lithofacies conforms to the sedimentary law.

In this modeling process, the sand and mudstone, as the first order facies, were first established. The sedimentary facies were simulated as the second order. In sedimentary facies simulation, distributary channel, channel edge and mouth bar are all simulated in sandstone facies. Therefore, the distribution of each type of sand body should be in sandstone facies. Due to the more argillaceous components, the distributary bay will be partially simulated in sandstone and mudstone facies. The distributary channel, channel edge and mouth bar were all simulated in sandstone facies, while the interdistributary bay was partially simulated in sandstone and mudstone facies because it contains more argillaceous components. Grain-size lithofacies, which can better reflect the reservoir quality, were tertiary facies, that is, the distribution of coarse sandstone, medium sandstone and fine sandstone was controlled and constrained on the basis of sedimentary facies model. Therefore, the distribution of each grain-size lithofacies is not only in the sandstone facies, but also in accordance with the sedimentary law of its spatial distribution. For example, in the channel sand body, coarse sandstone is often deposited in the middle and lower part of the river with strong hydrodynamics. Near the center of the river in the plane, to the edge of the river, the proportion of medium grained lithofacies and fine grained lithofacies gradually increases.

In the model established by traditional methods, the randomness of each lithofacies is strong, so it is difficult to ensure the rationality of the contact relationship between each lithofacies. This grain-size lithofacies modeling process has connotations of sedimentary geology, and the purpose of controlling grain-size lithofacies is achieved by constraining argillaceous content by sedimentary facies. By extracting the thickness distribution maps for each type of lithofacies in each layer of the model, we can analyze the sedimentary thickness and extent of coarse, medium and fine sandstones (see Figure 12). The distribution of grain-size lithofacies reflected in the model accords with the sedimentary law, that is, the continuous distribution of coarse sandstone-medium sandstone-fine sandstone is basically presented in the sandstone, and the changing process of lithofacies fully reflects the grain-size lithology characteristics of each sedimentary facies.

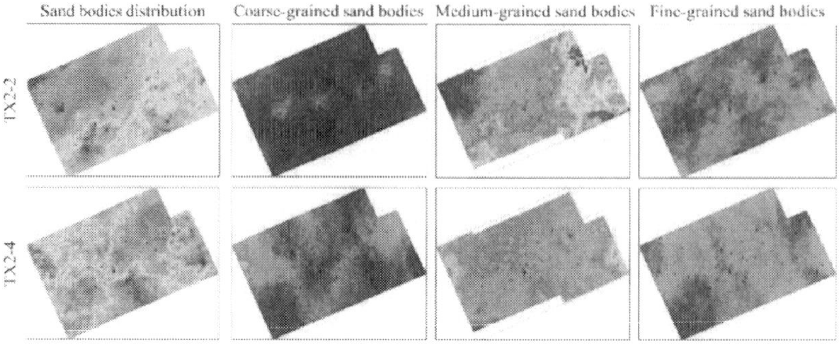


**Fig. 12.** Thickness distribution of different lithofacies models in each interval.

The thickness and average porosity of effective and high-quality reservoirs of TX2-2 and TX2-4 sand formations were extracted respectively from the established reservoir model (sandstone porosity greater than 3% is effective reservoir, and porosity greater than 4% is high-quality reservoir), and compared with the actual drilling situation of newly drilled well W1 (position as shown in Figure 1). In the model, the effective reservoir thickness of TX2-2 sand formation in well W1 is 32.00 m with an average porosity of 4.50%, and the high-quality reservoir thickness is 20.00 m with an average porosity of 4.90%. However, the effective reservoir thickness revealed by actual drilling is 39.69 m with an average porosity of 4.74%, and the high-quality reservoir is 35.29 m thick and has an average porosity of 4.88%. The anastomosis rate between the model and the well was achieved 85% (see Figure 13 and Table 1).

**Table 1.** Comparison of model and drilling parameters.

| Interval | Data | Effective reservoirs | | High-quality reservoirs | |
|---|---|---|---|---|---|
| | | Thickness/m | Average porosity | Thickness/m | Average porosity |
| TX2-2 | Model | 32.00 | 4.50% | 20.00 | 4.90% |
| | Well | 39.69 | 4.74% | 35.29 | 4.88% |
| TX2-4 | Model | 35.00 | 3.60% | 10.00 | 5.00% |
| | Well | 29.58 | 3.70% | 7.66 | 4.37% |

For tight clastic reservoirs, quantifying the three-dimensional distribution of high-quality reservoirs is the basis for efficient development of such reservoirs. The traditional sedimentary facies control modeling emphasizes the controlling effect of deposition relative to reservoir distribution, but the heterogeneity of the Xujiahe Formation gas reservoir in Xinchang area is strong, and it is difficult to control the change of internal reservoir and permeability properties with simple sedimentary facies. In this paper, a quantitative geological model of spatial distribution of grain-size lithofacies is established based on the qualitative analysis of tight reservoir which has more obvious control over the properties of reservoir and permeability. Based on this model, the spatial distribution of sand bodies with different qualities can be accurately described, and the distribution of relatively high-quality sand bodies in different intervals (such as medium lithofacies and coarse lithofacies in channel sand bodies) can be further quantitatively analyzed according to the sedimentary system. The lithofacies modeling method provided in this paper can provide a more accurate model basis for the quantitative characterization and effective development of high-quality reservoirs in tight sandstone gas fields.

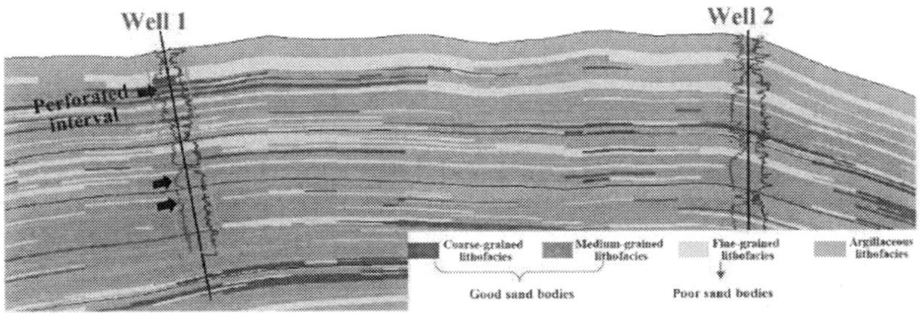


**Fig. 13.** Matching relationship between lithofacies model profile and new well.

## 8. Conclusions

(1) In the second member of Xichang area, a set of braided river delta is mainly developed in the middle-coarse-grained sediments with large thickness and wide planar distribution. Four sand body sedimentary types (or sedimentary facies) are further divided into distributary channel, channel edge, interdistributary bay and mouth bar, and the delta distributary channel is the main sedimentary facies.

(2) With the distribution of different types of sand bodies as training images, the sedimentary facies distribution model is established by multi-point geological statistics method, which can better reflect the overall sedimentary characteristics and evolution.

(3) The quantitative relationship between sedimentary facies and grain-size lithofacies can be linked by using argillaceous content as a "bridge". The spatial distribution probability of lithofacies is obtained by using 3D neural network clustering of argillaceous content and gamma-ray data.

(4) By controlling the types, ratios and boundaries of grain-size lithofacies and combining probabilistic bodies with cooperative constraints, the spatial distribution of grain-size lithofacies can be accurately simulated. The grain-size lithofacies model established in this paper complies with the depositional law in space and has more reasonable contact relations between various sand bodies.

(5) By comparing the thickness and average porosity of the effective reservoir and the high-quality reservoir encountered by the new well, the model anastomoses more than 85%.

## References

[1] Pytlak, L., Cross, D., Sachsenhofer, R. F., Bechtel, A., Linzer, H. G.: Gas accumulations in Oligocene–Miocene reservoirs in the Alpine Foreland Basin (Austria): evidence for gas mixing and gas degradation. International Journal of Earth Sciences 106, 2171-2188 (2017).

[2] Kordi, M.: Sedimentary basin analysis of the Neo-Tethys and its hydrocarbon systems in the Southern Zagros fold-thrust belt and foreland basin. Earth-Science Reviews 191, 1-11(2019).

[3] Capaldi, T. N., Horton, B. K., McKenzie, N. R., Stockli, D. F., Odlum, M. L.: Sediment provenance in contractional orogens: The detrital zircon record from modern rivers in the Andean fold-thrust belt and foreland basin of western Argentina. Earth and Planetary Science Letters 479, 83-97 (2017).

[4] Juárez-Arriaga, E., Lawton, T. F., Z, Y., Ocampo-Díaz, E., Stockli, D. F., Solari, L.: Sediment provenance, sediment-dispersal systems, and major arc-magmatic events recorded in the Mexican foreland basin, North-Central and Northeastern Mexico. International Geology Review 61, 2118-2142 (2019).

[5] Sharman, G. R., Hubbard, S. M., Covault, J. A., Hinsch, R., Linzer, H. G., Graham, S. A.: Sediment routing evolution in the North Alpine Foreland Basin, Austria: interplay of transverse and longitudinal sediment dispersal. Basin Research 30, 426-447 (2018).

[6] Yonkee, W. A., Eleogram, B., Wells, M. L., Stockli, D. F., Kelley, S., Barber, D. E.: Fault slip and exhumation history of the Willard Thrust Sheet, Sevier Fold-Thrust Belt, Utah: Relations to wedge propagation, hinterland uplift, and foreland basin sedimentation. Tectonics 38, 2850-2893 (2019).

[7] Lu, C., Liu, Z., Jia, H., Dai, Q., Li, M., Ren, M., Xia, S., Li, L., Wang, S.: The controls of

geomorphology and sediment supply on sequence stratigraphic architecture and sediment partitioning of the lacustrine rift basin in the Es-3 of Liuzan area, Nanpu Sag, Bohai Bay Basin, China. Australian Journal of Earth Sciences 65, 275-301 (2018).

[8] Silva, T. A., Girardclos, S., Stutenbecker, L., Bakker, M., Costa, A., Schlunegger, F., Lane, S. N., Molnar, P., Loizeau, J. L.: The sediment budget and dynamics of a delta-canyon-lobe system over the Anthropocene timescale: The Rhone River delta, Lake Geneva (Switzerland/France). Sedimentology 66, 838-858 (2019).

[9] Liu, J. L., Ji, Y. L., Yang, K. M., Song, H. Y., Liu, D. W., Wang, T. Y., Zheng, Z. Y.: Tectono-stratigraphy and sedimentary infill characteristics of Xujiahe Formation in western Sichuan foreland basin. Journal of China University of Petroleum (Edition of Natural Science) 39, 11-23 (2015).

[10] Meng, H., Zhong, D. K., Li, Z. P., Liu, Y. L., Zhao, J., Yang, Y. D.: Sequence stratigraphy and sedimentary facies of the fifth member of Upper Triassic Xujiahe Formation of the middle part of the West Sichuan Depression, South China. Journal of Stratigraphy 40, 278-288 (2016).

[11] Zhang, J. C., Zheng, H. R., Wang, G. L., Liu, Z. Q., Qi, Y. C., Huang, Z. W., Fan, X.: In-situ stresses, abnormal pore pressures and their impacts on the Triassic Xujiahe reservoirs in tectonically active western Sichuan basin. Marine and Petroleum Geology 122, 104708 (2020).

[12] Zhang, S. H., Tian, J., Ye, S. J., Wu, L., He, X. B., Zhuo, J. C., Fu, J.: The formation process of gas reservoirs in Xu2 Formation of Xinchang structural belt in Western Sichuan Depression. Natural Gas Industry 39, 17-22 (2019).

[13] Ma, X. J., Zhou, W., Tang, Y., Deng, H. C., Lei, T., Wang, J. J.: Timing of natural fractures formed in the gas reservoirs of the 2nd member of Xujiahe Fm in the Xinchang area, western Sichuan Basin. Natural Gas Industry 33, 15-19 (2013).

[14] Vander Vegt, H., Storms, J. E. A., Walstra, D. J. R., and Howes, N. C.: Can bed load transport drive varying depositional behaviour in river delta environments. Sedimentary Geology 345, 19-32 (2016).

[15] Shang, X. F., Li, M., Duan, T. Z.: Sedimentary system and sand bodies distribution of the second member of the Xujiahe Formation in the Xinchang area, Western Sichuan Depression, China. Interpretation 9(3), 927-944 (2021).

[16] Wang, M. C., Duan, T. Z., Ji, B. Y.: Research progress and application of multipoint statistics geological modeling technology. Journal of Palaeogeography 19(3), 557-566 (2017).

[17] Tahmasebi, P., Hezarkhani, A., Sahimi, M.: Multiple-point geostatistical modeling based on the cross correlation functions. Computational Geosciences 16(3), 779-797 (2012).

[18] Strebelle, S. B., Journel, A. G.: Reservoir modeling using Multiple-Point statistics. SPE New Orleans, Louisiana (2001).

[19] Wang, M. C., Duan T. Z.: Multipoint geostatistical facies modeling algorithm based on pattern vector distance. Acta Petrolei Sinica 39(8), 916-923 (2018).

[20] Demic, M.: Optimization of characteristics of elasto-damping elements of cars from the aspect of comfort and handing. International Journal of Vehicle Design 13(1), 29-46 (1992).

[21] Rudi, C., Paul M.B., V.: A fast quartet tree heuristic for hierarchical clustering. Pattern recognition 44(3), 662-677 (2011).

[22] Ogiesoba, O. C.: Application of thin-bed indicator and sweetness attribute in the evaluation of sediment composition and depositional geometry in coast-perpendicular subbasins, South Texas Gulf Coast. Interpretation 5, 87-105 (2016).

Journal of Physics: Conference Series 2594 (2023) 012050 doi:10.1088/1742-6596/2594/1/012050

# Numerical Simulation Study on Parameter Optimization of In-depth Fluid Diversion in Low-permeability Fractured Reservoir

**Wei Zhou1,2(✉), Daiyin Yin1,2**

1 Department of Petroleum Engineering, Northeast Petroleum University, Daqing 163318, China,
2 Key Laboratory of Enhanced Oil Recovery (Northeast Petroleum University), Ministry of Education, Daqing 163318, China
168001020037@stu.nepu.edu.cn

**Abstract**.Daqing Chaoyanggou Oil Field is a low-permeability fractured oil reservoir that has entered into the middle and high water-cut periods of waterflooding extraction. Now, the moisture content is 74.88% and the injected water flows along fractures. The moisture content of oil wells increases quickly, accompanied with relatively serious invalid waterflooding. The in-depth fluid diversion can be used to exploit residual oil in substrate effectively and thereby increase recovery efficiency of the low-permeability fractured oil reservoir. In this study, parameters of in-depth fluid diversion in low-permeability fractured reservoirs can be optimized by numerical simulation method. According to geologic characteristics and fracture development situations in the study area, a geologic model of the study area was constructed. A historical fitting of exploitation indexes was carried out to optimize the dosage, injection rate and injection time of control-displacement agent during in-depth fluid diversion. Results demonstrated that the optimal dosage, the optimal injection rate and the optimal injection time of the control-displacement agent were 0.1～0.15PV, 0.03～0.04PV/a and about 70% of comprehensive moisture content, respectively. This study can provide reference to residual oil recovery in low-permeability fractured reservoirs during middle and high water-cut periods.

## 1. Introduction

The reserves of low-permeability oil reservoirs in China account for 60%～70% of explored reserves[1-3], which will be an important continuous capacity of petroleum in China. The low-permeability reservoir is usually developed with abundant natural fractures[4-5]. The low-permeability fractured oil reservoir is mainly composed of substrate and fractures. It has typical two-pore dual-seepage tectonic characteristics and has significant differences between substrate and fractures in term of physical properties like permeability, porosity and saturation[6]. After water-flooding for a period, water injection is extremely easy to flow to the production well along fractures, thus increasing moisture content in the production well quickly and poor recovery of crude oils in the substrate[7,8]. These residual oils can be recovered by in-depth fluid diversion[9,10]. The control-displacement agent dosage, injection rate and injection time may influence the in-depth fluid diversion effect. In this study, a geologic model of the study area was constructed by numerical simulation and injection parameters of in-depth fluid diversion were optimized. Research results provide references to application of in-depth fluid diversion in oil fields.

Journal of Physics: Conference Series **2594** (2023) 012050 doi:10.1088/1742-6596/2594/1/012050

## 2. Construction of geologic model

A refined three-dimensional geologic model of Chao 45 Block was constructed by the Petrel geologic modeling software. The block area was 2.16km$^2$ and equidistant angular point coordinate grids were applied on the plane, with a step length of 20m × 20m. Parameters of the fracture system were set preliminarily as follows: porosity of fractures was 1/10 of that of the substrate; the permeability of east-to-west fractures was 7 times that of substrate; permeability of south-to-north fractures was 0.1 times that of substrate; and fracture density was 0.13 fractures/m. In the follow-up fitting process, permeability of fractures is adjusted locally according to dynamics of oil-water wells. The two-pore dual-seepage model was chosen by the Eclipse software for numerical simulation. The substrate system and fracture system were divided into two sets of grids. There were 47 sedimentation units longitudinally, which were divided into 94 simulation layers. The first 47 simulation layers were the substrate system and the rest 47 ones were fracture system. These 94 simulation layers were used for fitting of geological reserves, industrial volume, moisture content and oil yield, thus meeting the fitting requirements.

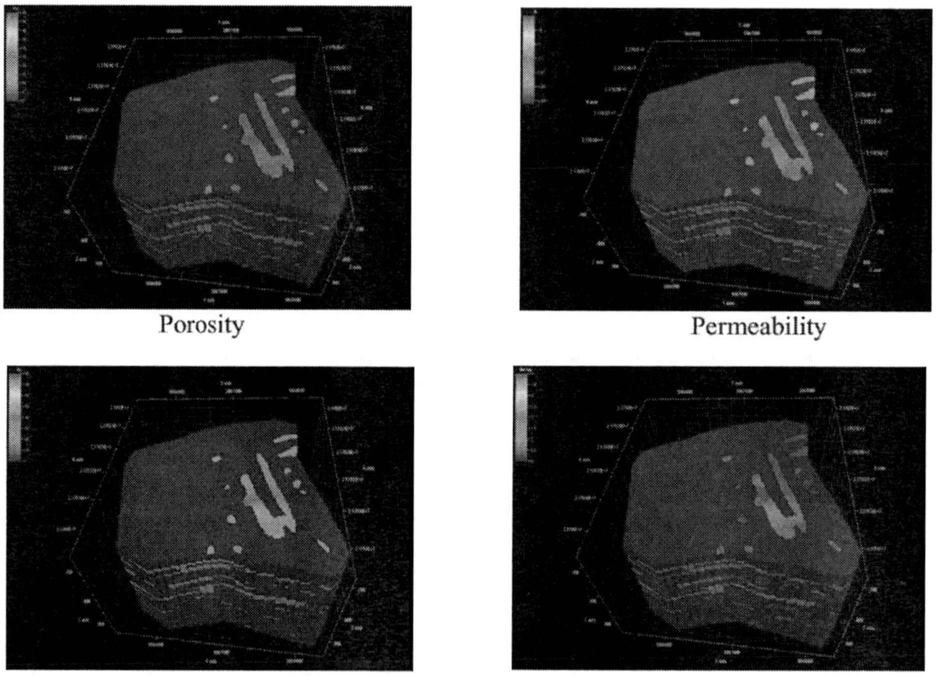

Porosity Permeability

Oil saturation Effective thickness

**Fig.1** Attribute models of the study area

## 3. Optimization of control-displacement agent dosage

The dosage of control-displacement agent was set 0.03PV, 0.05PV, 0.1PV, 0.15PV, 0.2PV, 0.25PV, 0.3PV and 0.4PV. After in-depth fluid diversion, waterflooding continued to 98% and the dosage of control-displacement agent was 0.04PV/a. The moisture contents and recovery efficiencies of 8 in-depth fluid diversion schemes and single waterflooding scheme were calculated by the model (Fig.2 and Fig.3).

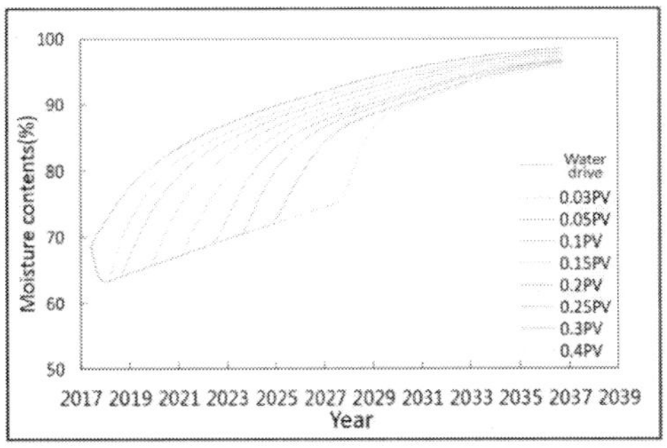


**Fig.2** Comparison of moisture contents under different dosage of control-displacement agent

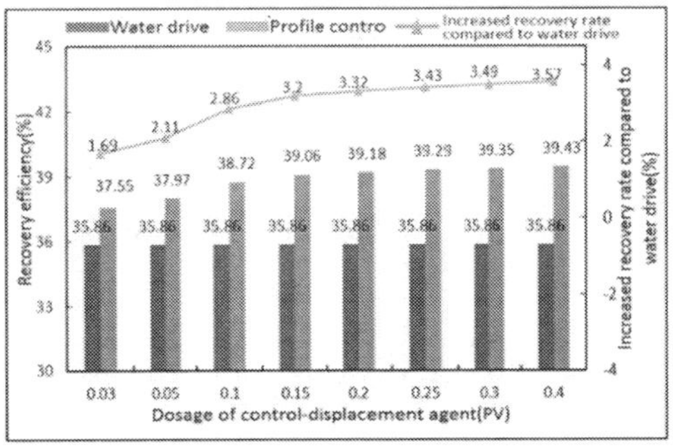


**Fig.3** Comparison of exploitation effects under different dosage of control-displacement agent

According to numerical calculation results, recovery efficiency increased gradually with the increase of dosage of control-displacement agent. After the dosage of control-displacement agent exceeded 0.15PV, the growth rate of recovery efficiency declined gradually. The recovery efficiency was 38.72% when the dosage was 0.1PV and it further increased by 0.34% to 39.06% when the dosage increased to 0.15PV. When the dosage continued to increase to 0.2PV, the recovery efficiency only increased by 0.12% compared with that at 0.15PV. According to observation of concentration field changes of control-displacement agent, the sphere of action expanded gradually when the dosage increased from 0.03PV to 0.15PV. Subsequently, the sphere of action expanded slightly. Combined with analysis of pressure analysis at bottom of wells, it became difficult to inject control-displacement agent after the dosage exceeded 0.15PV. Hence, the dosage of control-displacement agent was recommended 0.1~0.15PV with considerations to growth rate of recovery efficiency and injection capacity.

## 4. Optimization of injection rate of control-displacement agent

The dosage of control-displacement agent was set 0.1PV and the injection rate was set 0.02PV/a, 0.03PV/a, 0.04PV/a, 0.05PV/a and 0.06PV/a. After in-depth fluid diversion, waterflooding continued to 98%. The moisture contents and recovery efficiencies of 5 in-depth fluid diversion schemes were calculated by the model (Fig.4 and Fig.5).

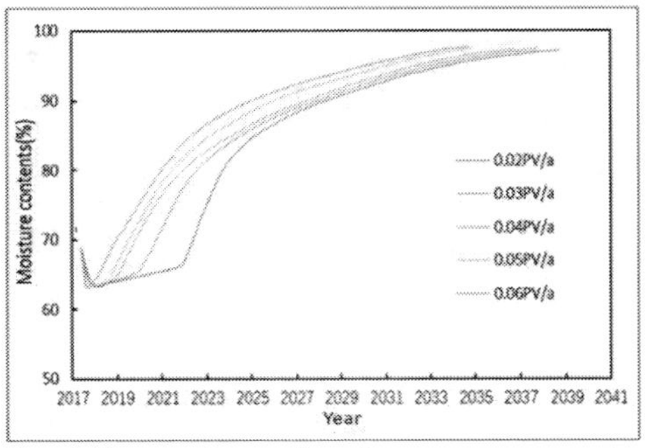


**Fig.4** Comparison of moisture content under different injection rates

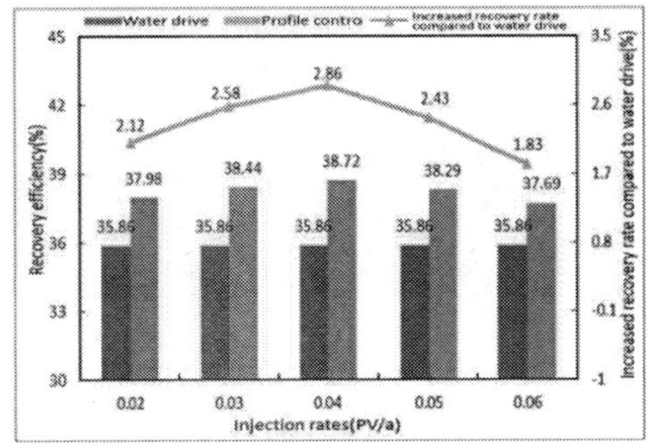


**Fig.5** Comparison of exploitation effect under different injection rates

According to numerical calculation results, the recovery efficiency was the highest (38.72%) when the injection rate was 0.04PV/a. When the injection rate was lower than 0.04PV/a, pressure in fractures increased slowly and moisture rate decreased to a small extent. The control-displacement agent failed to provide good blocking effect. When the injection rate was higher than 0.04PV/a, pressure at bottom of wells reached the formation fracture pressure and not all control-displacement agent was injected in, thus resulting in the low recovery efficiency. Hence, the injection rate of control-displacement agent was recommended 0.03~0.04PV/a with considerations to growth rate of recovery efficiency and injection pressure.

## 5. Optimization of injection rate of control-displacement agent

The dosage of control-displacement agent was set 0.1PV and the injection rate was set 0.04PV/a. Injection time was set at 50%, 60%, 70%, 80% and 90% of moisture contents. After in-depth fluid diversion, waterflooding continued to 98%[8-9]. The moisture contents and recovery efficiencies of 5 in-depth fluid diversion schemes were calculated by the model (Fig.6 and Fig.7).

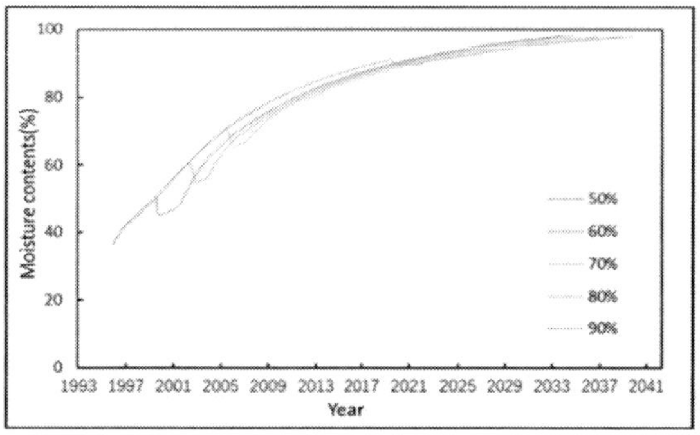


**Fig.6** Comparison of moisture content under different injection time

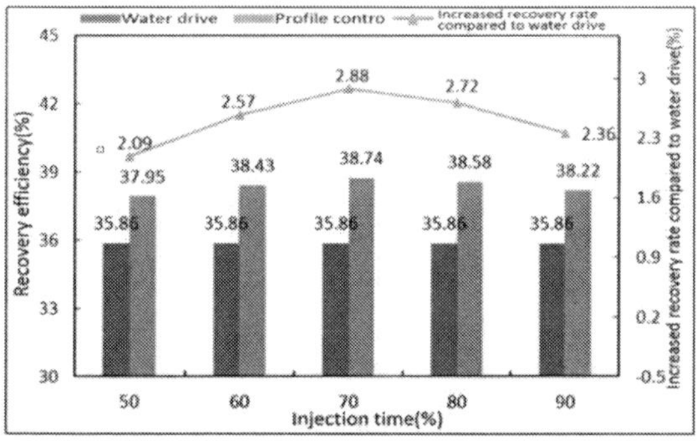


**Fig.7** Comparison of exploitation effect under different injection time

According to numerical calculation results, moisture content decreased more during in-depth fluid diversion when the control-displacement agent was injected earlier, but the moisture content increased quickly in the follow-up waterflooding stage. The highest recovery efficiency of in-depth fluid diversion was achieved when control-displacement agent was injected at 70% of moisture content. Earlier or delayed injection could decrease the recovery efficiency to some extents. This reflects that when the control-displacement agent was injected too early, there was small difference of water saturation between the substrate and fractures, which was disadvantageous for full development of fractures in control-displacement agent conductivity after in-depth fluid diversion. Meanwhile, gel molecules enter into pore channels of substrate and block some pore throat, forming a lot of residual oils. Hence, the growth rate of recovery efficiency decreased to some extent. If the control-displacement agent was injected too late, the water saturation of fractures was close to the maximum and fluid exchange between substrate and fractures declined accordingly. This was disadvantageous for improvement of water phase fluidity in fractures. After the injection time was delayed, the effective acting period of the oil-displacement system was shortened and the oil-displacement effect was deteriorated. With comprehensive considerations to above analysis and numerical simulation calculation results, it is recommended to inject control-displacement agent when the moisture content is 70%.

## 6. Conclusions

(1) With the increase of dosage of control-displacement agent, the recovery efficiency increases gradually, but its growth rate decreases after the dosage exceeded 0.15PV. Hence, the dosage of control-displacement agent is recommended 0.1~0.15PV with considerations to growth rate of recovery efficiency and injection capacity.

(2) When the injection rate is lower than 0.04PV/a, pressure in fractures increases slowly and the moisture content is decreased slightly. The control-displacement agent fails to provide good blocking effect. When the injection rate is higher than 0.04PV/a, pressure at bottom of wells reaches formation

fracture pressure and not all control-displacement agent is injected in, resulting in the low recovery efficiency. Therefore, the injection rate of control-displacement agent is recommended 0.03~0.04PV/a with considerations to growth rate of recovery efficiency and injection pressure.
(3) If the control-displacement agent is injected too early, there is small difference of water saturation between the substrate and fractures. Hence, fractures cannot guide flow of control-displacement agent fully after in-depth fluid diversion, thus decreasing growth rate of recovery efficiency. If the control-displacement agent is injected too late, the water saturation of fractures is close to the maximum and fluid exchange between the substrate and fractures declines accordingly and the improvement of water phase fluidity in fractures is deteriorated. Hence, it is recommended to inject control-displacement agent when the moisture content is 70%.

**References**

[1] XU T.(2015).Review on improving efficiency of recovery(EOR)in low permeability oil reservoir[J].Petrochemical Industry Technology,(11):97.
[2] TANG C.et al.(2011). Application of multi-functional complex gel low-permeability fractured oil reservoirs[J].Petrochemical Industry Application,2011(02):20-22,40.
[3] ZHANG Changyan.Causal Analysis and Treatment of Water Breakthrough of Low Permeability Reservoir with Complex Fracture System[J].Journal of Petrochemical Universities,2021, 34(05) :30-35.0
[4] WU Hongzhu, HAN Zhaohai, WANG Fenggang, et al.Research on high efficiency control and flooding technology of water channeling and flooding in fractured reservoirs[J].Energy Chemical Industry,2021(04):50-54.
[5] WEN Yuhuan, ZHOU Min, HUANG Yuchi, et al.Development Technology of Numerical Simulation in Crack Reservoirs[J].Journal of Chengde Petroleum College,2013(01):1-4,17.0
[6] Yin D , Zhou W.Mechanism of Enhanced Oil Recovery for In-Depth Profile Control and Cyclic Waterflooding in Fractured Low-Permeability Reservoirs[J]. Geofluids, 2021, 2021(1):1-9.
[7] BAI Baojun, LIU Xiange, LI Yuxiang.Advance in profile control through chemical water shutoff in chinese oil fields[J].Oil Drilling & Production Technology,1998(03):64-68.
[8] BAI Baojun, LIU Xiange, LI Yuxiang.Advance in profile control through chemical water shutoff in chinese oil fields[J].Oil Drilling & Production Technology,1998(03):64-68.
[9] Li Z, Yang G F, Song J M. Developing Deep Profile Control and Displacement Agent of Weak Gel for High Watercut Reservoirs[J]. Natural Gas Technology, 2009, 3(4): 46-48.
[10] ZHANG Yu. Study on Effect of Profile Controlling and Flooding System of Weak Gel on Gelling Properties[J].Natural Gas and Oil,2013,31(04):70-72.
[11] Zhou Yazhou;Yang Wenbin,Yin Daiyin.Experimental investigation on reservoir damage caused by clay minerals after water injection in low permeability sandstone reservoirs[J].Journal of Petroleum Exploration and Production Technology.2021.
[12] Daiyin Yin and Wei Zhou.Mechanism of Enhanced Oil Recovery for In-Depth Profile Control and Cyclic Waterflooding in Fractured Low-Permeability Reservoirs[J]. Geofluids. 2021
[13] Kai Liu, Daiyin Yin, Haibo Su.Transient transfer shape factor for fractured tight reservoirs: Effect of the dynamic threshold pressure gradient in unsteady flow[J].Energy Science & Engineering.2020,Vol.8(No.7)： 2566-2586.
[14] Dongqi Wang, Yin Daiyin, Junda Wang, Zhou Yazhou, Chengli Zhang. Influencing factors and microscopic formation mechanism of phase transitions of microemulsion system[J].Journal of Petroleum Exploration and Production Technology.2022,Vol.12(No.10)： 2735-2746.

Journal of Physics: Conference Series **2594** (2023) 012051 doi:10.1088/1742-6596/2594/1/012051

# Research on the Suitability of High-Density Leakage Plugging in Offshore High Temperature and High Pressure Wells

**Yan-jun LI[1,2], Shu-jie LIU[1,2], Jin YANG[1,3], Yu-hang ZHAO[1 (✉)], Yi-long XU[2], Hong ZHU[1], Xun ZHANG[1], Kai-die YANG[1]**

1 China University of Petroleum (Beijing), Beijing, China
2 Cnooc (China) Co.,LTD hainan Branch, Hainan, China
3 Cnooc South China Sea Oil and gas energy academician workstation
✉cup20220228@163.com

**Abstract**. A large number of fractured formations are developed in the high-temperature and high-pressure formations of the Yingqiong Basin in the South China Sea. During drilling, the risk of lost circulation is ex-tremely high, which will cause complex downhole accidents and seri-ously affect the drilling speed.In order to realize safe and efficient plugging of HPHT Wells, the following studies are carried out in this paper: Firstly, the characteristics of leakage, types of leakage and plug-ging and prevention mechanism of HPHT formation in Yingqiong Ba-sin are clarified based on core and indoor rock mechanics experiments. Secondly, a simulated plugging evaluation device with high tempera-ture resistance of 180 ℃ and high pressure resistance of 20Mpa was in-vented and applied to accurately evaluate the reservoir leakage behav-ior in Yingqiong Basin. Finally, a series of anti-high temperature plug-ging agents adapted to different loss types are systematically selected for the first time for different formation fracture types and the plugging effect is evaluated. In this study, through mechanism research, device research and development, accurate evaluation, the shaft wall pressure equivalent increased to 0.1g/cm3 without leakage. It provides technical support for the subsequent ultra-high temperature and high pressure stratum with weak leakage and narrow pressure window.

## 1. Introduction

Yingqiong Basin is a typical high-temperature and high-pressure basin in China's offshore waters. The resources of high-temperature and high-pressure natural gas are estimated to exceed 15 trillion cubic meters, accounting for 2/3 of the total re-sources[1]. In recent years, the discovered geological reserves in the high temperature and high pressure area account for 60% of the total discovered geological reserves in the western South China Sea, which has become the main growth point of natural gas reserves in the area[2-3].

However, the formation pressure system in Yingqiong Basin is complicated, the formation pressure of HPHT Wells is many steps, and the safe operation window is nar-row. The formation temperature is close to 200℃, the actual drilling pressure coefficient is 2.0 ~ 2.4, and the safe pressure window is 0 ~ 0.3. It can be seen that the deep HPHT Wells in Yingqiong Basin are facing harsh geological environment. Due to the narrow formation pressure window, various degrees of leakage occurred in many HPHT Wells in drilling, and the situation of well loss after overflow kill was present-ed. Practice has proved that high temperature and high pressure well lost circulation problem has the complexity and unpredictability of the field of high temperature and high pressure well plugging a success rate is low, and often appear many times after the collision and complex leakage phenomenon, the serious influence the drilling safety and efficiency, cause huge economic losses, so a high temperature and high pressure high density plugging fitment related research is particularly important[4-6].

The following problems generally exist in the process of high-temperature and high-pressure leak plugging: ① Lack of research on high-temperature and high-pressure leak mechanism and control law; ②Conventional evaluation methods for plugging have been established, but corresponding evaluation techniques have not been estab-lished for HPHT formations; ③The plugging material is relatively simple, easy to appear repeated leakage phenomenon, and the adaptability of the material to high temperature overpressure is difficult to verify.

In this paper, the characteristics and types of leakage in high temperature and high pressure stratum are summarized through rock mechanics experiments. At the same time, the plugging materials and evaluation methods are optimized based on the exist-ing plugging theory. Four kinds of accurate evaluation methods for HHT and HHP leakage behaviors were developed: leakage evaluation method for permeability HHT and HHP formation, leakage evaluation method for induced fracture HHT and HHP formation, diversion ability evaluation method for real irregular multi-branch leakage joints, and recovery trend evaluation method for fracture plugging slurry. A series of anti-high temperature plugging agents adapted to different fracture types and leakage types were systematically selected and their plugging effects were evaluated.

## 2. Mechanism of Plugging and Preventing Leakage in HTHP Stratum

### *2.1 Leakage Characteristics and Types*

Based on core sampling and rock mechanics experiments[7-8], the formation characteristics and potential leakage characteristics of XX-block in Yingqiong Basin are analyzed based on lithology, physical properties, cementation, porosity and permeability.

**Table 1.** Partial loss statistics of Yingqiong basin block

| Well number | Leakage zone | Lithology | Cementation condition |
|---|---|---|---|
| XX1-1-1 | 3936.20m | Quartz, a small amount of feldspar and dark minerals, clay, low total amount | Argillaceous cementation、loose layer |
| XX2-1-1 | 3256m | Quartz, calcite, a small amount of pyrite, feldspar, clay, high total | Sandy mudstone、Inlay cementation |
| XX1-1-2 | 4085m | Quartz, a small amount of feldspar, calcite, clay, high total amount | Filling metasomatic-inlay cementation |
| XX1-1-3 | 4103m | Quartz, a small amount of feldspar, calcite, clay, high total amount | Inlay cementation |
| XX1-1-4 | 4105~4138m | Quartz and calcite mainly, a small amount of feldspar, dolomite, The amount of clay is low | Inlay cementation |
| | 4329m | | Inlay cementation |
| XX1-1-5 | 4311m | Quartz, a small amount of feldspar, clay content is low | Inlay cementation |
| XX1-1-6 | 3868m | Quartz, secondary feldspar, slightly containing dark minerals | Marlaceous cementation、compact layer |
| YY1-1-1 | 2768m | Quartz, a small amount of feldspar, metamorphic rock, mica | Porous cementation |
| YY2-1-1 | 2524m | Quartz, a small amount of feldspar and dark minerals | argillaceous cementation、loose layer |

Table 1 summarizes the lithology and cementation of some lost formations in Yingqiong Basin, and makes a comparative analysis of the cementation types in their respective well blocks. The formations of YY1-1 well blocks are mostly pore cementation with less loss. The electron microscopy of their cores (left of Fig. 1) shows that there are more cements in their pores, which are easy to fill. The formation of well XX-1 is mostly cemented by pressure embedment, and there is widespread leakage phenomenon. Electron microscopy of the core (right of Fig. 1) shows that it is mostly in loose cementation state, and pore dissolution is common.

**Fig. 1.** Electron microscopic view of core

In addition, from the lithologic analysis of the well depth where the existing loss point is located, it is mainly fine siltstone and silt rock, with relatively poor cementation properties, and both have great loss potential, which is affected by the narrow density window, resulting in the occurrence of loss.

**Table 2.** Quantitative analysis of clay minerals around the leak horizon of the leak well in XX-block

| | | Clay mineral content % | | | | | | Illite/smectite mixed layer proportion % | |
|---|---|---|---|---|---|---|---|---|---|
| Well number | Well section | smectite | illite | kaolinite | chlorite | Illite/smectite mixed layer | Smectitech/loritemixed layer | Smectite layer | Illite layer |
| | | S | I | K | C | I/S | C/S | S | I |
| XX1-1-2 | 4032 | | 68 | 5 | 17 | 10 | | 10 | 90 |
| XX1-1-3 | 4090 | | 69 | 4 | 13 | 14 | | 10 | 90 |
| XX2-1-1 | 3496 | | 43 | 7 | 40 | 10 | | 10 | 90 |

By analyzing the clay minerals of sedimentary rock near the leakage layer, it can be seen that the main leakage layer in XX-block basically does not contain montmorillonite, and the illite content is high, most of which is more than 50%. The content of The Mixed layer of Aemon is between 10 and 20%, and it decreases rapidly with the increase of the depth. The illite layer is mainly in the mixed layer of Aemon. The content distribution of clay minerals (Table 2) shows that the risk of leakage is greatly affected by illite layer, and the expansion property of illite is not obvious when it meets water. However, the change of mechanical properties of illite itself after it meets water will reduce the overall strength of rock and increase the risk of leakage.

**Table 3.** Partial loss statistics of Yingqiong basin block

| Well number | Sand body | $K/10^{-3}\mu m^2$ | Φ/% |
|---|---|---|---|
| YY1-2-1 | H1 Ⅰa | 12.3~109.0 | 18.1~20.3 |
| YY1-2-1 | H1 Ⅱb | 4.0~90.1 | 17.1~18.9 |
| YY1-2-2 | H1 Ⅱb | 3.17~48.8 | 15.7~20.5 |
| YY1-2-3 | H1 Ⅰb | 6.66~49.1 | 16.5~18.1 |
| YY1-2-3 | H1Ⅲ | 5.1~6.5 | 16.4~17.2 |

A total of 24 cores were taken from four different sand bodies of YY1-2-1, YY1-2-2 and YY1-2-3 Wells. The permeability of the main reservoir in YY1-2 gas field is between 4.0 and 109.0mD, with an average permeability of 28.7mD. The porosity ranges from 15.7 to 20.5%, with an average of 17.9%.

*2.2 Plugging and preventing releakage mechanism*

Based on the traditional porous plugging theory, aiming at the shortcomings of the traditional plugging theory, the optimization of the following materials and evaluation methods is carried out to guide the development of high temperature resistance, compression and shear resistance materials and achieve the purpose of dynamic balance of plugging barrier.

*2.3 Existing stress cage plugging theory*
The stress cage theory (ideal filling theory) is based on the stacking efficiency of particles. When the cumulative volume fraction of particles is proportional to the square root of particle diameter, the particle material has the highest stacking efficiency, that is, it is believed that the material can form a dense sealing partition wall[9-11]. The theory is only suitable for temporary plugging, and the sealing barrier can not be formed in the deep part of the leakage due to the partial accumulation of the seam.

*2.4 Analysis of plugging and preventing leakage mechanism*
The existing leakage theory and control methods lack effective cognition and pertinence of stratum, leading to a series of leakage problems in engineering, such as "closing", "turning", "stopping", "spitting back" and "re-destruction". The lack of understanding of these problems in theory and practice will directly lead to the low construction rate of the sealing barrier or the risk of re-destruction.

This paper believes that the core reason for the traditional plugging system is the difference between the actual state of the fracture and the expected: after hours or even days of plugging operation, the second wing of the fracture produces dense mud cake, which will hinder the formation of subsequent solid barrier. Plugging operation in the risk of loss, the sooner the effect should be better. Therefore, the filtration control of plugging slurry is the prerequisite for the establishment of solid barrier. Too fast is too risky, resulting in the expansion of cracks and other problems, and too slow is difficult to escape. It is a better choice to control the filtration speed according to the downhole conditions and on-site conditions.

*2.5 Optimization of materials*
Based on closing reinforcement, wedged pressure and flipping the bridge plugging mechanism[12-14], developed a new type of resistance to high temperature and high strength rigid plugging material, through the indoor experiment to verify the protective mechanism of leakage, the results showed that the plugging type crack leakage, the common drilling fluid carrying sex good, big grain failed to reside, unable to form effective sealing barrier. The special plugging slurry can effectively make the large particles reside in the crack and further rapidly filter out to form a solid barrier. When plugging pore type loss, ordinary drilling fluid cannot cement sand and gravel because there is no dense solid barrier formed, but the solid barrier formed by special plugging slurry can cement sand and gravel to a greater extent, and the effect will be more obvious in preventing repeated leakage. It has been applied to a high temperature and high pressure exploration well, and achieved successful one-time plugging. After loss, the overall pressure equivalent of the wellbore is increased by 0.1g/cm$^3$.

## 3. An Accurate Evaluation Method for HTHP Formation Leakage Behavior

Loss and its evaluation of the development of technology are closely associated with the development of science and technology, restricted to the understanding of the mechanism and the mechanism for the loss, the early test evaluation method is relatively simple, some experiments and their results, and the actual differences and underground site conditions, test error is bigger, can't be real described the scene actual situation[15]. Nowadays, with the progress of technology, artificial seam plate, artificial pores (steel balls), sand bed, natural cracks and so on are gradually used to simulate underground cracks and pore morphology, respectively in static, dynamic and other conditions to evaluate the plugging performance of plugging materials[16-17].

Based on the existing evaluation methods, combined with the formation characteristics of Yingqiong Basin, the evaluation methods for leakage prevention and plugging were studied, and the following four accurate evaluation methods for leakage behavior of HPHT strata were established, which made up for the deficiency of traditional evaluation methods applied in HPHT strata.

*3.1 Loss evaluation method for permeable high temperature and high pressure formation*
The traditional method to simulate the plugging equipment of sand bed can not meet the plugging simulation evaluation under the condition of high temperature and high pressure.

Journal of Physics: Conference Series **2594** (2023) 012051 doi:10.1088/1742-6596/2594/1/012051

The high temperature and high pressure sand bed filtration apparatus consists of a porous and permeable leakage layer formed by filling sand beds in high temperature and high pressure pressure vessels. According to the selection of sand beds with different particle sizes, the pore sizes of different leakage channels are simulated. Through electric heating to the required experimental temperature, the external nitrogen cylinder is connected to the pressure plugging experiment. Fig. 4 (left) shows the design of filtration apparatus for high temperature and high pressure sand bed, and Fig. 4 (right) is the actual drawing. In this method, sand beds with different particle sizes can be used to simulate the plugging bearing capacity at different leakage rates. The bearing capacity can reach 20MPa and the temperature resistance can reach 180℃, and the experimental results can be avoided as far as possible due to equipment.

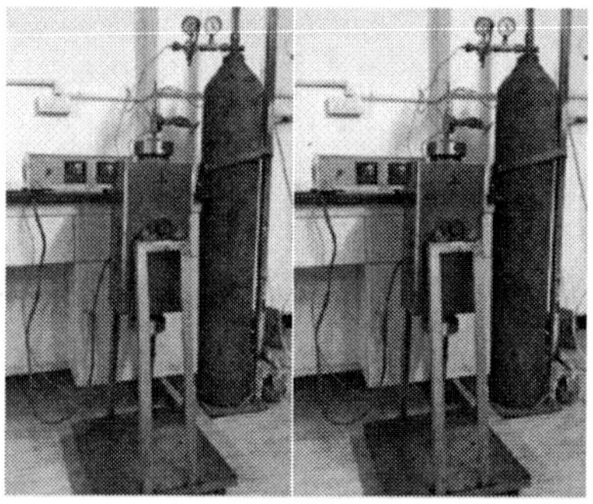

**Fig. 4.** High temperature and high pressure sand bed filtration apparatus

### *3.2 Loss evaluation method for induced fracture HHP formation*

The traditional method usually uses metal to make cracks, and it is impossible to judge the intrusion of plugging materials into cracks by simulating the destruction of cracks during disassembly.

The equipment for the first time to use natural rock formed natural cracks as the need to plug the cracks, can better simulate the formation of fractures, and crack width can be selected according to need, crack width 0.5-10mm can be selected, the length of 200mm. The plugging process can be heated and pressurized, and the simulated wellbore capacity is 3000ml, which can better simulate the plugging process and avoid the error of experimental results caused by the door sealing.

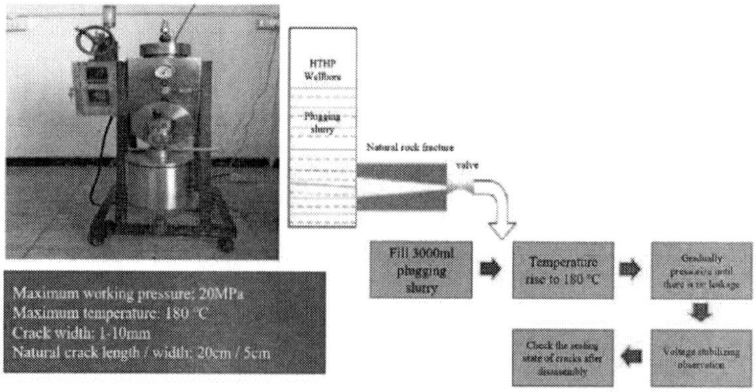


**Fig. 5.** Simulation of natural crack high temperature and high pressure plugging device

*3.3 Evaluation method for steering ability of real irregular multi-branch leaky joints*

The traditional method is usually one-way leakage, without considering the sealing situation of irregular (multi-branch, irregular) leakage, such as the plugging system does not have good steering ability, it will lead to inadequate sealing, resulting in the failure of plugging.

The design concept of the simulation fracture steering plugging evaluation instrument is to effectively evaluate the influence of plugging slurry on different types of fractures in the formation. The leak layer of plugging device adopts simulated cracks, and two simulated cracks of different sizes can be placed at the same time. In the upper part of the plugging slurry with pressure squeeze and test the plugging pressure, at the same time set up annular pressure, annular pressure is used to control the size of cracks and ensure the repeatability of the same particle size sand bed experiment, can more accurately test the plugging effect of plugging slurry on the leaking layer.

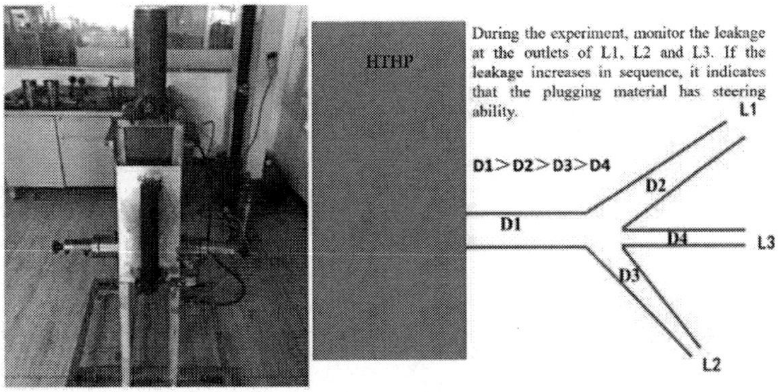


**Fig. 6.** Simulation fracture steering plugging evaluation instrument

*3.4 Evaluation method of flowback trend of plugging slurry*

The design concept of the simulated fracture flowback trend evaluation instrument is to effectively evaluate the influence of the reverse pressure on the pressure bearing capacity of the plugging layer after the successful sealing of the plugging layer by the plugging slurry and the pressure relief. Plugging meter leak layer of gravel filling can be used, can also be simulated crack, has three sets of pressure system, in addition to the regular on the top of squeeze pressure leak slurry plugging and test its plugging pressure ability, also increased the toroidal pressure and leak layer at the bottom of the pressure, toroidal pressure to cooperate to control leakage channel characteristics and leakage pore size and to ensure that the same particle size of sand bed experimental repeatability, Pressure at the bottom of the leaking layer is used to simulate the influence of reverse formation pressure on the plugging layer after the completion of plugging pressure relief, which can more accurately test the plugging effect of plugging slurry on the leaking layer.

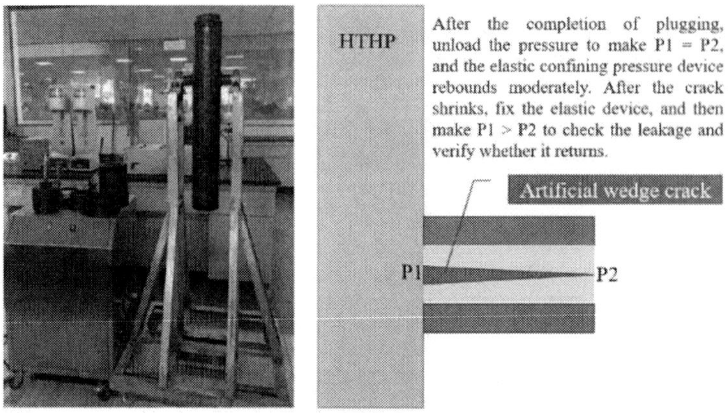


**Fig. 7.** Simulation leakage layer pressure system plugging instrument

## 4. Research and Development of Adaptation of Anti-high Temperature Plugging Agent

A series of plugging agents were selected according to different types of strata cracks to meet different plugging requirements under different conditions of strata cracks. At the same time, the plugging effects of plugging agents were evaluated through sand bed experiments, and the formulation requirements under the condition of optimal plugging effects were analyzed to provide reference and guidance for field construction[18-23].

### *4.1 STRH series plugging agent*

STRH series plugging agent is selected for the formation prone to leakage due to the development of small fractures. The plugging agent is based on THERM system and the basic formula is as follows: Fresh water + 1%prehydrated bentonite + 0.4%NaOH + 0.8%DISTEMP + 4%SMP + 4%SPNH+ 3%FT-1 + 3%EZCARB + 10%HCOOK + 0.3%$Na_2SO_3$ + barite.

**Table 4.** Partial loss statistics of Yingqiong basin block

| Sand tray | System | density / leakage / Temperature | 2.0 | 2.1 | 2.2 | 2.3 |
|---|---|---|---|---|---|---|
| 10D | THERM | 160℃ | 46.2 | 43.4 | 48.6 | 42.4 |
| | | 180℃ | 52.1 | 58 | 54.7 | 49.4 |
| | THERM+ 3%STRH | 160℃ | 19.5 | 20.8 | 26.4 | 18.6 |
| | | 180℃ | 21.3 | 23.4 | 21.7 | 19.9 |
| 5D | THERM | 160℃ | 38.2 | 36.4 | 42.6 | 39.7 |
| | | 180℃ | 39.5 | 40.6 | 38.1 | 36.8 |
| | THERM+ 3%STRH | 160℃ | 16.6 | 18.8 | 17.5 | 19.2 |
| | | 180℃ | 17.6 | 19.2 | 19.8 | 20.6 |

Experiments conducted in this paper (see Table 4 for data) show that adding 3%STRH can significantly reduce the loss.

### *4.2 BLN HT series plugging agent*

For medium and large fracture developed strata, the high-temperature resistant compound bridge PLUGGING BLN HT series plugging agent is selected. The plugging agent is based on THERM system and its basic formula is as follows: Fresh water + 1%prehydrated bentonite+ 0.4%NaOH + 0.8%DISTEMP + 4%SMP + 4%SPNH+ 3%FT-1 + 3%EZCARB + 10%HCOOK + 0.3%$Na_2SO_3$ + barite.

**Table 5.** BLN HT plugging agent sand bed test data

| System | density / leakage / Temperature | 1.8 | 1.9 | 2.0 | 2.2 |
|---|---|---|---|---|---|
| Base slurry | indoor temperature | 76.2 | 73.4 | 78.6 | 72.4 |
| | 180℃ | 82.1 | 88 | 84.7 | 89.4 |
| Base slurry+1% BLN-1 HT | indoor temperature | 70.5 | 69.8 | 68.4 | 65.6 |
| | 180℃ | 72.3 | 73.4 | 71.7 | 79.9 |
| Base slurry+2% BLN-1 HT | indoor temperature | 48.2 | 46.4 | 42.6 | 49.7 |
| | 180℃ | 59.5 | 50.6 | 58.1 | 56.8 |
| Base slurry+3% BLN-1 HT | indoor temperature | 46.6 | 48.8 | 47.5 | 49.2 |
| | 180℃ | 47.6 | 49.2 | 49.8 | 50.6 |
| Base slurry+4% BLN-1 HT | indoor temperature | 36.6 | 38.8 | 37.5 | 39.2 |
| | 180℃ | 27.6 | 29.2 | 29.8 | 30.6 |
| Base slurry+5% BLN-1 HT | indoor temperature | 16.6 | 18.8 | 17.5 | 19.2 |
| | 180℃ | 17.6 | 19.2 | 19.8 | 20.6 |

It can be seen from the experiment that when the density of base slurry +5% BLN-1 HT (containing high strength plugging agent) is 1.8 ~ 2.2, the plugging effect is the best, and it can plug 3-1mm cracks, and can form a sealing zone in the cracks, and no leakage.

MOSTRH series plugging agent for oil base drilling fluid was selected for weak stratum under high temperature and high pressure. The basic formula of the plugging agent is as follows: White oil /$CaCl_2$ aqueous solution + 0.8% primary emulsifier +1% auxiliary emulsifier + 1.2% wetting agent + 1.5% organic soil + 2.5%CaO + 5% fluid loss agent + barite.

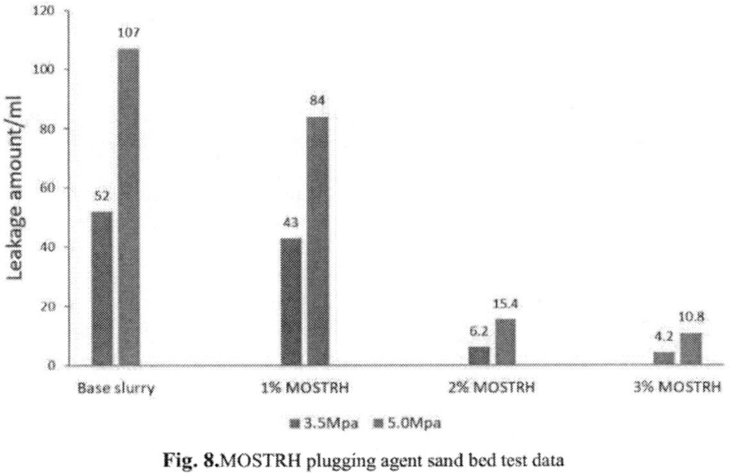


**Fig. 8.**MOSTRH plugging agent sand bed test data

Experimental data (Fig. 8) show that 2% MOSTRH can significantly reduce the loss of sand bed and disc, which has been applied to the development well of YY2-2 gas field.

## 5. Conclusions

1) Based on rock sample coring and rock mechanics experiments, the formation characteristics and potential leakage characteristics of XX-block in Yingqiong Basin are as follows: the formation of the

block has less leakage and more cement in the pore, which is easy to fill; When the formation of block is pressed and embedded cementation, there is usually leakage and loose cementation state, and pore dissolution is common. The cementation properties of fine siltstone and silt rock are poor and there is great potential of loss. The increase in illite content reduces the strength of the formation rock and increases the risk of loss.

2) Based on the existing evaluation methods and combined with the formation characteristics of Yingqiong Basin, the evaluation methods for leakage prevention and plugging were studied, and four accurate evaluation methods for leakage behavior of HPHT strata were established, which made up the deficiency of traditional evaluation methods applied in HPHT strata.

3) For the first time, a series of high-temperature resistant plugging agents adapted to different types of leakage were systematically selected and the plugging effect was evaluated: adding 3%STRH can significantly reduce the loss of leakage, which is suitable for the formation with micro and small fractures; Adding 5% BLN-1 HT (containing high strength plugging agent) can seal 3-1mm cracks, and can form a sealing zone in the cracks, and no leakage; The addition of 2% MOSTRH resulted in a significant loss reduction.

**References**

[1] Bo Wang. The temporary plugging fracturing cracks sealing and steering law research [D] China university of petroleum (Beijing), 2019.

[2] Cheng Han, Ming Luo, Zhongbao Liu, Yuhao Yang, Wenbiao Deng, Qingyuan Zhou. Ultra high temperature and high pressure drilling coring technology of semi submersible platform in Yingqiong Basin [J] Drilling and production technology, 2021,44 (04): 28-30.

[3] Cheng Han, Kaiwen Huang, Ming Luo, Xianyu Liu, Wenbiao Deng. Plugging technology of high temperature and high pressure wells in Yingqiong basin, South China Sea [J] Oil drilling technology, 2019,47 (06): 15-20.

[4] Chengyuan Xu. Deep fractured reservoir plugging material optimization theory and method [Z/OL]. (2020-10-15) [2022-01-21].

[5] Dorman-J, Gyory-L. Challanges And Experiance In The Applied Drilling Fluid Technology Of Ultra HTHP (High Temperture High Pressure) Wells[C]. PROCEEDINGS, 2006,137.

[6] Guanjie Zhang, Kongyou Wu, Caiwei Fan, Jingshou Liu, Guangjun Deng, Yujie Liu. Calculation and significance of rock mechanical parameters of deep reservoir in key areas of Yingqiong Basin [J] Journal of geology of colleges and universities, 2021,27 (05): 616-624.

[7] Hongwei Zhang, Fengjiang Zuo, Hongjun Li, et al. Drilling Fluid and Completion Fluid, 2015, 32(6):43-47.

[8] Jiang Wu, Yanjun Li, Wandong Zhang, Yuhao Yang. Research and application of leakage control technology in high temperature and high pressure formation [J] Oilfield chemistry, 2021,38 (02): 204-209.

[9] Jiaxiong He, Shenghong Chen, Hailing Liu, Shilin Liu. Analysis of oil and gas resource prospect and favorable exploration direction in Ying Qiong basin on the northern edge of the South China Sea [J] Natural Gas Geoscience, 2008 (04): 492-498.

[10] Mojtaba K M, Shervin T, Koroush T N, et al. A Brief introduction to high temperature and foam free water based drilling fluids[J]. SPE 180541,2016.

[11] M.A.TehraniA.Popplestone M-ISwaco. etc. Water-Based Drilling Fluid For HP/HTApplications[C]. SPE 105485.

[12] Mingmin He, Hongxing Li, Minghua Wang, Lita Guo, Chunli Yang, Shipei Zhang. Research on pressure bearing and plugging technology of cage envelope bridge [C] Proceedings of 2017 National Natural Gas academic annual conference, 2017:2508-2513.

[13] Song Deng, Honghai Fan, Jun Niu, Weiyan Ren, Yuhan Liu, Haoyu Diao, Zhaopeng Zhu. Study on safety density window of deepwater shallow drilling fluid [J] China offshore oil and gas, 2017,29 (04): 138-144

[14] Siyuan Lin, Yicheng Zhang, Xingqiang Zhong, Cheng Li, Jianglin Zhu, Donghui Luo. Fractured formation plugging packer evaluation technology research progress [J]. Journal of modern chemical industry, 2019 (6): 13. 1301-1304.

[15] WHITFILL D， HEMPHILL T. All lost-circulation material and systems are not created

ecpzal[ R] . SPE 843192003: 1-9.

[16] Weichu Yu, Changming Su, Xinran Huang, Chunyang Zhang. Journal of oil and gas technology,2008,30(05):116-117+380-381.

[17] Xijin Xing. Research on high-temperature high-density drilling fluid system in dongfang gas field, south China sea [J]. Drilling & production technology,2014,37(06):92-95.

[18] Xiaoyan Hu, Xu Wang, Lijun Zhang, et al. Progress in fine petrochemical industry,2016,17(06):8-12.

[19] Ya Liu. Research on formulation of high-temperature high-density oil-based drilling fluid [D]. Chongqing University of Science and Technology,2016.

[20] Xushen Li, Yingchao Zhang, Xibing Yang, Xuefeng Xu, Jianxin Zhang, Xiao Man. New understanding and progress of natural gas exploration in Yinggehai Qiongdongnan Basin [J] China offshore oil and gas, 2017,29 (06): 1-11.

[21] Yuhong Xie. Reservoir forming mechanism and resource prospect of high temperature and high pressure natural gas in the western South China Sea -- Taking Ying Qiong basin as an example [J] Petroleum drilling and production technology, 2016,38 (06): 713-722 .

[22] Yili Kang, Chengyuan Xu, Long Tang, Song Li, Daqi Li. Building a tough barrier around the well: Theory and method of lost circulation control [J] Petroleum exploration and development, 2014,41 (04): 473-479.

[23] Zhengquan Lu. Analysis of drilling fluid plugging materials and Discussion on leakage prevention and plugging technology [J] Scientific and technological innovation and application, 2019 (28): 157-158.

Journal of Physics: Conference Series **2594** (2023) 012052 doi:10.1088/1742-6596/2594/1/012052

# Uniform Acid Cleaning Technology and Its Application in Horizontal Wells with Open Hole Screen Tube

**Peifeng Jia1, 2, Chuanzhi Cui1(✉), Yingfei Sui1,Jiamin Ren2, Chao Lv2**

1 College of Petroleum Engineering, China University of Petroleum, Qingdao, China;
2. Petroleum Engineering Technology Research Institute of Sinopec Shengli Oilfield Company, Dongying, China.

Corresponding author. E-mail: ccz2008@126.com;

**Abstract**. After the completion of the open hole sieve tube in a horizontal well, it is necessary to replace the slurry and put it into production. Generally, circulating acid outside the tube cannot fully remove mud pollution. The acidification process needs to fully consider the heel effect of horizontal wells. Based on the horizontal pipe flow model and the characteristics of foam diversion, a mathematical model of acid washing suitable for open hole screen horizontal wells is established. Analysis suggests that lower flow rates and higher fluid viscosity during the acid washing process are more conducive to the uniform entry of the plugging solution into the formation. We have designed a uniform liquid mixing equipment with a flow blocking mechanism, and improved the acid washing construction process to achieve negative pressure reverse discharge of acid solution. This process can achieve uniform entry of the unblocking fluid into the formation after passing through the screen gap, and remove the blockage of the mud layer inside and outside the near screen tube. The horizontal wells using this process have a 30% increase in production capacity compared to oil wells in the same block, and the high yield validity period has been extended by 680 days.

## 1. Introduction

Sand production is very common in the process of developing unconsolidated sandstone reservoirs with horizontal wells, so it is necessary to control sand production in the formation at an early stage to ensure normal production of oil and gas wells[1,2]. The main harm is that the rod and pipe pump is damaged, which leads to the well lying, and the oil well is blocked by particle migration, which leads to the decline of production[3-6]. Pickling the production well section before the oil well is put into production can remove the near well plugging caused by the particles of the well entering fluid during the drilling and completion process, and delay the decline of oil well productivity. Conventional general acid cleaning process can not make the acid liquid enter the formation evenly, and the surrounding areas of some well sections can not be effectively acid cleaned, which will cause plugging damage near the well and affect the productivity of horizontal wells.

For horizontal well completion technology in unconsolidated sandstone, domestic scholars have conducted a lot of research. Wei Lihua et al. introduced the application and effect of open hole sand control completion technology for horizontal wells in Chengdao Oilfield, and proposed that acid washing after completion is helpful to remove near well plugging[7]. Wang Zenglin[8] pointed out that the integrated open hole sand control completion technology for horizontal wells has increased the discharge area, and the completion quality will be effectively improved by improving the perfection of oil wells through acid washing and other supporting technologies. Li Hongqian[9] and Xue Lina[10] analyzed and studied the influencing factors and technical parameters of horizontal well flushing technology. Zhang Jian[11] has studied the factors that cause the failure of the integrated technology of open hole sand control completion in horizontal wells, and believes that the well flushing technology needs to be optimized. In the process of sand control and production of horizontal wells, the general acid washing process with complex acid is usually used to avoid plugging of some screen sections after production due to drilling mud pollution, silty fine sand and mud plugging, so as to maximize the productivity of horizontal wells[12-14]. The existing general pickling process has the following problems: (1) The heel toe effect in the horizontal section and the formation heterogeneity lead to uneven acid distribution, which affects the pickling effect, and the productivity of the horizontal well section near the toe end is not fully developed. (2) The uneven acid entering into the formation resulted in no acid treatment near the well in some well sections and serious plugging after production. (3) Incomplete acid flowback in screen tube well section and formation deficit section will lead to screen tube corrosion and affect the effective period of sand control. In order to prolong the effective period of sand control for horizontal wells in unconsolidated sandstone formation and improve the productivity of single wells, it is necessary to analyze the main factors that affect the uniform acid entering into the formation, optimize the balanced acid washing reverse drainage technology on this basis, and improve the production effect of horizontal wells[15-19].

## 2. Mathematical model of near wellbore acid treatment for horizontal wells

### 2.1. *Mathematical model of foam diversion acidizing for horizontal wells*

Some text. The horizontal well section consists of n segments from the toe to the heel. During the flow of acid fluid from the heel to the toe, the acid fluid continuously enters the formation and the flow rate in the wellbore decreases, which is a process of variable mass flow[15,16]. Assumptions are as follows: the length of horizontal section of horizontal well is L, and the section radius is Rw; The horizontal section of the wellbore can be laminar flow or turbulent flow. The wellbore and reservoir flow are isothermal processes. There is no fluid inflow at the toe of the horizontal well. The horizontal section is one-dimensional axial movement.

Suction capacity of each segment:

$$\begin{cases} Q_w = \sum_{i=1}^{n} q_i = \sum_{i=1}^{n} \dfrac{2\pi L_i \cdot K_i \cdot \left[p_{\mathrm{wfi}} - p_{\mathrm{e}}\right]}{\mu_i \left[\ln \dfrac{R_{\mathrm{e}}}{R_{\mathrm{w}}} + S_i\right]} \\ p_{\mathrm{wfi+1}} = p_{\mathrm{wfi}} + \Delta p_{\mathrm{f}} + \Delta p_{\mathrm{ac}} \end{cases} \tag{1}$$

The friction pressure drop and Reynolds number are as follows:

$$\Delta p_{\mathrm{f}} = \frac{1}{2} f_i \frac{H_i}{D_{\mathrm{w}}} \rho u_i^2 \tag{2}$$

$$Re_i = \frac{\rho D u_i}{\mu} \tag{3}$$

$$f = \begin{cases} \dfrac{64}{Re_i} & Re_i \le 2300 \\ \dfrac{0.3164}{\sqrt[4]{Re_i}} & 2300 < Re_i < 10^5 \end{cases} \tag{4}$$

Kinetic energy loss:

$$\Delta p_{ac} = \rho\left(u_{i+1}^2 - u_i^2\right) \quad (5)$$

Kinetic energy loss: the horizontal wellbore model is coupled with the simplified two-dimensional reservoir flow model through the injection pressure, and the partial differential equation is discretized through the finite difference method, and the horizontal well acidizing process is numerically solved. The effects of formation permeability, acid viscosity, acid injection rate, wellbore radius and horizontal well section length on acid distribution and pressure distribution are studied. The uneven distribution of acid in the horizontal well section is caused by friction pressure drop and accelerated pressure drop during the flow of acid from heel to toe. The difference between the pressure in the heel wellbore and the formation pressure is large, resulting in a large amount of acid entering the formation. As shown in Figure 1, the calculation results show that the lower the reservoir permeability around the wellbore, the higher the viscosity of acid fluid and the lower the acid injection speed, the more uniform the acid fluid distribution along the way.

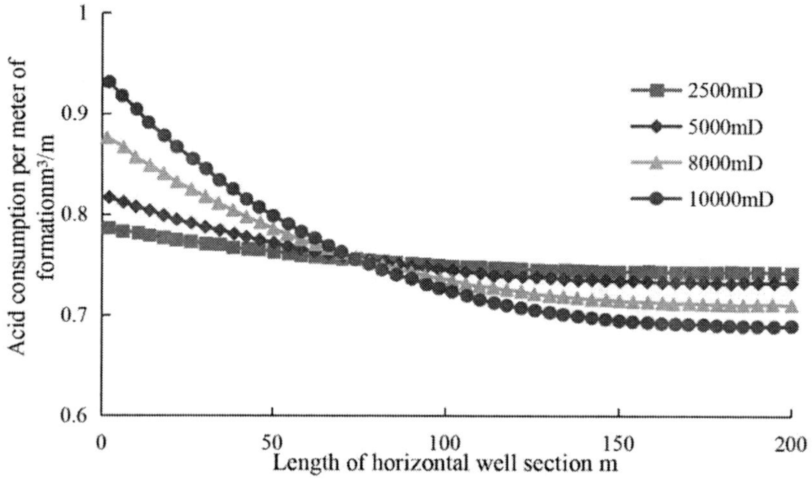


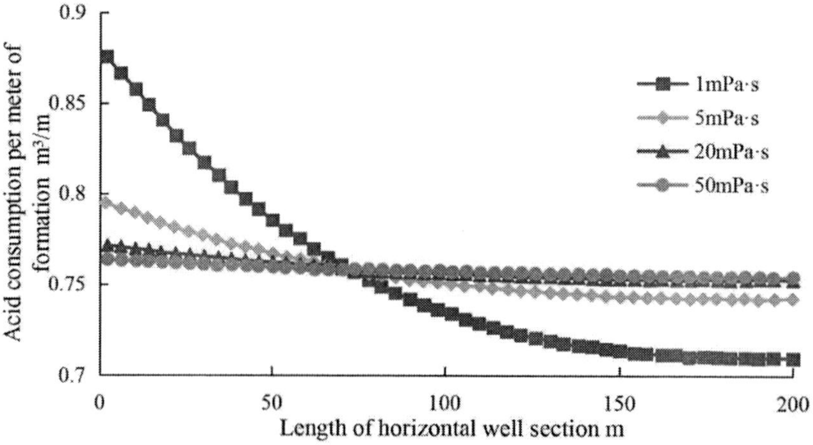

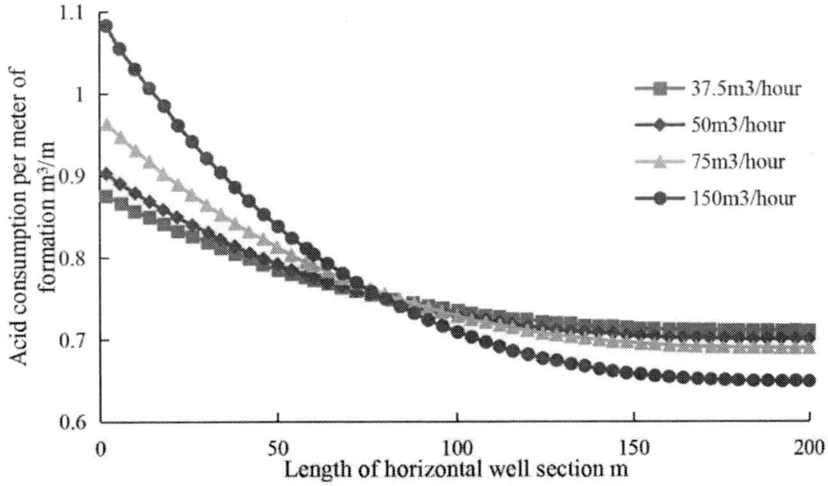


**Fig. 1 Effect of Permeability, Acid Viscosity and Acid Injection Rate on Near Well Sucked Acid Volume of Horizontal Well**

Based on the foam diversion mechanism, the foam flow characteristics are coupled to the variable mass flow equation of horizontal wells, and the mathematical model of foam diversion acidizing of horizontal wells is established[13,16]. The solution process is shown in Figure 2. Eq. (6) and (7) describe that foam blocks large pores, resulting in gas phase trap skin effect, which is shown by the reduction of formation permeability to acid. Meanwhile, density of foam acid in Eq. (9) μf Less than the density of pure acid solutionμi. That is to say, the foam acid cleaning process realizes the uniform entry of acid into the formation by reducing the formation permeability during acid feeding and the viscosity of acid liquid.

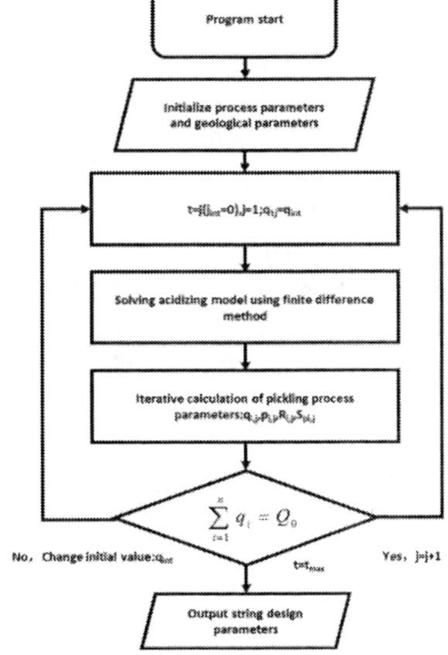


**Fig. 2 Calculation process of pickling model**

Front radius of foam in formation:

$$R_{fi}(t)=\sqrt{\frac{\int_0^t q_{fi}(t)\cdot\Gamma(t)dt}{\pi H_i\varphi_i(1-S_{wc})}+R_w^2} \tag{6}$$

Pseudo skin coefficient of gas trap:

$$S_{pi}(t)=\left(\frac{K_i}{K_{fi}}-1\right)\ln\frac{R_{fi}(t)}{R_w} \tag{7}$$

$$\frac{K}{K_f}=aK^b \tag{8}$$

Suction capacity of each segment:

$$q_i(t)=\frac{2\pi H_i\times K_i\times\left[p_{wfi}(t)-p_e\right]}{\mu_f\left[\ln\frac{R_e}{R_w}+S_i+S_{pi}(t)\right]} \tag{9}$$

Foam volume flow:

$$q_g(t)=\frac{T_e}{p_w(t)}\frac{z_g\times q_{gsc}p_{gsc}}{T_{gsc}} \tag{10}$$

Characteristic value of foam:

$$\Gamma(t)=\frac{q_g(t)}{q_1+q_g(t)} \tag{11}$$

Initial conditions:

$$\begin{cases}\Gamma=0.8 & t=0\\ p_w=p_e & t=0\end{cases} \tag{12}$$

### 2.2. *Calculation examples*

According to the above mathematical model and solution process of foam diversion acidizing for horizontal wells, the foam acidizing process is calculated by taking a horizontal well in Gudao Oilfield as an example. The main parameters are shown in Table 1. The bottom hole pressure change and acid distribution in horizontal well section of foam acidizing and conventional acidizing are compared, as shown in Figure 3. It can be found that compared with conventional acidizing, foam diverting acidizing distributes acid evenly, which is mainly because foam acid has a diverting effect, and its plugging effect on high permeability intervals is stronger than that on low permeability intervals, so that the acid fluid is evenly distributed in the horizontal wellbore.

**Table 1 Basic Data of Well Acidizing**

| Formation pressure (MPa) | Permeability ($10^{-3}\mu m^2$) | Foam acid viscosity (mPa·s) | Acid density ($kg/m^3$) | Well radius (m) | Total acid injection ($m^3$) | Length of horizontal well section (m) | Acidizing time (h) |
|---|---|---|---|---|---|---|---|
| 15 | 2 | 10 | 600 | 0.0875 | 150 | 200 | 4 |

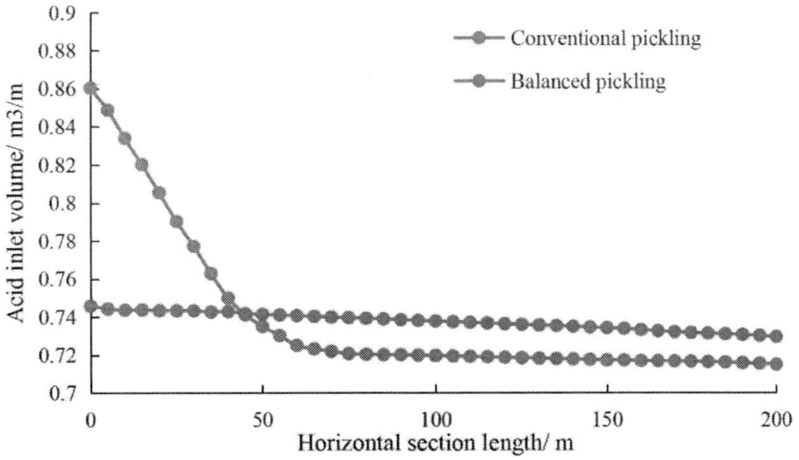


**Fig. 3 Distribution of Acid Solution in Horizontal Section**

## 3. Mathematical model of near wellbore acid treatment for horizontal wells

The foam acid cleaning process realizes the uniform injection string in the long horizontal well section, forms foam acid with acid liquid as the continuous phase and bubbles as the dispersed phase on the ground, continuously injects into the formation, and uniformly distributes acid throughout the well section. In the process of reverse drainage, in order to completely drain the acid liquid in the bottom layer, foam acid is used to realize the bottom hole negative pressure, so as to realize the rapid negative pressure backflow construction.

### *3.1. Design of uniform liquid dispensing string*

In order to ensure uniform pickling of long well section, uniform pickling string is designed and developed and equipped with uniform liquid preparation equipment, as shown in Figure 4. The principle is that a uniform injection channel is set on the injector, and the acid solution passing through the injection channel generates additional pressure drop, as shown in Formula 2-1 and 2-2. Couple the additional pressure drop to the mathematical model of foam pickling (Formula 1-1), and obtain the number of uniform injection channels n and the channel radius d through iterative calculation. The additional pressure drop reduces the flow rate of the wellbore fluid near the heel into the formation, further realizing balanced pickling. At the same time, the pipe string can meet the requirements of high pressure resistance and large displacement construction. At the same time, it strengthens the corrosion resistance of the tool, reduces the corrosion damage of the acid solution to the string, and has the functions of deep formation plugging removal and easy flowback of residual acid.

$$p_{ii} - p_{io} = \frac{2\left(f_{wo}^{2}\right)_{i} q_{si}^{2}}{\pi \rho_{wi} C_{di}^{2} d_{i}^{2}} \tag{13}$$

$$p_{ii} - p_{i(i-1)} = \frac{dp_i}{dz}\frac{L}{n} \tag{14}$$

### *3.2. Design process of balanced negative pressure pickling*

Nitrogen foam mixed drainage and unblocking on the ground can control the density of foam entering the well and adjust the bottom hole negative pressure value by controlling the flow rate of gas and liquid. The gradual reduction of foam density effectively prevented the damage of oil casing caused by sudden pressure drop, and the control of foam density effectively prevented the collapse of formation framework. The super negative pressure is formed at the bottom of the well through the optimization of the process flow, so that the blockage at the bottom of the well can be discharged more thoroughly.

(1) Well selection

The seepage capacity near the well is poor, and the permeability of the target layer is greater than $20 \times 10^{-3}$ μm$^2$ horizontal well, foam diversion acidizing process can be used if one of the following conditions is met: ① oil wells with long horizontal well section and large permeability difference; ② Oil wells with oil and water layers with rising water cut after acidification; ③ Repeated acidification wells with gradually poor acidification effect.

(2) Acidizing design

Data preparation includes: basic data of oil wells, reservoir data, conventional well testing data, recent production statistics, core analysis data, reservoir damage data, and early construction well history data.

Determine the acid system and additives, foaming agent concentration, acid dosage, foam slug design and dosage, displacement fluid dosage, foam diversion stage and pumping sequence.

(3) Construction evaluation and summary

The actual production increase (injection increase) after acidizing is used to evaluate the construction effect, and the acidizing diversion effect is evaluated based on the test results of the acidizing oil well production profile. Conformity rate of construction parameters and design parameters; Analysis and evaluation of continuity and safety of construction process and construction process effect.

## 4. Field application and effect

Since the maturity of this process, 85 wells of open hole screen horizontal well staged uniform negative pressure acid washing technology have been implemented in loose sandstone reservoirs of Gudao, Chengdong, Kenxi, Binnan and other oilfields. The development effect of typical wells is shown in Table 2. Through extensive field practice, it is believed that the oil wells with this process measure can maintain the productivity to the greatest extent, the productivity is 30% higher than that of similar oil wells, and the effective period of sand control is more than 180 days longer than that of similar oil wells.

Typical well example: open hole staged negative pressure acid cleaning of GOGD6-1P3 well.

Geological overview: permeability of Ng4-5 layer in GOGD 6-1P3 production interval $1999 \times 10^{-3}$ μm$^2$, the average effective porosity is 33.34%, the shale content is 10.72%, the average sand thickness is 5m, and the effective thickness is 2.3m. The viscosity of ground degassed crude oil is 1500mPa. s. The invasion of well fluid and mud during the drilling process of the well is easy to cause mud plugging, which can not effectively exert the oil well productivity. It is decided to adopt casing cementing jet plugging removal acidification technology to remove the pollution of drilling mud and cementing. The high liquid production period in this block is only 650 days, and the average liquid volume of horizontal wells is 17 m$^3$.

**Table 2 The development effect of typical wells**

| Well name | Production date | Period of validity (day) | Daily liquid production (m$^3$/day) | Daily oil production (t/day) | Water cut | Accumulated oil production |
|---|---|---|---|---|---|---|
| SXX154-P4 | 2015/8/30 | 1035 | 40.97 | 2.97 | 92.8 | 3073.95 |
| SXX154-P12 | 2016/1/30 | 882 | 37.68 | 4.87 | 87.1 | 4295.34 |
| GXXD9P61 | 2015/11/3 | 970 | 35.48 | 7.77 | 78.1 | 7536.9 |
| GXXD9P64 | 2015/3/27 | 1191 | 23.61 | 4.35 | 81.6 | 5180.85 |

| | | | | | | |
|---|---|---|---|---|---|---|
| GXXD9P63 | 2015/9/26 | 1008 | 22.19 | 5.26 | 76.3 | 5302.08 |
| CXX601-P148 | 2015/2/5 | 1241 | 27.82 | 13.93 | 49.94 | 17287.13 |
| CXX601-P151 | 2015/1/5 | 1272 | 16.74 | 9.97 | 40.46 | 12681.84 |
| WXX32-P11 | 2015/3/9 | 1209 | 16 | 8 | 50 | 9672 |
| CXX281-P31 | 2015/2/15 | 1231 | 22 | 4.8 | 78.4 | 5908.8 |
| CXX25-P1 | 2016/10/13 | 625 | 75 | 50 | 33.3 | 31250 |

Connect the positive circulation string, replace 7m$^3$ of foam acid, close the casing gate, lift and lower the string with a displacement of 300-500l/min, and extrude 17m$^3$ of foam acid, 9m$^3$ of high-quality completion fluid, shut in reaction 2h, and control pressure<15MPa. After the pressure diffusion is completed, connect the backwashing well and pressurize it with a pressure fan. When the pressure reaches 7MPa, the mixed discharge will start, and the high-quality completion fluid 27m$^3$ will be replaced by foam liquid with a density of 0.4~0.5g/cm$^3$, or the blowout will occur at the outlet. After continuing to displace the remaining completion fluid, stop the pressure fan, and then continue to backwash the well with formation sewage until the outlet pH is close to neutral. Wash the acid reactants to the ground, control the outlet when discharging the residual liquid, keep the outlet liquid volume stable, measure the pH value on paper, and observe the residual liquid at the outlet.

As shown in Figure 4, after the well has been put into production for plug removal and sand control by jet acidification, the average daily liquid production is 24.2m$^3$, the average daily oil production is 10.1t, and the average water content is 41.7%. At present, the well has been producing for 880 consecutive days with high liquid production, which exceeds the average value by 230 days. The cumulative oil production is 3600 tons, which is better than the oil wells in the same block that have not implemented this technology.

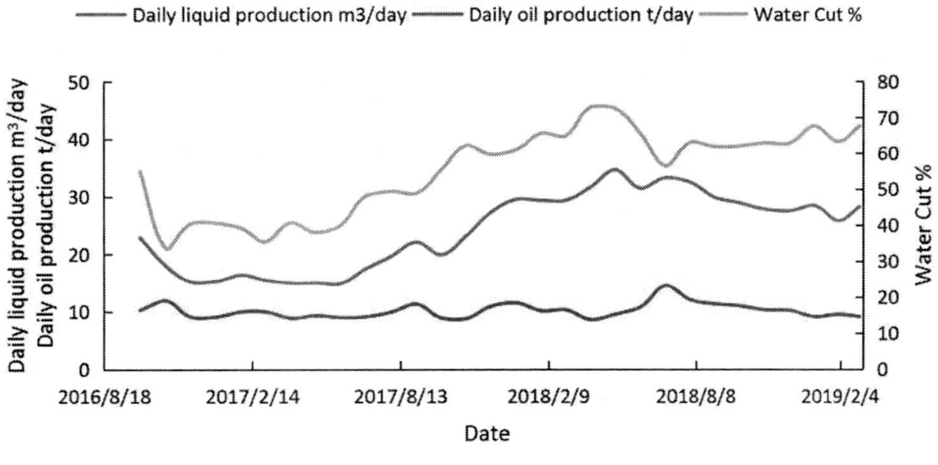


**Fig. 4 GOGD6-1P3 post measure production curve**

## 5. Conclusions

(1)Through theoretical research on acid washing of horizontal wells, a mathematical model of foam acid washing of horizontal wells is established. Through case calculation, it is believed that the difference in reservoir permeability will lead to the uneven entry of acid into the formation and affect the acid washing effect. Higher viscosity of acid solution and lower injection displacement are conducive to the uniform entry of acid into the formation.

(2)The proposed foam pickling process can effectively overcome heel toe effect and heterogeneity, make the acid liquid enter the horizontal formation evenly, and improve the pickling effect. The downhole uniform injection string is developed, and the additional differential pressure generated through the uniform flow channel can enable the acid to enter the formation evenly. This process has been applied to many unconsolidated sandstone blocks in Shengli Oilfield. The productivity of typical measure wells has been increased by 30%, and the effective period of high production has been extended by 680 days.

**References**

[1]Forrest J K. Horizontal gravel pack ing studies in a full-scale model wellbore[A]. SPE[C]. 20681, 1990.
[2]Collins D A, Nghiem L X, Agarwal R K and et al. Field-scale simulation of horizontal wells with hybrid grids. SPE[C]21218, 1991.
[3]Xiong Youming,Liu Mingli,Zhang Lin.Present status and development comments on horizontal well completion techniques in China[J].Petroleum Drilling Techniques, 2012, 40(1):1-6.
[4]Cui Chuanzhi.Study on Production Prediction for Horizontal Wells[D]. Beijing: China University of Geology,2005:7-44.
[5]Ren Minyan. Simulation of Outside Casing Gravel Packing Well Completion in Horizontal Wells[D]. Qingdao: China University of Petroleum,2012:12-26.
[6]Li Yanchao. Research on Productivity Prediction Method for Gravel-packed Horizontal Well Considering Sanding Influence[D]. Qingdao: China University of Petroleum, 2013:15-20.
[7]Wei Lihua, Ruan Hong, Wang Jin. Support technology of Open Hole Sand Control completion for horizontal wells in Chengdao Oilfield Journal of Oil and Gas Technology, 2007, 29(3): 302-304.
[8]Wang Zenglin, Zhang Quansheng. Support technology of well completion and oil production for horizontal wells inShengli Oilfield[J],Petroleum Geology and P Recovery Efficiency, 2008, 15(6):1-5.
[9]Li Hongqian, Liu Xisheng, Wang Hai. Discuss on Hole Cleaning in Horizontal Drilling[J]. Drilling Fluid & Completion Fluid, 1994,11(3):11-14.
[10]XUE Li-na.Research on Hole Opener Treatment of Horizontal Well[J]. Natural Gas Technology, 2007,1(4):31-34.
[11]Failure Analysis of Horizontal Open Hole Sand Control and Handling Measures[J]. China Petroleum Machinery,2014,42(1):121-124.
[12]Cheng L, Kam S I, Delshadm, el al. Simulation ofdynamic foam acid diversion processes[R]. SPE 68916,2001.
[13]Study on the flow-diversion characteristics of foam and its application[J]. Journal of Xi`an Shiyou University(Natural Science Edition), 2007,22(2): 100-102.
[14]RAO Fupei,DONG Yunlong, WU Jiesheng, et al. Technology of water control and well completion of horizontal wells in bottom water reservoir in Dagang Oilfield[J]. Oil Drilling & Production Technology, 2010, 32(3):107-109..
[15]Alakbari F. S., Mohyaldinn M. E., Muhsan A. S, et al. Chemical sand consolidation: from polymers to nanoparticles[J]. Polymers, 2020, 12(5): 1069.
[16]Kalgaonkar R., Chang F., Ballan A. N., et al. New advancements in mitigating sand production in unconsolidated formations[C]. SPE188043, 2017.
[17]Styward B., Wijaya R., Manalu D., et al. Sand conglomeration trial as an alternative to sand control case study from Mahakam Delta, Indonesia[C]. SPE191987, 2018.
[18]Ma C., Deng J., Dong X., et al. A new laboratory protocol to study the plugging and sand control performance of sand control screen[J]. Journal of Petroleum Science and Engineering, 2020, 184: 106548.
[19]Kuma P., Affeld C. J. Experience with chemical sand consolidation: case studies from west Africa[C]. SPE28267, 2018.

# Numerical Simulation Study of Creeping Rock Formation in Jin X Fracture Block of Jingqiu Oilfield

**Peng-wei LI1(✉) , Da-jun RU1,Jian-ning WANG1,Li-gong YANG1,Yong-sheng YE1,Bo ZHANG1,and Chao MA1**

1 Engineering Technology Research Institute of PetroChina Huabei Oilfield, Renqiu, Hebei

lpw31712056@163.com

**Abstract.** The Jin X fault block in Jing qiu Oilfield,as the main oil producing area of Huabei Oilfield.At present, there are a total of 141 wells drilled in this block, and 97 wells with damaged casing have been transformed;The casing transformation of wells has led to an increase in abandoned wells in this block, resulting in serious damage to the well network.The JinX fault block has a large number of damaged wells. The number of abandoned wells in this area has increased due to casing deformation, resulting in severe damage to the well network and high degree of casing damage. Among them,71 wells have been severely damaged due to casing deformation, such as scrapping, channeling, and pipe sticking, which have affected production, accounting for 73.2%. This has seriously affected the efficiency of oilfield production in the region.Previous statistics show that most of the casing deformation sections are in the paste salt formation, and the form of casing deformation and casing damage is mainly deformation. In order to investigate the mechanism of casing damage in the casing section, this paper establishes the creep equation of the casing layer in JinX block by conducting microstructure analysis of the casing layer and creep experiments in the casing layer, and then establishes a creep calculation model by using 3D numerical simulation technology, combining geological structure and well type conditions. The creep calculation model is then used to analyze the creep pattern under different conditions through simulation calculations, thus providing a reference for formulating effective casing loss prevention and control measures for the Jin X fault block.

## 1 Introduction

The JinX fault block in Jingqiu Oilfield is the main oil producing area of Huabei Oilfield. At present, there are a total of 141 wells drilled in this block, and 97 wells with damaged casing have been transformed; The casing transformation of wells has led to an increase in abandoned wells in this block, resulting in serious damage to the well network. Among them, 71 wells, accounting for 73.2%, have been severely damaged due to casing deformation, resulting in scrapping, channeling, and pipe sticking that have affected production. This area mainly exploits the oil layers of the Lower Tertiary Shahejie Formation, which are distributed in the 2900-3300m well section. The area has a large number of casing damage wells and a high degree of casing damage, which seriously affects the efficiency of oilfield production in the area. Preliminary statistics indicate that the deformation section of the casing is mostly between 2750-3000m, and the main form of casing damage is deformation. The lithology of this section mainly consists of grayish brown mudstone, gypsum shale, and salt rock interbedding, which is called gypsum salt rock layer. Casing deformation of gypsum salt rock section is mainly caused by plastic flow of rock generated by crustal stress [1-5].

In this regard, in order to explore the mechanism of casing damage in the gypsum salt layer of Jin X fault block in Jingqiu area, this paper established the creep constitutive equation of the gypsum salt layer of Jin X fault block by carrying out the microscopic fabric analysis of the gypsum salt layer and the creep experiment of the gypsum salt layer, and then used three-dimensional numerical simulation technology, combined with tectonics and well type conditions, established a creep geological model, and analyzed the creep laws under different conditions through simulation calculation, Thus providing reference for the development of effective casing damage prevention measures in the JinX fault block.

## 2 Determination of crustal stress magnitude and direction in Jin-X fault block

### *2.1 Determination of crustal stress*

a) Measurement of formation crustal stress by hydraulic fracturing

The on-site hydraulic fracturing test method is the most direct method for measuring the absolute stress in depth at present. It is based on the measured formation fracture pressure, instantaneous pump stop pressure, and fracture re tension pressure to back calculate the crustal stress. The basic idea is: after the formation fracturing, the pump is stopped instantaneously. At this time, the fracture will not expand forward, but still remain open. At this time, the pressure $p_s$ (instantaneous pump stop pressure) should be phase equilibrium with the minimum crustal stress of the vertical fracture,( $\sigma_h = p_s$ ). After momentarily stopping the pump and restarting it, the closed crack reopens. Since the pressure required to open( $p_r$ ) and close the crack ( $p_f$ )does not need to overcome the tensile strength of the rock compared to the formation fracture pressure( $S_t$ ), it can be approximately assumed that the tensile strength of the fracture layer is equal to the difference between these two pressures( $S_t = p_f - p_r$ ), i.e. The formation crustal stress can be back calculated by using three pressure( $p_f$ 、 $p_s$ 、 $p_r$ ) values that can be directly read from the fracturing pressure curve:

$$\begin{cases} \sigma_h = p_s \\ \sigma_H = 3\sigma_h - p_f - p_p + S_t = 3p_s - p_p - p_r \\ S_t = p_f - p_r \end{cases} \tag{1}$$

According to the tensile strength test of indoor rocks, the tensile strength of rocks is relatively small (about 2Mpa), therefore, when the hydraulic fracturing test is incomplete， $S_t = 2Mpa$ 。

b) Interpretation of formation crustal stress with logging data

According to the formation characteristics of different well depths and mechanical properties, the crustal stress values are different. For Jin X fault block, the crustal stress calculation model shown in the following formula is adopted in this study:

$$\begin{cases} \sigma_H = \left( \dfrac{\mu}{1-\mu} + \omega_1 \right)(\sigma_v - \alpha P_p) + \alpha P_p \\ \sigma_h = \left( \dfrac{\mu}{1-\mu} + \omega_2 \right)(\sigma_v - \alpha P_p) + \alpha P_p \end{cases} \quad (2)$$

In the equation: $\omega_1, \omega_2$ —tectonic stress coefficient; $\alpha$ — effective stress coefficient; $P_p$ — pore pressure; $\sigma_H, \sigma_h, \sigma_v$ — maximum and minimum horizontal crustal stress and overburden pressure; $\mu$ —poisson's ratio of strata。

c) Determination of longitudinal crustal stress profile

According to the principle of hydraulic fracturing testing crustal stress $p_f$ and stop pump pressure $p_s$ numerical substitution formula, Obtaining the maximum horizontal crustal stress of Jin X fault block $\sigma_H$ and minimum horizontal crustal stress $\sigma_h$ .The measured data is shown in Table 1.

**Table 1.** Record of Fracture Pressure Test for Well JinX-1

| Name of the well | depth (m) | Rupture pressure (MPa) | Maximum principal stress orientation |
|---|---|---|---|
| Jin X-1 | 2910 | 1.765 | 125° |
| | 3030 | 1.8782 | 125° |
| | 3155.8 | 1.801 | 125° |

Using trial calculation method and combining the rock mechanics parameter values of this section, the formula is introduced to obtain the structural stress coefficient of Jin-X fault block structure as follows:

$\omega_1$=0. 6927, $\omega_2$=0. 3122

According to the tectonic stress coefficient, the maximum and minimum crustal stress can be calculated, but the crustal stress calculated is for vertical wells. When calculating the crustal stress of inclined wells or directional wells, the crustal stress of vertical wells should be transformed accordingly. After the crustal stress transformation model is constructed, the transformed crustal stress components are:

$$\begin{aligned} \sigma_x &= \cos^2 a \left( \sigma_h \cos^2 \beta + \sigma_H \sin^2 \beta \right) + \sigma_v \sin^2 \alpha \\ \sigma_y &= \sigma_h \cos^2 \beta + \sigma_H \sin^2 \beta \\ \sigma_z &= \sin^2 a \left( \sigma_h \sin^2 \beta + \sigma_H \cos^2 \beta \right) + \sigma_v \cos^2 \beta \\ \tau_{xy} &= \cos a \sin \beta \cos \beta \left( \sigma_H - \sigma_h \right) \\ \tau_{xz} &= \cos a \sin a \left( \sigma_H \cos^2 \beta + \sigma_h \sin^2 \beta - \sigma_v \right) \\ \tau_{yz} &= \sin a \sin \beta \cos \beta \left( \sigma_H - \sigma_h \right) \end{aligned} \quad (3)$$

By substituting the tectonic stress coefficient into the above formula, the crustal stress gradient equation of Jin-X fault block is obtained as follows:

$$S_{max} = 0.024H - 10.555 \quad S_{min} = 0.0196H - 12.469 \quad (4)$$

The converted crustal stress gradient is:

$$S_{max} = 0.0228H - 3.5245 \qquad S_{min} = 0.0176H - 1.9043 \tag{5}$$

Combined with logging data, the formation crustal stress of Jin-X fault block reservoir is obtained, and the longitudinal crustal stress profile is drawn, as shown in Figure 1-1. It can be seen from the figure that the crustal stress of different layers are different. However, the three main crustal stress states are: overburden pressure>maximum horizontal crustal stress>minimum horizontal crustal stress. On this basis, the crustal stress corresponding well deviation is transformed, and the transformed crustal stress distribution is shown in Figure 1-2.

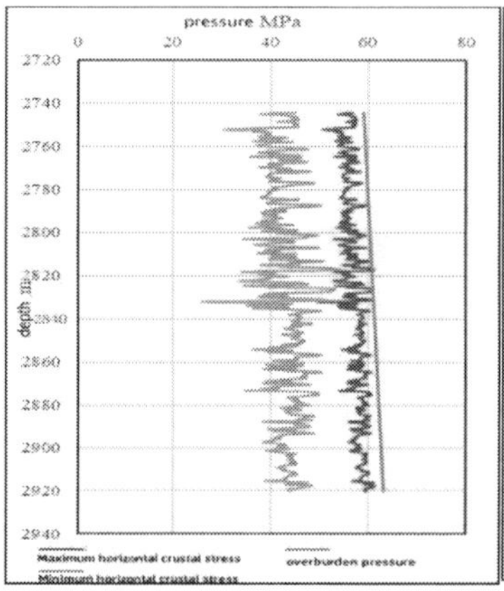


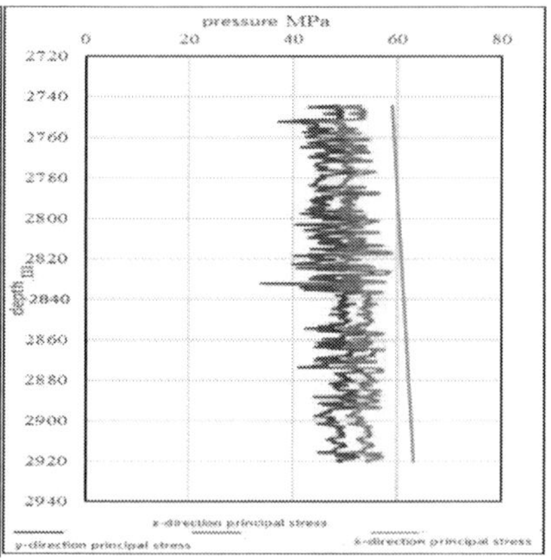


**Fig. 1-1.** crustal stress distribution diagram of Jin-X fault block **Fig. 1-2.** crustal stress distribution diagram after conversion

*2.2 Determination of crustal stress direction*

According to the dip logging of the fault block, the main stress direction in the middle is northwest, while the main stress direction in the east and west is northeast; The overall sand body distribution of Es2 section is in a northeast direction, with the sand body distribution direction in the middle of Es2 section 9 small layer being nearly east-west.

## 3 Creep testing and model determination of gypsum salt layer

Creep refers to the phenomenon where the deformation of a specimen increases with time under constant load. Under normal temperature and pressure, creep of rocks undergoes three stages: transient creep, steady-state creep, and accelerated creep, and ultimately begins to fail due to creep [6-10]. In oil extraction, the creep of rock mass often leads to the phenomenon of shrinkage and sticking when oil wells pass through salt rock formations, leading to serious drilling accidents.

### *3.1 Creep testing*

a) Whole rock mineral analysis and dissolution test

On site sampling and whole rock mineral analysis were conducted, with various mineral contents as follows:

Salt rock 100%, total clay 0%, anhydrite 0%, analcime 0%, quartz 0%, orthoclase 0%, plagioclase 0%, calcite 0%, dolomite 0%, siderite 0%, anhydrous mirabilite 0%, pyrite 0%. The mineral salt rock content of the salt rock layer in this block is 100%, and there are basically no other impurities. Conduct dissolution tests on the lithology and analyze its dissolution rate. Take 10g of salt rock sample and place it in a beaker. Take an appropriate amount of distilled water and immerse it in the salt rock for a dissolution test. After 3 minutes, the sample is completely dissolved. The sample and dissolution process are shown in Figure 2:

10g salt rock test sample dissolving dissolution completed (3 minutes)

**Fig.2.** Salt Rock Dissolution Test

b) Creep experiment

The rock samples used in the experiment were core samples from the drilled JinX block in the Huabei Oilfield. The indoor core column processing process involves first using a diamond core drill bit to drill a core on site Φ 50mm cylindrical sample, then flatten and polish both ends of the cylindrical sample to achieve a aspect ratio of 1.82.0 for the rock sample. Test process: The creep test is conducted using a stepped loading method to observe the creep behavior of salt rock samples at different stress levels. The axial pressure is loaded by gravity, and the confining pressure is controlled by oil pressure. The axial strain is automatically recorded by the data collection system, and data is sampled every 1 minute. The experiment is divided into two stages according to the loading stress level, with stable creep stage loaded under constant axial and confining pressure in each stage. At the final stage, the specimen was damaged, lasting for 90 hours. In the first stage, the axial pressure is 20.5MPa, the confining pressure is 15MPa, and the load is 23 hours; In the second stage, the axial pressure is 25.6MPa, the confining pressure is 15MPa, and the specimen is damaged after 24 hours of loading. Figure 3 shows the creep curves of each experimental stage, and Table 2 shows the creep rates of each stage of salt rock.

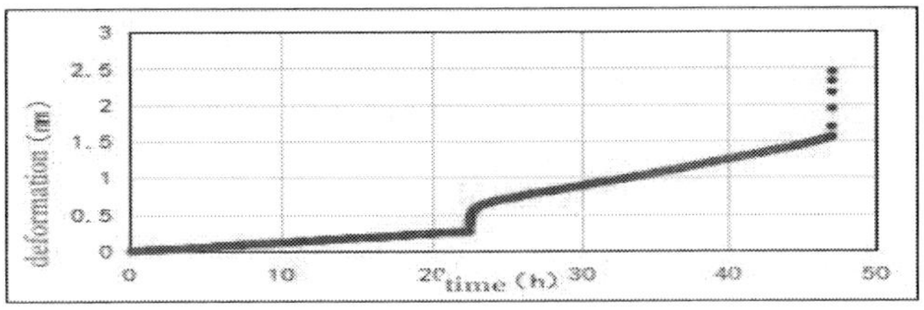

**Fig. 3.** Creep Curve of Salt Rock at Each Stage

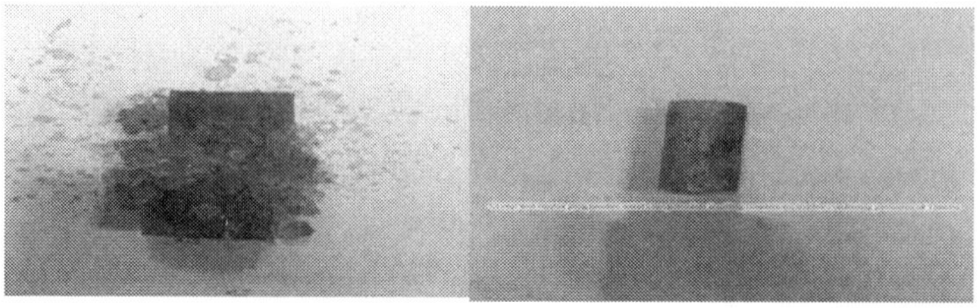

**Fig.4.** Core before and after the experiment

According to the test curve in Figure 3, it can be seen that although each subsequent creep stage is carried out on the basis of a certain creep strain history, each test stage still has its own transient creep stage and gradually enters the steady-state creep stage. With the changes in stress difference and confining pressure during the loading process, the creep of salt rock specimens exhibits completely different properties [11-16]. From the test curve, it can be seen that: (1) In the compression test of salt rock, both instantaneous strain and creep strain exist, and the entire test process conforms to the creep law of the rock. The rock is subjected to triaxial testing, and the test data is shown in Table 2. The confining pressure is constant, and the creep strain is determined by the stress difference (i.e. deviatoric stress); (2) Stable creep accounts for a large proportion of the entire creep process; (3) Deformation is plastic deformation, and after unloading the load, there is a slight recovery of strain, while most strains cannot be recovered. In the first stage, the stress difference is relatively small (5.5 MPa) and the steady-state creep rate is also small (only 0.0003/hr); In the second stage, the stress difference is relatively large (10.6MPa), and the steady-state creep rate increases (0.0008/hr); In step loading, as the stress difference increases, the steady-state creep rate of the specimen also increases, and the increase in stress difference can accelerate the creep rate of the salt rock. When the axial bias stress increases, the steady-state strain rate does not increase linearly. It can be seen that in step loading engineering, the steady-state creep rate of the specimen increases significantly with the increase of stress difference, and the increase of stress difference can accelerate the creep rate of salt rock.

**Table 2.** Creep Rate of Salt Rock at Each Stage

| | Phase 1 | Phase 2 |
|---|---|---|
| Axial compression(MPa) | 20.5 | 25.6 |
| confining pressure(MPa) | 15 | 15 |
| Stress difference(MPa) | 5.5 | 10.6 |
| Loading time(Hr) | 22.5 | 24.4 |
| Steady-state creep rate($Hr^{-1}$) | $3\times10^{-4}$ | $8\times10^{-4}$ |
| Total strain at this stage(%) | 0.561 | 3.280 |

*3.2 Determination of creep equation*

Based on a series of rock rheological tests, establish rheological equations that reflect the rheological properties of rocks, usually including empirical equation method and rheological model theory method. The former is based on the results of rock creep test, and empirical equation is established by regression fitting method of mathematical statistics; The latter is to idealize the medium into various models when studying the rheological properties of rocks. The models are composed of elements with basic properties (including elasticity, plasticity and viscosity). Through the series and parallel connection of these simple elements, some typical rheological models are obtained, and their equations are derived accordingly to establish the constitutive equation and characteristic equation of the model [17-21].

In this paper, FLAC3D finite difference calculation software is used for creep calculation, which provides the function of simulating the creep characteristics of materials and the change of material properties with time. This calculation option can achieve the following five creep material models: classical viscoelastic model; Exponential model; WIPP model; Viscoplastic model; Constitutive model of fractured salt rock. After conducting model comparison analysis, the WIPP model was used for salt rock creep calculation.

This article adopts a simplified WIPP creep model ( $\dot{\varepsilon} = A\sigma^{-n}$, $A = e^{-\frac{Q}{R}}$ ) .

Based on the data in Table 2, using CurveXpert mathematical regression software, the creep parameters of salt rock were determined as follows:

A= 2.055e-1MPa-1.5/yr-1， n= 1.5。 Therefore, the creep equation of the composite salt rock in the JinX fault block is determined as:

$$\dot{\varepsilon} = 2.055 \times 10^{-1} \sigma^{1.5} \tag{6}$$

## 4 FLAC3D software numerical simulation

*4.1 Establishment of creep calculation model*

In terms of model design, a certain thickness of creep rock layers in deep strata is taken as the research object. Due to the varying thickness of creep rock layers, a rock layer thickness of 50m is taken. Due to the axisymmetric deformation of the studied wellbore, the calculation area of the model is selected as 25m × 25m × One quarter of the 50m geometry is the calculation area, and the wellbore diameter is determined based on the actual situation. One fourth of the calculation domain has a free surface on the upper surface, which acts on the weight of the overlying rock layer, while the lower surface is constrained by vertical simple supports perpendicular to the surface. The front, rear, left, and right surfaces are all constrained in a normal direction, and hexahedral elements are used for element partitioning. Figures 5 and 6 are grid diagrams of the constructed model.

**Fig. 5.** Overall Grid Diagram

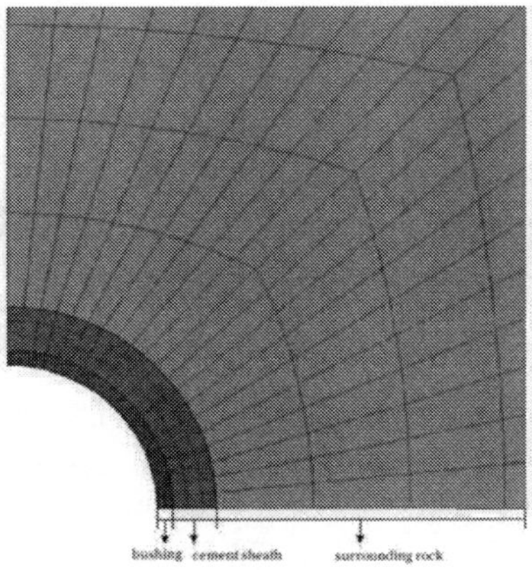


**Fig. 6.**Grid Details

For a single lithological interlayer, establish a grid area based on the actual rock layer thickness. Considering the different forces acting on the strata, the weight of the upper overlying strata can be simplified as an equivalent vertical load acting on the calculated surface of the strata.The equivalent overlying load value is determined by the formula $\sigma_z = \gamma h$ ， The schematic diagram of its mechanical model is shown in Figure 7. $\gamma$ Weighted average unit weight of overlying strata， h , the depth of the formation.

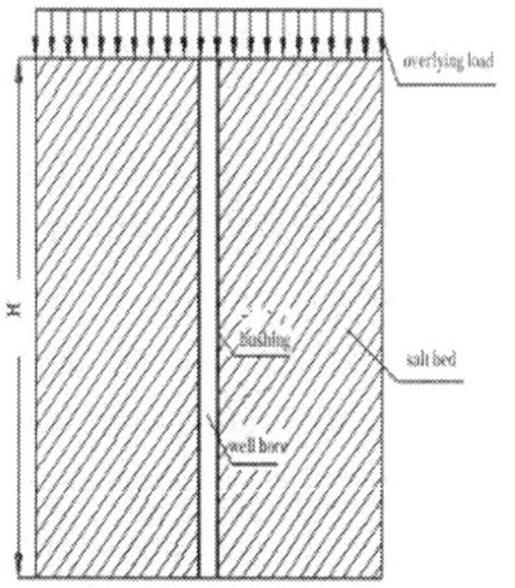


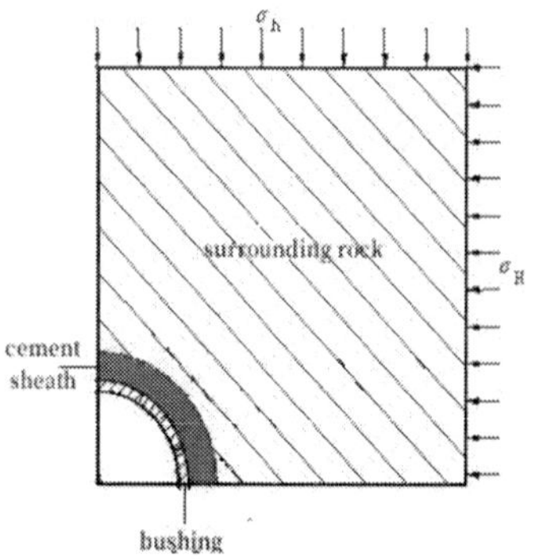


**Fig. 7.** Simplified mechanical schematic diagram for creep calculation
**Fig. 8.** Loading mode under non-uniform crustal stress

Due to the unknowability of the initial stress field, gravity field is often chosen as the uniform initial vertical stress field in numerical simulation analysis,Determined by formula calculation $\sigma_z = \gamma h$ , The horizontal stress field is calculated by $\sigma_x = K_0 \gamma h$ , among , $K_0 = \frac{v}{1-v}$ , $v$ ,Poisson's ratio.One quarter of the calculation domain has a free surface on the upper surface, which acts on the weight of the overlying rock layer, while the lower surface is constrained by vertical simple supports perpendicular to the surface. The front, rear, left, and right surfaces all have normal constraints. When calculating non-uniform crustal stress,The loading method is shown in Figure 8, and the maximum crustal stress is applied in the X and Y directions respectively $\sigma_H$ and minimum crustal stress $\sigma_h$ 。

*4.2 Selection of calculation parameters*

In creep analysis, the salt rock is analyzed using the power exponent constitutive model that comes with FLAC3D software and can analyze the creep law: $\dot{\varepsilon}_{cr} = A\sigma^{-n}$ ( A , n ,material parameters , $A = e^{-\frac{Q}{RT}}$ ) .Both the casing and cement sheath are made of pure elastic materials. According to the inversion results of the creep test of salt rock, the creep parameters of the salt rock formation are determined in Table 3.

During the drilling of the well in JinX block into the salt rock layer, due to the presence of falling fish, no logging data was obtained. In order to determine the elastic parameters of the salt rock layer, seismic layer velocity data was used, combined with the logging data of the upper layer of the well, to extract seismic characteristic parameters for feature compression. A relationship model between layer velocity and acoustic time difference was established, and the model was used to predict the acoustic time difference of the salt rock layer, And combined with the density of the formation, the elastic parameter of the salt rock formation is obtained as the Young's modulus $E = 11.758GPa$ ; poisson's ratio $\mu$ = 0.1747 , bulk modulus,6.024×103MPa, shear modulus,5.005×103MPa.

**Table 3.** Creep Parameters of Salt Rock

| parameterst ratum | A (MPa-nyr-1) | n | Elastic modulus (MPa) | Poisson's ratio |
|---|---|---|---|---|
| **Salt rock** | 2.055e-1 | 1.5 | 11757.8 | 0.1747 |

The parameters of the cement sheath are: bulk modulus 1.051 × 104MPa, shear modulus 8.9974 × 103MPa.

*4.3 Simulation Study on Formation Creep Pressure*

According to geological stratification data, after 2744m, it will encounter formations with strong fluidity such as salt layers, which pose a great threat to the casing. The most likely areas to cause complex situations such as sticking during drilling construction are also in this section of the formation. Therefore, based on the basic calculation conditions provided above and combined with the actual working conditions of the casing, FLAC3D software is used to simulate the changes in peristaltic pressure around the wellbore after drilling in non-uniform conditions, and mechanical analysis and judgment are conducted on the rationality of the use of the casing.

In practice, casing is often under the action of non-uniform crustal stress. When the casing is subjected to non-uniform loads, the bearing capacity of the casing is much lower than under uniform conditions. The greater the difference between the maximum external force acting on the casing (force in the 0 ° direction) and the minimum external force (force in the 90 ° direction), the more the strength of the casing decreases. When the external load exceeds the strength of the casing, failure occurs. It can be seen that under the action of non-uniform crustal stress, the external load generated by rock creep will do more harm to the casing.

According to the geological data of drilled and completed wells in JinX block, a 175.99m thick salt layer exists at 2744.15m~2920.14m. In drilling construction, when the wellbore size is 215.9mm and 241.3mm, 193.7mm, 177.8mm, and 139.7mm casing pipes are used. When the wellbore size is 215.9mm, the cement sheath thickness is 11.1mm, 19.05mm, and 38.1mm, respectively; When the wellbore size is 241.3mm, the thickness of the cement sheath is 23.8mm, 31.75mm, and 50.8mm, respectively. In the calculation, 25m from the model center is taken as the observation point. According to the actual situation, the overburden pressure of the model is set as 61.15MPa, the maximum crustal stress is 57.422MPa, and the minimum crustal stress is 43.078MPa. Then, simulate and calculate the stress curve at different depths within the salt rock section of different wells, taking 2930m as an example, as shown in Figures 9 to 11.

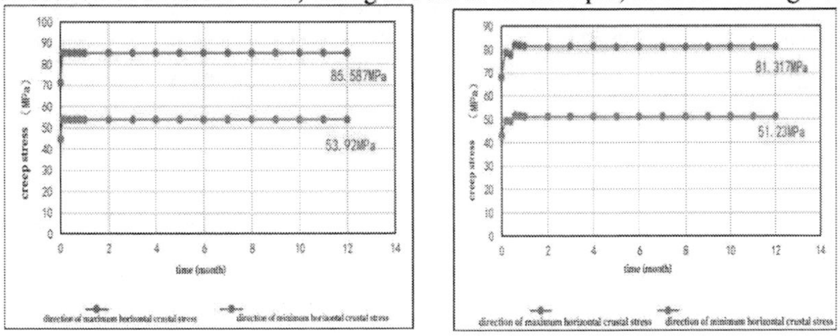


**Fig. 9.**Stress Time History Curve of Maximum and Minimum crustal stress Direction of 139.7mm Casing in 215.9mm (Left) and 241.3mm (Right) Holes

The non-uniform coefficient of stress on the stabilized casing calculated from the graph is Kp=0.63.

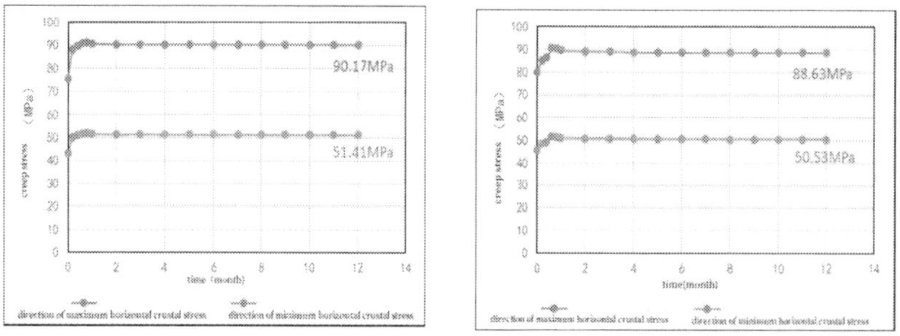


**Fig. 10.** Stress Time History Curve of Maximum and Minimum crustal stress Direction of 177.8mm Casing in 215.9mm (Left) and 241.3mm (Right) Holes

The non-uniform coefficient of stress on the stabilized casing calculated from the graph is Kp=0.57.

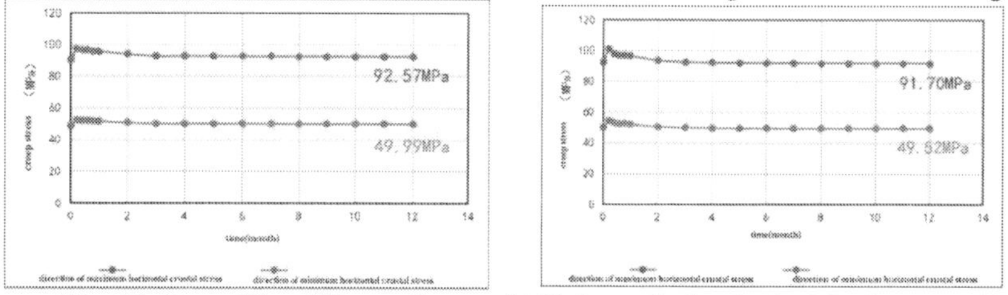


**Fig. 11.** Stress Time History Curve of 193.7mm Casing in the Maximum and Minimum crustal stress Directions of 215.9mm (Left) and 241.3mm (Right) Holes

The non-uniform coefficient of stress on the stabilized casing calculated from the graph is Kp=0.54.

Figures 9 to 11 show the stress time history curves in the maximum and minimum crustal stress directions of the casing wall. The overall trend of stress changes in the maximum crustal stress direction and the minimum crustal stress direction is "fast" first and then "slow". In about one month, the two began to stabilize. For different borehole sizes and casing sizes, the creep pressure of the casing in the direction of the maximum and minimum crustal stress was different. When the casing size of 215.9mm well is 139.7mm, 177.8mm and 193.7mm, the creep pressure in the maximum crustal stress direction is 85.59MPa, 90.17MPa and 92.57MPa, and the creep pressure in the minimum crustal stress direction is 53.92MPa, 51.41MPa and 49.99Ma. When casing size of 241.3mm well is 139.7mm, 177.8mm and 193.7mm, creep pressure in the direction of maximum crustal stress is 81.32MPa, 88.63MPa and 91.70MPa, and creep pressure in the direction of minimum crustal stress is 51.23MPa, 50.53MPa and 49.52MPa.

## 5 Conclusion

The JinX fault block area of Jingqiu Oilfield has a large number of casing damage wells, with a high degree of casing damage, mostly concentrated in the gypsum salt layer casing damage, which seriously affects the efficiency of oilfield production in the area.

Therefore, in order to explore the mechanism of casing damage in the gypsum salt rock section, this paper establishes the creep equation of the gypsum salt layer in JinX fault block by carrying out the microscopic fabric analysis of the gypsum salt layer and the creep experiment of the gypsum salt layer, and then uses three-dimensional numerical simulation technology, combined with tectonics and well type conditions, to establish a creep calculation model, and analyzes the creep law under different conditions through simulation calculation,Thus providing reference for the development of effective casing damage prevention measures in the JinX fault block.

### References

[1] Xu Peng,Yang Shengqi. Numerical simulation study on the mechanical properties of triaxial compression creep in composite rock formations[J]. Journal of Mining and Safety Engineering,2018,35(01):179-187.

[2] GHOLAMI R, RABIEI M, AADNOY B, et al. A methodology forwellbore stability analysis of drilling into presalt formations: a casestudy from southern Iran[J]. Journal of Petroleum Science and En-gineering, 2018, 167: 249–261.

[3] MOTA B, MATSCHEI T, SCRIVENER K. The influence of sodi-um salts and gypsum on alite hydration[J]. Cement and ConcreteResearch, 2015, 75: 53–65.

[4] Yuan Jiping. Calculation of external load of casing in salt rock formation[D]. Southwest Petroleum University, 2016.

[5] Zhai Xiaopeng, Lou Yishan et al. Influence of formation stiffness and relaxation time on casing external extrusion force in creeping rock formations[J].Journal of Petroleum,2015,36(10):1299-1304.

[6] Gholami R, Rasouli V , Aadnoy B , et al. Geomechanical and Numerical Studies of Casing Damages in a Reservoir with Solid Production[J]. Rock Mechanics and Rock Engineering, 2016, 49(4):1441-1460.

[7] Lu HF,Yao DOXI,Shen D,Wen L,Wang K. Numerical analysis of structural surface effects of laminated rock masses with compressive mechanical characteristics[J]. Journal of Anhui University of Technology (Natural Science Edition),2014,34(01):24-29.

[8] Lou Erbiao,Liu Hongtao,Li Ning,Yang Shangxiao,Zhou Bo,Cao Lihu,Zhao Li,He Renqing,Li Jianbo,Feng Chun. Analysis of creep mechanical properties of salt rock[A]. Chinese Society for Materials Research. Proceedings of China Materials Conference 2021[C]. China Society for Materials Research:China Society for Materials Research,2021:158-165.

[9] Jiang Tengjian, Zhao Lizai, Yu Jianxing. Study on creep mechanical properties and rheological model of sandstone[J]. Hydropower Energy Science,2021,39(03):124-128+133.

[10] Yang Shuai-Dong,Tan Wei-Jia,Fang Ying-Xue. Study on creep mechanical properties and long-term strength of quartz sandstone[J]. Hydroelectricity,2021,47(04):43-50.

[11] Dong ZK. Study of triaxial creep mechanics test and damage evolution law of salt rock[D]. Shijiazhuang University of Railways, 2020.

[12] LI Huan,Ngaha Tiedeu WILLIAM,Jaak DAEMEN,ZHOU Jun,MA Changkun. A power function model for simulating creep mechanical properties of salt rock[J]. Journal of Central South University,2020,27(02):578-591.

[13] Liu Y, Chen F, Zou Chenyang. Study on long-term strength of rocks based on creep mechanics test[J]. Hydropower,2019,45(12):113-116.

Journal of Physics: Conference Series **2594** (2023) 012054 doi:10.1088/1742-6596/2594/1/012054

# Study on Bionic Self-healing Cement Slurry System

**Jun ZHAO1, You-zhi ZHENG1, Jian YANG1, Fu-yun WANG1(✉), Yu HE1**

1 Engineering and Technology Research Institute of Southwest Oil & Gas Field Company,Chengdu, China

wang_fy@petrochina.com.cn
**Abstract.** Aiming at the technical problem of gas channing in annulus caused by long-term sealing performance failure of cementing cement ring, based on the technical idea of "prevent before plugging" and from the perspective of chemical bionics, a kind of bionic self-healing cement slurry system of dissolving viscosity was developed. The indoor evaluation results show that the cement stone has excellent physical properties. When the compressive strength is reduced by 15%, the elastic modulus of the cement stone is reduced by 41%, showing good toughness and playing the role of "first prevention". When the damaged cement stone is excited by hydrocarbon, a kind of soluble gel ester can be formed, and the cement stone skeleton can be re-bonded. Under the condition of methane curing, the recovery rate of compressive strength is 99%, the recovery rate of flexure strength is 58.2%, the recovery rate of gas permeability is 96.4%, and the recovery rate of liquid permeability is 100%, playing the role of "post-plugging". Therefore, it is concluded that the adhesive dissolving bionic self-healing cement technology can effectively realize the late integrity and realize the long-term sealing integrity after the cement ring is broken through self-healing, which provides a powerful technical support for preventing the abnormal annulus pressure, a hidden danger of oil and gas well production safety.

## 1. Introduction

Cementing cement of oil and gas well is a brittle material with "congenital defects" as well as building cement-based materials. The most commonly used measure to improve the mechanical properties of cement stone is to add toughening materials to modify the mechanical properties of cement stone, but under the combined action of bottom hole pressure and temperature, the integrity of cement ring will inevitably be damaged. If you can give the ability of self-healing after the formation of channeling channel after the failure of the cement ring in the later stage, you can effectively prevent the oil and gas channeling flow and restore the sealing ability of the cement ring. In the field of construction, concrete self-healing methods [1-6] include hydration gel penetration crystallization, liquid core fiber method, microbial method, microcapsule method, etc. There are great differences between cementing cement slurry and concrete in external environment, fluid medium, preparation conditions, interface conditions and performance requirements. Only in terms of preparation conditions, liquid core fiber method and microcapsule method are not suitable for cementing because high speed shear is required for cementing cement slurry.

With the development of cementing cement slurry technology, self-healing cementing technology has become one of the most effective technologies to prevent and control oil and gas channeling and improve the long-term sealing ability of cement ring [7-9]. According to the survey data, the self-healing technology of cementing cement ring started relatively late, only developed about 5 or 6 years ago, and most of the self-healing technology of the construction industry was used for reference. Relevant researches are advanced abroad and imitated in China, which are not mature [10-12].

## 2. Materials and methods

### *2.1. Experimental Material*

G grade oil well cement (Sichuan Jiahua Cement Factory). Bionic self-healing agent (self-developed). Oil well cement admixture (Enenos).

### *2.2. Experimental Instruments*

Atmospheric pressure thickener (Institute of Applied Technology, Shenyang Aerospace University); YA-300 electronic hydraulic testing machine; JSM-6490LV Scanning electron microscope (JEOL). PKZ-5000 electric bending testing Machine (Wuxi Jianyi Instrument Machinery Co., LTD.). Corrugated agitator; RTR-1000 triaxial rock mechanics test system (GCTS). DKS- II type core hole permeability tester (Changzhou Yi Yong Technology Co., LTD.).

### *2.3. Experimental methods*

The test mold with tested mechanical properties and permeability was maintained in a high temperature and high pressure curing device filled with methane gas, and a certain level of water was placed at the bottom of the test mold to maintain the humidity of the curing environment. Take out the cement stone that has been preserved to a certain age after cooling and cooling, continue to test the mechanical properties and permeability of cement stone, and calculate the recovery rate of mechanical properties and the reduction rate of permeability of cement stone. A self-developed method was used to fabricate seams before permeability test. Finally, the microstructure and internal morphology of the self-healing material and cement were observed and analyzed by optical microscope and scanning electron microscope (SEM).

## 3. Experimental Results and Discussion

### *3.1. Physical property test of self-healing materials*

The self-healing cement can effectively seal micro-cracks and micro-gaps in the cement ring. Its mechanism is that when the cement ring generates micro-cracks and micro-gaps and oil and gas enter, the material will expand or change from solid state to viscous flow state with certain fluidity under the excitation of oil and gas, which will close the channel, prevent channeling and restore the sealing ability of the cement ring. In order to investigate the response ability of self-healing materials to oil and gas, swelling tests were carried out in natural gas (composition and content are shown in Table 1) and diesel respectively. The experimental results are shown in Figure 1 and Table 1. As can be seen from Figure 1 and Table 1, the volume of the self-healing material does not expand in the oil and gas environment, but it shows strong viscosity dissolving ability. In the oil and gas environment, the phenomenon of "wire drawing" occurs, and new viscosity dissolving type crystals are generated, indicating that the self-healing cement ring can meet the performance requirements of the closed channeling channel.

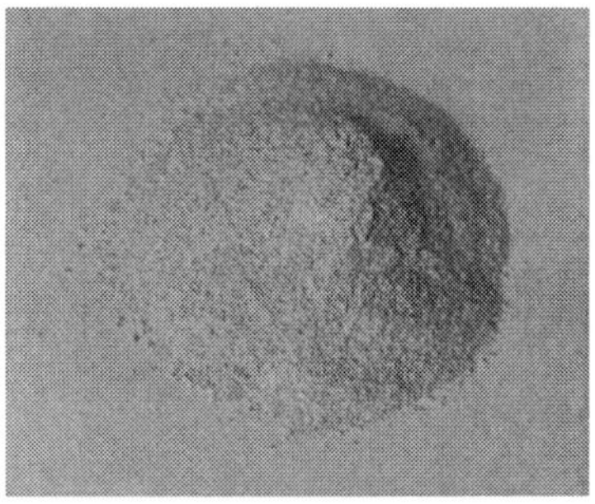

a) Bionic self healing material

b) Bionic self-healing material does not expand in contact with oil

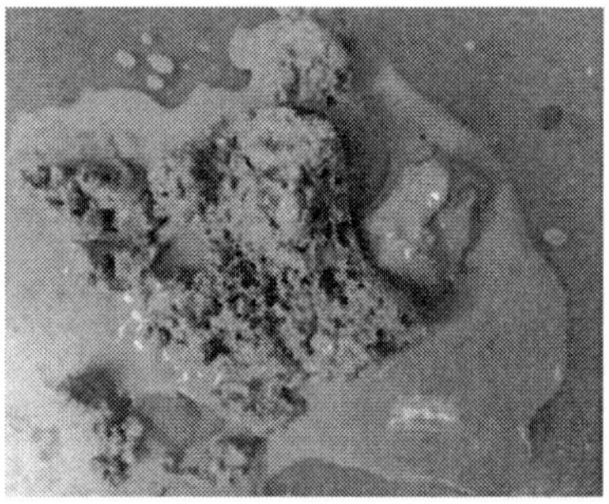

c) The material was dissolved in diesel oil for 7 days

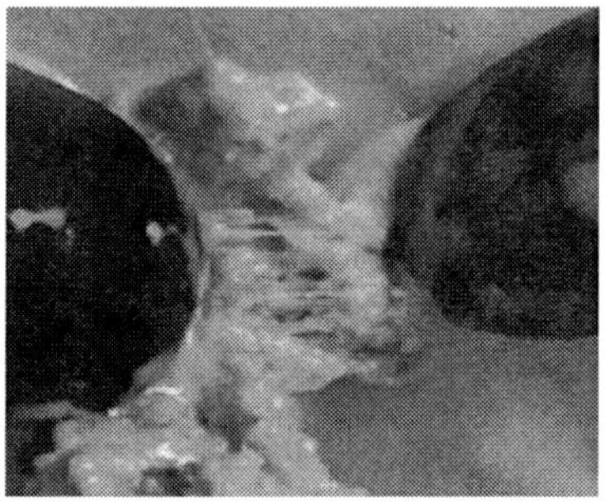

d) The material dissolved in diesel oil for 28d

**Fig. 1.** Self-healing material and swelling viscosity flow state

**Table 1** Quality changes of self-healing materials before and after curing in different media

| Curing medium | Experiment condition | Pre-curing mass (g) | After curing mass(g) | Mass increase rate(%) |
|---|---|---|---|---|
| natural gas | 30℃×0.15MPa×24h | 1.138 | 3.658 | 221.44 |
| diesel | 30℃×ordinary pressure×24h | 2.235 | 76.779 | 3335.3 |

*3.2. Evaluation of engineering performance of self-healing cement slurry system*

**Density Test.** The effect of adding 5%-25% (BWOC) biomimetic self-healing agent on the density of cement slurry system was further explored by keeping the cement slurry ratio of 0.44 unchanged. The test results are shown in Table 2.

**Table 2** Cement slurry density with different biomimetic self-healing agent dosage

| Self-healing agent dosage (%) | 0 | 5 | 10 | 15 | 20 | 25 |
|---|---|---|---|---|---|---|
| density (g/cm$^3$) | 1.90 | 1.88 | 1.86 | 1.84 | 1.82 | 1.80 |

As can be seen from Table 3, the cement slurry density decreases with the increase of the self-healing agent. The reason for this is that the density of self-healing agent is 1.90g/cm$^3$, which is lower than that of pure cement (3.15 g/cm$^3$), which is equivalent to reducing material in cement slurry. Therefore, the density of cement slurry decreases with the increase of self-healing agent under the condition of constant liquid-solid ratio. In practical application, the cement slurry density can be adjusted by admixture and water cement ratio.

**Thickening Time Test.** Self-healing materials with different dosages were thickened under low temperature, medium temperature, high temperature and ultra-high temperature respectively. The experimental results are shown in Table 3 and Figure 2.

**Table 3** Thickening time of different self-healing agent dosages at different temperatures

| serial number | temperature (℃) | Self-healing agent dosage(%) | Retarder dosage(%) | thickening time(min) |
|---|---|---|---|---|
| 1 | 50 | 10 | 0 | 239 |
| 5 | 90 | 10 | 0.6 | 230 |
| 8 | 130 | 10 | 2.5 | 220 |
| 11 | 170 | 10 | 4 | 356 |

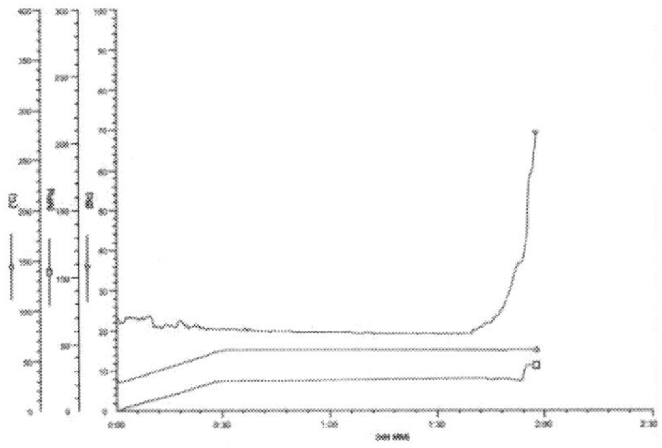

Thickening curve at 50℃

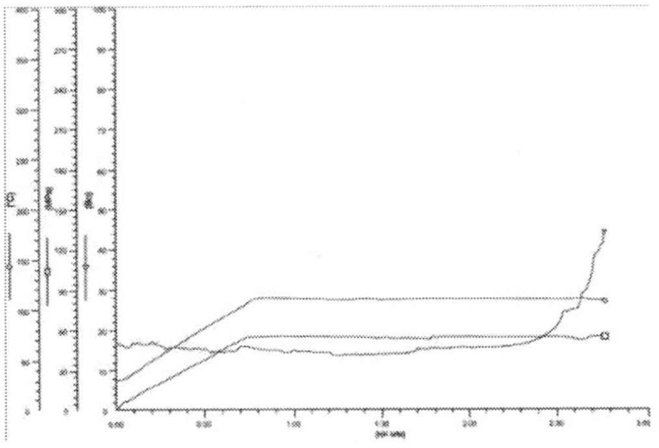

Thickening curve at 90℃

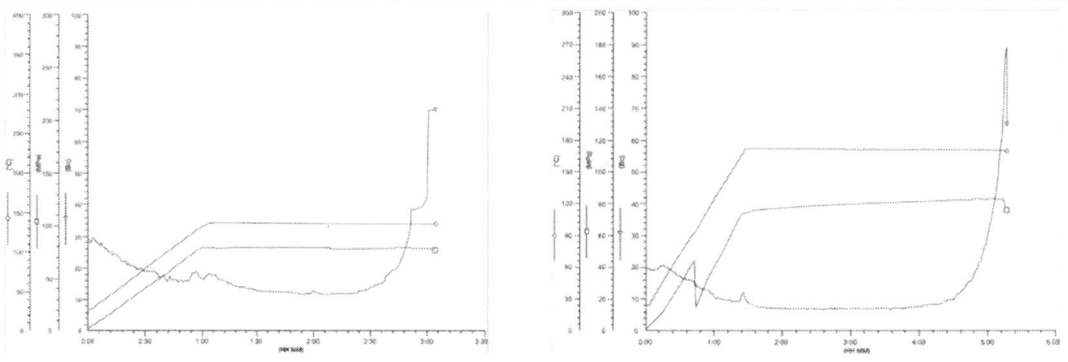

Thickening curve at 130℃ Thickening curve at 170℃

**Fig. 2.** Thickening curves of self-healing cement slurry at different temperatures

As can be seen from Table 4 and Figure 1, from low temperature to medium temperature to high temperature and then to ultra-high temperature, the self-healing cement slurry thickening time can be adjusted, the thickening curve is smooth, there is no bulge or core phenomenon, and the linearity of the thickening curve is not affected.

**Mechanical Property Test.** Select pure cement and self healing material added 10%, 20% cement stone for comparison. The results are shown in Table 4 and Figure 3.

**Table 4** Mechanical properties of cement with different dosages of self-healing agent

| Self-healing agent dosage(%) | 7d compressive strength(MPa) | 7d flexural strength(MPa) | 7d tensile strength(MPa) | Poisson's ratio | elasticity modulus(MPa) |
|---|---|---|---|---|---|
| 0 | 37.9 | 3.85 | 2.06 | 0.289 | 9363 |
| 10 | 35.2 | 4.05 | 2.72 | 0.145 | 6757 |
| 20 | 32.2 | 4.68 | 3.21 | 0.145 | 5957 |

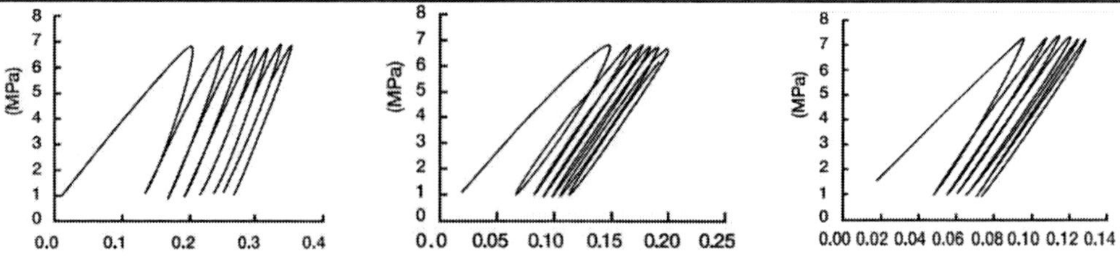


a) straight cement b) 10% self healing agent plus c) 20% self healing agent plus

**Fig. 3.** Cyclic stress-strain curves of cement with different dosage of self-healing agent

As can be seen from Table 5 and Figure 2, with the increase of dosage, the compressive strength and elastic modulus of low cement stone decrease by 15% and 41%, but the flexural strength and tensile strength of low cement stone increase significantly. In the stress-strain test under cyclic loading, the "residual deformation" of cement with self-healing material is significantly reduced and the mechanical properties are significantly improved. Therefore, the self-healing agent is a good admixture.

### *3.3. Research and development of self-healing performance evaluation device of cement slurry*

After many experiments and method improvement, the cement block module is directly split after plastic sealing with heat shrink tube before splitting and joint making, which can simulate the crack making of cement block well and facilitate the follow-up experiments. As shown in Fig. 4, the cracks after splitting are penetrable cracks, but there are no wide cracks after cutting on the surface of the cement module. As shown in Fig. 5, the permeability of the cement and the permeability data after healing can be more accurately tested during the permeability test of the cement.

**Fig. 4.** Direct splitting of plastic seal

**Fig. 5.** The penetrating joint on the surface of cement stone after plastic sealing splitting

In order to simulate the curing of self-healing cement stone in downhole high-temperature and high-pressure environment, methane gas curing device was made in this paper, as shown in Fig. 6. The device consists of a kettle body, a temperature control system and a heating system. The cement stone is put into the kettle body, and a certain pressure of methane is injected and heated to the required temperature for curing.

General view of maintenance device

Curing kettle body

Inside the curing kettle body

**Fig. 6.** Self-healing cement methane environmental curing device

### *3.4. Evaluation of healing performance of dissolving self-healing cement stone*

**Test Procedure of Mechanical Property Healing Test Experiment.** Test the compressive strength, flexural strength and splitting tensile strength of the cured stone cement module respectively; The tested cement was kept in methane gas at 90℃ for constant temperature curing and healing for 3 days、7 days、14 days and 28 days. The compressive strength, flexural strength and splitting tensile strength of cement at different curing ages were measured respectively. The recovery rate of cement with different strength was calculated respectively. The test results are shown in Table 5 to 7.

**Table 5** Compressive strength recovery rate of biomimetic self-healing cement

| sample name | compressive strength (MPa) | | | | | Maximum recovery rate (%) |
|---|---|---|---|---|---|---|
| | 0d | 3d | 7d | 14d | 28d | |
| straight cement | 37.9 | 0 | 1.3 | 2.5 | 3.4 | 8.97 |
| 10% self healing agent content cement stone | 34.2 | 5.2 | 12.6 | 18.4 | 30.7 | 89.7 |
| 20% self healing agent content cement stone | 30.9 | 8.5 | 15.6 | 21.3 | 29.2 | 94.5 |

**Table 6** Recovery rate of flexural strength of biomimetic self-healing cement

| sample name | Flexural strength (MPa) | | | | | Maximum recovery rate (%) |
|---|---|---|---|---|---|---|
| | 0d | 3d | 7d | 14d | 28d | |
| straight cement | 3.65 | 0 | 0 | 0 | 0 | 0 |
| 10% self healing agent content cement stone | 3.98 | 0 | 0 | 0.9 | 1.2 | 32.9 |
| 20% self healing agent content cement stone | 4.33 | 0 | 0.2 | 1.8 | 2.54 | 58.6 |

**Table 7** Recovery rate of split tensile strength of biomimetic self-healing cement

| sample name | Splitting tensile strength (MPa) | | | | | Maximum recovery rate (%) |
|---|---|---|---|---|---|---|
| | 0d | 3d | 7d | 14d | 28d | |
| straight cement | 2.06 | 0 | 0.25 | 0.42 | 0.63 | 30.5 |
| 10% self healing agent content cement stone | 2.58 | 0 | 0.68 | 1.59 | 2.04 | 79.1 |
| 20% self healing agent content cement stone | 2.72 | 0.32 | 0.87 | 1.86 | 2.31 | 84.9 |

Cement hydration is a long and continuous process, even pure cement itself also has a certain bionic self-healing function. From Table 5 to Table 7, it can be seen that the maximum recovery rate of compressive strength of pure cement is 8.97%, which reflects that pure cement itself has a certain self-healing ability, but the healing degree is not high. When 10% and 20% biomimetic self-healing agents were added, the recovery rate of compressive strength was more than 60%, and the maximum recovery rate reached 94.5%, which almost recovered to the level before splitting. The recovery rate of flexural strength is far less high than that of compressive strength. Even for low density cement with a recovery rate of 94.5%, the maximum recovery rate of flexural strength is only 58.6%. The recovery rate of splitting tensile strength of pure cement is 30.5%, and the recovery rate of splitting tensile strength of cement with 20% self-healing agent at 28d is 84.9%. This difference in recovery rate is mainly determined by the experimental method. After the fracture of flexural strength, only one crack can contact the oil-gas induced self-healing agent to produce self-healing, which is completely different from the hydrocarbon expansion self-healing agent studied by previous researchers. Instead of plugging micro-cracks by expansion to prevent the passage of oil, gas and water, oil-gas induced viscous fluid bonded cement cracks are generated. Therefore, there is a certain recovery of mechanical properties.

**Permeability Healing Test.** Experimental test steps: record the permeability K0 of the cement stone before and K1 of the cement stone after joint creation; The cement stone with permeability gap measured was kept in high purity methane gas (effective content 99.99%) for constant temperature curing and healing at 90℃ for 7d, 14d and 28d. The gas and liquid permeability of the cement stone at different curing ages were measured respectively, and the recovery of the permeability of the cement stone was calculated. The results are shown in Table 8 and Table 9.

**Table 8** Bionic self healing cement gas permeability measurement

| formula | pre-seam(mD) | 0d (mD) | 7d (mD) | 28d (mD) | Permeability recovery rate (%) |
|---|---|---|---|---|---|
| straight cement | 0.002 | 8.944 | 7.416 | 5.211 | 47.9 |
| Add 10% low density | 0.001 | 6.571 | 3.225 | 0.002 | 98 |
| Add 10% regular density | 0 | 6.543 | 3.221 | 0.003 | 97 |
| Add 20% regular density | 0 | 5.982 | 1.985 | 0.001 | 99 |

**Table 9** Permeability measurement of bionic self-healing cement fluid

| formula | pre-seam(mD) | 0d (mD) | 7d (mD) | 28d (mD) | Permeability recovery rate (%) |
|---|---|---|---|---|---|
| straight cement | 0.001 | 6.548 | 5.468 | 4.352 | 56.48 |
| Add 10% low density | 0 | 4.321 | 2.365 | 0 | 100 |
| Add 10% regular density | 0 | 4.225 | 1.587 | 0 | 100 |
| Add 20% regular density | 0 | 3.854 | 0.564 | 0 | 100 |

As can be seen from Table 8 and Table 9, both pure cement stone and self-healing cement stone are very dense. Before seam construction, the gas permeability of pure cement stone is only 0.002mD, and the liquid permeability is only 0.001mD. The gas permeability and liquid permeability of self-healing cement stone with more than 10% self-healing agent are 0. Although the pure cement stone has a certain self-healing function, the permeability reduction effect of pure cement stone after fracture making is little. Even after 28 days of methane curing, the gas permeability decreases from 8.944mD to 5.211mD, and the liquid permeability decreases from 6.548mD to 4.352mD. With the increase of self-healing agent addition, the permeability of self-healing cement gradually decreases to 0, and the permeability tested immediately after seam construction also gradually decreases. After three days of high-temperature and high-pressure methane curing, the permeability of self-healing cement slightly decreases. After 14 days of curing, the permeability decreases by more than 90% after 28 days of curing, and when the addition of self-healing agent exceeds 15%, the permeability of self-healing cement gradually decreases. The permeability of self-healing cement stone was reduced by 99% and the measured permeability was reduced by 100%. Therefore, the permeability of self-healing cement stone after curing in high temperature and high pressure methane environment for 28d is almost as good as that before seam construction, which can prevent oil and gas channeling flow.

As can be seen from the comparison between Table 8 and Table 9, whether it is pure cement stone or self-healing cement stone, the permeability tested in the liquid medium is lower than that tested in the gas medium. In particular, the permeability of the self-healing cement stone after self-healing is also 0, completely restoring the densification before the joint construction. On the one hand, this is because liquid viscosity is higher than gas, on the other hand, because the self-healing agent is more likely to react with methane in a wet environment to form new substances to heal cracks and prevent channeling.

*3.5. Action mechanism of bionic self-healing properties*

In order to reveal the mechanism of self-healing cement to complete self-healing under hydrocarbon exposure, firstly, optical microscope was used to observe the cracks of self-healing cement, and then SEM scanning electron microscope was used to observe the microscopic morphology of the section of self-healing cement before and after self-healing. The results are shown in Figures 7 and 8.

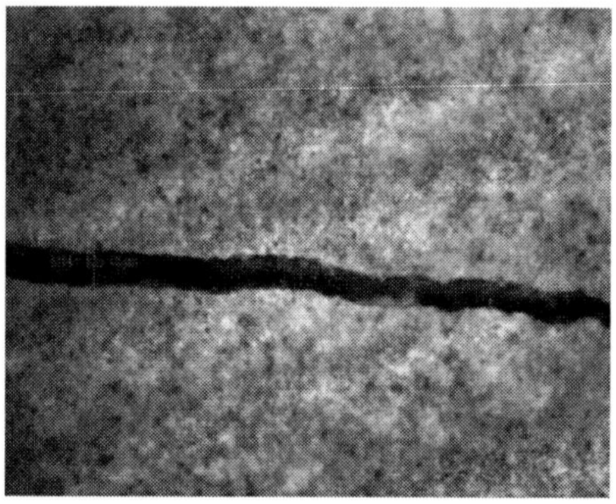

**Fig. 7.** Cement gap before self-healing

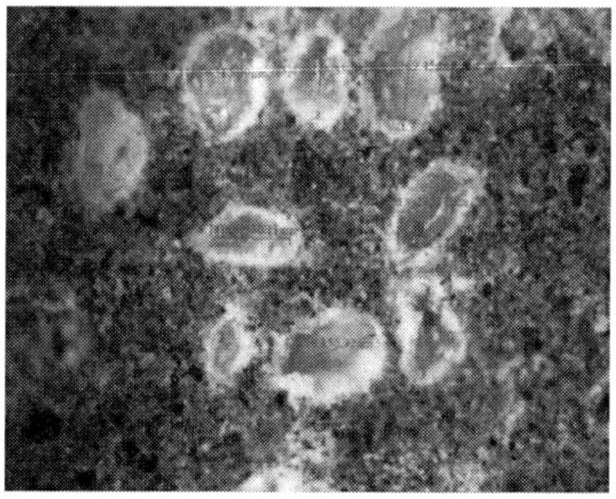

**Fig. 8.** Shape of cement section before self-healing

As can be seen from Fig. 7, the cracks formed by the splitting and joint creation of the cement stone after the thermoplastic process are more natural. Different from the regular cracks formed by wire cutting, they can more truly reflect the micro-cracks caused by the failure of the cement stone due to stress concentration in the later production and other operations of the cement ring underground. Fig. 8 shows that there are many irregular sac-like particles in the fracture section of the cement stone before healing. The wall of the sac-like particles is smooth and transparent but not stimulated, and they are distributed in the interior of the cement stone in an irregular sac-like shape.

The self-healing cement cracks after 3d curing with high concentration methane gas were observed by optical microscope, as shown in Fig. 9 and Fig. 10.

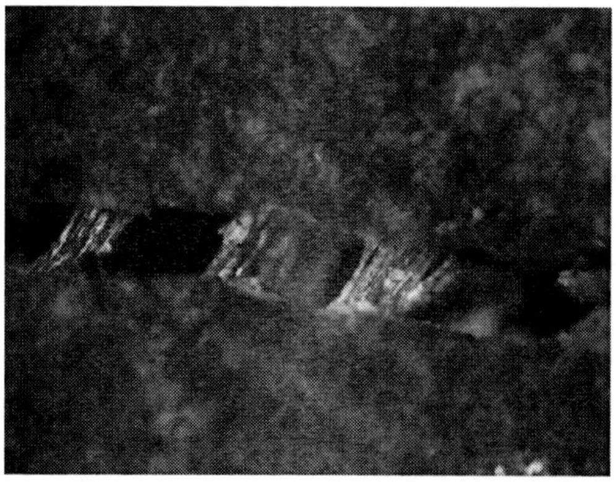

**Fig. 9.** Micro cracks of cement stone after 3d curing with high concentration methane

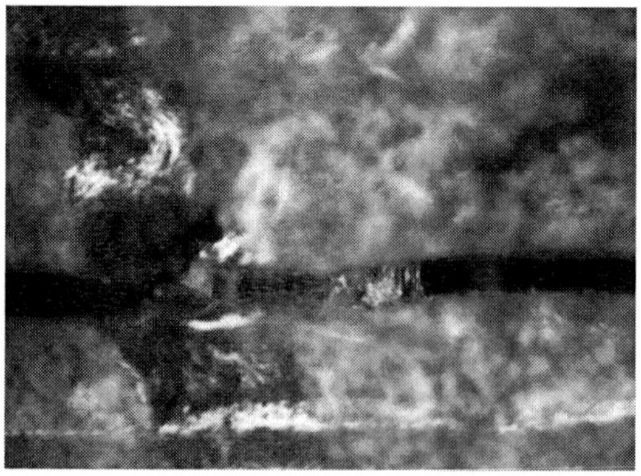

**Fig. 10.** Micro cracks of cement stone after curing with high concentration of methane for 7 days

As can be seen from Fig. 9, there are several viscous transparent gel objects in the micro-cracks of cement stone cured for 3d in high concentration methane gas to bond the cracks. The self-healing cement stone began to show self-healing properties. The self-healing material was excited by methane gas, and a new material was formed to bond the cracks of the cement stone. With the extension of reaction time, the transparent gel-like substance increased (Fig.10), and the crack began to heal.

SEM was used to observe the micro-morphology of the self-healing cement before and after self-healing, as shown in Fig. 11 and Fig. 12.

**Fig. 11.** Micromorphology of self-healing cement before healing

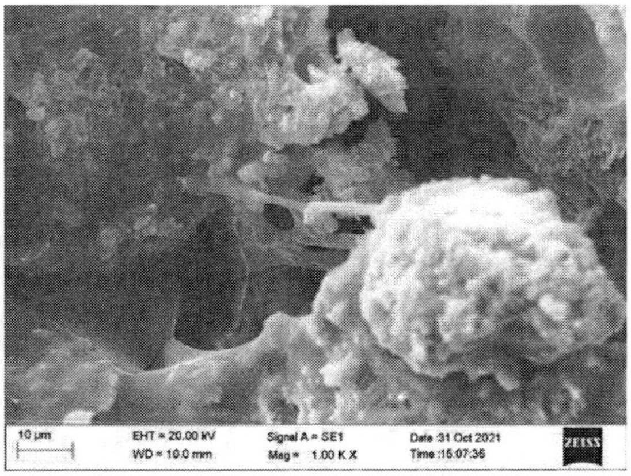


**Fig. 12.** Microstructure of self-healing cement

It can be seen from Fig. 11 and Fig. 12 that the self-healing material inside the self-healing cement stone exists in the crack in a "whiskle-like" or "sac-like" structure before self-healing. When the damaged cement stone is cured in methane gas, the self-healing material is automatically excited when it meets methane, and the self-healing agent reacts with methane to form a gel-like substance to connect the damaged cement skeleton. As the reaction continues, the generation of gel-like products gradually increases, until the cement skeleton is re-bonded, the mechanical properties of the cement stone are restored, the cement stone micro-cracks are blocked, and the gas channeling is prevented.

## 4. Conclusion and suggestion

Through the above research, the author comes to the following conclusions.

(1) Self-healing material has no adverse effect on the conventional properties of cement stone.

(2) The mechanical property of self-healing cement stone is significantly improved than that of pure cement stone, and the elastic modulus is reduced by 41% when the compressive strength is reduced by 15%, which plays the role of "first prevention" of gas channeling in the later stage of cement ring. The recovery rate of compressive strength can reach 96.1%, the reduction rate of gas permeability can reach 99%, and the reduction rate of liquid permeability can reach 100%, which plays the role of "post-plugging". Therefore, the bionic self-healing cement stone has the effect of "first prevent and then block" on the late gas channeling of cement stone.

(3) Through the analysis of electron microscopy, SEM and other testing means, it can be seen that the self-healing material evenly mixed in the cement stone reacts to produce a gelatinous ester substance under the excitation of oil and gas, which plays the role of the self-healing of the closed

channel, adhesive and cement stone skeleton, can prevent oil and gas channeling, and restore the sealing ability of the cement ring.

**Acknowledgments.** The project is supported by Key Research and Development Project of Sichuan Province: New Technology for Harmless Utilization of Oil-based cuttings after Oil Extraction in shale Gas Wells (212DYF1136), Research Project of petrochina Southwest Oil and Gas Field Company( No.20220302-02).

**References**

[1] Lu Wang. Preparation of self-healing gradient Mosi2-based antioxidant coatings by SAPS[D]. NWPU, 2018.

[2] Hai Yuan.Study on fracture healing property of asphalt mixture based on microwave heating[D]. Chongqing Jiaotong University, 2020.

[3] Da Shi. Effect of mineral admixtures on UHPC self-healing[D]. Hunan University, 2020.

[4] Zhang X, Zhu P, Aemail Z Z C C. A chemo-mechanically coupled continuum damage-healing model for chemical reaction-based self-healing materials[J]. International Journal of Solids and Structures, 2021: 111346.

[5] Manoj A A, Ramachandran R A, Menezes P L A P. Self-healing and superhydrophobic coatings for corrosion inhibition and protection[J]. International Journal of Advanced Manufacturing Technology, 2020, 106(5-6): 2119-2131.

[6] Pal S A C C, Mohanty I A C C, Panda I A C C. Self-healing Conventional Concrete Using Bacteria[J]. Lecture Notes in Civil Engineering, 2021, 75: 441-451.

[7] Zhenjie Yang, Bing Qi, Ani Liu, etc. Research progress on self-healing mechanism of microcracks in cement-based materials[J]. Oil drilling technology, 2009, 37(3): 124-128.

[8] Moroni A, Hoang-Kim A, Lio V, et al. Current augmentation fixation techniques for the osteoporotic patient[J]. Scandinavian journal of surgery: SJS: official organ for the Finnish Surgical Society and the Scandinavian Surgical Society, 2006, 95(2): 103-109.

[9] Moroni A, Rocca M, Faldini C, et al. Hydroxyapatite fully coated conic hip prosthetic stem: a long term animal study[J]. Annales chirurgiae et gynaecologiae, 1999, 88(3): 198-204.

[10] Baohui Zhao,Linglong Shi,Jianlong Zhou,etc.Application of self-healing anti-channeling cement technology to φ177.8mm liner cementing in well G103[C]//2015 Annual Drilling Technology Seminar and the 15th Oil Drilling Institute Director Conference, Bei Jing, Chinese Petroleum Society, 2015.

[11] Jianfeng Zhao． Research and development of self-healing technology of cement ring micro-fracture[J]． Science and technology guide，2010，(4)：288-291．

[12] Xiao Yao,Yalei Chen,Technical evaluation of cement ring self-healing materials at home and abroad[C]. Proceedings of 2012 National Cementing Research Conference.Petroleum Industry Press,2012.

Journal of Physics: Conference Series **2594** (2023) 012055 doi:10.1088/1742-6596/2594/1/012055

# Optimization Study of Injection-Production Parameters for $CO_2$ Enhanced Oil Recovery and Storage in Low Permeability Oil Reservoirs

**Ze-peng YANG1, Xin-wei LIAO1(✉), Xiao-chen WANG1**

1 School of Petroleum Engineering, China University of Petroleum Beijing, China
xinwei@cup.edu.cn

**Abstract**. To address the difficulty of water injection in the late stage of developing a certain low-permeability reservoir, $CO_2$ injection and fracturing parameter optimization were conducted. The Latin hypercube sampling and numerical simulation methods were used to study the five-spot well network model established, and evaluate the impact of injection and fracturing parameters on oil recovery and $CO_2$ storage using a comprehensive evaluation factor of oil displacement and storage. The results show that the injection rate is the main controlling factor affecting the oil recovery and storage effect, and the optimal injection rate is 6000m$^3$/d. There is a certain contradiction between the oil recovery and storage objectives at different stages of injection development, but after ten years of injection development, the objectives of oil recovery and storage are consistent. The optimal combination of ijection and fracturing parameters yields a recovery rate of 72.54% and a $CO_2$ storage amount of 101635.81 tons. This paper proposes that there is a certain contradiction between the oil recovery and storage objectives at different stages and evaluates schemes with different injection development periods by constructing a comprehensive evaluation factor for oil displacement and storage, which has certain reference value for optimizing parameters in the $CO_2$ injection development of low-permeability reservoirs.

## 1. Introduction

Low permeability reservoirs occupy a large proportion globally, characterized by poor reservoir properties where conventional water flooding is not effective, resulting in the problem of ineffective replenishment of reservoir energy during late-stage water injection, making it difficult for water to infiltrate while also preventing oil production [1,2]. $CO_2$ injection has been established as a mature and effective method for enhancing recovery after many years of mineral field experiments and theoretical analysis [3,4]. $CO_2$, with its excellent injection ability and unique oil displacement mechanism, has shown advantages in low permeability reservoir development [5-8]. At the same time, in the context of carbon neutrality, the use of $CO_2$ for enhanced oil recovery can not only increase microscopic oil recovery efficiency and improve reservoir development but also serve the purpose of carbon capture and storage [9-11]. $CO_2$ EOR technology, as an important CCUS technology in the oil and gas industry, is of significant importance to study the impact of development parameters on oil displacement and storage efficiency, and to optimize these parameters [12-15].

In this study, we established a five-spot well network for $CO_2$ injection and oil recovery in a low-permeability reservoir with direct air injection and horizontal well oil production, using a model comprising of production parameters including $CO_2$ injection rate, bottomhole pressure of production wells, and shut-in oil/gas ratio. Fracture parameters mainly included fracture half-length, width, height, permeability, and closure time. By using the Latin hypercube sampling method, the Pearson correlation coefficient, and the comprehensive evaluation factor, we analyzed the impact of injection and production parameters as well as horizontal well fracture parameters on the degree of oil recovery and $CO_2$ storage at different injection stages, in order to guide the optimization of $CO_2$ injection parameters in low-permeability reservoirs and site practices.

## 2. Method

### 2.1. Latin Hypercube Sampling

In order to comprehensively reflect the inherent law of all variable combinations and reduce the number of experiments as much as possible, a good experimental design is required when there are a large number of combinations of different parameter values and simulation times are long. To investigate the impact of different combinations of injection-production and fracture parameters on $CO_2$ EOR and storage, this study adopted the Latin Hypercube Sampling (LHS) method [16] for experimental design. LHS is a stratified sampling technique widely used in engineering practice, which uniformly divides the range of given random variables into hyperplanes and generates random values. It samples the same number of points in each interval, ensures that the sampling covers all parameter ranges evenly, and greatly improves sampling efficiency [17].

### 2.2. Pearson Correlation Coefficient

The Pearson Correlation Coefficient (PCC) [18] was used to measure the correlation between variables and the degree of recovery and storage rate. The calculation formula is as shown in Equation (1).

$$r_{XY} = \frac{\sum (X-\bar{X})(Y-\bar{Y})}{\sqrt{\sum (X-\bar{X})^2 \sum (Y-\bar{Y})^2}} \tag{1}$$

In the formula, X represents the variable and Y represents the corresponding index at a certain time. The value range of the correlation coefficient $r_{XY}$ is (-1, 1). The closer it is to -1, the greater the negative correlation degree, and the closer it is to 1, the greater the positive correlation degree. When the value is 0, it indicates complete irrelevance.

### 2.3. Comprehensive Evaluation Factor for EOR and Storage

In $CO_2$ EOR and storage, the main evaluation indicator for oil displacement effect is cumulative oil production, and the main evaluation indicator for storage effect is the quality of $CO_2$ underground storage. Considering the different magnitudes of evaluating EOR and storage effects, a dimensionless normalization process was adopted when constructing the objective function. The dimensionless oil recovery factor and $CO_2$ storage rate were defined as $f_1$ and $f_2$ (see Equation 2 and 3), respectively. The traditional method for evaluating both of these objectives involves transforming them into a single-objective problem through weighted summing methods in order to optimize parameters and evaluations. Based on this, a comprehensive evaluation factor for EOR and storage $F$ (see Equation 4) was constructed using a weighted linear summing method, which adjusts the weight $\omega_i$ to achieve a holistic evaluation that emphasizes different objectives [19,20].

$$f_1 = \frac{N_P - N_{P\min}}{N_{P\max} - N_{P\min}} \tag{2}$$

$$f_2 = \frac{m_{CO_2} - m_{CO_2\min}}{m_{CO_2\max} - m_{CO_2\min}} \tag{3}$$

$$F = w_1 f_1 + w_2 f_2 \tag{4}$$

In the formula: $N_P$ represents the cumulative oil production, $m^3$; $m_{CO_2}$ represents the buried mass of crude oil in the underground reservoir, $t$; $\omega_1$ and $\omega_2$ are the weights of the oil recovery and reservoir storage terms, $0 < \omega_1 < 1$ and $\omega_1 + \omega_2 = 1$.

## 3. Model Establishment and Scheme Design

### *3.1. The Establishment of a Mechanism Model*

According to the reservoir and fluid properties of a certain low-permeability oil reservoir, a mechanistic model is established with a size of 41×21×8, containing a total of 6888 grids. The horizontal step length is 20m, the vertical step length is 2m, the burial depth is 3400m, the initial formation pressure is 14MPa, the initial oil saturation is 0.69, the water saturation is 0.31, the formation temperature is 80.3℃, the average porosity is 7.46%, the average permeability is 5.36mD. In the center of the model, there is a horizontal production well with a horizontal section length of 650m, which is equipped with seven evenly spaced fractures, with a fracture spacing of 80m. Four gas injection wells located at the corners with the bottom perforation are also included in the model. The numerical model established is shown in Fig. 1.

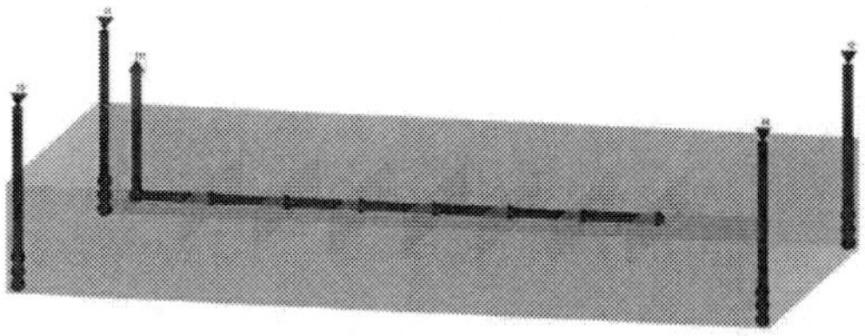

Fig. 1. Schematic diagram of five-point well pattern model.

The fluid components were merged into six pseudo-components. The critical parameters of the fluid model were obtained through phase equilibrium fitting (see Table 1). The recovery rates of injecting 1.2 times the pore volume of $CO_2$ under different displacement pressures were calculated by simulating capillary experiments. The relationship curve between the recovery rate and displacement pressure was plotted, and the minimum mixing pressure of the formation fluid was obtained as 15.5MPa (see Fig. 2).

Table 1. Reservoir fluid composition and critical parameters.

| Components | Molar Fraction, % | Critical Temperature, K | Critical Pressure, bar | Acentric Factor | Molecular Weight, kg/kg-mol |
|---|---|---|---|---|---|
| $N_2$ | 1 | 126.2 | 34.6 | 0.0377 | 28.013 |
| $CO_2$ | 0.68 | 304.19 | 73.82 | 0.228 | 44.01 |
| $H_2S$ | 1.78 | 373.53 | 89.63 | 0.0942 | 34.082 |
| $C_1$ | 47 | 190.56 | 45.99 | 0.012 | 16.043 |
| $C_2$to$C_6$ | 21.64 | 395.1276 | 41.1205 | 0.1659 | 50.4746 |

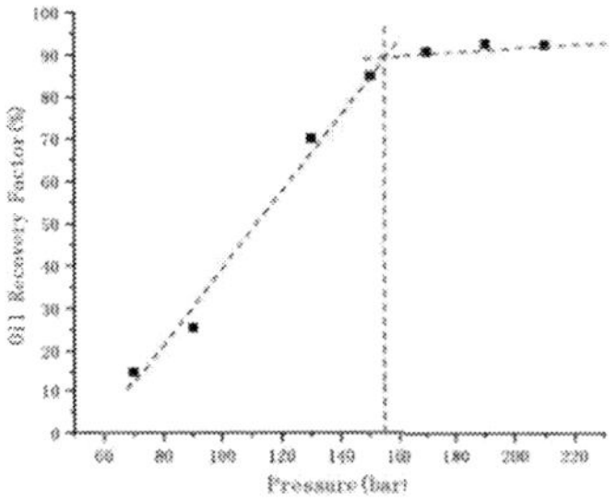


Fig. 2. Relationship curve between displacement pressure and recovery factor.

Journal of Physics: Conference Series **2594** (2023) 012055 doi:10.1088/1742-6596/2594/1/012055

### *3.2. LHS Scheme Design*

In this experiment, there were eight variables, including injection parameters such as injection rate, bottomhole flowing pressure of production well, and gas-oil ratio limit during shut-in, and hydraulic fracture parameters including fracture half-length, fracture width, fracture height, fracture permeability, and closure time. To simulate the gradual closure of fractures over time, the time-dependency of fractures was represented by multiplying the permeability coefficient by the initial fracture permeability. The permeability coefficient varied with time and is shown in Fig. 3. As each factor had a different effect on production and reserves, and the injection rate had the greatest impact, its influence was considered separately. The injection rate was divided into 20 intervals from 500 $m^3$/d to 10000 $m^3$/d for Latin hypercube sampling, and the sample space for each parameter is shown in Table 2.

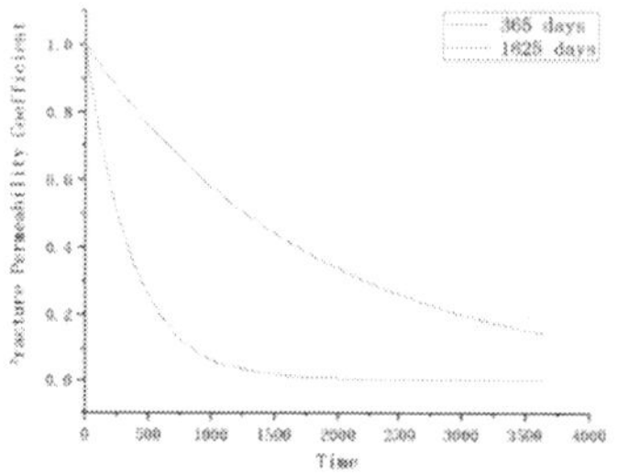


Fig. 3. Curves of permeability coefficient of fractures with closing time of 365 days and 1825 days respectively.

Table 2. Reservoir fluid composition and critical parameters.

| Parameter | Minimum value | Maximum value | Value |
|---|---|---|---|
| $CO_2$ injection rate, $m^3$/d | 500 | 10000 | 500 1000 1500 2000 2500 3000 3500 4000 4500 5000 5500 6000 6500 7000 7500 8000 8500 9000 9500 10000 |
| Bottomhole pressure of production wells, MPa | 7 | 12 | Algorithm control |
| Shut-in oil/gas ratio, $m^3/m^3$ | 500 | 1500 | Algorithm control |
| Fracture half-length, m | 50 | 120 | Algorithm control |
| Fracture width, m | 0.01 | 0.03 | Algorithm control |
| Fracture height, m | 2 | 8 | Algorithm control |
| Fracture permeability, mD | 500 | 5000 | Algorithm control |
| Fracture closure time, d | 365 | 3650 | Algorithm control |

## 4. Analysis of Simulation Results

### *4.1. Analysis of Influence Law of Gas Injection Rate*

The simulation results of Latin hypercube sampling for different injection rates show that before the injection rate reaches 6000 $m^3$/d, there was a significant positive correlation between the injection rate and recovery rate and $CO_2$ storage capacity. However, as the injection rate continues to increase, the recovery rate and $CO_2$ storage capacity showed fluctuations within a certain range due to the influence of other parameters (see Fig. 4). The impact of different injection rates on gas breakthrough time in production wells varied, with a general trend showing that the higher the injection rate, the faster the gas-oil ratio will increase, and the earlier the shut-in gas-oil ratio limit will be triggered. The number of wells opened and shut down after five, seven, and ten years under different injection rates are shown in Fig. 5.

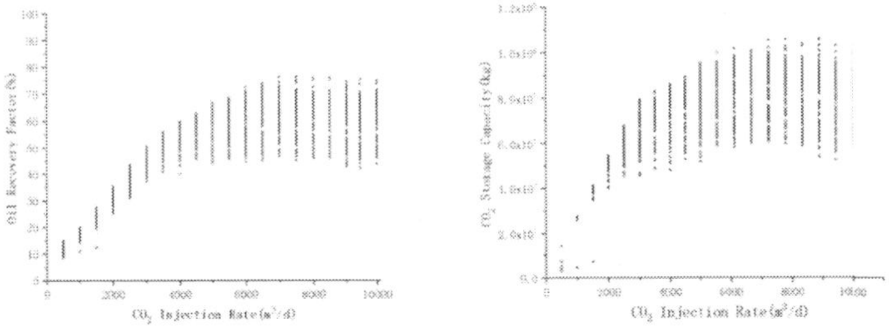

Fig. 4. Recovery factor and $CO_2$ storage capacity curve under different gas injection rates.

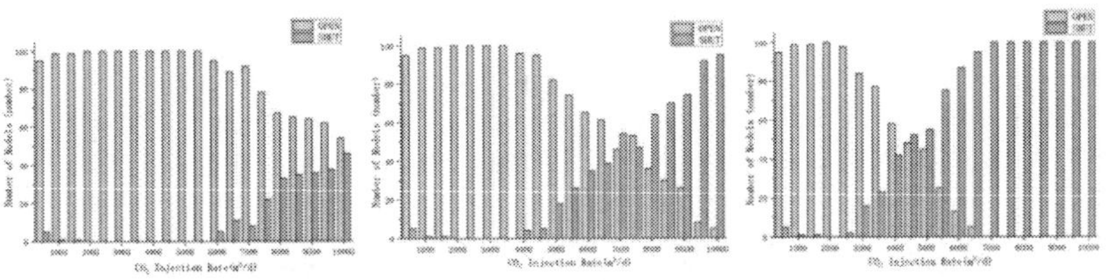

Fig. 5. Statistics of on-off wells of different schemes under different gas injection rates after five years (left), seven years (middle) and ten years (right) of gas injection development.

After five years of $CO_2$ injection development, some schemes with lower injection rates (500m$^3$/d, 1000m$^3$/d, and 1500m$^3$/d) experienced extensive gas breakthrough due to slow energy supplement in the reservoir, triggering well shut-in conditions under low wellbore flowing pressure and smaller shut-in gas-oil ratio restrictions. No shut-in situation was observed under injection rates from 2000 m$^3$/d to 5500 m$^3$/d, while an increasing number of shut-in schemes were observed as injection rates increased above 6000 m$^3$/d due to gas channeling induced shut-in. From seven years of $CO_2$ injection development, some shut-in incidents were observed from the injection rate of 4000 m$^3$/d, and the number of shut-in schemes gradually increased with increasing injection rates. After ten years of $CO_2$ injection development, all schemes with injection rates above 2000 m$^3$/d experienced varying degrees of gas channeling induced shut-in, and all schemes with injection rates above 7000 m$^3$/d caused massive gas channeling induced shut-in incidents.

Based on the analysis of different production efficiencies, $CO_2$ storage capacities, and gas channeling rules under different injection rates, the optimal injection rate is 6000 m$^3$/d under the maximum shut-in gas-oil ratio constraint, fracture parameters, and wellbore flowing pressure range. At the current reservoir pressure maintenance level, the annual injection rate is approximately 0.12HCPV/a. A further increase in injection rate will not result in a substantial increase in production and $CO_2$ storage, but instead will cause massive gas breakthrough, generating excessive and ineffective $CO_2$ circulation, which contradicts the purpose of utilizing $CO_2$ for oil displacement and storage development. Therefore, other parameter influence rules analysis and optimal design should be conducted under the ground single well injection rate of 6000m$^3$/d.

### 4.2. *Analysis of Influence Law of Other Parameters*

Under the condition of gas injection rate of 6000m$^3$/d, 100 sampling schemes with different parameter combinations were obtained using Latin hypercube sampling. The recovery rate, gas-oil ratio, and $CO_2$ storage capacity curves are shown in Fig. 6. The number of schemes that reach the well closure gas-oil ratio limit after five, seven, and ten years of gas injection development are 5, 35, and 87, respectively. There are differences in the impact of different injection and production parameter combinations on the recovery rate and $CO_2$ storage capacity at different time periods. Additionally, there are differences in the impact of the same objective function during different development stages. Pearson correlation coefficient is used to analyze the impact of injection and fracture parameters on the recovery rate and $CO_2$ storage capacity during different development stages.

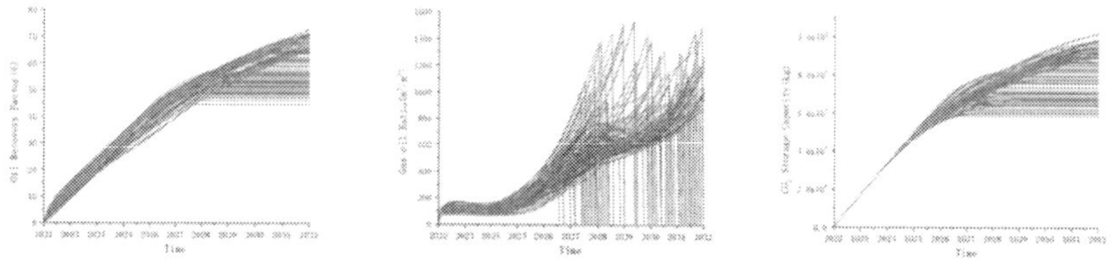

Fig. 6. Recovery factor (left), gas-oil ratio (middle) and $CO_2$ storage capacity(right) curve of Latin sampling scheme at gas injection rate of 6000m$^3$/d.

After one year of gas injection development, the Pearson correlation coefficients between each parameter and the recovery rate and $CO_2$ storage are shown in Fig. 7. The parameter that has the greatest impact on the recovery rate after one year of gas injection development is the bottomhole flowing pressure of the production well, which shows a negative correlation, while other parameters are positively correlated. At this stage, which is a rapid production increase period, $CO_2$ has not yet reached the bottom of the production well, and the main factor affecting production is the bottomhole flowing pressure of the production well. The greater the production pressure difference, the faster the production increase in the early stage, and the larger the fracture half-length, fracture height, fracture width. The parameters such as duration and permeability are, the greater the help in the early-stage production increase. For $CO_2$ storage, since the injection rate is the same, when there is no large-scale gas channeling and the well-closing condition is not triggered, the total storage amount is not much different. The impact rules of various parameters are completely opposite to the recovery rate. The main factor is the bottomhole flowing pressure of the production well, which is positively correlated with $CO_2$ storage, while other fracture parameters are negatively correlated with the storage amount. Therefore, the larger the bottomhole flowing pressure of the production well and the less developed the fractures are, the less degassing occurs during the initial development of the formation's crude oil, and the greater the $CO_2$ storage amount. The intersection diagram of the recovery rate and $CO_2$ storage after one year of gas injection development (see Fig. 8) is a Pareto frontier [21], which also shows that at this stage, the recovery rate and $CO_2$ storage present a contradictory state. Therefore, the optimization of injection and production, as well as fracture parameters becomes a multi-objective optimization problem.

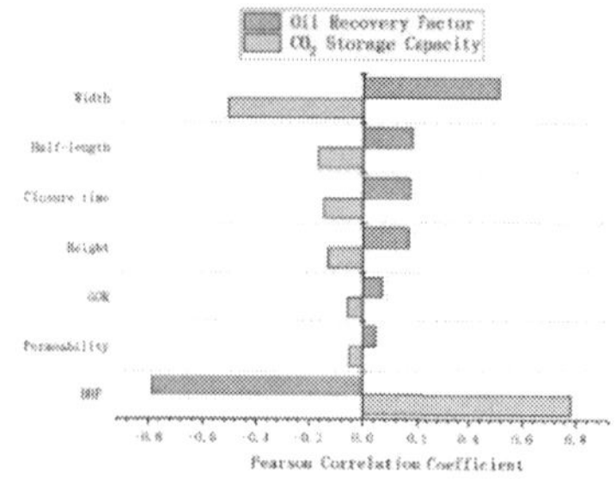


Fig. 7. The correlation coefficient between the parameters and the degree of recovery and $CO_2$ storage after one year of development.

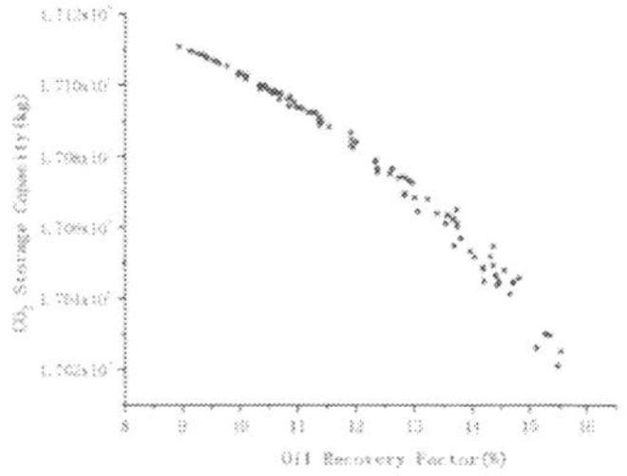


Fig. 8. Plot of intersection between recovery degree and $CO_2$ storage after one year of development.

After five years of gas injection, gas breakthrough began to appear in the Latin sampling scheme as shown in the gas-oil ratio curve (Fig. 6). The Pearson correlation coefficients of each parameter on the degree of recovery and the amount of $CO_2$ storage are shown in Fig. 9, with the duration of the fracture being the most important influencing factor, positively correlated with the degree of recovery and negatively correlated with the amount of $CO_2$ storage. The hydraulic fracturing fractures in horizontal wells gradually close after a certain period of time, and the duration of the action on the permeability coefficient of the fractures affects the permeability of the fractures. The longer the duration, the more favorable it is to improve the degree of recovery, but it will also increase the occurrence of gas breakthrough in production wells, which has a negative impact on $CO_2$ storage. At this time, the recovery rate and the amount of $CO_2$ storage began to show a trend where the degree of recovery and the amount of $CO_2$ storage were both optimal simultaneously, but there still existed some contradictions (see Fig. 10).

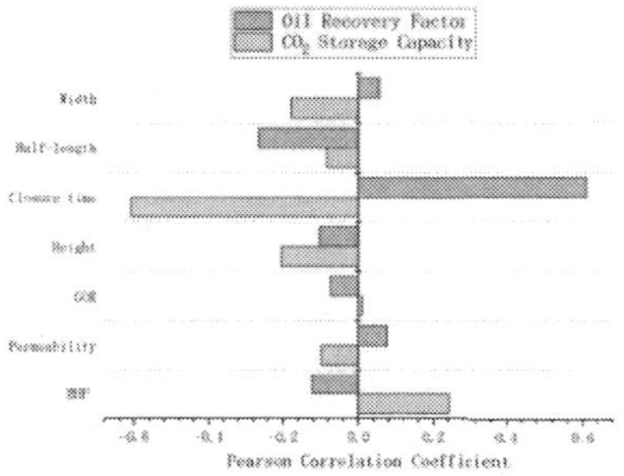


Fig. 9. The correlation coefficient between the parameters and the degree of recovery and $CO_2$ storage after five years of development.

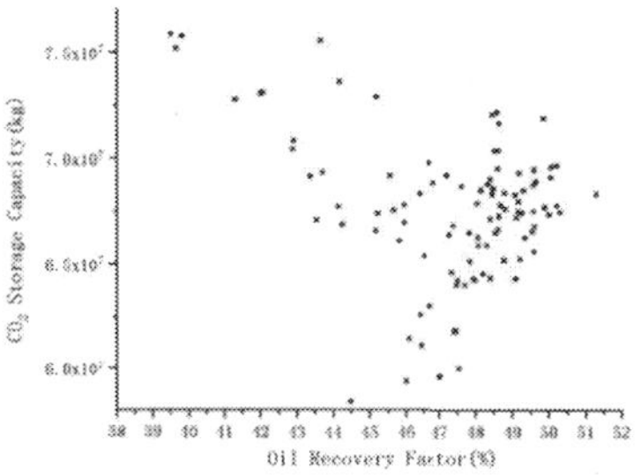


Fig. 10. Plot of intersection between recovery degree and $CO_2$ storage after five years of development.

After seven years of injection, 35 schemes had reached the gas-oil ratio limit. The Pearson correlation coefficient of each parameter on recovery and $CO_2$ storage is shown in Fig. 11. For recovery and $CO_2$ storage, the bottomhole producing pressure and gas-oil ratio are positively correlated, while the fracture parameters are negatively correlated. The bottomhole producing pressure and the duration of fractures have a more significant impact on $CO_2$ storage. Due to large-scale gas channeling, about one-third of the schemes were shut down, and the effect of the gas-oil ratio limit was reflected. A larger gas-oil ratio limit means a higher gas channeling tolerance, and production wells can continue producing at higher gas-oil ratios. At the same time, injection wells can continue to inject $CO_2$ to increase storage. The bottomhole producing pressure positively correlates with recovery and $CO_2$ storage, as the reservoir rock energy is supplemented after injection development. The pressure approaches the minimum miscibility pressure of the formation fluid, resulting in near-miscible displacement and a significant increase in oil recovery efficiency. The injected $CO_2$ displaces crude oil in the pore with higher efficiency and eventually achieves storage. Therefore, a higher bottomhole producing pressure at this stage positively correlates with recovery and storage. At this stage, there is a more obvious correlation between recovery and $CO_2$ storage (see Fig. 12).

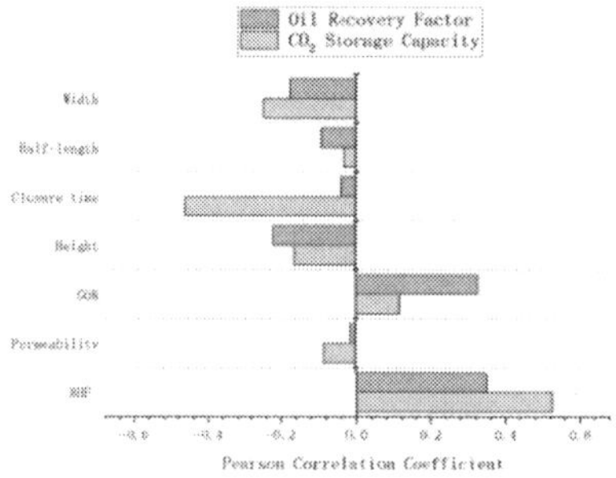


Fig. 11. The correlation coefficient between the parameters and the degree of recovery and $CO_2$ storage after seven years of development.

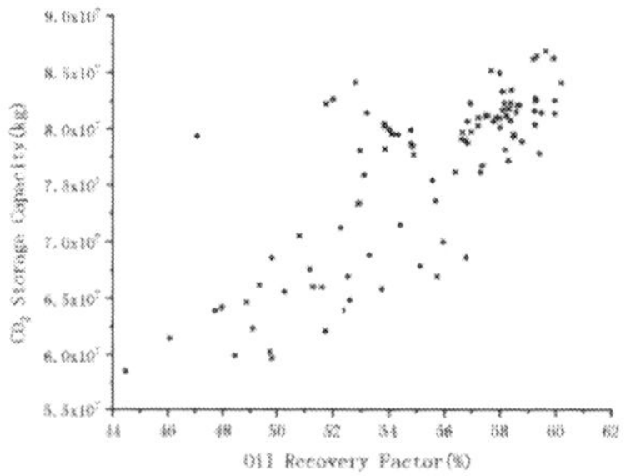


Fig. 12. Plot of intersection between recovery degree and $CO_2$ storage after seven years of development.

After ten years of injection, 87 schemes experienced large-scale gas channeling resulting in closed wells. The Pearson correlation coefficient of each parameter on recovery and $CO_2$ storage is shown in Fig. 13. The influence of each parameter on recovery and $CO_2$ storage did not change significantly. At this time, the linear relationship between recovery and $CO_2$ storage becomes more apparent, and there exist injection-production parameter schemes that simultaneously optimize both recovery and storage (see Figure 14).

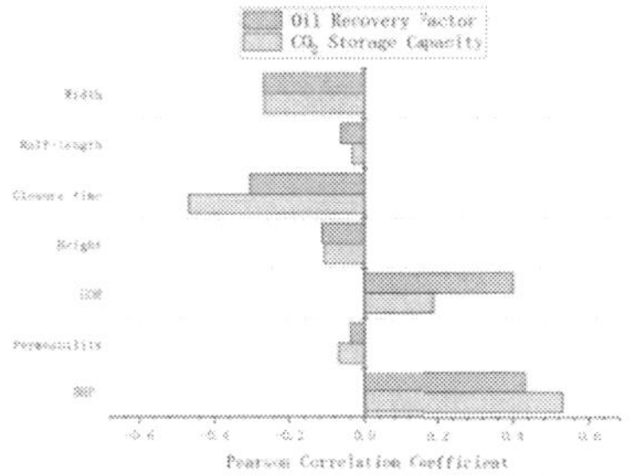


Fig. 13. The correlation coefficient between the parameters and the degree of recovery and $CO_2$ storage after ten years of development.

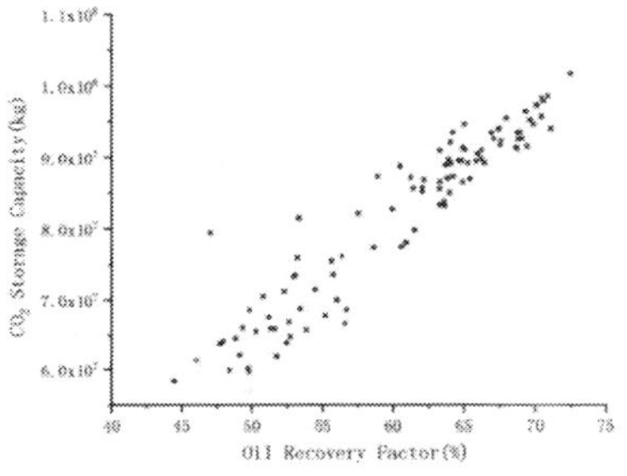


Fig. 14. Plot of intersection between recovery degree and $CO_2$ storage after ten years of development.

### *4.3. The Best Scheme of Lower Oil Displacement and $CO_2$ Storage In Different Periods*

Considering that many implementation conditions of $CO_2$ injection for oilfields and injection time often cannot meet the design requirements, there may be a lack of large-scale $CO_2$ fields or conditions for capturing fixed industrial waste gas for $CO_2$ capture around the oilfields. Moreover, after the above analysis, it is considered that there is a certain contradiction between the oil recovery effect and the $CO_2$ storage effect in different stages. Therefore, the phased evaluation of $CO_2$ injection has certain values.

After injecting gas for one, five, seven, and ten years, the comprehensive effect factor of enhanced oil recovery and $CO_2$ storage was utilized to evaluate different schemes. When enhanced oil recovery was emphasized, the weight of enhanced oil recovery was set to 0.9 and the weight of $CO_2$ storage was set to 0.1. When $CO_2$ storage was emphasized, the weight of enhanced oil recovery was set to 0.1 and the weight of $CO_2$ storage was set to 0.9. When enhanced oil recovery and $CO_2$ storage were equally important, both weights were set to 0.5. The evaluation results are shown in Table 3-6.

By utilizing the comprehensive evaluation factor of $CO_2$ flooding and storage, different schemes during various injection periods were evaluated. The results indicate that due to the different influences of injection-production parameters during different periods, there exists a certain contradiction between $CO_2$ flooding and storage targets. The same set of injection-production parameters cannot simultaneously satisfy the optimal $CO_2$ flooding effect and storage effect. After ten years of gas injection development, the goals of $CO_2$ flooding and storage were unified. The optimal combination of injection-production and fracturing parameters at an injection rate of 6000m$^3$/d is shown in Table 6, with a recovery rate of 72.54% and a $CO_2$ storage of 101635.8125 tons.

Table 3. Optimal scheme parameters of each target after one year of development.

| Target | Focus on oil displacement | Focus on storage | Equal emphasis on both |
|---|---|---|---|
| Best scheme serial number | 79 | 88 | 70 |
| $CO_2$ injection rate, $m^3/d$ | 6000 | 6000 | 6000 |
| BHP, MPa | 7 | 11 | 9.6 |
| Fracture width, m | 0.03 | 0.01 | 0.03 |
| Fracture closure time, d | 1788 | 1086 | 2157 |
| Shut-in oil/gas ratio, $m^3/m^3$ | 1322 | 812 | 1349 |
| Fracture height, m | 6 | 2 | 7 |
| Fracture permeability, mD | 2746 | 4655 | 1313 |
| Fracture half-length, m | 59 | 100 | 94 |
| $CO_2$ Storage Capacity, kg | 17025041.94 | 17110694.56 | 17064857.09 |
| Oil Recovery Factor, % | 15.53 | 8.93 | 13.72 |

Table 4. Optimal scheme parameters of each target after five years of development.

| Target | Focus on oil displacement | Focus on storage | Equal emphasis on both |
|---|---|---|---|
| Best scheme serial number | 95 | 90 | 65 |
| $CO_2$ injection rate, $m^3/d$ | 6000 | 6000 | 6000 |
| BHP, MPa | 84 | 75 | 88 |
| Fracture width, m | 0.02 | 0.03 | 0.03 |
| Fracture closure time, d | 3209 | 547 | 1322 |
| Shut-in oil/gas ratio, $m^3/m^3$ | 1155 | 593 | 1056 |
| Fracture height, m | 4 | 4 | 3 |
| Fracture permeability, mD | 1426 | 1962 | 1815 |
| Fracture half-length, m | 51 | 75 | 61 |
| $CO_2$ Storage Capacity, kg | 68246750.77 | 75543358.88 | 71832108.49 |
| Oil Recovery Factor, % | 51.29 | 43.65 | 49.85 |

Table 5. Optimal scheme parameters of each target after seven years of development.

| Target | Focus on oil displacement | Focus on storage | Equal emphasis on both |
|---|---|---|---|
| Best scheme serial number | 40 | 54 | 54 |
| $CO_2$ injection rate, $m^3/d$ | 6000 | 6000 | 6000 |
| BHP, MPa | 117 | 97 | 97 |
| Fracture width, m | 0.02 | 0.028 | 0.028 |
| Fracture closure time, d | 1898 | 1105 | 1105 |
| Shut-in oil/gas ratio, $m^3/m^3$ | 883 | 821 | 821 |
| Fracture height, m | 2 | 2 | 2 |
| Fracture permeability, mD | 513 | 4351 | 4351 |
| Fracture half-length, m | 90 | 104 | 104 |
| $CO_2$ Storage Capacity, kg | 84078497.15 | 86889396.83 | 86889396.83 |
| Oil Recovery Factor, % | 60.21 | 59.64 | 59.64 |

Table 6. Optimal scheme parameters of each target after ten years of development.

| Target | Focus on oil displacement | Focus on storage | Equal emphasis on both |
|---|---|---|---|
| Best scheme serial number | 65 | 65 | 65 |
| $CO_2$ injection rate, $m^3/d$ | 6000 | 6000 | 6000 |
| BHP, MPa | 88 | 88 | 88 |
| Fracture width, m | 0.03 | 0.03 | 0.03 |
| Fracture closure time, d | 1322 | 1322 | 1322 |
| Shut-in oil/gas ratio, $m^3/m^3$ | 1056 | 1056 | 1056 |
| Fracture height, m | 3 | 3 | 3 |
| Fracture permeability, mD | 1815 | 1815 | 1815 |
| Fracture half-length, m | 61 | 61 | 61 |
| $CO_2$ Storage Capacity, kg | 101635812.5 | 101635812.5 | 101635812.5 |
| Oil Recovery Factor, % | 72.54 | 72.54 | 72.54 |

## 5. Conclusions

(1) The development of $CO_2$ injection in low permeability reservoir can greatly enhance oil recovery and play the role of $CO_2$ storage at the same time.
(2) The injection rate has the greatest impact on the development of low-permeability reservoirs when injecting $CO_2$. When the injection rate reaches a certain level, gas channeling will occur and the effect of oil recovery will be greatly reduced. Therefore, there exists an optimal injection rate under certain operating conditions, and the closer the injection rate is to this optimal value, the better the development effect will be.
(3) In contrast, fracture parameters have a smaller impact compared to injection rate, wellbore flowing pressure and gas-oil ratio; when exploring its impact, it is better to avoid orthogonal or sampling with parameters that have greater influence, and classify different impact parameters before orthogonal analysis to reveal its impact law.
(4) The influence of injection and fracture parameters varies during different injection stages. In the early stage of injection, the lower the wellbore flowing pressure, the more favorable for the rapid increase of production; in the middle of the injection, the higher the wellbore flowing pressure, the more favorable for the $CO_2$ and reservoir fluid to form a mixed or near-mixed phase, which is conducive to improving the degree of recovery and $CO_2$ storage.
(5) There are some contradictions between the two objectives of oil displacement and $CO_2$ storage in different gas injection development periods. When the development objectives change, the injection-production parameters should be adjusted in time. In this study, utilizing a five-spot well network model of the selected research area in a ten-year gas injection process, the oil recovery and $CO_2$ sequestration objectives were achieved simultaneously with the optimal combination of injection-production and fracture parameters, resulting in a recovery rate of 72.54% and $CO_2$ sequestration of 101635.8125 metric tons.

## References

[1] Guo Ping, Li Shilun, Du Zhimin, et al. Evaluation of enhanced oil recovery by gas injection in low permeability reservoirs [J]. Journal of Southwest Petroleum Institute, 2002 (5): 46-50 +2.
[2] Han Dakuang. Discussion on EOR in deep development of high water cut oilfield [J]. Petroleum Exploration & Development, 1995 (5): 47-55 +98.
[3] Necmettin M. High Pressure Nitrogen Injection for Miscible/Immiscible Enhanced Oil Recovery[C]//SPE Latin American and Caribbean Petroleum Engineering Conference. OnePetro, 2003.
[4] Brock W R, Bryan L A. Summary Results of $CO_2$ EOR Field Tests, 1972-1987[C]//Low Permeability Reservoirs Symposium. OnePetro, 1989.

[5] Ghomian Y, Urun M, Pope G A, et al. Investigation of Economic Incentives for $CO_2$ Sequestration[C]//SPE Annual Technical Conference and Exhibition. OnePetro, 2008.
[6] Li Shilun, Sun Lei, Chen Zuhua, Li Jian, Tang Yong, Pan Yi. Further discussion on reservoir engineering concept and development mode of $CO_2$ flooding-EOR technology[J]. Reservoir Evaluation and Development, 2020, 10(3): 1-14.
[7] Liao Changlin, Liao Xinwei, Zhao Xiaoliang, et al. Technology of expanding swept volume in low permeability heterogeneous reservoir [J]. Petroleum Geology & Oilfield Development in Daqing, 2013, 32(5): 118-122.
[8] Li Shilun, Tang Yong, Hou Chengxi. Present situation and development trend of $CO_2$ injection enhanced oil recovery technology [J]. Reservoir Evaluation and Development, 2019, 9(3): 1-8.
[9] Gozalpour F, Ren S R, Tohidi B. $CO_2$ Eor and Storage in Oil Reservoir[J]. Oil & Gas Science and Technology, 2005, 60(3): 537-546.
[10] Ettehadtavakkol A, Lake L W, Bryant S L. $CO_2$-EOR and storage design optimization[J]. International Journal of Greenhouse Gas Control, 2014, 25: 79-92.
[11] Shen Pingping, Liao Xinwei. Carbon dioxide geological storage and enhanced oil recovery technology [M]. Carbon dioxide geological storage and enhanced oil recovery technology, 2009.
[12] He Houfeng, Hu Xuhui, Zhuang Yongtao, et al. Optimization of injection-production parameters of $CO_2$ flooding in low permeability reservoir and its application-Taking Block A of Shengli Oilfield as an example [J]. Petroleum Geology and Recovery Efficiency: 1-10.
[13] Ju Binshan, Yu Jinbiao, Lu Guangzhong, et al.Numerical simulation method and application of $CO_2$ displacement in low permeability reservoir [J]. Petroleum Geology and Recovery Efficiency, 2020, 27 (1): 126-133.
[14] Liao Guangzhi, He Dongbo, Wang Gaofeng, et al. Discussion on limit recovery ratio of carbon dioxide flooding under ultimate burial scenario [J]. Petro-level Exploration & Development, 2022, 49 (6): 1262-1268.
[15] Fan Lijun, Zhao Dongya, Liu Jiajia, et al. Joint optimization design and sensitivity analysis of oil displacement and storage in $CO_2$ well pattern [J]. Pe-trophy Engineering Construction, 2017, 43 (4): 7-12.
[16] Helton J C, Davis F J. Latin hypercube sampling and the propagation of uncertainty in analyses of complex systems[J]. Reliability Engineering & System Safety, 2003, 81(1): 23-69.
[17] Zhang F, Cheng L, Wu M, et al. Performance analysis of two-stage thermoelectric generator model based on Latin hypercube sampling[J]. Energy Conversion and Management, 2020, 221: 113159.
[18] Benesty J, Chen J, Huang Y, et al. Pearson correlation coefficient Noise Reduction in Speech Processing. 2. 1st[M]. Springer.
[19] Kovscek A R, Cakici M D. Geologic storage of carbon dioxide and enhanced oil recovery. II. Cooptimization of storage and recovery[J]. Energy Conversion and Management, 2005, 46(11): 1941-1956.
[20] Ma Xiaoshu, Li Yulong, Yan Lang. Comparison of traditional multi-objective optimization methods and multi-objective genetic algorithms [J]. Electric Drive Automation, 2010, 32 (3): 48-50 +53.
[21] Deb K. Multi-objective Optimisation Using Evolutionary Algorithms: An Introduction[J]. Multi-objective Evolutionary Optimisation for Product Design and Manufacturing, 2011: 3-34.

# Study on scaling mechanism of injected water in Yujiaping Oilfield

**Gang Liu1, Shuo Li1, Cheng Cheng1**

Water Injection Project Management Headquarters, Yan chang Oilfield Co., LTD., Yan 'an, 716000
547821232@qq.com

**Abstract.** Oilfield injected water scaling often has a great impact on production and development. In this study, the scaling mechanism of injection well was studied, and the scale samples were analyzed by X-ray diffraction and other methods to determine the elements and mineral composition of scale samples and finally determine the main components of scale samples, providing a basis for subsequent water quality treatment. The results show that scale formation in Yu jia ping oil area is mainly caused by corrosion, inorganic salt scale caused by chemical reaction and sediment carried by injected water. Among them, inorganic salt scale is the main scaling component, and its main reason is the changes of ions in the injected water under different chemical environments. After a series of chemical and physical reactions, some ions in the injected water will form insoluble, insoluble and slightly soluble inorganic salts. After long-term accumulation, it is easy to form scale, namely the so-called inorganic salt scale. Inorganic salt scale is generally composed of sulfate and carbonate, the more typical calcium carbonate, calcium sulfate, barium sulfate, etc.

## 1 Introduction

Scaling is an important problem in the process of oilfield exploitation. In recent years, although a lot of research work has been carried out in the field of sewage water quality treatment, there are still some outstanding problems, especially with the change of the quality and quantity of produced water and the aging of water treatment equipment, leading to some problems more prominent, such as substandard water quality, scale corrosion of water injection system, seriously affect the normal production and development of the oilfield. Through the investigation and study of oilfield water injection system scaling, it is found that injected water scaling can cause three consequences. Firstly, the impurities and corrosion products in the injected water can block the bottom of the injection well, leading to the decrease of water injection. Secondly, the precipitation of carbonate in the injected water causes scale formation in the injection pipe network and reduces the injection efficiency[1]. Third, the incompatibility of injected water and formation water clogs the near-wellbore area, resulting in increased injection pressure.

## 2 Basic theory of scaling

There are many types of scale, but usually oil field water contains only a few of them. The most common type of scale is carbonate scale, composed of $CaCO_3$, $MgCO_3$, $SrCO_3$, etc., which is easy to be removed by acidification and has relatively little harm. Sulfate scale, which is composed of $CaSO_4$, $BaSO_4$ and $SrSO_4$, is difficult to be removed by general methods, so it has great harm. Ferric scale is composed of $FeCO_3$, FeS, $Fe(OH)_2$ and $Fe_2O_3$. In fact, the general scale is not a single composition, often is mixed scale, but is mainly a certain scale.

People often say that scale refers to the slightly soluble or insoluble inorganic salts in water, they have a fixed lattice[2-5], and the simple scale compact, hard (such as $CaCO_3$, $BaSO_4$). The formation of scale mainly depends on whether the salt in the water is supersaturated or not and the growth process of salt crystals. At present, relatively mature scaling theories mainly include the following 3 kinds[6-9].

### *2.1 Incompatibility theory*

Mixing two chemically incompatible liquids (formation water with incompatible ions in different layers, formation water with surface water, and clear water with sewage) will result in unstable, precipitation-prone liquids because they contain different ions or different concentrations of ions. For example, when the formation water of the water type $NaHC0_3$ is mixed with the formation water of the water type $CaCl_2$, $CaCO_3$ scaling is likely to occur.

### *2.2 Theory of change of thermodynamic conditions*

A supersaturated solution will remain stable even with incompatible ions when downhole thermodynamic and dynamic conditions are constant. However, in the process of oil and well production, when the pressure drops, the temperature rises or the flow rate changes, the highly mineralized water is prone to scaling, and the opposite is true for calcium salts.

### *2.3 Adsorption theory*

Scaling is divided into three stages: precipitation, growth and deposition. Scale is a crystal structure. The surface of pipeline equipment is uneven, and it is a microscopic rough surface. Scale ions will be adsorbed on the wall surface, taking it as the crystallization center, and grow continuously to become dense and solid scale. In oilfield water injection system, the formation process of scale is often a mixed crystallization process, suspended particles in water can become crystal seeds, rough surface or other impurity particles can strongly catalyze the crystallization process, so that the solution at a low saturation will precipitate crystallization, the formation of scale.

## 3 Main influencing factors of scale formation

Factors that affect the formation of scale can be divided into internal and external causes. Internal causes are mainly scale forming ions, which are the material basis of scale formation. External factors are the conditions of scale formation, mainly including water salinity, temperature, system pressure, pH value, etc[10-14].

### *3.1 Temperature*

Most solid substances absorb heat when dissolved in water, according to the principle of equilibrium movement, when the temperature rises, the balance is conducive to moving in the direction of heat absorption, so the solubility of these substances increases with the increase of temperature, but calcium carbonate dissolution exothermic phenomenon, so its solubility decreases with the increase of temperature, which is the reason why our daily boiling water is prone to scale.

*3.2 pH value*

A large number of studies have shown that increasing the pH value of water is easy to produce carbonate crystallization. On the other hand, too low pH value will increase the corrosion of equipment and cause corrosion scaling. Reducing pH value increases the solubility of carbonate and weakens the trend of scale formation. In inorganic scale, this effect has a very obvious effect on calcium carbonate scale, but little effect on sulfate scale.

Reducing pH value will increase the solubility of carbonate and iron compounds, but will make the water more corrosive and corrosion problems. pH has little effect on the solubility of sulfate scale. This is not widely used for scale control. It only makes practical sense if a small change in pH can prevent scaling. Precisely control pH, which is often difficult in field production.

*3.3 Salinity*

The salinity of the water body, the concentration of inorganic salts, has a large effect on the solubility of carbonates and sulfates. The main influence is reflected in the salt effect. When the concentration of non-scaling ions increases within a certain range, the solubility of scaling inorganic salts will also increase; At the same time, it is also reflected in the co-ionic effect, that is, when the ions in a kind of scaling salt exist in other forms in the solution, it is easier to be extracted and scaled. Therefore, the decrease in salinity of the layer water after mixing with the injected fresh water may also cause scale formation.

**4 Analysis of scale samples in Yujiaping oil area**

The scale samples collected in Yujiaping oil area include the samples in the tubing of the 8507 wellbore and the samples in the water storage tank of the water injection station.

*4.1 X-ray Fluorescence spectrum Analysis (XRF)*

XRF is an instrument that can be used for rapid simultaneous determination of multiple elements. The XRF uses physical principles to detect the elements of a substance and can be used for qualitative and quantitative analysis[15-18]. That is, X-ray penetrates the internal electrons of atoms, and the outer electrons supplement to produce characteristic X-rays. According to the intensity of the characteristic X-rays of elements, the content information of each element can be obtained. This is the basic principle of X-ray fluorescence analysis. It can only measure elements, not compounds. But because XRF is a surface chemical analysis, the sample must meet many conditions to be accurate, such as smooth surface and uniform composition. The XRF analysis results of samples are shown in Table 1.

Table 1. XRF analysis results of scale sample and corrosion sample

| Elements | Scale sample inside tubing | Clean water storage tank scale sample |
|---|---|---|
| **Fe** | 70.9% | 96.7% |
| **Ca** | 20.2% | 0.952% |
| **Ba** | 2.51% | 1.09% |
| **Sr** | 1.82%. | 0.354% |
| **S** | 2.29% | 0.285% |

Note: Percentage content is for reference only (only indicates that scale sample contains these elements)

*4.2 X-ray diffractometer analysis*
The technique known as X-ray diffraction (XRD). A means of research used to obtain information about the composition of a material and the structure or morphology of atoms or molecules within the material by X-ray diffraction and analysis of its diffraction pattern. X-ray diffraction analysis is the main method to study the phase and crystal structure of materials. When a substance (crystal or non-crystal) is subjected to diffraction analysis, the substance is irradiated by X-ray to produce diffraction phenomena of different degrees. The substance composition, crystal type, intramolecular bonding mode, molecular configuration, conformation and so on determine the specific diffraction pattern of the substance. X-ray diffraction method has the advantages of no damage to the sample, no pollution, fast, high measurement accuracy, and can obtain a lot of information about the integrity of the crystal. Therefore, X-ray diffraction analysis as a modern scientific method of material structure and composition analysis, has gradually been widely[19-21] used in various disciplines of research and production.

Figure 1 shows the XRD analysis results of the samples in the tubing of the 8507 wellbore in Yujiaping oil area.

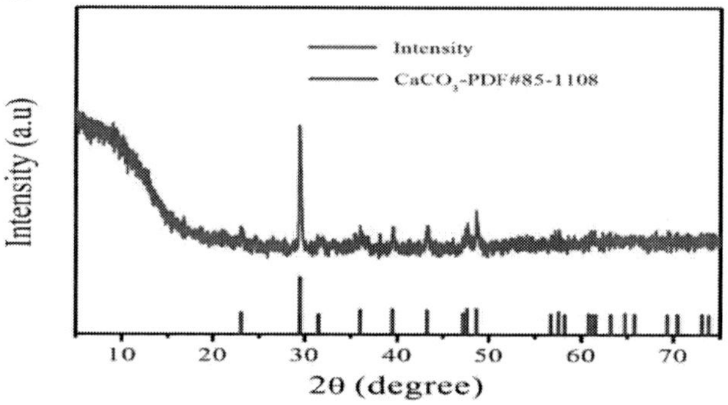


FIG. 1 XRD analysis results of scale samples in tubing

The samples in the tubing of wellbore 8507 in Yujiaping oil area are soluble in hydrochloric acid, and the results of acid dissolution test and scale prediction are consistent with the results of large-scale instrument analysis, that is, calcium carbonate is the main scale, followed by barium sulfate scale.

The XRD analysis results of the samples in the clean water storage tank of the water injection station are shown in Figure 2

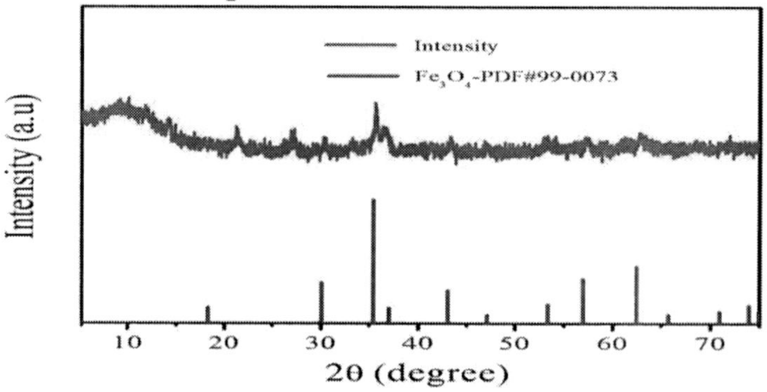


FIG. 2 XRD analysis results of scale samples in clear water storage tank of water injection station

The samples in the clear water storage tank of the water injection station were analyzed by XRD and EDX (FIG. 3). The samples were mainly rust (magnetite), calcium carbonate and amorphous clay deposits.

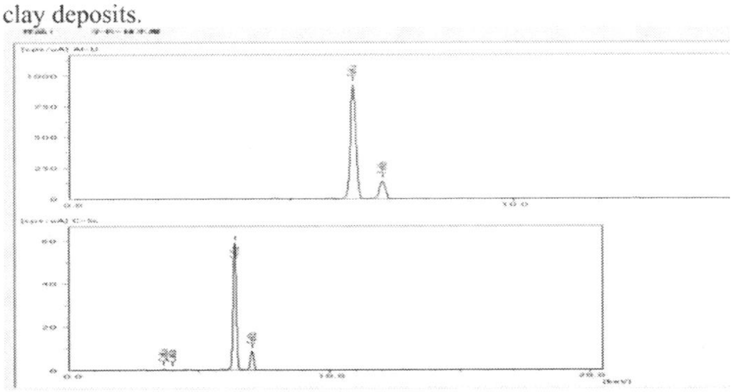

FIG. 3 EDX analysis results of scale samples in clear water storage tank of water injection station

**5 Analysis of scaling factors in Yujiaping oil area**

The results of sample analysis of scale formation in Yujiaping oil area show that calcium carbonate scale is the main type of scale formation, followed by barium sulfate scale, which is closely related to the composition of produced and injected water. The produced water in the oil area contains a large amount of scale forming $Ca^{2+}$, $Mg^{2+}$, $Sr^{2+}$ and $Ba^{2+}$. The content of anions such as $HCO_3{}_3^-$ and SO $_4^{2+}$is relatively small. The clear water also contains a small amount of HCO, CO $_3^{2-}$and $SO_4^{2-}$.

According to the solubility product principle, when the activity product of cationic and anion of an insoluble salt in water is greater than its solubility product, the water body is in a thermodynamic unstable state, and insoluble salt precipitates, that is, scale is produced. Specifically, it is the volume of $Ca^{2+}$ and $CO_3^{2-}$ activity product in water (how much ion content) that determines whether scale is formed. The reaction equation is as follows:

$$Ca^{2+} + CO_3^{2-} \rightarrow CaCO_3\downarrow;$$

$$Ca^{2+} + 2HCO_3^- \rightarrow CaCO_3\downarrow + H_2O + CO_2\uparrow.$$

When the water: $[Ca^{2+}] \times [CO_3^{2-}] > K_{sp\ CaCO3}$there will be a trend of calcium carbonate scaling;

Similarly, when: $[Ba^{2+}] \times [SO_4^{2-}] > K_{sp\ BaSO4}$ in water, there will be a tendency to produce barium sulfate scale, and the reaction equation is:

$$Ba^{2+} + SO_4^{2-} \rightarrow BaSO_4\downarrow$$

The reason why only calcium carbonate scale appears in carbonate scale, but no magnesium carbonate, strontium carbonate or barium carbonate is that the solubility product of calcium carbonate is equivalent to that of strontium carbonate and barium carbonate, while the solubility product of magnesium carbonate is larger. When $[CO_3^{2-}]$ is constant, $[Ca^{2+}]$ is much higher than the concentration (or activity) of $Mg^{2+}$, $Sr^{2+}$ and $Ba^{2+}$ in water, that is, calcium carbonate scale is preferentially formed. After the formation of calcium carbonate scale, $[CO_3^{2-}]$ has been consumed, and $[Ca^{2+}]$ in the water is much higher than the concentration of $Mg^{2+}$, $Sr^{2+}$ and $Ba^{2+}$, so the scale in Yujiaping oil area is dominated by calcium carbonate.

Similarly, in sulfate scale, the solubility product of calcium sulfate is larger, that of strontium sulfate is smaller, and that of barium sulfate is the smallest. When the concentration of sulfate is certain and the concentrations of $Sr^{2+}$ and $Ba^{2+}$ are similar, barium sulfate scale is preferentially formed according to the solubility product principle.

## 6 Conclusion

In this paper, the basic theory of scale formation in Yujiaping oil area is studied. It is believed that there are three main theories of scale formation at present: incompatibility theory, thermodynamic condition change theory and adsorption theory. Secondly, the influencing factors are analyzed: water salinity, temperature, system pressure, pH value and so on. Finally, X-ray fluorescence spectroscopy and X-ray diffractometer are used to study the elements and mineral composition of Yujiaping scale sample. The results show that: calcium carbonate scale is the main component, followed by barium sulfate scale, and the solubility product principle is used to explain.

## References

[1] WANG Jun, Ming Yunfeng. Research on anticorrosion and scale prevention technology of water injection system in Qianda Oilfield [J]. Industrial Water Treatment.2007,27(006):87-89
[2] Waard C De, Lots U Prediction of CO2 Corrosion of Carbon Steel [R]. A Working Party Report on Predicting CO2 Corrosion in Oil and Gas Industry, 1994:30~58.
[3] Warren P. Iverson. Research on the mechanisms of anaerobic corrosion [J]. International Biodeterioration and [3] Biodegradation 47, 2001:63 ~ 70.
[4] Iverson, W.P. Possible source of a phosphorus compound produced by sulfate-reducing bacteria that cause anaerobic corrosion of iron.Materials Protection, 1998:46 -- 49.
[5] Iverson, W.P.. Microbial corrosion of metals. Advances in Applied Microbiology 32, 1987:1 -- 36.]
[6] M B Kermani and A Morshed. Carbon dioxide corrosion in oil and gas production-A compendium[J]. Corrosion,2003,59(8):659 ~ 683.
[7] Khan M, Budair M O, Zubair S M. A parametric study of CaCO3scal-ing in AISI 316 stainless steel tubes [J]. Heat and Mass Transfer,2001,38 (1) : 115-121.
[9] Andritsos N, Kontopoulou M, Karabelas A, et al. Calcium carbonatedeposit formation under isothermal conditions [J]. The Canadian Journal of Chemical Engineering, 2009,74 (6) : 911-919.
[10] Shen J, Crosby C C. Insight into strontium and calcium sulfate scaling mechanisms in a wet producer [J]. Journal of Petroleum Technology, 1983,35 (7) : 1249-1255.
[11] Chen T, Neville A, Yuan M. Calcium carbonate scale formation -assessing the initial stages of precipitation and deposition [J].Jour -nal of Petroleum Science and Engineering, 2005,46 (3) : 185-194.
[12] Hasson D, Bramson D, Limoni-Relis B, et al. Influence of the flowsystem on the inhibitory action of CaCO3scale prevention additives[J]. Desalination, 1997,108 (1/2/3) : 67-79.
[13] Bian Chaofeng, Zhu Qijia, Chen Wu, et al. Chemical compatibility of injection water source and reservoir in oil field Research [J]. Chemical and Biological Engineering, 2006,23 (7) : 48-50.
[14] Li Lian-jiang, ZENG Qing-hui. Study on compatibility of injected water and reservoir in Shanghe Oilfield [J].Journal of Jianghan Petroleum Institute, 2003,25 (1) : 77-78.
[15] Li Jing, Sun Dongbai, Yang Dejun, Shenglin, Lu Minxu, Zhao Guoxian, et al. Corrosion Behavior of Tubing Steel under High Temperature and High Pressure CO2 [J]. Iron and Steel, 2001,36(6):48 ~ 51
[16] LI Jianping, Zhao Guoxian, HAO Shiming, Lu Minxu et al. CO2 Local Corrosion Rate of Common Tubing Steel [J]. Journal of Northeastern University (Natural Science Edition), 2004,25(11):1069 ~ 1071
[17] WANG Sheng-si. Review on detection techniques of sulfate Reducing Bacteria [J]. Chemical Industry of Oil & Gas, 1998(03): 192-193
[18] YU Dunyi, PENG Fangming, LV Zhanpeng et al. Analysis of casing corrosion in Well Wen 10-1, Zhongyuan Oilfield [J]. Chinese Journal of Corrosion and Protection,

1996,16(1):65 ~ 68
[19] Zhang Yunqiang, Liu Suping, Tao Lijiang et al. Comprehensive treatment of corrosion problem in Block Hu 19 of Zhongyuan Oilfield [J]. Corrosion and Protection in Petrochemical Industry, 2004,21(6):52 ~ 55
[20] NING Tingwei. Application and development of injected water fungicide in Shengli Oilfield [J]. Oilfield Chemistry, 1998,15(3):285 ~ 288
[21] Li Yuling. Corrosion and anticorrosion of water injection well. [21] Oil drilling and Production Technology [J]. 1982,5:69 ~ 72

Journal of Physics: Conference Series **2594** (2023) 012057 doi:10.1088/1742-6596/2594/1/012057

# Energy-load Matching and Shockwave Analysis of Electrical Explosion

**Zhong-xing HAN[1], Ben LIU[2], Yan-bao GUO[1(✉)], De-guo WANG[1(✉)], Tao LIN[1]**

[1] College of Mechanical and Transportation Engineering, China University of Petroleum (Beijing), Beijing, China

[2] Sinopec Research Institute of Petroleum Engineering, Beijing, China

gyb@cup.edu.cn; wdg@cup.edu.cn

**Abstract.** Electrical explosion shockwave is a technique for inducing formation resonance by the interruption of formation pressure balance. By matching complex nonlinear influencing factors, this paper optimizes the engineering parameters and estimates the influence range of the shockwave. It achieves the objective of efficient shockwave acquisition and amplitude control. The test apparatus rapidly discharges the stored energy of various capacitance-voltage schemes, causing the load of various diameter-length structures to generate shaped shockwaves. By characterizing the discharge circuit and the shockwave signal, the energy and load matching was studied. Extraction of shockwave signal characteristics for decomposition and reconstruction analysis. The matching results indicate that the 10 MW loop oscillation is generated within 5 μs of discharge time, with the load diameter structure getting a major effect on the deposited energy. The 50 MPa shaped shockwave is generated in a 100 mm diameter region. And the peak pressure is influenced primarily by the load length structure and voltage scheme. The optimal load structure is $D = 0.2$ mm and $L = 25$ mm, controlling 2 to 5 times the input power's enthalpy and increasing the voltage to create the shockwave more efficiently. The results of shockwave analysis indicate that the signal's time-domain characteristics correspond precisely with its propagation path. The wavelet packet analysis indicates that the shockwave possesses a broad frequency and a high peak. The detection signal is composed of the principal shock wave, the radiation pressure wave, and the resonance error. The results provide theoretical basis for engineering application.

## 1. Introduction

Wire electric explosion technology is a unique auxiliary technology to stimulate oil and gas well stimulation. Compared with conventional acidizing and hydraulic fracturing technology, electric explosion technology has the advantages of simple construction process, short construction period, and environmental protection. Since the 1960s, there has been much research on the basic theory and principle prototype [1-4].

The resistivity of Neutral Copper Gas is measured by the dark pause effect in a wire explosion [5]. Meanwhile, the weakening effects of water resistivity and phase state on underwater electric explosion are also studied [6,7]. The peak overpressure of the shock wave is discussed by conducting wire explosion under the different spatial arrangements and ambient pressure [8-10]. The magnitude and efficiency of the deposited energy are the fundamental reasons for experimental parameters' effects on the shockwave's peak overpressure in the process of wire electric explosion [11-14]. The preceding research focuses on a comprehensive analysis of the connection between discharge circuit properties and shock wave pressure.

In this paper, the dynamic effects of wire structure and energy storage on the performance of an electric explosion are analyzed further. In engineering practice, suitable wire structure and energy storage are selected. The wavelet packet is used to decompose and reconstruct the shock wave signal. This paper presents a method for measuring the shock wave of an electric explosion precisely.

## 2. Experimental Research

### *2.1 Test rig*

The underwater wire electric explosion test rig is shown in Figure 1. The main body of the test rig includes: Pulse power generator (HV DC power supply, Marx generator and trigger switch). Electrical pulse-shockwave energy conversion device (explosion bracket, electrode, consumable metal wire, cable and test tank). Signal acquisition device (Rogowski coil, high voltage divider, shockwave sensor, control module and oscilloscope).

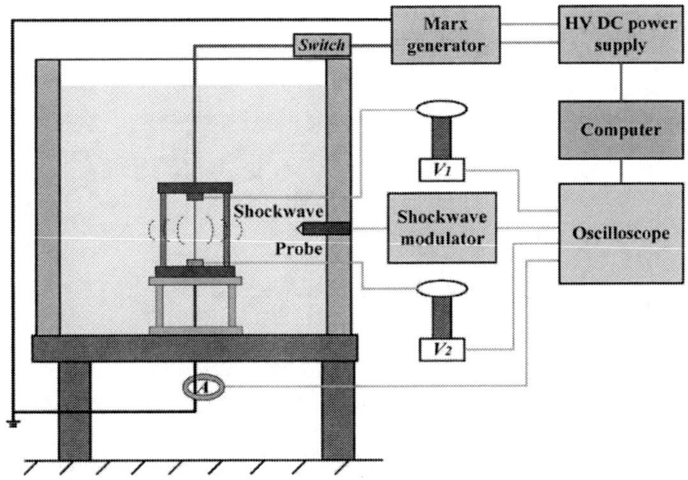


**Figure 1.** Schematic diagram of electric explosion test rig

The high-voltage DC power supply charges the Marx generator to store energy. The trigger switch transmits the pulse current to the energy conversion device. The pulse current drives the consumable Cu wire to make a rapid phase-change electrical explosion. Moreover, the electric explosion shockwave is diffused and propagated in the water.

### *2.2 Parameter design for the load structure and energy storage scheme*

Pure copper is used as the load metal wire in the development of overheated electric explosions. Under the influence of pulse current, the load rapidly goes through the phase transition from a solid state to a plasma state. The foundation of research into the compatibility of electric explosion load structure and energy scheme is the electric explosion-specific energy determined by the load's thermodynamic parameters. The thermodynamic parameters of copper listed in Table 1.

**Table 1** Thermodynamic parameters of the load [15]

| Material | $\rho$ / g·cm$^{-3}$ | $H_{melt}$ / kJ·mol$^{-1}$ | $H_{vaporize}$ / kJ·mol$^{-1}$ | $H_{total}$ / kJ·mol$^{-1}$ |
|---|---|---|---|---|
| Cu | 8.93 | 13.14 | 300.68 | 392.26 |

The load copper wire structure investigated in the experiment consists of five length and three diameter configurations. The influence of energy scheme should be excluded when studying the load structure. The amplification of the system's stored energy $E_{stored}$ to the wire's internal energy $E_{atom}$ represented as $K$. The wire's internal energy $E_{atom}$ approximate to thermodynamic parameter $H_{total}$. The driving energy of each group of loads under the structure matrix should be identical at a constant value $K$ of 3.5. The capacitance of the discharge test energy scheme is fixed at 30 μF to execute an overheated electric explosion. And the corresponding test voltage is converted according to Table 2.

**Table 2** Test parameters of load structure

| | $U$ / V | | | | |
|---|---|---|---|---|---|
| $(D\&L)$ / mm | $L_1 = 15$ | $L_2 = 20$ | $L_3 = 25$ | $L_4 = 30$ | $L_5 = 35$ |
| $D_a = 0.1$ | 1225 | 1415 | 1581 | 1733 | 1871 |
| $D_b = 0.2$ | 2450 | 2830 | 3163 | 3466 | 3743 |
| $D_c = 0.3$ | 3675 | 4245 | 4743 | 5199 | 5613 |

Voltage and capacitance two-variable schemes are investigated as energy storage in the experiment. The research should exclude the influence of load structure. Ensure that the load structure within the energy matrix is identical, with $D = 0.2$ mm and $L = 25$ mm. To compare the impact of the voltage and capacitance schemes, the amplification of driving energy goes from 2.01 to 4.69. $K_v$ and $K_c$ represent the amplification coefficients based on voltage and capacitance, respectively. The sequence test voltage and capacitance scheme are set according to Table 3 and Table 4.

**Table 3** Test parameters of voltage scheme

| $C$ / μF | $U$ / V | | | | |
|---|---|---|---|---|---|
| | $K_{v1} = 2.01$ | $K_{v2} = 2.55$ | $K_{v3} = 3.15$ | $K_{v4} = 3.81$ | $K_{v5} = 4.53$ |
| 30 | 2400 | 2700 | 3000 | 3300 | 3600 |
| 60 | 1700 | 1910 | 2122 | 2334 | 2546 |

**Table 4** Test parameters of capacitance scheme

| $U$ / V | $C$ / μF | | | | |
|---|---|---|---|---|---|
| | $K_{c1} = 2.01$ | $K_{c2} = 2.68$ | $K_{c3} = 3.35$ | $K_{c4} = 4.02$ | $K_{c5} = 4.69$ |
| 2400 | 30 | 40 | 50 | 60 | 70 |
| 3394 | 15 | 20 | 25 | 30 | 35 |

## 3. Results and Discussion

### *3.1 Influence of load structure on loop characteristics and shock wave peak*

Figure 2 illustrates the discharge circuit characteristics of a load electric explosion with different lengths and diameters, as determined by the experimental parameters in Table 2. The results indicate that the diameter plays a more important role on the discharge characteristics than its length. Using diameter as the main variable and length as the additional parameter, the discharge circuit's voltage and current curves are categorized into three groups. In the discharge loop, the oscillating ranges of voltage and current are represented by the yellow and green bands, respectively. And the black, red, and blue curves represent wire structures with three different diameters.

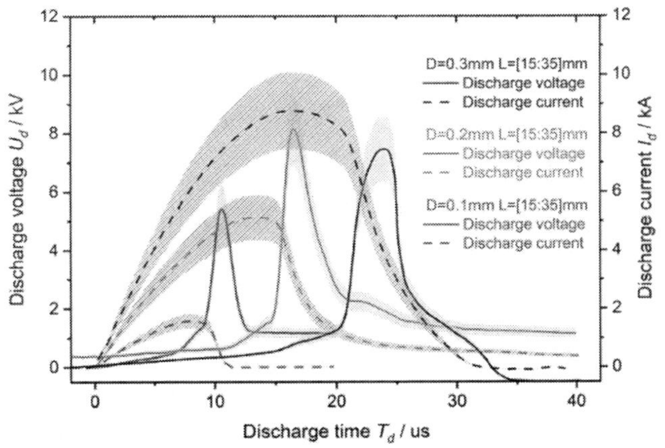


**Figure 2.** Circuit discharge characteristic curves of wires with different lengths and diameters

The overall analysis of the data reveals that the pulse voltage curve in the discharge loop is relatively steep, whereas the pulse current curve is relatively smooth. In general, the rising voltage curve lags the corresponding current curve. And the falling current curve crosses the rising voltage curve. The pulse current drives the loaded wire to store energy and deform. And the voltage rises significantly as its resistance increases. The wire is excited to generate an electric explosion when a certain amount of energy is accumulated. The wire transforms from the condensed state to the plasma state. The voltage curve falls behind the current curve during a dynamic shift in resistance value. At the crossing region of the curves, high-powered energy pulses are injected into the wires to generate shock waves.

The three data groups are then compared and analyzed in series. With the increase in wire diameter, the amplitudes of the discharge loop's pulse current and voltage are significantly amplified. However, the crossing region's duration remains approximately unaltered. Consequently, more energy is released when a wire with a larger diameter explodes electrically. Based on the characteristics of the circuit discharge shown in Figure 2, the ratio of the total deposited energy $E_{deposition}$ to the wire's internal energy $E_{atom}$ was calculated. The variation curves of peak pressure and specific deposited energy during an electric explosion with wires of various lengths and diameters are illustrated in Figure 3.

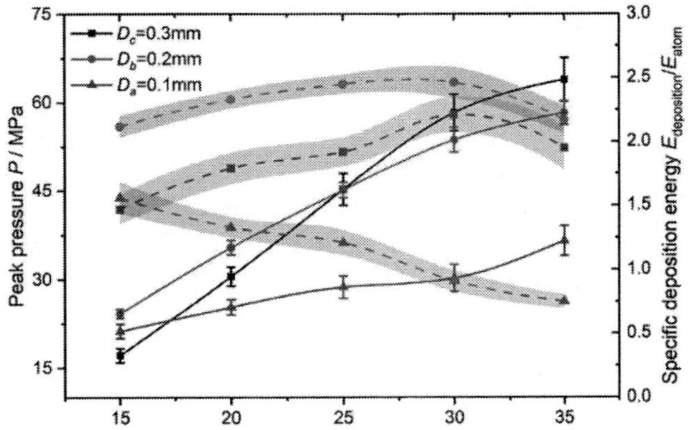


**Figure 3.** The peak pressure and specific deposited energy with various lengths and diameters

Figure 3 demonstrates that the peak pressure of a shock wave is reduced when the wire is too short or tiny. It might end in inefficient operations for downhole stimulation. The energy required rises and the conversion efficiency falls even though increasing the length and diameter of the wire can increase the peak pressure. It might discourage the commercialization and miniaturization of tools. The relatively long and thick wire should be chosen when the maximum shock wave peak is the design objective. But the late improvement effect is insignificant. The energy efficiency of selecting the optimal length-to-diameter ratio wire is increased when the target peak pressure of the working shock wave is determined. The results indicate that $D$ = 0.2 mm and $L$ = 25 mm is the best engineering-optimal group of wire structure parameters.

*3.2 Influence of energy storage scheme on shock wave peak value*

Under the optimal structural parameters, the wire electric explosion is driven by 2 ~ 4.7 times the specific energy in Table 3 and Table 4. Figure 4 shows the comparison curve of peak pressure after electric explosion for energy storage schemes with different capacitors and charging voltages.

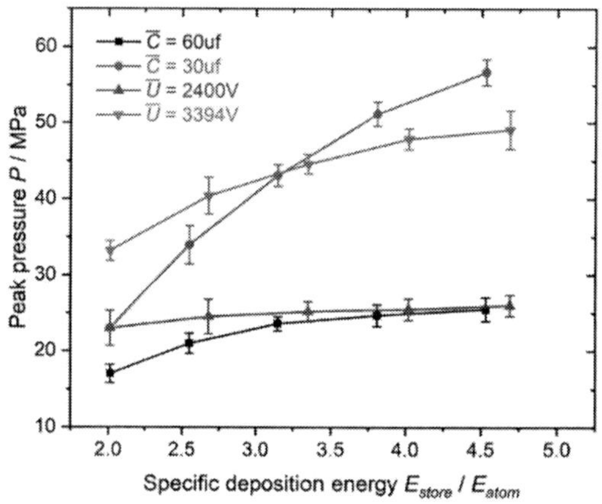


**Figure 4.** The comparison of peak pressure with different capacitors and charging voltages schemes

Four curves are interspersed and analyzed in Figure 4. As driving energy increases, the peak pressure of shock waves rises at varying rates under different voltage and capacitor energy storage schemes. The black and orange curves represent the peak pressure driven by 30 μF and 60 μF fixed capacitors in five voltage sequences. The blue and pink curves represent the peak pressure at the five capacitor sequences driven by fixed voltages of 2400 V and 3394 V, respectively. Comparative analysis of the two groups of results reveals the selection of energy scheme. The peak pressure growth rate is greater when the "fixed capacitance lifting voltage" scheme is utilized, as opposed to the "fixed voltage lifting capacitance" scheme. Boosting the shock wave by increasing the voltage and decreasing the capacitance when the total energy is equivalent.

The failure probability of the experiment is analyzed statistically excluding personal error which varies by the energy storage scheme's capacitance and charging voltage. The statistical results show that the failures mainly focus on energy storage schemes in larger or smaller tests. When the test fails under conditions of high specific energy, the discharge circuit's parameters are significantly lower than the normal values. And the capacitor has significant residual power. This failure may be caused by an uneven electrical explosion of the wire caused by the excess energy. The loop is truncated prematurely so that subsequent energy cannot continue to break down the wire. When the test fails under conditions of low specific energy, the discharge circuit's parameters exhibit characteristics of weak short circuit oscillation. The insufficient drive energy has transformed most of the wire into molten and overheated liquid copper. Consequently, the voltage-current waveform exhibits an oscillating state analogous to a short-circuit discharge.

*3.3 Shock wave analysis in time domain and frequency domain*

Under various energy schemes and load structures, shock wave pressure curves illustrate a similar trend. The primary distinction is the shock wave's peak pressure. Under the energy scheme ($U$ = 3000

V and $C = 30$ μF), the wire structure ($D = 0.2$ mm and $L = 25$ mm) is chosen to explode. The signal for the shock wave was detected 100 mm from the wire's center. Figure 5 depicts the segmented time domain characteristics and shock wave overpressure points.

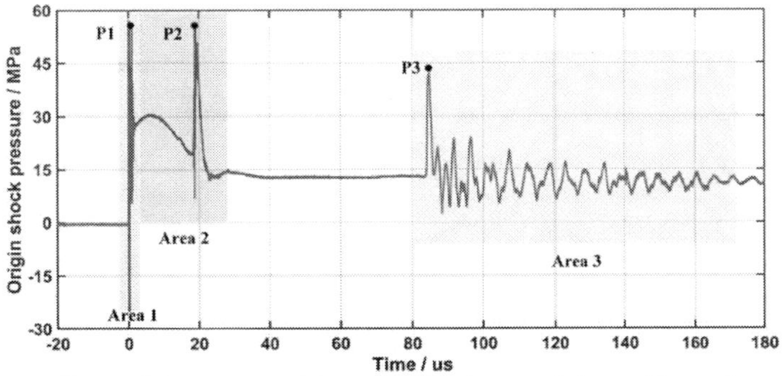


**Figure 5** the segmented time domain characteristics and key points

In shock wave research, the DB group wavelet is smooth and supportive. As the basis function, the db8 wavelet decomposes and reconstructs the original signal. This experiment's sampling frequency is 250 MHz, whereas the Nyquist frequency is 125 MHz. The signal is decomposed using a 10-layer wavelet packet, and the corresponding fundamental frequency band is 122 kHz. Figure 6 demonstrates the energy percentage statistics of each frequency band and the signal reconstruction outcomes.

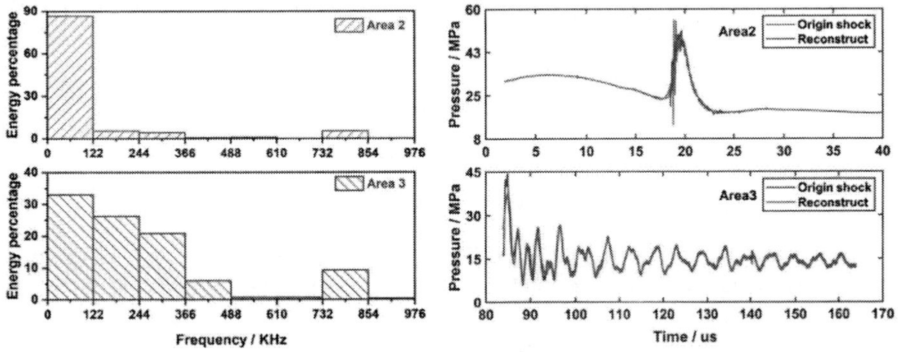


**Figure 6** The energy percentage statistics of each frequency band and the signal reconstruction

Combined with the results of Figures 5 and 6, the signal characteristics of the detected shock wave are separated into three areas. The first area occurs approximately within the first 2 μs. The detection signal exhibits high-frequency characteristics with the greatest amplitude being $P_1$ is 58 MPa. The signal results from the loop pulse current loading the sensor with electromagnetic interference. The travel speed should be light. The second area happens from 2 to 40 μs. The detection signal consists of a pulse signal with the greatest amplitude $P_2$ is 56 MPa. In conjunction with the rise of the pulse signal is a signal with a high interference frequency. 82% of the signal's energy is concentrated between 0 and 122 kHz. The third area lasts from 82 to 180 μs. The detection signal consists of a series of oscillating signals with the greatest amplitude $P_3$ is 42 MPa. The detection signal drifts by approximately 15 MPa. The amplitude of the oscillating signal decreases rapidly with time, whereas its period increases gradually. The signal's energy is primarily concentrated between 0 and 366 kHz. The original signal's characteristic regions Area2 and Area3 are subjected to trend analysis. Figure 7 illustrates the results of using the basic functions $f_n(t)$ and $g_n(t)$ for Sine and Gaussian signal fitting. $a_n$, $b_n$, $c_n$, $A_n$, $B_n$ and $C_n$ are the undetermined coefficients of the basic functions, respectively.

$$\begin{cases} f_n(t) = a_n \exp\left[-\left(\frac{t-b_n}{c_n}\right)^2\right] \\ g_n(t) = A_n \sin(B_n t + C_n) \end{cases} \tag{1}$$

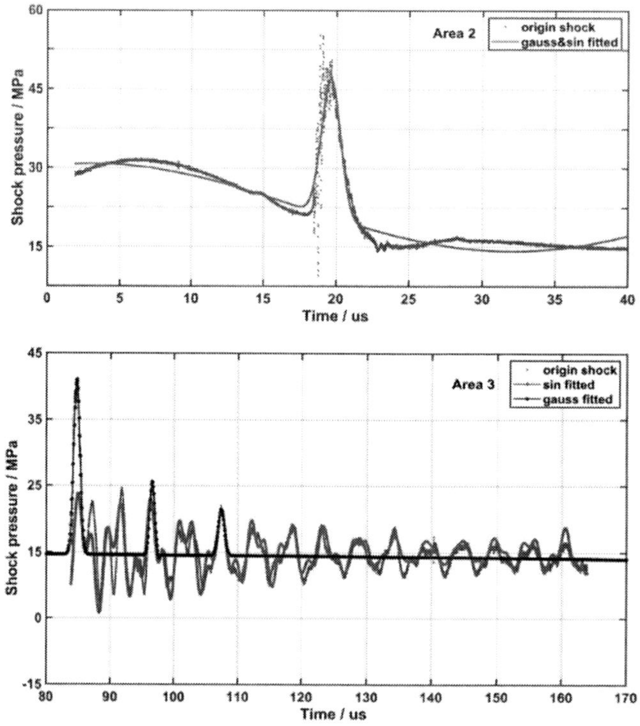


**Figure 7** The Sine and Gaussian signal fitting of Area2 and Area3

The Area2 and Area3 fitting results are $P_{Area2}(t)$ and $P_{Area3}(t)$, respectively. The desired shockwave signal was calibrated based on the fitting results. In addition, the causes of detection errors are assessed.

$$P_{Area2}(t) = f_0(t) + g_0(t) \tag{2}$$

$$P_{Area3}(t) = \sum_{i=1}^{5} f_i(t) + \sum_{j}^{4} g_j(t) \tag{3}$$

$$\begin{bmatrix} \boldsymbol{a} \\ \boldsymbol{b} \\ \boldsymbol{c} \end{bmatrix} = \begin{bmatrix} 0.70 & 0.39 & 0.01 & 0.16 & 0.10 & -3.16 \\ 1.92 & 8.48 & 9.76 & 9.65 & 10.74 & 108.7 \\ 9.07 & 6.84 & 2.07 & 5.56 & 6.75 & 397.8 \end{bmatrix}$$
$$\begin{bmatrix} \boldsymbol{A} \\ \boldsymbol{B} \\ \boldsymbol{C} \end{bmatrix} = \begin{bmatrix} 0.22 & 0.06 & 74.86 & 0.04 & 74.86 \\ 0.19 & 1.16 & 2.78 & 1.13 & 2.78 \\ 1.22 & 3.40 & 5.52 & -1.38 & -3.90 \end{bmatrix} \tag{4}$$

The Area2 signal is described and analyzed according to the fitting results. Signal components include the carrier, the high-frequency interference, and the Gaussian pulse. The carrier is a low-frequency sinusoidal signal caused by the influence of the electromagnetic field on the sensor. Which is induced by the relatively slow rise in current. The high-frequency interference signal begins at 18 μs and lasts above 1 μs. It caused by the influence of electromagnetic interference on the sensor. Which results from the relatively rapid drop in pulse current during the plasma discharge stage. The Gaussian pulse signal peaks around 19 μs. It originates from a high-energy radiation pressure wave during the wire electrical explosion. The radiation pressure and electromagnetic interference signal both travel at the speed of light. However, the radiation pressure is proportional to the energy output of the electrical explosion. Consequently, the waveform of the radiation pressure signal lags slightly behind that of the electromagnetic interference signal. The peak of the Gaussian pulse signal is the defining moment of the wire electrical explosion.

The Area3 signal is described and analyzed according to the fitting results. It is composed of an amplitude attenuation signal and an oscillation signal. The amplitude attenuation signal is fitted with a Gaussian of the fifth order. The Gaussian fit signal's local peak decays exponentially. The essential characteristic is the pressure attenuation of the shock wave front as it passes over the sensor. The oscillating signal is fitted with a Sine of the fourth order. The essential characteristic is the mechanical resonance of the detection system under shock wave action. The Sine fit signal represents the vibration mode of the detection device. The overall detection signal is the linear superposition of the shock wave front's attenuation and the detection device's resonance.

## 4. Conclusion

Based on comparative experimental analysis and pressure signal wavelet analysis, the following conclusions are drawn.

(1) The ideal energy conversion efficiency and shock wave peak pressure can be obtained when the wire structure and the driving energy are compatible. The optimal structure for a wire is $D$ = 0.2 mm and $L$ = 25 mm. The optimal amplification coefficient of driving energy is above 3.

(2) The polarized $D/L$ ratio is incapable of generating a stable electric explosion. When the total amount of stored energy remains constant, voltage regulation has a greater effect on shock waves than capacitance regulation.

(3) In the time domain, the shock wave of an electric explosion can be divided into three distinct areas. Each signal segment possesses varying degrees of broadband characteristics. The essence of the shock wave signal is a superposition of Gaussian and Sine signals. The signal's precision is determined by the mode of the detection system.

(4) The peak value of the signal fitting curve in the Area2 region corresponds to the radiation pressure wave, which marks the moment of the wire's electrical explosion. The Area3 signal frequency band 0~366 KHz carries a significant amount of energy. Consequently, it is crucial to consider the application of this shock wave frequency band.

**References**

[1] Kocis I, Kristofic T, Gajdos M, Horvath G, Jankovic S. Utilization of Electrical Plasma for Hard Rock Drilling and Casing Milling. SPE/IADC Drilling Conference and Exhibition2015. p. D021S12R02.

[2] Kocis I, Kristofic T, Gebura M, Horvath G, Gajdos M, Stofanik V, editors. Novel deep drilling technology based on electric plasma developed in Slovakia. 2017 XXXIInd General Assembly and Scientific Symposium of the International Union of Radio Science (URSI GASS); 2017 19-26 Aug. 2017.

[3] Krasik YE, Grinenko A, Sayapin A, Efimov S, Fedotov A, Gurovich VZ, et al. Underwater Electrical Wire Explosion and Its Applications. IEEE Transactions on Plasma Science. 2008;36(2):423-34.

[4] Patel K, Shah M, Sircar A. Plasma Pulse Technology: An uprising EOR technique. Petroleum Research. 2018;3(2):180-8.

[5] Bilbao L, Rodríguez Prieto G. Neutral Copper Gas Resistivity Measurements by Means of an Exploding Wire in Air. Applied Sciences. 2017; 7(8).

[6] Liu B, Wang D, Guo Y. Influence of water conductivity on shock waves generated by underwater electrical wire explosion. Physics Letters A. 2018;382(1):49-54.

[7] Han R, Wu J, Zhou H, Wang Y, Ding W, Ouyang J, et al. Effects of water states on the process of underwater electrical wire explosion under micro-second timescale pulsed discharge. The European Physical Journal Plus. 2020;135(1):50.

[8] Ageev PG, Ageev NP, Pashchenko AF, Kasilov VP, Ganiev SR, Kurmenev DV. Experimental Study of Plasma-Impulse Impact: Intensity of Pressure Pulsations in the Medium Processed. Journal of Machinery Manufacture and Reliability. 2019;48(2):184-9.

[9] Rososhek A, Efimov S, Maler D, Virozub A, Krasik YE. Shockwave generation by electrical explosion of cylindrical wire arrays in hydrogen peroxide/water solutions. ApPhL. 2020;116(24):243702.

[10] Smirnov AP, Zhekul VG, Taftai EI, Khvoshchan OV, Shvets IS. Experimental Study of Pressure Waves upon the Electrical Explosion of Wire under the Conditions of Elevated Hydrostatic Pressure. Surface Engineering and Applied Electrochemistry. 2020;56(2):192-200.

[11] Grinenko A, Efimov S, Fedotov A, Krasik YE, Schnitzer I. Efficiency of the shock wave generation caused by underwater electrical wire explosion. JAP. 2006;100(11):113509.

[12] Li L, Qian D, Zou X, Wang X. Effect of Deposition Energy on Underwater Electrical Wire Explosion. IEEE Transactions on Plasma Science. 2018;46(10):3444-9.

[13] Liu B, Wang D, Guo Y. Effect of Circuit Parameters and Environment on Shock Waves Generated by Underwater Electrical Wire Explosion. IEEE Transactions on Plasma Science. 2017;45:2519-26.

[14] Liu Y, Ren Y-J, Liu S-W, Lin F-C, Liu Y, Shen Y-K. Comparison and analysis of shockwave characteristics between underwater pulsed discharge and metal wire explosion. PhPl. 2020;27(3):033503.

[15] Chase MW, Jr. NIST-JANAF THERMOCHEMICAL TABLES, Fourth Edition, Monograph 9 (Part I and Part II). 9. Fourth Edition ed1998. p. pp. 1005.

# Synthesis and Characterization of a Cationic Micro-crosslinking Polymer and its Application as a Fluid Loss Reducer in Water-based Drilling Fluids

**Zhong-yi Wang[1,2], Jin-sheng Sun1[*], Shui-xiang Xie[3], Xian-bin Huang[1,2]**

1Key Laboratory of Unconventional Oil & Gas Development (China University of Petroleum (East China)), Ministry of Education, Qingdao, Shandong, China
2School of Petroleum Engineering, China University of Petroleum (East China), Qingdao, Shandong, China
3China National Petroleum Group Safety and Environmental Protection Technology Research Institute Co, China

20160048@upc.edu.cn

**Abstract.** In addition to flat rheology to deal with the wellbore stabilization problem caused by narrow safety density window when drilling deepwater wells, upgrading the mud cake quality by fluid loss reducer and plugging to stabilize the wellbore is also an important measure to deal with this problem. For the existing fluid loss reducers for deepwater water-based drilling fluids, it is difficult to balance the plugging performance. In this study, a cationic micro-crosslinking polymer was synthesized as a fluid loss reducer by reversed-phase emulsion polymerization. N-isopropylacrylamide (NIPAM) and acrylamide (AM) were chosen as the main body of the synthesis to enhance the hydrophilicity of the products. Methacryloyloxyethyl trimethyl ammonium chloride (DMC) was selected to provide cationic groups to enhance the residency of the products in the formation. The molecular structure of ENAD was characterized using infrared spectroscopy (FT-IR), nuclear magnetic resonance hydrogen spectroscopy (NMR) and X-ray photoelectron spectroscopy (XPS), and the molecular structure of the product was as expected. The thermal stability of ENAD was analyzed by TGA, and the initial thermal decomposition temperature was 283 °C. The filtration effect of ENAD in BF under different media was evaluated. The experimental results show that ENAD can withstand temperature up to 150℃ and has good filtration performance. 16h after aging at 150℃, the API filtration loss (FLAPI) is 8.8mL, the sand bed intrusion depth is 4.4cm, the high temperature and high pressure filtration loss (FLHTHP) is 32mL, the high temperature and high pressure filtration loss of plugging 10μm sand disc is 92mL. After comparing with sulfonated asphalt (FT), it is found that ENAD has better filtration performance. It is found that ENAD has better performance in reducing filtration loss. The mode of interaction between ENAD and the formation and its own mechanism of filtration loss reduction were analyzed by zeta potential analysis and SEM. With the increase of ENAD addition, the absolute values of zeta potential of BF before and after aging decreased from 48.7 and 32.1 to 43.3 and 27.1, respectively. ENAD enhances the interaction force with bentonite particles through its own adsorption properties. The strength of the mud cake is further strengthened by thermal deformation properties. It also enhances the force between the polymer and the formation through electrostatic gravitational adsorption, which further enhances the retention capacity of the polymer in the formation.

Compared with conventional fluid loss reducers, ENAD has superior filtration effect and can be used as a fluid loss reducer for deepwater water-based drilling fluids.

## 1. Introduction

There are 22 sedimentary basins in the South China Sea, with a total basin area of $100\times10^4$ km$^2$. 6 basins in the northern part of the South China Sea, with an area of $37\times10^4$ km$^2$, and 16 basins in the southern part of the South China Sea, with an area of $63\times10^4$ km$^2$[1]. The 22 basins have $226.3\times10^8$t of oil reserves and $15.84\times10^8$m$^3$ of natural gas reserves, accounting for about 1/3 of China's total oil and gas reserves. However, the South China Sea oil and gas fields are concentrated in the shallow waters of the northern part of the South China Sea, and some of them are jointly developed with foreign enterprises[2]. This indicates that China's development of oil and gas resources in the South China Sea is relatively low, especially the development of deepwater oil and gas resources in the South China Sea[3]. Deepwater oil and gas resources have great potential for development, and will certainly be called a key area for oil and gas resources development in the future[4]. As offshore oil and gas production continues to climb, oil and gas drilling operations are gradually moving from shallow water areas to deeperwater areas. However, compared to land-based oil and gas drilling operations, deepwater drilling fluids face serious wellbore stability issues.

Deepwater areas are significantly lower compared to the overburden pressure to which onshore formations are subjected, showing high activity, undercompaction, poor cementation, easy collapse, swelling and dispersion, but little change in pore pressure, which makes the gap between it and the rupture pressure small, resulting in a narrower window of safe density. Coupled with the complex geology of deepwater shallow formations, this exacerbates the instability of shallow formations and places more stringent requirements on the wellbore stability performance of drilling fluids[5]. In response to the strict wellbore stability requirements under the narrow safety density window, many high performance deepwater water-based drilling fluids have been developed, among which filter loss reducers as the core treatment agents of high performance drilling fluid systems have been intensively studied[6]. ULTRADRILL drilling fluid is a high-performance deepwater water-based drilling fluid system introduced by Schlumberger MI SWACO, and the core material was developed by Arvind Patel et al[7]. Among them, POLYPAC UL, a low-viscosity polyanionic cellulose, is applied to this drilling fluid system as a filtration loss reducer, which reduces filtration loss by enhancing the aggregation of bentonite particles. Baker Hughes' K. Morton et al[8]. developed PERFORMAX, a high-performance water-based drilling fluid system for deepwater drilling, in which the filter loss reducer MAX-TROL is a sulfonated resin that acts as a high-temperature and high-pressure filter loss reducer through its own plugging effect. CNOOC has introduced a soil-phase free, strongly inhibited, high performance deepwater water-based drilling fluid system HEM, which was mainly developed by Jiansheng Luo and Huike Li[9]. The filtration loss reducer PF-FLOTRO is a kind of temperature-resistant starch, which reduces the filtration loss by increasing viscosity and lifting cut. The API filtration loss of the system was only 6.5 mL after aging at 160℃ for 16 h, which has good performance in filtration loss reduction. Among the above filter loss reducers, cellulose and starch mostly achieve the effect of filter loss reduction through viscosity increase and aggregation of bentonite particles[10]. The resin material mainly reduces the filtration loss by cross-linking and plugging[11]. It is necessary to develop a new type of filter loss reduction agent for deepwater water-based drilling fluids that can simultaneously seal and stabilize the wellbore and improve the quality of the mud cake to reduce filtration loss.

To address the above related issues, a cationic micro-crosslinking polymer was synthesized by reversed-phase emulsion polymerization in this study. N-isopropylacrylamide (NIPAM), acrylamide (AM), and methacryloyloxyethyl trimethyl ammonium chloride (DMC) were used as the main body of the synthesis, and N, N-methylenebisacrylamide (MBA) was used as the cross-linking agent. The molecular structural properties of ENAD were characterized by FT-IR, NMR and XPS analyses. The thermal stability of ENAD was analyzed by TGA. The filtration loss reduction effect of ENAD in BF on different media was evaluated. The mode of interaction between ENAD and the formation and its own filter loss reduction mechanism were analyzed by zeta potential analysis and SEM.

## 2. Materials and Methods

### *2.1. Materials*
N-Isopropyl acrylamide (NIPAM, 99wt% with stabilizer MEHQ), acrylamide (AM, 98wt%), methacryloyloxyethyl trimethyl ammonium chloride (DMC, 75wt% aqueous solution with stabilizer MEHQ), azo diisobutyronitrile (AIBN, 98wt%), N, N-methylene bisacrylamide (MBA, 99wt%), oleic acid sorbitol ester (Span80, pharmaceutical grade), polyoxyethylene sorbitan monooleate (Twen80, pharmaceutical grade), sodium dodecyl sulfate (SDS, 92.5wt%-100.5wt%) anhydrous sodium carbonate (Na2CO3, 99.5wt%), alkaline alumina (75wt%) were purchased from Shanghai Aladdin Biochemical Technology Co. Sulfonated asphalt (FT, industrial grade), purchased from Shandong Deshunyuan Petroleum Technology Co. Bentonite (industrial grade), purchased from China Huai'an Group. 5# white oil (industrial grade), purchased from Shenzhen ZhongRunTong Chemical Co.

### *2.2. Experimental Methods*

*2.2.1. Synthesis of ENAD.* 100mL of deionized water was added to the beaker, and 12g NIPAM, 6g AM, 2g DMC and 0.02g MBA were added to the deionized water respectively under the action of magnetic stirring, and the inner phase was obtained by stirring well. A certain amount of Twen80 was added to the internal phase and stirred for 10 minutes using a magnetic stirrer until the emulsifier was fully dissolved.

125mL 5# white oil was added to another beaker, 0.02g AIBN and a certain amount of Span80 were added and stirred for 10 minutes using a magnetic stirrer until the emulsifier was fully dissolved to get the outer phase. After washing the high-speed shear emulsifier, pour the external phase into a large beaker. Set the speed of high-speed shear emulsifier to 2000rpm, pour the inner phase during shearing, and shear for 10 minutes to obtain the emulsion.

The emulsion was poured into a three-mouth flask, and the three-mouth flask was placed in a water bath, and after passing nitrogen gas for 20 minutes, the water bath was turned on, and after the water bath reached 65°C, the reaction was kept stirring for 7h to obtain the cationic micro-crosslinking polymeric filter loss reducing agent ENAD.

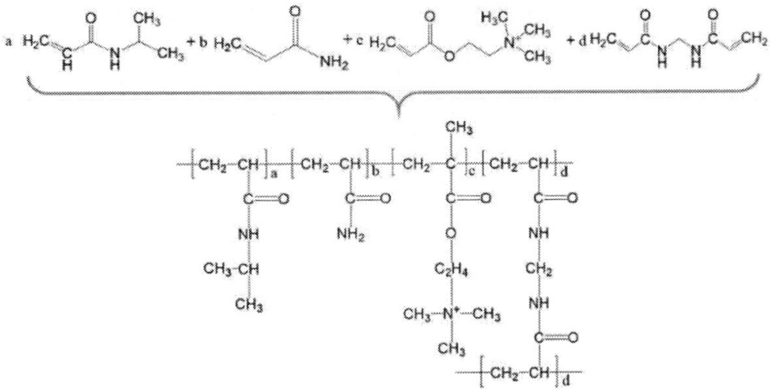

**Fig. 1. Synthesis process and molecular structure of ENAD**

*2.2.2. Characterization.* The infrared spectra of ENAD were tested by using an infrared spectrometer (IRTRacer-1000, Shimadzu, Japan). The filter loss reducer emulsion was broken using ethanol and washed by centrifugation several times, and then the sample was lyophilized for 24 h using a freeze dryer to obtain a white powder sample of the filter loss reducer. The sample was mixed with KBr and ground using an agate mortar, and the sample was prepared by using the KBr compression method (2 MPa, 3 min). FT-IR spectra of ENAD were measured in the wave number range of 500-4000 $cm^{-1}$ at room temperature with a resolution of 4 $cm^{-1}$.

The TGA of ENAD was tested by using a thermogravimetric analyzer (TGA2, Mettler, USA). The powder samples after breaking the emulsion were placed in an oven at 105°C for 24 hours. The test was started according to the instrument's program, with a set temperature range of 30°C-800°C, a ramping rate of 10°C/min and a protective gas of N2. TGA graphs of ENAD were obtained.

The NMR hydrogen spectra of ENAD were tested by nuclear magnetic resonance spectrometry (Bruker Analytik, Rheinstetten, Germany) using D2O as solvent. For the test, approximately 3 mg of the post-breakout powder sample was dissolved in heavy water (TMS internal standard, 0.03% volume fraction) and the signal was collected over 16 cycles.

The surface elements of ENDA were determined by X-ray photoelectron spectroscopy (XPS, axis ultra, Kratos, UK). A monochromatic X-ray source with an aluminum target was used. The acceleration voltage was 15 kV, the beam was 500 μm and the number of scans was 10.

*2.2.3 Performance evaluation.* In this study, 4wt% bentonite mud was used as the test base fluid (BF) because most water-based drilling fluids (WBDF) are prepared with 2-4wt% bentonite to achieve increased viscosity and reduced filtration[12].

A 4wt% bentonite mud configuration: add 400mL distilled water in a high stirring cup, then add 16g bentonite and 1.2g Na2CO3 in distilled water, stir at high speed (6000r/min) for 20min, stop twice to scrape off the clay adhering to the wall of the cup, place in a closed container for 24h. get 4wt% bentonite mud, performance of 4wt% bentonite mud must comply with GBT 16783.1-2014 standard.

Rheological and filtration performance. Bentonite-based dispersions play an essential role in the formulation of water-based drilling fluids (WBDF). To study the effect of ENAD on BF performance, the FT was chosen as the counterpart. The rheological and filtration performance of ENAD and FT in BF with different concentrations (2wt%, 3wt%) were tested before and after hot rolling at 150°C for 16h. The rheological parameters (θ600, θ300, θ6, θ3) of BF were tested with a ZNN-D6B viscometer (ZNN-D6B, Tongchun Petroleum Instruments, Qingdao, China), and the filtration (FL) of BF was tested with a medium-pressure filtration meter (SD6A, Tongchun Petroleum Instruments, Qingdao, China). The apparent viscosity (AV), plastic viscosity (PV), yield value (YP) and API filtration (FLAPI) of BF were calculated according to the API standard.

$$AV=0.5\cdot\theta_{600} \tag{1}$$
$$PV=\theta_{600}-\theta_{300} \tag{2}$$
$$YP=0.5\cdot(\theta_{300}-PV) \tag{3}$$
$$FL_{API}=FL\cdot 2 \tag{4}$$

Where AV is the apparent viscosity (mPa·s), PV is the plastic viscosity (mPa·s), YP is the yield value (Pa), and $FL_{API}$ is the API filtration.

Sand bed plugging performance. To study the plugging performance of ENAD on different permeability formations, BF containing 3wt% ENAD and 3wt% FT were prepared as test muds[13, 14]. After 16h hot rolling at 150℃, the sand bed of 40-60 mesh rang is used as the medium. The amount of filtration or the depth of intrusion into the sand bed (0.69 MPa, 30 min) was tested using a medium-pressure sand bed filtration meter (FA, Hengtai Da Mechanical and Electrical Equipment Co., Ltd., Qingdao, China).

High temperature and high pressure (HTHP) filtration performance. During drilling fluid circulation, a high temperature and pressure environment exists at the bottom of the well [15-18]. To investigate the filtration performance of ENAD under high temperature and high pressure environment, the high temperature and high pressure filtration meter (GGS71-B, Qingdao Tongchun, Qingdao, China) was used to test the high temperature and high pressure filtration (FLHTHP, 3.5 MPa, 30 min) of BF containing ENAD and FT after aging. Test conditions: temperature (25°C,150°C), 3wt% addition.

$$FL_{HTHP}=FL\cdot 2 \quad (5)$$

Plugging performance of artificial ceramic cores under HTHP. To further investigate the filtration performance of ENAD under high temperature and high pressure environment[19], the ceramic cores plugging filtration (3.5 MPa, 30 min) of ENAD after aging at 150°C with 3wt% addition was tested using a high temperature and high pressure ceramic cores plugging apparatus (GGS71-B, Qingdao Tongchun, Qingdao, China). The porosity of the artificial ceramic cores used for testing was 10μm.

*2.2.4. Mechanistic Analysis.*To determine whether the EMNC surface has positive charge properties brought by cationic monomers. Testing the zeta potential of BF containing 0 wt%-1wt%-2wt%-3wt% ENAD before and after hot rolling at 150°C for 16h. The surface electrical properties of ENAD are judged from the change in potential. Zeta potential was measured with a zeta potential analyzer (Nano ZS, Malvern, UK).

The mud cakes and artificial ceramic cores obtained from the experiments under different conditions were processed. The microscopic morphology of the mud cake and artificial ceramic core surfaces were observed using scanning electron microscopy. The surfaces of the mud cakes and artificial ceramic cores were rinsed with deionized water, dried and sprayed with gold, and then the microscopic morphology of the surfaces were observed with a scanning electron microscope (EVO-15/LS, Zeiss, Germany).

## 3. Results and Discussion

### *3.1. Characterization of ENAD*

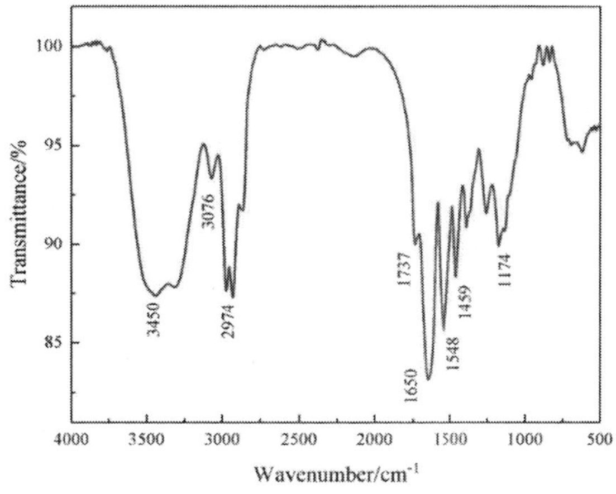


**Fig. 2. FT-IR of ENAD.**

The FT-IR of ENAD are shown in Figure 2. The N-H bond stretching vibrations of the amide groups on NIPAM, MBA and AM caused a peak of 3450 $cm^{-1}$[20]. The stretching vibration peaks of $-CH_3$ on

NIPAM and DMC appear at wavelength 3076 $cm^{-1}$[21]. The wavelength corresponding to the $-CH_2-$ stretching vibrational peak in the DMC, MBA and polymer main chain of ENAD is 2974 $cm^{-1}$. The stretching vibration of the C=O bond of ENAD produces 2 peaks at wavelengths 1737 $cm^{-1}$ and 1650 $cm^{-1}$[20]. The peaks at wavelengths 1548 $cm^{-1}$, 1459 $cm^{-1}$ and 1174 $cm^{-1}$ are the peaks of the C-N bond stretching vibration on ENAD[20]. FT-IR results show that ENAD contains functional groups specific to various monomers.

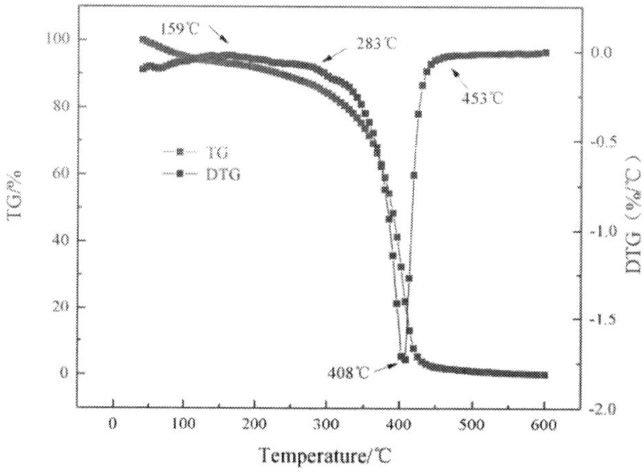


**Fig. 3. TG of ENAD.**

The TG results of ENAD are shown in Figure 3. Given that the hydrophilic group in ENAD adsorbs part of the moisture in the air, the TG curve of ENAD decreases slowly when the temperature is 30~159°C, and the mass loss is 2.2% at this time. When the temperature is 159°C~283°C, the TG curve is smooth and ENAD basically does not decompose. the decomposition of functional groups and main chains on the side chains of ENAD molecular chains causes the TG curve to drop sharply from 283°C to 453°C. At this stage, the decomposition rate is the fastest at 448°C, and the mass loss is 9.6% at 283°C. After 453°C, the TG curve of ENAD drops more slowly because the main chain of ENAD was basically degraded by heat. In general, ENAD has good thermal stability and its thermal degradation temperature is after 283°C, with a mass loss of 9.6% at 283°C.

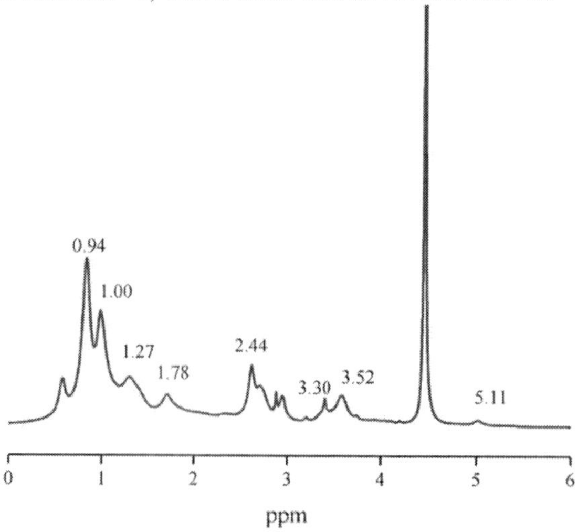


**Fig. 4. NMR for ENAD.**

The NMR hydrogen spectrum of ENAD is shown in Figure 4. The methyl group ($-CH_3$) in the NIPAM molecular chain appears at 1.00 ppm[22]. The methyl group ($-CH_3$) attached to the propylene group in the DMC molecular chain appears at 1.27 ppm. The methylene (-$CH_2$-) of the main chain of ENAD copolymer appeared at 1.68 ppm[22]. The hypo methyl (-CH-) in the molecular chain of MBA appears at 2.44 ppm. Generally, the molecular structure of the target product is consistent with the pre-designed one.

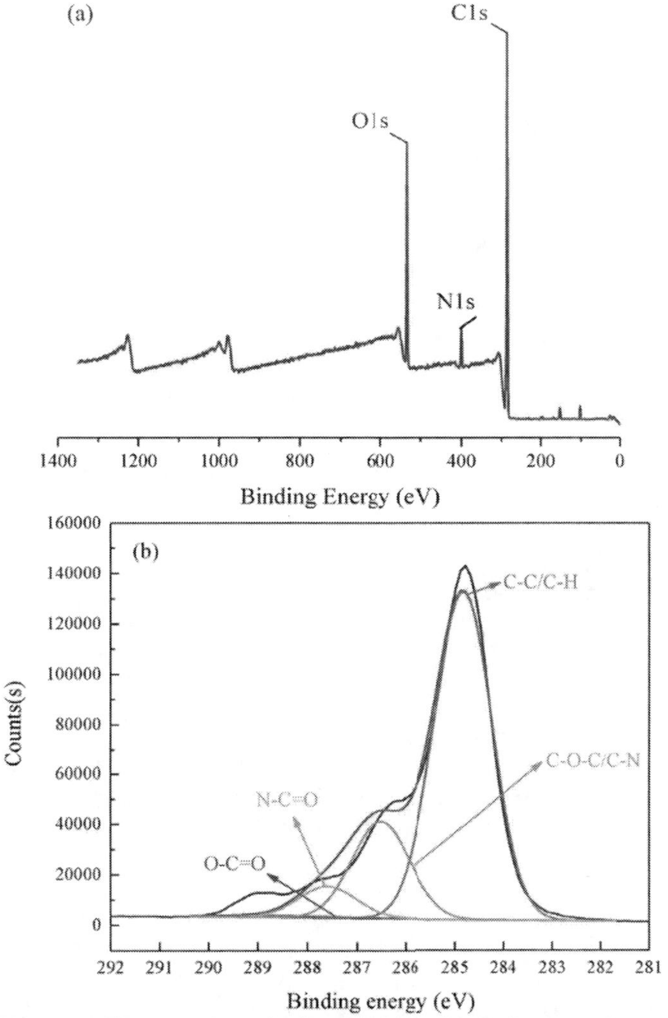


**Fig. 5. XPS for ENAD, (a) Elemental analysis of ENAD (b) C-element absorption spectrum of ENAD.**

The XPS spectra of ENAD are shown in Figure 5(a). ENAD exhibits a characteristic peak of C1s at a binding energy of 285 eV[23]. The binding energy at 400 eV exhibits the characteristic peak of N1s[23]. The binding energy at 532 eV exhibits the characteristic peak of O1s[23]. The C1s XPS spectra and fitted peaks of ENAD are shown in Figure 5(b). the C1s peaks of ENAD split into four peaks at the binding energies of 284.8 eV, 287.6 eV and 286.5 eV. They correspond to C-C/C-H, N-C=O and C-O-C/C-N, respectively[24]. Generally, the molecular structures of the target products were consistent with the predesigned ones.

### *3.2. Performance of ENAD*

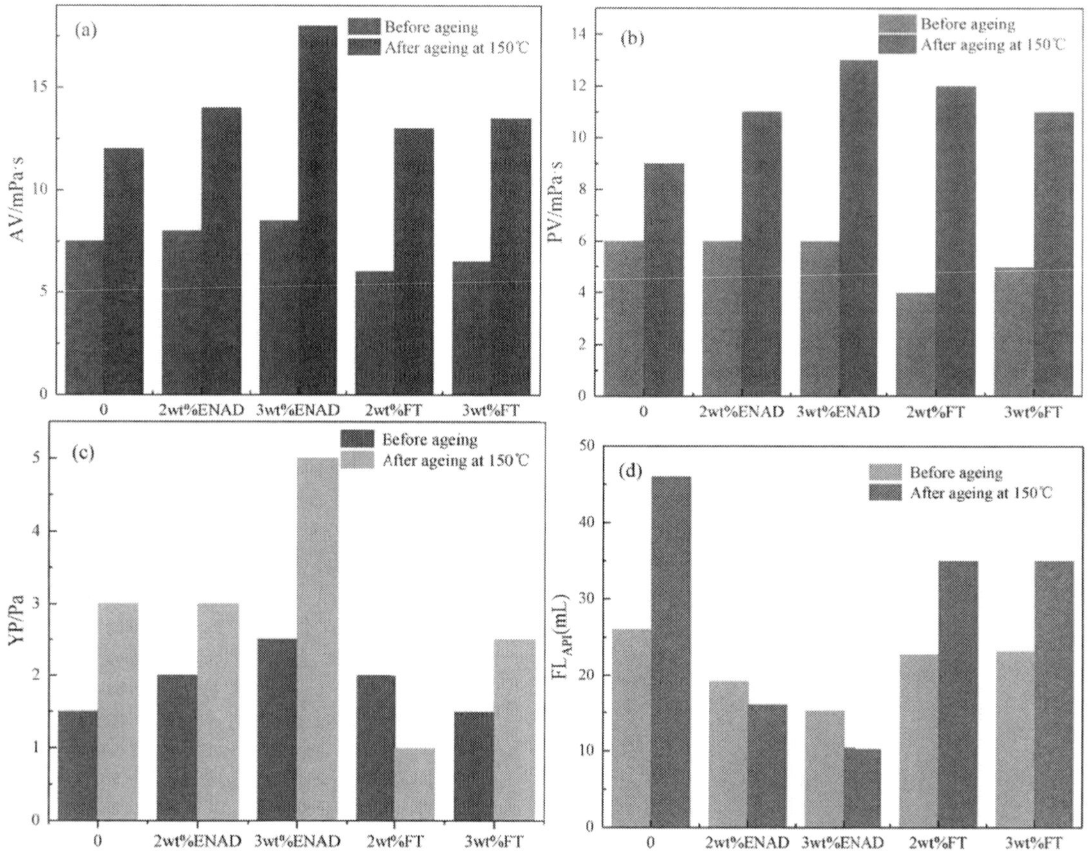


**Fig. 6. Rheological and filtration performance of BF under different test conditions. (a) AV performance of BF, BF with 2-3wt%ENAD and BF with 2-3wt%FT. (b) PV performance of BF, BF with 2-3wt%ENAD and BF with 2-3wt%FT. (c) YP performance of BF, BF with 2-3wt%ENAD and BF with 2-3wt%FT. (d) $FL_{API}$ of BF, BF with 2-3wt%ENAD and BF with 2-3wt%FT**

BF is a dispersion system consisting of bentonite particles as the dispersed phase and water as the dispersing medium[25]. Due to the lamellar structure of bentonite particles and their own negative charge properties. The filtration and rheological characteristics of the dispersion system leading to the whole bentonite-based slurry are influenced by the additives. Apparent viscosity (AV), plastic viscosity (PV) and yield value (YP) are commonly used to express the rheological characteristics of drilling fluids. Among them, PV can reflect the strength of internal mechanical friction, which is influenced by solid and liquid phases. YP depends on the structural strength of the spatial network formed by the clay particles and polymer chains, while AV is a combined expression of PV and YP[26]. API filtration ($FL_{API}$) is used to express the filtration characteristics of drilling fluids[27].

As shown in Fig. 6(b) and Fig. 6(c), the PV was maintained constant at 6mPa·s before aging when the addition of ENAD in BF was increased from 0 to 2.0wt% and 3.0wt%. And when the addition of ENAD in BF was increased from 0 to 2.0wt% and 3.0wt%, YP increased from 1.5 mPa-s to 2.5 mPa-s before aging. Due to the lamellar structure of bentonite, ENAD is dispersed between the bentonite lamellar layers, and the unique micro-crosslinking structure enhances the structural strength of the dispersed medium, which further enhances the resistance to the flow of water molecules between the dispersed phases[28]. This in turn leads to a small increase in YP. The PV increased from

9mPa·s to 11mPa·s and 13mPa·s when the addition of ENAD in BF was increased from 0 to 2.0wt% and 3.0wt% after 16 h of hot rolling at 150°C. This phenomenon is due to the thermal dispersion effect of clay particles on the one hand, the PV of BF appears to be enhanced compared to that before aging[29]. On the other hand, it is due to the enhanced dispersion of ENAD after thermal rolling, the molecular chains of ENAD stretch, the micro-crosslinking structure expands, and the liquid-liquid flow friction is enhanced, leading to an increasing PV[30]. At this point, the YP increased from 3mPa·s to 5mPa·s when the addition of ENAD in BF increased from 0 to 2.0wt% and 3.0wt%. This is due to the fact that the dispersion of ENAD further enhances the structural strength of the dispersion medium, which in turn leads to an increase in YP. However, ENAD did not cause substantial changes in the rheological characteristics of BF, indicating its good compatibility.

As shown in Figure 6(d), the API filtration of BF before hot-rolling decreased from 26 mL to 19.2 mL and 15.2 mL when the addition of ENAD in BF was increased from 0 to 2.0wt% and 3.0wt%. It shows that ENAD can reduce the filtration of BF more effectively at room temperature compared to FT. The API filtration of BF decreased from 46 mL to 16 mL and 8.8 mL when the addition of ENAD in BF was increased from 0 to 2.0wt% and 3.0wt% after hot rolling at 150°C for 16 h. The API filtration of BF decreased from 46 mL to 35 mL and 35 mL when the addition of FT in BF increased from 0 to 2.0wt% and 3.0wt%. It shows that ENAD is able to reduce the filtration of BF more effectively after hot rolling compared to FT. This is due to the fact that ENAD can enhance the strength of the cake by micro-crosslinking the structure. The cation adsorption better matches the filter cake pore throat and resides in the filter cake pore space to plug it[31]. This reduces the filtration loss of BF. In summary, ENAD has a more effective filtration loss reduction effect than FT.

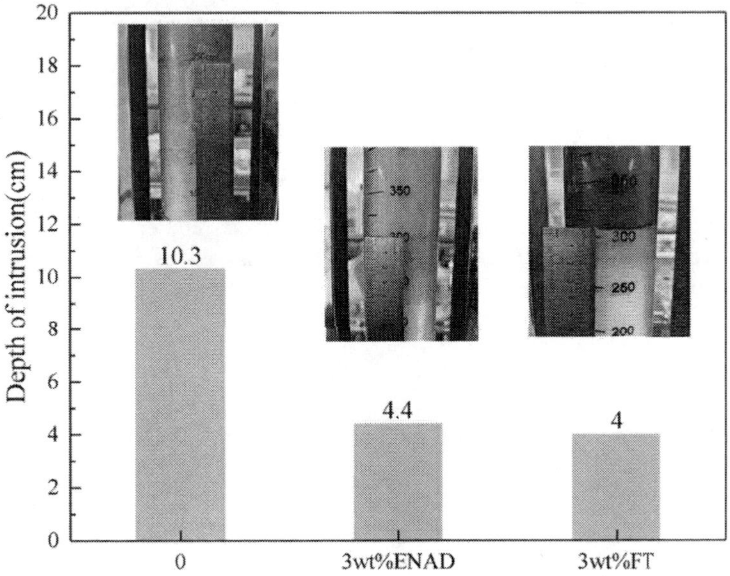


**Fig. 7. Depth of intrusion of BF, BF with 3wt%ENAD and BF with 3wt%FT on 40-60 mesh sand bed**

As shown in Figure 7, the intrusion depth of BF was 10.3 cm in 30 min when the mesh range of the sand bed was 40-60 mesh. It shows that the filtration of BF in drilling encounters with permeable formations is high without additives. The depth of BF intrusion into the 40-60 mesh sand bed was reduced to 4.0 cm by adding 3wt% FT to the BF. The depth of BF intrusion into the 40-60 mesh sand bed was reduced to 4.4 cm by adding 3wt% ENAD to the BF. It shows that ENAD not only can effectively improve the quality of mud cake, but also can effectively plug the hole joints of sand bed stacking. To a certain extent, it reduces the fluid loss of BF in drilling encounter with similar strata, and ENAD has good effect of reducing filtration loss for permeable formation.

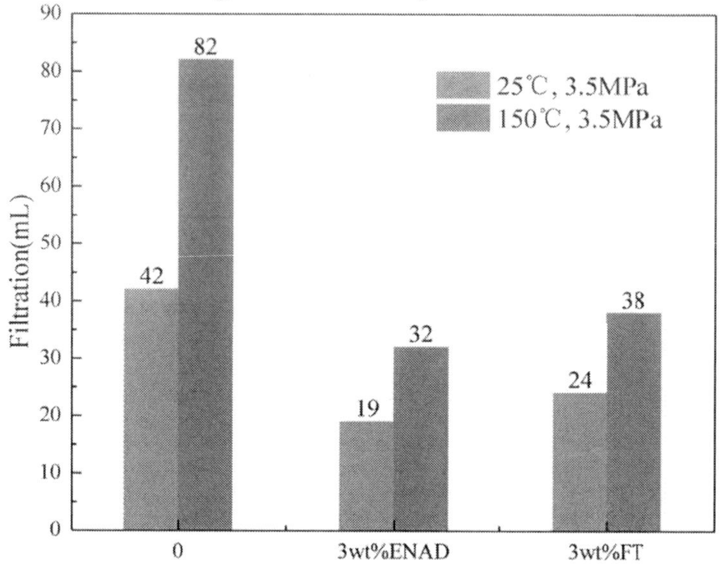


**Fig. 8. High-temperature and high-pressure filtration performance of BF, BF with 3wt%ENAD and BF with 3wt%FT under different test conditions**

**Table. 1. Filtration reduction rates of ENAD and FT**

| Test cases | Reduction rate (25℃， 3.5MPa) | Reduction rate (150℃， 3.5MPa) |
|---|---|---|
| BF+3wt%ENAD | 54.76% | 60.98% |
| BF+3wt%FT | 42.86% | 53.66% |

As shown in Figure 8, the filtration of BF was 43 mL at room temperature. The filtration of BF with 3wt% ENAD was 19 mL and the filtration of BF with 3wt% FT was 24mL. The filtration was 83 mL for BF, 32 mL for BF with 3wt% ENAD, and 38 mL for BF with 3wt% FT when tested at high temperature (150°C). From the test results, it can be seen that both ENAD and FT can effectively reduce the filtration of BF at both room temperature and high temperature conditions. Among them, ENAD has a better filtration loss reduction effect.

As can be seen from Table 1, the filtration reduction rate of ENAD for BF was 54.76% and that of FT for BF was 42.86% under the test at room temperature. The filtration reduction rate of ENAD for BF was 60.98% and that of FT for BF was 53.66% under the high temperature (150°C) condition test. FT insoluble in water can play a part in filling the pore throat and cracks to play a plugging role. Moreover, FT has the physical property of softening point by itself, softening and deforming under certain temperature and pressure, thus plugging cracks. So after warming up, the filtration reduction rate of FT has increased instead. For ENAD, the same phenomenon of increased filtration reduction rate after warming is also observed. This is due to the stretching of the molecular chains of the micro-crosslinking structure after warming, which enhances the heat deformation ability of ENAD and improves the adsorption space of ENAD. Compared with FT, ENAD has better filtration loss reduction performance.

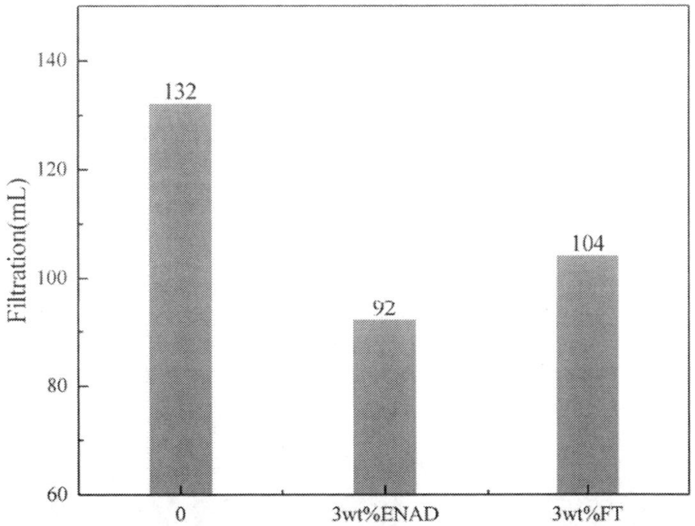


**Fig. 9. Artificial ceramic core plugging performance of BF, BF with 3wt%ENAD and BF with 3wt%FT**

As shown in Figure 9, the plugging performance of BF under different additive conditions was evaluated using 10 μm artificial ceramic cores as the medium. The filtration was 132 mL for BF tested under high temperature and high pressure (150°C, 3.5 MPa), 104 mL for BF with 3wt% FT and 92 mL for BF with 3wt% ENAD. Compared with FT, ENAD can more effectively reduce the filtration of BF under the condition of 10μm artificial ceramic core as the medium, and has better filtration loss reduction performance. In the face of different media, ENAD still has good filtration loss reduction performance. It also proved that ENAD can effectively plug the formation pores and joints through the interaction with bentonite.

*3.3. Mechanistic Analysis of ENAD*

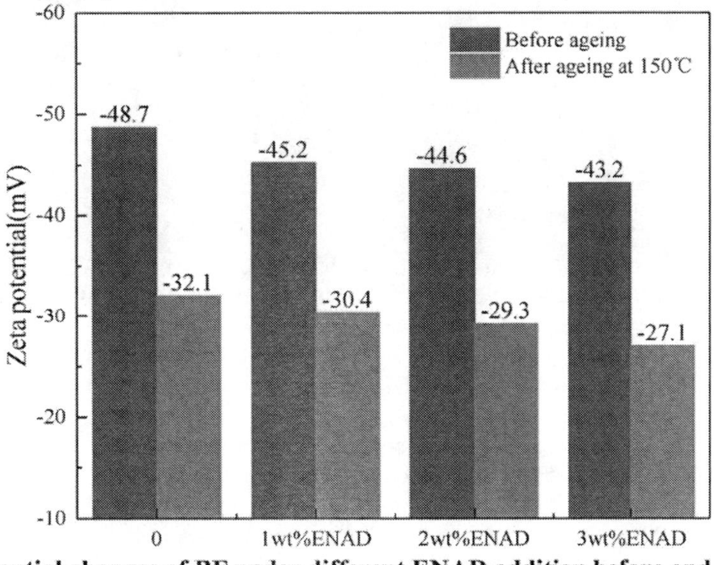


**Fig. 10. Zeta potential changes of BF under different ENAD addition before and after hot rolling**

As shown in Figure 10, the absolute value of the negative zeta potential of the BF before and after the heat roll decreases with the increasing amount of ENAD addition. Before hot rolling, the absolute value of negative zeta potential kept decreasing as the ENAD addition increased from 0wt% to 3wt%, and the zeta potential increased from -48.7 mV to -45.2 mV, -44.6 mV, and -43.2 mV, respectively. After hot rolling, the zeta potential increased from -32.1 mV to -30.4 mV, -29.3 mV, and -27.1 mV, respectively, with the increase of ENAD addition from 0wt% to 3wt%, and the absolute value of negative zeta potential kept decreasing. This indicates that ENAD with cationic groups is able to exert a certain adsorption effect with the negatively charged bentonite particles. This action neutralizes a part of the charge on the surface of the bentonite particles. It leads to a reduction of the diffuse double electric layer and affects the dispersion of the bentonite particles. However, the absolute value of the negative zeta potential of the BF with 3wt% ENAD added before and after aging was not reduced to a great extent. It indicates that although ENAD has adsorption effect on bentonite particles, the effect on the dispersibility of bentonite particles is not serious.

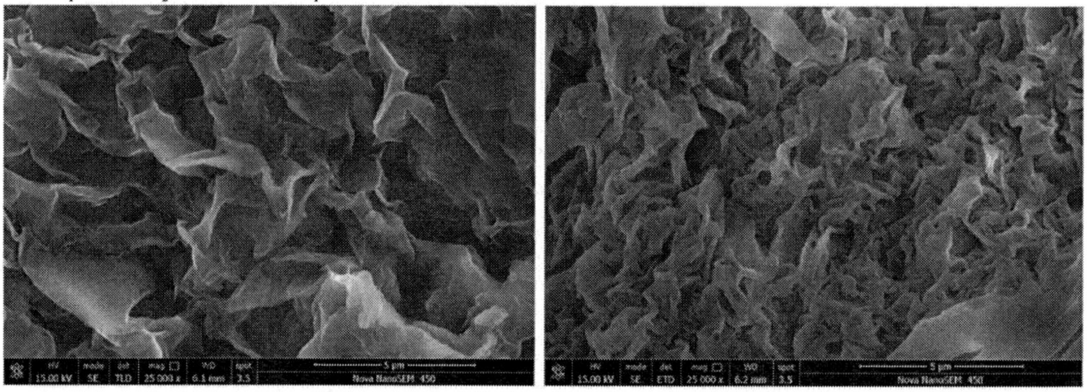

**Fig. 11. (a) SEM images of the HTHP filter cake formed by the original BF and (b) the HTHP filter cake formed by the BF containing 3wt% ENAD.**

**Fig. 12. (a) SEM images of the HTHP sand disc by the original BF and (b) the HTHP sand disc by BF containing 3 wt% ENAD.**

The HTHP filter cake formed by BF and BF containing 3 wt% ENAD was observed using SEM (150°C, 30 min, 3.5 MPa). As shown in Fig. 11(a), the filter cake surface of BF showed flaky aggregation of clay particles due to irreversible flocculation of clay particles induced by high temperature. And a large number of micro cracks and micro porosity were observed on its surface, indicating that the weak force between clay particles made it difficult to form a dense plugging layer. As shown in Figure 11(b), the particle size of clay particles was significantly reduced and the formed mud cake was more dense after adding ENAD, which reduced the filtration of fluid. It indicates that after the adsorption of ENAD on the surface of bentonite particles, the micro-crosslinking structure of ENAD itself can effectively enhance the strength of the mud cake.

Artificial ceramic cores plugged by BF and BF containing 3wt% ENAD were observed using SEM (150°C, 30 min, 3.5 MPa). The artificial ceramic cores after breaking were observed as shown in Figure 15(a). It was found that a small amount of bentonite particles existed in the cracks and pores of the artificial cores after plugging them with BF, but it was difficult to reside in them to form an effective plugging layer. The presence of particles was basically not observed on the surface of the artificial ceramic core. After using BF containing 3 wt% ENAD to plug the artificial ceramic cores, a large number of clay particles can be found on the surface of the artificial ceramic cores, forming a dense plugging layer in a large and extensive manner. A large number of particles were also observed to reside and accumulate in the fractures of the ceramic cores, forming a dense plugging layer through pore filling and particle bridging.

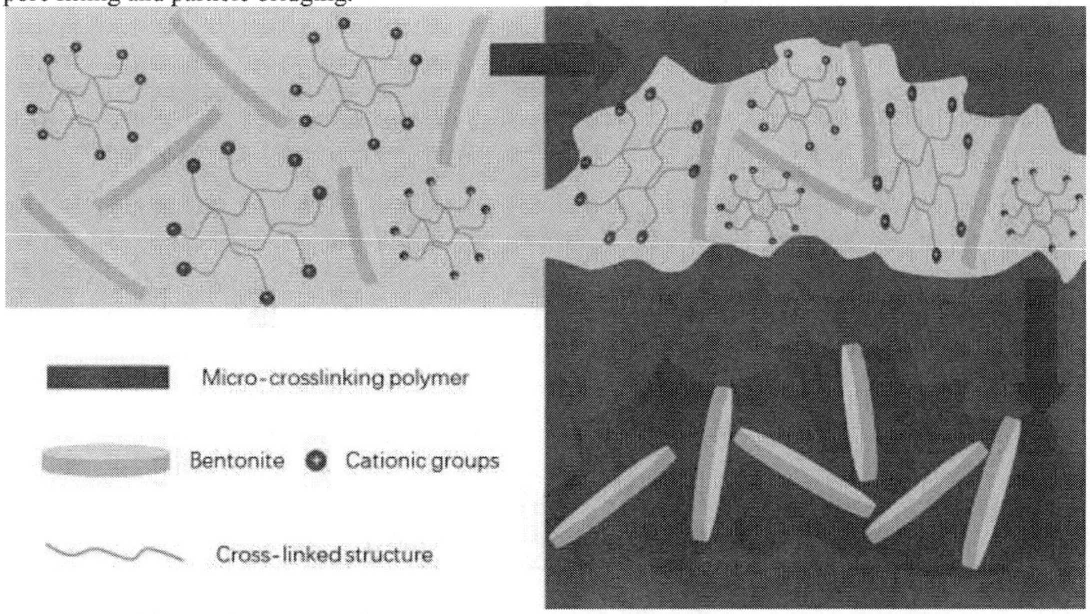


**Fig. 13. Schematic diagram of ENAD filtration loss reduction mechanism**

Based on the above facts and test results, the plugging mechanism of ENAD can be illustrated schematically in Figure 13. When ENAD is in BF, it has good mating properties and is uniformly dispersed in the BF system together with bentonite particles. Under the positive pressure differential during circulation, ENAD acts in two ways during the drilling process. On the one hand, ENAD is able to adsorb with bentonite particles, and its micro-crosslinking structure can assist bentonite particles to form a dense mud cake on the surface of the wellbore, which effectively reduces the filtration loss of BF. On the other hand, ENAD is able to enter the formation with bentonite particles. The electrostatic gravitational force between the cationic groups and the negative charge of the formation enhances the particle's ability to reside in fractures and pores. It also forms a more compact plugging zone with bentonite particles through its own thermal deformability. It fills the pores between different shaped particles with a certain degree of self-adaptability. In short, ENAD enhances mud cake strength by adsorption of micro-crosslinking structure and enhances formation residence by electrostatic gravitational force acting together. While forming a surface dense mud cake on the surface of the wellbore, it is also able to effectively plug microfractures and microporosity at different scales. It further strengthens the strength of the plugging layer with good filtration loss reduction and plugging performance.

## 4. Conclusion

In this paper, a cationic micro-crosslinking polymer ENAD was synthesized by reversed-phase emulsion polymerization. The physicochemical properties of ENAD were characterized by FT-IR, TGA/DSC, NMR, XPS and particle size analysis. FT-IR, NMR and XPS indicated that the molecular structure and functional groups of ENAD were in accordance with the molecular design. TGA indicated that ENAD had good thermal stability with an initial thermal decomposition temperature of 283℃. The rheological and filtration properties of ENAD were analyzed before and after aging at 150 ℃ for 16 h using BF as the test dispersion system and FT as the contrast ratio. The results showed that in the case of 16 h after aging at 150 ℃. The API filtration loss of BF containing 3 wt% ENAD was 8.8 mL. The medium pressure sand bed intrusion depth of BF containing 3 wt% ENAD was 4.4 cm. the high temperature and high pressure filtration loss of BF containing 3 wt% ENAD was 32 mL. The high temperature and high pressure 10 μm sand disc filtration loss of BF containing 3 wt% ENAD was 92 mL. The performance of ENAD was all due to FT. This indicates that ENAD can still maintain positive filtration loss reduction performance in BF at high temperature. The filter loss reduction control mechanism of ENAD was investigated by zeta potential analysis and SEM analysis. Specifically, ENAD enhances the strength of the mud cake and the residency in the formation through a special micro-crosslinking structure and cationic adsorption synergy. This cationic micro-crosslinking polymer can be used as a filter loss reducer for BF in deepwater drilling operations.

**Acknowledgments.** The project is supported by National Natural Science Foundation of China (52288101) and Key R & D Program of Shandong Province (No.2020ZLYS07)

## References

[1] XU H, SUN J, LIAO J, et al. Bioherm petroleum reservoir types and features in main sedimentary basins of the South China Sea [J]. Journal of Earth Science, 2012, 23(6): 828-41.
[2] SUN Q, WU S, CARTWRIGHT J, et al. Shallow gas and focused fluid flow systems in the Pearl River Mouth Basin, northern South China Sea [J]. Marine Geology, 2012, 315: 1-14.
[3] ZHU W, ZHONG K, LI Y, et al. Characteristics of hydrocarbon accumulation and exploration potential of the northern South China Sea deepwater basins [J]. Chinese Science Bulletin, 2012, 57: 3121-9.
[4] ZHANG G, QU H, CHEN G, et al. Giant discoveries of oil and gas fields in global deepwaters in the past 40 years and the prospect of exploration [J]. Journal of Natural Gas Geoscience, 2019, 4(1): 1-28.
[5] ALCáZAR-VARA L A, CORTéS-MONROY I R. Drilling fluids for deepwater fields: An overview [J]. Recent Insights in Petroleum Science and Engineering, 2018, 71.
[6] REN X, LIU R, MA Z. Experimental investigation of a low-molecular-weight polymer coating agent for deep-sea oil and gas drilling [J]. Journal of Petroleum Exploration and Production Technology, 2021, 11(7): 2953-62.
[7] PATEL A D, STAMATAKIS E, DAVIS E, et al. High performance water based drilling mud and method of use [Z]. Google Patents. 2004
[8] MORTON E K, BOMAR B B, SCHILLER M W, et al. Selection and evaluation criteria for high-performance drilling fluids; proceedings of the SPE Annual Technical Conference and Exhibition, F, 2005 [C]. OnePetro.
[9] HUAIKE L, TIE G, LEI G, et al. High performance water-based drilling fluids—A high efficiency muds achieving superior shale stability while drilling deepwater well with HPHT in South China Sea [J]. Science, 2019, 7(4): 98-103.
[10] ZHONG H, KONG X, CHEN S, et al. Preparation, characterization and filtration control properties of crosslinked starch nanospheres in water-based drilling fluids [J]. Journal of

Molecular Liquids, 2021, 325: 115221.
[11] PU L, XU P, XU M, et al. Lost circulation materials for deep and ultra-deep wells: A review [J]. Journal of Petroleum Science and Engineering, 2022: 110404.
[12] XIE B, TING L, ZHANG Y, et al. Rheological properties of bentonite-free water-based drilling fluids with novel polymer viscosifier [J]. Journal of Petroleum Science and Engineering, 2018, 164: 302-10.
[13] YU L, DONG M, DING B, et al. Emulsification of heavy crude oil in brine and its plugging performance in porous media [J]. Chemical Engineering Science, 2018, 178: 335-47.
[14] CHENG K, LIU Y, QI Z, et al. Laboratory Evaluation of the Plugging Performance of an Inorganic Profile Control Agent for Thermal Oil Recovery [J]. Energies, 2022, 15(15): 5452.
[15] XU Z, SONG X, LI G, et al. Gas kick simulation in oil-based drilling fluids with the gas solubility effect during high-temperature and high-pressure well drilling [J]. Applied Thermal Engineering, 2019, 149: 1080-97.
[16] AGWU O E, AKPABIO J U, EKPENYONG M E, et al. A comprehensive review of laboratory, field and modelling studies on drilling mud rheology in high temperature high pressure (HTHP) conditions [J]. Journal of Natural Gas Science and Engineering, 2021, 94: 104046.
[17] MAO H, YANG Y, ZHANG H, et al. A critical review of the possible effects of physical and chemical properties of subcritical water on the performance of water-based drilling fluids designed for ultra-high temperature and ultra-high pressure drilling applications [J]. Journal of Petroleum Science and Engineering, 2020, 187: 106795.
[18] LIU J-P, ZHANG X-F, ZHANG W-C, et al. Carbon nanotube enhanced water-based drilling fluid for high temperature and high salinity deep resource development [J]. Petroleum Science, 2022, 19(2): 916-26.
[19] HUANG X-B, SUN J-S, HUANG Y, et al. Laponite: A promising nanomaterial to formulate high-performance water-based drilling fluids [J]. Petroleum Science, 2021, 18: 579-90.
[20] ZHENG H, ZHANG S, YANG C, et al. Simultaneous removal of Ni (II) and Cr (VI) from aqueous solution by froth flotation using PNIPAM-CS intelligent nano-hydrogels as collector [J]. Journal of Molecular Liquids, 2021, 342: 117551.
[21] SRIDHAR S P, SOMAN A K, SANKARAN T S, et al. Temperature responsive poly (N-isopropylacrylamide-block-styrene) block copolymer coatings with tunable hydrophilicity [J]. Surfaces and Interfaces, 2020, 21: 100800.
[22] GöKTAŞ M, DENG G. Synthesis of poly (methyl methacrylate)-b-poly (N-isopropylacrylamide) block copolymer by redox polymerization and atom transfer radical polymerization [J]. Indonesian journal of chemistry, 2018, 18(3): 537-43.
[23] SINGH V, RAWAT K S, MISHRA S, et al. Biocompatible fluorescent carbon quantum dots prepared from beetroot extract for in vivo live imaging in C. elegans and BALB/c mice [J]. Journal of Materials Chemistry B, 2018, 6(20): 3366-71.
[24] ALEGRE C, BUSACCA C, DI BLASI A, et al. Electrospun carbon nanofibers loaded with spinel-type cobalt oxide as bifunctional catalysts for enhanced oxygen electrocatalysis [J]. Journal of Energy Storage, 2019, 23: 269-77.
[25] SUN J-S, HUANG X-B, LV K-H, et al. Methods, technology status and research progress for improving the high temperature stability of water-based drilling fluids [J]. Journal of China University of Petroleum (Natural Science Edition), 2019, 43(05): 73-81.
[26] HUANG X-B, SUN J-S, JIANG G-C, et al. Development of modified fatty acid cutting agent and its application [J]. Journal of China University of Petroleum (Natural Science Edition), 2019, 43(03): 107-12.
[27] LI H, HUANG X-B, SUN J-S, et al. Improving the anti-collapse performance of water-based drilling fluids of Xinjiang Oilfield using hydrophobically modified silica nanoparticles with cationic surfactants [J]. Petroleum Science, 2022.
[28] LIU Y, LI Y, LIU Y, et al. Three-Dimensional Polymer Nanofiber Structures for Liquid Contamination Adsorption [J]. ACS Applied Nano Materials, 2022, 5(4): 5640-51.

[29] MOLZ F, PARR A D, ANDERSEN P, et al. Thermal energy storage in a confined aquifer: Experimental results [J]. Water Resources Research, 1979, 15(6): 1509-14.
[30] DAI X, ZHANG X, WEN J, et al. Research progress on high-temperature resistant polymer separators for lithium-ion batteries [J]. Energy Storage Materials, 2022.
[31] XIA X, GUO J, FENG Y, et al. Hydrophobic associated polymer "grafted onto" nanosilica as a multi-functional fluid loss agent for oil well cement under ultrahigh temperature [J]. RSC advances, 2016, 6(94): 91728-40.

Journal of Physics: Conference Series **2594** (2023) 012059 doi:10.1088/1742-6596/2594/1/012059

# Exploration and Application of Acoustic Emission Detection Technology in Leakage Detection of the Bottom Plate of Crude Oil Metal Storage Tanks

**Yu-fang LIU1(✉) , Li-xin SHI1, Qi-er SA2,Tian-ning ZHANG1,Yan-qiang CHAI2, Xiao-feng SI1, YU-meng Wang1,and Bing -jun LI3**

1Engineering Technology Research Institute of Huabei oilfiel d Company,2 Huabei Oilfield Erlian Branch,3 Huabei Oilfield Production Plant 5

cy5_lyfang@petrochina.com.cn

**Abstract.** The use of acoustic emission detection technology can conduct online and non-stop detection and evaluation of crude oil metal storage tanks, achieving the integrity detection and evaluation of storage tanks. In order to explore the applicability of acoustic emission detection technology in the leakage detection of the bottom plate of crude oil metal storage tanks, this project carried out online acoustic emission detection of crude oil metal storage tanks, magnetic flux leakage detection after tank opening and cleaning, and compared and analyzed the results of two detection technologies. The adaptability of online acoustic emission detection technology was analyzed and explored. It was found that acoustic emission technology can qualitatively evaluate the corrosion of tank bottom plates, but cannot accurately locate corrosion defects. For tanks with severe corrosion that require maintenance, it is still necessary to open the tank and use magnetic flux leakage and other detection techniques to determine the location of defects. Through comparative analysis of acoustic emission testing technology and magnetic flux leakage testing technology, the results obtained from acoustic emission testing technology are conservative compared to the actual results. It is recommended to upgrade the evaluation of corrosion degree for the acoustic emission evaluation results of crude oil metal storage tanks.

## 1. Introduction

Since the 14th Five Year Plan period, with the increasing production of major oil fields in China and the increasing consumption of petroleum, the petrochemical industry has flourished, and the application scale of crude oil metal storage tanks has become increasingly widespread. At the same time, with the increase of the online service cycle of crude oil metal storage tanks, deformation of the tank body, corrosion of the tank wall, leakage of the bottom plate, failure of safety accessories, and settlement of the tank body also occur from time to time, laying significant hidden dangers for the safe production and operation of station storage tanks [2-4]. As an important production facility in oil field stations, metal storage tanks can cause serious economic losses, environmental pollution, and casualties in the event of safety accidents. How to scientifically and effectively solve the safe operation of storage tanks is an urgent need for enterprise tank integrity management [5-7]. Corrosion, as an important factor in crude oil metal storage tanks, requires key protection, and regular inspection of tanks is a necessary means to ensure tank safety. Understanding the corrosion status of tanks can prevent accidents such as leaks. At present, the commonly used methods for tank inspection are mainly open tank inspection, which requires the tank to be stopped, the medium inside the tank to be emptied, impurities inside the tank to be removed, and some positions even require polishing operations. Then, magnetic flux leakage testing, eddy current testing, magnetic particle testing, radiographic testing, and other methods are used to determine the corrosion situation and leakage location [8-9]. Although this type of detection method can visually display the defect situation, the detection workload is large, requiring operations such as shutdown, tank pouring, and tank cleaning, and the tank detection process takes a long time. Acoustic emission detection technology, as a technical means for online detection of metal storage tanks, can not only ensure the normal operation of the tank without the need for shutdown or cleaning of the tank floor, greatly reducing the workload of inspection due to tank opening, but also accurately determine the specific location of tank floor corrosion leakage [10-12]. Establishing a integrity management system for large crude oil storage tanks is of great significance to oil and gas storage and transportation enterprises. One is that integrity management not only effectively implements general inspection and maintenance operations, but also advocates the use of advanced detection technology to accurately predict the potential hazards of crude oil storage tanks, achieving the goal of zero equipment accidents. The second is to optimize the inspection and maintenance cycle by using risk based inspection techniques, which can reduce production costs while achieving intrinsic safety of tank operation [13-14].

This article introduces a case study of a technical service company using acoustic emission testing technology to conduct online inspection of crude oil metal storage tanks in Inner Mongolia. This case is based on risk assessment by analyzing the possible defects and severity of defects in the storage tank, selecting the online detection method of acoustic emission testing technology to further grasp the integrity information of the storage tank, providing the most direct basis for integrity evaluation, and further deciding whether to stop production and repair the storage tank. Propose a reasonable maintenance cycle based on the predicted results, and minimize economic expenses as much as possible while ensuring safety.

## 2. Acoustic Emission Technology in Tank Leakage Detection Method

Acoustic emission is a physical phenomenon. Due to external force or internal force, material or structure will be deformed or broken, and strain energy will be released in the form of stress wave in an instant. Factors such as material fracture, crack propagation, stress redistribution, friction, and collision may all release this energy [15]. The principle of acoustic emission detection technology is to use the instrument to detect this strain energy, infer the acoustic emission source by analyzing the acoustic emission signal, and evaluate its active defects [16].

For atmospheric pressure metal storage tanks, acoustic emission signals can propagate through both oil and material surfaces simultaneously, and the signal strength gradually decreases with increasing distance. Under the condition of excluding external interference, the pressure of the storage tank is loaded within the acceptable range of the storage tank, and the corrosion layer in the corrosion thinning area inside the tank begins to change, resulting in corresponding acoustic emission signals [17]. The sensors fixed on the outer wall will receive this dynamic signal and send it to the hardware and software in the form of an electrical signal. Through data analysis and processing, the corrosion and leakage situation of the storage tank will be ultimately determined. Among them, corrosion of storage tanks often occurs at the bottom of the tank, and acoustic emission technology is often used to detect the corrosion of the bottom of atmospheric storage tanks. Zhang Shuxin et al. [18] evaluated the corrosion status of the bottom plate of crude oil storage tanks using acoustic emission detection technology and verified it with the results of magnetic flux leakage during tank opening. The study found that the local corrosion degree of the tank bottom plate was highly correlated with the event localization statistics of acoustic emission detection, while the correlation degree with impact statistics was relatively low; The actual corrosion of the tank bottom plate is more severe than the evaluation results of online acoustic emission testing. Chen Jianfei et al. [19] from Shengli Oilfield applied the German VALLEN AMSY-5 acoustic emission instrument to detect the bottom of atmospheric storage tanks in use in the oilfield. They analyzed the on-site inspection situation and measured data, and found that acoustic emission parameters can better predict the corrosion status of the tank bottom. However, since acoustic emission obtains dynamic information, it is necessary to reduce the interference of other noise when sensors obtain information. At the same time, Jiang Linlin et al. [20] of China petroleum engineering Technology Research Institute conducted corrosion evaluation on the bottom plates of four storage tanks using acoustic generation technology, and Weng Xiang et al. [21] of Xinjiang Oilfield conducted corrosion evaluation and detection on the bottom plates of one floating roof oil storage tank, all of which achieved good results.

The acoustic emission online detection technology can qualitatively evaluate the corrosion of the bottom plate of metal storage tanks, but it is conservative compared to the actual results. The evaluation results should be appropriately corrected and the corrosion degree rating should be upgraded [22-23]. The process of acoustic emission detection of storage tanks is shown in Figure 1.

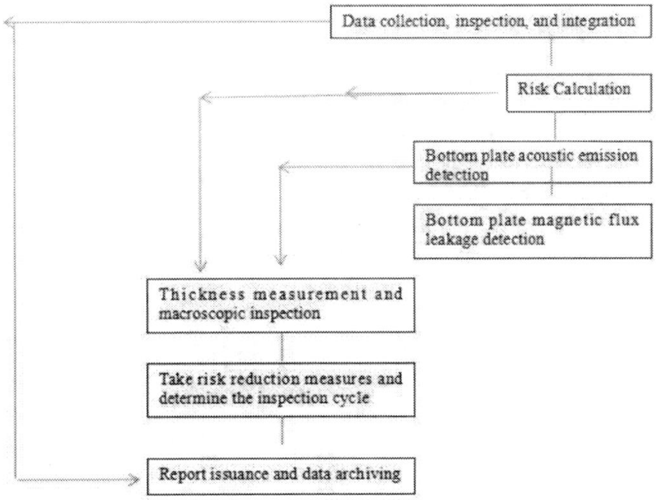


**Fig.1.** Acoustic Emission Testing Tank Flow Chart

## 3. Application of leak detection technology based on acoustic emission technology

### *3.1 Overview of storage tanks*

The storage tank for acoustic emission testing in this inspection belongs to one crude oil metal storage tank of North China Oilfield Erlian Branch. The on-site photos of the storage tank are shown in the figure. The crude oil metal storage tank has a capacity of 5000m3 and was put into use in 1989. The basic information is shown in Table 1.

Journal of Physics: Conference Series **2594** (2023) 012059 doi:10.1088/1742-6596/2594/1/012059

**Fig.2.** Site Photo of Crude Oil Metal Storage Tank

Table 1 Basic information of storage tanks

| Basic information | specification parameters |
|---|---|
| Tank material: | carbon steel |
| Medium type | crude oil |
| year of construction | 1989 |
| Size (diameter × Height, meters) | 24×11m |
| Foundation type | concrete |
| Tank capacity | 5000m3 |
| Tank roof type | dome |
| Working temperature | 25℃ |
| Nominal wall thickness | 11mm |
| Operating liquid level | 80% |

### *3.2 Acoustic emission detection process*

The acoustic emission detection adopts the SAMOS-32 acoustic emission detection and analysis system of American Physical Acoustics Company, paired with an integrated acoustic emission probe (response bandwidth of 20khz-100khz). The sensor is coupled and fixed with a vacuum grease and magnetic suction seat between the sensor and the tank wall, and polished at a distance of 30-50 cm from the tank bottom φ A 30mm area with a metallic luster is used to install acoustic emission sensors. Six sensors are uniformly set up (1-6 in Figure 3) and connected using cables. After the liquid level of the storage tank reaches the online detection height of acoustic emission (80% of the tank capacity), the inlet and outlet valves, agitators, heaters, and other interference detection devices are closed. The stability time of the storage tank is 2 hours.

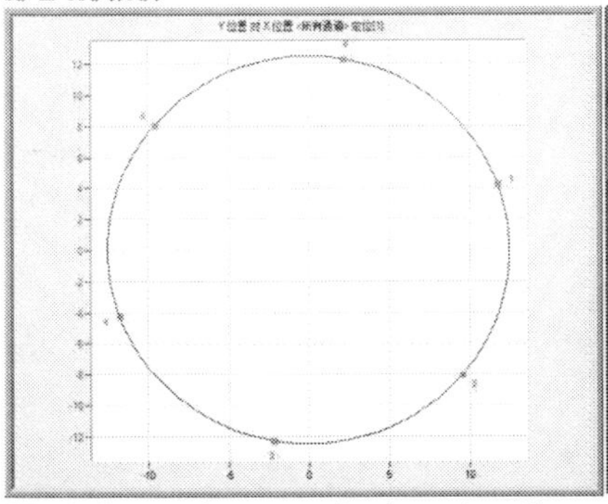

**Fig.3.** Sensor layout and orientation diagram **Fig.4.** On site inspection diagram

### *3.3 The Analysis of Acoustic Emission Testing Data*

A total of 1 storage tank was tested in this batch. During the inspection period of the tank, it was disturbed by periodic signals and continuously collected for 2 hours. During the collection process, the temperature of the tank wall was 22 degrees Celsius, the local temperature was around 25 degrees Celsius, and the temperature of the outer iron sheet of the insulation layer reached around 40 degrees Celsius in the direct sunlight area. The results obtained through the analysis of the acoustic emission detection and analysis system are shown in Figure 5-7.

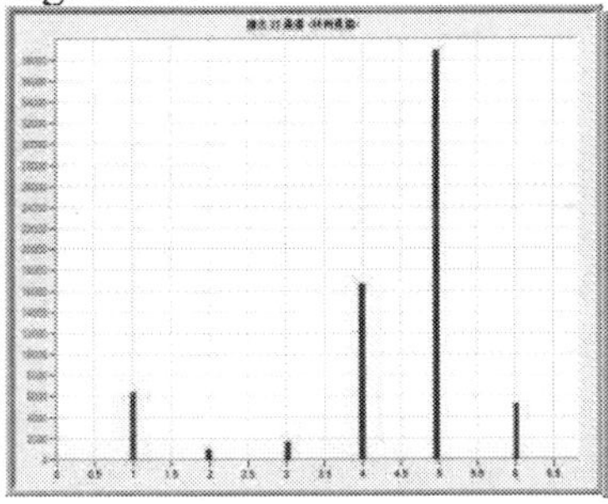
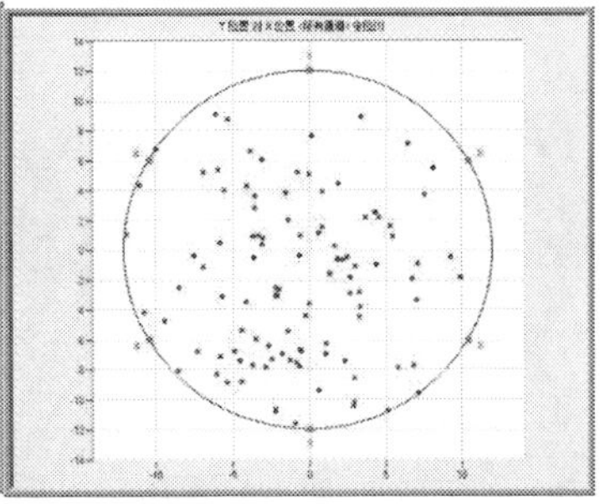

**Fig.5.** Comprehensive rating of corrosion condition **Fig.6.** Impact Time Accumulation Chart

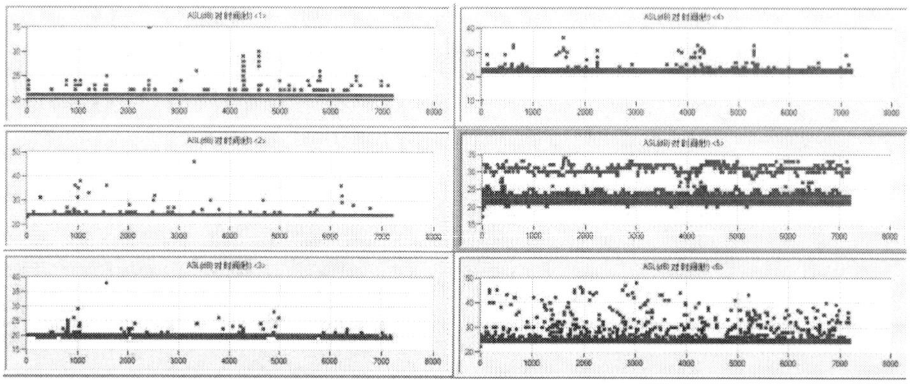

**Fig.7** ASL Time Scatter Plot

The collected data is processed and monitored, including human interference noise, environmental noise, mechanical interference, and condensation droplets. The corresponding noise data is filtered based on signal characteristics. During the calibration process, it is ensured that each sensor effectively receives the signal and reaches a certain amplitude, meeting the requirements for on-site use. At the same time, it can be determined whether the connection sequence of each channel is correct. As shown in Figure 6, the cumulative impact time graph of the acoustic emission detection plane and the ASL time scatter plot shown in Figure 7 can be obtained. From the graph, it can be determined that the impact data of channels 4 and 5 present a high level. Based on this, Analyze the risk of corrosion leakage in the vicinity of this area.

## 4. Open can magnetic flux leakage detection

After the tank was opened and cleaned, magnetic flux leakage testing was conducted on the key suspected areas of acoustic emission. The results showed that 17 defects with an equivalent depth exceeding 40% were found on the middle bottom plate, with a maximum equivalent depth of 85%. No defects with an equivalent depth exceeding 40% were found on the edge bottom plate.

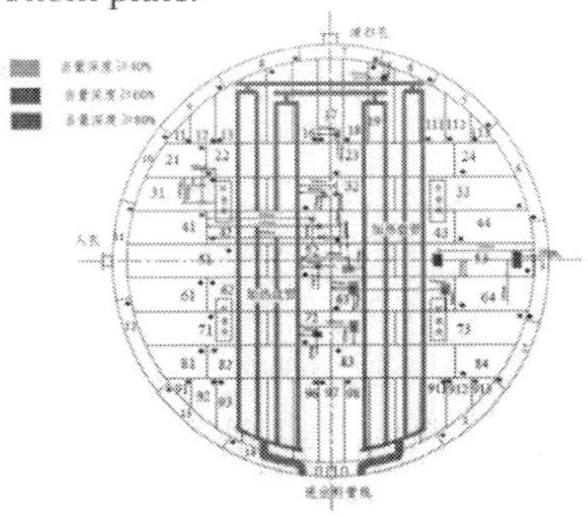

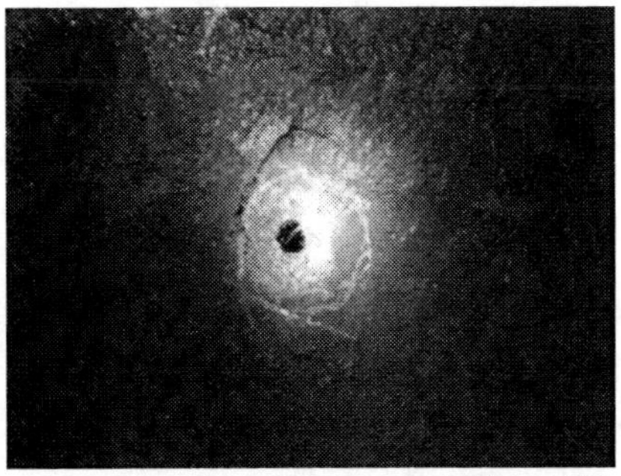

**Fig.8.** Schematic diagram of testing parts **Fig.9.** Corrosion defect map of the middle plate

## 5. Comparison of Acoustic Emission Testing and Magnetic Flux Leakage Testing Results

By comparing and analyzing the results of acoustic emission testing and magnetic flux leakage testing, it can be concluded that the leakage of the bottom plate of the storage tank and the activity of the sound source are rated as Level II and Level V, respectively. This means that the possibility of leakage is small but the corrosion is severe. Therefore, a plan should be formulated as soon as possible to arrange maintenance, and dynamic monitoring must be carried out for its use ($\leq$ 1 year). After opening the tank for inspection, it was found that there was already 1 meter deep oil sludge inside the tank. According to the analysis of magnetic flux leakage testing results, it can be seen that the bottom plate of the storage tank is corroded by 26% to 100%, with one perforation point. Due to the 1 meter thick oil sludge covering the bottom plate, no crude oil leakage occurred. So analysis shows that acoustic emission detection technology can obtain qualitative results of corrosion on the bottom plate of storage tanks, and compared to the actual situation, the detection results are relatively conservative.

## 6. Conclusion

(1) Acoustic emission technology can qualitatively evaluate the corrosion of tank bottom plates, but it cannot accurately locate corrosion defects (positioning errors are often measured in meters). It can only comprehensively evaluate and grade the severity of tank bottom plate corrosion, and provide suggestions for the next steps to be taken. For tanks with severe corrosion that require maintenance, it is still necessary to open the tank and use magnetic flux leakage and other detection techniques to determine the location of defects. The significance of acoustic emission testing lies in the overall evaluation of the severity of tank bottom plate corrosion, identifying tanks that require priority maintenance from a large number of tanks, and assisting managers in formulating maintenance plans.
(2) Through comparative analysis of acoustic emission testing technology and magnetic flux leakage testing technology, the results obtained from acoustic emission testing technology are conservative compared to the actual results. It is recommended to upgrade the evaluation of corrosion degree for the acoustic emission evaluation results of crude oil metal storage tanks.

**References**

[1] Fan Guotao,Zhao Yongtao. Acoustic Emission Data Analysis of Tank Settlement[J]. Journal of Physics: Conference Series,2021,2025(1).

[2] SUMMA V. In-service above ground storage tanks inspection robotics technology update[J]. The Inspectioneering Journal， 2008， 14 （2）： 11-19.

[3] SCHEMPF H， CHEMEL B， EVERETT N. Neptune： aboveground storage tank inspection robot system[J]. IEEE Robotics & Automation Magazine， 1995， 2 （2）： 9-15.

[4] CRUZ A C， RIBERIO M S. In service inspection robotized tool for tanks filled with hazardous liquids — Robtank Inspec[J]. Industrial Robot， 2005， 32 （2）： 157-162.

[5] SATTAR T P ， LEON-RODRIGURZ H ， SHANG J ， et al. Automated NDT of floating production storage oil tanks with a swimming and climbing robot[C]. London ： 8th International Conference on Climbing and Walking Robots and their Support Technologies for Mobile Machines， 2005： 935-942.

[6] SIEGER T. PETROBOT ： opening up the oil ， gas ， and petrochemical markets for robotics： 610401[R]. Rijswijk： Shell Global Solutions International BV， 2016： 17-23.

[7] Zhao Jie, Liu Yanfeng, Peng Xiwen, Ju Zichen, Wang Yuanyuan, Yang Chao. Analysis of hidden danger management schemes for atmospheric storage tanks in gathering and transportation systems [J]. Petroleum and Chemical Equipment, 2023,26 (01): 129-132

[8] Wang Shi, Deng Jin, Xing Shu, Wang Yanlong, Du Liang. Application of online acoustic emission technology in tank bottom leakage detection [J]. Petrochemical Equipment Technology, 2022,43 (06): 32-35+40+5-6
[9] Hu Huasheng, Sun Jie. Application of Typical Corrosion Detection for the Bottom Plate of Atmospheric Storage Tanks for Hazardous Chemicals [J]. Equipment Supervision, 2022 (03): 35-38
[10] Qin Changning Research on the Identification Method of Tank Bottom Corrosion State Based on Acoustic Emission [D]. Shenyang University of Technology, 2022
[11] Zhang Hanqi,Liu Tao,Lu Jiyun,Lin Renbang,Chen Changhao,He Zifan,Cui Shengming,Liu Zhixiang,Wang Xiaowei,Liu Bo,Xiong Ke,Wu Qi. Static and ultrasonic structural health monitoring of full-size aerospace multi-function capsule using FBG strain arrays and PSFBG acoustic emission sensors[J]. Optical Fiber Technology,2023,78.
[12] Ma Yuefei,Liu Meiyu,Yang Lu,Dai Peng,Fan Junwei. Damage analysis of concrete-filled square stainless steel columns based on acoustic emission and Markov chain methods[J]. Engineering Structures,2023,281.
[13] Hu Jia,Yu Yonghua,Yang Jianguo,Jia Haichao. Research on the generalisation method of diesel engine exhaust valve leakage fault diagnosis based on acoustic emission[J]. Measurement,2023,210.
[14] Wang Chun,Li XinRu,Cheng LuPing,Jiang DongPing,Xiong ZuQiang,Lei BinBin,Zhang PanLong,Zhan ShuaiFei. Mechanical properties and acoustic emission characteristics of deep hard coal after segmented high-temperature treatment.[J]. Scientific reports,2023,13(1).
[15] Dong Longjun,Zhang Yihan,Bi Shuijin,Ma Ju,Yan Yihao,Cao Heng. Uncertainty investigation for the classification of rock micro-fracture types using acoustic emission parameters[J]. International Journal of Rock Mechanics and Mining Sciences,2023,162.
[16] Xu Xiangbo,Jin Zuquan,Yu Yong,Li Ning. Damage source and its evolution of ultra-high performance concrete monitoring by digital image correlation and acoustic emission technologies[J]. Journal of Building Engineering,2023,65.
[17] Barile Claudia,Pappalettera Giovanni,Paramsamy Kannan Vimalathithan, Casavola Caterina. A Neural Network Framework for Validating Information–Theoretics Parameters in the Applications of Acoustic Emission Technique for Mechanical Characterization of Materials[J]. Materials,2022,16(1).
[18] Zhang Shuxin, Sun Bingbing, Luo Xiaowu, Zhou Huiping, Weng Xiang, Zhu Yongbin, Luo Jinheng. Online Acoustic Emission

Testing and Verification of Crude Oil Tank Bottom Plate [J]. Petroleum Pipe and Instrument, 2022, 6 (02): 75-78

[19] Chen Jianfei, Sheng Hua, Qiu Dongquan, Wang Anquan, Xu Xiaoguang. Application of acoustic emission in the detection of atmospheric storage tanks in oil fields [J]. Inner Mongolia Petrochemical, 2011,37 (19): 28-31

[20] Jiang Linlin, Li Lingjie, Su Bihuang, Wang Zhitao, Zhang Yanjun. Application of acoustic emission technology in corrosion detection of tank bottom plates [J]. Corrosion and Protection, 2021,42 (02): 56-59+77

[21] Weng Xiang, Cheng Mengmeng, Luo Xiaowu, Deng Liyuan. Research on the adaptability of online acoustic emission detection for defects in the bottom plate of crude oil storage tanks [J]. Petroleum Industry Technical Supervision, 2021,37 (12): 21-24

[22] Huang Xin,Wang Tong,Luo Yanbin,Guo Jiaqi. Study on the Influence of Water Content on Mechanical Properties and Acoustic Emission Characteristics of Sandstone: Case Study from China Based on a Sandstone from the Nanyang Area[J]. Sustainability,2022,15(1).

[23] Deng Huang-Shi,Fu He-Lin,Zhao Yi-Bo,Shi Yue,Huang Xianquan. Using acoustic emission parameters to study damage and fracture characteristics of concrete with different pour intervals cold joints[J]. Theoretical and Applied Fracture Mechanics,2022,122.

[24] Author, F.: Article title. Journal 2(5), 99–110 (2016).

[25] Author, F., Author, S.: Title of a proceedings paper. In: Editor, F., Editor, S. (eds.) CONFERENCE 2016, LNCS, vol. 9999, pp. 1–13. Springer, Heidelberg (2016).

[26] Author, F., Author, S., Author, T.: Book title. 2nd edn. Publisher, Location (1999).

[27] Author, F.: Contribution title. In: 9th International Proceedings on Proceedings, pp. 1–2. Publisher, Location (2010).

[28] LNCS Homepage, http://www.springer.com/lncs, last accessed 2016/11/21.

# Effect of Gas Injection on the Enhancement of Condensate Recovery in a Gas Condensate Reservoir

**A Esmaeili[1]*, L E Garzón Rojas[1], A CRodrigues Nogueira[1], R Simões Angélica[1], S P Aranha Da Paz[1], M Dehghani[1], C Guerrero[1]**

[1]Universidade Federal do Pará (UFPA), Belem, Brazil

esmailyabdollah@gmail.com

**Abstract.** So far, many methods have been used to increase production from gas condensate reservoirs, but in general, all these methods can be placed in two groups:1) Methods of maintaining pressure or preventing pressure drop; 2) Creating chemical changes in reservoir fluids to improve production mechanisms due to capillary forces.Also, many of the methods used have realized both of the above purposes and at the same time while maintaining the reservoir pressure (or preventing pressure drop) also change the chemical physical properties of the reservoir. One of these methods is gas injection. What causes severe changes in the performance of condensate gas reservoirs as soon as the pressure reaches the dew point pressure is the change of the reservoir fluid from a single-phase gas state to a two-phase gas condensate state. In order to accurately study the fluid behavior of one of condensed gas reservoirs, the reservoir fluid was simulated by software, four samples of nitrogen gas, carbon dioxide, methane and separator gas, were injected into the reservoir and then the effect of changing injection rate, injection pressure and injection duration on gas condensate recovery has been investigated by these injection gases.

## 1. Introduction

The study of condensate gas reservoirs is one of the challenging cases in the field of petroleum engineering. Typically, production from these reservoirs without pressure stabilization results in the extraction of a large amount of gas with a relatively small amount of condensate. By reducing the reservoir pressure and reaching the dew point pressure, natural production from the reservoir usually leads to an increase in the saturation of gas condensate in the reservoir. This increase in condensate saturation causes two major problems; First, it causes the relative permeability of gas to decrease and the amount of gas produced on the surface also decreases; Second, these condensates are stuck in the reservoir and can no longer be easily removed, and in some cases, it causes it to be completely irremovable.Figure 1 shows a common characteristic of gas condensate fluids.

In order to choose the best scenario of increased harvest that leads to the least wastage of condensate in the reservoir and increase the recovery factor, different production and injection scenarios have been investigated. Three types of gas, carbon dioxide, nitrogen, and exhaust gas from the separator have been used for injection. The results showed that returning the exhaust gas from the separator is the best scenario to solve the problem. It was also observed that the best time to carry out the processes of increased harvesting is at the same time as the production starts and before the pressure drops below the dew point of condensate gas.

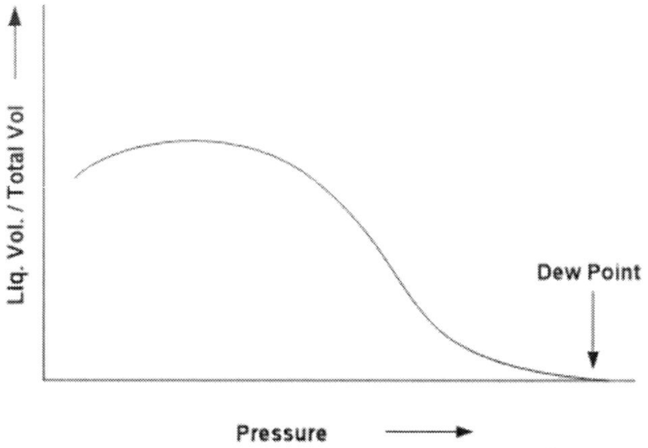


**Figure 1.** Liquid drop out of gas condensate

## 2. Effects of the Types of Injection Gas

In condensate gas reservoir, when the pressure of the reservoir reaches below the pressure of the dew point of the gas, as a result of reverse condensation, the gas in the reservoir turns into a liquid. Accumulation of condensate in the reservoir, especially near the well wall, causes blockage of the fluid passage, reduces gas mobility, and thus reduces the ability to produce from the well. The formation of these valuable condensates inside the reservoir means their loss, because it is no longer possible to produce these condensates. The gas injection process is carried out for two purposes: fully maintaining the pressure and preventing a sharp drop in pressure. Maintaining full pressure is one of The best methods to increase and maximize the amount of recovery Condensate is in this method, the gas produced is re-injected into the reservoir and only gas condensate is produced from the reservoir and this gas is produced. It is injected into the reservoir as a cycle and produced from it be made The point here is that this method is very expensive.

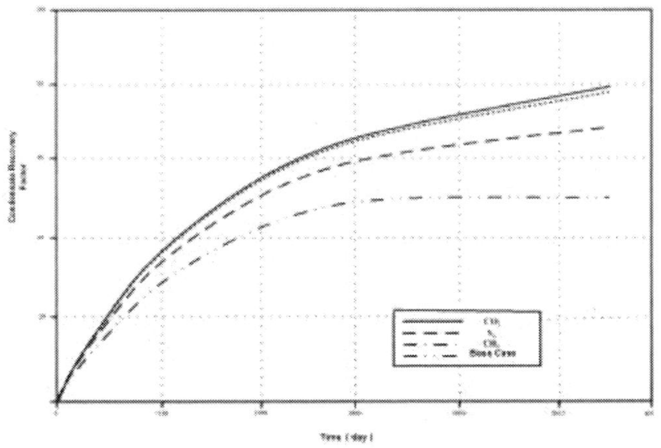


**Figure 2** - Types of Injection Gas and its Effects

## 3. Effect of Injection at Different Reservoir Pressure

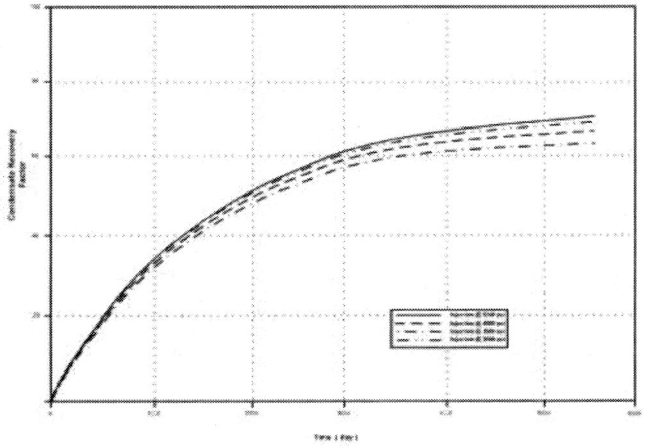

Figure 3 - Effect of Injection at Different Reservoir Pressure

## 4. Effect of Injection Rate

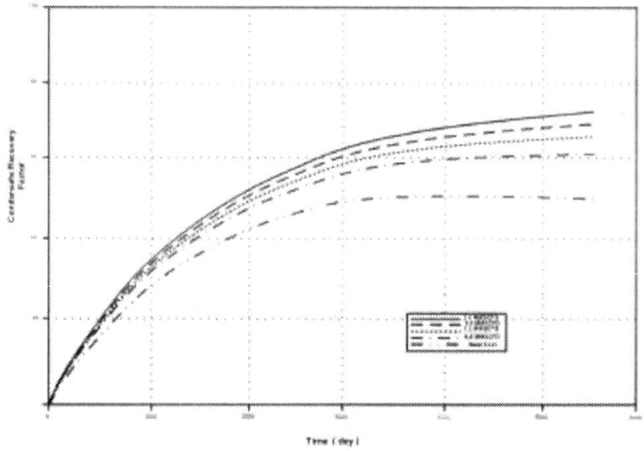

**Figure 4** - Effect of Injection Rate

## 5. Effect of Injection Duration

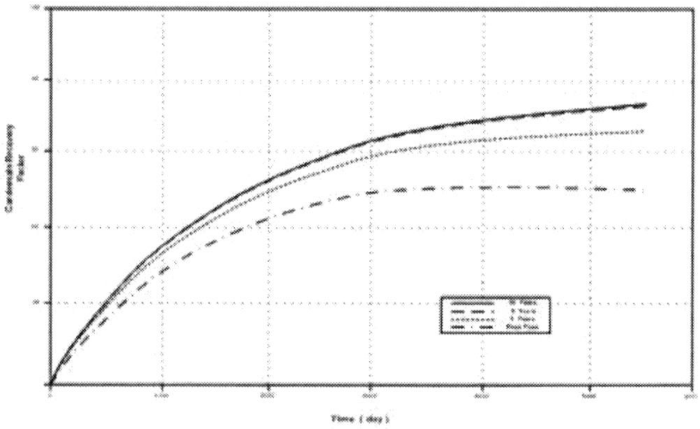

**Figure 5** - Effect of Injection Duration

## 6. Conclusions

Due to the rapid pressure drop in the reservoir and reaching less than the dew point pressure reservoir, condensate forms in the reservoir and falls into the trap. as a result Most of these precious liquids remain in the reservoir and cannot be produced. As a result, there is a need to maintain reservoir pressure with different methods and the fluid moves to the production well and does not liquid in the reservoir be made. In this study, injection of different gases with different pressures and flow rates injection has been investigated. The most optimal injection scenario of each gas It is introduced to the reservoir in each section. In this section to summarize Different gas injection scenarios are discussed and the optimal injection scenario for each gas It has been determined and compared to the most suitable option for this reservoir to become clear.

## 7. References

[1]. D.R Mc Cord & Assoc., "Reservoir Engineering and Geological Study and Analysis for Fracture Operation", vol. 2, 2018.
[2]. D.R.Mc Cord & Assoc., "Fracture Study of Asmari Reservoir", 2019.

[3]. "Log Interpretation Charts", Schlumberger Well Services, 2011.
[4]. "Schlumberger Log Interpretation Principles", Schlumberger Educational Services, 2015.
[5]. Tiratsoo, E.N., "Oil Field of the World", Gulf Publishing Company, 2012.
[6]. Sylvain j. Pirson, "WELL LOG ANALYSIS", Prentice-Hall, 2016.
[7]. Thompson & Wright, "OIL PROPERTY EVALUATION", Colorado School of Mines, 2011.
[8]. James R. Jorden & Frankl. Campbell,"Well Logging", Society of Petroleum Engi-neers, 2019.
[9]. "Petroleum Measurement Tables", ASTM, 2021.

**Acknowledgments**

The authors wish to express their sincere thanks and gratitude to Universidade Federal do Pará for their helps and encouragements in connection with this study. We are grateful for all their assistances.

# The need for an environmental strategy in the oil, gas and petrochemical industries

**L E Garzón Rojas[1]*, A Esmaeili[1], A C Rodrigues Nogueira[1], M Dehghani[1], C Guerrero[1]**

[1]Universidade Federal do Pará (UFPA), Belém, Brazil

lauraestefaniagarzon28@gmail.com

**Abstract.** The different stages of oil exploration, extraction, and processing, as well as its down-stream industries and the activities of refineries and petrochemical exploitation, leave multiple effects on the environment. For certain, as a result, soil, air, water (including surface and underground), organisms, plants, trees and even humans are affected by these effects. Hence, the fundamental importance of sustainable development and attention to the needs of the future generation, as well as the relevance of preventing the re-duction of environmental quality, encourages and requires oil industry and environmental experts to think of solutions and provide operational-strategic solutions to control and reduce the adverse environmental effects caused by these industries. Indeed, it is necessary to diversify control strategies and include environmental considerations in the planning of the oil, gas, and petrochemical industries, but the current situation is far from ideal. This article first briefly presents some environmental challenges and problems of the oil, gas and petrochemical industries and then describes the need for an environmental strategy to overcome these issues.

## 1. Introduction

The environment in which we coexist interrelates several physical, biological, and chemical factors that can affect us as a species, as well as influence other living organisms present on the planet. The equilibrium presents in the different terrestrial and marine ecosystems, the biodiversity of fauna and flora, and the multiplicity of water sources such as aquifers, seas, rivers, and marshes, are considered indispensable elements of the natural environment and each of them contributes effectively to preserving appropriate conditions for life [1]. In fact, the living nature of our environment is a part of our life. Nevertheless, most of the responsibility for the imbalance of these conditions lies in the hands of humans, who play an important role in the accelerating destruction of nature due to the uncontrolled exploitation of natural resources. Nowadays, the current environmental crises around the world have awakened new initiatives for the preservation and protection of natural resources, as well as campaigns for the diffusion of environmental culture, acquiring a special position in relation to energy activities in the world.

The reality is that enhanced measures are required to prevent and control the harmful effects that may occur in the environment, effects that were not previously considered relevant, but which require that scientists, governments and society in general work to align technical-scientific development and environmental protection, seeking their com-patibility through sustainable strategies that promote the continuation of life for future generations. On the other hand, the oil, gas and petrochemical industries have generated employment for many people and, therefore, play a key role in today's social and economic life, supply investments of more than US$511 billion in 2020, however, many of the processes involved the use of chemical substances that can have negative effects on the environment and on health [2]. Thus, it is essential to continue devising strategies to eliminate or reduce these risks to an acceptable level and to continue generating not only economic benefits but also scientific, social and environmental benefits that can be sustainable.

Following the definition of strategy, it can be deduced that it is necessary to generate guidelines that allow overcoming the current situation to the desired one. The industrial sector continues growing worldwide, as well as the energy demands and environmental impacts of these advancements as evidenced by the 34 percent of total net anthropogenic greenhouse gas (GHG) emissions coming from the energy sector in 2019 according to the Intergovernmental Panel on Climate Change (IPCC). Therefore, the concern for environ-mental protection becomes a priority within the criteria and standards of sustainable industrial development of oil, which, being a very valuable resource for the production of energy and various derivatives, must invest in the supervision of each of the extraction phases, contributing to the elimination of several sources of environmental pollution that are sometimes irreparable [2].

Contamination of the environment with petroleum substances can rarely happen naturally in some cases, for example, in some oil-rich regions of the world, where the petroleum resources are found near the earth's surface, it may come out from underground as a boiling spring and pollute the surrounding environment with this substance. The major occurrence of pollution cases is caused by human activities, since oil rich areas are usually located in seabeds and coastal areas, during extraction, oil spills in different extraction routes can reach the waters of the sea or the ocean, as well as the air with the volatile hydrocarbon's evaporation and the accumulation or pollutants of the ecosystems [3].

Creating oil pollution is done in the way of transferring this substance to be delivered to another destination, which can be caused by leakage from pipes or caused by ocean tankers carrying oil, conndicating with last statistics made by ITOPF-2022, where topics such as lost as a result of tanker incidents are evaluated, that for the period 1970-2022, it was approximately 5.88 million tons, and for the 2022 was approximately 15,000 tonnes. Environmental contamination by petroleum hydrocarbon, leaked in the various forms of extraction and transportation is one of the significant concerns of the recent world, although it happens mostly unintentionally, but in many cases, it is a consequence of human negligence [1].

Due to the physical-chemical nature of oil, its entry into our environment can cause a lot of issues, many of which are known today, and many others might be known when the consequences become apparent in the future, not only for humans but also for other bio-tic components of the ecosystem [4]. When crude oil reaches the surface of the earth or water, its lighter parts, which are also very toxic, such as benzene and toluene, quickly evaporate and pollute the air in that area, but the heavier parts of petroleum compounds remain in the environment for a long time. It remains and pollutes the earth's soil and the water of the seas and oceans as well as their shores. Therefore, although oil and its compounds are necessary for life, carelessness in its production and transportation and contamination of the environment with these substances can take life from humans and other creatures on this planet.

During the last decades several environmental debates have been highlighted which have allowed the development of new chemical and physical methods that seek to remediate pollution generators such as burying, evaporation, dispersion and others, however the high cost of these technologies has driven to innovation in the decomposition of pollutants by ecological methods such as bioremediation [1].
There are several causes that can generate the entry of millions of litters of crude oil into the seas due to different types of maritime accidents, such as oil spills, ship groundings, fires, among others; these spills of hundreds of thousands of litters of oil residues into the environment can be caused by the movement of land, sea and air vehicles, as well as by the activities of hundreds of refineries and related industrial centres. Meanwhile, the development programs of oil, gas and petrochemical

industries have doubled the environmental challenges of oil, gas and petrochemical industries; due to the development of legislation and the signing of some protocols, agreements and conventions related to maritime accidents and pollution, the number of accidents and incidents has been gradually decreasing, however the marine environment is still polluted by ships all over the world [5].

## 2. Pollution of the Marine Environment by Petroleum Products

The toxic effects of pollutants, especially petroleum substances, on aquatic organisms depend on the physical and chemical factors of the environment, the type, composition, quality, and amount of petroleum substances in the aquatic environment, as well as the various aquatic species that are exposed to petroleum substances [6]. Several studies indicated that the highest short-term damage was caused by light oils, as well as the effects of petroleum substances on aquatic life, show that mammals and fish are not the only creatures that are destroyed by hydrocarbons, some petroleum compounds act in a harmful way on plankton, crabs, shrimp, among other small species, even generating permanent damage in their evolution, diversification and sometimes massive death [7].

Species such as seabirds that swim on the surface of the water are strongly affected by oil spills, as their feathers quickly become stained with oil and they lose their vital characteristics, such as the ability to fly, hunt, adapt to changes in body temperature and move from one environment to another [8]. The marine creatures, from the smallest microbes to the largest whales, are not immune to pollution and these toxic compounds can destroy them; either by superficial contact or by the entry of various petroleum com-pounds into the digestive system of birds and other species, causing serious injury and ultimately death [9].

The existence of rich oil aquifers in the bed and margin of the North Sea of Iran, which is internationally known as the Caspian Sea, and the extensive exploitation of these areas caused since the last century a large oil pollution in the region according to the report of the Caspian Sea Environment Bureau, being taken protection meaty as the treaty gives UNEP, 2003. Due to the geomorphological characteristics of the sea, most of these pollutions occur towards the south, where the oil pollutants accumulate together with the chemical contamination caused by the rivers flowing into the sea, being these ecological problems influenced by rising waters and environmentally destructive activities, which are extremely dangerous for the coastal ecosystems. In summary, the environmental problem caused by the unrestrained activities of the co-inhabitants of the Mazandaran Sea margin has caused ecosystemic degradations along the Mazandaran Sea margin [10].

The Persian Gulf accounts for 60% of the global oil tankers travel, where every year a million tons of crude oil enters as a result of oil operations, exploration, transport, sea collisions, and the emptying of the balance water of vessels, becoming the main marine pollution hazard, exposing this sea to physical and chemical de-structuring of the habitats, the Persian Gulf being one of the locations most affected by oil spills [11]. The bio-diversity of this coastal environment, as mangrove forest, is fundamental in the conservation of the biological balance, being an area of high production of nutrients that are discharged into the seas through the rivers and biodiversity including corals, sponges and fish, which are being affected by the destructive effects resulting from oil activities. It is necessary to know the effect of the pollutants, especially in the food chains, by testing the living organisms, after contact with these oil pollutants [7].

Although, the Persian Gulf is one of the most varied aquatic environments in the world, the presence of huge oil and gas reserves in this area has led to the expansion of drilling operations, extraction, refining, construction of oil facilities, loading and transport in supertankers and increased pollution of the marine environment, these hydrocarbons are the main pollutants of the environment for biological systems [11].

## 3. Some Problems of Environmental Strategy of Oil, Gas and Petrochemical Industries

The It can be affirmed that the most polluting element of the environment today is humans. The presence of humans in many regions of the world has caused the disruption of order, harmony and biodiversity in those regions, leading to habitat degradation, biodiversity decline, climate change and pollution accumulation, as reported by the WHO European Centre for Environment and Health in 2022 [12]. The anthropogenic influence generated by the institutions that exploit subway resources has stronger environmental implications than other industries; the exploitation of subway resources requires more precision and planning because it is mainly carried out in less penetrated areas that may be more vulnerable to some type of environmental degradation, seeking to preserve the existing natural state and not to destroy it.

Within the oil and gas industry, deforestation is caused not only by the process of oil drilling but also by the infrastructure needed during the subsurface resource search and exploration phase is a common area of concern as the cutting of thousands of trees be-comes inevitable to facilitate access to these deposits; although companies are obliged to invest capital in the reforestation of these areas, this process is still slow and the impact on climate change is evident, requiring quicker and more effective actions [13]. The formation of oil layers in less destroyed areas means that the slightest change in the environment has a greater impact; however, it is important to monitor each stage of the pro-cess, including the transfer of the oil to the refineries from the extraction points, since in most cases, due to the passage of pipelines through desert areas and bad weather, there is a possibility that these pipelines may be damaged [14].

Furthermore, with the presence of huge offshore drilling rigs and oil fields, water pollution together with deforestation, soil erosion, interruption of water flows and overexploitation of natural resources in these areas, as well as the domestic waste generated, make this activity one of the most dangerous due to the lack of control and the spread of contaminated waste from the installation of drilling, as well as noise, vibration and other consequences that can have a great impact on the ecosystems. In this type of contamination generated in the exploitation phase may trigger the vertical migration of contaminants to aquifers and the sea, when leaving the subsoil associated compounds overflow and migrate to surrounding areas, being populations of diverse species such as fish affected by oil contamination of water, and birds that feed on these infected fish can even die, causing an alteration of the life cycle and the food chain, which also affects human life [15].

With the growth of human knowledge, a reliable alternative has been introduced for al-most everything in terms of fuel, however, no reliable alternative has been provided for the environment. Destroying the environment is easier than planting trees or protecting natural resources. Although good protective and monitoring measures have been formed in the last few years, there is a long way to reach the desired situation.

The current demand for energy supply following sustainable and environmentally friendly initiatives is a booming issue, this type of alternative fuels has many advantages, however, although they are less polluting than conventional fuels, their continued use involves some consequences such as contamination and microbiological corrosion, in addition to the limitations in infrastructure, adaptation and resources required for their mass implementation [16]. The scientific development of these alternative fuel sources has advanced significantly in technical terms, nevertheless, the environmental and conservation aspect continues to be only superficially addressed, and it is necessary that the protection, monitoring as well as control measures be more rigorously specified and implemented, regardless of the type of energy source.

The use of toxic and flammable solvents is one of the most important hazardous problems in the chemical industries; having harmful effects not only for the health of workers, but also the possibility of the entry of these solvents into the environment, which can cause the spread of contaminants that require high-cost processes for their elimination and control [17]. In general, the problems caused by the separation of solvents from products, the complete recycling of these solvents and environmental hazards were among the factors that turned researchers' attention to the use of green solvents in order to use them to absorb the emissions of greenhouse gases, like CO2 from natural gas [18].

Chemical industries have a key role in the social and economic life of societies, not only by generating employment and an average global sale of chemical products valued at 3,669 billion euros in 2019, but also by the influence of their advances in the update of new alternatives in the transformation towards a toxic-free environment. However, this same industry has contributed negatively by using several chemical substances during its processes that can have harmful effects on the environment or human health, so eliminating or reducing these risks to an acceptable level is a very important issue [19]. Now-adays, several protocols, model for safety standards worldwide and legal restrictions have been developed in the stages of use, transfer, purification or disposal of chemical substances, contributing to the reduction of the release of hazardous substances and their possible risks; since this is a human health issue, it is essential to generate higher visibility of these risks and their containment measures to achieve safe and sustainable standards for the promotion of innovation as well as for the protection of the environment [20].

In brief, some of the environmental challenges of the oil industry can be summarized as follows: 1- Lack of attention to research and applied studies 2- Insufficient implementation of new technologies 3- Failure to use qualified and efficient experts 4- Lack of interaction and structural communication and coherent between the sub-sectors 5- Inattention to the risk of environmental hazards of industries 6- Weakness of environmental laws, guidelines and standards 7- Absence of strategic vision and long-term planning.

## 4. Suggested Solutions

As is known, crude oil and its products, either during extraction or during land and sea transportation, as a result of incidents and accidents cause soil, water or the environment in general. The presence of petroleum hydrocarbons on the surface of the sea and on land is considered a serious threat to the ecosystem, because they are extremely harmful and carcinogenic to the biodiversity; their occurrence can cause fire, underground water pollution and air pollution, for this reason, the cleaning of these pollutions should be done as soon as possible to have a safe and risk-free environment.

There are many standard and common methods for cleaning and decontamination of petroleum materials, which are limited due to high cost and low efficiency. For example, biological cleaning of petroleum materials are processes that convert toxic compounds into non-toxic and safe substances; this action takes place as a result of the metabolic activities of microorganisms that are able to use petroleum materials as their source of energy and carbon. The supply of this technology can be very useful, because it is able to convert the toxic compounds of petroleum substances into non-toxic substances without causing disruption in the natural environment. In comparison with other physical and chemical cleaning technologies such as burning and burying oil sludge, the biological method is much cheaper and more affordable.

In recent years, the oil industry has taken great steps to control oil spills and develop cleaner fuels with less sulfur to prevent air pollution in cities, but these strategies have to be more implemented to impact positively in the control of the problems caused by cli-mate change. Carbon, as one of the elements of oil, is one of the main elements of green-house gases that destroy the cooling system of the planet Earth. The goal of the Kyoto and UE Protocols is to reduce the level of carbon emissions and with that the reaction of most oil companies is to announce an increase in efforts to find natural gas that has a less polluting effect than oil. The second step of these companies is to try to find alternative types of energy such as solar energy to be used at different stages of exploration, production, infrastructure, and transportation.

Certainly, some of the solutions facing the oil industry can be summarized as follows: 1- Attention to environmentally friendly technologies: for example, the construction of environmentally save double-walled oil tankers to transport oil, in order to reduce the possibility of pollution due to oil spills. 2- Adaptation and localization of environmental laws, standards and guidelines: one of the serious needs is to review and adapt the general laws and international standards for different industrial areas of the specificities of the countries and to determine and compile the criteria, regulations and standards appropriate for each region. 3- Development of a strategic environmental assessment with the objective of including integrated biophysical, economic, social and political considerations in development policies, programs and plans. 4- Conducting training courses and promotional activities: long-term attitude should be taught to managers and employees of industries through courses and explanatory meetings. 5- Management of environmental crises: Although the observance of environmental laws leads to the control of pollutants and pollution, but the occurrence of unexpected incidents in the field of polluting industries, in the absence of management of emergency situations, can increase the depth of the resulting disaster several times. Incidents such as marine accidents for oil tankers can be a good example in this field. 6- Necessity of corrective measures for old units: As mentioned, some oil industry units were built in the past, at that time, compliance with environmental standards was not very important, as well as the outdated technologies.

But now, due to the undeniable role of these units in the country's industries, there is no way to stop their activities, but also, it is not possible to ignore the pollution from these materials with this justification. Therefore, by properly managing these pollutions and drafting new laws and amending the pollutions, the pollution burden of these industries should be reduced or updated and thus faced with their adverse effects. 7- Creation of a center for decision-making and policy research within companies: taking into account that many of the challenges in the environmental field of the oil and related industries are structural and are caused by the way of management and planning in this field; the creation of a center of knowledge whose mission is policy research, decision-making, strategic and pathological studies in this field and to analyse the management processes that affect the environment and to propose and provide solutions in this field together with technical studies. In addition to the above, these think tanks provide the ability to establish communication between the oil industry and research and academic centers.

8- Use of natural gas: The share of gas as a clean fuel in the consumption basket of fossil fuels should increase more than before. In this regard, the natural gas refining capacity and the number of areas covered by the gas supply network should be increased. 9- Implementation of plans for the development and improvement of the quality of refinery products: with the implementation of these plans, the quality of the products will be im-proved in such a way that the amount of cyclic (aromatic) compounds in the produced gasoline will be reduced to the level of acceptable international standards, and with the construction of new units for desulfurization of oil and gas and white oil, the amount of sulfur in these products will decrease. 10- Crude oil associated gas collection: Crude oil associated gas collection plans take a way to ensure that no gas is flared, thus, the global oil industry needs to implement strategies to avoid polluting the environment. 11- Elaboration of the strategy for sustainable development of fossil fuels by a committee of experts according to compilations of established protocols, which will be part of the strategy for sustainable development of industries and mines. 12- Use of crude oil processing mobile devices, these facilities are built and operated in order to prevent environmental pollution and oil wastage. 13- Providing fuel to nomadic and isolated areas: In order to prevent the destruction of forests and pastures, extensive measures should be taken to supply gas to villages in forested areas. For example, the implementation of rural com-pressed natural gas project will have very good results.

## 5. Conclusion

Despite the determination to reduce the dimensions and consequences of environmental pollution in recent years, these efforts have been insufficient and sometimes inefficient and ineffective. All of these concerns can be solved when the conditions for achieving the following prerequisites are created: 1- Creating an atmosphere of understanding and common understanding of the problems and issues between the officials of the Ministry of Oil and the Environmental Protection Organization 2- The partnership of all officials on the necessity of creating an effective and strategic interaction mechanism between effective institutions in this field 3- Agreement on the necessity of compiling an environmental strategy for the oil, gas and petrochemical industries 4- Determining the pre-requisites for compiling an environmental strategy for the oil, gas and petrochemical industries 5- Emphasis on creating active and strong centers for research and continuous intellectual feeding of workers and managers. 6- Applying scientific supervision and having the necessary legal and organizational powers to continuously monitor the development and implementation of the environmental strategy of the oil, gas and petrochemical industries. Since today, short-term planning and decisions are not the answer to environmental problems and challenges caused by industries, all environmental measures should be planned and followed in the form of long-term environmental strategies. The ecological modification suffered during these last decades of negative industrial impacts is a main aspect to be considered at the moment of exploiting and extracting resources, understanding that there is no way to deconstruct these consequences, however, the new exploration requires higher investments of time and money to restore and prevent dam-ages to the environment.

## 6. References

[1].Fowzia A, ANM Fakhruddin. A Review on Environmental Contamination of Petroleum Hydrocarbons and its Biodegradation. Int J Environ Sci Nat Res. 2018; 11(3):63-69. DOI: 10.19080/IJESNR.2018.11.555811.

[2].International Labour Office, 2022. The future of work in the oil and gas industry: Opportunities and challenges for a just transition to a future of work that contributes to sustainable development (Geneva, 28 November–2 December 2022), Sectoral Policies Department, Geneva, ILO.

[3].Das, N. & Chandran, P., 2010. Microbial Degradation of Petroleum Hydrocarbon Contaminants: An Overview. SAGE-Hindawi Access to Research Biotechnology Research International Volume 2011, Article ID 941810, 13 pages doi:10.4061/2011/941810

[4].National Research Council (US) Committee on Oil in the Sea: Inputs, Fates, and Effects. Oil in the Sea III: Inputs, Fates, and Effects. Washington (DC): National Academies Press (US); 2003. 2, Understanding The Risk. Available from: https://www.ncbi.nlm.nih.gov/books/NBK220690/

[5].Kozanhan, M.K., 2019. Maritime Tanker Accidents and Their Impact on Marine Environment. Scientific Bulletin of Naval Academy, Vol. XXII 2019, pg. 324-342. doi: 10.21279/1454-864X-19-I1-047

[6].Bakke, T., Klungsøyr, J. and Sanni, S., 2013. Environmental impacts of produced water and drilling waste discharges from the Norwegian offshore petroleum industry, Marine Environmental Research, Volume 92, Pages 154-169, ISSN 0141-1136, doi: https://doi.org/10.1016/j.marenvres.2013.09.012

[7].Norouzi, E.L., and Mohsenpour, R., 2020. Investigation of oil pollution on aquatic animals and methods of its prevention. Journal of Aquaculture & Marine Biology. 9(5): 160–165.

[8].Environment and Climate Change Canada, 2003. Catalogue number CW69-4/85-2003E-IN ISBN 0-662-34255-0. https://www.hww.ca/en/issues-and-topics/oil-pollution-and-birds.html Date of access: 22/04/2023

[9].Wells, P. G. 2001. Oil and seabirds—the imperative for preventing and reducing the continued illegal oiling of the seas by ships. Marine Pollution Bulletin 42(4): 251–252.

[10].Caspian Environment Programme, 2002. Transboundary Diagnostic Analysis For The Caspian Sea. Volume 2: 1-132.

[11].Hassanshaian, M., Amirinejad, N. and Askarinejad, M., 2020. Crude oil pollution and biodegradation at the Persian Gulf: A comprehensive and review study. J Environ Health Sci Eng. Dec, 18(2): 1415–1435. doi: 10.1007/s40201-020-00557-x

[12].A health perspective on the role of the environment in One Health. Copenhagen: WHO Re-gional Office for Europe; 2022. Licence: CC BY-NC-SA 3.0 IGO.
[13].World Rainforest Movement, 2013. Oiled forests – the case of Ecuador. WRM Bulletin 196 3 December 2013. Available in: https://www.wrm.org.uy/bulletin-articles/oiled-forests-the-case-of-ecuador
[14].Bravo, E., 2007. Los Impactos de la Explotacion Petrolera en Ecosistemas Tropicales y la Biodiversidad. INREDH.
[15].Cordes EE, Jones DOB, Schlacher TA, Amon DJ, Bernardino AF, Brooke S, Carney R, DeLeo DM, Dunlop KM, Escobar-Briones EG, Gates AR, Génio L, Gobin J, Henry L-A, Herrera S, Hoyt S, Joye M, Kark S, Mestre NC, Metaxas A, Pfeifer S, Sink K, Sweetman AK and Witte U (2016) Environmental Impacts of the Deep-Water Oil and Gas Industry: A Review to Guide Management Strategies. Front. Environ. Sci. 4:58. doi: 10.3389/fenvs.2016.00058
[16].Lee J.S, Ray R.I, Little B.J. 2010. An assessment of alternative diesel fuels: microbiological contamination and corrosion under storage conditions. Biofouling. Aug;26(6):623-35. doi: 10.1080/08927014.2010.504984
[17].ILO, International Labour Organization, 2021. Exposure to hazardous chemicals at work and resulting health impacts: A global review International Labour Office. Geneva. ISBN: 978-9-22-034219-0
[18].Adeyemi, I., Abu-Zahra, M.R.M., Alnashef, I., 2017. Novel Green Solvents for CO2 Capture, Energy Procedia, Volume 114, Pages 2552-2560, ISSN 1876-6102, https://doi.org/10.1016/j.egypro.2017.03.1413.
[19].Bonzini M, Leso V, Iavicoli I. Towards a toxic-free environment: perspectives for chemical risk assessment approaches. Med Lav. 2022 Feb 22;113(1):e2022004. doi: 10.23749/mdl.v113i1.12748. PMID: 35226649; PMCID: PMC8902740.
[20].COM, 2020. 667.
[21].Journal of Foraminiferal Researcher-Volume22,24,25 Number 1,2,3,4 - 1992,1994,1995
[22].Michael Tanzer. "Oil and the Gulf Crisis." in Beyond the Storm: A Gulf Crisis Reader. New York: Olive Branch. 1991. P. 265
[23].Reynolds, K., 2015. Are post-spill environmental studies becoming the norm? INTERSPILL, Amsterdam.

**Acknowledgments**
The authors wish to express their sincere thanks and gratitude to Univesidade Federal do Pará for their helps and encouragements in connection with this study. We are grateful for all their assistances.

# Research and Application of Intelligent Water Source Well Air Defense Pumping and Dry Burning Technology

**Bao-yuan GAO1,2(✉) ,Zhao-hui LI1,2,Wei-shi ZHENG1,2**

1 Drilling & Production Engineering Technology Research Institute, CNPC Chuanqing Drilling Engineering Company Limited, Sichuan, China
2 National Engineering laboratory for Low-permeability oil and gas field exploration and development, Shaanxi, China
gcy_gby@126.com

**Abstract.** The surface water injection technology for developing low permeability reservoirs comes from the water supply of water source wells. The groundwater resources in oil fields are limited, with high water consumption and a decrease in water level, lacking effective monitoring methods. In recent years, there have been frequent occurrences of unsafe waste of electrical energy due to dry burning during pumping in water source wells, which has seriously affected the normal water injection of oilfield formations. In order to prevent dry burning of water source wells by pumping air, research on intelligent air defense dry burning technology for water source wells has been carried out. The automatic pump stop alarm at low water level and the automatic pump start at high water level avoid the phenomenon of dry burning during pumping in the water source well, ensure the normal operation of the water source well, and ensure sufficient water injection from the source. Reduce employee labor intensity and achieve the goal of optimizing human resources. Associate the start and stop status of the water source well with the tank position of the water injection station to achieve automatic water distribution and inventory, achieve digital management and fine management, and achieve the goal of reducing costs and increasing efficiency, which has promotion value.

## 1 Preface

With the expansion of oilfield development scale, there are more and more oilfield water injection. Due to the limited groundwater resources, the water level drops, and the lack of effective monitoring means, the static level can only be observed through the water outlet pipe during each workover. When the water is drained, the power cannot be cut off in time, resulting in the phenomena of empty pumping and dry burning of water source wells, which leads to a large number of water source wells' submerged pumps being burnt out and wasting a large amount of electric energy, which seriously leads to the water supply of water source wells being stopped. Due to the reduction of water injection into the ground, the formation pressure and the oil production rate are reduced, It has seriously affected the completion of crude oil production. In order to ensure the normal water supply of the water source well, control the operation of the submersible pump, and meet the actual needs of the oilfield production and operation, the intelligent remote monitoring system for the water level of the water source well has been studied [1]. The intelligent remote monitoring of the water level, outlet pressure, outlet flow and electric parameters of the submersible pump of the water source well has been carried out, and the existing data resources of the water source well and the intelligent achievements have been rationally utilized, which has played the role of low water level early warning and high water level automatic pump start, It avoids the phenomenon of empty pumping and dry burning of water source wells [2], optimizes the working system of water source wells, ensures the normal operation of water source wells and energy conservation and consumption reduction [3], and ensures the sufficient water injection in the oilfield from the source [4]-[5].

## 2 Main equipment configuration of water supply system of water source well

The main equipment of the water supply system of the water source well includes: the upper computer of the engineer station, the remote monitoring center, the intelligent remote monitoring device (composed of the Siemens programmable controller PLC and its expansion module, the human-machine interface touch screen, the soft starter, the low-voltage electrical apparatus and the remote control terminal RTU, etc.) [6], the submersible pump, the outlet pressure gauge, the outlet flow meter, the radio frequency admittance sensor, the single-phase valve, the gate valve, the outlet pipeline, and the water storage tank, Water source wells, etc. [7].

## 3 Design of intelligent remote monitoring system for water level of water source well

The intelligent remote monitoring system for water level of water source wells is mainly composed of two parts: the remote monitoring device for water level of water source wells and the remote monitoring center [8].

### *3.1 Composition of remote monitoring device for water level of water source well*

The remote monitoring device for water level in the water source well includes the monitoring part for water level, outlet pressure, outlet flow and three-phase electrical parameters of the submersible pump, the part for soft starter to start the submersible pump and the part for remote control terminal RTU to collect and upload data [9].

### *3.2 Working principle of water level monitoring of water source well*

An intelligent monitoring device for water level of water source well is designed, which is simple in installation, reliable in performance, and can measure water level up to 914 meters deep with a small error of 1 ‰.

The RF admittance sensor probe is connected to the RF admittance transmitter probe cable, and the reference cable is connected to one end of the wellhead ground water outlet line. The RF admittance sensor probe installed parallel to the wall of the water outlet line connected to the submersible pump constitutes an electrolytic capacitor with the submersible pump water outlet line. The RF admittance sensor probe is the positive electrode of the electrolytic capacitor, and the water outlet line connected to the submersible pump is the negative electrode of the electrolytic capacitor, Water as electrolyte. When the height of the electrolyte water changes, that is, the water level of the water source well changes, the capacitance value also changes.

The working process of the radio frequency admittance device to measure the water level: the standard sine signal generating circuit sends out a 100kHZ radio frequency signal, and the interference components are filtered by the filtering circuit to obtain a relatively pure 100kHZ radio frequency signal. The capacitance driving circuit is used to improve the measuring range of the water level. The transformer bridge measurement circuit converts the water level signal into a voltage signal, and the measured voltage signal is divided into two channels: including the water level signal and the water hanging in the RF admittance sensor signal; Provide synchronous signal at sampling time. The synchronous signal is processed and the sampling time is given. At this time, the measured signal is sampled to obtain a pure water level signal, thus eliminating the influence of water hanging on the RF admittance sensor probe. The detected water level signal is sent to the programmable control system for A/D conversion and data processing through the low-pass filter circuit to eliminate some interference, and the actual water level is displayed through the display circuit. The actual water level is compared with the set water level up and down, and the signal for controlling the submersible pump is given. The purpose of low water level warning and high water level automatic pump start is achieved, and the phenomena of empty pumping and dry burning of water source wells are avoided [10].

### *3.3 Design of remote monitoring center*

The RTU module of the remote control terminal in the water level remote monitoring device uploads the water level signal collected by the radio frequency admittance sensor of the water source well, the water outlet pressure signal collected by the pressure gauge, the water outlet flow signal collected by the flowmeter, the electrical parameter signal of the submersible pump collected by the multi-function electric energy meter and the start and stop signal to the digital network through the special network of the oil field [11], After being processed by the firewall, it will be sent to the monitoring switch, which will further send the processed data to the central control room computer and the upper computer of the engineer station of the monitoring center through the network cable to realize the intelligent remote monitoring of the water level of the water source well, data storage, and network release [12].

## 4 Field application

More than 30 wells have been successfully applied in the oilfield, improving the accuracy of real-time monitoring of production data of water source wells, realizing remote real-time monitoring of water level, water pressure, water flow and electric parameters of electric submersible pump, playing the role of low water level warning and high water level automatic pump start, solving the problems of empty pumping and dry burning of water source wells, improving the management efficiency of water source wells, optimizing the working system, and ensuring the normal operation of water source wells, From the source, enough water is injected into the oilfield [13].

## 5 Conclusion

(1) From the perspective of cost performance, the radio frequency admittance sensor has certain advantages in measuring the water level of the water source well, such as low cost, reliable performance, large measuring range of water level, small error, simple installation method, and the installation only needs to be parallel to the water outlet line.
(2) The radio frequency admittance sensor is used to measure the water level of the water source well. The power carrier technology and anti-interference technology are applied in the transmitter circuit, and the collected data is accurate and reliable.
(3) Combined with the intelligent management of the oilfield, it has realized the unified and refined management, analysis and judgment of the necessary production data of the water level, water yield, outlet pressure, three-phase electricity of the submersible pump and other parameters of the water

source well, optimized the operation system of the water source well, and ensured the normal operation of the water source well and energy conservation and consumption reduction.
(4) After the remote start and stop of the water source well is achieved, the probability of errors during manual start and stop operations is effectively reduced, the rate of equipment damage is reduced, the frequency of well repairs is reduced, and the goal of cost reduction and efficiency increase is achieved[14].
(5) Measure the return flow rate, achieve automatic water inventory, and control the interval of well opening to achieve the goal of fine management[15].

## References

[1]Ghidaoui M. S., Zhao M., Mcinnis D. A., et al. A Review of Water Hammer Theory and Practice[J]. Applied Mechanics Reviews, 2018, 58(1): 49.
[2]Zhao Xiaolong, Yu Shichun, Li Yongqing, et al. R&D and Design of Water Source Well Management System [J]. Petrochemical Automation, 2015:5（51）： 60-62.
[3]Gao Baoyuan, He Yijun. Research on the application of energy conservation protection and remote monitoring devices for water source wells [J]. Automation and Instrumentation, 2015:15（02）:10-11.
[4]Zhang Qian, Wang Linping, Wang Man, et al. Application of energy conservation protection in water source wells of Changqing Oilfield [J]. Energy Conservation Technology, 2014:32（03）:272-274.
[5]Liu Liqiang, Wang Yaping, Qi Ranran, et al. Intelligent control system for water source wells [J]. Energy Saving Technology, 2014:32 (03): 272-274
[6]Li Ye, Yuan Xuemin, Zhao Zhu. Selection and application of online analytical instruments [J]. Petrochemical Automation, 2012,48 (04): 70-73.
[7]Gao Xikui, Zhu Weidong, Cheng Mingxiao. Online Analytical System Engineering Technology [M]. Beijing: Chemical Industry Press, 2014:
134-135.
[8]Zhang Haifeng, Wang Huan, Yu Hui, et al. Application of intelligent protection and remote monitoring system for water source wells in oilfield digital management [J]Journal of Hefei University, 2012:22 (02): 44-46,84.
[9]Wu Bo. Research on wireless network centralized monitoring system for water source wells [J]. Gansu Science and Technology, 2005:34 (03): 119-120.
[10]You Tao, Diao Haisheng, Zhu Wentao, et al. Any realization of remote control and interface display of oilfield water source wells [J]. PetroChina and Chemical Standards and Quality, 2009 (04): 247.
[11]Kang Haipan, Xu Yonggao, Wang Xin. Data transmission mode of oilfield water source wells [J]. Oil and Gas Field Surface Engineering, 2012:31 (12): 68.
[12]Rong Zhenyu. Embedded oilfield water source well network remote measurement and control system [J]. Instrument technology and sensor, 2008 (03): 70-72.
[13]Xie Lichun, Lei Zhanxiang, Zhao Jinling, et al. Development and application of intelligent control instrument for automatic protection of water source wells [J]. Petroleum Machinery,2007,35（11）:73-74.
[14]Kim S. H. Design of surge tank for water supply systems using the impulse response method with the GA algorithm[J]. Journal of Mechanical Science and Technology, 2018, 24(2): 629-636.
[15] Zielke W. Frequency dependent friction in transient pipe flow[J]. Trans.asme J.basic Engng, 2019, 90(3): 414.

# Research and Application of Digital Monitoring System for Pulse Current Cathodic Protection of Oil Pipeline

**Bao-yuan GAO1,2(✉), Zhao-hui LI1,2, Tan Huan1,2**

1 Drilling & Production Engineering Technology Research Institute, CNPC Chuanqing Drilling Engineering Company Limited, Sichuan, China
2 National Engineering laboratory for Low-permeability oil and gas field exploration and development, Shaanxi, China
gcy_gby@126.com

**Abstract.** The transportation of crude oil mainly relies on long-distance buried oil pipelines, which are generally made of steel pipes. Oil leakage often occurs due to corrosion, causing environmental pollution and unsafe accidents. A digital monitoring system for pulse current cathodic protection of oil pipelines was studied, with a focus on the composition and functions of the digital monitoring system for pulse current cathodic protection of oil pipelines. In 2022, the system was applied on two oil pipelines in Tongzhai operation area of the Third Oil Production Plant of Changqing oil field. The system realized the pulse current cathodic protection of oil pipelines to prevent corrosion. At the same time, it can digitally monitor the pulse current cathodic Conservation status status and adjust the working state of the potentiostat, and cooperate with the pulse current cathodic protection monitoring expert system for auxiliary analysis, which has received good economic and social benefits.

## 1 Preface

Currently, the transportation of crude oil in oil fields mainly relies on long-distance buried oil pipelines, generally made of steel pipes. Oil leakage from pipes often occurs, resulting in environmental pollution and unsafe accidents. Cathodic protection measures for oil pipelines are effective means to prolong service life and reduce operation failure of oil pipelines. The cathodic protection of oil pipeline is relatively economical when it is implemented in a wide range of fields, but it needs to be managed and maintained by a special person, and it needs to have a stable and reliable uninterrupted DC power supply and implement uninterrupted monitoring and control [1]. At present, the monitoring and control of cathodic protection devices for oil pipelines are generally carried out by pipeline inspectors or maintenance workers on a regular basis. With the continuous improvement of the modern management level of the oil field, it is urgent to replace the unattended [2].
To this end, a remote monitoring system for cathodic protection of oil pipelines has been developed [3]. The purpose of cathodic protection of oil pipelines to prevent corrosion and remote monitoring and

control of cathodic protection constant potential control and corrosion potential of oil pipelines in the duty room has been achieved, and good economic benefits have been received.

## 2 Composition of remote monitoring system

Cathodic protection is to use DC power supply to apply cathodic polarization current to the protected metal to make the metal surface cathodic polarization. When the polarization potential is greater than or equal to the open circuit potential of the corrosion battery, the electrode potential of the cathode and anode of the metal corrosion battery is equal to or more negative, and the protected metal stops corrosion. Generally, the protection potential is between -0.85V and -1.15V (measured with copper sulfate reference electrode), which can stop metal corrosion. When the protection potential is greater than -0.85V, the pipeline cannot be protected; Less than the minimum protection current When the cathodic polarization is more negative than about -1.15V, the hydrogen evolution reaction begins. The process of realizing this principle for cathodic protection of oil pipelines is that the negative pole of the potentiostat power supply is connected with the oil pipeline, and the positive pole of the potentiostat is connected with the auxiliary anode and the deep well grounding anode, thus forming a protection circuit. The cathodic protection method of oil pipeline has the advantages of large output power, long-distance anode configuration, continuous adjustable output current, large protection range, no restriction of environmental resistivity, good economy and long service life of protection device, which makes it more and more important in anti-corrosion engineering.
The cathodic protection remote monitoring system of oil pipeline is composed of online cathodic protection corrosion inhibitor and remote monitoring center [4], [5].

## 3 Design of remote monitoring system

### *2.1 Design of online corrosion inhibitor*

The online corrosion protection instrument is mainly composed of the instrument monitoring cathodic protection current, constant potential and reference potential, constant potential instrument and remote control terminal RTU module [6]. The potentiostat is used to output continuous and constant DC voltage and current, with high operating efficiency; The monitoring instrument is used to display the cathodic protection current, constant potential and reference potential locally, and upload the collected cathodic protection current, constant potential and reference potential data to the remote control terminal RTU module through RS485 interface; The RTU module then transmits the received cathodic protection current, constant potential, reference potential and start-stop signal data to the monitoring center through a special network to realize remote monitoring and achieve the purpose of safe and efficient operation of the oil pipeline without corrosion for a long time [7].

### *2.2 Main technical parameters*

(1) Cooling mode: fan temperature-controlled air cooling. When the internal temperature of the power supply is higher than 60 ℃, the fan starts to operate and cool down;
(2) Environmental conditions: temperature range - 30~+60 ℃, relative humidity of air ≤ 85%;
(3) Input voltage: 220V (1 ± 10%), 50HZ;
(4) Output voltage: 0~60VDC;
(5) Rated output current: 30A;
(6) Interface communication: RJ45 interface communication.

### *2.3 Design of remote monitoring center*

As the control terminal, the remote monitoring center is provided with human-machine interface by the monitoring software running on the computer. It receives the cathodic protection current, constant potential, reference potential and remote start/stop data from the remote control terminal RTU of the cathodic protection online corrosion protection instrument through the private network. It can display the cathodic protection current, constant potential, reference potential and remote start/stop data in a visual way on the computer of the remote monitoring and control center, and analyze and judge the cathodic protection abnormalities, The power supply of the online corrosion inhibitor can be turned off remotely, and the online corrosion inhibitor can be started remotely under normal conditions. The SCADA system of the remote monitoring center has high communication rate and good stability, and has significant application advantages [8]. This is of great significance to ensure the stability and reliability of the cathodic protection system of the entire oil pipeline.

## 4 Field application

The independently developed remote monitoring system for cathodic protection of oil pipelines has been applied on the oil pipelines of the Third Oil Production Plant of Changqing Oilfield. The system realizes online detection of the cathodic protection status of oil pipelines, and can monitor and adjust the constant potential working status of the corrosion inhibitor at any time through remote monitoring, so that the entire cathodic protection system is in the best working status. More than one year of operation practice shows that cathodic protection of oil pipelines has been realized to prevent corrosion, and the scientific, accurate and timely management of cathodic protection of oil pipelines has been improved, and good economic benefits have been received [9].

## 5 Conclusion

(1) The cathodic protection of oil pipeline is realized to prevent corrosion. The constant potential corrosion inhibitor is used to overcome the disadvantages of the traditional silicon rectifier corrosion inhibitor, such as low operating efficiency, vibration noise, heat and waste of electric energy.
(2) The remote monitoring system for cathodic protection of oil pipeline adopts humanized operation and clear charts, which can directly and timely check the condition of each monitoring point, so that the fault can be eliminated in time. This has positive significance for improving the cathodic protection level of oil pipelines, reducing pipeline corrosion and extending the service life of pipelines [10].
(3) The casing of oil and water wells adopts pulse current cathodic protection technology, which sends current into the corroded well section outside the casing through pulse, continuously providing protection to the casing and slowing down the corrosion rate of the casing. Compared with traditional technology, the protection depth of the casing is extended to more than three times, allowing deeper wells to receive uniform protection throughout the entire line [11].
(4) It can remotely monitor the operating conditions of cathodic protection in oil pipelines, keeping the protection potential of the pipelines constant within the control range, achieving the best protection effect, solving the major problem of on-site management not keeping up, and reducing the risk of corrosion in oil pipelines [12].
(5) The efficiency of pulse current cathodic protection for oil pipelines has been improved by 60% compared to direct current cathodic protection. The potential distribution of the protective sleeve is more uniform, the total current demand is lower, and the consumption of anode materials is lower, saving anti-corrosion costs [13].

**References**

[1]Jin Haifeng Design and application of cathodic protection technology in petrochemical facilities [J]. Oil and Gas Field Surface Engineering, 2009, 28 (9): 37-38.
[2]Wang Yanxia. Application of cathodic protection technology for oil and water wells in oil production plants [J]. Industry, 2016, 1 (01): 121-121.
[3]Zhang Xinhua, Wang Yan. Application of cathodic protection monitoring system in oilfield stations [J]. Petrochemical Corrosion and Protection, 2008,25 (2): 59-61.
[4]Wang Feng. Forced current cathodic protection and intelligent monitoring system [J]. Oil and gas field surface engineering, 2008,27 (7): 68,70.
[5]Zhao Peng, Liu Chenglong, Chen Liang. Application of SCADA system in oilfield production [J], China New Technology and New Products, 2014,15:8-8.
[6]Gao Xikui, Zhu Weidong, Cheng Mingxiao. Online analytical system engineering technology [M]. Beijing: Chemical Industry Press, 2014:134-135.
[7]Li Ye, Yuan Xuemin, Zhao Zhu. Selection and application of online analytical instruments [J]. Petrochemical Automation, 2012,48 (04): 70-73.
[8]Xu Qing, Zhou Haobin, Yuan Sen, et al. Design of wireless data acquisition system for oil well casing pulse current cathodic protection based on GPRS [J]. Comprehensive Corrosion Control, 2015 (12): 42-43.
[9]Wang Chengyu, Chen Yuchen, Hou Yun, et al. Circuit breaker remote monitoring system based on Web and LabVIEW [J]. Chemical Automation and Instrumentation, 2014,41 (06): 632-635.
[10]Wu Chunxue, Li Hongkui. Oilfield development and application of computer measurement and control system [M]. Beijing: Petroleum Press, 2001.
[11]BALLA E. System Identification Modelling and Ime Based Pid Control of Impressed Current Cathodic Protection Sys-tems ].Massey University,2018,6(5):2055-2074.
[12]PARSAM H. Simulation of Cathodic Protection Potential Distribution on Oil Well Casings D].Journal of Petroleum Science and Engineering,2019,72(4):215-219.
[13]METWALLY I.Factors Affecting Pulsed-cathodic Protection Effectiveness for Deep Well Casings [J].Anti-corrosion Methods and Materials,2017,56(4):196-205.

Journal of Physics: Conference Series **2594** (2023) 012064 doi:10.1088/1742-6596/2594/1/012064

# Study on Surfactants for Replenish Energy before Fracturing in Tight Reservoir

**Chun-tian Liu1, Liang-liang Ma2, Zheng-mao Wang3, Shou-liang Lu1, Zhi-rui Song1, Xin-xin Li1, Ying Wang1**

1. Exploration and Development Research Institute of Daqing Oilfield
2. Daqing Oilfield Development Department
3. Petro China Oil&Gas and New Energy Company

dqpilct@163.com

**Abstract**. Tight reservoir in the periphery of Daqing oilfield is abundant. However, due to tight lithology, poor physical properties, and insufficient formation energy, water injection is not effective, development effect is not ideal, and utilization degree is low. Therefore, it is proposed to implement energy replenishment before fracturing, integral fracturing, and energy storage after fracturing. In the energy replenishment stage, surfactants with good compatibility with the formation and strong oil washing ability are selected as injection medium to play the dual role of energy replenishment and oil washing, and improve the recovery rate of tight sandstone reservoirs. Surfactants are evaluated by established method; LH anionic surfactant and KR anionic non-ionic surfactant are preferred. Under alkali free condition, interfacial tension of two surfactants maintain at order of 10-3. Recovery degree of surfactants in natural cores reach more than 15%. Field application conducted in 30 wells in Block P, including 27 vertical wells and 3 horizontal wells. Oil is seen in vertical wells at 7.8% flow back rate. The average daily liquid production of the 20 wells that have been pumped is 7.5 tons per day, with a daily oil production of 1.3 tons, and a cumulative oil increase of 223.8 tons per well. The average daily oil increase of a single well in the three months after fracturing a qualified oil well is 5.2 tons, 1.6 times and 2.2 times that of dessert fracturing and overall fracturing in the same period. The average daily liquid production of 3 horizontal wells is 24.2 tons, and the daily oil production is 9.6 tons. The test results show that the overall design of the fracturing energy supplement scheme is reasonable, and the selected surfactant can supplement the formation energy while improving the oil washing efficiency, which will provide technical support for exploring new ways to improve oil recovery in tight reservoirs.

## 1. Introduction

The peripheral tight oil reservoirs in Daqing Oilfield are rich in reserves and have great development potential. Due to tight lithology, poor physical property, insufficient formation energy, water injection is not effective. Although a lot of adjustment and reconstruction measures have been taken, the development effect is still not ideal, single well productivity is low, overall recovery factor is low. How to supplement formation energy efficiently and reasonably is the difficulty and key point to improve recovery efficiency of tight sandstone reservoir. Currently, the most commonly used methods for supplementing formation energy include water injection and gas injection. Due to the small pores, high capillary pressure, and high irreducible water saturation in tight sandstone reservoirs, water injection is very likely to cause water lock damage to the reservoir and affect development effectiveness; Gas flooding is prone to viscous fingering, leading to gas channeling, which cannot effectively improve oil recovery [1-10]. Surfactants can improve oil washing efficiency by reducing oil-water interfacial tension, changing the wettability of rock surface and emulsifying crude oil, etc. They have been widely studied and applied in medium, high permeability, and low permeability reservoirs, the study of applicability in tight reservoir is more and more extensive, but the report of practical application is relatively few [11-18].

For the preferred P block in Daqing tight reservoir, field tests were carried out to transform the development mode with the core of reconstructing well patterns, reconstructing fracture patterns, and artificial reservoirs. Energy replenishment before fracturing, overall fracturing, and energy storage after fracturing were implemented to achieve maximum reservoir transformation, maximum fracture controlled reserves, and effective utilization of tight reserves. Through laboratory experiments, two kinds of surfactants with excellent performance are selected as energy replenishing medium to inject solution into oil wells or wells. After soaking and oil-water permeation, they played the dual roles of energy-replenishing and oil-washing, enhanced oil recovery in tight sandstone reservoirs.

## 2. Surfactant optimization

According to basic characteristics of Daqing periphery reservoir, five different types of surfactants are selected, including LH anionic surfactant, KR, XW, HL anionic non-ionic surfactant, ZT biosurfactant, etc. The performance evaluations of these surfactants are carried out.

### *2.1. Experimental Conditions*

Oil and water: water from well of target blocks. Use them after filtration. Crude oil from target oil layers. Use them after dehydration.

Experimental temperature: determined as 90 ℃ based on the actual situation of the target block.

Instrument: Rotating droplet interface tension meter; Homogenizer; Brinell viscometer; American TA rheometer; High temperature oven; High temperature and high pressure saturated oil device; Infiltration device.

At present, no industrial standard has been established for evaluation of fracturing displacement surfactant. Combined with evaluation standards of current composite flooding systems, an evaluation method is determined.

### *2.2. Interfacial tension*

Interfacial tension between surfactant system and crude oil is evaluated by rotating drop method. KR, XW, HL and ZT Surfactant are directly formulated with water to form a salt free system. The interfacial tension of four systems are tested, that is, surfactant concentrations are 0.3%, 0.2%, 0.1% and 0.05% respectively. LH Surfactant form a salt-added system. The interfacial tension of four systems is detected, that is, the surfactant/salt concentrations are (0.3%, 1.2%), (0.3%, 0.6%), (0.1%, 1.2%), (0.1%, 0.6%) respectively.

Four detection points of KR and LH can reach ultra-low interfacial tension respectively. There is no ultra-low interfacial tension process at 0.05% of HL surfactant. Four detection points of XW and ZT surfactants cannot reach the order of magnitude of $10^{-3}$ (Table 1).

**Table 1. Minimum interfacial tension of surfactants**

| | System number | 1# | 2# | 3# | 4# |
|---|---|---|---|---|---|
| KR | Surfactant concentration (wt%) | 0.05 | 0.1 | 0.2 | 0.3 |
| | Minimum interfacial tension($mN\cdot m^{-1}$) | 0.0008 | 0.0006 | 0.0014 | 0.0008 |
| LH | Surfactant concentration (wt%) | 0.1 | 0.1 | 0.3 | 0.3 |
| | KCl concentration (wt%) | 0.6 | 1.2 | 0.6 | 1.2 |
| | Minimum interfacial tension($mN\cdot m^{-1}$) | 0.0063 | 0.0011 | 0.0014 | 0.0014 |

### 2.3. *Emulsification*

Emulsification experiment is carried out with homogenizer. Put crude oil and surfactant system into a plugged colorimetric tube according to the volume of 1:1, homogenize it with a homogenizer at speed of 10000r/min for 1min. Place it vertically in a constant temperature oven at oil layer temperature. Record total volume of emulsion, upper phase volume, middle phase volume and lower phase volume until the volumes of upper phase and lower phase do not change.

Test emulsifying capacity of each surfactant at concentration 0.3%. Other two sets of emulsification experiments are carried out simultaneously. After emulsification, oil phase viscosity is measured for 1h and 24h by rheometer.

According to emulsification test results, after surfactant interacts with crude oil, water evolution rate is KR<LH2<HL<LH1< XW from fast to slow, and the viscosity does not increase significantly, which can meet on-site injection requirements (Table 2).

**Table 2.** Oil phase viscosity and water evolution of emulsified crude oil with different surfactants

| Surfactant type | | KR | HL | LH1 | LH2 |
|---|---|---|---|---|---|
| Oil phase viscosity (mPa·s) | Place for 1H | 23.8 | 21.4 | 22.0 | 25.3 |
| | Place for 24H | 20.7 | 22.7 | 20.5 | 21.6 |
| Rate of precipitation (%) | Place for 1H | 60 | 71 | 91 | 90 |
| | Place for 6H | 85 | 100 | 100 | 98 |
| | Place for 24H | 100 | 100 | 100 | 100 |

### 2.4. *Imbibition*

At present, the main development method for tight reservoirs is volume fracturing, and after fracturing, fractures become the main seepage channel. However, there are still a large amount of residual oil in the matrix rock that cannot flow. Imbibition of this part of crude oil into the oil flow channel will significantly improve oil recovery, which is a key technology for developing tight reservoirs. The oil evolution rate of the imbibition experiment reflects the ability of different systems to extract crude oil from the core.

Natural cores are used to carry out imbibition experiments to detect the effect of each surfactant product at 0.3% concentration.

According to experimental results, oil precipitated by LH surfactant system is relatively large, and oil separation rate can reach more than 30%. Mainly because this surfactant is petroleum sulfonate, it has good mutual solubility with crude oil of Daqing ultra-low permeability reservoir, has strong emulsification carrying capacity, and 1.2% salt is added to the system to form appropriate salinity conditions, which promotes the interaction between surfactant and crude oil. Oil separation rate of KR surfactant is more than 15% (Table 3).

**Table 3.** Results of natural core imbibition experiments with different surfactant systems

| Surfactant type | Active agent/KCl concentration(wt%/wt%) | 48h Oil precipitation rate (%) | 180h Oil precipitation rate (%) |
|---|---|---|---|
| LH1 | 0.3/1.2 | 71.3 | 61.5 |
| LH2 | 0.3/1.2 | 49.5 | 32.0 |
| KR | 0.3 | 6.3 | 15.6 |
| HL | 0.3 | 2.2 | 4.5 |

## 3. Field test

Field tests were conducted in Zone P, and 30 oil and water wells were selected, 240m×100m rectangular well pattern, with an oil recovery rate of 0.01% and a recovery degree of 3.09%.

The number of development layers of a single well is 6.7, the effective thickness of a single well is 16.5 m, the permeability is 1 mD, the viscosity of crude oil is 4.4 mPa·s, and oil saturation is 50%. The reservoir sensitivity is relatively strong and water sensitive.

### *3.1. Test scheme*

The integrated associated fracturing fluid is used to meet the performance requirements of fracturing fluid (glide water, clean water, sand-carrying fluid), which has the following advantages: 1.less gel breaking residue in the system; 2.The gel breaker contains surfactant components that have an oil displacement effect; 3.By adjusting the proportioning at different stages, meet the sand carrying requirements of different sand ratios; 4.simple liquid dispensing without special equipment.

LH surfactant is a petroleum sulfonate type that requires the addition of potassium chloride to fully exert the oil washing effect. However, potassium chloride reduces the viscosity of the sand-carrying fluid and affects the sand carrying capacity. Therefore, clean water + LH surfactant + potassium chloride is used as a pre slug in oil wells to supplement energy. At the same time, potassium chloride can play an anti-swelling role. Considering the formation adsorption loss, in order to achieve the optimal imbibition effect, the optimized injected LH surfactant concentration is 0.3%. Subsequently, inject sand carrying fluid for volume fracturing.

KR surfactant can achieve ultra-low interfacial tension without adding salt, and has little impact on the viscosity of the sand carrying fluid. Therefore, this system can be injected into water wells together with the sand-carrying fluid, and the whole process of fracturing can be enhanced.

### *3.2. Test effect*

All 27 vertical wells in the test area have seen oil, with a flowback rate of 7.8%. Currently, 20 wells have been pumped, with an average daily liquid production of 7.5 tons per well and a daily oil production of 1.3 tons, with a cumulative increase of 223.8 tons per well. 11 wells that have been pumping for more than 90 days have an average daily liquid production of 8.3 tons, daily oil production of 2.3 tons, water content of 64.9%, flowback rate of 25.6%, and cumulative oil increase of 4631 tons (Fig.1).

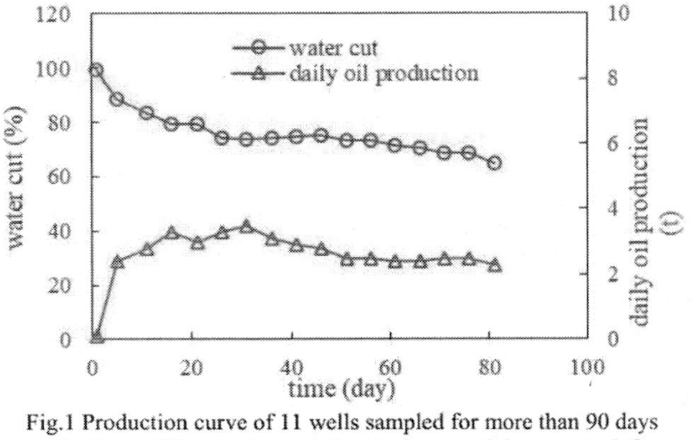


Fig.1 Production curve of 11 wells sampled for more than 90 days

Compared with previous single well sweet spot fracturing and integrated fracturing, the oil increase intensity of qualified oil wells has increased by 0.27t/d·m and 0.34t/d·m respectively. The average daily oil increase of a single well in 3 months after fracturing is 5.2 tons, which is 1.6 times and 2.2 times that of sweet spot fracturing and integrated fracturing in the same period (Table 4).

The average daily liquid production of 3 horizontal wells in the test area is 24.2 tons, and the daily oil production is 9.6 tons. It is estimated that the accumulative oil increase in the test area during ten years is $13.69\times10^4$t, and the recovery degree is increased by 11.8%.

**Table 4.** Field Test Results of Qualified Oil Wells in Block P

| | Initial daily oil production intensity (t/d·m) | Average daily liquid production in the first three months (t/d) | Average daily oil production in the first three months (t/d) |
|---|---|---|---|
| Integral fracture network | 0.14 | 8.6 | 2.4 |
| Single well fracture pattern | 0.21 | 8.5 | 3.2 |
| Fracturing recharge | 0.48 | 11.2 | 5.2 |

## 4. Conclusion

Based on the characteristics of oil water, mineral composition, temperature, and other characteristics of tight oil layers in the periphery of Daqing, two surfactants, LH and KR, are selected through laboratory experiments. The indoor natural core imbibition degree can reach 15% and 30%, respectively. It is proved that surfactants can improve the flow of oil phase in tight rock cores by reducing the interfacial tension between oil and water, modifying the wettability of rocks, and emulsifying and dispersing, thereby improving oil recovery.
Field tests were carried out in 30 oil and water Wells in P zone. In the production well, clean water +0.3%LH surfactant +1.2% potassium chloride are used as the pre-slug to achieve the best imbibition effect of LH surfactant, improve the efficiency of oil washing while replenishing energy, and then inject the sand carrying liquid. KR surfactant is used in the injection well and injected together with the sand carrying fluid to enhance the energy throughout the fracturing process. Both injection methods have achieved good results. The average daily oil increase of a single well that meets the standard three months after fracturing is 5.2 tons, which is 1.6 times and 2.2 times that of dessert fracturing and overall fracturing in the same period. The production improvement effect is significant. This technology has broad application prospects.

**References**

[1] Fu Jing, Yao Bowen, Lei Zhengdong et al. Technology status of enhanced oil recovery in ultra-low permeability tight reservoirs in North America. Journal of Southwest Petroleum University (Science & Technology Edition), 2021, 43(05):166-183.
[2] Da Yinpeng, Xue Xiaojia, Liu Ming, et al. Energy enhancement mechanism of energy storage and re-fracturing in ultra-low permeability oilfield. Science Technology and Engineering, 2021, 21(33):14139-14146.
[3] Huang Ning. A new development model of volume fracturing and imbibition in ultra-low permeability reservoir. Chemical Industry Management, 2020(03): 218.
[4] Kang Yili, Tian Jian, Luo Pingya, et al. Technical bottleneck and development strategy of enhanced oil recovery in tight reservoirs. Acta Petrolei Sinica, 20, 41(04):467-477.
[5] Zhou Fujian, Su Hang, Liang Xingyuan, et al. Integration technology of efficient fracture-network reconstruction and enhanced oil recovery in tight oil reservoirs. Petroleum Exploration and Development, 2019, 46(05): 1007-1014.
[6] Zhang Junfeng, Bi Haibin, Xu Hao, et al. New progress in exploration and development of tight oil in foreign countries and its reference significance. Acta Petrolei Sinica, 2015, 36 (2):127-137.
[7] Wang Xiuyu, Ju Mingshuang, Yang Wensheng, et al. Law and mechanism of dynamic

imbibition and displacement in tight reservoir. Petroleum Geology and Recovery Efficiency, 2019, 26 (3):92-98.
[8] Wu Zhongbao, Zeng Qian, Li Jin, et al. A new way of water injection huff and puff in volumetric reservoir to effectively supplement formation energy development. Petroleum Geology and Recovery Efficiency, 2017, 24 (5):78-83.
[9] Gao Tao, Zhao Xisen, Dang Hailong, et al. Mechanism and application of waterflooding in tight reservoirs of Yanchang Oilfield. Special Oil and Gas Reservoirs, 2018, 25 (4):134-137.
[10] Zhang Benyan, Dang Wenbin, Wang Shaopeng, et al. CO2 injection to enhance oil recovery of Chang 8 tight sandstone reservoir in Honghe Oilfield, Ordos Basin. Oil & Gas Geology, 2016, 37 (2):272-275.
[11] Su Yubin, Lin Guanyu, Han Yue. Effect of surfactant on spontaneous imbibition displacement in tight sandstone reservoir. Fault-block Oil & Gas Field, 2017, 24 (5):97-100.
[12] Shen Anqi, Liu Yikun, Qiu Xiaohui, et al. Study on enhancing imbibition recovery of tight reservoir with surfactant. Oilfield Chemistry, 2016, 33 (4):696-699.
[13] Wu Wei. Enhanced Oil Recovery technology by emulsion surfactant flooding in ultra-high temperature medium and low permeability reservoirs. Petroleum Geology and Recovery Efficiency, 2018, 25 (2):72-76.
[14] Hu Xiaodong, Hou Mingming, Hu Wenting. Research progress on Synthesis and application of anionic gemini surfactants for oil field. Petroleum and natural gas chemical industry, 2011, 33 (4): 494-498.
[15] Feng Xiaoye, Xi Lihui. Application of Gemini Surfactants in recent years. Petrochemical applications, 2019, 10:58-61.
[16] DAI C L, WANG X K, LI Y Y, et al.Spontaneous Imbibition Investigation of Self-Dispersing Silica Nanofluids for Enhanced Oil Recovery in Low-Permeability Cores[J].Energy & Fuels, 2017, 31(3):2663-2668.
[17] XU D R, BAI B J, WU H R, et al. Mechanisms of imbibition enhanced oil recovery in low permeability reservoirs: Effect of IFT reduction and wettability alteration [J]. Fuel, 2019, 244: 110-119.
[18] ALVAREZ J O, SCHECHTER D S. Altering wettability in Bakken shale by surfactant additives and potential of improving oil recovery during injection of completion fluids[C].SPE Improved Oil Recovery Conference. Society of Petroleum Engineers, 2016.

Journal of Physics: Conference Series **2594** (2023) 012065 doi:10.1088/1742-6596/2594/1/012065

# Removal of Amphiphilic Polymer from Tertiary Oil Recovery Wastewater Using Insoluble Cyclodextrin Polymer

**Yuan LI[1](✉) and Ruian ZHANG[1]**

1 Party school (Training Center) of Sinopec Shengli Petroleum Administration Co., Ltd. , Dongying, Shandong, China

lyfly_01@163.com

**Abstract**. The wastewater produced by amphiphilic polymer flooding is seriously emulsified and difficult to treat. In recent years, insoluble β-cyclodextrin (β-CD) polymers were widely employed in removal and utilization of organics with specific sizes and shapes. In this paper, a new type of host β -CD insoluble polymer (HCDP) was synthesized using β -CD with tetrafluoro-phthalonitrile (TFPN) as the crosslinker. A kind of guest amphiphilic polymer (GAP) was synthesized and the inclusion performance of HCDP to GAP was studied by viscosity method. Besides, The inclusion complexes of HCDP and GAP was regenerated using washing method and the inclusion capacity of the regenerated CDP was tested for five cycles. The results showed that GAP's apparent viscosity increased dramatically at about 800 mg/L and it showed linear relationship (R-square was 0.99902) with polymer concentration below 400mg/L, so GAP's concentration can be checked by viscosity method. HCDP can rapidly adsorb GAP in aqueous solutions and reach the adsorption equilibrium in about 20 minutes in the process of inclusion, and the maximum equilibrium adsorption amount of HCDP to GAP was about 150 mg/g. HCDP showed a strong recycling capacity. After five cycles of regeneration, the adsorption capacity did not significantly reduced, and the adsorption capacity remained above 140 mg/g. All the results prove that the new type of β-CD insoluble polymer can be used to encapsulate amphiphilic polymers in EOR wastewater, the study provides a new way for the resource utilization of amphiphilic polymers in EOR wastewater.

## 1. Introduction

Water resources are very valuable ecological resources and the protection of water bodies receives increasing attention internationally and domestically [1]. During the enhanced oil recovery (EOR) process, a lot of chemicals are employed to improve the swept volume such as polymers and oil washing efficiency such as surfactant and alkali [2-4]. Composite oil displacement technology has also been widely applied, such as SP flooding, AP flooding and ASP flooding, etc [5-6]. Chemical flooding can enhance oil recovery, but the produced liquid contains residual chemicals and is commonly emulsified as oil in water emulsion, which is difficult to handle with conventional methods and restricts the wide application of EOR technology[7-10]. The treatment of EOR produced fluid has also attracted large quantities of researchers' interest, mainly including the demulsification process and water treatment process. As to the treatment processes for the chemical flooding wastewater, several methods were employed include coagulation/flocculation, oxidation, biodegradation, membrane filtration or combined method [11-15].

Amphiphilic polymers (AP) have good viscosity and solubilization properties, which can greatly improve oil recovery [16]. However, the wastewater produced by AP flooding is seriously emulsified and difficult to treat. How to treat AP flooding wastewater, especially how to realize the resource utilization of residual AP in wastewater is an urgent problem to be solved.

Many materials (such as chitosan, zeolite, polymer, cyclodextrin, etc.) have been made into adsorbent and can adsorb organic pollutants. Among many adsorbents, cyclodextrin polymers have attracted researchers' interest because of their cheap, easy to prepare and green characteristics. In recent years, water-insoluble β-cyclodextrin (β-CD) polymers are widely employed in organic's removal and utilization, because the insoluble polymers as the host can encapsulate specific sizes of guest organic matter [1, 17-21].

In this paper, tetrafluoro-phenylene dinitrile and β-Cyclodextrin is used as raw material for cross-linking polymerization β-Cyclodextrins containing the original cavity structure and can form host and guest inclusion complexes with many organic compounds. In this study, the removal performance of amphiphilic polymers in wastewater by insoluble β-cyclodextrin polymer was determined, which provides a reference for the practical application of cyclodextrin chemistry in the field of wastewater treatment.

## 2. Experimental

### 2.1. Synthesis of Host Polymer

According to the synthesis method in literature [18], tetrafluoro-phenylene dinitrile is used as a cross-linking agent. β-CD and tetrafluoro-phthalonitrile (TFPN) were crosslinked and polymerized to synthesize a new type of β-CD insoluble polymer (HCDP), the reaction equation is shown in Fig. 1.

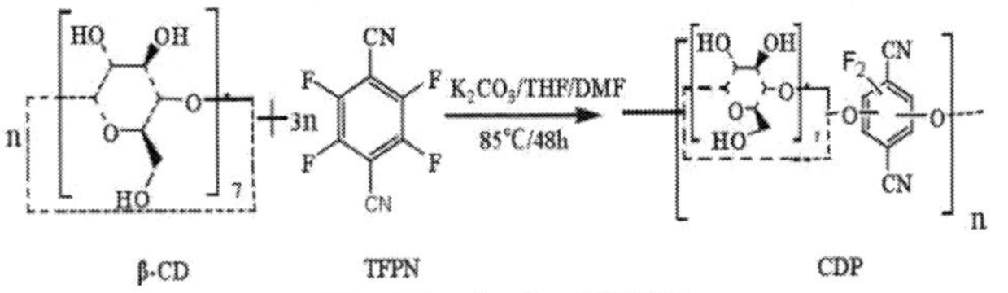


**Fig. 1** Synthesis of HCDP

### 2.2. Synthesis of Guest Polymer

In this paper, the synthesis of guest amphiphilic polymer (GAP) mainly uses KPS-SHS system (the molar ratio was 1:1) as the initiator of the reaction, and sodium dodecyl sulfate as the surfactant of solubilizing hydrophobic monomer by free radical micelle polymerization as described in literature [16]. The reaction equation is shown in Fig. 2.

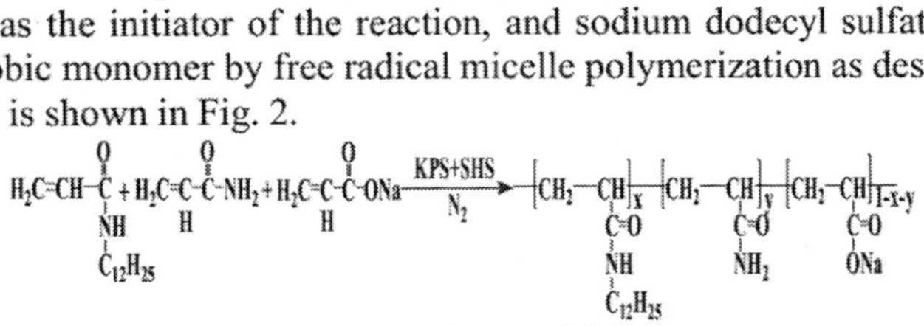


**Fig. 2** Synthesis of GAP

### 2.3. Viscosity Measurements of GAP

Deionized water was employed to dissolve GAP at room temperature for 4~6 h to ensure GAP was fully dissolved, and then GAP solutions was stayed for 24 h at 45 ℃ to remove air bubbles. NDJ-5S viscometer was employed to measure GAP's apparent viscosity, the measurement temperature was 45 ℃ and the stirring speed was 6 rpm.

### 2.4. Inclusion capacity of HCDP to GAP

The inclusion properties of HCDP to GAP was systematically studied by viscosity method. 25 mL GAP solution was placed in a 100 mL conical flask and add 5 mg HCDP was added into the conical flask at a rotating speed of 300 rpm. After a certain time of inclusion, the viscosity of the inclusion system was immediately measured to determine the remaining GAP concentration and the adsorption capacity of HCDP to GAP.

## 3. Results and Discussions

GAP's apparent viscosity was shown in Fig. 3 and it can be seen that GAP's apparent viscosity increased dramatically at about 800 mg/L because extensive intermolecular association occurred due to the hydrophobic association of hydrophobic groups lead to network structures in GAP solutions. The critical aggregation concentration (CAC) of GAP was about 800 mg/L and thus GAP had good thickening properties above 800 mg/L. But below 800 mg/L, GAP's apparent viscosity showed certain linear relationship with GAP concentration.

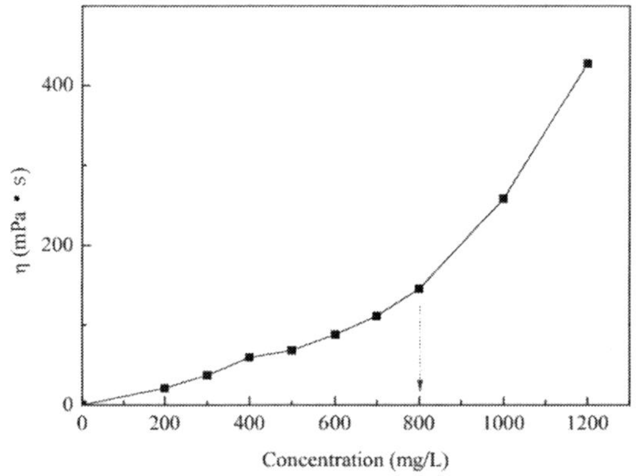


**Fig. 3** The apparent viscosity of GAP solutions.

Fig. 4 showed that GAP's apparent viscosity had linear relationship (R-square was 0.99902) with polymer concentration below 400mg/L. As in GAP's dilute solutions, the polymer concentration can be checked by the viscosity method.

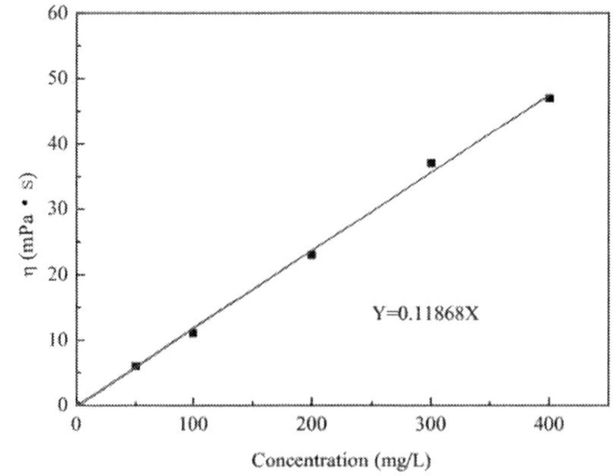


**Fig. 4** The apparent viscosity of GAP solutions as a function of concentration below 400mg/L.

Fig. 5 showed the infrared spectrum of TFPN, β-CD and HCDP. The peak of 2935 $cm^{-1}$and 1025 $cm^{-1}$ represent the stretching vibrations of the C-H and C-O-C bonds of methylene, which are the common characteristic peak of HCDP and β-CD. The results showed that HCDP synthesized was the expected product and the holes of β-CD were retained in HCDP.

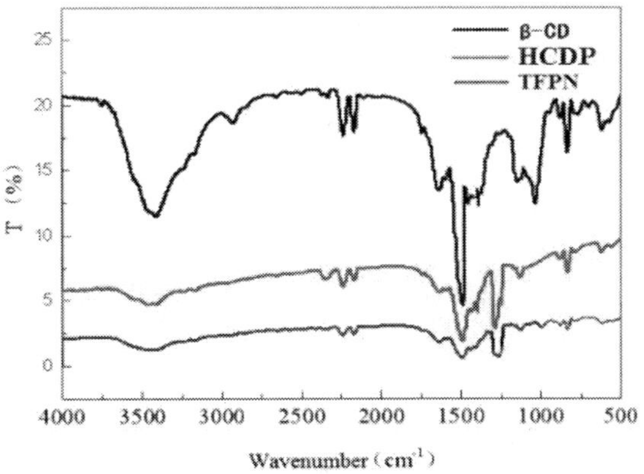


**Fig. 5** The infrared spectrum of TFPN, β-CD and insoluble β-CD polymer.

Fig. 6 showed GAP's apparent viscosity as a function of inclusion time using HCDP as the adsorption material. In the process of inclusion of GAP, HCDP can rapidly adsorb and reach the adsorption equilibrium in about 20 minutes. The adsorption capacity of HCDP to GAP was showed in Fig. 7. Similarly, the adsorption capacity of HCDP to GAP increased rapidly with the increasing inclusion time and can reach absorption equilibrium in about 20 min. The equilibrium adsorption amount was about 150 mg/g.

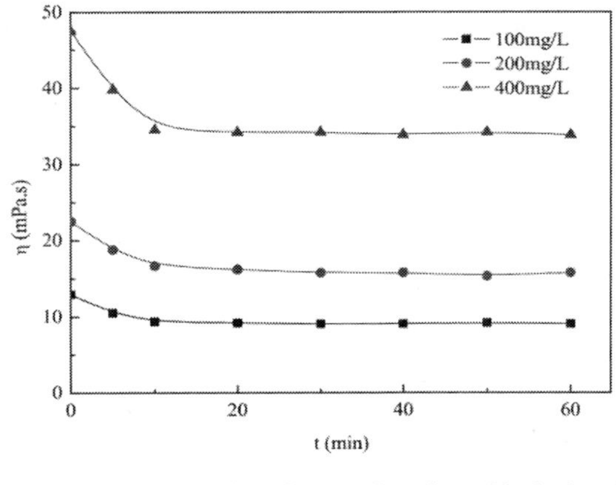


**Fig. 6** GAP's apparent viscosity as a function of inclusion time.

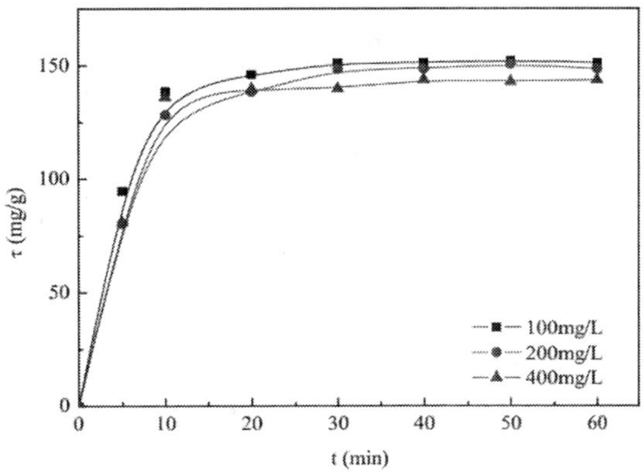


**Fig. 7** The adsorption capacity of HCDP to GAP as a function of inclusion time.

The inclusion complexes of HCDP and GAP was regenerated using washing method with methanol and hydrochloric acid mixed solution for three times, and then dried 45 ℃ for 1 h. The regenerated HCDP had good inclusion capacity to GAP was shown in Fig. 8. After five cycles, the adsorbent still had good adsorption capacity, and the adsorption capacity of HCDP for the fifth time can still reach 142 mg/g, indicating that the new HCDP prepared is a highly efficient adsorbent that can be recycled.

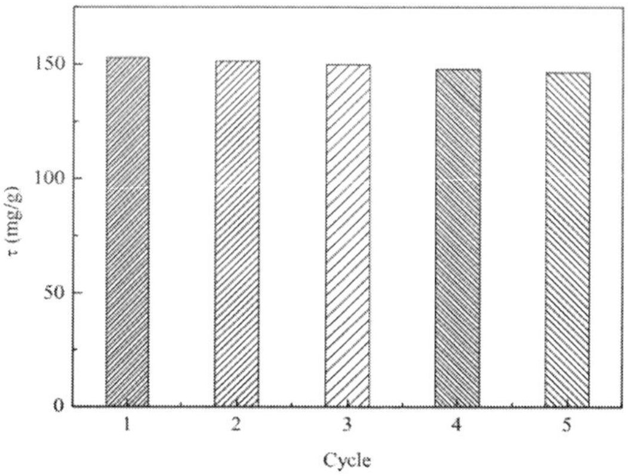


**Fig. 8** The adsorption capacity of HCDP to GAP after regeneration.

## 4. Conclusions

A new type of host β-Cyclodextrin insoluble polymer (HCDP) was synthesized using β-CD with tetrafluoro-phthalonitrile (TFPN) as the crosslinker. Guest amphiphilic polymer (GAP) was synthesized and the inclusion performance of HCDP to GAP was studied.

The results showed that GAP's apparent viscosity increased dramatically at polymer concentration of about 800 mg/L. GAP's apparent viscosity showed linear relationship (R-square was 0.99902) with polymer concentration below 400mg/L so the polymer concentration can be checked by the viscosity method. HCDP can rapidly adsorb GAP in aqueous solutions and reach the adsorption equilibrium in about 20 minutes in the process of inclusion, and the maximum equilibrium adsorption amount of HCDP to GAP was about 150 mg/g. HCDP showed a strong recycling capacity. After five cycles of regeneration, the adsorption capacity did not significantly reduced, and the adsorption capacity remained above 140 mg/g.

The results prove that the new type of β-CD insoluble polymer can be used to encapsulate amphiphilic polymers in EOR wastewater, the study provides a new way for the resource utilization of amphiphilic polymers in tertiary oil recovery wastewater.

## References

[1] Morin-Crini, N., Crini, G. Environmental applications of water-insoluble β-cyclodextrin-epichlorohydrin polymers. Prog. Polym. Sci. 38(2), 344-368 (2013)

[2] Alvarado, V., Manrique, E. Enhanced Oil Recovery: An Update Review. Energies 3(9), 1529-1575 (2010)

[3] Sheng, J.J., Leonhardt, B., Azri, N. Status of Polymer-Flooding Technology. J. Can. Petrol. Technol. 54(2), 116-126 (2015)

[4] Liu, Q., Dong, M., Ma, S., et al. Surfactant enhanced alkaline flooding for western Canadian heavy oil recovery. Colloid Surface A 293(1-3), 63-71(2007)

[5] Liu, H., Zhang, Y., Li, Y., et al. Influence on emulsification in binary flooding of oil displacement effect. J. Disper. Sci. Technol. 37(1), 89-96(2016)

[6] Hou, J., Liu, Z., Dong, M., et al. Effect of Viscosity of Alkaline/Surfactant/ Polymer (ASP) Solution on Enhanced Oil Recovery in Heterogeneous Reservoirs. J. Can. Petrol. Technol. 45(11), 27-33(2006)

[7] Deng, S., Yu, G., Jiang, Z., et al. Destabilization of oil droplets in produced water from asp flooding. Colloid Surface A 252(2-3), 113-119(2005)
[8] Wang, B., Wu, T., Li, Y., et al. The effects of oil displacement agents on the stability of water produced from ASP (alkaline/surfactant/polymer) flooding. Colloids & Surfaces A 379(1-3), 121-126(2011)
[9] Yan, L., Cui, Z., Song, B., et al. Improving performances of double-chain single-head surfactants for SP flooding by combining with conventional anionic surfactants. Journal of Dispersion Science & Technology 39(1), 130-137 (2018)
[10] Majeed, T., Kamal, M.S., et al. A review on foam stabilizers for enhanced oil recovery. Energy & Fuels, 35(7), 5594-5612 (2021)
[11] Fang, S., Duan, M., Long, W, et al. Treatment of produced liquid from polymer flooding by using cationic surfactant. Separation Science & Technology 49(3), 432-437(2014).
[12] Nie, C., Xu, L., Gu, D., et al. Towards efficient demulsification of produced water in oilfields: solar step directional degradation of polymer on interfacial film of emulsions. Energy & Fuels 30(11), 9686-9692(2016).
[13] Pi, Y., Zheng, Z., Bao, M., et al. Treatment of partially hydrolyzed polyacrylamide wastewater by combined Fenton oxidation and anaerobic biological processes. Chemical Engineering Journal 273, 1-6(2015)
[14] Liu, J., Ren, J., Xu, R., et al. Biodegradation of partially hydrolyzed polyacrylamide by immobilized bacteria isolated from HPAM-containing wastewater. Environmental Progress & Sustainable Energy 35(5), 1344-1352(2016)
[15] Zhang, B., Yu, S., Zhu, Y., et al. Application of a polytetrafluoroethylene (PTFE) flat membrane for the treatment of pre-treated ASP flooding produced water in a Daqing oilfield. Rsc Advances 6(67), 62411-62419(2016)
[16] Xu, B., Zhang H., Bian, H. Delayed Crosslinking Amphiphilic Polymer Gel System with Adjustable Gelation Time Based on Competitive Inclusion Method. Polymers-Basel 11, 381(2019)
[17] Crini, G. and Morcellet, M. Synthesis and applications of adsorbents containing cyclodextrins. Journal of Separation Science 25(13), 789-813(2015)
[18] Alsbaiee, A., Smith, B.J., Xiao, L., et al. Rapid removal of organic micropollutants from water by a porous β-cyclodextrin polymer. Nature 529(7585), 190-194(2015)
[19] Kong, J., Xu, B. Synthesis of Insoluble β-Cyclodevrin Polymer Microspheres for Removal of Organics in Aqueous Solution. International Core Journal of Engineering 4(1), 120-124(2018)
[20] Choi J., Jeong D., Cho E., et al. Pentynyl Ether of β-Cyclodextrin Polymer and Silica Micro-Particles: A New Hybrid Material for Adsorption of Phenanthrene from Water. Polymers 9(1), 10(2017)
[21] Celebioglu A., Ipek S., Durgun E., et al. Selective and Efficient Removal of Volatile Organic Compounds by Channel-type Gamma-Cyclodextrin Assembly through Inclusion Complexation. Ind. Eng. Chem. Res. 56(5), 7345-7354(2017)

# Development and Application of The Graft Coated Viscosity Reducer

**Fei Zhou(✉), Xue-liang Li, Di Yao, Yi Sun, Wei Dong, Lei Wang**

Liaohe Oilfield Company, Panjin, Liaoning 124010, China

66739885@qq.com

**Abstract** The graft coated viscosity reducer was developed for the high cost of mixing light oil. The viscosity reducing mechanism is that the polar hypertonic agents in thSe viscosity reducer dissociate the net structures of the resin and asphalt and then form eutectic grafting with these dissociated macromolecules when the heavy oil in the water was broken or in the solution of associating state. The heavy oil takes shape of small particles while the resin is out of structure and not easy to get agglomerated. The released resin and the coating agent for the viscosity reducer form a layer of film on the surface of small oil particles to prevent the resin and asphalt from network structuring again. The heavy oil particles are not easy to be in coalescence concerning electrical repulsion and solution association among the coating in between. The heavy oil particles are dispersed in the medium to achieve the stabilization of viscosity reduction. For example, the viscosity of the crude oil from Liaohe Oilfield is 87.2mPa•s after adding the agent at 50℃ (the original viscosity was 4919mPa•s). The viscosity reducer did not corrode the pipelines and wellbores and had no effect on the dehydration of united stations. The construction was carried out by continuously dripping the viscosity reducer into the 16 wells of Liaohe Oilfield. The viscosity all decreased to below 100mPa•s afterwards. The current, load and other parameters all decreased along with the application. So this method could take the place of mixing light oil.

## 1 Introduction

Heavy oil is rich in resin and asphaltene, it has the characteristics of high viscosity and high density, which seriously restricts the development and exploitation of heavy oil. Reducing the viscosity of heavy oil and improving the fluidity of heavy oil are of great significance for solving the exploitation and gathering of heavy oil[1]. The most widely used chemical viscosity reduction of heavy oil is emulsification viscosity reduction method. The factors affecting emulsification viscosity reduction of heavy oil are divided into chemical factors and physical factors. Chemical factors refer to emulsion systems: crude oil, water and emulsifiers; physical factors refer to external factors, including oil-water volume ratio, temperature, mixing method, salt content, etc[2]. There are two main lifting processes for heavy oil recovery in Liaohe Oilfield : viscosity reduction by blending light oil and electric heating. The cost of the two processes is high and the annual cost is over 100 million yuan. Therefore, it is necessary to propose a new heavy oil viscosity reduction technology[3], reduce mining costs. In this paper, according to the properties of heavy oil in Liaohe Oilfield, a graft coated viscosity reducer was developed and its performance was evaluated.

## 2 Viscosity reduction mechanism of the graft coated viscosity reducer

The mechanism of graft coated viscosity reducer can be understood from two aspects[4]: First, the contact between the viscosity reducer solution and the heavy oil can reduce the oil-water interfacial tension. Therefore, after stirring at a certain temperature, the oil is dispersed in a granular form in its aqueous solution to form emulsion-like liquid. The molecules in the viscosity reducer are adsorbed around the oil droplets, and form a directional unimolecule protective film, prevent the oil droplet repolymerization. In the process of emulsion liquid[5,6] flow, the friction force of liquid flow on pipe wall and intramolecular friction force can be weakened. Second, due to the wetting effect of the viscosity reducer aqueous solution[7], the flow resistance is significantly reduced, that is, a layer of aqueous solusion water film is adsorbed on the pipe wall, so that the friction between the crude oil and the pipe wall becomes the friction between the aqueous solution and the pipe wall, to achieve the purpose of a significant decrease in flow resistance.

Before the gum asphaltenes form aggregates, the reticulate framework[8,9] of asphaltenes is first dissociated by the polar high permeability additives in the viscosity reducer. The heavy oil forms small particles, and the gum loses its structure and is not easy to agglomerate.

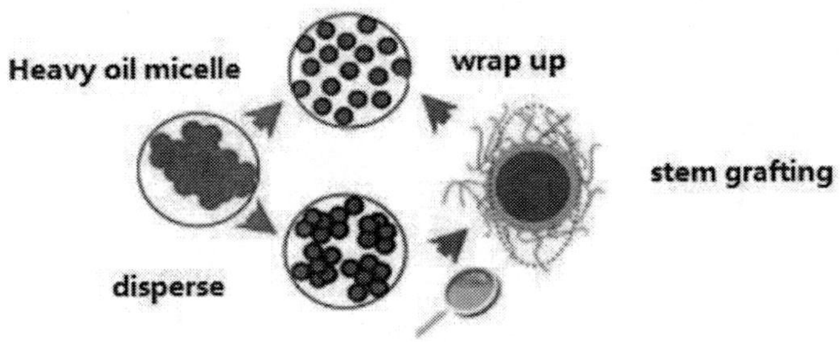


Fig. 1 Viscosity reduction mechanism diagram

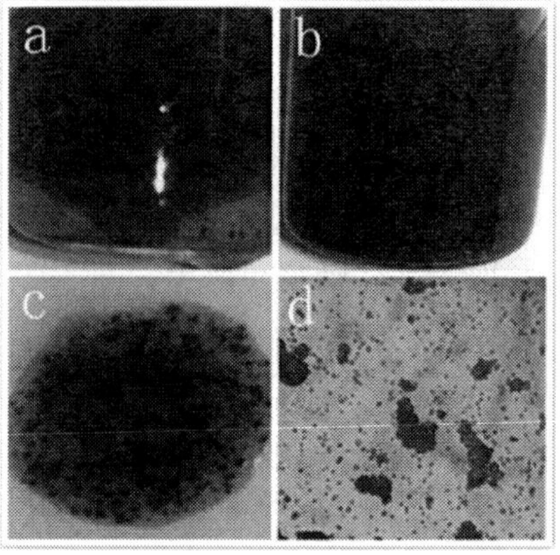


Fig.2 Comparison of effects before and after the graft coated viscosity reducer added
(a.Unadded heavy oil b.Emulsifying heavy oil with additives c.Macro photo after emulsification d.Micro photo after emulsification)

It can be seen from Figs.1 and Figs.2. The released gum and the surface active components in the viscosity reducer form a film on the surface of the small particles to prevent the asphaltene from accumulating again to form a reticulate framework, so as to achieve the effect of stably reducing the viscosity of heavy oil[10]. The formed heavy oil dispersion system makes the small particles present an ' tumbling flow ' state during the flow process and reduces the internal friction. To achieve the purpose of heavy oil viscosity reduction.

## 3 Experiment part

### *3.1 Experimental instruments and samples*

Experimental instruments: BROOKFIELD rotational viscometer, constant temperature water bath, electronic balance, beaker, color comparison tube.

Experimental samples: heavy oil in the fifth operation area of Shuguang Oil Production Plant of Liaohe Oilfield, field water samples.

### *3.2 Preparation of Medicament*

2% Ethylene vinyl acetate copolymer + 2% dichloroethane + 1% polyacrylate + the rest is light oil. Mix the above substances into a flask, macromolecular grafting agent was prepared by fully stirring 30-60min at normal pressure 50℃ water bath.

Sodium heavy alkyl benzene sulfonate 25% ( solid content is 30 %) + AES( aliphatic alcohol ether sodium sulfate) 55% ( concentration is 20%) + OP-10 ( alkyl phenyl polyoxyethylene) 10% ( concentration is 20%) + polyethylene glycol 8% + 0.5% sodium dodecyl benzene sulfonate + 1.5% sodium hydroxide ( concentration is 20%). The above agents are fully mixed and stirred to obtain a coated viscosity reducer.

### *3.3 Experimental method*

In the experiment, 50mL long and thin beaker was used. First, macromolecular graft agent was added and stirred, and then coated viscosity reducer was added. The viscosity can be determinated after mixing evenly.

*3.3.1 Emulsion preparation and apparent viscosity determination method*

The preparation and determination of apparent viscosity of heavy oil O/W system were carried out in accordance with the provisions of the industry standard Q/SY 118-2013 5.2.

*3.3.2 Viscosity reduction effect evaluation method*

Viscosity reduction rate calculation method:

$$P = \frac{\mu_0 - \mu_1}{\mu_0} \times 100 \quad \cdots\cdots\cdots\cdots\cdots\cdots\cdots\cdots\cdots\cdots\cdots \text{(1-1)}$$

In the formula: P - viscosity reduction rate of heavy oil, %;

μ0 - viscosity of heavy oil before adding viscosity reducer, mPa • s;

μ1 - viscosity of heavy oil after adding viscosity reducer, mPa • s.

The greater the viscosity reduction rate, the better the viscosity reduction effect of the viscosity reducer on heavy oil[11,12].

**4 Result and discussion**

In this paper, the effects of additive agent concentration, time, temperature and water content on viscosity reduction were investigated.

*4.1 Effect of additive concentration on viscosity reduction effect*

According to the test steps, the viscosity values under the conditions of 500mg/L, 1000mg/L, 1500mg/L and 2000mg/L were tested respectively. The results are as follows:

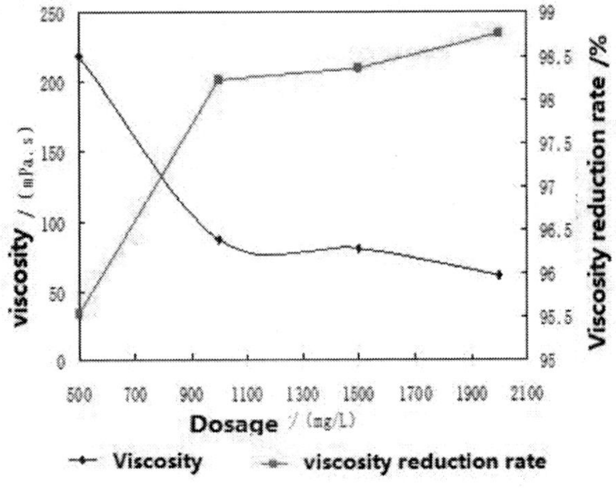


Fig. 3 Effect of additive dosage on viscosity

According to Figure 3, when the additive dosage is 2000mg/L, 1500mg/L, 1000mg/L, the apparent viscosity is below 300mPa·s. The viscosity reduction rate is all achieve above 90%, and the static viscosity is below 200 mPa·s, which meets the requirements of wellbore liquid lifting. When the additive dosage is 500 mg/L, although the viscosity is below 600 mPa·s, the static viscosity is greater than 1500 mPa·s, which cannot meet the requirements of the standard Q/SY 118-2013, so the preferred additive dosage is 1000 mg/L.

### *4.2 Effect of time on viscosity reduction*

According to the test steps, the viscosity values with additive dosages of crude oil of 500mg/L, 1000mg/L and 2000mg/L were tested respectively, and the stability of viscosity values within 2h was determined. The results are as follows :

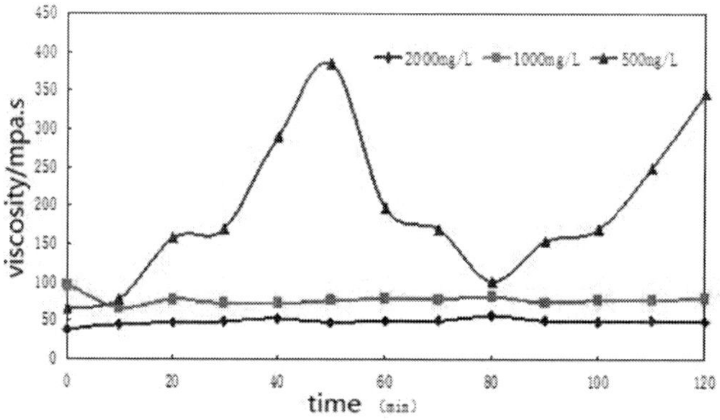


Fig. 4 Effect of time on viscosity

It can be seen from figure 4 that when the additive dosage is 2000 mg/L and 1000 mg/L, the viscosity can maintain good stability within 2h. However, when the dosage was 500mg / L, the viscosity value fluctuates greatly and the emulsion liquid is unstable. This is due to the fact that 2000mg/L and 1000mg/L of the viscosity reducer can be fully adsorbed at the oil-water interface and form a dense film. And this layer of film is not damaged with the passage of 2h time, showing the stability of crude oil emulsion liquid, more than 1000mg/L additive dosage is the optimal selection.

### *4.3 Effect of temperature on viscosity reduction*

The changes of viscosity at different temperatures under the condition of 1000mg/L additive dosage were determined. The results are as follows :

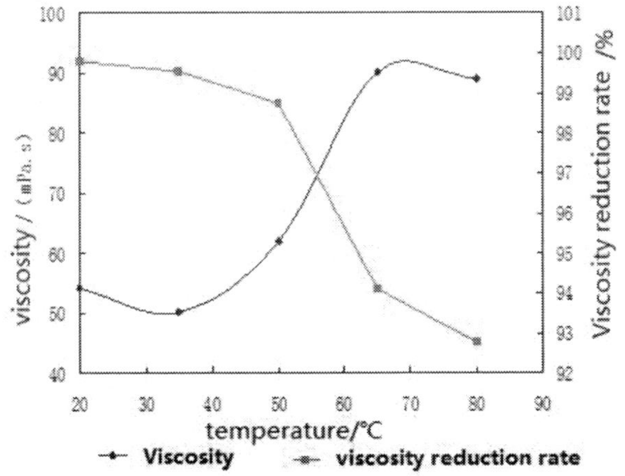


Fig. 5 Effect of temperature on viscosity

As shown in Fig.5, the viscosity-temperature curve is ' weird'.This is because at 80 ℃, due to the high temperature, the temperature determines the viscosity, the viscosity is small. When it reaches 60-70℃, the viscosity is determined by both the viscosity reducer and the temperature. Some small asphaltene particles wrapped by the viscosity reducer are destroyed, and the viscosity increases. When the temperature continues to decrease, the viscosity reducer plays a major role, and the viscosity begins to decrease again.

*4.4 Effect of water content on viscosity reduction effect*

The changes of viscosity under different water content were measured under the condition of 1000mg/L additive dosage. The results are as follows :

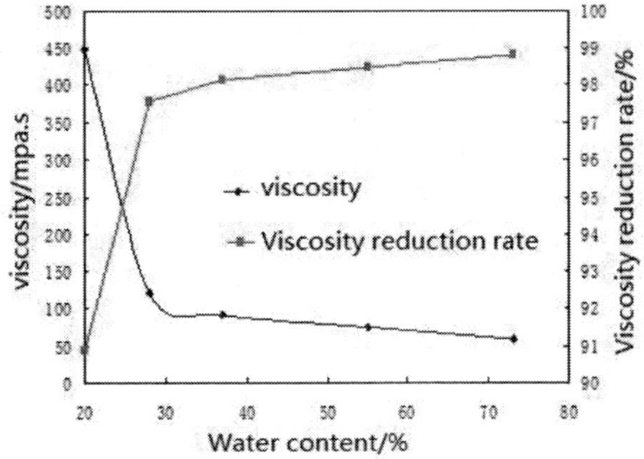


Fig.6 Effect of water content on viscosity

It can be seen from figure 6 that when the water content is less than 20%, due to too little water content, oil and water cannot reach the emulsification condition and cannot be emulsified. When the water content is 73%, there is a small amount of stratification and the excessive water content is meaningless to the actual production. Therefore, the water content of the field test well is the best effect between 20% and 70%.

### *4.5 Compatibility assessment*

Viscosity reducer 1000mg/L, while doing blank experiments. Demulsifier was added according to the fifth operation area water treatment process, and the demulsification temperature was 90℃. Its effect on water treatment is measured. The experimental results are shown in Table 1.

Table 1. Compatibility experiment assessment

| number | crude oil water content (%) | the amount of dehydration (mL) | Crude oil water content after dehydration (%) | emulsion splitter(mg/L) | flocculating agent(mg/L) | coagulant aid(mg/L) | Suspension content(mg/L) |
|---|---|---|---|---|---|---|---|
| 1 | 40~90 | 82 | <5 | 100 | -- | -- | 9 |
| 2 | 40~90 | 82 | <5 | 100 | -- | -- | 9 |

It can be seen from this that the dehydration amount of the colorimetric tube with and without the viscosity reducer is equal, both of which are 82mL, and the water content in the upper oil is less than 5%, indicating that the viscosity reducer has no effect on demulsification. Therefore, the addition of viscosity reducer will not affect the dehydration of the combined station.

## 5 Conclusion

A viscosity reducer suitable for Liaohe heavy oil was developed and compared. The optimum concentration of the graft coaded viscosity reducer is 1000mg/L. The viscosity of heavy oil is below 100mPa·s at 50℃. The viscosity reduction rate is achieved more than 95 % and stable. Reaching oil well lifting conditions. And the viscosity first increases and then decreases with the increase of temperature. It has the best viscosity reduction effect when the water content is between 20% and 70%. It has no effect on the water treatment work of the joint station.

## References

[1]Jun-biao Gu, Long-she Shen. Application of Decreasing Viscosity Technique of Heavy Oil on Crude Production[J]. Liaoning Chemical Industry,2003,32(10):40-43.
[2]Gui-sheng Cui. Study on Viscosity Reduction Method and Mechanism of Heavy Oil Emulsification[D]. East China : China University of Petroleum,2009.
[3]Min Huang,Fang-tian Li,Zu-hua Shi. Viscosity Reducer of Heavy Oil. DJH-1[J]. Oilfield Chemistry,2000,17(2):137-139.
[4]Bing Qin. Study on The Relationship Between Structure and Properties of Heavy Oil Emulsified Viscosity Reducer[D]. Bei Jing:Petrochemical science academy,2001.

[5]Feng-ying Zhang,Jian Li,Lin Zhu,etc. Research Progress of Heavy Oil Oil-soluble Viscosity Reducer[J]. Special Oil and Gas Reservoir,2006,13(2):16-20.
[6]Bao-ming Wei. Metal Corrosion Theory and Application[M]. BeiJing: Chemical Industry Publishing House,1984:3-5.
[7]Wen-jie Liang. Petrochemistry[M]. ShanDong DongYing:University of Petroleum Press,1993:64-65.
[8]Xian-qing Han. Principles of Enhanced Oil Recovery. BeiJing: Petroleum Industry Publishing House,1993:131-169.
[9]Guo-ran Liu. Heavy Oil Emulsification Viscosity Reduction Technology[J]. Special Oil and Gas Reservoir,1995,2(1):57-59.
[10]LANGEVIND, POTEAUS, HéNAUTI,et al. Crude oil emulsion properties and their application to heavy oil transportation [J]. Oil & Gas Science and Technology-Rev IFP, 2004, 59(5): 511-521
[11]PARIDAR S，NAZAR A，KARIMI Y. Experimental evaluation of asphaltene dispersants performance using dynamic light scattering[J]. Journal of Petroleum Science and Engineering，2018，163: 570 — 575
[12] AWUAL M R，HASAN M M，ISLAM A，et al． Optimization of an innovative composited material for effective monitoring and removal of cobalt(II) from wastewater[J]. Journal of Molecular Liquids，2019，298: 112035.

# Development and Application of Polymer Plugging Remover for Polymer Injection Well in Changqing Oilfield

**Xiang-hui WAN1,2 (✉), Wen-xue JIANG1,2, Yang XU1,2**

1 ChuanQing Drilling Co. Research Institute of Drilling and Producing Engineering, Xi'an,China

2 National Engineering Laboratory for Exploration and Development of Low-Permeability Oil & Gas Fields, Xi'an,China

superproeric@qq.com

**Abstract.** Water injection is a conventional method to maintain reservoirs pressure and improve production in low permeability oilfield. To decrease high water-cut and further enhance oil recovery, Polymers, such as polymer microsphere, preformed polymer gel, cross-linked polyacrylamide, are extensively used to control injection profile in Changqing oilfield. However, due to the low permeability and narrow pore throats of tight reservoirs, pressure in some water flooding wells increases rapidly after the polymer was injected, negatively affecting the energy supplement of the well cluster. A polymer plugging remover, composed of oxidizer, acid and chelating agent to relieve blockage, was developed with viscosity reduction and structure decomposition. The release period is more than 48 hours, which is conducive to deep-seated plugging removal. After treatment, the average diameter of polymer residue is 39-254nm and the dissolution rate is 15.76-53.42%. The permeability of the core sample plugged with polymer has increased significantly before and after treatment by polymer plugging remover, especially, it recovered by 28.67% for microsphere polymer. Five field tests have been completed successfully in the Jiyuan oilfield of Changqing. Before treatment, conventional acidification was performed in the pilot wells for two times at least, which is proved invalid. However, after polymer degraded by the plugging remover, the injection rate increased by 16.2m$^3$/d, the pressure dropped by 3.4MPa and the validity period has exceeded 130 days on average. The polymer plugging remover provides an alternative way to increase injectability when dealing with high pressure and under-injection wells caused by polymer flooding.

## 1. Introduction

Changqing Oilfield is a typical reservoir, namely low porosity, low permeability and low richness. Water injection is an important method to maintain and restore reservoir energy and to improve the recovery rate in tight reservoir [1]. In response to the rapid increase in water content ratio in production well during the injection, the polymer flooding was performed to control water cut and improve sweep efficiency. The profile controlling materials mainly include polymer microspheres [2], PPG [3], cross-linked polyacrylamide [4], surfactants [5], of which polymer accounts for more than 90% [6]-[7]. Since 2017, the number of polymer injected well in Changqing Oilfield has reached 4,000 wells per year,

and it has changed from single well injection to general injection through water injection station [8]. Profile control agents could retain in porous media by chemisorptions and mechanical trapping. With the increase of polymer injection volume and the heterogeneity of the reservoir properties, 10 percent of water wells currently exhibit high injection pressure and fail to reach the desired injection rate; as a result, intermittent injection is adopted instead. The number of polymer clogged wells is still increasing, which seriously affects the water drive efficiency.
866 treatments in water injection wells were conducted to enhance injectivity in Changqing Oilfield in 2019. Though it takes effect to some extent that pressure decreases and injection increases, the proportion of polymer injection wells was as high as 77% in 78 invalid wells in statistics. It is common that low water flooding efficiency and short validity period present in the post-treatment wells [9]-[11]. In response to the above problems, the polymer plugging remover technique is proposed to relieve the deep-seated plug and enhance injection for the high-pressure and under-injection wells, which provides a high efficiency solution for exploitation of the block.

## 2. Experimental Materials and Instruments

### *2.1. Materials*
Polymer degradation agent (oxidizer), sustained-release plugging remover (acid and chelating agent), core sample from Changqing Oilfield, polymer microspheres, single-phase gel (a preformed particle gel), and expandable particles from sample collection in the oilfield.

### *2.2. Instruments*
ICS-5000 ion chromatography (Japan), Nano C laser particle size analyzer (USA), HTHP acidification core flow analyzer (USA)

## 3. Analysis of Polymer Blockage

According to Chang 8 core sample analysis of Jiyuan oilfield, the average porosity is 10.6% and the average permeability is 0.85md. The median radius is 0.26μm and the diameter of pore throat of dominant channel is 0.3-1.9 μ m [12]. Taking Block A as an example, 151 wells have been conducted profile control treatment in the past five years. As a result, the average monthly decline has decreased from 1.9% to 1.0% after profile control agent, mainly polymer microspheres, was injected. Based on the statistics of 16 wells, the average injection pressure increased from 13.9 MPa to 17.6 MPa and the average water absorbing thickness increased from 9.7 to 15.1 m, which indicates that water injection profile is improved. However, the injection pressure rises rapidly, for example, 11 out of 16 wells are under-injection and the pressure are as high as 19 MPa.

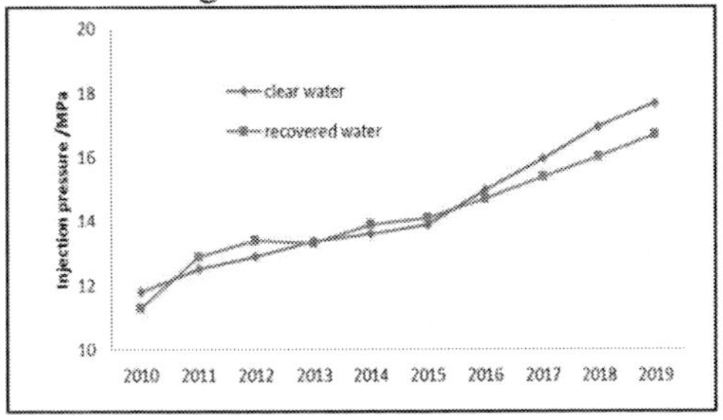


**Fig. 1.** Change of water injection pressure over the time
The Nano C laser particle size analyzer was used to test the particle size of the profile control agents in Changqing Oilfield. The results are shown in the Fig. 2 and Fig.3.

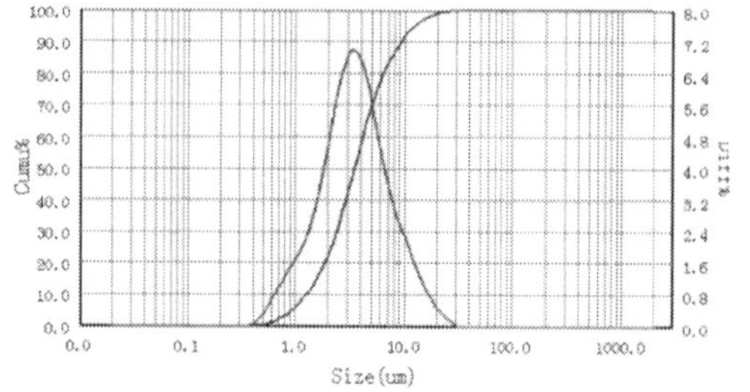


**Fig. 2.** Particle size of peg( single-phase gel)

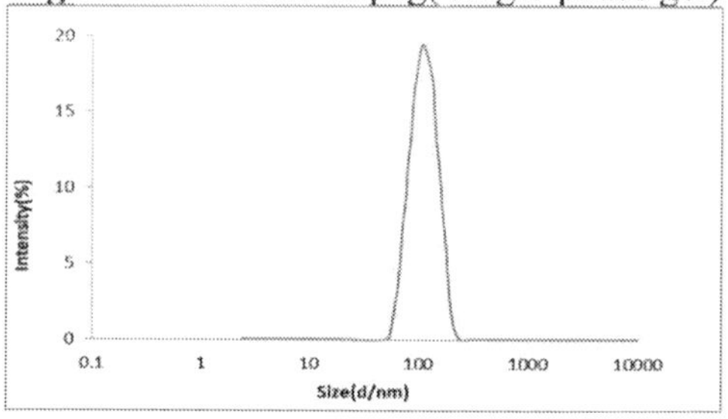


**Fig. 3.** Particle size of polymer microspheres

The particle size distribution of PEG ranges from 0.5μm to 50μm, and that of polymer microspheres is between 100nm to 400nm which can expand with time. The size of expandable particles ranges from 0.5mm to 8mm. According to the 1/3 bridging principle, the above profile control agents could lead to plugging and cause rapid pressure rise in the well. The profile control materials jamming from the near wellbore area to deep reservoir are as follows: expandable particles, PEG single-phase gel, and polymer microspheres.

## 4. Experiment and Result

The polymer plugging remover for injection wells is composed of polymer degradation additives and sustained-release plugging remover agent, which are processed with inverse emulsion by the mass ratio of 1:3.

The reactive components are released slowly through demulsification by temperature and acid reaction. The catalyst component, released by the polymer degradation additives, can activate, decompose and dissolve the polymer chain. $H^+$, ionized from sustained-release decompose agent, can clean and dissolve the inorganic plug, enlarge the capillary radius of channel. As a result, the pressure of polymer injection well is reduced. Preferably, the ratio of ionized $H^+$ to the effective content of catalyst components is 2:1.

### *4.1. Release Performance Test*

Under the conditions of acid rock reaction in a certain temperature, the release rate can be determined by the hydrogen ion's concentration in the experiment. Natural marble block, ranging from 3 to 5 mm, is used as the reactant at 60 ℃ . The weight loss of marble is easily converted to the molar concentration of hydrogen ion based on the chemical reaction formula. The test results are shown in Fig. 4.

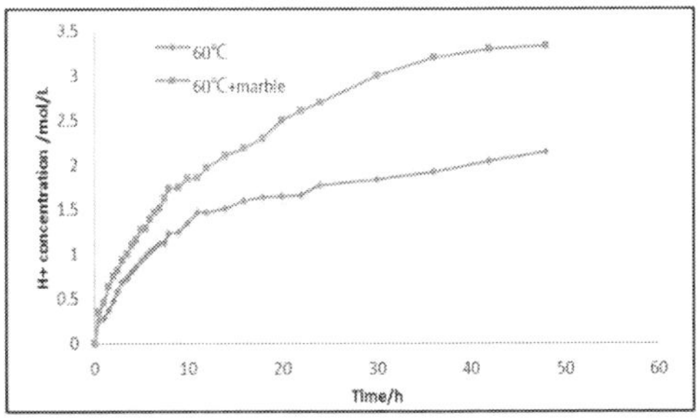


**Fig. 4.** Results of hydrogen ion concentration released by plugging removal fluid

The results show that the full release time of the remover is more than 48 h under dual factors of reactant and temperature, while the release efficiency is only 64.2% for the single factor of temperature. Because the average content of calcite in reservoir interstitial material is 4.6% in Jiyuan oilfield, the operation distance of the remover will be further, which is beneficial to the deep-seated blockage in the formation. As a result, shut-in time should be prolonged for more than 24 hours after the treatment.

### *4.2. Particle Size Test of Polymer after Degradation*

First, 20g natural marble was added to a reaction vessel in a reheating furnace, then 100ml polymer plugging remover was added. At last, 10g polymer microspheres, single phase gel and expandable agent were added separately to the vessel under different reaction time at 70 C. The average particle size of the flooding agent was measured by laser particle size analyzer at different times. The results are showed in Fig. 5.

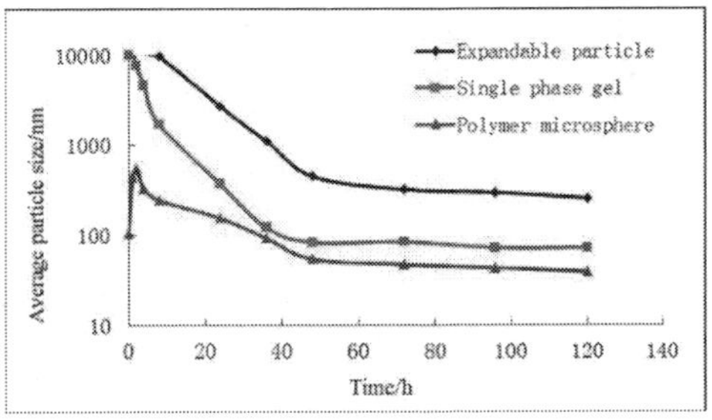


**Fig. 5.** Particle size of flooding agent after degradation by polymer plugging remover

The results show that the particle diameters of the three kinds of profile control agents were basically constant after 48 hours. After breakdown, the average particle size of the expandable granular agent was 254nm while the single-phase gel flooding agent was 72nm and the polymer microsphere was 39nm. At the same time, the once single-phase solution turned into separate layers obviously. The final particle size of the polymer flooding agent was smaller than the average median diameter of the reservoir, which is 520nm. If water injection is continued, the residue will migrate towards deeper formation and blockage can be relieved. It can also be discharged back to the surface by the swabbing pump.

### *4.3. Degradation Test*

The polymer microspheres, single-phase gel, and expandable particles were treated with ethanol and dried. Two kilograms of sample were put into 100ml of remover at 70℃ for different time, then were filtered and dried at 105℃. The dissolution ratio of the polymer agent was tested compared with a blank sample, the results are shown in Fig. 6.

.

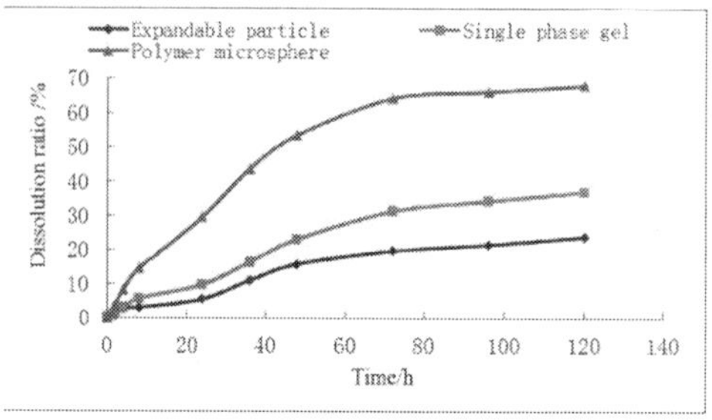


**Fig. 6.** Degradation of profile control agent by plugging removal fluid

From the results, it can be seen that the remover has a certain degradation capacity to the three kinds of polymer. The degradation rate of expandable particles, single-phase gel, and expandable particles is 15.76%, 23.10% and 53.42% respectively in 48h, indicating the remover has a certain breakdown ability for polymer materials. The three kinds of polymer have distinct dissolution rates, which are closely relate to their synthesis techniques and inherent properties, such as molecule weight, micro structure, solubility, particle size distribution and specific surface.

### *4.4. Dynamic Core Flow Test*

Dynamic core flow test is conducted with HTHP core flow instrument, the core flow test is designed to evaluate the compatibility between the polymer plugging remover and the formation. The selected sample core is from Yanchang series of Jiyuan oilfield; the temperature is 70 C.

The experiment procedure is as follows: forward displace the core sample with formation water to calculate the permeability $K_1$; forward displace the profile control agent for 2 PV and keep the core sample in static state for 120h; forward displace the core sample with formation water to calculate the permeability $K_2$; forward displace the polymer plugging remover and keep the core sample in static state for 12 h; forward displace the core sample with formation water to calculate the permeability $K_3$; calculate the damage ratio of profile control agent and the polymer plugging remover. The experimental results are shown in Table 1.

**Table 1.** Core flooding of profile control agents and plugging remover

| Series | Polymer | NO. | Gas permeability, $10^{-3}\mu m^2$ | porosity , % | damage ratio by profile modifier, % | damage ratio after treated with remover, % |
|---|---|---|---|---|---|---|
| Chang 8 | Expandable particles | 8 | 1.057 | 11.33 | 82.92 | 37.54 |
| Chang 8 | Expandable particles | 10 | 0.759 | 10.43 | 81.66 | 29.85 |
| Chang 8 | single-phase gel | 14 | 0.943 | 9.65 | 94.51 | 24.68 |
| Chang 8 | single-phase gel | 18 | 0.733 | 8.47 | 95.48 | 17.22 |
| Chang 8 | polymer microspheres | 11 | 0.683 | 8.24 | 95.75 | -24.7 |
| Chang 8 | polymer microspheres | 9 | 0.828 | 8.31 | 94.84 | -32.64 |

The results show that the core damage rate of polymer microspheres, single-phase gel and expandable particles is 95.30%, 95.99%, and 82.29% respectively at 70℃, indicating they can modify profile and plug water drive channel. After plugging of the core by the polymers, the blockage was relieved to varying degrees once the remover was displaced. The damage ratio of polymer microspheres after treatment was -28.67%, while it was 20.95% and 33.70% for single phase gel and expandable particles separately. The test results show that the remover is the most effective for the microsphere breakdown. The plugging materials are degraded and the core is expected to be stimulate, which indicates that the remover has good adaptability to the reservoir.

## 5. Field Application

The remover has been applied to five wells in Jiyuan Oilfield. All of the candidate wells had been treated for more than two times with other working fluid. Based on the reservoir thickness, injection pressure, polymer type and total injection volume, the period of under-injection, the designed plugging remover volume is 80m$^3$ and displacement volume is 50m$^3$. The displacement rate is 400-600L/min and the shut-in time is 72 hours. 130 m$^3$ produced liquid is swabbed before water injection starts. After the treatment, the average pressure decreased is 3.4MPa, and the daily injection increased by 16.2m$^3$/d. Up to now, the average cumulative injection increased by 2024.6m$^3$, and the average valid period has exceeded 130 days. All the wells being treated are still in good performance that will continue to be tracked in the following days. In a word, increased injectivity has been achieved for polymer flooding wells in low permeability oilfield.

D127 well, after three times of hydraulic fracturing and acidizing treatments, the injection pressure was maintained at 16.1MPa in the early times. Due to the higher water cut of the corresponding production well, the compound profile control agent, polymer microsphere and single-phase gel, were injected in August 2019, and the pressure increased rapidly to 19.4MPa. After acidizing and acid fracturing for two times, the injectivity fail to reach the desired with a valid period less than 3 months. When the water pressure reaches 21.2MPa, the maximum limit of the injection infrastructure, the desired injection rate is 20m$^3$/d but the actual rate is as less as 3m$^3$/d. After polymer plug removal, the water injection pressure decreases by 3.6 MPa and the daily injection rate increase by 27m$^3$, thus the energy of well group is fully supplemented. Well analysis showed that profile control polymer was mainly blocked in the middle or deep in the wellbore, conventional acid was consumed too fast in the near well zone. Thus, the removal of polymer plugging by acid was poor, resulting in invalid attempt or short valid period. The polymer plugging remover makes up for the deficiency of conventional fluid by its own advantages. At the same time, it is analyzed that the small particle size polymer residue which cannot be completely breakdown will be discharged back to the surface by flowback or swab, so as to reduce its secondary plugging to the deep formation, which is beneficial in prolonging the effective period and improving the injectivity.

Table 2 Performance evaluation of wells after treatment

| well | Before the treatment | | | | After the treatment | | | | Daily increase rate of injection m$^3$/d | Cumulative increase of injection volume m$^3$/d |
|---|---|---|---|---|---|---|---|---|---|---|
| | Tubing pressure MPa | Casing pressure MPa | injection allocation rate m$^3$/d | Actual injection rate m$^3$/d | Tubing pressure MPa | Casing pressure MPa | injection allocation rate m$^3$/d | Actual injection rate m$^3$/d | | |
| D127 | 21.2 | 21.1 | 20 | 3 | 17.6 | 17.6 | 30 | 30 | 27 | 3051 |
| Y95 | 19.2 | 19.2 | 20 | 5 | 16.2 | 16.2 | 20 | 20 | 15 | 1854 |
| D221 | 19.1 | 19.1 | 20 | 16 | 15.4 | 15.3 | 20 | 20 | 4 | 568 |
| Y47 | 19.5 | 19.4 | 20 | 8 | 15.9 | 15.8 | 30 | 30 | 22 | 2648 |
| G19 | 19.3 | 19.3 | 20 | 7 | 16.2 | 16.1 | 20 | 20 | 13 | 2002 |
| sum | 17.8 | 17.8 | 30.1 | 17.2 | 15.8 | 16.3 | 31.0 | 28.4 | 11.2 | |

## 6. Conclusion

(1) Through the particle size analysis of profile control polymers and pore throat of the low-permeability reservoir, profile controlling polymers can cause serious blockage in the water injection wells and decrease of water injectivity.

(2) The release time of the polymer plugging remover is more than 48 hours under the formation condition. It can greatly degrade and reduce the polymer particle size, dissolve polymer residues and increase the core permeability, which has a good adaptability to low permeability reservoirs.

(3) The polymer plugging remover has greatly improved the injectivity of five repeated under-injection and high-pressure wells. The average decreased pressure is 3.4MPa, the daily increased injection is 16.2m$^3$/d, and the average validity period has exceeded 130 days, and it continues to be effective.

## References

[1] Liu Guangfeng, Lei Yan, Xin Meng, et al. Study on water flooding improvement technology in southwest Chang 8 reservoir of block L1 in Jiyuan Oilfield. Petrochemical Industry Application 35(12), 34-38(2016).

[2] Yuqin Jia, Haien Yang, Chen Cheng, et al, Field Application and Performance Evaluation of Polymer Microsphere Profile Control in Low Permeability Oil Reservoir. In: International Petroleum Exhibition & Conference, Abu Dhabi(2019).
[3] Abdulmohsin Imqam; Baojun Bai, et al. Preformed Particle Gel Propagation Through Super-K Permeability Sand and Its Resistance to Water Flow During Conformance Control. In: SPE/IATMI Asia Pacific Oil & Gas Conference and Exhibition, Nusa Dua(2015).
[4] R. S. Seright; Mac Seheult; Todd Talashek. Injectivity Characteristics of EOR Polymers. SPE Res Eval & Eng 12 (05), 783–792(2009).
[5] Lirio Quintero, Michael Deighton, Henry Nguyen. Surfactant Enhanced Waterflood SEW in Medium and Light Crude Oil Reservoirs In: the Offshore Technology Conference Asia, Offshore Technology Conference, Houston, Texas, USA (2018).
[6] Xue Fangfang, Ma Bo, Chu Liguo, et al. Detection Method of Nanoscale Polymer Microspheres in Oilfield Injection and Production Fluids. Journal of Liaoning University of Petroleum & Chemical Technology37(1), 165-168(2020).
[7] Jia Yuqin, Yang Haien, Zhang Tao, et al. Adaptability Analysis of Polymer Microsphere Flooding in Low Permeability Reservoir and Reservoir Selection Journal of Liaoning University of Petroleum & Chemical Technology, 37(1), 38-40(2017).
[8] He Haizhou,Zhang Dexin, Liu Chunlin, et al. Adaptability of polymer microsphere flooding in low permeability reservoir: a case study of block A in Changqing Oilfield. Petrochemical Industry Technology 26(2), 118(2019).
[9] Peng Chong, Li Jibiao,Zheng Jiangang, et al. Performance Evaluation of Dimeric Surfactant Compound System and Its Application in Longdong Oilfield. Oilfield Chemistry 36(4), 657-661(2019).
[10] Wang Erzhen, Wang Yong, Song Zhaojie, et al. Application of Long-term On-line Injection Enhancement Technology in Changqing Jiyuan Oilfield Oilfield Chemistry 36(2), 262-266(2019).
[11] Deng Zhiying, Zhang Suiwang, Song Zhaojie, et al. Study and application of online diversion acidizing technology in injection wells in ultra-low permeability reservoirs. Oil & Gas Geology40(2), 430-435(2019).
[12] Zhou Dan, Gu Liang, Li Taiyu, et al. Study on multi-component composite micro-emulsion micellar liquid injection technology in Chang 8 reservoir of Jiyuan oilfield. Petrochemical Industry Application36(9), 19-23(2017).

# Study on Nozzle Structure of Hydraulic Jet Fracturing

**Meng CAI 1,2, Xin-ying1,2 CHENG(⊠), Xiu-hong CHEN 1,2, Chong-jia ZHU 1,2and Yan ZHANG 1,2**

1 Daqing Oilfield Production Technology Institute, Daqing City, Heilongjiang Province, China
2 Heilongjiang Provincial Key Laboratory of Oil and Gas Reservoir Stimulation, Daqing City, Heilongjiang Province, China
chengxinyingi@petrochina.com.cn

**Abstract.**In the process of hydraulic jet fracturing and perforation, fracturing fluid carrying sand is injected through annulus to fracture the pay zone. The numerical simulation analysis with the software of ANSYS-Fluent was applied aiming at the problem of low quality of fracturing fluid through the nozzle jet. By means of the analysis and comparison to some important structural parameters of the nozzle, the optimum jet quality through the nozzle was obtained. The research results showed. When the nozzle outlet diameter is from 4.5mm to 4.8mm; the length of the column is from 7mm to 12mm and the inlet radius of curvature is 9mm, the optimum jet quality of the nozzle can be obtained. The jet quality through the nozzle can be improved effectively through the analysis and optimization of the structure of the nozzle, which provides effective technical support for the hydraulic jet fracturing technology.

## 1 Introduction

In the process of hydraulic jet fracturing, the nozzle is a key component of the jet formation process, and the quality of the nozzle directly affects the quality of the jet [1]. The quality of the jet used for sandblasting includes jet velocity, jet density and pressure drop loss [2]. Under the same flow condition, the difference of nozzle structure can lead to the difference of jet quality. In order to judge the quality of nozzle structure, it is very important to find out the criterion to measure nozzle structure. Therefore, in order to improve the jet quality, it is necessary to study the new nozzle and optimize the main structural parameters of the nozzle (outlet diameter, straight column section length, inlet curve radius) to obtain the optimal jet quality[3-5].

## 2. Nozzle structure and parameters

The overall structure of the nozzle is simple, and the external design is stepped or equal diameter, so that it can be installed in the nozzle. The shape of the flow channel inside the nozzle is mainly a shrinkage curve, as shown in Figure 1. The structural parameters of the nozzle mainly include the diameter of the nozzle inlet, the diameter of the outlet, the horizontal length, the diameter of the inlet contraction Angle/inlet curve and the length of the straight column section, as shown in Figure 1. Different structural parameters correspond to different jet mass. Through simulation research, the structural parameters with the highest jet mass can be found [4-6].

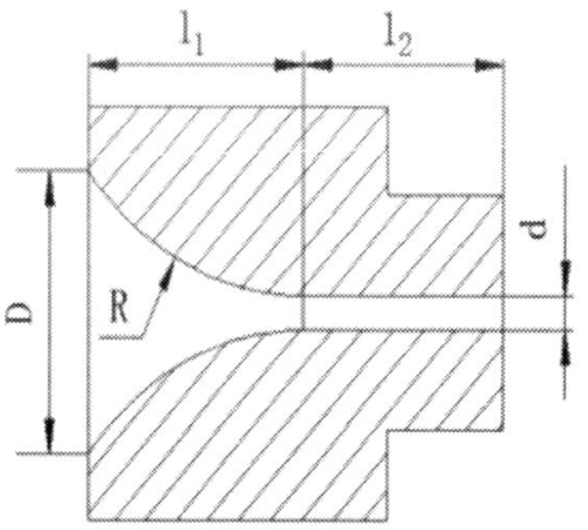


**Fig. 1.** Structure diagram of curved nozzle

## 3. Optimization analysis of nozzle parameters

In the process of sand blasting, the inlet diameter of the nozzle has little influence on the jet mass. Therefore, under the condition of a certain gun wall thickness, the outlet diameter of the nozzle d, the length of the straight column section l2 and the inlet curve radius R determine the jet performance of the nozzle. Therefore, the above three parameters are optimized and analyzed respectively[5-8].

### *3.1 Influence of outlet diameter on jet performance*

In the perforating process, the minimum jet velocity of the nozzle is 230m/s; when the inlet displacement is 1m3/min and the number of nozzles is 4, the minimum diameter of the nozzle outlet should be 4.8mm without considering other geometric parameters. When the diameter of the nozzle outlet is less than 4.8mm, whether higher efflux effect can be obtained, the outlet diameter of 2mm, 3mm, 4mm and 4.8mm can be taken respectively for simulation analysis and calculation[9-10].

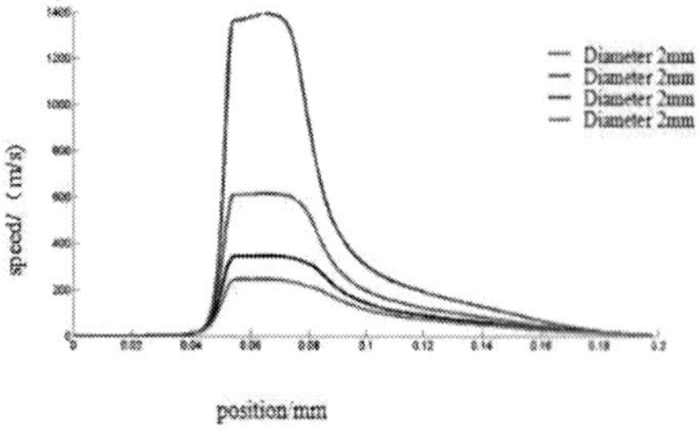


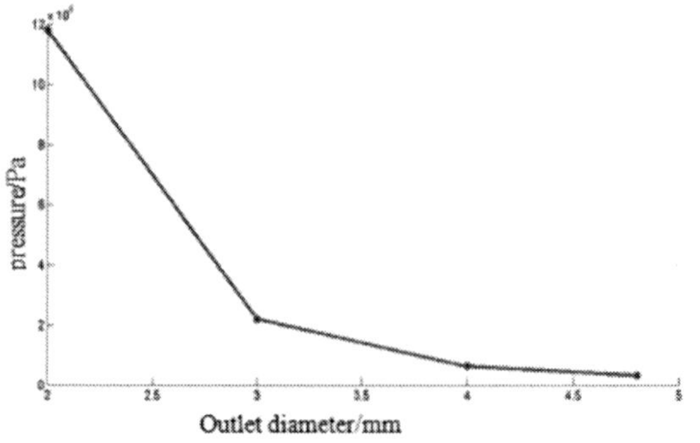


**Figure.2-a.** Influence of outlet diameter on jet velocity **Figure.2-b.** Influence of outlet diameter on pressure difference

According to Figure 2-a, the nozzle aperture shrinks after the liquid enters the nozzle, and the liquid velocity increases to the maximum value within a short time. When the liquid exits the nozzle, the axis velocity gradually decreases. When the outlet diameter of the nozzle changes, a larger jet velocity can be generated, but the equivelocity nucleus of the outlet diameter of the smaller nozzle has a smaller length and a faster attenuation, indicating that the jet density is relatively poor. Meanwhile, according to Figure 2-b, the pressure difference is inversely proportional to the outlet diameter. The smaller the outlet diameter, the greater the pressure difference, the more pressure energy is converted into kinetic energy, and the greater the pressure drop loss will inevitably be.

### *3.2 Influence of straight column length on jet performance*

As can be seen from Figure 3-a, the velocity variation trend of the nozzle in the straight column section of different lengths is basically the same, but the axial velocity attenuation of the jet becomes slower and slower as the length of the straight column section increases, and the jet density increases. At the same time, it can be seen from Figure 3-b that the longer the straight column section is, the greater the pressure drop will be. This is because when other structural parameters remain unchanged, the distance that the fluid flows through the straight column section of the jet increases with the increase of the length of the straight column section, thus increasing the pressure drop at the straight column section. Therefore, the design of this value should be combined with the gun wall thickness limit and pressure drop loss requirements.

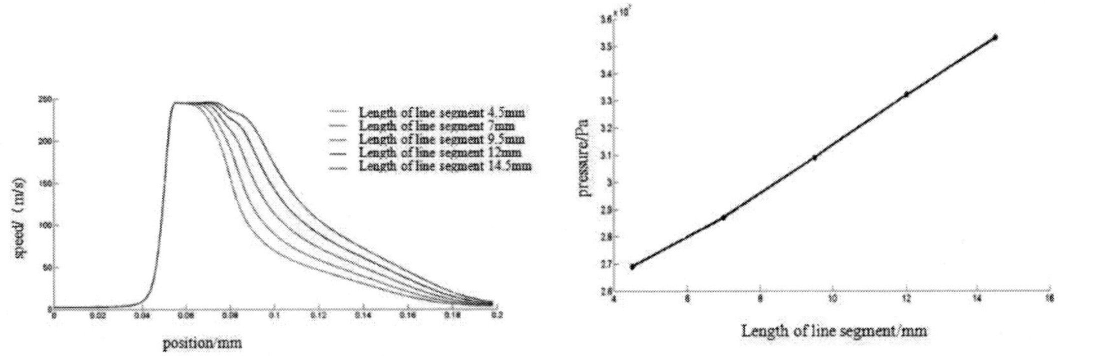


**Figure.3-a.** Effect of length of straight column segment on jet velocity **Figure.3-b.** Effect of length of straight column segment on pressure difference

### *3.3 Influence of inlet curve radius on jet performance*

According to Figure 4-a, the changing trend of jet velocity relationship of nozzles with different inlet curve radii is basically the same. However, it can be seen from Figure 4-b that the pressure difference at the inlet and outlet of the nozzle is different with different curve radii. Under the five radii (r=6-10mm), the pressure drop first decreases and then increases with the increase of the radius. The pressure difference is the minimum when the radius is 9mm. Therefore, when other parameters remain unchanged, a value near the radius of the entry curve segment of about 9mm is established as the optimal value.

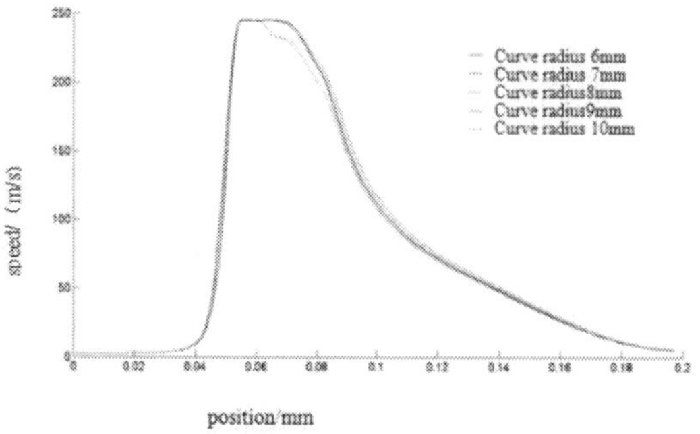


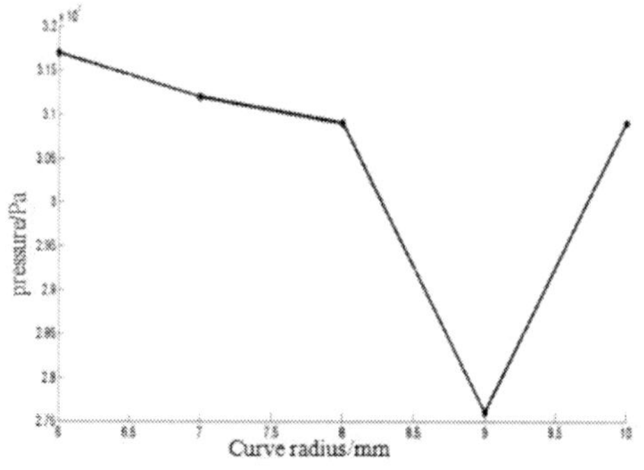


**Figure.4-a.** Influence of inlet curve radius on jet velocity **Figure.4-b** Influence of inlet curve radius on pressure difference

### *3.4 Comprehensive optimization of nozzle parameters*

The above simulation analysis is obtained after fixing two parameters and optimizing the third parameter[5]. In order to obtain the optimal jet quality, the parameters can be designed in combination. In the process of perforating, the pressure difference of the nozzle should not be too large. Therefore, the outlet diameter d≥4.5mm, the length of the straight column section l2≤12mm, and the curve radius of the inlet section R=9mm can be selected. The jet velocity is related to the diameter of the nozzle outlet, so the outlet diameter d≤4.8 is selected. The density of the jet is related to the diameter of the nozzle outlet and the length of the straight column section, so the outlet diameter d≥4mm and the length of the straight column section l2≥7mm can be selected. In the actual design, l2 of the straight column section should also be combined with the wall thickness limit of the spray gun and the importance of the pressure drop loss and the density of the jet. Inlet diameter has little influence on jet quality. In order to reduce the processing difficulty during design, inlet diameter D=2d is selected.

## Conclusion

(1) The smaller the outlet diameter of the nozzle, the higher the jet velocity, the greater the pressure drop loss and the worse the jet density; The longer the length of the straight column section of the nozzle, the greater the pressure drop loss and the better the density of the jet. With the increase of the curvature radius of the nozzle inlet passage, the pressure drop loss decreases first and then increases, and reaches the minimum when the curvature radius is 9mm. The inlet diameter of the nozzle has little effect on the jet mass of the nozzle.

(2) Structural optimization design results of hydraulic injection annulus sand fracturing nozzle: diameter of nozzle outlet ranges from 4.5mm to 4.8mm, length of straight column section ranges from 7mm to 12mm, inlet curvature radius R=9mm, inlet diameter D=2d.

## References

[1] Fu Gangdan, Li Xianwen, Ren Yong, Wang Zaiqiang, GUI Jie. Laboratory experimental study on optimization of hydraulic blasting perforation parameters [J]. Special Oil & Gas Reservoirs,2011,18(3):97-99.

[2] Yu Yong, Yang Biao, Du Baotan, et al. Application of hydraulic blasting perforation in oil and gas field development [J]. Special Oil and Gas Reservoirs, 2002,9(4):56-58.
[3] He Feng, Xie Junshi, Yang Jinglong. Influence of nozzle internal flow path profile on jet flow field. Chinese Journal of Applied Mechanics,2001,18(4):114-120
[4] Wei Dejun, Cao Zhongwen, Li Ping. Effect of Nozzle Shape on Spraying Ability [J]. Liaoning Chemical Industry,2006,35(7):92-95. (in Chinese)
[5] NIU Wei-yi. Numerical Simulation of Influence of flow tube Diameter on Separation Efficiency of cyclone [J]. Chemical Technology and Engineering,2015,8(1):45-48.
[6] DENG J X，FENG Y H，DING Z L，et a1．Wear be． havior of ceramic nozzles in sand blasting treatments [J]．Journal of the European Ceramic Society，2003 (23)：323—329．
[7] NANDURI M，TAGGART D G，KIM T J．The effects of system and geometric parameters on abrasive water jet nozzle wear J 1. International Journal of Machine Tools&Manufacture，2002，42(5)：615-623．
[8 ] NANDURI M，TAGGART D G，KIM T J．A study of nozzle weal-in abrasive entrained water jetting environ— ment[J]．Journal of Tribology，2000，122(2)：465-471．
[9 ] Cui Lu, Huang Shaopu, Kang Wenquan, et al. Research on erosion of multi-jet hydraulic injection tools under perforating conditions [J]. Oil Drilling and Production Technology,2018,40(5):620-625.
[10 ] Huang Zhongwei, Li Gensheng, Tian Shouzeng, et al. Analysis of downhole tool wear law in hydraulic jet multistage fracturing [J]. Journal of Chongqing University (Natural Science Edition), 2014,37 (5) : 77-82.

Journal of Physics: Conference Series **2594** (2023) 012069 doi:10.1088/1742-6596/2594/1/012069

# Experimental Study of Condensate Vaporization into Bubbles under Different Operating Parameters

**Ce Wang1, Yu-ling Lü1(✉), Chao Song2, Qing-wei Feng3, Xiao-xuan Chen1, Li-min He1**

1 College of Pipeline and Civil Engineering, China University of Petroleum (East China), Qingdao, Shandong, China

2 China Petroleum Pipeline Engineering Corporation (CPPE), Chengdu Branch, Chengdu, Sichuan, China

3 Xianhe Oilfield, Sinopec Shengli Oilfield Branch, Dongying, Shandong, China

lylchuyun@163.com

**Abstract.** Produced water (PW) is a natural by-product of the oil and gas production process, where crude oil is dispersed in the form of micron-sized emulsified droplets in the produced water due to mixing between crude oil and water caused by equipment, piping and valve shear action. Flotation is the main method for treating emulsified oil droplets in oilfield produced water. The conventional flotation technology suffers from a series of problems such as large reflux ratio and poor flotation effect, thus limiting its further application. In this paper, bubbles are prepared by vaporization of low boiling point condensate through theoretical analysis and experimental study. In this paper, the law of bubble generation from condensate solution at different temperatures after decompression and release is investigated. The results demonstrate that the median particle size of gas bubbles generated from condensate is related to the condensate content, mixing pressure, release pressure, and temperature. The median particle size can reach 9.65 μm when mixed at 2.0% condensate content, 0.2 MPa mixing pressure, and 35℃, and released at atmospheric pressure, which is much smaller than the size of gas bubbles generated by dissolved air flotation. The condensate bubbles as alkane congeners are more likely to adhere to the oil droplets in the extracted water, which is conducive to improving the air flotation oil removal efficiency of oily wastewater.

## 1 Introduction

Air flotation separation, as a physical oil removal technology, is widely used in oilfield wastewater treatment processes. The principle is to inject or try to generate a large number of uniform and dispersed tiny bubbles into the water to be treated, which are used as a medium for mutual adhesion with impurities such as oil droplets or suspended particles in the water to form a less dense polymeric substance that floats on the surface of the water body in order to accelerate the separation process between liquid and liquid, liquid and solid, etc. in the water [1]. Compared with other oil removal techniques, air flotation has the following advantages: (1) shorter residence time and higher loading rate; (2) less sensitivity to oil density and viscosity; (3) smaller footprint for compact design; and (4) better separation of small and light particles [2]. In addition, flotation has been widely used in other industries such as paper deinking [3], oil sands processing [4], grease recovery [5], microbial separation [6], and plastics recovery [7].

Bubble manufacturing technology is one of the main factors affecting the air flotation effect, the smaller the diameter of bubbles and the higher the number density, the better the air flotation effect, and the traditional microbubble manufacturing technology can be divided into three main categories: dissolved gas precipitation bubbles, induced gas manufacturing bubbles, and electrolytic precipitation bubbles [8]. For bubble fabrication techniques, many researchers have conducted related studies. Takahashi T. et al [9] conducted a fundamental study of dissolved gas at 0.1 MPa - 0.5 MPa pressure and bubble precipitation at atmospheric pressure and obtained empirical equations for calculating bubble diameter; Bowonder B. and Kunmar R. [10] investigated the process of bubble fabrication by porous discs and found that for The number of effective sites for bubble formation is a function of the surface tension, viscosity and density of the liquid and the gas flow rate; Ketkar D. R. et al [11] investigated the effect of several shapes of the electrode surface and the variation of the current density on the diameter of the electrolytic precipitation bubbles.

In order to further improve the air flotation separation efficiency and reduce the operating cost, experts and scholars at home and abroad have continuously explored the bubble fabrication technology, among which the air flotation process combining CTour process and jet mixer proposed by ProSep, Norway, injects oily wastewater, condensate, and flocculant into the jet mixer at the same time, and precipitates bubbles under the pressure plunge of the mixer [12]. Sabey et al [13] named the process CTour, which investigated the effects of dissolved gas in water, free gas in water, solid particles, and oil-water interfacial tension on the performance of the CTour system and concluded that these parameters do not affect the performance, but the structural design of the injection and mixing system and the operating conditions can have a significant effect on the oil removal efficiency. Statoil conducted an offshore test in the statfjord field, where the use of the CTour process resulted in a 70% improvement in oil droplet removal efficiency, as well as effective removal of corrosion inhibitor and hydrocarbon components from the produced water [14]. Thus, in this paper, the CTour air flotation method was also adopted to study the pattern of bubble generation from condensate solution at different temperatures after decompression and release.

The key to the oil condensate removal process is to produce small enough bubbles to meet the air flotation separation requirements under air flotation conditions, and the size distribution of the bubbles during the air flotation process can have an important effect on the oil removal efficiency. The reason is that the size of oil droplets and bubbles affects the collision efficiency, which in turn affects the air flotation efficiency. Smaller bubbles have a larger specific surface area for the same volume and can provide more attachment area for flocs [15], and reducing the bubble size during flotation can significantly improve the air flotation separation efficiency [16], N. Ahmed et al [17] studied the effect of bubble size on the particle removal efficiency of air flotation process by controlling the impeller speed of a multiphase flow pump to change the bubble size. When the bubble diameter was reduced from 655 μm to 75 μm, the air flotation efficiency was improved by a factor of 100. Similarly, Yoon R. H. [18] showed that reducing the bubble size during flotation can increase the probability of copolymerization of bubbles with fine particles. Yan et al [19] showed that reducing the bubble size

improves the efficiency of oil removal by air flotation, one of the reasons being that smaller bubbles increase the relative surface area available for droplet adhesion, thus improving efficiency. In addition, smaller bubbles also lead to an increase in collision frequency, which is a key factor in capturing oil droplets [20]. In addition, according to Lim et al [21], the decrease in gas-liquid contact angle and the increase in bubble size lead to a decrease in oil adhesion efficiency. From the above studies, it is clear that bubble size is related to many factors, and for the vaporization of condensate into bubbles process, temperature and pressure are the key to affect the bubble size, and there is no systematic study on the effect of different operating parameters on bubble size at home and abroad.
Therefore, this paper investigates the size of bubbles generated by condensate at different operating parameters, which is the key to determine the effectiveness of air flotation treatment. To this end, a bubble observation system with adjustable parameters is built in this paper, and the bubble size distribution of condensate at different temperature and pressure is analyzed by measuring the bubble size under different operating parameters using microscopic photography.

## 2 Theoretical Analysis

When the condensate solution passes through the pressure reducing valve, the pressure is suddenly reduced from the mixing pressure $p_1$ to the release pressure $p_2$, and the condensate droplets vaporize in this process, forming a large number of bubbles. The formation of bubbles can be divided into three processes, (1) the pressure around the condensate droplet decreases continuously, when the pressure is lower than the bubble point pressure, the intermolecular forces can not bind their own molecules to achieve vaporization. (2) The gas molecules dispersed in the solution agglomerate to form gas molecular clusters; (3) The gas molecular clusters closely pile up and expand to form micro bubbles.
Let the surface energy of the cluster consisting of n gas molecules be $w_0$.
Let the process of forming molecular clusters from individual gas molecules uniformly dispersed in the solvent be an isothermal process, then the change in chemical potential of this process is:

$$\mathrm{d}G = nV_\mathrm{m}\mathrm{d}p \quad (1)$$

Changes in the chemical potential for the formation of gas molecular clusters because the solvent is a non-pressurizable fluid.

$$G_2 - G_1 = -(p_1 - p_2)nV_\mathrm{m} \quad (2)$$

Thus the free energy change during the conversion of solvent molecules in high-pressure condensate solution to gas molecular clusters by decompression is obtained as follows.

$$\Delta F_1 = -(p_1 - p_2)nV_\mathrm{m} + w_0 \quad (3)$$

For a certain impetus ($p_1$-$p_2$), Assuming that the expansion process in which the gas molecular aggregates expand instantaneously and gain enough space to become a gas state is a thermodynamic equilibrium process, then:

$$G_2 = G_3 = G_4 \quad (4)$$

So it can be deduced that:

$$\Delta F_2 = 0 \quad (5)$$

The surface energy of the system changes during the formation of microbubbles by tightly clustered gas molecules:

$$w_1 = 4\pi\sigma r^2 - w_0 \tag{6}$$

External work of gas expansion:

$$w_2 = \left(p_g - p_2\right) 4\pi r^3 / 3 \tag{7}$$

$$\Delta F_3 = w_1 - w_2 = 4\pi\sigma r^2 - w_0 - 4\pi r^3 \left(p_g - p_2\right)/3 \tag{8}$$

Then the change in the free energy of the system during the formation of the bubble is:

$$\Delta F = \Delta F_1 + \Delta F_2 + \Delta F_3 = -\left(p_1 - p_2\right) n V_m + 4\pi\sigma r^2 - 4\pi r^3 \left(p_g - p_2\right)/3 \tag{9}$$

From equations (2) and (4), we get:

$$G_4 - G_1 = G_3 - G_1 = -\left(p_1 - p_2\right) n V_m \tag{10}$$

Let the gas be an ideal gas, and the solution is an ideal solution in all states, then:

$$p_g = nkT \tag{11}$$

$$\mathrm{d}G = nkT\mathrm{d}\ln p \tag{12}$$

Integration of equation (12) gives:

$$G_4 - G_1 = nkT \ln\left(p_g / p_1\right) \tag{13}$$

where $p_g$ can be expressed as:

$$p_g = 2\sigma / r + p_2 \tag{14}$$

From equation (10) and (13), we get

$$nkT \ln\left(p_g / p_1\right) = -\left(p_1 - p_2\right) n V_m \tag{15}$$

Substituting equation (14) and (15) into equation (9), we can get the free energy change of condensate depressurization into bubble process:

$$\Delta F = \left(p_2 + 2\sigma / r\right) \ln\left(\frac{p_2 + 2\sigma / r}{p_1}\right) + \frac{4}{3}\pi\sigma r^2 \tag{16}$$

Therefore, the free energy change of the condensate decompression bubble formation process is related to the release pressure $p_2$, the surface tension $\sigma$ of the liquid, the bubble radius $r$, and the mixing pressure $p_1$.

## 3 Experimental

### *3.1 Experimental Equipment*

In order to accurately grasp the distribution pattern of condensate bubbles under different operating parameters, a bubble observation process was built. The condensate is released through the mixing tank and bubbles are generated in the observation sample tank, which is photographed and observed using a high-speed camera. The whole process is shown in Fig. 1.

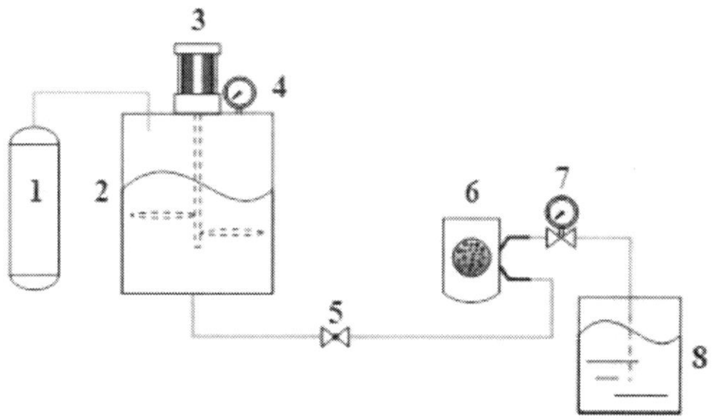


1 - Nitrogen cylinder; 2 - Mixing tank; 3 - Stirrer; 4 - Pressure gauge; 5 - Regulating valve; 6 - Observation sample tank; 7 - Back pressure valve with pressure gauge; 8 - Recovery drum.

**Fig. 1. Schematic diagram of bubble preparation and observation system**

The main function of the mixing pressure tank is to mix the condensate with water thoroughly. The mixing tank is equipped with an agitator with a speed of 700 r/min, which allows the condensate to be completely dispersed in the water. The air inlet is connected to the nitrogen tank to control the mixing pressure in the tank. In order to accurately control the experimental temperature, the use of digital display silicone rubber heating tape, which is composed of nickel-chromium alloy wire and insulating materials, designed with high power density, fast heating, high thermal efficiency, long service life, can be directly wound on the surface of the mixing pressure tank, good contact, uniform heating, simple installation, safe and reliable, adjustable temperature range of 0-150 ℃, temperature control accuracy of plus or minus 1 ℃, fast heating speed, and a digital display The temperature controller can display the temperature in real time. The regulating valve is used to control the flow rate, so that the observation sample tank appears continuous and stable bubbles, the observation area is a circle with a diameter of 40mm, which is convenient for dynamic observation of bubbles.

A backpressure valve with a pressure gauge is installed behind the observation tank to measure and adjust the release pressure in real time. A high-speed camera (Hot-Shot 1280) equipped with a 300x microscope head (Mitutoyo) was used to take pictures of bubble size distribution at a frame rate of 1000 fps and 1280*512 pixels, and an LED light source (Olympus LFP-60WP) with 60W power and adjustable brightness was used to ensure sufficient light intake. At different shutter speeds the field of view is clear and bright.

### *3.2 Experimental Materials and Methods*

The gas chromatograph was used to analyze the condensate used, and the phase diagram of the experimental condensate was obtained by Aspen HYSYS simulation as shown in Fig. 2. The bubble point pressure at each experimental temperature is given in Table 1, from which it can be seen that the condensate can be vaporized during the decompression and release process at the experimental temperature.

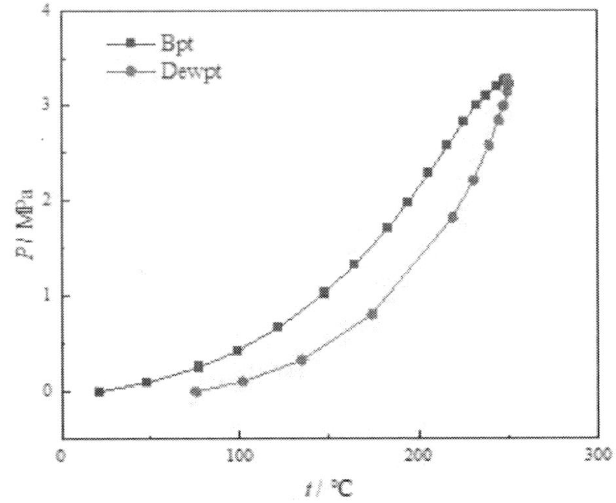


**Fig. 2. Phase diagram of condensate**

**Table 1.** Bubble point pressure table for condensate at experimental temperature

| Temperature *t* / ℃ | Bubble point pressure *P* / MPa |
|---|---|
| 25 | 0.0139 |
| 30 | 0.0310 |
| 35 | 0.0481 |
| 40 | 0.0651 |
| 45 | 0.0822 |

Each experiment controls the stirrer speed constant, adds nitrogen to control the experimental pressure, turns on the heating tape to control the experimental temperature, and after 30 minutes of heating, opens the regulating valve for the process of releasing bubbles by decompression.

Currently, the methods for measuring the bubble size include conductivity probe method, fiber optic probe method, photoelectric capillary method, laser diffraction particle size meter method[22], atomic force microscopy method[23] and image analysis method[24], and image analysis method, as one of the widely used methods for bubble particle size measurement, can directly obtain visible images and analyze the average diameter of bubbles, and the advantage of image analysis method is that it can measure the diameter as small as The advantage of the image analysis method is that it can measure microbubbles as small as 0.8 μm in diameter[25], so this paper uses the image analysis method to analyze the bubble size distribution. After the bubbles in the observation sample tank were stabilized, the bubble photographs were started to be taken, and the experiment was repeated three times for each set of operating parameters, and more than 15 photographs were selected for processing and analysis in each experiment to eliminate the effect of chance. Fig. 3a shows the bubble photos taken by high-speed camera, with the frame rate selected as 1000fps and shutter speed selected as 1/3000s. At the gas-liquid phase interface, the light is reflected due to the incident angle exceeding 48.7°, and the light passing through the center of the bubble can pass through smoothly. Therefore, the bubble captured by high-speed camera shows a black circle with a bright spot in the middle, and the image is processed using ImageJ image processing software as shown in Fig. 3b to clearly label the bubble for bubble particle size statistics. The median particle size $d_{50}$ was used to characterize the average bubble diameter.

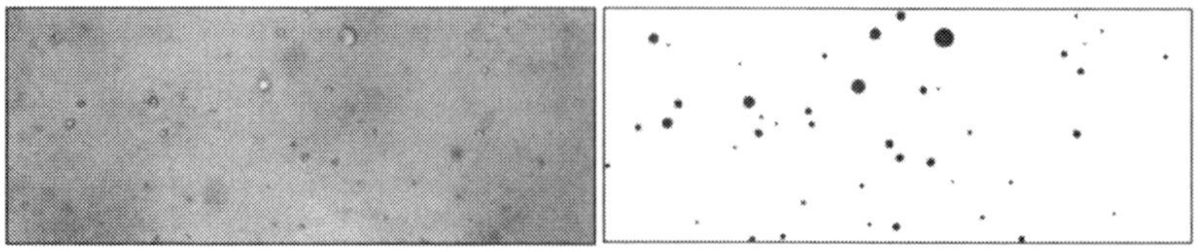

(a) High-speed camera to take bubble photos (b) ImageJ processed bubble photo

**Fig. 3. Comparison of high-speed camera photos and processed photos**

In order to obtain accurate bubble size distribution, the number of bubbles analyzed in each group of experiments was between 1000 and 3000.The experimental conditions are shown in Table 2.

**Table 2. Experimental operation parameters**

| Parameters | Values |
|---|---|
| Mixing pressure $P_1$ / MPa | 0.10,0.15,0.20,0.25 |
| Release pressure $P_2$ / MPa | 0,0.01,0.02,0.03 |
| Temperature $t$ / ℃ | 25,30,35,40,45 |
| Condensate content $C$ / % | 0.5,1.0,1.5,2.0,2.5 |

## 4 Results and Discussion

### *4.1 Effect of Condensate Content on Bubble Particle Size*

In order to investigate the effect of the variation of condensate content on the bubble distribution, experiments were conducted with condensate contents of 0.5% ~ 2.5%, and the bubble distribution was analyzed using micrographs. Fig. 4 shows the variation of bubble size at different condensate contents when the mixing pressure is 0.1 MPa, the release pressure is atmospheric pressure, and the temperature is 25°C. From the bubble photos, it can be seen that when the condensate content reaches 1.5% before, the number of bubbles is small and there are a small number of large bubbles, and the overall effect is not good, but after reaching 1.5%, the changes of bubble number as well as bubble particle size gradually stabilize, which is mainly because when the condensate is less, there are fewer nucleation locations that can produce bubbles, so the number is smaller, and when the condensate content increases, the condensate in the effluent droplets are more and become the sites of bubble generation after the sudden pressure drop, so the number of bubbles will increase.

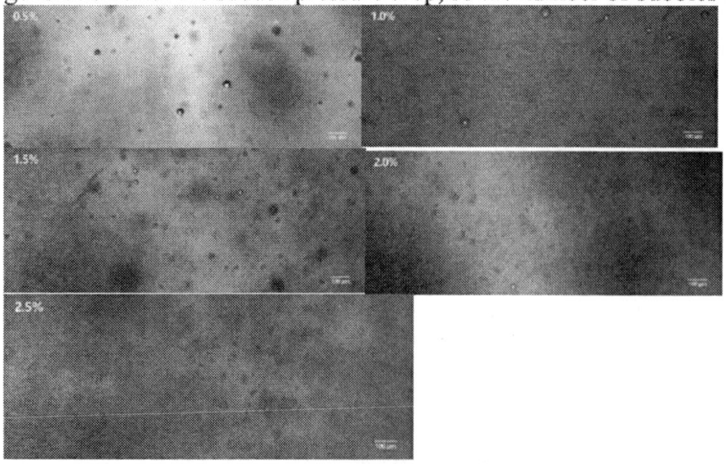


**Fig. 4. Photographs of bubbles taken at different condensate contents**

The particle size of the photographed bubbles was counted using ImageJ, and the diameter of each bubble was analyzed to obtain the probability density distribution of the bubble particle size dg at different condensate contents, as shown in Fig. 5. The median particle size $d_{50}$ at different condensate contents is shown in Fig. 6. In the range of condensate content selected for the experiment, the peak particle size of bubbles was basically unchanged. When the condensate content reached 2.5%, the number of bubbles larger than 30 μm increased, which was due to the increase of collision and aggregation between bubbles as the number of bubbles increased, so the subsequent experiments were conducted under the condition that the condensate content was 2.0%.

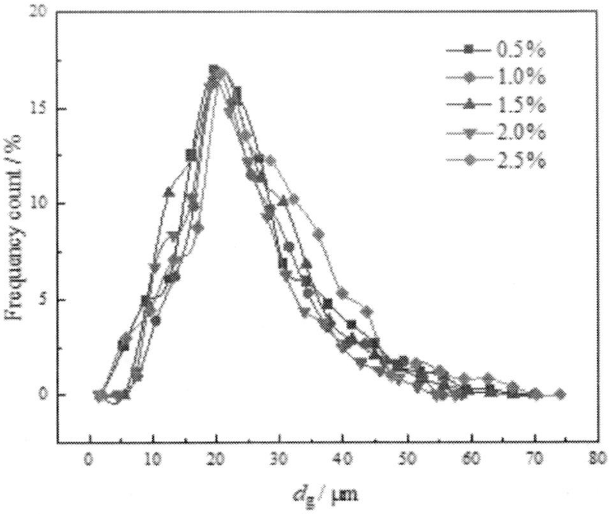


**Fig. 5. Probability density distribution of bubbles at different condensate contents**

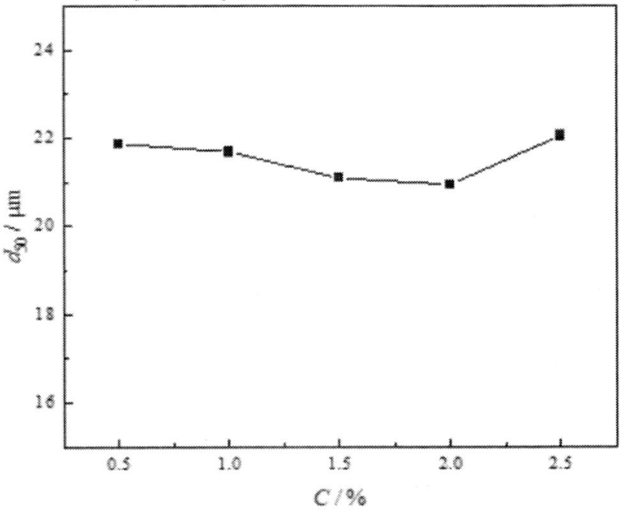


**Fig. 6. Variation pattern of the average diameter of bubbles with condensate content**

### *4.2 Effect of Temperature on the Particle Size of Bubbles*

Temperature mainly affects the vaporization effect of condensate, and Fig. 7 shows the variation of bubble particle size with temperature when the mixing pressure is 0.1 MPa and the release pressure is atmospheric pressure. The temperature range from 25℃ to 45℃ is selected as the temperature variation range. From the probability density distribution graph, it can be seen that the peak point gradually shifts to the left with the increase of temperature and the proportion of small bubbles gradually increases, and at 40 ℃, the bubble particle size gradually increases and large bubbles are generated due to the increase of temperature.

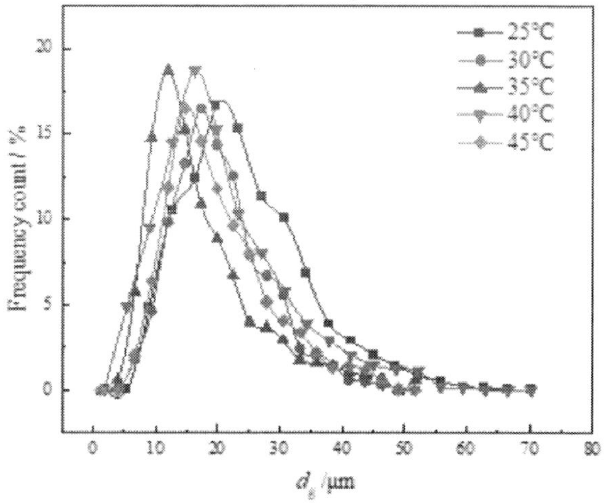


**Fig. 7. Probability density distribution of bubbles at different temperatures**

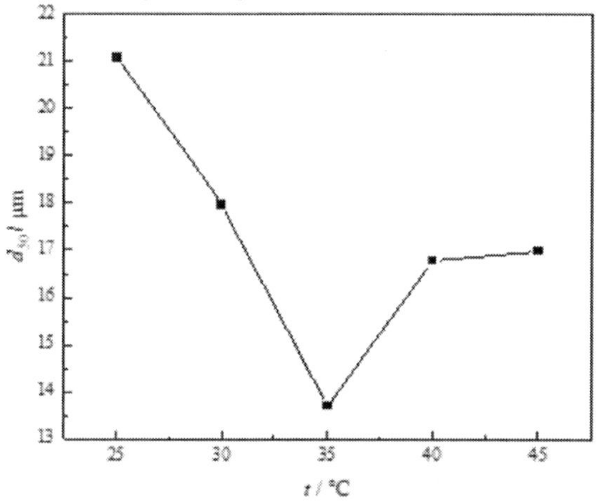


**Fig. 8.** The variation pattern of the average diameter of bubbles with temperature

According to the bubble median particle size variation curve (Fig. 8), it can be concluded that the average bubble diameter gradually decreases with the increase in temperature below 35°C. This is when the surface tension decreases due to the increase in temperature, and according to Mandrea et al[26], it was demonstrated that the bubble diameter is proportional to the surface tension of water as follows.

$$r = \frac{2\sigma}{p_g - p_2} \tag{1}$$

Therefore, the median particle size of bubbles decreases as the temperature increases. However, with the increasing temperature, the Brownian motion of microbubbles intensifies, which leads to the increase of the aggregation between bubbles and the continuous collision and aggregation between bubbles, so that the bubble diameter increases at 40 and 45°C.

### *4.3 Effect of Mixing Pressure on Bubble Size*

When the temperature is kept constant, the mixing pressure is too small will lead to vaporization of condensate in the mixing tank and produce fewer bubbles. Fig. 9 shows the variation of bubble particle size with the mixing pressure when the release pressure is atmospheric pressure and the temperature is 35°C. With the increase of pressure, the number of bubbles increases because the increase of mixing pressure causes more condensate components to precipitate out to form small bubbles after the sudden pressure drop, and the bubble distribution is more dense. According to the thermodynamic analysis, when the mixing pressure increases, the free energy change of the condensate decompression bubble formation process decreases, which is more favorable to this process.

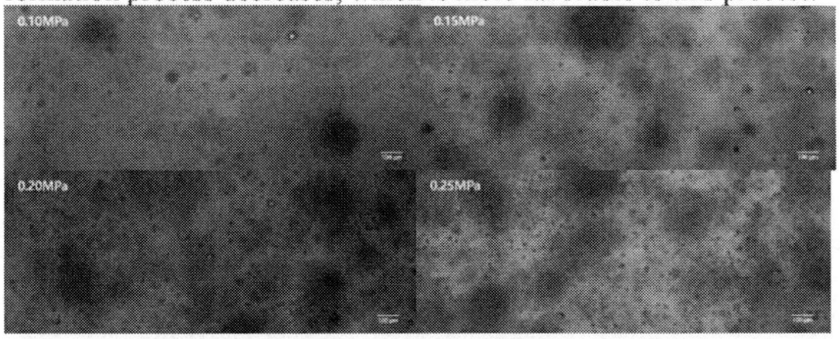


**Fig. 9. Photographs of bubbles taken at different mixing pressures**

The particle size distribution of bubbles at different pressures is shown in Fig. 10. With the increase of mixing pressure, the bubble size gradually shifts to the direction of small diameter, the particle size distribution is more concentrated, the large bubbles gradually decrease, and the proportion of bubbles smaller than 20 μm increases.

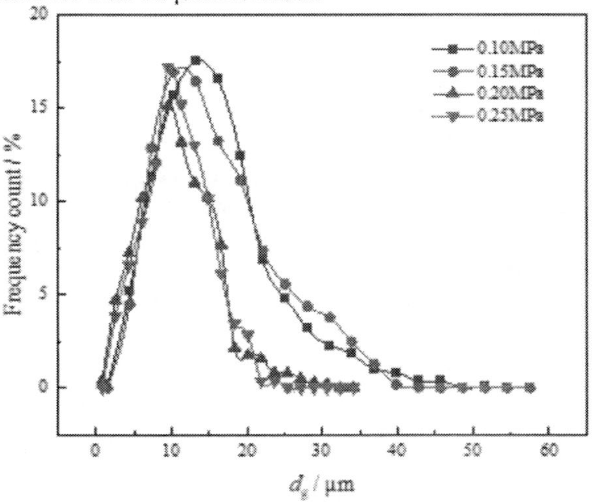


**Fig. 10.** Probability density distribution of bubbles under different mixing pressure

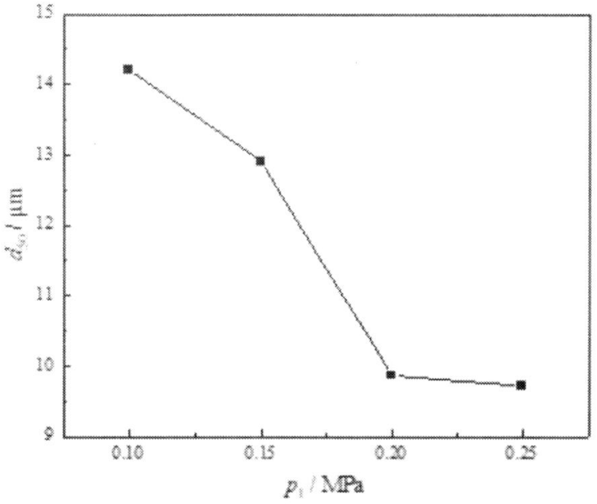


**Fig. 11.** Variation of median particle size of air bubbles with mixing pressure

The variation of the median bubble size at different mixing pressures is shown in Fig. 11, where the median bubble size decreases as the mixing pressure increases. From the thermodynamic analysis, it can be concluded that bubbles must cross an energy barrier to form in solution, and the radius of bubble formation and the energy required to form bubbles decrease significantly as the mixing pressure increases. Therefore, a higher mixing pressure is more favorable to obtain small bubbles.

### *4.4 Effect of Release Pressure on the Size of Bubbles*

The release pressure is the pressure in the observation sample tank, in which the pressure is controlled by a backpressure valve. The release pressure has an important effect on the energy barriers to be overcome in the bubble forming process, as well as the bubble diameter size. Fig. 12 shows the variation of bubble particle size with release pressure for a mixing pressure of 0.2 MPa and a temperature of 35°C. As the release pressure increases, the number of bubbles tends to decrease. According to the thermodynamic analysis of the bubble formation process of condensate vaporization, increasing the release pressure increases the amount of free energy change of the process, which makes the bubble formation process need to overcome a larger energy barrier.

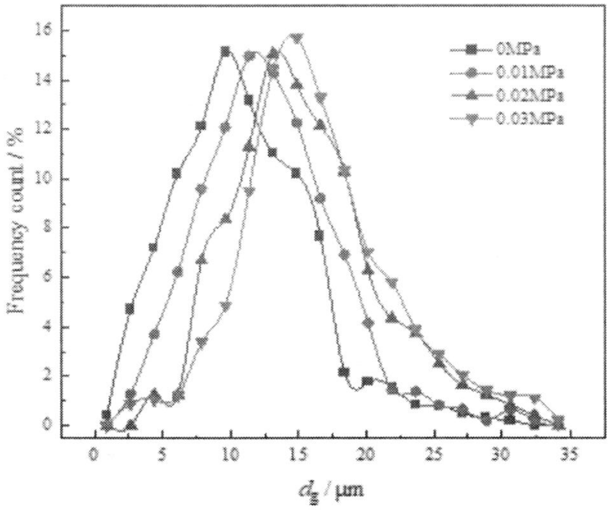


**Fig. 12. Probability density distribution of bubbles at different release pressures**

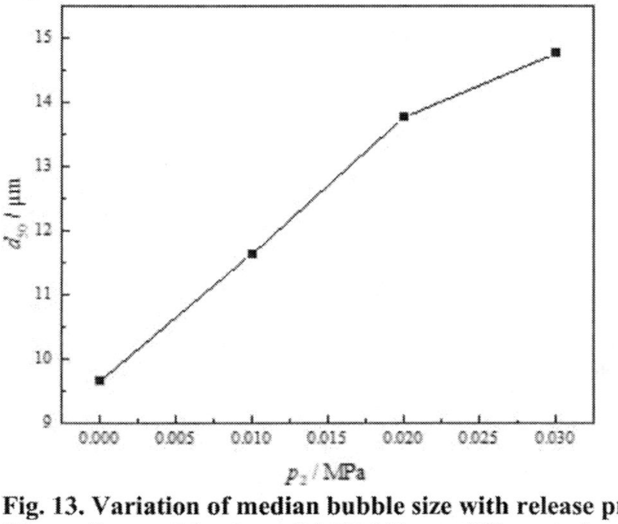


**Fig. 13. Variation of median bubble size with release pressure**

The median particle sizes of the bubbles at different release pressures were extracted and plotted in Fig. 13. The median particle size of bubbles increases with the increase of release pressure, which is the same rule as that obtained from the bubble diameter calculation equation (17). After bubbles are generated, they keep expanding in order to adapt to the increase of ambient pressure until their own internal pressure equilibrates with the external pressure. From the thermodynamic analysis, the increase in release pressure raises the energy barrier required to generate bubbles, and at the same time, the pressure difference of the condensate depressurization process into bubbles decreases, which is not conducive to the bubble generation.

## 5 Conclusion

In this paper, an experimental study of the vaporization effect of condensate at different parameters is investigated using a combination of experimental and theoretical methods, aiming to provide support for the application of condensate to the air flotation process. A bubble observation system, which can accurately measure the bubble size, was built to study the size and number of bubbles produced by condensate under different operating parameters, and to demonstrate the feasibility of condensate application in the air flotation process. The effects of condensate content, temperature, mixing pressure, and release pressure on the bubble size distribution were studied. Among them, temperature, mixing pressure and release pressure have a large influence on bubble distribution. The higher the temperature, the higher the mixing pressure and the lower the release pressure, the smaller the diameter of the bubbles obtained. Compared with the bubble diameter obtained by dissolved air flotation, the bubble diameter obtained by condensate vaporization is smaller, which is more favorable to the air flotation process. The entire air flotation process involving condensate needs to be studied in future research in order to achieve efficient oil removal.

**Acknowledgments.** The project is supported by National Natural Science Foundation of China (Number 52174067), the opening fund of Shandong Provincial Key Laboratory of Oil & Gas Storage and Transportation Safety, and the Fundamental Research Funds for the Central Universities.

**Abbreviations**

$C$——Condensate volume content, %
$d_{50}$——Bubble median particle size, μm
$\Delta F$、$\Delta F_1$~$\Delta F_2$——The process of bubble formation and the energy change of the system in each process, J
$G$、$G_1$~$G_4$——Process of bubble formation and the chemical potential of each state, J
$k$——Boltzmann's constant
$n$——Number of gas molecules
$P$——Pressure, kPa
$p_g$——Internal pressure of bubble, Pa
$p_1$——Condensate and water mixing pressure, Pa
$p_2$——Observe the release pressure in the sample tank, Pa
$r$——bubble radius，m
$T$——Absolute temperature, K
$t$——Temperature, ℃
$V_m$——Volume of gas molecules, $m^3$
$w_0$——Surface energy of gas molecular clusters, J
$w_1$——System surface energy change, J
$w_2$——Gas expansion does external work, J
$\sigma$——Surface tension of liquids, N/m

**References**

[1]Wang, C., Lü, Y., Song. C., et al. Separation of emulsified crude oil from produced water by gas flotation: A review. Science of The Total Environment, 157304(2022).
[2]Piccioli, M., Aanesen, S. V., Zhao, H., et al. Gas Flotation of Petroleum Produced Water: A Review on Status, Fundamental Aspects, and Perspectives. Energy & Fuels, 34(12): 15579–15592 (2020).
[3]Tsatsis, D. E., Papachristos, D. K., Valta, K. A., et al. Enzymatic deinking for recycling of office waste paper. Journal of Environmental Chemical Engineering, 5(2): 1744–1753 (2017).
[4]Wang, C., Liu, Q., Ivey, D. G., et al. Bi-wetting property of oil sands fine solids determined by film flotation and water vapor adsorption. Fuel, 197: 326–333 (2017).
[5]Pintor, A. M. A., Vilar, V. J. P., Botelho, C. M. S., et al. Oil and grease removal from wastewaters: Sorption treatment as an alternative to state-of-the-art technologies. A critical review. Chemical Engineering Journal, 297: 229–255 (2016).

[6]Mathiazakan, P., Shing, S. Y., Ying, S. S., et al. Pilot-scale aqueous two-phase floatation for direct recovery of lipase derived from Burkholderia cepacia strain ST8. Separation and Purification Technology, 171: 206–213 (2016).
[7]Nguyen, T. D., Al Tahtamouni, T. M., Huong, P. T., et al. Selective flotation separation of ABS/PC from ESR plastic wastes mixtures assisted by ultrasonic catalyst/H2O2. Journal of Environmental Chemical Engineering, 7(5): 103354 (2019).
[8]Hoseini, S. M., Salarirad, M. M., Alavi Moghaddam, M. TPH removal from oily wastewater by combined coagulation pretreatment and mechanically induced air flotation. Desalination and Water Treatment, 53(2): 300–308 (2015).
[9]Takahashi, T., Miyahara, T., Mochizuki, H. Fundamental Study of Bubble Formation in Dissolved Air Pressure Flotation. Journal of Chemical Engineering of Japan, 12(4): 275–280 (1979).
[10]Bowonder, B., Kumar, R. Studies in bubble formation - IV: bubble formation at porous discs. Chemical Engineering Science, 25(1): 25–32 (1970).
[11]Ketkar, R., Mallikarjunan, R., Venkatachalam, S. Electroflotation of quartz fines. International Journal of Mineral Processing, 31(1): 127–138 (1991).
[12]Knudsen, L., Hjelsvold, M., Frost, K., et al. Toward Zero Environmental Impact of the Produced Water[A]. All Days[C]. Aberdeen, United Kingdom: SPE, SPE-83994-MS (2003).
[13]Sabey, B. CTour – Simultaneous Removal of Dispersed and Dissolved Hydrocarbons in Produced Water. Annual-PW-Seminaor, 1–11 (2006).
[14]Grini, P., Hjelsvold, M., Johnsen, S. Choosing Produced Water Treatment Technologies Based on Environmental Impact Reduction[A]. OnePetro, (2002).
[15]Li, H., Hu, L., Xia, Z. Impact of Groundwater Salinity on Bioremediation Enhanced by Micro-Nano Bubbles. Materials, Multidisciplinary Digital Publishing Institute, 6(9): 3676–3687 (2013).
[16]Bui, T., Nam, S-N., Han, M. Micro-Bubble Flotation of Freshwater Algae: A Comparative Study of Differing Shapes and Sizes. Separation Science and Technology, Taylor & Francis, 50(7): 1066–1072 (2015).
[17]Ahmed, N., Jameson, G. The effect of bubble size on the rate of flotation of fine particles. International Journal of Mineral Processing, 14(3): 195–215 (1985).
[18]Yoon, R-H. Microbubble flotation. Minerals Engineering, 6(6): 619–630 (1993).
[19]Yan, S., Yang, X., Bai, Z., et al. Drop attachment behavior of oil droplet-gas bubble interactions during flotation. Chemical Engineering Science, 223: 115740 (2020).
[20]Robinson, D. Oil and gas: Treatment and discharge of produced waters offshore. Filtration + Separation, 50(2): 20–23 (2013).
[21]Lim, W., Lau, E., Poh, P. Analysis of attachment process of bubbles to high-density oil: Influence of bubble size and water chemistry. Journal of the Taiwan Institute of Chemical Engineers, 68: 192–200 (2016).
[22]Tasaki, T., Wada, T., Fujimoto, K., et al. Degradation of methyl orange using short-wavelength UV irradiation with oxygen microbubbles. Journal of Hazardous Materials, 162(2): 1103–1110 (2009).
[23]Wang, W., Li, K., Ma, M., et al. Review and perspectives of AFM application on the study of deformable drop/bubble interactions. Advances in Colloid and Interface Science, 225: 88–97 (2015).
[24]Oliveira, C., Rodrigues, R., Rubio, J. A new technique for characterizing aerated flocs in a flocculation–microbubble flotation system. International Journal of Mineral Processing, 96(1): 36–44 (2010).
[25]Zhang, W-H., Zhang, J., Zhao, B., et al. Microbubble Size Distribution Measurement in a DAF System. Industrial & Engineering Chemistry Research, American Chemical Society, 54(18): 5179–5183 (2015).
[26]Mândrea, L., Oprina, G., Chihaia, O., et al. Theoretical and Experimental Study of Gas Bubbles Behavior. International Journal of Modeling and Optimization, 7(3): 145–151 (2017).

# Failure Mechanism and Control Measures of Cu-$CuSO_4$ Electrode in Saline Soil Environment

**Chun-yu LI(✉) , Xin-yuan FENG, Peng-cheng HE, Ben-quan LI,** Chun-bao LI

Drilling & Production Technology Research Institute of PetroChina Qinghai Oilfield Company

lchunyqh@petrochina.com.cn

**Abstract.** Saline soil environment is widely distributed in western China, especially in Qaidam Basin. High salinity soil in Qaidam Basin has higher corrosion risk to metal pipelines. The Cu-$CuSO_4$ electrode is used as the long-term reference electrode in the cathodic protection system of oil and gas pipeline. In saline soil environment, the Cu-$CuSO_4$ reference electrode generally has a problem of excessive drift, which seriously affects the normal operation of cathodic protection system. The failure mechanism of the Cu-$CuSO_4$ reference electrode in saline soil environment was demonstrated: soil ion pollution and drain drying caused potential drift. In view of the main factors affecting Cu-$CuSO_4$ reference electrode drift in saline soil environment, measures such as adding packing material and setting up maintenance device were developed, and the field application effect was good. It provides a technical idea for prolonging the life of cathodic protection system in saline soil environment.

## 1. Introduction

Cathodic protection is widely used in corrosion protection of oil and gas transmission pipelines, which plays an important role in resisting external corrosion and stray current interference. Long - term reference electrode is an important auxiliary device of forced current cathodic protection system. It is mainly used for calibrating output parameters of cathodic protection instrument and long-term monitoring of cathodic protection potential[1]。 With the gradual popularization and application of intelligent cathodic protection facilities, the long-term reference electrode is more and more widely used as the reference electrode for measuring cathodic protection parameters of intelligent test pile.

Cu-$CuSO_4$ long-term reference electrode is the main long-term reference electrode used in cathodic protection system of onshore oilfield pipeline. Its potential accuracy affects the accuracy of cathodic protection potential measurement. And tis effective life is very important to maintenance workload and cost.

The saline soil is widely distributed in Qaidam Basin, and the main landforms are desert, salt marsh and grassland. The forced current cathodic protection system is adopted in all the oil and gas transmission pipelines in Qaidam Basin, and the reference electrode system is Cu-$CuSO_4$ reference electrode. Cu-$CuSO_4$ reference electrode is widely used in general soil and freshwater environment [2], but its application in saline soil environment, especially in wet salt lake lakeside and saline soil with high groundwater level, needs further study. It is found that the long term reference electrode of cathodic protection system generally has the problems of short effective time and large potential drift.

It is difficult to set and evaluate the parameters of the cathodic protection instrument, and also has an adverse effect on the efficiency of pipeline cathodic protection.

In this paper, the main reason of failure of Cu-$CuSO_4$ long-term reference electrode and its failure mechanism in saline soil environment are discussed and analyzed. The measure of prolonging the effective life of the reference electrode has been developed and applied in the construction and maintenance of the cathodic protection system of oilfield pipeline with good results.

## 2. Saline soil environment and its influence

Saline soil is defined as soil with soluble salt content greater than 0.3%. Qaidam Basin is located in the extreme arid desert environment of Qinghai-Tibet and Xin-jiang, and the saline soil exists widely. Except for a few desert environments with low salt content, the salt content in other environments is more than 0.3%, up to 80%. According to the salt content, it is classified into strong saline soil and super saline soil environment. According to the type, it includes chlorine saline soil and sulfuric acid saline soil, of which chlorine saline soil accounts for the main part.

The oil and gas transmission pipeline in Qaidam Basin connects the main oil and gas production area and consumption area. It has the characteristics of wide distribution, long distance and complex environment along the route. Saline soil is widely distributed along the pipeline, and the soil environment involves salt lake lakeside, salt marsh, arid Gobi, desert, etc. Soil salt content generally ranges from 0.5% to 45%, and chloride ion content ranges from 0.01% to 32.76% [3]. The soil moisture content is generally low and the soil is dry, but the groundwater level is higher in the salt lake lakeside, low-lying saline-alkali land and some grassland areas. Some of the pipes are even soaked in salt water. Under the influence of snowmelt water replenishment in some areas, the changes of dry and wet soil are obvious.

According to the detection and evaluation results of the pipeline cathodic protection system, the 20 long-lasting reference electrodes used in the cathodic protection system of 10 oil and gas transmission pipelines generally have the problem of negative drift, and the specific potential drift is shown in Table 1. 15 long-lasting reference electrode had a potential drift greater than 50mv, the maximum was -556mv, only 3 had a potential drift less than 50mv, the minimum was -15mv, one long-lasting reference electrode was damaged, and the potential drift oscillated between -1.14v and -2.8V.

In the forced current cathodic protection system, the long-term reference electrode is mainly installed on the potentiostat or the intelligent test pile, which is used for the calibration of the output potential of the potentiostat and the measurement of the pipeline potential. In the process of daily detection and inspection, the conventional forced current cathodic protection pipeline is only tested on the power point. According to the on/off potential difference of the pipeline cathodic protection system during regular inspection, the pipeline cathodic protection state is judged and the cathodic protection instrument parameters are adjusted. The accurate protection potential and long-term reference electrode drift are only tested during regular inspection or special inspection. The drift of long-term reference electrode potential brings great challenges to the accuracy of cathodic protection apparatus output and the effectiveness of pipeline cathodic protection. On the one hand, the output potential dis-played by the potentiostat is the potential difference between the zero-position line potential (pipeline potential) and the long-term reference electrode potential. When the potential of the long-term reference electrode shifted, the potentiostat showed that the potential changed accordingly, and the displayed value could not reflect the potential difference between the pipeline and the earth. This results in a difference between the output potential and the potential to be set, resulting in over-protection or under-protection of the pipeline. On the other hand, the equipotential parameters uploaded in the intelligent cathode-protection pile test are the difference between the pipeline potential and the long-term reference electrode potential. When potential drift occurs at the long-acting reference electrode, the uploaded cut-off potential is not consistent with the actual protection potential, which is more likely to cause mis-judgment of the pipeline protection potential.

**Table 1.** Statistics of drift of long-acting reference electrode

| serial number | location | Difference (portable reference electrode) /mv |
|---|---|---|
| 1 | Pipeline A-1# | -344 |
| 2 | Pipeline A-2# | -294 |
| 3 | Pipeline A-3# | -14 |
| 4 | Pipeline B | -271 |
| 5 | Pipeline C-1# | -556 |
| 6 | Pipeline C-2# | -60 |
| 7 | Pipeline C-3# | -88 |
| 8 | Pipeline D-1# | -165 |
| 9 | Pipeline D-2# | -129 |
| 10 | Pipeline D-3# | -166 |
| 11 | Pipeline D-4# | -253 |
| 12 | Pipeline D-5# | -33 |
| 13 | Pipeline D-6# | -233 |
| 14 | Pipeline D-7# | -369 |
| 15 | Pipeline E | -1140~-2800 |
| 16 | Pipeline F | -15 |
| 17 | Pipeline G | -271 |
| 18 | Pipeline H | -89 |
| 19 | Pipeline I | -84 |
| 20 | Pipeline J | -220 |

## 3. Failure mechanism of long-term reference electrode

### *3.1. Common Failure Causes*

According to relevant research and engineering practice, the following problems exist in the use of Cu-$CuSO_4$ long-term reference electrode:

(1) Ion contamination: the long-term reference electrode is directly in contact with the soil, and the potential is measured through ion exchange. As a result, the ions in the soil, especially the small volume of Cl-, will inevitably enter the long-acting reference electrode. $CuSO_4$ solution is polluted to change the electrode system in the electrode, thus leading to the drift of electrode potential [4].

(2) Leakage of electrode solution: when the soil around the electrode is dry or the temperature changes greatly, the copper sulfate solution in the electrode will leak out of the electrode. The $CuSO_4$ solution in the electrode was reduced continuously. When the copper rod in the electrode is exposed too much or cannot contact with electrolyte [5], the long-term reference electrode potential will change significantly [6]. This is more evident in the arid western regions [7].

(3) Increase of internal resistance: with the infiltration of soil ions into the reference electrode or the leakage of the solution in the electrode, oxidation reaction will occur on the surface of the copper rod. This changes the original pure copper-copper sulfate electrode system and leads to potential changes [8]. Improper selection or contamination of leaky materials that also provide ion channels will hinder ion penetration rates. This increases the contact resistance of the reference electrode [9].

(4) Other factors: If the sealing of the long-term reference electrode and the cable connection line is not strict, external damage will lead to corrosion or even wire break-ing at the joint, affecting the measurement effect. Meanwhile, high soil contact resistance in dry environment will also affect the accuracy of reference electrode potential [10].

### *3.2. Failure mechanism in saline soil environment*

The saline soil is widely distributed in Qaidam basin. According to the soil test results, the main characteristics are as follows: the salt content in extremely dry soil such as Gobi and desert is relatively low, generally less than 10%; The salt content in low-lying wet or dry wet alternating soil is high, generally in 10%-45% directly, and even up to 80%. The morphology of the failed long-term reference electrode of cop-per sulfate is shown in Figure 1. The solution in the electrode has dried up and precipitated blue copper sulfate crystal and yellow green copper chloride crystal at the bot-tom. This indicates that a large number of chloride ions in soil infiltrated into the reference electrode during the use of the long-acting reference electrode. The penetration rate and concentration of chloride ions into the long-acting reference electrode cavity are affected by the following factors: the concentration of chloride ions in soil, the degree of soil moisture, soil properties and the change of soil wetting and drying. A large number of salt lakes and low - lying salt - alkali areas are distributed in Qaidam Basin. These locations have high soil salt content and high water table, and some of the pipelines are even immersed in saturated salt water. Chloride ion contamination was the highest.

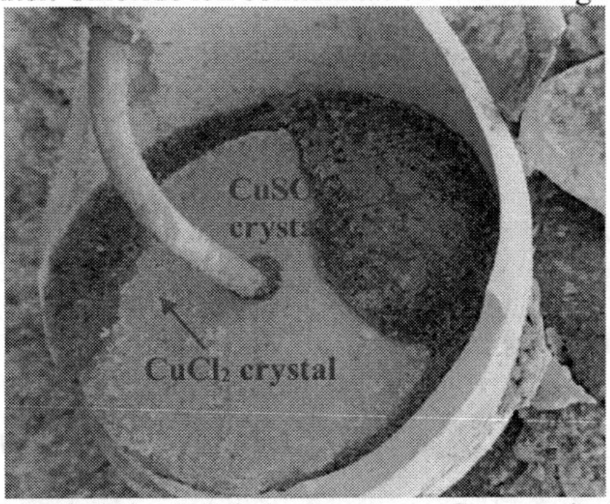


**Fig. 1** Failed Cu-$CuSO_4$ long-acting reference electrode

In arid soil, the failure mechanism of the long-term reference electrode is shown in FIG. 2 (a), where the osmotic pressure in the electrode is much higher than that in the soil. Electrolyte seepage pressure is maximum [11]. With the continuous seepage of electrolyte in the long-term reference electrode, the copper rod is exposed or even suspended. Problems such as oxidation, electrode and electrolyte separation occur, leading to potential drift or failure of long-term reference electrode.

In moist soil environment, the failure mechanism of the long-term reference electrode is shown in Figure 2 (b). Different from the conventional soil solution, the con-centration of soluble ions such as $Cl^-$ in the saline soil solution is high. Driven by the concentration difference, Cl- plasma will penetrate into the electrode. This changes the original Cu-$CuSO_4$ electrode system, and the reference electrode potential shifts.

In the alternating wet and dry soil environment, the long-term reference electrode is faced with the following conditions: the electrolyte seepage when the soil is dry, and the chloride plasma infiltration into the polluted electrolyte when the soil is wet. The problem of reference electrode potential drift occurs most quickly.

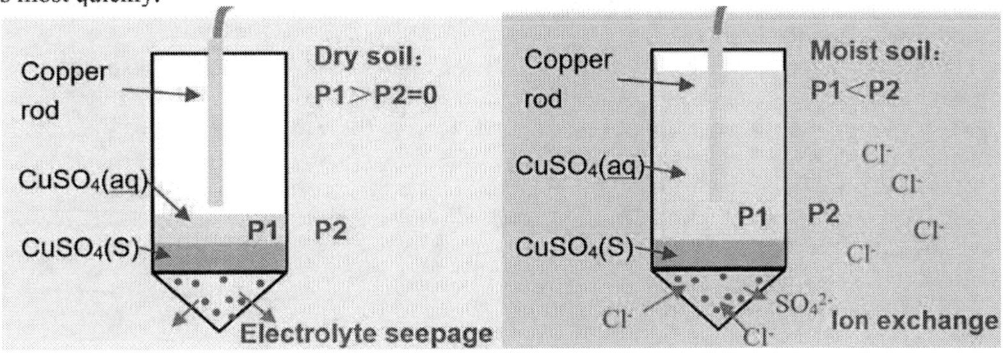

Fig. 2 Schematic diagram of long-acting reference electrode failure mechanism in saline soil environment. (a) Dry soil environment (b) Moist soil environment

## 4. Optimization and application

In the saline soil environment of Qaidam Basin, the main problems facing the long-lasting reference electrode are electrolyte leakage caused by drying and electrolyte system change caused by ion pollution. According to the actual content of chloride ions in soil, Ag-AgCl electrodes can also be used to reduce soil $Cl^-$ ion pollution in humid environment. But that doesn't solve the problem of electrolyte leakage and other ion contamination of the soil in dry environments. At the same time, the difference of electrode potential will also lead to the inconvenience of production, such as the existence of two sets of potential systems lead to misoperation and other problems and the use of Cu-$CuSO_4$ portable electrode test potential conversion problems.

In order to slow down the leakage rate of long-term reference electrode electrolyte, relevant researches on electrode materials and electrolyte have been carried out at home and abroad. In terms of electrode leakage materials, graphite junction cop-per/saturated cupric sulfate reference electrode [12], ceramic leakage material long-term reference electrode [13], composite material long-term reference electrode [14] and other Cu-$CuSO_4$ long-term reference electrodes have been produced and evaluated. Gel materials have also been used to assist in reducing the leakage rate of reference electrodes, including gel electrolyte [15], gel sandwich ceramic shell [16], gel cladding [17], etc. The laboratory evaluation effect is good, but the long-term application effect has not been evaluated [18]. Although the above long-term reference electrode can slow down the leakage of electrolyte in theory and evaluation, there are still the following problems: first, the structure of the electrode is relatively complex, the pore size of the leakage material is small, and the anti-blocking ability is insufficient; Second, the complex design increases the internal resistance of the electrode. In field use, it is also necessary to solve the interaction between leakage rate and electrode internal resistance [1]. Third, there are few field applications of these electrodes, and the practical application effect is not clear. Fourth, the ability to resist soil chloride ion infiltration pollution is not clear.

A replaceable long-acting reference electrode [19] was also designed to replenish and replace the reference electrode electrolyte to avoid drying out the long-acting reference electrode and soil ion contamination. However, the structure of the electrode is complex, requiring high operation and maintenance, and the later maintenance workload and cost are high.

In the saline soil environment of Qaidam Basin, the Cu-$CuSO_4$ electrode can be optimized from two aspects of moisturizing and mitigating ion pollution, and its effective life can be extended. In practice, the following measures are adopted to pro-long the effective life of the long-acting reference electrode:

(1) Add the reference electrode filler and set the curing device (as shown in Figure 3). Sacrificial anode packing material was transplanted, and the optimized packing material was 45% hydrated $CaSO_4$ and 40% bentonite and15% $Na_2SO_4$. The bentonite expands when wet and keeps humidity to reduce the electrolyte leak-age rate. $Na_2SO_4$ and $CaSO_4$ can extract enough $SO_4^{2-}$ ions for conducting current. At the same time, the closely bonded filler can also block and delay the penetration rate of the ion reference electrode in soil.

The additional maintenance device can be watered and maintained regularly. Especially in the dry soil environment, the combination of bentonite in the filler can effectively prolong the electrode wet environment, reduce the electrolyte seepage of long-term reference electrode, and reduce the grounding resistance, thus extending the effective life.

(2) The electrode leakage material is preferred. The ceramic material with reason-able porosity was selected as the reference electrode leakage material, and the penetration area was set reasonably to slow down the electrolyte leakage rate of the electrode, so as to extend the effective life of the electrode.

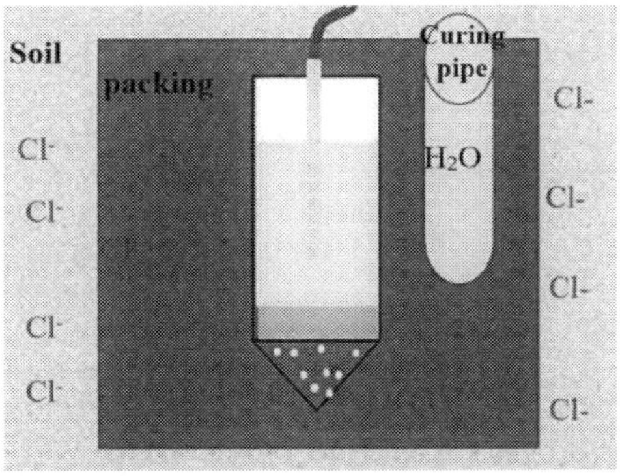


Fig. 3 Schematic diagram of optimized long-acting reference electrode

In the process of pipeline cathodic protection system construction and repair, the long-term reference electrode is replaced according to the optimized method. As shown in Figure 4, the reference electrode body is made of pore ceramic material, which is coated and buried in the filler. At the same time, the maintenance pipe is arranged, one end is buried in the filler, and a point is exposed to the ground for watering and maintenance.

At present, the replaced long-acting reference electrodes all work well, and the drift of all long-acting reference electrodes is less than 10mv, which meets the application requirements of pipeline cathode-protection system.

Fig. 4 Installation and curing device of long-acting reference electrode

## 5. Conclusions

(1) The long-term reference electrode potential in Qaidam Basin The long - term reference electrode in pipeline cathodic protection system generally has the problems of short effective life and excessive potential drift.

(2) Electrolyte leakage and soil ion pollution caused by drought were the main causes of potential drift failure of $Cu$-$CuSO_4$ long-term reference electrode.

(3) Add electrode filler for long-term reference electrode can extended its useful life.

(4) Set the maintenance device and maintain it regularly can keep the reference electrode wet for a long time.

(5) The porous ceramic electrode can be selected to ensure the conductivity and reduce the electrolyte leakage.

## References

[1] Dong Liang, Wu Tong, Wu Fang Yun, et al. Progress of Permanent Copper/Saturated Copper

Sulfate Reference Electrode[J]. Corrosion & Protection, 2020, 41(6):1-6.

[2] Li Zili, Xie Yuehui, Hao Hongna, et al. Research Progress of Measurement Methods for Cathodic Protection Potential of Buried Pipelines[J]. Corrosion & Protection, 2012, 33(1):55-59.

[3] Li Chunyu, Wang Jie, Cao Junfeng, et al. Pipeline Anticorrosive Coating Quality and Pipe-line Corrosion Analysis in Salty Soil Environment at Qaidam Basin[J]. Oil-Gasfield Surface Engineering, 2019, 38(6):73-79.

[4] Pawe L S, Lope Z R, Ondak E. Chemical and Environmental in Fluences on Copper/Copper Sulfate Reference Electrode Half Cell Potential[J]. Materials Performance, 1998,37(5):24-29.

[5] Yang Xuemei. Research on new technology of oil and gas Pipeline Corrosion and Protection[D]. Tian Jin: Tianjin University,2007.

[6] Sirotads, Ulikhin A N,Sham shetdinova N K. The Causes of Failure of Reference Electrodes[C]// Corrosion2014. Houston, Tx:Naceinterna-Tional,2014:3800.

[7] Ruan Jinghong, Liu Guo, Wang Xiuyun, et al. Failure Analysis and Countermeasures of Cathodic Protection System for Tahe Oilfield Pipelines[J]. Corrosion & Protection, 2012.261-264.

[8] Wong Yongji. Discussion on internal resistance of copper sulfate reference electrode and development of graphite junction[J]. Oil & Gas Storage and Transportation. ,1983,2(6):6-15.

[9] Xu Naixin. Factors affecting the accuracy of reference electrode[J]. Corrosion & Protection, 1995, 16(3):142-145.

[10] Li Zhengjun. Applicability of Cathodic Protection Probe in Western Typical Soil Environments[J]. Corrosion & Protection. 2018,39(9):719-721.

[11] Jiang Guohui, Han Shuqiang, Gao Guoping, et al. Defects and Improvement of Long-term Cu/CuSO4 Reference Electrode in Offshore Environments[J]. Total Corrosion Control. 2014, 28(2):50-51.

[12] Wong Yongji. Development of long - lasting copper sulfate reference electrode [J]. Physical Testing and Chemical Analysis（Part B:Chemical Analysis）, 1987,23(4):198-199.

[13] Dunn D S, Brossia C S. Semi-permanent Reference Electrode:US7459067[P].2008.

[14] Xu Lianjun. Long acting copper sulfate reference electrode:201120151510.0[P].2012.

[15] Lin Cunguo, Xu Likun. Copper/saturated copper sulfate gel reference electrode: 200610068457.1[P].2007.

[16] Liu Yang. Development of a new long life reference electrode [D]. Harbin: Harbin Engineering University,2006.

[17] Lin C G, Xu L K, Liu Y. Study of the Application of Gel Electrolyte in the Reference Electrode of Cu/CuSO4[J]. Corrosion Science and Technology, 2008,7(3):179-181.

[18] YIN Nana, LI Mingqian, DAI Xin, et al. Preparation and Properties of Electrographic Gel-stype All-solid Ag/AgCl Reference Electrode[J]. Chinese Journal of Inorganic Analytical Chemistry, 2018,8(3):58-62.

[19] MA Xiaoliang, CHANG Guichuan, DING Runhua, et al. Study on cyclic reference electrode [J]. Natural Gas and Oil, 2014,32(1):77-79.

Journal of Physics: Conference Series **2594** (2023) 012071 doi:10.1088/1742-6596/2594/1/012071

# Progress and Suggestions for Fracturing Technologies of Tight Oil in Daqing Oilfield

**Xian-jun Wang1,2, Yu Sun1,2(✉) , Wei Wang1,2, Hong-tao Zhang1,2, Xiang Gao1,2**

1.Daqing Oilfield Production Technology Institute, Daqing,Heilongjiang
2.Heilongjiang Provincial Key Laboratory of Oil and Gas Reservoir Stimulation, Daqing Oilfield, Daqing, Heilongjiang
475734077@qq.com

**Abstract:** Daqing oilfield is rich in tight oil reserves, and it's a realistic field for increasing the efficiency of Daqing oilfield. After 11 years of technical exploration and practice, based on the idea of integration of geological engineering, Daqing oilfield has formed two technical series ,which are volumetric fracturing with tight spacing in horizontal well and network fracturing in vertical well , and realizes three-dimensional reservoir transformation. However, with the exploration and development, the reservoir physical properties gradually deteriorate, and there are some problems such as unclear stimulation mechanism and difficult production improvement in fracturing stimulation. Therefore, based on the introduction of the progress of tight oil fracturing technology in Daqing oilfield, the challenges faced by tight oil fracturing in Daqing oilfield are analyzed. According to the geological characteristics of tight oil reservoir in Daqing oilfield, three concepts are proposed ,which respectively are horizontal well fracturing changing from further close cutting to uniform fracture initiation and balanced distribution of sand fluid,vertical well fracturing changing from large displacement, large seam fracturing to moderate displacement, moderate scale multi-branch seam fracturing , fracturing concept changing from "trading initial production for benefit" to "cumulative production for benefits". And it puts forward 5 relevant technical research directions. It has certain guiding significance for assisting the efficient development of tight oil in Daqing oilfield.

## 1 Introduction

The successful exploration and development of tight oil in the Bakken Formation in the Williston Basin and the Eagle Ford Formation in the Western Gulf Basin have triggered a global boom in tight oil exploration, and tight oil has become a key field to replace conventional energy exploration and development [1-7]. Tight oil resources in China are abundant and have huge exploration and development potential. At present, many tight oil reserves have been discovered in Ordos Basin, Songliao Basin, Junggar Basin and other large-scale tight oil reserves. Fuyu and Gaotaizi tight oil reservoirs are mainly developed in the north of Songliao Basin, with great resource potential and wide distribution [8-11], which are realistic fields for the benefit and reservoir increase of Daqing oilfield. During the 13th Five-Year Plan period, Daqing Oilfield took the integration of geological engineering as the idea, and developed two technical series of horizontal well close cut volume fracturing and vertical well fracture network fracturing, which realized the three-dimensional transformation of reservoir. At present, nearly 20 development test zones have been opened up in Fuyu reservoir, and good production and construction results have been achieved [12]. However, there are still some technical bottlenecks, such as the difficulty in understanding the stimulation mechanism, the gradual deterioration of reservoir quality, and the difficulty in fracturing to increase production, which need to be solved. The monitoring results of field show that there are obvious unbalanced fluid intake and expansion phenomena of each cluster fracture in the stage multi-cluster fracturing[13-14]. And Cramer et al. [15] found that not every perforation could be opened in the fracturing process due to differences in perforation phase and quality, etc. How to make each cluster fracture expand evenly in the fracturing stage has become an urgent problem to be solved.Therefore, this paper summarizes the fracturing technology progress of tight oil in Daqing oilfield , analyzes the existing technical problems and challenges, proposes three ideas,respectively are "Horizontal well fracturing changes from further close cutting to uniform fracture initiation and balanced distribution of sand and fluid, large displacement and large seam vertical wells fracturing changes to moderate displacement, and moderate scale multi-branch seam fracturing , fracturing concept by 'first-timer in benefit' to 'tired in economic benefit ".Then this paper gives five research direction,including the whole crack effectively support horizontal well fracturing, and vertical well multistage temporary plugging fracturing, fracturing and energy integration technology, postfrac flowback reasonable system research and high efficiency refracturing of horizontal wells , in order to realize the efficient development of tight oil in Daqing oil field and realize benefit construction.

## 2 Geological overview of tight oil in Daqing Oilfield

Daqing tight oil is mainly distributed in the central and southern part of Daqing placanticline, Sanzhao Sag, Longhupao terrace and Qijia-Gulong Sag [10]. It is mainly controlled by the provenance system in the south and the north, and three sedimentary environments, namely, winding river, delta plain and delta front, are developed. The buried depth is between 1700m and 2600m, and the reservoir lithology is mainly siltstone. The pore throat is dominated by micro - and nano-scale pores and throats, and the pore diameter is mainly below 10μm, and the throat diameter is mainly less than 4μm [16].

Influenced by continental deposition, the heterogeneity of reservoir is obvious. Several sand beds are developed longitudinally, one or two sand beds are developed in one sand bed group, and each sand bed is separated by 8-10m mudstone. The single layer is thin and the lateral oil layer is discontinuous. The thickness of the single layer sandstone in the Fuyu oil layer is 2~4m, and the sand body is lenticular and short banded with poor continuity .

The physical properties of the reservoir are poor and the natural fractures are not developed. Daqing tight oil reservoirs generally have a porosity of 2%~15% and a permeability of 0.02~ 0.5md, which is a mesoporous tight reservoir. The oil saturation is low, 40%~50%,and the viscosity of formation crude oil is relatively large, generally within 4~8mPa·s. The stress difference of the reservoir is 5~7MPa, the brittleness is not obvious, and the natural fractures are not developed.

Compared with tight oil reservoir at home and abroad, tight oil in Daqing Oilfield has the characteristics of low gas-oil ratio, poor fluidity, thin sandstone thickness, large interlayer span, longitudinal sand and mud interbed development, transverse discontinuity and no natural fracture development. Therefore, a single well type cannot be used for effective development, and the experience of Marine tight oil development such as the Bakken Formation in the Williston Basin cannot be copied. Therefore, it is necessary to carry out targeted fracturing technology to achieve effective production of tight oil.

## 3 Progress of tight oil fracturing technology in Daqing Oilfield

Since 2011, the direction of oil and gas exploration in Daqing Oilfield has changed from conventional oil and gas to unconventional oil and gas.According to the integrated working idea of "pre-exploration first, exploration technology, increase production and reservoir, evaluation and follow-up, open up experimental areas, and explore effective development methods" [17], and drawing on domestic and foreign tight oil development experience, two technical series of horizontal well close cut volume fracturing and vertical well fracture network fracturing, which are suitable for tight oil characteristics in Songliao Basin, have been gradually formed, and three-dimensional reservoir transformation has been realized. It provides strong technical support for the construction of tight oil production capacity blocks.

### *3.1 Progress of horizontal well fracturing technology*

Daqing Oilfield began to explore volume fracturing technology for tight oil horizontal wells in 2011. With the gradual deterioration of reservoir physical properties, the theoretical research and field practice understanding are gradually improved, and the reform concept is constantly updated. The cluster spacing, in-section fracture placement, temporary plugging and diversion technology, proppant and fracturing fluid system in the experimental zone have experienced different development stages.

The first stage: 2011-2014. In 2011, we began to explore the volume fracturing technology for horizontal wells, gradually forming the "modular" design technology, and realizing three-dimensional transformation from "block" to “body”. In 2014, the core technology was further refined to form the "eight-optimized" volumetric fracturing technology. Field tests were conducted in typical blocks, using the composite plug fracturing process, with clusters ranging from 50 to 100m, and 1~4 clusters within the stage, using the guar fracturing fluid system.The proppant changed from the original all-coated sand to a mixture of coated sand and quartz sand. The fracturing effect of large cluster spacing fracture network was initially observed, but the underground microseismic monitoring showed that there were a large number of unused reservoirs between the fractures, the reserve loss was large, and the production difference between wells was large, so the technical effect was not fully reflected.

The second stage: 2014-2017. In 2015, we began to reduce the spacing of clusters and increase the scale of fracturing. Field tests were carried out. Composite bridge plug fracturing technology was adopted, with the spacing of clusters ranging from 50 to 20m, and 2~4 clusters within the stage. The combination of variable viscous slippery water + guar gum and slippery water + association fluid was used, and quartz sand tailing ceramsite was used as proppant. Small cluster spacing and large-scale fracturing had significant oil increase effect, but the increase of fracture scale lead to the increase of construction cost.

The third stage:2018-2019. In order to both increase production and reduce cost, in 2018, we began to focus on developing small cluster spacing close cut volume fracturing technology and supporting efficiency and cost reduction technology, and carried out field tests in typical blocks with cluster spacing of 10-20m, and optimized 5~6 clusters with temporary plugging and turning once in the section to improve construction efficiency. The fracturing fluid and proppant system has changed from slippery water + association fluid carrying quartz sand tailing ceramsite to variable viscous slippery water carrying different particle size quartz sand combination, to reduce raw material cost. The fracturing process of soluble plug was preliminarily tested.

The fourth stage: Since 2020, centering on the development concept of "increase production and reduce cost", volume fracturing technology has been formed with the core of "close cutting in plane and longitudinal through-layer transformation" to increase production, and "variable viscous slippery water, temporary plugging in the section and soluble bridge plug" to reduce cost. On the plane, the principle of strengthening class I reservoir and optimizing class II reservoir placement is adopted. The cluster spacing of class I is 5-10m, the cluster spacing of class II is 8-10m, and the sand intensity is 4-6t/m. Vertically, it breaks through the traditional fracturing thinking of "mudstone interlayer as shielding layer" and adopts the through-layer fracturing technology to realize three-dimensional production, forming two sets of technologies: "fracturing mud layer to find sand layer" for Fuyu reservoir and "one seam to penetrate multiple layers" for Gaotaizi reservoir.At the same time, the hydraulic fracturing fluid system of variable slimy water is improved to achieve "low cost, low damage, low friction, high sand carrying", and the highest continuous sand carrying ratio is 45%, realizing the upgrading of fracturing fluid technology in tight reservoirs.The multi-cluster temporary blocking steering technology in the section is improved. When 5-6 clusters are placed in the segment, the optimal temporary blocking turn is 1 time, and when 7-8 clusters are placed in the segment, the optimal temporary blocking turn is 2 times, so as to improve the overall degree of transformation in the section. The innovative ball-seat soluble plug technology is formed, which can withstand 120℃ and bear 70MPa of pressure. The bridge plug can be dissolved quickly after fracturing, realizing rapid production after fracturing and saving the cost of drilling plug.

During the 13th Five-Year Plan period, under the condition of gradual deterioration of reservoir physical properties, close cut volume fracturing technology was integrated and applied to 60 wells, and the initial daily oil production of single well after fracturing was 14.7t, which achieved good results. Compared with the end of the 12th Five-Year Plan, the initial daily oil production of a single well after fracturing is increased by 1.26 times, and the comprehensive construction cost of the same scale is reduced by 13%.

### *3.2 Progress of vertical well fracturing technology*

In 2011, Daqing Oilfield began to explore fracture-mesh fracturing technology. Induced fractures and fracture-mesh systems are formed by slug injection of slickwater and clean water with large displacement, which greatly improved the contact area between fractures and reservoirs.In 2015- 2017, it began to be applied on a large scale, and gradually formed the vertical well fracture network fine fracturing technology.In 2018, it began to apply in typical blocks. According to the characteristics of thin interbedded development, it carried out longitudinal fine stratification and large-scale fracture network fracturing reconstruction in plane, and improved the flow conductivity of the fracture network system through the combination of multi-particle size quartz sand support.Several rounds of field tests were carried out to adjust the sand adding strength of different types of reservoirs. After several adjustments, the sand adding strength of Class I reservoir increased to 19.5m3/m, and the daily oil production increased by 23%. After three adjustments, the sand adding strength of Class II and III reservoirs increased to 15-17m3 /m, and the daily oil production increased by only 7.1%. The design ideas of "improving class I, fracturing class II and controlling class III" for vertical wells are determined, and the reconstruction strength design principles of tight oil vertical wells are formed.Fracture network fracturing technology is currently the main technology for the economic and effective development of peripheral tight reservoirs. By the end of the 13th Five-Year Plan period, 746 wells have been applied in total, and the initial daily oil increase per well after fracturing is 4.5t, with an average cumulative oil increase per well of 1410t, which is more than 5 times of the conventional process, and the stage input-output ratio is 1:1.45.

In view of the lack of natural fractures in tight oil reservoirs and the difficulty of forming complex fracture networks, branch fracture fracturing technology tests were carried out in tight oil reservoirs in 2019.Multiple branch fractures are formed on both sides of the main fracture by temporarily plugging and turning inside the fracture for two or three times, and new lateral fractures are opened, so as to control the fracture belt length, increase the fracture bandwidth, tap the potential lateral remaining oil, increase the oil drainage area, and improve the reservoir development effect.In 2019, comparative tests of branch fracture fracturing and fracture-mesh fracturing were carried out in typical blocks, and class II and III reservoirs accounted for 75% of the test wells. Compared with fracture-mesh fracturing, the oil production intensity of branch fracture fracturing was increased by 30%, which initially achieved good results, and clarified the direction for production improvement of this type of reservoir. In 2020, further tests were carried out in Block P, and 6 wells were constructed with an initial average daily oil production of 5.8t. Among them, 2 wells were put into operation for one year with an average cumulative oil production per well of 1284t, showing obvious production improvement.

In order to solve the problems of thin monolayer thickness, weak stress shielding (1-3MPa), large fracture height extension, and serious ineffective reconstruction in vertical wells, longitudinal fracture height control fracturing technology tests were carried out in tight oil reservoirs in 2020. At present, a total of 33 wells have been fractured. The initial daily oil production is 4.1t, and the cumulative oil production for per well is 829t in 272 days.

## 4 Challenges

First, it is difficult to understand the mechanism of fracturing stimulation. The intensive cutting and fracturing of horizontal wells has achieved good results, but the interference of stress shadow is enhanced after small spacing and dense placement of fractures, and it is difficult to identify fracture morphology. The influence of perforation parameters on fracturing construction and post-fracturing effect is still unclear. The influence of fracturing fluid viscosity, displacement, fracture roughness and proppant particle size on effective propping is not clear enough. The research on stimulation mechanism and sand production mechanism after fracturing started late, and the understanding of energy increase imbibition and reservoir damage is unclear. All of these seriously restrict the targeted fracturing design.

Second, the reservoir is highly heterogeneous, and the precise fracturing design is difficult. Tight oil reservoirs in Daqing Oilfield are characterized by longitudinal misstacking, non-concentration, thin single sand body thickness, and transverse discontinuity. The differences in pore throat characteristics, start-up pressure, stress anisotropy and other microscopic characteristics lead to large differences in fracturing effects between adjacent blocks and adjacent wells in the same blocks, which makes it difficult to form regular reconstruction understanding and ensure the effectiveness of reconstruction.

Thirdly, the reservoir quality deteriorates gradually, and it is difficult to increase production by fracturing. With the deepening of exploration and development, tight oil development shifts from Class I to Class II. In 2020, the proportion of Class I tight oil reservoirs decreases by 11.4 percentage points. Numerical simulation of long-term productivity of horizontal Wells shows that long-term cumulative production of class II reservoirs gradually slows down with the decrease of cluster spacing. There is an urgent need to develop technologies that can effectively develop Class II reservoirs.

Fourth, fracturing fluid and tools are difficult to meet the needs of reconstruction. The cost of tight oil fracturing has decreased year by year through the adoption of new technologies, optimization of solutions, and optimization of fracturing fluids. However, due to the difficulty of development and the large scale of fracturing, the low cost demand is still not met. Currently, the oil field is in the exploration and development stage of "low oil price, post-reservoir and unconventional". As the development objects become more and more complex and the difficulty of economic and effective development becomes more and more difficult, the demand for technology tends to the direction of "low cost, high efficiency, green and intelligent". The fracturing fluid system and fracturing tools need to continuously tackle the key problems and improve the performance to meet the development needs.

## 5 Next Steps

According to the technical challenges faced by Daqing Oilfield, it is necessary to further emancipate the mind and change the fracturing philosophy to meet the needs of effective utilization and reconstruction of Class II reservoirs and realize beneficial development.

### *5.1 With the goal of even fracture initiation and balanced distribution of sand and liquid, the effective support technology for the whole fracture of horizontal wells is tackled*

As the cluster spacing of horizontal wells decreases further, the induced stress and interfracture interference effect of artificial fractures gradually increase, and the sand fluid of each cluster in the horizontal wells is unbalanced and some clusters cannot be fully reconstructed. Long-term cumulative production gradually slows down with the decrease of cluster spacing. In order to realize the benefit development of tight oil, horizontal well fracturing should be transformed from further close cutting to even fracture initiation and balanced distribution of sand and liquid. The goal is to fully transform each cluster in the horizontal well to ensure that each cluster has a contribution to productivity and improve the transformation effect without increasing the construction cost.

The test results of optical fiber clustering at home and abroad show that the phenomenon of crack initiation and uneven extension exists in the intensive cutting of multiple clusters in the segment, and the sand and liquid of each cluster in the segment enters unevenly, and some clusters have no capacity contribution (Figure 1, 2). This indicates that the horizontal well production will not increase significantly with further close cutting. Therefore, it is necessary to explore the effective support technology for the whole fracture of horizontal wells to ensure the uniform initiation of fractures in each cluster and improve the production of a single well. The first is to optimize the perforating parameters. Adopting fixed plane, slope perforating, etc. to ensure the uniform initiation of fractures in each cluster. The second is to optimize the number of clusters in the segment. The joint parameters are improved by combining the numerical model and productivity test results to ensure the uniform expansion of clusters. The third is to optimize the fracturing fluid system, develop the low-cost and high suspension fracturing fluid, thus to ensure the uniform sand flow of each cluster, and improve the effective support fracture length. The fourth is to optimize the proppant combination and study the law of proppant fracture migration to ensure the effective support of the whole fracture system.

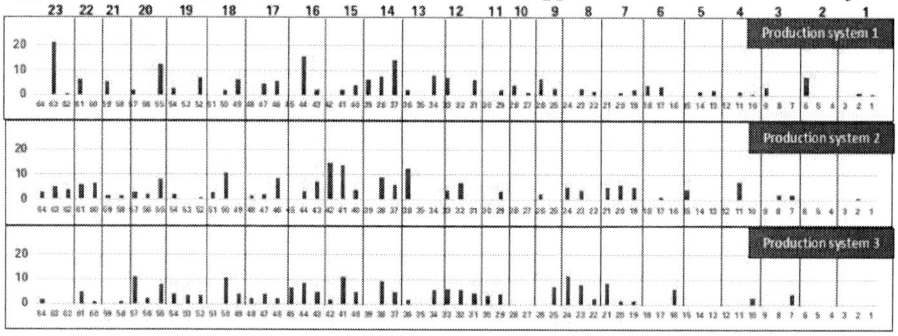


Fig. 1 Results of fiber cluster productivity test in well A

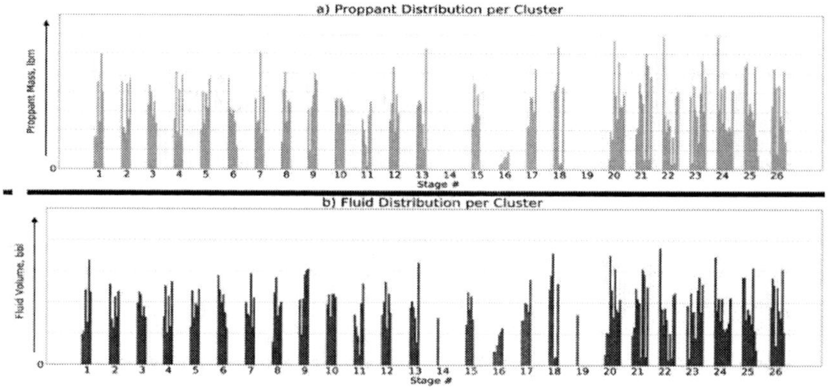


Fig. 2 Results of HFTS-2 fiber cluster sand solution test in North America

*5.2 The multi-stage temporary plugging and diversion fracturing technology of vertical wells is studied with the aim of increasing production and reducing cost at a moderate displacement and scale*

The thickness of the vertical single layer of Daqing tight oil is thin, and the stress and lithology occlusion are weak. The large displacement construction results in the high expansion of the longitudinal fractures, and the sand addition strength of the reservoir is only 30% of the design, and the ineffective reconstruction is serious. Due to the lack of development of natural fractures on the plane, it is difficult to form complex fracture network, and the cost of large-scale fracturing is high.

Therefore, the large-displacement and large-scale fracture network fracturing of vertical wells should be transformed to the multi-branch fracture fracturing with moderate displacement and moderate scale, so as to improve the reconstruction degree and reduce the construction cost.
Therefore, it is necessary to tackle the multi-stage temporary plugging and diversion fracturing technology of vertical well. Through manual intervention of stress field, the vertical height of fractures is controlled to increase the effective fracture length of reservoir. On the plane, a moderate scale of multi-stage temporary plugging and turning in fractures can improve the complexity of fractures, increase the contact area between fractures and reservoirs, realize three-dimensional deployment of reservoirs, and further reduce cost and increase efficiency.

*5.3 Tackling three key fracturing technologies with the goal of improving the cumulative production*
In unconventional reservoirs such as shale oil in North America, natural fractures are developed, the distance of oil and gas seepage from matrix to fracture is short, and the initial production of high-intensity reconstruction is high. A well in Midland Basin produced 7791m3 of oil in the first month, and 445~763m3 of oil a year later. Therefore, the development of unconventional oil and gas in North America does not pursue long-term conductivity,while using initial production for benefits, to recover cost quickly. But the natural fractures in Daqing tight oil reservoir are not developed, the reservoir seepage capacity is poor, the initial production is low, and the number of wells with oil production ≦10t in the first month after pressure is 75%. For example, the oil production of horizontal wells in Daqing X block in the first month is only 210t (Figure 3). Therefore, in order to realize the benefit development of Daqing tight oil, the concept of fracturing should be transformed from "initial production for benefit" to "cumulative production for benefit". The long-term conductivity of fractures should be pursued, and the effective period of wells should be extended by addressing key issues such as fracturing and energy enhancement integration technology, reasonable flowback system after fracturing and efficient re-fracturing technology of horizontal wells.

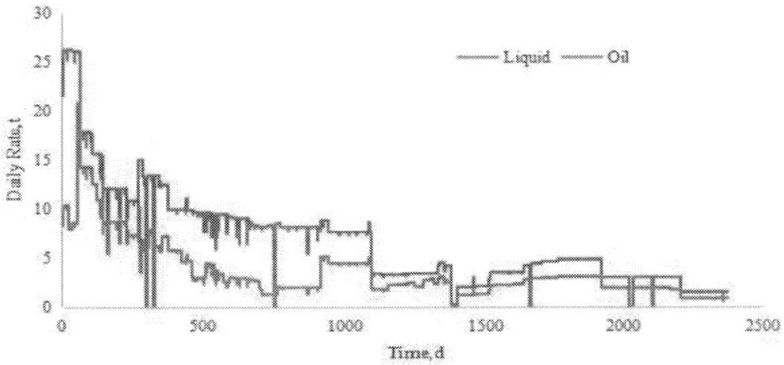


Fig. 3 Production curve of a horizóntal well in X Block

First, fracturing and energy enhancement integration technology should continue to carry out research on . According to the differences in reservoir physical properties and crude oil properties of different blocks, the adaptability of different energy enhancement systems (such as N2, CO2 ,etc.) with the reservoir are evaluated, meanwhile, the performance of upgrading tools is improved to effectively increase reservoir energy and delay production decline. The goal of "controlling decline and increasing EUR" can be achieved through a single transformation of fracturing and energy enhancement.
Second, reasonable backflow and production system need be optimized to improve long-term production effect. Currently, forced flowback, controlled pressure flowback and closed well flowback have their advantages and disadvantages. It is necessary to form a personalized flowback system based on reservoir microscopic characteristics, seepage mechanism and fracturing fluid performance, so as to improve fracturing fluid flowback efficiency, control proppant backflow effectively, avoid sand

production, ensure post-pressure effect and improve accumulative production.
Third, for fractured low-producing wells, the production of old wells can be restored through repeated fracturing methods such as replenishing new fractures, fracturing new and old fractures, fracturing new and old fractures and increasing energy. However, due to the large scale of refracturing, there are problems such as low efficiency of double sealing and single clamping, long construction cycle and poor targeting of temporary plugging in the whole well. In the next step, we can study the conbination technology of double sealing and single clamping string and temporary plugging steering, with 3~4 clusters within the stage and temporary plugging for 1~2 times to improve construction efficiency. Meanwhile, we can explore and carry out wellbore reconstruction processes such as sleeve in casing and expansion tube to recreate new boreholes and improve the effect of repeated reconstruction, thus to revitalize the old well resources, prolong the life cycle of the well and improve the cumulative production.

## 6 Conclusion

(1) Through continuous innovation and development, Daqing Oilfield has formed a set of fracturing technology to meet the fracturing needs of tight oil. However, with the continuous in-depth development, the fracturing technology and supporting tools need to be improved.
(2) Strengthen mechanism research and big data analysis. The research on initiation and propagation mechanism of artificial fractures should be strengthened. On the basis of geological re-understanding, we should carry out big data analysis, establish learning curve, and improve fracturing scheme design. Through mechanism and data driving, fracturing effects can be further improved .
(3) Accelerate the integration of geological engineering research. A large platform of integration of geological engineering can be established. The integrated model of geological engineering can be improved through detailed geological understanding, and the reservoir heterogeneity and stress shadow can be fully considered to achieve a reasonable match between fracture and sand body, fracture and stress, reconstruction volume and production, so as to improve the pertinence of project design.
(4) Field monitoring tests should be strengthen. The distributed optical fiber monitoring technology or hawk-eye TV logging technology can be used to monitor the initiation and propagation of fractures in real time, identify the fracture morphology, judge the distribution of sand fluid in each cluster, and know the production of a single cluster, so as to guide the next program optimization and improve the effective reconstruction degree of the whole well.

## References

[1] Du Jinhu, He Haiqing, Yang Tao, etc. Progress and challenges of tight oil exploration in China [J]. China Petroleum Exploration, 2014, 19 (1): 1-9.

[2] Jia Chengzao, Zheng Min, Zhang Yongfeng. China's unconventional oil and gas resources and prospects for exploration and development [J]. Petroleum Exploration and Development, 2012, 39 (2): 129-136.

[3] Hu Wenrui. Geology-engineering integration-a necessary way to realize profitable exploration and development of complex reservoirs[J]. China Petroleum Exploration，2017,22(1)：1-5.

[4] Su Hao, Lei Zhengdong, Zhang Diqiu, etc.Volume fracturing parameters optimization of horizontal well in tight reservoir [J].Lithologic Reservoirs,2018,30(4):140-148.

[5] ZOU Caineng,TAO Shizhen,Yang Zhi,et al.New Advance in Unconventional Petroleum Exploration and Research in China[J]. Bulletin of Mineralogy,Petrology and Geochemistry,2012，31 (4)：312-322.

[6] SUN Longde，ZOU Caineng，JIA Ailin，et al. Development characteristics and orientation of tight

oil and gas in China[J] Petroleum Exploration and Development, 2019, 46(6): 1015-1026.

[7] QIU Zhen, LI Jianzhong, WU Xiaozhi, et al. Exploration status, main geologic characteristics and their differences of tight oil between America and China[J]. Lithologic Reservoirs, 2015, 27(4): 119-126.

[8] Wang Faxian.Refracturing techique and its field test for the horizontal well in the tight oil[J]. Petroleum Geology&Oilfield Development in Daqing, 2018, 37(4): 171-174.

[9] Jin Chengzhi,He Jian,Lin Qingxiang,et al.Fracturing stimulation based on geology-engineering integration to tight oil reservoirs in Block Fang 198-133,northern Songliao Basin.China Petroleum Exploration, 2019, 24(2): 218-225.

[10] Cui Baowen, Meng Qi'an, Bai Xuefeng, et al. Petroleum exploration progress and suggestions for North Songliao Basin[J]. Petroleum Geology&Oilfield Development in Daqing, 2018, 37(3): 1-9.

[11] Wang Yuhua. Analysis and countermeasure on the exploration situation of Daqing Oilfield[J]. Petroleum Geology&Oilfield Development in Daqing, 2019, 38(5): 23-33.

[12] Wang Guangyun, Wang Fenglan, Zhao Bo, et al. Exploration and development situation and development strategy of Daqing Oilfield Company[J]. China Petroleum Exploration, 2021, 26(1): 55-73.

[13] MILLER C K, WATERS G A, RYLANDER E I, Evaluation of production log data from horizontal wells drilled in organic shales[R]. SPE 144326, 2011.

[14] ZHU D, HILL D, ZHANG S. Using temperature measurements from production logging/downhole sensors to diagnose multistage fractured well now profile[D]. Texas: Texas A&M university, 2018.

[15] CRAMER D, FRIEHAUF K, ROBERTS G, et al. Integrating DAS, treatment pressure analysis and video-based perforation imaging to evaluate limited entry treatment effctiveness[R]. SPE 194334, 2019.

[16] Lin Tiefeng,Kang Dejiang,Jiang Lina.Geological characteristics and exploration potential of the tight oil in Fuyu oil reservoirs of North Songliao Basin[J].Petroleum Geology & Oilfield Development in Daqing ,2019, 38(5): 94-100.

[17] Cui Baowen, Meng Qi'an, Bai Xuefeng, et al. Horizontal well techniques and their exploration practices in the tight oil reservoirs in North Songliao Basin[J].Petroleum Geology&Oflfield Development in Daqing, 2014, 33(5): 16-22.

Journal of Physics: Conference Series **2594** (2023) 012072 doi:10.1088/1742-6596/2594/1/012072

# Study on Mechanical Calculation Method of Injection-production String in Changqing Gas Storage

**Chang-ling WANG1(✉), Rui-qi WANG2, Zhao DONG3 Xu-mei LI1, De-hua MA1, Xiao-ming WANG1**

1 Oil and Gas Technology Research Institute of Changqing Oilfield Company of CNPC

2 Changqing Branch of CNPC Logging Company

3 No. 1 Oil Production Plant of Changqing Oilfield Company of CNPC

1458734964@qq.com

**Abstract.** on the basis of previous studies, According to the working conditions of injection and production wells in gas storage,this paper selects the mathematical model of the downhole string suitable for Changqing gas storage under the effects of piston, axial force, expansion, temperature and spiral bending. The mechanical analysis and calculation of injection-production string under simulated working condition and actual working condition are carried out by using corresponding mechanical analysis software, and the parameters such as the axial load, triaxial Factor of safety, minimum absolute triaxial Factor of safety, string elongation, string shortening are given , according to the calculation results, the safety strength of injection-production string is checked, and the load safety criterion for production and operation of string is given.Field Test and calculation of injection-production string are carried out .According to the actual test results, the deformation rule of the injection production string under the actual gas injection condition is summarized, and compared with the results calculated by the software, which verifies the accuracy of the calculating software and forms the method of the strength check of the injection and production string. It can provide effective evaluation means for the safe operation of injection and production string in gas storage. The safety failure of the injection and production pipe string is reduced, the safety and efficient operation of the gas storage is ensured.

## 1 Introduction

The mechanical analysis and deformation calculation of injection-production string in gas storage are complicated because of the joint action of many forces. Therefore, it is very important to establish a suitable calculation model for accurately calculating the magnitude, direction, elongation, Factor of safety, safe range of load, etc. At the same time, the suitable calculation software should be used to calculate and analyze the forces on the string. In addition, proper on-site testing is also necessary.

Only if the calculation model, software calculation and on-site testing are consistent, can a perfect mechanical analysis and safety strength check method of injection-production string be formed.

**2 Selection and establishment of mechanical calculation model for injection and production well string**[1,2] .

The external factors affecting the stress and deformation of the injection-production string mainly include: gravity, buoyancy, fluid pressure inside and outside the string[3,4], fluid flow viscosity, temperature, constraint mode at the bottom packer, completion operation sequence, etc. Through literature research, the pipe string is analyzed and calculated according to the types of stress and deformation. The corresponding mechanical calculation model is selected and established.

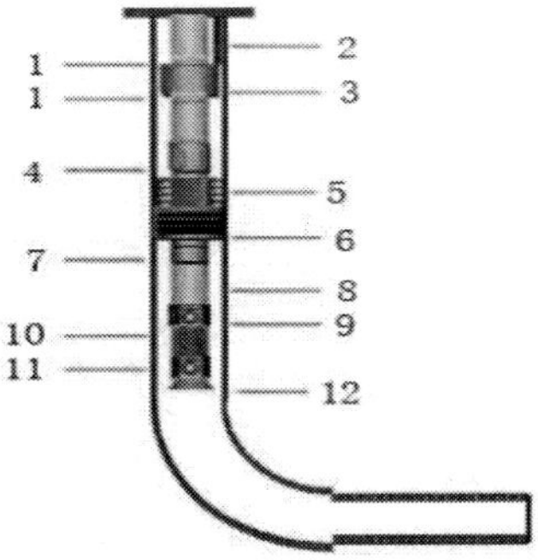


Fig.1Schematic diagram of completionstring of injection-production well

1 - flow nipple 2 - hydraulic control pipeline 3 - downhole safety valve 4 - sliding sleeve 5 - anchor tool 6 - hydraulic permanent packer 7 - milling extension tube 8 - 1 oil tube 9 - plug working tube 10 - screen tube 11 - working cylinder of suspension pressure gauge 12 - rabbet

*2.1 Analysis of axial force effect*

Due to the hydraulic effect, the effective axial force and the straight real axial force need to be considered for the axial force of the injection-production string[5,6]. Based on previous studies, this paper determines the calculation method of the straight real axial force in the test process.

Load calculation:

Set the packer as the coordinate origin,positive upward, positive axial pressure and negative tension. From the literature, it can be determined that the equivalent force at any point $z$ on the string is $F_\omega$ (unit: N), then the true axial force (unit: N) can be obtained as:

$$F_\tau(z)=F_\omega(z)-P_i(z)A(i)+P_o(z)A_o \quad (1)$$

Where: $P_i(z)$ is the pressure at the oil pipe $z$ , unit: MPa; $P_o(z)$ Is the pressure $z$ outside the oil pipe, unit: MPa; $A(i)$ Is the internal circular load area of the tubing, $A(i)=\pi\left(\frac{d_i}{2}\right)^2$ in mm2; $A_o$ ,Is the external circular load area of the oil pipe, $A(o)=\pi\left(\frac{d_o}{2}\right)^2$ in mm$^2$.

Stress calculation:

The axial stress $F_\tau(z)$ generated by the true axial force is:

$$\sigma_F(z) = -\frac{F_\tau(z)}{A_o - A_i}$$ , unit: MPa (2)

Deformation calculation:

According to Hooke's law, the corresponding axial deformation $\sigma_F(z)$ can be determined from the axial stress. The axial strain is:

$$\varepsilon_F(z) = \frac{d\Delta L_F(z)}{dz} = \frac{\sigma_F(z)}{E} = -\frac{F_\tau(z)}{E(A_o - A_i)}$$ (3)

Therefore, the axial deformation is:

$$\Delta L_F(z) = \Delta L_F(z_o) - \frac{1}{E(A_o - A_i)}\int_{z_o}^{z} F_\tau(z)dz$$ , unit: m (4)

Based on similar derivation, other mechanical calculation models such as expansion effect, temperature effect, piston effect and spiral bending effect can be obtained.

Total effect force analysis:

Total stress and deformation of injection-production string:

The total axial load is:

$$\sum F = F_v + F_\tau + F_P + F_T + F_M$$ , unit: KN (5)

The total axial force is:

$$\sum \sigma = \sigma_v + \sigma_F + \sigma_p + \sigma_T + \sigma_M$$ unit: $MPa$ (6)

The total axial deformation is:

$$\sum \Delta L = \Delta L_v + \Delta L_F + \Delta L_p + \Delta L_T + \Delta L_M$$ unit: $m$ (7)

$\sum \Delta L$ -Total axial deformation $\Delta L_V$ - piston effect axial deformation $\Delta L_F$ - axial force effect axial deformation $\Delta L_T$ - temperature effect axial deformation $\Delta L_P$ - expansion effect axial deformation $\Delta L_M$ - spiral bending effect axial deformation

*2.2 Analysis, calculation and result analysis of stress on injection-production string*

According to the stress analysis and the selection of mechanical model, combined with the domestic and international investigation of the mechanical analysis software, Landmark Wellcat mechanical analysis software is used for the stress analysis and safety strength check of the injection-production string.

Simulated calculation:

Calculation conditions: take 139.7mm oil pipe as an example, use the maximum injection-production parameters to simulate the calculation (main working conditions: injection 1200000m3/d, production 2000000m3/d,The gas injection pressure is 30MPa, the wellhead temperature is 20 ℃, and the packer depth is 2935.18m).

Simulate the main process of calculation:

Brief description of calculation process: input relevant design parameters (such as injection, gas production, gas injection pressure, etc.) - software analysis and calculation - output calculation results (such as axial load, triaxial safety factor, etc.)

Some input and output data interfaces:

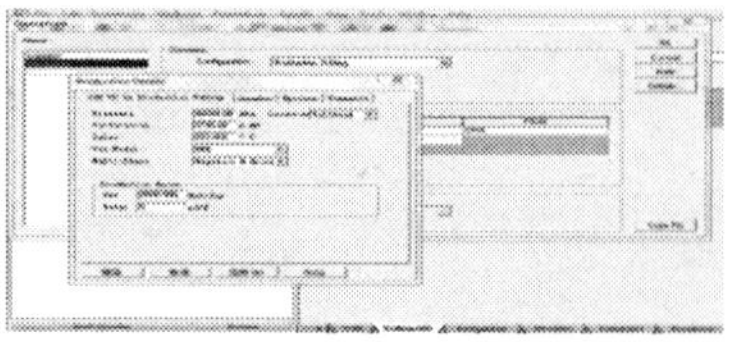

Fig.2 Production parameter parameter condition)

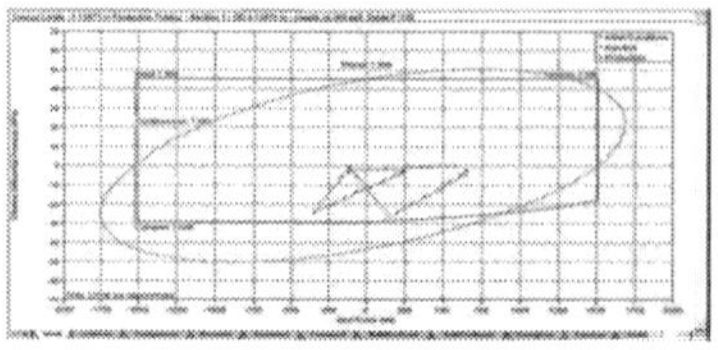

Fig.3 safety range diagram setting table (mining

*2.3 Analysis of calculation results*

Under the injection condition, the tubing length is 0.624m due to axial force effect and piston effect, 0.710m due to expansion effect, 1.334m due to temperature effect, and the overall length of the tubing string changes to 0m.

Under production conditions, the tubing is shortened by 1.894m due to axial force effect and piston effect, the tubing string is lengthened by 0.689m due to expansion effect, the tubing string is lengthened by 1.205m due to temperature effect, and the overall length of the tubing string changes by 0m.

It can be seen from the table of minimum absolute safety factor that the minimum absolute triaxial safety factor is 2.054 under injection conditions. Under the mining condition, the minimum absolute triaxial safety factor is 2.92. The load of injection and production conditions are within the safe range.

Calculate with actual production parameters:

Actual production parameters: 139.7mm tubing, 2935.18m depth of packer, 480000m3/d gas injection volume, 25.42 MPa gas injection pressure (pressure variation range: 25.42-30.54 MPa), 43.78 ° C temperature (temperature variation range: 43.78-60.20 ° C)

Calculation results:

Under the injection condition, the tubing length is 1.008m due to axial force effect and piston effect, the tubing length is 0.524m due to expansion effect, the tubing length is 0.484m due to temperature effect, and the overall length of the tubing string changes to 0m.

Under the injection condition, the minimum absolute triaxial safety factor is 2.228, and the load under the injection condition is within the safe range.

## 3 Test and analysis of injection-production string

In order to verify the accuracy of the calculation software, Jingping * - * - * well was selected for on-site detection of the axial and radial deformation of the production string[7,8].

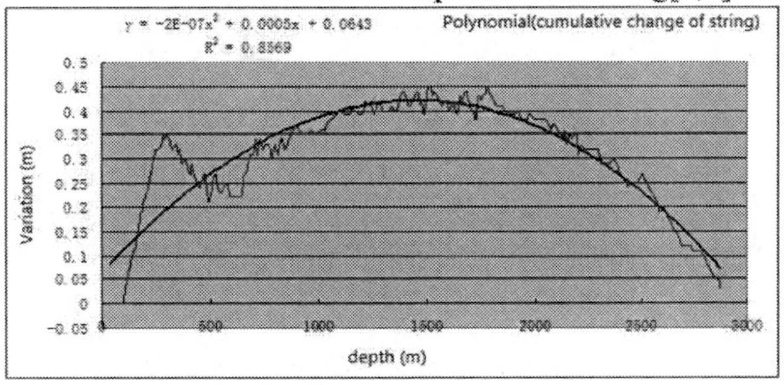


Fig.4 Variation and depth of the production string

From the change curve, the tensile deformation of 1 to 25 pipe strings is relatively obvious, and the tensile deformation of each pipe string is relatively uniform, and the axial shrinkage of 25 to 55 oil pipes occurs. From 55 to 253 oil pipes, the overall shape of the pipe string changes in a parabola, to close to 0, indicating that the tensile deformation of the pipe string is controlled by the wellhead and permanent packer (2896.93m). The overall trend of string change is to stretch first and then compress, and the total amount is approximately 0.

The test results show that the injection-production string will deform under production conditions. The deformation rules of the string are summarized as follows:

Axial deformation law: with the increase of depth, the pipe string is stretched first and then compressed[9,10]. The change trend of the overall length of the pipe string is close to 0 m.

Radial deformation law: the upper well section shrinks (0-1850 m) and the lower well section expands.

## 4 Conclusion

4.1 In this paper, the mathematical model of the downhole pipe string suitable for Changqing gas storage is selected and determined under the piston effect, axial force effect, expansion effect, temperature effect and spiral bending effect through the analysis of the stress on the pipe string.

4.2 The strength check of the injection-production string of Changqing gas storage is calculated by using a professional calculation software. The mechanical calculation method of injection and production well string is formed.

4.3 The simulation calculation results of the maximum injection-production condition parameters show that the safety factor of the injection-production well in Changqing Gas Storage is high under both gas production and gas injection conditions, which can achieve safe production.

4.4 Through the method of actual measurement, the deformation of the injection-production string under production conditions is measured, and the deformation law of the string is summarized.

4.5 The change trend of the overall length of the pipe string is close to 0 m according to the software calculation analysis and actual measurement. The two conclusions are consistent. It shows that the software calculation method can be used to calculate the safety factor of gas storage pipe string.

## Reference

[1]Zhou Guangwei;Qian Changzhao;Chen Changping : Research on the Approximate Calculation Method of the Fundamental Frequency and Its Characteristics on a Tensioned String Bridge, Processes，2022-01-08

[2]Xiangyu Yan;Hong Hu;Zhihua Chen;Yan Yang: An Improved Mathematical Calculation Method for Beam String Structure Based on Static Equilibrium Principle, International Journal of Steel Structures, 2020-05-28

[3]Beivydas Edmundas :A SIMPLIFIED CALCULATION METHOD FOR SYMMETRICAL LOADING OF A SINGLE-SPAN COMPOSITE STRING STEEL STRUCTURE, Engineering Structures and Technologies, 2019-10-14

[4]Wei-kai YI :Mechanical analysis of multi-cycle cyclic alternating load of injection-production string in gas storage, China and foreign energy 2023,28(03) ; 58-622

[5]Ming-fei LI, Bing-hui KANG , Qiao WANG , etc.: Mechanical analysis and safety evaluation of telescopic completion string under typical working conditions, petroleum drilling and production technology 2022,44(01)37-423

[6]Xiao-lei, SHI Wen-jun HUANG , Deli GAO: Mechanical analysis of high temperature and high pressure string considering buckling and friction, petroleum machinery 2020,48(11)4

[7]Geng TANG,hang WANG , Yu-fei LI: Completion string mechanical analysis method and application based on experimental testing, petrochemical applications 2020, 39(06)24-285

[8]Jiandong DING , Zhang-hua LIAN , Yi-ran DING , etc.: Static and dynamic safety evaluation and software development of gas storage injection and production string, oil drilling and production technology 2018,40(02) : 215-2217

[9]Kai-qiang LIU , Yin-ping CAO , Hao YANG: Research status and development trend of horizontal well string mechanics, petrochemical application 2016,35(02)10

[10]Den PAN: Mechanical study of expandable casing [ J ] . Natural gas industry, 2004,24(9) ; 54-5613

Journal of Physics: Conference Series **2594** (2023) 012073 doi:10.1088/1742-6596/2594/1/012073

# Research on Refracturing Technology of Ultra-low Permeability Fractured Oilfield

**Ming-guo LI1(✉), Gao-liang XUAN1, Wei HAN1, Hong-ni ZHANG1, Mei-heng DUAN1, Xue-ping ZHOU1**

1 Oil and Gas Engineering Research Institute, PetroChina Jilin Oilfield Company, Songyuan, China
1844054030@qq.com

**Abstract**. The development of ultra-low permeability reservoir in Daqingzi area, Qian an. With a long period of mining, the number of refractured wells is increasing year by year, but there is no benefit after fracturing. There is an urgent need for low-cost fracturing techniques to make refracturing more cost-effective. In this paper, on the basis of fully understanding reservoir physical properties, in-depth analysis of the major problems facing the refracturing of Daqingzi oilfield, and analysis of refracturing potential. Through field experiment, post-frac analysis and effect evaluation, the refracturing technology route suitable for this area is gradually explored. In this field experiment, the problem of underinjection was solved through the analysis and targeted transformation of injection wells. The average daily injection capacity of a single well was increased, injection pressure was reduced, the formation pressure of surrounding producing wells was restored to the original formation pressure, and the refracturing time of producing wells was reached, and the individual refracturing design of producing wells was carried out. In addition, post-frac analysis and effect evaluation were carried out. The results show that the injection-production relationship of the well group is improved obviously after the measures, the effect of increasing fluid and oil is remarkable, and the average fracturing cost per well is reduced. The new technology of holistic fracturing of injection and producing wells centered on injection well is defined. It provides technical support for increasing production and controlling cost of ultra-low permeability reservoir.

## 1. Introduction

With the development of Jilin oilfield for a long time, the focus of refracturing is on the ultra-low permeability reservoir. This kind of reservoir has poor physical property and large well network. It is difficult to establish an effective injection-production relationship in a block by fracturing only at the end of a single production well and injection well, which has many problems such as high cost and low formation energy. This paper will focus on this kind of reservoir to solve the contradiction of reservoir development step by step[1-20].

Journal of Physics: Conference Series **2594** (2023) 012073 doi:10.1088/1742-6596/2594/1/012073

## 2. Analysis of problems and potential of refracturing in Daqingzi Oilfield

The physical property of the Jinyuan area in Daqingzi oilfield is relatively good, the average permeability is 5-8md, and the injection and production are sensitive. Affected by the distribution of remaining oil and the unclear understanding of waterline, the small-scale refracturing of old producing wells in the early stage has a high water-cut, and the refracturing of old producing wells is not an effective means to improve production in this area.

The average permeability in Qianyuan area of Daqingzi oilfield is less than 3md, and it is difficult to establish the injection-production relationship. There are some problems such as the injection well cannot be injected, low formation energy of producing well and high fracturing cost. In the early stage, self-storage fracture network fracturing was carried out in producing wells. After fracturing, high yield fluid and high water-cut , which is effective, but it has no benefit due to high cost. Small displacement fracturing without sand was carried out in injection well, long-term injection increase was not realized, and injection-production relationship was not established with the well. There is an urgent need for low-cost fracturing techniques to make refracturing more cost-effective.

## 3. Refracturing technology route centered on injection wells

There are three major reconstruction problems that have not been solved in fracturing oil and injection well. The project team sorted out the development conflict and the effect of the measures, and then gradually identified that the effective renovation of the injection well is the key factor to improve the economic efficiency of refracturing, in Qianyuan area of Daqingzi oilfield. And it began to explore the new technology of holistic fracturing of injection and producing wells centered on injection well (see Fig. 1), and gradually solved the three major reconstruction problems in Qianyuan area of Daqingzi oilfield[10-11,16-18,20-22].

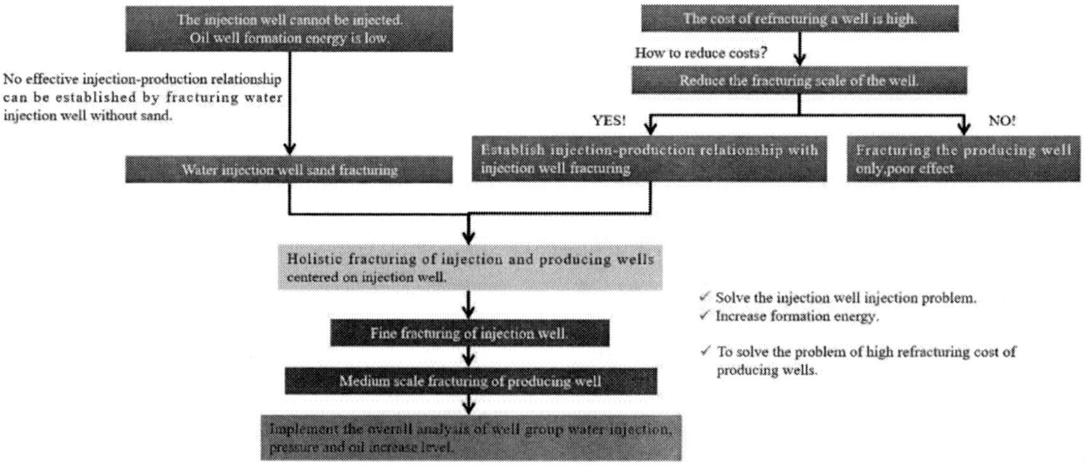


**Fig. 1.** Roadmap of "holistic fracturing of injection and producing wells centered on injection wells"

### *3.1. Overall planning of injection wells*

Block selection: Considering the water absorption capacity of small layer, injection capacity of small layer, block recovery degree and formation energy, the blocks with water absorption index less than $10m^3/MPa$, injection capacity of single well less than $30m^3$, recovery degree less than 15% and pressure retention level less than 80% are preferred.

Injection well optimization: Considering the complete well condition evaluation, injection conditioon tracking and underinjection cause analysis, the wells with normal well condition, daily actual injection distribution less than 10%, and underinjection due to zone difference or near-well pollution and blockage are preferred.

Individual design of injection well: Considering the vertical characteristics and horizontal plane contradictions of reservoir comprehensively, the reservoir is classified, and then the reconstruction countermeasures of each layer are defined.

### *3.2. Overall planning of producing wells*

Determination of refracturing time for producing well: Monitor the water absorption of small injection layer, the improvement of water injection and the pressure of adjacent producing well. When the injection well reaches the daily injection distribution level, the water injection layer achieves balanced injection and the pressure of the main control producing well recovers, the refracturing time of producing well will be reached.

Individual design of producing well: Combined with the understanding of fluid increase level after fracturing of injection well and the law of block effect, the producing well in the dominant position is suitable for fracturing with less sand and the producing well in the weak position is suitable for fracturing with more sand.

Post-frac effect evaluation: Summarize the actual daily production effect, conduct production test analysis, and finally determine the technical adaptability.

## 4. Experimental scheme and effect

### *4.1. Experimental scheme*

According to the above technical route, Block C of Daqingzi oilfield is selected as the test Block. The water absorption index, injection capacity and recovery degree of Block C are all at a low level. By comparing the formation pressure in the north and south development zones of Block C, it is clear that the pressure holding level in the north of Block C is only 50%, so it is selected as the key area.

Four water injection wells in the main development area with good well condition and serious underinjection due to formation reasons are selected as test objects. According to the longitudinal (sonic time difference, permeability, oil saturation) and plane contradictions, the reservoir is divided into three types, and the reconstruction requirements and designed penetration ratio of each type of reservoir are defined:

Class I reservoir (interpreted as oil reservoir with acoustic time difference greater than 235, permeability greater than 4, and oil saturation greater than 50%): Small-scale fracturing with designed penetration ratio of 30%;

Class II reservoirs (interpreted as same layer of oil and water, acoustic time difference of 220-235, permeability of 2-4, oil saturation of 40%-50%): medium-scale fracturing, designed penetration ratio of 40%;

Class III reservoirs (interpreted as strong or extremely strong waterlogged, acoustic time difference 210-225, permeability 1-2, oil saturation 30%-40%): large-scale fracturing with designed penetration ratio of 50%.

The following Table 1 gives conditions, fracturing scale and parameters of four injection wells.

**Table 1.** Conditions, fracturing scale and parameters of four injection wells in Block C experimental area

| Well | Pressure (MPa) | Standard daily injection (t) | Cause analysis of underinjection | Specific reason | Layer number | Thickness (m) | Reservoir classification | Transformation demand | Displacement ($m^3$/min) | Sand content ($m^3$) | Fluid volume ($m^3$) | Sand strength |
|---|---|---|---|---|---|---|---|---|---|---|---|---|
| A1 | 18 | 0 | Stratigraphic | Near- | qn1 | 9.8 | I | Small scale | 4 | 17 | 163 | 1.7 |

| | | | | | | | | | | | | |
|---|---|---|---|---|---|---|---|---|---|---|---|---|
| | | | cause | hole plugging | : 11.12 | | | | | | | |
| | | | | | qn1 : 9 | 2 | II | Medium scale | 4 | 8 | 121 | 4 |
| | | | | | qn1 : 1.8 qn2 : 24 | 6.4 | III | Large scale | 5 | 30 | 138 | 4.8 |
| A2 | 17.8 | 0 | Stratigraphic cause | Poor reservoir physical property | qn1 : 11.12 | 7.6 | I | Small scale | 4 | 14 | 147 | 1.8 |
| | | | | | qn1 : 9 | 1.6 | II | Medium scale | 4.5 | 10 | 135 | 6.3 |
| | | | | | qn1 : 7 | 3.2 | I | Small scale | 4 | 10 | 116 | 3.1 |
| A3 | 17.5 | 5 | Stratigraphic cause | Near-hole plugging | qn1 : 11.12 | 5.4 | I | Small scale | 5 | 13 | 157 | 2.4 |
| | | | | | qn1 : 8.9 | 4.8 | II | Medium scale | 5 | 16 | 189 | 3.3 |
| | | | | | qn2 : 24 | 2.2 | III | Large scale | 4 | 13 | 167 | 5.9 |
| A4 | 18 | 0 | Stratigraphic cause | Near-hole plugging | qn1 : 12.15 | 4.8 | I | Small scale | 5 | 10 | 148 | 2.1 |
| | | | | | qn1 : 7.9 | 7.2 | II | Medium scale | 5 | 22 | 261 | 3.1 |
| | | | | | qn2 : 24 | 2.2 | III | Large scale | 5 | 12 | 145 | 5.5 |
| **Average single small layer** | | | | | | **4.8** | | | **5** | **14** | **157** | **3.0** |

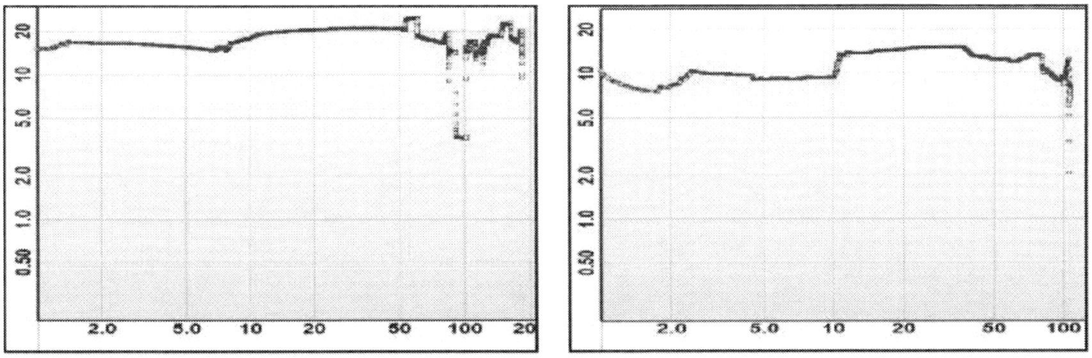


**Fig. 2.** Comparison of net pressure between self-storage energy three-stage and pre-slickwater at the same displacement

The changes of injection well injection and formation energy were tracked. Six months after injection, the average daily injection capacity of a single well was increased by 25m$^3$, the injection pressure was decreased by 0.6-1.8MPa, and the formation pressure of surrounding producing wells recovered quickly. When the pressure holding level of producing wells in the dominant position was increased from 50% to 100%, the time for refracturing of producing wells was reached. Weak response producing wells are preferred. The three-stage fracturing technology of "fracturing-networking-supporting" is used to improve the net pressure level (see Fig. 2). The designed penetration ratio of producing wells in the dominant orientation is 50%, and the designed penetration ratio of producing wells in the weak orientation is 70%-80%, and the matching degree between fractures and well pattern is improved. The following Table 2 gives Conditions, fracturing scale and parameters of three production wells.

**Table 2.** Conditions, fracturing scale and parameters of three producing wells around the injection wells

| Well | Location with well | Layer number | Thickness (m$^3$) | Actual parameter | | | |
|---|---|---|---|---|---|---|---|
| | | | | Displacement (m$^3$/min) | Single-layer sand content (m$^3$) | Single-layer slickwater liquid quantity (m$^3$) | Single layer guanidine gum liquid quantity (m$^3$) |
| B1 | Weak bearing | qn1: 7.8.9/12 | 9.8 | 4-5.5-4 | 18 | 295 | 183 |
| B2 | Weak bearing | qn1: 7/11、12 | 10.4 | 4-5.2-4 | 18 | 300 | 218 |
| B3 | Dominant bearing | qn1: 7,9/12.14 | 19 | 4-5.5-4 | 26 | 400 | 380 |

### *4.2. Effect evaluation after refracturing*

After well test analysis, artificial fractures meet the reconstruction demand (see Fig. 3), and the average oil increase per well is 2t, and the cumulative oil increase is expected 420t, well input-output ratio 1.7 , achieving a qualitative breakthrough. The following Table 3 gives fracturing cost and output-input ratio of three producing wells.

**Table 3.** Fracturing cost and output-input ratio of three producing wells around the injection well

| Well | Half-length (m) | Conductivity (md.m) | Skin | Original daily production (t) | Post-fracture production | | | Fracturing cost (Ten thousand yuan) | Input - output ratio |
|---|---|---|---|---|---|---|---|---|---|
| | | | | | Early daily production (t) | Stable daily production (t) | Accumulative oil is expected (t) | | |
| B1 | 182 | 470 | -6.5 | 1.3/0.6 | 14.2/3.1 | 7.1/3.1 | 361 | 46.5 | 1.40 |
| B2 | 145 | 442 | -3.2 | 4.9/1.2 | 10/3.4 | 7.4/2.4 | 480 | 42.4 | 2.04 |
| B3 | 120 | 515 | -2.8 | 2.3/1.2 | 15.2/4.4 | 11.3/2.8 | 520 | 49.9 | 1.88 |

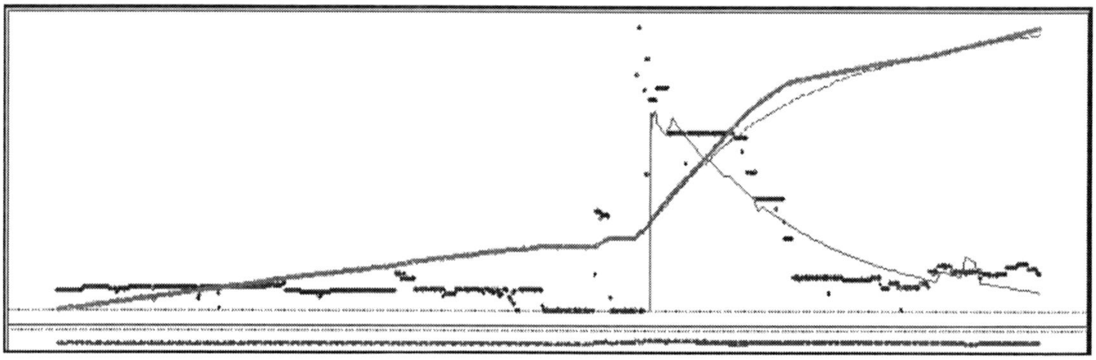

**Fig. 3.** Historical production fitting diagram of well B1

After fracturing the injection wells of the three well groups in the experiment, a cumulative increase of 45,000 m$^3$ was injected, and the cumulative increase of oil in the wells was 936t. At present, the overall daily increase of oil in the well groups was 6t (see Fig. 4). Compared with other well groups in the same block, the daily injection allocation and daily production of measure well group are on the rise, which is stronger than the development level of other well groups and improves the injection-production relationship as a whole.The following Table 4 gives production comparison of well groups.

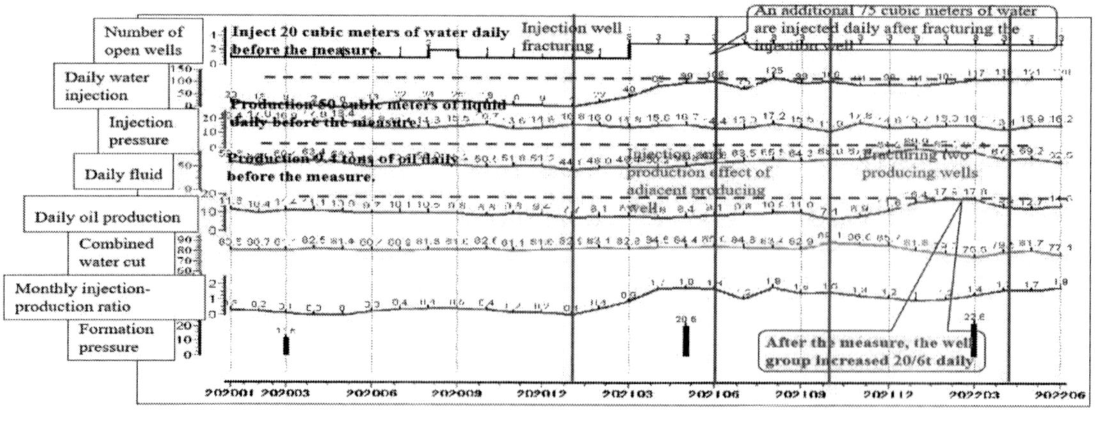


**Fig. 4.** Comprehensive dynamic curve of 3 injection wells +13 producing wells

**Table 4.** Production comparison of well groups in Block C

| Well | Standard daily injection (t) | Actual daily injection (t) | 202209 | | | comparison | | | classification |
|---|---|---|---|---|---|---|---|---|---|
| | | | Daily fluid production (t) | Daily oil production (t) | Water cut | Daily fluid production (t) | Daily oil production (t) | Water cut | |
| A1 | 50 | 40.6 | 18.4 | 3.6 | 80.4 | 3.9 | 0.5 | 1.5 | rise |
| A3 | 20 | 19.1 | 31.6 | 6.6 | 78.9 | 1.8 | 3.2 | -9.6 | rise |
| A2 | 45 | 38.5 | 23.9 | 4.2 | 82.3 | 9.3 | 1.1 | 3.4 | rise |
| O1 | 20 | 4.1 | 6.6 | 0.6 | 91.3 | 2.6 | 0.1 | 4.3 | stable |
| O2 | 0 | 0 | 8.9 | 0.7 | 90.0 | 5.9 | 0.3 | 4.7 | stable |
| O3 | 30 | 28.6 | 18.8 | 5.3 | 71.7 | -3.0 | 0.2 | -4.9 | stable |
| O4 | 50 | 32.3 | 21.9 | 11.0 | 47.7 | -10.9 | 0.4 | -20.1 | stable |
| D1 | 40 | 33.2 | 18.2 | 7.8 | 57.1 | -19.2 | -6.9 | -3.5 | decline |
| D2 | 0 | 0.0 | 16.9 | 5.0 | 70.3 | -4.3 | -3.8 | 11.9 | decline |
| A4 | 0 | 0.0 | 20.1 | 5.4 | 73.4 | -11.0 | -6.9 | 12.8 | decline |
| D3 | 30 | 14.6 | 3.8 | 68.8 | 55.5 | -16.7 | 59.9 | -1.1 | decline |
| D4 | 50 | 35.9 | 10.3 | 3.5 | 66.5 | -3.0 | -1.5 | 3.3 | decline |
| D5 | 10 | 1.7 | 11.0 | 3.8 | 65.2 | -10.9 | -5.7 | 8.6 | decline |
| **total** | **345** | **249** | **210.4** | **126.3** | **40.0** | **-55.4** | **41.0** | **-27.9** | |

## 5. Conclusion

(1) Through sorting out the development of contradictions, the effect of measures, it is clear that there are three major reconstruction problems in Qianyuan area of Daqingzi oilfield such as no injection of injection well, low formation energy of producing well and high fracturing cost.

(2) In order to solve the three major reconstruction problems, a new technical route of "holistic fracturing of injection and producing wells centered on injection well" is proposed to improve the injection-production relationship of blocks.

(3) At the injection well end, the standard of well selection and individual design is formed. In the process, it is necessary to combine the vertical and plane contradictions of the well to clarify the fracturing reconstruction requirements of each small layer, optimize the fracturing parameters of the well individually, and ensure the uniform water absorption of each layer.

(4) At the production well end, the standard of determining the refracturing time and individual design is formed. By combining formation energy changes, optimize the fracturing parameters of the well individually, and improve the matching degree between fractures and well patterns.

(5) Through field experiments and post-pressure evaluation, the results show that the new technology of "holistic fracturing of injection and producing wells centered on injection well" improves the injection-production relationship of well group, and the effect of increasing fluid and oil is obvious. It provides technical support for increasing production and controlling cost of ultra-low permeability oilfield.

## References

[1] Wang Anlun, Chen Yinghe,Wei Jianguang, Li Jiangtao,Zhou Xiaofeng. Experimental study on the mechanism of five point pattern refracturing for vertical & horizontal wells in low permeability and tight oil reservoirs[J]. Energy,2023,272.

[2] Wang Youjing, Song Xinmin. Comprehensive Characterization Integrating Static and Dynamic Data for Dynamic Fractures in Ultra-Low Permeability Reservoirs: A Case Study of the Chang 6 Reservoir of the Triassic Yanchang Formation in the Ordos Basin, China[J].

Minerals,2022,12(10).
[3] Wang Daobing, Dahi Taleghani Arash, Yu Bo, Wang Meng, He Chunming. Numerical Simulation of Fracture Propagation during Refracturing[J]. Sustainability,2022,14(15).
[4] Lei Sheng, Pan Yuan, Huang ZiYi, Yao ZhenHua, Zhou YuHui, Wang WeiXiao, Li Su, Lei YuQi. Progress and Enlightenment of Refracturing Technology for Shale Gas in North America[J]. Frontiers in Energy Research,2022.
[5] Cui Ping. Practice and cognition of effective water-flooding development mode in Gaotaizi block of ultra-low permeability reservoir[J]. E3S Web of Conferences,2022,358.
[6] Liu Xiaoqiang, Qu Zhanqing, Guo Tiankui, Huang Wenjie. Numerical simulation of stress field distribution before re-fracturing treatment[J]. IOP Conference Series: Earth and Environmental Science,2021,814(1).
[7] Guo Yintong, Wang Lei, Chang Xin, Zhou Jun, Zhang Xiaoyu. Study on Fracture Morphological Characteristics of Refracturing for Longmaxi Shale Formation[J]. Geofluids,2020,2020.
[8] Oleg Yurievich Smetannikov, Yuriy Aleksandrovich Kashnikov,Sergey Gennadievich Ashikhmin, Artem Eduardovich Kukhtinskiy. Numerical model of fracture growth in hydraulic re-fracturing[J]. Fracture and Structural Integrity,2019,13(49).
[9] Science - Geoscience; Recent Studies from University of North Dakota Add New Data to Geoscience (Refracturing: Well Selection, Treatment Design, and Lessons Learneda Review)[J]. Science Letter,2019.
[10] Stephen Mraz. Refracturing Revitalizes Old Oil Wells[J]. Machine Design,2017.
[11] "Kash", Sudhendu,Jbeili, Terry. Beating the decline through refracturing[J]. World Oil,2015.
[12] Chen Qi, Zheng Jiadan. Numerical simulation of refracturing and fracture steering in low permeability reservoir[J]. Journal of Guangdong Institute of Petrochemical Technology,2022,32(03):69-72.
[13] Zhou Haijun. Research on refracturing technology of low efficiency horizontal well in A Oilfield[J]. Exploration engineering in western China,2022,34(01):106-108+119.
[14] Zhang Wenkang, Yan Yajie, Cao Wei. Study on temporary blocked steering and re-fracturing Technology of low permeability horizontal Wells[J]. Chemical management,2021(36):197-198.DOI:10.19900/j.cnki.ISSN1008-4800.2021.36.095.
[15] Da Yinpeng, Xue xiaojia, Liu Ming, Shi Peize, Qu Hongyan, Zhou Fujian. Mechanism of energy storage and re-fracturing in ultra-low permeability oil fields[J]. Science, Technology and Engineering,2021,21(33):14139-14146.
[16] Wang Yin, Luo Pan, Wang Qiang. Influencing factors and productivity prediction of refracturing Wells: A case study of Dadonggou Area in Fuxian County[J]. Chemical design communication,2021,47(09):30-31.
[17] He Dailan. Development status of old well refracturing technology[J].Technology and industry,2021,21(01):264-268.
[18] Meng Xiangcan. Refracturing stimulation technology in Block A of Daqing Zijing Oilfield[J]. Oil and gas well testing,2020,29(05):44-49.DOI:10.19680/j.cnki.1004-4388.2020.05.008.
[19] Liu Lu, Sun Qi, Guo QingHua. Research and application of horizontal well refracturing technology in low permeability oilfield[J]. Petrochemical technology,2020,27(10):34-35.
[20] You Yuting. Analysis of well and layer selection method and influencing factors of refracturing effect[J]. China Petroleum and Chemical standards and quality,2020,40(04):7-8.
[21] Xiao Yu, Sun Yaoyao,He Jingcong. Application effect evaluation of injection well microfracturing technology in an oilfield in Bohai Sea[J]. Offshore oil,2018,38(03):36-38.
[22] Han Chunlin. Research on effective fracturing technology of injection well in low permeability reservoir[J]. Chemical design communication,2018,44(02):231.

# Application of Remote Wireless Intelligent Layered Water Injection Technology in Dagang Oilfield

**Zu-chang SONG(✉), Yang LIU , Yong-tao ZHANG, Ji-jun YANG, Ying-ying WANG, Min LI**

Dagang Oilfield Branch of China Petroleum and Natural Gas Co. Ltd., Tianjin, China

songzchang@petrochina.com.cn

**Abstract**. A remote wireless intelligent layered water injection technolo-gy is proposed to address the problems of traditional layered water in-jection well testing in oil fields, such as the inability to achieve continu-ous dynamic monitoring of injection parameters, high failure rate, and low success rate of testing and commissioning. The technology consists of a host computer client, a server, a surface controller, and an under-ground intelligent water distributor. The underground intelligent water distributor takes the injected water as the carrier, and uses the fluctua-tion of the water injection pressure in the tubing for signal transmission to realize two-way communication between the ground controller and the downhole intelligent water distributor. The underground intelligent water distributor integrates high-precision pressure gauges, controllable water nozzles, and calculation and control modules to form an intelli-gent control system, realizing the reception of ground down transmitted commands or data, the calculation and control of water injection flow rate of water nozzles, and the storage and upload of underground data. Remote wireless intelligent layered water injection technology has been applied in 43 wells, with a success rate of 100%. The maximum depth is 3897m, and the maximum wellbore inclination is 56.07 °. This tech-nology realizes remote control of water injection wells, intelligent au-tomatic measurement and adjustment, real-time data monitoring, and automatic early warning functions, laying a foundation for the con-struction of "Digital Intelligence" oilfield.

## 1. Introduction

Dagang Oilfield is a typical complex fault block oilfield with multiple layers, strong heterogeneity, deep reservoir burial, high reservoir temperature, and prominent interlayer contradictions [1]. After more than 50 years of development, it is currently in the stage of high water content development, with a comprehensive water content of 90% [2]. Improving development efficiency under high water content conditions puts forward higher requirements for layered water injection technology [3]. Bridge eccentricity and bridge concentric injection are the main injection processes in Dagang Oilfield, accounting for 76%[4]. The bridge injection technology utilizes cable carrying instruments for flow measurement and adjustment. The packer can also be inspected step by step using specialized electric sealing instruments [5]. The water distributor adopts a bridge type flow passage, effectively eliminating interference between different water injection layers during measurement and adjustment, ensuring the effectiveness of layered water injection [6]. However, adjusting the water injection volume and sealing inspection requires the use of cable lifting instruments, which results in heavy periodic testing tasks and low work efficiency, which cannot meet the needs of fine development of water drive reservoirs. The main issues are as follows:

Unable to achieve continuous dynamic monitoring of injection parameters. Firstly, specialized teams, professionals, and instruments are required for measurement and adjustment, and it is not possible to conduct measurement and adjustment at any time. Secondly, the flow measurement and adjustment are carried out while testing, and the relevant data can only reflect the status during the testing process. After the completion of the survey, only the total water injection volume and pressure values can be recorded at the wellhead, and flow and pressure data for each layer cannot be obtained. The layered water volume and pressure fluctuations caused by the influence of strata, pipe columns, and tools in the later stage cannot be monitored in a timely and effective manner[7]..

High failure rate and low success rate of testing and debugging. The measuring and adjusting instrument is inserted through a cable and connected to the underground water distributor, and the opening size of the water nozzle is changed to achieve the purpose of adjusting the water injection volume. The testing process is greatly influenced by factors such as instruments, wellbore, and oil reservoir, resulting in a low success rate. During the testing and adjustment process, the torque of the water distributor is high, and the phenomenon of not being able to open the water nozzle often occurs. Secondly, Dagang Oilfield is located in the Bohai Bay and belongs to the coastal oilfield. The well inclination is large, and the problem of scaling in the pipe string after long-term water injection is common. It is difficult to lower the instruments, and the phenomenon of not lowering the instruments in place also occurs. It is necessary to implement pipeline work in advance, increase the operating procedures, increase labor intensity, and affect the success rate of measurement and adjustment[8]..

In order to solve the above problems, a remote wireless intelligent layered water injection technology is proposed. This technology is composed of host computer client, server, surface controller and downhole intelligent water distributor, and integrates technologies such as computer network communication, piezoelectric control, artificial intelligence control and waveform pulse signal transmission. The separate injection data and water injection status of each layer can be remotely monitored from the host computer client, without the need for well test trucks and personnel to arrive at the well site, technicians can easily complete the measurement, allocation and sealing of each layer through the host computer client in the office Separator seal inspection work. The field test application shows that this technology is not affected by the ground environment of the well site, weather changes, wellbore, time and other factors, and has significant advantages such as high measurement and adjustment efficiency, high fractional injection pass rate, high measurement and adjustment success rate, and low cost[9]..

## 2. Technical Solutions

The remote wireless intelligent layered water injection technology consists of a surface part and an underground part[10].. The ground part mainly includes the upper computer client, server, remote wireless communication network, electric control cabinet, and ground controller; the downhole part mainly includes two-way anchors, packers, intelligent water distributors, and ball seats. As shown in Figure 1.

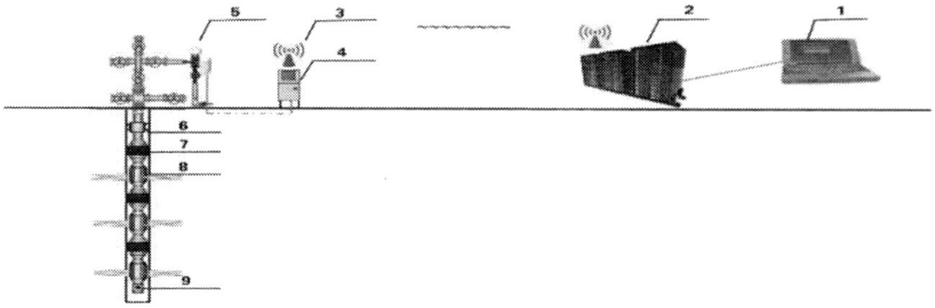


1. Host computer client 2. Server 3. Remote wireless communication network 4. Electric control cabinet 5. Ground controller 6. Two-way anchor 7. Packer 8. Intelligent water distributor 9. Ball seat

**Figure 1.** Schematic diagram of the overall technical solution

When layered parameter testing and flow adjustment are required, a measurement and adjustment command is issued through the host computer client, and the command is sent to the wellhead ground controller through the ground wireless communication network. The surface controller controls the action of the electric control valve according to the control instructions transmitted remotely and according to the wave code code specification, sends wave code information to the underground, and controls the switch and opening of the electric control valve at the same time to realize constant current control and constant pressure control[11].. The downhole intelligent water distributor controls the opening of the faucet according to the received control instructions to realize flow control. Downhole nozzle front pressure, nozzle rear pressure, nozzle opening, flow and other data are transmitted to the ground controller in the form of wave codes, and the ground controller analyzes the downhole wave code commands and sends the analyzed data back to the host computer client, thereby realizing Hierarchical parameter admission.

## 3. Key Technology

### *3.1. Long-distance Wireless Communication Network*

The long-distance wireless communication network is the channel to realize the transmission of data information between the server and the wellhead surface controller, mainly using VPDN technology[12].. Install a customized SIM card that supports VPDN technology on the industrial wireless router inside the electric control cabinet, and build a virtual dedicated channel on the public network that is free from external interference, and can safely access the internal data resources of the enterprise network[13]. The wireless communication network connects the ground controller with the data center server (internal network of the oil field), and the host computer client uses the user name and password to log in and access the server, ensuring the safety and reliability of the remote transmission data. As shown in Figure 2.

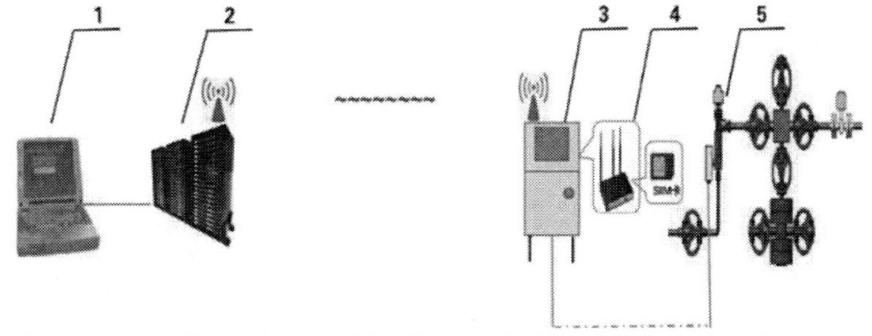


Host computer client 2. Server 3. Electric control cabinet 4. Industrial wireless router 5. Ground controller

**Figure 2.** Schematic diagram of information transmission between ground controller and server

The RJ45 network port of the ground controller is connected to the network port of the wireless industrial router through a network cable; install a customized SIM card (supporting VPDN technology) in the industrial wireless router to realize the connection between the router and the server in the data center; deploy the upper computer software on the server, and then Realize the remote transmission function of encrypted data from the well site data to the data center via the RJ45 network port and industrial wireless router.

### *3.2. Ground Controller*

#### *3.2.1. Structure and Working Principle*

The surface controller is the central mechanism connecting the downhole intelligent water distributor and the host computer terminal, and is installed on the wellhead water injection process[14].. The main functions include: first, according to the collected pressure information and flow information, perform constant current control and constant pressure control on the electric control valve; second, according to the control instructions transmitted remotely, convert the computer control information To control the action of the valve, send wave code control instructions to the downhole; the third is to collect the ground flow, process pressure, water injection pressure, temperature, valve opening, layered flow, and layered pressure data; the fourth is to receive the wave code sent back from the downhole information. It is mainly composed of flowmeter, communication module, chip main control circuit, process pressure gauge, water injection pressure gauge and electric control valve. As shown in Figure 3.

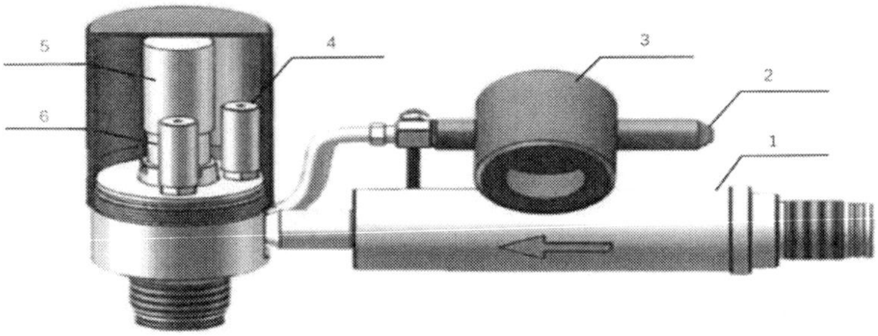


1. Flow meter 2. Communication module 3. Chip main control circuit 4. Process pressure gauge 5. Electric control valve 6. Water injection pressure gauge

**Figure 3.** Structure diagram of ground controller

The ground controller is a device that integrates testing instruments and control systems. The system integrates the sensors of the flow meter, pressure gauge, and valve controller into a local area network system, and then transmits the data to the microprocessor for control processing. The network communication module establishes bidirectional communication between the upper computer processing terminal and the remote management system. The microprocessor will collect data for processing and analysis, control the opening and closing of the electric control valve, and achieve constant current control and constant voltage control. According to the remotely transmitted control instructions, control the action of the electric control valve, send wave code information to the underground, and analyze the collected underground data before transmitting it to the upper computer client.

#### *3.2.2. The Main Technical Parameters*

Flow control accuracy 0.2 %
Flow control range 0.1-50 $m^3/h$
Pressure control accuracy 0.2 %
Pressure control range 0-60 MPa
Water injection temperature ≤80 ℃
Working pressure ≤35 MPa
Operating voltage 24 VDC
Power Consumption 0.5W
Degree of protection IP65
Installation method Welding

### *3.3. Intelligent Water Distributor*

#### *3.3.1. Structure and Working Principle*

The intelligent water distributor is used for underground data collection, data transmission, control command execution, command conversion and transmission, and electric adjustment of water nozzles. It is directly connected to the pipe string and goes down the well together with the pipe string, with one placed in each layer. Mainly composed of upper connector, battery, sensor, data storage module, chip control module, motor, controllable water nozzle, lower connector, water outlet, etc. As shown in Figure 4.

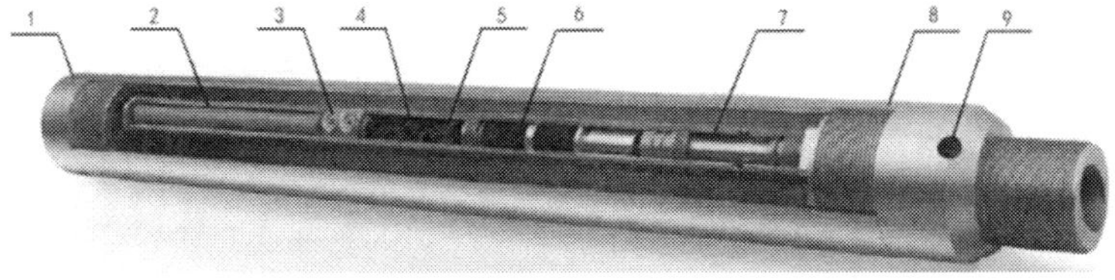


1. Upper connector 2. Battery 3. Sensor 4. Data storage module 5. Chip control module
6. Motor 7. Controllable water nozzle 8. Lower joint 9. Water outlet
**Figure 4.** Schematic diagram of intelligent water distributor structure

The intelligent water distributor adopts high-performance batteries and low energy consumption high torque motors, which can meet normal use for more than 3 years. Sensors can collect data such as pressure in front of the downhole nozzle, pressure behind the nozzle, nozzle opening, flow rate, etc., and transmit them to the surface controller as needed. The intelligent underground water distributor adopts a rotary controllable water nozzle, with a water nozzle opening of 0-16 positions, each with 5 small positions. The water nozzle opening control accuracy is high. According to the actual water quality, scaling characteristics, and water injection well flow rate of the oilfield, corresponding water nozzles and anti-corrosion technologies are matched to effectively prevent the impact of water nozzle corrosion and damage on water injection. The water distributor controls the opening of the water nozzle based on the received control instructions, achieving flow control. At the same time, based on the pressure before and after the water nozzle, calculate the injection pressure difference, and automatically adjust the water nozzle according to the injection pressure difference and corresponding flow rate to maintain a stable injection pressure difference, so that the injection volume can be relatively stable within the designed injection volume. Effectively solve the changes in injection water volume caused by changes in formation pressure, overcome the defects of fixed mode injection, and solve problems such as interlayer differences and water channeling caused by insufficient or excessive injection.

One click sealing inspection function, which sends sealing inspection instructions on the ground, and the water distributors in each layer of the underground achieve automatic sealing inspection. The sealing inspection sequence is from the bottom layer to the top layer, and after the entire well sealing inspection is completed, all the water distributors return to the opening before the sealing inspection. 30 minutes later, the sealing layer water distributor will upload the sealing result curve: layer level+sealing result.

### *3.3.2. The Main Technical Parameters*

Working pressure 35 MPa
Working temperature 150 ℃
Total tool length 1240 mm
Maximum outer diameter 114 mm
Minimum inner diameter 44 mm
Working time >3 years
Connecting threads 27/8TBG
Applicable well depth ≤ 4000 m
Initial state off

### *3.4. Wave Code Communication Technology*

Wave code communication technology is used for data and instruction transmission between ground controllers and underground intelligent water distributors. Wave code communication technology uses water inside the oil pipe as the carrier and uses pressure waves to transmit commands and data, achieving long-distance wireless transmission of data from the surface to the underground and underground to the surface. The wellhead uses the electric control valve switch of the ground controller to control the pressure inside the pipe string to generate specific pressure pulses, which are transmitted to the underground intelligent water distributor through the oil pipe. The intelligent water

Journal of Physics: Conference Series **2594** (2023) 012074 doi:10.1088/1742-6596/2594/1/012074

distributor located underground detects the pressure pulse signal, analyzes it, and executes corresponding action instructions. The underground intelligent water distributor controls the wellhead water injection pressure and flow rate through the opening and closing of the water nozzle to generate specific pulse signals, and the ground controller interprets the underground wave code instructions. The wave code encoding rules are shown in Figure 5.

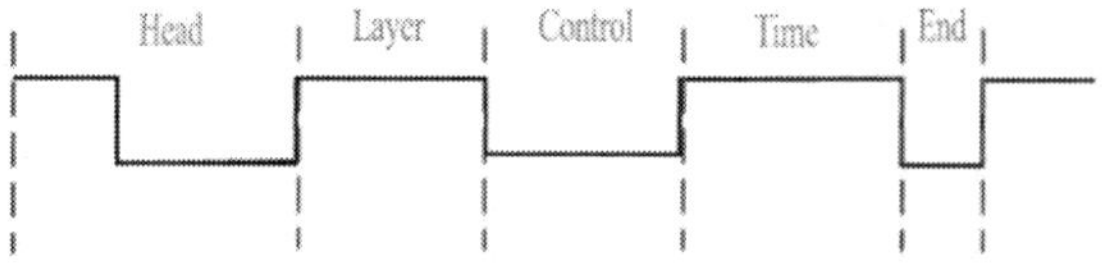


**Figure 5.** Schematic diagram of wave code encoding rules

Wave code communication technology ensures that data transmission and testing operations are not affected by wellbore structure, scaling, corrosion, and other factors of the downhole string, achieving powerful functions such as testing operations with fluid channels and layered water injection in highly deviated wells.

## 4. On site application

### *4.1. Check the sealing performance of the packer*

Taking a three stage separate injection well in Dagang Oilfield of PetroChina as an example. The well depth is 3206m, and the maximum inclination is 29.2 °. It adopts 1 bidirectional anchor, 3 packers, 3 intelligent water distributors, and 1 ball socket. In April 2023, the sealing performance of the injection packer for this well was tested, and the specific testing results are as follows:

The upper computer client sends a one click sealing inspection command to the ground controller, which converts the command into wave code information and sends it to the underground intelligent water distributor. The intelligent water distributor automatically completes the sealing inspection work according to instructions and transmits the sealing inspection results back to the ground controller. The sealing results are analyzed by the ground controller and then transmitted to the server through the ground wireless communication network. The upper computer client monitors the sealing process in real-time. The analysis results show that the sealing test results of the three packers in the entire well are all sealed. As shown in Figures 6, 7, and 8.

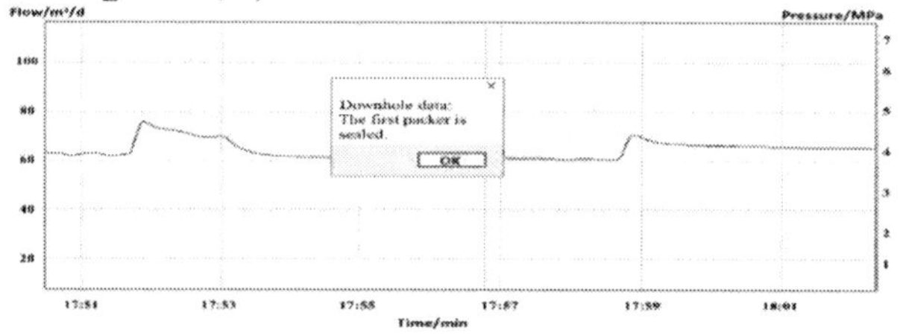


**Figure 6.** Analysis Results of Sealing Inspection of the 1st Packer

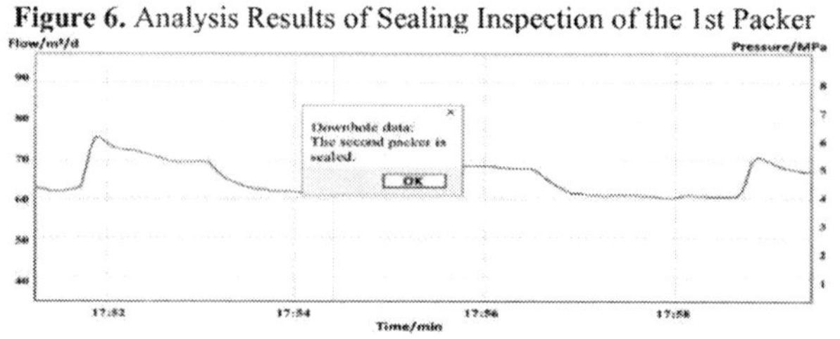


**Figure 7.** Analysis Results of Sealing Inspection of the 2nd Packer

Journal of Physics: Conference Series **2594** (2023) 012074 doi:10.1088/1742-6596/2594/1/012074

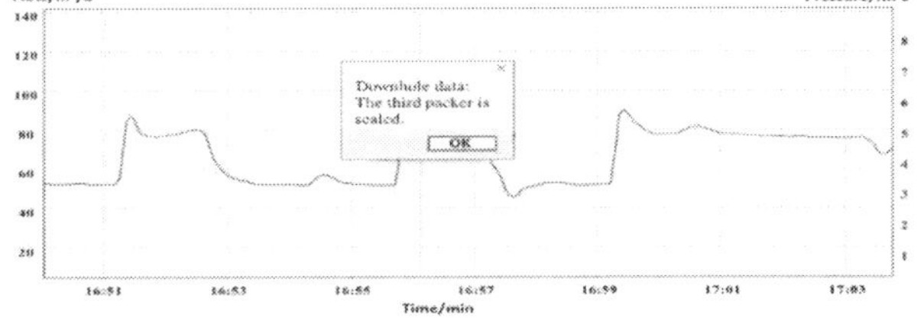


**Figure 8.** Analysis Results of Sealing Inspection of the 3rd Packer

### 4.2. *Flow Testing and Adjustment*

Taking one three-stage separate injection well from Dagang Oilfield as an example. The well depth is 3206m, and the maximum inclination is 29.2 °. It adopts 1 bidirectional anchor, 3 packers, 3 intelligent water distributors, and 1 ball socket. In April 2023, flow testing and allocation was conducted according to injection requirements. The entire well is injected with 130m3/d, with 50m3/d for layer one, 50m3/d for layer two, and 30m3/d for layer three. The specific deployment of traffic testing is as follows:

The testing and debugging instructions are sent to the ground controller through the upper computer client, with injection of 50m3/d for layer one, 50m3/d for layer two, and 30m3/d for layer three. The instructions are encoded by the ground controller and sent to the underground intelligent water distributor. The water distributors in each layer control the opening of the water nozzle according to the instructions, achieving the goal of remote control and flow adjustment. The intelligent water distributor can accurately adjust and record the water injection volume of each layer, and all three sub injection layers meet the injection requirements. The required water injection volume for layer one is 50.0m3/d, and the actual water injection volume is 49.92m3/d, with an error of 0.2%. The required water injection volume for layer 2 is 50.0m3/d, and the actual water injection volume is 50.16m3/d, with an error of 0.3%. The required water injection volume for layer three is 30.0m3/d, and the actual water injection volume is 30.48m3/d, with an error of 1.6%. The actual water injection volume of the entire well is 130.56 m3/d, with an error of 0.4%. The flow test and adjustment results are shown in Figure 9.

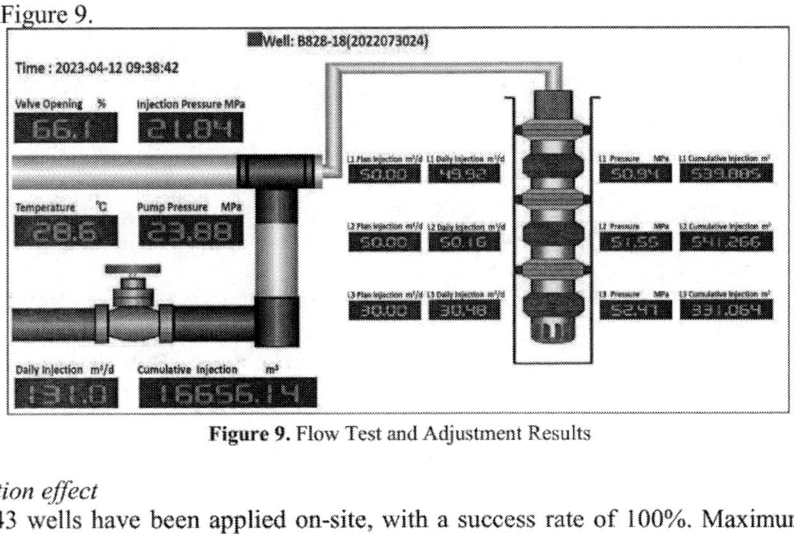


**Figure 9.** Flow Test and Adjustment Results

### 4.3. *Application effect*

Up to now, 43 wells have been applied on-site, with a success rate of 100%. Maximum well depth: 3897m, maximum well inclination: 56.07 °, maximum reservoir temperature: 110 °, maximum validity period: 1493 days. Through automatic measurement and adjustment of stratified flow rate, the qualification rate of separate injection in water injection wells has been maintained at over 90% for a long time. Not only does it achieve real-time continuous monitoring of parameters such as pressure and flow in each layer, but it also achieves remote fine allocation of layered water injection volume and monitoring of the sealing condition of the packer. At the same time, it has the ability to replay and query historical data, providing more detailed and accurate data support for dynamic adjustment of oil reservoirs. This technology reduces the investment in testing and debugging teams and equipment, solves the problem of insufficient testing and debugging team and equipment strength, and saves 3.4 million yuan in testing and debugging costs annually. Compared with existing bridge injection technology, it is not affected by factors such as well site environment, weather changes, and time, and has significant advantages such as high testing efficiency, high success rate, and low management cost.

## 5. Conclusion

(1) This study proposes a remote wireless intelligent injection technology to achieve digital and intelligent management of oil fields, improve water well management level, reduce management and development costs of old oil fields, and lay the foundation for the construction of "digital and intelligent" oil fields.
(2) This technology applies a multi equipment integration solution, and a system comprehensively completes functions such as water allocation, layered flow testing, layered sealing inspection, wireless data transmission, ground data collection, and remote control of water injection wells,
(3) The ground controller uses an electric control valve for constant flow control and constant pressure control, ensuring stable water injection and reducing the problem of over or under injection of water wells caused by pressure fluctuations.
(4) VPDN technology can achieve encrypted data transmission, greatly improving data security.
(5) This technology is not limited by the ground environment and wellbore of the well site, reducing the difficulty of the injection separation process, enhancing the automation control performance, and has strong adaptability for deep and inclined wells.

## References

[1] Li Hong. Reservoir types, development characteristics and development methods of complex fault-block oilfields [J]. Test Production Technology, 1997, 18(3): 8.
[2] Zhao Ming, Cai Mingjun, Pan Hong. Countermeasures and Practices of Fine Water Injection Development Technology in Dagang Oilfield [C]. //Proceedings of the Eighth Bohai Bay Oil and Gas Field Exploration and Development Technology Symposium. 2016:465-471.
[3] Liu He, Pei Xiaohan, Luo Kai, Sun Fuchao, Zheng Lichen, Yang Qinghai. Current status and development trend of separate layer water injection technology for oil and gas field development in China [J]. Petroleum Exploration and Development, 2013, 40(6): 733- 737.
[4] Song Zuchang, Liu Yang, Gai Xubo, Lai Xueming, Li Qiaowei, Dou Shoujin, Wu Lian, Zhang Le. Bridge-type concentric separate injection technology and its application in deep inclined wells [J]. Petroleum Mining Machinery, 2013 ,042(007):62-65.
[5] Song Zuchang, Liu Yang, Gai Xubo, et al. Research and Application of Continuously Adjustable Bridge Eccentric Layered Water Injection Technology [J]. Petroleum Mine Machinery, 2016(3):4.
[6] Liu Y , Wang X , Song Z , et al. Research and Application of Novel Bridge Concentric Zonal Water Injection Technology[C]// International Petroleum Technology Conference. 2021.
[7] Yi Chen. Introduction of Zonal Water Injection Technology in Offshore Oilfields[J]. World Journal of Innovation and Modern Technology,2023,6(1).
[8] Chen Yi. Introduction of Zonal Water Injection Technology for Highly-deviated Well in Offshore Oilfields[J]. Journal of Research in Science and Engineering,2023,5(2).
[9] Cheng Hanlie,Yang Dong,Lu Changsong,Qin Qiang,Cadasse David. Intelligent Oil Production Stratified Water Injection Technology[J]. Wireless Communications and Mobile Computing,2022,2022.
[10] Wan Juye,Zhuang Yuan,Huang Yuhan,Qian Yejian,Qian Lijun. A review of water injection application on spark-ignition engines[J]. Fuel Processing Technology,2021,221.
[11] Wu Zhijun,Xie Wei,Yu Yang,Li Liguang,Deng Jun. Comparison of spray characteristics of

gasoline and water-in-gasoline mixture at elevated fuel temperature conditions[J]. Fuel,2021,304.

[12] Zhang Xu,Zhu Ruibin,Li Gaofeng,Jiang Yan,Xiong Wei,Zhao Fei,Wang Jianning,Lin Chunqing,Liu Qilun,Li Ran. Application of wireline intelligent separate-layer water injection technology in Huabei Oilfield[J]. Journal of Physics: Conference Series,2021,1983(1).

[13] Dib Ghady,Haberschill Philippe,Rullière Romuald,Revellin Rémi. Thermodynamic investigation of quasi-isothermal air compression/expansion for energy storage[J]. Energy Conversion and Management,2021,235.

[14] Sun Xin,Nan Rong Li,Shu Bo Zhao,Wang Hang,Zhu Li Juan,Han Li Hong. Corrosion Behavior for Casing in Producing Well with Water Injection Technology during Long-Term Service[J]. Materials Science Forum,2020,5930.

# Study on Optimal Design of Constant Current Water Distributor

**Jing-qi WANG 1,2, Guang-lei GAO 1,2, Yin TONG 1,2, Qing-feng YUE 1,2, Chao WANG 1,2 (✉)**

1 Daqing Oilfield Production Technology Institute, Daqing Oilfield Limited Company,Daqing,China
2 Heilongjiang Provincial Key Laboratory of Oil and Gas Reservoir
443696466@qq.com

**Abstract**. Daqing Field has entered the late stage of ultra-high water cut development. The number of subzones of water injection wells is increasing year by year, and the number of zones of high-efficiency annual survey has exceeded 240,000. The testing capacity has reached a bottleneck, the contradiction among test efficiency, test team and qualified rate of water injection is becoming more and more obvious. The constant flow water distributor can keep the water injection quantity of the applied layer basically unchanged under the frequent fluctuation of water injection pressure, the water injection zone will be changed from repeated adjustment to little adjustment or even no adjustment, greatly reducing the testing workload and improving the testing efficiency. In order to further improve the performance of constant-current water distributor, the stress simulation of valve spool of constant-current water distributor was carried out. Based on the mechanical analysis of the valve core of constant flow distributor, the force acting on the valve core of constant flow distributor is obtained by numerical simulation, and the accuracy of the simulation results is verified by laboratory experiments. According to the requirements of starting pressure difference and spring stiffness of constant current water distributor, the constant current water distributor is optimized and perfected. The experimental results show that the error is less than 5% , which meets the requirements of field application. This study provides a basis for further improving the design of constant-current water distributor. Up to now, 392 wells have been tested with constant-current distributor, the effect of constant-current is good, and the average efficiency of single well is increased by 28.3%.

## 1. Introduction

Daqing oil field into the extremely high aquifer period, the relationship between the injection and mining is more complex, inter-layer, intra-layer and plane contradictions are more prominent, layer division needs to be more detailed, water distribution accuracy, measurement cycle and other work requirements are higher, the test workload greatly increased[1-6].Constant flow water distribution technology can realize constant flow water distribution, which can not only reduce the test and allocation layer, but also limit the flow fluctuation of the limiting layer, reducing the test workload and greatly improving the test efficiency. Therefore, in order to improve the constant flow effect of the constant flow dispenser, a simulation study on force of valve core in water distributor with constant flow rate was carried out to meet the requirements of field injection.

## 2. Structure Principle

The constant current water distributor is mainly composed of fishing rod, pressure cap, cam cavity, cam, water distributor body, spring, valve core, water nozzle, screen plug, etc. the structure is shown in Fig 1.

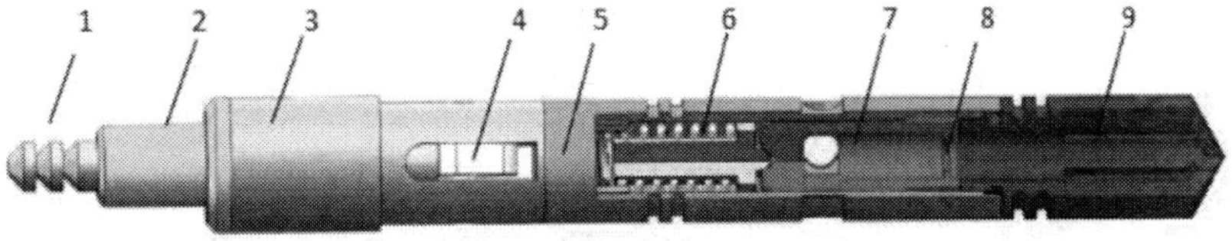


**Fig. 1.** Schematic diagram of the structure of the constant flow water distributor

1 - fishing rod; 2 - pressure cap; 3 - cam cavity; 4 - cam; 5 - main body of water distributor; 6 - spring; 7 - valve core; 8 - water nozzle; 9 - screen plug

The constant flow principle of the constant flow distributor is: when the injected water flows through the constant flow distributor, the injection pressure and the spring force act on the two ends of the valve core at the same time, making the spool move back and forth with the change of the pressure difference between the two ends; when the injection pressure increases or the formation pressure decreases, the pressure difference at both ends of the valve core increases, the valve core moves to the left and compresses the spring. At this time, the water outlet area decreases and the flow returns to the original set value; When the water injection pressure decreases or the formation pressure increases, the differential pressure decreases and the valve core moves to the right. At this time, the water outlet area increases and the flow returns to the original set value[7-10]. The working principle is shown in Fig 2.

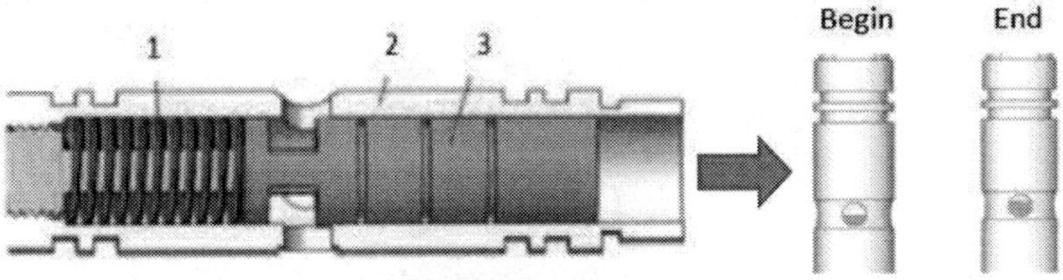


a.Spool section view b.Position diagram before and after spool movement

**Fig. 2.** Working principle diagram of constant current water distributor

1 - spring; 2 - main body; 3 - valve core

## 3. Spool Force Analysis

### *3.1. Stress Condition of Valve Cor*

The valve core of the constant flow distributor is subjected to two forces, based on the cross-section A of the valve core, one is the force Fw from the fluid to the valve core, the other is the spring force Ft, and the two forces follow the movement of the valve core.and its stress condition is shown in Fig 3. The valve core is in equilibrium under the action of the two forces, so as to ensure the constant flow at the outlet of the constant flow water distributor and achieve the constant flow effect. Through numerical simulation, the force of fluid on the valve core when the valve core is in the equilibrium position is obtained, and then the elastic value provided by the spring is obtained.

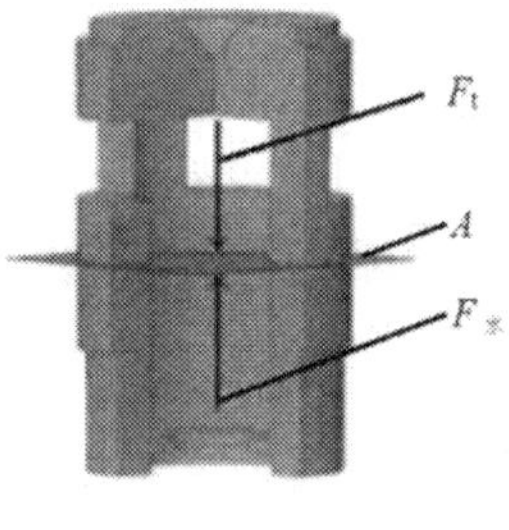


**Fig. 3.** Schematic diagram of the force of the spool

### *3.2. Internal Flow Field Characteristics of Constant Flow Water Distributor*

The three-dimensional geometric model of the fluid domain of the constant flow water distributor is established in the ratio of 1:1. In order to ensure the flow characteristics of the flow field, the water outlet and inlet are lengthened, and the rest are modeled and analyzed according to the actual situation under the working state of the valve core. The schematic diagram of the fluid domain of the constant flow water distributor is shown in Fig 4.

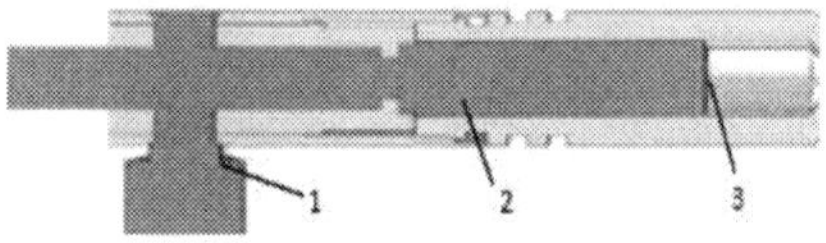


**Fig. 4.** Schematic diagram of the fluid area of the constant flow water distributor

1 - water outlet; 2 - fluid domain; 3 - water inlet

The grid independence is calculated for the fluid domain of the constant flow distributor. When the number of grids divided is about $61.8\times10^4$, as the number of grids increases, the calculation error is less than 1% and no longer increases.Considering the calculation time and calculation results comprehensively, the number of grids selected to be divided is $61.8\times10^4$ [11], and the grid division is shown in Fig 5.

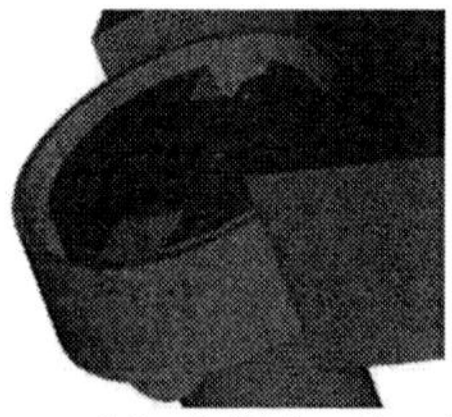

**Fig. 5.** The grid division diagram of the fluid domain of the constant flow distributor

### *3.3. Force Analysis of Spool*

There are 6 force-bearing surfaces of the spool in the fluid domain, as shown in Fig 6.

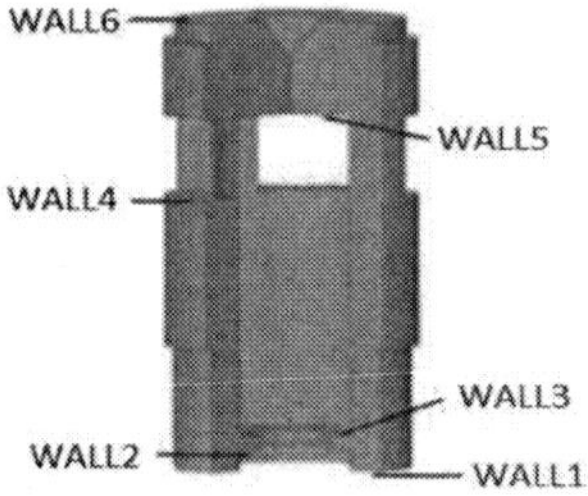


**Fig. 6.** Schematic diagram of valve core stress surface position

Journal of Physics: Conference Series **2594** (2023) 012075 doi:10.1088/1742-6596/2594/1/012075

Among them, the force surfaces of WALL1, WALL2, and WALL5 give upward force to the valve core, and the force surfaces of WALL3, WALL4, WALL6 give downward force to the valve core. The shape of each force surface is shown in Fig 7, and the area data of each force surface as shown in Table 1.

a.WALL1 b.WALL2 c.WALL3 d.WALL4 e.WALL5 f.WALL6

**Fig. 7.** Shape of each stress surface of valve core

**Table 1.** Area data of each stress surface of valve core

| Stress surface | WALL1 | WALL2 | WALL3 | WALL4 | WALL5 | WALL6 |
|---|---|---|---|---|---|---|
| Area ($mm^2$) | 146 | 75 | 75 | 123 | 194 | 211 |

The area of upward pressure on the valve core is equal to the area of downward pressure on the valve core. Taking the daily injection volume of $30m^3$, the valve core is 3mm at the water outlet (the initial position is 2mm), and the pressure at the water outlet is zero as an example, the pressure cloud diagram of the fluid domain is simulated, as shown in Fig 8.

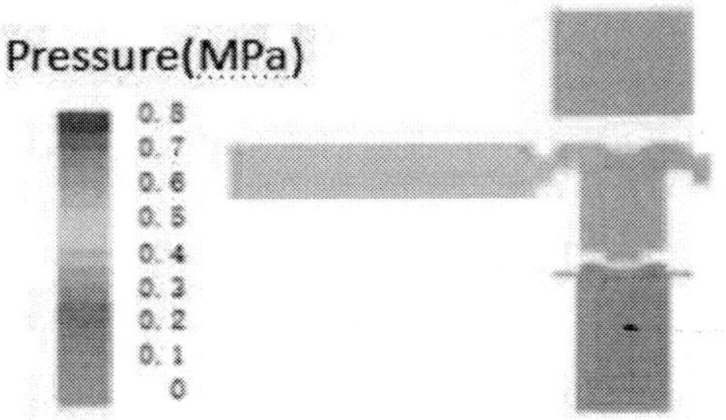


**Fig. 8.** The pressure cloud diagram of the fluid domain of the constant flow water distributor

In order to analyze and calculate the force on the spool, it is necessary to further analyze the pressure on each force surface of the spool.The pressure of each surface of WALL1~WALL6 is simulated, and the pressure cloud diagram of each surface is shown in Fig 9, and the pressure and pressure data are shown in Table 2.

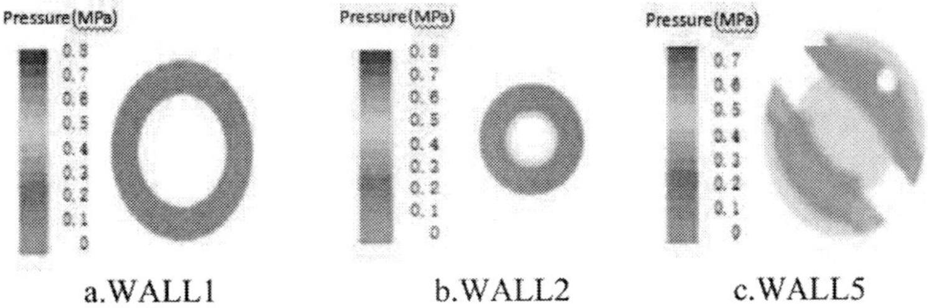


a.WALL1 b.WALL2 c.WALL5

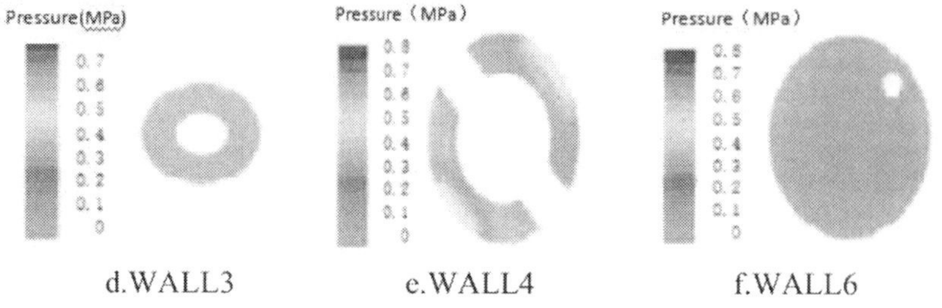


**Fig. 9.** Pressure nephogram of each stress surface of valve core

**Table 2.** Pressure and pressure data of each stress surface of valve core

| Stress surface | WALL1 | WALL2 | WALL3 | WALL4 | WALL5 | WALL6 | Total |
|---|---|---|---|---|---|---|---|
| Pressure (MPa) | 0.70 | 0.68 | 0.22 | 0.25 | 0.30 | 0.26 | |
| Pressure (N) | 63.210 | 32.171 | -10.415 | -17.838 | 38.499 | -37.720 | 67.907 |

Note: The negative sign in the table represents the direction

It can be seen from the table that when the valve core position is 3mm and the daily injection volume is $30m^3$, the total force of the valve core is 67.907N, and the direction is the compression spring direction.At this time, 67.907N of elastic force needs to be provided by the spring force to make the valve core reach a balanced state.

**Table 3.** Force data of spring when valve core is at different positions

| Spool position (mm) | 2.0 | 2.5 | 3.0 | 3.5 | 4.0 | 4.5 | 5.0 | 5.5 |
|---|---|---|---|---|---|---|---|---|
| Spool displacement (mm) | 0 | 0.5 | 1.0 | 1.5 | 2.0 | 2.5 | 3.0 | 3.5 |
| Spring force (N) | 65.438 | 66.003 | 67.907 | 69.810 | 71.317 | 72.118 | 73.765 | 76.631 |

## 4. Indoor and Field Tests

### *4.1. Verification of Simulation Results*

The accuracy of the simulation results is verified by indoor experiments. Fix the valve core at different positions from the water outlet, change the inlet pressure of the constant flow water distributor to make the flow constant at the daily injection volume of $30m^3$, and compare the experimental results with the simulation data, as shown in Table 4. It can be seen from the data in the table that the simulated inlet pressure is very close to the experimental inlet pressure, and the average error is within 5%, which proves the accuracy of the simulation method.

**Table 4.** The inlet pressure data table when the valve core is at different positions from the water outlet

| Spool position (mm) | 2.0 | 2.5 | 3.0 | 3.5 | 4.0 | 4.5 | 5.0 | 5.5 |
|---|---|---|---|---|---|---|---|---|
| Experimental pressure (MPa) | 0.60 | 0.62 | 0.62 | 0.63 | 0.64 | 0.86 | 1.36 | 5.80 |
| Simulated pressure (MPa) | 0.593 | 0.596 | 0.599 | 0.641 | 0.650 | 0.890 | 1.470 | 5.500 |
| Error% | 1.17 | 3.87 | 3.39 | -1.75 | -1.56 | -3.49 | -8.09 | 5.17 |

### *4.2. Calculation of Water Nozzle Area*

The water nozzle in the constant flow distributor has an incompletely contracted thin-walled small hole structure, and its flow calculation formula is [12]:

$$Q = C_{\mathrm{d}} A \sqrt{\frac{2\Delta P}{\rho}} \tag{1}$$

In the formula: Q—daily injection volume, $m^3$;
Cd—flow coefficient;
A—Water nozzle area, $mm^2$;
$\Delta P$—pressure difference between inlet and outlet, MPa;
$\rho$—The density of water, $kg/m^3$.
In field applications, the pressure difference between the inlet and outlet of the constant flow distributor is required to be within the range of 0.5 to 1.0 MPa. Formula (1) can be used to calculate under different constant flow conditions.The water nozzle area of the constant flow water distributor when the starting pressure difference is 0.5 and 1.0MPa respectively.

### *4.3. Value of Spring Stiffness*

The rectangular spring can effectively prolong the service life of the constant current water distributor [13]. Because the rectangular spring with an outer diameter of 11mm is applicable in the constant current water distributor, the stiffness value of the rectangular spring meeting the service conditions of the constant current water distributor is limited, specifically: 2.6N/mm, 3.1N/mm, 4.1N/mm, 4.9N/mm and 7N/mm.

### *4.4. Results of Laboratory Experiments*

According to the numerical simulation results of the force on the valve core of the constant flow water distributor and the requirements of the starting pressure difference and spring stiffness of the constant flow water distributor, the constant flow water distributor is optimized and the indoor experiment is carried out. Taking the experimental results of the constant flow water distributor with a daily injection volume of $30m^3$ as an example, the experimental data are shown in Table 5 and the test curve is shown in Fig 10.

**Table 5.** Indoor experimental results of a constant flow water distributor with a daily injection volume of $30m^3$

| Experimental pressure (MPa) | 0.5 | 1.0 | 1.5 | 2.0 | 2.5 | 3.0 | 3.5 | 4.0 | 4.5 | 5.0 |
|---|---|---|---|---|---|---|---|---|---|---|
| Daily injection ($m^3$) | 28.8 | 29.2 | 28.9 | 29.4 | 30.5 | 30.7 | 31.5 | 30.2 | 31.3 | 30.9 |
| Flow error% | 4.00 | 2.67 | 3.67 | 2.00 | 1.67 | 2.33 | 5.00 | 0.67 | 4.33 | 3.00 |

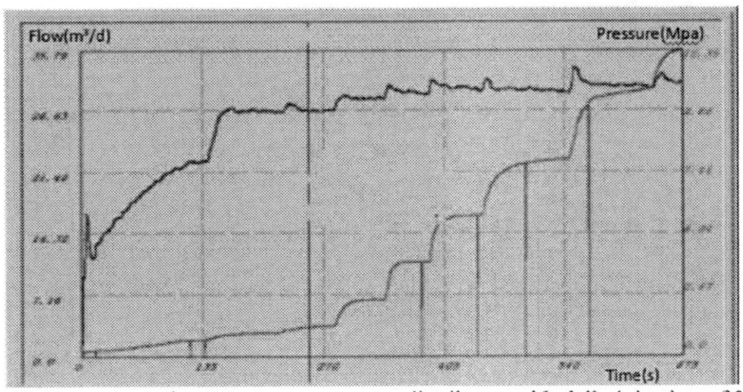


**Fig. 10.** Test curve of constant current water distributor with daily injection of $30m^3$

It can be seen from the experiment that the constant current performance of the optimized constant current water distributor is improved. Under different injection pressure difference, the flow error is less than 5%, which meets the requirements of field application.

### *4.5. Field Test Results*

Up to now, a total of 392 wells have been tested in the field, the success rate of the field test is 96% , the longest service life is more than 13 months, and the average constant flow error of single well is less than 20% , which meets the requirement of field water injection, the average efficiency of single well measurement and adjustment is increased by 28.3% .

## 5. Conclusion

(1)Through the force analysis of the valve core of the constant flow water distributor, the internal flow field characteristics of the constant flow water distributor are simulated and studied, and the force model of the valve core of the constant flow water distributor is established to obtain the constant flow state when the valve core is in different positions.The force condition of, and the comparison between experimental results and simulation results verifies the accuracy of the simulation method.
(2)According to the numerical simulation results of the force on the valve core of the constant flow water distributor, according to the existing rectangular spring stiffness value and the starting pressure difference requirements of the constant flow water distributor, more suitable springs and nozzles are selected for the constant flow water distributor of different dosages to improve the constant current effect of the constant current distributor can extend the service life of the constant current distributor.
(3)Through indoor experiments, it is verified that the constant flow performance of the optimized constant flow water distributor is improved.
(4)Through extensive on-site applications, the accuracy of the simulation method and the feasibility of constant flow water injection technology have been further verified.
(5)In the next step, the constant flow water distributor with more injection will be optimized, and field experiments will be carried out to further verify the applicability of the simulation study on the stress of the valve core of the constant flow water distributor.

**References**

[1] Wang Feng Shan,Jia De Li,Zhang Shu Jin, et al. Identification-Free Adaptive Separate Layer Water Injection Control Based on Expert Amendment [J].Applied Mechanics and Materials. Volume 44-47 , Issue 44-47 . 2010. PP 1470-1475.
This reference has two entries but the second one is not numbered (it uses the 'Reference (no number)' style.
[2] Yuhai Cui,Rui Xia,Yongqiang Kong.Modelling and structure optimisation on throttle tube of pre-throttle water distributor [J]. nternational Journal of Modelling, Identification and Control. Volume 36 , Issue 3 . 2020. PP 248-255.
[3] JIA Deli,PEI Xiaohan,LIU He,et al.Research on the Flow Control Strategy of Water Distributor in Water Injection Well [C].The thirty-three th Chinese control conference,2014.
[4] Liu Yikun, Bi Yongbin, Sui Xinguang. Prediction of oilfield development indexes in the later stage of high water cut [J]. Daqing Petroleum Geology and development,27 (1): 58-60(2008).
[5] Zhu Zhenkun, Zhao Mingxin, Ma Qiang, et al. Development of a downhole non-contact intelligent dispenser[G]. Oil Production Engineering Research Institute of Daqing Oilfield Co., Ltd. Oil Production Engineering Collection 2016 First Series. Beijing: Petroleum Industry Press, 1-4(2016).
[6] Li Lidong. Plug type multi parameter tester [G]//Oil Production Engineering Research Institute of Daqing Oilfield Co., Ltd. oil production engineering anthology, Volume 1, 2016. Beijing: Petroleum Industry Press, (1): 5-9(2016).
[7] Wu Jianming, Xie Mingzheng, Yin Sihua, et al. Characteristics analysis of constant flow and

quantitative water nozzle[J]. Oil and Gas Field Surface Engineering,26(8):31-32(2007).
[8] Wang Haiying, Zhang Xuan. Application of latent analysis eccentric constant flow plug in oil production engineering [J]. Petroleum Instrument,23 (5): 78-79(2009).
[9] Luan Yuntian.The application of constant current plugging device in layered water injection technology[J].Inner Mongolia Petrochemical Industry, (6):102-104(2014).
[10] Zhang Yun. Development and application of new constant current water distribution plug [J]. Petroleum pipe and instrument, (2): 70-73(2018).
[11] Zong Heming, Han Yi, Wang Jingqi, et al. Simulation research on the internal flow field of the constant current plugging device[G]//Oil Production Engineering Research Institute of Daqing Oilfield Co., Ltd. Oil Production Engineering 2019 First Series. Beijing: Petroleum Industry.
[12] Feng Jiansheng. Analysis of measurement principle and flow calculation problems of orifice flowmeter [J]. Coal mine safety,45 (10): 218-221(2014).
[13] Qu Yanzhe, Zhang Guilian. Stability of cylindrical spiral spring with rectangular cross-section of spring wire[J].Journal of Harbin Institute of Technology,34(1):113-115(2002).

# Structural Optimization Design of Throttle Valve Core for Intelligent Layered Polymer Injection String Controlled By Cable

**Ying-ying Wang(✉) , Jian-guang Chen, Zhi-xue Qiao, Zhao-yun Yang, Zu-chang Song, Hai-chao Yang**

Oil production technology research institute of PetroChina Dagang Oilfield, Tianjin China
33515098@qq.com

**Abstract**. Optimization Design of Throttling Valve Core Structure for Wired Intelligent Layered Polymer Injection String Abstract: Wired intelligent layered polymer in-jection technology can improve the coverage of polymer flooding, alleviate inter-layer conflicts in reservoir development, and improve crude oil recovery. It is an important component of tertiary oil recovery. However, the layered polymer in-jection technology currently used in most domestic oilfields is still the method of fishing and changing the injection nozzle for testing and adjustment. The alloca-tion efficiency of injection volume testing for each layer is slow, with large er-rors, and it is impossible to monitor polymer injection parameters throughout the process. To this end, research has been carried out on the cable controlled intelli-gent layered polymer injection string. By optimizing a highly efficient and low viscosity loss streamlined layered polymer injection throttling device with contin-uously adjustable gaps, the adjustment stroke is shorter and the measurement and adjustment efficiency is higher. Using finite element simulation, the low shear throttle control unit was studied and optimized. By moving the polymer core up and down slightly in the axial direction, the relative gap between the polymer core and the working cylinder was changed to achieve flow adjustment. At the same time, it is necessary to complete the direct reading measurement and adjustment process to form a single pipe layered polymer injection measurement and adjust-ment integration technology. The cable controlled intelligent layered polymer in-jection matching technology connects with the ground intelligent control system through cables to achieve real-time monitoring and online control of polymer flow, pressure, temperature, and other parameters, improving monitoring accura-cy and efficiency, and providing technical support for the construction of the "Digital Intelligence Oilfield".

## 1. Introduction

Layered polymer injection process is an important means of tertiary oil recovery, which can increase the coverage of polymer flooding, alleviate interlayer conflicts, and improve crude oil recovery rate [1]. The layered polymer injection process in Dagang Oilfield has gone through three stages of development: initiation, development, and integration. The intelligent layered polymer injection technology is still in the initial stage of exploration and research [2]. At present, the main problems faced by the layered polymer injection process are: the technology is relatively backward, and the field is still dominated by general pipe string and oil sleeve surface polymer injection; Low success rate of testing and debugging: During the process of fishing and plugging testing and debugging, it is easy to encounter obstacles and obstacles, resulting in a low success rate; The difficulty of pipe string sealing inspection is high, and the success rate of mechanical sealing inspection and setting is low, with potential accidents; Unable to meet the needs of reservoir dynamic monitoring [3]. Therefore, based on the bridge eccentricity+measurement and adjustment linkage polymer injection process, and drawing on the idea of intelligent layered water injection technology, research on the supporting technology of intelligent layered polymer injection with cables is carried out. This technology is a key means to achieve online intelligent control of layered polymer injection wells, and to achieve remote monitoring and control of the base and wellhead. The intelligent layered polymer injection technology with cables mainly includes two parts: ground control and downhole tools. The ground is controlled by ground control equipment, data transmission system, and intelligent layered polymer injection software control system. The core equipment is an integrated intelligent polymer mixer, which integrates flow meters and regulating cores [4].

## 2. Research and Experiment on Underground Polymer Flow Measurement Technology

Polymer solution downhole measurement technology is one of the core technologies of layered injection intelligent measurement and commissioning technology [5]. The main downhole permanent flowmeters used in stratified injection Wells are vortex street, electromagnetic, differential pressure and ultrasonic. Differential pressure flowmeters calculate flow rate by throttling differential pressure, which causes viscosity loss and is not suitable for polymer flow measurement. The flow test of vortex flowmeter requires Reynolds number to be in the range of 5×103 ~ 7×104, in which the linearity is the best when Reynolds number is in the range of 2×104 ~ 7×106 [6]. However, when vortex flowmeter is used for downhole polymer measurement, the Reynolds number of polymer is less than 5000 based on polymer viscosity calculation commonly used at present [7]. To solve this problem, indoor tests of vortex flowmeter and electromagnetic flowmeter are carried out to select suitable polymer downhole measuring equipment [8].

### *2.1. Indoor selection test of flow meter*

Ground tests were conducted on vortex flow meters and electromagnetic flow meters, using polymer solutions from XX polymerization station. The experimental process is as follows: Two types of flow meters are connected in series in a pipeline, and the performance of the flow meter is determined by changing the flow rate passing through the pipeline, observing the error between the flow meter test results and the basic flow rate.

#### *2.1.1. Test conditions*

XX aggregation station; 19 million molecular weight; Concentration 1500mg/L

#### *2.1.2. Test process*

Connect two types of flow meters in series in a pipeline, change the process reference flow rate from 5-80 m3/d, and observe the flow meter test results.

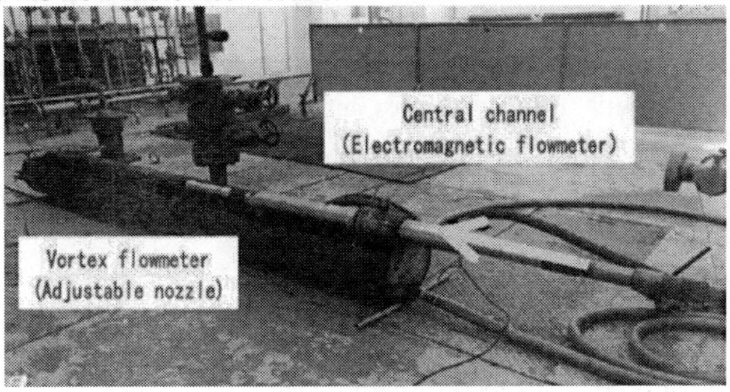


**Fig. 1.** Ground Series Test of Vortex Flowmeter and Electromagnetic Flowmeter

### 2.1.3. Experimental results

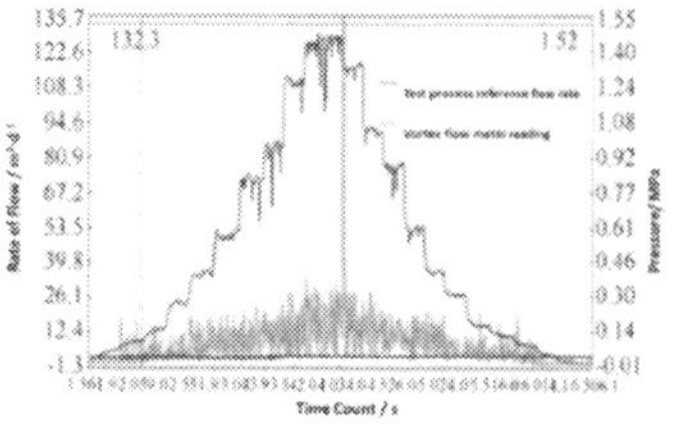

**Fig. 2.** Flow rate variation curve of vortex flowmeter test

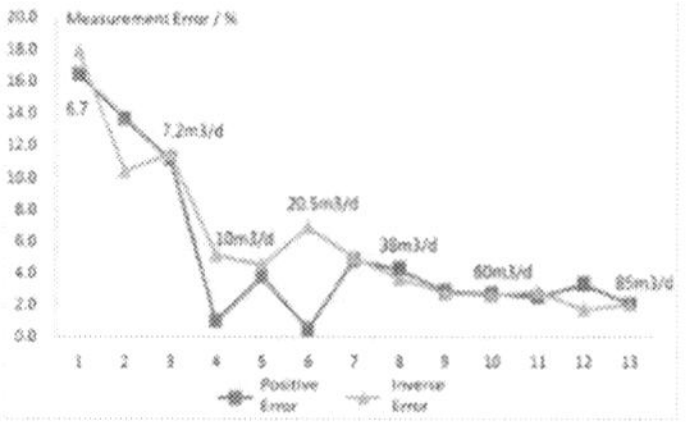

**Fig. 3.** Flow rate variation curve of electromagnetic flowmeter test

(1) When the flow rate changes during the vortex flowmeter test process, the vortex flowmeter reading does not follow the change and is not suitable for polymer flow measurement;

(2) Electromagnetic flow meters are suitable for polymer flow measurement with a wide range and can maintain high measurement accuracy.

## 2.2. Electromagnetic flow meter measurement accuracy test

The working principle of electromagnetic flow meters is to measure the flow rate of conductive liquids based on Faraday's law of electromagnetic induction. Different flow rates generate different induction electric types in a magnetic field. In principle, electromagnetic flow meters are suitable for polymer flow measurement because they are a flow rate measurement method [9]. According to the actual underground working conditions [10], two types of underground electromagnetic flow meters were selected for the experiment: one is a central channel electromagnetic flow meter with an internal diameter of 46mm, and the flow rate of each layer is obtained by decreasing layer by layer. The advantage of this flowmeter is that it has a wide applicable flow range and is not easily clogged; The disadvantage is that obtaining single-layer traffic requires a gradual decrease in accuracy, especially for small traffic. Another type is a small diameter electromagnetic flowmeter, with a diameter of 12mm, which ensures direct testing of layered polymer flow under the condition of a center of 46mm. The advantage of the second type of flowmeter is that it can achieve single-layer direct testing, ensuring a central channel, and high testing accuracy.

### 2.2.1. Multiple indoor tests were organized for two types of electromagnetic flow meters:

(1) Electromagnetic flow meter with inner diameter of 12 and 17mm (single layer direct testing), range of 5-30m3/d

(2) Electromagnetic flow meter with an inner diameter of 25mm (single layer direct testing), with a range of 5-50m3/d

(3) 46mm electromagnetic flow meter for internal well passage (measuring single layer with decreasing method), with a range of 5-120m3/d

— ***Note: Test conditions of 19 million and concentration of 1500mg/L***

### 2.2.2. Experimental conclusion:

(1) The central channel electromagnetic flowmeter is suitable for polymer flow measurement, with a wide range and can maintain high accuracy above 10m3/d. When measuring above 10m3/d, the polymer testing accuracy is within+5%, and when measuring below 10m3/d, the polymer testing error is within 20%. Meanwhile, it was found through experiments that when the flowmeter calibrated in water is directly used for polymer measurement, the overall measurement value is relatively small and requires secondary calibration.

On the right side of the red line are the calibration results in an aquatic environment, with testing errors within+5%. When the calibrated central channel electromagnetic flowmeter is directly used for polymer measurement, the maximum test error is nearly 25%, and secondary calibration is required in the polymer environment.

(2) The internal diameter of a small diameter electromagnetic flowmeter has a significant impact on the polymer flow range, and the measurement range of a small diameter electromagnetic flowmeter is much smaller than that of a water medium. Indoor experiments have shown that the electromagnetic flow meter with an inner diameter of 12mm has a testing range of 5-30m3/d under polymer conditions. After exceeding 30m3/d, the flow fluctuation is significant, but the average value is in good agreement with the actual value.

## 3. Research and Experiment on the Throttling Core of Intelligent Injection Dispenser

Polymer solution is a kind of pseudoplastic fluid of non-Newtonian fluid [11]. At low flow rate or at rest, it appears viscous because of their entanglement and high viscosity [12]. However, when the flow rate increases, these relatively scattered chain shaped particles exhibit a phenomenon of shear thinning due to the shear stress between the flow layers [13]. When the polymer solution in the injection well passes through the water injection nozzle, it will undergo shear degradation, causing adhesive damage, and is also prone to blockage [14]. Therefore, it is necessary to conduct special research on the injection joint nozzle.

### *3.1. Optimization of Layered Polymer Injection Throttling Valve Core Structure*

At present, the commonly used throttling structures in domestic oil fields include [15]: annular groove structure and shuttle structure, both of which achieve throttling by adjusting the length of the core entering and exiting the chamber.

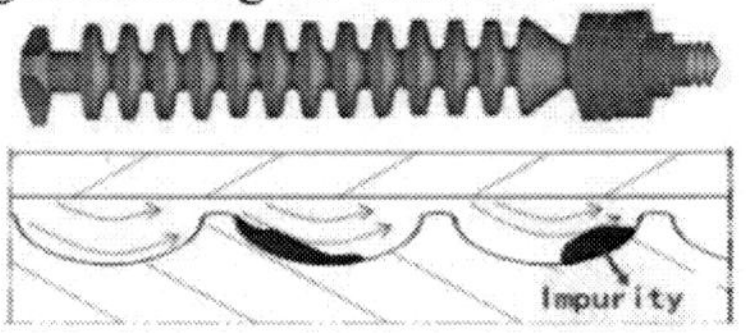


**Fig. 4.** Annular groove structure throttle valve core

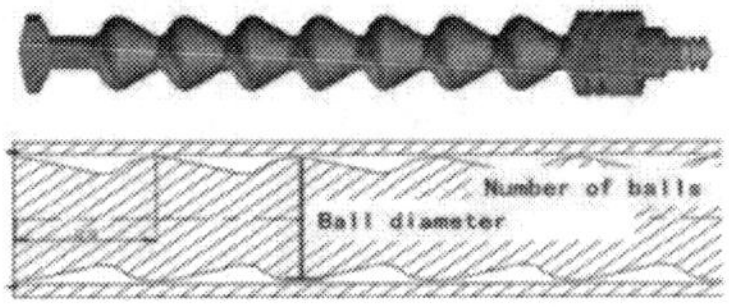


**Fig. 5.** Shuttle shaped ball structure throttle valve core

On the basis of these two structures（Fig. 6）, the shuttle shaped throttle core is optimized to be a streamline throttle core [16]: it is mainly to change the smooth plane structure of the chamber into an independent ring groove structure, each core is located in a ring groove separately, and moving half a vertical well distance can achieve full sealing to full opening [17]. Compared to before, the travel distance of the core is greatly shortened, and the testing time is correspondingly shortened to within 15 minutes.

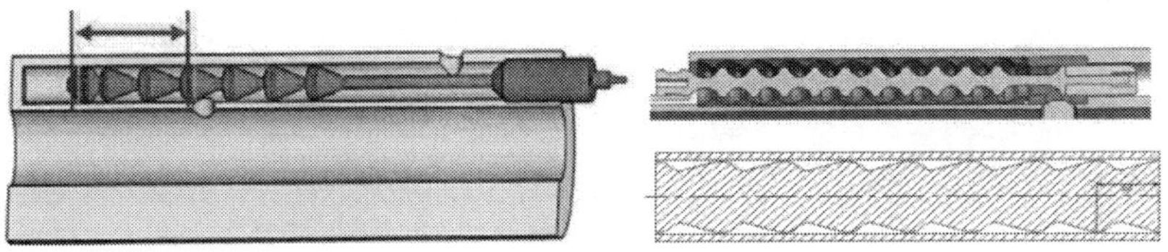

**Fig. 6.** Structure Diagram of streamline Core

**Table 1.** Technical parameters of throttle cores with different structures

| Serial Number | Tool Structure | Technical Parameter | | |
|---|---|---|---|---|
| | | Length/mm | Inner diameter/mm | Inner diameter of offset hole/mm |
| 1 | Annular groove structure | 1640 | 46 | 25 |
| 2 | Shuttle structure | 1750 | 46 | 20 |
| 3 | Streamline structure | 1160 | 46 | 20 |

### 3.2. *Throttling core performance test*

The throttling core flow channel is a complex structure, making it difficult to provide reasonable evaluation results in theory [18]. Currently, there are mainly two methods in China: indoor inspection evaluation and numerical simulation [19].

#### 3.2.1. *Indoor testing*

Indoor experimental evaluation: A detection process design and a polymer adhesion detection device were designed.

Indoor hydraulic characteristics tests were conducted on three types of throttling cores, using polymer solutions from XX polymerization station. Compared with the throttling pressure difference and viscosity loss rate, the streamline structure has a low viscosity loss rate and a large throttling pressure difference, which meets the design standard requirements [20]. During the detection process, real-time adjustments are made to the various parameters of the core using direct manual methods. Based on the continuously adjusted data, the changes in the main parameters of the adjustable throttle core, such as viscosity loss, displacement, and throttle pressure difference, are obtained

Test conditions: Injection volume of polymer flooding reagent is 30 m3/d, injection concentration is 2500 mg/L, and molecular weight of polymer flooding reagent is 25 million.

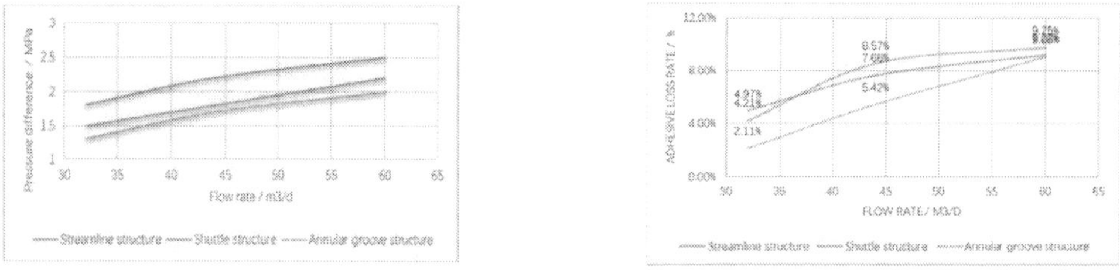


**Fig. 7.** Comparison of hydraulic characteristics of throttle cores with different structures

#### 3.2.2. *Evaluation of indoor test results:*

From the comparison of indoor fishing success rate, hydraulic characteristic parameters throttle pressure difference, and viscosity loss rate, streamline and shuttle shaped throttle valves have higher fishing success rate and better performance.

**Table 2.** Indoor test data table for throttle cores with different structures

| Process type | One time fishing success rate/% | Throttling pressure difference/MPa | Adhesive loss rate/% |
|---|---|---|---|
| Streamline | 100 | 1.3 | 7.9 |
| Shuttle shaped | 87 | 1.12 | 8.5 |
| Circular groove | 18 | 0.03 | 12 |

#### *3.2.3. Numerical Simulation*

In order to obtain the throttling pressure drop caused by different streamline throttling core sub structures when HPAM solution flows through, the large-scale finite element engineering calculation software [21] was used to conduct numerical simulation under different number of slots. The finite element model of streamline runner, the mesh division and boundary conditions of the calculation model, and the structure diagram of the intelligent polymer dispenser are designed.

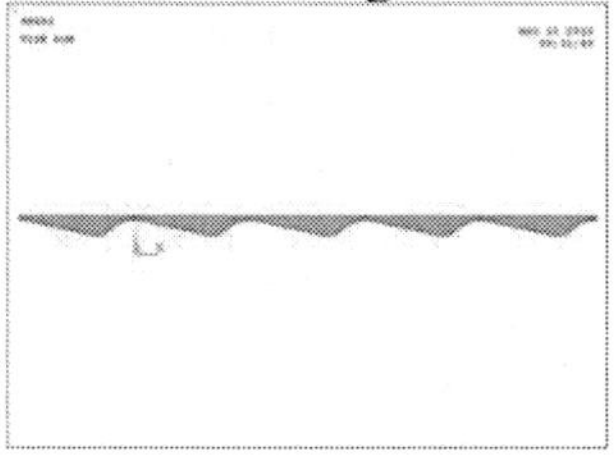

**Fig. 8.** Five shuttle shaped flow channel finite element models

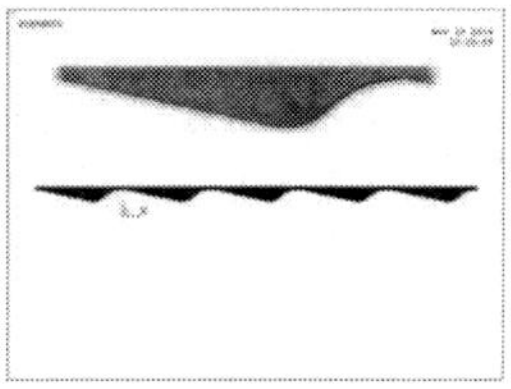

**Fig. 9.** Mesh division and boundary conditions of computational models

The simulation software was used to simulate the changes in viscosity loss rate and throttle pressure difference under different groove numbers and flow rates.

**Fig. 10.** Change diagram of throttle pressure difference under different displacement and adjustment clearance

— *Empirical conclusion*

a) The viscosity loss rate of valve core varies with different groove numbers; The viscosity loss rate of valve cores with different flow rates and the same number of grooves also varies;

b) The viscosity loss rate is approximately proportional to the increase of the number of grooves;

c) As the number of grooves increases, the adhesion loss rate gradually increases, and the maximum value does not exceed the performance index of 10%, which meets the optimization design requirements of the valve core.

After multiple numerical simulations, the relationship between throttle pressure difference and injection volume, ball number, and ball diameter was plotted into a curve chart. The use of a chart can quickly determine the required core diameter and quantity for a certain flow range.

## 4. On site pilot test

From 2019 to 2022, three wells were tested for remote control of layered polymer injection technology, avoiding the use of testing and debugging vehicles and personnel, reducing testing and debugging costs. The annual cost of testing and debugging was saved by 32000 yuan per well, and the annual cost of testing and debugging was saved by 96000 yuan for three wells; Improving the accuracy of layered polymer injection and improving the effect of polymer flooding, corresponding to an increase of 514 tons of oil in 10 beneficiary wells, is approximately 1.8 million yuan.

**Table 3.** Production Parameter Data of Field Test Wells

| Well number | Assignment time | Well depth/m | Number of segments/segment | Injection/$m^3$/d | Actual injection/$m^3$/d |
|---|---|---|---|---|---|
| AAA-1 | 2018.11 | 1841 | 3段 | 25/25/30 | 27/28/30 |
| AAA-2 | 2020.12 | 1266 | 2段 | 40/40 | 43/37 |
| AAA-3 | 2021.4 | 1317 | 3段 | 20/30/30 | 21/29/28 |

Starting from 2019, pilot testing was carried out in a certain oil production unit's AAA-1 well, which is a three stage layered polymer injection well. Three sets of intelligent polymer allocators and cable packers were designed for the pipe string. During construction, conduct pre well testing on the water distributor, thread the cable through the packer on the ground, and perform assembly inspection on the tools. When entering the well through a cable packer, inspect and install an armored cable protector at the tubing coupling. After running all the tubing strings, remove the blowout preventer, pass the cable through the wellhead, and install the wellhead. After completing the well, use intelligent polymer injection software on site for testing and sealing, and open the water nozzle of the second graded polymer; Press the pump truck to about 5MPa and read the flow data of the secondary aggregator; Observe the final pressure change of the 1 and 3 grading aggregators to determine the sealing condition of the downhole packer. After passing the sealing inspection, conduct testing and adjustment; Flow verification; Separate each grading aggregator and calibrate it with the wellhead flow meter; Let the aggregator fully open, read the flow rate of each layer, and prioritize adjusting the opening of the control layer to achieve the injection volume. Monitor the flow, pressure, and temperature data of each layer daily through an intelligent layered polymer injection software system; Compare the deviation of wellhead flow rate and regularly send polymer switch commands.

## 5. Conclusion

(1) The intelligent layered polymer injection matching technology with cables is connected to the ground intelligent control system through cables, achieving real-time monitoring and online control of polymer flow, pressure, temperature and other parameters, improving monitoring accuracy and efficiency, and providing technical support for the construction of "Digital Intelligence Oilfield".

(2) The control system adjusts the injection volume of layers based on real-time layer flow rate and injection allocation plan, ensuring that the entire process of layered polymer injection is controlled.

(3) The downhole pressure and flow parameters provide data support for the optimization of polymer flooding development plan.

(4) Remote intelligent testing and adjustment can greatly reduce the testing cost of layered injection in polymer injection wells without moving the pipe string and avoiding construction.

(5) It can maintain the qualification rate of polymer injection well allocation for a long time and achieve precise injection allocation throughout the entire polymer injection cycle.

## References

[1] De Miranda Diego Alves,Rauber Willian Kévin,Vaz Miguel,Alves Marcus Vinícius Canhoto,Lafratta Fernando Humel,Nogueira André Lourenço,Zdanski Paulo Sergio Berving. Analysis of numerical modeling strategies to improve the accuracy of polymer injection molding simulations[J]. Journal of Non-Newtonian Fluid Mechanics,2023,315.

[2] Ahmed Shehzad,Hanamertani Alvinda Sri,Alameri Waleed,Al-Shalabi Emad W.,Hashmet Muhammad Rehan. Experimental investigation of flow diversion and dynamic retention during polymer flooding in high salinity fractured carbonates using CT imaging[J]. Geoenergy Science and Engineering,2023,221.
[3] Zhang Jiyang, Liu Renyong, Guo Wei, Xu Xiaoyu, Zhu Jiahui, Wang Ruihua, Wang Fang Research and Application of Cable Controlled Intelligent Layered Oil Production Technology [J] Petroleum Mining Machinery, 2022,51 (05): 35-41
[4] Yang Lingzhi, Zhou Zhiping, Yang Haien, Li Falong, Hu Gaixing Digital injection technology for pre installed cables of underground flexible composite pipes [J] Petroleum Drilling Technology, 2022,50 (06): 120-125
[5] Wang Erhu, Zhang Jie Research on the Application of Fine Injection Production Technology in Water Drive Reservoirs [J] Cleaning the World, 2022, 38 (06): 114-116
[6] Xu Lu Optimization and experimental evaluation of secondary sand control screen tube for polymer flooding [J] Technical Supervision of the Petroleum Industry, 2022, 38 (02): 46-49
[7] Zhang Zhiqun Analysis of Factors Influencing Polymer Flooding Effect [J] Chemical engineering and Equipment, 2021, (10): 73-74
[8] Wang Zhiguo,Wang Zaiqiang,Huang Wei,Dou Yihua,Cao Yinping. Progress in mechanical analysis of CO2 flooding injection string[J]. IOP Conference Series: Earth and Environmental Science,2021,651(3).
[9] Guolin Jing,Shan Tang,Xiaoxiao Li,Huaiyuan Wang. The analysis of scaling mechanism for water-injection pipe columns in the Daqing Oilfield[J]. Arabian Journal of Chemistry,2017,10(S1).
[10] Wen Houzhen Research on the Mechanics Theory, Analysis Method, and Application of Fine Layered Injection Production String [D]. Northeast Petroleum University, 2016
[11] Du Liang The application of layered polymer injection technology in oil wells of Class II oil reservoirs [J] Oil and Gas Field Surface Engineering, 2014,33 (10): 117
[12] Xin Xu. Analysis and Calculation of the Flow Pressure in the Multipoint Steam Injection String of a Horizontal Well[A]. Information Engineering Research Institute, USA. Proceedings of 2014 5th International Conference on Information Technology for Manufacturing Systems (ITMS 2014) [C]
[13] Yun Wang,Jun Li,Guang Qiang Cao. Material Optimization of Injection and Production String in Su4 Underground Gas Storage[J]. Advanced Materials Research,2014,3246(962-965).
[14] Juquanli Mechanical Analysis of Layered Water Injection Packer String [D]. Xi'an University of Petroleum, 2014
[15] Xiang Yong, Shi Zanrong, Tan Chaoqin, Wang Hui, Lu Kaidong, Yang Liping, Deng Han Development and application of a software for analyzing the stress on the string of layered water injection wells [J] China Petroleum and Chemical Industry, 2013, (11): 51-53
[16] Hou Zhen Wen,Zun Ce Wang,Feng Xia Lv. The Mechanical Model and Boundary Condition of Separated Zone Water Injection String ' s Axial Force[J]. Applied Mechanics and Materials,2012,2034(229-231).
[17] Michael Zettlitzer,Fabian Moeller,Daria Morozova,Peter Lokay,Hilke Würdemann. Re-establishment of the proper injectivity of the CO 2 -injection well Ktzi 201 in Ketzin, Germany[J]. International Journal of Greenhouse Gas Control,2010,4(6).
[18] Liu Huajun, Wang Zhiqing, Sun Lei, Li Chunyun, Liu Ruoxu, Yang Yushan Research and Application of Layered Steam Injection String [J] Journal of Changjiang University (Self Science Edition) Science and Engineering Volume, 2007, (02): 216-218+348
[19] Mingchao Gang,Dexin Yuan,Jing You,Weiyi Xie,Rui Wang,Hanxiang Wang,Wenxiu Zhao. Mechanical Analysis of Separate Layer Injection String during Operation Condition and its Application[P]. Proceedings of the 2016 International Forum on Mechanical, Control and Automation (IFMCA 2016),2017.
[20] Dekui Xu,Chongjiang Liu,Xingliang Song,Zhenkun Zhu,Hangpeng Cao,Xiaoyu Xu,Lili

Wang,Junli Liu,Chao Wang. Study on Application of Intelligent Separated Layer Water Injection Technology in Daqing Oilfield[P]. DEStech Transactions on Computer Science and Engineering,2018.

## AUTHOR INDEX